A DICTIONARY OF
MODERN
ENGLISH USAGE

# A DICTIONARY OF
# MODERN
# ENGLISH USAGE

BY

## H. W. FOWLER

SECOND EDITION

*revised by*

SIR ERNEST GOWERS

OXFORD
AT THE CLARENDON PRESS

*Oxford University Press, Ely House, London W. 1*

GLASGOW   NEW YORK   TORONTO   MELBOURNE   WELLINGTON
CAPE TOWN   IBADAN   NAIROBI   DAR ES SALAAM   LUSAKA   ADDIS ABABA
DELHI   BOMBAY   CALCUTTA   MADRAS   KARACHI   LAHORE   DACCA
KUALA LUMPUR   SINGAPORE   HONG KONG   TOKYO

FIRST EDITION 1926
SECOND EDITION 1965
REPRINTED WITH CORRECTIONS
1965 (TWICE), 1966
1968 (WITH CORRECTIONS)
1970, 1972

PRINTED IN GREAT BRITAIN

# PREFACE TO THE REVISED EDITION

'It took the world by storm' said *The Times*, in its obituary notice of H. W. Fowler, about *The King's English*, published by him and his younger brother Frank in 1906. That description might have been more fitly applied to the reception of *A Dictionary of Modern English Usage* which followed twenty years later, planned by the two brothers but executed by Henry alone. This was indeed an epoch-making book in the strict sense of that overworked phrase. It made the name of Fowler a household word in all English-speaking countries. Its influence extended even to the battlefield. 'Why must you write *intensive* here?' asked the Prime Minister in a minute to the Director of Military Intelligence about plans for the invasion of Normandy. '*Intense* is the right word. You should read Fowler's *Modern English Usage* on the use of the two words.'[1] Though never revised, the book has kept its place against all rivals, and shown little sign of suffering from that reaction which commonly awaits those whose work achieves exceptional popularity in their lifetime.

What is the secret of its success? It is not that all Fowler's opinions are unchallengeable. Many have been challenged. It is not that he is always easy reading. At his best he is incomparable. But he never forgot what he calls 'that pestilent fellow the critical reader' who is 'not satisfied with catching the general drift and obvious intention of a sentence' but insists that 'the words used must . . . actually yield on scrutiny the desired sense'.[2] There are some passages that only yield it after what the reader may think an excessive amount of scrutiny—passages demanding hardly less concentration than one of the more obscure sections of a Finance Act, and for the same reason: the determination of the writer to make sure that, when the reader eventually gropes his way to a meaning, it shall be, beyond all possible doubt, the meaning intended by the writer. Nor does the secret lie in the convenience of the book as a work of reference; it hardly deserves its title of 'dictionary', since much of it consists of short essays on various subjects, some with fancy titles that give no clue at all to their subject. What reporter, seeking guidance about the propriety of saying that the recep-

---

[1] *The Second World War*, v. 615.          [2] S.V. ILLOGICALITIES.

tion was held 'at the bride's aunt's', would think of looking for it in an article with the title 'Out of the Frying-Pan'?

There is of course more than one reason for its popularity. But the dominant one is undoubtedly the idiosyncrasy of the author, which he revealed to an extent unusual in a 'dictionary'. 'Idiosyncrasy', if we accept Fowler's own definition, 'is peculiar mixture, and the point of it is best shown in the words that describe Brutus: "His life was gentle, and the elements So mixed in him that Nature might stand up And say to all the world This was a man." One's idiosyncrasy is the way one's elements are mixed.'[1] This new edition of the work may therefore be suitably introduced by some account of the man. The following is based on a biographical sketch by his friend G. G. Coulton published in 1934 as *Tract XLIII* of the Society for Pure English.

He was born in 1858, the son of a Cambridge Wrangler and Fellow of Christ's. From Rugby he won a scholarship to Balliol, but surprisingly failed to get a first in either Mods. or Greats. After leaving Oxford he spent seventeen years as a master at Sedbergh. His career there was ended by a difference of opinion with his headmaster, H. G. Hart (also a Rugbeian). Fowler, never a professing Christian, could not conscientiously undertake to prepare boys for confirmation. Hart held this to be an indispensable part of a housemaster's duty. Fowler was therefore passed over for a vacant housemastership. He protested; Hart was firm; and Fowler resigned. It was, in Fowler's words, 'a perfectly friendly but irreconcilable' difference of opinion. Later, when Hart himself had resigned, Fowler wrote to Mrs. Hart that though Sedbergh would no doubt find a new headmaster with very serviceable talents of one kind or another, it was unlikely to find again 'such a man as everyone separately shall know (more certainly year by year) to be at once truer and better, gentler and stronger, than himself'.

Thus, at the age of 41, Fowler had to make a fresh start. For a few years he lived in London, where he tried his hand as an essayist without any great success, and attempted to demonstrate what he had always maintained to be true—that a man ought to be able to live on £100 a year. In 1903 he joined his brother in Guernsey, and in 1908, on his fiftieth birthday, married a lady four years younger than himself. The brothers did literary work together. Their most notable productions were a translation of Lucian and *The King's English*. The great success of the latter pointed the road they were to follow in future.

When war broke out Henry was 56. He emerged from retirement to

[1] s.v. IDIOSYNCRASY.

take part in the recruiting campaign. But he found himself more and more troubled by the thought that he was urging others to run risks which he would himself be spared. So he enlisted as a private in the 'Sportsmen's Battalion', giving his age as 44. His brother, aged 45, enlisted with him. Their experiences are fully told in letters from Henry to his wife, now in the library of St. John's College, Cambridge. It is a sorry story, summarized in a petition sent by the brothers to their commanding officer in France in February 1916.

[Your petitioners] enlisted in April 1915 at great inconvenience and with pecuniary loss in the belief that soldiers were needed for active service, being officially encouraged to mis-state their ages as a patriotic act. After nine months' training they were sent to the front, but almost immediately sent back to the base not as having proved unfit for the work, but merely as being over age—and this though their real ages had long been known to the authorities. . . . They are now held at the base at Étaples performing only such menial or unmilitary duties as dish-washing, coal-heaving and porterage, for which they are unfitted by habits and age. They suggest that such conversion of persons who undertook purely from patriotic motives the duties of soldiers on active service into unwilling menials or servants is an incredibly ungenerous policy. . . .

This petition secured Fowler's return to the trenches, but not for long. Three weeks later he fainted on parade, and relegation to the base could no longer be resisted. This seemed the end. 'By dinner time', he wrote to his wife shortly afterwards, 'I was making up my mind to go sick and ask to be transferred to a lunatic asylum.' This drastic measure proved unnecessary, for in a few days he was to go sick in earnest. He was sent back to England, and after some weeks in hospital was discharged from the Army, having spent eighteen dreary months in a constantly frustrated attempt to fight for his country.

After their discharge the brothers returned to Guernsey, but the partnership only lasted another two years; Frank died in 1918. In 1925 Henry and his wife left the island to live in a cottage in the Somerset-shire village of Hinton St. George. There he remained until his death in 1933, occupied mainly with lexicographical work for the Clarendon Press and on the book that was to make him famous. An exceptionally happy marriage ended with the death of his wife three years before his own. The unbeliever's memorial to her was, characteristically, a gift of bells to the village church.

The most prominent element in Fowler's idiosyncrasy was evidently what the Romans called *aequanimitas*. He knew what he wanted from life; what he wanted was within his reach; he took it and was content. It pleased him to live with spartan simplicity. Coulton quotes a letter he wrote to the Secretary of the Clarendon Press in reply to an offer to pay the wages of a servant. Fowler was then 68 and the month was November.

> My half-hour from 7.0 to 7.30 this morning was spent in (1) a two-mile run along the road, (2) a swim in my next-door neighbour's pond—exactly as some 48 years ago I used to run round the Parks and cool myself in (is there such a place now?) Parson's Pleasure. That I am still in condition for such freaks I attribute to having had for nearly 30 years no servants to reduce me to a sedentary and all-literary existence. And now you seem to say: Let us give you a servant, and the means of slow suicide and quick lexicography. Not if I know it: I must go my slow way.

So he continued to diversify his lexicography with the duties of a house-parlourmaid and no doubt performed them more scrupulously than any professional.

He has been described by one who had been a pupil of his at Sedbergh as 'a man of great fastidiousness, (moral and intellectual)', and he is said to have shown the same quality in his clothes and personal appearance. Coulton compares him to Socrates. Though not a professing Christian, Fowler had all the virtues claimed as distinctively Christian, and, like Socrates, 'was one of those rare people, sincere and unostentatious, to whom the conduct of life is *ars artium*'.

Such was the man whose idiosyncrasy so strongly colours his book. The whimsicality that was his armour in adversity enlivens it in unexpected places; thus by way of illustrating the difficulty there may be in identifying a phenomenon he calls 'the intransitive past participle', he observes that 'an angel dropped from heaven' has possibly been passive, but more likely active, in the descent. The simplicity of his habits has its counterpart in the simplicity of diction he preaches. The orderly routine of his daily life is reflected in the passion for classification, tabulation, and pigeon-holing that he sometimes indulges beyond reason. Above all, that uncompromising integrity which made him give up his profession rather than teach what he did not believe, and to go to the battlefront himself rather than persuade younger men to do so, permeates *Modern English Usage*. That all kinds of affectation and humbug were anathema to his fastidious mind is apparent on almost every page. Perhaps it was this trait that made him choose, as his first literary enterprise, to try to introduce to a wider public the works of that archetypal debunker, Lucian.

Much of *Modern English Usage* is concerned with choosing the right word, and here the need for revision is most evident, for no part of 'usage' changes more quickly than verbal currency. To a reader forty years after the book was written it will seem to be fighting many battles that were won or lost long ago. 'Vogue words' get worn out and others take their place. 'Slipshod extensions' consolidate their new positions. 'Barbarisms' become respected members of the vocabulary. 'Genteel-isms' and 'Formal words' win undeserved victories over their plainer rivals. 'Popularized technicalities' proliferate in a scientific age. Words unknown in Fowler's day—*teenager* for instance—are now among our hardest worked.

Articles on other subjects have better stood the test of time, but many call for some modernization. One or two have been omitted as no longer relevant to our literary fashions; a few have been rewritten in whole or in part, and several new ones added. About those that deal with 'grammar' in the broadest sense something needs to be said at greater length.

There were two sides to Fowler as a grammarian. In one respect he was an iconoclast. There was nothing he enjoyed debunking more than the 'superstitions' and 'fetishes' as he called them, invented by peda-gogues for no other apparent purpose than to make writing more difficult. The turn of the century was their heyday. Purists then enjoyed the sport of hunting split infinitives, 'different to's', and the like as zestfully as today they do that of cliché-hunting. The Fowlers' books were a gust of common sense that blew away these cobwebs. It was refreshing to be told by a grammarian that the idea that *different* could only be followed by *from* was a superstition; that to insist on the same preposition after *averse* was one of the pedantries that spring of a little knowledge; that it is better to split one's infinitives than to be ambiguous or artificial; that to take exception to *under the circumstances* is puerile; that it is nonsense to suppose one ought not to begin a sentence with *and* or *but* or to end one with a preposition; that those who are over-fussy about the placing of the adverb *only* are the sort of friends from whom the English language may well pray to be saved; that it is a mistake to suppose that *none* must at all costs be followed by a singular verb; that it is futile to object to the use of *to a degree* in the sense of *to the last degree*; that to insist on writing *first* instead of *firstly* is pedantic artificialism; and that to forbid the use of *whose* with an inanimate antecedent is like sending a soldier on active service and insisting that his tunic collar shall be tight and high. If writers today no longer feel the burden of fetters such as these they have largely the Fowlers to thank.

On the other hand, Fowler has been criticized—notably by his famous contemporary Jespersen—for being in some respects too strict and old-fashioned. He was a 'prescriptive' grammarian, and prescriptive grammar is not now in favour outside the schoolroom. Jespersen, the 'grammatical historian', held that 'of greater value than this prescriptive grammar is a purely descriptive grammar which, instead of acting as a guide to what should be said or written, aims at finding out what is actually said or written by those who use the language'[1] and recording it objectively like a naturalist observing the facts of nature.[2] Fowler, the 'instinctive grammatical moralizer' (as Jespersen called him and he welcomed the description), held that the proper purpose of a grammarian was 'to tell the people not what they do and how they came to do it, but what they ought to do for the future'.[3] His respect for what he regarded as the true principles of grammar was as great as was his contempt for its fetishes and superstitions. He has been criticized for relying too much on Latin grammar for those principles. In part he admitted the charge. 'Whether or not it is regrettable', he said, 'that we English have for centuries been taught what little grammar we know on Latin traditions, have we not now to recognize that the iron has entered into our souls, that our grammatical conscience has by this time a Latin element inextricably compounded in it, if not predominant?'[4] At the same time he had nothing but contempt for those grammarians whom he described as 'fogging the minds of English children with terms and notions that are essential to the understanding of Greek and Latin syntax but have no bearing on English'.[5]

The truth is that the prime mover of his moralizing was not so much grammatical grundyism as the instincts of a craftsman. 'Proper words in proper places', said Swift, 'make the true definition of a style.' Fowler thought so too; and, being a perfectionist, could not be satisfied with anything that seemed to him to fall below the highest standard either in the choice of precise words or in their careful and orderly arrangement. He knew, he said, that 'what grammarians say should be has perhaps less influence on what shall be than even the more modest of them realize; usage evolves itself little disturbed by their likes and dislikes'. 'And yet', he added, 'the temptation to show how better use might have been made

[1] *Essentials of English Grammar*, p. 19.
[2] *Enc. Brit.*, s.v. GRAMMAR.
[3] *SPE Tract XXVI*, p. 194.
[4] Ibid.
[5] s.v. CASES 2.

of the material to hand is sometimes irresistible.'[1] He has had his reward in his book's finding a place on the desk of all those who regard writing as a craft, and who like what he called 'the comfort that springs from feeling that all is shipshape'.

He nodded, of course. Some of his moralizings were vulnerable even when he made them; others have become so. Some revision has been necessary. But no attempt has been made to convert the instinctive grammatical moralizer into anything else. In this field therefore what has been well said of the original book will still be true of this edition: 'You cannot depend on the Fowler of *Modern English Usage* giving you either an objective account of what modern English usage *is* or a representative summary of what the Latin-dominated traditionalists would *have it be*. *Modern English Usage* is personal: it is Fowler. And in this no doubt lies some of its perennial appeal.'[2]

Anyone undertaking to revise the book will pause over the opening words of Fowler's own preface: 'I think of it as it should have been, with its prolixities docked. . . .' He cannot be acquitted of occasional prolixity. But his faults were as much a part of his idiosyncrasy as his virtues; rewrite him and he ceases to be Fowler. I have been chary of making any substantial alterations except for the purpose of bringing him up to date; I have only done so in a few places where his exposition is exceptionally tortuous, and it is clear that his point could be put more simply without any sacrifice of Fowleresque flavour. But the illustrative quotations have been pruned in several articles, and passages where the same subject is dealt with in more than one article have been consolidated.

Only one important alteration has been made in the scope of the book. The article TECHNICAL TERMS, thirty pages long, has been omitted. It consisted of definitions of 'technical terms of rhetoric, grammar, logic, prosody, diplomacy, literature, etc., that a reader may be confronted with or a writer have need of'. The entries that are relevant to 'modern English usage' have been transferred to their alphabetical places in the book. For the rest, the publication of other 'Oxford' books, especially the *COD* and those on English and classical literature, has made it unnecessary to keep them here. The eight pages of French words listed for their pronunciation have also been omitted; a similar list is now appended to the *COD*.

[1] s.v. THAT REL. PRON. I.
[2] Randolph Quirk in *The Listener*, 15 March 1958.

I have already referred to the enigmatic titles that Fowler gave to some of his articles, and their effect in limiting the usefulness of the book as a work of reference. But no one would wish to do away with so Fowleresque a touch; indeed, I have not resisted the temptation to add one or two. I hope that their disadvantage may be overcome by the 'Classified Guide' which now replaces the 'List of General Articles'. In this the articles (other than those concerned only with the meaning, idiomatic use, pronunciation, etc., of the words that form their titles) are grouped by subject, and some indication is given of their content wherever it cannot be inferred from their titles. This also rids the body of the book of numerous entries inserted merely as cross-references.

E. G.

# ACKNOWLEDGEMENTS

I GRATEFULLY record my obligation to all those who have contributed to this edition with suggestion, criticism, and information; they are too many for me to name them all. I must also be content with a general acknowledgement to the many writers (and their publishers) whom I have quoted, usually because they said what I wanted to say better than I could myself.

As an amateur in linguistics, I am especially indebted to those who have readily responded to my appeals for expert guidance, particularly to Mr R. W. Burchfield, Mr G. V. Carey (who contributes an article on the use of capitals), the late Dr R. W. Chapman, Professor Norman Davis, Mr P. S. Falla, and Professor Randolph Quirk. Mr Peter Fleming will recognize his own hand in more than one article. Mr D. M. Davin, who has been in charge of the work for the Clarendon Press, has been infinitely helpful. To Mr L. F. Schooling my obligation is unique. He not only started me off with a comprehensive survey of what needed to be done, but has shared throughout in every detail of its execution, fertile in suggestion, ruthless in criticism, and vigilant in the detection of error.

ACKNOWLEDGMENTS

# PREFACE TO THE FIRST EDITION

I think of it as it should have been, with its prolixities docked, its dullnesses enlivened, its fads eliminated, its truths multiplied. He had a nimbler wit, a better sense of proportion, and a more open mind, than his twelve-year-older partner; and it is matter of regret that we had not, at a certain point, arranged our undertakings otherwise than we did.

In 1911 we started work simultaneously on the *Pocket Oxford Dictionary* and this book; living close together, we could, and did, compare notes; but each was to get one book into shape by writing its first quarter or half; and so much only had been done before the war. The one in which, as the less mechanical, his ideas and contributions would have had much the greater value had been assigned, by ill chance, to me. In 1918 he died, aged 47, of tuberculosis contracted during service with the B.E.F. in 1915–16.

The present book accordingly contains none of his actual writing; but, having been designed in consultation with him, it is the last fruit of a partnership that began in 1903 with our translation of Lucian.

H. W. F.

# ACKNOWLEDGEMENTS IN THE FIRST EDITION

I cannot deny myself the pleasure of publicly thanking Lt-Col. H. G. Le Mesurier, C.I.E., who not only read and criticized in detail the whole MS. of this book, but devised, at my request, a scheme for considerably reducing its bulk. That it was not necessary to adopt this scheme is due to the generosity of the Clarendon Press in consenting to publish, at no high price, an amount much greater than that originally sanctioned.

On behalf of the Press, Mr. Frederick Page and Mr. C. T. Onions have made valuable corrections and comments.

The article on *morale* has appeared previously in the *Times Literary Supplement*, that on *only* in the *Westminster Gazette*, and those on Hyphens, Inversion, Metaphor, Split infinitive, Subjunctives, and other matters, in *SPE* Tracts.

H. W. F.

# CLASSIFIED GUIDE TO THE DICTIONARY

THE articles listed in this Guide are classified according as they deal with (I) what may for convenience be called 'usage', that is to say points of grammar, syntax, style, and the choice of words; (II) the formation of words, and their spelling and inflexions; (III) pronunciation; and (IV) punctuation and typography. The Guide does not include any articles that are concerned only with the meaning or idiomatic use of the title-words, or their spelling, pronunciation, etymology, or inflexions.

## I. USAGE

**absolute construction.** ('*The play being over,* we went home.')

**absolute possessives.** ('Your and our(s) and his efforts.')

**abstractitis.** Addiction to abstract words.

**adjectives misused.**

**ambiguity.** Some common causes.

**Americanisms.**

**analogy.** As a literary device. As a corrupter of idiom.

**archaism.**

**avoidance of the obvious.** In choice of words the obvious is better than its obvious avoidance.

**basic English.**

**battered ornaments.** An introduction to other articles on words and phrases best avoided for their triteness.

**cannibalism.** For instance the swallowing of a *to* by another *to* in 'Doubt as to whom he was referring'.

**cases.** The status of case in English grammar. Some common temptations to ignore it. References to other articles on particular points.

**cast-iron idiom.** More on the corruption of idiom by analogy.

**-ce, -cy.** Differences in meaning between words so ending, e.g. *consistenc(e)* (*y*).

**cliché.**

**collectives.** A classification of nouns singular in form used as plurals.

**commercialese.**

**compound prepositions and conjunctions.** *Inasmuch as, in regard to,* etc.

**didacticism.** Showing itself in attempts to improve accepted vocabulary etc.

**differentiation.** Of words that might have been synonyms, such as *spirituous* and *spiritual*; *emergence* and *emergency*.

**double case.** Giving references to other articles which illustrate the making of a single word serve as both subjective and objective.

**double passives.** E.g. 'The point is sought to be avoided.'

**elegant variation.** Laboured avoidance of repetition.

**elision.** Of auxiliaries and negatives: *I've, hasn't,* etc.

**ellipsis.** Leaving words to be 'understood' instead of expressed, especially parts of *be* and *have,* of *that* (conj.) and of words after *than.*

**enumeration forms.** The proper use of *and* and *or* in stringing together three or more words or phrases.

**-er and -est.** Some peculiarities in the use of comparatives and superlatives

**ethic.** For the 'ethic dative'.

**euphemism.**

**euphuism.**

**false emphasis.** Sentences accidentally stressing what was not intended to be stressed.

**false scent.** Misleading the reader.

**feminine designations.** Their use.

**fetishes.** References to articles on some grammarians' rules misapplied or unduly revered.

**foreign danger.** Foreign words and phrases misused through ignorance.

**formal words.** Deprecating their needless use.

**French words.** Their use and pronunciation.

**fused participle.** The construction exemplified in 'I like you pleading poverty.'

**gallicisms.** Borrowings from French that stop short of using French words without disguise, e.g. 'jump to the eyes'.

**generic names and other allusive commonplaces.** *A Jehu, Ithuriel's spear*, and the like.

**genteelisms.**

**gerund.** Its nature and uses. Choice between gerund and infinitive in e.g. *aim at doing, aim to do.*

**grammar.** The meaning of the word and the respect due to it.

**hackneyed phrases.** The origin and use of the grosser kind of cliché.

**hanging-up.** Keeping the reader waiting an unconscionable time for verb or predicate.

**haziness.** Shown in overlappings and gaps.

**headline language.**

**hyperbole.**

**hysteron proteron.** Putting the cart before the horse.

**-ic(al).** Differentiation between adjectives with these alternative endings.

**-ics.** *-ic* or *-ics* for the name of a science etc.? Singular or plural after *-ics?*

**idiom.** Defined and illustrated.

**illiteracies.** Some common types.

**illogicalities.** Defensible and indefensible.

**incompatibles.** Some ill-assorted phrases of similar type: *almost quite, rather unique*, etc.

**incongruous vocabulary.** Especially the use of archaisms in unsuitable setting.

**indirect object.**

**indirect question.**

**-ing.** Choice between the *-ing* form and the infinitive in such sentences as 'Dying at their posts rather than surrender(ing)': 'doing more than furnish(ing) us with loans.'

**intransitive past participle.** As a grammatical curiosity in e.g. 'fallen angels'.

**inversion.** Its uses and abuses.

**-ion and -ment**
**-ion and -ness**
**-ism and -ity** ⎱ Differentiation in meaning between nouns from the same verb with these different endings.

**irrelevant allusion.** The use of 'hackneyed phrases that contain a part that is appropriate and another that is pointless or worse', e.g. to 'leave (severely) alone'.

**italics.** Their proper uses.

**jargon.** Distinguishing argot, cant, dialect, jargon, and other special vocabularies.

**jingles.** Supplements the article repetition of words or sounds.

**legerdemain.** Using a word twice without noticing that the sense required the second time is different from that of the first.

**letter forms.** Conventional ways of beginning and ending letters.

**literary critics' words.**

**literary words.**

**litotes.** A variety of **meiosis.**

**long variants.** E.g. *preventative* for *preventive*; *quieten* for *quiet.*

**love of the long word.**

**-ly.** Ugly accumulation of adverbs so ending.

**malapropisms.**

**meaningless words.** *Actually, definitely, well*, etc.

**meiosis.** Understatement designed to impress.

**membership.** Use of *-ship* words for *members, leaders*, etc.

**metaphor.**

**misapprehensions.** About the meaning of certain words and

**phrases,** e.g. *leading question, prescriptive right.*

**misquotations.** Some common examples.

**names and appellations.** Conventional ways of speaking to and of relations and friends.

**needless variants.** Of established words.

**negative mishandlings.** Especially those that lead one to say the opposite of what one means.

**noun-adjectives.** As corrupters of style.

**novelty hunting.** In the choice of words.

**number.** Some problems in the choice between singular and plural verbs.

**object-shuffling.** Such as 'Instil people with hope' for 'instil hope into people'.

**officialese.**

**oratio obliqua, recta.**

**out of the frying pan.** Examples of a writer's being faulty in one way because he has tried to avoid being faulty in another.

**overzeal.** Unnecessary repetition of conjunctions, prepositions, and relatives.

**pairs and snares.** Some pairs of words liable to be confused.

**paragraph.**

**parallel sentence dangers.** Damaging collisions between the negative and affirmative, inverted and uninverted, dependent and independent.

**parenthesis.**

**participles.** On the trick of beginning a sentence with a participle. Also giving references to other articles on participles.

**passive disturbances.** On the impersonal passive (*it is thought* etc.). Also giving references to other articles on the passive.

**pathetic fallacy.**

**pedantic humour.**

**pedantry.**

**perfect infinitive.** 'I should (have) like(d) to have gone.'

**periphrasis.**

**personification.** E.g. using *crown* for *monarch, she* for *it*.

**phrasal verbs.** Their uses and abuses.

**pleonasm.** Using more words than are required for the sense intended.

**poeticisms.**

**polysyllabic humour.**

**popularized technicalities.** Including 'Freudian English'.

**position of adverbs.** Common reasons for misplacing them.

**preposition at end.**

**preposition dropping.** ('Eating fish Fridays'; 'going places' etc.)

**pride of knowledge.** Showing itself disagreeably in the choice of words.

**pronouns.** Some warnings about their use.

**quasi-adverbs.** Adjectival in form (*preparatory, contrary*, etc.).

**quotation.** Its uses and abuses.

**repetition of words or sounds.**

**revivals.** Of disused words.

**rhyming slang.**

**rhythm.**

**Saxonism and anti-Saxonism.**

**semantics.**

**sentence.** What is a sentence?

**sequence of tenses.**

**Siamese Twins.** Such as *chop and change*; *fair and square*.

**side-slip.** A few examples of sentences that have gone wrong through not keeping a straight course.

**slipshod extension.** Of the meaning of words, and consequent verbicide.

**sobriquets.**

**sociologese.**

**split infinitive.**

**stock pathos.**

**sturdy indefensibles.** Examples of ungrammatical or illogical idiom.

**subjunctive.** Modern uses of a dying mood.

**superfluous words.** Some that might be dispensed with.

**superiority.** Apologizing for the use of homely phrases.

**superstitions.** Some outworn grammatical pedantries.

**swapping horses.** Three sentences gone wrong, one through failure to maintain the construction of the opening participle, and the others through failure to remember what the subject is.

**syllepsis and zeugma.** Defined and distinguished.

**synonyms.**

**tautology.** Especially on the use of the 'abstract appendage'.

**-tion words.** Addiction to *position* and *situation* and similar abstract words.

**titles.** Changing fashion in the designation of peers.

**to-and-fro puzzles.** Sentences that leave the reader wondering whether their net effect is positive or negative.

**trailers.** Specimens of sentences that keep on disappointing the reader's hope of coming to the end.

**-ty and -ness.** Differentiation between nouns with these alternative endings.

**u and non-u.**

**unattached participles.**

**unequal yokefellows.** A collection (from other articles) of varieties of a single species: *each . . . are*; *scarcely . . . than* and others.

**unidiomatic -ly.** Against 'the growing notion that every adjective, if an adverb is to be made of it, must have a *-ly* clapped on to it'.

**verbless sentences.**

**vogue words.**

**vulgarization.** Of words that depend on their rarity for their legitimate effect, e.g. *epic.*

**walled-up object.** Such as *him* in 'I scolded and sent him to bed.'

**Wardour Street.** The use of antique words.

**word patronage.** Another manifestation of the attitude described in **superiority.**

**working and stylish words.** Deprecating, with examples, 'the notion that one can improve one's style by using stylish words'.

**worn-out humour.** Some specimens.

**worsened words.** Such as *imperialism, appeasement, academic.*

# II. WORD FORMATION, INFLEXION, AND SPELLING

## A. GENERAL

**ae, oe.** *Medi(a)eval, (o)ecumenical.*

**analogy** (2). As an influence in word-making.

**aphaeresis.**

**apocope.**

**back-formation.** E.g. *diagnose, burgle.*

**barbarisms.** Unorthodox word-formation.

**curtailed words.** Including acronyms.

**didacticism.** Deprecated in the spelling of familiar words.

**eponymous words.** Some familiar examples.

**facetious formations.**

**feminine designations.** Ways of forming them.

**hybrids and malformations.** Developing the article **barbarisms.**

**new verbs in -ize.**

**onomatopoeia.**

**portmanteau words.** *Motel, Oxbridge,* etc.

**reduplicated words.** *Hugger-mugger* etc.

**spelling points.** Spelling reform. Double or single consonants? References to articles on particular points of spelling. Some special difficulties.

**true and false etymology.** Some examples of words whose looks belie their origin.

## B. WORD BEGINNINGS

**a-, an-** (= not).

**aero-, air-.**

**bi-.** As in *bi-monthly.*

**brain- compounds.**

**by-, bye-.**

**centi-, hecto-.**

**co-.**

**de-, dis-.**

deca-, deci-.
demi-.
em- and im-, en- and in-. As
  alternative spellings in some words.
ff. For capital F in proper names.
for-, fore-.
hom(oe) (oi) o-.
in- and un-. Choice between in
  negative formations.

non-.
para-.
re-.
self-.
semi-.
super-.
tele-.
vice-.
yester-.

## C. WORD ENDINGS

-able, ible.
-al nouns. Their revival and invention deprecated.
-atable.
brinkmanship. For the *-manship*
  suffix.
-ce, -cy. As alternative ways of
  forming nouns.
-ed and 'd. *Tattoed* or *tattoo'd* etc.
-edly. Distinguishing the good and
  the bad among adverbs so formed.
-ee.
-eer.
-en and -ern. Adjectives so ending.
-en verbs from adjectives. Distinguishing between the established
  and the dubious.
-er and -est. Or *more* and *most* for
  comparative and superlatives.
-ey and -y. *Horsey* or *horsy* etc.
-ey, -ie, and -y in pet names.
  *Auntie, daddy*, etc,
forecast. Past of *-cast* verbs.
-genic.
-iana.
-ion and -ment.  ⎫
-ion and -ness.   ⎬ As alternative ways
-ism and -ity.   ⎭ of forming nouns.

-ist, -alist, -yist. *Agricultur(al)ist,*
  *accompan(y)ist,* etc.
-ize, -ise. Choice between z and s in
  verbs so ending.
-latry.
-less.
-lily. Formation of adverbs from
  adjectives in *-ly*.
-logy.
-or (and *-er*) as agent terminations.
-our and -or in *colo(u)r, hono(u)r,*
  etc.
-phil(e).
-re and -er. In *cent(re)(er)* etc.
-some.
suffragette. For the *-ette* suffix.
-t and -ed. *Spoilt* or *spoiled* etc.
-th nouns. Deprecating the revival
  of obsolete or the invention of new.
-ty and -ness. As alternative ways
  of forming nouns.
-ular.
-valent.
-ward(s).
warmonger. For the *-monger* suffix
-wise, -ways.
-worthy.
-xion, -xive. Or *-ction, -ctive.*

## D. PLURAL FORMATIONS

-ae, -as. Of words ending a.
-ex, -ix. Of words so ending.
-ful. *Handful* etc.
Latin plurals.
o(e)s. Of words ending -o.
-on. Of words so ending.
plural anomalies. Of words ending
  -s in the singular. Of compound

words. Of words ending -y. References to other articles on plurals of
particular words or terminations.
-trix. ⎫
-um.  ⎬ Of words so ending.
-us.  ⎭
x. As French plural.

## E. MISCELLANEOUS

be (7). *Ain't I, Aren't I.*
centenary. Words for the higher
  anniversaries (*tercentenary* etc.).

dry. Spelling (i or y) of derivatives
  of monosyllables in *-y*.
-fied. *Countrified* or *countryfied* etc.

**M.P.** Singular and plural possessive forms.

**mute e.** Retained or omitted in inflexions and derivatives of words so ending (*lik(e)able, mil(e)age,* etc.).

**-o-.** As a connecting vowel (*Anglo-Indian, speedometer,* etc.)

**one word or two.** Giving references to articles on the writing of e.g. *altogether, all together, anyrate, any rate, into, in to.*

**-our, -or-.** E.g. in *colo(u)rist, honourable.*

**possessive puzzles.** Of proper names ending s and other difficulties. Use of 's as a bare plural.

**singular s.** Vagaries of words ending s in the singular.

**-s-, -ss-, -sss.** The writing of e.g. *focus(s)ed, mis(-)spell, mistress-ship.*

**-ved, -ves.** Words ending f making v in inflexions.

**verbs in -ie, -y, and -ye.** Their inflexions.

**y and i.** Choice between in such words as *cipher, gypsy.*

**-z-, -zz-.** *Buz* or *buzz* etc.

## III. PRONUNCIATION

**-ade, -ado.**

**arch(e)(i)-.**

**-ciation.**

**didacticism.** Illustrated and deprecated.

**diphth-.**

**false quantity.** An expression to be banished from any discussion of English pronunciation.

**French words,** and other foreign words.

**Greek g.** Soft or hard?

**homophone.**

**-ies and -ein.**

**-ile.**

**-in and -ine.**

**-ise.**

**Italian sounds.**

**-ite.**

**Latin phrases.**

**-lived.**

**noun and adjective accent.** ⎫
**noun and verb accent.** ⎬ Differences in pronunciation indicating different parts of speech.
**participles** (5). ⎭

**pn-.**

**pronunciation.** (1) Some recent trends. (2) Silent t. (3) Silent h. (4) ă or ah in e.g. *pass* and ŏ or aw in e.g. *loss.* (5) ŏ or ŭ in e.g. *comrade.* (6) Long u. (7) er or ur in e.g. *demurring.* (8) al- followed by consonant. (9) -ough-. (10) Some proper names curiously pronounced.

**ps-.**

**pt-.**

**quad-.**

**quat-.**

**re-.**

**received pronunciation,** or 'standard English'.

**recessive accent.**

**retro-.**

**-th and -dh.** Plurals of words ending -th.

**u and non-u.**

**-valent.**

**wh.**

## IV. PUNCTUATION AND TYPOGRAPHY

**æ and œ.** Use of the ligatures.

**capitals.**

**diaeresis.**

**hyphens.** A general article containing also references to articles on particular points.

**italics.**

**misprints** to be guarded against.

**period in abbreviations.** *Mr.* or *Mr*? *Rev.* or *Rev*?

**stops.** Comma. Semicolon. Colon. Full stop. Exclamation mark. Question mark. Inverted commas. Parenthesis symbols.

# KEY TO PRONUNCIATION

## VOWELS

ā ē ī ō ū ōō (*mate, mete, mite, mote, mute, moot*)
ă ĕ ĭ ŏ ŭ ŏŏ (*rack, reck, rick, rock, ruck, rook*)
(The light vague *er* sound often given to short vowels in
unstressed syllables, and the *i* sound often given to unstressed
*e*, are not separately distinguished.)
ār ēr īr ōr ūr (*mare, mere, mire, more, mure*)
ar er or (*part, pert, port*)
ah aw oi oor ow owr (*bah, bawl, boil, boor, brow, bower*)

## CONSONANTS
### of which the value needs defining

ch (*child, each*: not as in *chaos, champagne, loch*)
dh (dhăt, mŭ′dher, = *that, mother*)
g (*gag, get*: not as in *gentle*)
j (jŭj = *judge*)
ng (*singer*: not as in *finger, ginger*)
ngg (fĭ′ngger = *finger*)
s (saws = *sauce*: not as in *laws*)
th (*thinketh*: not as in *this, smooth*)
zh (rōōzh, vĭ′zhn, = *rouge, vision*)
For h, r, w, in ah, ar &c., ow, owr, see Vowels

# ABBREVIATIONS, SYMBOLS, ETC.

a., adjective
aa., adjectives
adj., adjective
adv., adverb
advl, adverbial
APD, Authors' and Printers' Dictionary
arch., archaic
A.V., Authorized Version
c., century
cc., centuries
cf. ( = *confer*), compare
COD, Concise Oxford Dictionary
conj., conjunction
CUP, Cambridge University Press
DNB, Dictionary of National Biography
E, English
e.g. ( = *exempli gratia*), for instance
ellipt., elliptical
*Enc. Brit.*, Encyclopaedia Britannica
Evans, E's Dictionary of Contemporary American Usage
F, French
Gk, Greek
Gram., Grammar

i.e. ( = *id est*), that is
indic., indicative
ind. obj., indirect object
L, Latin
Lit., Literature
lit., literally
MS., manuscript
MSS., manuscripts
n., noun
NEB, New English Bible
nn., nouns
obj., object
OED, Oxford English Dictionary
OF, Old French
OID, Oxford Illustrated Dictionary
opp., as opposed to
OUP, Oxford University Press
part., participle present
pers., person
pl., plural
p.p., past or passive participle
pr., pronounce
pref., prefix
prep., preposition
pron., pronoun
refl., reflexive
rel., relative
R.V., Revised Version

sc. ( = *scilicet*), to wit
s.f. ( = *sub finem*), near the end
sing., singular
Skeat, S's Etymological Dictionary
SOED, Shorter Oxford English Dictionary
SPE, (Tracts of the) Society for Pure English
subj., subjunctive
s.v. ( = *sub voce*), under the (specified) word
TLS., Times Literary Supplement
U.K., United Kingdom
U.S., United States of America
usu., usually
v., vb, verb
var., variant
vol., volume
wd, word
Webster, W's New International Dictionary

/, placed between separate quotations
[ ], containing words that are not part of the quotation

Small capitals refer the reader to the article so indicated, for further information

**a, an.** 1. *A* is used before all consonants except silent h (*a history, an hour*); *an* was formerly usual before an unaccented syllable beginning with h and is still often seen and heard (*an historian, an hotel, an hysterical scene, an hereditary title, an habitual offender*). But now that the h in such words is pronounced the distinction has become anomalous and will no doubt disappear in time. Meantime speakers who like to say *an* should not try to have it both ways by aspirating the h. *A* is now usual also before vowel letters that in pronunciation are preceded by a consonantal sound (*a unit, a eulogy, a one*). Before letters standing for abbreviations or symbols the choice is usually determined by the sound of the letter, not of the word it represents, e.g. *an R.A., an M.P.*; but that is the sort of thing about which we ought to be allowed to do as we please, so long as we are consistent.

2. The combinations of *a* with *few* and *many* are a matter of arbitrary but established usage: *a few, a great many, a good many*, are idiomatic, but *a many* is now illiterate or facetious and *a good few* is colloquial; *a very few* is permissible (in the sense some-though-not-at-all-many, whereas *very few* means not-at-all-many-though-some), but *an extremely few* is not; see FEW.

3. *A, an,* follow instead of preceding the adjectives *many, such,* and *what* (*many an artist, such a task, what an infernal bore!*); they also follow (i) any adjective preceded by *as* or *how* (*I am as good a man as he; knew how great a labour he had undertaken*), (ii) usually any adjective preceded by *so* (*so resolute an attempt deserved success; a so resolute attempt* is also English, but suggests affectation), and (iii) often any adjective preceded by *too* (*too exact an,* or *a too exact, adherence to instructions*). The late position should not be adopted with other words than *as, how, so, too*; e.g. in *Which was quite sufficient an indication | Can anyone choose more glorious an exit? | Have before them far more brilliant a future|*, the normal order (*a quite* or *quite a sufficient, a more glorious, a far more brilliant*) is also the right one.

4. *A, an,* are sometimes ungrammatically inserted, especially after *no* adj., to do over again work that has already been done; so in *No more signal a defeat was ever inflicted* (*no* = not a; with this ungrammatical use cf. the merely ill-advised arrangement in *Sufferred no less signal a defeat*, where *no* is an adverb and *a* should precede it; see 3 above). Other examples of the mistake are: *The defendant was no other a person than Mr. Benjamin Disraeli* (*no other* = not another). | *Glimmerings of such a royally suggested even when not royally edited an institution are to be traced* (*even . . . edited* being parenthetic, we get *such a royally suggested an institution*).

**a-, an-,** not or without. Punctilious word-making requires that these should be prefixed only to Greek stems; of such compounds there are some hundreds, whereas Latin-stemmed words having any currency even in scientific use do not perhaps exceed half a dozen. There are the botanical *acapsular* and *acaulous*, the biological *asexual* and *acaudate*, and the literary *amoral*. This last being literary, there is the less excuse for its having been preferred to the more orthodox non-moral. *Amoral* is a novelty whose progress has been rapid. In 1888 the OED called it a nonce-word, but in 1933 full recognition had to be conceded. These words should not be treated as precedents for future word-making.

**abbreviations.** See CURTAILED WORDS.

**abdomen.** The orthodox British pronunciation is *ăbdō'men*, giving the *o* the same value as in the Latin word, though doctors, the chief users of the word, often say *ab'dŏmen*, which is standard in America.

**abetter, -or.** *-er* is the commoner general form, *-or* the invariable legal one.

**abide.** For *a*. in its current sense (*abide by* = keep) *abided* is usual, but in its archaic sense of remain or dwell it makes *abode* only.

**-able, -ible,** etc. 1. Normal use of *-able* as living suffix. 2. Choice between *-able* and *-ible* (or *-uble*). 3. Negative forms of adjectives in *-ble*. 4. *-ble* words of exceptional form or sense.

**1.** Normal use of -able as living suffix. The suffix *-able* is a living one, and may be appended to any transitive verb to make an adjective with the sense *able*, or *liable*, or *allowed*, or *worthy*, or *requiring*, or *bound, to be ——ed*. If the verb ends in mute *-e*, this is retained after soft c or g (*pronounceable, manageable*) and generally dropped after other consonants (*usable, forgivable*), but on this see MUTE E. Verbs ending in -y preceded by a consonant change y into i (*justifiable, triable*), but not when preceded by a vowel (*buyable, payable*). Verbs with the Latin-derived ending *-ate* that have established adjectives drop the *-ate* (*demonstrable, abominable, alienable, appreciable, calculable, expiable, execrable*, etc.); and new adjectives from such verbs should be similarly formed, but for possible exceptions see -ATABLE.

**2.** Choice between -able and -ible (or -uble). The *-ible* form is the natural one for words derived from Latin verbs ending *-ĕrĕ* or *-īrĕ*, making adjectives in *-ibilis* (*dirigible, audible*). Otherwise *-able* is the normal form and should be used unless there is a well-established *-ible* form for the word, or it belongs to a set that form their adjectives that way; for instance *perceivable* and *prescribable* should not be substituted for *perceptible* and *prescriptible* and the established *convertible* should be decisive for preferring *avertible* to *avertable*. On the other hand adjectives in *-able* may be formed even from those verbs whose established representatives end *-ible* when the established word has to some extent lost the verbal or contracted a special sense. Thus a mistake may be called *uncorrectable*, because *incorrigible* has become ethical in sense; *solvable* may be preferred because *soluble* has entered into an alliance with *dissolve*; a law must be described as *enforceable* to disclaim any relationship between that passive-sense adjective and the active-sense *forcible*; and *destroyable by dynamite* may seem less pedantic than *destructible by* because *destructible* tends to be purely adjectival. The existence of a single established *-ible* word of a more or less technical kind need not be allowed much weight; e.g. *fusible* does not suffice to condemn *confusable, diffusable*, and *refusable*.

**3.** Negative forms of adjectives in -ble. The adjectives in *-ble* being required with especial frequency in negative contexts, the question often arises whether the negative form of any particular word should be made with *in-* or *un-*. The general principle is (*a*) that negatives from *-ble* words other than those in *-able* have *in-* (or *ig-, il-, im-, ir-*); the only exceptions are words already beginning with the prefix *im-* or *in-* (*impressible, intelligible*), and (*b*) negatives from words in *-able* ordinarily have *un-*, but there are numerous exceptions with *in-* (e.g. *improbable, inestimable*). These latter have a tendency, no doubt due to the greater familiarity of *un-*, to develop an alternative negative form with that prefix (e.g. *approachable, surmountable*). See IN- and UN-.

**4.** -ble words of exceptional form or sense. The normal formation and sense of adjectives in *-able* have been explained in paragraph 1; and adjectives in *-ible* have the same ordinary range of sense. There are, however, large numbers of words, and certain usages, that do not conform to this simple type, and to some of them (*a reliable man*, *perishable articles*, *dutiable goods*, *feedable pasture*, an *unplayable wicket*, an *actionable offence*, *payable ore*, *unwritable paper*, and others) exception is often taken. The advocatus diaboli who opposes their recognition has the advantage of an instantly plausible case that can be put clearly and concisely: we do not rely a man, nor perish articles, nor play a wicket; therefore we have no right to call a man unreliable, and so with the rest. An answer on the same pattern would be that neither do we dispense a man, yet our right to call him indispensable is not questioned. But it is better to go on broader lines, sacrificing the appearance of precision and cogency, and point out that the termination -*ble* has too wide a range in regard both to formation and to sense and the analogies offered by the -*ble* words are too various and debatable to allow of the application of cut-and-dried rules. The words and usages to which exception is taken should be tested not by the original Latin practice, nor by the subsequent French practice, nor by the English practice of any particular past period, even if any of these were as precise as is sometimes supposed, but by what inquiry may reveal as the now current conception of how words in -*ble* are to be formed and what they may mean. In determining that conception we cannot help allowing the incriminated words themselves to count for something. It may seem unfair that *reliable* should itself have a voice in deciding its own fate; but it is no more unfair than that possession should be nine points of the law. The existence of the still more modern *payable ore*, *playable wicket*, *unwritable paper*, has in the same way its value as evidence; the witness-box is open to the prisoner. Apart, however, from this special proof that the current conception of -*ble* is elastic, it is easy to show that at the present stage of its long history and varied development it could not be rigid. In the first place the original formation and meaning of many common words containing

it are obscured by the non-existence in English of verbs to which they can be neatly referred (*affable*, *amenable*, *delectable*, *feasible*, *plausible*, and many others). Secondly, there are many common words in which the sense of -*ble* either is (as sometimes in Latin), or (which is as much to the point) seems to be, not passive but active (*agreeable*, *capable*, *comfortable*, *hospitable*, *viable*, etc.). Thirdly, -*ble* is often appended, or (which is as much to the point) seems to be appended, to nouns instead of to verbs (*actionable*, *companionable*, *fashionable*, *seasonable*, *unexceptionable*, etc.). To take a single example in detail, no one but a competent philologist can tell whether *reasonable* comes from the verb or the noun *reason*, nor whether its original sense was that can be reasoned out, or that can reason, or that can be reasoned with, or that has reason, or that listens to reason, or that is consistent with reason. The ordinary man knows only that it can now mean any of these, and justifiably bases on these and similar facts a generous view of the termination's capabilities; *credible* meaning for him worthy of credence, why should not *reliable* and *dependable* mean worthy of reliance and dependence? *Durable* meaning likely to endure, why should not *payable* and *perishable* mean likely to pay and perish?

In conclusion, a small selection follows of words in -*ble*, other than those already mentioned, that illustrate the looser uses of the termination; the paraphrases are offered merely by way of accommodating each word to what is taken to be the current conception of -*ble*: *accountable*, liable to account; *answerable*, bound to answer; *appealable*, subject to appeal; *available*, that may avail; *bailable*, admitting of bail; *chargeable*, involving charge; *clubbable*, fit for a club; *conformable*, that conforms; *conversable*, fit for conversing; *demurrable*, open to demur; *jeepable*, capable of being traversed by a jeep; *impressionable*, open to impressions; *indispensable*, not admitting of dispensation; *knowledgeable*, having or capable of knowledge; *laughable*, providing a laugh; *marriageable*, fit for marriage; *merchantable*, fit for the merchant; *objectionable*, open to objection; *operable*, capable of being operated on; *peaceable*, inclined to peace; *personable*, having person or presence; *pleasurable*, affording pleasure; *practicable*, adapted for practice; *profitable*, affording profit; *proportionable*, showing proportion; *revertible*, liable to reversion; *risible*, adapted for provoking laughter; *sizable*, having size; *skatable*, fit for skating; *unconscionable*, not according to conscience.

**ablutions** seems to be emerging from the class of PEDANTIC HUMOUR, which is its only fitting place outside religious ceremonial, to claim serious recognition as a FORMAL WORD. This should not be conceded. Though we have prudishly created unnecessary difficulty for ourselves by denying to the word *lavatory* its proper meaning, we still have *wash-place* and do not need monstrosities like *a. facilities*, *a. cubicles*, and *mobile a. centres*.

**abolishment, abolition.** See -ION AND -MENT.

**aborigines.** The word being still usually pronounced with a consciousness that it is Latin (i.e. with -*ēz*), the sing. *aborigine* (-*nĕ*) is felt to be anomalous and avoided or disliked; the adj. *aboriginal* used as a noun is the best singular.

**above.** *The passage quoted a.*; *the a. quotation*; *the a. is a quotation*. There is ample authority, going back several centuries, for this use of *a.* as adverb, adjective, or noun, and no solid ground for the pedantic criticism of it sometimes heard.

**abridg(e)ment.** For spelling see JUDGEMENT.

**absence.** For *conspicuous by a.* see CONSPICUOUS.

**absolute construction.** Defined by the OED as 'standing out of the usual grammatical relation or syntactical construction with other words', it consists in English of a noun or pronoun that is not the subject or object of

any verb or the object of any preposition but is attached to a participle or an infinitive, e.g. *The play being over*, we went home. / Let us toss for it, *loser to pay*.

1. The insertion of a comma between noun and participle in the absolute use is indisputably wrong. It arises from the writer's or the compositor's taking the noun, because it happens to stand first, for the subject of the main verb; and it puts the reader to the trouble of readjusting his notion of the sentence's structure. *The King having read his speech from the throne, their Majesties retired* is the right form; but newspaper writing or printing is so faulty on the point that it would be likely to appear as *The King, having read his* etc. Thus: *By mid-afternoon Lock, having taken seven wickets for 47, it was all over.* / *The House of Commons, having once decided against the capital penalties, it was declared impossible that there could be another execution for forgery.* The temptation to put a comma in this position is so strong that one may be found even in the rubric of a ceremonial service, presumably prepared with scrupulous care: *Bath King of Arms, having bowed first to those Knights Grand Cross who have been installed previously and then to those who are not to be installed, they thereupon sit in the seats assigned to them.*

2. The case in this construction is the subjective; e.g. *There being no clear evidence against him, and he* (not *him*) *denying the charge, we could do nothing.* There is little danger of the rule's being broken except where a pronoun stands as a complement. Though no one would write *me being the person responsible*, the form *the person responsible being I* is likely to be shrunk from; *me* should not be used except colloquially; *myself* is usually possible, but not always. The formula *whom failing* (= or in default of him) should be either *who failing* or *failing whom*; the justification of *failing whom* is that *failing* has, like *during* etc., passed into a preposition, and *whom failing* is a confusion between the two right forms.

3. The construction may be elliptical,

with the participle omitted: *He a scholar, it is surprising to find such a blunder.* But it cannot be used without a noun or pronoun: *he* should be inserted before the participles in *It was his second success of the day, having won the Royal Winter Fair Trophy earlier.* / *The formal garden was conceived by the sixth earl, but, dying in 1844, it was left to his son to complete it.* See UNATTACHED PARTICIPLES.

4. The following example of one absolute construction enclosed in another is a pretty puzzle for those who like such things: *To the new Greek Note Bulgaria replied by a Note which was returned to the Bulgarian Foreign Minister,* Greece, it being declared, not wishing *to enter into any bargaining.* It is clear enough that this will not do; it must be changed into (*a*) *it being declared that Greece did not wish,* or (*b*) *Greece not wishing, it was declared, to* . . .; but why will it not do? Because the absolute construction 'it being declared' cannot, like the 'it was declared' of (*b*), be parenthetic, but must be in adverbial relation to the sentence. Knowing that, we ask what 'it' is, and find that it can only be an anticipatory *it* (see IT) equivalent to 'that Greece did not wish'; but the consequent expansion 'Greece, that Greece did not wish being declared, not wishing' makes nonsense.

**absolute possessives.** Under this term are included the words *hers, ours, theirs,* and *yours,* and (except in their attributive-adjective use) *his, mine,* and *thine.* The ordinary uses of these need not be set forth here though it is perhaps worth remarking that the double possessive of such constructions as *a friend of mine, that facetiousness of his,* is established idiom. See OF 7. But a mistake is often made when two or more possessives are to be referred to a single noun that follows the last of them: the absolute word in -s or -ne is wrongly used in the earlier place or places instead of the simple possessive. The correct forms are: *your and our and his efforts* (not *yours and ours*); *either my or your informant*

*must have lied* (not *mine*); *her and his mutual dislike* (not *hers*); *our without your help will not avail* (not *ours*). There is no doubt a natural temptation to substitute the wrong word; the simple possessive seems to pine at separation from its property. The true remedy is a change of order: *your efforts and ours and his; my informant or yours; our help without yours.* It is not always available, however; *her and his mutual dislike* must be left as it is.

**abstractitis.** The effect of this disease, now endemic on both sides of the Atlantic, is to make the patient write such sentences as *Participation by the men in the control of the industry is non-existent* instead of *The men have no part in the control of the industry; Early expectation of a vacancy is indicated by the firm* instead of *The firm say they expect to have a vacancy soon; The availability of this material is diminishing* instead of *This material is getting scarcer; A cessation of dredging has taken place* instead of *Dredging has stopped; Was this the realization of an anticipated liability?* instead of *Did you expect you would have to do this?* And so on, with an abstract word always in command as the subject of the sentence. Persons and what they do, things and what is done to them, are put in the background, and we can only peer at them through a glass darkly. It may no doubt be said that in these examples the meaning is clear enough; but the danger is that, once the disease gets a hold, it sets up a chain reaction. A writer uses abstract words because his thoughts are cloudy; the habit of using them clouds his thoughts still further; he may end by concealing his meaning not only from his readers but also from himself, and writing such sentences as *The actualization of the motivation of the forces must to a great extent be a matter of personal angularity.*

The two quotations that follow are instructive examples of the difficulties that readers may find in following the meaning of writers suffering from this disease. The first is English and its subject is the way in which business men arrive at decisions; the second is American and its subject is the testing of foods specially designed for use in certain types of military aircraft, or possibly in space-ships.

1. *Whereas the micro-economic neoclassical theory of distribution was based on a postulate of rationality suited to their static analysis and institutional assumptions, we are no longer justified in accepting this basis and are set the problem of discovering the value premises suited to the expectational analysis and the institutional nature of modern business. The neo-classical postulate of rationality and the concept of the entrepreneur as the profit maximizing individual, should, I think, be replaced by a sociological analysis of the goals of the firm in relation to its nature as an organization within the socio-political system.*

2. *Strangeness of samples has been shown to lead to relative rejection of products in the comparative absence of clues to a frame of reference within which judgement may take place. Variation in clues selected by judges as a basis for evaluation lead to greater inter-judge disagreement. Addition of a functional (utilitarian) basis for judgement tends to reduce relative importance of product physical characteristics as a basis for judgement. In the absence of any judgemental frame of reference reduction in the number of product physical attributes apparent to the judge appears to reduce operation of bases for rejection and increase homogeneity of judgement between subjects; inter-sample discrimination is also reduced.* See also PERIPHRASIS, MEMBERSHIP, TAUTOLOGY, and -TION WORDS.

**abysmal, abyssal.** The first is the word for general use (*abysmal ignorance, degradation, bathos*); *abyssal*, formerly used in the same way, has now been appropriated as a technical term meaning of the bottom of the ocean or of a depth greater than 300 fathoms.

**Academe** properly means *Academus* (a Greek hero); and its use as a poetic variant for *academy*, though sanctioned

by Shakespeare, Tennyson, and Lowell, is a mistake; but *the grove of A.* (Milton) means rightly The Academy.

**Academy.** *The A., the Garden, the Lyceum, the Porch, the Tub,* are names used for five chief schools of Greek philosophy, their founders, adherents, and doctrines: *the A.,* Plato, the Platonists, and Platonism; *the Garden,* Epicurus, the Epicureans, and Epicureanism; *the Lyceum,* Aristotle, the Aristotelians, and Aristotelianism; *the Porch,* Zeno, the Stoics, and Stoicism; *the Tub,* Antisthenes, the Cynics, and Cynicism.

**accent(uate).** In figurative senses (draw attention to, emphasize, make conspicuous, etc.) the long form is now much the commoner; in literal senses (sound or write with an accent), though either will pass, the short prevails; and the differentiation is worth encouraging.

**acceptance, acceptation.** The words, once used indifferently in several senses, are now fully differentiated. *Acceptation* means only the interpretation put on something (*the word in its proper acceptation means love; the various acceptations of the doctrine of the Trinity*), while *acceptance* does the ordinary work of a verbal noun for *accept* (*find acceptance,* be well received; *beg* or *ask* one's *acceptance of,* ask him to accept; cf. *ask his acceptation of,* ask how he understands; *cards of acceptance,* accepting an invitation; *acceptance of persons,* favourable regard; *acceptance of a bill,* drawee's accepting of responsibility; *endorses my acceptance of the terms,* agrees with me in accepting them; cf. *endorses my acceptation of them,* agrees with my view of their drift).

**accept of.** In all senses of *accept* other than that of accepting a bill of exchange etc. *accept of* was formerly almost as widely used as the simple verb; this was still so when letter A of the OED was published in 1888. It has since fallen into disuse and is becoming an ARCHAISM, though it has lingered long enough for the COD

(1964) to record it as still permissible 'with a slight suggestion of formality or condescension'.

**access, accession.** There are probably, in modern usage, no contexts in which one of these can be substituted for the other without the meaning's being modified. But the wrong one is sometimes carelessly or ignorantly chosen. With regard to arriving, *accession* means getting there, *access* opportunity of getting there; accordingly *accession to the throne* means becoming sovereign, *access to the throne* opportunity of approaching the sovereign. We can say *His access to fortune was barred*, or *His accession to fortune had not yet taken place*, but not the converse. The idea of increase, often present in *accession*, is foreign to *access*; *an access of fury, fever, joy, despair*, etc., is a fit or sudden attack of it, which may occur whatever the previous state of mind may have been, whereas *an accession of* any of them can only mean a heightened degree of the state that already existed; *our forces have had no accession*, have not been augmented in numbers, *have had no access*, have not been able to enter.

**accessary, accessory.** The words, though they have separate histories, are often confused. The following distinction was favoured by the OED (1888). *Accessary* involves the notion of complicity or intentional aid or consent, and is accordingly used only where that notion is applicable, i.e. chiefly (as a noun) of persons and (as an adjective) of persons or their actions (*he was an accessary, if not the principal*; *the accessaries also were punished*; *this course has made us accessary to the crime*; *was guilty of accessary action*). *Accessory* has no such implication of consent, and, though it includes the notion of contributing to a result, emphasizes especially the subordinate nature of the contribution; it is applied chiefly to things (*the accessory details of the picture*; *that is only an accessory*, an unessential feature; *the accessories*, the not indispensable accompaniments).

Unfortunately this useful differentiation has been blurred by the encroachments of *accessory* on the province of *accessary*. *Accessory before (or after) the fact* is now the more usual spelling.

**accidence.** See GRAMMAR.

**acclimatize, -imate, -imatization, -imatation, -imation.** *Acclimatize, acclimatization*, are the forms for which general usage has decided in Britain, though in U.S. the shorter form is sometimes used for the verb. Some British writers wish to retain the others with reference to the process when brought about by natural as opposed to human agency; but it is doubtful whether the words are in common enough use for the differentiation to gain currency; and, failing differentiation, it is better that the by-forms should perish.

**accommodation** has long been a FORMAL WORD for rooms in a hotel etc. It has recently been pressed into service to meet the inconvenience of our having no single word to cover *house, flat*, and *lodgings*, and is worked hard in that capacity by housing authorities. *Accommodation unit* seems to have been killed by the ridicule that greeted its first appearance, but the cliché *alternative accommodation*, meaning somewhere else to live, remains as an unhappy legacy of the general post that marked the early days of the second world war. See also ALTERNATIVE.

**accompan(y)ist.** See -IST.

**accomplice, accomplish.** The OED gives the pronunciation with -ŏm-, not -ŭm-, as the established one for both words, though 'the historical pronunciation' of *accomplish* was with -ŭm-. This ruling is still followed by the dictionaries and, on the whole, in usage, though -ŭm- is sometimes heard. See PRONUNCIATION 5.

**accord, account.** The phrases are *of* one's *own accord, on* one's *own account*; *of* one's *own account* is a confusion. See CAST-IRON IDIOM.

**according as.** There is a tendency to repeat the phrase (like BETWEEN), with a mistaken idea of making the construction clearer, in contexts where the repetition is not merely needless, but wrong. For instance, the second *according as it* should be omitted in *The big production will be harmful or the reverse, according as it can command the Government to insure it a monopoly in all circumstances, or according as it works with the knowledge that, if it abuses its trust, the door is freely open to the competing products of other countries.*

The error is at once apparent if the clause (for it is in fact a single clause) is reduced to its simplest expression—(will be harmful or the reverse) according as it is irresponsible or responsible; no one would write *or according as it is responsible*; the temptation comes in long sentences only, and must be resisted. *Or according as* is legitimate only when what is to be introduced is not, as in the quotation, the necessarily implied alternative or the other extreme of the same scale, but another scale or pair of alternatives. Man attains happiness or not *according as he deserves it or not* (right), *according as he deserves it or does not deserve it* (right), *according as he deserves it or according as he does not deserve it* (wrong), *according as he deserves it or according as he can digest his food* (right).

**account.** Unlike *regard*, and like *consider*, this verb does not in good modern usage admit of *as* before its complement; *I account it a piece of good fortune*; *you are accounted wise* or *a wise man*.

**accumulative.** The word, formerly common in various senses, has now given place to *cumulative* in most of them, retaining in ordinary use only the sense given to accumulating property, acquisitive.

**ace.** See TOP, ACE, CRACK.

**achieve** implies successful effort. Its use in *on achieving the age of 21* is unsuitable and in *officers achieving redundancy* is absurd.

**acid test.** See POPULARIZED TECHNICALITIES and HACKNEYED PHRASES.

**acknowledge(ment).** For -dg(e)-ment see JUDGEMENT.

**acoustic.** Pronunciation varies between -ow- and -oo-; the latter is perhaps commoner, and is preferred by the OED. In its favour is the adoption from French, the sound of Greek *ov* in the more recent English pronunciation of Greek, and the general impression that the value of *ou* in outlandish words is oo; in favour of -ow- is the older English pronunciation of Greek, and the preponderating value of *ou* in English. *Acoo'stic* is recommended.

**acquaintanceship** is a NEEDLESS VARIANT for *acquaintance*.

**acronym.** See CURTAILED WORDS.

**act** vb. In the sense *behave like*, the word, once used as freely as *play* (*act the lover*, *act the child*), has fallen into disuse. Even *play* in this sense is now rarely used apart from certain phrases (e.g. *play the fool*; *play the man*); *act like a* is the usual expression.

**act, action.** The distinction between the two words is not always clear. The natural idea that *act* should mean the thing done, and *action* the doing of it, is not even historically quite true, since *act* represents the Latin noun *actus* (which is very close to *actio* in sense) as well as the Latin participle *actum*; but, even if not true, it has influence enough to prevent *act* from being commonly used in the more abstract senses. We can speak only of the *action*, not the *act*, of a machine, when we mean the way it acts; and *action* alone has the collective sense, as in *his action throughout* (i.e. his acts or actions as a whole) *was correct*. There are also other senses in which there is obviously no choice open. In contexts that do admit of doubt, it may be said generally that *action* tends to displace *act*. If we were making the phrases for the first time now, we should probably prefer *action* in *Through God will we do great acts*, *The Acts of the Apostles*, *By the* *act of God, Be great in act as you have been in thought, I deliver this as my act and deed.* This tendency, however, is by no means always effective; it is immaterial, for instance, whether we say *we are judged by our acts* or *by our actions*; there is no appreciable difference between *it was an act*, and *it was an action, that he was to regret bitterly.* And in certain contexts *act* more than holds its ground: (1) in the sense deed of the nature of; *it would be an act* (never action) *of folly, cruelty, madness, kindness, mercy*, etc.; similarly in the sense deed characteristic *of*; *it was the act* (rarely *action) of a fool* (cf. *the actions of a fool cannot be foreseen*, where the sense is not *characteristic deed*, but simply *deed*). On the other hand, when for *of folly* or *of a fool* etc. *foolish* etc. is substituted, *action* is at least as common as *act—a cruel, kind, foolish, noble, base, action* or *act*. (2) In the sense instant of doing: *caught in the act, was in the very act of jumping.* (3) In antithesis to *word, thought, plan*, etc., when these mean every word, each thought, a particular plan, rather than speech, thinking, planning: *faithful in word and act* (but *in speech and action); innocent in thought and act* (but *supreme in thought and action); the act was mine, the plan yours* (but *a strategy convincing in plan, but disappointing in action*).

**activate, actuate.** *Activate* was marked *obs.* in the original OED, but has since been recalled to life as a technical term of chemistry and physics, used especially of promoting the growth of bacteria in sewage and of making substances radioactive. It should not be allowed to become a POPULARIZED TECHNICALITY and displace *actuate* (= to set a machine in motion or to prompt a person to action). *He was activated by the best possible intentions* will not do.

**actuality.** See LITERARY CRITICS' WORDS.

**actually.** See MEANINGLESS WORDS.

**acuity, acuteness.** See -TY AND -NESS.

**adapt(at)ion.** The OED gives examples of *adaption* from Swift and Dickens, but the longer form alone is now in general use. For *adapt(er)(or)* see -OR.

**ad captandum** 'for catching (the common herd', *vulgus*). Applied to unsound specious arguments. *An a. c. presentation of the facts.*

**addle, addled.** The adjectival use of *addle* as in *an addle egg, his brain is addle*, is correct, and was formerly common; but to prefer it now to the usual *addled* is a DIDACTICISM. It still prevails, however, in compounds, as *addle-pated, addle-brained.*

**-ade, -ado.** Pronunciation. Most of the -ade words have anglicized their ending into *-ād—arcade, brocade, cascade, cavalcade, esplanade, fusillade, serenade*, etc. A few retain *-ahd* as their only pronunciation, e.g. *aubade, ballade, charade, façade, glissade. Promenade* shows a curious reluctance to follow the lead of *esplanade. Promenahd* is still usual, but as long ago as 1933 the SOED recognized *-ād* as an alternative. *Accolade* seems to have crossed the boundary but not yet settled down on the other side; the COD gives *-ād* first with *-ahd* as an alternative; with *pomade* it is the other way about. The *-ado* words have been having similar experiences. *Barricado, gambado*, and *tornado* are now *-ādo* only; for *bravado* the COD still gives *-ahdo* only, and prefers that pronunciation for *desperado*. The more exotic words such as *amontillado, avocado*, INCOMMUNICADO, and *Mikado* remain *-ahdo* only. For *-ada* words see ARMADA and CICADA.

**adequate.** For unidiomatic use see INADEQUATE.

**adhere, adhesion.** The established phrase *give* one's *adhesion to* a policy, party, leader, etc., means to declare one's acceptance of, and describes a single non-continuous act. *Adhere to* is narrower; it is not used, by good writers at least, in the corresponding sense *accept* or *declare acceptance of,* but only in that of remaining constant to.

**adjacent.** *A very good maiden over from Benaud contained a loud shout for a catch behind the wicket. This one certainly turned, and May was certainly very adjacent.* *Adjacent,* says the OED, means 'not necessarily touching, though this is by no means excluded'. We cannot therefore accuse this reporter of using the word incorrectly, whatever we may think of the playfulness that prompted him to prefer it to the monosyllable *near* or *close*.

**adjectivally, adjectively,** etc. *Adjectivally and substantivally* are preferable to *adjectively* and *substantively*. First, the words *adjective* and (in the grammatical sense) *substantive* are now regarded as nouns. So far as they are still used as adjectives, they are felt to be nouns used attributively; adverbs formed directly from them therefore cause uneasiness. Secondly the adjectives *adjectival* and *substantival* are of such frequent occurrence in modern grammar that it is natural to form the adverbs from them, especially since the former has an even wider currency as a polite substitute for some more expressive but less printable word (*He threatened to knock my adjectival block off* ), cf. EPITHET. Thirdly adverbs from the other part-of-speech names correspond to *adjectivally,* not to *adjectively—adverbially, pronominally, verbally,* etc., not *adverbly* etc.

**adjectival nouns.** See NOUN ADJECTIVES.

**adjectives misused.** 'An adjective', says the OED, 'is a word standing for the name of an attribute which being added to the name of a thing describes the thing more fully and definitely, as a *black coat.*' Adjectives, then, ought to be good friends of the noun. In fact, as has been well said, they have become its enemies. They are often used not to 'describe the thing more fully and definitely' but rather to give it some vague and needless intensification or limitation; as if their users thought that the noun by itself was either not impressive enough or too

stark, or perhaps even that it was a pity to be content with one word where they might have two. *The operation needs considerable skill and should be performed with proper care.* / *Effective means of stopping the spread of infection are under active consideration and there is no cause for undue alarm.* The adjective-noun pairs in these sentences are typical of the worser kind of present-day writing, especially business and official. It is clear that *considerable, proper, effective,* and *active* are otiose and *undue* is absurd; their only effect is to undermine the authority of the nouns they are attached to.

*It is my hope that this year concrete and positive steps will be taken to achieve progress towards the union of Africa.* The speaker may perhaps be pardoned for feeling that *steps* needed reinforcing by an adjective; a step may be short or tottery, though it is true that steps of that kind are not likely to 'achieve progress'. He might reasonably have said *decisive* or *definitive.* He saved himself the trouble of thinking of a suitable adjective by putting in a couple of clichés. One may perhaps walk up concrete steps but one cannot 'take' them, and any step must be positive unless indeed it is a step backwards; the speaker cannot have thought it necessary to warn his hearers against thinking that that was what he meant.

The habit of propping up all nouns with adjectives is seen at its worst in those pairs in which the adjective is tautological, adding nothing to the meaning of the noun; such are *grateful thanks, true facts, usual habits, consequent results, definite decisions, unexpected surprise,* and scores of others commonly current. Constant association with an intensifying adjective deprives a noun of the power of standing on its own legs. Thus *danger* must always have its *real, part* its *integral,* and *crisis* its *grave* or *acute,* and *understatements* must be *masterly.* The only hope for a noun thus debilitated is for the combination to be recognized as a cliché and killed by ridicule; there are signs for instance that in this way *test* is ridding itself of *acid* and *moment* of *psychological*. See HACKNEYED PHRASES.

It is convenient, though sometimes confusing, that adjectives when used attributively may denote relationship, not quality; a *male nurse* is a nurse who is a male, but a *sick nurse* is not a nurse who is sick; nor did the old phrase a *mad doctor* mean a doctor who was mad. But this free-and-easy property of adjectives is no excuse for failing to choose the most fitting one for use in the ordinary way as a qualifier; for instance the weather may be *hot* or *cold* and commodities may be *dear* or *cheap*, but temperatures and prices are more suitably described as *high* or *low*.

**adjust.** *It is argued that . . . this enables the prostitute and her client to adjust to society.* This 'elliptical intransitive' use of *a.* is said by the OED to be obsolete, and no later example is given of it than 1733. Modern idiom required the reflexive pronoun to be expressed—*to adjust themselves to society*—until the old construction was revived as a term of psychology.

**administratrix.** For pl. see -TRIX.

**admission, -ittance, -issible, -ittable.** Of the nouns, *admission* is used in all senses (*No admittance except on business* is perhaps the only phrase in which the substitution of *admission* would be noticed), while *admittance* is confined to the primary sense of *letting in*, and even in that sense tends to disappear. *You have to pay for admission* is now commoner than *for admittance*, and so with *What is needed is the admission of outside air*; *admission 2s.6d.* is now the regular form; on the other hand *Such an admittance* (instead of *admission*) *would give away the case* is now impossible.

The difference between the adjectives is that *admissible* is the established word, and *admittable*, though formerly current, is now regarded as merely made for the occasion, and used only when the connexion with *admit* is to be clear; this is chiefly in the predicate, as *Defeat is admittable by anyone without dishonour*.

**admit. 1.** *Admit of,* formerly used for *admit* in several senses, is now restricted to the sense *present an opening* or *leave room for*, and to impersonal nouns usually of an abstract kind as subject: *His veracity admits of no question* (but not *I can admit of no question*); *A hypothesis admits by its nature of being disputed* (but not *he admits of being argued with*); *A jet air-liner does not admit of careless handling*.

**2.** *Admit to. Grey then admitted to his financial manipulations.* One may either *confess* one's misdeeds or *confess to* them, but if *admit* is used idiom will not tolerate *to*. See CAST-IRON IDIOM.

**adopted, adoptive.** The anomalous use of *adopted* with *parents, father, mother,* etc., is to a certain extent excused by such allowed attributive uses as *the condemned cell*; that is the cell of the condemned, and the adopted father is the father of the adopted. Similarly *divorced* is applied equally to the successful petitioner and the unsuccessful respondent. But while *condemned* and *divorced* save a clumsy periphrasis, *adopted* saves only the trouble of remembering *adoptive*.

**adumbrate.** See FORMAL WORDS.

**advance(ment).** There are no contexts in which *advancement* can be substituted for *advance* without damage to or change in the sense; in the following sentence *advance* should have been written: *It will not be by the setting of class against class that advancement will be made.* It is true that both words can be used as verbal nouns of *to advance*; but *advance* represents its intransitive and *advancement* its transitive sense; *the advance of knowledge* is the way knowledge is advancing, whereas *the advancement of knowledge* is action taken to advance knowledge. Apart from this verbal-noun use with *of* following, and from a technical sense in law, *advancement* has only the sense of preferment or promotion, never the more general one of progress.

**adventurous, venturesome, adventuresome, venturous.** Usage has decisively declared for the first two and against the last two. *Adventuresome* and *venturous*, when used, are due to either ignorance or avoidance of the normal.

**adverse.** Unlike *averse*, this can be followed only by *to*; *Politicians who had been very adverse from the Suez-Canal scheme* is wrong.

**advert.** See ARCHAISM.

**advertise.** Not *-ize*; see -ISE, -IZE.

**advocate.** Unlike *recommend, propose, urge*, and other verbs, this is not idiomatically followed by a *that*-clause, but only by an ordinary or a verbal noun. In *Dr. Felix Adler advocates that close attention shall be paid to any experiments*, either *urges* should be substituted for *advocates*, or *that* and *shall be paid* should be omitted or give place to *the paying of*.

**-ae, -as,** in plurals of nouns in -a. Most English nouns in -a are from Latin (or latinized Greek) nominative feminine singular nouns, which have in Latin the plural ending *-ae*. But not all; e.g. *sofa* is from Arabic; *stanza* and *vista* are from Italian; *subpoena* is not nominative; *drama* and *comma* are neuter; *data, strata, stamina,* and *prolegomena* are plural; and with all such words -ae is impossible. Of the majority, again, some retain the Latin -ae in English either as the only or as an alternative plural ending (*formulae* or *-las, lacunae* or *-nas*), and some have always *-as* (*ideas, areas, villas*). The use of plurals in -ae therefore presents some difficulty to non-latinists. For most words with which -ae is possible or desirable the information is given in their dictionary places; for the principle of choice when both -ae and -as are current see LATIN PLURALS 1, 3.

**æ, œ.** These ligatures (see DIGRAPH), of which the pronunciation is identical (ē), are also in some founts of type so much alike that compositors often use one for the other and unlearned readers have their difficulties with spelling increased. It seems desirable that in the first place all words in common enough use to have begun to waver between the double letter and the simple e (as *pedagogy* now rarely *pae-* or *pæ-*, *medieval* still often *-aeval*, *ecumenical* still usually *oe-* or *œ-*, *penology* now rarely *poe-* or *pœ-*) should be written with the e alone, as *phenomenon* now is; and secondly, in words that have not yet reached or can for special reasons never reach the stage in which the simple e is acceptable, ae and oe should be preferred to æ and œ (*Caesar, gynaecology, paediatrics, homoeopathy, diarrhoea, archaeology, Boeotian, Oedipus*; the plurals and genitives of Latin first-declension nouns, as *sequelae, Heraclidae, aqua vitae*). This is in fact the present tendency of printers. In French words like *chef-d'œuvre* the ligature œ must obviously be kept; whether it is kept or not in *manoeuvre*, where the pronunciation is anomalous, is of no importance.

**aeon, æon, eon.** The first form is recommended; see Æ, Œ.

**aerate, aerial** are no longer written with a DIAERESIS, and now that the common pronunciation of the new noun *aerial* is indistinguishable from *Ariel*—slovenly perhaps but curiously appropriate—the old adjective can hardly fail to conform; and with *aeroplane* (and other *aero-* compounds) pronounced as though they began in the same way as *aircraft*, we shall probably soon give up all attempt to pronounce *aer* in any of its compounds in the disyllabic way we suppose the Greeks to have pronounced it.

**aero-, air-.** The two *aero*-compounds still in popular use—*aeroplane* and *aerodrome*—are unlikely to maintain themselves much longer against pressure from America, where *air-* has always been the favoured prefix. *Aerodrome* is already giving way to *airport, airfield,* and *airstrip*; and *aircraft* (formerly collective but now often used for a single machine) and even *airplane* are increasingly used.

**aery, aerie, eyry, eyrie.** The victory of the last form over the other three seems to have been undeserved. According to Skeat and the OED, it and *eyry* are due to a theory of the derivation (from *ey*, M.E. for *egg*; *eyry* = *eggery*) that is known (though the ultimate origin of *aery* is doubtful) to be wrong. Of the alternative pronunciations recognized by the dictionaries ($\bar{a}'r\bar{i}$, $\bar{e}'r\bar{i}$, and $\bar{i}'r\bar{i}$) the first is preferred.

**aesthet(e)(ic).** The adjective, which means etymologically *concerned with sensuous perception*, was introduced into English to supply *sense of beauty*, with an adjective, and used in such contexts as *a. principles*, *from an a. point of view*, *an a. revival occurred*, *a. considerations do not appeal to him*. By a later extension it was given the meaning *professing or gifted with this sense* (*I am not a.*; *a. people*), thus providing an adjective for the noun *aesthete*. This was a much later introduction; the OED's first quotation is 1881 and it is significant that its first definition, beginning 'One who professes a special appreciation of what is beautiful', was changed by the SOED some 50 years later to 'One who professes a superior appreciation of what is beautiful'. The word is less used now than it was at the end of the 19th c., but the opposite of an *aesthete*, according to the COD, is still a *hearty* in English university use. The adjective is less in place when given the meaning *dictated by* or *approved by* or *evidencing this sense* (*a very a. combination*; *aesthetically dressed*; *a. chintzes and wallpapers*; *flowers on a table are not so a. a decoration as a well-filled bookcase*); and still less so when it is little more than a stilted substitute for *beautiful* (*that green is so a.*; *a not very a. little town*).

**affect, effect.** These verbs are not synonyms requiring differentiation, but words of totally different meaning, neither of which can ever be substituted for the other. *Affect* (apart from other senses in which it is not liable to confusion with *effect*) means have an influence on, produce an effect on, concern, effect a change in: *effect* means bring about, cause, produce, result in, have as result. *These measures chiefly a. the great landowners. It does not a. me. It may seriously a.* (i.e. injure) *his health. A single glass of brandy may a.* (alter for better or worse the prospects of) *his recovery. A single glass of brandy may e.* (bring about) *his recovery. This will not a.* (change) *his purpose. This will not e.* (secure) *his purpose.*

**affinity.** The prepositions normally used after this are, according to context, *between* and *with*. When the sense is less *relationship* or *likeness* than *attraction* or *liking*, *to* or *for* are sometimes used instead of *with*. This should not be done: in places where *with* is felt to be inappropriate the truth is that *affinity*, which properly describes a reciprocal relationship only, has been used of a one-sided one, and should itself be replaced by another word. Cf. *sympathy with* and *for*.

**affirmative.** See NEGATIVE.

**afforce.** *There is no suggestion that either House should be afforced for the purpose, as the House of Lords . . . is afforced by the addition of judges of the highest degree.* As long ago as 1888 the OED described as obsolete all the uses of *a.* except 'to reinforce or strengthen a deliberative body by the addition of new members, as a jury by skilled assessors or persons acquainted with the facts', and its supporting quotations refer only to the practice of 'afforcing' juries in the Middle Ages. In the COD the word is not given. Its use in the above quotation cannot escape the suspicion of being a REVIVAL prompted by PRIDE OF KNOWLEDGE.

**afford.** The modern use of *can't afford to* in the sense of *daren't* makes for confusion. *Can we afford to do this?* asks a politician about a popular proposal, meaning have we the money to do it? *Can we afford not to do it?*

retorts another, meaning dare we face the consequences of not doing it? The two arguments are not in the same plane and will never meet.

**à fond.** See FRENCH WORDS It should be remembered that *à fond* and *au fond* mean different things, *à fond* to the bottom, i.e. thoroughly, and *au fond* at bottom, i.e. when one penetrates below the surface.

**a fortiori.** Introducing a fact that, if one already accepted is true, must also and still more obviously be true. *It could not have been finished in a week; a. f. not in a day.*

**after.** English novelists, rashly trying to represent Irish characters as speaking in their native idiom, almost always betray their ignorance of its subtleties. Their commonest mistake is their wrong use of the expression *I'm after doing so-and-so*. It does not mean *I want to do* or *I am about to do*. It means *I have just done*.

**aftermath.** *Our own generation can be proud of what it has done in spite of war and its aftermath.* The use of *aftermath* in the sense of an unpleasant consequence of some event is firmly established, and only a pedant can object to it on the ground that the word in its primary sense (a second growth of grass in a season after the first has been cut) is beneficent rather than unpleasant. But the metaphor is not yet dead enough to tolerate the use with it of an incongruous epithet such as *violent*. See METAPHOR 2 C.

**afterward(s).** *Afterward*, once the prevalent form, is now obsolete in British use, but survives in U.S.

**age.** For synonymy see TIME.

**aged.** One syllable in *aged* 21 etc. and *an aged horse* (i.e. more than 6 years old); two syllables in *an aged man* etc.

**agenda.** *What emerged from the Commonwealth Conference was not a cut-and-dried agenda.* Although *agenda* is a plural word, it is pedantry to object to the common and convenient practice of thus treating it as a singular one. If a singular is needed for one *item of*

the agenda there seems no escape from that rather cumbrous phrase; *agendum* is pedantic and *agend* obsolete.

**aggrandize(ment).** The accent of the verb is on the first and of the noun on the second syllable. See RECESSIVE ACCENT.

**aggravate, aggravation.** For many years grammarians have been dinning into us that to *aggravate* has properly only one meaning—to make (an evil) worse or more serious—and that to use it in the sense of *annoy* or *exasperate* is a vulgarism that should be left to the uneducated. But writers have shown no less persistence in refusing to be trammelled by this admonition. The OED, which calls the usage 'fam.', gives examples that date back to 1611 and include quotations from Richardson and Thackeray. They have their distinguished followers today. *But Archbishop Tenison, though much out of favour with the Queen, outlived her in the most aggravating manner* (G. M. Trevelyan). / *He had pronounced and aggravating views on what the United States was doing for the world* (Graham Greene). / *Then he tried to be less aggravating* (E. M. Forster). / *Syngman Rhee has made it plain that he will go to any lengths to aggravate the communists into a renewal of the fighting* (*The New Statesman*). It is time to recognize that usage has beaten the grammarians, as it so often does, and that the condemnation of this use of *aggravate* has become a FETISH. After all, the extension from aggravating a person's temper to aggravating the person himself is slight and natural, and when we are told that Wackford Squeers pinched the boys in aggravating places we may reasonably infer that his choice of places aggravated both the pinches and the boys.

**ago.** If *ago* is used, and the event to be dated is given by a clause, it must be by one beginning with *that* and not *since*. The right forms are: *He died 20 years ago* (no clause); *It is 20 years since he died* (not *ago*); *It was 20 years ago that he died.* The following examples are wrong; the tautology *ago since* is naturally commoner, but is equally wrong, in sentences like the second, where a parenthesis intervenes: *It is barely 150 years ago since it was introduced.* | *It is seven years ago, when the Calder Hall station was begun, since a start was made with turning nuclear power to peaceful purposes.* For similar mistakes see HAZINESS.

**agree.** The normal uses of *agree* are as an intransitive verb often with a preposition (*a. with*, concur with a person, *a. to*, consent to a project, *a. on*, decide something by mutual consent). Its use as a transitive verb in the last sense, without *on*, was said by the OED in 1888 to be applicable only to discrepant accounts and the like, but it is now much wider, especially in the p.p. *An agreed statement was issued after the meeting.* | *The committee has power to agree its own procedure.* | *It proved impossible to agree a price.* That is unexceptionable. But the same cannot be said of the encroachment of a transitive *agree* on the province of *agree to*. *The chairman has not yet agreed the draft circular.* | *The use of tear-gas was agreed by the Commissioner of Police.* | *There is ample evidence that the petitioner agreed the course of action taken by the respondent.* Here *agree* usurps the place of some more precise, and therefore better, word such as *approve, sanction, confirm, condone.*

**agricultur(al)ist.** See -IST.

**aim.** In the article CAST-IRON IDIOM it is observed that in the secular conflict between idiom and analogy, analogy perpetually wins; it is for ever successful in recasting some piece of cast iron. This is what has happened to *aim*. Until recently it was possible to say with confidence that in Britain this verb in the metaphorical sense of purpose or design or endeavour was idiomatically followed by *at* with the gerund, not by *to* with the infinitive. *He aimed at being* (not *he aimed to be*) *the power behind the throne.* Even in 1933 the OED Supp. gave only one British example (from Thomas

Hardy) of *aim to*. But the analogy of *purpose, try, intend*, which take the infinitive, reinforced by the general use of that construction with *aim* in America, is proving too strong; and it is unlikely that eyebrows were raised by any members, however purist, of the audience that in 1958 heard a Minister at the Annual Congress of the Conservative party say *What we aim to do is to widen the whole field of house purchase.*

**ain't.** See BE 7.

**air-.** See AERO-.

**aisle** has escaped from its proper meaning of the lateral division of a church separated from the nave by pillars, and is commonly applied also to the central passage-way of the nave, and indeed to any passage-way between seats in a church, corruptly replacing *alley*, says the OED. In America it has strayed even further, and is used of what in England would be called gangways in a theatre or railway carriage

**aitch-bone.** *H-bone, edge-bone, ash-bone*, and other forms, are due to random shots at the etymology. *Aitchbone*, though it does not reveal the true origin of the word (L *natis* buttock, with loss of n- as in *adder* etc.), suggests no false one and corresponds to the pronunciation.

**-al nouns.** When a noun in *-al* is given in its alphabetical place with a simple reference to this article, the meaning is that its use is deprecated. There is a tendency to invent or revive unnecessary verbal nouns of this form. The many that have passed into common use (as *trial, arrival, refusal, acquittal, proposal*) have thereby established their right to exist. But when words of some age (as *revisal, refutal, retiral, accusal*) have failed to become generally familiar and remained in the stage in which the average man cannot say with confidence off-hand that they exist, the natural conclusion is that there is no work for them that cannot be adequately done by the more ordinary verbal nouns in -ion (*revision*) and

-ation (*refutation, accusation*) and -*ment* (*retirement*). When there is need on an isolated occasion for a verbal noun that shall have a different shade of meaning from those that are current (e.g. *accusal* may suggest itself as fitter to be followed by an objective genitive than *accusation*; cf. *the accusal of a murderer, the accusation of murder*), or that shall serve when none already exists (there is, e.g., no noun *beheadment*), it is better to make shift with the gerund (*the accusing, the beheading*) than to revive an unfamiliar *accusal* or invent *beheadal*. The use of rare or new -al nouns, however, is due only in part to a legitimate desire for the exactly appropriate form. To some writers the out-of-the-way word is dear for its own sake, or rather is welcome as giving an air of originality to a sentence that if ordinarily expressed would be regarded as commonplace; they are capable of writing *bequeathal* for *bequest*, *agreeal* for *agreement*, *allowal* for *allowance*, or *arisal* for *arising*. Except for this dislike of the normal word, we should have had *account* instead of *recountal* in *Of more dramatic interest is the recountal of the mission imposed upon Sir James Lacaita*, and *to recount these* in *But this is not the place for a recountal of these thrilling occurrences*; cf. *retiral* in *There were many retirals at the dissolution. Referral, surprisal, supposal, decrial*, may be mentioned among the hundreds of needless -al words that have been actually used.

**à la.** The sex of the person whose name is introduced by this does not affect the form, *la* agreeing not with it but with an omitted *mode*: *à la reine*; *à la* (not *au*) *maître-d'hôtel*; *a Home-rule Bill à la* (not *au*) *Gladstone. Au* with adjectives, as in *au naturel, au grand sérieux* (cf. *à la française* etc.), is not used in English except in phrases borrowed entire from French.

**alarm, alarum.** *Alarum* is by origin merely a variant of *alarm*, and the two nouns were formerly used without distinction in all senses. Later *alarum* was restricted to the senses of alarm-signal, warning-signal, or clock or other apparatus that gives these. This being a clear and useful differentiation, it is to be regretted that it should have not been maintained: *alarum* survives only for the clock (even there fighting a losing battle with *alarm clock*) and in the jocular use of old stage-direction *alarums and excursions*. The use of *alarm* for the air-raid warning was the death-blow to *alarum*.

**albeit,** i.e. *all be it* (*that*), or, in full, *all though it be that*, was classed as an ARCHAISM in the first edition of this book. It has since been picked up and dusted and, though not to everyone's taste, is now freely used, e.g. *It is undeniable that Hitler was a genius, a. the most evil one the modern world has known.*

**ale, beer.** Both words are more than 1,000 years old, and seem originally to have been used as synonyms for the liquor made from fermented malt. They were distinguished when *beer* was appropriated to the kind brewed with an infusion of hops, first imported in the 16th c. This distinction has now disappeared; *beer* has become a generic word comprising all malt liquors except stout and porter, though brewers still call some of their products *ales*, especially with a distinguishing adjective, e.g. *pale, brown, rustic, audit*. In ordinary use, as at table, *beer* is the natural word; *ale* has a flavour of GENTEELISM.

**alibi** is a Latin word meaning *elsewhere* which has been adopted by British Law as a name for the defence against a criminal charge that seeks to establish the accused's innocence by proving that he was in some other place than the scene of the crime at the time when it was committed. It is a useful word—indispensable indeed in its proper place—with a precise meaning. That it should have come to be used as a pretentious synonym for *excuse* is a striking example of the harm that can be done by SLIPSHOD EXTENSION. Perhaps the vogue of detective stories is responsible for the corruption. So many of them rely on an alibi for their plot that ignorant readers think that the word will do for any means of rebutting a charge. Not that the ignorant are now the only offenders; the following quotations are taken respectively from the title of a serious book, a review in the TLS, a speech by a Trade Union leader, and a speech by a Cabinet Minister. *The SS: Alibi of a Nation.* | *It is doubtful whether an audience even of Mr. Wilson's coevals and juniors could be stimulated by a tract like this to anything but an extension of that bored mistrust of their elders which already provides so many of them with an alibi.* | *If people decide not to run after they have been nominated, that must not be used as an alibi to discredit the elected representatives of the people.* | *We are certainly not prepared to use the temporary petrol shortage as an alibi for not building roads.* The mischief is that, if this goes on, we shall be left without a word for the true meaning of *alibi*.

**alien.** The prepositions after the adjective are *from* and *to*. *From* is the earlier usage, and represents the commoner Latin construction, though *alienus* with the dative is also good Latin. There is perhaps a slight preference for *from* where mere difference or separation is meant (*We are entangling ourselves in matters a. from our subject*), and for *to* when repugnance is suggested (*cruelty is a. to his nature*). But this distinction is usually difficult to apply, and the truth seems rather that *to* is getting the upper hand of *from* in all senses (cf. DIFFERENT, AVERSE).

**align(ment), aline(ment).** The OED pronounces for the spelling with *ne* and against that with *gn*. On the verb it says 'As *line* is the Eng. spelling of Fr. *ligne* and *ligner*, there is no good reason for retaining the unetymological *g* in the derivative'; and on the noun 'The Eng. form *alinement* is preferable to *alignment*, a bad spelling of the Fr.' Usage as clearly pronounces for the *-gn-* forms; in the OED quotations *gn* is just four times as numerous as *ne*. The claims of usage and etymology

are often hard to decide between (cf. RHYME). No one would propose to correct *admiral*, *aisle*, or *cretin*, back into conformity with Arabic *amir*, Latin *ala*, or Latin *christianus*, though the insertion of *d* and *s*, and the omission of *h*, are 'unetymological'; and on the other hand unnoticed corrections of words taken from French (as *scholastic*, respelt on Greek σχολή after being taken directly from *scolastique*) are innumerable. But *align* is not only the established form; it is also more correct than *aline*. Correction on Latin analogies (*adlineare*, *allineare*) could only give *alline*; and *aline* regarded as a purely English formation would have no meaning, *a-* in the sense *to* not being a recognized formative element. *Alline(ment)* seems defensible but inexpedient, *aline(ment)* indefensible, and *align(ment)* unobjectionable. The last is now given first place by the COD, and in the expression the *nonaligned countries* that is always the spelling.

**-(al)ist.** For such alternative forms as *agriculturist* and *agriculturalist*, see -IST.

**alkali.** The pronunciation is usually -*lī*, but sometimes -*lĭ*. The plural should be and usually is *alkalis*; but the -*lĭ* pronunciation, suggesting *alkaly* as the singular, has produced a by-form *alkalies*.

**allegory.** For *a*. and *parable*, see SIMILE AND METAPHOR.

**alleluia.** See HALLELUJAH.

**allerg(ic)(y).** These words were invented early in the 20th c. by the medical profession to describe the condition of a patient who is unusually sensitive to some substance. They soon found favour as POPULARIZED TECHNICALITIES, no doubt because they filled a gap usefully. In its medical sense *allergy* is usually applied to the peculiar reaction of an individual to something ordinarily harmless. Similarly in its extended meaning it was used to suggest an instinctive and perhaps

# alliteration

unreasonable dislike. *I am allergic to Dr. Fell* says in 6 words what it takes the famous rhyme 24 to say. *I am a. to* provides a convenient middle course between the stilted *I have an instinctive antipathy to* and the colloquial *I have a thing about*. But, like most popularized technicalities, these are becoming VOGUE WORDS, apt to displace common words that might be more suitable, such as *dislike, repugnance, aversion*.

**alliteration.** The purposive use in a phrase or sentence of words beginning with or containing the same letter or sound. *After life's fitful fever*; In a *summer season* when *soft* was the *sun*. The much-quoted line of Charles Churchill *Apt alliteration's artful aid* is not as good an example of alliteration as it looks, since only the first two *a*'s have the same value.

**all of.** *All of the ministers were present.| In all of the book there is no better chapter.* This intrusive *of* was said by the OED in 1888 to be a comparatively modern construction, and rare except with pronouns (*all of whom, all of us,* etc.). It has since made headway, especially in U.S., but in Britain *all the ministers* and *in all the book* are still regarded as preferable.

**allow of.** This is undergoing the same limitation as ADMIT OF, but the process has not gone so far; *Jortin is willing to allow of other miracles, A girl who allows of no impertinent flattery,* are hardly felt to be archaic, though *of* would now usually be omitted. The normal use is, and sense, however, are the same as those of *admit of*.

**all right.** The words should always be written separate; there are no such forms as *all-right* or *allright,* and *alright* (on the analogy of *already* and *altogether*), though described by the OED Supp. as a frequent spelling of *all right,* is still regarded as a vulgarism, inadmissible not only in those obvious cases where the two words are completely independent, as in *The three answers, though different, are all*

right, but also where they may be regarded as forming a more or less fixed phrase. So: *The scout's report was 'All right'* (i.e. all is right). | *Is he all right? | 'Will you come for a walk?' 'All right.' | All right, you shall hear of this again. | Oh, I know them apart all right.*

The reason [says Onions] why the prejudice against *alright* is so strong seems, however unconsciously, to be the recognition of the colloquial levity of the phrase, and an objection to its literary pretence to be a good grammatical adverb which its condensed form would ensure. . . . Even if we are prepared to admit *alright* for some uses, here at all events (*That's quite all right | I'm quite all right*) we should spell it out, the stress being full and even. It will almost inevitably establish itself in the long run, but it is to be hoped that the example of *already* will be closely followed and that it will be restricted to adverbial uses such as *The difficulty can be got over alright.* Even here it is at present barely justifiable since the vocalic value of *all* is usually retained and no marked differentiation of meaning has yet taken place (SPE Tract XVIII).

Now that in colloquial use *all right* has been virtually put out of business by O.K. (q.v.) the progress of the composite form towards recognition may be slower than Onions thought.

**all that.** See THAT, ADJ. AND ADV. I.

**allusion, allude.** I. For pronunciation see PRONUNCIATION 6.

2. The words are much misused. An allusion is a covert or indirect reference, in which the application of a generality to the person or thing it is really aimed at, or the identification of something that the speaker or writer appears by his words to have in mind but does not name, is left for the hearer or reader to make; it is never an outright or explicit mention. *Allude to* has the same limitations. Examples of the legitimate use are: *We looked at each other wondering which of us he was alluding to*; *Though he never uses your name, the allusion to you is obvious*; *He is obscure only because he so often alludes* (or *only owing to his frequent allusions*) *to contemporary events now forgotten.*
The misuse is seen in: *When the*

speaker happened to name Mr. Gladstone, the allusion was received with loud cheers. | The lecturer then alluded at some length to the question of strikes.

It may be added that *allude to* is often chosen, out of foolish verbosity, when the direct *mean* would do better; so *When you said 'some people cannot tell the truth', were you alluding to* (did you mean) *me?*; but this is an abuse rather than a misuse.

**ally**, n. and v. These words and their inflected forms used all to be pronounced with the accent on the second syllable (*alī', alī'z, alī'd*). But the noun, both singular and plural, is now commonly accented on the first syllable, and so is the p.p. of the verb when used attributively (*the allied forces*).

**almanac**. The only modern spelling, though the variant with *k* is maintained by *Whitaker's Almanack*.

**almoner**. The OED prefers *ăl'mŏner* to *ahm-*. But in the most familiar use of the word, *hospital a.*, *ahm-* is now commoner, perhaps by analogy with *alms*, and the COD puts it first. See PRONUNCIATION 8.

**almost**. *He was in a state of almost collapse*. Although the OED denies to *almost* the status of an adjective, it recognizes this usage, which it describes as 'qualifying a sb. with implied attribute' and quoting *An almost quaker | His almost impudence*. Cf. the similar use of NEAR. For *almost quite* see INCOMPATIBLES, and for the curtailed form *most* see MOST(LY) 2.

**alone**. The adverbial use of *alone* with *not* in place of the usual *only* (*more might be done, and not alone by the authorities, but . . .*) is a survival, and like other archaisms is to be avoided. In ordinary writing it is one of the thousand little mechanical devices by which 'distinction of style' is vainly sought; the following passage shows it in characteristic surroundings: *Recourse to porters, whose limited loads are carried on the head, savours more of operations in the West African bush than on the Indian frontier, so that not*

alone in the region passed through, but in its transport will our latest little war wear an interesting and unusual aspect. This censure does not apply to the adjectival use (*It is not youth alone that needs . . .*), in which *not* and *alone* are usually separated. For *let alone* see LET.

**à l'outrance**. The French phrase is *à outrance* or *à toute outrance*, never *à l'outrance*. Those who use French phrases to suggest that they are at home with French should accordingly be careful to write *à outrance*. For those who use them merely as the handiest way of expressing themselves the form that is commoner in English (*à l'o.*) is as good as the other, and it does not lay them open to the charge of DIDACTICISM. So with *double entendre* for French *double entente*, *nom de plume* for *nom de guerre*, and others.

**alright**. See ALL RIGHT.

**also**. The word is an adverb, and not a conjunction; nevertheless, it is often used in the latter capacity where *and* (*also*), *but also*, or *as well as*, would be in place. In talk, where the informal stringing on of afterthoughts is legitimate, there is often no objection to this (*Remember your passport and money; also the tickets*); and the deliberate afterthought may appear similarly in print (*The chief products are hemp and cigars; also fevers*). But it is the writer's ordinary duty to settle up with his afterthoughts before he writes his sentence, and consequently the unassisted *also* that is proper to the afterthought gives a slovenly air to its sentence.

*Great attention has been paid to the history of legislation, also* [and also] *to that of religion. | We are giving these explanations gently as friends, also* [and] *patiently as becomes neighbours. | 'Special' is a much overworked word, it being used to mean great in degree, also* [but also] *peculiar in kind. | Mr. Sonnenschein's volume will show . . . Shakespeare's obligations to the ancients, also* [as well as] *the obligations of modern writers to Shakespeare. Also* is now much in demand to introduce a VERBLESS SENTENCE in the guise of an afterthought. *I am a product of long corridors and the noise of wind under the tiles. Also of endless books.*

**alternate(ly), alternative(ly). 1.** *Alternative* (offering a choice) had formerly also the sense now belonging only to *alternate* (by turns); now that the differentiation is complete, confusion between the two is even less excusable than between DEFINITE and DEFINITIVE. Examples of the wrong use of one for the other are: *The journey may be made by road or alternately by rail.* | *Alternatively they sat and walked in the moonlight talking of this and that.*
**2.** There are also difficulties about the correct use of the noun *alternative*, and for dealing with these it is necessary to realize clearly the word's different senses. These are now distinguished, with illustrations that may serve to show what is idiomatic and what is not:
(*a*) A set, especially pair, of possibilities from which one only can be selected; this is often practically equivalent to *choice. The only a. is success or* (not *and*) *death. We have no a. in the matter.*
(*b*) Either of such pair or any one of such set. *Either a. is, any of the aa. is, both aa. are, all the aa. are, intolerable. The aa., the only aa., are success and* (not *or*) *death.*
(*c*) The second of such pair, the first being in mind. *We need not do it; but what is the a.? We must do it; there is no* (not *no other*) *a. The* (not *the other*) *a. is to die.*
(*d*) Some other of such a set, one at least being in mind. *If we decline, what are the aa.? The only a. is to die. You may say* lighted *or* lit *or* alight; *there is no other a. The only aa. to it are gas and* (not *or*) *candles.*
**3.** *Alternative* implies a choice, and should not be used where there is none, as a pretentious synonym for *other, new, fresh, revised,* and the like. This SLIPSHOD EXTENSION is very common especially in official writing, perhaps because of the baneful influence of the cliché *alternative accommoda-*

*tion.* (See ACCOMMODATION.) *The meeting for the 24 Oct. has been cancelled; an alternative date will be announced as soon as possible* (fresh). | *The sale of the premises is due to be completed by the 31 Dec., when vacant possession will have to be given. In the meantime your directors have alternative premises in view* (other). | *The programme cannot be carried out on account of illness and alternative arrangements are being made* (fresh or revised). | *There will be full consultation in the coalfields about securing alternative work for the men displaced by the closing of the pits* (other).
**4.** The notion that because it is derived from Latin *alter* (one or other of two) *alternative* cannot properly be used of a choice between more than two possibilities is a FETISH.

**although.** See THOUGH.

**altogether.** Confusion between this and *all together* is not uncommon. *Until at last, gathered altogether again, they find their way down to the turf.* | *France . . . must take her place at the head of the group of States in Western Europe which, though now impoverished, would altogether form an important element in the safeguarding of liberty. All together* should have been used in each of these examples. *Altogether* is right only in the senses *entirely* or *on the whole.*

**amateur.** The best pronunciation is *ă'mătŭr,* the next best *ămătŭr'*; it is high time that vain attempts at giving the French *-eur* should cease, since the word is in everyday use. Cf. LIQUEUR.

**ambience.** The adjective *ambient* has been in use for centuries, mostly as a poeticism, occasionally as a scientific word; but the noun *ambience* was unknown to the compilers of Letter A of the OED in 1888. It has now become a prime favourite of journalists and critics, usually as a pretentious synonym for *surroundings, environment, milieu, atmosphere,* and the like. *In Anderston, scheduled for slum clearance and redevelopment, a Glasgow man has taken an ordinary pub and added a buttery* *where the cooking is excellent and the ambience like high-grade Soho. | It is a nice paradox that, while the BBC ethos may favour a crusade fervently concerned with questions of social justice, the ambience of ITV might be more suitable to a strategy based on simple considerations of material well-being. | Beyond the selfishness and stupidity, the terrible ordinariness of people of fashion . . . remains the general agreeable ambience of an upper class, and to this ambience Proust remained faithful all his life. | If ever an overture needed a resonant ambience in which its various elements might harmoniously fuse, it is Tannhäuser's.* The French form *ambiance*, says the OED Supp., is used for the arrangement of accessories to support the main effect of a piece; so the writer of the last example might perhaps say that he chose the word for this specialized sense.

**ambiguity.** The word as here used includes not only those ambiguities that leave the reader genuinely puzzled which of two interpretations is right, but also those in which one of these, probably the more natural grammatically, is clearly not what the writer meant. The fault of this kind of writing is not so much obscurity as clumsiness. Like that described in the article FALSE SCENT, it misleads the reader only momentarily, if at all, but makes him think the writer a fool for not being able to say what he means. As the purpose of this dictionary is to help writers to express themselves clearly and accurately, and the causes of ambiguity are innumerable, reference might perhaps here be made to its contents *passim*. But a few of the more frequent of these causes deserve special notice.

'Of all the faults found in writing', wrote Cobbett in 1818, 'the wrong placing of words is one of the most common, and perhaps it leads to the greatest number of misconceptions.' Or, in the words of another old grammarian, the golden rule of writing is 'that the words or members most nearly related should be placed in the sentence as near to one another as possible, so as to make their mutual relation clearly appear'. Few books since published on grammar and usage have omitted to enliven an arid subject by stock examples of this fault, such as the well-worn one of the advertisement of a piano for sale by a lady going abroad in an oak case with carved legs. Nevertheless, carelessness of this kind is still so common that it may be worth while to give some modern examples by way of warning. *No child shall be employed on any weekday when the school is not open for a longer period than four hours. | Please state from what date the patient was sent to bed and totally incapacitated by your instructions. | To ask the Minister of Agriculture if he will require eggs to be stamped with the date when they are laid by the ſarmer. | The Committee considered again the recommendation that —— soap be used throughout the hospital because Mr. G. had objected to it at the Medical Committee. | No force was used beyond that necessary to put an end to the uproar by the stewards. | I can recommend this candidate for the post for which he applies with complete confidence. | We had a leisurely breakfast and recalled the old tales of the city and the love story of Hero and Leander, who swam the Hellespont, and spent a great deal of time just looking at the beautiful view.* All these ambiguities can be removed by rearrangement to 'bring the words or members most nearly related as near to one another as possible'.

Some other common causes of ambiguity are:

The use of a word with two meanings in a context in which it makes for uncertainty or absurdity. *Much conversation was going on about him. | | Certain remedies are available for this condition. | The meeting had been over-looked by the police on duty. | Miss Pickhill grasped the pince-nez which hung from a sort of button on her spare bosom.*

The use of pronouns in such a way that the reader cannot be sure of their antecedents. See PRONOUNS 3.

Uncertainty whether *shall* or *will*, or *should* or *would*, is used as a simple auxiliary or as implying volition or obligation. *It is of interest to note that, even as a tiro in politics, he should have taken his own line.*

Failure to make clear the field of operation of a word or phrase, e.g. an adverb *He needs more suitable companions*; / or a participle *His dogs might sometimes come to resemble the frightened and exhausted rabbit who in the end walks towards the stoat seeking to devour him*; / or a relative or other subordinate clause *I accused him of having violated the principles of concentration of force which had resulted in his present failures*; / *The child has not yet learnt to express a thought fully and clearly, so that only one meaning is conveyed*; / or a genitive *The case will be reheard before a full Bench: Counsel will argue both the merits and the jurisdiction of the Court*; / or a negative *The letter was not sent because of information received.* See also BECAUSE 2; EVEN; HYPHENS; NEGATIVE MISHANDLING; ONLY; POSITION OF ADVERBS; UNATTACHED PARTICIPLES; WITHOUT.

Ambiguities may sometimes be removed by punctuation, but an attempt to correct a faulty sentence by inserting stops usually betrays itself as a slovenly and ineffective way of avoiding the trouble of rewriting. 'It may almost be said that what reads wrongly if the stops are removed is radically bad; stops are not to alter meaning but merely to show it up' (*The King's English*).

**ambivalen(t)(ce).** These terms were invented by psychoanalysts early in the 20th c. to mean the coexistence in one person of opposing emotional attitudes towards the same object, or the simultaneous operation in the mind of two irreconcilable wishes. The words are new, but not the condition they describe: Catullus when he wrote *odi et amo* and Ovid when he wrote *video meliora, proboque; deteriora sequor* were suffering from ambivalence. The discovery of so imposing a word for so common a condition quickly led to its becoming a POPULARIZED TECHNI-

CALITY, with the usual consequence that it now has to do a great deal of work, of which some at least would be better done by other words. Typical examples of its fashionable use are: *A move to the country has an ambivalent character. On the one hand . . . it is a sort of challenge. . . . On the other hand . . . it really is cheaper in the country. | The Soviet attitude towards us is ambivalent. On the one hand they do not wish to miss the opportunity of exploiting the Middle Eastern situation. . . . On the other hand they do not wish to lose the advantages of peaceful coexistence.* It will be a pity if the homely expression *mixed feelings*, which served us well and long, is wholly displaced by this usurper.

**amen.** '*Ahmen* is probably a comparatively modern Anglican innovation of about a hundred years' standing. Roman Catholics, one is glad to note, on the whole retain the English *āmen*' (D. M. Low in *Essays and Studies*, 1960).

**amend.** See EMEND.

**amenity.** The OED gives only the pronunciation -*mĕn*- but -*mēn*- is now more usual. *Amenities* has become something of a VOGUE WORD, especially in housing, for what conduces to human pleasure or convenience, and can mean anything from indoor sanitation to a distant prospect. See OFFICIALESE.

**America(n).** The use of *America* for *the United States* and *American* for *(citizen) of the U.S.* is open to as much and as little objection as that of *England* and *English(man)* for *Great Britain (and Northern Ireland)*, *British*, and *Briton* (see ENGLAND). It will continue to be protested against by purists and patriots, and will doubtless survive the protests. 'If it's any comfort to those who resent it, the usage is founded on a lazy disinclination to pronounce the longer name rather than arrogance, and it has no official sanction' (Evans).

**Americanisms.** It was a favourite

theme of Mencken that England, now displaced by the United States as the most powerful and populous English-speaking country, is no longer entitled to pose as arbiter of English usage. 'When two-thirds of the people who use a certain language decide to call it a *freight* train instead of a *goods* train, they are "right"; then the first is correct usage and the second a dialect.' We are still far from admitting this claim, but in fact are showing signs of yielding to it in spite of ourselves. The close association of the two countries in the second world war and the continued presence of members of the U.S. Air Force among us have done much to promote American linguistic infiltration; and more is being done by the popularity of American products for stage, cinema, radio, television, and comic strip, and the apparent belief of many English entertainers that to imitate American diction and intonation is a powerful aid to slickness. In vocabulary this infiltration is notorious; in grammar and idiom it is more subtle but hardly less significant. Our growing preference for PHRASAL VERBS over simple ones with the same meaning (*meet up with*, *lose out on*), the use of the plain SUBJUNCTIVE without an auxiliary in such a sentence as *he is anxious that the truth be known*, the effects of HEADLINE LANGUAGE, especially as an eater-up of prepositions (*world food production* for *production of food in the world*), the obliteration of the distinction between SHALL and WILL that the few who understood it used to consider the hall-mark of mastery of the niceties of English idiom, the foothold gained by the American *I don't have* at the expense of the English *I haven't got* (see DO 2), the victory of *aim to do* over *aim at doing*, the use of *in* instead of *for* in such a phrase as *the first time in years*, the progress made by DUE TO towards the status of a preposition and of LIKE towards that of a conjunction—such things as these, trifling in themselves, are cumulatively symptoms of surrender by the older competitor to the younger and more vigorous.

In vocabulary we accept Americanisms if they have the appeal of novelty and aptness or fill a gap for us; thus we have recently admitted *baby-sitter*, TEENAGER, KNOW-HOW, and GIMMICK as readily as our fathers welcomed *stunt*, *blurb*, and *rubberneck*. When it is a question of adopting the meaning given in America to a word we use differently, we often put up more resistance, but not always: since the war we have taken to using *assignment* and *executive* in the American senses; *fix* for *mend* and *mad* for *angry* are common. *Radio* is displacing *wireless* and *mail* encroaching on *post* and the British workman when on strike seems now to find *scab* a more satisfactorily expressive word than *blackleg*. *Gun* has been enlarged to include a revolver and an automatic, and *date* to cover an *engagement*, and the convenience of such expressions as *commuter* for one who travels daily to and from work and *reservation* for *advance booking* is irresistible. *Luggage* has partly yielded to *baggage*, but only if we are travelling by sea or air. Occasionally we have taken back words that went over with the Pilgrim Fathers and were forgotten by those they left behind; the reappearance among us of the now superfluous verb *loan* is evidence that we do not always show sound judgement in doing this. But some of the long-standing differences in vocabulary seem so firmly rooted as to be unshakable. There is no question of merit in it: we may indeed maintain that *lift* is a better word than *elevator* and *motor car* than *automobile*, but we cannot deny that *fall* is prettier than *autumn* and *sidewalk* more sensible than *pavement*.

The following is a list of some of the commoner of the persisting differences.

*Clothes*

| English | American |
|---|---|
| Boot | Shoe |
| Bowler | Derby |
| Braces | Suspenders |
| Galoshes | Rubbers |
| Nappy | Diaper |

| English | American | English | American |
|---|---|---|---|
| Pyjamas | Pajamas | Waste-paper-basket | Waste-basket |
| Sock suspenders | Garters | *Legal and official* | |
| Vest | Undershirt | Bylaw | Ordinance |
| Waistcoat | Vest | Inland Revenue | Internal Revenue |
| *Commercial* | | Prison van | Patrol wagon |
| Chemist | Druggist | Witness-box | Witness stand |
| Commercial traveller | Travelling salesman | Ticket-of-leave | Parole |
| Draper's shop | Dry goods store | *Sports and games* | |
| Fancy Goods | Notions | Cannon (billiards) | Carom |
| Hire purchase | Buying on time | Draughts | Checkers |
| Hoarding | Billboard | Fruit machine | Slot machine |
| Ironmonger | Hardware dealer | Nine-pins | Ten-pins |
| Market gardener | Truck farmer | Pack (of cards) | Deck |
| Men's hairdresser | Barber | Shooting | Hunting |
| Note (paper money) | Bill | Touch-lines | Side-lines |
| Post (vb) | Mail | *Transport* | |
| Press-cutting agency | Clipping service | Bonnet (of car) | Hood |
| | | Boot (of car) | Trunk |
| Receptionist (hotel) | Desk clerk | Coach (railway) | Car |
| | | Engine-driver | Engineer |
| Shop | Store | Goods train etc. | Freight train etc. |
| Shop-walker | Floor-walker | Gradient | Grade |
| Sleeping partner | Silent partner | Guard (of passenger train) | Conductor |
| Slot machine | Vending machine | Level crossing | Grade crossing |
| Street vendor | Peddler | Lorry | Truck |
| Timber (sawn) | Lumber | Pavement | Sidewalk |
| *Food* | | Permanent way | Roadbed |
| Beetroot | Beets | Petrol | Gas(oline) |
| Biscuits and small cakes | Crackers and cookies | Points (railway) | Switch |
| Chicory | Endive (and vice versa) | Railway | Railroad |
| | | Return ticket | Round-trip ticket |
| Corn | Grain | Roundabout | Traffic circle or rotary |
| Maize | Corn | Saloon (car) | Sedan |
| Scone | Biscuit | Shunt | Switch |
| Sweets | Candy | Signal-box | Switch-tower |
| Tin | Can | Silencer (of car) | Muffler |
| Treacle | Molasses | Single ticket | One-way ticket |
| Undercut | Tenderloin | Sleeper | Cross-tie |
| *Household* | | Subway | Underpass |
| Cooker | Oven | Terminus | Terminal |
| Cookery book | Cookbook | Tram | Street-car |
| Cupboard | Closet | Underground railway | Subway |
| Drawing-pin | Thumb-tack | | |
| Dust-bin | Garbage can | Van | Delivery truck |
| Flat | Apartment | Wing (of car) | Fender |
| Jug | Pitcher | *Miscellaneous* | |
| Paraffin | Kerosene | Aluminium | Aluminum |
| Tap | Faucet | Autumn | Fall |
| Veranda | Porch | Bank holiday | Legal holiday |

| English | American |
| --- | --- |
| Caretaker | Janitor |
| Council school | Public school |
| Cutting (newspaper) | Clipping |
| Dust cart | Garbage truck |
| Ex-Service man | Veteran |
| Fanlight | Transom |
| Friendly Society | Fraternal order |
| Holiday | Vacation |
| Lift | Elevator |
| Perambulator | Baby carriage |
| Pig breeding | Hog raising |
| Private soldier | Enlisted man |
| Public school | Private school |
| Queue | Line |
| Rear (family etc.) | Raise |
| Scribbling-block | Scratch-pad |

**amid, amidst.** 1. Both are LITERARY WORDS, subject to the limitations of their kind.

2. As to the alternative forms, the OED states that 'There is a tendency to use *amidst* more distributively, e.g. of things scattered about, or a thing moving, in the midst of others'. This may have been true at that time (1888) but the only distinction that can now be hazarded is that *amid* has dropped out of ordinary use even more than *amidst*, and is therefore felt to be inappropriate in many contexts that can still bear *amidst* or *in the midst of*. When we find *amid* in a passage of no exalted or poetical kind (*A certain part of his work . . . must be done amid books*), our feeling is that *amidst* would have been less out of place, though *among* would have been still better.

**among, amongst.** There is certainly no broad distinction either in meaning or in use between the two. The OED illustrates under *amongst* each of the separate senses assigned to *among*; it does, however, describe *amongst* as 'less usual in the primary local sense than *among*, and, when so used, generally implying dispersion, intermixture, or shifting position'. Such a distinction may be accepted on authority, but can hardly be made convincing by quotations even on the liberal scale of the OED. It is remarkable, at any rate,

that one of the forms should not by this time have driven out the other (cf. *on* and *upon*, *although* and *though*, *while* and *whilst*, *amid* and *amidst*). The survival of both without apparent differentiation may possibly be due to the unconscious desire for euphony or ease; few perhaps would say *amongst strangers* with *among* to hand; *amongst us* is easier to say than *among us*. Some confirmation of this is found by comparing the ratio borne by initial vowels to initial consonants after *amongst* with the corresponding ratio after *among*; reckoned upon the 19th-c. quotations in the OED, this ratio is four times as high for *amongst* as for *among*. Though the total number of quotations is too small to justify the belief that this proportion prevails generally, it probably indicates a tendency. It may be said with some confidence that (1) *among* is the normal word, (2) *amongst* is more usual before vowels, but (3) before *the*, which so commonly follows as easily to outnumber all other initials, the two forms are used quite indifferently. See also BETWEEN 1.

**amoral, non-moral.** See A-, AN-.

**amphibious** (living both on land and in water) was applied during the second world war to operations in which land forces are carried into action in ships and landing-craft covered by naval forces and put ashore in territory held by the enemy. No new principle of warfare was involved, but a new word may have seemed to be called for by the complexity of modern equipment and the invention of machines that are in fact amphibious. The word is inadequate, since air forces also normally take part in such operations, but an attempt to recognize this fact by coining the absurdity *triphibious* happily proved abortive.

**amphibol(og)y.** A statement so expressed as to admit of two grammatical constructions each yielding a different sense. *Stuff a cold and starve a fever* appears to be two sentences containing separate directions for two maladies, but may also be a conditional sentence meaning If you are fool enough to stuff a cold you will produce and have to starve a fever.

**ample,** as an attributive adjective in the sense *plenty of* (*he has a. courage; there is a. time; a. opportunities were given*), seems, when the OED was published in 1888, to have been used only with nouns denoting immaterial or abstract things. But it is now often attached to nouns that, like *butter, oil, water, coal,* denote substances of indefinite quantity; *We have a. water for drinking, There is a. coal to carry us through the winter.* This extension is natural and unexceptionable.

**amuck,** the familiar spelling, due to popular etymology, but going back to the 17th c. and well established, should be maintained against the DIDACTICISM *amok.*

**anacoluthon.** A sentence in which there is wrongly substituted for the completion of a construction something that presupposes a different beginning. *Can I not make you understand* that *if you don't get reconciled to your father* what is to happen to you? (the *that* construction requires a statement, not a question, to complete it). *Pliny speaks of divers engaged in the strategy of ancient warfare, carrying tubes in their mouths and so* drew *the necessary air down to their lungs.* (*Drew* presupposes a preceding *who carried* instead of *carrying.*) See also NOMINATIVUS PENDENS.

**analogy.** 1. As a logical resource. 2. As an influence in word-formation. 3. As a corrupter of idiom.

1. As a logical resource. The meaning of analogy in logic is inference or procedure based on the presumption that things whose likeness in certain respects is known will also be found alike or should be treated as alike, in respects about which knowledge is limited to one of them. It is perhaps the basis of most human conclusions, its liability to error being compensated for by the frequency with which it is the only form of reasoning available; but its literary, not its logical, value is what we now have to do with. Its literary merits need not be pointed out to anyone who knows the Parables, or who has read the essays of Bacon or Montaigne, full of analogies that flash out for the length of a line or so and are gone: *Money is like muck, not good unless it be spread.* What does need pointing out is unfortunately its demerit—the deadly dullness of the elaborate artificial analogy favoured by those who think it necessary to write down to their audience and make their point plain to the meanest capacity. Specimens fully bearing out this generalized description are too long to quote; but the following gives a fair idea of the essential stupidity of these fabricated analogies, against which no warning can be too strong. Let it be read and compared with the money that is like muck: *The Government are playing the part of a man entrusted with the work of guarding a door beset by enemies. He refuses to let them in at once, but provides them with a large bag of gold, and at the same time hands them out a crowbar amply strong enough to break down the door. That is the Government's idea of preserving the Union and safeguarding the integrity of the United Kingdom.*

2. As an influence in word-formation. In the making of words, and in the shape that they take, analogy is the chief agent. Wanting a word to express about some idea a relation that we know by experience to be expressible about other ideas, we apply to the root or stem associated with it what strikes us as the same treatment that has been applied to the others. That is, we make the new word on the analogy of the old; and in ninety-nine cases we make it right, being all old hands at the job; for each of us, in the course of a day, makes large numbers of words he has never seen in a dictionary or grammar, words for which his only warrant is merely an unconscious conviction that they are analogical. Nearly every inflexion is an instance; we are taught, perhaps, that the past of *will* is

*would,* or the plural of *ox* is *oxen,* but not that the past of *inflate* is *inflated* or the plural of *book* is *books;* those forms we make when we want them by analogy, and generally make them right. Occasionally, however, we go wrong: *The total poll midway in December was 16,244 so that upwards of half the electors were abstentients.* The writer wanted a single word for persons guilty of abstention, and one too that would not, like *abstainer,* make us think of alcohol. *Dissension* came into his head as rhyming with *abstention;* if that yields, said he, *dissentients,* why should *abstention* not yield *abstentients?* But the correspondence between *abstention* and *dissension* is not quite so close as he assumed; if he had remembered *dissentire* and *abstinere,* analogy would have led him to *abstinents* instead of to *abstentients* and perhaps eventually to the conclusion that there was no way out but to write *half the electors abstained.* That is a live newspaper instance of the fallibility of analogy, and dead specimens may be found in any etymological dictionary—dead in the sense that the unsoundness of their analogical basis excites no attention as we hear the words. Who thinks of *chaotic, operatic, dilation,* and *direful,* as malformations? Yet none of them has any right to exist except that the men who made them thought of *eros* as a pattern for *chaos, drama* for *opera, relate* for *dilate,* and *dread* for *dire,* though each pattern differed in some essential point from the material to be dealt with; the objection to some of the words is explained in the separate articles. These malformations, it is true, have now all the rights of words regularly made. They have prospered, and none dare call them treason; but those who try their luck with *abstentients* and the like must be prepared to pass for ignoramuses. See also HYBRIDS AND MALFORMATIONS and (talking of *ignoramuses,* for which false analogy has been known to substitute *ignorami*) LATIN PLURALS.

**3.** As a corrupter of idiom. That is the capacity in which analogy chiefly requires notice in this book. If by

*idiom* we mean the way in which it comes natural to an Englishman to word what he wishes to express, and if by *an idiom* we mean any particular combination of words, or pattern of phrase, or construction, that has become habitual with us to the exclusion of other possibilities, we find the pranks played by analogy upon idiom and idioms are innumerable. Each of the following extracts exhibits an outrage upon idiom, and each is due to the assumption that some word or phrase may be treated in a way that would be legitimate if another of roughly the same meaning had been used instead; that other is given in brackets, sometimes with alternative forms: *Men enjoy to play under him* (like to play *or* enjoy playing). | *He gives an example to all* (sets an example *or* gives a lead). | *Many opposition 'experts' appear to demur from the change* (demur to *or* dissent from). | *After the captain had had a word to the bowler* (had a word with *or* said a word to). | *Nothing is to be gained by blinking at it* (blinking it *or* winking at it). | *They showed no inclination to throw in the sponge* (throw up the sponge *or* throw in the towel). | *The proposals are generally identical to the policies of the opposition* (similar to *or* identical with). | *He smiled away awkward questions but answered frankly to most* (replied to most *or* answered most). | *His boundaries were hit to all parts of the compass* (all parts of the field *or* all points of the compass). | *The broadcasting authorities can be trusted not to derogate the supremacy of Parliament* (impair the supremacy *or* derogate from the supremacy). | *The double task was performed only at the expenditure of laborious days and nights* (by the expenditure *or* at the cost). | *Questions in which an intimate acquaintance of Scottish jurisprudence was particularly useful* (acquaintance with *or* knowledge of).

These are all casual lapses; each reveals that the writer is not a master of idiomatic English, but they are not caused by any widespread misapprehension of the meaning of particular

words. There are, however, words
whose sufferings under analogy are
more serious, so that the unidiomatic
substitute tends to supplant the true
English; some of these, dealt with in
separate articles, are: *as well as* (=
besides; see WELL); ANTICIPATE (=
expect); ADVOCATE (= recommend);
DUE TO (= owing to); FOIST (= fob);
FEASIBLE (= possible); HARDLY 2 (=
no sooner); HOPE (= expect); IN
ORDER THAT (= so that); INSTIL (=
inspire); OBLIVIOUS (= insensible);
POSSIBLE (= able); PREFER (= wish
rather); REGARD 3 (= consider);
SUPERIOR 3 (= better); VARIOUS (=
several). See also the general articles
OBJECT-SHUFFLING, SLIPSHOD EXTEN-
SION, and CAST-IRON IDIOM.

**anastrophe.** A term of rhetoric
meaning the upsetting, for effect, of
such normal order as preposition
before noun or object after verb. 'No
war or battle's sound Was heard *the
world around*'. '*Me* he restored and
*him* he hanged'.

**analysis.** 'But what good came of it
in the ultimate analysis?' Quoth little
Peterkin. That is how this famous
question would now be phrased by
the many people who find *in the end* or
*at last* too bald for them, and have
made a cliché out of *in the ultimate* (or
*final*) *analysis*, a POPULARIZED TECH-
NICALITY from chemistry.

**anchoret, anchorite.** The OED
states that the two forms were then
equally common. The first has the
two advantages of representing the
Greek original (ἀναχωρητής) more
closely, and corresponding better with
the surviving feminine *anchoress* (*an-
choritess* being now archaic). Nevertheless *anchorite* is now more usual
than *anchoret* and *anchoress* is rare.

**anchovy.** Formerly usually *ănchō'vĭ*;
now by RECESSIVE ACCENT *ă'nchŏvĭ* has
become commoner.

**anchylose, ankylose.** The right form
would be *ancylose*. The established
one was until recently *anchylose*, with

*h* inserted to indicate the *k* sound;
these irregular devices are regrettable,
since it is a matter of absolute indifference whether the hard sound is
preserved or not, while the inserted *h*
puts the Greek scholar off the track.
The form with *k*, always an occasional
variant, secures the sound more certainly and does not deceive the
scholar; and now that it has been
adopted for the disease *ankylostomiasis*
(hook-worm) it will probably supersede *ch* in the other derivatives of the
same Greek word.

**and.** 1. *And which.* 2. Bastard enumeration. 3. Commas in enumeration.
4. Omission of *and* in enumerations.
5. *And* beginning a sentence.

1. For *and which* see WHICH WITH
AND OR BUT.

2. Bastard enumeration. There is
perhaps no blunder by which hasty
writing is so commonly defaced at
present as the one exemplified in *He
plays good cricket, likes golf and a
rubber of bridge.*
The test for this now prevalent
slovenliness is fortunately very simple:
there must be nothing common to two
or more of the items without being
common to all. In the *He plays*
example the word *likes* is common
to the golf and bridge items, but has
no relation to the cricket item. In the
following examples what is common to
some but not all items is in roman type
and corrections are added in brackets.
But it may be said in general that insertion of the missing *and*, from which
ignorant writers shrink consciously or
unconsciously, is usually attended with
no more damage to euphony than that
repetition of essential words by the
fear of which ELEGANT VARIATION, in
all its distressing manifestations, is
produced; there is nothing to offend
the ear in *He plays good cricket, and
likes golf and a rubber of bridge.*
*Hence* loss of *time, of money, and sore
trial of patience* (correct to *time and
money*). | It was *terse, pointed, and a
tone of good humour made it enjoyable*
(correct to *terse and pointed*). | *His
workmen* are *better housed, better fed,*

*and get a third more in wages* (correct to *housed and fed*). | *For* it *fails to include many popular superstitions, does not evidence any great care in its composition, and its arrangement is amateurish* (correct to *and is amateurish* in arrangement).

A few bad enumerations are added for which carelessness rather than a wrong theory seems responsible. *The centuries during which the white man kidnapped, enslaved, ill-treated, and made of the black a beast of burden (and made a beast of burden of*; or *and ill-treated the black and made him a beast of burden*). | *Many of these authoresses are rich, influential, and are surrounded by sycophants who* . . . (*and surrounded*).| *It is true he has worked upon old material, has indulged in no more serious research than a perusal of the English and French chronicles of the age and most of the modern works dealing with the subject* (*material and indulged*).

**3.** *And* and commas in enumerations. See STOPS, COMMA (B).

**4.** Omission of *and* in enumerations. *Manchester has a tradition for fine journalism, fine libraries, fine music.* | *Barristers, civil servants, politicians who had learnt how to examine a text closely* . . . *would be able to tear the heart of a brief or a minute.* | *There is always the delicate music with its suggestions of passion, alarm, tragedy.* | *It is ridiculous that students of English should never come into contact with the great prose, pregnant with matter, of Hobbes, Clarendon, Burke, Gibbon, Berkeley, Hume, Macaulay, Carlyle, Bradley.* | *There were few people who had not heard of the practical eccentric, purposeful joker, fabulous showman.* | *Lectureships, television, journalism, are the kind of jobs that command respect.* | This omission of the customary *and* between the last two items of an enumeration (a form of the rhetorical figure known as ASYNDETON) is evidently becoming fashionable. Perhaps it originated in unconscious imitation of HEADLINE LANGUAGE, where space cannot be spared for *ands*. Perhaps it is meant to suggest that the enumeration is not exhaustive: the writer could

add more if he chose. Perhaps the staccato effect is thought to introduce the same kind of liveliness as the VERBLESS SENTENCE. As with the verbless sentence, any effectiveness the device may have depends on its being used sparingly enough to be unobtrusive.

**5.** *And* beginning a sentence. That it is a solecism to begin a sentence with *and* is a faintly lingering SUPERSTITION. The OED gives examples ranging from the 10th to the 19th c.; the Bible is full of them.

**and/or.** The ugly device of writing *x and/or y* to save the trouble of writing *x or y or both of them* is common and convenient in some kinds of official, legal, and business documents, but should not be allowed outside them.

**anent,** apart from its use in Scottish law-courts, where it is in place, is chiefly met with in letters to the press; that is, it is a favourite with unpractised writers who, on their holiday excursions into print, like to show that they possess gala attire. See WARDOUR STREET. *Anent* is often found in the company of dubious syntax and sense, as in: *Sir, Your remarks today on the result of the Canadian election anent the paragraph in the Philadelphia Record is, I am glad to see, the first sign of real appreciation of* . . . (*is* should be *are*; and he is not glad that it is the *first* sign).

**aneurysm,** is preferable to the now more usual *aneurism.* The y is due to Greek εὐρύς wide; the false form suggests the totally different sense *nervelessness*; cf. MESEMBRIANTHEMUM.

**angina.** OED gives preference to *ǎ′njǐna,* but the long *i* has now prevailed; the COD gives no other pronunciation.

**angle** n., **angle** v., **angler.** The noun *angle* (fish-hook) is now scarcely used. The verb is chiefly a stylish synonym (see WORKING AND STYLISH WORDS) for *fish.* Perhaps it is to give a veneer of respectability to our actions that we choose the stylish word when we *angle for votes*, though we are content to *fish for compliments* or *in troubled waters*. *Angler* on the other hand can be put to good use, since *fisher* is archaic and *fisherman* is mostly applied to one who makes his living with nets. But the taint of archaism attaching to *angle* has affected *angler* and, though we still have *anglers' clubs*, it is now more natural to say of one skilled in the art *he is a good fisherman* than *a good angler*. For *angle* in the sense of *point of view* see POINT.

**Anglo-Indian.** For ambiguous meaning see MULATTO 4.

**animalcula.** See LATIN PLURALS 3.

**another.** 1. For *one a.* see EACH 2. 2. *This was quite a. thing from coldness.* / *Its people seen as a. kind from us.* This use of *from* instead of *than* with *a.* in its sense of *different* is condemned by the OED as catachrestic, but these quotations show that it may occasionally be found in good writers today. Cf. the use of *than* instead of *from* with *different*. See THAN 8.

**antagonize.** As a synonym for *oppose*, *resist*, *neutralize*, *counteract*, the word is recognized in dictionaries and seems to be sometimes used in America (*The Democrats have given notice to antagonize this and all other Bills for* ...). But in Britain it retains this sense only as a term of physics; its ordinary meaning is to provoke to opposition (*It assumes infallibility and warns off critics in a tone of determination sufficient to antagonize the man who approaches its findings with an open mind*). With us therefore the verb has parted company with the nouns *antagonist* and *antagonism*, which still mean opponent and opposition. To antagonize someone is to make him your antagonist, perhaps unintentionally, as in the example given above. In this sense the verb, little more than a needless variant if used for *oppose*, is obviously valuable; it conveys an idea not otherwise expressible in one word.

**antetype.** See -TYPE

**anticipate.** 1. Meaning. The use of *a.* as a synonym for *expect*, though very common, is a SLIPSHOD EXTENSION. The element of *forestall* present in *a.* ought to have been preserved and is still respected by careful writers. LOVE OF THE LONG WORD has led to the habit of writing, e.g., *We anticipated this crisis* when what is meant is that we foresaw it; the proper meaning is that we took steps beforehand to meet it.

2. Unidiomatic use. This is the natural result of 1. *Exhibitions of feeling were, of course*, anticipated *to take place on Monday.* / *This book, which we repeat, might have been* anticipated *to contain a manifesto of the aims of the young intellectuals of America, proves to be* ... / *A noteworthy act, which may be* anticipated *to have far-reaching effects in the future.* The OED has 9 separately numbered definitions, and 35 quotations, for *anticipate*. None of the definitions and none of the quotations, suggest the possibility of such a use as is seen in all the above newspaper extracts; the writers have thought their sentences with the homely *expect*, which would have served perfectly, and then written them with the FORMAL WORD *anticipate* ANALOGY has duped them into supposing that since it vaguely resembles *expect* in sense it must be capable of the same construction.

**anticlimax.** Annulment of the impressive effect of a climax by a final item of inferior importance. *The rest of all the acts of Asa, and all his might and the cities which he built, are they not written in the book of the chronicles of the kings of Judah? Nevertheless in the time of his old age he was diseased in his feet.*

**anti-novel** etc. After the second world war the prefix *anti* came into use as a label for anything that flout tradition and convention. 'In an anti-novel', writes a reviewer of one of them, 'action there is of a sort, but the characters buzz about sluggishly like winter flies.' 'The old novel', write another, 'was about the nobility and heroism of man; the anti-novel prefers ignobility.' The typical anti-hero is a feckless creature without any of the adventurousness of his picaresque archetype; if he has any adventures they are likely to be of a sordid kind— 'plenty of bed and booze'. Speculating on the reason for the contemporary passion for being 'anti' Sir Charles Snow has diagnosed it as 'an expression of that nihilism which fills the vacuum created by the withdrawal of positive directions for living, whether religious or humanist. In my happier nightmares', he adds, 'I see myself attending an anti-play with an anti-audience after a dinner prepared by an anti-cook'.

**anti-saxonism.** See SAXONISM

**antithesis.** Such choice or arrangement of words as emphasizes a contrast. *Crafty men contemn studies; simple men admire them; and wise men use them.*

**antitype.** See -TYPE.

**antonomasia.** A figure whereby an epithet or other phrase stands for a proper name (e.g. *the Iron Duke*) or conversely the use of a proper name to express a general idea (e.g. *a Solomon*). For examples see SOBRIQUETS and GENERIC NAMES.

**anxious.** The objections made to it in the sense *eager* (*to* hear, improve, go, etc.) as a modernism, and in the sense *likely to cause anxiety* (*It is a very a. business; You will find her an a. charge*) as an archaism, are negligible; both are natural developments, the first is almost universally current, and the second is still not infrequent.

**any.** 1. Compounds. *Anybody, anything, anyhow, anywhere*, are always single words; so also the now obsolescent adverb *anywise* (but *in any wise*). For *anyone, any one*, see EVERY ONE. *Any way* is best written as two words (*I cannot manage it any way*) except when it means *at all events, however that may be, at any rate* (*Anyway, I can endure it no longer*); at *any rate*, not *at anyrate*. *Any time* is

sometimes written as one word in U.S. but is always two in Britain. *Anyplace* (= anywhere) is U.S. only.

2. *He is the most generous man of anyone I know.*
This common idiom, which looks illogical (*of all I know* being the logical form) may be defensible as a development or survival of the archaic type *Caesar, the greatest traveller, of a prince, that had ever been*, where *of* means *in the way of*, and we should now write *for* instead. But it is more likely to be merely a loose use of *anyone* as though it were a plural. In any case those who dislike the illogical and wish to avoid such a sentence as *Surroundings that are the most gently beautiful of any English cathedral* can always write *more gently beautiful than those of any other.*

3. *Then, for the first time, she paid any attention to my existence.* This use of *any* (= some) is idiomatic only in negative or interrogative statements, express or implied. *Have you any bananas? No we haven't any bananas.* But *yes we have some bananas.* If the sentence quoted had run *This was the first time that she* etc. it would have had enough negative implication ('she had not paid any before') to justify *any*; but as written, with *for the first time* in parenthesis, it becomes uncompromisingly positive, and *any* will not do.

**aphaeresis.** The loss of an initial letter, syllable, etc. *Special* was formerly *especial, adder naddre*, and *cute acute*. Cf. SYNCOPE, APOCOPE.

**aphetic.** An adj. now often used instead of the rare *aphaeretic* and meaning 'resulting from aphaeresis or aphesis'—the latter a comparatively modern word to express gradual and unintentional aphaeresis as in *squire* (esquire), *'shun* (attention).

**aphorism, apophthegm,** etc. The word *aphorism*, meaning literally a definition or distinction, is of medical origin; it was first used of the Aphorisms of Hippocrates, who begins his collection with one of the most famous of all famous sayings *Art is long; life* *is short*. The word has come to denote any short pithy statement containing a truth of general import (OED). It is distinguished from the *axiom*, which is a statement of a self-evident truth, and also from the *theorem*, which is a demonstrable proposition in science and mathematics; the aphorism concerns itself with life and human nature, and its truths are incapable of scientific demonstration. The *epigram*, on the other hand, though concerned with life like the aphorism, and possessing its terse and pointed form, is lacking in general import; it is not the statement of a general truth. Other words with much the same significance as aphorism are *maxim, apophthegm, saying*; and although these have sometimes been distinguished from one another —the maxim being defined as containing not only a general truth but also advice and admonition, and the apophthegm or saying as an aphorism or maxim expressed in speech—these distinctions of the rhetoricians are not of much importance and have seldom been observed in the current usage of the words. *Sentence* and *gnome* are old names for the aphorism which have fallen out of use. (Adapted from the Introduction to *A Treasury of English Aphorisms* by Logan Pearsall Smith, 1943.)

**apocope.** The loss of a final letter or syllable or more. *My, curio, cinema*, were formerly *mine, curiosity, cinematograph*. Cf. APHAERESIS and SYNCOPE.

**apology.** It is perhaps the disuse of *apology* in the sense of 'justification, explanation or excuse of an incident or course of action' (OED), as Socrates and Newman used it, that has led to ALIBI's unseemly occupation of the vacant place.

**apophthegm.** This has been the usual spelling in Britain since Dr. Johnson, for the sake of etymology, preferred it to the older *apothegm*, still current in U.S. But even in Britain few people attempt the difficult task of sounding the *ph*, and the pronunciation *ap'othem* is recognized by all dictionaries, by

most as the only one. For meaning see APHORISM.

**aposiopesis.** Significant breaking off so that the hearer must supply the unsaid words. *If we should fail* ——. *Oh, go to* ——*!* Pronounce *ăposīopē'sis*.

**a posteriori.** Working back from effects to causes, i.e. inductively. *God's in his heaven—all's right with the world* is an a posteriori inference if it means The world is so clearly good that there must be a god in heaven; but an a priori inference if it means that since we know there is a god, the state of the world must be right. See also INDUCTION, DEDUCTION.

**apostrophe.** See POSSESSIVE PUZZLES, and -ED AND 'D, for some points concerning its use.

**appal(l).** The single *l* is now standard spelling, but *-lled* etc., see -LL-, -L-, 1.

**a(p)panage.** Either form will do; *appa-* is perhaps commoner in general, and *apa-* in learned use. Sometimes ignorantly used for *appendage*. *Appanage* meant originally provision made for the younger children of monarchs by way of territorial or other jurisdiction; hence generally any dependency or perquisite, and, by further extension, any natural accompaniment or attribute.

**apparatus.** Pl. *-uses*. See LATIN PLURALS. Pronounce *-ātus* not *-ahtus*.

**apparent(ly).** 1. Either pronunciation (*-ãrent, -ãrent*) is legitimate, though the latter is more usual; see FALSE QUANTITY. 2. For commas before and after *apparently* see THEREFORE. 3. *Heir apparent*. See HEIR.

**appeared** is liable to the same misuse as *seemed* with the PERFECT INFINITIVE.

**appeasement.**                See WORSENED WORDS.

**appendix.** Pl. *-dices* (-sēz) or *-dixes* (-ĕz); see -EX, -IX, 4.

**applicable.** The stress should be on the first syllable. See RECESSIVE ACCENT.

**appraisal** was formerly a literary word (OED's quotations are from Coleridge, De Quincey, and M. Arnold) but under U.S. influence it is now replacing *appraisement* in Britain as a word of general use in political, military, and business affairs; and in these unstable times *reappraisal* (sometimes agonizing) has inevitably followed.

**appreciate** is overworked as a synonym of *understand* or *recognize* in official and business letters, especially in the phrase *It is appreciated that* to introduce a recognition of the point of view of the person addressed or *you will appreciate that* to ask him to recognize the writer's. This tendency to wander can be restrained by using the word only with a noun as its object: *I appreciate your great patience* but not *I appreciate that you have had a lot to put up with*; there the writer can choose between *recognize, know, realize,* and *admit*.

**apprehend -ension, comprehend -ension.** So far as the words are liable to confusion, i.e. as synonyms of *understand(ing)*, the *ap-* forms denote the getting hold or grasping, and the *com-* forms the having hold or full possession, of what is understood. What is *beyond my apprehension* I have no cognizance of; what is *beyond my comprehension* I am merely incapable of fully understanding. To *apprehend* a thing *imperfectly* is to have not reached a clear notion of it; to *comprehend* it *imperfectly* is almost a contradiction in terms. *I apprehend that A is B* advances an admittedly fallible view; *I comprehend that A is B* states a presumably indisputable fact.

**approximately** is much in use as a FORMAL WORD for *about*; if there is a difference it is that *a.* suggests a more careful calculation. *With near approach to accuracy* is the OED's definition. *Very approximately* should therefore imply a nearer approach to accuracy than *approximately* alone, instead of the opposite—*very roughly*—as it is always intended to do. But as it is universally understood in the sense intended it might be classed as a STURDY INDEFENSIBLE.

**a priori.** Working forward from known or assumed causes to effects, i.e. deductively. For an example see A POSTERIORI. Pronounce *ā prīo'rī*. See LATIN PHRASES. See also INDUCTION, DEDUCTION.

**apropos** is so clearly marked by its pronunciation as French, and the French construction is, owing to *à propos de bottes*, so familiar, that it is better always to use *of* rather than *to* after it if any preposition is used; *apropos what you were saying* is, however, good colloquial idiom. Probably *to* is partly accounted for by some confusion with *appropriate* (*His interpolation of stories that were not always strictly apropos to the country through which we were passing*).

**apt, liable, likely, prone, calculated.** Followed by *to* with the infinitive in the senses *having an unfortunate tendency* (apt), *exposed to a risk* (liable), the first two of these words are so near in meaning that one of them (*liable*) is often used where the other would be better. It may perhaps be laid down that *apt* is the right one except when the infinitive expresses not merely an evil, but an evil that is one to the subject. This is so, and therefore *liable* is right, in: *We are l. to be overheard* (being overheard is an evil to us); *The goods are l. to suffer.* It is not so, and therefore *apt* is the only word, in: *Curses are a. to come home to roost* (the evil is not to the curses, but to the curser); *Damage is a. to be done*; *Difficulties are a. to occur*; *Lovers' vows are a. to be broken.* It is usually not so, and therefore *apt* is usually the right word, in: *He is a. to promise more than he can perform* (but *liable*, if the evil suggested is the shame he feels); *Cast iron is a. to break* (but *liable*, if we are sorry for the iron and not for its owner).

Since *liable* is apt to encroach, and *apt* is liable to suffer neglect, the best advice is never to use *l.* till *a.* has been considered and rejected.

In British usage *apt* always implies a general tendency; for a probability arising from particular circumstances *likely* is the word. *He is apt to be late for his engagements*, but *He is likely to be late for his engagement if he does not start soon*. In the second sentence *apt* would be idiomatic in U.S. but not in Britain. *Prone* has the same significance as *apt*, but differs in being suitable only for use with persons, as in the expression *accident-prone*, a product of the motor-car age. The natural meaning of *calculated* is designed to produce a certain result (*a story calculated to make the reader's hair stand on end*), often in the negative of unsuccessful designs (*a disguise not calculated to deceive anyone*), and sometimes ironically (*a proposal calculated to have the most unexpected results*). See also LIABLE and INAPT.

**Arab, Arabian, Arabic.** With exceptions for a phrase or two, such as *gum arabic*, the three adjectives are now differentiated, *Arab* meaning of the Arabs, *Arabian* of Arabia, and *Arabic* of the language or writing or literature of the Arabs. So we have *an Arab chief, child, girl, horse, league, sheikh*; *Arab courage, fanatics, fatalism, philosophy, traditions*; *the Arabian desert, fauna, flora, gulf, and nights*; *Arabian gales*; *the Arabic numerals*; *an Arabic word*; *Arabic literature, writing*. *Arab* and *Arabian* can sometimes be used indifferently; thus *an Arab village* is one inhabited by Arabs; if it happens to be in Arabia it is also *an Arabian village*, and may be called by either name; *the Arab war* is one with Arabs; *the Arabian war* is one in Arabia; and the two may be one. Also *Arabian* may still be used instead of *Arab* of what belongs to or goes back to the past, as *Arabian conquests, monuments, philosophy, records*.

**arbiter, arbitrator.** These words, originally synonymous and still sometimes so treated, have developed a differentiation that should be respected. An arbiter makes decisions of his own accord and is accountable to no one for them, as a dictator may be the arbiter of a people's fortunes or a dressmaker the arbiter of fashion. An arbitrator decides an issue referred to him by the parties and is accountable if he fails to act judicially or to observe a procedure prescribed by statute. An arbiter acts arbitrarily; an arbitrator must not.

**arbo(u)r.** In Britain the bower is spelt with *u* and the mechanical term without.

**arc,** vb. For inflexions see -C-, -CK-.

**arch,** adj. For meaning see JOCOSE.

**arch-, arche-, archi-.** Though the prefix *arch-* (chief etc.) is pronounced *arch* in all words except *archangel* and its derivatives, the longer forms are always pronounced *arki*: so *archbishop* (-ch-), but *archiepiscopal* (-k-); *archdeacon* (-ch-), but *archidiaconal* (-k-). The ch is hard in *archetype, archimandrite, Archimedes, architectonic, architrave*.

**archaism.** A certain number of words through the book are referred to this article, and such reference, where nothing more is said, is intended to give warning that the word is dangerous except in the hands of an experienced writer who can trust his sense of congruity. Even when used to give colour to conversation in historical romances, what Stevenson called *tushery* is more likely to irritate the reader than to please him. Archaic words thrust into a commonplace context to redeem its ordinariness are an abomination. More detailed remarks will be found in the general articles INCONGRUOUS VOCABULARY, REVIVALS, and WARDOUR STREET. Particular words under which the question of archaism is discussed are *anent, aught, bounden, chide, choler(ic), confidant* (n.), *derring-do, except* (conj.), *howbeit, parlous, perchance, save* (prep.), *surcease*; and a few specimens of those for which the mere reference to this article or to WARDOUR STREET has been thought sufficient are *advert, bashaw, belike, betwixt, broider(y), certes, fortune* (vb.), *peradventure, quoth(a)*, and *whit*.

**archetype.** See -TYPE.

**area.** For synonyms see FIELD.

**are, is.** When one of these is required between a subject and a complement that differ in number (*these things . . . a scandal*), the verb must follow the number of the subject (*are*, not *is*, *a scandal*). See NUMBER I.

**argot.** See JARGON.

**arguing in a circle.** The basing of two conclusions each upon the other. That the world is good follows from the known goodness of God; that God is good is known from the excellence of the world he has made.

**argumentum ad —.** *a. a. hominem*, one calculated to appeal to the individual addressed more than to impartial reason; *a. a. crumenam* (purse), one touching the hearer's pocket; *a. a. baculum* (stick) or *argumentum baculinum*, threat of force instead of argument; *a. a. ignorantiam*, one depending for its effect on the hearer's not knowing something essential; *a. a. populum*, one pandering to popular passion; *a. a. verecundiam* (modesty), one to meet which requires the opponent to offend against decorum.

**arise,** in the literal senses of getting up and mounting, has given place, except in poetic or archaic use, to *rise*. In ordinary speech and writing it means merely to come into existence or notice or to originate *from*, and that usually (but cf. *new prophets a. from time to time*) of such abstract subjects as *question, difficulties, doubt, occasion, thoughts, result, effects*.

**Aristotele'an, Aristote'lian.** The latter has prevailed. Cf. MEPHISTO-PHELIAN, HERCULEAN.

**arithmetical.** See PROGRESSION.

**armada.** The OED gives *armáda* only, but *armahda*, perhaps because of our present tendency to give a foreign value to the vowels in any word that looks foreign (see PRONUNCIATION I), is now usual.

**arms** (weapons). The singular is late, rare, and best avoided. Instead of describing a particular pattern of rifle or sword as 'a beautifully balanced arm', it is worth while to take the trouble to remember *weapon*. *Firearm* is the statutory term for the types of weapon, defined in the Firearms Act, that it is illegal to possess without a certificate, but it is not in general use, and we do well to sacrifice the exhaustive brevity of *The report of a firearm was heard* and risk ambiguity with *gun*, inaccuracy with *pistol*, or extravagance with *pistol, rifle, or gun*—unless, of course, we have the luck to hit upon *shot*. The only sense in which the singular is idiomatic is in such phrases as *arm of the service* (cavalry, infantry, or artillery) or *the air arm* (cf. *the secular arm*).

**army and navy.** This, the familiar order (e.g. *Army and Navy Club, Stores*), is rightly corrected in toasts, public speeches, and the like, into *Navy and Army*; but where precedence is not in question it is both needless and impossible to get the correction accepted.

**around** is, in British use, a variant of *round* disappearing until recently but now coming back under American influence. It is still the normal form in certain combinations, as *a. and about*, (*the air*) *a. and above* (*us*), all *a.* (*are signs of decay*); and it can be used without being noticeable in a few of the senses of *round*, as *seated a. the table, diffuses cheerfulness a. her, spread destruction a.* But it is hardly possible to say *winter comes a., all the year a., win one a., drinks all a., sleep the clock a.* American usage is quite different; among the examples in an American dictionary are the following, all of which are still unnatural for an Englishman: *He went through, but I ran a.; He turned a.; The earth turns a. on its axis; Go a. to the post office; The church a. the corner*.

**arouse** The relation of this to *rouse* is like that of *arise* to *rise*; that is, *rouse* is almost always preferred to it in the literal sense and with a person or animal as object. *A.* is chiefly ᵘˢᵉᵈ with the senses 'call into existence, occasion', and with such abstracts as *suspicion, fears, cupidity, passions,* as object of the active or subject of the passive: *This at once aroused my suspicions; Cupidity is easily aroused. Rouse* would be more suitable in *I shook his arm, but failed to a. him.*

**arrive.** For the absolute sense 'reach success or recognition' see GALLICISMS.

**art.** For the broad distinction between *a.* and *science* see SCIENCE.

**artiste. 1.** The word is applicable to either sex; *artists and artistes* as a phrase for male and female performers is a mere blunder.

**2.** In the sense 'professional singer, dancer, or other public performer', *artiste* serves a useful purpose; it is best restricted to this use, in which it conveys no judgement that the performance is in fact artistic. If it is desired to intimate that a cook, tailor, hairdresser, etc., or an artiste, makes his occupation into a fine art, *artist,* and not *artiste,* is the right word: *He is quite an artist; What an artist!*

**as. 1.** *Equally as.* **2.** Causal *as* placed late. **3.** *As to.* **4.** *As if, as though.* **5.** *As,* relative pronoun. **6.** Omission of *as.* **7.** *As =* in the capacity of. **8.** *As from* etc. **9.** Case after *as.* **10.** *As a fact.* **11.** *As well as.* **12.** *As concern(s)* etc. **13.** *As and when, as such,* and *as per.*

**1.** For the illiterate *equally as (good* etc.) see EQUALLY.

**2.** Causal *as* placed late, *as* meaning since, because, seeing that, for, etc. To causal or explanatory *as*-clauses, if they are placed before the main sentence (*As he only laughed at my arguments, I gave it up*) there is no objection. The reverse order (*I gave it up, as he only laughed at my arguments*) is, except when the fact adduced is one necessarily known to the hearer or reader and present to his mind (*I need not translate, as you know German*), unpleasant to anyone with a literary ear. All good writers instinctively avoid it; but, being common in talk, it is much used in print also by those who have not yet learnt that composition

is an art and that sentences require arrangement. The first passage quoted below suggests the kind of writer most liable to this weakness; the second and third, with their successive *as*-clauses, exhibit the total lack of ear that accompanies it: *One is pleased to find that Mr. P. Gannon still maintains his form, as he won the Open Challenge cup in face of such opponents as . . . | The Sunningdale man, indeed, put up a most strenuous fight, as his 154 equalled the total of . . . Mr. Carlisle's golf, however, was much more consistent than Mr. Gannon's, as to his two 77's Mr. Gannon opposed an 82 and a 72. | They strongly advocate a diminution of the petroleum duty, as it would lead to a great increase of work, it being largely used for industrial purposes, as coal is scarce here.*

**3.** *As to.* This has a legitimate use— to bring into prominence at the beginning of a sentence something that without it would have to stand later (*As to Smith, it is impossible to guess what line he will take*); it has, for instance, been wrongly omitted in: *Whether the publishers will respond to Sir Charles Stanford's appeal or not it is too early to speak with any confidence.* A spurious imitation of legitimacy occurs when *as to* introduces what would even without it stand at the head of the sentence, as in: *As to how far such reinforcements are available this is quite another matter*; omit 'as to' and 'this', and the order is unaffected; the writer has chosen to get out of the room by a fire-escape when the door was open. *As to* is of course idiomatic when *to* is part of an infinitive of result or purpose and *as* refers back to a preceding *so. He so acquitted himself as to please everybody.*

Apart from this, *as to* is usually either a slovenly substitute for some simple preposition (*Proper notions as to* [of] *a woman's duty.* | *She had been sarcastic as to* [about] *his hunting.* | *Piece of business as to* [upon] *which Dr. Thorne had been summoned.* | *Bantered himself as to* [on] *his own want of skill.* | *The manufacturer complains that everything as to* [concerning] *the future is left to the whim of the Board of Trade*); or it is entirely otiose (*The only points on which the Government found fault were* [*as to*] *the Permanent Settlement and* [*as to*] *the system on which* . . . | *Asked* [*as to*] *what effect the arrest of the players would have on the American tour, Mr. Yeats said* . . . | *With the idea of endeavouring to ascertain* [*as to*] *this.* | *Doubt is expressed* [*as to*] *whether this will affect the situation*).

As might be expected, those who put their trust in a phrase that is usually either vague or otiose are constantly betrayed by it into positive bad grammar: *Unless it has some evidence* as to whom *the various ideas belong* (i.e. as to to whom). | *A different dance* according as to whether *the child is a male or a female* (i.e. according as the child is). | *It is open to doubt* as to what extent *individual saving prevails,* and *No two people seem to agree* as to what extent *it is one country and one race* (i.e. as to to what extent). | *It is not quite clear* as to what *happened* (This implies the ungrammatical *As to what happened is not clear* instead of the normal *What happened is not clear*). | *The question does not relate solely to the possibility of Mr. Whitaker accepting appointment, but also* as to whether *any more suitable candidate can be suggested* ('relate as to'?—relativity would seem to be as upsetting to Priscian as to Euclid). The popular favourites: *The question as to whether, The doubt as to whether,* may almost be included among the ungrammatical developments, since *the* doubt or question demands an indirect question in simple apposition (*The question whether, The doubt whether*); in such forms as *Doubts are expressed as to whether,* the 'as to' is not incorrect, but merely repulsive. *An interesting question therefore remains as to how far science will provide us with the power* may lawfully be written; *The interesting* etc. may not. See also QUESTION 6.

**4.** *As if, as though.* These should ordinarily be followed by the 'past' form of the conditional, and not by the present indicative (*would,* not *will; could,* not *can; did,* not *does; was* or *were,* not *is; had gone,* not *has gone;*

knew, not *knows*). The full form of the incorrect *It is scanned curiously, as if mere scanning* will *resolve its nature* is *It is scanned curiously,* as it would be scanned *if mere scanning* would *resolve its nature;* and the omission of *it would be scanned* leaves *would resolve* unchanged. *As though,* about which the same demonstration is not possible, is precisely equivalent to *as if*, and is subject to the same rule; and the rule applies to the still more elliptical use of either form in exclamations (*As if,* or *As though, you didn't* [not *don't*] *know that!*) as well as to the use with an expressed main sentence. The tendency to put the verb in a present tense is especially common after *it looks* or *seems.* Here there may be the justification that the clause gives a supposed actual fact; but sometimes the supposition is admittedly false: *It looks as if the Party is bringing pressure on Dr. Adenauer.* | *To the observer from without it seems as if there has been some lack of stage-management.* | *It is not as though a sound liquor is supplied. Is* and *has* may pass in the first and second quotations, but the last cries out for *is* to be changed to *was* or *were.*

**5.** *As,* relative pronoun. The distinction between *as* the relative pronoun, capable of serving as subject or object of a verb, and *as* the relative adverb, not capable of so serving, must be grasped if a well-marked type of blunder is to be avoided. Examples of the blunder are: *The ratepayers have no direct voice in fixing the amount of the levy,* as is possessed *by the unions.* | *With a speed of eight knots,* as has been found practicable, *the passage would occupy five days.* | *There were not two dragon sentries keeping ward,* as *in magic legend* are usually found *on duty.* If these sentences, the faultiness of which will probably be admitted at sight, are examined, it will be seen that for each two cures are possible. One is to substitute for *as* an undoubted relative pronoun (*such as* or *which*) and write, *such as is possessed; which has been found practicable; such as are usually found.* The other is to insert a missing subject or object (*as one is* *possessed*; as it *has been found practicable* to steam; *as* dragons *are usually found*). Either method of correction suggests the same truth, that the adverbial *as* in these sentences is not a relative pronoun, and has been wrongly treated as one. The fact is that when *as* is used as a relative pronoun the antecedent is never a simple noun that has already been expressed (which must be represented by an ordinary relative—*such as, which, who, that*), but a verb or verbal notion, or a previously unexpressed noun, that has to be gathered from the main sentence. Thus we cannot say *To affect poverty, as is now often affected* (i.e. which poverty is affected); but we can say *To affect poverty, as is now often done* (i.e. which affecting is done). If this test is applied to the incorrect sentences above, it will be found that each antecedent of the supposed relative pronoun is of the illegitimate kind, a simple expressed noun—*voice, speed, sentries*. It may perhaps occur to the reader that a legitimate substitute for *as is possessed by the unions* in the first example would be *as the unions have*, and that *as* then represents *which voice* and consequently invalidates our rule. That it seems to do so, however, is owing to a peculiarity of the verb *have*. The *ratepayers possess no voice, as the unions* do; *the ratepayers exercise no voice, as the unions do*; *the ratepayers have no voice, as the unions* (not *do*, but) *have*. *Have* (in the sense *possess*), being never represented like other verbs by DO, is allowed, when used where *do* would be substituted for any other verb, to take the construction proper to *do*; *as the unions possess it* or *do*; *as the unions exercise one* or *do*; but *as the unions have* simply.

On the other hand, failure to recognize that *as* is a relative pronoun sometimes produces mistakes of a different kind: *Epeiros, as it is well known, was anciently inhabited by* . . . (*as* = which fact, and *it* is therefore impossible unless *as* is omitted).

**6.** Omission of *as*. *But it is not so much as a picture of time as a study of humanity that Starvecrow Farm claims*

*attention* (*as as a study* has been too much for even a literary critic's virtue); cf. the omissions of *to* in the *as to* quotations in the last paragraph of 3 above; these are examples of CANNIBALISM. For the omission of *as* after *regard* see REGARD.

**7.** *As* = in the capacity of. When this is used, care must be taken to avoid the mistake corresponding to what is called the UNATTACHED PARTICIPLE; we can say *He gave this advice as leader of the opposition*, or *This advice was given by him as leader*, *he* and *him* supplying the point of attachment; but we cannot say *The advice which he tendered to the Peers was given as leader of the opposition*. The writer of the following has fallen into this trap through being in too great a hurry with his *that*: *I should like to say that as a social worker in this field, without exception, no contraceptive technique had been applied by any of the young women who came to me for assistance.*

**8.** *As from* etc. An agreement, especially one for an increase of pay, is often said to come into effect *as from* a certain date. That is reasonable if it is retrospective: the agreement is to be treated as if it had been made earlier. Otherwise the *as* is superfluous. That curious compound preposition sometimes used, *as and from*, is merely silly, whatever the date may be.

**9.** Case after *as*. It is a matter of no great practical importance, case being distinguishable only in a few pronouns, and these pronouns occurring so seldom after *as* that most of the examples given in illustration will have an artificial air; but some points may be noticed. (*a*) Sometimes a verb is to be supplied; the right case must then be used, or the sense may be spoilt; *You hate her as much as I* implies that I hate her; *You hate her as much as me* implies that you hate me. (*b*) *As* is never to be regarded as a preposition; the objective case after it, when right, is due either to the filling up of an elliptic sentence as in (*a*) or to causes explained in (*c*) and (*d*); *When such as her die, She is not so tall as me*, may pass colloquially but not in prose. (*c*) The phrases *such—as he* etc., *so—as he* etc., may be treated as declinable compound adjectives (cf. German *was für ein*), which gives *Such men as* he *are intolerable* but *I cannot tolerate such men as him*; *Never was so active a man as he* but *I never knew so active a man as him*; to ban this construction and insist on writing *he* always, according to the (*a*) method, seems pedantic, though *he* is always admissible. (*d*) In many sentences the supplying of a verb supposed to have been omitted instead of repeated, as in (*a*), is impossible or difficult, and the case after *as* simply follows that of the corresponding noun before *as*; *as* is then equivalent to *as being* (cf. Greek ὡς ὤν, ὡς ὄντα, etc.); so *I recognized this man as him who had stared at me*; *You dressed up as she*, *You dressed yourself up as her*, *I dressed you up as her*, *You were dressed up as she*; *The entity known to me as I*, *The entity that I know as me*. Similar questions arise with *than* and are discussed in THAN 6.

10. *As a fact*; see FACT.

11. *As well as*; see WELL.

12. For *as concern(s)*, *regard(s)*, *follow(s)*, see CONCERN, FOLLOW.

13. For *As and when* see IF AND WHEN, for *as per* see COMMERCIALESE, and for *as such* see SUCH.

**ascendancy, ascendant.** 1. Spelling. Though *-ancy* is not much commoner than *-ency*, it is better as corresponding to *ascendant*, which is much commoner than *ascendent*. 2. Usage and meaning. *The ascendancy of*, *Have an* or *the ascendancy over*, *be in the ascendant*, are the normal phrases. In the first two *ascendant* for *ascendancy* would be archaic; in the third, which is less detached than the others from its astrological origin, *ascendancy* would be wrong, and when used (*It is not recorded what stars were in the ascendancy when Winston Churchill was born*. / *Jimmie's better angel was in the ascendency*) is attributable to ignorance. Both words mean domination or prevailing influence, and not upward tendency or rising prosperity or progress.

**ashen.** See -EN ADJECTIVES.

**Asian, Asiatic.** As a colourless adjective for *Asia*, *Asian* is the older. *Asiatic*, which later largely superseded it, originally had a contemptuous nuance, especially when applied to a literary style, 'prose somewhat barbarously rich and overloaded' (M. Arnold). (For this the word *Asianic* has since been coined, defined by the OED Supp. as 'characterized by the florid and inflated style of Asiatic Greeks in the three centuries before Christ'.) Modern preference for *Asian* over *Asiatic* (*Asian Relations Conference*, the *Afro-Asian Group*) is a mid-20th-c. REVIVAL. *Asiatic* never wholly got rid of its taint, and the reversion to *Asian* was no doubt prompted by the West's recognition of the expediency of being polite to the East (cf. the disappearance of *Chinaman* and the conversion of *Eurasian* into *Anglo-Indian*). And so, though we used to speak of *Asiatic cholera*, it was natural in 1957 to speak of *Asian influenza*.

**aside**, written as one word, does not mean on each side (*We sat five aside in the suburban train*; *They were playing three aside*); *a side* must be written; cf. £500 *a year* (not *ayear*). *Aside from* (= apart from) is good U.S. idiom but not yet at home in England.

**aspirant** is a victim of the RECESSIVE ACCENT. In 1888 the OED gave the pronunciation as *aspir'ant* (occasionally *as'pĭrant*). The COD now puts *as'pĭrant* first.

**assay, essay, vbs.** A differentiation has been established by which *assay* is confined to the sense *test*, and *essay* to the sense *attempt*; the OED says: 'Except as applied to the testing of metals, *assay* is now an archaic form of *essay*.' *Essay* itself has by this time the dignity attaching to incipient archaism; but the distinction should be observed.

**asset.** The great popularity that this word now enjoys as a saver of trouble to those who have not time to choose between *possession*, *gain*, *advantage*, *resource*, and other synonyms is modern, and an effort should be made to keep it within bounds. Most of those who use it are probably unaware that, though now treated as plural, *assets* was once (cf. *riches*) a singular. The *-s* is not a plural termination, French *assez* (enough) being its source; and the original sense of the word is what suffices or should suffice to meet liabilities. The false singular originates in uses like *The chances of a dividend depend upon the realization of two assets, one a large debt, and the other.* ... It is now firmly established, and not without value if reasonably used, but the following quotations show how tempting it is as a thought-saver: *Her forehand drive—her most trenchant a.* (stroke). | *It was Mr. John Ball who showed us that the experience of years is an incalulable a. when it comes to the strain of a championship* (invaluable). | *Nor is it every doctor who despises club practice; many find it a very handy a.* (source of income). | *As sound a head as that of his Reverence is a distinct a. to society* (of value).

**assign.** Except *assignation* (*-sĭg-*) derivatives and allied words (*-able, -ee, -or, -ment,* etc.) are pronounced *-sīn-*. *Assignment* in the sense of a task assigned to someone is an Americanism that took root in Britain during the second world war.

**assist,** in the sense be present (at a performance etc.), is now a GALLICISM; in the sense help (to potatoes etc.), it is a GENTEELISM.

**assonance.** See REPETITION s.f. As a term of prosody it means the rhyming of one word with another in the accented vowel and those (if any) that follow, but not in the consonants, e.g. *devil* and *merit*.

**assume, presume.** Where the words are roughly synonymous, i.e. in the sense *suppose*, the object-clause after *presume* expresses what the presumer really believes, till it is disproved, to be true; that after *assume* expresses what the assumer postulates, often as a confessed hypothesis. It is possibly this distinction that has led to the curious variation in the adverbs ordinarily used—*assumedly* (as has in fact been supposed by someone as a hypothesis) and *presumably* (as anyone might naturally suppose from the evidence).

**assure, assurance.** These words have never found general acceptance in the sense of paying premiums to secure contingent payments, though they are used by insurance offices and agents, and so occasionally by their customers, especially of the ordinary life policy. One insures *against* something but is assured *of* it. Hence *Life Assurance* where the definite sum is a certainty on maturity or death, but *Term Insurance* where what is covered is the risk of the insured person's dying within a specified period, with no return of premium or benefit if he survives it. Apart from such technical use, *insure* and *insurance* hold the field.

**astronomical.** The use of this word in the sense of so vast as to be comparable with numbers and distances observed by astronomers is now enough to be unknown to the OED and even to its 1933 Supp. It is now so well established as to be in danger of becoming a VOGUE WORD, and so of being unsuitably used. When applied to numbers and distances it is a reasonable metaphor, though so hyperbolical that it is often felt to need a qualifying *almost*. But the metaphor is strained when put to such a use as *In recent years there has been an almost astronomical rise of public interest in architecture.* Here the writer has made matters worse with his *almost*, which takes us away from the metaphorical and into the actual, and suggests that the same system of measurement can be applied to the distances of the stars and the volume of public interest in art.

**asyndeton.** The omission, for effect, of conjunctions by which words or sentences would in normal speech be connected. *I came, I saw, I conquered.* See also AND 4.

**at, in.** According to the OED *at* is used 'of all towns except the capital of our own country and that in which the speaker dwells, if of any size' If this means that *in* is not used of other towns, the exception has become too narrow. We now speak as readily of being *in* any town as of being *at* it; if there is any difference it is that *in* conveys a suggestion of being physically within and *at* does not. *New College is* in *Oxford.* | *The new term* at *Oxford begins this week.*

**-atable.** In spite of the general rule given under -ABLE for the formation of adjectives in -able from verbs in -ate, the short form with -at- omitted would be disagreeably pedantic in many cases where either the verb itself is little used in literature, or the dropping of -at- amounts to disguising the word (as in disyllables, *create*, *vacate*, etc.), or the -able adjective is likely to be very seldom used, or confusion with another word might result. Thus *incubate*, at least in the sense in which its -able adjective is likely to be wanted, is a technical rather than a literary word; *inculcable* is not instantly recognized as from *inculcate*; *inculpable* is both likely to be understood as not culpable, and unlikely to be often wanted; and accordingly *incubatable*, *inculcatable*, *inculpatable*, are less objectionable than *incubable*, *inculcable*, and *inculpable*. The practice should be to use -atable where the shorter form is felt to be out of the question. Simple reference to this article under any word means that -atable is better.

**ate,** past of *eat*. The U.S. pronunciation *āt* is beginning to edge out the British *ĕt*.

**atom, molecule, nucleus, proton, neutron, electron.** To the mere literary man without scientific knowledge, the relations of these words to each other are puzzling, and not easy to learn, even in an elementary way, from dictionaries, especially as dictionaries have had difficulty in keeping up with the discoveries and terminology of nuclear physics. A short explanation may therefore be useful.

We now know only too well that Democritus, the exponent of the atomic theory nearly 2,500 years ago, was

right when he maintained that matter should be regarded not as continuous but as composed of discrete particles. He called these σώματα άτομα (uncuttable bodies), and *atom* has been the name for them ever since. According to modern theory the atom, so small that some hundred million of them placed side by side would measure only about an inch, may be compared to the solar system on a minute scale. Its *nucleus*, about one hundred thousandth part of the whole, and composed of two types of particles, *protons* and *neutrons*, corresponds to the sun. Circulating round it in orbits like the planets are bodies known as *electrons*. The protons in the nucleus and the electrons circulating round it are electrically charged, positively and negatively respectively. There are the same number of electrons as of protons and they carry charges of the same magnitude; and so, neutrons being uncharged, the atom as a whole is electrically neutral.

An atom is the smallest specific unit of a chemical element, and the elements are distinguished from one another by the different numbers of protons in the nuclei of their atoms. A *molecule* is the smallest specific unit of a chemical compound, and is therefore a combination of atoms.

**atonal, dodecaphonic, serial, twelve-tone** are all used as descriptive terms for certain types of modern music, notably that of Schönberg and his followers. But there is an essential difference, not always respected, between *atonal* and the others. *Atonal*, the only one of the four old enough to appear in the OED Supp., and there defined as 'a style of composition in which there is no conscious reference to any scale or tonic', is a purely negative term, vague and intrinsically subjective, applicable to any music that does not give a feeling of key. The others are positive terms for rigid methods of composition which, treating all twelve notes of the octave as of equal and independent value (hence *dodecaphonic* and *twelve-tone*), prescribe certain rules governing the sequence in which they should be used (hence *serial*). That music so composed will naturally be atonal is incidental; Schönberg himself disapproved of the use of the word as a descriptive label for his music.

**attempt.** For *was attempted to be done* etc. see DOUBLE PASSIVES.

**attic.** See GARRET.

**attraction.** A tendency less commonly operative in English (except in mere blunders) than in Latin and Greek, by which a word is changed from the correct case, number, or person, to that of an adjacent word. *When* him [*whom*] *we serve 's away*; *The wages of sin is death*. And (as a blunder) *I have always been surprised at how little the members of one part in the House of Commons knows what happens in the other*. See NUMBER 4.

**audience, audition.** In their meaning of a formal interview these words are differentiated: *audience* has long been used for one given by a monarch or other high personage for the transaction of business; *audition* was adopted at the beginning of the 20th c. for one given by an impresario to an aspiring performer. *Audition* is now being used as a verb. There seems no more reason to take exception to this than to the use as verbs of *commission* and *petition*, unless it be that one is new and the others established by centuries of usage.

**au fond.** See À FOND.

**aught.** See WARDOUR STREET. For *a. I know* is the only phrase in which the word is still current in ordinary speech, and even there *all* is displacing it.

**autarchy, autarky.** *Autarchy*, absolute sovereignty, *autarky* (sometimes misspelt -*chy*) self-sufficiency, are from different Greek words: this accounts for the difference in spelling, which should be preserved.

**authentic, genuine.** The distinction commonly drawn between the words is by no means universally observed, especially when either is used without special reference to the other; but, when a distinction is present to the mind, *authentic* implies that the contents of a book, picture, account, or the like, correspond to facts and are not fictitious, and *genuine* implies that its reputed author is its real one: *a genuine Hobbema*; *An authentic description*; *The Holbein Henry VIII is both authentic and genuine* (represents Henry as he was, or is really a portrait of him, and is by Holbein). The artificial character of the distinction, however, is illustrated by the fact that, *genuine* having no verb of its own, *authenticate* serves for both.

**authoress** is a word that has always been disliked by authoresses themselves, perhaps on the ground that sex is irrelevant to art and that the word implies disparagement of women's literary abilities. They had good reason to be sensitive: as recently as 1885 a Savoy audience was invited to applaud the sentiment that 'that singular anomaly the lady novelist' never would be missed. When the OED was published the authoresses seemed to be getting their way; for the opinion is there given that *authoress* 'is now only used when sex is purposely emphasized; otherwise in all the senses, and especially the last [sc. a female literary composer], *author* is now used of both sexes'. But *authoress* is dying hard. Seventy years after the OED's pronouncement a book reviewer still finds it natural to say *The authoress does not much discuss the quality of acting in Garrick's theatre*, and a High Court judge to say *The wife was in several respects the authoress of her own wrong because of her nagging*. See also FEMININE DESIGNATIONS.

**authoritarian, authoritative.** The differentiation is complete: -*arian* means favourable to the principle of authority as opposed to that of individual freedom; -*ative* means possessing due or acknowledged authority; entitled to obedience or acceptance (OED).

**autocar, autocycle, automobile.**
In spite of our general disposition to
borrow words from America, we have
firmly rejected the U.S. *automobile* in
favour of our *motor car, motor,* or now
more generally *car,* though we preserve *automobile* in the title of the
Association and Club, and *Autocar* in
that of the periodical. *Autocycle* remains in the *A. Union,* but the vehicles
it is interested in are in common parlance either *motor bicycles* or *mopeds.*

**automation.** *Mechanization,* enabling
machines to do work formerly done by
human muscles, produced the industrial revolution. *Automation,* enabling
machines to do work formerly done by
the human brain, is said to be producing another, not less epoch-making. It is a pity that the word for
something so important (with its back-
formation verb *automate* now inevitably appearing) should be a
'BARBARISM'. But *automatism* was not
available, having been given a different
job; and it cannot be denied that the
words chosen have practical advantages over the more orthodox but
more cumbrous *automatize* and *auto-
matization* that might have been
formed from the Greek verb αὐτοματίζω.
Besides, we do not want any more
NEW VERBS IN -IZE than we can help.

**avail,** vb. The constructions *be availed
of, avail of,* illustrated in the following
quotations are wrong: *If economical
means of transport are availed of* (made
use of). | *No salvage appliances could
have been availed of in time* (used). |
*A desire to avail of the quieter phase of
national emotion* (take advantage of).
The normal construction is *avail* one-
*self of* (*I shall a. myself of your kind
offer*). From this are wrongly evolved
('with indirect passive, especially in
U.S.' says the OED, but with examples from English newspapers too)
such forms as *The offer was availed of*;
the absurdity of this is patent as soon
as the method is applied to similar
reflexive verbs; because we can say
*They busy themselves in politics, You
should rid yourself of cant, Many devote
themselves to religion,* we do not infer

that *Politics are busied in, Cant should
be ridded of, Religion is often devoted to,*
are English; as little does *avail* one*self
of the offer* justify *the offer was availed
of. Available,* which perhaps encourages the use of these bad constructions,
lends them no real support; its original
sense was *that can avail,* and what it
would suggest is rather *avail the offer*
than *avail of the offer.*

**avant-garde.** *Eugene Ionesco, one of
the avant-garde of French playwrights.* |
*Though he read the accepted authors, he
was also attracted by the avant-garde of
writing in his time.* | *In television . . . a
conscientious critic in the prime of life
can see everything, avant garde and
rearguard, and all that lies between.* | *It
might seem odd to many a 78-year-old
composer to be represented in a Festival
of this kind at all, so avant-garde is the
International Society of Contemporary
Music in its trends.* In 1888 the OED
said that *avant-garde* had been replaced by the aphetic form *vanguard,*
and had become archaic or obsolete.
The 1933 Supp. did not disturb this
verdict. But the critics have since
decided that *vanguard* will no longer
do; *avant-garde* has become for them
the inevitable word, to serve the same
purpose in the middle of the 20th
century as *fin de siècle* did at the end of
the 19th. It seems, however, that they
are now beginning to feel the need of a
word for what is even more avant
garde than avant garde itself, and
*vanguard* is coming back into its own.
*Extreme vanguard music* is the description applied by a music critic to a
concert displaying 'mere belligerence
towards the condition of music' given
by a group of young composers under
dadaist influence.

**avenge, revenge, vengeance.**
*Avenge* and *vengeance* are one pair,
*revenge* verb and noun another. The
distinction between the two pairs is
neither very clear nor consistently observed. This is natural, since the same
act done under the same conditions
may be described either as *vengeance*
or as *revenge* according to the point of
view from which it is regarded. It may

averse44avoidance

be said, roughly, that *vengeance* is
redressing the balance by an offender's
being made to suffer something more
or less equivalent to his offence, while
*revenge* is the satisfying of the offended
party's resentment by the same means;
one act may effect both, but it will be
spoken of by one name or the other
according to context. It is in harmony
with this that the subject of the active
verb *revenge* is ordinarily the wronged
party, and its object either himself or a
wrong done at least indirectly to him,
while the subject of *avenge* is ordinarily
a disinterested party, and its object
another person or a wrong done to
another. Exceptions are numerous,
but choice may be assisted by
remembering the general principle
that personal feeling is the thing
thought of when *revenge* is used, and
the equalization of wrongs when
*avenge* or *vengeance* is used.

**averse, aversion.** To insist on *from*
as the only right preposition after
these, in spite of the more general use
of *to* (*What cat's averse to fish?*—Gray.
*He had been averse to extreme courses.*—
Macaulay. *Nature has put into man an
aversion to misery.*—Locke) is one of
the pedantries that spring of a little
knowledge. If *averse* meant originally
turned away, and *a* is Latin for away,
this did not prevent even the writers
of classical Latin from using the dative
after *aversus*; nor, if it had, need it
prevent us, to whom the original and
physical sense is not usually present,
from using after *averse* and *aversion*
the preposition that best fits their true
sense.

**avocation,** originally a calling away,
an interruption, a distraction, is used
in Britain, though not in U.S., as
a stylish synonym for *vocation* or
*calling*, with which it is properly in
antithesis. This has destroyed a useful
differentiation; but *vocation* has retained the idea of a 'calling'; a man is
fortunate if his avocation, in the sense
of his occupation, is also his vocation.

**avoidance of the obvious** is very
well, provided that it is not itself obvious; but, if it is, all is spoilt. Expel

*eager* or *greedy* or *keen* from your
sentence in favour of *avid*, and your
reader wants to know why you have
done it. If he can find no better
answer than that you are attitudiniz-
ing as an epicure of words for whom
nothing but the rare is good enough,
or, worse still, that you are painfully endeavouring to impart some
much needed unfamiliarity to a platitude, his feeling towards you will
be something that is not admiration.
The obvious is better than obvious
avoidance of it. *Nobody could have
written 'Clown' who had not been (as
Mr. Disher is known to be) an avid
collector of pantomime traditions and
relics. | Everything is just in a state of
suspended animation, and the House,
instead of being in its usual bustle on
account day, is devoid. | Lord Lans-
downe has done the Liberal Party a good
turn by putting Tariff Reform to the
front; about this there can be no* dubiety.,
*If John never 'finishes' anything else, he
can at least claim by sheer labour to have
completed over* five-score *etchings.* There
are some who would rather see *eager*
and *empty* and *doubt* and *a hundred* in
those sentences than *avid*, *devoid*,
*dubiety*, and *five-score*; and there are
some who would not; the examples are
typical enough to sort tastes.
Several words can be thought of that
have been through this course. Starting as variants for the business word,
some with their own special and useful
functions, they have been so seized
upon by those who scorn to talk like
other people as to become a badge by
which we may know them; after which
they pass into general use by the side
of the words to which they were preferred, giving the language pairs of
useless synonyms that have lost whatever distinction there may once have
been between them. Such words are
*cryptic, dual, facile, forceful,* and *Gallic*
as used without the justification of
special meanings instead of *obscure,
double, easy, forcible,* and *French*; on all
of these comment will be found in
their separate articles. A few examples
of the uses deprecated are: '*A sensible
young man, of rough but mild manner,*

*and very seditious'; this description, excepting the first clause, is somewhat cryptic.* | *The combination of cricket and rowing 'blues' is very rare; the late J. W. Dale was the last Cambridge man to earn the dual distinction.* The 'dual event' is perhaps already, and will surely soon be, upon us. | *The reunion of a Labour and Socialist World Conference has not proved to be so facile to arrange as it appeared.* | *'I blame the working of the Trade Board' said Mr. Newey, forcefully, 'for keeping wages at an artificial figure.'*

**avouch, avow, vouch.** The living senses of the three words are distinct; but, as a good deal of confusion has formerly prevailed between them, the dictionaries are not very helpful to inquirers, providing as they do quotations under each for senses that now belong only to one of the others; it is therefore worth while to state roughly the modern usage. *Avouch,* which is no longer in common use, means guarantee, solemnly aver, prove by assertion, maintain the truth or existence of, vouch for (*A miracle avouched by the testimony of . . .*; *Millions were ready to avouch the exact contrary*; *Offered to avouch his innocence with his sword*). *Avow* means own publicly to, make no secret of, not shrink from admitting, acknowledge one's responsibility for (*Think what one is ashamed to avow*; *Avowed himself my enemy*; *Avowed his determination to be revenged*; *Always avows,* and cf. in the contrary sense *disavows, his agents*). *Vouch* is now common only in the phrase *vouch for,* which has taken the place of *avouch* in ordinary use, and means pledge one's word for (*Will you vouch for the truth of it?*; *I can vouch for his respectability*).

**await, wait.** *Await* is always transitive, but *wait* is not always intransitive. *I am awaiting to hear your decision* is not English; *I await,* and *I wait, your decision* are equally good though the latter is less usual. See WAIT.

**awake, awaken, wake, waken.** *Awake* has past *awoke,* rarely *awaked,* and p.p. *awaked* sometimes *awoken* and rarely *awoke*; *wake* has past *woke,* rarely (and that usually in transitive sense) *waked,* and p.p. *waked,* rarely *woke* or *woken*; *awaken* and *waken* have *-ed.*

Distinction between the forms is difficult, but with regard to modern usage certain points may be made: (1) *Wake* is the ordinary working verb (*You will wake the baby*; *Something woke me up*; *I should like to be waked at 7.30*; *Wake the echoes*), for which the others are substituted to add dignity or formality, or to suit metre, or as in (3) or (5) below. (2) *Wake* alone has (and that chiefly in *waking*) the sense *be* or *remain awake* (*Sleeping or waking*; *In our waking hours.*) (3) *Awake* and *awaken* are usually preferred to the others in figurative senses (*When they awoke,* or *were awakened,* to *their danger*; *This at once awakened suspicion*; *The national spirit awoke,* or *was awakened*; *A rude awakening*). (4) *Waken* and *awaken* tend to be restricted to the transitive sense; *when he wakens* is rarer for *when he wakes* than *that will waken him* for *that will wake him.* (5) In the passive, *awaken* and *waken* are often preferred to *awake* and *wake,* perhaps owing to uncertainty about the p.p. forms of the latter pair; *it wakened me* is rare for *it woke* or *waked* me, but *I was wakened by it* is common for *I was waked* or *woke* or *woken by it*; see also the alternative forms in (3) above. (6) *Up* is very commonly appended to *wake,* rarely to *waken,* and never to *awake* and *awaken.*

**away.** 1. For *once and away* see ONCE. 2. For *cannot away with* see ARCHAISM. 3. *Way* as an abbreviation of *away* in such compound adverbs as *w. above, w. back, w. below, w. up* is a common Americanism that has not yet taken root here; it is surprising, and perhaps significant of a change in our attitude towards it, to find such sentences as these in *The Times* and *The T.L.S.*: *Education begins way back in the home. At a time when tension between India and Pakistan, for reasons way above the understanding of those killed or killing,* *is rising to danger point. | That betraying adjective 'strange' places him . . . way back on the rickety rationalist Olympus of Edward Gibbon.*

**awfully.** See TERRIBLY.

**axiom.** See APHORISM.

**ay, aye.** The word meaning *yes* is pronounced ī, and the poetic word meaning *ever* is pronounced ā; but which spelling corresponds to which pronunciation has been disputed. The nautical *Ay, ay, sir* is usually written thus; and *aye* is the usual spelling now for *ever*; on the other hand *the ayes have it* is so written, though -*es* may there be intended for the plural termination. *Ay* (ī) yes, *aye* (ā) ever, seem now established (as is recognized by the COD) though the authority of the OED was on the other side.

**azure.** Under the influence of the speak-as-you-spell movement (see PRONUNCIATION 1) the conventional pronunciation ă'zhĕr is giving place to one that gives full value to the *u*, recognized at least as an alternative by most modern dictionaries.

# B

**babe, baby.** In the primary sense *baby* is now the regular form, and *babe* archaic, poetic, or rhetorical. In figurative use, *babe* implies guilelessness, innocence, or ignorance, and *baby* unmanliness. The slang use of *baby* for a girl friend is an Americanism that has made little headway with us, but we owe to that country the indispensable *baby-sitter*.

**bacchanal, bacchant(e).** The nouns *bacchanal* and *bacchant* are used of both males and females, but with a tendency to be restricted to males; *bacchante* is used of females only. *Bacchant* is always pronounced ba'kănt; *bacchante* is băkă'nt, bă'kănt, or băkă'nti. *Bacchant* has plural *bacchants* or *bacchantēs*; *bacchante* has *bacchantes* (-ts or -tīz).

**bacillus.** Pl. *bacillī*; see -US, and

LATIN PLURALS. For the meaning see MICRO-ORGANISMS.

**back formation.** A dictionary definition of the term is: The making from a supposed derivative (as *lazy, banting*) of the non-existent word (*laze, bant*) from which it might have come. It is natural to guess that the words *scavenger, salvage,* and *gridiron* are formed from verbs *scavenge* and *salve* and a noun *grid,* and consequently to use those verbs and noun if occasion arises. Those who first used them, however, were mistaken, and were themselves making the words out of what they wrongly took for their derivatives; similarly *banting* is not formed from *bant,* but is the name first of a man, and then of his system, out of which the verb to *bant* was made by back formation. This will perhaps sufficiently explain the references made here and there to this article. Some back formations are not generally recognized as such, and have the full status of ordinary words, e.g. *diagnose* (from *diagnosis*), *drowse* (from *drowsy*), *sidle* (from *sideling* = *sidelong*), *grovel* (from *grovelling,* an adverb), *resurrect* (from *resurrection*), and BURGLE (from *burglar*). But when first invented they are generally felt to be irregular and used only ignorantly, slangily, or jocosely; so *buttle, chinee,* ENTHUSE, *frivol, liaise, locomote, maffick,* ORATE, PROCESS (go in procession), *revolute.* Other articles that may be looked at are BRINDLE, FILTRATE, GRIDIRON, GUERILLA, and SALVE.

**back of** as a preposition meaning behind is an American, not a British, idiom, unless it is to be regarded as a preposition in the phrase *b. of beyond.*

**background.** *Tell me your background* has become the fashionable way of saying *tell me all about yourself.* Within reason this is a proper and useful extension of a meaning defined by the OED as that which lies at the back of or beyond the chief object of contemplation. *Egypt is the b. of the whole history of the Israelites. | The successful candidates included boys and girls of* *very different bb.* But it has become too popular as a VOGUE WORD, saving its user the trouble of thinking of a more precise word such as *origin, cause, reason, explanation, qualifications,* and others. *The activities of a few shop stewards provide the b. of the trouble* (are the cause). / *Your b. seems more suitable for employment in an engineering establishment* (qualifications).

**backlog.** To the OED this is only known as a large log placed at the back of a fire (chiefly U.S.). That this would remain comparatively unburnt while the rest of the fire was consumed was presumably the reason for the now very popular metaphorical use of the word (esp. U.S.) in the sense of arrears. What the Telephone Department of the G.P.O. calls *List of waiting applicants* the American Telephone and Telegraph Company calls *Backlog of held orders.*

**backward(s).** As an adverb either form may be used; as an adjective *backward* only.

**bade.** See BID.

**baggage.** Englishmen travel by land with *luggage* and by sea and air with *baggage.* Americans, more sensibly, travel everywhere with *baggage.*

**bail out, bale out.** In the sense of securing someone's release from custody by guaranteeing his reappearance, the spelling is always *bail.* The OED says that the same spelling should be used for emptying a boat of water; *bale* is 'erroneous' because the derivation is from French *baille,* bucket. But, perhaps owing to an instinct for DIFFERENTIATION, popular usage prefers *bale* both for this and for making a parachute descent from an aircraft in an emergency.

**balance,** in the sense *rest* or *remainder,* is, except where the difference between two amounts that have to be compared is present to the mind, a SLIPSHOD EXTENSION. We may fairly say 'you may keep the balance', because the amount due and the amount that more than covers it suggest com-

parison; but in 'the b. of the day is given to amusement' such a comparison between amounts is, though not impossible, farfetched, and the plain word (*rest,* or *remainder*) is called for. For IMBALANCE see that article.

**ballad(e).** A *ballad* was originally a song as accompaniment to dancing; later any simple sentimental song especially of two or more verses each to the same melody, e.g. Jonson's *Drink to me only.* A separate modern use is as the name of simple narrative poems in short stanzas, such as *Chevy Chase.*
  A *ballade* is an elaborate poem consisting of three eight (or ten) line stanzas and a four (or five) line envoy, all on three (or four) rhymes only in the same order in each stanza, and with the same line ending each stanza and the envoy. An old French form, revived in France and England in the 19th c.

**ballyrag, bullyrag.** The etymology is unknown; the second form is probably due to a supposed connexion, described by the OED as 'unlikely', with *bully;* the first form is far more common, and seems preferable.

**balmy, barmy.** *Barmy,* not *balmy,* as it is sometimes misspelt, is the slang word for crazy. It is the adjective of *barm,* the froth that forms on the top of fermenting malt liquor; the OED gives 17th-c. examples of its use with *brain* and *head* in the sense of frothy, excited. *Balmy* is the adjective of *balm,* fragrant salve, the stuff about which Jeremiah asked whether there was none in Gilead. See HACKNEYED PHRASES.

**balusters, banisters.** Originally the latter was a corruption of the former: both meant uprights supporting a rail. They have become differentiated in two respects: *banisters* are commonly wood and indoors and *balusters* stone and outside; and *banisters* is used of the complete structure of uprights and what they support and *balusters* is not; *balustrade* is the word for that.

**banal, banality.** These were imported from France by a class of writers whose jaded taste relished novel or imposing jargon. In French they have had a continuous history and a natural development from their original to their present sense; in English they have not, and though they have been with us for very many years are not yet wholly naturalized. With *common, commonplace, trite, trivial, mean, vulgar, truism, platitude,* and other English words, to choose from, we should confine *banal* and *banality,* since we cannot get rid of them, to occasions when we want to express a contempt deeper than any of the English words can convey. We have not even agreed about the pronunciation. *Bä′năl, bă′năl, bănal′,* and *bănahl′* all have some dictionary recognition. The fully anglicized *bă′nal* is recommended.

**banisters.** See BALUSTERS.

**baptist(e)ry.** The -tery form is the better.

**bar.** B. *sinister,* popularly supposed to be a symbol of illegitimacy, is strictly incorrect, *bend* or *baton sinister* being the true term. It is, however, so familiar that to correct it, except where there is real need for technical accuracy, is pedantic; see DIDACTICISM.

**barbarian, barbaric, barbarous.** The differences in usage among the three adjectives are roughly, and setting aside special senses of the first, as follows. *Barbarian,* as an adjective, is now regarded as an attributive use of the noun *barbarian;* i.e. it is used only in such ways as would be admissible for nouns like *soldier* or *German* (*a soldier king* or *people; German ancestry* or *thoroughness*), and means consisting of barbarians, being a barbarian, belonging to or usual with barbarians. So we have *barbarian tribes, hosts, frankness, courage; a barbarian king, home, empire; barbarian man* (the human race as barbarians); *the barbarian world.*

The other two words are ordinary adjectives, but differ in their implica-

tions. *Barbaric* is used with indulgence, indifference, or even admiration, and means of the simple, unsophisticated, uncultured, unchastened, tasteless, or excessive kind that prevails among barbarians. We speak of *barbaric taste, finery, splendour, costume, gold, hospitality, simplicity, strength, health.*

*Barbarous,* on the other hand, always implies at least contempt, and often moral condemnation; it means that is unfit for or unworthy of or revolts or disgraces or would disgrace the civilized: *barbarous ignorance, speech, customs, style, words, cruelty, treatment, tyranny.*

**barbarism, barbarity, barbarousness.** The three nouns all belong to the adjective *barbarous,* but the first two are now (putting aside intentional archaism and metaphor) clearly distinguished. *Barbarism* means uncivilized condition, grossly uncultivated taste, an illiterate expression, or a word of the kind described in the next article; *barbarity* means grossly cruel conduct or treatment, or a grossly cruel act; *barbarousness* may be substituted for either of the others where the sense *quality* or *degree* is to be given unmistakably: *They live in barbarism; The barbarism,* or *barbarousness, of his style; 'Thou asketh' is a barbarism; He treats prisoners with barbarity; The barbarity,* or *barbarousness, of the decree is irrelevant; Unheard-of barbarities.*

**barbarisms.** The word is here used only in the last of the senses assigned to it in the preceding article: it has the meaning the Greeks gave it of a word formed in an unorthodox way, and it conveys no more pejorative an implication than that it is the sort of thing one might expect from a foreigner. Even so, *barbarisms* may seem a hard word to fling about, apt to wound feelings, though it may break no bones; perhaps it would be better abstained from, but so too would the barbarisms themselves. What after all is a barbarism? It is for the most part some word that, like its name, is apt to wound feelings—the feelings, however, of much fewer persons. They will only be those who have not merely, like the Eton boys of a former generation, 'a profound conviction that there are such languages' as Greek and Latin, but a sufficient acquaintance with and love of those languages to be pained by outrages upon their methods of word-formation. In this era of democracy it can hardly be expected that the susceptibilities of so small a minority should be preferred to the comfort of the millions, and it is easier for the former to dissemble their dislike of barbarisms than for the latter to first find out what they are and then avoid them.

There are unfortunately two separate difficulties, both serious. We may lack the information that would enable us to decide whether any particular word is or is not a barbarism. This is indeed obtainable from a competent philologist; but life is not long enough to consult a competent philologist every time one of the hundreds of dubious words confronts us; nor yet is it long enough for an *ad hoc* course of Latin and Greek grammar. And then, even if the philologist has been consulted, or the course gone through, what can we do about it? A barbarism is like a lie; it has got the start of us before we have found it out, and we cannot catch it. It is in possession, and our offers of other versions come too late.

That barbarisms should exist is a pity. To expend much energy on denouncing those that do exist is a waste. To create them is a grave misdemeanour; and the greater the need of the word that is made, the greater its maker's guilt if he miscreates it. A man of science might be expected to do on his great occasion what the ordinary man cannot do every day, ask the philologist's help; that the famous *eocene-pleistocene* names were made by 'a good classical scholar' (see Lyell in DNB) shows that word-formation is a matter for the specialist. For further discussion see HYBRIDS AND MALFORMATIONS.

**baritone.** See BARYTONE.

**bark, barque.** The two forms have become differentiated, *bark* being a poetic synonym for ship or boat, and *barque* the technical term for a ship of special rig.

**barmy.** See BALMY.

**baronage, barony, baronetage, baronetcy.** The forms in -*age* are collectives, meaning all the barons, all the baronets, list or book of barons, etc. Those in -*y* are abstracts, meaning rank or position or title of a baron or baronet.

**baroque, rococo.** The term *baroque* (misshapen pearl) was applied, at first contemptuously but later respectfully, to a style of architecture that originated in Rome in the early 17th c. and showed an audacious departure from the traditions of the Renaissance. Asymmetry of design, luxuriance of ornament, strange or broken curves or lines, and polychromatic richness were its main features. The word was later extended to the other visual arts of the baroque period, which is generally regarded as having lasted to the middle of the 18th c. The best known early exponents are Borromini in architecture, Bernini in sculpture, and Rubens in painting.

*Rococo* (rock-work) is sometimes treated as synonymous with baroque, but is more properly confined to a later development of it, especially in France, lighter and more fanciful, and with ornament even less related to structure. The characteristics of baroque are grandeur, pomposity, and weight; those of rococo are inconsequence, grace, and lightness. Baroque aims at astounding, rococo at amusing.

It has become fashionable to apply the word *baroque* to literature and music also. What it is intended to mean when so used has been the subject of much argument and little agreement, except, it seems, that the poetry of Crashaw and the music of Vivaldi are typically baroque. Such definitions of baroque poetry as have been attempted, as for instance that its marks are 'instability, mobility, metamorphosis and the domination of decor', are unlikely to give the ordinary reader confidence that he will recognize it with certainty. The application of the term to music is even more recent. It started in Germany, and its propriety is not universally accepted. 'It is clear that a word professing to embrace such various products as the later madrigals and Handel's operas, Peri's melodrama and Bach's Art of Fugue, has no justification beyond mere convenience' (Grove's *Dictionary of Music*, 1954 ed.).

This quest for the meaning of *baroque* when applied to the non-visual arts is the more difficult because those so using the word do not always make it clear, even, one suspects, to themselves, whether to them it is a chronological term, meaning any work produced during the baroque period, or a descriptive one, meaning a work with baroque characteristics, whenever produced, or a mixture of the two, meaning a work with baroque characteristics, produced during the baroque period. The Muse of History seems now to have been infected by her sisters of Poetry and Music. From an important historical work recently published a reviewer quotes *b. absolutism, b. raison d'état*, and *an adventure in b. power politics*, and he adds, 'There is a strong case for persuading historians to use the epithet *baroque* with restraint, for it is fast becoming as devoid of meaning as *feudal*'.

**barrage.** Originally a dam, especially on the Nile. The meaning was extended during the first world war to the curtain of intense gunfire known as an *artillery b.* and during the second to the anti-aircraft device known as a *balloon b.* Its more recent use for a rapid and noisy discharge of questions or other interjections at a meeting must be due to NOVELTY HUNTING; the old metaphor *volley* is more apt.

**barytone, -ritone.** The first is the correct spelling etymologically, and the invariable one for the Greek grammatical term; the second is the one in musical use. The -y- is the normal English transliteration of the original Greek *v*, which has been changed to

-i- in the derivative Italian *baritono*. The prevailing though not invariable English practice in adopting words at second hand is to undo such intermediate changes and transliterate the originals consistently. It would have been justifiable to take the Italian *baritono* whole; but when we anglicized the ending we ought to have followed the ordinary English method of transliteration.

**basal, basic.** These unEnglish-looking adjectives, neither of which existed before the 19th c., were manufactured as adjuncts to certain technical uses of the noun *base* in botany, chemistry, and architecture, where *fundamental* would have been misleading. But a tendency then arose to allow these upstarts to supplant *fundamental*, with its 500-year tradition, in those general and figurative contexts in which they are unnecessary and incongruous. No doubt those who started the fashion hoped that what they had to say would be more convincing if seasoned with a pinch of the up to date and scientific. The native element of *basal* and *basic* is seen in: *The elytra have a basal gibbosity*; *The basal portion of the main petiole*; *Its capital resting on its basic plinth*; *Basic salts, phosphates, oxides*. *Basal* in a figurative sense has now dropped out of favour, but *basic*, with the help of such established phrases as *b. dyes, industry, petrol allowance, slag*, and *steel*, and now BASIC ENGLISH, is in a fair way to win its victory over *fundamental*.

**basalt.** The OED gives preference to the stress on the second syllable, but it is now usually on the first, and the COD puts that pronunciation first.

**based upon.** See UNATTACHED PARTICIPLES.

**Basic English** is the name given by C. K. Ogden to a device for 'debabelizing' language propounded by him in 1929. Many others have tried to do the same thing by constructing an artificial language for universal use; J. M. Schleyer for instance who invented *Volapük* in 1879 and L. L. Zamenhof who invented *Esperanto* in 1887. But the likelihood that any such language will in fact be universally used has remained very remote indeed. As Ogden has said, 'an artificial language still awaits a millennium in which conversion shall cease to be confined to a few thousand enthusiasts'. His approach is different. Instead of inventing a new language he would use English, with its vocabulary drastically limited and its grammar simplified, as an international auxiliary language. He had of course no thought that basic English ever should or could supersede literary English; his aim was to provide the peoples of all countries, nearly a quarter of whom already have some knowledge of English, with a vehicle of communication with one another in the ordinary affairs of life.

The vocabulary consists of 850 words chosen as capable of doing, alone or in combination, all the essential work of 20,000. It is claimed for the system that anyone who masters it will have at his command 'idiomatic English with no literary pretensions but clear and precise at the level for which it was designed'. It is obvious that ordinary prose, when translated into a language with so limited a vocabulary, cannot retain its elegance or its rhetorical appeal, and Basic English has suffered from being an easy target for ridicule by those who misunderstand its object. To criticize it on this account is unreasonable; it should be judged only by its aptitude to achieve the purpose put thus by an enthusiastic advocate, Sir Winston Churchill: 'It would certainly be a grand convenience for us all to be able to move freely about the world and to be able to find everywhere a medium, albeit primitive, of intercourse and understanding'. But Ogden's system, after a promising start, has languished— 'bedevilled by officials' said its inventor. It remains to be seen whether anything will come of the variant called *Essential World English* with which, thirty years later, Lancelot

Hogben has tried to give fresh life to the conception.

**basis.** *Basis* is a frequent component of PERIPHRASIS, much used by those who think simple adverbs too bald and would rather write, e.g., *on a provisional b.* than *provisionally*.

**bas-relief, bass-relief, basso-relievo, basso-rilievo.** The first form is French, the last Italian, and the other two are corruptions; the plural of the third is *basso-relievos*, and of the fourth *bassi-rilievi*. It is recommended to use the first and pronounce it *bǎ'srilēf*.

**bathetic, bathotic.** These are made in imitation of *pathetic*, *chaotic*; but *pathetic* (from παθητικός) is not analogous, and *chaotic* is itself irregular. An adjective for *bathos* is, however, almost a necessity to the literary critic, and the OED states that *bathetic* is 'A favourite word with reviewers'; it is the better of the two and has rightly prevailed. *Bathotic* is called a nonce-word by the OED; it appeared once in 1863 and has not been seen since.

**battalion** has plural *battalions*, and not even in poetic style *battalia*. *Battalia* is a singular word (It. *battaglia*) meaning battle array; but being archaic, and often preceded by *in* (*Friedrich draws out in battalia.*—Carlyle), it is taken as meaning battalions.

**battered ornaments.** On this rubbish-heap are thrown, usually by a bare cross-reference, such synonyms of the ELEGANT-VARIATION kind as *alma mater*, *daughter of Eve, gentle sex,* and *Emerald Isle*; such metonymies as *the buskin* or *cothurnus* and *the sock* for tragedy and comedy; such jocular archaisms as *consumedly* and VASTLY; such foreign scraps as *dolce far niente*, HOI POLLOI, and CUI BONO?; such old phrases as *in* DURANCE *vile* and *suffer a sea-change*; such adaptable frames as *where* ——*s most do congregate* and *on* —— *intent*; and such quotations, customarily said with a wink or written instead of one, as *Tell it not in Gath* or *own the soft impeachment*. The title of the article, and their present company, are as much comment as is needed for most of them; but a few words will be found elsewhere on those that contain a word in small capitals; and other articles from which the list may be enlarged are: CLICHÉ; FACETIOUS FORMATIONS; GALLICISMS 5; HACKNEYED PHRASES; INCONGRUOUS VOCABULARY; IRRELEVANT ALLUSION; QUOTATION; OUT-HEROD; POPULARIZED TECHNICALITIES; SOBRIQUETS; STOCK PATHOS; VOGUE WORDS; WARDOUR STREET; WORKING AND STYLISH WORDS; WORN-OUT HUMOUR; and ZEUGMA.

**bay.** For *b.* and *gulf* see GULF.

**bay, bow, -window.** A *bay-w.*, so named because it makes a bay in the room, is one that projects outwards from the wall in a rectangular, polygonal, or semicircular form; *bow-w.*, though often loosely applied to any of these shapes, is properly restricted to the curved one.

**-b-, -bb-.** Monosyllables ending in b double it before suffixes beginning with vowels if the sound preceding it is a short vowel, but not if it is a long vowel or a vowel and r: *cabby, webbed, glibbest, bobbed, shrubbery*; but *dauber, barbed*. Words of more syllables (e.g. *rhubarb, sillabub,, hubbub, Beelzebub, cherub*) are few, and it will suffice to mention *cherubic* (so spelt), and *hobnob* (*-bbed, -bbing*).

**be.** 1. Number of the copula. 2. *Be* and *were*, subjunctives. 3. *Be*+adverb +participle. 4. Elliptical omissions. 5. Confusion of auxiliary and copulative uses. 6. Case of the complement. 7. Forms.

1. For the number of the verb between a subject and a complement of different numbers (*The wages of sin* is *death*; *The only obstacle* are *the wide ditches*), see NUMBER 1.

2. For use and abuse of *be* and *were* as subjunctives (*If an injunction* be *obtained and he* defies *it*; *It* were *to be wished*), see SUBJUNCTIVES.

3. For mistaken fear of separating *be* from its participle etc. (*If his counsel* still is *followed*; *The right* wholly to be

*maintained*), see POSITION OF ADVERBS.

4. *He is dead, and I alive*; *I shall dismiss him, as he ought to be.* For such forms see ELLIPSIS 1, 3.

5. Confusion of auxiliary and copulative uses. In *The visit was made* we have *was* auxiliary; in *The impression was favourable* we have *was* copulative. It is slovenly to make one *was* serve in both capacities, as in *The first visit* was *made and returned, and the first impression of the new neighbours on the Falconet family highly favourable*; *was* should be repeated after *family*—though, if *created* had stood instead of *highly favourable*, the repetition would have been unnecessary.

6. Case of the complement. The rule that the complement must be in the same case as the subject of the copula (*You believed that* it *was* he; *You believed* it *to be* him) is often disregarded in talk (*It wasn't me*), but should be observed in print, except where it would be unnatural in dialogue. The temptation in its simplest forms is rare, but may occur; Meredith, for instance, writes *I am she, she* me, *till death and beyond it*, where the ungrammatical *me* is not satirically intended. This should not be imitated.

7. Forms. Those that require notice are (*a*) *an't, ain't*, and (*b*) the singular subjunctives. (*a*) *A*(*i*)*n't* is merely colloquial, and as used for *isn't* is an uneducated blunder and serves no useful purpose. But it is a pity that *a*(*i*)*n't* for *am not*, being a natural contraction and supplying a real want, should shock us as though tarred with the same brush. Though *I'm not* serves well enough in statements, there is no abbreviation but *a*(*i*)*n't I?* for *am I not?* or *am not I?*; for the *amn't I* of Scotland and Ireland is foreign to the Englishman. The shamefaced reluctance with which these full forms are often brought out betrays the speaker's sneaking fear that the colloquially respectable and indeed almost universal *aren't I* is 'bad grammar' and that *ain't I* will convict him of low breeding. (*b*) The present subjunctive has *be* throughout (*Be I fair or foul*; *If*

*thou be true*; *Be it so*), the form *beest*, originally indicative but used for a time as second singular subjunctive, being obsolete. The singular of the past subjunctive is *were*, *wert*, *were* (*If I were you*; *Wert thou mine*; *It were wise*), *were* for the second person being obsolete.

**bean.** When the young person of today says *I haven't a bean*, or, more probably, *Actually I'm positively beanless*, he is echoing an expression, *Not worth a bean*, that can be found in Langland and earlier. The persistence of the old symbol of worthlessness (cf. *peppercorn*) in our slang term for lack of money contrasts oddly with the instability of those we use for money in possession. *Lolly*, current when this article was written (1959), had within living memory a long series of predecessors including *chink*, *dibs*, *dough*, *needful*, *oof*, *ready*, *rhino*, *spondulicks*, and *tin*.

**bear,** vb. See FORMAL WORDS. For p.p. see BORN(E).

**beat.** The old p.p. *beat*, still the only form in *dead-beat*, lingers colloquially also in the sense *worsted*, *baffled* (*I'm absolutely beat*), but it is unlikely that this is the origin of the expressions *beat generation* and *beatnik*. They probably come from *beat*=rhythm, and reflect passion for jazz rather than despair at the state of the world.

**beau-ideal.** If the word is to be used it should be pronounced *bō-īdē′al*, and written without accent. But neither in its only French sense of ideal beauty, nor in its current English sense of perfect type or highest possible embodiment of something, is there any occasion to use it, unless as a shoddy ornament. The English sense is based on the error of supposing *idéal* to be the noun (instead of the adjective) in the French phrase; and the English noun *ideal*, without *beau*, is accordingly the right word to use, unless *flower*, *perfection*, *paragon*, *pattern*, *pink*, or some other word, is more suitable.

**beautiful.** *But the home b. needs other growing greenery when the festive season arrives.* | *THE BED BEAUTIFUL. To see the English bed of supreme beauty you must take train . . .* Such vulgarizing adaptations of Bunyan have upon readers the effect described in IRRELEVANT ALLUSION.

**because. 1.** After such openings as *The reason is*, *The reason why . . . is*, the clause containing the reason should not begin with *because*, but with *that*. For examples see REASON 3, and for similar overlappings see HAZINESS.
**2.** *Because* following a negative clause is often a cause of ambiguity. Does the *because* clause disclaim a reason why a thing was done or does it give a reason why a thing was not done? *He did not oppose the motion because he feared public opinion.* Does this mean that it was not fear of public opinion that made him oppose, or that fear of public opinion made him refrain from opposing? There is a similar ambiguity in the following quotations, and the fact that the reader is unlikely to choose the wrong alternative is no excuse for offering it to him (see AMBIGUITY). *He said he is taking no inferiority complex into his discussions with Mr. Kruschev because the United States is second best in the missile field.* | *The witness said that the case was not brought before the committee because of the incident the night before.*

**bedizen.** The OED allows both *ī* and *ĭ*, but prefers the *ī*, and states that 'all English orthoepists' do so. But popular usage apparently does not; the COD now gives short *i* only.

**befal(l), befel(l).** The second l should be kept; see -LL-, -L- 3.

**beg.** Such expressions as *b. to state*, *b. to acknowledge*, *b. to remain* can claim an honourable origin in that this use of *beg* (i.e. *beg leave to*) was prompted by politeness. The OED quotes from Chatham *Lett. Nephew* 'There is likewise a particular attention required to contradict with good manners; such as begging pardon, begging leave to doubt and such like phrases.'

But the beg-phrases of COMMERCIALESE now serve merely to introduce a flavour of stiffness and artificiality into what should be spontaneous and friendly.

**begging the question.** See MIS-APPREHENSIONS, and PETITIO PRINCIPII, of which it is the English version.

**begin. 1.** Past tense *began,* formerly also (and still rarely) *begun.* **2.** For *It was begun to be built* etc. see DOUBLE PASSIVES. See also COMMENCE.

**behalf** and **behoof** are liable to confusion both in construction and in sense. *On his* etc. *behalf,* or *on behalf of all* etc., means *as representing him, all,* etc. (*I can speak only on my own behalf*; *Application was made on behalf of the prosecutor*); *on* is the normal preposition; the phrase does not mean, except additionally and by chance, for the advantage of, and it is still in common use. *Behoof,* now an archaism, means simply advantage. *For* or *to his* etc. *behoof,* or *for* or *to the behoof of all* etc., means *for* or *to the advantage of him, all,* etc. (*For the behoof of the unlearned*; *To the use and behoof of him and his heirs*; *Taking towns for his own behoof*); *for* and *to* are the prepositions.

**beholden, beholding.** As past participle of *behold, beholden* is now obsolete except in poetry. In the sense bound by gratitude (which it got when *behold* could still mean hold fast) it is still in use, though archaic by the side of *obliged. Beholding* in that sense is an ancient error due to ignorance of how *beholden* got its meaning, and is now obsolete.

**behoof.** See BEHALF.

**behove, behoove.** The first spelling is the better; indeed the second is virtually obsolete except in U.S., where it is preferred. As to pronunciation, the OED says 'Historically it rimes with *move, prove,* but being now mainly a literary word, it is generally made to rime with *rove, grove,* by those who know it only in books'.

**belike.** See ARCHAISM and WARDOUR STREET.

**belittle.** The OED says 'The word appears to have originated in U.S.; whence is recent English use in sense 3', which is 'depreciate, decry the importance of'. (Sense 1 is 'to make small' and sense 2 is 'to cause to appear small by contrast, to dwarf'.) It is only in sense 3 that we have adopted the word, but we have done this so enthusiastically that we are inclined to forget the old-established words, of which we have a large supply suitable for various contexts and shades of meaning—*cry down, decry, depreciate, deride, disparage, lower, make light of, minimize, poohpooh, ridicule, run down, slight.*

**belles-lettres** survives chiefly in publishers' circulars, library catalogues, and book reviews, its place having been taken elsewhere by *literature* (sometimes *pure literature*) used in a special sense; that sense is, as defined by the OED, 'Writing which has claim to consideration on the ground of beauty of form or emotional effect'. Like other words that require a speaker to attempt alien sounds (such as the ending *-etr* is), *belles-lettres* can never become really current. Its right to live at all, by the side of *literature,* depends on the value of a differentiation thus expressed by the OED: 'But it is now generally applied (when used at all) to the lighter branches of literature or the aesthetics of literary study'; i.e. *Paradise Lost* is literature rather than belles-lettres, though *The Essays of Elia* is both. This restricted application, however, itself needs defence, *b.* properly including the epic as much as the toy essay, just as *literature* does. We could in fact do very well without *b.,* and still better without its offshoots *belletrist* and *belletristic.* See also LITERATURE.

**belly** is a good word now almost done to death by GENTEELISM. It lingers in speaking of animals, ships, and aeroplanes and in proverbs and phrases, to which the second world war added one or two—the *soft underbelly of* the Axis, the *bellyaching* of officers who questioned their orders. But on the whole the process goes on, and the road to the heart lies less often through the *b*. than through the stomach or the tummy. The slaying of the slayer attempted by *tummy*, though half-hearted and facetious, illustrates the vanity of genteel efforts; a perpetual succession of names, often ending in nursery ineptitudes, must be contrived. *Stomach* for *belly* is a specially bad case, because the meaning of *stomach* has to be changed before it can take the place of *b*. in many contexts. The tendency, however, is perhaps irresistible. See EUPHEMISM.

**beloved**, when used as a past participle (*b. by all; was much b.*), is disyllabic (*-ŭvd*); as a mere adjective (*dearly b. brethren; the b. wife of*), or as a noun (*my b.*), it is trisyllabic (*-ŭvĕd*); the first of these rules is sometimes broken in ignorance of usage, and the second for the sake of the emphasis attaching to what is unusual. Cf. *blessed, cursed*.

**below, under.** There is a fairly clear distinction between the prepositions, worth preserving at the cost of some trouble. But the present tendency is to obscure it by allowing *under* to encroach; and if this continues *b*. will seem more and more stilted, till it is finally abandoned to the archaists. The distinction is that *b*., like its contrary *above* (cf. also the Latin *infra* and *supra*), is concerned with difference of level and suggests comparison of independent things, whereas *under*, like its contrary *over* (cf. also the Latin *sub* and *super*), is concerned with superposition and subjection, and suggests some interrelation. The classes b. us are merely those not up to our level; those u. us are those that we rule. *B. the bridge* means with it higher up the stream; *u. the bridge*, with it overhead. Contexts in which *b*. is both right and usual are *b. par, b. the belt*. Contexts in which *u*. has encroached are *men b. 45, b.* one's *breath, no one b. a bishop, incomes b. £500*. Contexts in which *u*. is both right and usual are *u. the sun, the sod, the table, the*

circumstances, the Stuarts, tyranny, protection, one's *wing*, one's *thumb*, *a cloud*. Cf. also BENEATH.

**beneath** has still one generally current sense—too mean(ly) or low for (*He married b. him; It is b. contempt; It would be b. me to notice it*). Apart from this it is now little more than a poetic, rhetorical, or emotional substitute for *under(neath)* or *below*.

**Benedick,** not *Benedict*, is the spelling in *Much Ado*, and should always be the spelling when the name is used generically for a confirmed or captured bachelor; but *Benedict* is often used (*Penalize the recalcitrant Benedicts by putting a heavy tax upon them*) either (and probably) in ignorance, or on the irrelevant ground that Shakespeare might have done well to use the more etymological form in -ct.

**benign, benignant, malign, malignant.** The distinction between the long and short forms is not very clear, nor is any consistently observed. But it may be said generally that *benign* and *malign* refer rather to effect, and *benignant* and *malignant* to intention or disposition: *Exercises a benign* or *malign influence; A benignant* or *malignant deity*. An unconscious possessor of the evil eye has *a malign* but not *a malignant look*; discipline is benign rather than benignant, indulgence benignant rather than benign. The difference is the same in kind, though less in degree, as that between *beneficent, maleficent*, and *benevolent, malevolent*. It is to be noticed, however, (1) that the impulse of personification often substitutes the -*ant* forms for the others, e.g. as epithets of *destiny, chance*, etc.; (2) that the distinction is less generally maintained between *benign* and *benignant* than between the other two (e.g. *of benign appearance* is common, where *benignant* would be better); (3) that nevertheless in medical use as epithets of diseases, morbid growths, etc., the forms are *benign* (as would be expected) and *malignant* (contrary to the rule). This use of *malignant* is perhaps a stereotyped example of the personifying tendency, which *benign* escaped because *benignant*, a comparatively recent formation, did not exist when the words were acquiring their medical sense. See also MALIGNANCY.

**bereaved, bereft.** The essential principle is perhaps that *bereaved* is resorted to in the more emotional contexts, *bereft* being regarded as the everyday form (cf. BELOVED). The result in practice is that (1) *bereft* is used when the loss is specified by an *of*-phrase, and *bereaved* when it is not, the latter naturally suggesting that it is the greatest possible (*Are you bereft of your senses?*; *The blow bereft him of consciousness*; *A bereaved mother*; *Weeping because she is bereaved*); but (2) *bereaved* is sometimes used even before *of* when the loss is that of a beloved person (*A mother bereft*, or *bereaved*, *of her children*; *Death bereft*, or *bereaved*, *her of him*). See -T AND -ED.

**beseech.** *Besought* is the established past and p. participle, though *beseeched*, on which the OED comment is merely 'now regarded as incorrect', still occurs, probably by inadvertence, and Milton has *beseecht*.

**beside(s).** The forms have been fully differentiated in ordinary modern use, though they are often confused again in poetry, and by those who prefer the abnormal or are unobservant of the normal. (1) *Beside* is now only a preposition, *besides* having all the adverbial uses; *besides* would have been normal in *And what is more, she may keep her lover beside*. / *We talked of thee and none beside*. (2) *Beside* alone has the primary prepositional senses 'by the side of' (*Sat down beside her*; *She is an angel beside you*), 'out of contact with' (*beside oneself*, *the question*, *the mark*, *the purpose*). (3) *Besides* alone has the secondary prepositional senses 'in addition to', 'except'; it would have been normal in *Other men beside ourselves*. / *I have no adviser beside you*. For *besides* = 'as well as' see WELL.

**bespeak.** The p.p. form *bespoke* perhaps lingers only, beside the now usual *bespoken*, as an attributive adjective meaning made to order (*bespoke goods*, *boots*, etc.) in contrast with *ready-made*, and even this is now old-fashioned.

**bestir** is now always used reflexively (*must b. myself*), and never, idiomatically, as an ordinary transitive verb; *stirred* should have been used in *The example of the French in Morocco has bestirred Italy into activity in Africa*.

**bet.** Both *bet* and *betted* are in idiomatic use as past tense and p.p. *He bet me £5 I could not*; *They betted a good deal in those days*; *I have bet £500 against it*; *How much has been bet on him?*; *The money was all betted away*. These examples, in which it will probably be admitted that the form used is better than the other, suggest that *bet* is preferred in the more usual connexion, i.e. with reference to a definite transaction or specified sum, and *betted* when the sense is more general.

**bête noire.** See FRENCH WORDS. Those who wish to use the phrase in writing must not suppose, like the male writer quoted below, that the gender can be varied: *From the very first, and for some reason that has always been a mystery to me, I was his* bête noir.

**bethink** has constructions and meanings of its own, and can never serve as a mere ornamental substitute for *think*, as in *They will bethink themselves the only unhappy on the earth*.

**better.** The idiomatic phrase *had better* requires care. See HAD 1.

**better, bettor.** See -OR. According to the OED the tendency was towards *-or*. But that was in 1888, and the tendency seems now to be the other way; the COD prefers *-er*.

**betterment.** The use of the word in general contexts, apart from its technical application to property, is an example of SAXONISM. *The late Lady Victoria devoted her entire life to the b. of the crofters and fishermen*. If the writer had been satisfied with the English for *betterment*, which is *improvement*, he would not have been blinded by the unusual word to the fact that he was writing nonsense; the lady's effort was not to better or improve the crofters, but their lot.

**between** is a sadly ill-treated word; the point on which care is most necessary is that numbered 6. 1. *B* and *among*. 2. *B. you and I*. 3. *B. each, every*. 4. *B. . . . and b.* 5. *Difference b.* 6. *B. . . . or* etc.

1. *B.* and *among*. The OED gives a warning against the superstition that *b.* can be used only of the relationship between two things, and that if there are more *among* is the right preposition. 'In all senses *between* has been, from its earliest appearance, extended to more than two. . . . It is still the only word available to express the relation of a thing to many surrounding things severally and individually; *among* expresses a relation to them collectively and vaguely: we should not say *the space lying among the three points* or *a treaty among three Powers.*' But the superstition dies hard. Seventy years after those words were written the following sentence cannot escape suspicion of being under its influence: *The peaceful, independent, and self-governing status of Cyprus is conditional on the continuance of cordial relations among Britain, Greece, and Turkey.*

2. *B. you and I*, which is often said, perhaps results from a hazy recollection of hearing *you and me* corrected in the subjective. See I for fuller discussion.

3. *B. each, every*. *B.* may be followed by a single plural (*b. two perils*) as well as by two separate expressions with *and* (*b. the devil and the deep sea*). Its use with a single expression in which a distributive such as *each* or *every* is supposed to represent a plural is very common (Ruskin has *b. each bracket*); but its literal absurdity offends the purists and it is best avoided. *A batsman who tried to gain time by blowing his nose b. every ball* (after every ball). *The absence of professional jealousy that must exist in future b. each member of our profession* (b. the members, or, if emphasis is indispensable, b. each member . . . and the rest).

4. *B. . . . and b.* The temptation to repeat *b.* with the second term in long sentences must be resisted; *B. you and b. me* is at once seen to be absurdly wrong; the following is just as bad: *The claim yesterday was for the difference b. the old rate, which was a rate by agreement, and b. the new, of which the Water Board simply sent round a notice*. See OVERZEAL.

5. *Difference b. B.*, used after words like *difference*, seems to tempt people to put down for one of the terms the exact opposite of what they mean: *My friend Mr. Bounderby would never see any difference b.* leaving *the Coketown 'hands' exactly as they were and requiring them to be fed with turtle soup and venison out of gold spoons* (for *leaving* read *refusing to leave*). | *There is a very great distinction between a craven truckling to foreign nations and* adopting *the attitude of the proverbial Irishman at a fair, who goes about asking if anybody would like to tread on the tail of his coat* (read *avoiding* for *adopting*).

6. *B. . . . or* etc. In the commonest use of *b.*, i.e. where two terms are separately specified, the one and only right connexion between those terms is *and*. But writers indulge in all sorts of freaks; the more exceptional and absurd of these, in which *against, whereas,* and *to* are experimented with, are illustrated in: *It is the old contest b. Justice and Charity, b. the right to carry a weapon oneself against the power to shelter behind someone else's shield* (here ELEGANT VARIATION has been at work; to avoid repeating *between . . . and* is more desirable than to please the grammarian). | *He distinguishes b. certain functions for which full and rigorous training is necessary, whereas others can very well be discharged by me who have had only the limited training* (read *and others that can*). | *Societies with membership b. one thousand to five thousand*. These are freaks or accidents; the real temptation, strong under certain circumstances, is to use *or* for *and*; *They may pay in money or in kind* is wrongly but naturally converted into *The choice is b. payment in money or in kind*. So *Forced to choose b. the*

*sacrifice of important interests on the one hand* or *the expansion of the Estimates on the other.* | *Our choice lies b. forfeiting the liberties of two million West Berliners* or *the survival of the rest of us.* These again are simple, requiring no further correction than the change of *or* to *and.* Extenuating circumstances can be pleaded only when one or each of the terms is compound and has its parts connected by *and*, as in: *The question lies b. a God and a creed*, or *a God in such an abstract sense that does not signify* (read *b. a God and a creed on the one hand*, and *on the other a God in such* etc.). | *The conflict, which was previously b. the mob and the Autocracy, is now b. the Parliament and the King* or *the Parliament and the Bureaucracy* (this means that the question now is whether Parliament and King, or Parliament and Bureaucracy, shall rule, and this way of putting it should be substituted: *The conflict was previously b. mob and Autocracy; but the question* etc.)

**betwixt.** See ARCHAISM. But the phrase *betwixt and between*, meaning an intermediate position, survives as a colloquial cliché.

**beverage.** See PEDANTIC HUMOUR, and WORKING AND STYLISH WORDS.

**beware** is now not inflected, and is used only where *be* would be the verb part required with *ware* meaning cautious, i.e. in the imperative (*B. of the dog!*), infinitive (*He had better b.*), and present subjunctive (*Unless they b.*). *Bewaring, I beware* or *bewared, was bewared of*, etc., are obsolete.

**bi-** prefixed to English words of time (*bi-hourly, bi-weekly, bi-monthly, bi-quarterly, bi-yearly*) gives words that have no merits and two faults: they are unsightly hybrids, and they are ambiguous. To judge from the OED, the first means only two-hourly; the second and third mean both two-weekly, two-monthly, and half-weekly, half-monthly; and the last two mean only half-quarterly, half-yearly. Under these desperate circumstances we

can never know where we are. If it were not for *bicentenary*, which lacks a vernacular equivalent, there would be no reason why all the *bi-* hybrids should not be allowed to perish, and the natural and unambiguous *two-hourly* and *half-hourly, fortnightly* and *half-weekly, two-monthly* and *half-monthly, half-yearly and half-quarterly, two-yearly* and *half-yearly*, of which several are already common, be used regularly in place of them and the words (*biennial, bimestrial*) on which they were fashioned; these latter have now almost become ambiguous themselves from the ambiguity of the misshapen brood sprung of them. They cause confusion in the most surprising places. *An annual bulletin is our first aim; but biennial issues may become possible if the Association enlarges as we hope.* (From a bulletin issued by the International Association of University Professors of English.) *Biannual*, probably invented to stand to *biennial* as *half-yearly* to *two-yearly*, is sometimes confused with and sometimes distinguished from it. *Half-yearly* is the right word.

**bias** makes preferably *-ased, -asing*. See -S, -SS-.

**bicentenary, bicentennial.** See CENTENARY.

**biceps, triceps.** If plurals are wanted, it is best to say *-cepses*, the regular English formation; not *-cipites* (the true Latin), both because it is too cumbrous, and because Latin scholars do not know the words as names of muscles. But *biceps* as a plural, originating as a mere blunder (cf. FORCEPS), is common and may oust others. See LATIN PLURALS 4.

**bid.** 1. In the auction sense the past and p.p. are both *bid* (*He bid up to £10; Nothing was bid*).
2. In other senses, the past is usually spelt *bade* and pronounced *băd* (or now increasingly *bād*); the p.p. is *bidden*, but *bid* is preferred in some phrases, especially *Do as you are bid*.
3. *Bid* one go etc. has been displaced in speech by *tell* one *to go* etc., but lingers in literary use. The active is

usually followed by infinitive without to (*I bade him go*), but the passive by *to* (*He was bidden to go*).

**4.** For the use of the noun as a space-saver see HEADLINE LANGUAGE.

**bide.** Still current in Scotland (*bide a wee*), but in England, apart from archaism and poetic use, the word is now idiomatic only in *b.* one's *time*, and its past in this phrase is *bided*.

**biennial.** See BI-.

**big, great, large.** The differences in meaning and usage cannot be exhaustively set forth; but a few points may be made clear. See also SMALL. Roughly, the notions of mere size and quantity have been transferred from *great* to *large* and *big*; *great* is reserved for less simple meanings, as will be explained below. *Large* and *big* differ, first, in that the latter is more familiar and colloquial, and secondly, in that each has additional senses—*large* its own Latin sense of generous (cf. LARGESS), and *big* certain of the senses proper to *great*, in which it tends to be used sometimes as a colloquial and sometimes as a half-slang substitute. It will be best to classify the chief uses of *great* as the central word, with incidental comments on the other two.

**1.** With abstracts expressing things that vary in degree, *great* means *a high degree* of (*g. care, ignorance, happiness, tolerance, charity, joy, sorrow, learning, facility, generosity, comfort*). *Big* is not idiomatic with any of these; and though *large* is used with *tolerance, charity,* and *generosity,* it is in a special sense—broad-minded or prodigal. With words of this kind that happen themselves to mean size or quantity (*size, quantity, bulk, magnitude, amount, tonnage*) *large* and *big* are sometimes used, though neither is as idiomatic as *great,* and *big* is slangy or at best colloquial.

**2.** *Great* may be used not to imply size, but to indicate that the person or thing in question has the essential quality of his or its class in a high degree; so *a g. opportunity, occasion,*

*friend, landowner, majority, schoolmaster, shot*(shooter), *nuisance, stranger, brute, fool, haul, race* (contest), *undertaking, success, linguist, age.* Here *large* could be substituted with *landowner, majority, haul,* and *undertaking,* but merely because a large quantity of land, votes, fish, or money is involved; *big* could stand with the same four on the same ground; it is increasingly used also with most of the others. This is unfortunate; *a great fool* should mean a very foolish fool, and *a big fool* one whose stature belies his wits.

**3.** *A great* has the meaning *eminent, of distinction,* and *the g.* the meaning *chief, principal, especial* (*a g. man; g. houses; a g. family; the g. advantage,* or *thing, is*); and from these comes the use of *great* as a distinctive epithet (*the g. auk; G. Britain; Alexander the G.; the Lord G. Chamberlain*), with the idea of size either absent or quite subordinate. In these senses *large* cannot be used, though it would stand with many of the same words in a different sense (*a g. family* has distinguished, but *a l. family* numerous, members). Here again the substitution of *big* for *great* makes for confusion; *a big man* should refer to the man's size, or be extended only (as in *the big men of the trade*; cf. *large* with *landowner* etc. in 2) to express the quantity of his stock or transactions. But thanks no doubt to the fondness of modern journalism for short words and snappy phrases *big* is now so often used as a distinctive epithet instead of *great,* even when difference of size is not the salient point of distinction, that it is losing its slangy flavour. *The big race, a big occasion, The Big Five* (banks).

**4.** Finally, *great* does sometimes mean of remarkable size—the sense that it has for the most part resigned to *large* and *big*; but it is so used only where size is to be represented as causing emotion. *Large* and *big* give the cold fact; *great* gives the fact coloured with feeling; e.g. *He hit me with a great stick* is better than *with a large* or *big stick,* if I am angry about its size; but in *Perhaps a big* or *large stick might do* it would be impossible to substitute *great*. Similarly *Big dogs are better out of doors*, but *I am not going to have that great dog in here*; *His feet are large* or *big*, but *Take your great feet off the sofa*. *What a great head he has!* suggests admiration of the vast brain or fear of the formidable teeth it probably contains, whereas *What a large head he has!* suggests dispassionate observation.

**bilateral, unilateral, multilateral.** These words, firmly established in the jargon of physiology and diplomacy, especially in the phrase *u. disarmament*, are in danger of becoming POPULARIZED TECHNICALITIES and driving out the old-fashioned words *two-sided*, *one-sided*, and *many-sided*. That their use should have led a much respected newspaper to insert so surprising a headline as *BILATERAL TRIANGLE IN SOUTH AMERICA* is the greater reason for avoiding them.

**billet doux.** Pronounce *bǐ'lǐdōo'*. The plural is *billets doux*, but should be pronounced *bǐ'lǐdōo'z*.

**billion.** It should be remembered that this word does not mean in American use (which follows the French) what it means in British. For us it means the second power of a million, i.e. a million millions (1,000,000,000,000); for Americans it means a thousand multiplied by itself twice, or a thousand millions (1,000,000,000), what we call a *milliard*. Since *billion* in our sense is useless except to astronomers, it is a pity that we do not conform.

**bill of rights.** This term, when used in U.S., ordinarily refers not to the famous statute passed by the English Parliament in 1689 but to the Amendments to the U.S. Constitution adopted in 1791 to prevent the Federal Government from encroaching on the liberties of the people.

**bipartisan.** This unlovely substitute for *two-party* had made enough progress by 1933 to be recognized by the OED Supp., whose earliest example is dated 1920. It seems to have come to stay, on both sides of the Atlantic. Perhaps it could claim in self-defence that it contains an implication not necessarily present in *two-party*, namely that of agreement between the two.

**bishopric.** See SEE.

**bitumen.** Of the alternative pronunciations *bitu'men* and *bit'umen* the OED originally preferred the first, but the second is now more usual; the COD gives no other. See RECESSIVE ACCENT.

**bivouac.** Participles *-cked, -cking*; see -C-, -CK-.

**black(en).** See -EN VERBS.

**blame.** *I am to b.* is an illogicality long established as idiomatic. *Don't b. it on me* is a colloquialism not yet recognized by the dictionaries, a needless variant of *don't b. me for it*, and not to be encouraged.

**blank verse.** Strictly, any unrhymed verse; but in ordinary use confined to the five-foot iambic unrhymed verse in which *Paradise Lost* and the greater part of Shakespeare's plays are written.

**-ble.** See -ABLE.

**blended, blent.** *Blended* is now the everyday form (*carefully blended teas*; *he successfully blended amusement with instruction*); but *blent* survives as a participial adjective in poetic, rhetorical, and dignified contexts (*pity and anger blent*).

**blessed, blest.** The attributive adjective is regularly disyllabic (*blesséd innocence*; *what a blesséd thing is sleep!* *the blesséd dead*; *every blesséd night not a blesséd one*), and so is the plural noun with *the*, which is an absolute use of the adjective. But the monosyllabic pronunciation is sometimes used in verse, or to secure emphasis, or in archaic phrases; the spelling is then *blest: our blest Redeemer*; *that blest abode*; *Blest Pair of Sirens*. The past tense, past participle, and predicative adjective are regularly monosyllabic the spelling is usually *blessed* in th past tense, *blest* in clearly adjectival contexts, and variable (usually *blessed*) in the p.p. (*He blessed himself*; *God has blessed me with riches*; *He is blessed*, or *blest*, *with good health*, *in his lot*, etc.; *Blessed*, or *blest*, *if I know*; *Those who win heaven*, *blest are they*; *It is twice blest*); in the beatitudes and similar contexts, however, *blessèd* is usual. *Blessèd* sometimes makes *-est*; see -ER AND -EST 4.

**blessedness.** For *single b.*, see WORN-OUT HUMOUR.

**blink.** *That is the dark side, and nothing is to be gained by blinking at it.* To condone an offence is to *wink at it.* But to refuse to recognize an unpleasant fact is to *blink it*, not to *blink at it.* See ANALOGY.

**blithesome** is a NEEDLESS VARIANT of *blithe*; see -SOME.

**blond(e).** The practice now usual is to retain the *-e* when the word is used either as noun or as adjective of a woman or the lace and drop it otherwise (*the blonde girl*; *she is a blonde*; *she has a blond complexion*; *the blond races*).

**bloom, blossom.** Strictly *bloom* n. and v. refers to the flower as itself the ultimate achievement of the plant, and *blossom* n. and v. to the flower as promising fruit. The distinction is perhaps rather horticultural than literary or general; at any rate it is often neglected. But *The roses are in bloom*, *The apple-trees are in blossom*, and other uses, confirm it; and in figurative contexts, the *blooming-time* or *bloom* of a period of art is its moment of fullest development, when its *blossoming-time* or *blossom* is already long past.

**bloom** (in the foundry). See REVIVALS.

**blue.** The OED derives the slang verb (= squander) from *blow*, implying that *blue* comes from a misunderstanding of the past tense *blew.* But the forms are now always *blue*, *blued*.

**bluebell.** In the south this is the wild hyacinth, *Scilla nutans*; in the north, and especially in Scotland, it is another name for the *harebell*, *Campanula*

*rotundifolia*, with fewer, larger, and thinner-textured flowers than the other.

**boatswain.** The nautical pronunciation (*bō'sn*) has become so general that to avoid it is more affected than to use it.

**bodeful** is a modern stylish substitute for *ominous*; 'very frequent in modern poets and essayists' said the OED in 1888, but its popularity has since waned. See WORKING AND STYLISH WORDS, and SAXONISM.

**bog(e)y, bogie.** The OED prefers *bogy* for the bugbear, and *bogie* in rolling-stock. The name of the imaginary golfing colonel is usually spelt *Bogey*.

**bona fide(s).** *Bona fide* (pronounce *bōnă fīdĕ*) is a Latin ablative meaning *in good faith.* Its original use is accordingly adverbial (*Was the contract made bona fide?*); but it is also and more commonly used attributively like an adjective (*Was it a bona fide contract?*). In this attributive use the hyphen is admissible, but not usual; in the adverbial use it is wrong. *Bona fides* is the noun; the mistake is sometimes made by those who know no Latin of supposing *fides* to be the plural of *fide*: *The fact that Branting accepted the chairmanship of the Committee should be sufficient evidence of its* bona-fide. / *His bona fides were questioned.*

**bond(s)man.** The two forms are properly distinct, *bondsman* meaning a surety and being connected with the ordinary *bond* and *bind*, and *bondman* meaning a villein, serf, or man in bondage, and having (like *bondage*) nothing to do with *bond* and *bind*. But *bondsman* is now rare in its true sense, and on the other hand is much more used than *bondman* in the sense proper to the latter.

**bond washing and dividend stripping.** Most of us are familiar with these terms, but few know much more about them than that they are devices for the legal avoidance of taxation. In the course of the duel provoked by them between the tax avoider and the legislature they have developed a protean variety of detail, but their essence remains the same. In their original and simplest form they were collusive transactions by which a person liable to high rate of surtax would avoid liability by selling investments cum dividend and buying them back at a lower price after the dividend had been paid to the purchaser; in this way he converted what would have been taxable income in his hands into a nontaxable gain. The other party to the deal would be either a tax-exempt body (e.g. a charity) or someone (e.g. a dealer in securities) who, unlike the ordinary taxpayer, was taxable on his gains from transactions in securities and so could set off his loss on resale against his liability. Thus, provided that the difference between the two prices, with incidental expenses, did not exceed the amount of the dividend, the only loser would be the Revenue.

**bonne bouche.** The meaning of the phrase in French is not that which we have given it; but variation of meaning or form is no valid objection to the use of a phrase once definitely established; see À L'OUTRANCE.

**born(e).** The p.p. of *bear* in all senses except that of birth is *borne* (*I have borne with you till now*; *Was borne along helpless*); *borne* is also used, when the reference is to birth, (*a*) in the active (*Has borne no children*), and (*b*) in the passive when *by* follows (*Of all the children borne by her one survives*); the p.p. in the sense of birth, when used passively without *by*, is *born* (*Was born blind*; *A born fool*; *Of all the children born to them*; *The melancholy born of solitude*).

**botanic(al).** The -ic form is 'now mostly superseded by *botanical*, except in names of institutions founded long ago, as "The Royal Botanic Society" ' —OED. See -IC(AL).

**both.** 1. *Both . . . as well as.* 2. Re-

dundant *both*. 3. Common parts in *both . . . and* phrases.

1. *Both . . . as well as.* To follow *b*. by *as well as* instead of *and*, as is often done either by inadvertence or in pursuit of the unusual, is absurd; how absurd is realized only when it is remembered that the *as well* of *as well as* is itself the demonstrative to which the second *as* is relative, and can stand in the place occupied by *both* instead of next door to *as*. *The metrostyle will always be of exceeding interest, b. to the composer as well as to the public.* | Which *differs from* who *in being used b. as an adjective as well as a noun.* In these examples either omit *both* or read *and* for *as well as*; *as well*, it will be seen, can be shifted into the place of *both* if the object is to give timely notice that the composer or the adjective is not the whole of the matter.

2. Redundant *both*. The addition of *both* to *equal(ly)*, *alike*, *at once*, *between*, or any other word that makes it needless, is at least a fault of style, and at worst (e.g. with *between*) an illogicality. In the examples, *both* should be omitted, unless the omission of the other word(s) in roman type is preferable or possible: *If any great advance is to be* at once b. *intelligible and interesting.* | *We find* b. *Lord Morley and Lord Lansdowne* equally *anxious for a workable understanding.* | *The International Society is not afraid to invite comparisons* between *masters* b. *old and new.* See also FALSE EMPHASIS 2.

3. Common parts in *both . . . and* phrases. Words placed between the *both* and the *and* are thereby declared not to be common to both members; accordingly, *He was* b. *against the Government and the Opposition* is wrong; the right arrangements are (*a*) *he was b. against the Government and against the Opposition*, (*b*) *he was against b. the Government and the Opposition*, preferably the latter.

**bother.** See POTHER.

**bottleneck** does not seem ever to have been used in its literal sense. The OED (1888), which gives some fifty other *bottle-* compounds, makes no mention of it. Nor is it so used today, but it is much in demand figuratively. In the OED Supp., where the first example of its use is dated 1896, it is defined as 'a narrow or confined space where traffic may become congested'; and this, with good reason, has remained a common use. During the second world war it was a favourite way of describing any obstruction that impeded the flow of production or the supply of some needed commodity; in this sense it has been overworked; popular new metaphors usually are (cf. CEILING, TARGET, and see METAPHOR), and has brought deserved derision on those who have done violence to the metaphor by speaking of the biggest *b*. when they mean the most constrictive, or describing a universal shortage as a *world-wide b.*, or demanding that *bb*. should be ironed out.

**bounden** is still used in *bounden duty* though not in *in duty bound*. It is also used alternatively with *bound* as the p.p. of *bind* in the sense *beholden* (*I am much bounden*, or *bound, to you*); but the whole verb, including the p.p., is a mere ARCHAISM in this sense.

**bounteous, -iful.** See PLENTEOUS.

**bourgeois**, a French word meaning 'a member of the mercantile or shop-keeping class of any country' (OED), should have been one that commanded respect for solid worth. In fact, as C. S. Lewis has pointed out, it has had the unfortunate experience of being applied first by the class above the bourgeoisie to mean 'not aristocratic, therefore vulgar' and then by the class below it to mean 'not proletarian, therefore parasitic, reactionary'.

As the name of a printing type the word is anglicized and pronounced *bĕrjois'*.

**bourn(e).** There are two words, which were originally *burn* and *borne*, but are now not distinguished, consistently at any rate, either in spelling or in pronunciation. The first, which survives in place-names (e.g. *Bournemouth, Eastbourne*) and retains in Scotland its

original form *burn*, means a stream; but outside Scotland is now applied as a current word only to the streams of the chalk downs, full in winter and dry in summer; it serves in poetry as an ornamental synonym for *brook*. The second means properly a boundary (from French *borne*) as in *The undiscovered country from whose borne No traveller returns*, but is used almost solely, with a distorted memory of that passage, in the sense of destination or goal. The OED prefers *bourn* stream, and *bourne* goal, and the differentiation would be useful.

**brace,** n. (= two). See COLLECTIVES 5.

**brachylogy.** Irregular shortening down of expression. *Less sugar, This is no use,* and *A is as good or better than B,* are brachylogies for Less *of* sugar, This is *of* no use, and A is as good *as* or better than B; the first is established as idiomatic; the second is at worst a STURDY INDEFENSIBLE; the last is still regarded by many as illegitimate.

**brackets.** See STOPS, and for use in the sense of *class*, see GROUP.

**brain(s),** in the sense of wits, may often be either singular or plural, the latter being perhaps, as the OED suggests, the familiar and the former the dignified use. Some phrases, however, admit only one or the other, e.g. to have a tune on the *b.*, but (although the physical object is otherwise always *b.*) to blow one's *bb.* out.

**brain compounds.** The OED (1888) lists a large number of *b.*-compounds of which few are in common use today. For instance, the jilted heroine no longer gets *b. fever*, as she was prone to do in the Victorian novel, and the poet's *b. brat* of 1630 is now, more politely, a *b. child*. In 1932 the OED Supp. added, among others, the now popular *b. wave* (originally a telepathic message but now a 'sudden inspiration or bright thought') and *b. storm* ('a succession of sudden and severe paroxysms of cerebral disturbance'), sometimes, it is 'said, used in U.S. as a verb for a process of inducing the birth of a collective *b. child*. *Bb. trust*, *b.-washing*, and *b. drain* are still more recent. The first was originally applied to the body of advisers appointed by Franklin Roosevelt after his first election as President, and was later adopted in Britain as the name for broadcast discussions by small groups of polymaths. *B.-washing* came from America as a response to the need for a convenient word to describe certain modern methods of indoctrination, especially those employed by the Chinese on prisoners of war in Korea. The phrase itself is as old as Shakespeare. 'It's monstrous labour', said Octavius at the carouse on Pompey's Galley, 'when I wash my brain and it grows more foul'. *B. drain* is not, as might be thought, the final stage of b. washing. It is a compendious expression of concern at the tendency of promising young British scientists to go and work in America.

**brake, break,** nn. The words meaning (1) bracken, (2) thicket, (3) lever, (4) crushing or kneading or peeling or harrowing implement, (5) steadying-frame, though perhaps all of different origins, are spelt *brake* always. The word that means checking-appliance is now invariably *brake*, but *break* occasionally occurred in the 19th c. owing to a probably false derivation from *to break* (the OED refers it to no. 3 above, which it derives from OF *brac* = F *bras* arm). The word meaning horse-breaker's carriage-frame, and applied also to a large wagonette, and now preserved in the type of motor-car that has taken the place of the wagonette as a *shooting b.* varies; *brake* is certainly more usual for the latter. The word meaning fracture etc. is always *break*.

**branch.** For synonymy see FIELD.

**bran(d)-new.** The spelling with -d is right (fresh as from the furnace); but the d is seldom heard, and often not written.

**brash.** '*The Quiet American*', Graham Greene's scathing denunciation of the United States' *brash and clumsy political warfare against communism in SE. Asia.* / *One feels that the late Tommy Handley, in his brash proletarian way, really invented 'One-upmanship' in his power to master every situation.* / *Bagehot had no enthusiasm whatever for 'democracy', which he equated with the brash and vulgar American republic.* / *Did Mr. Butler resent the brash zeal with which Mr. Nabarro rushed into the delicate negotiations about making dripfeed oil heaters safer?* This adjective, called by the OED 'obs. or dial.' in 1888, but persisting as a U.S. colloquialism, has been taken up by British journalists and given so much work to do that the dictionaries are finding it hard to keep abreast. Definitions they have tried include *active*, *bold*, *callow*, *cheeky*, *forward*, *hard*, *harsh*, *hasty*, *impetuous*, *insolent*, *quick*, *rash*, *rough*, *saucy*, *sharp*, *sudden*, *tactless*, and *tart*. That is a lot to pack into one monosyllable. No wonder the word is popular.

*Brash* has several other meanings. In the sense of assault it is obsolete, but the noun meaning a slight feverish attack or an eruption of water is still current; so is the collective word meaning fragments of disintegrated rock or hedge-clippings. The OED leaves open the question with which of these words, if any, the now popular adjective is connected.

**brass tacks.** See RHYMING SLANG.

**brave** in the sense of fine or showy is an ARCHAISM, and in the sense of worthy a GALLICISM; *make a b. show*, however, is fully current, and Miranda's *b. new world* has been given a fresh lease of life by Aldous Huxley.

**braze.** See REVIVALS.

**break.** 1. For p.p. see BROKE(N). 2. For spelling of nouns see BRAKE, BREAK.

**breakdown.** This POPULARIZED TECHNICALITY from statistics, having the charm of novelty, is apt to make those who use it forget the danger that its literal meaning may intrude with ludicrous effect. *A complete b. of our exports to dollar countries is unfortunately impossible.* | *The houses should be broken down into types.* | *The b. of patients by the departments under whose care they were before discharge should be strictly followed.* | *Statistics of the population of the United States of America, broken down by age and sex.* This use of *b.* is established and unexceptionable, but humdrum words like *classify* are safer.

**breakthrough.** This military metaphor has become a VOGUE WORD since the second world war, applied especially to some signal achievement in scientific research. It is an apt metaphor, and has no need to be bolstered, as it often is, by adjectives such as *major.* But, like all vogue words, it is being overworked.

**breeches** etc. The singular noun and its derivatives (*breechloader*, *breeching*, etc.) have *-ēch-* in pronunciation; *breeches* the garment has always *-ĭch-*, and the verb *breech* (put child into bb.) followed this, but the modern practice of putting small boys into breeches as soon as they can stand has made the word obsolete. 'Breeches is a double plural, the form *breek* being itself plural; as *feet* is from *foot* so is *breek* from *brook*.'—Skeat.

**brevet,** n. and v. Pronounce *brĕ'vĕt*, not *brĕvĕ't*; the past and p.p. are accordingly *breveted*, see -T-, -TT-.

**brier, briar.** For the word meaning thorny bush, the spelling *brier* is nearer the original and preferable; the name of the pipe-wood is an entirely different word; it is also best spelt *brier*, but *briar* is more usual.

**brilliance, -cy.** See -CE, -CY.

**brindle(d), brinded.** The original form *brinded* is archaic, and should be used only in poetry. *Brindled*, a later variant of it, is now the ordinary adjective, and *brindle*, a BACK-FORMATION from this, and convenient as a name for the colour and the dog, should be used only as a noun.

**brinkmanship.** *American brinkmanship has not led to British panic.* Thus *The Times*, whose leader-writers are never shy of giving currency to a neologism if they think it useful. To add the suffix *-ship* to a noun is a recognized way of producing a compound that means 'the qualities or character associated with, or the skill or power of accomplishment of, the person denoted by the noun' (OED), and in most of these compounds that noun ends *-man*—*statesmanship, horsemanship, seamanship,* etc. The conceit of making FACETIOUS FORMATIONS by treating *-manship* as the suffix was invented by Stephen Potter with his *Gamesmanship, Lifemanship,* and *Oneupmanship,* and he has had many imitators; *brinkmanship* is said to have been coined by Adlai Stevenson. Few of these pleasantries are likely to prove more than jocular and transitory slang, but *brinkmanship* is evidently felt to supply the need for a word denoting the qualities or character associated with one whose conduct of his country's foreign policy puts anxious spectators in mind of a man precariously balancing himself on the edge of a precipice.

**Britain, British, Briton.** For the relation of these to *England, English(man),* see ENGLAND.

**Briticism,** the name for an idiom used in Great Britain and not in America, is a BARBARISM; it should be either *Britannicism* or *Britishism,* just as *Hibernicism* or *Irishism* will do, but not *Iricism. Gallicism* and *Scot(t)icism* cannot be pleaded, since *Gaulish* and *Scotch* are in Latin *Gallicus* and *Scot(t)icus,* but *British* is *Britannicus.* The verbal critic, who alone uses such words, should at least see to it that they are above criticism.

**Britisher** 'was an American term that had a currency in the U.S. in the late 18th and 19th cc. but is practically never heard today. The ordinary American . . . accords the Irish separate recognition but all other male inhabitants of the British Isles are

*Englishmen* to him' (Evans). If this is so, it is time that British writers reconciled themselves to relinquishing the word in its convenient function of announcing that the user of it is American.

**broad, wide.** Both words have general currency; their existence side by side is not accounted for by one's being more appropriate to any special style. What difference there is must be in meaning; yet how close they are in this respect is shown by their both having *narrow* as their usual opposite, and both standing in the same relation, if in any at all, to *long*. Nevertheless, though they may often be used indifferently (*a b.* or *a w. road*; *three feet w.* or *b.*), there are (1) many words with which one may be used but not the other, (2) many with which one is more idiomatic than the other though the sense is the same, (3) many with which either can be used, but not with precisely the same sense as the other. These numbered points are illustrated below. The explanation seems to be that *wide* refers to the distance that separates the limits, and *broad* to the amplitude of what connects them. When it does not matter which of these is in our minds, either word does equally well; if the verges are far apart, we have a w. road; if there is an ample surface, we have a b. road; it is all one. But (1) backs, shoulders, chests, bosoms, are b., not w., whereas eyes and mouths are w., not b.; *at w. intervals*, *give a w. berth*, *a w. ball*, *w. open*, in all of which *b.* is impossible, have the idea of separation strongly; and *w. trousers*, *w. sleeves*, *w. range*, *w. influence*, *w. favour*, *w. distribution*, *the w. world*, where *b.* is again impossible, suggest the remoteness of the limit. Of the words that admit *b.* but refuse *w.* some are of the simple kind (*b. blades, spearheads, leaves*; *the b. arrow*), but with many some secondary notion such as generosity or downrightness or neglect of the petty is the representative of the simple idea of amplitude (*b. daylight*, *B. Church*, *b. jests*, *b. farce*, *b. hint*, *b. Scotch*, *b. facts*, *b. outline*).

(2) Some words with which one of the two is idiomatic, but the other not impossible, are: (preferring *broad*) *expanse*, *brow*, *forehead*, *lands*, *estates*, *brim* (though when the brim is very broad the hat becomes a *wide-awake*), *mind*, *gauge*; (preferring *wide*) *opening*, *gap*, *gulf*, *culture*.

(3) Some illustrations of the difference in meaning between *broad* and *wide* with the same word; the first two may be thought fanciful, but hardly the others: *A w. door* is one that gives entrance to several abreast, *a b. door* one of imposing dimensions; *a w. river* takes long to cross, *a b. river* shows a fine expanse of water; *a w. generalization* covers many particulars, *a b. generalization* disregards unimportant exceptions; a page has *a b. margin*, i.e. a fine expanse of white, but we allow *a w. margin* for extras, i.e. a substantial difference between the certain and the possible costs; *a w. distinction* or *difference* implies that the things are very far from identical, but *a b. distinction* or *difference* is merely one that requires no subtlety for its appreciation; *a b. view* implies tolerance but a *w. view* scope only.

**broadcast.** For past tense see FORE-CAST.

**broccoli** (not *-oco-*, nor *-lo*) is the best spelling. The word is an Italian plural, and is generally used collectively like *spinach* etc.; but if *a* or the plural is wanted, *a broccoli*, *two broccolis*, are the forms.

**brochure, pamphlet.** The introduction of the word *b.* in the 19th c. was probably due to misconception of the French uses. In French *b.* is used where the French *p.* (chiefly applied to scurrilous or libellous or violently controversial pamphlets) is inappropriate. The sense 'a few leaves of printed matter stitched together' has always belonged in English to *p.*, though it has by the side of this general sense the special one (different from the French) 'p. bearing on some question of current interest (esp. in politics or theology)'. 'Dans sa brochure appelée en anglais *pamphlet*', quoted in French dictionaries from Voltaire, gives us the hint that the two words had the same meaning. But because of the special sense of *pamphlet*, *brochure* has now found a useful place in our language to denote a commercial pamphlet, e.g. of a travel agency.

**broider(y).** See ARCHAISM, POETIC-ISMS.

**broke(n).** The form *broke*, now obsolete or a blunder in most senses, lingered until recently as p.p. of *break* = dismiss the service (*he was broke for cowardice*) and is still idiomatic in the slang phrase (*stony*) *broke*.

**brood.** *Those who did not die young frequently got chubby, but you needn't brood about that now.* One may b. *on* or *over* something, but not *about* it. See CAST-IRON IDIOM.

**brow.** *In the sweat of thy brow* is a MISQUOTATION (*face*).

**brusque,** though formerly so far naturalized as to be spelt *brusk* and pronounced *brŭsk*, is now usually pronounced *brŏŏsk*.

**brutal, brute, brutish.** *Brutal* differs from *brute* in its adjectival or attributive use, and from *brutish*, in having lost its simplest sense 'of the brutes as opposed to man' and being never used without implying moral condemnation. Thus, while *brute force* is contrasted with skill, *brutal force* is contrasted with humaneness. In torturing a mouse, a cat is *brutish* and a person *brutal*.

**buck.** See HART.

**buffalo.** Pl. -*oes*; see -O(E)S 1.

**buffet.** The OED pronounces this *bŭ´fĭt* in the sense *sideboard* or *cupboard*, and as French in the sense *refreshment bar*. But we ordinarily follow the U.S. practice of calling both *boofā* and reserving *bŭfĭt* for the blow.

**bulk.** *The bulk of* in the sense of *most*, the greater part, the majority (*We have disposed of the bulk of the surplus stocks*) can be supported by good authority

going back over 200 years; but one of those synonyms is generally preferable, especially when the idea of large size inherent in *bulk* makes that word unsuitable.

**bumble-bee, humble-bee.** See NEEDLESS VARIANTS. Neither form, however, though there is no difference of meaning, is a mere variant of the other; they are independent formations, one allied with *boom*, and the other with *hum*. The first form has now virtually driven out the second.

**bunkum, buncombe.** The first spelling is the prevalent one, often shortened to *bunk*. The second, from an American place-name, is the original; but the word is equally significant with either spelling, and no purpose is served by trying to re-establish the less usual.

**bur, burr.** The word meaning prickly seed-vessel etc. is usually, and might conveniently be always, *bur*; the word describing northern pronunciation is always *burr*; in all the other words, which are less common, *burr* is usual and might well be made universal.

**burden, burthen.** The second form is, even with reference to a ship's carrying capacity, for which *burden* is now often used, a NEEDLESS VARIANT; and in other uses it is an ARCHAISM.

**bureaucrat** etc. The formation is so irregular that all attempt at self-respect in pronunciation may as well be abandoned. We must be content to accept the popular pronunciations *bū´rŏkrăt, bŭrŏ´krăsĭ*, and *bŭrŏkră´tic*.

**burgh, burgher.** *Burgh* is the Scottish equivalent of the English *borough* and is so pronounced. *Burgher*, an archaism, is pronounced *ber´ger*.

**burgle.** See BACK-FORMATION. A verb being undoubtedly wanted, and words on the pattern of *burglarize* being acceptable only when there is no other possibility (see NEW VERBS IN -IZE), it is gratifying that *burgle* has outgrown its early facetiousness and become generally current.

**burlesque, caricature, parody, travesty.** For *b*. as a type of stage-performance see COMEDY ETC. In wider applications the words are often interchangeable; a badly conducted trial, for instance, may be called a *b.*, *a c.*, *a p.*, or *a t.*, *of justice*; a perverted institution may be said, without change of sense, to *b.*, *c.*, *p.*, or *t.*, *its founder's intentions*; and, the others having no adjectives of their own, the adjective *burlesque* can serve them, as well as its own noun, in that capacity (*a b. portrait, poem*, etc.). Two distinctions, however, are worth notice: (1) *b.*, *c.*, and *p.*, have, besides their wider uses, each a special province; action or acting is burlesqued, form and features are caricatured, and verbal expression is parodied; (2) *travesty* differs from the others both in having no special province, and in being more used than they when the imitation is intended to be an exact one but fails. See also HUMOUR, WIT, etc.

**burnt, burned.** *Burnt* is the usual form, especially in the p.p.; *burned* tends to disappear, and is chiefly used with a view to securing whatever impressiveness or beauty attaches to the unusual; see -T AND -ED.

**burr.** See BUR.

**burst, bust.** In the slang expressions *b. up, b.-up, go a b., on the b.*, etc., the spelling *bust* is established and should be used by those who use the phrases. So too *bronco-buster* and *block-buster*.

**business, busyness.** The second form, pronounced *bi′zĭnĭs*, is used as the simple abstract noun of *busy* (the state etc. of being busy) so as to distinguish it from the regular *business* with its special developments of meaning.

**but.** 1. Case after *b.* = except. 2. Redundant negative after *b.* 3. Illogical *b.* 4. Wheels within wheels. 5. *B.* . . . *however*. 6. *B. which* and *b.* beginning a sentence.

1. Case after *but* = except. The question is whether *b.* in this sense is a preposition, and should therefore

always take an objective case (*No one saw him but me*, as well as *I saw no one but him*), or whether it is a conjunction, and the case after it is therefore variable (*I saw no one but him*, i.e. but I did see *him*; *No one saw him but I*, i.e. but *I* did see him). The answer is that *but* was originally a preposition meaning outside, but is now usually made a conjunction, the subjective case being preferred after it when grammatically needed. Out of a large collection of examples of *but* followed by an inflected pronoun, 95% showed the conjunctional use; *Whence all b. he* (not *him*) *had fled* exemplifies, in fact, the normal modern literary use though the OED says that the prepositional use is equally correct. The fact seems to be that *all but him* is used (*a*) by those who either do not know or do not care whether it is right or not—and accordingly it is still good colloquial—, and (*b*) by the few who, being aware that *b.* was originally prepositional, are also proud of their knowledge and willing to air it—and accordingly it is still pedantic-literary. It is true that the conjunctional use has prevailed owing partly to the mistaken notion that *No one knows it b. me* is the same sort of questionable grammar as *It is me*; but it has prevailed, in literary use, and it is in itself legitimate; it would therefore be well for it to be universally accepted.

2. Redundant negative after *but*. But (now rare), *but that* (literary), and *but what* (colloq.), have often in negative and interrogative sentences the meaning *that* . . . *not*. But just as *I shouldn't wonder if he didn't fall in* is often heard in speech where *didn't fall* should be *fell*, so careless writers insert after *but* the negative already implied in it. Examples (both wrong): *Who knows b. that the whole history of the Conference might* not *have been changed?* | *Who knows but what agreeing to differ may* not *be a form of agreement rather than a form of difference?*

For similar mistakes, see HAZINESS and NOT 4.

3. Illogical *but*. A very common and exasperating use of *but* as the ordinary

adversative conjunction is that referred to in NOT 5 and more fully illustrated below. A writer having in his mind two facts of opposite tendency, and deciding to give them in two separate and complete sentences connected by *but*, forgets that the mere presence of the opposed facts is not enough to justify *but*; the sentences must be so expressed that the total effect of one is opposed to that of the other. He must not be seduced into throwing in an additional circumstance in one (usually the second) of his sentences that will have the unintended effect of neutralizing the contrast: *In vain the horse kicked and reared, b. he could not unseat his rider* (if the kicking was in vain, the failure to unseat involves no contrast; either *in vain* or *but* must be dropped). / *Pole was averse to burning Cranmer, b. it was Mary who decided that his recantation was not genuine and that he must die* (The fact in contrast with Pole's averseness is Cranmer's having to die; this may be given simply—*but Cranmer was burnt*, or with additional details—*it was decided* etc., or even *Mary decided* etc.—, as long as the opposition between the sentences remains; but *it was Mary who decided* at once makes the second sentence harmonious instead of contrasted with the first; correct to *and it was Mary who decided*, or *but Mary decided*). / *The task is not easy, but Mr. A's production never looked like being equal to it* (Mr. A's failure is in harmony, not in contrast, with the difficulty mentioned in the first part of the sentence. Either substitute *and* for *but* or read *but Mr. A might have made more of it than he did*). / *It is in no spirit of hostility to the Committee of Union and Progress that these lines are written; but it is a sincere appeal to the men of courage and goodwill at Salonica to strive to set their house in order* (Either omit *but*, or convert the two sentences into one by writing *but in sincere appeal*; we then have the correct form *It is not black, but white* instead of the incorrect *It is not black, but it is white*).

Since less excusable blunders than these, due to gross carelessness, occa-sionally occur (e.g. *It is not an evergreen, as is often represented; b. its leaves fall in the autumn, and are renewed in the spring*), it may be well to put down the right and wrong types in the simplest form: (right) *It is not black, but white: It is not black; it is white: It is not black; but it is nearly black*: (wrong) *It is not black, but it is white: It is not black, b. it is nearly white.*

4. Wheels within wheels. A few examples will show the disagreeable effect produced when, inside one of the contrasted sentences connected by *but*, an internal contrast, also indicated by *but*, is added: *But he did not follow up his threats by any prompt action against the young king, b. went off to Germany to conclude the campaign against his brother Lewis of Bavaria. B. on arriving in Bavaria he did not strike down his enemy, b. made a six months' truce with him. / You have come not to a scattered organization, b. to an organization which is in its infancy, b. which is yet real./ I gazed upon him for some time, expecting that he might awake; b. he did not, b. kept on snoring. / The reformers affirm the inward life, b. they do not trust it, b. use outward and vulgar means./ There was a time when golf was a Scottish speciality, b. it has followed Scottish precedents in spreading over the whole country south of the Tweed. B. we are glad that it is a Scot who has ventured to blame golf.*

5. *But . . . however* is perhaps always due to mere carelessness: *If any real remedy is to be found, we must first diagnose the true nature of the disease; b. that, however, is not hard. / B. one thing, however, had not changed, and that was . . . / The enemy's cavalry withdrew with losses, b. they returned, however, reinforced by . . .*

6. For *but which* see WHICH WITH AND OR BUT, and for the superstition about beginning a sentence with *but* or *and* see AND 5.

**buz(z).** See -Z-, -ZZ-.

**by,** prep., owing to the variety of its senses, is apt to be unintentionally used several times in the same sentence;

when the uses are parallel and the repetition intentional (*We can now travel by land, by sea, or by air*), monotony is better than the ELEGANT VARIATION (*by land, on the sea, or through the air*) often affected. But *by land, sea, or air* is better than either, and such accidental recurrences of *by* as here shown are slovenly (cf. OF): *The author's attempt to round off the play* by *causing Maggie to conquer* by *making John laugh* by *her poor joke about Eve was not worthy of him.* | *Palmerston wasted the strength derived* by *England* by *the great war* by *his brag.*

**by, bye, by-.** The spelling, and usage in regard to separating the two parts, hyphening them, or writing them as one word, are variable. As the noun and adjective are merely developments of the adverb or preposition, it would have been reasonable to spell always *by*; but *bye* for the noun is too firmly established in some uses to be abolished (e.g. *a bye* at cricket and *drawing* or *playing off a bye* in a competition), and it would be convenient if the noun were always *bye*, and the adjective and adverb always *by*. But *by the by* is usually so written, though the second *by* is a noun, perhaps on the analogy of *by and by*, where both are adverbs. *By(e)-law* is probably a corruption of the obsolete *byrlaw*, the local custom of a township; it is often given an *e*, though the OED puts *by-law* first. *Bye-bye* is also unconnected with *by*: as a nursery word it is a sound intended to lull a child to sleep (cf. *lullaby*) and as a farewell it is a clipping of *good-bye*, which is itself a contraction of *God be with you*.

As to the hyphening of *by* as a prefix, the authorities, as is usual with hyphens, are not agreed. If we choose the COD as our guide we shall find that we may write as single words *bygone*, *bypass*, *bypath*, *byplay*, *bystander*, *by-street*, and *byword*, but we should use a hyphen in *by-blow*, *by-election*, *by-lane*, *by-law*, *by-name*, *by-product*. If, as suggested in the article HYPHENS, unnecessary hyphens ought to be avoided, all the members of the second class,

except *by-election* with its awkward conjunction of vowels, might surely be treated as eligible for promotion to the first.

**by and large,** in its popular sense of more or less, on the whole, has wandered even further from its original meaning than most POPULARIZED TECHNICALITIES. It is a nautical term for sailing alternately close to the wind and with the wind abeam or aft. The phrase seems to have settled down comfortably to its new job, and is doing no harm to anyone except to sailors, who are annoyed at the ignorance that this use of it betrays.

# C

**cabal, cabbalist(ic), cabbala,** etc. These are the right spellings, though the words are from the same source. *Cabal*, as a word for a secret faction is earlier than the 'Committee for Foreign Affairs' of Charles II with which it is usually associated; that the initial letters of the names of the members of that body formed the word was a coincidence.

**cable(gram).** The verb *cable* (transmit, inform, etc., by cable) is both convenient and unobjectionable; *cablegram* is not only a BARBARISM, but a needless one, since *cable* (cf. *wire* vb. and n.) serves perfectly as a noun also in the sense submarine telegram.

**cachet** (except in the medical sense) is mainly a LITERARY CRITICS' word (*bears the c. of genius* etc.) and should not be allowed to extrude native words; *stamp, seal, sign manual*, are usually good enough for English readers. See FRENCH WORDS; and, for synonymy, SIGN.

**cachinnate, -ation, -atory.** See POLYSYLLABIC HUMOUR.

**caddie, caddy.** The golf-attendant has *-ie*; the tea-box has *-y*.

**caddis** is preferable to *caddice*

**cadre**, being an established military technicality, should be anglicized in sound and pronounced *kah'der*, in plural *kah'derz*; the French pronunciation is especially inconvenient in words much used in the plural.

**café** is naturalized in the sense coffee-house or restaurant—so thoroughly naturalized indeed as to be *kăf* to many road-transport workers. In the sense coffee it is a FRENCH WORD.

**caffeine.** See -IES, -EIN, where the pronunciation *kăfēn'* is recommended.

**cagey** is an American colloquialism of recent date (it is not to be found in the 1928 Webster or the 1933 OED Supp.) that has won popularity in Britain also. It is used as a synonym of *wary* or *close*, applied especially to one who, for fear of being indiscreet, is uncommunicative or evasive when asked for information

**calcareous, -rious.** The first form is definitely wrong, the ending being from Latin -arius, which gives -arious or -ary in English; but it is so firmly established that a return to the correct but now obsolete second form is out of the question.

**calculate.** 1. *C.* makes *calculable*; see -ABLE 1. 2. The American colloquialism is an example of SLIPSHOD EXTENSION; the sense I consider-as-the-result-of-a-calculation passes into the simple sense I consider. *We shall win, I c., by a narrow majority* shows the normal use, the assumption at least being that the numbers have been reckoned and compared. *We shall be in time, I c.* is (according to British usage) correct if the time wanted and the time at disposal have been worked out in detail, but wrong if it is a mere general expression of sanguineness. This use of *calculate* has never found favour in Britain, and is now said to be passing out of use in U.S. *Figure* is sometimes used there in the same loose way, and is open to the same criticism. See also RECKON.

**calculus.** The medical word has pl. *-lī*; the mathematical, *-luses*. See -US.

**calibre.** This spelling, and the pronunciation *kă'liber*, are now established, though the French pronunciation lingered for the human quality after it had disappeared for the internal diameter. See -RE AND -ER.

**caliph** is the spelling and *kă'lĭf* the pronunciation put first by the OED, which states, however, that 'orientalists now favour *Khalîf*'; see DIDACTICISM.

**calligraphy** etc. should not be altered to *calig-*. Greek compounds are made either with καλλι- from κάλλος beauty, or with καλο- from καλός beautiful. Choice is therefore between *calligraphy* and *calography*; and as the actual Greek compounds were καλλιγραφία etc., *calligraphy* is obviously right.

**callus.** Pl. *-uses*; see -US. The word is often wrongly spelled *callous* from confusion with the adjective.

**calmative**, being queer both in pronunciation (*kă'lmatĭv*, not *kah'matĭv*) and in formation (there is no Latin word for *-ative* to be attached to), should be left to the doctors as a technical term, if even they have a use for it beside *sedative*.

**caloric, calorie.** *Caloric*, originally a name for the supposed form of matter to which the phenomena of heat and combustion were once ascribed, and later for heat generally, is now little used except in its compounds, such as *calorimeter*. *Calorie* is a term for any one of several thermal units, especially for the unit of heat or energy produced by any food substance. Like *caloric* in its day, *calorie* seems to be a temptation to those given to PEDANTIC HUMOUR, who now find the same satisfaction in saying *intake of calories* instead of *eating* as their predecessors did in saying *caloric* instead of *heat*.

**cambric.** Pronounce *kā-*.

**camellia.** The spelling with -ll- is quite fixed, and the mispronunciation *-mē-* now so prevalent as to be justified by usage.

**camelopard** does not contain the word *leopard*, and should be neither spelt nor pronounced as if it did. Pronounce *kamĕ'lopard*, if there should be occasion to use this now archaic or facetious name for *giraffe*.

**Canaan(ite).** The pronunciation *kā'-nyan* is a quite justifiable escape from the difficult and unEnglish *kā'nă-ăn*; *kā'nă-ăn* passes into *kā'năyăn*, and that into *kā'nyan*. The pronunciation *ka'năn*, the ordinary clerical one, is a simpler evasion.

**canard** should be anglicized, and have the d of the sing. and the ds of the pl. sounded.

**candelabrum.** Pronounce *-ābrum*. The pl. *-bra* is still preferred to *-brums*; the false sing. *-bra* with pl. *-bras* should not be used.

**candid** is a WORSENED WORD. It originally meant kindly, uncensorious; Jane Austen so uses it. The change came during the 19th c. 'To be candid, in *Middlemarch* phraseology, meant to use an early opportunity of letting your friends know that you did not take a cheerful view of their capacity, their conduct or their position.' Today *I must be candid* is invariably a warning of unpleasantness to follow.

**canine.** The pronunciation *kā'nīn* (not *kanī'n*, nor *kā'nīn*) is both the commonest and the best. *Feline, bovine, asinine, leonine*, are enough to show that RECESSIVE ACCENT is natural; and, if *kă* is due to dread of FALSE QUANTITY, it is certainly not worth fighting for on that ground.

**cannibalism.** That words should devour their own kind is a sad fact, but the guilt is perhaps less theirs than their employers'; at any rate the thing happens: *As to which additional commodities the guaranteed price should be applied, Mr. Gaitskell said that such a fundamental issue should be discussed between the next Labour government and the industry.* To has swallowed a *to*. | *Harvey, McDonald, and O'Neill make up as powerful a trio of batsmen known in modern test cricket. As* has swallowed an *as*. | *It is more or less—and certainly more than less—a standardized product.*

*More* has swallowed a *more*. | *Although the latter were overwhelmingly superior in numbers, the former had the advantage of being under one control, and that of Napoleon himself. That* has swallowed a *that*; the full form would be 'and that control the control of', which gives 'and that that of'; but this cannibal may perhaps be thought to have consumed himself rather than another of his kind. | *The most vital problem in the etymological study of English place-names is the question as to what extent personal names occur in place-names. To* has swallowed a *to*, as its way is when employed by *question-*AS-*to* writers.

**cannibalize,** meaning to use one of a number of similar machines to provide spare parts for the others, is one of the more felicitous of the NEW VERBS IN -IZE.

**cannon.** 1. For plural see COLLECTIVES 3. 2. As the natural name for the thing, *c.* is passing out of use (though revived for the armament of military aircraft) and giving place to *gun*, which is now the regular word except when context makes it ambiguous.

**cañon, canyon.** The second is recommended. Pronounced *kă'nyon*.

**cant.** For meaning and use see JARGON.

**cantatrice,** on the rare occasions when it is used, is ordinarily pronounced as Italian (*-ēchă*), sometimes as French (*-ēs*); *singer*, or *vocalist*, should be preferred when it is not misleading; other English substitutes, as *songstress, female singer*, are seldom tolerable, but *soprano* etc. may be a way out. See FEMININE DESIGNATIONS.

**canton(ment).** The noun *canton* is usually *kan'tŏn*, sometimes *kăntŏ'n*. The verb is in civil use *kăntŏ'n*, but in military use generally *kantoo'n*. The noun *cantonment*, which is military only, is generally *kantoo'nment*.

**canvas(s).** The material is now always spelt *-as*; so also the verb meaning to line etc. with *c.*; for the plural of the noun, and for *canvas(s)ed* etc. in this sense, see -s-, -ss-. The verb meaning to discuss, ask for votes, etc., has always -ss; so also has the noun meaning the process etc. of canvassing in this sense.

**caoutchouc.** Pronounce *kow'chook̄*.

**caper** (herb). See SINGULAR S.

**capita, caput.** See PER.

**capitalize, -ization, ist.** Accent the first, not the second syllables; see RECESSIVE ACCENT.

**capitals.** Apart from certain elementary rules that everyone knows and observes, such as that capitals are used to begin a new sentence after a full stop, to introduce a quotation, and for proper names and those of the days and months, their present use is almost as anarchic as that of HYPHENS. Uniformity is lacking not only in practice but also in precept: no two sets of style rules would be found to agree in every respect. Rather than add to the confusion by attempting to prescribe a new code, this article will reproduce, with small adaptations approved by the author, the advice given by Mr. G. V. Carey in the publications of the Cambridge University Press *Punctuation* and *Mind the Stop*:[1]
  The use of Capitals is largely governed by personal taste, and my own, while not favouring seventeenth-century excess, happens to favour even less the niggardliness now sometimes apparent. The printed page that is starved of capitals suffers not merely in appearance (to my eye at any rate) but also in function, for denial of capitals to well-known bodies, institutions, officials and the like militates against ready reference. The following suggestions, while claiming no higher status than that, aim at being at least systematic, logical, and unambiguous. Appended in brackets are alternative usages that commend themselves to some.
  (i) Capitals are appropriate for full titles of persons, ranks, offices, institutions, countries, buildings, books, plays, etc., whether general or particular, singular or plural—and for the *whole* of such titles.
    King of the Belgians, Dukes of Burgundy, Admiral of the Fleet, a Prime Minister, the Home Secretary, Bishops of Durham, the Attorney-General, the Royal Society, the North-West Frontier, the Western Powers, St. Paul's Cathedral.
  [Some prefer lower case when the titles are used generally or in the plural:
    dukes of Burgundy, a prime minister, bishops of Durham, an attorney-general.
  Against this practice is the fact that it cannot be consistently followed without occasional ambiguity: for example, an admiral of the fleet is not necessarily identical with an Admiral of the Fleet, nor a foreign secretary with a Foreign Secretary. Another recent tendency in certain quarters is to capitalize only half of a title, as in the following, which surely qualify as titles by now:
    south-east Asia, the western Powers, the welfare State.]
  (ii) When only part of a title is given the practice of retaining a capital for the 'particular' but dropping it for the 'general' is fairly common and, though any form of hair-splitting is liable to cause more trouble than it is worth, usually harmless. Thus:
    the Duke (also 'the Dukes' if they have been specified), the Bishop, the Admiral, the Cathedral.
  But
    kings, princes, dukes, a bishop, a cathedral.
  [On the other hand, the number of anti-capitalists who would have lower case for both sets of the above is far from negligible.]
  (iii) In reference to institutions, bodies, etc., it is desirable, when their names are repeated in shortened form, to retain the capital for the shortened title. For example:
    The Commissioners of Inland Revenue reconsidered the matter. . . . Eventually the Commissioners agreed to . . .

---

[1] The publishers are indebted to the Syndics of the Cambridge University Press for permission to reprint these extracts.

The Church Assembly met yesterday at ... One of the most important problems considered by the Assembly was ...

A representative of the R.S.P.C.A., who attended, said that information had reached the Society that ...

[It would be not unusual to find, on their recurrence, *the commissioners, the society,* though possibly not *the assembly.* Which again implies inconsistency; whereas the retention of capitals makes also for easier reference and avoids possible ambiguity.]

(iv) With the fairly numerous words that have a more restricted (or concrete) sense as well as a commoner (or abstract) sense, the invariable use of lower case for the latter and capital for the former—irrespective of singular/plural or particular/general distinctions—is recommended. Such words, amongst many others, are:

state (= condition, circumstances) State (= organized community);
power (= strength etc.) Power (= powerful nation);
government (= the function of governing) Government (= the body that governs);
minister, ministry (religious) Minister, Ministry (political);
church (= the building) Church (= the body);
east, west, western, etc. (directional) East, West, Western, etc. (ethnological);
underground (below ground, generally) Underground (of railways).

[The particular/general criterion already referred to is often applied here: e.g. *the Government,* but *a government* and *governments; the Minister,* but *a minister* and *ministers*—all in the political sense. This is surely to draw the dividing-line in the wrong place. The same applies to the common practice of reverting to lower case when such a word is used adjectivally, for example, *a government appointment;* to which *a Government appointment* is to be preferred.]

(v) There are a few words normally requiring an initial capital that, when applied to common objects, exchange the capital for a small letter. Most of them are place-names (or adjectives formed therefrom) that have in course of time lost their purely local association. I have in mind 'india-rubber', 'brussels sprouts', 'roman type', 'french windows', 'venetian blinds'; and readers will no doubt be able to supply plenty of others.

(vi) There is (or was until lately) a convention in certain quarters of printing the names of streets etc. as 'Regent-street', 'Portland-place', 'Shaftesbury-avenue', 'Berkeley-square', and so on. Why 'street', 'road', etc., should in this one connexion fail to conform with the normal rule for the use of capitals in full titles has always been a mystery to me; and the insertion of the hyphen seems peculiarly gratuitous. The streets themselves are not labelled in this way, nor do their names generally so appear on advertisements, buses, Underground stations, notepaper headings, visiting-cards, and the like. The thing seems quite pointless; and when we get, for instance, 'Charing Cross-road' —I have seen it printed even 'Charing cross-road'—the result is a silly ambiguity. There are signs that this irritating practice is beginning to decline, so let us hope that 'Regent Street' and 'Shaftesbury Avenue' will soon be the invariable form and that we may say farewell to 'Leicester-square'.

If this whole topic should seem to have been unduly laboured, some responsibility belongs to the journal that I have always taken, and still take for the most part, as the soundest guide to modern usage. *The Times* in its treatment of capitals, especially under the headings (i) to (iii) above, now continually bewilders me. In the same context it will print, repeatedly, *Civil Service* and *Civil servant;* the *East* and *Middle East* appear to be legitimate, but not the *West; East* and *West Germany* have now usually become *east* and *west Germany* (a first step towards reunification?); one recent leading article spoke of

... the Middle East or south-east Asia ... complaints against France in North Africa, the Dutch in West Indies, and the South African Government in south-west Africa. Again, from a parliamentary report: No formal request has been made by the Governor to the bishop for the archdeacon's removal. If there is method here, it is hard to discern it. Let it be repeated: the employment of capitals is a matter not of rules but of taste; but consistency is at least not a mark of bad taste.

**carafe.** The chief use of *c.* was formerly as a slightly genteel word for the water-bottle that stood on every washhand-stand with an inverted tumbler over it. For that purpose the word has fallen into disuse with the article of furniture on which this kind of *c.* used to stand; but the word is still common, without any taint of genteelism, for the vessel in which light table wines are often served.

**caravanserai, -sera, -sary.** The first spelling is the best. It should be pronounced *-rī* in spite of Fitzgerald's rhyming it with *day* in *The Rubaiyat.*

**carcass, -ase.** The -ss form stands first in the Oxford dictionaries; the verb is always so spelt but *-ase* must be at least as usual for the noun.

**careen.** The only meanings known to the dictionaries up to the middle of the 20th c. were nautical. In the transitive sense (the commonest) a ship was careened when she was turned over on her side to clean or repair her hull; in the intransitive sense a ship careened when she heeled over under sail. But the word had appeared] in U.S. fiction in contexts in which it was clearly intended to have a quite different meaning—that of rapid movement. *Henry heard the Ford taxi coming out of the side street and saw it careening up on to the dock.* The taxi was conveying gangsters from a bank they had robbed to a boat they hoped to escape in, and as the bank's alarm siren was already sounding we may presume that the author's intention was to indicate very rapid movement indeed. Why this innovation should have proved attractive to writers of fiction we can only guess.

Those who so use it cannot claim the support of Meredith's description of the huge bulk of Prince Lucifer careening o'er Afric's sands, for it is clear that the word there means *leaning.* Perhaps its popularity comes from a latent suggestion of simultaneous careering and careening—of a pace so wild that the vehicle sways from side to side and takes its corners on two wheels. That explanation seems to be favoured by the first lexicographer to take note of the new meaning, with the definition 'to lurch or toss from side to side' (Webster).

**caricature.** See BURLESQUE.

**caries** is a Latin singular meaning decay. For pronunciation see -IES, EIN.

**carillon.** The anglicized pronunciation *kărĭ'lŏn* (or *kă'rĭlŏn*) should be preferred to the hybrid sound *-ilyon.*

**cark(ing).** The verb is practically obsolete, and the adjective, surviving only as a stock epithet of *care*, should be let die too.

**carnelian.** See CORNELIAN.

**carotid.** Pronounce *kărŏ'tĭd*; see FALSE QUANTITY.

**carousal, carousel.** These words are often confused. *Carousal* (with the stress on the second syllable) is an -AL NOUN formed from *carouse*—unnecessarily since *carouse* is itself a noun as well as a verb—and means a drinking-bout; it is said to be derived from the German *gar aus*, 'no heel taps'. *Carousel* (with the stress on the last syllable), sometimes spelt *carrousel*, means a tournament; in U.S. it is used for what we call a merry-go-round. It comes from the Italian *corosello*, which, according to Skeat, is a corrupt form of *garosello* = rather quarrelsome. 'No doubt', he adds, '*garoselo* was turned into *corosello* by confusion with *coricello*, a little chariot or car . . . owing to the use of chariots at such festivities.'

**carpet.** *On the c.* (under discussion). Now that the sense required for *c.*, viz. tablecloth, is obsolete, we make the phrase serve a different purpose as a slang expression for *under reprimand*; and if we must use a GALLICISM for *under discussion* we leave *tapis* untranslated.

**carrel.** See REVIVALS.

**carte, quart(e),** in fencing. The first spelling, still the commonest except in technical books following French authorities, should be preferred if only as keeping the pronunciation right.

**cartel,** in the old senses (a challenge, or an agreement for the exchange of prisoners) is pronounced *kar'tel*; in the modern sense of manufacturers' combine it represents German *Kartell*, and was accordingly disposed at first to accent the last syllable, but has now moved far enough towards the old pronunciation to give equal stress to both.

**case.** There is perhaps no single word so freely resorted to as a trouble-saver, and consequently responsible for so much flabby writing. The following extract from a legal treatise, in which the individual uses are comparatively justifiable, shows how the word now slips off the pen even of an educated writer: *In the majority of cc. where reprisals have been the object, the blockade has been instituted by a single State, while in cc. of intervention several powers have taken part; this is not, however, necessarily the c.*

To obviate the suspicion of an intolerant desire to banish it from the language, let it be admitted that *case* has plenty of legitimate uses, as in: *In your c. I would not hesitate*; *A bad c. of blackmailing*; *I am only putting a c.*; *Circumstances alter cc.*; *In c. of fire, give the alarm*; *Take brandy with you in c. of need*; *The plaintiff has no c.*; *What succeeds in one c. may fail in another*; *Never overstate your c.*; *In no c. are you to leave your post*; *It would be excusable for a starving man, but that was not your c.*; *There are seven cc. of polio.*; *In any c. I will come.*

The bad uses are due sometimes to the lazy impulse to get the beginning of a sentence down and to let the rest work itself out as it may, and some-

times to a perverted taste for long-windedness, PERIPHRASIS, or ELEGANT VARIATION. It will be seen that *in the case of*, the worst offender, can often be simply struck out (brackets are used to show this), and often avoided by the most trifling change, such as the omission of another word (also bracketed). Many examples are given, in the hope that any writer who has inspected the misshapen brood may refuse to bring more of them into the world: *Older readers will, at least (in the c. of) those who abhor all Jingoist tendencies, regret that the authors have . . .* | *He has used this underplot before in (the c. of) 'The Fighting Chance'.* | *That he could be careful in correcting the press he showed in (the c. of) the 'Epistle to John Driden'.*| *In the cc. above noted, when two or more handlings of the same subject by the author exist, the comparison of the two usually suffices to show how little vamping there is in (the c. of) the latter.* | *(In the c. of) Pericles (, the play) is omitted.*| *(In the c. of) cigars sold singly (they) were made smaller.* | *(In the c. of Purvey his) name was first mentioned in connexion with Bible translation in 1729* (Purvey's). | *There are many (cc. of) children who have lost their parents and (of) parents who do not know the whereabouts of their children.* | *Mr. Mintoff has demanded full employment in the c. of all dockyard employees* (for all dockyard employees). | *This is the first disaster in the c. of a Viscount air-liner* (to a Viscount air-liner). | *In the c. of no poet is there less difference between the poetry of his youth and that of his later years* (No poet exhibits less). | *Even in the purely Celtic areas only in two or three cc. do the first bishops bear Celtic names* (only two or three of the first bishops bear). | *That in all public examinations acting teachers in every c. be associated with the Universities* (teachers be always associated). | *In many (cc. of) largely frequented buildings, as much dust as this may be extracted every week.* | *His historical pictures were (in many cc.) masterly* (Many of his).

The ELEGANT VARIATIONist, as was implied above, is in clover with *case*; *instance* provides him with one of those doubles that he loves to juggle with, and *be the case* enables him to show his superiority to the common mortal who would tamely repeat a verb; we conclude with a few of his vagaries: *Although in eight cc. the tenure of office of members had expired, in every instance the outgoing member had been re-elected.* | *Thunderstorms have in several cc. occurred, and in most instances they have occurred at night.* | *In thirty-two cc. there are Liberal candidates in the field, and in eleven instances Socialists supply the third candidate.* | *There are four cc. in which old screen-work is still to be found in Middlesex churches, and not one of these instances is so much as named.* | *This Conference will lay a foundation broader and safer than has hitherto been the c.* (been laid). | *It is not often worth while harking back to a single performance a fortnight old; but this is not the c. with the Literary Theatre Club's production of Salome* (but it is worth while).

**casein.** For pronunciation see -IES, -EIN.

**cases.** 1. General. 2. The status of case. 3. Temptations.

1. *General.* The sense of case is not very lively among English-speakers because, very few words having retained distinguishable case-forms, it is much more often than not unnecessary to make up one's mind what case one is using for the purpose of avoiding solecisms. Mistakes occur chiefly, though not only, with (*a*) the few words having case-forms, mostly personal pronouns, and (*b*) the relative pronouns. Accordingly, necessary warnings, with illustrations and discussion, are given in the articles HE, I, LET, ME, SHE, THAN, THAT REL. PRON. 6, THEY 4, US 1, WE 1, WHAT 3, WHO AND WHOM 1, 2. To those warnings the reader is referred for practical purposes, and the present article can be devoted to a confession of faith in case as an enduring fact, a miscellaneous collection of quotations showing that it cannot quite be trusted to take care of itself, and a glance at the conditions that make mistakes most likely.

2. *The status of case.* Is case, then,

a notion permanently valuable and inevitably present, or can we, and may we as well, rid our minds of it? We know that grammarians are often accused, and indeed often guilty, of fogging the minds of English children with terms and notions that are essential to the understanding of Greek and Latin syntax, but have no bearing on English. We know that the work done by the classical case-endings has been in large part transferred in English to two substitutes: the difference between the nominative and the accusative (or subject and object) is indicated mainly by the order in which it arranges its words; and the dative, ablative, locative, and such cases, are replaced by various prepositions. We know that English had once case-forms for nouns as well as pronouns, and that nevertheless it found them of so little use that it has let them all disappear. We know that, if the novelists are to be trusted, the uneducated find the case-endings even of pronouns superfluous. 'Me and my mate likes ends' said the ruffian who divided the rolypoly between himself and his ally and left their guest the hiatus; he had no use for *I*, even when the place to be filled was that which belongs to the subject, and the instinct of case, if it exists untaught, might have been expected to operate. We know, lastly, that not everyone who has learnt grammar enough to qualify as journalist or novelist is quite safe on his cases when the test is a little more severe than in *Me and my mate*.

Is the upshot that case is moribund, that our remaining case-forms are doomed to extinction, that there is behind them no essential notion or instinct of case itself, that no fuss whatever need be made about the matter, that the articles of which a list was given above are much ado about nothing, and that the right policy is to let the memory of case fade away as soon as we can agree whether *I* or *me*, *she* or *her*, *who* or *whom*, is to be the survivor of its pair? Possibly it is; SUBJUNCTIVES are nearly dead; case too may be mortal; but that fight to a finish between *I* and *me* and the other pairs will be a lengthy affair, and for as long as it lasts the invisible cases will have their visible champions to muster round. Meanwhile let me confess my faith that case visible and invisible is an essential of the English language, and that the right policy is not to welcome neglect of its rules, but to demand that in the broadcasts, the newspapers, and the novels, from which most of us imbibe our standards of language, they should be observed.

3. *Temptations.* A. First in frequency and deadliness comes the personal pronoun in a place requiring the objective case followed by a relative that must be subjective. Examples: *Three years of dining are a preliminary for* he *who would defend his fellows.* / *Should not a Christian community receive with open arms* he *who comes out into the world with clean hands and a clean heart?* / *They came to fight in order to pick up the challenge of* he *who had said 'Our future lies on the water'.* / In these the temptation has been to regard *he-who* as a single word that surely cannot need to have the question of case settled twice over for it; and hazy notions of something one has heard of in classical grammar called relative attraction perhaps induce a comfortable feeling that one will be safe whether one writes *he* or *him*. That is a delusion; neither relative attraction nor inverse attraction (the right term here) is a name to conjure with in modern English grammar, though the textbooks can muster a Shakespearian and Miltonic example or two; in modern grammar they are only polite names for elementary blunders. All the foregoing examples should have *him* instead of *he*.

B. The next temptation is to assume, perhaps from hearing *It is me* corrected to *It is I*, that a subjective case cannot be wrong after the verb to *be*. *I saw a young girl gazing about, somewhat open-mouthed and confused, whom I guessed* (correctly) *to be* she *whom I had come to meet.* / *It is not likely that other and inferior works were done at the same time by an impostor pretending to be* he. / In these examples

it is not *to be* that decides the case of *she* and *he*; it is *whom* and *impostor*, and *her* and *him* must be substituted.

C. It is hard not to sympathize with the victims of the next trap. *One comes round again to the problem of Kant—* he, *too, a cosmopolitan like Goethe.* / *It is sad to look in vain for a perambulator in Nursemaids' Walk, and to discover only one solitary person, and* he *a sentry, on the steps of the Albert Memorial.* Appositions such as 'him, too, a cosmopolitan', 'and him a sentry', do sound as if one was airing one's knowledge of the concords. Well, perhaps it is better to air one's knowledge than one's ignorance of them; but the escape from both is to be found in evading the pronoun (*another cosmopolitan,* or *also a cosmopolitan*) or sacrificing the apposition (*and he was a sentry*).

D. The invisibility of case in nouns and pronouns tempts us to try sometimes whether they may not be made to serve two masters, as in Oliver Cromwell's *What you call a gentleman and is nothing else,* and St. Paul's *Eye hath not seen, nor ear heard, neither have entered into the heart of man,* the things *which God hath prepared for them that love him.* Modern examples are: *A plan which I have often tried and has never failed me.* / *It gave a cachet of extreme clericalism to the Irish party which it does not deserve, but must prejudice it not a little in the eyes of English radicalism.* / *Yet the coal is there in abundant quantities, and there is nothing the world* wants *so much or* can *be dispensed with such handsome profit to those who produce it.* In the verse from *Corinthians, things* has to serve *seen-and-heard* as object, and *have entered* as subject. I *Cor.* ii. 9 is the reference, and a glance at the R.V. with its *which* in italics shows that the Revisers did not regard its grammar as passable. The NEB translators avoid the issue by recasting the sentence so as to confine *things* to the subjective case. The last example has the peculiarity that the word whose case is in question, viz. *that,* not only has no distinguishable cases, but is not on show at all; but the sentence is un-

grammatical unless it is inserted twice
—*nothing* that *the world wants so much*,
or that *can be dispensed*. See also THAT
rel. pron. 6, WHAT 3, and WHICH 2.

E. Another trap is the compound
subject or object; when instead of a
single pronoun there are a pronoun
and a noun to be handled, the case
often goes wrong where if the pronoun
had been alone there would have been
no danger. *By that time Mr. Mac-
donald will be in possession of the
decision of the Conservative Party, and
it will then be for* he and his advisers
*to take a decision*. Even the divider of
the rolypoly, who can easily be believed to have said *Me and my mate
likes*, would never have said *Me likes*;
still less could we have had here *It will
be for he to take*.

F. *Let Gilbert's future wife be* whom
*she may*. This example is a little more
complicated, but of a kind that not infrequently presents itself. The temptation is to look before and after, and
doubt in which direction the governing
factor is to be found. We first, perhaps,
put aside the error of supposing that *be*
requires a subjective, i.e. *who*, and remember that *let* puts *wife* in the objective, which raises a presumption that
the same case will follow, i.e. *whom*.
But then it perhaps occurs to us that
the part to be played by *who*(m) is that
of complement to *may* (*be*), which
ought to be in the same case as *she*.
In this difficulty the last resource is to
write the sentence in full, *Let Gilbert's
wife be her who she may be*; and the
insertion of the omitted *her* having
provided the first *be* with the objective
complement that it requires, we find
ourselves able to write *who* as the subjective complement required by the
second *be*. *Who* is in fact the grammatical English; cf. WHOEVER.

**cast(e).** *Caste* is sometimes wrongly
written for *cast* in certain senses less
obviously connected with the verb *cast*
—mould, type, tendency, hue, etc.
The confusion is the more natural
since *cast* was formerly the prevalent
spelling for the hereditary class also;
but the words are now differentiated,

and *cast* is the right form in such contexts as: *reflections of a moral c.,
heroines of such a c., a man of the c.
of Hooker and Butler, my mind has a
melancholy c., his countenance was of the
true Scottish c., a strongly individual c.
of character, their teeth have a yellow-
ish c.*

**caster, -or.** The word meaning fine
sugar, pepperbox, etc., and swivelled
furniture wheel, should strictly be
*caster*, the first group being from the ordinary sense of *cast* (throw) and the last
from an obsolete one (veer). But *-or*,
probably due to confusion with other
*castors*, is now usual.

**cast-iron idiom.** Between IDIOM and
ANALOGY a secular conflict is waged.
Idiom is conservative, standing in the
ancient ways, insisting that its property
is sacrosanct, permitting no jot or tittle
of alteration in the shape of its phrases.
Analogy is progressive, bent on extending liberty, demanding better reasons
than use and wont for respecting the
established, maintaining that the matter is what matters and the form can go
hang. Analogy perpetually wins, is for
ever successful in recasting some piece
of the cast iron, and for that reason no
article in this book is likely to be
sooner out of date in some of its examples than this. Idiom as perpetually
renews the fight, and turns to defend
some other object of assault. 'I doubt
that it ever happened', 'He is regarded
an honest man', 'He was quoted to
have said', 'A hardly won victory',
'Hanker to learn the truth', 'The state
to which we have arrived', 'With a
view of establishing himself'—all these,
says Idiom, are outrages on English;
correct them please to 'I doubt *whether*
it ever happened', 'He is regarded *as* an
honest man', 'He was quoted *as having*
said', 'A *hard*-won victory', 'Hanker
*after learning* the truth', 'The state *at*
which we have arrived', 'With a view
*to* establishing himself'. But why?
retorts Analogy. Is not to *doubt* to *be
unconvinced*? Is not *regarding considering*? Is not *quoting* what a man has
said *reporting* it? Is not *-ly* the adverbial ending, and is not *won* to be

modified by an adverb? Is not *hankering* the same as *longing*, and is to have *arrived* any different from to have *come*? And if *in view of* is English, why should *with a view of* be unEnglish? Away with such hair-splittings and pedantries! When one word is near enough to another to allow me to use either, I propose to neglect your petty regulations for the appurtenances proper to each.

Not that Analogy, and those whom it influences, are offenders so deliberate and conscious as this description of them might seem to imply; they treat *regard* like *consider* not because they choose to flout the difference that Idiom observes, but because it comes natural to them to disregard distinctions that they have not noticed. In ANALOGY 2 it has been pointed out that Analogy has very important functions to perform apart from waging its war upon Idiom; and therefore the admission that this book is wholly partisan in that war need not be interpreted as a condemnation of analogy always and everywhere. The Analogy that wars against Idiom is unsound or hasty or incomplete analogy.

The cast-iron nature of idiom may now be illustrated by a few phrases, shortened down to the utmost, in which some change that to the eye of reason seems of slight importance has converted English into something else: He did it on his own ACCORD; CONTENTED himself by saying; LEST the last state becomes worse than the first; Is to a great MEASURE true; Had every MOTIVE in doing it; The RESENTMENT I feel to this Bill; We must RISE equal to the occasion; Fell SHEERLY down; Stood me in splendid STEAD; Guests came by THE hundreds; You need not BROOD about that now; He ADMITTED to his financial manipulations; It is not PROBABLE to happen. Discussion or actual quotations for these lapses will be found under the words in capitals; and a few articles that have special bearing on the present subject are: AIM; CLAIM; DOUBT(FUL); FACT; IN ORDER THAT; OBLIVIOUS; PLEASURE; PREFER 3; REGARD 2; RESORT; SUCH 1; THAT, CONJ. 2; UNIDIOMATIC -LY; and VIEW.

**castle.** *C. in the air* is English; *c. in Spain* is a GALLICISM. Both mean a visionary project.

**casualty** means an accident. Its extention to include the victims of a casualty (*Casualties were taken to hospital*) is recent but established.

**casuistic(al).** The OED has four quotations for each form; of the -ic four, three are later than the 18th c., of the -ical four only one; from which it would seem that -ic is the modern choice; see -IC(AL).

**catachresis.** Wrong application of a term, use of words in senses that do not belong to them. The popular uses of *chronic* = severe, *alibi* = excuse, *mental* = weak-minded, and *mutual* = common, are examples.

**catacomb.** Pronounce -*ōm*.

**catchup.** See KETCHUP.

**category** is a philosophical term with a narrower meaning than *class*, but under the influence of LOVE OF THE LONG WORD it is used freely as a synonym of the simpler one. For the sake of precision it would be better if *category* were used by no one who was not prepared to state (1) that he does not mean *class*, and (2) that he knows the difference between the two; see WORKING AND STYLISH WORDS, and POPULARIZED TECHNICALITIES.

**Catholic.** It is open to Roman Catholics to use *C.* by itself in a sense that excludes all but themselves; but it is not open to others to use it instead of *Roman Catholic* without implying that no other Church has a right to the name of *C.* Neither the desire for brevity (as in *the C. countries*) nor the instinct of courtesy (as in *I am not forgetting that you are a C.*) should induce anyone who is not Roman C. to omit the *Roman*. The words should not be hyphened.

**catholic(al)ly.** Both forms are rare, and consequently no differentiation has been established; *a catholicly* and *a catholically minded person* may mean either one of wide sympathies etc. or one inclined to Catholicism; today *catholic minded* is more likely to be used than either.

**catsup.** See KETCHUP.

**cattle.** The distinguishing names by age and sex are sometimes puzzling to those who do not live in the country. They are: *Bull* an uncastrated male reared for breeding; *Bullock* or *steer* a castrated male reared for beef; *Calf* an animal of either sex not more than a year old; *Cow* a female that has calved; *Heifer* a female that has not yet calved; *Ox* an obsolete term on the farm; formerly applied to castrated males that did the draught work later assigned to horses, and later still to tractors.

**cause.** *The main cause of the higher price of meat in France is due to the exclusion of foreign cattle.* The main cause is the exclusion; the price is due to the exclusion; out of two rights is made a wrong. See HAZINESS for this type of blunder; with *reason* it is still commoner than with *cause*. See REASON 3.

**causerie.** Informal essay or article in newspaper or broadcast, especially on literary subjects and as one of a series. Named after Sainte-Beuve's *Causeries du Lundi* (Monday talks), a series of weekly criticisms in the *Constitutionnel* and *Moniteur* newspapers.

**causeway, causey.** Either form is correct, the first being not a false spelling of the second, but a shortening of *causey-way*. *Causeway*, however, has virtually ousted *causey* (except in some local names), so that those who use the latter are naturally taken for pedants protesting against an error that is, after all, not an error; see DIDACTICISM.

**cavalcade** means a procession of people on horseback. It has been trying to keep up with the times by extending its meaning to cover first processions in which no horses take part and then, under the lead of Noel Coward, pageants generally. The word has become too popular, and is often used where one of the older words— *procession, display, pageant*—would be more suitable, as in the announcement of a forthcoming *cavalcade of motor-scooters ridden by historical characters from Nero to Napoleon*. But it is better to accept the fact that *cavalcade*, like many other words, has broken its etymological bounds than to try to drive it back with such a barbaric weapon as *motorcade*, said to have been invented by the Salvation Army and shyly adopted by *The Times* with apologetic inverted commas. *The people of London will have a chance to see a 'motorcade' when, in an open car, President Eisenhower and Mr. Macmillan lead a procession of cars from London Airport to the American Embassy.* This word seems however to have since been making progress, especially in America.

**caveat.** In Latin the *a* of the first syllable is short, and the word is often so pronounced in the phrase *caveat emptor*. Otherwise the established pronunciation is *cāveat*. See FALSE QUANTITY.

**-c-, -ck-.** When a word ending in c has a suffix beginning with a vowel added to it, the hard c is preserved before the native suffixes -ed, -er, -ing, and -y, and indicated by the addition of k (*mimicked, bivouacker, trafficking, panicky*), but not before the classical suffixes -ian, -ism, -ist, -ity, -ize (*musician, criticism, publicist, electricity, catholicize*). For the inflexions of the verb *to arc* (short-circuit) the OED Supp. prefers *arcked, arcking* to *arced, arcing* but all the quotations give c only, and this is more usual, though the sound is hard.

**cease** is rapidly giving way to *stop*, as *cast* has given way to *throw*; it is no longer the word that presents itself first, except in a few well-established partnerships such as *cease fire* and, less firmly, *cease work*. We substitute it for *stop* when we want our language to be dignified; it is now poetic, rhetorical, formal, or old-fashioned, though not sufficiently so to have such labels attached to it in dictionaries. No effort should be made to keep words of this kind at work; they should be allowed to go into honourable retirement, from which the poets and the rhetoricians can summon them at need. For the guidance of anyone who still wants to use the word it is worth remarking that *cease to trouble, cease from troubling*, and *cease troubling* are equally idiomatic. See FORMAL WORDS.

**-ce, -cy.** Among the hundreds of words corresponding to actual or possible adjectives or nouns in -ant or -ent, large numbers now present no choice of form: no one hesitates between *avoidance, forbearance, admittance, magnificence, coincidence*, or *intelligence*, and a form in -cy; nor between *buoyancy, constancy, vacancy, agency, decency*, or *cogency*, and a form in -ce. But in many cases it may easily happen that one has doubts which is the right form, or whether one is as good as the other, or whether both exist but in different senses: *persistence* or *persistency*? *competence* or *competency*? *consistence* or *consistency*?

When there is doubt about a word not given in its place in this book, and again when one is given without further comment than *See* -CE, -CY, it is to be presumed that either -ce or -cy may be used; but three generalities may be added. First, that short words favour -cy, and longer ones -ce; it was not by design, but by a significant accident, that all the -cy words given above as having no -ce alternatives were metrical matches for *buoyancy*. Secondly, that many words tend to use the -ce form in the singular, but -*cies* rather than -*ces* in the plural, e.g. *irrelevance*, but *irrelevancies*. And thirdly, that euphony often decides, in a particular context, for one or the other ending. Of the first point a good illustration is provided by *frequency* and *innocence*; formerly both endings were common

for each, but now from the shorter adjective *frequent* -ce is almost obsolete, and from the longer *innocent* -cy is an archaism preserved by Bible texts. On the second it may be added that words used concretely in the plural meaning specimens of the quality etc. (*truculencies* = truculent phrases, *irrelevancies* = irrelevant points, *inadvertencies* = acts of inadvertence) partly account for the peculiarity, seeing that when there is differentiation it is -cy, not -ce, that tends to the concrete, as in *emergency* = event that emerges compared with *emergence* = the emerging. And on the third point convincing examples will be found under TRANSPARENCE.

Articles in which differentiation between the two forms is recorded as existing or recommended are CONSISTENCE, DEPENDENCE, EMERGENCE, INDEPENDENCE, INDIFFERENCE, PERMANENCE, PERSISTENCE; *Residency* and *Excellency* are forms chiefly used in special senses while the -ce forms do the general work; and under COMPLACENCY the -cy form is recommended for differentiation not from *complacence*, but from *complaisance*.

**cee-spring, C-spring.** The second form is perhaps better; cf. A-line, H-iron, L-joint, S-piece, T-square, U-bolt, V-neck, Y-track.

**ceiling, floor.** These words had a great vogue during and after the second world war, especially in OFFICIALESE, as picturesque terms for the upper and lower limits of the permissible. There is no harm in that: if rationing and other restrictions on liberty are necessary we ought to be grateful to those who would gild the pill for us by substituting the brightness of metaphor for the drabness of words used literally. The trouble is that these metaphors are too fascinating; they are worked to death in all sorts of incongruous contexts. We are accustomed to think of ceilings as pretty stable things; constant talk about raising them and lowering them and extending them and waiving them and abolishing them and restoring them and so on sounds rather silly, and in the end these well-meant attempts to brighten the official vocabulary may do more harm than good. Sometimes the ceilings turned out to be floors too, which was very puzzling. *The effect of this announcement is that the total figure for 1950–1 of £410 m. can be regarded as a floor as well as a ceiling.* See METAPHOR 1.

**cello.** Pl. -*os*; see -O(E)S 6. Being now much commoner than *violoncello,* it does right to shed its apostrophe.

**Celt(ic), K-.** The spelling *C-* is the established one, and no useful purpose seems to be served by the substitution of *K-,* though the pronunciation *s-,* once general, seems now, perhaps under the influence of DIDACTICISM, to be giving place to *K-.*

**centenary, centennial,** nn. meaning hundredth anniversary. *Centenary,* the usual British form, has the disadvantage that the notion of years is not, except by modern development, contained in it; this, however, is true also of *century,* and need not count for much. *Centennial,* chiefly used (as a noun) in America, has the disadvantage that it gives a less convenient pattern for forming the names of higher anniversaries on. That these are sometimes wanted is a reason for maintaining *centenary.*

The shots made at these higher names often result in monstrosities. Of these *bicentenary,* which might have been *ducenary,* and *tercentenary* (*trecenary*), must be taken as established; *quatercentenary* and *quincentenary* have put in an appearance, and are unlikely to be ousted by *quadringenary* and *quingenary.* Nor is anyone, however passionate a supporter of etymological correctness, likely to try to introduce *sescenary, septingenary, octingenary,* or *nongenary*; ANALOGY will insist on *cent* coming into the word somehow; *sexcentenary* and *octingentenary* have already been used. To *millenary* no exception can be taken. The pronunciations *sentē'nari* (not *se'ntĕ-,* as in U.S.)

and *mĭlē'nărĭ* (desirable in itself for distinction from *millinery*) would suit best others that it might be necessary to use. But the advice given here to those who are at a loss for an -*enary* word is to content themselves with -*hundredth anniversary.*

**centi-, hecto-.** In the metric system *centi-* denotes division, and *hecto-* multiplication, by a hundred; cf. DECA-, DECI-, and KILO-, MILLI-.

**cento** (composition made up of scraps from other authors). Pl. -*os,* see -O(E)S 6. The pronunciation is *sĕ-*; the word is Latin, but is often mispronounced *chĕ-* as if Italian; the Italian is *centone,* and the French *centon.*

**centre.** *Supreme authority was at last centred in a single person.* | *The organ, and an entirely new system of church music centring round it.* Which is the better preposition, *in* or *round*? To reject idiom because it does not make sense if literally construed is to show oneself ignorant of the genius of the language. But when, as here, the idiomatic expression (*centre in* or *on*) happens also to be the logical one, there is nothing to be said for preferring the illogical *centre round,* as though *centre* and *gather* were synonymous. As a noun *centre* has its own precise meaning and should not be used as a GENTEELISM for *middle.*

**centrifugal, centripetal.** The stress should be on the second syllable See RECESSIVE ACCENT.

**century.** The first century of the Christian era consisted of the years 1 to 100, the second of years 101 to 200, and so on. It follows that each century contains one year only (the last) beginning with the number that names it; the first 99 begin with a number lower by one. For the curiously different Italian reckoning see TRECENTO.

**-cephalic.** Pronounce -*sĕph-.* Compounds (*brachy-, dolicho-, hydro-,* etc.) accent the -al-.

**cerement** is disyllabic (*sērm-*). Cf. REREDOS.

**ceremonial, ceremonious,** aa. *Ceremonial* means connected with or constituting of or fit for a ceremony (i.e. a piece of ritual or formality) or ceremonies (*the -al law*; *a -al occasion*; *for -al reasons*; *-al costume*). *Ceremonious* means full of or resulting from ceremony, i.e. attention to forms (*why be so -ous?*; *-ous people*; *-ous politeness*). In these examples the termination not used could hardly be substituted, even with change of meaning. But with some words *-al* and *-ous* are both possible, though with different significance: *a -ous court* is a sovereign's court in which ceremony is much observed; *a -al court* would be a judicial court set up to regulate ceremonies; a visitor may make *a -ous entry* into a room, but an army *a -al entry* into a town that has capitulated.

**certes.** Pronounce *ser'tĕz*. See AR-CHAISM.

**certitude** is now restricted to the single sense of absolute conviction or feeling quite sure; *certainty* can, but often does not, mean this also, and the use of *c.* may therefore obviate ambiguity.

**cervical** is pronounced *servī'kl* by purists, but *ser'vĭkl* is commoner (*As the sabre true cut cleanly through his cervical vertebrae*). See on *doctrinal* in FALSE QUANTITY; the Latin for neck is *cervix, -īcis*.

**cess.** See TAX.

**ceteris paribus.** Pronounce *sē'terĭs pă'rĭbus* in spite of its inconsistency with *et cĕtera*. See LATIN PHRASES.

**chagrin.** The pronunciation *shăgrē'n* is preferred in Britain, *shăgrī'n* in U.S.

**chain reaction.** See REACTION.

**challenging.** *The test series of 1962/3 has begun on an entertaining and c. note./ The plan is a c. attempt to make use of idle acres. Challenging* has become a VOGUE WORD, used with a freedom that must often leave readers wondering who is challenging whom.

**cham.** Pronounce *kăm*. The word is an obsolete form of *Khan* (ruler) and is seldom used except as a sobriquet of Dr. Johnson (*The Great Cham*).

**champaign, champagne.** According to the OED these words should be distinguished not only in spelling but also in pronunciation: *shampā'n* for the wine and the province it comes from and *cha'mpān* for the tract of open country.

**chance,** vb, as a synonym for *happen* (*it chanced that . . .*; *I chanced to meet him*) stands in the same relation to it as CEASE to *stop*.

**Change,** in *on Change,* is not an abbreviation of *Exchange*, and should have no apostrophe.

**Chanticleer.** See SOBRIQUETS.

**chanty, sh-,** sailors' hauling-song The pronunciation *sh-* is accounted for by the supposed derivation from the French *chantez*, sing ye. Formerly spelt *ch*, but since Sir Richard Terry popularized the *Sea Shanties* it is now always *sh*.

**chap, chop,** jaw or cheek. In *lick one's cc., fat-cc., c.-fallen*, both spellings are common; in *Bath c., chap* only is used, and in *the cc. of the Channel, chops* only.

**chaperon.** The addition of a final e is wrong. Pronounce *shă'pĕrōn*.

**chap** (man) is a CURTAILED WORD. *Chapman* formerly meant a merchant of substance, like those who used to provide Solomon with gold, and the 'chapmen riche and therto sad and trew' of the *Man of Lawes Tale*, but later was applied only to pedlars.

**char-à-banc** is a word still in popular use, in spite of the competition of *motor-coach*; the spelling *charabanc* (plural *-s*) and pronunciation *shă'rabang* should be accepted. See FRENCH WORDS; the French spelling in the singular is *char-à-bancs*.

**character** is a valuable and important word with several well-marked senses. The worst thing that can happen to such a word is that it should be set to do inferior and common work which could be more suitably done by meaner words and has to be done so often that the nobler word is cheapened by familiarity. *Character*, like *case* and other good words, now occurs a hundred times as a mere element in PERIPHRASIS for once that it bears any of its independent senses. The average writer can perhaps not be expected to abstain from the word for the word's sake; but, if he realizes that at the same moment that he degrades the word he is making his sentence feeble and turgid, he will abstain from it for his own sake. A few slightly classified examples of the abuse are therefore added.

(*a*) *C.* is used with adjectives as a substitute for an abstract-noun termination, *-ness*, *-ty*, etc.: *The very full c. of the stage-directions indicates . . .* (fullness) / *On account of its light C., Purity and Age ——'s whisky is a whisky that will agree with you* (lightness. But this is the kind of literature in which such idioms are most excusable). / *Unmoved by any consideration of the unique and ancient c. of the fabric* (uniqueness and antiquity).

(*b*) A simple adjective *x* is watered into *of a x character*; the right water for such solutions, which are bad in themselves when not necessary, is *kind*; but the simple adjective is usually possible: *Employment of a patriotic c.* (patriotic employment). / *There is no unemployment of a chronic c. in Germany.* / *The attention which they receive is of a greatly improved c.* / *His influence must have been of a very strong c. to persuade her.* / *Payments of the c. in question* (of this kind; or such payments).

**character, characteristic.** For synonymy see SIGN.

**char(e).** The form *chare* (part. *charing*, pron. *-ār-*) was said by the OED to be the usual one. This was doubtful even then, and *charwoman* later established *char* as a CURTAILED WORD for her and *charring* for what she does. EUPHEMISM has since been at work and changed her first into *charlady* and then into *daily help*. For the impersonal noun

(a job of housework) U.S. *chore* has now replaced *char*.

**charge.** 1. *All dogs in c. of servants must be kept on a lead.* 'Who leads which?' asks a popular columnist, commenting on this notice in a London square. There was a time when the notice would have been unambiguous: *A nurse in c. of children*, says the OED, is a modern way of saying *children in c. of a nurse*. Having created this confusion, we are trying to get rid of it by inserting a *the* when *c. of* relates to the custodian—*dogs in the c. of servants*; *children in the c. of a nurse*.

2. *I don't think anyone could ever charge me of being a fellow-traveller.* To *charge* someone *of* something is an unidiomatic construction on the ANALOGY of *accuse*. A charge (n) *of* being but to charge (vb) *with* being.

**charivari.** Pronounce *shărĭvar'ĭ*. The word looks Italian but is in fact French.

**chastise** is never spelt with z; see -IZE, -ISE. Pronounce the noun *chas'- tĭzment*. See RECESSIVE ACCENT.

**cheap(ly).** See UNIDIOMATIC and-LY DEAR.

**check and checker.** See CHEQUE and CHEQUER.

**(check)mate.** *Mate* is the usual form in chess, and *checkmate* in figurative use.

**cheerful, cheery.** The latter has reference chiefly to externals—voice, appearance, manner, etc. Resignation may be cheerful without being cheery; and a person may have a cheerful, but hardly a cheery, spirit without his neighbours' discovering it. The cheerful feels and perhaps shows contentment, the cheery shows and probably feels it.

**cheque**, though merely a variant of *check*, is in British usage clearly and usefully differentiated from it with the sense bank-draft, *check* in this sense being chiefly American.

**chequer, checker.** The first spelling is very much commoner in Britain for both the noun and the verb. The U.S. game *checkers* is the British *draughts*.

**cherub, cherubic.** *Cherub* has pl. *cherubim* chiefly when *the Cherubim* are spoken of as a celestial order; *cherubims* is wrong; in figurative use *cherubs* is usual. *Cherubic* is pronounced *chĕrōō′bic*.

**chevalier d'industrie.** The expression is not, as one might think, the French for *tycoon*; it means *adventurer*.

**chevaux de frise.** See FRENCH WORDS. *Cheval de frise* is now rare; *chevaux de frise* is treated either as sing. or as pl. (*a wall with a c.d.f.*, or *a wall with c.d.f.*). So called because first employed in Friesland as a device for stopping cavalry charges.

**chevy.** See CHIVY.

**chiaroscuro.** Pronounce *kyar′-oskoor′ō*.

**chiasmus.** When the terms in the second of two parallel phrases reverse the order of those in the first to which they correspond. If the two phrases are written one below the other, and lines drawn between the corresponding terms, those lines make the Greek letter chi, a diagonal cross:

I cannot            dig

to beg            I am ashamed.

**chick(en).** *Chicken* is the original and still the ordinary form, *chick* serving as a diminutive used chiefly of an unhatched or unfledged bird, the young of small birds, or (endearingly, in pl.) children. For pl. of *chicken* see COLLECTIVES 3.

**chide** stands to *scold* as *proceed* to *go*. Past *chid*, p.p. *chid(den)*.

**chiefest,** formerly common, is now felt to be an unnatural form, and used only as a deliberate ARCHAISM.

**childish, childlike.** The distinction

drawn is so familiar that *childish* is in some danger of being restricted to the depreciatory use that is only one of its functions, while *childlike* is applied outside its sphere. The face, for instance, that we like a child to have should be called not a childlike, but a childish face; the rule that *childish* has a bad sense is too sweeping, and misleads. *Childish* used of adults or their qualities, and *childlike* (which should always be so used), have the opposite implications of blame and approval. *Childish* means 'that ought to have outgrown something or to have been outgrown', and *childlike* 'that has fortunately not outgrown something or been outgrown'. *Childish simplicity* in an adult is a fault; *childlike simplicity* is a merit; but *childish simplicity* may mean also simplicity in (and not as of) a child, and convey no blame. *Childish enthusiasm* may be either a child's enthusiasm or a man's silly enthusiasm; *childlike enthusiasm* is only that of a man who has not let his heart grow hard.

**childly.** See REVIVALS.

**chilli** is the right spelling for the capsicum pod (unconnected with *Chile*); pl. *chillies*.

**chill(y).** The form *chill* (as adj.) is only a LITERARY WORD, *chilly* being that in general use.

**chimera, -aera, -æra.** The first spelling is best. See Æ, Œ. The pronunciation *kĭm-* is preferable to *kīm-*.

**Chinaman** has acquired a derogatory flavour and is falling into disuse in its old sense, though oddly revived in cricket jargon for an off-break (to a right-hand batsman) bowled by a left-hand bowler. For Chinese nationals *Chinese* is now used in the singular as well as the plural. *Chinee* for *Chinaman* is a BACK-FORMATION from *Chinese* pl., and, being felt to be irregular, was always rare except as jocular slang; even for that purpose it was superseded by *Chink*.

**chiropodist** is a BARBARISM which no longer accurately represents the occupation. Treatment of the hands is now the concern of the *manicurist*; the *c.* is left with the feet only. But he shows little disposition to recognize this pruning of his functions by accepting the more suitable title of *pedicurist*. The dictionaries give the pronunciation *kǐro-*, but the popular *shǐro-* is common.

**chivalry** etc. The pronunciation *sh-*, instead of *ch-*, though based on a mistake, is now established. Of the adjectives *chivalrous* and *chivalric* the second should be either let die as a NEEDLESS VARIANT or restricted to the merely grammatical function of representing the phrase 'of chivalry', as in *the chivalric ages*.

**c(h)ive.** Spell with the *h* and so pronounce.

**chivy, che-.** The *-i-* certainly gives the prevailing sound, and has now become the accepted spelling (sometimes *chivvy*) except for *Chevy Chase*.

**chlorine.** For pronunciation see -IN AND -INE.

**chloroform.** The pronunciation *klǒro-* seems to be gaining on *klōro-*, formerly standard, and must not be condemned merely because it is a FALSE QUANTITY.

**choir, quire.** The first spelling, which goes back little further than the 18th c., neither bears its pronunciation on its face nor represents the French or the Latin forms well, and is therefore inferior. But attempts to restore *quire*, though it still appears in the Prayer Book, would be futile; it must be content with its meaning as a measure of paper.

**choler(ic).** *Choler*, except when used historically with reference to the four humours, is now a mere ARCHAISM; *choleric*, however, has survived it, and is preferable in some contexts to *irascible*, *quick-tempered*, etc.; pronounced *kǒ'lerik*.

**chop, cutlet.** A chop is cut from the loin and includes a rib; a cutlet is cut from the neck, or may be a small piece

of meat from any part and include no bone: it is derived from French *côtelette* (diminutive of *côte*, rib) and has no connexion with English *cut*.

**chorale.** Pronounce *kǒrah'l*. As to spelling, the *-e* is strictly incorrect, but both usual and convenient, obviating confusion with the adj. *choral*; cf. LOCALE and MORALE.

**c(h)ord.** There are two words that are written *chord*. One of them, that used in Harmony, is from *accord* and has no connexion with *cord*. The other (*touch the right chord*; *the chord of an arc*; *the vocal chords*; *the spinal chord*) is the same as *cord*, but has had its spelling corrected after the Greek original. It is well to remember that in the four phrases mentioned *chord* means simply string; but the spelling *cord*, which would have been legitimate and avoided confusion in any of them, is ruled out by custom except in the last two.

**Christian name.** See FORENAME.

**chronic** in the illiterate use for bad, intense, severe, (*the weather has been c.*; *that was a c. fight last night*), is a SLIPSHOD EXTENSION. See POPULARIZED TECHNICALITIES.

**chrysalis** has pl. *chrysalises*, *chrysalids*, or *chrysalides* (*krǐsǎ'lǐdēz*); the first should be made the only form.

**chute.** See SHOOT.

**-ciation.** Nouns in *-ation* from verbs in *-ciate* have, if they follow their verbs, the very unpleasant combination of two neighbouring syllables with the *-sh-* sound (*ĕmāshiǎ'shn* from *emaciate*). The alternative pronunciation *-sǐāshn*, sometimes recognized by the OED (e.g. in *association*), avoids the bad sound, and is legitimate on the analogy of *denunciation*, *pronunciation*, *annunciation*, of which all might have had, and the last has in *annunciate*, a verb in *-ciate* as well as that in *-ounce*. Words in *-tiation* (as *initiation*) can perhaps hardly be treated in the same way, except those that, like *negotiation*, have alternative forms with *-c-* for *-t-*;

*nĭgōsĭā'shn* seems possible, but not *propĭsĭā'shn*.

**cicada, cicala, cigala.** The first is the original Latin word taken into English (pronounce *-kā-* or *-kah-*); the second is Italian (*-kah-*); the third is the French *cigale* with termination assimilated in English to the others (*-gah-*). The first is recommended.

**cicatrice, cicatrix.** The first, pronounced *sĭ'katrĭs* and in pl. *sĭ'katrĭsĕz*, is the English word. The second, pronounced *sĭkā'trĭks* and in pl. *sĭkătrī'-sēz*, is the Latin in surgical and other scientific use.

**cicerone.** Pronounce *chĭchĕrō'nĭ* but *sĭsĕrō'nĭ*, no doubt by assimilation to *Cicero*, is common; even *sĭsĕro'n* may sometimes be heard; pl. *ciceroni*, pronounce *-nē*.

**cinema, cinematograph, kin-.** The *cin-* forms are obviously more handy for words in constant popular use, and we were right to accept them heartily. There is indeed very little in any of the objections once made to them. The points are: (1) *c* or *k*?; (2) the syllable accented; and (3) the curtailed form of *cinema*. (1) English *c* for Greek *k*, far from being wrong, is normal; cf. *catholic, cenotaph, Circe, colon, cubic, cycle*. It may be regrettable that, since the scientific words *kinetic* and *kinematic* are abnormally spelt, the connexion of *cinematograph* with them is obscured; but that is their fault, not its. (2) The vowel sounds and the syllable accents will be found justified in the article FALSE QUANTITY. The chief objection—to misplacing in *sĭ'-nĕma* the stress of the Greek *kĭnē'ma*—falls to the ground when it is remembered that *cinema* is not the Greek word *kinema* at all, but a curtailed form of *cinematograph*, whose second syllable is bound to be *-nĕ-* in popular speech. (3) Curtailing is an established habit, no worse in *cinema* than in the schoolboy's *prep.*, our ancestors' *mob*, or our own *fridge, polio*, and *telly*. *Cinema* itself has been curtailed to provide a useful prefix, *cine-camera, cine-projector*, etc. See CURTAILED WORDS.

**Cingalese.** See SINHALESE.

**cinq(ue).** The five on dice etc. is pronounced *sĭngk*, and best spelt *cinque*. *Ace, deuce, trey* (*-ā*), *cater* (kā-), and *sice* (sīs), are the others of the series. In *Cinque Ports* the pronunciation is the same.

**cinquecento.** Pronounce *chĭngkwĭ-chĕ'ntō*; for meaning see TRECENTO.

**cinq(ue)foil.** Pronounce *sĭ'ngkfoil*. The OED puts the longer form first.

**cipher** is the spelling preferred by the dictionaries, but *cypher* is usual for the secret codes.

**circuit(ous).** Pronounce *ser'kĭt*, but *serkū'ĭtus* (not *ser'kĭtus*).

**circumbendibus.** See FACETIOUS FORMATIONS.

**circumlocutional, -nary, -utory.** Though an adjective is often wanted for *circumlocution*, the practice being now so common, only the last (the best of a bad lot) has won any favour and is perhaps preferable to the formal *periphrastic*. But *roundabout* will often serve. See also PERIPHRASIS.

**circumstance.** The objection to *under the cc.*, and insistence that *in the cc.* is the only right form, because what is round us is not over us, is puerile. To point out that *round* applies as much to vertical as to horizontal relations, and that a threatening sky is a c. no less than a threatening bulldog (*Under the circumstances I decided not to venture*), might lay one open to the suspicion of answering fools according to their folly. A more polite reply is that 'the cc.' means the state of affairs, and may naturally be conceived as exercising the pressure under which one acts. *U. t. cc.* is neither illogical nor of recent invention (1665 in OED), and until the grammarians started telling us it was wrong was far more often heard than *i. t. cc.* The OED, far from hinting that either form is incorrect, assigns them different functions: 'Mere situation is expressed by "*in* the circumstances", action affected is performed "*under* the circumstances".'

Or, as the sociological jargon of today would put it, What is done *under the cc.* is influenced by *environmental factors.*

**cirrus** has pl. *cirrī*; see -US.

**cither(n),cittern,gittern,zither(n).** Only the last is still in common use. When the forms are distinguished *cither* is the general word including the ancient cithara and its more modern representatives, *zither(n)* is appropriated to the Tyrolese instrument, and *cithern, cittern, gittern,* all mean the one common in the 16th and 17th cc.; *cittern* and *gittern* might well be dropped as NEEDLESS VARIANTS.

**city** in common usage is applied to any large and important town, but in Great Britain is strictly an honorary title held by ancient custom (especially in the case of episcopal sees) or granted by Royal Charter, and not necessarily implying any greater powers of local government than those of a borough. Cities vary in size and importance from Birmingham with a population of over a million to Wells with one of under 7,000, and in antiquity from Oxford, which has been a city from time immemorial, to Cambridge, which was made one in 1951. The most famous, the City of London, is also one of the smallest, with an area of one square mile and a resident population of about 5,000.

**clad.** See CLOTHE.

**claim**, vb. The primary meaning of *c.* is to demand recognition of a right: *He claimed to be the next-of-kin.* A natural extension is to use *claim* instead of *say* etc. in the sense of claiming credence for an improbable assertion (*He claims to have seen a flying saucer*) or claiming recognition of a successful achievement (*The Union claims that the response to its strike call was complete*). But more questionable extensions have followed. There is no doubt a vigour about *claim*—a pugnacity almost—that makes such words as *allege, assert, contend, declare, maintain, profess,*

*represent, say,* and *state* seem tame by comparison. That may explain, but does not justify, its present use as a VOGUE WORD, impoverishing our vocabulary by elbowing out those weaker rivals, with their various shades of meaning. In the following examples, which anyone could double from almost any day's newspapers, other words are suggested that might suitably have been used, if only to give *claim* a rest. *The State Department claims that discrimination is being shown against the American film industry* (declares). | *There are those who claim that NATO has an aggressive purpose* (assert). | *The police took statements from several people who claimed that they had seen the gunmen* (said). | *He claimed that the story he had told had never varied* (maintained). | *From the right, Mr. Brown claimed that the Labour Party must not split over defence policy* (? urged). The excursions of *claim* must surely have reached their limit in the Hawaiian newspaper headline *OAHU BARMAID CLAIMS RAPE.*

**clandestine** according to the OED should be pronounced *klănde′stĭn* and this is recommended although the RECESSIVE ACCENT has been at work, and some recent dictionaries give *klăn′destĭn.*

**clari(o)net.** The two forms denote the same instrument, but *-inet* is now in general use. The stress is usually on the last syllable as in the French word, *clarinette,* from which *c.* comes.

**classic(a!).** These adjectives, in their senses of relating to the classics and conforming to the rules of Greek and Latin antiquity, are distinguished rather by suitability to different contexts than by difference of meaning. *Classical* is the usual word, and it would perhaps never be noticeably the wrong one, even where *classic* is more idiomatic (e.g. we can say, if we choose, *This is classical ground*). On the other hand, there are many combinations in which *classic* would sound ridiculous; *classic education, classic allusions,* are impossible. *Classic*, however, apart from being used in the plural as a noun meaning the general body of Greek and Latin literature, has its own separate meaning of outstandingly important or authoritative. *St. Andrews is the classic home of golf.* | *Rylands v. Fletcher was cited as the classic case.* | *The Derby is one of the five classic flat races.*

**clause.** 1. Grammar. It conduces both to clearness and to brevity if the word in its grammatical sense is applied only to what is sometimes called a *subordinate c.*, and never either to a complete sentence or to the framework of the sentence, which is often called the *main* or *principal c.*, but may equally well be called *main sentence*. The definition of a c., then, should be 'subordinate words including a subject and predicate, but syntactically equivalent to a noun or adjective or adverb'; in this book the word is always to be understood thus.
2. Law. Legal documents are ordinarily divided into *cc*. Those of a parliamentary Bill are rechristened *sections* when the Bill becomes an Act.

**clear(ly).** See UNIDIOMATIC -LY.

**cleave**, split, has past tense *clove* or *cleft* or *cleaved*, p.p. *cloven* or *cleft* or (arch.) *clove*, which attach themselves capriciously to different objects: *cloven-footed, cloven hoof, cleft palate, cleft stick, clove hitch.*

**cleave**, stick, has past tense *cleaved* or (arch.) *clave*, p.p. *cleaved*.

**clench, -inch.** The spellings are so far differentiated as to be generally applied thus: we *clench* our hands, jaws, teeth, and object held; we *clinch* an argument or a bargain; we usually *clench*, but sometimes *clinch*, a rivet, rope, or nail (always when using the nail to construct a *clinker*, or *clincher-built* boat). Boxers go into *clinches*, and the fact or statement that settles an argument is a *clincher*.

**clerk.** The pronunciation *-erk*, normal U.S., is now occasionally heard in Britain instead of the long-established *-ark*, sometimes facetiously, sometimes seriously, perhaps by infection from America or from an excessive respect for spelling.

**clever** is much misused, especially in feminine conversation, where it is constantly heard in the sense of learned, well read, bookish, or studious; a woman whose cleverness is apparent in all she does will tell you that she wishes she was c., that she cannot read c. books (meaning those of the graver kind), and that Mr. Jones must be a very c. man, for he has written a dictionary. But in fact ignorance and knowledge have no relation to cleverness, which implies ingenuity, adroitness, readiness, mental or manual quickness, wit, and other qualities incompatible with dullness, but not with ignorance or dislike of books.

**clew, clue.** The words are the same, but the more recent *clue* is now established in the usual sense of idea or fact that may lead to a discovery, while *clew* is retained in the nautical sense, and in the old-fashioned sense skein or ball of thread or yarn from which the usual sense of *clue* has been developed. The word has sprung into favour in the slang expressions of ignorance *I haven't a clue* | *I'm absolutely clueless*. Perhaps this is due to the popularity of crossword puzzles and detective stories.

**cliché** means a stereotype; in its literary sense it is a word or phrase whose felicity in a particular context when it was first employed has won it such popularity that it is apt to be used unsuitably and indiscriminately. Some observations on how clichés are born and the harm they do will be found under HACKNEYED PHRASES; and many examples of different kinds of *cc*. are given in that article and others, especially BATTERED ORNAMENTS, IRRELEVANT ALLUSION, METAPHOR, SIAMESE TWINS, and VOGUE WORDS.
The word is always used in a pejorative sense, and this obscures the truth that words and phrases falling within the definition are not all of a kind. There are some that always deserve the stigma—those threadbare and facetious ways of saying simple things and

those far-fetched and pointless literary echoes which convict their users either of not thinking what they are saying or of having a debased taste in ornament. A few obvious specimens are *durance vile, filthy lucre, sleep the sleep of the just, tender mercies, own the soft impeachment, suffer a sea change, leave no stone unturned*. There are others that may or may not deserve to be classed with them; that depends on whether they are used mechanically, taken off the peg as convenient reach-me-downs, or are chosen deliberately as the fittest way of saying what needs to be said. To take one or two examples from the many hundreds of words and phrases that it is now fashionable to brand as clichés, writers would be needlessly handicapped if they were never allowed to say that something was a *foregone conclusion*, or *Hobson's choice*, or a *white elephant*, or that someone was *feathering his nest* or *had his tongue in his cheek* or *a bee in his bonnet*. What is new is not necessarily better than what is old; the original felicity that has made a phrase a cliché may not be beyond recapture. The enthusiasm of the cliché-hunters is apt to run away with them. The professional writer of today is less in need of being warned against using clichés than of being reminded of the advice once given by J. A. Spender:

'The hardest worked cliché is better than the phrase that fails. . . . Journalese results from the efforts of the non-literary mind to discover alternatives for the obvious where none are necessary, and it is best avoided by the frank acceptance of even a hard-worn phrase when it expresses what you want to say.'

Or, to quote Clutton Brock:

'A writer may of course attain to a familiar metaphor in his own process of expression, but if he does, if it is exactly what he has to say, then it will not seem stale to the reader. . . . If an image forces itself upon a writer because it and it alone will express his meaning, then it is his image, no matter how often it has been used before.'

Those well-worn clichés that are now as zestfully pilloried as split infinitives were by an earlier generation are not the kind most likely to tempt a writer into the lazy acceptance of a prefabricated phrase. They are readily recognizable, and present themselves without disguise for deliberate adoption or deliberate rejection. More insidious are those apparently innocent phrases that, almost unnoticed, are on the way to becoming clichés. They slip past the barrier without scrutiny. Thus (to take a few examples of the many that might be given) a writer may say 'CLIMATE of opinion' without asking himself whether he means anything more than 'opinion', or 'within the FRAMEWORK of' instead of a bare preposition, or that something is 'grinding to a halt' which is doing no more than slowly stopping; or he may find that 'in this day and age' has thrust aside the plain word 'today', which was all that he needed, or that 'in the ultimate analysis' has done the same to 'in the end'.

Clichés are plentiful in the linguistic currency of politics, domestic and international. They too, however happy in their original application, soon lose any semantic value they may once have had, and become almost wholly emotive. That, for instance, has been the fate of *self-determination, appeasement, power politics, parity of esteem, underprivileged classes, victimization*, and innumerable others, including *democracy* itself, the classic example of a Humpty-Dumpty word. Even those admirable recent coinages *cold war, iron curtain, peaceful coexistence*, and *wind of change* are now so near to clichés as to offer themselves as substitutes for thought. It has been said by one who ought to know that 'When Mr. Khrushchev says *peaceful coexistence* he means almost precisely what we mean by *cold war*.'

**clientele** should be written without italics or accent; the fully anglicized pronunciation *klĭentēl* is favoured by the Oxford Dictionaries, but *klĕontāl* (or *-ĕl*) is still probably more usual.

**climacteric.** The old pronunciation was *klĭmăktĕ'rĭk*, which stands first in the OED; but *klīmă'kterĭk* (see RECESSIVE ACCENT) is now commoner; it is preferred by the COD and is likely to prevail.

**climate, clime.** 1. *Clime* differs from *climate* not only in being mainly a poetic word but also in meaning; essentially it means a tract of country. Some reference to the climate may be implied but not necessarily, and it never means, like *climate*, weather conditions alone.

2. A figurative use of *climate*, apparently unknown until the middle of the 20th c., has now forced its way into the dictionaries. 'Trend or attitude of community or era, character of something' is the COD's definition. It has produced a CLICHÉ, *climate of opinion*, so popular as to encourage the hope that it will soon be worked to death. Anyone tempted to use it would be wise to ask himself whether *opinion* by itself would not do just as well, as it certainly would in this typical example: *It may look as if the c. of o. in the House had hardened against going into the Common Market.* Cf. FRAMEWORK.

**climax.** A Greek word meaning *ladder*. In rhetoric, arrangement of a series of notions in such an order that each is more impressive than the preceding. (1) *Eye hath not seen,* (2) *nor ear heard,* (3) *neither have entered into the heart of man,* / the things which God hath prepared; three progressive stages of strangeness. In popular use it is generally applied to the culmination only.

**close.** See GENTEELISM.

**close(ly).** See UNIDIOMATIC -LY.

**closure, gag, guillotine, kangaroo.** The first is the name given to a provision by which debate in the House of Commons can be cut short in spite of the wish of the minority to continue it; the closure is brought into operation by a motion That the Question be now put. See also REVIVALS.

*Gag* is the word used, chiefly by the closured party, to describe the ordinary closure or its developments, the guillotine and the kangaroo.

The *guillotine,* or closure by compartments, is thus defined in the *Ency. Brit.*: 'The guillotine means that the House decides how much time shall be devoted to certain stages of a measure, definite dates being laid down at which the closure shall be enforced and division taken.'

The *kangaroo,* or kangaroo closure, is a further development. The guillotine having the disadvantage that the limited time may be wasted on minor matters and none be left for important ones, the Chairman of Committees is empowered to select the amendments that shall be debated, the unselected ones being voted on without debate.

**clothe** has *clad* beside *clothed* both as past and p.p. While *clothed,* however, is suitable to all contexts (except where *dressed* is preferable as less formal). *clad* is (1) intolerably archaic in effect as past and only slightly less so as p.p., and 2) never used absolutely, but always with some specification of the kind of clothing (cf. *ironclad*). Accordingly, *clad* cannot be substituted in *You were fed and clothed at my expense*; *He clothed himself hurriedly*; *When he was clothed he admitted us.* But *clothed* can be substituted in any of the following phrases, which are selected as favourable for the use of clad: *Lightly, well, insufficiently, clad*; *He clad himself in shining armour*; *Clad with righteousness*; *Hills clad with olives*; *Clad in blue.*

**clothes.** The old pronunciation is *klōz,* with ample authority from rhymes in 17th-c. and 18th-c. poets, including Shakespeare (*Then up he rose, and donned his clothes*). But this is often deliberately abstained from in the mistaken belief (once supported by the OED but abandoned by its successors) that it is 'vulgar or careless', and, unless the articulation of the *th* is found too difficult, it is likely to disappear under the influence of the speak-as-you-spell movement. See PRONUNCIATION 1.

**clue.** See CLEW.

**co-.** There are three ways of writing *cooperate* (coop-, co-op-, coöp-), and two of writing *copartner* (cop-, co-p-). The DIAERESIS is less in favour here than in America; moreover it is possible only in some words (those in which *co-* is followed by a vowel), whereas the hyphen is possible in all. But it should be recognized that hyphens in the middle of words are no ornament, and admittance should be refused to all that cannot prove their usefulness. In the classified list given below of the commoner words beginning with *co-* together or *co-*complementary, the spelling printed is to be taken as standard. But see HYPHENS, where it is suggested that many words now usually given a hyphen would be better without.

**1.** In some words the hyphen is never used: *coadjutor, coagulate, coalesce, coalition, coerce, cognate, cohabit, cohere, coincide, coition.*

**2.** Many are either so common or so analysable at a glance that the hyphen, though sometimes used, is entirely superfluous: *coeducational, coefficient, coequal, coessential, coeval, coexecutor, coexist, coextensive, coheir, coinstantaneous, cooperate, coopt, coordinate, coparcenary, copartner.*

**3.** Some are used and seen only by the learned, who may be expected to know them at a glance without hyphens: *coacervation, coadunate, coaxial, cosecant, coseismal, cosine, cotangent.*

**4.** Some always have the hyphen apparently by way of a (*sic*), or announcement that the spelling is intentional: *co-religionist, co-respondent.*

**5.** Some, if no hyphen is used, tend to fall at the first glance into wrong syllables and so perplex: *co-belligerent, co-latitude, co-pilot, co-signatory, co-tenant, co-tidal, co-trustee, co-worker.*

**6.** When a writer believes himself to be making a new word, he naturally uses the hyphen: *my co-secretary, their co-authorship*, etc.

**coal.** 1. *Haul*, and *call, over the cc.* are both in use, though the former is perhaps commoner. The reference is to the burning of heretics. 2. *Coal-tit*

is a better spelling than *cole-*, since the latter obscures the connexion with *c*.

**coastal** is a BARBARISM, the *-oa-* showing at once that *-al* has been added to an English and not a Latin word. The properly formed adjective *costal* having been put to another purpose, we showed a greater regard for the niceties of word-making when we used the noun attributively, as in *coastguard, coastline,* and *coast waiter*, than when we invented the term *Coastal Command* which has ensured for that adjective a firm place in our vocabulary.

**cocaine.** The pronunciation *kokā'n*, stigmatized by the OED (in 1893) as vulgar, is now the only one, as the COD recognizes. Cf -IES, -EIN.

**coccyx.** Pronounce *kŏ'ksĭks*.

**Cockaigne** is properly the name of a luxurious Utopia; the use of it for London as the home of Cockneys is a mistake or a pun.

**cockle.** *The cc. of the heart* is of some age (quoted from 1671) but of disputed origin; such phrases are best not experimented with, but kept, if used at all, to their customary form and context (*rejoice, warm, the cc. of the heart*).

**cock's-comb, cockscomb, coxcomb.** The first for the comb of a cock, the second for the fool's cap and the plants, and the third for the fop.

**coco(a), coker.** *Cacao* and *coco*, independent words, have corrupted each other till the resulting *cocoa* is used always for the drink and often for the coco(a)-nut palm; *coker*(*nut* etc.) is a shop spelling devised to obviate the confusion. *Coco-nut, coco fibre*, etc., are still used, though the *-a* more often appears; they should be kept in existence if possible, and *cocoa* be restricted to the drink and the powder from which it is made; the uncrushed seeds and the tree are still usually spelt *cacao*.

**codex** has pl. *codicēs*; see -EX, -IX.

**codify.** The pronunciation *kŏ-* would accord with the general tendency to prefer short vowels in such forms seen in *gratify, pacify, ratify, edify, specify, verify, vilify, vivify, modify*; whether a similar list on the other side could be made is very doubtful. But in *codify* kō- is in fact winning, no doubt because of the influence of *code*.

**cog.** The phrase *cogged dice* is due to a misunderstanding of the old *to cog dice*, which meant not to load them, but to cheat in throwing them; *loaded* should be used.

**cognate.** When a noun that is the object of a verb does not express the external person or thing on which the action is exercised (the direct object) but rather supplements the verb adverbially, it is called the *cognate*, or the *internal*, or the *adverbial, object* or *accusative*:

  is playing *bridge* (cognate);
  I hate *bridge* (direct);
  lived a good *life* (internal or cognate);
  spent his *life* well (direct);
  looked *daggers* (adverbial or cognate).

In the last example *daggers* is a metaphor for a look of a certain kind, and therefore cognate with the verb.

**cognizance, cognizant, cognizable, cognize.** *Cognize* alone has the *-g-* always sounded. Of the four, *cognizance* is the word from which the others have sprung, and it had for some time no *-g-* to be sounded. The introduction of the *-g-* has affected pronunciation, and *kŏg-* is now common in all, though *kŏn* may still be heard in the first three. *Cognizable* is usually stressed on the first syllable, whether the *g* is pronounced or not. For synonyms of *-nce*, see SIGN.

**coincidence.** *The long arm of c.* is a HACKNEYED PHRASE. Varying its form, endowing it with muscles, making it throw people about, and similar attempts at renovation, only make matters worse: *The author does not strain the muscles of coincidence's arm to bring them into relation.* | *Nor does Mrs. Moberly shrink from a use of 'the long arm' quite unnecessarily.* | *The long arm of c. throws the Slifers into Mercedes's Cornish garden a little too heavily.*

**colander, cullender.** Both are pronounced kŭ′lender; the first spelling, which is nearer the Latin stem (cf. *percolate*), is also more frequent in the 19th-c. quotations in the OED and is now general.

**col-, com-, con-.** See PRONUNCIATION 5.

**cold war.** This expression, happily invented by Walter Lippman in 1947 to describe the relations then existing between the United States of America and the Soviet Union, has, with all its merits, two disadvantages. One is that it is becoming a CLICHÉ, and so, like all clichés, tending to bemuse thought. The other is that the metaphor places politicians in a dilemma when they try to follow it up. Are they to advocate a rise or a fall in the temperature of the cold war? The latter would seem to intensify an already alarming crisis; the former to bring the hot war a stage nearer. This problem has to be dodged by doing violence to the metaphor. *In an atmosphere of 'live and let live' for a generation the cold war may well become less turbulent.*

**collaborator.** See WORSENED WORDS.

**collectives.** The word is applied to many different things. What is common to all is that a noun singular in form is used with a plural implication. For instance *flock* (a number of sheep or parishioners) is one kind of collective, and *flock* (woollen waste) is another. The first may be treated as singular or plural (*His f. was attacked by wolves*; *His f. was without a pastor* or *were unanimous in disapproval*), and can itself be given a plural form with the ordinary difference in meaning from the singular (*shepherds tending their flocks*). The other *flock* (woollen waste) can be used in either the singular form (with a singular verb) or the plural (with a plural verb) without any difference of meaning (*A mattress of flock* or *flocks*; the *flock has, the flocks have, not been disinfected*). But the word *collective* is applied to both, as well as to many equally dissimilar kinds of noun.

Collectives may be roughly divided into the following groups.

1. Nouns denoting a whole made up of similar parts, such as *committee, crew, firm, orchestra, soviet.* These are also called *nouns of multitude.* See NUMBER 6.

2. Nouns that make no plural form but are used as both singular and plural, e.g. *counsel* (= barrister), *deer, grouse, salmon, sheep, trout,* or, in a few cases, as plural only, e.g. *cattle.*

3. Nouns that make plural forms in the normal way but whose singular form may also be used as a plural, sometimes with little if any difference of meaning, e.g. *cannon, duck, fish,* but more often with some special implication. For instance *hair, straw,* and *timber* are used for the mass, *hairs, straws,* and *timbers* for the particular; *chicken, lamb, cabbage,* and *potato* are on the dinner table, *chickens, lambs, cabbages,* and *potatoes* are in their natural state; *shot* and *game* are for the pellets and their objectives, *shots* and *games* for those words in their other senses; *elephant* and *lion* are in the bush, *elephants* and *lions* in the zoo; *fruit* is the current word and *fruits* the archaic.

4. Names of materials used for a collection of things made from them, e.g. *china, linen, silver.*

5. Words of number or amount that when used after definite or indefinite numerals have the singular instead of the plural form (*six brace of grouse; a few hundredweight of coal*); and so with *dozen, score, hundred,* and occasionally with *fathom* and *pound.*

6. Abstract singulars used instead of concrete plurals, e.g. *accommodation* (= rooms or lodgings), *kindling* (= pieces of wood), *royalty* (= royal persons), *pottery* (= earthenware articles). For some abuses of this liberty see MEMBERSHIP.

7. Even nouns denoting substances of indefinite quantity such as *butter* and *water* are classed as collectives.

For the question whether collectives should be given singular or plural verbs see NUMBER 6.

**college.** Although some schools (notably Eton and Winchester) have an ancient right to be called *cc.,* the indiscriminate assumption of the name by schools that are no more colleges than others contented with the ordinary title is a sad degradation and obscuring of the word's meaning. To speak of sending a boy to college when what is meant is merely to school is an abuse of words that should be resisted, even though it is too late to ask the self-styled 'colleges' to consider whether it is for their real dignity to use *c.* in the same way as our grandfathers were rebuked by the OED for using *academy.* See WORKING AND STYLISH WORDS.

**collusion** etc. The notion of fraud or underhandedness is essential to collusion, and the following is a misuse: *The two authors, both professors at Innsbruck, appear to be working in c.* The supposed arrangement is merely that their periods shall not overlap; *in collaboration* will therefore not do; if *in concert* will not, the thing must be given at length.

**colon.** See STOPS.

**colossal** in the sense not of enormous (as in *c. folly* etc.), but of indescribably entertaining or delightful, is a Germanism not deserving adoption even by advertisers tiring of FABULOUS. The similar use of IMMENSE, though we do not name it *honoris causâ,* and its freshness has faded, is at least of native development.

**colour** makes *colourable, colourist,* but *coloration, decolorize* are the more favoured spellings for those words. See -OUR- AND -OR-.

**colourful.** The figurative use of this word was originally described by the OED as rare, but this comment was withdrawn in the 1933 Supp., which gives numerous examples of its use in the sense of *gay, vivid, striking, picturesque, bright, vivacious,* etc. *It is a colourful life to say the least. | So movingly and colourfully does he tell his tale.| Hampshire is to me a bundle of memories,* all colourful. The word has earned a rest, especially from the critics.

**columnist,** meaning a journalist who writes chattily about people and events, is an Americanism naturalized in England in the second quarter of the 20th c.

**combat.** Pronunciation is still hesitating between *kum-* and *kom-*. See PRONUNCIATION 5. The COD puts *kŭm* first. Part. and p.p. *-ating, -ated*; see -T-, -TT-.

**come** in such a phrase as *He will be twenty-one come Sunday* is a subjunctive ('Let Sunday come') as it is in *Come one come all*. This is called by the OED arch. and dial., and when used today in serious prose it has a WARDOUR STREET flavour. *The festivalgoer can, come the autumn, compare his notes with Burney's.*

**come-at-able, get-at-able.** Write with the hyphens. *C.* was made as long ago as the 17th c., but, except in *g.* (1799), the experiment has not been successfully repeated, and probably will not be.

**comedian, tragedian,** have, in the sense *actor*, the feminines *comedienne, tragedienne,* best pronounced *komē̆'-diē̆'n, trajē̆'diē̆'n,* and written without accents; see FRENCH WORDS 2. The words are properly applicable also to the writers of comedies and tragedies, and their usurpation by the actors leaves us without distinctive words for the former; we have to be content with *playwright* or *dramatist.* But the introduction of *comedist* and *tragedist* for the writers is a remedy worse than the disease; we cannot begin now to talk of *the Greek comedists and tragedists,* for instance.

**comedy, farce, extravaganza, burlesque.** As species of drama, the four are distinguished in that *comedy* aims at entertaining by the fidelity with which it presents life as we all know it, *farce* at raising laughter by the outrageous absurdity of the situations or characters exhibited, *extravaganza* at

diverting by its fantastic nature, and BURLESQUE at tickling the fancy of the audience by caricaturing plays or actors with whose style it is familiar. In U.S. the last 'has a special meaning—a theatrical entertainment featuring coarse comedy and dancing. Of late years striptease . . . has become so indispensable a part of all burlesque shows that *striptease* and *burlesque* are now almost synonymous' (Evans).

**comic(al).** The broad distinction, sometimes obscured by being neglected, is that that is *comic* of which the aim or origin is comedy, and that is *comical* of which the effect, whether intended or not, is comedy. A *comic actor* is merely one who acts comedy; a *comical actor,* one who makes the audience laugh. *Comic hesitation* is that in which the hesitator is playing the comedian; *comical hesitation,* that in which observers find comedy, whether the hesitator meant them to or was unconscious of them. Accordingly, *comic* is the normal epithet (though *comical* may be used, in a different sense) with *actor, opera, scene, relief, song, singer, paper; comical* is normal (subject to the converse reservation) with *face, effect, expression, deformity, earnestness, attempt, terror, hesitation, fiasco.* There is some tendency (*the attempt was comic in the extreme; The disaster had its comic side*) to use *comic* where *comical* is the right word. This may possibly be a sign that *comical* is on the way to become archaic and obsolete, a process likely to be helped by the colloquial use of *comic* as a noun. This would be regrettable; for the difference of meaning is fairly definite and of real use. But some of the publications called *Comics* are neither comic nor comical.

**comity,** from Latin *cōmis* courteous (though pronounced *kŏm-*), means courtesy, and *the c. of nations* is the obligation recognized by civilized nations to respect each other's laws and usages as far as their separate interests allow. It has nothing to do with Latin *cŏmes* companion, and phrases based on this false derivation (*obtain admittance to the c. of states*; *entered into the c. of nations*; *a useful member of the civilized c.*), and implying the sense *company, association, league, federation,* etc., are wrong.

**comma.** See STOPS.

**commando.** Pl. *-os*; see -O(E)S 6. The word was originally applied to irregular military units used by the Boers against the Africans, and afterwards against the British in the South African war. In the second world war it was revived in England as a name for specially trained parties of troops sent to raid enemy-occupied territory.

**commence(ment).** The writers who prefer *ere* and *save* to *before* and *except* may be expected to prefer *c.* to *begin-(ning)* in all contexts. *Begin* or *start* is the word always thought and usually said, but it is translated sometimes before it is said, and often before it is written, into *c.*, which is described by the OED as 'precisely equivalent to the native *begin*'. It is a good rule never to do this translation except when the simpler word is felt to be definitely incongruous; see FORMAL WORDS. In official announcements *c.* is appropriate; the playbill tells us when the performance will *c.*, though we ask each other when it *begins* or *starts*. It is suitably used in Acts of Parliament and *Commencement* is a technical term at Cambridge and some other universities for the ceremonial conferring of degrees; in U.S. it is also used for what we call *Speech-day*. The grave historical style also justifies *c.*, and historians' phrases, such as *c. hostilities,* keep their form when transferred to other uses, though we *begin,* and do not *c.,* a quarrel; similarly we *c. operations,* but merely *begin* or *start dinner.* As against the precise equivalence mentioned above, it should be observed that *begin* has, owing to its greater commonness, more nearly passed into a mere auxiliary than *c.*; and from this it results (1) that *begin,* not *c.,* is even in formal style the right word before an infinitive; in *The landholders commenced to plunder indis-*

*criminately,* anyone can perceive that *began* would be better; (2) that *c.* retains more than *begin* the positive sense of initiative or intention, and is especially out of place with an infinitive when this sense is absent, as in *Even the warmest supporters of the Chancellor of the Exchequer must be* commencing to feel *that he should give some slight consideration to . . .* The use of *commence* with a predicative noun (e.g. *to commence author*) is now an ARCHAISM.

**commendable** is one of the few four-syllable *-able* adjectives that have resisted the RECESSIVE ACCENT. Pronounce *commen'dable.*

**commentator** is not, as might be supposed, a new word coined for the needs of broadcasting. It is an old one (= writer of a commentary) revived for a new purpose.

**commercialese.** Some words and phrases characteristic of the way of writing called pejoratively *commercialese* are as follows: *Ult.* (last month), *inst.* (this month), *prox.* (next month), *even date* (today), *favour* or *esteemed favour* (a customer's letter), *to hand* or *duly to hand* (received), SAME (it), *your goodself* (you), BEG (a meaningless prefix used before verbs of all kinds), PER (by), *advise* (tell), *as per* (in accordance with), RE (about), *be in receipt of* (receive), *enclosed please find* (I enclose), *and oblige* (please), *assuring you of our best attention at all times* (a formal ending). For instance a letter written in typical commercialese might begin *Your esteemed favour of even date to hand and we beg to thank your goodself for same.*

It can no doubt be pleaded in extenuation of commercialese that much of it originated in a wish to treat the customer with almost obsequious respect. But it has become an artificial jargon, and few would now disagree with the verdict of the Departmental Committee on the Teaching of English in England:

We have no hesitation in reporting that *Commercial English* is not only objectionable to all those who have the purity of the language at heart but also contrary to the true interests of commercial life, sapping its vitality and encouraging the use of dry, meaningless formulae just where vigorous and arresting English is the chief requisite. Further, its sweeping condemnation by the leading business firms of the country demonstrates that, whatever its origin may have been, *Commercial English* now continues to retain its hold upon commercial schools and colleges solely through the influence of an evil tradition and of the makers of text-books to whom such a tradition is of commercial value. The large business houses however are giving a lead which must in time have its effect upon the commercial community generally, and it is our confident hope that *Commercial English* will presently become one of the curiosities of dead and forgotten speech.

That was written in 1921. The internal revolt against commercialese has undoubtedly had some effect, especially in the larger firms, but by no means as yet to the extent that the Committee hoped.

**commiserate.** *The late Emperor Francis Joseph, who commiserated* with *the imperial bird for that it had but a single head.* The orthodox use of *c.* is transitive, and the OED gives no quotation showing *with.* But the ANALOGY of *sympathize* and *condole* has since got the better of idiom, and *commiserate with* cannot be denied recognition, as the COD admits. See CAST-IRON IDIOM.

**commitment, committal.** In nearly all senses the two forms are interchangeable, but *-tal* gains ground while *-ment* loses it. The sense 'being committed to doing something', however, belongs almost only to *-ment,* the sense 'perpetration of an offence' almost only to *-tal,* and the sense 'sending to prison or for trial at a superior court' only to *-tal.*

**committed** in the sense of *biased* or *prejudiced* is a word greatly loved by literary critics. *Such loose thinking does not mean that Christian historians have not got a good deal to say which their less committed colleagues will ignore at their peril.* Sometimes authors are goaded into flinging back a word that one of them has called 'that pretentious current favourite term of the self-appointed proprietors of politics,

philosophy, and culture'. *Your readers should not be obliged to tolerate so-called reviews which read like the hysterical effusions of committed parties with bad consciences or with blind obsessions preventing them from seeing the real purpose of an author's publication.* | *Reserved in future for more delicate and non-controversial aspects of literary criticism, your reviewer may well feel glad to make room for less committed commentators.* In international relations the word has established itself (as an alternative to *neutral* or *non-aligned*) in the expression *non-committed countries.*

**committee,** in the original sense of person to whom something is committed (esp. now the care of a lunatic), is pronounced *kŏmĭtē'.*

**commonplace, platitude, triviality, truism.** All these words are often used as terms of reproach in describing the statements made by a speaker or writer; but none of them is identical in sense with any other, and if they are not to be misused a rough idea at least of the distinctions is necessary. It is something to remember that no one should welcome *platitude, triviality,* or *truism* in the strict sense, as a description of a statement of his own, whereas it may be a merit in a statement to be a *commonplace* or a *truism* in its loose sense.

A *commonplace* is a thing that, whether true or false, is so regularly said on certain occasions that the repeater of it can expect no credit for originality; but the commonplace may be useful. It was formerly used in the sense of a notable saying, without any implication of triteness. Hence the *Commonplace Book* for recording such sayings.

A *platitude* is a thing the stating of which as though it were enlightening or weighty convicts the speaker of dullness; a platitude is never valuable. The word is misused in: *It is a p. that the lack of cottages is one of the chief of the motive forces which drive the peasantry to the towns.* In U.S. platitudinous remarks are aptly termed *bromides.*

A *triviality* is a thing the saying of which as though it were adequate to the occasion convicts the speaker of silliness; a triviality is never to the purpose.

A *truism* in the strict sense (to which it might be well, though perhaps now impossible, to confine it) is a statement in which the predicate gives no information about the subject that is not implicit in the definition of the subject itself. *What is right ought to be done*; since the *right* is definable as *that which ought to be done*, this means *What ought to be done ought to be done*, i.e. it is a disguised identical proposition, or a truism. *It is not well to act with too great haste*; *too great haste* being haste greater than it is well to act with, the sentence tells us no more, though it pretends to, than anyone who can define *too great haste* knew before the predicate *is not well* was added. But *What is right pays*, or in other words *Honesty is the best policy*, is not a truism either in the strict sense (since it makes a real statement and not a sham one) or in the loose sense (since its truth is disputable); nor is *It is not well to act in haste* a truism of either kind. Both statements, however, are commonplaces, and often platitudes.

A *truism* in the loose sense is a thing that, whether in point or not, is so indisputably true that the speaker is under no obligation to prove it, and need not fear contradiction. This sense is a SLIPSHOD EXTENSION; the writer who describes his principle as a *t.* in order to justify his drawing conclusions from it would do better to call it an *axiom*; and the critic who depreciates some one else's statements as *tt.*, not in the strict sense, but meaning merely that they are too familiar to be of value, should call them *platitudes* or *commonplaces*.

**common sense.** There is no reason why this should not be written as one word, but it rarely is. When written as two it should not be hyphened except when it is used attributively. *The philosophy of common sense*; *The common-sense philosophy*. See HYPHENS.

**communal.** The OED gives *komū′nal* preference over *kŏ′mŭnal*, but RECESSIVE ACCENT has prevailed and the latter is now recognized as the established pronunciation; the COD admits no other. See FALSE QUANTITY (on *doctrinal*).

**commune.** The noun is pronounced *kŏ′mŭn*. In the verb *komū′n* is now usual (see NOUN AND VERB ACCENT), but *kŏ′mŭn* is still heard.

**commuter.** This useful Americanism is now thoroughly naturalized. It meant originally the holder of a rail season (U.S. commutation) ticket travelling daily between his home in the country and his work in town. Its meaning has been widened to the extent of including those who use other means of transport for the same purpose, but not yet far enough to make it an apposite choice, even jocular, in *There is every reason to think that the Foreign Office would function more efficiently if the activities of Britain's diplomatic commuters were curbed.*

**comparative(ly).** Timid writers who shrink from positive statements have a bad habit of using *comparatively* and *relatively* to water down their adjectives and adverbs, forgetting that those words can properly be used only when some comparison is expressed or implied. *There were many casualties but comparatively few were fatal* is a proper use of *comparatively*, meaning as it does that the proportion of deaths to total casualties was low. But to say *casualties were comparatively few*, meaning that there were not very many of them is a misuse of the word. Politicians and officials are specially given to using these words (and UNDULY) as a measure of protection instinctively taken against the notorious danger of precision in politics. See also FEW.

**comparatives.** For misuses, see -ER AND -EST, MORE, and THAN.

**compare.** 1. In the sense suggest or state a similarity is regularly followed by *to*, not *with*; in the sense examine or set forth the details of a supposed

similarity or estimate its degree, it is
regularly followed by *with,* not *to. He
compared me to Demosthenes* means that
he suggested that I was comparable to
him or put me in the same class; *He
compared me with Demosthenes* means
that he instituted a detailed comparison or pointed out where and how far
I resembled or failed to resemble him.
Accordingly, the preposition in each
of the following is the one required by
idiom: *Witness compared the noise to
thunder; The lecturer compared the
British field-gun with the French; Compared with,* or *to, him I am a bungler*
(this is a common type in which either
sense is applicable).
After the intransitive verb (*a boiled
mullet cannot c. with a baked one*), and
after *in comparison, with* alone is possible.
  2. With *compare,* as with LIKE (see
that article s.f.), it is easy to slip
into the mistake of comparing two
things that are not comparable:
*There is no one to be compared with his
influence* (read *him in influence*). | *Dry-
den's prose, which is meant to be popular,
loses nothing of its value by being compared with his contemporaries* (read *with
that of his contemporaries*).

**compass.** For synonymy see FIELD.

**compendium.** Pl. *-ums, -a;* see -UM.

**compensate. 1.** *To compensate* is
defined by Skeat as 'to requite suitably'. The implication is that what is
paid by way of compensation is given
ex gratia, and not in discharge of any
legal obligation. Perhaps it is a relic
of feudalism that what is paid by the
Crown as the purchase price of property compulsorily acquired from its
subjects is still called *compensation.*
  2. *He is a very shy fellow with women,
endlessly compensating for all he is
worth: at worst insanely narcissistic,
intensely feminine.* This intransitive use
of *c.* is a POPULARIZED TECHNICALITY
from psychology.
  3. *Compensate* was formerly *kompe'-
nsāt* but is now *ko'mpensāt;* see RECES-
SIVE ACCENT. For *compensatory* the
COD still accents the second syllable,

but *compensā'tory* is a strong rival
and may prevail.

**competence, -cy.** Neither has any
sense in which the other cannot be
used; but the practice seems to be
growing, desirable for the sake of
differentiation, of confining *-cy* to
the sense of modest means and preferring *-ce* for the senses *ability* and legal
capacity; see -CE, -CY.

**complacence, -cy.** There is no distinction that can be called established;
the second form is now much commoner, and is less liable to confusion
with *complaisance* (see foll.); *complacence* might be dropped as a NEEDLESS
VARIANT; see -CE, -CY.

**complacent, -ency, complaisant,
-ance. 1.** The two sets have clearly
differentiated meanings, but are often
confused; it would help to obviate this
confusion if the more easily distinguished pronunciation of the second
set (*kŏmplīză'nt, -ă'ns,* not *komplā'znt,
-ā'zns*) were made invariable, and
if *complacency* were always preferred
to *-acence* (see prec.).
  2. He is *complacent* who is pleased
with himself or his state, or with other
persons or things as they affect him;
the word is loosely synonymous with
*contented.* He is *complaisant* who is
anxious to please by compliance, service, indulgence, or flattery; the word
is loosely synonymous with *obliging.*
The wrong choice has been made in
each of these sentences: *He owed such
funds as he possessed to French compla-
cency.* | *He has nothing more to expect
from the complacency of the authorities.*|
*The display of the diamonds usually
stopped the tears, and she would remain
in a* complaisant *state until* . . .
  Note, 1924. I wrote the above in
1913, fortified by the OED descriptions (dated 1893) of *complacence,
-acency,* and *-acent,* in the senses
proper to *complaisance, -aisant,* as respectively '*Obs.*', '? *Obs.*', and '? *Obs.*'
It is a curious illustration of the changing fashions in words that I have since
collected a dozen newspaper examples

of *complac-* words wrongly used for *complais-*, and none of the contrary mistake. It looks as if some journalists had forgotten the existence of *complais-* and the proper meaning of *complac-*.

Note, 1957. Perhaps because there are fewer things justifying complacency today than there used to be, *complacent* has intensified its pejorative colour and is now generally used as the suitable adjective for those who are given to WISHFUL thinking. *Complaisant* is no longer intruding on *complacent's* territory, but is tending to disappear, leaving its work to be done by *obliging*.

**complement. 1.** As a term of grammar *c.* means that which completes, or helps to complete, the verb, making with it the predicate. This (A) is the widest sense of the word, not excluding e.g. the direct object of a transitive verb, or adverbs. It is possibly the most reasonable application of the term; it is also the least useful, and the least used. (B) Often the direct object is excluded, but all other modifications or appendages of the verb are called complements; a sense found convenient in schemes of sentence analysis, but too wide to be precise and too narrow to be logical. (C) A further restriction admits only such words or phrases as are so essential to the verb that they form one notion with it and its meaning would be incomplete without them; thus in *He put his affairs in order* the verb *put* is essentially incomplete without its complement *in order*, whereas in *He replaced the volumes in order* a new detail merely is added by the adverb *in order* to the complete verb *replaced*; some verbs are in their nature incomplete, e.g. the auxiliaries, and, in *must go*, *go* is the complement of *must*. A serviceable use, especially if it were established as the only one. (D) Lastly, in the narrowest sense, *c.* is applied only to the noun or adjective predicated by means of a copulative verb (*be, become*, etc.) or a factitive verb (*make, call, think*, etc.) of the subject *He is a fool*; *He grew wiser*; *He was*

made *king*) or of the object (Call no man *happy*); in such examples as the last, the complement is called an *objective* or an *oblique c.* A sense frequent in Latin grammars.

**2.** In the verb *-ent* is clearly sounded if not given the main accent; in the noun it is neither accented nor clearly sounded; see NOUN AND VERB ACCENT.

**complete,** vb, in one sense is a FORMAL WORD for *finish*; both are now threatened by the dubious newcomer FINALIZE. In another sense (to make complete by supplying what was missing) it is a normal working word.

**complex** (e.g. inferiority c., Electra c., Oedipus c., etc.) is a term of psychoanalysis meaning a group of repressed emotional ideas responsible for an abnormal mental condition. It is a familiar example of a POPULARIZED TECHNICALITY turning into a VOGUE WORD. OED Supp. quotes from *Athenaeum 1919* 'A Complex is now a polite euphemism for a bee in one's bonnet'. It is popularly used in contradictory senses: sometimes, like ALLERGY, to indicate repulsion (*Muriel's losing her sex complex . . . she is getting tangled up with some man*), sometimes attraction (*A fond aunt with a commiseration complex*). In its best-known association, *inferiority c.*, its meaning is almost universally misunderstood: to the popular mind the natural manifestation of an *i. c.* is diffidence; to the psychoanalyst it is aggressiveness.

**compliment.** The pronunciation varies as with COMPLEMENT.

**complin(e).** 'The final *e* is modern and unhistorical'—OED. But it seems to have established itself. Pronounce *kom'plĭn*.

**compose.** See COMPRISE.

**composure.** *The composure of the language of the characters makes us feel that the speaker took the whole sense for granted.* This use of *composure* in the sense of *composition* is an ARCHAISM. The only extant meaning of *composure* is calmness of demeanour.

**compound prepositions, conjunctions,** etc. A selection of these is: *as to* (AS 3);| INASMUCH AS; *in* CONNEXION *with*; IN ORDER THAT or *to*; *in relation to*; IN SO FAR *as*, *that*; IN THAT; *in the* CASE *of*; *in the* INSTANCE *of*; *in the matter of*; *in the* NEIGHBOURHOOD *of*; *in the region of*; *of the character of*; *of the* NATURE *of*; *of the order of*; *on the* BASIS *of*; *owing to* (*in spite of*) *the* FACT *that*; *so far as* . . . *is* CONCERNED *relative to*; *with a* VIEW *to*; *with reference to*; *in* or *with* REGARD *to*; *with* RESPECT *to*. And one or two specimens of their sorry work are: *At least 500,000 houses are required, and the aggregate cost is in the region of £400,000,000.* | *Sir Robert Peel used to tell an amusing story of one of these banquets,* in the case of *which he and Canning were seated on opposite sides of Alderman Flower.* | *If I have a complaint to proffer against Mr. Bedford, it certainly is, except perhaps in the case of 'Monna Vanna', not in the matter of the plays to which he has refused a licence, but in regard to a few of the plays which he sanctioned.* | *We have arranged a competition in regard to imitating Pepys's style.*

But so much has been said on the subject, and so many illustrations given, elsewhere (see PERIPHRASIS, and the words in small capitals in the list above) that nothing but a very short general statement need be made here. Of such phrases some are much worse in their effects upon English style than others, *in order that* being perhaps at one end of the scale, and *in the case of* or *as to* at the other; but, taken as a whole, they are almost the worst element in modern English, stuffing up what is written with a compost of nouny abstractions. To young writers the discovery of these forms of speech, which are used very little in talk and very much in print, brings an expansive sense of increased power; they think they have acquired with far less trouble than they expected the trick of dressing up what they may have to say in the right costume for public exhibition. Later they know better, and realize that it is feebleness instead of power

that they have been developing; but by that time the fatal ease that the compound-preposition style gives (to the writer, that is) has become too dear to be sacrificed.

**comprehend.** See APPREHEND.

**comprise.** *Many of the airs used by Handel in his operas have been arranged by Sir Thomas Beecham to c. a ballet suite.* | *The Government of the Federation and the three territories which c. it.* | *The four submarines comprising the nuclear deterrent.* This lamentably common use of *comprise* as a synonym of *compose* or *constitute* is a wanton and indefensible weakening of our vocabulary. For the distinction between *comprise* and *include* see INCLUDE.

**comptroller, cont-.** The first spelling is not merely archaic, but erroneous, being due to false association with *count*(F *conter* f. L *computare*). Government officials bearing that title are now ordinarily *controllers*, but statute or antiquity has established *comptroller* for some of them, e.g. *C. and Auditor-General, Deputy Master and C. of the Mint, C. of the Royal Household*, and *C.-General of the Patent Office*.

**comrade.** For the pronunciation (*kŏ-* or *kŭ-*) see PRONUNCIATION 5.

**concept** is a philosophical term. The philosophers have not been allowed to keep it to themselves, and its extended use is sometimes legitimate; but the substitution of it for the ordinary word *conception* in such a context as the following is due to NOVELTY-HUNTING: [a caricature has been described] *Now this point of view constantly expressed must have had its influence on popular concepts.* See POPULARIZED TECHNICALITIES. The fate of such words is often to be put to menial work, and *concept* is no exception, as in the advertisement *A new c. in make-up, blessing your skin with its incredible beauty benefits.*

**concern. 1.** In (*so far*) *as concerns* or *regards*, the number of the verb (which is impersonal, or has for its unexpressed subject 'our inquiry' or some such phrase) is invariable; the change to plural, as in the quotation that follows, is due, like *as* FOLLOW, to misapprehension: *Many of these stalks were failures, so far as* concern *the objective success.*

2. The idiomatic use of *so far as . . . concerned*, e.g. *so far as I am concerned*, meaning *so far as I have any say in the matter*, or *for all I care* (*That's all right s. f. a. I a. c.*; *You may go to the devil s. f. a. I a. c.*) is often improperly, and even absurdly, extended to serve as one of the least excusable types of COMPOUND PREPOSITIONS. *The punishment does not seem to have any effect so far as the prisoners are concerned* (on the prisoners). / *The months of January, February, and part of March 1963 were disastrous as far as the building industry of this country was concerned* (for the building industry). / *The girl is entirely unknown as far as the larger cinema audiences are concerned* (to the larger audiences).

3. For the unidiomatic omission of the verb in such phrases as *so far as concerns* see FAR 4.

**concernment** has no senses that are not as well, and now more naturally and frequently, expressed by the noun *concern*; the substitution of the latter was censured as affectation in the 17th c., but the boot is now on the other leg, and *c.* should be dropped as a NEEDLESS VARIANT.

**concession** is a word widely used in British domestic politics for the practical recognition by Authority that a citizen, or a body of citizens, with a grievance has an unanswerable case, e.g. for the restoration of rights or possessions expropriated during an 'emergency' or, like the 'extra-statutory cc.' granted by the Board of Inland Revenue, for waiving a liability that causes indefensible hardship in particular circumstances. It is an unfortunate word for this purpose; a flavour of royal condescension is still latent in it, suggesting that what to one party may seem an act of bare justice

is regarded by the other as one of grace and favour, and so widening the gulf that is supposed to exist between 'Us' and 'Them'. But the word is too convenient to permit of any hope of its disuse.

**conciseness, concision.** The first is the English word familiar to the ordinary man; *concision* is the LITERARY CRITICS' WORD, more recent in English, used by writers under French influence, and often requiring the reader to stop and think whether he knows its meaning: *The writing of verse exacts concision, clear outline, a damming of the waters at the well-spring.* See -ION AND -NESS.

**concomitance, -cy.** The second is now a NEEDLESS VARIANT; see -CE, -CY.

**concur.** For the pronunciation of inflected forms see PRONUNCIATION 7.

**condign** meant originally *deserved*, and could be used in many contexts, with *praise* for instance as well as with *punishment*. It is now used only with words equivalent to *punishment*, and means deservedly severe, the severity being the important point, and the desert merely a condition of the appropriateness of the word; that it is an indispensable condition, however, is shown by the absurd effect of: *Count Zeppelin's marvellous voyage through the air has ended in c. disaster.*

**condition.** The use of *c.* as a transitive verb is condemned by some purists as 'not English', but in fact the word has been so used for more than 400 years, and for more than 300 in the sense most common today—to govern as a condition. The OED quotes M. Arnold *Limits we did not set condition all we do.* Its use in the sense of to put into good condition, or make responsive to suggestion, especially of horses and dogs, prisoners of war and deviationists, is comparatively modern, and is being widened both by the invention of air-conditioning and by advertisers' adopting it as a word of appeal, *These cigarettes are CONDITIONED!* The similar use of *recondition* is also recent (the earliest example in OED Supp. is dated 1920) but is now established, especially of buildings, ships, and motor cars and other machinery.

**conditional clauses.** See SUBJUNCTIVE.

**conduce.** *A sore throat did not c. him to make a major effort.* This should be either *induce him to make* or *c. to his making*; *conduce* as a transitive verb is obsolete.

**conductress.** See FEMININE DESIGNATIONS.

**conduit.** Pronounce *kŭ'ndĭt*. See PRONUNCIATION 5.

**confection.** The French dressmaking term properly means no more than a piece of attire not made to measure; but, being applied chiefly to fashionable wraps etc., it is sometimes misunderstood as expressing in itself (like *creation*) the speaker's exclamatory admiration.

**confederacy, -eration.** See FEDERATION.

**confer(r)able.** Of the verbs in *-fer* accented on the last syllable, two form adjectives in *-ble* of which the spelling and accent are fixed (*pre'ferable* and *tra'nsferable*). The others, for which various forms have been tried (*confer*, *confe'rrable*; *defer*, none; *infer*, *i'nferable* and *infe'rible* and *infe'rrable* and *infe'rrible*; *refer*, *re'ferable* and *refe'rrable* and *refe'rrible*), should be made to follow these two; *inferable* and *referable* are doing so, but the dictionaries would still have us say *conferr'able*.

**confidant, -ante, -ent.** *Co'nfident* was in use as a noun meaning confidential friend or person to whom one entrusted secrets long before the other forms were introduced; but it is now an ARCHAISM, and to revive it is pedantry. *Confidant* is masculine and *confidante* feminine; they are indistinguishable in pronunciation, and accent the last syllable.

**conform(able).** *He is as anxious as anyone to conform with the laws and*

*spirit of the game.* Idiom demands *conform to*, not *with*. For such formations as *conformable* see -ABLE 4.

**congeries.** The pronunciation favoured by the dictionaries is *-jĕ'rĭēs*, but see -IES, -EIN.

**conjunction.** A word whose function is to join like things together, i.e. a noun or its equivalent with another noun or its equivalent, an adjective etc. with another, adverb etc. with adverb etc., verb with verb, or sentence with sentence. The relation between the things joined is shown by the particular conjunction chosen (*but*, *and*, or *nor*; *if*, *although*, or *because*; *that* or *lest*; *since* or *until*). Some conjunctions, in joining two sentences, convert one into a dependency of the other, or of a clause in it, and are called *subordinating* or *strong cc.*, the others being *co-ordinating* or *weak* (strong— I hate him *because* he is a Judas: weak —I hate him; *for* he is a Judas). Many words are sometimes conjunctions and sometimes adverbs (*therefore*, *so*, *however*, *since*, etc.); and such words as *when* and *where*, though often in effect cc., are more strictly described as relative adverbs with expressed or implied antecedent (*I remember the time when*, i.e. at which, *it happened*; *I will do it when*, i.e. at the time at which, *I see fit*).

**conjunctive (mood)** is a term that had much better be dropped. The forms denoted by *c.* and *subjunctive* are the same, and *subjunctive* is the much better known name. *C.* might have been useful in distinguishing uses if it had been consistently applied; but it means sometimes the forms however used (*subjunctive* then being a division under it restricted to the subordinate uses), sometimes the forms when used as main verbs (*subjunctive* then being a division parallel to it restricted as before), and sometimes merely the forms when used as main verbs of conditional sentences (*subjunctive* then being, very unreasonably, the name for all uses, dependent or independent, and *c.* a division under it). This is hopeless confusion; *c.* should be given up, *subjunctive* be used as the name of the forms whatever their use, and the differences of function be conveyed by other words (*dependent, conditional, optative,* etc.).

**conjure** in the sense *beseech* is pronounced *konjoor'*, in other senses *kŭ'njer*.

**conjuror, -er.** Although in the OED 19th-c. quotations *-or* is five times as common as *-er* for the juggler, the latter is put first by the SOED and other more recent dictionaries; the partner in an oath is always *-or* with the stress on the second syllable.

**connexion, -ction. 1.** The first is the etymological spelling; see -XION. But the second is now more common, and standard U.S.
**2.** *In c. with* is a formula that every one who prefers vigorous to flabby English will have as little to do with as he can; see COMPOUND PREPOSITIONS and PERIPHRASIS. It should be clearly admitted, however, that there is sometimes no objection to the words; this is when they are least of a formula and *c.* has a real meaning (*Buses run i. c. w. the trains*; *The isolated phrase may sound offensive, but taken i. c. w. its context it was not so*). In the prevalent modern use, however, it is worn down into a mere compound preposition, with vagueness and pliability as its only merits. The worst writers use it, from sheer love of verbiage, in preference to a single word that would be more appropriate (*The three outstanding features i. c. w.* [i.e. of] *our 'Batchworth Tinted', as sample set enclosed, are as follows*). The average writer is not so degraded as to choose it for its own sake, but he has not realized that when *i. c. w.* presents itself to him it is a sign that laziness is mastering his style, or haziness his ideas. Of the examples that follow, the first two are characteristic specimens of compound-prepositional periphrasis: *The special difficulty in Professor Minocelsi's case arose* i. c. w. *the view he holds relative to the historical value of . . .* (Prof. M. was specially hampered by

his views on). | *Regulations* with regard to *the provision of free places* i. c. w. *secondary education* (Regulations for providing free places in secondary schools). | *The general secretary expressed his disgust* i. c. w. *the award, and said his executive would have to consider what further steps should be taken* i. c. w. *pursuing their claim* (Disgust at the award: steps to pursue their claim).

**connoisseur.** Pronounce *kŏnĭser'*; the modern French spelling (*-nai-*) should not be used.

**connote, denote.** Both mean to signify, but with a difference. A word denotes its primary meaning—its barest adequate definition; it connotes the attributes commonly associated with it. For instance *father* denotes one that has begotten; it connotes male sex, prior existence, greater experience, affection, guidance, etc.: *ugly* denotes what is unpleasing to our sight; it connotes repellent effect, immunity from the dangers peculiar to beauty, disadvantage in the marriage market, etc. Connote is sometimes used (loosely says the COD) in the sense of denote, i.e. 'mean', but the differentiation is worth preserving.

**conscience.** See SAKE.

**consecutive.** A c. clause is a subordinate clause that expresses the consequence of the fact stated in the sentence on which it depends; and a c. conjunction, in English *that* corresponding to a preceding *so* or *such*, is the word joining such a clause to the sentence (He was so angry *that* he could not speak). Some grammarians, however, would call this example an *adverbial clause of degree*, and confine the term *consecutive* to such a sentence as *He was angry, so that we were glad to get away*.

**consensus** means unanimity, or unanimous body, of opinion or testimony. The following quotation, in which it is confused with *census*, is nonsense: *Who doubts that if a consensus were* *taken, in which the interrogated had the honesty to give a genuine reply, we should have an overwhelming majority?* The misspelling *concensus* is curiously common.

**consequential** is a word severely restricted in its application by modern idiom; it is unidiomatic in several of the senses that it might have or has formerly borne.

1. Where doubt can arise between it and *consequent*, the latter should always be used when the sense is the simple and common one of *resulting*, and *-ial* be reserved for that of *required for consistency with something else*. Thus *In the consequent confusion he vanished*, but *The consequential amendments were passed. Consequential confusion* is not English; *the consequent amendments* is, but means not (as with *-ial*) those necessitated by one previously accepted, but those that resulted from (e.g.) the opposition's hostility or the discovery of a flaw. The right use is seen in *A good many of these undiscussed changes were only consequential alterations* and the wrong one in *The anomalies that flow from present law, and the consequential state of confusion created in the public mind on particular occasions, are evidence that the Act is a bad one*. But the following sentence (in which *consequent* would be better, but either is possible) shows that the line is sometimes hard to draw: *Yet whilst he washes his hands of the methods of the Albert Hall, with its consequential campaign of resistance and its cry of 'no servant tax', he declares that the Bill must not be passed*.

2. *C.* does not mean of consequence; a c. person may or may not be important; all we know is that he is self-important; *Mr. C. bustled about, feeling himself the most c. man in the town* would not now be English.

3. *C.* does not now mean having great consequences. For *so desperate and so c. a war as this* there should be substituted *a war so desperate and so pregnant with consequences*.

4. *C. damage* is damage that is consequent on an act but not the direct and immediate result of it.

**conservative.** The use of this word as an epithet, in the sense of moderate, safe, or low, with *estimates, figure*, etc., originating in U.S., is now firmly established in Britain also—too firmly, indeed, for it promotes forgetfulness of those simpler words, which would generally serve as well or better.

**conservatoire, conservatory.** The French, German, and Italian musical institutions are best called by their native names—*conservatoire, conservatorium, conservatorio*. In England the use of *conservatory* in this sense has disappeared, and such an institution is called an *Academy*, a *College*, or a *School, of Music*.

**consider.** *Salvat considers Bruce Tulloh and Gordon Pirie, if he runs in this event, as his most dangerous rivals.* | *Though they suddenly feared the African, they felt a repulsion sharp as their own fear at the idea that they might some day be protected from him by Dr. Verwoerd, whom they c. as an evil force.* Idiom demands either *whom they c. an evil force* etc. or *whom they regard as an evil force* etc. Under REGARD 3 numerous examples are given of the use of that word without *as*, on the analogy of *consider* (e.g. *He regards the public interest taken in the current rivalry of Democratic candidates for the nomination responsible*). The quotations given above seem to show that *regard* is now beginning to have its revenge by corrupting the idiom of *consider*.

**considerable.** 1. *C.* in the sense *a good deal of* is applied in British use only to immaterial things (*I have given it c. attention*). The use with material things is an Americanism; the following are from definitions in two American dictionaries: *Silk fabric containing c. gold or silver thread.* | *Certain pharmaceutical preparations similar to cerates, but containing c. tallow*. British idiom requires *a considerable amount of*.

2. *C.* is a flabby adjective, a favourite resource of flabby thinkers who feel a need to give their nouns the prop of an adjective but have not the courage to use a more virile one. In such sen-

tences as *There are c. difficulties in this proposal | He is a man of c. ability* the nouns are more effective if allowed to stand on their own legs. See ADJECTIVES MISUSED.

**considerateness, consideration.** *Consideration,* so far as it is comparable with *-ateness,* means thought for others, while *-ateness* means the characteristic of taking such thought; see -ION AND -NESS. Sometimes therefore it does not matter which is used (*He showed the greatest -ateness* or *-ation*; *Thanks for your -ateness* or *-ation*). But more often one is preferable: *His -ateness is beyond all praise*; *I was treated with -ation*; *He was struck by the -ateness of the offer.*

The official cliché *under consideration,* now, it seems, so well worn as to need reinforcement by an adjective such as *active, earnest, serious,* is apt, like all clichés, to make its users forget the simple way of saying what needs to be said. *It has not yet been decided when the meeting will take place, but dates under very serious consideration are . . .* Why not just *The most likely dates are . . .?*

**consist.** *C. of* introduces a material, and *c. in* a definition or statement of identity; we must not say *the moon consists in green cheese* (no one would), nor *virtue consists of being good* (many do). ELEGANT VARIATION between the two is absurd: *The external world consisted, according to Berkeley,* in *ideas; according to Mr. Mill it consists* of *sensations and permanent possibilities of sensation. Of* is wrong in *The most exceptional feature of Dr. Ward's book undoubtedly consists of the reproduction of photographs.*

**consistence, -cy.** See -CE, -CY. The *-cy* form is now invariable in the noun that means being consistent, i.e. not inconsistent (*-cy is an overrated virtue*). In the noun meaning degree of thickness in liquids usage varies; *A -ce something like that of treacle,* and *Mud varying in -cy and temperature,* are both from T. H. Huxley; it would be well if *-ce* could be made the only form in this

sense, as *-cy* in the other. It is sometimes doubtful now whether freedom from inconsistency is meant or metaphorical solidity; among the OED quotations are: *Reports begin to acquire strength and -ce*; *A vague rumour daily acquiring -cy and strength.* The removal of such doubt would be one of the advantages of the limitation proposed above for *-ce.* But the present trend seems to be towards using *-cy* for all purposes.

**consistory.** The difficulty of pronouncing this word with the accent on the first syllable has checked the influence of RECESSIVE ACCENT and *consis'tory* is probably commoner. Cf. GLADIOLUS and LABORATORY.

**console,** bracket, organ unit, etc. Accent the first syllable (*kŏ'nsōl*).

**consols** may be pronounced with the stress on either the first or second syllable. Most dictionaries prefer the second; the COD gives no other.

**conspectus.** Pl. *-uses*; see -US.

**conspicuous.** *C. by absence* is a CLICHÉ so overworked as a tinsel way of saying *absent* (*Even in the examination for the M.D., literary quality and finish is often c. by its absence*) that it would be a bold writer who now used it even in a context as apt as that for which it was coined by Tacitus when he said about a funeral at which portraits of distinguished relatives were displayed *Praefulgebant Cassius atque Brutus eo ipso quod effigies eorum non videbantur.*

**constable.** Pronounce *kŭn-.* See PRONUNCIATION 5.

**constitution(al)ist.** See -IST.

**construct, construe, translate,** with reference to language. To *translate* is to reproduce the meaning of a passage in another language, or sometimes in another and usually a more intelligible style of the same language. To *construe* is to exhibit the grammatical structure of a passage, either by translating closely or by analysis, and so it is often tantamount to *translate* or *interpret.* A sense of *construe* formerly common is that in which *construct* is taking its place (Prohibit *should not be constructed*, or *construed, with an infinitive*). The older pronunciation of *construe* (for which *conster* was long the prevalent form) is *kŏ'nstroō*; the *kŏnstroō'* now often heard is no doubt due to the NOUN-AND-VERB-ACCENT tendency.

**constructive. 1.** In legal and quasi-legal use is applied to an act that, while it does not answer to the legal definition of what constitutes that act (*c. trust, notice, malice*, etc.), is seen when the true construction is put upon its motive or tendency to be equivalent to such an act.
  **2.** A use of *c.* that has become very popular is as an antonym of *destructive*. It is specially associated with criticism: authors and politicians whose works and deeds meet with disapproval protest indignantly that they would have welcomed *c.* criticism, meaning perhaps praise. The exhortation to be *c.* is becoming a parrot-cry; and the word is often used as a cliché that contributes nothing to the sense: *The Foreign Office proposes to put forward a c. suggestion for dealing with the situation in Laos.*

**consuetude.** Pronounce *kon'swĭtūd*.

**consumedly.** See TERRIBLY.

**consummate. 1.** Pronounce the adj. *kŏnsŭ'măt*, the verb *kŏ'nsumăt*; see PARTICIPLES 5A. **2.** See -ATABLE.

**consummation.** For *a c. devoutly to be wished* see HACKNEYED PHRASES.

**consumption, consumptive.** That popular usage should have dropped these words in favour of *Tuberculo(sis)(us)* and then taken to the abbreviation *T.B.* is no doubt an example of the workings of EUPHEMISM (cf. *V.D.*). See also PHTHISIS.

**contact.** The use of *c.* as a verb (*get into touch with*) gave no little offence when it first appeared here from America. But convenience has prevailed over prejudice, and the diction-aries now give it full recognition: after all, it is an ancient and valuable right of the English people to turn their nouns into verbs when they are so minded. Now that the word has settled down as a verb, we shall no doubt get into the habit of accenting the second syllable. See NOUN AND VERB ACCENT.

**contemporary.** *The Progressive Party is the only political party in South Africa which is contemporary; all the others, including the Liberal Party, are genuine antiques.* If we are to believe the OED, which tells us that the adjective *c.* means 'living or existing or occurring together in time', this sentence is manifest nonsense. The fact is that *c.* is being more and more used in the sense not of contemporary with some other specified person or event but with the user of the word—a needless synonym for present day. *It is not difficult to see reflected in their choice either their own sentimental attachment to the past or ruthless preference for the c.* | '*Strip the Willow*' *has scraps and glimmers of c. meaning;* '*The Duchess of Malfi*' *has not.* From this it is a small step to give it the meaning of modern, up-to-date, abreast of the times, almost as good indeed as that much-favoured adjective AVANT-GARDE. If this misuse goes on, *c.* will lose its proper meaning altogether, and no one who reads, say, that *Twelfth Night* is to be produced with contemporary incidental music will think the sentence capable of any other meaning than that the music will be by a living composer.

**contemptible, contemptuous.** *Mr. Sherman, speaking in the Senate, called the President a demagogue who contemptibly disregarded the Government, because President Wilson, speaking at Columbia yesterday, said an International Labour Conference would be held at Washington, whether the Treaty was ratified or not.* Mr. Sherman probably meant, and not improbably said, *contemptuously*. What is done contemptuously may be regarded as contemptible by the person to whom it is done: the right word may depend on the point of view. Either might have been used in *The judges described the increase of their salaries provided for in the Bill before Parliament as contemptuous*; i.e. to the judges the offer seemed so contemptible that Parliament must have been contemptuous in making it. See PAIRS AND SNARES.

**content**, v. *C.* one*self with* (not *by*) is the right form of the phrase that means not go beyond some course; the following are wrong: *We must c. ourselves for the moment* by *observing that from the juridical stand-point the question is a doubtful one.* | *The petition contents itself* by *begging that the isolation laws may be carried out.*

**content(ment).** The two forms now mean practically the same, *contentment* having almost lost its verbal use (*The contentment of his wishes left him unhappy*) and meaning, like *content*, contented state. *Contentment* is the usual word, *content* surviving chiefly in *to heart's content* and as a poetic or rhetorical variant.

**content(s)**, what is contained. The OED says 'The stress *conte'nt* is historical, and still common among the educated'. But the stress *co'ntent* due, no doubt, to the wish to differentiate from *content* = contentment, has now become the usual pronunciation in both singular and plural.

**contest.** 1. Pronounce the noun *kŏ'ntĕst*, the verb *kŏntĕ'st*; see NOUN AND VERB ACCENT.
2. The intransitive use of the verb (*Troops capable of contesting successfully against the forces of other nations*; cf. the normal *contesting the victory with*) is much rarer than it was, and is better left to *contend*.

**context.** See FRAMEWORK.

**continental.** 'Your mother,' said Mr. Brownlow to Mr. Monks in *Oliver Twist*, 'wholly given up to continental frivolities, had utterly forgotten the young husband ten years her junior.' This use of *continental* reflects the common belief in England that the Continent, especially France, offers unwonted opportunities for gaiety and self-indulgence. It persists in such expressions as *c. Sunday, c. cabaret*, now not necessarily in the pejorative sense intended by Mr. Brownlow but suggesting either envy or reprobation, or a mixture of both, according to the taste of the user. Such feelings toward what we suppose to be the continental way of life have no doubt changed with the mellowing of Victorian prudery, but are unlikely to disappear so long as we are not allowed to gamble where we please or to drink whenever we are so disposed.

**continual, continuous.** That is *-al* which either is always going on or recurs at short intervals and never comes (or is regarded as never coming) to an end. That is *-ous* in which no break occurs between the beginning and the (not necessarily or even presumably long-deferred) end.

**continuance, continuation, continuity.** *Continuance* has reference to *continue* in its intransitive senses of last, go on; *continuation* to *continue* in its transitive senses of prolong, go on with, and (in the passive) be gone on with. Choice between the two is therefore open when the same sense can be got at from two directions; *We hope for a -ance of your favours* means that we hope they will continue; *We hope for a -ation of them* means that we hope you will continue them; and these amount to the same thing. But the addition that continues a tale or a house is its *-ation*, not its *-ance*, and the time for which the pyramids have lasted is their *-ance*, not their *-ation*; we can wait for a *-ation*, but not for a *-ance*, *of hostilities*; and, generally speaking, the distinction has to be borne in mind. *Continuity*, though occasionally confused with *continuance*, is less liable to misuse, and it is enough to say that its reference is not to *continue*, but to *continuous*. The *continuity man* or *girl* in the film studio, who joins up scenes so as to make a continuous story, is rightly so called. For *solution of continuity* see POLYSYLLABIC HUMOUR.

**contradictious, -tory.** The meanings given to *contradicting, captious, cavilling, cantankerous, quarrelsome,* do not belong to *contradictory*; if either word is to be used, it must be *-tious*; but this, though not in fact a new word, is always used with an uneasy suspicion that it has been made as a stopgap, and it is better to choose one of the many synonyms.

**contrary.** 1. The original accent (*kŏntrār′ĭ*) lingers (1) with the uneducated in all ordinary uses of the adjective (not, perhaps, in *the c.*); (2) with most speakers in the jocose or childish *c.* for *perverse* or *peevish* (*Mary, Mary, quite contrary*), and in *contrariness, -ly,* used similarly; (3) with many speakers in *contrariwise,* especially when it either represents *on the c.* rather than *in the c. manner,* or is used playfully.

2. *On the c., on the other hand.* The idiomatic sense of *o. t. o. h.* is quite clear; except by misuse (see below) it never means *far from that,* i.e. it never introduces something that conflicts with the truth of what has preceded, but always a coexistent truth in contrast with it. The following two examples should have *o. t. c.* instead of *o. t. o. h.*: *It cannot be pleaded that the detail is negligible; it is, o. t. o. h., of the greatest importance.* | *The object is not to nourish 10,000 cats by public charity; it is, o. t. o. h., to put them to sleep in the lethal chamber.* An example of the right use is: *Food was abundant; water, o. t. o. h., was running short.*

The use of *o. t. c.* is less simple; it may have either of the senses of which *o. t. o. h.* has only one; i.e. it may mean either *on the other hand* or *far from that*; but if it stands first in its sentence it can only mean *far from that.* Thus *Food was abundant; o. t. c., water was running short* is impossible; but *Food was abundant; water, o. t. c., was running short* is not incorrect, though *o. t. o. h.* is commoner and, with a view to future differentiation, preferable. If *o. t. c.* is to stand first, it must be in such forms as *Food was not abundant; o. t. c., it was running short.*

**contrary, converse, opposite.** These are sometimes confused, and occasionally precision is important. If we take the statement *All men are mortal,* its contrary is *Not all men are mortal,* its converse is *All mortal beings are men,* and its opposite is *No men are mortal.* The contrary, however, does not exclude the opposite, but includes it as its most extreme form. Thus *This is white* has only one opposite, as *This is black,* but many contraries, as *This is not white, This is coloured, This is dirty, This is black*; and whether the last form is called the *contrary,* or more emphatically the *opposite,* is usually immaterial. But to apply *the opposite* to a mere contrary (e.g. to *I did not hit him* in relation to *I hit him,* which has no opposite), or to the converse (e.g. to *He hit me* in relation to *I hit him,* to which it is neither contrary nor opposite), is a looseness that may easily result in misunderstanding. The temptation to go wrong is intelligible when it is remembered that with certain types of sentence (*A exceeds B*) the converse and the opposite are identical (*B exceeds A*).

**contrast.** 1. Pronounce the noun *kŏ′ntrahst,* the verb *kŏntrah′st*; see NOUN AND VERB ACCENT. 2. The transitive use of the verb with one of the contrasted things as subject, in the sense *be a c.* to or *set off by c.,* was formerly common, but in modern writing is either an archaism or a blunder; *with* should always be inserted. The use meant is seen in: *The sun-tinged hermit and the pale elder c. each other.* | *Monks whose dark garments contrasted the snow.* | *The smooth slopes are contrasted by the aspect of the country on the opposite bank.*

**controversy.** Accent the first syllable. See RECESSIVE ACCENT.

**contumac(it)y.** See LONG VARIANTS.

**contumely.** The possible pronunciations, given here in order of merit, are no less than five: *kŏ′ntūmlĭ, kŏ′ntūmē′lĭ, kŏ′ntūmĭlĭ, kontū′mĭlĭ, kontū′mlĭ.* The well-known line *The oppressor's wrong, the proud man's c.* does much to kill the last two, which are irreconcilable with it, and to encourage the first, which seems, to those whose knowledge of metre is limited, to fit blank verse better than the second or third (the COD's choice). But any of the first three has to overcome popular dislike of a stressed syllable followed by more than one unstressed, especially if there are three. See RECESSIVE ACCENT.

**conundrum.** The derivation is unknown, but the word is not Latin, and its only plural is *-ums*.

**conversation(al)ist.** See -IST.

**convers(e)(ely)(ant). 1.** Pronounce *kŏ'nvers, kŏnver'slī, kon'versant.* **2.** For the sense of *the converse,* see CONTRARY, CONVERSE, OPPOSITE.

**convict.** Pronounce the noun *kŏ'nvĭkt,* the verb *kŏnvĭ'kt;* see NOUN AND VERB ACCENT. *Convict* and *convince* come from the same Latin verb, and their meanings originally overlapped. Now *convict* is used only in the sense of prove guilty and *convince* only in that of persuade by argument or evidence. But *conviction* still serves both.

**convolvulus.** Pl. *-uses;* see -US.

**convoy.** *C.,* like ALLY, shows a disposition to drop the difference in pronunciation between the noun and the verb (see NOUN AND VERB ACCENT) and to make both *kon'voi.* For verb inflexions see VERBS IN -IE etc. 4.

**cooperate, coopt, coordinate.** For *co-o-, coö-, coo-* see CO-.

**copulative.** Copulative verbs are such as, like the chief of them, *be,* link a complement to the subject (He *is* king; we *grow* wiser); among them are included the passives of factitive verbs (This *is considered* the best). For copulative conjunctions see DISJUNCTIVE.

**coquet(te)** etc. The noun is now always *-ette,* and is applied to females only. The verb, formerly *coquet* only, is often now *-ette,* and will no doubt before long be *-ette* only; the accent, and the influence of *-etting, -etted,*

*-ettish,* will ensure that. The noun *coquetry,* for which *kō'kĭtrĭ* is now the standard pronunciation, may perhaps change similarly to *kŏkĕ'trĭ.*

**cord, chord.** For uses in which the spelling is doubtful, see C(H)ORD.

**cordelier.** Pronounce *kordĕlēr'.*

**core.** 'Rotten at the core' (*heart*) is a MISQUOTATION.

**co-respondent** etc. See CO-.

**cornelian, car-.** The first is right (from French *corneline*), and the second, standard in U.S., is due to mistaken etymology.

**cornucopia.** Pl. *-as,* not *-ae.*

**corolla.** Pl. *-as;* see -AE, -AS.

**corona.** Pl. *-ae;* see -AE, -AS.

**corona(l)(ry).** The circlet is *kŏ'rŏnal.* The anatomical terms are *kŏrō'nal* and *kŏ'rŏnărĭ,* but, now that the second is a familiar word, the first is unlikely to hold out against *kŏ'rŏnal,* already recognized as an alternative by some dictionaries.

**corporal, corporeal,** aa. Neither is now a common word except in particular phrases. *Corporal* means *of the human body,* and is common in *-al punishment;* it is also occasionally used with *deformity, beauty, defects,* and similar words, instead of the usual *personal* or *bodily. Corporeal* means *of the nature of body, material, tangible;* so our *-eal habitation* (the body), the *-eal presence of Christ in the Sacrament.*

**corps.** Pronounce in sing. *kōr,* but in pl. (though the spelling is the same) *kōrz.*

**corpulence, -cy.** There is no difference; *-ce* is recommended; *-cy* should be dropped as a NEEDLESS VARIANT.

**corpus.** Pl. *corpora;* see -US.

**corral.** Pronounce *kŏră'l.*

**correctitude,** a modern formation, ascribed by the OED Supp. to an association of *correct* with *rectitude,* has a narrower meaning than *correctness:*

it is used always of correctness of conduct or behaviour generally, with an implication of conscious rectitude.

**correspond** in the sense of be similar etc. takes *to*; in the sense of communicate by letter it takes *with*.

**corrigendum.** Pl. (much commoner than the sing.) *-da*; see -UM.

**cortège.** By English people the word is rarely used except of a funeral procession.

**cortex.** Pl. *-ices* (*-ĭsēz*); see -EX, -IX.

**costume.** The verb, now rarely used except in p.p., is *kŏstū'm*. This pronunciation is preferred by the OED for the noun also; but *kŏ'stūm* (see NOUN AND VERB ACCENT) is now commoner and is put first by the COD.

**cot(e).** The word for *bed* is, or was, Anglo-Indian, is unconnected with the other words, and is always *cot*. The poetic word for *cottage*, and the word for *shelter* (usually seen in compounds, as *sheep-c.*), represent allied but separate old-English words; *cot* is now invariable in the sense *cottage*, and *cote* usual in the sense *shelter*; the latter is usually pronounced *cōte*, but not in its commonest combination *dovecote*.

**cothurnus.** Pl. *-nī*; see -US. As a word for *tragedy*, *c.* is a BATTERED ORNAMENT.

**cottar, cotter, cottier.** The words are clearly distinguished from *cottager* by being applicable not to any one who lives in a cottage, but to peasants doing so under certain conditions of tenure. As compared with each other, however, there is no differentiation between them that is of value; it is merely that the *-tar*, *-ter* forms are more used of the Scottish variety, and *-tier* of the Irish. It would be well if *cottar* were made the sole form, *cotter* left to the word meaning pin or bolt, and *cottier* abandoned.

**couch,** bed, sofa, etc. *C.* may be used as a general word, often with a poetic flavour, for anything that is lain on, bed, lair. But it is commonly differentiated: a *couch* has the head end

only raised, and only half a back, suggesting a session with the psychiatrist; a *sofa* has both ends raised, and the whole of the back, and is, or should be, made for comfort. A *settee* does not differ from a sofa except that it may be without arms. An *ottoman* has neither back nor arms. Nor has a *divan*; its distinctive feature, until the coming of the *divan bed*, was that it was a fixture against a wall.

**couch,** the weed. The OED prefers the pronunciation *kowch*, and describes *kōōch* as that of the southern counties only.

**couchant.** Pronounce *kow'chănt*.

**could.** For such forms as *Could he see you now* see SUBJUNCTIVES.

**co(u)lter.** Spell with -u- and pronounce *ō*.

**council, -sel, -cillor, -sellor.** A board or assembly, and the meeting of such a body, has always *-cil*, and a member of it is *-cillor*. The abstract senses *consultation, advice, secret* (*keep one's c.*) belong to *-sel*, and one who gives advice is, as such, a *-sellor*, though he may be a *-cillor* also; *my -cillors* are the members of my (e.g. the king's) council; *my counsellors*, those who advise me officially or otherwise. But the *Counsellors of State* appointed in England to perform the royal duties during the sovereign's absence from the country are, unconformably, so spelt. *Counsel* has also the semi-concrete sense of the person or persons (never *counsels*) pleading for a party to a lawsuit (formally designated *Mr. X of Counsel*); the use is originally abstract, as when *All the wealth and fashion* stands for *all the rich* etc. *people*, or as though *advice* were said for *adviser(s)*.

**countenance, face, physiognomy, visage.** *Face* is the proper name for the part; *countenance* is the face with reference to its expression; *physiognomy*, to the cast or type of features. (For the facetious use of PHYSIOGNOMY see that article.) *Visage* is now a LITERARY WORD, used ornamentally for *face* without special significance.

**counterpoint** (adj. *contrapuntal*). See HARMONY etc.

**coup.** Pronounce *kōō*; pl. *coups*, pron. *kōōz*.

**course.** 1. There are three common uses of *of course* as an adjunct to a statement of fact: the polite ('I am sure you must be intelligent enough to know this already, but I may as well mention it'), the disdainful ('You must indeed be an ignoramus to have to ask that question'), and the showing-off ('I am so well informed that I naturally have such knowledge at my finger-tips'). This last variety, as the herald of an out-of-the-way fact that one has just unearthed from a book of reference, is a sad temptation to journalists: *From this marriage came Charles James Fox; his father was, o. c., created Baron Holland in 1763.* | *Milton o. c. had the idea of his line from Tacitus.* | *He is, o. c., a son of the famous E. A. Sothern, of 'Lord Dundreary' fame.* See SUPERIORITY.
2. The habit of some printers of automatically encasing adverbial phrases in commas may have disastrous results if applied to o. c.: *The navigating officer made no change, of course, until a collision was inevitable.*

**court martial.** See HYPHENS and PLURAL ANOMALIES.

**courteous, courtesy,** are variously pronounced *ker-* and *kor-*; the first is recommended.

**cousin.** See RELATION.

**Coventry.** The OED prefers *kŏ-* to *kŭ*; and *kŏ-* seems now to be established, though *kŭ-* is still given as an alternative by some dictionaries. The same is true of *Covent Garden.* See PRONUNCIATION 5.

**coverlet, -lid.** Both forms are old; the first is better, the ending almost certainly representing French *lit* bed, and not English *lid*.

**covert,** n. The *-t* is now so seldom sounded, and is so often omitted even in writing, that what distinction remains between *covert* and *cover* in the sense of a shelter for game may be said to be valueless. But *covert*, with the *t* sounded, survives in a few combinations such as *c. coat, wing cc.*

**coward(ly).** The identification of coward and bully has gone so far in the popular consciousness that persons and acts in which no trace of fear is to be found are often called *coward(ly)* merely because advantage has been taken of superior strength or position. Such action may be unchivalrous, unsportsmanlike, mean, tyrannical, and many other bad things, but not cowardly; cf. the similar misuse of DASTARDLY. For the adverb see -LILY.

**cowslip.** The true division is cow and slip, not cow's and lip; and the usual pronunciation with s, not z, is accordingly right.

**coxcomb, cocks-.** See COCKSCOMB.

**coyote.** Some dictionaries give the pronunciation as *koi-ō'tĕ* and others as *kŏ-yō'tĕ*, a difference of opinion that is no doubt explained by the fact that, according to Webster, the correct Spanish pronunciation is '*kō-yō'tā*, almost *koi-ō'tā*'. (In western America the first syllable is *kī-*.) All dictionaries agree in permitting a disyllabic alternative (*-ōt*).

**crack.** See TOP, ACE, CRACK.

**cramp,** as an adjective meaning crabbed or hard to understand (*c. words, terms, style*), narrow (*a c. corner*), niggling (*writes a c. hand*), has now had its senses divided between *crabbed* and *cramped*, and the use of it is an affectation.

**cranesbill, crane's-bill.** The apostrophe and hyphen are better dispensed with in established words of this type; cf. COCKSCOMB.

**cranium.** Pl. *-ia*; see -UM.

**crape, crêpe.** The first is used for the black mourning material and the second for any other crêpe fabric.

**crayfish, craw-.** The first is the British form, the second 'now used chiefly in U.S.' (OED). A corruption of Fr. *ecrevisse*.

**creative** is a term of praise much affected by the critics. It is presumably intended to mean original, or something like that, but is preferred because it is more vague and less usual (cf. *seminal*). It has been aptly called a 'luscious, round, meaningless word', and said to be 'so much in honour that it is the clinching term of approval from the schoolroom to the advertiser's studio'.

**credence, credit.** Apart from the isolated phrase *letter of credence* (now usually *credentials*) and the concrete ecclesiastical sense table or shelf, *credence* has only one meaning—belief or trustful acceptance. The use seen in *Two results stand out clearly from this investigation . . .; neither of them gives any credence to these assertions . . .* is a mere blunder; *give credence to* means believe, simply; *support* or *credibility* is the word wanted. *Credit*, on the other hand, is rich in meanings, and it is a pity that it should be allowed to deprive *credence* of its ewe lamb; *credence* would be better in *Charges like these may seem to deserve some degree of credit*, and in *To give entire credit to whatever he shall state*. Even *give credit* (*to*) has senses of its own (*I give him credit for knowing better than that*; *No credit given on small orders*), which are all the better for not being confused with the only sense of *give credence to* (*One can give no credence to his word*).

**credible.** See VIABLE.

**crenel(le), crenellated.** Spell *crenel* and pronounce *krĕ'nĕl*.

**creole.** See MULATTO 3.

**cretin.** The OED gives only *krē'tin*, but *krĕ-* has since made headway and is put first by some dictionaries.

**crick, rick (wrick),** whether identical in origin or not, are commonly used in slightly different senses: *crick* for a cramp or rheumatic pain, especially of the neck or back, *rick* for a sprain or strain, especially of a joint. *Wrick* is

the preferable spelling (cf. *wrinkle* and *crinkle*, *wrack* and *rack*), but has fallen into disuse.

**cringe** makes *-ging*; see MUTE E.

**crisis.** Pl. *crises* (*krī'sēz*). The proper meaning of the word is a state of affairs in which a decisive change for better or worse is imminent. Used loosely for any awkward, dangerous, or serious situation it is a SLIPSHOD EXTENSION.

**criterion.** Pl. *-ia*; see -ON 1. For synonymy see SIGN.

**critique** is in less common use than it was, and, with *review*, *criticism*, and *notice*, ready at need, there is some hope of its dying out, except so far as it may be kept alive by the study of Kant.

**crochet, croquet,** make *-eting*, *-eted*, pronounced *krō'shiing*, *krō'kiing*, *krō'-shĕd*, *krō'kĕd*.

**crooked.** See -ER AND -EST 2. A stick that is not straight is a *krŏŏkĕd* stick; one provided with a crook is a *krŏŏkt* stick.

**crosier, -zier.** The OED prefers -s-.

**cross-section.** This has become a VOGUE WORD, often used in the sense of *sample* because it looks more scientific. (See POPULARIZED TECHNICALITIES.) A *c.-s.* is a transverse cut across an object so as to expose its inner layers, as a tree trunk when cut across will expose the lines that mark its growth. A *sample* is a small separated part of something that will serve to show the quality of the mass. A *c.-s. of public opinion* in a particular locality (a favourite use of the term) can only properly be so called if the number of samples taken from each section of the community is in the same proportion to the total samples as the number of people in that section is to the whole community. But a *c.-s. of opinion* is a convenient way of indicating that the samples composing it are random and diverse, and it is perhaps unreasonable to expect any greater exactitude in the use of the expression.

**crow.** The past is now usually *crowed* (*They crowed over us*; *The baby crowed loudly*; *The cock crowed*, or *crew*, *at dawn*); *crew* is used always in *the cock crew* when there is reference to the N.T. passage (it is preserved in the N.E.B.), and alternatively with *crowed* when *cock* is the subject in other connexions.

**crown.** *The C.* is often used as a phrase for the king or queen regarded not as a person, but as a part of the constitution. It does not follow that pronouns appropriate to *king* can be used after it, as in these absurdities: *The incontestable fact that the C. nowadays acts, and can only act, on the advice of his Ministers.* | *The people of this country are little likely to wish to substitute for this* [rule by Cabinet] *rule by the C., for whom the experiment would be most fraught with peril.*

**crucial** properly means something that finally decides between two hypotheses—*decisive, critical*. A word with so fine a point deserves better treatment than the SLIPSHOD EXTENSION that uses it merely as a synonym for *important*.

**cryptic** might usefully be reserved for what is purposely equivocal, like the utterances of the Delphic oracle, and not treated as a stylish synonym for *mysterious, obscure, hidden*, and other such words. See WORKING AND STYLISH WORDS and AVOIDANCE OF THE OBVIOUS.

**cubic(al).** *Cubic* is the form in all senses; *cubical* only in that of shaped like a cube. So *-ic measure, content, foot, equation*; but *a -ical box* or *stone*. *Cubic*, however, is sometimes used in the sense of cube-shaped, and always of minerals crystallizing in cubes, as *-ic alum, saltpetre*. See -IC(AL).

**cui bono?** As generally used, i.e. as a pretentious substitute for To what end? or What is the good?, the phrase is at once a BATTERED ORNAMENT and a blunder. The words mean To whom was it for a good?, i.e. Who profited by it or had something to make out of it?,

i.e. If you want to know who brought it about ask yourself whose interest it was that it should happen. Those who do not want it in this sense should leave it alone. The following is an amusing attempt to press the correct translation of the Latin into the service of the ordinary pointless use: *We have had repeated occasion of late to press the question 'Cui bono?' in relation to the proposal to force the Government to a creation of peers. We must ask it again, in reference to the scandal of yesterday. What is the good of it? Who stands to gain?* See MISAPPREHENSIONS.

**culinary.** Pronounce *kū′lĭnărĭ*. The word is a favourite with the POLYSYLLABIC HUMOURIST.

**cullender.** See COLANDER.

**cult,** as now used, dates only from the middle of last century; its proper place is in books on archaeology, comparative religion, and the like. Its unsuccessful attempt to oust *worship* in general use has left it with a suggestion of something new-fangled or cranky.

**cultivated, cultured,** are the past participles of verbs that the OED defines in the same way: 'lit. to till or to produce by tillage, fig. to improve and develop by education and training'. *Cultivate* is fully alive in both senses; *culture* (v.) is now rarely used except in its past participle. There it has lost its literal meaning but retained its figurative one; it has also acquired a new one by being used as a noun to describe micro-organisms developed in prepared media. In their figurative sense *cultivated* and *cultured* are used synonymously and it would be a useful differentiation if this duty were left to *cultured*, and *cultivated* confined to its literal meaning. But this is perhaps less likely to happen now that culture (n.) has acquired the wider meaning of 'the civilization of a people, esp. at a certain stage of its development or history' (OED Supp.), and has been tainted by association with *kultur*, defined by the OED Supp. as 'civilization as conceived by the Germans, esp.

in a derogatory sense as involving race-arrogance, militarism and imperialism'.

**cum(m)in.** The OED prefers *cumin*: but, besides the service done by the second *m* in keeping the pronunciation of a not very common word steady, the spelling of *Matt.* xxiii. 23 (*cummin* in all versions) is sure to prevail in a word chiefly used with reference to that passage.

**cumulative.** See ACCUMULATIVE.

**cumulus.** Pl. *-lī*; see -US.

**cuneiform.** The slovenly pronunciation *kū'nĭform*, not uncommon, should be avoided, and to this end *kūnē'ĭform* is preferable to the more difficult *kū'nĕĭform*; cf. CONTUMELY.

**cup.** For 'cups that cheer' see HACKNEYED PHRASES.

**curaçao, -çoa.** Spell *-çao*; pronounce *kūr'ăsō*.

**curate.** For 'the curate's egg' see HACKNEYED PHRASES, and WORN-OUT HUMOUR.

**curator.** Pronounce *kūrā'tor* except in the Scots-law use (ward's guardian), in which it is *kūr'ător*.

**curb, kerb.** The second is a variant merely, not used in U.S. but in Britain now much commoner than *curb* in the sense footpath-edging; the *kerb market* where dealings are carried on in the street after the Stock Exchange is closed is always so spelt. The same spelling seems likely to prevail in the closely allied senses fender, border, base, framework, mould. For the bit and in the sense check n. or v., *curb* is invariable.

**curriculum.** Pl. *-la*; see -UM.

**cursed, curst.** The adjective *cursed* is disyllabic except sometimes in verse; the form *curst* is chiefly used either to show that the rare monosyllabic pronunciation is meant (esp. in verse), or to differentiate the archaic sense ill-tempered.

**curtailed words.** Some of these establish themselves so fully as to take the place of their originals or to make them seem pedantic; others remain slangy or adapted only to particular audiences. A few specimens of various dates and status have here been collected as possibly useful to those who have, or wish to have, views on the legitimacy of curtailment: *ad*(vertisement); *amp*(ere); *bra*(ssiere); (omni)-*bus*; *cab*(riolet); (violon)*cello*; *consols* (consolidated annuities); *co-op*(erative society or store); *cox*(swain); *deb*(utante); *demob*(ilize); *exam*(ination); (in)*flu*(enza); *fridge*(refrigerator); *gym*(nasium); *gen*(eral information); *homo*(sexual); *incog*(nito); *knicker*(bocker)*s*; *marg*(arin)*e*; *mike* (microphone); *mod*(eration)*s*; *op*(us); *pant*(aloon)*s*; *para*(graph); (tele)*phone*; *photo*(graph); (aero)*plane*; *polio*(myelitis); *pop*(ular); *pram* (perambulator); *prefab*(ricated house); *prep*(aration); *pro*(fessional); *prom*(enade concert); *pub*(lic house); *quad*(rangle, ruplet); *recap*(itulate); *rev*(olution)*s*; *scrum*(mage); *spec*(ulation); *sub*(altern, marine, scription, stitute); *taxi*(meter cab); *telly* (television); *turps* (turpentine); *vet*(erinary surgeon); *viva* (voce); *zoo*(logical gardens).

Another way of forming curtailed words is to combine initial letters, a method now so popular, especially in America, that a word—*acronym*—has been coined for it. The first world war produced a few—*Anzac* (Australian and New Zealand Army Corps), *Dora* (Defence of the Realm Act), *Wrens* (Women's Royal Naval Service), and the second a great many; among them *Asdic* (Allied Submarine Detection Investigation Committee), *Cema* (Council for Encouragement of Music and the Arts, now the Arts Council of Great Britain, which is not amenable to this treatment), *Ensa* (Entertainments National Service Association), *Fany* (First Aid Nursing Yeomanry), *Fido* (Fog Investigation Dispersal Operation), *Naafi* (Navy Army and Air Force Institutes), *Octu* (Officer Cadet Training Unit), *Pluto* (Pipe Line under the Ocean), *Radar* (Radio Detection and Ranging), *Remc* (Royal Mechanical and Electrical Engineers), *Shaef* (Supreme Headquarters

Allied Expeditionary Force), *Wracs* (Women's Royal Army Corps), and *Wrafs* (Women's Royal Air Force). The process has since continued: the United Nations Organization itself is *Uno* and some of its branches have lent themselves to this form of abbreviation, e.g. *Unrra* and *Unesco*. The Western Alliance has given us *Gatt* (General Agreement on Tariffs and Trade), *Nato* (North Atlantic Treaty Organization), and *Shape* (Supreme Headquarters of the Allied Powers in Europe); the initials of the organization most frequently referred to and with the most cumbrous title (The Organization for European Economic Cooperation) are unfortunately unpronounceable. But European economic cooperation on a smaller scale has produced *Benelux* (Belgium, Netherlands, Luxembourg), and *Efta* (European Free Trade Area). Some of our trade unions admit of this labour-saving device, e.g. *Aslef* (Association of Locomotive Engineers and Firemen), *Natsopa* (National Society of Operative Printers and Assistants) and *Nalgo* (National Association of Local Government Officers). *Neddy* for the National Economic Development Council and *Nicky* for the National Incomes Commission were irresistible, and we always welcome any opportunity of giving pet names of this sort to the new inventions of atomic or electronic science, e.g. *Hector* (Heated Experimental Carbon Thermal Oscillator Reactor) and *Ernie* (Electronic Random Number Indicating Equipment).

**curts(e)y, courtesy.** *Courtesy* is archaic and affected for *curtsy*; *curtsy* n. and v. (*curtsied, curtsying*) is better than *curtsey*, which involves *curtseyed*; see VERBS IN -IE etc. 2, 6.

**customs.** See TAX.

**cybernetics,** from the Greek κυβερνήτης helmsman, is a name invented in America for a 'new science' which, to quote from a broadcast on the Third Programme, 'deals with the processes of control and communication and endeavours to predict human and animal behaviour by the construction of models (or machines) which will imitate these activities in a concrete way'. 'The most recent models incorporate complex and subtle mathematical concepts of probability, even of the probability of probabilities. But it is still by no means certain that they have yet bridged the gap between the biological or psychological and the physical' (*The Listener*, 30 Nov. 1956).

**cycle.** For *c.* as a time-word see TIME.

**cyclopaedia, -dic.** For *-pae-, -pæ-, -pe-* see Æ, Œ. The longer forms *encyclo-* are in themselves better, and *encyclopaedia*, being common in titles, is also the prevalent form; but *cyclopaedic* (not *-ical*) is often used for the adjective; cf. *accumulate* and *cumulative*.

**cyclopean, -pian.** The first (*sīklopē'an*) is more usual than the second (*sīklō'pĭan*); but neither is wrong.

**cyclop(s).** The form recommended for the singular is *cyclops*, and those who are content to make it serve as a plural also will avoid the dilemma of having to choose between the faintly absurd *cyclopses* and the faintly pedantic *cyclō'pēs*.

**cynic(al).** As an adjective, *cynic* is used only in the sense 'of the ancient philosophers called cynics' (except in the technical terms *cynic year, cynic spasm*), and the word that describes temperament etc. is *cynical*; see -IC(AL).

**cypher.** See CIPHER.

**Cyrenaic.** See HEDONIST.

# D

**daemon, dæ-.** Write *dae-*; see Æ, Œ. This spelling, instead of *demon*, is used to distinguish the Greek-mythology senses of supernatural being, indwelling spirit, etc., from the modern sense of devil.

**dail.** Some dictionaries give the pronunciation *doil*, others *dawl*. The true sound, difficult for an Englishman, lies between the two.

**dais.** The OED recognizes no other pronunciation than *dās*: 'Always a monosyllable in French and English where retained as a living word; the disyllabic pronunciation is a "shot" at the word from the spelling.' But the shot has struck home. The SOED admitted *dā′is* as an alternative, and some modern dictionaries give this alone.

**damning,** in the sense *cursing*, is pronounced without the *n*; in the sense *fatally conclusive* the *n* was formerly sounded but is rarely heard now.

**danger.** It is curious that *in d. of life* and *in d. of death* should mean the same. *Life* is the only exception to the idiom that requires *in d. of* to be followed by the peril, not by what is exposed to it.

**dare.** 1. *Dare* and *dares*. 2. *Durst*. 3. *Dare say*. 1. *Dare* as 3rd pers. sing. pres. indic. is the idiomatic form instead of *dares* when the infinitive depending on it either has no *to* or is 'understood'; this occurs chiefly, but not only, in interrogative and negative sentences. Thus *dares*, though sometimes used in mistaken striving after correctness, would be contrary to idiom in *Dare he do it?*; *He dare not!*— *Yes, he dare*; *He dare do anything*; *No one dare oppose him.*
   2. *Durst*, which is a past indicative and past subjunctive beside *dared*, is obsolescent, and nowhere now required, like *dare* above, by idiom; the contexts in which it is still sometimes preferred to *dared* are negative sentences and conditional clauses where there is an infinitive either understood or having no *to* (*But none durst*, or *dared to*, or *dared, answer him*; *I would do it if I durst*, or *dared*).
   3. *Dare say* as a specialized phrase with the weakened sense *incline to*

*think, not deny, admit as likely*, or ironically in the same sense as the slang *says you* (cf. the unweakened sense in *I dare say what I think*, *Who dare say it?*, *He dared to say he*, or *that he, would not do it*). This is much commoner in the first person than in the others and has certain peculiarities: (*a*) even when not parenthetic (*You, I d. s., think otherwise*), it is never followed by the conjunction *that* (*I d. s. it*, not *that it, is a mere lie*); (*b*) it is never *dare to say* in direct speech, and the *to* is rare and better avoided in indirect speech also (*He dared say the difficulty would disappear*; *I told him I dared say he would change his mind*; *He dares say it does not matter*); (*c*) to avoid ambiguity, it is sometimes written as one word (*I dare say she is innocent*, I am sure of it; *I daresay she is innocent*, I can believe it); but this device is useless as long as it is not universally accepted, and it cannot be applied to the indirect *dares* and *dared*; it is simpler to avoid *I dare say* in the unspecialized sense wherever it can be ambiguous.

**darkling** is an adverb formed with the now forgotten adverbial termination *-ling*, and is a poetic word meaning in the dark (*Our lamps go out and leave us d.*; *The wakeful bird sings d.*). By a natural extension it is also used as an attributive adjective (*Like d. nightingales they sit*; *They hurried on their d. journey*). But since it has nothing to do with the participial *-ing* it does not mean growing dark etc.; from the mistaken notion that it is a participle spring both the misuses of the word itself and the spurious verb *darkle*, which never won any real currency, as the analogous *grovel* did, and may be allowed to die unregretted.

**dash.** See STOPS.

**dastard(ly).** The essential and original meaning of the words is the same as that of COWARD(LY), to the extent at least that both pairs properly connote want of courage; but so strong is the false belief that every bully must be a coward that even acts requiring great courage (such as a political assassination that exposes the perpetrator to the risk of immediate lynching) are described as dastardly if carried out against an unsuspecting or helpless victim. The true meaning is seen in 'A laggard in love and a dastard in war'. The words should at least be reserved for those who do avoid all personal risk.

**data** is a Latin plural (*The d. are*, not *is, insufficient.* | *What are the d.?* | *We have no d.*); the singular, comparatively rare, is *datum*; *one of the data* is commoner than *a datum*; but *datum-line*, is used for a line taken as a basis. Latin plurals sometimes become singular English words (e.g. *agenda*, *stamina*) and *data* is often so treated in U.S.; in Britain this is still considered a solecism, though it may occasionally appear. *This d. and that obtained from radio star scintillation will be analysed at the Jodrell Bank Experimental Station*. Pronounce *dāta*.

**date.** For *d., epoch*, etc., see TIME. *Date* in the sense of engagement or appointment (*make a d. with*), called by the OED Supp. U.S. colloq., has had a cordial welcome in Britain, especially where the two persons concerned are young people of opposite sexes, but is still at best colloquial. Sometimes the verb, used transitively, is made to suffice. *I've dated her for tomorrow.*

**daughter-in-law.** See -IN-LAW.

**davits.** Pronounce *dă-* not *dā-*.

**day and age.** *In this d. and a. there are misdemeanours that are more serious than any felonies.* | *The book will be devoured nostalgically by the middle-aged. But let them not forget the young, whom, in this d. and a., it will profit even more.* | *In a d. and a. when the scientists can split the atom and send rockets to the moon . . . it ought to be possible to have a supply of electricity that could not fail over so wide an area.* | *The bath in that d. and a. was still a mark of social distinction.* | *The field in which he works demands no less than statesmanship, and, in this d. and a.,* a little more. | *An outworn boomerang device . . . used by schoolboys as a substitute for intellectual assurance and by statesmen at their peril (and ours) in this d. and a.* These quotations, all taken from ordinary journalistic reading over a period of a year or so show how popular this cliché has become. Its literal meaning is not easily grasped, and the nuance that is supposed to distinguish it from some more commonplace expression such as *in these days* is not always clear to the reader or even perhaps to the writer.

**-d-, -dd-.** Monosyllables ending in d double it before suffixes beginning with vowels if the sound preceding it is that of a single short vowel, but not if it is a long vowel or a doubled vowel or a vowel and r: *caddish, redden, bidding, trodden, buddy*; but *deaden, breeder, goodish, plaided, harden*. Words of more than one syllable follow the rule for monosyllables if their last syllable is accented or is itself a word in combination (*forbidding, bedridden*), but otherwise do not double the d (*nomadic, wickedest, rigidity, periodical*).

**D-day,** though generally understood as referring only to the 6 June 1944, the day of the Allied invasion of Normandy, is in fact a general military term for the date on which any important operation is planned to begin.

**dead letter,** apart from its Pauline and post-office uses, is a phrase for a regulation that has still a nominal existence but is no longer observed or enforced; the application of it to what was never a regulation but has gone or is going out of use, as quill pens, top hats, steam locomotives, etc., or to a regulation that loses its force only by actual abolition (*identity cards are now a d. l.*), is a SLIPSHOD EXTENSION. Capital punishment is a *d. l.* in Belgium, where it remains a legal penalty though never inflicted, but cannot properly be so described in a country where it has been abolished by law.

**deadline** originally meant the line round a military prison beyond which a prisoner was liable to be shot. Its

**deal** 120 **decad(e)**

recent extension to serve for any limit beyond which it is not permissible to go (especially the time within which a task must be finished) is useful, although, like all popular new metaphors, it breeds forgetfulness of common words (e.g. *limit*) that might sometimes be more suitable. Cf. CEILING.

**deal, n.** 1. The use of *a d.* instead of *a great* or *good d.*, though as old as Richardson and Johnson (the Shakespearian *what a deal!* can hardly be adduced), has still only the status of a colloquialism, and should be avoided in writing even when the phrase stands as a noun (*saved him a d. of trouble*), and still more when it is adverbial (*this was a d. better*). 2. *A d.* in the sense of a piece of bargaining or give-and-take, though still colloquial in such phrases as *a square d., a raw d.*, was recognized by the OED Supp. as 'now in general English use and applied to international as well as interparty agreements'. It entered history in the name (*The New Deal*) given by President Franklin Roosevelt to his social and economic reforms in 1932.

**deal, v.** *It deals very shrewdly, but also with warm conviction, of the war between the poet and the laggard following who cry everlastingly for a New Thing.* The writer, after his parenthesis, continues as though he had begun *He treats,* with which *of* would be idiomatic. *Deals* needs *with.* See CAST-IRON IDIOM.

**dean, doyen,** though originally the same word, meaning the senior member of a community, have become differentiated: *dean* is the title of an ecclesiastical or academic officer; *doyen* (pr. *dwah'yen*) is a title of respect for the senior member of any community, especially of a diplomatic corps.

**dear(ly).** 1. With the verb *love, dearly* is now the regular form of the adverb, and *dear* merely poetic; but with *buy, sell, pay, cost,* etc., *dear* is still idiomatic, and the tendency born of mistaken grammatical zeal to attach an UNIDIOMATIC -LY should be resisted. 2. Between *dear* and *expensive* there

are differences of nuance that deserve more respect than they get. *E.* should mean merely that a thing costs a lot of money, *d.* that it costs more than it is worth. A collar-stud at half a crown may be dear but not expensive; a motor car at £5,000 expensive but not dear. Similarly *inexpensive* suggests a good bargain, *cheap* an inferior article.

**dearth, lack.** *I think it of interest to point out* what a singular d. of information exists *on several important points.* | *The headmaster* showed a considerable l. of cooperation *with the governing body.* Read *how little we know; did not cooperate.* For this favourite journalistic device see ABSTRACTITIS.

**debark(ation)** are natural shortenings of the better established *disemb-* (cf. *debus, detrain*) and may supersede them.

**debouch(ment).** Though the dictionaries still favour *deboosh*, the anglicized *debowch* is becoming common, and will probably establish itself.

**debris, dé-.** Write without accent and pronounce *dĕ'brē.* The word is thoroughly naturalized though it has not anglicized the pronunciation of its second syllable.

**debunk.** See DE-, DIS-.

**début, débutant(e).** *Début* can only be pronounced as French, and should not be used by anyone who shrinks from the necessary effort. There is no reason why *debutant* should not be written without accent, pronounced *dĕ'būtant*, and treated like *applicant* etc. as of common gender. But in its commonest use we prefer the noncommittal course of shortening it colloquially to *deb.*

**deca-, deci-.** In the metric system, *deca-* means multiplied, and *deci-* divided, by ten; *decametre*, 10 m., *decimetre*, $\frac{1}{10}$ m.; so with *gramme, litre,* etc.

**decad(e)** is now always spelt with the final *e*, and the pronunciation *dĕ'kād* is now more usual than *dĕ'kăd.*

**de- and dis-.** The function of these prefixes is to form a compound verb with the sense of undoing the action of the simple one, and they are an invaluable element in the language, constantly providing us with useful new words such as (to take one or two recent examples) *debunk*, *debus*, *de-ice*, *degauss*, *demob*, and *derate*. But the device should be used with discrimination; it is dangerously easy for a writer to invent a new word of this kind to save himself the trouble of thinking of an existing antonym. 'At the present rate of distortion of our language', it has been said, 'it looks as if we should soon be talking about *black* and *disblack*, *good* and *disgood*.'

There are, however, two reasons for coining a *de-* or *dis-* word that cannot be ascribed to indolence. One is where the action to be undone is expressed by a word used in a special sense that would not be conveyed by any existing antonym; this justification can be pleaded for inventing, e.g., *derequisition*, *decontaminate*, *derestrict*, rather than using such words as *restore*, *cleanse*, and *free*. Even on behalf of so unprepossessing a word as *desegregation* (*The line of absolute resistance to desegregation of the schools in Virginia has suddenly crumbled*) it can be argued that only with its help could the writer make his meaning immediately clear without circumlocution.

The other good reason for such coinages is to provide words of intermediate meaning, as we have long had *dispraise* for something between *praise* and *blame*, and *disharmony* for something between *harmony* and *discord*. Thus economists, who are specially given to these experiments, have invented *disincentive*, *disinflation*, *diseconomy*, and *dissaving* as words with meanings less positive than the antonyms *deterrent*, *deflation*, *extravagance*, and *spending*. These words have not had a universal welcome, but no one can deny that they spring from a laudable desire to make language more precise. And the writer who said that the gradualness with which the Commonwealth has evolved has 'shielded the British from the bruising shocks of disimperialism' can claim to have invented a pregnant word.

**decided, decisive.** *Decisive* is often used loosely where *decided* is the right word, just as DEFINITIVE is a common blunder for *definite*, and DISTINCTIVE an occasional one for *distinct*. A *decided* victory or superiority is one the reality of which is unquestionable; a *decisive* one is one that decides or goes far towards deciding some issue; a *decided* person is one who knows his own mind, and a *decided* manner that which comes of such knowledge; a *decisive* person, so far as the phrase is permissible at all, is one who has a way of getting his policy or purpose carried through. The two meanings are quite separate; but, as the *decided* tends to be *decisive* also, it gets called so even when decisiveness is irrelevant. Examples of the wrong use are: *The serjeant, a decisive man, ordered . . . | A decisive leaning towards what is most simple. | It was not an age of decisive thought. | Poe is decisively the first of American poets.* The following suggests a further confusion with *incisive*: *The Neue Freie Presse makes some very decisive remarks about the Italian operations at Preveza.*

**decimate** meant originally to kill every tenth man as a punishment for cowardice or mutiny. Its application is naturally extended to the destruction in any way of a large proportion of anything reckoned by number, e.g. a population may be said to be decimated by a plague. But undue advantage is taken of this latitude by a journalist who applies the word to the virtual extermination of rabbits by myxomatosis; and anything that is expressly inconsistent with the proper sense (*A single frosty night decimated the currants by as much as 80%*) must be avoided. A startling example of what this may lead to was given by a contributor to correspondence in *The Times* on the misuse of the word LITERALLY. 'I submit the following' he wrote, 'long and lovingly remembered from my "penny dreadful" days: *Dick, hotly pursued by the scalp-hunter, turned in his saddle,* *fired, and literally decimated his opponent.'* See SLIPSHOD EXTENSION.

**declarant, declaredly, declarative, declaratory.** Pronounce -*ā̆rant*, -*ā̆rēdlĭ*, -*ărătĭv*, -*ărătorĭ*; for the second see -EDLY.

**declinal, declination, declinature,** in the sense courteous refusal (*The declinals were grounded upon reasons neither unkind nor uncomplimentary.* / *Yuan persists in his declination of the Premiership.* / *The reported declinature of office by the Marquess of Salisbury*) are three unsatisfactory attempts to provide *decline* with a noun. It is better to be content with *refusal* (if *declining* will not do) modified, if really necessary, by an adjective. See -AL NOUNS.

**decorous.** Pronunciation has not yet settled down between *dĕkōr'us* and *dĕk'ŏrus*; *decorum* pulls one way and *decorate* the other. Most authorities prefer *dĕkōr'us* but *dĕk'ŏrus* (now put first by the COD) may win. See RECESSIVE ACCENT.

**dedicated.** *He is that rara avis a dedicated boxer.* The sporting correspondent who wrote this evidently does not see why the literary critics should have a monopoly of this favourite word of theirs, though he does not seem to think that it will be greatly needed in his branch of the business.

**deduction.** See INDUCTION, DEDUCTION.

**deem** is an indispensable word in its legal sense of assuming something to be a fact which may or may not be one. (*Any person who has not given notice of objection before the prescribed day will be deemed to have agreed.*) For its use as a stilted word for think see WORKING AND STYLISH WORDS.

**deep(ly).** See UNIDIOMATIC -LY.

**deer.** See COLLECTIVES 2.

**defect.** For 'the defects of his qualities' see HACKNEYED PHRASES. Pronounce *dĕfĕkt'*, but see PRONUNCIATION 1.

**defective, deficient.** The differentiation tends to become complete, *defective* being associated more and more with *defect*, and *deficient* with *deficit*. That is deficient of which there is either not enough or none, that is defective which has something faulty about it; so *deficient quantity, revenue, warmth, means; defective quality, condition, sight, pronunciation, boots; a defective chimney, valve, manuscript, hat.* With some words quantity and quality come to the same thing; for instance, *much* or *great insight* is the same as *deep* or *penetrating insight*; consequently a person's insight may be described either as *defective* or *deficient*. Again, deficiency in or of a part constitutes a defect in the whole; thus milk may be *defective* because it is *deficient* in fatty matter. Lastly, either word may sometimes be used, but with a difference of meaning from the other; *deficient water* or *light* is too little water or light; but *defective water* is impure and *defective light* is uncertain; similarly, *a defective* differs from *a deficient supply* in being irregular or unreliable rather than insufficient in the aggregate. This useful distinction was blurred in the old Mental Deficiency Acts which, though bearing that title, dealt with persons termed *mental defectives*. But both expressions have now been abolished as statutory terms; *mental deficiency* has become *subnormality* and is included, with *mental illness*, in the generic term, *mental disorder.* See also I.Q. and PSYCHOPATHIC.

**deficit.** The pronunciation *dĕfĭ'sĭt* is wrong; the OED prefers *dē'fĭsĭt* to *dē'fĭsĭt* and it is now usual; the Latin quantity (*dēficit = is lacking*) is no guide. See FALSE QUANTITY.

**definite, definitive. 1.** Confusion between the two, and especially the use of *definitive* for *definite*, is very common. Many writers seem to think the words mean the same, but the longer and less usual will be more imposing, and mistakes are made easy by the fact that many nouns can be qualified by either, though with different effects. Putting aside exceptional senses that have nothing to do with the confusion (as when *definitive* means of the defining kind), *definite* means defined, clear, precise, unmistakable, etc., and *definitive* means having the character of finality. Or, to distinguish them by their opposites, that is *definite* which is not dubious, vague, loose, inexact, uncertain, undefined, or questionable; and that is *definitive* which is not temporary, provisional, debatable, or alterable. *A definite offer* is one of which the terms are clear; *a definitive offer* is one that must be taken or left without haggling. *Definite jurisdiction* is that of which the scope or the powers are precisely laid down, and *definitive jurisdiction* is that from which there is no appeal; either word can be applied, with similar distinctions, to *answer, terms, treaty, renunciation, statement, result,* etc. But with many words to which *definite* is rightly and commonly applied (*a definite pain, accusation, structure, outline, forecast*) *definitive* either is not used except by mistake for *definite*, or gives a meaning rarely required (e.g. *a definitive forecast* means, if anything, one that its maker announces his intention of abiding by). The following examples show wrong uses of *definitive*: *We should be glad to see more definitive teaching.* | *The Bill has not yet been drawn up, and the Government are not responsible for 'forecasts', however definitively they may be written.* | *The definitive qualities of jurisprudence have not often found so agreeable an exponent as the author of these essays.*

2. *Definite* and *definitely* are overworked words. In the nineteen-twenties young people (and some older ones) forgot the word *yes*: an affirmative reply was always *definitely* and a negative one usually *definitely not*. That fashion has waned, but the words have still a habit of intruding where they are not wanted. *Investigations have shown that there is a definite need for accommodation for patients who are convalescent.* | *This has caused two definite spring breakages to* loaded vehicles. | *This is definitely harmful to the workers' health.* See MEANINGLESS WORDS.

**defrayal.** See -AL NOUNS.

**degree.** The phrase *to a d.*, however illogical it seems as a substitute for *to the last degree* (cf. *to the* NTH), is at least as old as *Cecilia* (*He really bores me to a d.*) and *The Rivals* (*Your father, sir, is wrath to a d.*), and objection to it is futile. Though called by the COD colloquial, it appears occasionally in serious writing today. *As for the style, it is cliché-ridden to a d.* | *All his other films have been heavy and humourless to a d.*

**deism, theism.** Though the original meaning is the same, the words have been so far differentiated that *deism* is understood to exclude, and *theism* (though less decidedly) to include, belief in supernatural revelation, in providence, and in the maintenance of a personal relation between Creator and creature.

**delectable.** In ordinary use (except by advertisers) the word is now ironical only; i.e. it is to be taken always, as *precious* is sometimes, to mean the opposite of what it says; even in this sense it is becoming rare. In poetry, sometimes in fanciful prose, and in *the d. mountains*, it retains its original sense; so in *Of all the fleeting visions which I have stored up in my mind I shall always remember the view across the plain as one of the most d.*

**deliberative.** The sense not hasty in decision or inference, which was formerly among those belonging to the word, has been assigned to *deliberate* by modern differentiation; the use of *d.* in that sense now instead of *deliberate* (*All three volumes are marked by a cautious and d. tone, that commends them to thoughtful men*) is to be classed with the confusions between ALTERNATIVE, DEFINITIVE, and *alternate, definite*.

**delusion, illusion.** It cannot be said that the words are never interchangeable; it is significant of their nearness in meaning that *illusion* has no verb in common use corresponding to *delude* (*illude* having almost died out), and *delusion* has none corresponding to *disillusion* (*undeceive* and *disillusion* being used according as the delusion has been due to deceit practised on the victim or to his own error). Nevertheless, in any given context one is usually better than the other; two distinctions are here offered:

1. A d. is a belief that, though false, has been surrendered to and accepted by the whole mind as the truth, and so may be expected to influence action. An i. is an impression that, though false, is entertained provisionally on the recommendation of the senses or the imagination, but awaits full acceptance and may be expected not to influence action. We labour under dd., but indulge in ii. The dd. of lunacy, the ii. of childhood or of enthusiasm. Delusive hopes result in misguided action, illusive hopes merely in disappointment. That the sun moves round the earth was once a d., and is still an i. The theatre spectator, the looker at a picture or a mirror, experience i.; if they lose consciousness of the actual facts entirely, the i. is complete; if the spectator throws his stick at the villain, or the dog flies at his image, i. has passed into d.

2. The existing thing that deludes is a d.; the thing falsely supposed to exist, or the sum of the qualities with which an existing thing is falsely invested, is an i. Optimism (if unjustified) is a d.; Heaven is (if non-existent) an i. If a bachelor dreams that he is married, his marriage is an i.; if he marries in the belief that marriage must bring happiness, he may find that marriage is a d. A mirage, or the taking of it for a lake, is a d.; the lake is an i. What a conjuror actually does—his real action—is a d.; what he seems to do is an i.; the belief that he does what he seems to do is a d. The world as I conceive it may for all I know be an i.; and, if so, the world as it exists is a d.

**demagogic etc.** For pronunciation see GREEK G.

**demand.** We d. something *of* or *from* a person; we make a d. *on* him.

**demean.** There are two verbs. One, which is always reflexive, means to conduct oneself or behave, and is connected with *demeanour* and derived from old French *demener* (*He demeans himself like a king*). The other, which is usually but not always reflexive (*I would not d. myself to speak to him*; *A chair which it would not d. his dignity to fill*), means to lower or debase. This seems to be the product of a confusion between the first verb and the adjective *mean*, and, though it is occasionally found as a normal word in good authors, it is commonest on the lips of the uneducated or in imitations of them, and is best avoided except in such contexts.

**demesne.** 1. *Dĕmē´n* used to be the favoured pronunciation, but *dĕmā´n*, said by the OED to be 'in good legal and general use and historically preferable', is now probably commoner. 2. *Demesne, domain.* The two words are by origin the same, but in technical use there are several distinctions between them that cannot be set forth here. In the wide general sense of sphere, region, province, the established form is *domain*, and the use of *demesne* is due to NOVELTY-HUNTING.

**demi.** The only English words in ordinary use today in which the prefix *demi-* survives are *demigod*, *demijohn*, and *demisemiquaver*. We now use *half-* or *semi-*, or, for words of Greek origin, *hemi-*. The variant *demy* (pr. *dĕmī´*) remains for the size of paper and the scholar of Magdalen College Oxford, who used to be given half a fellow's allowance.

**demise, devise.** Both are legal terms for the assignment of property, but *devise* is now limited to realty passing by will (personalty is *bequeathed*), and *demise* has also a special application to the transfer of sovereignty on the death or abdication of a monarch (*d. of the*

*Crown*). *Demise* is, by transfer, sometimes used for the occasion that caused it—a FORMAL WORD for *death*.

**demoniac(al).** The adjectives are not clearly differentiated; but there is a tendency to regard *-acal* as the adjective of *demon*, so that it is the form chosen when wickedness is implied, and *-ac* as the adjective of the noun *demoniac*, so that it is chosen to convey the notion of the intensity of action produced by possession (*demoniacal cruelty, demoniac energy*). Pronounce *-ŏ′nĭăk, -ī′ăkl*.

**de′monstrate, de′monstrator, de-mo′nstrable, demo′nstrative.** The accents are those shown with *de′-monstrable* as an alternative for the third. See RECESSIVE ACCENT.

**demote** is an importation from America recent enough for the supporting English quotations in the OED Supp. to mark it as a colloquialism by putting it in inverted commas. Its convenience is likely to establish it in our vocabulary; etymologically it is no less respectable than *promote*.

**demur.** For the pronunciation of *demurrer* etc. see PRONUNCIATION 7.

**Denmark.** For 'something rotten in the state of D.' see IRRELEVANT ALLUSION.

**denote.** See CONNOTE.

**dénouement.** The clearing up, at the end of a play or tale, of the complications of the plot. (The Aristotelian word is λύσις, loosing, that follows the δέσις, tying). A term often preferred to the English *catastrophe* because in popular use that word has lost its neutral sense.

**dent, dint,** are variants of the same word, meaning originally a blow or the mark caused by one. Both words retain the latter sense; the former survives only in the phrase *by dint of*.

**denunciate.** See LONG VARIANTS.

**department.** For synonymy see FIELD.

**depend.** The construction illustrated below, in which *it depends* is followed by an indirect question without *upon*, is common, but slovenly: '*Critics ought to be artists who have failed.' Ought they? It all depends who is going to read the criticism, and what he expects to learn from it.* Even less defensible (except in the colloquialism *it all*—or *that*—*depends*) is the wholly unsupported use of *depend. Those captains whose choice, while more or less of a formality, has depended from match to match.*

**dependable.** For such formations (that can be depended *upon*), see -ABLE 4.

**dependant, -ent.** In Britain the noun has *-ant*, the adj. *-ent*. In U.S. *-ent* is usual for both.

**dependence, -ency.** The first is now used in all the abstract senses (*a life of -ce; no -ce can be put upon his word; the -ce of the harvest on weather*), and *-cy* is virtually confined to the concrete sense of a thing that depends upon or is subordinate to another, especially a dependent territory. See -CE, -CY.

**depositary, -tory.** The first is properly applied to the person or authority to whom something is entrusted, and *-tory* to the place or receptacle in which something is stored; and the distinction is worth preserving, though in some contexts (*a diary as the d. of one's secrets; the Church as the d. of moral principles*) either may be used.

**depot.** Write without accents or italics, and pronounce *dĕ′pō* (U.S. *dē′pō*).

**deprecate** (do the reverse of *pray* for) and its derivatives *-cation, -catory*, often appear in print, whether by the writer's or the compositor's blunder, in place of *depreciate* (do the reverse of *praise*) and its derivatives *-ciation, -ciatory*: *Mr. Birrell's amusing deprecation of the capacity of Mr. Ginnell to produce a social revolution in Ireland. | The self-deprecatory mood in which the English people find themselves.*

**deprival.** See -AL NOUNS.

**deprivation.** Pronounce either *dĕprī-* or *dēprĭ-*, not *dĕprī-* nor *dēprī-*.

**derby** etc. Unlike the English county, town, earl, and horse-race, which are pronounced *darbi*, the *derby hat* (U.S. for *bowler*) and the shoes called *derbies* are pronounced *dĕr-*. The slang word for handcuffs is spelt, as it is pronounced, *darbies*.

**derelict, dereliction,** in modern usage are differentiated: *derelict* = abandoned, especially of a ship; *dereliction*, especially of duty = reprehensible neglect.

**derisory.** The OED definitions (dated 1894) make no distinction between this and *derisive*, being almost in the same words for both. About the meaning of *derisive* (conveying derision, deriding) there is no doubt; and if *derisory* means precisely the same it may well be regarded as a NEEDLESS VARIANT, so clearly is *derisive* now in possession. But, by the sort of differentiation seen in MASTERFUL and *masterly*, a distinct sense has been given to *derisory*; it was recognized by the OED Supp. and is now common. As *derisive* means conveying derision, so *derisory* means inviting or worthy only of derision, too insignificant or futile for serious consideration; it is applied to offers, plans, suggestions, etc. As Larousse illustrates the use of *dérisoire* by 'proposition derisoire', the new sense may be a GALLICISM, but it would be a natural enough development in English, the word being no longer needed in the sense now nearly monopolized by *derisive*. If the differentiation is to be satisfactory, *derisory* should, like *masterful*, be no longer recognized in its former sense. See also RISIBLE. The following quotation gives the modern sense unambiguously: *They will not cover the absence of those supplies from the Ukraine and Roumania which were promised to the people and have only been forthcoming in derisory quantities.*

**derring-do.** This curious word, now established as a 'pseudo-archaic noun' (COD) meaning desperate courage, is traced to a misinterpreted passage of Chaucer, in which Troilus is described as second to none 'In dorryng don that longeth to a knyght', i.e. 'in daring (to) do what belongs to a knight'. Spenser, a lover of old phrases, apparently taking it for a noun, as if the line meant 'in bold achievement, which is a knightly duty', made *derrynge do* in this sense a part of his regular vocabulary. (The spelling is due to a misprint in Lydgate, who adopted the phrase.) The derivation is a surprise; but, if Spenser did make a mistake, it does not follow that modern poetical writers, under the lead of Scott, should abstain from saying 'deeds of derring-do'; the phrase is part and parcel of an English that is suited to some occasions.

**descant.** The noun as a technical term of music (*dĕ′skant*) and the verb meaning to talk at large (*dĕskant′*) are from the same O.F. word meaning to sing, though they have wandered a long way apart.

**descendable, -ible** (heritable). The dictionaries prefer the second; see -ABLE 2.

**description.** The less this is used as a mere substitute for *kind* or *sort* (*no food of any d.*; *crimes of this d.*; *every d. of head-covering*), the better. *It is not an avocation of a remunerative description—in other words it does not pay.* Mr. Micawber's second thoughts were best. See WORKING AND STYLISH WORDS.

**desert, dessert.** Only the noun meaning an arid waste is accented on the first syllable; the noun meaning what one deserves, the noun meaning the fruit course and the verb meaning to abandon or abscond are all accented on the second.

**desiderate** (feel the want of, 'think long for') is a useful word in its place, but is so often misplaced that we might be better without it. Readers, outside the small class that keeps up its Latin, do not know the meaning of it, taking it for the scholar's pedantic or facetious form of *desire*. Writers are often in the same case (see the sentence quoted below; we do not d. what we can preserve), and, if they are not, are ill-advised in using the word unless they are writing for readers as learned as themselves: *In this she acts prudently, probably feeling that there is nothing in the Bill that could prevent her, and those like-minded, acting as benevolently towards their servants as before, and so preserving the 'sense of family unity' she so much desiderates.*

**desideratum.** Pl. *-ta*; see -UM. Pronounce *dĕsĭdĕrā'tum*.

**desire** is a GENTEELISM for *want*.

**desolated,** as polite exaggeration for *very sorry* etc., is a GALLICISM.

**despatch.** See DISPATCH.

**desperado.** Pronounce *-ahdō*, see -ADE-, -ADO. Pl. *-oes*, see -O(E)S 1.

**desperation** never now means, as formerly, mere despair or abandonment or loss of hope, but always the reckless readiness to take the first course that presents itself when every course seems hopeless.

**despicable.** Pronounce *dĕ'spĭkăbl*, not *dĕspĭ'kabl*; see RECESSIVE ACCENT.

**despoil.** *Sir, Not only are portable wirelesses now common in Kensington Gardens but they even d. Kew Gardens, the only quiet place left in London. Despoil* means 'rob', with an implication that the robbing is done pretty thoroughly. *Spoil* may be used, with a touch of archaism, in the same sense, as the Children of Israel were said to have spoiled the Egyptians when they borrowed expensive jewellery and clothes from them and then decamped. But only *spoil* has the meaning of 'mar the enjoyment of'.

**destine.** (*Who was*) *destined to be* etc., when it means no more than *who has since become* or *afterwards became*, is a BATTERED ORNAMENT.

**detail.** As military terms both noun and verb are *dĕtāl'*. In ordinary usage the noun is *dē'tāl*; the verb is rare

except in its p.p., and that is always *dē'tailed*.

**deter.** Pronounce the participle *dĕtŭr'ĭng*, but the adj. *detĕ'rent*; see PRONUNCIATION 7.

**determinately, determinedly.** The first is used of what has definite limits; the sense with determination, in a resolute way, does not belong to it at all, though some writers use it (*Thurlow applied himself -ately to the business of life*) as an escape from the second. A better escape is to use *resolutely, firmly, with determination*, or some other substitute. The objection to *determinedly*, which is very general, is perhaps based on reluctance to give it the five syllables that are nevertheless felt to be its due (see -EDLY). An example or two will illustrate the ugliness of the word: *In causes in which he was heart and soul convinced no one has fought more -edly and courageously* (with greater determination and courage). | *Cobbett opposed -edly the proposed grant of £16,000.* | *However, I -edly smothered all premonitions.* | *He is -edly opposed to limited enfranchisement.*

**determinism.** See FATALISM.

**detour, dé-.** Write without accent and italics. The pronunciations *dā'toor* in Britain and *dĕ'toor* in U.S. are probably now more usual than *dĭtoor'* preferred by the OED.

**deus ex machina** (pronounce *mā'kĭnä*). The 'machine' is the platform on which, in the ancient theatre, the gods were shown suitably aloft and from which they might descend to take a hand in settling the affairs of mortals. Hence the phrase means the providential intervention of a person or event to solve a difficulty in the nick of time in a play or novel.

**Deuteronomy.** The accentuation *dūtĕrŏ'nŏmĭ* is better than *dū'tĕrŏnŏmĭ*, which is impossible for the ordinary speaker; cf. GLADIOLUS. See RECESSIVE ACCENT.

**device.** For synonymy see SIGN.

**devil**, n. *Devil's advocate* is very dangerous to those who like a picturesque phrase but dislike the trouble of ascertaining its sense. Often it is taken to mean a tempter, and in the following example the not unnatural blunder is made of supposing that it means a whitewasher, or one who pleads for a person who either is or is supposed to be wicked: *Because the d.'s a. always starts with the advantage of possessing a bad case, Talleyrand's defender calls forth all our chivalrous sympathy.* The real *d.'s a.*, on the contrary, is one who, when the right of a person to canonization as a saint is being examined, puts the devil's claim to the ownership of him by collecting and presenting all the sins that he has ever committed; far from being the whitewasher of the wicked, the d.'s a. is the blackener of the good. And in this other example the writer referred to is in fact devil's advocate in 'the rest of his book', and something quite different ('God's advocate', say) in 'an early chapter': *He tries in an early chapter to act as 'devil's advocate' for the Soviet Government, and succeeds in putting up a plausible case for the present régime. But the rest of his book is devoted to showing that this Bolshevist case is based on hypocrisy, inaccuracy, and downright lying.* See MISAPPREHENSIONS.

**devise.** See DEMISE, DEVISE.

**deviser, -sor.** *Devisor* is the person who devises property, and is in legal use only; *-er* is the agent-noun of *devise* in other senses; see -OR.

**devolute**, though an old verb in fact, has been dormant for three centuries, and is to be regarded rather as a BACK-FORMATION from *devolution* than as a REVIVAL; it is unnecessary by the side of *devolve*, which should have been used in (with *on* for *to*): *The House will devise means of devoluting some of its work to more leisured bodies.*

**dexter.** See SINISTER.

**dext(e)rous.** The shorter form is recommended.

**d(h)ow** is included by the OED among 'words erroneously spelt with dh'; *dow* was common down to 1860, but the *h* is now too firmly fixed to be dislodged.

**diabolic(al).** Roughly, *-ic* means of, and *-ical* as of, the devil: *Horns, tail and other -ic attributes*; *He behaved with -ical cruelty.* See -IC(AL).

**diaeresis.** The pronouncing of two successive vowels as separate sounds and not as a single vowel or diphthong, or the mark over the second (¨) sometimes used to indicate such separation, as in *Chloë, Danaë.* (A peculiarity in the French use of the mark may be mentioned by way of warning; in such words as *aiguë, ciguë*, the mark means not that the e is separate from the u, but that ue is not silent as in *fatigue* and *vogue*, but forms a distinct syllable.) The mark, when used in English, should not be regarded as a permanent part of any word's spelling, but kept in reserve for occasions on which special need of it is felt; cf. Æ, Œ, and CO-. In fact it is in English an obsolescent symbol; it has been dropped from the *aer-* words (see AERATE), and we now prefer, unreasonably, to use a hyphen for this purpose; we write *co-op*, not *coöp*, to distinguish the cooperative store from the place to keep hens in. But we still write *Boötes* and *naïve* (if we prefer the disyllabic pronunciation, see NAIF etc.), and if we took to writing *zoölogy* we should perhaps hear that word less often mispronounced.

**dialect.** For *d., patois, vernacular,* etc., see JARGON.

**dialectal, -ic, -ical.** The natural adjective for *dialect* would be *-ic* or *-ical*, and both forms were once used as such, besides serving as adjectives to the noun *dialectic*. But to avoid confusion *dialectal* has been formed and has found acceptance, so that we now speak of *dialectic(al) skill*, but *dialectal words* or *forms* if we prefer an adjectival form to the more ordinary use of the noun attributively (*dialect words*).

**dialogue** is not necessarily the talk of

two persons; it means conversation as opposed to monologue, to preaching, lecturing, speeches, narrative, or description; see DUOLOGUE.

**diarchy, dy-.** Spell *di-*. *D.* is to *monarchy* as *dibasic*, *dicotyledon*, *dioxide*, and *disyllable* are to *monocotyledon*, *monoxide*, *monosyllable*, and the other *mono-* words. *Monologue* and *dialogue* are not a relevant pair, *dialogue* having nothing to do with Gk. *di-* two-.

**diastole.** Pronounce *dīǎ'stolǐ*.

**dichotomy,** a technical term of logic, astronomy, and botany, has become a POPULARIZED TECHNICALITY and for many writers is now the inevitable word for any division, difference, cleavage, etc., between two things. *The French Revolution, however, deepened the d. between radical and conservative.* There was no need to look beyond *cleavage*.

**dicky.** See RHYMING SLANG.

**dictate.** Accent the noun (usu. pl.) *dǐ'ktāt(s)*, the verb *dǐktā't*; see NOUN AND VERB ACCENT.

**dictatress, -trix.** See FEMININE DESIGNATIONS.

**dictionary, encyclopaedia, lexicon.** A *d.*, properly so called, is concerned merely with words regarded as materials for speech; an *e.* is concerned with the things for which the words are names. But since some information about the thing is necessary to enable the words to be used rightly, most dictionaries contain some matter that is strictly of the encyclopaedic kind; and in loose use *d.* comes to be applied to any encyclopaedia that is alphabetically arranged. *Lexicon* means the same as *d.*, but is usually kept to the restricted sense, and is moreover rarely used except of Greek, Hebrew, Syriac, or Arabic dd. See also GLOSSARY.

**didacticism.** 'No mortal but is narrow enough to delight in educating others into counterparts of himself'; the statement is from *Wilhelm Meister*. Men, especially, are as much possessed by the didactic impulse as women by the maternal instinct. Some of them work it off *ex officio* upon their children or pupils or parishioners or legislative colleagues, if they are blest with any of these; others are reduced to seizing casual opportunities, and practise upon their associates or upon the world, if they can, in speech or print. The speaker who has discovered that *Juan* and *Quixote* are not pronounced in Spain as he used to pronounce them as a boy is not content to keep so important a piece of information to himself; he must have the rest of us call them *Hwan* and *Kēhōtā*; at any rate he will give us the chance of mending our ignorant ways by doing so. The orientalist whom histories have made familiar with the *Khalif* is determined to cure us of the delusion, implanted in our childish minds by hours with some bowdlerized *Arabian Nights*, that there was ever such a being as our old friend the Caliph. Literary critics saddened by our hazy notions of French do their best to improve them by using *actual* and *sympathetic* as though their meanings were the same as those of *actuel* and *sympathique* (see GALLICISMS). Dictionary devotees whose devotion extends to the etymologies think it bad for the rest of us to be connecting *amuck* with *muck*, and come to our rescue with *amok*. These and many more, in each of their teachings, teach us one truth that we could do as well without, and two falsehoods that are of some importance. The one truth is, for instance, that *Khalif* has a greater resemblance to Arabic than *Caliph* and *Kēhōtā* to the way Cervantes spoke than *Kwiksot*. Is that of use to anyone who does not know it already? The two falsehoods are, the first that English is not entitled to give what form it chooses to foreign words that it has occasion to use; and the second that it is better to have two or more forms coexistent than to talk of one thing by one name that all can understand. If the first is not false, why do we say *Germany* and *Athens* and *Brussels* and *Florence* instead of

*Deutschland* and the rest, or allow the French to insult us with *Londres* and *Angleterre*? That the second is false not even our teachers would deny; they would explain instead that their aim is to drive out the old wrong form with the new right one. That they are most unlikely to accomplish, while they are quite sure to produce confusion temporary or permanent; see MAHOMET for a typical case.

Seriously, our learned persons and possessors of special information should not, when they are writing for or speaking to the general public, presume to improve the accepted vocabulary or pronunciation. When they are addressing audiences of their likes, they may naturally use, to their heart's content, the forms that are most familiar to both parties; but otherwise they should be at pains to translate technical terms into English. And, what is of far greater importance, when they do forget this duty, we others who are unlearned, and naturally speak not in technical terms but in English, should refuse to be either cowed by the fear of seeming ignorant, or tempted by the hope of passing for specialists, into following their bad example without their real though insufficient excuse. See also POPULARIZED TECHNICALITIES.

**differ,** in the sense be different, exhibit a difference, is followed only by *from*, not by *with*. In the sense have a difference of opinion, express dissent, dispute, it is followed usually by *with*, but sometimes by *from*.

**difference.** *There is all the d. in the world between deceiving the public by secret diplomacy and carrying on the day-to-day business of negotiation from the housetops*. Why, certainly; but was it worth while to tell us so obvious a fact? If the writer had put in a *not* before either *deceiving* or *carrying*, he would have told us both something of value and what he meant. For other examples of this ILLOGICALITY see BETWEEN 5.

**different. 1.** That *d*. can only be followed by *from* and not by *to* is a SUPERSTITION. *To* is 'found in writers of all ages' (OED), and the principle on which it is rejected (You do not say *differ to*; therefore you cannot say *d. to*) involves a hasty and ill-defined generalization. Is it all derivatives, or derivative adjectives, or adjectives that were once participles, or actual participles, that must conform to the construction of their parent verbs? It is true of the last only; we cannot say *differing to*, but that leaves *d*. out in the cold. If it is all derivatives, why do we say *according, agreeably,* and *pursuant, to instructions*, when we have to say *this accords with*, or *agrees with*, or *pursues, instructions*? If derivative adjectives, why *derogatory to, inconceivable to*, in contrast with *derogates from, not to be conceived by*? If ex-participle adjectives, why do *pleases, suffices, neglects, me* go each its own way, and yield *pleasant to, sufficient for*, and *negligent of, me*? The fact is that the objections to *d. to*, like those to AVERSE *to*, SYMPATHY *for*, and COMPARE *to*, are mere pedantries. This does not imply that *d. from* is wrong; on the contrary, it is 'now usual' (OED); but it is only so owing to the dead set made against *d. to* by mistaken critics. For *Different than* see THAN 8.

**2.** *Different* has a prominent place in the jargon of advertisers. Every day we hear or read the slogan that so-and-so's cigarettes, soap, beer, chocolates, etc. are 'different'. Why do the advertisers attribute so great a potency to this word? It is not necessarily a virtue to be different; that depends on what one is different from, and how; and this we are not told. Perhaps the appeal of the word is to human nature's perpetual craving for change. The OED Supp., refusing to this use of the word any higher status than slang, defines it as 'out of the ordinary, special, *recherché*'. It must be losing its appeal, for advertisers are finding that it needs invigorating; we are now told that some of their wares are *super-different*, which is, it seems, only one degree below FABULOUS.

**differential** has various special uses as both adjective and noun in several branches of science. It is also a useful adjective in the sense of relating to or depending on a difference, especially with reference to rates of pay. The expression *differential rates of wages*, properly used, is a compendious way of saying that the differences in the rates are differential differences, that is to say they depend on differences in other things, e.g. the skill needed for the job. *Differential rents* are rightly so called; they depend on differences in the tenants' incomes. But then the rot sets in. *Differentials*, properly applied as a noun to these differential differences, is increasingly used, under the influence of LOVE OF THE LONG WORD, as an imposing synonym for differences of all sorts. No more than *difference* was meant for instance in *For a long time this territory was run under a paternalist régime, but on the principle that, in so far as political power was devolved, it was shared by all three races on an equal basis—the assumption being that the numerical differential was cancelled out by the differential of relative skills.* Perhaps the rot might be stopped if everyone were to bear in mind that Ophelia did not say *You must wear your rue with a differential*, nor did Wordsworth write *But she is in her grave, and O the differential to me.*

**differentiation.** In dealing with words, the term is applied to the process by which two words that can each be used in two meanings become appropriated one to one of the meanings and one to the other. Among the OED's 18th-c. quotations for *spiritual* and *spirituous* are these two: *It may not here be improper to take notice of a* wise and spiritual *saying of this young prince. | The Greeks, who are a* spirituous and wise *people*. The association of each with *wise* assures us rather startlingly that a change has taken place in the meaning of *spirituous*. It and *spiritual* have now been appropriated to different senses, and it would be difficult to invent a sentence in which one would mean the same as the

other; in other words, differentiation is complete. In a living language such differentiation is perpetually acting upon thousands of words. Most differentiations are, when fully established, savers of confusion and aids to brevity and lucidity, though in the incomplete stage there is a danger of their actually misleading readers who have not become aware of them when writers are already assuming their acceptance. When complete they improve the language by making it a finer instrument, just as the opposite process—SLIPSHOD EXTENSION—blunts it. Differentiations become complete not by authoritative pronouncements or dictionary fiats, but by being gradually adopted in speaking and writing; it is the business of all who care for the language to do their part towards helping serviceable ones through the dangerous incomplete stage to that in which they are of real value. There are many references through the book to this article. The matter is simple in principle, the difficulty being in the details; and all that need be done here is to collect, with some classification, a few differentiated words, those about which information is given in their dictionary places being printed in small capitals.

A. Words completely and securely differentiated: *adulteration* and *adultery*; *can* and *con*; *catch* and *chase*; *cloths* and *clothes*; *coffer* and *coffin*; *coign* and *coin*; *conduct* and *conduit*; *convey* and *convoy*; *costume* and *custom*; *courtesy* and CURTSY; *cud* and *quid*; *dam* and *dame*; *defer* and *differ*; DIVERS and *diverse*; EMERGENCE and *emergency*; PRONOUNCEMENT and *pronunciation*.

B. Words fully differentiated, but sometimes confounded by ignorant or too learned writers: ACCEPTANCE and *acceptation*; ALTERNATE and *alternative*; CONJURE' and *con'jure*; CONTINUANCE and *continuation*; DEFINITE and *definitive*; *distinct* and DISTINCTIVE; ECONOMIC and *economical*; ENORMOUSNESS and *enormity*; ESPECIAL(LY) and *special(ly)*; EXCEEDING(LY) and *excessive(ly)*; FATEFUL and fatal; *historic* and HISTORICAL; IMMOVABLE and *irremovable*; *informer* and *informant*; *intense* and INTENSIVE; LEGISLATION and *legislature*; LUXURIANT and *luxurious*; MASTERFUL and *masterly*; OLYMPIAN and *Olympic*; PRECIOSITY and *preciousness*; *rough* and ROUGHEN; *slack* and SLACKEN; STOREY and *story*; *transcendent* and TRANSCENDENTAL; TRIUMPHAL and *triumphant*; *uninterested* and DISINTERESTED.

C. Words in which an incipient or neglected differentiation should be encouraged: ASSAY and *essay* (vbs); CORRECTITUDE and *correctness*; DEAR and *expensive*; DEFECTIVE and *deficient*; *derisive* and DERISORY; EGOISM and *egotism*; FALSEHOOD, *falseness*, and *falsity*; *loose* and *loosen* (-EN verbs); OBLIQUENESS and *obliquity*; OPACITY and *opaqueness*; SPRINT and *spurt*; TRICKSY and *tricky*.

D. Words in which a desirable but little recognized differentiation is here advocated: APT and *liable*; CONSISTENCE and *consistency*; INCLUDE and *comprise*; INQUIRY and *enquiry*; PENDANT, *pennant*, and *pennon*; *proposal* and PROPOSITION; SPIRT and *spurt*; THAT and *which*; UTILIZE and *use*; WASTE and *wastage*.

E. Words vainly asking for differentiation: SPECIALITY and *specialty*.

F. Differentiated forms needlessly made: SPIRITISM for *spiritualism*; *stye* for STY; *tyre* for TIRE.

**digest.** Pronounce the noun *dī'jĕst*, the verb *dĭjĕ'st*; see NOUN AND VERB ACCENT.

**digit** has technical uses in arithmetic and other sciences; as a mere substitute for *finger*, it ranks with PEDANTIC HUMOUR.

**digraph, diphthong, ligature.** There is much confusion in the use of these words.

*Ligature* (binding together) is a typographical term for the union of two letters as when *fi* is printed *fi* or *oe œ*. Pairs of vowels when so printed are often wrongly called diphthongs.

*Diphthong* (two-sounding) is a term

for a compound vowel sound produced by combining two simple ones. Diphthongal sounds in English are *ī*, *ū*, *oi*, and *ou* (as in *loud*). Some phoneticians, but not all, would add *ā* and *ō*. It is immaterial whether the sound is represented by one letter or two; the *ī* in *idol* and the *ū* in *duty* are no less diphthongs than the *ei* in *eider* and the *ue* in *due*. But pairs of vowels which, though combined, are pronounced with a simple vowel sound (e.g. head, feet, boot, soup), are digraphs (see below) not diphthongs.

*Digraph* (two-writing) is a term for any two written consonants expressing a sound not analysable into two, as ph, dg, ch; or any two written vowels expressing a vowel sound, whether simple or compound, that is pronounced in one syllable, as (simple) ea in *beat* or *head*, ee in *heed*, ui in *fruit*, (diphthong) oi, eu, ou (in *out*). Digraphs therefore include all diphthongs except those that are written as single letters (see above) and are consequently often supposed not to be diphthongs.

**dike, dyke.** The first is the better form.

**dilatation, -lation, -latator, -lator, -later.** The forms **dilation, dilator** are wrongly formed on the false ANALOGY of *calculation, -lator*, etc. in which *-at-* represents the Latin 1st-conj. p.p. stem; in *dilate*, unlike *calculate*, the *-at* is common to the whole Latin verb, of which the adj. *latus* (wide) is a component. In surgical use the correct *-latation* and the incorrect *-lator* prevail.

**dilemma.** The use of *d.* merely as a finer word for difficulty when the question of alternative does not arise is a SLIPSHOD EXTENSION. The word is a term of logic, meaning an argument that forces an opponent to choose between two alternatives both unfavourable to him: he is presented with an *argumentum cornutum* or is on the horns of a dilemma, either of which will impale him. *Dilemma* should be used only when there is a pair, or at any rate a definite number, of lines that might be taken in argument or action and each is unsatisfactory. See POPU-LARIZED TECHNICALITIES.

**dim.** For 'dim religious light' see IRRELEVANT ALLUSION.

**diminutive** has a valuable technical sense in grammar; in general use (*a d. child, pony, apple, house, nose*) it used to be a favourite of the POLYSYLLABIC HUMOURIST and a faint aura of facetiousness still hangs about it.

**diocese.** The spelling *diocess* is now obsolete but the pronunciation is usually weakened to *-ĕs* or *-ĭs*. For *d.*, *bishopric*, and *see*, see SEE.

**diphth-.** *Diphtheria, diphthong*, and their derivatives, are sometimes misspelt, and very often mispronounced, the first -h- being neglected; *dĭfth-* is the right sound, and *dĭpth-* a vulgarism.

**diploma.** The pl. is now always *-mas*.

**diplomat(ist).** The shorter formation, standard in U.S., is increasingly used in Britain.

**direct(ly).** 1. The right adverb if straight is meant is *direct*; *directly* should be confined to the meaning immediately. *Go directly direct to Paris.* See UNIDIOMATIC -LY.
2. The conjunctional use of *directly* (*I came d. I knew*) is quite defensible, but is chiefly colloquial.

**directress, -trix.** See FEMININE DESIGNATIONS. Neither has found much favour as feminine of *director*, but *-trix* has a use in geometry (pl. *-trices*, see -TRIX).

**direful** is a POETICISM for *dire* in sense, and in formation is based on a false analogy (*dreadful*).

**dis-.** See DE- and DIS-.

**disassemble.** See DISSEMBLE, DIS-ASSEMBLE

**disassociate** is a NEEDLESS VARIANT of *dissociate*.

**disbalance.** See IMBALANCE.

**disc.** See DISK.

**discernable, -ible.** The second has driven out the first; see -ABLE 2.

**disciplinary.** The OED recognizes only *dĭ'sĭplĭnărĭ*; but this strains the principle of the RECESSIVE ACCENT almost to breaking-point, and *dĭsĭplĭ'-nărĭ*, admitted by the COD as an alternative, is here recommended.

**discomfit.** There is a tendency to use this in too weak or indefinite a sense (*Bell, conscious of past backslidings, seemed rather discomfited*). It is perhaps mistaken sometimes for the verb belonging to the noun *discomfort*. It has nothing to do with that; its primary meaning is overwhelm or utterly defeat.

**discontent.** For 'the winter of our d.' see IRRELEVANT ALLUSION.

**discrete** (separate, abstract, etc.) should be accented *dĭ'skrēt*, not *dĭs-krē't*; the first is both natural in English accentuation (cf. the opposed adj. *concrete*), and useful as distinguishing the word from the familiar *discreet*.

**discuss**, used with wine, food, etc., as object, may be classed with WORN-OUT HUMOUR.

**disfranchise**, not *-ize*; see -IZE, -ISE. An older and better word than *disenfranchise*.

**disgruntle(d).** 'Now chiefly U.S.' was the OED's comment, repeated by the SOED in 1933. If it was true, the verb was only visiting America: it was used in England 300 years ago, and its p.p., though not the rest of it, has come back to claim a useful place in our vocabulary.

**disgustful** was formerly common in the sense disgusting, but has been so far displaced by that word as to be a NEEDLESS VARIANT and is now only used jocularly.

**disillusion(ize).** It is a pity that there should be two forms of the verb.

The first is recommended; *disbud, discredit, disfigure, dismast,* give sufficient support for the use of *dis-* before a noun in the sense deprive or rid of; *-ize* is the refuge of the destitute and should be resorted to only in real destitution (see NEW VERBS IN -IZE); and the verbal noun is undoubtedly *disillusionment.*

**disinterested** means free from personal bias. Its use in the sense of uninterested (caring, like Gallio, for none of these things) was called by the OED 'qy. obs.'; but this comment was withdrawn in the 1933 Supp. and modern examples were given. This revival has since gathered strength; the examples that follow are drawn from a variety of sources: a cabinet minister, a gossip columnist, a book reviewer, and a dramatic critic. *I hope this will excuse me from the charge of being disinterested in this matter.* / *The doctors seem curiously disinterested in what happens to the patient afterwards.* / *The Italian campaign had started, owing to the disinterest of Washington, as a secondary campaign.* / *Through a fog of involved plot-making . . . one finally stumbles . . . on the philosophical kernel of the play. But by then a fair share of the audience may have justifiably succumbed to exasperation or disinterest.* A valuable DIFFERENTIATION is thus in need of rescue, if it is not too late.

**disjunctive.** Conjunctions implying not combination but an alternative or a contrast (as *or, but*) are so called, the others (as *and*) being *copulative.* The distinction is of some importance in determining the number of a verb after a compound subject; see NUMBER 2, 3.

**disk, disc.** 'The earlier and better spelling is disk' (OED). But in eleven of the sixteen examples given by OED Supp. of the modern uses of the word the spelling is *disc;* and that is now the popular form, e.g. *disc harrow, disc jockey,* and *slipped disc.*

**dispatch, des-.** The OED gives good reasons for preferring *dis-,* and the COD still puts it first; but *des-* continues a strong rival See also FORMAL WORDS.

**dispel** means to drive away in different directions, and must have for object a word to which that sense is applicable (*darkness, fear, cloud, suspicions*), and not, as in the following sentence, a single indivisible thing: *Lord C. effectually dispelled yesterday the suggestion that he resigned because he feared . . .* He might dispel the suspicion, or repel the suggestion, suspicion being comparable to a cloud, but suggestion to a missile.

**disposal, disposition.** In some contexts there is no choice (*His -ition is merciful; The -al of the empty bottles is a difficulty*); in some the choice depends upon the sense required. *The -ition of the troops* is the way they are stationed for action, and is general's work; *The -al of the troops* is the way they are lodged etc. and is quartermaster's work. When doubt arises, it is worth while to remember that *-ition* corresponds to dispose, and *-al* to dispose of. So *The -ition of the books is excellent* (they are excellently disposed, i.e. arranged), but *The -al of the books was soon managed* (they were soon disposed of, i.e. either sold or got out of the way); *The -ition of the body is stiff* (it is stiffly disposed, i.e. arranged), but *The -al of the body proved impossible* (it could not be disposed of, i.e. destroyed or concealed). *The testamentary -ition of property* (the way it is disposed or arranged by will) and *The testamentary -al of property* (the way it is disposed of or transferred under a will), describing the same act from different points of view, are naturally used without much discrimination. The same is true of *at* one's *-al* or *-ition;* but in this formula *-al* is now much commoner, just as *You may dispose of the money as you please* is now commoner than *You may dispose it.*

**disputable.** Accent *dĭ'spūtăbl,* not *dĭspū'tăbl;* see RECESSIVE ACCENT.

**disseise, -ze, disseisin, -zin.** Spell *-se, -sin* and pronounce *sēz*; see SEISE.

**dissemble, disassemble.** *They were engaged in dissembling their machines.* Now that the old word *disassemble* has been revived, there is no excuse for putting *dissemble* to this novel use.

**dissemble, dissimulate.** There is no clear line of distinction between the two. *Dissemble* is the word in ordinary use, and the other might have perished as a NEEDLESS VARIANT, but has perhaps been kept in being because, unlike *dissemble*, it is provided with a noun (*dissimulation*), and a contrasted verb (*simulate*), and is more convenient for use in connexion with these.

**dissoluble, dissolvable.** 1. Pronounce the second *dizo'lvabl* and the first *dĭ'sŏlōōbl* or *dĭsŏ'lūbl*; the word is hesitating between the rival attractions of *so'luble* and *dis'solute*. See RECESSIVE ACCENT.
2. *Dissoluble* is the established word, and may be used in all senses; but *dissolvable* often represents *dissolve* when it means make a solution of in liquid (*sugar* is *-vable* or *-uble in water*), and sometimes in other senses (*a Chamber -uble* or *-vable at the Minister's will*); see -ABLE 2.

**distil(l).** The modern form is *-il* (*-ill* in U.S.); see -LL-, -L-.

**distinction,** as a LITERARY CRITICS' WORD, is, like *charm*, one of those on which they fall back when they wish to convey that a style is meritorious, but have not time to make up their minds upon the precise nature of its merit. They might perhaps defend it as an elusive name for an elusive thing; but it seems rather to be an ambiguous name for any of several things, and it is often doubtful whether it is the noun representing *distinctive* (markedly individual), *distinguished* (nobly impressive), *distingué* (noticeably wellbred), or even *distinct* (concisely lucid). A few quotations follow; but the vagueness of the word cannot be brought out without longer extracts than are admissible, and the reader of reviews must be left

to observe for himself: *His character and that of his wife are sketched with a certain d.* / *She avoids any commonplace method of narration, but if she achieves a certain d. of treatment in the process, she detracts enormously from the interest of her story.* / *The book is written with a d. (save in the matter of split infinitives) unusual in such works.* / *Not only is distinctness from others not in itself d., but distinctness from others may often be the very opposite of d., indeed a kind of vulgarity.* / *Despite its length, an inclination to excessive generalization, and an occasional lack of stylistic d. verging upon obscurity, this book is a remarkable piece of literary criticism.*

**distinctive** means 'serving or used to discriminate', 'characteristic', 'so called by way of distinction'. But it is often misused (cf. DEFINITIVE, ALTERNATIVE) for *distinct*: *The refugees at length ceased to exist as a d. people.* / *Distinctively able and valuable.* On the other hand *distinctively* would have been the appropriate word in *The Swiss name of Edelweiss will be given to the village, the houses having the high-pitched roofs and other features of distinctly Swiss architecture.* Sometimes too *d.* is misused for *distinguished*: *During a long public life he served the interests of his class well in many d. positions.* / *Mr. K., Mr. R. B., Miss J. S. M., . . . and a number of other d. people.*

**distinctly,** in the sense really quite, is usually the badge of the superior person indulgently recognizing unexpected merit in something that we are to understand is not quite worthy of his notice: *The effect as the procession careers through the streets of Berlin is described as d. interesting.* / *Quite apart from its instructive endeavours, the volume is d. absorbing in its dealing with the romance of banking.*

**distrait, -te.** See FRENCH WORDS. Use *-ait* (*-ā*) of males and of things (*expression, air, mood, answer*, etc.), and *-aite* (*āt*) of females.

**disyllable, diss-.** The first is better; the double *s* is due to French, in which it served the purpose of indicating the hard sound (s, not z); the prefix is *di-*, not *dis-*.

**diurnal** should no longer be used in the sense of daily, i.e. recurring every day, though that was formerly one of its possible meanings. In modern use, (1) when opposed to *nocturnal* it means by day, (2) when opposed to *annual* etc. it means occupying a day.

**divan.** See COUCH. Pronounce *dĭvăn'*.

**divers(e).** The two words are the same, but differentiated in spelling, pronunciation, and sense. *Divers* (now mostly archaic or jocular), implying number, is accentuated on the first syllable, and *diverse*, implying difference, on the second; cf. *several* and *various*, each of which has both senses without differentiation.

**dividend stripping.** See BOND WASH-ING.

**do.** 1. *Did* subjunctive. 2. *Do have.* 3. *Do* as substitute.
  1. For *did* as in *Did I believe it, it would kill me* see SUBJUNCTIVES.
  2. *Do have.* Protests used to be common against the use of *do* as an auxiliary to *have*. Perhaps these were due to resistance to Americanisms, for in U.S. *Do you have a match? I don't have a match* are idiomatic where our own idiom requires *Have you (got) a match? I haven't (got) a match.* In British idiom *do have* is permissible only subject to two limitations: that *have* is used in a sense different from that of physical possession, and that habit or repetition or general practice is implied. *Do you have coffee?* means for us is it our habit to drink it; *have you coffee?* means is there any to make the drink with. *Do* when so used is performing its usual function as a colourless auxiliary in negative and interrogative sentences and as an emphasizing auxiliary in affirmative sentences. *People under 21 do not have the vote* (i.e. it is not our custom to give them one): *Do we have to pay for admission?* (i.e. is it the practice to make

a charge). If the sentence is affirmative—*People under 21 do have the vote | We do have to pay for admission*—the *do* implies that the assertion is contrary to expectation, and is analogous to the emphatic use of *do* in such sentences as *Do have a drink | You do have some odd ideas.* But these limitations on the use of *do have* apply only to the present tense; *did have* is as idiomatic as *had* in all senses of *have* in the past, and contains no implication of habit or custom. *I believe, although I do not have a copy of the speech, that I am quoting correctly* is idiomatic in America but not in England. *I believed, although I did not have a copy of the speech, that I was quoting correctly* is idiomatic in both countries. The American idiom is, however, encroaching on ours (see AMERICANISMS) and may prevail. This, says C. S. Lewis, is regrettable. 'The language which can with the greatest ease make the finest and most numerous distinctions of meaning is the best. . . . It was better to have the older English distinction between "I haven't got indigestion" (I am not suffering from it at the moment) and "I don't have indigestion" (I am not a dyspeptic) than to level both, as America has now taught most Englishmen to do, under "I don't have".' For this use of *got* see GET 1.
  3. *Do* as substitute. The use of *do*, whether by itself or in conjunction with *as, so, it, which*, etc., instead of a verb of which some part has occurred previously, is a convenient and established idiom; but it has often bad results.
  (*a*) *They do not wish to see the Act of 1903 break down, as break down it is bound to do*; omit either *break down* or *do.*
  (*b*) *Great Britain is faithful to her agreements when she finds an advantage in doing so. | The ratepayers have no voice, as the unions do, in fixing the amount of the levy. Do* etc. must not be substituted for a copulative *be* or for *have* in the sense of possess (see AS 5). Read *in being so* and *as the unions have.*
  (*c*) *As to the question whether sufficient*

*is known as to the food of birds, the author feels bound to reply that we do not. | Although nothing is said as to Cabinet rank being associated with the two offices, it may be assumed that both do so. | The title of 'Don' is now applied promiscuously throughout Spain very much as we do the meaningless designation of 'Esquire'. | Some of them wrote asking to be reinstated, which we did. | If imposition of the federation by force is ruled out, as all the members of the Commission do . . . | A large number had been grudgingly supported by relatives who would now cease to do so. | Why was it not pushed to a victorious conclusion in the House of Lords, where the party had the power to do so?* Unless the subject and the voice of *do* will be the same as those of the previous verb, it should not be used. But transgression of this rule is of varying gravity; sometimes it results in flagrant blunders, as in the first two examples, and sometimes merely in what, though grammatically defensible (since *do so* means strictly *act thus*) is nonetheless an offence against idiom.

(*d*) *The dissolution which was forced upon the country was deliberately done so as to avoid giving an advantage to the Unionists. | The ambassador gave them all the assistance which the Imperial nature of his office made it obligatory upon him to do so. | We have got to make a commission in the Territorial Force fashionable, the right thing for every gentleman to do. | To inflict upon themselves a disability which one day they will find the mistake and folly of doing.* In these examples *do* is in grammatical relation to a noun (*dissolution, assistance, commission, disability*) that is only a subordinate part of the implied whole (*the forcing of a dissolution, the giving of assistance, the holding of a commission, the inflicting of a disability*) to which alone it is in logical relation; we do not *do* a dissolution, a commission, etc. These sentences, in which *do* is a transitive verb meaning *perform*, are not genuine examples of the substitute *do*; but the mistakes in them are due to the influence of that idiom.

**docile.** The OED pronounces *dō'sīl* or *dŏ'sīl*, with preference to the first, which has since become the only current pronunciation in Britain. See -ILE.

**doctor.** See PHYSICIAN.

**doct(o)ress.** See FEMININE DESIGNATIONS.

**doctrinal.** In Britain the pronunciation *dŏktrī'nal* seems to be making strong headway against *do'ktrīnal* which is preferred by most dictionaries and is standard U.S. See FALSE QUANTITY and RECESSIVE ACCENT.

**document.** It is sometimes forgotten that the word includes more than the parchments or separate papers to which it is usually applied; a coin, picture, monument, passage in a book, etc., that serves as evidence, may be a d., and the following remark on 'Documents illustrative of the Continental Reformation' is absurd: *It is a collection not only (as the title implies) of dd., but also of passages from books and letters.* The phrase *human d.* is more than a mere metaphor and underlies the new meaning given to *documentary* (too recent to have appeared in the OED Supp.) for a film or television programme presenting an activity of real life with imagination but without fictional colouring, sometimes using professional actors, sometimes not.

**dodecaphonic.** See ATONAL.

**doe.** See HART.

**dogma.** Pl. *-mas*; see LATIN PLURALS.

**doily, doiley, doyly.** The first is the OED spelling and the usual one now. It is an EPONYMOUS WORD.

**dolce far niente.** See BATTERED ORNAMENTS.

**domain.** For synonymy see FIELD. See also DEMESNE.

**Domesday, dooms-.** *D. Book* is spelt *Domes-* but pronounced *dōomz-*; elsewhere the spelling is *dooms-*.

**domestic,** n. If one of the two words in *domestic servant* has to serve for the whole, the differentiating one, *domestic*, is clearly the more suitable, and was in fact originally the one used. A change of fashion brought *servant* into favour and *domestic* became almost a GENTEELISM. But the taint of subservience that is supposed to attach to *servant* forced that word in turn out of currency, first in U.S. and then in Britain, and *the master* and *the mistress* likewise disappeared. OED Supp. gives an American quotation dated 1818: *Servants, let me here observe, are called 'helps'. If you call a servant by that name they leave without notice.* In Britain, after a temporary reversion to *domestic*, we have fallen back on the American word, but neither that EUPHEMISM nor the attempt of the *Institute of Houseworkers* to give the occupation a new look seems to have had any marked effect on its present unpopularity.

**domesticity.** The OED pronounces *dō-*; see FALSE QUANTITY. But the short *o* must now be at least as common and will probably prevail. The COD puts it first.

**donate.** The OED's comment 'chiefly U.S.' is still true, but the word is freely used in Britain as a FORMAL WORD for give. It is a BACK-FORMATION from *donation*.

**donation** has escaped the taint of formality that still attaches in Britain to *donate*, and serves to mark a useful distinction between a single gift (*donation*) and one repeated periodically (*subscription*).

**double case.** *An ex-pupil of Verrall's . . . cannot but recall the successive states of mind that he possessed—or, more truly, possessed him—in attending Verrall's lectures.* Here *that* is first objective and then subjective; see CASES 3 D, THAT rel. pr. 6, WHAT 3, and WHICH 2.

**double construction.** *They are also entitled* to prevent *the smuggling of alcohol into the States,* and to reason-able assistance *from other countries to that end.* 'Entitled to prevent [infin.] . . . and to assistance [noun]' is a change of a kind discussed in SWAPPING HORSES.

**double entendre** is the established English form, and has been in common use from the 17th c.; the modern attempt to correct it into *double entente* suggests ignorance of English rather than knowledge of French; cf. À L'OUTRANCE. See FRENCH WORDS.

**double negatives.** See NEGATIVE MISHANDLING.

**double passives.** *The point is sought to be evaded*: monstrosities of this kind, which are as repulsive to the grammarian as to the stylist, perhaps spring by false analogy from the superficially similar type seen in *The man was ordered to be shot.* But the forms from which they are developed are dissimilar: They ordered the man *to be shot,* but They seek *to evade* the point; whereas *man* is one member of the double-barrelled object of *ordered, point* is the object not of *seek* at all, but of *evade.* It follows that, although *man* can be made subject of the passive *was ordered* while its fellow-member is deferred, *point* cannot be made subject of the passive *is sought,* never having been in any sense the object of *seek.*

To use this clumsy and incorrect construction in print amounts to telling the reader that he is not worth writing readable English for; a speaker may find himself compelled to resort to it because he must not stop to recast the sentence he has started on, but writers have no such excuse. Some of the verbs most maltreated in this way are *attempt, begin, desire, endeavour, hope, intend, propose, purpose, seek,* and *threaten* (commonest of all perhaps is *fear* in such a sentence as *all the passengers are feared to have been killed* or *feared killed*; so common as to qualify as a STURDY INDEFENSIBLE). A few examples follow: *Now that the whole is attempted to be systematized. The mystery was assiduously, though vainly, endeavoured to be discovered.*

*The darkness of the house (forgotten to be opened, though it was long since day) yielded to the glare. | A process whereby a tangle of longlasting problems is striven to be made gradually better. | A new definition of a drunkard was sought to be inserted into the Bill.* In legal or quasi-legal language this construction may sometimes be useful and unexceptionable: *Diplomatic privilege applies only to such things as are done or omitted to be done in the course of a person's official duties. | Motion made: that the words proposed to be left out stand part of the Question.* But that is no excuse for admitting it to literary English.

**doubletalk,** formerly applied in America to a particular kind of jocular nonsense, is now mostly used for a way of speech that deliberately employs words not as symbols but as excitants of emotion. It is a weapon of propaganda generally associated with the Communist ideology, but by no means confined to it. By using doubletalk the same thing may be called either *tyranny* or *democracy,* either *aggression* or *liberation,* either *rebellion* or *self-determination.* In Orwell's *Nineteen Eighty-four* this language, called by him *Newspeak,* was an instrument of government designed eventually to produce *doublethink,* 'the power of holding two contradictory beliefs in one's mind simultaneously and accepting both'.

**doubt(ful).** It is contrary to idiom to begin the clause that depends on these with *that* instead of the usual *whether,* except when the sentence is negative (*I do not doubt . . .; There is no doubt . . .; It was never doubtful . . .*) or interrogative (*Do you doubt . . .?; Is there any doubt?; Can it be doubtful . . .?*). *Whether* (or *if*) is used to imply that doubt exists, *that* (or *but that*) to imply that it does not. The mistake against which warning is required is the use of *that* in affirmative statements. It is especially common (probably from failure to decide in time between *doubt* and *deny* or *disbelieve, doubtful* and *false*) when the clause is placed before

*doubt(ful)* instead of in the normal order. *Whether* should have been used in: *I must be allowed to doubt that there is any class who deliberately omit . . . | Mr. Mikoyan thought the present position a good one and doubted that it would be altered. | That the movement is as purely industrial as the leaders claim may be doubted. | That his army, if it retreats, will carry with it all its guns we are inclined to doubt.* An even worse solecism is the use of *doubtful* without any conjunction: *It is more than doubtful they have learned the lessons which the war should have inculcated.* Nor can the use of *that* be tolerated as a far from elegant variation: *It is very doubtful whether it was ever at—and still more doubtful that it came from— . . .*

**doubtless, no doubt, undoubtedly,** etc. *Doubtless* and *no doubt* have been weakened in sense till they no longer convey certainty, but either probability (*You have doubtless* or *no doubt heard the news*) or concession (*No doubt he meant well enough*; *It is doubtless very unpleasant*). When real conviction or actual knowledge on the speaker's part is to be expressed, it must be by *undoubtedly, without (a) doubt,* or *beyond a doubt* (*He was undoubtedly guilty*).

**dour.** Pronounce to rhyme with *moor,* not with *hour.*

**douse, dowse.** The latter is the only spelling of the verb concerned with the divining-rod (so *dowser, dowsing-rod,* etc.). The other verb or verbs are spelt both ways; *-use* is commoner, and is to be encouraged for the sake of DIFFERENTIATION.

**dow.** See DHOW.

**dower, dowry.** The two words, originally the same, are differentiated in ordinary literal use, *dower* being the widow's life share of her husband's property, and *dowry* the portion brought by a bride to her husband; but in poetic or other ornamental use *dower* has often the sense of *dowry*; and either is applied figuratively to talents etc.

**down.** See UP AND DOWN.

**doyen.** See DEAN.

**dozen.** See COLLECTIVES 5.

**drachm, drachma, dram.** *Drachm* was formerly the prevalent form in all senses; but now the coin is almost always *drachma* (pl. *-s* or *-e*), the small draught of alcoholic liquor is always *dram*, and *dram* is not uncommon even where *drachm* is still usual, in apothecaries' and avoirdupois weight. Pronounce *drachm* drăm, and *drachma* dră′kma.

**draft, draught,** etc. *Draft* is merely a phonetic spelling of *draught*, but some differentiation has taken place. *Draft* has ousted *draught* in banking, and to a great extent in the military sense detach(ment); it is also usual in the sense (make) rough copy or plan (a good *draftsman* is one who drafts Bills well, a good *draughtsman* one who draws well). In all the other common senses (game of *dd.*, air-current, ship's displacement, beer on d., beast of d., haul of fish, dose, liquor), *draught* is still the only recognized British form; in U.S. *draft* is much more widely used.

**dragoman.** The pl. is correctly *-mans*, and usually *-men*; the word is a corruption of an Arabic word meaning interpreter, and has no connexion with English *man*. Insistence on *-mans* is DIDACTICISM.

**dream.** The ordinary past and p.p. is *dreamt* (*-ĕmt*); *dreamed* (*-ēmd*) is preferred in poetry and in impressive contexts. See also -T and -ED.

**drib(b)let.** *Driblet* is both the usual and (f. obs. vb *drib*+*-let*) the more correct form.

**drink** has past tense *drank*, p.p. *drunk*; the reverse uses (*they drunk*, *have drank*) were formerly not unusual, but are now blunders or conspicuous archaisms.

**droll.** For synonymy see JOCOSE.

**dromedary.** The abnormal pronunciation *drŭm-*, put first in the OED, continues to resist the influence of the spelling, though the COD has relegated it to second place.

**drunk(en).** The difference, as now established, is complex. In the sense now the worse for drink, *drunken* is strictly the correct attributive adjective and *drunk* the predicative; *I saw a drunken man*; the man I *saw was drunk*. But *drunk* is now increasingly used colloquially as an attribute, and even as a noun. *There were a lot of drunk people* (or *drunks*) *about*. *Drunken* is, however, still the only admissible form for the meaning given to drink or symptomatic etc. of drunkenness (*A lazy -en lying ne'er-do-weel*; *His -en habits*). It may be used predicatively also, but only in the sense *given to drink* (cf. *He was -en and dissolute* with *He was drunk and incapable*); *He was -en yesterday* is contrary to modern idiom.

*Drunk* is fertile of synonyms. The formal *inebriated* is now little used, and the formal *intoxicated* only in formal contexts; polite alternatives are *under the influence* and *the worse for liquor*. The genteelisms *tipsy* and *half-seas-over* are outmoded, and so are the slang *boozed*, *oiled*, *primed*, *screwed*, and *squiffy*. Slang terms still current include, in ascending order of severity, *lit up*, *merry*, *tiddly*, *tight*, *bottled*, *canned*, *pickled*, *sloshed*, *sozzled*, *plastered*, *blind*, *blotto*, and *out*.

**dry** etc. The spelling in some derivatives of *dry* and other adjectives and verbs of similar form (monosyllables with *y* as the only vowel sound) is disputable. The prevalent forms for the words derived from the adjective are *drier*, *driest*, *drily* (sometimes *dryly*), *dryness*, *dryish*. The word derived from the verb—*dryer* or *drier*—has become much commoner owing to the invention of machines called by that name for drying hair, grain, laundry, and other things. The makers of them are not yet agreed about the spelling, but *drier* seems to be the more favoured.

# dual(istic) 141 due

These inconsistencies, however regrettable, are beyond remedy; and the case is not much better with many of the other similar words. See FLIER, SKIER, SLY, SPRY, and VERBS in -IE 6.

**dual(istic).** *Dual* has recently come into favour for certain special purposes, e.g. *d. control* (of motorcars and aircraft), *d. purpose* (of cattle), *d. personality* (of schizophrenics) and *d. carriage-way*. But in their original meanings both words are of the learned kind, and better avoided when such ordinary words as *two, twofold, twin, double, connected, divided, half-and-half, ambiguous*, will do the work: *The skirt was dual* (divided), *and rather short.* | *Dual* (double) *ownership.* | *The dual* (connected) *questions of 'abnormal places' and a minimum wage would bring about a deadlock.* | *The Government is pleased with the agitation for electoral reasons, but does not desire it to be too successful; the reason for this dualistic* (ambivalent) *attitude is that* . . . For the term of philosophy see MONADISM.

**dub. 1.** As a transitive verb *d.* has several meanings. The oldest is to confer a knighthood by a tap on the shoulder with a sword, later extended, often jocularly, to inventing a nickname (*Mr. G. is now dubbed the Young Pretender by some of the right-wing revisionists*). The most recent is to re-record the sound-track of a film, especially when substituting a different language for the original. The origin of this usage is obscure; it may be a further extension of the old meaning, or it may be an abbreviation of *double*.

**2.** The noun *dubbin(g)* should be spelt with *-g*; it is from *dub* smear with grease, and parallel to *binding, seasoning*, etc.

**dubiety.** Pronounce *dūbī'ĕtĭ*; see FALSE QUANTITY. For *d.* and *doubt*, see WORKING AND STYLISH WORDS.

**duck.** For pl. see COLLECTIVES 3.

**due.** Has *due to*, using the weapon of ANALOGY, won a prescriptive right to be treated as though it had passed, like OWING TO, into a compound preposition? May we now regard as idiomatic such sentences as *Due to the great depth of water, air pressures up to 50 lb. per square inch will be necessary.* | *Rooks, probably due to the fact that they are so often shot at, have a profound distrust of man.* | *Due to last night's rain play will be impossible before lunch*? Or must we say that, although in these quotations *owing to* would stand, *due*, which must, like ordinary participles and adjectives, be attached to a noun, and not to a notion extracted from the sentence, is impossible; that it is not the pressures, the rooks, and play that are due but the force of the pressure, the timidity of the rooks, and the absence of play?

The prepositional use of *owing to* is some 150 years old, but of a similar use of *due to* there is not a vestige in the OED (1897); in the 1933 Supp. it is said to be 'frequent in U.S. use', and in 1964 the COD tersely dismisses it as 'incorrect'. The original edition of the present Dictionary said this: 'It is now as common as can be, though only, if the view taken in this article is correct, among the illiterate; that term is here to be taken as including all who are unfamiliar with good writers, and are consequently unaware of any idiomatic difference between *Owing to his age he was unable to compete* and *Due to his age he was unable to compete*. Perhaps the illiterates will beat idiom; perhaps idiom will beat the illiterates; our grandsons will know.' Now, when this usage is still 'as common as can be' and is freely employed by BBC announcers, it seems clear that idiom, though still resisting stoutly, is fighting a losing battle. The offending usage has indeed become literally part of the Queen's English. *Due to inability to market their grain, prairie farmers have been faced for some time with a serious shortage of sums to meet their immediate needs.* (Speech from the Throne on the opening of the Canadian Parliament by Elizabeth II, 14 Oct. 1957.)

**dul(l)ness, ful(l)ness.** The spelling is not fixed, but it is best to use -*ll*-, as in all other words in which -*ness* follows -*ll* (*drollness, illness, shrillness, smallness, stillness,* etc.); see -LL-, -L-, 4.

**duologue** is a bad formation, but is now past resisting. There are indeed difficulties in the way of making a good one; *dyologue,* which would be better only in one respect, is indistinguishable in sound from *dialogue*; *dilogy* is used for another purpose; moreover it conflicts with *trilogy* and *tetralogy*; *dittologue* suggests *ditto*; *biloquy* after *soliloquy* would have been less bad than *duologue* after *monologue,* but has not been given a chance.

**durance, duress(e).** 1. *Durance* now means only the state of being in confinement, is a purely decorative word, and is rare except in the cliché *in durance vile. Duress* means the application of constraint, which may or may not take the form of confinement; it is chiefly in legal use, with reference to acts done under illegal compulsion, and is commonest in the phrase *under duress.*
2. Spell *duress* and pronounce *dūrĕs'*; the spelling with a terminal *e* is obsolete.

**Dutch.** See NETHERLANDS.

**duteous, dutiful.** The second is the ordinary word; *duteous* (a rare formation, exactly paralleled only in *beauteous*) is kept in being beside it by its metrical convenience (six of the seven OED quotations are from verse), and when used in prose has consequently the air of a POETICISM; see also PLENTEOUS.

**dutiable.** For such forms see -ABLE 4.

**duty.** See TAX.

**dwell,** in the sense have one's abode, has been ousted in ordinary use by *live,* but survives in poetic, rhetorical, and dignified use as well as in official use as *dwellings.*

**dyarchy.** See DIARCHY.

**dye** makes *dyeing* as a precaution against confusion with *dying* from *die*; cf. *singeing* but *impinging.* See VERBS IN -IE etc. 7, and MUTE E.

**dynamic(al).** Both words date from the 19th c. only, and -*ic* has virtually superseded -*ical* except as the adjective of *dynamics* (-*ical principles*; *an abstract* -*ical proposition*). See -IC(AL).

**dynamo.** Pl. -*os*; see -O(E)S 5.

**dynast.** Pronounce *dĭn*- rather than *dīn*-.

# E

**each.** 1. Number of, and with, *e.* 2. *Each other.* 3. *Between e.*
1. Number. *E.* as subject is invariably singular, even when followed by *of them* etc.: *E. of the wheels has 12 spokes* (not *have*). When *e.* is not the subject, but in apposition with a plural noun or pronoun as subject, the verb (and complement) is invariably plural: *The wheels have 12 spokes e.*; *the wheels e. have 12 spokes* (this latter order is better avoided); *the wheels are e. 12-spoked.* But the number of a later noun or pronoun, and the corresponding choice of a possessive adjective, depend upon whether *e.* stands before or after the verb, and this again depends on the distributive emphasis required. If the distribution is not to be formally emphasized, *e.* stands before the verb (or its complement, or some part of the phrase composing it), and the plural number and corresponding possessive are used: *We e. have our own nostrums* (not *his own nostrum,* nor *our own nostrum*); *They are e. of them masters in their own homes.* If the distribution is to be formally insisted on, *e.* stands after the verb (and complement) and is followed by singular nouns and the corresponding possessives: *we are responsible e. for his own vote* (also sometimes, by confusion, *e. for our own votes,* and sometimes, by double confusion, *e. for our own vote*). The following forms are incorrect in various degrees: *Brown, Jones, and*

echelon

*Robinson e. has a different plan. | You will go e. your own way. | They have e. something to say for himself. | E. of these verses have five feet. | They e. of them contain a complete story. | We are master e. in his own house. | Guizot and Gneist, e. in their generation, went to school to the history of England to discover . . . | The People's Idols mount, e. his little tub, and brazen-throated, advertises his nostrum, the one infallible panacea.* It is said that in the hymn-lines 'Soon will you and I be lying *E.* within our narrow bed' *our* has been substituted for the original *his*; if so, the corrector, offended by *his* of the common gender, failed to observe that he was restricting the application to married couples. See also NUMBER 11.

2. *Each other* is now treated as a compound word, the verb or preposition that governs *other* standing before *e.* instead of in its normal place: *they hate e. o., they sent presents to e. o.,* are usually preferred to *e. hates the other(s), they sent presents e. to the other(s).* But the phrase is so far true to its origin that its possessive is *e. other's* (not *others'*), and that it cannot be used when the case of *other* would be subjective: *a lot of old cats ready to tear out e. other's* (not *others'*) *eyes; we e. know what the other wants* (not *what e. o. wants*). Some writers use *e. o.* only when no more than two things are referred to, *one another* being similarly appropriated to larger numbers; but this differentiation is neither of present utility nor based on historical usage. The old distributive of two as opposed to several was not *e.,* but *either*; and *either other,* which formerly existed beside *e. o.* and *one another,* would doubtless have survived if its special meaning had been required.

3. *Between e.* For such expressions as 'three minutes b. e. scene' see BETWEEN 3.

**earl.** See TITLES.

**earthen, earthly, earthy.** *Earthen* is still in ordinary use (see -EN ADJECTIVES) in the sole sense made of earth (either soil or potter's clay). *Earthly* has two senses only: (1) belonging to

this transitory world as opposed to heaven or the future life, and (2) in negative context, practically existent or discoverable by mortal man. *Earthy* means of the nature, or having an admixture, of earth (soil, dross, gross materialism). *An earthen mound, rampart, earthenware. Earthly joys, grandeur; the earthly paradise; their earthly pilgrimage; is there any earthly use, reason,* etc.?; *for no earthly consideration*; cf. the slang *he hasn't an earthly* (i.e. chance). *An earthy precipitate formed in a few minutes; the ore is very earthy; an upright man, but incurably earthy in his views and desires.*

**eas(il)y.** *Easy* as an adverb, instead of the normal *easily,* survives only as a vulgarism and in a few phrases, mostly colloquial: *stand easy, easy all, take it easy, easy come easy go, easier said than done.*

**easterly, northerly, southerly, westerly.** Chiefly used of wind, and then meaning east etc. or thereabouts, rather from the eastern etc. than from the other half of the horizon; otherwise only of words implying either motion, or position conceived as attained by previous motion: *an easterly wind; took a southerly course; the most easterly outposts of western civilization.* But *south* (not *southerly*) *aspect; the eastward* (not *easterly*) *position; the west* (not *westerly*) *end of the church; western* (not *westerly*) *ways of thought.*

**eat.** The past is spelt *ate* (sometimes *eat*) and pronounced *ĕt,* but the U.S. pronunciation *āt* is encroaching.

**ebulli(ent)(tion).** Pronounce -*bŭl*-not as in *bull.*

**echelon.** *With another General Election at least four years away, there seems no desperate hurry for Mr. Macmillan to turn his attention to the higher echelons of party management.* An *echelon* is a particular kind of formation (it might be roughly described as 'staggered') of troops, ships, or aircraft. To use it, as it is apparently used here, for any kind of graded organization is a SLIPSHOD

EXTENSION attributable presumably to NOVELTY HUNTING, which destroys the special significance of the word in the same way as that of ALIBI, for instance, is being destroyed by its use for any kind of excuse or defence.

**echo.** Pl. *echoes*; see -O(E)S 1.

**economic(al).** The nouns *economics* and *economy* having nearly parted company (though Political Economy, like the Queen's Proctor, impedes actual divorce), it is convenient that each should have its own adjective. Accordingly *-ic* is now associated only with economics, and *-ical* only with economy; an *economic rent* is one in the fixing of which the laws of supply and demand have had free play; an *economical rent* is one that is not extravagant. In practice the first generally means a rent not too low (for the landlord), and the second one not too high (for the tenant). In *The question of economical help for Russia by sending her goods from this country*, the wrong word has been chosen.

**-ed and 'd.** When occasion arises to append the *-ed* that means *having* or *provided with* so-and-so to words with unusual vowel terminations (-a, -i, -o, etc.), it is best to avoid the bizarre appearance of *-aed* etc. and to write '*d*: *one-idea'd, mustachio'd; the wistaria'd walls; a rich-fauna'd region; long-pedigree'd families; the fee'd counsel; subpoena'd witnesses; mascara'd eye-lashes; a shanghai'd sailor; ski'd mountaineers.* Even with familiar words in *-o*, as *halo* and *dado*, the apostrophe is perhaps better; and *ideaed, pedigreeed, shanghaied* and such words are deliberately avoided because they look absurd.

**edge.** For *e.-bone* see AITCH-BONE; for *edgeways, -wise*, see -WISE, -WAYS.

**edifice.** See WORKING AND STYLISH WORDS.

**editress.** See FEMININE DESIGNATIONS.

**-edly.** An apology is perhaps due for 'setting out a stramineous subject' at the length this article must run to; but some writers certainly need advice on it (*Women and girls stayed their needles while the Liberal leader's wife and daughter chatted informedly with them*), and few writers have time for the inductive process required, in default of perfect literary instinct, to establish sound rules.

Experiments in unfamiliar adverbs of this type (as *embarrassedly, boredly, mystifiedly, biassedly, painedly, awedly*) lay the maker open to a double suspicion: he may be NOVELTY-HUNTING (conscious, that is, of a dullness that must be artificially relieved) or he may be putting down the abnormal in the belief that it is normal and so betraying that his literary ear is at fault.

The following is offered as a fairly complete list of the standard words; there are some hundreds of others to which there is no objection, but these will suffice to test doubtful forms by. The list is in three parts, first adverbs from adjectives in *-ed*, secondly adverbs from adjective-noun compounds in *-ed*, and lastly adverbs from true past participles.

It will probably be admitted by everyone that the list is made up wholly of words known to be in the language already and not having to be manufactured for some special occasion with doubts about their right to exist. Most readers will admit also that, while it is physically possible to say any of those starred without allowing a separate syllable to the *-ed-*, the only ones actually so pronounced by educated persons are those with two stars; *fixedly*, for instance, demands its three syllables, and *unconcernedly* its five.

1. From adjectives in *-ed*: *belatedly, benightedly, conceitedly, crabbedly★, crookedly★, dementedly, deucedly★, doggedly★, jaggedly★, learnedly★, nakedly, raggedly★, ruggedly★, sacredly, stiltedly, wickedly, wretchedly★.*

2. From adjective-noun compounds in *-ed*: *-bloodedly* (*cold-b.* etc.), *-fash-ionedly*★★ (*old-f.* etc.), *-handedly* (*open-h.* etc.), *-headedly* (*wrong-h.* etc.),

-heartedly (warm-h. etc.), -humoured-
ly** (good-h. etc.), -mindedly (absent-
m.), -naturedly** (ill-n. etc.), -sidedly
(lop-s. etc.), -sightedly (short-s. etc.),
-spiritedly (low-s. etc.), -temperedly**
(ill-t. etc.), -windedly (long-w. etc.),
-wittedly (slow-w. etc.).

3. From true past participles (including some with corresponding negative
or positive forms in equally or less
common use, which need not be mentioned): abstractedly, admittedly, ad-
visedly*, assuredly*, avowedly*, col-
lectedly, confessedly*, confoundedly,
connectedly, constrainedly*, consumedly*, contentedly, cursedly*, decidedly,
dejectedly, delightedly, deservedly*, designedly*, devotedly, disappointedly,
disinterestedly, disjointedly, dispiritedly,
distractedly, excitedly, fixedly*, guard-
edly, heatedly, hurriedly**, jadedly,
markedly*, misguidedly, perplexedly*,
pointedly, professedly*, repeatedly,
reputedly, resignedly*, restrainedly*,
rootedly, statedly, unabatedly, un-
affectedly, unconcernedly*, undaunt-
edly, undisguisedly*, undisputedly,
undoubtedly, unexpectedly, unfeign-
edly*, unfoundedly, uninterruptedly,
unitedly, unreservedly*, unwontedly.

The upshot is that, among the
hundreds of adverbs in -edly that may
suggest themselves as convenient
novelties, (a) those that must sound
the e are unobjectionable, e.g. ani-
matedly, offendedly, unstintedly; (b) of
those in which the e can (physically)
be either sounded or silent none (with
the exception of those in classes (c)
and (d) below) are tolerable unless
the writer is prepared to have the e
sounded; thus the user of composedly,
confusedly, dispersedly, pronouncedly,
absorbedly, and declaredly, will not
resent their being given four syllables
each, and they pass the test; but no one
will write experiencedly, accomplish-
edly, boredly, skilledly, or discouragedly,
and consent to the ed's being a distinct
syllable; they are therefore ruled out;
(c) hurriedly suggests that such forms
as palsiedly, worriedly, variedly, fren-
ziedly, and studiedly (from verbs in
unaccented -ȳ) are legitimate; (d) words
in unaccented -ure, -our, or -er, seem

to form passable adverbs in -edly without the extra syllable, as measuredly,
injuredly, perjuredly, labouredly, pam-
peredly, bewilderedly, chequeredly; most
two-starred words in the second part
of the standard list answer to this
description; (e) few if any from verbs
in -ȳ, or from those in -ble, -cle, etc., as
triedly, satisfiedly, troubledly, puzzledly,
are endurable.

These conclusions may be confirmed
by comparing some couples of possible words. Take dementedly and
derangedly, open-handedly and open-
armedly, admittedly and ownedly, dispiritedly and dismayedly, delightedly
and charmedly, disgustedly and dis-
pleasedly. The reason why the first
of each couple seems natural and the
second (except to novelty-hunters) unnatural is that we instinctively shrink
from the ed syllable (archaic when
phonetics allow the e to be silent)
except in established words; charmedly
as a disyllable is felt to flout analogy,
and as a trisyllable is a bizarre mixture
of the archaic and the newfangled.

**educate** makes educable (see -ABLE 1).

Those who like to remind their
audiences on school speech days that
education is a drawing out and not a
putting in are not quite right in their
etymology. It is true that the Latin
word educere, the primary meaning of
which is to draw out, was also used in
the sense of to rear, but that verb
would have given us eduction. Education is derived from a different word,
educare, the primary meaning of which
is to bring up, to educate.

**education(al)ist.** See -IST.

**-ee.** Apart from its rare use as a
diminutive (coatee, bootee) and its
seemingly arbitrary appearance in
some words (bargee, grandee, goatee,
settee), this suffix is most commonly
used for the indirect object of a
verb, especially in legal terms: lessee,
vendee, trustee, referee are persons to
whom something is let, sold, entrusted, or referred. Being originally
an adaptation of the French p.p. é, it
also serves in some words for the

direct object (EMPLOYEE, *trainee, examinee*). Such words as *refugee, debauchee, absentee*, being derived in this way from French reflexive verbs, where subject and object are the same, have the appearance of agent-nouns; and this no doubt accounts for a modern tendency to make new agent-nouns by using the suffix *-ee*. But we already have at least three suffixes for that purpose (*-er, -or*, and *-ist*) and to use one whose natural meaning is the opposite is gratuitously confusing. The unskilled workers used to 'dilute' skilled workers in time of war should have been called *diluters* instead of *dilutees*; the skilled were the dilutees. For *escapee* see ESCAPE.

**-eer.** This suffix is the anglicized form of French *-ier* (Latin *-arius* or *-iarius*) meaning one who is concerned with. It has not necessarily a derogatory implication; it has none in *auctioneer, mountaineer, musketeer, scrutineer, volunteer*, and others. But the purpose of coining new words in *-eer* has usually been to give them a contemptuous flavour, suggesting that the person so described either does good things badly (*sonneteer, sermoneer*, etc.) or bad things well (*racketeer, profiteer*, etc.).

**effect,** vb. See AFFECT.

**effective, effectual, efficacious, efficient.** The words all mean having effect, but with different applications and certain often disregarded shades of meaning. *Efficacious* applies only to things (especially now to medicines) used for a purpose, and means sure to have, or usually having, the desired effect. *Efficient* applies to agents or their action or to instruments etc., and means capable of producing the desired effect, competent or equal to a task. *Effectual* applies to action apart from the agent, and means not falling short of the complete effect aimed at. *Effective* applies to the thing done or its doer as such, and means having a high degree of effect.
*An efficacious remedy; a drug of known efficacy.*
*An efficient general, cook; efficient*

*work, organization; an efficient car. Efficient cause* is a special use preserving the original etymological sense 'doing the work'; that which makes a thing what it is. The efficiency of an engine is the ratio of useful work it does to the energy it expends.
*Effectual measures; an effectual stopper on conversation. Effectual demand* in economics is demand that actually causes the supply to be brought to market.
*An effective speech, speaker, contrast, cross-fire; effective assistance, cooperation. An effective blockade, effective capital, effective membership, effective strength* (of a military unit), preserve a now less common sense 'not merely nominal or potential but actual'.

**e.g.** is short for *exempli gratiâ*, and means only 'for instance'. Non-latinists are apt to think that it does not matter whether *e.g.* or *i.e.* is used; so *Mr —— took as the theme of his address the existence of what he called a psychic attribute, e.g., a kind of memory, in plants.* Italics, and a following comma, are unnecessary, but not wrong. Both abbreviations should be reserved for footnotes or very concise writing; in open prose it is better to write *for example* or *for instance*; *namely* or *that is to say*. See also I.E.

**ego(t)ism.** The two words are 18th-c. formations. Etymologically, there is no difference between them to affect the sense, but *egoism* is correctly and *egotism* incorrectly formed—a fact that is now of no importance, since both are established. *Egotism* used to be the more popular form, and (perhaps consequently) restricted to the more popular senses—excessive use of *I* in speech or writing, and self-importance or self-centredness in character. But *egoism* now shows signs of ousting *egotism* even in these senses; it is also used in metaphysics and ethics as a name for the theory that a person has no proof that anything exists outside his own mind (now more usually *solipsism*), and for the theory that self-interest is the foundation of morality. However arbitrary the differentiation may be, it serves a useful purpose if it can be maintained. But the adj. *egotistic* seems now to be threatened by the POPULARIZED TECHNICALITY *egocentric*.

**egregious.** The etymological sense is simply eminent or of exceptional degree (*e grege*, out of the flock, as Horace calls Regulus *egregius exsul*). The use of the word has been narrowed in English till it is applied only to nouns expressing contempt, and especially to a few of these, such as *ass, coxcomb, liar, impostor folly, blunder, waste*. The *e. Jones* etc. is occasionally used in the sense *that notorious ass Jones*; and with neutral words like *example e.* is the natural antithesis to *outstanding*—an *outstanding example of fortitude, an e. example of incapacity*. Reversion to the original sense, as in the following, is mere pedantry: *There is indeed little aforethought in most of our daily doings, whether gregarious or egregious.*

**eirenicon, ir-.** Usually spelt *eir-*, and pronounced *īrē'nikŏn*. As it is chiefly in learned use, it is odd that the spelling should be anomalous. *Irenicum* would be the latinized and normally transliterated form; *irenicon* the normally transliterated Greek form; *eirenikon* the Greek written in English letters. All these have been rejected for the now established mixture *eirenicon*.

**either.** 1. The pronunciation *ī-*, though not more correct, has almost wholly displaced *ē* in England, though not in U.S.
2. The sense each of the two, as in *the room has a fireplace at e. end*, though more naturally expressed by *each*, cannot be considered unidiomatic.
3. The sense any one of a number (above two), as in *e. of the angles of a triangle*, is loose; *any* or *any one* should be preferred.
4. The use of a plural verb after *e.*, as in *if e. of these methods are successful*, is a very common grammatical blunder.
5. *Either . . . or.* In this alternative formula *e.* is frequently misplaced. This should be avoided in care-

ful writing, but is often permissible colloquially. There are two correct substitutes for *You are e. joking or have forgotten*. Some writers refuse one of these, *You e. are joking or have forgotten*, on the ground that it looks pedantic; but there is no such objection to the other, *E. you are joking or you have forgotten*. In conversation, however, the incorrect form is defensible because a speaker who originally meant to say (*are*) *forgetting* (corresponding to *are joking*) cannot, when he discovers that he prefers *have forgotten*, go back without being detected (as a writer can) and put things in order. Some examples follow of the slovenliness that should not be allowed to survive proof-correction.

*. . . unless it sees its way to do something effective e. towards keeping the peace or limiting the area of conflict. | Their hair is usually worn e. plaited in knots or is festooned with cocks' feathers. | It is not too much to say that trade unions e. should not exist, or that all workers should join compulsorily. | The choice before the nations will be e. that of finding a totally different and far better method of regulating their affairs, or of passing rapidly from bad to worse.*

*Either . . . or* is sometimes not disjunctive, but equivalent to both . . . and or alike . . . and: *The continuance of atrocities, the sinking of the Leinster, the destruction of French and Belgian towns and villages, are a fatal obstacle either to the granting of an armistice or to the discussion of terms*. In such cases, *alike* (or *both*) *. . . and* should be preferred, or else proper care should be taken with *either*; 'an obstacle to either granting an armistice or discussing terms' would do it.

**eke,** adv. is an ARCHAISM sometimes used by way of PEDANTIC HUMOUR.

**eke out.** The meaning is to make something, by adding to it, go further or last longer or do more than it would without such addition. The proper object is accordingly a word expressing not the result attained, but the original supply. You can eke out your income or (whence the SLIPSHOD EXTENSION) a scanty subsistence with odd jobs or by fishing, but you cannot eke out a living or a miserable existence. The first quotation below illustrates the right use, and the others the wrong ones.

*Mr. Weyman first took to writing in order to e. o. an insufficient income at the Bar. | These disconsolate young widows would perforce relapse into conditions of life at once pitiful and sordid, eking out in dismal boarding-houses or humble lodgings a life which may have known comfort. | Dr. Mitford eked out a period of comparative freedom from expense by assisting the notorious quack, Dr. Graham.* A legitimate extension is the use of *eke out* in the sense of making a thing go further not by adding to it but by using it sparingly, to dole out. *The accumulated gossip could be eked out for weeks at his own tea-parties. | A man the very thought of whom has ruined more men than any other influence in the nineteenth century, and who is trying to e. o. at last a spoonful of atonement for it all.*

**elaborateness, elaboration.** See -ION AND -NESS.

**elder, -est.** These forms are now almost confined to the indication of mere seniority among the members of a family; for this purpose the *old-* forms are not used except when the age has other than a comparative importance or when comparison is not the obvious point. Thus we say *I have an elder* (not *older*) *brother* in the simple sense a brother older than myself; but *I have an older brother* is possible in the sense a brother older than the one you know of; and *Is there no older son?* means Is there none more competent by age than this one? Outside this restricted use of family seniority, *elder* and *eldest* linger in a few contexts such as *elders* meaning persons whose age is supposed to demand the respect of the young, and as the title of lay officers of the Presbyterian Church, the *elder brethren* of Trinity House, the *elder hand* at piquet, and *elder statesman*.

**electric(al).** See -IC(AL). The longer form, once much the commoner (the OED quotes *electrical shock, battery, eel,* and *spark,* never now heard), survives only in the sense *of* or *concerning electricity* (e.g. *e. department, knowledge, Trade Union*), and is not necessarily preferred even in that sense except where there is danger that *electric* might mislead.

**electrocute, -cution.** This word does not claim classical paternity; if it did, it would indeed be a BARBARISM. It is merely a PORTMANTEAU WORD formed by telescoping *electro-* and *execution,* and, as it is established, protest is idle.

**electron.** See ATOM etc. Pl. *-ns;* see -ON 2.

**eleemosynary.** Seven syllables: *ĕlĕē-mŏ′zĭnărĭ.*

**elegant variation.** It is the second-rate writers, those intent rather on expressing themselves prettily than on conveying their meaning clearly, and still more those whose notions of style are based on a few misleading rules of thumb, that are chiefly open to the allurements of elegant variation. Thackeray may be seduced into an occasional lapse (*careering during the season from one great dinner of twenty* covers *to another of eighteen* guests—where, however, the variation in words may be defended as setting off the sameness of circumstance); but the real victims, first terrorized by a misunderstood taboo, next fascinated by a newly discovered ingenuity, and finally addicted to an incurable vice, are the minor novelists and the reporters. There are few literary faults so widely prevalent, and this book will not have been written in vain if the present article should heal any sufferer of his infirmity.

The fatal influence is the advice given to young writers never to use the same word twice in a sentence —or within 20 lines or other limit. The advice has its uses; it reminds any who may be in danger of forgetting it that there are such things as pronouns, the substitution of which relieves monotony. The official would have done well to remember it who writes: *Arrangements are being made to continue the production of these houses for a further period, and increased numbers of these houses* (them) *will therefore be available.* The advice also gives a useful warning that a noticeable word used once should not be used again in the neighbourhood with a different application. This point will be found fully illustrated in REPETITION; but it may be shortly set out here, a kind providence having sent a neatly contrasted pair of quotations: (A) *Dr. Lebbé* seriously *maintains that in the near future opium-smoking will be as* serious *as the absinthe scourge in France*; (B) *The return of the Nationalists to Parliament means that they are prepared to treat* seriously *any* serious *attempt to get Home Rule into working order.* Here (A) would be much improved by changing *serious* to *grave*, and (B) would be as much weakened by changing *serious* to *real.* The reason is that the application of *seriously* and *serious* is in (A) different (the two being out of all relation to each other) and in (B) similar. *I am* serious *in calling it* serious suggests only a vapid play on words; *we will be* serious *if you are* serious is good sense; but the rule of thumb, as usual, omits all qualifications, and would forbid (B) as well as (A). A few examples follow of the kind of repetition against which warning is needed, so as to bring out the vast difference between the cases for which the rule is intended and those to which it is mistakenly applied: *Meetings at which they* passed *their time* passing *resolutions pledging them to resist.* | *A debate which took wider ground than that* actually *covered by the* actual *amendment itself.* | *We much* regret *to say that there were very* regrettable *incidents at both the mills.* | *The figures I have* obtained *put a very different complexion on the subject than that generally* obtaining. | *Doyle drew the* original *of the outer sheet of Punch a*s *we still know it; the* original *intention*

*was that there should be a fresh illustrated cover every week.*

These, however, are mere pieces of gross carelessness, which would be disavowed by their authors. Diametrically opposed to them are sentences in which the writer, far from carelessly repeating a word in a different application, has carefully not repeated it in a similar application. The effect is to set readers wondering what the significance of the change is, only to conclude disappointedly that it has none: *The Bohemian Diet will be the second Parliament to elect* women *deputies, for Sweden already has several* lady *deputies.* | *There are a not inconsiderable number of* employers *who appear to hold the same opinion, but certain* owners—*notably those of South Wales—hold a contrary view to this.* | *Mr. John Redmond has just now a path to* tread *even more thorny than that which Mr. Asquith has to* walk. What has Bohemia done that its females should be mere women? Are owners subject to influences that do not affect employers? Of course they might be, and that is just the reason why, as no such suggestion is meant, the word should not be changed. And can Mr. Asquith really have taught himself to walk without treading? All this is not to say that *women* and *employers* and *tread* should necessarily be repeated— only that satisfactory variation is not to be so cheaply secured as by the mechanical replacing of a word by a synonym. The true corrections are here simple, (1) *several* alone instead of *several women* (or *lady) deputies*, (2) *some* alone instead of *certain employers* (or *owners*), (3) *Mr. Asquith's* instead of *that which Mr. Asquith has to tread* (or *walk*); but the writers are confirmed variationists—nail-biters, say, who no longer have the power to abstain from the unseemly trick.

Before making our attempt (the main object of this article) to nauseate by accumulation of instances, as sweetshop assistants are said to be cured of larceny by cloying, let us give special warning against two temptations. The first occurs when there are successive phrases each containing one constant and one variable. The variationist fails to see that the varying of the variable is enough, and that the varying of the constant as well is a great deal too much; he may contrive to omit his constant if he likes, but he must not vary it: *There are 466 cases; they consist of 366 matrimonial suits, 56 Admiralty actions, and 44 Probate cases* (strike out *suits* and *actions*; but even to write *cases* every time is better than the variation). / *The total number of farming properties is 250,000; of these only 800 have more than 600 acres; 1,600 possess between 300 and 600 acres, while 116,600 own less than eight acres apiece* (if *while* is changed to *and*, *possess* and *own*, which anyhow require not *properties* but *proprietors*, can be dropped; or *have* can be repeated). / *At a higher rate or lower* figure, *according to the special circumstances of the district* (omit *rate*). / *It was Tower's third* victory, *and Buxton's second* win (drop either *victory* or *win*).

The second temptation is to regard *that* and *which* as two words that are simply equivalent and (the variationist would say *and which*) exist only to relieve each other when either is tired. This equivalence is a delusion, but one that need not be discussed here; the point to be observed in the following quotations is that, even if the words meant exactly the same, it would be better to keep the first selected on duty than to change guard: *He provides a philosophy* which *disparages the intellect and* that *forms a handy background for all kinds of irrational beliefs* (omit *that*). / *A scheme for unification* that *is definite and* which *will serve as a firm basis for future reform* (omit *which*). / *A pride* that *at times seemed like a petty punctilio, a self-discipline* which *seemed at times almost inhuman in its severity* (repeat *that*).

And now the reader may at length be turned loose among dainties of every kind; his gorge will surely rise before the feast is finished. *It is stated that 18 rebels* were killed: *8 French soldiers* lost their lives. / *There are four cases in which old screen-work is still to be found*

*in Middlesex churches, and not one of these* instances *is so much as named.* / *In 32 cases there are Liberal candidates in the field, and in all* instances *so-called Socialists supply the third candidate.* / *Dr. Tulloch was for a time* Dr. Boyd's *assistant, and knew* the popular preacher *very intimately, and the picture he gives of* the genial essayist *is a very engaging one.* / *Rarely does the 'Little Summer' linger until* November, *but at times its stay has been prolonged until quite late in* the year's penultimate month. / *The addressee of many* epistles *in the volumes of* 'Letters *of Charles Dickens'.* / *The export trade of the U.S. with the Philippines has increased by nearly 50%, while that of the U.K. has decreased by* one-half. / *Curiously enough, women* played *the male* parts, *whilst men* were entrusted with *the female* characters. / *France* is *now* going through *a similar experience* with regard to *Morocco to that which* England had to *undergo* with reference to *Egypt.* / *While I feel quite* equal to the role *of friendly and considerate employer, I do not feel* adequate to the part *of a special Providence.* / *Were I an artist, I could* paint *the Golf Links at Gaya and call it 'A Yorkshire Moor'; I could* depict *a water-way in Eastern Bengal and call it 'The Bure near Wroxham'; I could* portray *a piece of the Punjab and call it 'A Stretch of Essex'.* / *Other county championship matches are at* Bath, *where Somersetshire and Gloucestershire* play, *at Fylde, where Lancashire* meet *Durham, and Gosforth, where Yorkshire* tackle *Northumberland.* Difficulties begin to arise *whenever we attempt to compare the estimates* obtained by one method *with those* deduced by another technique. / *He may thus be said to have had the troops' respect to the same extent as* Marlborough *but with all his virtues he never found his way to their hearts as did* John Churchill *with all his failings.* / *The other candidates* [at a by-election a year after the general election of 1959] *are Mr. G. P. . . . who contested Monmouth* last year, *Mr. F. E., who contested Stroud* in the general election, *and Dr.*

K. D. P., *who contested Cheltenham in 1959.* / *Not only should an agreement be* come to, *but it has always been certain that it will be* arrived at. / *They spend a few weeks longer in their winter home than in their summer* habitat. / *It is interesting and satisfactory that* a Wykehamist and an Oxonian *should be succeeded by* an Oxonian and Wykehamist. It will also be interesting and satisfactory to anyone who has lasted out to this point to observe that this skilled performer, who has brought off a double variation (reversing the order of the titles, and stripping the second Wykehamist of his article), has been trapped into implying by the latter change that the successor is one man and the predecessor(s) two.

For elegant variation of *said* in dialogue see INVERSION s.f.

**elegy.** In the strict sense a song or poem of mourning (old synonym *threnody*) and properly applied in English to such pieces as *Lycidas*, *Adonais*, and *Thyrsis* or, by a slight extension, to a poem such as Gray's *Elegy in a Country Churchyard*, inspired by a general sense of the pathos of mortality. But since the favourite ancient metre for such pieces was the elegiacs so named on that account, a natural reaction caused anything written in elegiacs to be called an elegy, whatever its subject, and the name was extended to cover any short poem, irrespective of metre, that was of the subjective kind, i.e. was concerned with expressing its author's feelings. The present tendency is to restrict the word to its original sense.

**elemental, elementary.** The two words are now clearly differentiated. *Elemental* refers to 'the elements' either in the old sense of earth, water, air and fire, or as representing the great forces of nature conceived as their manifestations (or, metaphorically, the human instincts comparable in power to those forces); *elementary* refers to elements in the more

general sense of simplest component parts or rudiments. *Elemental* fire, strife, spirits, passion, power: *elementary* substances, constituents, facts, books, knowledge, schools. The -al form is often wrongly chosen by those who have not observed the differentiation, and think that an occasional out-of-the-way word lends distinction to their style; so: *The ever growing power of the State, the constant extension of its activities, threaten the most* elemental *liberties of the individual.* / *Responsible government in Canada was still in its most* elemental stage.

**elevator,** by the side of the established English *lift*, is a cumbrous and needless Americanism; it should at least be restricted to its established meanings in grain-storage, aeronautics, and anatomy.

**elfish.** See ELVISH.

**eliminate, -ation. 1.** The essential meaning (etymologically 'turn out of doors') is the expulsion, putting away, getting rid, or ignoring, of elements that for some reason are not wanted; the verb does not mean to extract or isolate for special consideration or treatment the elements that *are* wanted, as in *He would e. the main fact from all confusing circumstantials*, and in *Hypotheses of the utmost value in the elimination of truth.* See POPULARIZED TECHNICALITIES.

**2.** The verb makes *eliminable* (see -ABLE 1).

**elision.** A dictionary definition is 'The suppression of a letter (especially a vowel) or a syllable in pronunciation'. We all do this when using a pronoun (or sometimes a noun) with an auxiliary verb, or an auxiliary verb with *not* (*I'm, you'll, they've, didn't, sha'n't, won't, spring's here, John's won*, etc.), unless we want to emphasize the elided word by articulating it. These elisions are not ordinarily reproduced in print except when it records the spoken word, as in dialogue. Or rather they used not to be, though some old writers of the chattier sort would allow themselves an occasional *'tis* or *'twas*. Today we may find them in every variety of prose, from the journalistic to the academic. Some of the writers given to this practice keep their elisions in reserve and bring one out only now and then, as if to enliven their matter with a flash of spontaneity. Others use them all the time, especially the negatived auxiliaries. Possibly this may eventually become standard usage. There would be nothing revolutionary about it; we long ago took to printing the possessive *-es* as *'s*, and our conversion of *can not* into a single word may be significant of a trend towards further telescoping of our negatived auxiliaries. Meanwhile the printing of these elided forms in serious prose will no doubt continue to grate on some old-fashioned people. They seem to be intended, like the VERBLESS SENTENCE, to create a pleasant sense of intimacy between writer and reader. For some readers they may. On others, until we have got used to them, they may obtrude themselves as a gimmick. A friendly conversational atmosphere is an excellent thing, but digs in the ribs and slaps on the back, however well intentioned, are seldom the best way of promoting one.

**ellipsis.** 1. *Be* and *have*. 2. Second part of compound verb. 3. With change of voice. 4. *That* (conj.). 5. After *than*. 6. With inversion. 7. *That* (rel. pron.).

*Ellipsis* means the omission from a sentence of words needed to complete the construction or the sense. That the reader may at once realize the scope of the inquiry, a few ellipses of miscellaneous types are first exhibited:
*The ringleader was hanged and his followers ∧ imprisoned.* / *The evil consequences of excess of these beverages is much greater than ∧ alcohol.* / *Mr. Balfour blurted out that his own view was ∧ the House of Lords was not strong enough.* / *No state ever has ∧ or can adopt the non-ethical idea of property.* /

*The House of Lords would have really revised the Bill, as no doubt it could be ∧ with advantage.* / *Not only may such a love have deepened and exalted, and ∧ may ∧ still deepen and exalt, the life of any man, but* . . .

When a passage would, if fully set out, contain two compound members corresponding to each other, how far may the whole be shortened by omitting in one of these members ('understanding', in grammatical parlance) a part that is either expressed in the other or easily inferable from what is there expressed? Possible varieties are so numerous that it will be better not to hazard a general rule, but to say that the expressed can generally be 'understood', and the inferable can be in specially favourable circumstances, and then proceed to some types in which mistakes are common.

1. Ellipsis of parts of *be* and *have*. Not only the expressed part can be understood, but also the corresponding part with change of number or person: *The ringleader was hanged and his followers imprisoned*; *He is dead, and I alive*; *The years have passed and the appointed time come*; *Later the bomb was covered by the incoming tide and the disposal men forced to suspend work.* These are permissible; not all that is lawful, however, is expedient, and the licence is not to be recommended outside sentences of this simple pattern. With the intervening clause in the following quotation it is clearly ill-advised: *A number of stumbling-blocks have been removed, and the road along which the measure will have to travel straightened out*; it should be observed that it is the distance of *straightened* from *have been*, and not the change of number in the verbs, that demands the insertion of *has been*.

2. Ellipsis of second part of compound verb. Only the expressed part can be understood; *No State can or will adopt* would be regular, but *No State has or can adopt* is (however common) an elementary blunder. The understanding of an infinitive with *to* out of one without *to* (*A standard of public opinion which ought and we be-*

*lieve will strengthen the sense of parental responsibility*) is equally common and equally wrong; insert after *ought* either *to strengthen* or *to*. Wrong too is the understanding of an infinitive *be* out of a preceding past participle *been* in *These usually came from constituencies which have never been won by Labour and probably never will.*

3. Ellipsis with change of voice. Even if the form required is identical with that elsewhere expressed, the word should be repeated if the voice is different; as in *Though we do not believe that the House of Lords would have really revised the Bill, as no doubt it could be ∧ with advantage.* Still less can the passive *managed* be supplied from the active *manage* in *Mr. Dennett foresees a bright future for Benin if our officials will manage matters conformably with its 'customs', as they ought to have been ∧.* And with these may be classed the leaving us to get *to be* out of the preceding *to* in *If the two lines are to cross, the rate of loss ∧ reduced to zero, and a definite increase in the world's shipping to be brought about . . .*; or, even worse, leaving us to get *be* out of a preceding *go*, as in *He said the country needed a national compact that restrictive practices would go and the strike weapon ∧ hung on the wall.*

4. Omission of *that* (conjunction). Though this is strictly speaking not an ellipsis, but rather an exercise of the ancient right to abstain from subordinating a substantival clause (*And I seyde his opinioun was good*—Chaucer), it may conveniently be mentioned here. Three examples will suffice to show the unpleasantness of ill-advised omission, and to suggest some cautions: *Sir,—I am abashed to see ∧ in my notice of Mr. Bradley Birt's ' "Sylhet" Thackeray' ∧ I have credited the elder W. M. Thackeray with sixteen children.*/ *Mr. Balfour blurted out that his own view was ∧ the House of Lords was not strong enough.*/ *I assert ∧ the feeling in Canada today is such against annexation that . . .* The first illustrates the principle that if there is the least room for doubt where the *that* would come, it should be expressed and not under-

stood. The second leads us to the rule that, when the contents of a clause are attached by a part of *be* to such words as *opinion, decision, view,* or *declaration* (a very common type), *that* must be inserted. At the same time it illustrates the motive that most frequently causes wrong omissions—the sensible reluctance to make one *that*-clause depend on another; but this is always avoidable by other, though often less simple, means. The third involves a matter of idiom, and reminds us that, while some verbs of saying and thinking can take or drop *that* at will, many have a strong preference for one or the other use (see THAT, CONJ. 2); *assert* is among those that habitually take *that*.

5. Ellipsis after *than* is extremely common, and so various in detail as to make the laying down of any general rule impossible. The comparative claims of brevity on the one hand, and on the other of the comfort that springs from feeling that all is shipshape, must in each case be weighed with judgement. It will be best to put together a few examples, ranging from the more to the less obvious, in which doubts whether all is right with the sentence obtrude themselves. *The evil consequences of excess of these beverages is much greater than ∧ alcohol* means than the evil consequences of excess of alcohol are great. How are we to compress that? Shall we (*a*) omit *are great*? Yes, everyone does it; (*b*) omit *the evil consequences of excess of*? No, no one could do it but one who could also write, like this author, *consequences is*; (*c*) retain all this? No—waste of words; (*d*) shorten to *those of excess of*? Yes, unless the knot is cut by writing *than with alcohol.* | *That export trade is advancing with greater rapidity than our trade has ever increased*; i.e. than any rapidity with which ours has increased; shorten to *more rapidly than our trade ever has.* | *The proceedings were more humiliating to ourselves than I can recollect in the course of my political experience*; i.e. than I can recollect any proceedings being humiliating; shorten to *any that I can recollect.* | *The interpretations are more uniformly admirable*

*than could, perhaps, have been produced by any other person*; i.e. than any would have been admirable that . . . ; alter to *than what could*, though the misplacing of *perhaps*, which belongs to the main sentence, will cry all the louder for correction.

6. Ellipsis complicated by INVERSION. In questions, and in sentences beginning with *nor* and certain other words, inversion is normal, the subject standing after instead of before the auxiliary (*Never did I hear*, not *Never I heard*). When a sentence or clause thus inverted has to be enlarged by a parallel member of the kind in which ellipsis would naturally be resorted to, difficulties arise. *Why is a man in civil life perpetually slandering and backbiting his fellow men, and is unable to see good even in his friends?* The repetition of *is* without repeating *why* and the subject is impossible; in this particular sentence the removal of the second *is* would solve the problem as well as the re-insertion of *why is he*; but repetition is often the only course possible. *Not only may such a love have deepened and exalted, and may still deepen and exalt, the life of any man of any age, but* . . . The inversion has to be carried on; that is, *not only*, and the subject placed after *may*, must be repeated if *may* is repeated; and, *may* being here indispensable, nothing less will do than *not only* (with *and* omitted) *may it still deepen*.

7. For ellipsis of *that* (rel. pron.), and of prepositions governing it, see THAT, REL. PRON. 3 and 4.

**else.** The adverb *e.* has come so near to being compounded with certain indefinite pronouns and words of similar character (*anybody, everyone, little, all*, etc.) that separation is habitually avoided, and e.g. *Nobody is ignorant of it e.* is unidiomatic; correspondingly, the usual possessive form is not *everyone's* etc. *e.*, which is felt to be pedantic though correct, but *everyone else's*. With interrogative pronouns the process has not gone so far. Though *What e. did he say?* is the normal form, *What did he say e.?* is unobjectionable; corre-

spondingly, *who else's* may be used colloquially, but *whose else* has maintained its ground; and of the forms *Who else's should it be?, Whose e. should it be?, Whose should it be e.?*, the last is perhaps the best.

**elus(ive)(ory), illus(ive)(ory).** It is convenient for the avoidance of confusion that *elusive* has become the normal adjective of *elude* and *illusory* that of *illude* while *elusory* and *illusive* are falling into disuse. That is elusive which we fail, in spite of efforts, to grasp physically or mentally; *the elusive ball, half-back, submarine; elusive rhythm, perfume, fame; an elusive image, echo, pleasure.* That is *illusory* which turns out when attained to be unsatisfying, or which appears to be of more solid or permanent value than it really is; *illusory fulfilment, success, victory, possession, promises.*

The elusive mocks its pursuer, the illusory its possessor.

**elvish, elfish.** It is no longer true, as was said in the OED, that 'the older form *elvish* is still the more usual'; and *elfin* is more usual than either. For some similar questions see -VE.

**em- and im-, en- and in-.** The words in which hesitation between e- and i- is possible are given in the form recommended.

*embed, empanel, encage, encase, enclose* etc., *encrust* (but *incrustation*), *endorse, endorsement* (but *indorsation*), *endue, enfold, engraft, engrain* (but *ingrained*), *ensure* (in general senses), *entrench, entrust, entwine, entwist, enwrap; insure* (in financial sense), *insurance, inure* (but *enure* for the legal word), *inweave.* See also IM-.

Tenacious clinging to the right of private judgement is an English trait that a mere grammarian may not presume to deprecate, and such statements as the OED's *The half-latinized* enquire *still subsists beside* inquire will no doubt long remain true. See IN-QUIRE. Spelling, however, is not one of the domains in which private judgement shows to most advantage, and the general acceptance of the above forms on the authority of the OED, undisturbed by the COD, would be a sensible and democratic concession to uniformity.

**embargo.** Pl. *-os*; see -O(E)S 6.

**embryo.** Pl. *-os*; see -O(E)S 4.

**emend(ation).** The words are now confined to the conjectural correction of errors in MSS. or printed matter; they are not used, like *amend(ment)*, of improvement or correction in general.

**emergence, emergency.** The two are now completely differentiated, *-ce* meaning emerging or coming into notice, and *-cy* meaning a juncture that has arisen, especially one that calls for prompt measures, and also (more recently) the presence of such a juncture (*in case of -cy*). See -CE, -CY. In a world that is seldom without an emergency of one sort or another, the word has become a much used EUPHEMISM, especially by politicians and officials, for unpleasant possibilities that they shrink from referring to more bluntly.

**emotive, emotional.** *Emotional* means not only given to emotion but also, like *emotive*, appealing to the emotions. But modern usage has differentiated the words usefully by assigning *emotive* to the cause and *emotional* to the effect.

**employee, employé.** In 1897 the OED gave *employé* as the normal form and labelled *employee* as 'rare exc. U.S.', but that comment was withdrawn in the 1933 Supp. *Employee* has now rightly prevailed as a good plain word with no questions of spelling and pronunciation and accents and italics and genders about it; and in the COD *employé* is now given the italics that brand it as foreign. See also -EE.

**enable.** The ordinary meaning of this word is to empower (someone to do something). Its use with the thing to be done as its object, in the sense of to make possible, to facilitate, was said to be obsolete by the OED, but a revival led to the cancellation of this comment in the OED Supp., and the word now appears in very respectable company with this meaning. *A Royal Commission reported that this enabled substantial tax evasion. | To be sumptuous in display was expected of any prominent manager in the late Victorian and Edwardian theatre, and the economics of the day enabled it.*

**-en adjectives.** The only adjectives of this type still in ordinary natural use with the sense made of so-and-so are *earthen, flaxen, hempen, wheaten, wooden,* and *woollen*; we actually prefer *earthen vessels, flaxen thread, hempen rope, wheaten bread, wooden ships,* and *woollen socks,* to *earth vessels, flax thread, hemp rope, wheat bread, wood ships,* and *wool socks.* Several others (*ashen, brazen, golden, leaden, leathern, oaken, oaten, silken, waxen*) can still be used in the original sense (consisting of ashes, made of brass, etc.) with a touch of archaism or for poetic effect, but not in everyday contexts: *the ashen relics in the urn* but *an ash deposit in the stokehole; the brazen hinges of Hell-gate,* but *brass hinges do not rust; a golden crown* in hymns and fairy-stories, but *a gold crown* in an inventory of regalia; *a lead pipe,* but *leaden limbs; a leathern jerkin,* but *a leather pouch; silken hose,* but *silk pyjamas; an oaken staff,* but *an oak umbrella-stand; an oaten pipe,* but *oat cake; the comb's waxen trellis,* but *wax candles.* The chief use of the *-en* adjectives is in secondary and metaphorical senses—*ashen complexion, brazen impudence, golden prospects, leaden skies, leathern lungs, silken ease, waxen skin,* and the like. When well-meaning persons, thinking to do the language a service by restoring good old words to their rights, thrust them upon us in their literal sense where they are out of keeping, such patrons merely draw attention to their clients' apparent decrepitude—apparent only, for the words are hale and hearty, and will last long enough if only they are allowed to confine themselves to the jobs that they have chosen.

**en- and in-.** See EM- AND IM-.

**enclave.** The anglicized pronunciation *ĕn'klāv* (or *ĕnklāv'*) is now established.

**encomium.** Pl. usually *-ms*; see -UM.

**endeavour.** *A somewhat ponderous jibe has been endeavoured to be levelled at the First Lord of the Admiralty because he . . .* For this use of *endeavour*, with which *somewhat* is in perfect harmony, see DOUBLE PASSIVES. See also FORMAL WORDS.

**ended, ending.** *Statistics for the six months* (*ended*) (*ending*). If the terminal date is in the future *ending* is always used and *ended* ordinarily when it is past. But it is pedantic to object to *ending* for a past date on the ground that a present participle cannot suitably be used of a past event. If the reference is to the initial date the word is always *beginning*, never *begun*, whatever the date.

**endemic, epidemic.** An endemic disease is one habitually prevalent in a particular place; an epidemic disease is one that breaks out in a place and lasts for a time only. To insist that, when the disease is of animals, the words must be *enzootic* and *epizootic* is PEDANTRY.

**endorse.** You can endorse, literally, a cheque, driving licence, or other paper, and, metaphorically, a claim or argument. But to talk of endorsing material things other than papers in the sense of expressing approval of them, 'cracking them up', is an example of unsustained metaphor (see METAPHOR, 2A). The OED Supp. says this is U.S., and the COD, referring to its use in advertising (*So-and-so endorses our pills*), calls it vulg.

**end-product** is a term of chemistry, meaning the final product of a process that involves by-products. Figuratively used it has become a POPULAR-IZED TECHNICALITY claiming as its victims *consequence, result, outcome, upshot,* etc. *The e.-pp. of these dialectics are . . . the roots of all our discontents.* / *The radical—over-reckless —decision to rush independence to the Congolese by next June . . . is simply the e. p. of the process inevitably set in motion by Britain in West Africa on the morrow of Indian independence.* Perhaps the vogue of *e. p.* is partly due to what used to be called 'Television's most popular Panel Game *What's My Line?*' in which 'Is there an end-product?' was a stock question.

**enforce.** Adj. *-eable*; see -ABLE 2. *They were prepared to take action with a view to enforcing this country into a premature and vanquished peace.* / *Until Mr. Marples enforces motorists to obey the highway code he can look for little cooperation from pedestrians.* This use of *e.* for *force* or *compel* or *drive*, with a person as object, though common two or three centuries ago, is obsolete; today we force a person into peace (or to obey), or enforce peace (or obedience) on a person. See NOVELTY-HUNTING and OBJECT-SHUFFLING.

**England, English(man).** The incorrect use of these words as equivalents of *Great Britain* or *The United Kingdom, British, Briton,* is often resented by other nationals of the U.K., like the book-reviewer who writes of Lord Cherwell's 'dedication to the service of Britain, which, in the annoying way foreigners have, he persisted in calling "England"'. Their susceptibilities are natural, but are not necessarily always to be deferred to. For many purposes the wider words are the natural ones. We speak of the *British Commonwealth*, the *British Navy, Army,* and *Air Force* and *British trade*; we boast that *Britons* never never never shall be slaves; we know that Sir John Moore sleeps in a grave where a *Briton* has laid him, and there is no alternative to *British* English if we want to distinguish our idiom from the American. But it must be remembered that no Englishman, or perhaps no Scotsman even, calls himself a Briton without a sneaking sense of the ludicrous, or hears himself referred to as a BRITISHER without squirming. How should an Englishman utter the words *Great Britain* with the glow of emotion that for him goes with *England*? His sovereign may be Her *Britannic* Majesty to outsiders, but to him is Queen of *England*; he talks the *English* language; he has been taught *English* history as one continuous tale from Alfred to his own day; he has heard of the word of an *Englishman* and aspires to be an *English* gentleman; and he knows that *England* expects every man to do his duty. 'Speak for *England*' was the challenge flung across the floor of the House of Commons by Leo Amery on 2 Sept. 1939. In the word *England*, not in *Britain* all these things are implicit. It is unreasonable to ask forty millions of people to refrain from the use of the only names that are in tune with patriotic emotion, or to make them stop and think whether they mean their country in a narrower or wider sense each time they name it.

**english**, vb. See REVIVALS and SAXONISM.

**enhance.** *Spain felt that the war could not touch her, but that, on the contrary, while the rest of Europe was engaged in mutual destruction, she would be materially enhanced.* A dangerous word for the unwary. Her *material prosperity* may *be enhanced*, but *she* cannot *be enhanced* even *in material prosperity*, though a book may *be enhanced in value* as well as have *its value enhanced. E.* (*and be enhanced*) with a personal object (or subject) has long been obsolete. See CAST-IRON IDIOM.

**enjoin.** It is questionable whether the construction with a personal object and an infinitive (*The advocates of compulsory service e. us to add a great army for home defence to . . .*) conforms to modern idiom though the OED quotes Steele, *They injoined me to bring them something from London*, and Froude, *The Pope advised and even enjoined him to return to his duties*; and the COD admits this construction without comment. The ordinary modern use is *e. caution* etc. *upon* one, not *e.* one *to* do or be. It is confusing that in legal use in Britain (and more widely in U.S.) the meaning of *e.* (to prohibit, especially by injunction) is precisely the opposite of its ordinary one.

**enormous, enormity.** The two words have drifted so far apart that the use of either in connexion with the limited sense of the other is unadvisable. *Enormous sin* and *The impression of enormity produced by the building* are both etymologically possible expressions; but use of the first lays one open to suspicion of pedantry, and of the second to suspicion of ignorance. *The enormity of the destruction suffered during the war. Enormousness* is not a pretty word, but the writer could have found a way out by writing *vastness* or *enormous extent*.

**enough and sufficient(ly). 1.** In the noun use (= adequate amount), the preference of *s.* to *e.* (*have you had s.?*; *s. remains to fill another*) may almost be dismissed as a GENTEELISM; besides being shorter, *e.* has the grammatical advantage of being a real noun.

**2.** In the adjective use (*is there e.,* or *s., butter?*) *s.* has the advantage of being a true adjective, while *e.* is only a quasi-adjective; *a s. supply* is possible, and *an e. supply* is not. In spite of the fact, however, that *s.* is always and *e.* only sometimes available, *e.* is to be preferred as the more natural and vigorous word wherever mere amount can be regarded as the only question: *is there e. butter*, or *butter e., for the week?*; *he has courage e. for anything*. But where considerations of quality or kind are essential, *s.* is better; compare *for want of s. investigation* with *there has been investigation e.*; the first implies that the investigation has not been thorough or skilful, the second that the time given to it has been excessive.

**3.** In the adverbial use, neither word suffers from a grammatical handicap, *e.* being as true an adverb as *sufficiently*. Choice is dictated (often without the chooser's knowledge) in part by the feeling that a plain homely word, or a formal polysyllable, is appropriate (*he does not idle e.*; *he does not indulge s. in recreation*), and in part by the limitation of *e.* pointed out above to mere amount or degree (*the meat is not boiled e.*; *he does not s. realize the consequences*); often, however, *e.* is so undeniably more vigorous that it is worth while to help it out with *clearly*, *fully*, *far*, *deeply*, etc., rather than accept the single word *s.*; compare *he has proved his point clearly e.* with *he has s. proved his point*.

**enquire, enquiry, in-.** See INQUIRE.

**enrol(l).** Spell *enrol*, *enrolment*, but *-lling* etc.; see -LL-, -L-.

**ensure, insure, assure.** For *e.* and *i.*, see EM- AND IM-; for *e.* and *a.* and for *i.* and *a.*, see ASSURE.

**entail.** In spite of the increasing tendency to differentiate (see NOUN AND VERB ACCENT) the OED's ruling that the noun, like the verb, has the accent on the last syllable has not been changed by its successors. But lawyers, the chief users of it, ordinarily accent the first.

**enterprise,** not *-ize*; see -IZE, -ISE.

**enthral(l).** Spell *enthrall*; see -LL-, -L-.

**enthuse.** See BACK-FORMATION. A word of U.S. origin which in neither country has emerged from the stage of slang, or at best colloquial.

**entitled** means having a right (*to* do something) or a just claim (*to* some advantage); it does not mean bound (*to* do) or liable (*to* a penalty); but it is sometimes badly misused: *Germany has suffered bitterly, is suffering bitterly, and Germany is* entitled to suffer *for what she has done.* | *If these people choose to come here* [into court] *and will not learn our heathen language, but prefer their gibberish or jargon, I consider they* are entitled to pay *for it*.

**entity.** The word is one of those regarded by plain people, whether readers or writers, with some alarm and distrust as smacking of philosophy. Its meaning, however, is neither more nor less recondite than that of the corresponding native word, which no one shies at; *e.* is *being*, and *an e.* is *a being*. The first or abstract sense (existence) is comparatively rare; *e. is better than nonentity* means the same as *it is better to be than not to be*. In the second or concrete sense (something that exists), *an e.* differs only so far from *a being* that *being* is not used, except by philosophers, of things that are non-sentient or impersonal; a plant or a stone or a State may be called an *e.*, but is not, outside of philosophy, called a being, and even the philosophers prefer *ens*. *E.* therefore has a right to its place even in the popular vocabulary.

**entourage.** Pronounce *ontoorahzh'*, 'an extreme example of obstinate refusal to take English pronunciation, probably because it is still confined to literary use'.

**entresol.** See FLOOR.

**entrust.** Modern idiom allows only two constructions: to e. (a task, a charge, a secret) *to* someone; to e. (someone) *with* a task etc., and the latter construction is less natural than the former if what is entrusted is a material object, e.g. money. The verb no longer means to put trust in simply (that is to *trust*, not *e.*), nor to commission or employ or charge *to* do (for which those verbs, or again to *trust*, will serve). The obsolete uses are seen in: *King Edward* entrusted him *implicitly, and invariably acted upon his advice.* | *By victory the fighting men have achieved what their country* has entrusted them to do. See CAST-IRON IDIOM.

**enumeration forms.** One of the first requisites for the writing of good clean sentences is to have acquired the art of enumeration, that is, of stringing together three or four words or phrases of identical grammatical value without going wrong. This cannot be done by blind observance of the rule of thumb that *and* and *or* should be used only once in a list. It will suffice here to illustrate very shortly the commonest type of error: *The introductory paragraph is sure, firm, and arouses expectancy at once. | If he raises fruit, vegetables, or keeps a large number of fowls.| A matter in which the hopes and fears of so many of My subjects are keenly concerned, and which, unless handled with foresight, judgment, and in the spirit of mutual concession, threatens to . . .* (Cabinet English, presumably; certainly not the King's English).

The matter will be found fully discussed under AND 2 and 4; OR is liable to corresponding ill treatment; and a particular form of bad enumeration is set forth in the article WALLED-UP OBJECT. See also STOPS, COMMA B.

**enunciate** makes *-ciable*; see -ABLE 1.

**enure.** See INURE, and EM- AND IM-.

**envelope** n. The French spelling (*-ppe*) has long gone, and the French pronunciation should no longer be allowed to embarrass us, but give way to *ě'nvělōp*; all the more now that the verb *envelop*, from its frequency in military bulletins, has become popular instead of merely literary.

**-en verbs from adjectives.** It being no part of most people's business to inquire into such matters, the average writer would probably say, if asked for an offhand opinion, that from any adjective of one syllable an *-en* verb could be formed meaning to make or become so-and-so. That, at any rate, was roughly the position taken up by one party to a newspaper controversy some years ago on the merits of *quieten*. A very slight examination shows it to be remote from the facts; *-en* cannot be called a living suffix. There are on the one hand some 50 *-en* verbs whose currency is beyond question; on the other hand as many adjectives may be found that, though they look as fit as the 50 or turning into verbs by addition of

*-en*, no one would dream of treating in that way. Some of them are allowed to become verbs without the *-en* (*lame, wet, blind, foul*); others have to go without a cognate verb (*grand, wise, sore*); others have their beginning operated on instead of their end (*large* and *enlarge, fine* and *refine, new* and *renew, plain* and *explain, strange* and *estrange, dense* and *condense*); and the despotism of usage is still clearer when it is noticed that we can say *moisten* but not *wetten, quicken* but not *slowen, thicken* and *fatten* but not *thinnen* and *leanen, deafen* but not *blinden, sweeten* but not *souren, sharpen* but not *blunten, cheapen* but not *dearen, freshen* but not *stalen, coarsen* but not *finen*. Between the two sets of adjectives whose mind is made up, some taking and some refusing *-en*, there are a few about which questions may arise. With some the right of the *-en* verb to exist is disputable, and with others the undoubted existence of two verbs (e.g. *loose* and *loosen*), one being identical with the adjective and one having *-en*, raises the question of differentiated senses; and some remarks may be offered on each. The following is the list, thought to be fairly complete, of the ordinary *-en* verbs, not including anomalous ones like *strengthen*, nor any whose right to exist is dubious: *blacken, brighten, broaden, cheapen, coarsen, darken, deaden, deafen, deepen, fasten, fatten, flatten, freshen, gladden, harden, harshen, lessen, liken, lighten, loosen, louden, madden, moisten, quicken, redden, ripen, roughen, sadden, sharpen, shorten, sicken, slacken, smarten, soften, steepen, stiffen, straighten, straiten, sweeten, tauten, thicken, tighten, toughen, weaken, whiten, widen, worsen.*

The debatable words are:

*black* and *blacken*: the second is the wider word used for most purposes, *black* being confined to the sense put black colour upon, besides being only transitive except when used with *out*, meaning to suffer a momentary loss of consciousness. You black boots, glass, someone else's eye, or your own face, and black out a passage as censor or a building to prevent light inside it from being seen outside. You blacken a character; stone blackens or is blackened with age.

*brisken*: not in OED originally, but included in the 1933 Supp. with quotations from 1799. Not in current use.

*dampen*: old in English, but 'now chiefly U.S.' (OED) and a SUPER-FLUOUS WORD.

*fat* and *fatten*: the first is chiefly archaic, kept alive by *the fatted calf*, but also survives as a business word in cattle-breeding circles; *fatten* is the ordinary word.

*glad* and *gladden*: *gladden* is now the ordinary word, but *to glad* is still in poetical use, and is familiar in Moore's

'I never nursed a dear gazelle
To glad me with its soft black eye ...'

*greaten*: 'now archaic' (OED); but a word formerly much used.

*liven*: a modern word used (generally with *up*) as a more colloquial synonym of *enliven*, and also intransitively.

*loose* and *loosen*: the broad distinction is that *loose* means undo or set free (opposite to bind), and *loosen* means to make looser (opposite to tighten).

*mad* and *madden*: *mad* was formerly much used, especially as intransitive in the sense act madly. This is now obsolete, so that 'far from the madding crowd', which is an example of it in Gray's line 'Far from the madding crowd's ignoble strife', is perhaps generally taken in the mutilated form to which Thomas Hardy gave currency to mean far from the distracting crowd. The only present function of *to mad* is to supply a poetical synonym for *madden*, which has suffered from wear and tear as a trivial exaggeration for annoy.

*olden*: this had a vogue during the 19th c. in the sense make or become older in looks or habits, and was an especial favourite of Thackeray's; but, with *to age* well established, it is a SUPERFLUOUS WORD and has rightly fallen into disuse.

*quiet* and *quieten*: *quiet* as a verb dates from 1440 at least, and appears in the Prayer Book, Shakespeare, Burke, and Macaulay, besides many good minor writers; it is both transitive and intransitive. For *quieten*, perhaps the only -*en* verb from an adjective of more than one syllable, the most authoritative name quoted by the OED is Mrs. Gaskell. Its inflexions (*quietened, quietening*, etc.) are ugly, and it should have been dismissed as a SUPERFLUOUS WORD instead of being allowed to usurp the place of the older one. The favour now shown to it is perhaps attributable not so much to a preference for it over *quiet* as to ignorance that *quiet* is a verb.

*right* and *righten*: *to right* is established, and *righten* (called 'rare' by the OED, though used occasionally from the 14th c. on) is a SUPERFLUOUS WORD.

*rough* and *roughen*: both are in full use, with some idiomatic differentiation, though often either will do; see ROUGH(EN).

*slack* and *slacken*: as *rough(en)*; see SLACK(EN).

*smooth* and *smoothen*: the OED gives numerous examples of *smoothen*, each of which, however, makes one wonder afresh why on earth (except sometimes *metri gratiâ*) the writer did not content himself with *smooth*; *smoothen* had clearly had a vogue in the early 19th c., but is now a SUPERFLUOUS WORD.

*stout* and *stouten*: *stout* occurs only in special senses, and is archaic; *stouten* also is now rare.

*white* and *whiten*: *to white* is perhaps only used in echoes of 'whited sepulchres' and of 'as no fuller on earth can white them'.

*worsen*, though some writers may shy at it and reluctantly prefer *deteriorate* in the intransitive sense, is quoted from Milton, George Eliot, and others, and is now common and unexceptionable.

**environs.** The OED recognizes the two pronunciations ĕnvīr'onz, ĕ'nvironz, in that order. So does the COD, but there can be little doubt that the first has now prevailed, perhaps because *environment* and *environmental*, now much used in psychology and sociology, demand the long *i*.

**envisage.** *It is not envisaged at the moment that any reports should be published. E.* is a 19th-c. importation from France now enjoying a vogue as a

FORMAL WORD for *face, confront, contemplate, imagine, intend, recognize, realize, view, visualize,* and *regard.* In the sense of forming a mental picture of something that may exist in the future (*In making these proposals I am envisaging a completely modernized railway system*), e. may no doubt express its user's intention better than any of the other words, but such occasions must indeed be rare in comparison with those in which it is a pretentious substitute for one of them.

**epic.** The proper use of this word is of a narrative poem celebrating the achievements of some heroic person of history or tradition and suggesting the Homeric or Miltonic muse. It has been debased by SLIPSHOD EXTENSION, especially at the hands of film producers, who seem to use it as a catchword to describe any story told on the screen that they hope the public will find exciting. *FABULOUS HOLLYWOOD EPICS!* And (to quote *The Times*) 'it is the present fashion to label as "epic" any play whose anti-hero in the jargon of the time, heroes being "old hat", meanders through countless untidy little scenes of squalid life'. Gossip writers also like to use it when they feel the need of a change from FABULOUS. *Mr. A. O. walked briskly over to our table at the end of his epic Dorchester party at three o'clock this morning.* See VULGARIZATION.

**epicene.** Having no real function in English grammar, the word (n. and adj.) is kept alive chiefly in contemptuous use, implying physical as well as moral sexlessness; for this purpose it is better suited than *common* or *neuter* owing to their familiarity in other senses.

**Epicurean.** See HEDONIST.

**epidemic.** See ENDEMIC.

**epigram.** Literally 'on-writing'. Four distinct meanings, naturally enough developed. First, now obsolete, an inscription on a building, tomb, coin, etc. Second (inscriptions being often in verse, and brief), a short poem, and especially one with a sting in the tail.

Third, any pungent saying. Fourth, a style full of such sayings. See also APHORISM.

**epithet** is suffering a VULGARIZATION that is giving it an abusive imputation. This was jocular in origin: *epithet* was used instead of a dash to stand for an adjective too unseemly to be printed. OED Supp. quotes *If you make the Varsity boat 'easy all' with your epithetted clumsiness, it is ten-and-sixpence.* The corruption has now gone so far that an American broadcaster giving a serious talk can describe the remark '*Politician*' *is a job description, not an epithet* as 'a neat saying'.

**epoch, epoch-making.** Under TIME, the meaning of the word *epoch* is explained. If an epoch were made every time we are told that a discovery or other event is epoch-making, our bewildered state of ceaseless transition from the thousands of eras we were in yesterday to the different thousands we were in today would be pitiful indeed. But luckily the word is a blank cartridge, meant only to startle, and not to carry the bullet of conviction. Cf. UNIQUE and UNTHINKABLE.

**eponymous words.** The following are some of the words now in common use that entered the language as the names of people with whom the things or practices they stand for were associated:

*Banting*: William B. (1797–1878), London cabinet-maker and dietician.

*Bloomers*: Amelia Bloomer (1818–94), American feminist.

*Bowdlerize*: Thomas Bowdler (1754–1825), expurgator of Shakespeare and Gibbon.

*Boycott*: Capt. Charles Cunningham B. (1832–97), land agent in Co. Mayo.

*Bradshaw*: George B. (1801–53), engraver and printer.

*Braille*: Louis B. (1809–52), French inventor.

*Brougham* (carriage): Lord B. and Vaux (1778–1868).

*Bunsen* (burner etc.): Professor R. W. B. (1811–99).

*Burke* (stifle): William B. (1792–1829) murderer and resurrectionist.

*Cardigan*: 7th Earl of C. (1797–1868).

*Clerihew*: E. Clerihew Bentley (1875–1956).

*Derby* (horse race): 12th Earl of Derby (1776–1834).

*Derrick*: A 17th c. hangman.

*Doily*: the Doyly family (late 17th c.), linen-drapers.

*Galvanize*: Luigi Galvani (1737–98), Italian physiologist.

*Greengage*: Sir William Gage (c. 1725).

*Guillotine*: Dr. J. I. Guillotin (1738–1814), French physician.

*Hansard*: Luke H., printer (1752–1828).

*Hansom*: Joseph Aloysius H. (1803–82), architect.

*Macadam*: John Loudon M. (1756–1836), surveyor.

*Macintosh*: Charles M. (1766–1843).

*Mae West*: Miss M. W. (b. 1892), film actress.

*Mansard* (roof): François M. (1598–1666), architect.

*Mesmerize*: F. A. Mesmer (1733–1805), Austrian physician.

*Morse*: Samuel M. (1791–1872), American inventor.

*Pinchbeck*: Christopher P. (1670–1732), clockmaker.

*Plimsoll* (line): Samuel P. (1824–98), member of parliament.

*Pullman* (car): George M. P. (1831–97), American industrialist.

*Quisling*: Vidkun Q. (1887–1945), Norwegian traitor.

*Sadism*: Count (usually called Marquis) de Sade (1740–1814).

*Sam Browne* (belt): General Sir S. B. (1824–1901).

*Sandwich*: 4th Earl of S. (1718–92).

*Shrapnel*: General Henry S. (1761–1842).

*Silhouette*: Etienne de S. (1709–67).

*Spoonerism*: Rev. W. A. Spooner (1844–1930).

*Watt*: James W., engineer (1736–1819).

*Wellington* (boot): 1st Duke of W. (1769–1852).

**equable.** The quality indicated is complex—not merely freedom from great changes, but also remoteness from either extreme, a compound of uniformity and moderation. A continuously cold climate or a consistently violent temper is not e.; nor on the other hand is a moderate but changeable climate or a pulse that varies frequently though within narrow limits.

**equal.** 1. *The navy is not e. in numbers or in strength to* perform *the task it will be called upon to undertake; perform* should be *performing*; see GERUND 3, and ANALOGY.

2. *This work is* the e., if not better than anything *its author has yet done*: e. lends itself particularly to this variety of UNEQUAL YOKEFELLOWS. Read *is e. to, if not better than, anything. . . .*

**equally as.** 1. The use of *as* instead of *with* in correlation with *equally* (*Hermes is patron of poets equally as Apollo*) is a relic of the time when *equally with* had not been established and writers were free (as with many other correlative pairs) to invent their own formulae.

2. The use of *equally as* instead of either *equally* or *as* by itself is an illiterate tautology, but one of which it is necessary to demonstrate the frequency, and therefore the danger, by quotation. These should be corrected by using *equally* alone where a comparison is not expressed within the sentence, and *as* alone where it is. *The labour crisis has furnished evidence equally (as) striking.* | *A practice in some respects equally (as) inequitable.* | *Surely actors' working conditions are* (*equally*) *as important as those of administrative staff.* | *The opposition are* (*equally*) *as guilty as the government.* But sentences in which *equally as* is used to signify precise numerical equality may need more radical alteration: *The expansion of the Canadian labour force will be equally as large as in the last five years* (will be the same as in the last five years). Whether the following can pass may be left to the

reader to consider: *As one British yachtsman said ruefully this afternoon,* Sceptre *would like heavier winds, but there was no assurance that such conditions would not suit* Columbia *equally as well.* He will certainly have no hesitation in passing *How is he to devise a picture which satisfies us equally as a pattern and as an illusion of life?*

**equat(e)(ion).** 'The not distinguishing where things should be distinguished', said John Selden, 'and the not confounding where things should be confounded, is the cause of all the mistakes in the world.' Anyone expressing the same sentiment today would probably say 'the equating of things that ought not to be equated' and so on; *equate* is now the VOGUE WORD for treat as equivalent, fail to distinguish between, confound. Examples: *He equates Hitler and Nasser, and was determined that here would be no Munich.* / *Would you accept an equation between the democratic platform in U.S. and socialism in England?* / *He had those pale, blue, rather shifty eyes that unobservant people often equate with dishonesty.* This is a modern extension of the meaning of a technical term of astronomy and mathematics, harmless enough and indeed valuable if used with discretion. But we are seeing too much of it. For *personal e.* see PERSONAL.

**equerry.** In the circles where such people are found they are called *ĕkwĕ′rĭs.* Elsewhere the established pronunciation is *ĕ′kwerĭ,* and the OED gives it precedence, though it explains that, as against *ĕkwĕ′rĭ,* it probably owes its victory to the word's being popularly confused with *equus* horse, *equine,* etc.; see TRUE AND FALSE ETYMOLOGY. The RECESSIVE ACCENT tendency, however, would perhaps in any case have prevailed.

**equivalence, -cy.** There appears to be no sort of differentiation; the four-syllabled word is now much commoner, and the five-syllabled might well be let die. See -CE, -CY.

**equivocation** (in logic). A fallacy consisting in the use of a word in different senses at different stages of the reasoning. If we conclude from Jones's having a thick head (i.e. being a dullard) that he is proof against concussion, we take *thick head* to mean first dull brain and afterwards solid skull, which is an equivocation.

**era.** For synonymy see TIME.

**-er and -est, more and most.** Neglect or violation of established usage with comparatives and superlatives sometimes betrays ignorance, but more often reveals the repellent assumption that the writer is superior to conventions binding on the common herd. The remarks that follow, however, are not offered as precise rules, but as advice that, though generally sound, may on occasion be set aside.

1. The normal *-er* and *-est* adjectives. 2. Other common *-er* and *-est* adjectives. 3. *-er* and *-est* in adverbs. 4. Adjectives tolerating *-est* but not *-er.* 5. Stylistic extension of *-er* and *-est.* 6. Emotional *-est* without *the.* 7. Superlative in comparisons of two. 8. Comparatives misused. 9. Superlatives misused.

1. The adjectives regularly taking *-er* and *-est* in preference to *more* and *most* are (*a*) all monosyllables (*hard, sage, shy,* etc.); (*b*) disyllables in -y (*holy, lazy, likely,* etc.), in -le (*noble, subtle,* etc.), in -er (*tender, clever,* etc.), in -ow (*narrow, sallow,* etc.); (*c*) many disyllables with accent on the last (*polite, profound,* etc.; but not *antique, bizarre,* or the predicative adjectives *afraid, alive, alone, aware*); (*d*) trisyllabic negative forms of (*b*) and (*c*) words (*unholy, ignoble,* etc.).

2. Some other disyllables in everyday use, not classifiable under terminations, as *common, cruel, pleasant,* and *quiet* (but not *constant, sudden,* etc.) prefer *-er* and *-est.* And many others, e.g. *awkward, brazen, buxom, crooked,* can take *-er* and *-est* without disagreeably challenging attention.

3. Adverbs not formed with *-ly* from adjectives, but identical in form with their adjectives, use *-er* and *-est*

naturally (*runs faster, sleeps sounder, hits hardest, hold it tighter*); some independent adverbs, as *soon, often*, do the same; -*ly* adverbs, though comparatives in -*lier* are possible in archaic and poetic style (*wiselier said, softlier nurtured*), now prefer *more wisely* etc.

4. Many adjectives besides those described in 1 and 2 are capable in ordinary use, i.e. without the stylistic taint illustrated in 5 and 6, of forming a superlative in -*est*, used with *the* and serving as an emphatic form simply, while no one would think of making a comparative in -*er* from them: *in the brutalest, civilest, timidest, cheerfullest, cunningest, doggedest, damnablest, manner*. The terminations that most invite this treatment are -*ful*, -*ing*, -*able*, -*ed*, and -*id*; on the other hand the very common adjective terminations -*ive*, -*ic*, and -*ous*, reject it altogether (*curiouser and curiouser* is a product of Wonderland). Though it is hard to draw a clear line between this use and the next, the intent is different. The words are felt to be only slightly less normal, and yet appreciably more forcible, than the forms with *most*; they are superlatives only, and emphasis is their object.

5. As a stylistic device, based on NOVELTY-HUNTING, and developing into disagreeable mannerism, it was once the fashion to extend the use of -*er* and -*est* to many adjectives normally taking *more* and *most*, and the reader got pulled up at intervals by *beautifuller, delicater, ancientest, diligentest, delectablest, dolefuller, devotedest, admirablest*, and the like. The trick served Carlyle's purpose, but grew tiresome in his imitators and is no longer popular. The extreme form of it is shown in the next paragraph.

6. The emotional -*est* without *the*. For *In Darkest Africa* it may perhaps be pleaded that the title was deliberately chosen for its emotional content. But such a sentence as *Mlle Nau, an actress of considerable technical skill and a valuable power of exhibiting deepest emotion* is so obviously critical and unemotional that it shows fully the VULGARIZATION of a use that is appropriate only to poetic contexts. In so analytic a mood the critic should have been content with *deep emotion*. If he had been talking descriptively, he might have gone as far as 'she exhibited the deepest emotion'; but not unless he had been apostrophizing her in verse as 'deepest emotion's Queen', or by whatever lyric phrase emotion (and not analysis) might have inspired, should he have dared to cut out his *the* and degrade the idiom sacred to the poets. Not that he is a solitary or original sinner; half the second-rate writers on art and literature seem to think they have found in this now hackneyed device an easy way of exhibiting intense but restrained feeling.

Other examples: *The problem is not one of Germany alone; many of the other States which were in the Central Alliance are* in worst plight *for food, so far as can be gathered.* | . . . *addressed the Senate, declaring that* widest diversity *of opinion exists regarding the formation of a League of Nations.* | *An extraordinary announcement is made tonight, which is bound* to stir profoundest interest *among all civilized peoples, and to mark a really new epoch in the story of democracy.* | *But Stoddard did not strike the local note, whereas Stedman could tell of Stuyvesant and the* "*Dutch Patrol*" in pleasantest fashion *and in accordance with the very tone of the Irving tradition.* | *Mr. Vanderlip is, therefore,* in closest touch *with the affairs of international finance.*

If the reader will be good enough to examine these one by one, he will certainly admit this much—that such superlatives are, for better or worse, departures from custom, and that in each sentence a change from '――est' to '(a) very ――' or 'the most――' or 'the ――est' would be a return to normal English. If he will next try to judge, from all the specimens taken together, what effect is produced by this artifice, it may be hoped, though less confidently, that he will agree with the following view. The writers have no sense of congruity (see INCONGRUOUS VOCABU-

LARY), and are barbarically adorning contexts of straightforward business-like matter with ill-judged scraps of more exalted feeling; the impression on sensitive readers is merely that of a queer simulated emotionalism.

7. **Superlative in comparisons of two.** They were forced to give an answer which was not their real answer but only the nearest to it of two alternatives. This use of -est instead of -er where the persons or things compared are no more than two should normally be avoided; the raison d'être of the comparative is to compare two things, and it should be allowed to do its job without encroachment by the superlative. Nearer should have been used in the example given. But exceptions must be admitted. Use of the comparative instead of the superlative would be pedantry in such phrases as Put your best foot foremost; May the best man win; Get the best of both worlds; and who would wish thus to weaken Milton's Whose God is strongest, thine or mine?

8. Certain illogicalities to which the comparative lends itself may be touched upon. For Don't do it more than you can help see HELP. Better known than popular is cured by resolving better into more well. It is more or less—and certainly more than less—a standardized product is a case of CANNIBALISM, one of the necessary two mores having swallowed the other. Unwise striving after double emphasis accounts for He excelled as a lecturer more than as a preacher, because he felt freer to bring more of his personality into play, and for Were ever finer lines perverted to a meaner use? In the first (a mixture of freer to bring his and free to bring more of his) the writer has done nothing worse than give himself away as a waster of words; the second is discussed under ILLOGICALITIES.

9. In superlatives, the fairest of her daughters Eve is still with us: see ILLOGICALITIES. And here is a well contrasted pair of mistakes; the first is of a notorious type (for examples see NUMBERS), and the second looks almost as if it were due to misguided avoidance of the supposed danger; read have for has in the first, and has for have in the other: In which case one of the greatest and most serious strikes which has occurred in modern times will take place. / Houdin was a wonderful conjuror, and is often reckoned the greatest of his craft who have ever lived.

**-er and -or.** 1. The agent termination -er can be added to any existing English verb; but with many verbs the regular agent-noun ends in -or and that in -er is occasional only; with others both forms are established with or without differentiation of sense; see -OR.

2. When -er is added to verbs in -y following a consonant, y is ordinarily changed to i (occupier, carrier); but y is retained between a vowel and -er (player, employer, buyer). But monosyllabic verbs in -y are capricious in this respect. See DRY.

**Erastianism.** See JANSENISM.

**ere.** See INCONGRUOUS VOCABULARY, and VULGARIZATION. Ere can still be used with effect if the atmosphere is right (It is so easy to talk of 'passing emotion' and to forget how vivid the emotion was ere it passed), but in all ordinary prose before should be preferred; the following quotations show the fish out of water at its unhappiest: The iniquitous anomaly of the plural voter will be swept away ere we are much older. / As many people may be aware, Christmas books are put in hand long ere the season with which they are associated comes round. / In the opinion of high officials it is only a matter of time ere the city is cleared of the objectionable smoke pollution evil.

**ergo** (Latin for therefore) is archaic or obsolete in serious use, but still serves the purpose of drawing attention facetiously to the illogical nature of a conclusion: He says it is too hot for anything; ergo, a bottle of Bass. See PEDANTIC HUMOUR.

**Eros.** We talk erotics more than we did, and there is an Eros that all Londoners have seen; so the name has a future before it, and its pronunciation matters. The Greek word, in English mouths, is ē'rōz, but the OED Supp., putting ēr'ŏs first, reflects the Londoners' choice. It is no worse than *Sŏcrates, Nēro, Plāto,* and many other classical names that we pronounce with a FALSE QUANTITY.

**erratum.** Pl. *-ta*; see -UM.

**ersatz.** This word for a substitute (impliedly inferior) for the real thing was borrowed from Germany at the time of the first world war. We had no real need of it—we have our own *artificial, imitation, mock, sham,* and *synthetic*—and the vogue it enjoyed for a time must have been due to that NOVELTY HUNTING which also prompts the costumier to reject the well worn words and advertise his imitation mink as *mink simulation.*

**erst, erstwhile.** See INCONGRUOUS VOCABULARY and LATE.

**escalate.** *Undeclared war between India and China has escalated sharply during the past few days.* | *The President chose to begin with one of the less drastic options, leaving open the possibility of escalation to more violent ones. Escalate,* a recent BACK-FORMATION from *escalator* (both originally U.S.) was not needed; *escalade* (n. and vb.) has long been in similar metaphorical use. But *-ate* is likely to drive *-ade* out; it has the advantage of novelty and a more native look, and a moving staircase provides a more up-to-date metaphor than a scaling ladder.

**escap(ee) (ism) (ist).** *Escapism* and *escapist,* for those who would escape from reality into fantasy, are words too recent to be in the OED Supp.; they are no doubt a natural product of the atomic age. Cf. WISHFUL thinking. *Escapee,* whose French form (see -EE) is said to be due to its having been originally applied to French convicts from New Caledonia escaping to Australia, is a SUPERFLUOUS WORD that should not be allowed to

usurp the place of *escaper.* One might as well call deserters *desertees.*

**esoteric.** See EXOTERIC, EXOTIC.

**especial(ly).** 1. (*E*)*special(ly).* 2. *Especially* with inversion. 3. *Especially as.* 4. *More especially.*

1. (*E*)*special(ly).* The characteristic sense of the longer adjective and adverb is pre-eminence or the particular as opposed to the ordinary, that of the others being limitation or the particular as opposed to the general. There is, however, a marked tendency in the adjectives for *especial* to disappear and for *special* to take its place. It may be said that *special* is now possible in all senses, though *especial* is still also possible or preferable in the senses (*a*) exceptional in degree, as *My especial friend is Jones*; *He handles the matter with especial dexterity*; *Oxford architecture receives especial attention,* (*b*) of or for a particular person or thing specified by possessive adjective or case, as *For my* or *Smith's especial benefit*; *For the especial benefit of wounded soldiers.* In the adverbs the encroachments of the shorter form are more limited; a writer may sometimes fall into saying *The reinforcements arrived at a* specially *critical moment,* where *an especially* would be better, but it is as little allowable to say *The candidates,* specially *those from Scotland, showed ability* as it is to say *Candidates must be* especially *prepared* or *An arbitrator was* especially *appointed.* Two examples follow of *especially* used where *specially* is clearly meant; in both the sense is not to an exceptional degree, but for one purpose and no other: *The founder of the movement will be* especially *interviewed for the ten o'clock news.* | *Agreeable features of the book will be the illustrations, including a number of reproductions of prints* especially *lent.*

2. *Especially* with inversion. The word is a favourite with victims of this craze (see INVERSION): *Springs of mineralized water, famous from Roman times onwards; especially did they come into renown during the nineteenth century.* | *Mr. Campbell does not recognize a change of opinion, but*

*frankly admits a change of emphasis;* especially is he *anxious at the present time to advance the cause of Liberal Evangelism.*

**3.** *Especially as.* It is worth notice that of the causal *as*-clauses discussed in AS 2 some types intolerable in themselves are made possible by the insertion of *especially* before *as*: *I shall have to ask for heavy damages, as my client's circumstances are not such as to allow of quixotic magnanimity.* As by itself is, as usual, insufficient to give the remainder of the sentence the fresh push-off that the introduction of an unforeseen consideration requires; but *especially* inserted before *as*, by bespeaking attention, prevents the tailing off into insignificance that would have ruined the balance.

**4.** *More especially* is a common form of TAUTOLOGY. *Some showers may develop more especially in the eastern counties. Especially* is quite capable of doing the job by itself.

**espionage.** *Es′piŏnij, -ahzh, -āj* and *espi′ŏnij* all have some dictionary authority. Preference is recommended in that order.

**esplanade.** Pronounce *-ād* not *-ahd* (unlike *promenade*). See -ADE, -ADO.

**esquire** was originally a title of function; the esquire was the attendant of the knight and carried his gear. It later became a title of rank, intermediate between *knight* and *gentleman*, the right to which is still defined by law in a way that to modern ideas is in some respects curious. Barristers are esquires (at any rate after they have taken silk; there seems some doubt about the outer bar), but solicitors are never more than mere gentlemen; justices of the peace are esquires but only while they are in the Commission. A class of esquires that must by now be of considerable size is that of the eldest sons of knights and their eldest sons and so on in perpetuity. But the impossibility of knowing who is an esquire and who not, combined with a reluctance

to draw invidious distinctions, has deprived *esquire* of all significance. and it looks as though one odd product of the Century of the Common Man might be to promote the whole adult male population to this once select and coveted status.

**essay.** **1.** For *e.* and *assay*, vv., see ASSAY.

**2.** The verb is accented on the last syllable; the noun, in its now commonest sense of a kind of literary piece, on the first. But in the wider and now rare sense of an attempt the old accent on the last may still be heard; that it was formerly so accented is evident from lines like *Whose first essay was in a tyrant's praise.* | *This is th' essay of my unpractis'd pen.* | *And calls his finish'd poem an Essay.* See NOUN AND VERB ACCENT.

**essence and substance, essential-(ly) and substantial(ly).** The words started in life as Latin philosophical terms translating the Greek οὐσία (lit. being) and ὑπόστασις (lit. underneathness). The meaning of the Greek words was practically the same, 'true inwardness' being perhaps the nearest equivalent in native English, but the second was substituted by later Greek philosophers for the first as used by earlier ones; similarly in Latin *substantia* was a post-Augustan synonym for Cicero's *essentia*. It is therefore natural that *essence* and *substance*, *essential(ly)* and *substantial(ly)*, should on the one hand be sometimes interchangeable, and on the other hand develop, like most synonyms, on diverging lines with differentiations gradually becoming fixed. It may be said roughly that *s.* has moved in the direction of material and quantity, *e.* in that of spirit and quality. The strictly philosophical or metaphysical uses are beyond the scope of this book; but some examples of the words in popular contexts may serve to show how they agree and disagree.

**1.** Examples in which either is possible, sometimes with and sometimes

without change of sense, or with degrees of idiomatic appropriateness:
*God is an essence* (or less often *a s.*), i.e. a self-existent being. | *I can give you the substance of what he said* (or less often *the e.*, implying the cutting out of all superfluous details). | *But he took care to retain the substance of power* (or less usually *the essence*, or archaically *the substantials*, or quite well *the essentials*). | *The essence of morality is right intention, the substance of it is right action* (the words could not be exchanged in this antithesis, but in either part by itself either word would do; *the e.* is that without which morality would not be what it is, *the s.* is that of which it is made up). | *Distinguish between the mere words of Revelation and its substance* (or *e.*). | *They give in substance the same account* (or *in essence* rarely, or *substantially* or *essentially*). | *The treaty underwent substantial modifications* (or *e.*, but *s.* means merely that they amounted to a good deal, *e.* that they changed the whole effect). | *Desire of praise is an essential part of human nature* (or *s.*, but if *e.*, human nature without it is inconceivable, and if *s.*, human nature is appreciably actuated by it). | *There is an essential difference* (or *s.*; the latter much less emphatic). | *All parties received substantial justice* (or rarely *e.*, which implies much less, if any, ground for dissatisfaction).

**2.** Examples admitting of *essen-* only:
*The essence of a triangle is three straight lines meeting at three angles.* | *What is the essence of snobbery?* | *Time is the essence of the contract.* | *Kubla Khan may be called essential poetry.* | *The qualities essential to success.* | *This point is essential to the argument.* | *An essentially vulgar person.*

**3.** Examples admitting of *substan-* only:
*Butter is a substance.* | *Parting with the substance for the shadow.* | *There is no substance in his argument.* | *A man of substance.* | *A cloth with some substance in it.* | *His failure to bring any substantial evidence.* | *A substantial meal.* | *A substantially built house.*

**essential, necessary, requisite.** The words so far agree in the sense in which they are all commonest, i.e. 'needed', that in perhaps most sentences containing one of them either of the others could be substituted without serious change of meaning. It often does not matter whether we say 'the e.' or 'the n.', or 'the r., qualities are courage and intelligence only'. They have reached the meeting-point, however, from different directions, bringing each its native equipment of varying suitability for various tasks. For instance, in *We can hardly say that capital is as r. to production as land and labour* the least suitable of the three has been chosen, the word wanted to class the relation of land and labour to production being the strongest of all, whereas *r.* is the weakest.

If we call something *e.* we have in mind a whole that would not be what it is to be or is or was if the part in question were wanting; the e. thing is such that the other thing is inconceivable without it. *E.* is the strongest word of the three.

When we call something *n.*, we have in mind the irresistible action of causality or logic; the n. thing is such that the other cannot but owe its existence to it or result in it. *N.* doubles the parts of *indispensable* and *inevitable*.

When we call something *r.*, we have in mind merely an end for which means are to be found; the r. thing is that demanded by the conditions, but need not be the only thing that could satisfy their demands, though it is usually understood in that sense. The fact that r. has no negative form corresponding to *une.* and *unn.* is significant of its less exclusive meaning.

For a trivial illustration or two: Bails are r., but neither e. nor n., for cricket; not e., for it is cricket without them; not n., for their want need not stop the game. In the taking of an oath, religious belief is e., but neither n. nor r.; the unbeliever's oath is no oath, but the want of belief need not prevent him from swearing, nor will belief help him to swear. In this book the alphabetical arrangement is unessential, but not unnecessary, and very requisite; the dictionary without it would be a dictionary all the same, but the laws of causality make the publishers demand and the writer supply alphabetical order, and without it the purpose would be very badly served.

**establishment.** *Their enthusiasm is tempered by the coolness of the E. 'Space travel', said the Astronomer Royal recently, 'is utter bilge'. | Some of the blame must go to the Theatrical E. Their influence on the theatre is disastrous in every way. | Sir Oliver musters the women's colleges, 23 heads of houses, 4 heads of halls, and 9 professors. If anything, a vote for Mr. Macmillan is a vote against what many dons regard as the University E. | In his career he has shown enviable ability to drop bricks without disaster, to keep the respect of the E. but retain the liberty to hold advanced opinions. | The fusty E. with its Victorian views and standards of judgement must be destroyed. | Today the membership list of the Yacht Club is studded with E. names | Mr. Crossman has suggested that it is the duty of the Labour Party to provide an 'ideology for non-conformist critics of the E.'* The use of *establishment* illustrated by these quotations leapt into popularity in the nineteen-fifties. It was started by the use of *Establishment* by certain writers as a pejorative term for an influence that they held to be socially mischievous. In their choice of a word for the object of their attack they have been too successful. *Establishment* has become a VOGUE WORD, and whatever meaning they intended to give it, never easy to gather with precision from their various definitions, has been lost beneath the luxuriance of its overgrowth. As one of them has sadly recorded: 'Intended to assist inquiry and thought, this virtuous, almost demure, phrase has been debauched by the whole tribe of professional publicists and vulgarizers who today imagine that a little ill-will entitles them to comment on public affairs. Corrupted by them, the *Establishment* is now a harlot of a

phrase. It is used indiscriminately by dons, novelists, playwrights, poets, composers, artists, actors, dramatic critics, literary critics, script writers, even band leaders and antique dealers, merely to denote those in positions of power whom they happen to dislike most' (Henry Fairlie in *The Establishment*, Anthony Blond, 1959).

**estate.** *The three estates of the realm*, i.e. the Lords Spiritual, the Lords Temporal, and the Commons, is often wrongly applied to Sovereign, Lords, and Commons. The use of the phrase being now purely decorative, and the reader being often uncertain whether the user of it may mean Sovereign and Parliament, or Parliament, or all bishops and all peers and all electors, it is perhaps better left alone. *The third e.* is a phrase often used for the French BOURGEOISIE before the Revolution; *the fourth e.* is a jocular description of the Press as one of the powers that have to be reckoned with in politics, and the *fifth e.* would now be a fitting title for the Trades Union Congress.

**esteem, estimat(e)(ion).** 1. *Estimate* makes *estimatable* and *esteem estimable*. 2. For *success of esteem* see GALLICISMS. 3. The sense of a judgement formed by calculation or consideration belongs to *estimate* and not to *estimation*, which means not the judgement itself, but the forming of it. The tendency described in LONG VARIANTS often leads writers astray, as in: *Norwegians can only wish that the optimistic estimation of Mr. Ponting of the British minefields at Spitzbergen will come true.*
4. The use of *in my* etc. *estimation* as a mere substitute for *in my* etc. *opinion* where there is no question of calculating amounts or degrees, as in *The thing is absurd in my e.*, is illiterate. *Tories love discussion: they cannot have too much of it. But they think it is going too far to translate words into action. That is not, in their e., playing the Parliamentary game.*

**estop** is a useful word so long as it is restricted to its proper sense; to give it a wider one betrays either ignorance or pedantry. The proper (legal) sense is (in the passive) 'to be precluded by one's own previous act or declaration from alleging or doing something'. Two quotations will show (*a*) the right and (*b*) the wrong use: (*a*) *No one defended more joyously the silencing of Mr. Asquith last July, and Mr. Maxse is estopped from complaining, now that his own method has been applied to himself*; (*b*) *The road winds along the side of a barren mountain till it appears to be estopped by a high cliff.*

**etc.** is invaluable in its right place—lexicography for instance. But to resort to it in sentences of a literary character (*His faults of temper etc. are indeed easily accounted for*) is amateurish, slovenly, and incongruous: *A compliment of this kind is calculated to increase their enthusiasm, courage, etc., to do their utmost.* | *The Covenanted Civil Service with its old traditions and its hereditary hatred of interlopers, be they merchants, journalists, doctors, etc.* On the other hand, in the contexts to which it is appropriate, it is needless purism to restrict its sense to what the words could mean in Latin, i.e. (*a*) and *the rest* as opposed to and *other things*, (*b*) *and* the like as opposed to *or* the like, (*c*) and other *things* as opposed to *persons*; the first restriction would exclude *His pockets contained an apple, a piece of string, etc.*; the second would exclude '*Good*', '*fair*', '*excellent*', *etc.*, *is appended to each name*; the third would exclude *The Duke of A, Lord B, Mr. C, etc., are patrons. Et hoc genus omne* is a phrase on which the literary man who finds himself sorely tempted to 'end with a lazy etcetera', but knows he mustn't, sometimes rides off not very creditably.

The reasonable punctuation with *etc.* is to put a comma before it when more than one term has preceded, but not when one term only has: *toads, frogs, etc.*; but *toads etc.* For the difference between *etc.* and *et seq.* see SEQ.

**ethic(al), ethics.**  1. *ethic dative.* 2. *ethic, ethics.*  3. *ethics*, number. 4. *ethics, morals.*  5. *ethical, moral.*

1. *ethic dative.* *Ethic* has now been almost displaced as an adjective by *ethical*. It is still used occasionally, but is noticeably archaic; the only exception to this is in the *ethic dative*. This, in which the word means emotional or expressive, is the name for a common Greek and Latin use in which a person no more than indirectly interested in the fact described in the sentence is introduced into it, usually by himself as the speaker, in the dative, which is accordingly most often that of the first personal pronoun. Thus, in *Quid mihi Celsus agit?*, the word *mihi* (lit., to or for me) amounts to a parenthetic 'I wonder'. The construction was formerly English also: in *He that kills me some six or seven dozen of Scots at a breakfast* the word *me* amounts to a parenthetic 'Just fancy!'

2. *ethic, ethics.* Of the two nouns the second is the one for ordinary use. It means the science of morals or study of the principles defining man's duty to his neighbours, a treatise on this, or a prevailing code of morality (*Ethics is*, or *are, not to be treated as an exact science*; *That is surely from the Ethics*, i.e. Aristotle's; *Our modern ethics are not outraged by this type of mendacity*). *Ethic* in any of these senses has a pedantic air; it is chiefly in technical philosophic use, and its special meaning is a scheme of moral science (*The attempt to construct an ethic apart from theology*).

3. For the grammatical number of *ethics* see -ICS 2.

4. *ethics, morals.* The two words, once fully synonymous, and existing together only because English scholars knew both Greek and Latin, have now so divided their functions that neither is superfluous. They are not rivals for one job, but holders of complementary jobs; *ethics* is the science of morals, and *morals* are the practice of ethics; *His ethics may be sound, but his morals are abandoned.* That is the broad distinction; the points where confusion

arises are three: (a) sometimes those who are talking about morals choose to call them ethics because the less familiar word strikes them as more imposing; (b) there is an impression that ethics is somehow more definitely than morals disconnected from religion; (c) the distinction is rather fine between the sense of *ethics* as a prevailing code of morals, and morals themselves; but, though fine, it is clear enough.

5. *ethical, moral*. It is in the nature of things that the dividing line between the adjectives should be less clear than with the nouns. For, if ethics is the science of morals, whatever concerns morals evidently concerns ethics too, and is as much ethical as moral; and vice versa. Nevertheless, we talk of *a moral*, but not *an ethical*, *man*, and we perhaps tend more and more to talk of *the ethical* rather than *the moral basis* of society, education, and so forth. At the same time, since *immoral* is popularly associated with sexual immorality, *unethical* has come into vogue in U.S. as an adjective for the conduct of the man who is immoral in other ways, especially in violating the accepted code of a profession or business. In England too we avoid the word *immoral* for the same reason, but should probably say *dishonest*, or, if we do not want to use so harsh a word, *unscrupulous*. This may not be true for long; the American use of *unethical* is gaining currency here. Perhaps that is because the words *right* and *wrong* and *good* and *wicked* seem now to be old-fashioned.

**ethos.** Pronounce *ē'thŏs*. It means the characteristic spirit informing a nation, an age, a literature, an institution, or any similar unit. (*The Liberal Party, it has been said, has now an ethos but no real policy*.) In reference to a nation or state, it is the sum of the intellectual and moral tendencies manifested in what the Germans called the nation's Kultur; like Kultur, it is not in itself a word of praise or blame, any more than *quality*.

**-ette.** For the uses of this suffix see FEMININE DESIGNATIONS s.f. and SUFFRAGETTE.

**euphemism** means (the use of) a mild or vague or periphrastic expression as a substitute for blunt precision or disagreeable truth. The heyday of euphemism in England was the mid-Victorian era, when the dead were *the departed*, or *no longer living*, pregnant women were *in an interesting condition*, novelists wrote *d——d* for damned and *G—d* for God, bowdlerized editions of Shakespeare and Gibbon were put into the hands of the young and trousers were *nether garments*, or even, jocosely but significantly, *unmentionables* or *inexpressibles*. We are less mealy-mouthed now, though still more given to euphemisms than our Continental neighbours; the notice *Commit no nuisance* or *Decency Forbids* was even in our own day sometimes used for the injunction put more bluntly in France as *Défense d'uriner*. But euphemism is a will-o'-the-wisp for ever eluding pursuit; each new word becomes in turn as explicit as its predecessors and has to be replaced. The most notorious example of the working of this law is that which has given us such a plethora of names for the same thing as *jakes, privy, latrine, water-closet, w.c., lavatory, loo, convenience, ladies, gents, toilet, powder-room, cloaks*, and so on, endlessly. There are of course—or were before the publication of *Lady Chatterley's Lover*—some words, now a small and rapidly diminishing number, too tainted by bawdy and ribaldry to be usable, and for these polite synonyms must be found. But delicacy becomes absurdity when it produces such an anticlimax as is contained in *Pathological tests suggest that she had two blows on the head, was strangled and probably assaulted*. 'It is a pity', said a President of the Probate Divorce and Admiralty Division in 1959, 'that plain English is not used about these matters in divorce proceedings. When I say plain English I mean that, so far as I know, ever since the tablets of stone were translated into English in the English version of the Bible, *adultery* has been the word, not *misconduct* or *intimacy* or any other paraphrase of it.'

In the present century euphemism has been employed less in finding discreet terms for what is indelicate than as a protective device for governments and as a token of a new approach to psychological and sociological problems. Its value is notorious in totalitarian countries, where assassination and aggression can be made to look respectable by calling them *liquidation* and *liberation*. In Western democracies too use is made of the device of giving things new names in order to improve their appearance. Thus what were at first called crudely Labour Exchanges and Distressed Areas are now *Employment Exchanges* and *Development Areas*: the poor are the *lower income brackets* or the *underprivileged ˉclasses*: poor-law relief is *national assistance*; those who used to be known as backward and troublesome children are now *maladjusted*; ladies once termed mistresses (itself a euphemism for the earlier *concubines* and *paramours*) are *unmarried wives*; insanity is now *mental disorder*; lunatic asylums are *mental hospitals*; criminal lunatics are *Broadmoor patients*, and every kind of unpleasant event that might call for action by the government is discreetly referred to as an EMERGENCY. The same device is used to give a new look to an old occupation. Thus charwomen have become *dailies*, gaolers *prison officers*, commercial travellers *sales representatives*, and RATCATCHERS *rodent operators*. Dustmen, naturally resenting our wounding habit of emphasizing their unlikeness to dukes, are now *refuse collectors* or *street orderlies*; boardinghouses have been rechristened *guest houses*; many butchers call themselves *purveyors of meat* and at least one a *meat technologist*; hairdressers are *tonsorial artists* and undertakers *funeral furnishers* or *directors*, or (U.S.) *morticians*.

**euphuism.** The word is often ignor-

antly used for euphemism with which it is entirely unconnected. It is named from Lyly's *Euphues* (i.e. The Man of Parts), fashionable in and after the 16th c. as a literary model, and it means affected artificiality of style, indulgence in antithesis and simile and conceits, subtly refined choice of words, preciosity. It is, unlike euphemism, a word which no one but the literary critic is likely to need. A single example of the common misuse will suffice: *While a financial euphuism christened railway construction a 'transformation of capital', and not an expenditure.* See POPULARIZED TECHNICALITIES and PAIRS AND SNARES.

**Eurasian.** See MULATTO 1 and 4.

**evaluate.** *Henderson, whose services to the Labour Party have never been properly evaluated, preferred loyalty to the party's principles, and thereby saved the Labour Party from complete disaster.* Evaluate is a term of mathematics meaning to find a numerical expression for; hence, more generally, to express in terms of the known. Its use as a synonym of *value* in that word's sense of to have a high opinion of, to esteem, can only be attributed to the septic influence of LOVE OF THE LONG WORD. An example of its correct use is . . . *a statement by the Colonial Secretary of minimal changes in the mathematics of the new Northern Rhodesian constitution — mathematics which are excusably beyond the capacity of the ordinary citizen to evaluate.*

**evasion, evasiveness.** The latter is a quality only; in places where quality, and not practice or action, is the clear meaning, *evasion* should not be used instead of it. The right uses are shown in *His evasion of the issue is obvious; he is guilty of perpetual evasion*; but *the evasiveness of his answers is enough to condemn him.* See -ION AND -NESS.

**eve.** *On Christmas E., on the E. of St. Agnes, on the e. of the battle, on the e. of departure, on the e. of great developments.* The strict sense of *e.* being the evening or day before, the first two phrases are literal, the last is metaphorical, and the two others may be either, i.e. they may mean *before* (either with an interval of days or weeks, or with a night intervening) or actually on the same day. Nevertheless, despite the risk of ambiguity, they are all legitimate; what is not legitimate is to use the word in its metaphorical sense and yet remind the reader of the literal sense by some turn of words that involves it, as in *The most irreconcilable of Irish landlords are beginning to recognize that we are on the e. of the dawn of a new day in Ireland.* See METAPHOR 2 B.

**even.** 1. Placing of *e.* 2. *E. so.*

1. Placing of *e.* It will be seen in POSITION OF ADVERBS that their placing is a matter partly of idiom and partly of sense; *e.* is one of those whose placing is important to the sense. *Even I did not see him on Monday* implies that I was more likely to see him than anyone else was. *I did not e. see him on Monday* implies that I had expected not only to see him but also to speak to him. *I did not see e. him on Monday* implies that he was the person I expected to see. *I did not see him e. on Monday* implies that Monday was the day on which I expected to see him.

2. *Even so.* This is a phrase that has its uses; it often serves as a conveniently short reminder to the reader that the contention before him is not the strongest that could be advanced, that deductions have been made, that the total is net and not gross. But some writers become so attached to this convenience that they resort to it (*a*) when it is a convenience to them and an inconvenience to their readers, i.e. when it takes a reader some time to discover what exactly the writer means by it, and (*b*) when either nothing at all or one of the everyday conjunctions would do as well. The following passages are none of them indefensible, but all exemplify the ill-judged *e. so,* used (when it conveys too much) to save the writer trouble, or (when it conveys too little) to gratify his fondness for the phrase: *Just at present the Act is the*

*subject of misconceptions and misrepresentations, some of which can only be dissipated by actual experience of its working. It may be that,* e. so, *the people will dislike the Act* (even after experience). / *I hope it won't come to this; but,* e. so, *bridge-players will continue to take their finesses and call it just the luck when they go down* (even if it does). / *It is natural that France should be anxious not to lose on the swings what she gains on the roundabouts, and she has some reason for nervousness as to the interaction of commerce and politics.* E. so, *she will do well not to be over-nervous* (But). / *If the absent are always wrong, statesmen who have passed away are always gentlemen. But,* e. so, *we were not prepared for this tribute to those statesmen who fought for Home Rule in 1886 and 1893* (omit e. so).

**evensong** etc. See MORNING.

**event.** *In the e. that the Suez Canal should suddenly become blocked.* This is an Americanism; in Britain idiom prefers *in the e. of . . . becoming.*

**eventuality, eventuate.** The words are much used in OFFICIALESE and also to be found in flabby journalese; some characteristic specimens are: [*Disinterested management is advocated*] *as a second string to the bow of temperance reformers, a provision for the eventuality of the people refusing to avail themselves of the option of veto.* / [*That the Territorial Force is on the eve of a breakdown*] *is very far from the case, however dear such an eventuality might be to the enemies of the voluntary system.* / *The bogies that were raised about the ruin did not eventuate, yet employers still want the assistants to work for long hours.* / *And why did not that policy eventuate?*

**ever** is often used colloquially as an emphasizer of *who, what, when,* and other interrogative words, corresponding to such phrases as *who in the world, what on earth, where* (can he) *possibly* (be?). When such talk is reproduced in print, *ever* should be a separate word —*what ever* etc., not *whatever* etc. It may even be used by itself as a colloquialism. *Did you ever?* (hear such a thing). For *e.* in letters see LETTER FORMS.

**ever so** (*though it were ever so bad* etc.). See NEVER SO.

**everyday** (adj.). One word.

**every one.** 1. *Every one, everyone.* 2. Number of pronoun after *e.*

1. *Every one, everyone. The . . . drawings are academical in the worst sense of the word; almost everyone of them deserves a gold medal.* In this sentence the making of the two words into one is undoubtedly wrong; this should only be done where *everybody* might be substituted. That is never true when, as here, things and not persons are meant, nor yet when, as here, a partitive *of* follows; in either of those cases it is agreed that the words should be kept separate. Unfortunately there is not the same agreement on the corresponding rule that when *everybody* can be substituted *everyone* should be so written. The question cannot be decided for *everyone* by itself; the parallel *anyone*, *no one*, and *someone* must be taken into account; of these *no one* alone is fixed, and that is always two words, owing to the natural tendency to pronounce *noone* noon. On the side of one word we have (*a*) the fact that all the four words, when they mean anybody etc., have only one accent instead of the two that are heard when they mean any single etc., (*b*) the general usage of printers, based on this accentuation, with all except *no one*. On the side of two words we have (*a*) consistency, since the others thus fall into line with *no one*, Mahomet-and-the-mountain-fashion, (*b*) escape from the mute e before a vowel inside a single word in *someone*, which is undesirable though not unexampled, (*c*) the authority of the OED, which gives precedence in all four to the separation. A very pretty quarrel. The opinion here is that the accent is far the most important point, that *anyone* and *everyone* and *someone* should be established, and that *no-one*

is the right compromise between the misleading *noone* and the inconsistent *no one*. The rules would then be these: (1) *Anyone, everyone, no-one,* and *someone*, in the sense anybody, everybody, etc.; (2) *any one, every one, no one, some one*, each with two accents, in other uses.

The foregoing advice, given in 1926, now represents the standard practice, except that *no one* has more backing than *no-one*, and is recommended.

2. Number of pronoun after *everyone* (*E. had made up their minds*; *E. then looked about them silently*); on this question see NUMBER 11.

**evidence**, vb. To evidence something is to be the proof, or serve as evidence, of its existence or truth or occurrence. To say that you evidence care is a wrong use of the word if what is meant is that you are behaving carefully, but it is a right use if what is meant is that your appearance shows that someone has been careful to see you are properly turned out. It will be seen that *show* or *exhibit* could take the place of *e.* rightly used, but also that they would stand where *e.* could not. Writers with a preference for the less common or the more technical-looking word are sometimes trapped by the partial equivalence into thinking that they may indulge their preference by using *e.* instead of *show.* A right and a wrong sentence will make the limitation of meaning clearer; it must be borne in mind, however, that that definition does not pretend to cover all senses of *e.,* but only those in which it is in danger of misuse.

Right use: This work *of Mr. Phillipps, while it bears all the marks of scholarship, bears also the far rarer impress of original thought, and* evidences *the power of considering with an unusual detachment a subject which . . .*
Wrong use: Mr. Thayer evidences a remarkable grasp *of his material, and a real gift for the writing of history.*

**evince** has lost most of its meanings: the only one remaining is that of indicating a quality or state of mind. *His*

*speech evinced a reluctance to carry his argument to its logical conclusion.*
But those who like a full-dress word better than a plain one continue to use and sometimes to misuse it. The writer of the first of the quotations below, in putting *evince* next door to *evident*, surely evinces a fondness for it that borders on foolishness. The writer of the second must have been unaware that, though either a person or an attitude can e. an emotion, neither a person nor an emotion can e. an attitude; an attitude is nothing if not visible, and what is evinced is inferable but not visible. The writer of the third has fallen into the trap of OBJECT SHUFFLING, and written *evince* where he meant *evoke. Both the Tories and the Labour Party evinced an evident anxiety to stir up trouble on the labour unrest in the railway world. | The Opposition welcomed the Bill on first reading, did not divide against it on second reading, and have, on the whole, only evinced a legitimately critical attitude in Committee. | A prewar built period style residence with four bedrooms, for which £6,750 is asked, evinced much less interest than another at the same figure on the opposite side of the road.*

**ex-.** For such patent yet prevalent absurdities as *ex-Lord Mayor, ex-Chief Whip, ex-Tory Solicitor-General* (except in another sense than its writer means), see HYPHENS 6; and for alternatives, LATE.

**exactly, just.** *E. what has happened or what is about to happen is not yet clear. | Just how the words are to be divided.* This now familiar idiom, in which *e.* or *j.* is prefixed to an indirect question, is a modern development. The *e.* or *j.* sometimes adds point, but is more often otiose, and the use of it becomes with many writers a disagreeable mannerism. See also JUST.

**exceeding(ly) and excessive(ly).** The difference is the same as that between *very great* or *very much* and *too great* or *too much.* It is not inherent in the words, nor very old, *excessive(ly)*

having formerly had both meanings; but it is now recognized by most of those who use words carefully, and is a useful DIFFERENTIATION. It follows that *I am excessively obliged to you* is not now standard English, and that *I was excessively annoyed* should be said in repentant and not, as it usually is, in self-satisfied tones. A passage in which a good modern writer allows himself to disregard the now usual distinction may be worth giving: *I have said that in early life Henry James was not 'impressive'; as time went on his appearance became, on the contrary, excessively noticeable and arresting.*

**except.** *Except the Lord keep the city, the watchman waketh but in vain.* This use of *e.* as a conjunction governing a clause, i.e. as a substitute for the *unless* or *if . . . not* of ordinary educated speech, is now either an ARCHAISM resorted to for one or other of the usual reasons, or else an illustration of the fact that old constructions often survive in uneducated talk when otherwise obsolete. In the following quotation, archaism, chosen for one of the less defensible reasons, is the explanation: *But, e. the matter is argued as a mere matter of* amour propre *how is it possible to use such high-flown language about a mere 'change of method'?*

**excepting** as a preposition has one normal use. When a possible exception is to be mentioned as not made, the form used, instead of *not except,* is either *not excepting* before the noun or *not excepted* after it: *All men are fallible except the Pope; All men are fallible, not excepting the Pope* or *the Pope not excepted.* Other prepositional uses of *excepting* are unidiomatic; but the word as a true participle or a gerund does not fall under this condemnation: *He would treble the tax on brandy excepting only,* or *without even excepting, that destined for medicine.* An example of the use deprecated is: *The cost of living throughout the world, excepting in countries where special causes operate, shows a tendency to keep level.*

**exception.** *The e. proves the rule,* and phrases implying it, are so constantly introduced in argument, and so much more often with obscuring than with illuminating effect, that it is necessary to set out its different possible meanings, viz. (1) the original simple legal sense, (2) the secondary rather complicated scientific sense, (3) the loose rhetorical sense, (4) the jocular nonsense, (5) the serious nonsense. The last of these is the most objectionable, though (3) and (4) must bear the blame for bringing (5) into existence by popularizing an easily misunderstood phrase; unfortunately (5) is much the commonest use. See POPULARIZED TECHNICALITIES.

1. 'Special leave is given for men to be out of barracks tonight till 11.0 p.m.'; 'The exception proves the rule' means that this special leave implies a rule requiring men, except when an exception is made, to be in earlier. The value of this in interpreting statutes is plain. 'A rule is not proved by exceptions unless the exceptions themselves lead one to infer a rule' (Lord Atkin). The formula in full is *exceptio probat regulam in casibus non exceptis.*

2. We have concluded by induction that Jones the critic, who never writes a kindly notice, lacks the faculty of appreciation. One day a warm eulogy of an anonymous novel appears over his signature; we see that this exception destroys our induction. Later it comes out that the anonymous novelist is Jones himself; our conviction that he lacks the faculty of appreciation is all the stronger for the apparent exception when once we have found out that, being *self*-appreciation, it is outside the scope of the rule—which, however, we now modify to exclude it, saying that he lacks the faculty of appreciating *others.* Or again, it turns out that the writer of the notice is another Jones; then our opinion of Jones the first is only the stronger for having been momentarily shaken. These kinds of exception are of great value in scientific inquiry, but they prove the rule not when they are seen to be exceptions, but when they have been shown to be

either outside of or reconcilable with the principle they seem to contradict.

3. *We may legitimately take satisfaction in the fact that peace prevails under the Union Jack, the Abor expedition being the exception that goes to prove the rule.* On the contrary, it goes to disprove it; but no more is meant than that it calls our attention to and heightens by contrast what might otherwise pass unnoticed, the remarkable prevalence of peace.

4. 'If there is one virtue I can claim, it is punctuality.' 'Were you in time for breakfast this morning?' 'Well, well, the exception that proves the rule.' It is by the joint effect of this use and 3 that the proverb comes to oscillate between the two senses Exceptions can always be neglected, and A truth is all the truer if it is sometimes false.

5. It rained on St. Swithin, it will rain for forty days; July 31 is fine and dry, but our certainty of a wet August is not shaken, since today is an exception that (instead of at one blow destroying) proves the rule. This frame of mind is encouraged whenever a writer, aware or unaware himself of the limitations, appeals to the use described in (2) without clearly showing that his exception is of the right kind: *That the incidence of import duties will be affected by varying conditions, and that in some exceptional cases the exporter will bear a large share of it, has never been denied; but exceptions prove the rules and do not destroy them.* / *The general principle of Disestablishing and Disendowing the Church in Wales will be supported by the full strength of Liberalism, with the small exceptions that may be taken as proving the rule.*

**exceptionable, exceptional, unex-.** The *-able* and *-al* forms, especially the negatives, are often confused by writers or compositors. *Exceptional* has to do with the ordinary sense of *exception,* and means out of the common; *exceptionable* involves the sense of *exception* rarely seen except in *take exception to* and *open to exception*; it means the same as the latter phrase, and its nega-

tive form *unexceptionable* means offering no handle to criticism. The usual
mistake is that shown in: *The picture is in unexceptional condition, and
shows this master's qualities to a marked
degree.*

**excessive(ly).** See EXCEEDING(LY).

**excise.** See TAX.

**exclamation mark.** See STOPS.

**executive.** Apart from its general use
for one of the three branches of
government, of which the others are
the *legislative* and the *judicial, e.* is, in
Britain, the name given to one of the
three general classes of civil servants of
which the others are the *administrative*
and the *clerical.* In America it means a
high officer with important duties in a
business organization, and this meaning, outside the civil service, has now
become common in Britain also.

**executor.** In the special sense
(testator's posthumous agent) pronounce *ĕkzĕ'kūtor*, in other senses
*ĕ'kzĕkūtor.* For the feminine of *executor* see -TRIX.

**exemplary.** *My experience today is e.
of the attitude of agents.* The ordinary
meaning of *e.* is serving as a model or
a warning (*e. behaviour, e. punishment*).
There is more than a touch of archaism
in its use in the sense of typical.

**exercise.** For the cliché *object of
the e.* See OBJECTIVE.

**exigence, -cy.** *-cy* is now the commoner form; *-ce* has no senses in which
*-cy* would be unsuitable, while *-ce*
sounds archaic in some; it would be
well if *-cy* became universal; see -CE
AND -CY.

**exist.** See SUBSIST, EXIST.

**-ex, -ix.** Naturalized Latin nouns in
*-ex* and *-ix* (genitive *-icis*) vary in the
form of the plural. The Latin plural
is *-ices* (*-ĭsēz* or *-īsēz*), the English *-exes*
(*-ĕksēz*); some words use only one of
these, and some both. See LATIN
PLURALS.

   **1.** Words in purely scientific or tech-

nical use (*codex, cortex, murex, silex,*
etc.) are best allowed their Latinity;
to talk of *codexes, cortexes, murexes,* and
*silexes,* is to take indecent liberties with
palaeography, physiology, ichthyology,
and geology, the real professors
of which, moreover, usually prefer
*-ices.*
   **2.** Latin words borrowed as trade
names (*duplex, lastex, perspex, pyrex,
triplex,* etc.) are for the period of their
lives English; if a plural is needed for
any of them it should be *-es.*
   **3.** Words that have become the established English for an object (e.g. *ilex*)
use *-exes; under the shade of the ilices*
shows ignorance of English more conspicuously than knowledge of Latin;
cf. -US and -UM. The question whether
the Latin in preference to the native
names (e.g. of *ilex* for *holm-oak*) should
be encouraged or prevented is a separate one, to be decided for the individual word.
   **4.** There are some words, however,
whose use is partly scientific and partly
wider, e.g. *apex, appendix, index,
matrix, vertex, vortex;* of these both
plurals are used, with some tendency,
but no more, to keep *-xes* for popular
or colloquial and *-ices* for scientific or
formal contexts: *The line just avoids the
apexes of the hills,* but *The shells have
their apices eroded. | Six patients had
their appendixes removed,* and *I hate
books with appendixes,* but *The evidence
is digested in five appendices. | The
volume is rounded off by splendid
indexes,* but *Integral, fractional, and
negative indices. | A heap of old stereo-
type matrixes,* but *Some of the species
of whinstone are the common matrices of
agate and chalcedony. | Arrange the
trestles with their vertexes alternately
high and low,* but *In the vertices of
curves where they cut the abscissa at
right angles. | Whirlpools or vortexes
or eddies,* but *The vortices of modern
atomists.* There is thus considerable
liberty of choice; but with most words
of this class the scientific use, and
consequently the Latin plural, is much
commoner than the other; *index* (other
than in the mathematical sense) is the
principal exception.

**5.** For a fuller discussion of the plurals of words ending -TRIX see that article.

**ex officio.** When used as an adjective, the words should be hyphened: *I was there* ex officio, but *the* ex-officio *members of the committee.* See HYPHENS.

**exordium.** Pl. *-ms* or *-ia*; see -UM.

**exoteric and exotic** (pron. *ĕxōtĕ'ric* and *ĕxŏ'tic*), of the same ultimate derivation, have entirely different applications. That is exoteric which is communicable to the outer circle of disciples (opp. *esoteric*); that is exotic which comes from outside the country (opp. *indigenous*); *exoteric doctrines*; *exotic plants. Exotic* is now a VOGUE WORD in the sense of *outlandish*, to which it is preferred because it has a more learned appearance.

**ex parte,** when used as an adj., should be hyphened: *speaking ex parte,* but an *ex-parte statement*; see HYPHENS.

**expect.** Exception is often taken to the use of *expect* to mean suppose, be inclined to think, consider probable. This extension of meaning is, however, so natural that it seems needless purism to resist it. *E.* by itself is used as short for e. to find, e. that it will turn out that; that is all: *I e. he will be in time; I e. he is there by this time; I e. he was there; I e. you have all heard all this before; Mr.* ——'s *study is scholarly and thorough, and has had a good deal of expansion, we e., since it took the* —— *Essay Prize*, i.e. if the facts ever happen to come to our knowledge, we shall be surprised if they are not to that effect. The OED remarked that the idiom was 'now rare in literary use'. That was owing to the dead set made at it; but it is so firmly established in colloquial use that it is not surprising that the period of exile seems to have been short.

**expectant.** The EUPHEMISM *e. mother* for pregnant woman is modern; it entered the statute book in 1918. But *e.* has long been similarly used with other nouns in legal phraseology, e.g. *e. heir.*

**expectorate, -ation** are GENTEELISMS that in America were once used as the established words for *spit* and *spitting*, and in Britain occasionally preferred to them as more polite, especially in public notices deprecating the habit. In both countries they have proved an exception to the rule (see EUPHEMISM) that, when a genteelism has outgrown its gentility and become itself the plain rude word for the rude thing, a new genteelism has to be found for it. In this case recognition of the vanity of genteelism has worked in reverse; the old rude word has been restored and *expectorate* (apart from its use in medicine) relegated to the vocabulary of the POLYSYLLABIC HUMORIST. Perhaps this is because the practice, less tolerated now that an urge to chew is usually satisfied with gum instead of tobacco, is no longer thought to deserve a cloak of gentility. Strictly, the words have different meanings. To *spit* is to eject saliva etc. from the mouth; to *expectorate* involves hawking up phlegm etc. from lower down.

**expediency, -ce.** The form first given is now much commoner in all surviving senses; there is no incipient differentiation, and it is desirable that the now rare *-ce* should be abandoned. See -CE, -CY.

**expensive.** See DEAR.

**expletive.** The OED gives the pronunciations *ĕ'ksplĭtĭv, ĕksplē'tĭv*, in that order. The noun use (oath or other interjection) being frequent and popular, and the adjective use (serving to fill out) literary and especially grammatical, the two pronunciations might well have been made use of for DIFFERENTIATION; but *-plētiv* seems to be winning the day.

**explicit and express.** With a certain class of nouns (e.g. *declaration, testimony, promise, contract, understanding, incitement, prohibition*), either adjective can be used in the general sense of definite as opposed to virtual or tacit or vague or general or inferable or implied or constructive. One may nevertheless be more appropriate than the other. That is explicit which is set forth in sufficient detail; that is express which is worded with intention. What is meant by calling a promise explicit is first that it has been put into words and secondly that its import is plain; what is meant by calling it express is first, as before, that it has been put into words, and secondly that the maker meant it to bind him in the case contemplated. This second element in the meaning of *express* is now generally present in it where it is roughly synonymous with *explicit*, but has come into it by accident. An express promise was by origin simply an expressed promise, i.e. one put into words, *express* being a Latin participle of the kind seen in *suspect* = suspected, *subject* = subjected, and many others. When its participial sense ceased to be appreciated, it was natural that the familiar adjectival sense (*for the express purpose of*; *express malice is when one with a sedate deliberate mind and formed design doth* . . .) should influence its meaning; the idea of special intention is now almost invariably conveyed by *express* when it is preferred to *explicit*. See also IMPLICIT.

**explore every avenue.** See HACKNEYED PHRASES.

**exposé** is an unwanted GALLICISM; *exposition* will serve in one of its senses and *exposure* in the other.

**exposition** in the sense public show of goods etc. is a GALLICISM (or Americanism) for exhibition, especially a large one, often international.

**ex post facto.** This is the established spelling; but the person who knows the Latin words is worse off with it in this disguise than one who does not; it should be *ex postfacto* (*ex* on the footing of, *postfacto* later enactment). The ordinary rule of HYPHENS would then be applied, and we would say *It is undesirable to legislate ex postfacto*, but *ex-postfacto legislation is undesirable*. E. legislation is, for instance, the making of an act illegal after it has

been committed; but what is referred to in *facto* is not the 'doing' of the action but the 'enacting' of the law.

**express,** adj. See EXPLICIT.

**express,** vb. *Both men afterwards expressed themselves perfectly satisfied.* Insert *as* after *themselves*. There is no authority for to express oneself satisfied etc.; at any rate the OED has no acquaintance with it; and it certainly requires the support of authority, whereas no such support is needed for the use with *as*. The fact is that ANALOGY is being allowed to confuse *express* with *declare* just as *regard* is wrongly given the construction of *consider*.

**extant** had formerly the same sense as existent or existing, and was as widely applicable. Its sense and its application have been narrowed till it means only still in existence or not having perished at the present or any given time, and is applied almost exclusively to documents, buildings or monuments, and customs. *E. memory, the e. generation, the e. crisis, e. States*, are unlikely or impossible phrases, and *the e. laws* would be understood only of such as were on record but not in operation, of laws as documents and not as forces. The pronunciation recommended is *ĕ′kstant*, but *ĕksta′nt* is not uncommon, especially when used predicatively.

**extemporaneous(ly) and extemporary, -ily,** are cumbersome words; *extempore* (4 syll.) is seldom unequal to the need. See LONG VARIANTS.

**extend.** *E.* for *give* or *accord* has been a septic influence in journalism. It might have been natural English; you e. your hand literally, and from that through extending the hand of welcome to the metaphorical extending of a welcome is a simple enough passage. But native English did not go that way, perhaps because *give* and *accord* were already in constant use, one for everyday and the other for more formal contexts. *E.* in this sense has done its development in America, and come to us full-grown via the newspapers—a bad record. To e. a welcome is tolerable because of its obviousness as a metaphor; but the extending of a hearty reception, sympathy, congratulations, a hearing, a magnificent send-off, and the like, should if possible be barred (in America a congregation 'extends a call' to the reverend gentleman of its choice); we have still *give, accord,* and *bestow* to choose between, with *offer* and *proffer* to meet the demand for other shades of meaning. The following quotation shows an application in which even the notion of friendliness inherent in the metaphor has disappeared: *Being promptly deported by the German police, he appealed to the Foreign Office for redress, but Lord Salisbury informed him in a characteristically pointed official dispatch that he could see no grounds whatever for taking exception to the treatment which had been extended to him.*

Two points are to be observed in regard to the above advice: (*a*) The condemnation does not touch such sentences as *You should e. to me the same indulgence,* where the metaphor may be different, and the meaning 'widen it so as to include me as well as someone else'; (*b*) it is not maintained that *e.* has never had the sense of give or accord in native English—it had in the 16th–18th centuries—but only that the modernism does not descend direct from the native use; having been re-imported after export to America, it is now ill at ease in the old country.

**extent.** In the phrase *to . . . extent, e.* should not be qualified by adjectives introducing any idea beyond that of quantity; *to what, to any, to some, to a great* or *vast* or *enormous* or *unknown* or *surprising, e.,* but not *Some of the girls even go* to the man-like e. of *holding meetings in the Park to discuss their grievances.*

**extenuate.** 1. *E.* makes *-uable*; see -ABLE 1. 2. The root meaning being to thin down or whittle away, the proper

object of the verb in its sense of make excuses for is a word expressing something bad in itself, as *guilt, cowardice, cruelty,* and not a neutral word such as *conduct* or *behaviour.* These latter, though neutral in themselves, are often converted by context into unmistakable words of blame, and are then legitimate objects of *e.* Hence the misapprehension arises that it can always govern them, and consequently that the meaning of excuse belongs to the verb. In fact it belongs to the combination between the verb and an object meaning something blamable, though that object need not always be expressed; the common phrase *extenuating circumstances,* for instance, means circumstances extenuating the guilt of the guilty person. From this misapprehension comes the further error of supposing that you can extenuate, i.e. make excuses for, a person. In such cases etymology is of value.

**exterior, external, extraneous, extrinsic.** Etymologically the four differ only in the formative suffixes used, and there is no reason why any of them might not have acquired the senses of all; *outside* is the fundamental meaning. It will be best to take them in pairs.

1. *exterior* and *external.* That is exterior which encloses or is outermost, the enclosed or innermost being interior. These opposites are chiefly applied to things of which there is a pair, and with conscious reference, when one is spoken of, to the other: the exterior court is one within which is an interior court; the exterior door has another inside it; exterior and interior lines in strategy are concentric curves one enclosing the other; and the exterior surface of a hollow ball, but not of a solid one, is a legitimate phrase.

That is external which is without and apart or whose relations are with what is without and apart, that which is within being internal. *The external world* (situated outside us), *external evidence* (derived from sources other than the thing discussed), *external remedies* (applied to the outside of the body), *external debt* and *relations* (having a sphere of operation outside the country concerned).

In many phrases either *exterior* or *external* may be used, but usually with some difference of underlying meaning; e.g. the exterior ear is thought of as the porch of the interior ear, but the external ear is the ear as seen by the outsider. Again, a building's exterior features and external features are different things, the former being those of its outside only, and the latter all, whether of outside or inside, that can actually be seen. Similarly, with the nouns, *exterior* has the definite narrow material meaning of the outside, as opposed to the inside of a building or the inner nature of a person, while *externals* includes all about a person that reveals him to us, his acts and habits and manner of speech as well as his features and clothes.

2. *extraneous* and *extrinsic*. That is extraneous which is brought in, or comes or has come in, from without. A fly in amber, a bullet in one's chest, are extraneous bodies; *extraneous aid, interference, light, sounds*; extraneous points are questions imported into a discussion from which they do not naturally arise.

That is extrinsic which is not an essential and inherent part of something, essential properties being intrinsic. A florin's intrinsic value is what the metal in it would have fetched before it was coined; its extrinsic value is what is added by minting. A person's extrinsic advantages are such things as wealth and family interest, while his courage and talent are intrinsic advantages.

It is worth notice that *extrinsic* is now rare, being little used except when a formal contrary is wanted for the still common *intrinsic*. *Extraneous* on the other hand exists on its own account; it has no formal contrary, *intraneous* being for practical purposes non-existent, and must make shift with *internal, intrinsic, indigenous, domestic, native*, or whatever else suits the particular context.

**extraneous.** See EXTERIOR.

**extraordinary.** The OED gives precedence to the five-syllable pronunciation (-*trŏr*-) over the six (-*trăŏr*-). So does the COD. But the speak-as-you-spell movement is making for the longer, and the second syllable is often carefully enunciated by those who speak in public or over the air. See PRONUNCIATION 1.

**ex(tra)territorial(ity).** The forms were once used quite indifferently, but the longer is now more usual. To the classical latinist it seems the only reasonable one, since *extra*, and not *ex*, is the classical Latin for outside of; and this is perhaps a stronger consideration than the saving of a syllable. It would certainly be better to have one spelling only, and *extra-* is recommended, especially as *exterritorial* now suggests a former member of the Territorial Army.

**extricate** makes -*cable*; see -ABLE 1.

**extrinsic.** See EXTERIOR.

**-ey and -y in adjectives.** The adjectival suffix is -*y*, not -*ey*. Weak spellers are often in doubt whether, when -*y* is appended to nouns in MUTE E (as *race*), the *e* is to be dropped or kept. With the very few exceptions given below, it should be dropped (*racy*, not *racey*). A selection of -*y* adjectives from nouns in mute -*e* will suffice to show the normal formation, and this is followed by another list, containing words of the kind in which the bad speller goes wrong. He often does so because he conceives himself to be making a new, or at least hitherto unprinted, word, and is afraid of obscuring its connexion with the noun if he drops the *e*—a needless fear. The safe -*y* adjectives are: *bony, breezy, briny, chancy, crazy, easy, fleecy, fluky, gory, greasy, grimy, hasty, horsy, icy, lacy, mazy, miry, nervy, noisy, oozy, prosy, racy, rosy, scaly, shady, shaky, slimy, smoky, snaky, spicy, spiky, spongy, stony, wavy, wiry*. The shaky -*y* adjectives are: *caky, cany, fluty, gamy, homy,* *liny, mity, mousy, nosy, pursy, sidy, stagy, tuny, whity.*

The exceptions referred to above are:

1. When an adjective in *-y* is made from a noun in *-y*, *e* is inserted to part *y* from *-y*: *clayey*, not *clayy*.

2. *Hole* makes *holey*, to prevent confusion with *holy* = hallowed.

3. Adjectives from nouns in *-ue* ($\overline{oo}$) retain the *e*: *gluey* and *bluey*, not *gluy* or *bluy*.

**eye**, vb. For the present participle (*eying* or *eyeing*) see VERBS IN -IE etc. 7.

**-ey, -ie, -y, in pet names.** (By pet name is meant a name used affectionately, familiarly, or jocosely instead of the ordinary one for some person, animal, or thing.)

It would be idle to attempt to prescribe the spelling of proper names of this kind. The right to spell one's children's names as one pleases is one of the few privileges left to the individual in the Welfare State. Moreover, the *-ey* etc. terminations seem to be going out of favour; pet names are being cut down to monosyllabic size— *Al, Art, Don, Ern, Ken, Les, Perce, Reg, Ron, Sid,* and so on. But an examination of others may be not without interest. The most established type of all (*baby, daddy, granny*) has *-y*; it would be a simplification if *-y* could be made universal; but *-ie* is preferred in Scotland (*laddie, lassie, caddie*); the retention of mute *-e*, giving *-ey* (*dovey, lovey,* etc.) is more defensible than in the adjectives made with *-y* (see -EY AND -Y IN ADJECTIVES); and generally variety seems unavoidable.

In the list the recommended form stands first or alone; the principle has been to recommend plain *-y* wherever usage is not thought to be overwhelmingly against it; the addition of another ending in brackets means that that form is perhaps commoner, but not so much so as to make the recommended one impossible. Some of the words included (*booby, caddy, collie, coolie, nippy, nosy, puppy, toddy,* and perhaps others) are not in fact formations of the kind in question, but being mistakable for them are liable to the same doubts. There is some tendency when a word is much used in the plural (*frillies, goodies, johnnies, kiddies, kilties, sweeties*) to think that *-ie* must be the singular termination. Adjectives like *comfy* are given here because the *-y* is an ending of the kind we are concerned with, and not the adjective suffix.

*aunty* (*-ie*); *baby*; *billy* = cooking-can; *blacky*; *bobby* = policeman; *booby*; *bookie* = bookmaker; *buddy*; *bunny*; *caddie* = golf-attendant; *caddy* = tea-box; *cissy*; *clippie*; *collie*; *comfy*; *cookie* = cake; *coolie*; *corbie* = crow; *daddy*; *darky* (*-ey*); *deary* (*-ie*); *doggy* (*-ie*); *ducky*; *fatty*; *frilly*; *Froggy*; *girlie*; *goody*; *goosey* (*y*); *granny*; *hanky*; *hoodie* = crow; *hubby*; *johnny* (*-ie*); *kiddy*; *kilty* (*-ie*) = Highland soldier, *kitty* (*-ie*) = kitten; *kitty* = pool; *laddie, lassie*; *lovey-dovey*; *mammy*; *missy* (*-ie*); *mounty* (*-ie*) = Canadian Mounted Policeman; *mummy* = mother; *nancy*; *nanny*; *nappy*; *nippy*; *nosy* (*-ey*); *nicy*; *nighty*; *nunky* = uncle; *nursey*; *piggy* (*-ie*); *pinny*; *puppy*; *sawney*; *shimmy* = chemise; *slavey*; *sonny*; *sweety*; *teddy* = bear or boy; *toddy*; *tommy* = bar, gun, rot, or Atkins; *tummy*; *wifie*.

For the plurals of *-ey* nouns see PLURAL ANOMALIES 4.

**eyot.** Pronounce *āt*; the OED calls it 'a more usual variant of *ait*', and 'an artificial spelling'.

**eyrie.** See AERY.

# F

**fable.** See SIMILE AND METAPHOR.

**fabulous** means mythical, legendary, but was long ago extended to do duty as an adjective for something that is real but so astonishing that you might think it was legendary if you did not know better. It has become fabulously popular as a term of eulogy or allure. To advertisers, it would seem, *f.* is the word of paramount appeal, more potent even than DIFFERENT, and that is saying a lot. *You'll look lovelier every day with f. pink ——— | FOLLIES STRIPTEASE: London's only challenge to Paris. Thirty f. girls. | TAKE YOUR PICK, featuring new surprises and f. prizes. | Win a f. trip to Paris. | THE MOUSETRAP, Seventh f. year. | Ludmilla Teherina—the face above the f. legs. | This f. machine embodies refinements found in no other typewriter.* | Journalists too, especially gossip columnists, have a great affection for the word. *Among the 350 guests was Mr. B., also a f. party giver. | The Princess's room opens on to a f. view. | Fifth Avenue: row after row of f. stores. | I stood with an array of Cunard executives to greet Mr. Hicks and Lady Pamela after their f. wedding. | Like I said, Texas is a f. place at Christmas time. Real f. | This most f. and exciting art sale of the century. | It contains f. photographs of life in Red China. | Zimbalist, producer of f. Hollywood epics.* There are, however, signs, at least in the advertisement world, that the vogue of *f.* is passing its zenith. *Fantastic* and *sensational* and *stupendous* are beginning to contest its leadership.

*Fabulous* (often contracted to *fabs*) and *fantastic* are also in that long succession of words which boys and girls use for a time to express high commendation and then get tired of, such as, to go no farther back than the present century, *topping, spiffing, ripping, wizard, super, posh, smashing*.

**facetiae**, in booksellers' catalogues, is, like *curious*, a euphemism for erotica; the following extract from such a catalogue was vouched for by the *Westminster Gazette*: 'FACETIAE. 340—Kingsley (C.) Phaethon; or Loose Thoughts for Loose Thinkers, 2nd ed., 8vo, boards, 1s., 1854'.

**facetious.** For synonymy see JOCOSE.

**facetious formations.** A few specimens may be collected in groups illustrating more or less distinct types.

*Pun or parody*: anecdotage; godwottery; goluptious; judgematical; sacerdotage.

*Mock mistakes*: mischevious; splendiferous; underconstumble.

*Popular etymology, real or supposed*: highstrikes (hysterics); jawbation (jobation); trick-cyclist (psychiatrist).

*Mock Latin*: crinkum-crankum; hocus-pocus; high-cocalorum; holusbolus; snip-snap-snorum.

*Portmanteau words*: brunch (breakfast lunch); chortle (snort chuckle); galumph (gallop triumph); SMOG (smoke fog); squarson (squire parson); and perhaps hokum (hocus-pocus bunkum) and NATTER (nag chatter). See PORTMANTEAU for more.

*Incongruous Greek or Latin trimmings to English words*: absquatulate; bardolatry; circumbendibus; fistical; omnium gatherum; squandermania.

*Irreverent familiarity*: crikey (Christ); gorblimey (God blind me); gosh (God).

*Onomatopoeia, obvious or obscure*: belly-flopper; bubblyjock; collywobbles; crackerjack; gobbledygook; rumbustious; whiz-bang.

*Long and ludicrous*: antigropelos; discobulate; galligaskins; hornswoggle; panjandrum; skedaddle; skulduggery; spondulicks; slubberdegullion; spiflicate; tatterdemalion; transmogrify.

For facetious formations in *-manship* see BRINKMANSHIP.

**facile.** For pronunciation of the adjective see -ILE. The Latin adverb in the phrase *facile princeps* is *fă'sĭli*. The adjective's value as a synonym for easy or fluent or dextrous lies chiefly in its depreciatory implication. A f. speaker or writer is one who needs to expend little pains (and whose product is of correspondingly little import); a f. triumph or victory is easily won (and comes to little). Unless the implication in brackets is intended, the use of *f.* instead of its commoner synonyms (*a more economical and f. mode; with a f. turn of the wrist*) is ill-judged and usually due to AVOIDANCE OF THE OBVIOUS.

**facilitate.** *The officer was facilitated in his search by the occupants.* We f. an operation, not the operator.

**fact** is well equipped with idiomatic phrases. We have *in f., in point of f., as a matter of f., the f. is,* and *the f. of the matter is,* all unquestionably established. It is a pity that the invention *as a f.* (of which no example is recorded in the OED) should have been thrust upon us in addition to all these. But that is no great matter now, since all have been superseded by the inevitable *actually.* (See MEANINGLESS WORDS.) Perhaps the reason why we have all these ways of emphasizing that we are telling the truth is that we are an empirical people and feel that our reader or listener would like to be assured at the outset that we are concerned with facts and actualities, not with theories and surmises.

The PERIPHRASES *owing to the f. that* and *in spite of the f. that* seldom have any advantage over the simple conjunctions *because* and *although.*

**factious, factitious, fictitious.** Though the words are not synonyms even of the looser kind, there is a certain danger of confusion between them because there are nouns with which two or all of them can be used, with meanings more or less wide apart. Thus *factious rancour* is the rancour that lets party spirit prevail over patriotism; *factitious rancour* is a rancour that is not of natural growth, but has been deliberately created to serve someone's ends; and *fictitious rancour* is a rancour represented as existing but in fact imaginary. A party cry has a *factious value,* a silver coin a *factitious value* (cf. *extrinsic,* see EXTERIOR etc.), and a bogus company's shares a *fictitious value.*

**factor** is one of those words (cf. INVOLVE) which are so popular as thought-saving reach-me-downs that all meaning is being rubbed off them by constant use. A *f.* is something that contributes to an effect (*The Rent Act was an important f. in the result of the by-election*), but it is made to serve for such words as *circumstance, component, consideration, constituent, element, fact, event* in contexts where its true meaning is only faintly present if at all.

**factotum.** Pl. *-ms;* see -UM.

**facund.** See FECUND, FACUND.

**faerie, faery.** Pronounce *fā'ĕrĭ.* 'A variant of *fairy.* In present usage, it is practically a distinct word, adopted either to express Spenser's peculiar modification of the sense, or to exclude various unpoetical or undignified associations connected with the current form *fairy.*'—OED. The distinction should be respected by all who care for the interests of the language and not only for their own momentary requirements. To say *Faerie* when one merely means *Fairyland* in trivial contexts is VULGARIZATION.

**faience.** The use in English of a foreign 'general term comprising all the various kinds of glazed earthenware and porcelain'—the whole definition given in the OED—is hard to divine. Most of those who read the word are disappointed to find, on appeal to a dictionary, that it means nothing more specific. Originally it was applied to the majolica pottery made at Faenza in Italy, a town famous for medieval pottery.

**fail. 1.** For *a failed harvest, coup, stockbroker,* etc., see INTRANSITIVE PAST PARTICIPLES.
**2.** *failing* (= in default of) is a participle developed through the absolute construction into a preposition: 'if' or 'since so-and-so fails' means the same as 'in case of' or 'on the failure of so-and-so'. Either the absolute or the prepositional use is grammatically legitimate, but not a mixture of the two; the form 'whom failing' familiar in companies' proxy notices is such a mixture; strictly it should be either 'failing whom' (preposition and objective) or 'who failing' (absolute and subjective).
**3.** *Fail* is one of the words apt to cause the sort of lapse noticed in NEGATIVE MISHANDLING in which the writer, by not sorting out his negatives, stumbles into saying the opposite of what he means: *New Year's Day is a milestone which the least observant of us can hardly fail to pass unnoticed.*

**fain** is now archaic in the general sense glad, eager, willing (*f. by flight to save themselves*), but survives, though with a touch of archaism, with *would* (*I would f. die a dry death*), and also in the sense willing to make shift with (*He would f. have filled his belly with the husks the swine did eat*).

**fair(ly). 1.** For *bid f., fight* or *hit* or *play f., f. between the eyes* etc., *speak* one *f.*, see UNIDIOMATIC -LY.

**2.** For the avoidance of ambiguity it should be remembered that *fairly* has the two oddly different senses of utterly (*I was f. beside myself*) and moderately (*a f. good translation*), and that the context does not always make it clear which is meant. Cf. QUITE.

**fairy, fay, fey. 1.** For *Fairyland* and *Faerie*, see FAERIE.

**2.** *Fairy* and *fay*. The difference is not in meaning but merely in appropriateness to different contexts; *fairy* being now the everyday form, *fay* should be reserved for occasions demanding the unusual. The spelling *fey* is only for the adjective of different derivation meaning conscious of doom.

**faithfully. 1.** For *yours f.* see LETTER FORMS.

**2.** In *promise f., f.* is a colloquial substitute for *definitely, explicitly, expressly, emphatically,* or *solemnly*.

**3.** *Deal f. with* is a phrase of biblical sound and doubtless of puritan origin, now used for the most part jocularly in the sense show no lenity—one of the idioms that should not be spoilt by over-frequent use.

**fall. 1.** For *is fallen, fallen angel*, etc., see INTRANSITIVE P.P.

**2.** The noun *f.* as a synonym for the ordinary *autumn* is either an AMERICANISM, a provincialism, or an ARCHAISM; as the last, it has its right and its wrong uses; as either of the others, it is out of place in Britain except in dialogue. That is a pity. As was said in *The King's English, fall* is better on the merits than *autumn* in every way, and we once had as good a right to it as the Americans, but we have chosen to let the right lapse.

**fallacy, fallacious.** The meaning of *fallacious* is inviting a wrong inference, misleading. It and *fallacy* are sometimes ignorantly used as LONG VARIANTS of *false* and *falsehood*. *A fallacious report that the course was flooded kept many people away.* No doubt the report was misleading, but that was because it was untrue, not because it contained a fallacy. *Lord B. states that unemployment reached its greatest heights under the Labour government in 1931. The repetition of this old fallacy* (falsehood) *cannot be allowed to go unchallenged.*

A fallacy in logic is 'an argument which violates the laws of correct demonstration. An argument may be fallacious in *matter* (i.e. misstatement of facts), in *wording* (i.e. wrong use of words), or in the *process of inference*. Fallacies have, therefore been classified as: I. Material, II. Verbal, III. Logical or Formal.'—*Encycl. Brit.* Some types of fallacy are of frequent enough occurrence to have earned names that have passed into ordinary speech, and serve as a short way of announcing to a false reasoner that his conscious or unconscious sophistry is detected. Such are *arguing in a circle, equivocation, begging the question* (*petitio principii*), *ignoratio elenchi, argumentum ad hominem* etc., *non sequitur, post hoc ergo propter hoc, false analogy, undistributed middle*, all of which will be found alphabetically placed in this dictionary.

**false analogy** in Logic is the unfounded assumption that a thing that has certain attributes in common with another will resemble it also in some attribute in which it is not known to do so; e.g. that of a pair of hawks the larger is the male, on the ground that other male animals are larger than female; or that *idiosyncracy* is the right spelling because many other words ending in the sound *-krăsĭ* are spelt with *-cy*.

**false emphasis. 1.** *That being so, we say that it would be shameful if domestic servants were the only class of employed* *persons left outside the scheme of State Insurance.* What the writer means is that it would be shameful for servants to be left out when all other employees are included. What he says is that it would be shameful for nobody except servants to be excluded—which is plainly neither true nor his contention. The disaster is due to his giving too emphatic a place to a subordinate, though important, point; what is shameful is the servants' exclusion, not the inclusion of anyone or everyone else.

2. An especially common form of false emphasis (already touched on in BOTH 2) is the use of the emphatic word *both* (which means one as well as the other, or in one case as well as in the other) in places where that full sense is either unnecessary or impossible, instead of *the two*, *they*, or nothing at all. The point is clear if the two sentences (*a*) *Both fought well*, and (*b*) *To settle the matter both fought*, are compared. In (*a*) emphasis is wanted; not only one fought well, the other did too; but in (*b*) of course one did not fight without the other's fighting, since it takes two to make a fight; the needless *both* makes the reader wonder whom else they both fought. Obvious as the mistake is, it is surprising how often it occurs in sentences little more abstruse than (*b*): *Both men had something in common* (with whom? With each other; then why not *the two*, or *the men*, or *the two men*, or simply *they*?). | *Lord Milner had fixed these prices because the Food Controller and the Board of Agriculture* both *disagreed as to what they should be, and he had at least the wisdom to fix a price that they both disliked* (the first *both* is needless and misleading; the second is right). The next instance is at once more excusable and more fatal, both for the same reason: that hard thinking is necessary to get the thing disentangled: *This company has found that the men they employ in America can be depended on to produce a minimum of 40% more output than the men they employ abroad, and yet these men both in America and elsewhere may be of the same race and nationality*

*at birth.* Here the point is not that America and the other country are in some matter alike, but that the difference between the employee in the one and the employee in the other, wherever they may have been born, is constant. *Both*, inserted where it is, hopelessly disguises this; read *The men employed in America may be of the same race and nationality at birth as those employed elsewhere.*

**falsehood, falseness, falsity.** DIFFERENTIATION has been busy with the three, but has perhaps not yet done with them. At present *A falsehood* is a lie; *falsehood* is lying regarded as an action, but it is also a statement or statements contrary to fact or the truth. *Falseness* is contrariety to fact regarded as a quality of a statement, but it is also lying and deception regarded as an element in character. *Falsity* is interchangeable with *falseness* in its first but not in its second sense. In the following examples the word used is, except where an alternative is shown, the only one of the three consistent with modern usage: *That is a falsehood*; *You told a falsehood*; *He was convicted of falsehood*; *Truth would be suppressed together with falsehood*; *Truth exaggerated may become falsehood*; *The falseness*, or *falsity*, *of this conclusion is obvious*; *A falseness that even his plausibility could not quite conceal.*

**false quantity.** The phrase should be banished from the discussion of how to pronounce English words. The use of it betrays the user's ignorance that standard English teems with what are in one sense or another false quantities. Its implication is that, with some limitations or other, the sound of vowels in English words derived from Greek and Latin is decided by the sound in the words from which they come. But these limitations are so variously conceived that mere mention of false quantity is valueless. Take a score of words about the pronunciation of which opinions differ, and on which classical quantities might be expected to throw light; the classical quantities are marked where they matter, and accents are added when acceptance of the classical quantity would naturally result in a particular stress: amēnity, appărent, cănī'ne, commū'nal, corō'-nary, dēficit, doctrī'nal, ēquable, glă'-dĭŏlŭs, ĭdyl, inter'nĕcine, mĕticulous, pătriot, salī'vary, Salonī'ca, sēmaphore, sīmian, subsīdence, trībunal, vertī'go. It will be clear from this list that adherence to classical quantity may operate singly or doubly, i.e. on the sound of a vowel only, or through it on the word's balance, and that the secondary is much more noticeable than the primary effect; the difference between pătriot and pātriot, appărent and appārent, is slight, but that between doctrī'nal and dŏ'ctrĭnal, vertī'go and ver'tĭgo, is very great. How little weight is to be attached to classical quantity as an argument merely for one vowel sound against another will be plain from another score of examples, some of them actual Latin words, in which the unquestioned pronunciation is a false quantity: ăgent, ălien, bŏnă fīde, cōmic, corrōborate, dēcent, ēcho, ēthics, et cētera, fastīdious, ĭdĕa, jocōsity (and all in -osity), mīlitary, mĭnor, mītigate, ŏdour, pathĕtic (and most in -etic), sēnile, sōlitary, varĭety (and all in -iety). It is useless to call out 'false quantity' to someone who says cănine or ĭdyl or trību'nal or amĕnity when he can answer you with ăgent, fastĭdious, mĭnor, or ĕcho. The simple fact is that, in determining the quality of a vowel sound in English, classical quantity is of no value whatever; to flout usage and say Sōcrates is the merest pedantry.

In its secondary effect, as an influence in selecting the syllable in English words that shall bear the stress, classical quantity is not so negligible. A variation of stress being a much more marked thing than a vowel difference, the non-latinist's attention is arrested when a neighbour whom he credits with superior knowledge springs doctrī'nal upon him, and doctrī'nal gets its chance. Whether doctrī'nal is right is another question; the superior-knowledged one knows that *doctrina*

has a long i; but has he satisfied himself that a long i, not in *doctrina* but in *doctrinalis*, i.e. with no stress on it, has any right to affect the stress of *doctrinal*? Or again, has the Greek scholar who knows kīnēma and objects to cĭnēma reflected that cinema does not represent *kinema* itself, but is a shortening of *ci'nemă'tograph*, which again has passed through French and indeed been there 'assembled' on its way from Greek to English? If he has, he will probably hold his peace. In many words, such as *canine*, *saline*, *vertigo*, the latinist's first thoughts (kănī'n, sălī'n, vertī'gō) do not call for similar rethinking; but he has still to reckon with the RECESSIVE-ACCENT tendency, which has as good a right to a voice in the matter as his erudition, and will fight hard and perhaps victoriously, as it has for kă'nīn, sā'līn, and ver'tĭgo.

After all deductions, however, a small province is left in which the false-quantity principle may fairly reign; if *clematis* is pronounced klĕmă'tĭs and *enema* inē'ma, what has been done is this: in Greek words adopted without modification, a syllable that in the original is neither long in quantity nor stressed has been wantonly made the stressed syllable in English; they should be klĕ'mătĭs and ĕ'nĕmă. But on such disputes as those between protă'gonist and prō'tagō'nist, cŏ'mmunal and commū'nal, i'nternē'cine, and inter'nĕcine, să'līvary and salī'vary, mă'rītal and marī'tal, cer'vĭcal and cervī'cal, ă'nthropoid and anthrō'-poid, its decision is not final; it is not judge, but a mere party to the suit.

**false scent.** The laying of false scent, i.e. the causing of a reader to suppose that a sentence or part of one is taking a certain course, which he afterwards finds to his confusion that it does not take, is an obvious folly—so obvious that no one commits it wittingly except when surprise is designed to amuse. But writers are apt to forget that, if the false scent is there, it is no excuse to say they did not intend to lay it; it is their business to see that it is not there, and this requires more care than might be supposed. The reader comes to a sentence not knowing what it is going to contain. The writer knows. Consequently what seems to the writer, owing to his private information, to bear unquestionably only one sense may present to the reader, with his open mind, a different one. Nor has the writer even the satisfaction of calling his reader a fool for misunderstanding him, since he seldom hears of it; it is the reader who calls the writer a fool for not being able to express himself.

The possibilities of false scent are too miscellaneous to be exhaustively tabulated; the image of the reader with the open mind, ready to seize every chance of going wrong, should be always present to the inexperienced writer. A few examples, however, may suggest certain constructions in which special care is necessary: *It was only after Mr. Buckmaster, Lord Wodehouse, and Mr. Freake, finding that they were unable to go, that the England team as now constituted, but with Major Hunter in the place of Captain Cheape, was decided on.* The writer knew that *after* was to be a preposition governing *Mr. B.* etc. *finding*; but the reader takes it for a conjunction with a verb yet to come, and is angry at having to reconsider. Such things happen with the FUSED PARTICIPLE. / *Four years, the years that followed her marriage, suffice Lady Younghusband for her somewhat elaborate study, 'Marie Antoinette: Her Early Youth, 1770–1774'.* The reader does not dream of jumping over Lady Y. to get at the owner of *her* (marriage) till *1770–1774* at the end throws a new light on the four years. See PRONOUNS for more such false scent. / *The influences of that age, his open, kind, susceptible nature, to say nothing of his highly untoward situation, made it more than usually difficult for him to cast aside or rightly subordinate.* Only the end of the sentence reveals that we were wrong in guessing that the influences and his nature etc. to be parts of a compound subject.

In the first of the examples given

above the absurd comma after *Mr. Freake* helps to lead the reader astray. He may be no less deceived by the absence of a necessary comma. *However the enterprise may turn out in the end to have been not without its lessons.* Unless a comma follows *however* the reader can hardly fail to be momentarily misled. But it is a bad habit to rely on commas to prevent the laying of a false scent. The first example should have been written *after Mr. B.* etc. *had found* and the last *the enterprise may however*.

**fan(atic).** 1. Pronounce *fănă'tĭk.* The word has lost its fully adjectival use. We say *I call a man fanatical* (or *a f.*), but not simply *f.*) *who* ... See -IC(AL).
2. The obsolete abbreviation *fan* was revived in U.S. in the 19th c. for a keen and constant spectator of a sport, especially baseball; and with the coming of the cinema and television the word has won immense popularity on both sides of the Atlantic for enthusiastic admirers of some kinds of entertainment and some people who entertain them; and, in *fan-mail*, for the letters in which that admiration is expressed.

**fantasia.** *fahntahzē'ah, făntah'zĭa* were the pronunciations given by the OED, the former being the Italian one, appropriate for the musical term, the latter the popular one for the word in its transferred senses. But both have since been superseded, the second wholly and the first largely, by the fully anglicized *făntā'zia*.

**far.** 1. *Farther, further.* 2. (So) *f. from.* 3. *F.-flung.* 4. *As* and *so f. as.* 5. *So f. as, so f. that.*
1. For *farther, further,* see FARTHER.
2. (So) *f. from. So far from 'running' the Conciliation Bill, the Suffragettes only reluctantly consented to it.* This idiom is a curious, but established, mixture between 'Far from running it they consented to it reluctantly' and 'They were so far from running it that they consented to it reluctantly'. It is always open, however, to those who dislike illogicality to drop the *so* in the

shorter form—'Far from running it they consented to it reluctantly'. But it is waste labour to tilt against STURDY INDEFENSIBLES. Writers who use this one should be careful not to leave the participle in the air (see UNATTACHED PARTICIPLES) as in *So far from increasing the subsidy given to this theatre, it should be radically reduced.*

3. *Far-flung.* The emotional value of this, though perhaps lessening as our *f.-f.* empire melts away, is sometimes reckoned so high as to outweigh such trifling matters as appropriateness: *Set against all its* [the war's] *burden of sorrow and suffering and waste that millions of men from f.-f. lands have been taught to know each other better.* The lands are distant; they are not far-flung; but what matter? *F.-f.* is a signal that our blood is to be stirred; and so it is, if we do not stop to think. *He is already popular, even in the remotest parts of this f.-f. constituency.*

4. *As* and *so f. as. As* or *so f. as x* cannot be used as short for *as far as x goes* or *so far as concerns x*; in the following examples *concerns, regards, is concerned, goes,* etc., should have been inserted where omission is indicated: *As far as getting the money he asked for* ∧, *Mr. Churchill had little difficulty.* | *The result was that the men practically met with a defeat so far as* ∧ *obtaining a definite pledge in regard to their demands.* For misuse of the phrase *so far as x is concerned* see CONCERN 2.

*As* or *so far as,* regarded as a compound preposition, is followed primarily by a word of place (*went as far as York*); secondarily it may have a noun (which may be an infinitive or gerund) that expresses a limit of advance or progress (*He knows algebra as far as quadratics; I have gone so far as to collect,* or *as far as collecting, statistics*). But when the purpose is to say not how far an action proceeds, but within what limits a statement is to be applied, as in the examples at the beginning of this section, *as* and *so far as* are not prepositions, but conjunctions requiring a verb. The genesis of the misuse may be guessed at thus:

*I have gone as far as collecting statistics* (right). *As far as collecting statistics you have my leave to proceed* (correct, but unnatural order). *As far as collecting statistics he is competent enough* (cf. *knows algebra as far as quadratics*; defensible, but better insert *goes*; the Churchill sentence quoted is just below this level). *As far as collecting statistics, only industry is necessary* (impossible).

5. *So far as, so far that. His efforts were so far successful* (a) *as they reduced,* or (b) *as to reduce,* or (c) *that they reduced, the percentage of deaths.* The (b) and (c) forms mean the same, and their interpretation is not in doubt: he reduced the percentage, and had that success. The meaning of (a) is different: if you want to know whether and how far he succeeded, find out whether and how far he reduced the percentage; perhaps he did not reduce it, and therefore failed. But the (a) form is not infrequently used wrongly instead of (b) or (c). *The previous appeal made by M. Delcassé was so far successful as the Tsar himself sent orders to comply* (read *that* for *as*; the sending of orders clearly took place, and such sending is not a variable by which the degree of success could be measured).

**farce.** See COMEDY. The connexion with the etymological sense (stuffing) lies in the meaning 'interpolation' the farce having originated in interludes of buffoonery in religious dramas.

**farouche.** The meaning, simply sullen-mannered from shyness (*cheval f., cheval qui craint la présence de l'homme* —Littré), is obscured by association with ferocious; according to the OED 'the connexion is untenable'; see TRUE AND FALSE ETYMOLOGY.

**farther, further.** The history of the two words appears to be that *further* is a comparative of *fore* and should, if it were to be held to its etymology, mean more advanced, and that *farther* is a newer variant of *further*, no more connected with *far* than *further* is, but affected in its form by the fact that *further,* having come to be used instead of the obsolete comparative of *far* (*farrer*), seemed to need a respelling that should assimilate it to *far*. This is intended as a popular but roughly correct summary of the OED's etymological account. As to the modern use of the two forms, the OED says: 'In standard English the form *farther* is usually preferred where the word is intended to be the comparative of *far*, while *further* is used where the notion of *far* is altogether absent; there is a large intermediate class of instances in which the choice between the two forms is arbitrary.'

This seems to be too strong a statement: a statement of what might be a useful differentiation rather than of one actually developed or even developing. The fact is surely that hardly anyone uses the two words for different occasions; most people prefer one or the other for all purposes, and the preference of the majority is for *further*. Perhaps the most that should be said is that *farther* is not common except where distance is in question, and that *further* has gained a virtual monopoly of the sense of moreover, both alone and in the compound *furthermore*. The three pairs of quotations following are selected for comparison from the OED stores.

1. Comparative of *far*: *If you can bear your load no farther, say so.*—H. Martineau. *It was not thought safe for the ships to proceed further in the darkness.*—Macaulay.

2. No notion of *far*: *Down he sat without farther bidding.*—Dickens. *I now proceed to some further instances.*—De Morgan.

3. Intermediate: *Punishment cannot act any farther than in as far as the idea of it is present in the mind.*—Bentham. *Men who pretend to believe no further than they can see.*—Berkeley

On the whole, though differentiations are good in themselves, it is less likely that one will be established for *farther* and *further* than that the latter will become universal. In the verb, *further* has the field virtually to itself.

**fascist** etc. The Italian words—*fasci-*

*sta* pl. *-ti, fascismo*—are pronounced (roughly) *fahshē'stah, -tē, -ē'smo*. In English *făshism* and *făshist* are more usual than *făsism* and *făsist* and are preferred by the COD.

**fatalism, determinism.** The philosophical distinction between the words cannot here be more than roughly suggested, and is itself more or less arbitrary. *F.* says: Every event is pre-ordained; you cannot act as you will, but only in the pre-ordained way. *D.* says: You can act (barring obstacles) as you will; but then you cannot will as you will; your will is determined by a complex of antecedents the interaction of which makes you unable to choose any but the one course. That is, *f.* assumes an external power decreeing irresistibly every event from the greatest to the least, while *d.* assumes the dependence of all things, including the wills of living beings, upon sequences of cause and effect that would be ascertainable if we were omniscient. The difference between the two views as practical guides to life is not great; one assures us that what is to be will be, the other that whatever is cannot but be; and either assurance relieves us of responsibility; but those are called determinists who decline to make assumptions (involving the ancient notion of Fate) about an external directing will.

Such, very roughly, is the difference between the two theories; but the popular distinction today is not between the names of two contrasted theories, but between the name of an abstract philosophy and that of a practical rule of life. *D.* is the merely intellectual opinion that the determinist or fatalist account of all that happens is true; *f.* is the frame of mind that disposes one at once to abandon the hope of influencing events and to repudiate responsibility for one's actions; *d.* is regarded as a philosophy, and *f.* as a faith.

**fateful.** *Will the Irish question, which has been fateful to so many Governments, prove one of the explosive forces which will drive the Coalition asunder?*

Correct to *fatal*. NOVELTY-HUNTING, the desire to avoid so trite a word as *fatal*, is responsible for many *fatefuls*; cf. FORCEFUL. There was a reason good enough for inventing *fateful*, in the restriction of the older *fatal* to a bad sense; *fateful* can mean big with happy fate as well as with unhappy. But to use *fateful*, as in the quotation, where *fatal* is the right word is to renounce the advantage gained by its invention, and to sacrifice the interests of the language to one's own momentary desire for a gewgaw. See PAIRS AND SNARES.

**father-in-law.** See -IN-LAW.

**fathom.** See COLLECTIVES 5.

**fault.** 1. *At* or *in f*? Strictly *I am at f.* means I am puzzled, and *I am in f.* means I am to blame. But the distinction is subtle (a hound that is *at* f. may also be *in* f.) and seems to be disappearing; *at* f. is now more usual in both senses.
2. *Fault* as a transitive verb in the sense of to find f. with. The OED gives 16th-c. examples of this but calls it rare. It is enjoying a revival, established in tennis, where a server may be faulted by the umpire as a bowler may be no-balled in cricket, and now put to more serious use by leader writers and literary critics. *The details of the scheme, to which the party does not stand committed, may be faulted, but the objects of the reform command a wide measure of agreement.* / *This tendency to fault Herodotus for not going about his business like a 19th-c. classicist is also apparent in the introduction.* There seems no reason why exception should be taken to this addition to the large number of nouns that we make serve as verbs, except by those to whom every novelty in the English language is offensive.

**faun, satyr,** are the Latin and the Greek names for woodland creatures, half beast and half man in form, half beast and half god in nature. Horse's tail and ears, goat's tail and horns, goat's ears and tail and legs, budding

horns, are various symbols marking not the difference between the two, but that between either of them and man. The faun is now regarded rather as the type of unsophisticated and the satyr of unpurified man; the first is man still in intimate communion with Nature, the second is man still swayed by bestial passions.

**fauna, flora,** are singular nouns used as collectives, not plurals like *carnivora* etc. Their plurals, rarely needed, are *faunas* and *floras*, or *faunae* and *florae*. They are Latin goddess names made to stand for the realms of animals and of plants, especially those to be found in any given district.

**fay.** See FAIRY 2.

**feasible.** With those who feel that the use of an ordinary word for an ordinary notion does not do justice to their vocabulary or sufficiently exhibit their culture (see WORKING AND STYLISH WORDS), *f.* is a prime favourite. Its proper sense is practicable, 'capable of being done, accomplished, or carried out' (OED). That is, it means the same as *possible* in one of the latter's senses, and its true function is to be used instead of *possible* where that might be ambiguous. A thunderstorm is *possible* (but not *f.*). Irrigation is *possible* (or *f.*). A counter-revolution is *possible*; i.e. (a) one may happen for all we know, or (b) we can if we choose bring one about; but, if (b) is the meaning, *f.* is better than *possible* because it cannot properly bear sense (a) and therefore obviates ambiguity.
The wrong use of *f.* is that in which, by SLIPSHOD EXTENSION, it is allowed to have also the other sense of *possible*, and that of *probable*. This is described by the OED as 'hardly a justifiable sense etymologically, and . . . recognized by no dictionary'. It has, however, become very common; in all the following quotations, it will be seen that the natural word would be either *probable* or *possible*, one of which should have been chosen: *It seems f. that Lord Folkestone might, without further in-*

*vestigation, have offered a journalistic fledgling some minor job on the paper, but that he should . . . discuss Harris taking over the editorship becomes absurd . . . | Witness said it was quite f. that if he had had night binoculars he would have seen the iceberg earlier. | We ourselves believe that this is the most f. explanation of the tradition.*

**feast.** For 'f. of reason' see HACK-NEYED PHRASES.

**feature,** vb. The use of this in cinema announcements instead of *represent* or *exhibit* started in America and did not take long to establish itself in England. The OED originally gave no meaning that supported this use, but the 1933 Supp. admitted 'to make a special feature of, to exhibit as a prominent feature in a dramatic piece'. Popular usage now sometimes turns it upside-down and speaks not only of a per-formance featuring a performer but also of a performer featuring a per-formance. *A fine display by Graveney featured the Gloucestershire innings.* A later stage of this rake's progress is the intransitive use of the word in the sense of 'figure'.*'Busman's Honeymoon' was a play in which Wimsey featured. | Linguistic difficulties will not feature among the road mines on the path to the Free Trade Area.*

In the world of theatre and film the shine has now worn off *feature*, and a new word has had to be found. Popu-lar favourites are no longer *featured*: it would be almost insulting. They are *starred*, or at least *co-starred*.

**fecund, facund. 1.** The OED gives preference to *fĕcund* over *fē-*, but the latter is now more usual; the COD gives no other.

**2.** The literary critic who writes of *The fecund Walpole and the facund Wells,* fishing up the archaic *facund* for the sake of the play on words (see PARONOMASIA), may impress his readers by his cleverness but is unlikely to give them a clear idea of the difference in the art of the two novelists that corre-sponds to the difference in the vowels.

Both words are from Latin adjectives, *fecundus* meaning prolific and *facundus* meaning eloquent. The English words, when applied to writers or speakers, merge in the sense fluent or luxuriant, which may be the reason why only one has remained in use.

**federation, confederation, con-federacy.** The OED says: '*Confeder-acy* now usually implies a looser or more temporary association than *con-federation*, which is applied to a union of States organized on an intentionally permanent basis. . . . In modern politi-cal use, *confederation* is usually limited to a permanent union of sovereign States for common action in relation to externals. . . . The United States of America are commonly described as a *Confederation* (or confederacy) from 1777 to 1789; but from 1789 their closer union has been considered a "federation" or federal republic.' / [On *federation*] 'Now chiefly *spec.* the for-mation of a political unity out of a number of separate States, so that each retains the management of its internal affairs.'

It is certainly true that *confederacy* implies a looser and more temporary association than the other two. NATO, for instance, is a confederacy. But the word is falling into disuse because it is tainted by the sinister meaning given to *confederate*. Whether there is, or ever has been, any precise difference in the ways in which *confederation* and *federation* are used (apart from the Federals and Confederates of the American Civil War) may well be doubted.

**feel.** *The Committee f. that Mr. X must share the responsibility for this un-fortunate occurrence.* 'To feel', says the COD, defining the sense in which the word is here used, 'is to have a vague or emotional conviction that.' That is no way for a committee to record a grave conclusion, with its suggestion that they are guided by intuition rather than by the evidence. Officials, perhaps from a misplaced modesty that shrinks from positive assertion, are too fond of announcing conclusions in this namby-pamby fashion.

**feldspar**, not *felspar*. The first part is German *Feld* field, not *Fels* rock.

**felicitate.** See FORMAL WORDS.

**fellow** and hyphens. See HYPHENS for the principles that it is suggested should decide between e.g. *fellow man*, *fellow-man*, and *fellowman*. Usage, however, is far from observing those principles with *f*. All the combinations of *f.* with a noun (except *f.-feeling*, for which see below) would be best written as two separate words without hyphen, and they all are sometimes so written. But, owing to the mistaken notion that words often used in juxtaposition must be hyphened, the more familiar combinations are so often seen with the hyphen that they now look queer and old-fashioned without it. *F.-feeling*, which is more of a true compound than the rest, would be better written *fellowfeeling*, but this also is ordinarily given the hyphen.

Those who are not afraid of seeming old-fashioned can write all the items of the following list except *fellowship* and *f.-feeling* as two separate words, and this is recommended. But where a hyphen is here inserted, it is usual: *f. author*, *f. Christian*, *f.-citizen*, *f.-commoner*, *f.-countryman*, *f. crafts-man*, *f.-creature*, *f. executor*, *f.-feeling* or *fellowfeeling*, *f. heir*, *f. lodger*, *f.-man*, *f.-officer*, *f. passenger*, *fellowship*, *f. sinner*, *f.-soldier*, *f. subject*, *f. sufferer*, *f. traveller*, *f. worker*.

**felspar.** See FELDSPAR.

**female, feminine, womanly, womanish.** The fundamental difference between *female* and *feminine* is that the first is wider, referring to the sex, human or not, while the other is limited to the human part of the sex. This would leave it im-material in many contexts which word should be used; and yet we all know that, even in such contexts, one and not the other is nearly always idiomatic.

A female is, shortly put, a she, or, put more at length, a woman-or-girl-or-cow-or-hen-or-the-like. The noun use is the original; but, like all nouns, the word can be used attributively, and through the attributive use this noun has passed into an adjective. The female sex is the sex of which all members are shes; that is the attributive use; passing to, or rather towards, the full adjectival use, we say so-and-so is female, meaning that it is of or for the female sex. Beyond that point as an adjective *female* has not gone. *Feminine*, on the other hand, is not a noun that has gone part way to complete adjectivehood; it has been an adjective all its life, and means not merely of or for women, but of the kind that characterizes or may be expected from or is associated with women. This means that there are two factors in choosing between *female* and *feminine*, (*a*) that of the difference between all sex and human sex, and (*b*) that of the difference between the noun-adjective and the true adjective.

The result is this: when the information wanted is the answer to the question *Of* (or *for*) *which sex?*, use *female*, provided that the context sufficiently indicates the limitation to humankind; when the question is *Of what sort?*, use *feminine*. So we get *female ruler, cook, companion, Paul Pry*, but *feminine rule, cookery, companionship, curiosity*; *female attire, organs, children, servants, screws*; *the female ward* of a prison; *female education* is the education provided for (of course, human) females, while *feminine education* is that which tends to cultivate the qualities characteristic of women. *Feminine* is the epithet for beauty, features, arguments, pursuits, sympathy, weakness, spite, and the like. The feminine gender is the one that includes nouns resembling women's names; a man may be called feminine, but not female, if he is like women.

*Womanly* is used only to describe qualities peculiar to (a) good women as opposed to men (*w. compassion, sympathy, intuition*, etc.) or (b) developed women as opposed to girls (*w. beauty, figure, experience*). *Womanish*, on the other hand, has become a pejorative word for those qualities of timidity, frivolity, and instability which men used to think, until modern war taught them better, were always characteristics of the other sex.

**female, woman.** *F.* in its noun use is sometimes convenient as a word that includes girls as well as women, and sometimes as including non-human as well as human f. creatures. Where such inclusion is not specially desired, to call a woman a female is resented as impolite; the noun has acquired the same kind of disparaging overtone as INDIVIDUAL. It is not reasonable to extend this resentment to the adjective use of *female*; but this is what probably accounts for the apparent avoidance of the natural phrase *f. suffrage* and the use of the clumsy *woman suffrage* instead. As with *f. education* (for which see the preceding article), *f. suffrage* is short for the suffrage of (of course, human) f. creatures, i.e. women. In the first edition of this book the hope was expressed that when the way women were going to vote came to be a common theme of discussion it would be called the *female vote* and not the *woman vote*; for to turn *woman* into an adjective with *female* ready made would be mere perversity. In the event we have been neither impolite enough to call it the *female* vote (though we still speak of the *male vote*) nor perverse enough to turn *woman* into an adjective; we have evaded the dilemma by calling it the *women's vote*.

**feminine designations.** There are several possible ways of indicating the female sex in an occupational or agent noun. If the male counterpart ends in *-man* or *-master* this will naturally be turned into *-woman* or *-mistress* as in *horsewoman, policewoman, schoolmistress, postmistress*. Words borrowed from the French usually carry with them their feminine forms (*masseuse, confidante*), but not always: an *artiste* for instance is not a woman painter. Latin words used in the law also keep their vernacular inflexions (*executrix, testatrix*, etc.), and so does at least one Greek word, *heroine*. Another common device is to use *lady, woman*, or *girl* attributively, as we speak of a *lady doctor*, a *woman teacher*, and a *girl student*. Here there is a risk of ambiguity: the first word in such compounds may be used not adjectivally but as a genitive: a *woman-hater* is not a woman who hates nor is a *lady-killer* a murderess. See HYPHENS.

These are, however, secondary ways of forming feminine designations; the standard way has always been the use of the inflexion *-ess*. This has been applied not only to occupational terms such as *abbess, actress, governess* (the feminine equivalent of *governor* in its old sense of tutor-guardian), *hostess, seamstress, shepherdess, sorceress*, and many others, but also to agent-nouns denoting more casual activities such as *adulteress, adventuress, benefactress, foundress, murderess, patroness, temptress*, and the like.

Feminine designations seem now to be falling into disuse. 'The agent-nouns in *-er* and the sbs. indicating profession etc.', said the SOED in 1933, 'are now treated as of common gender whenever possible.' With some of them we have no choice. A woman ambassador cannot be an *ambassadress*, for that is the title given to the wife of an ambassador; nor can we pray for Elizabeth our Queen and *Governess*; nor can a woman mayor be a *mayoress*, for that is a separate office to be filled by one of her relatives. But such cases are few, and can have little to do with the curious fact that we are dropping these words at the very time when it might be thought, as was said in the first edition of this book, that 'with the coming extension of women's vocations, feminines for vocation-words are a special need of the future'. Perhaps the explanation of this paradox is that it symbolizes the victory of women in their struggle for equal rights; it reflects the abandonment by men of those ideas about women in the professions that moved Dr Johnson to his rude remark about women preachers. Modern woman justifiably resents any such implications. *She was a professional sculptor (she hated the word sculptress as a concession to feminism) who had studied her art in Paris and been a pupil of Rodin.*

We have not thought it necessary to coin new words such as *barristress, councilloress, solicitress, secretaress*; we are content to say *my Lord* to a woman sitting in a judicial capacity that would entitle a man to be so addressed, and we have not given distinctively feminine titles to women officers in the Services. Even where feminine designations are available we do not make much use of them. *Cateress, dictatress* (and -*ix*), *directress, doctress, editress, inspectress, inventress, oratress, paintress, presidentess, professoress, tailoress, tutoress*, all recorded in the dictionaries, many of them of impressive antiquity, are either wholly neglected or allowed to die gradually. (For AUTHORESS see that article.) Even the words ending -*man* are sometimes preferred to their -*woman* counterparts, especially if the first part of the compound is of generic rather than particular significance. We say *airwoman* and *horsewoman* but not *chairwoman* (perhaps it sounds too like charwoman) or *alderwoman*, and we do not hesitate to treat such words as *craftsman, houseman* (in the hospital sense), *sportsman, spokesman* as of common gender unless there is need to distinguish the sexes, as when we are asked to vote for the *sportsman* and *sportswoman* of the year. Some -*ess* designations of course survive. It is natural that words like *abbess, actress, deaconess, stewardess, wardress*, should do so, for in those occupations women have duties peculiar to their sex, but it is less easy to explain why *hostess, manageress*, and *waitress* offer a stout resistance to the general trend, and *proprietress* and *instructress*, and perhaps *poetess*, are more tenacious of

life than most of the old -*ess* words. One or two of these we have indeed brought back into use for special purposes such as *conductress*, if what she conducts is a vehicle not an orchestra. But what we usually do today, if we must have a feminine designation, is to use *woman* attributively, or sometimes, playfully and ignorantly (see SUFFRAGETTE), the suffix -*ette. Conductorette* had a short life, and few will wish a longer one to the more recent coinages *announcerette* (by a gossip columnist), *drum-majorette* (by the U.S. entertainment world), and *stewardette* (by a shipping line). Any capacity to amuse that *undergraduette* may once have had has now waned, though the word, wearing the label 'joc' or 'slang', has made its way into the dictionaries. In serious use *usherette* alone seems to have established itself.

**feminineness, feminism**, etc. The words on record in the OED are: *feminacy, feminality, femineity, feminicity, feminility, feminineness, femininism, femininity, feminism, feminity*. Of these *feminacy, feminality, feminicity*, and *feminility*, may be put out of court as mere failed experiments except to the extent that *feminalities* has acquired a special meaning as 'the sort of knick-knacks that women like to have about them' (OED). *Femineity, -ineness, -inity*, and -*ity*, remain as competitors for the sense of woman's nature and qualities, none of them perceptibly differentiated in meaning. *Feminineness* is a word that does not depend on usage or dictionary-makers for its right to exist; it can, of course, be used; *femininity* and *feminity* are both as old as the 14th c. and have been in use ever since; of the two, -*inity* is the more correct form, and has found greater favour, though -*ity* is more euphonious and manageable, and is as justifiable as e.g. *virginity*; *femineity* is the 19th-c. formation, needless beside the others.

*Feminism* and *feminism* should have meanings different both from the above and from each other. *Feminism* should mean (*a*) an expression or idiom peculiar to women, and (b) the tendency in a man to feminine habits. The first sense is in fact that in which it is now used; the second has been taken over by *effeminacy*. *Feminism* (with *feminist* attached) should mean faith in woman, advocacy of the rights of women, the prevalence of female influence. This sense, now well established, is novel enough not to be recorded in the OED (1901) but the 1933 Supp. gives examples dating from 1895.

**femur.** Pl. *femurs* or *femora*; see LATIN PLURALS.

**ferae naturae.** *The law applies only to animals f. n.*; *Rabbits are f. n.*; *Rabbits are among the f. n.* The first two sentences show the correct, and the third the wrong use of the phrase, and the three together reveal the genesis of the misuse. *F. n.* is not a nominative plural, but a genitive singular, and means not 'wild kinds', but 'of wild kind', and it must be used only as equivalent to a predicative adjective, and not as a plural noun. Pronounce *fērē nātū'rē*. See LATIN PHRASES.

**feral, ferial.** Although the Latin adj. *feralis* means pertaining to funeral rites (whence the rare English word *feral* with the same meaning) the ordinary use of *feral* in English is as an adjectival form of the Latin *ferus*, meaning wild, undomesticated (see preceding article). It has no connexion with *ferial*, from Latin *feriae*, holidays, an ecclesiastical term for something (e.g. a form of service) appertaining to an ordinary day as opposed to a festal or fast day.

**fer(r)ule.** The cap or ring for a stick has two *r*s, and is also spelt *ferrel*; the teacher's implement (now in allusive use only) has one *r*, and is also spelt *ferula*. The two words are of separate origins.

**fertile.** The OED gives precedence to *-ĭl*; but *-īl* is now usual except in U.S. See -ILE.

**festal, festive.** Both words point to feast or festival, but the reference in *-al* is more direct; a person is in festal mood if there is a festival and he is in tune with it, but he may be in festive mood even if he is merely feeling as he might if it were a festival. *A festal day*; *in festal costume*; *a festive scene*; *the festive board*. The distinction is not regularly observed, but, such as it is, it accounts for the continued existence of the two words. There is something of the same difference between *festival* and *festivity* or *festivities*.

**fetid, foetid.** The OED prefers *fĕ'tĭd* as spelling and pronunciation, but in the latter respect custom has not followed it; *fē-* is now usual. The Latin original is, correctly spelt, *fētidus*.

**fetishes,** or current literary rules and conventions misapplied or unduly revered. Among the more notable or harmful are: SPLIT INFINITIVE; FALSE QUANTITY; avoidance of repetition (see ELEGANT VARIATION); the rule of thumb for WHICH WITH AND OR BUT; a craze for native English words (see SAXONISM); pedantry on the foreign pronunciation of foreign words (see DIDACTICISM and FRENCH WORDS 2); the notion that ALTERNATIVES cannot be more than two, that NONE must have a singular verb and that RELIABLE, AVERSE *to*, and DIFFERENT *to*, are marks of the uneducated; the rule of thumb for *and* and *or* in ENUMERATION FORMS; the dread of a PREPOSITION AT END; the idea that successive metaphors are mixed METAPHOR; the belief that common words lack dignity (see FORMAL WORDS). See also SUPERSTITIONS.

Pronunciation of *fetish*. The OED gives *fĕt-* precedence over *fēt-*, but the latter has since made headway and some modern dictionaries put it first. Though it has the air of a mysterious barbarian word, it is in reality the same as *factitious*, and means (like an idol, the work of men's hands) a made thing.

**fetus.** See FOETUS.

**feverish.** It is strange that hardly anyone now asks the sensible question 'Is he f.?'; it must always be the absurd one 'Has he got a temperature?' Perhaps this is an example, more than ordinarily foolish, of the working of EUPHEMISM. But to have a

few

fey

temperature in the sense of having one that is above normal is firmly established colloquially as a STURDY INDEFENSIBLE (cf. to a DEGREE), and to correct those who use the phrase is pedantry.

**few.** 1. *Comparatively f.* 2. *Fewer number.* 1. As will be seen from the newspaper extracts below, ugly combinations of *comparative(ly)* with *a few* and *few* are common. There is no possible objection to putting the adverb *comparatively* before the adjective *few*, as in *Comparatively few people are in the secret*; that is a normal construction not requiring comment apart from a warning against the frequent misuse of *comparatively* when no true comparison is expressed or implied (see COMPARATIVE). But *a comparatively few* is quite another matter, and so is *the comparative few*. The extracts now follow: *The one beneficial treatment for such men could not be obtained excepting for* a comparatively few. / *Discussion in and out of the House has reduced these to* a comparatively few points. / *The* comparative few *who take season tickets seldom travel every day*.

It is remarkable in the first place that of an idiom now enjoying such a vogue no trace whatever should appear in the OED's quotations either for *few* or for *comparative(ly)*; the explanation is doubtless that people of literary discernment, and even the writers of books in general, recoil from such a monstrosity, or used to do so. It is indeed easier to call it a monstrosity than to prove it one, because *a few* is itself an anomalous phrase, and therefore analogies for its treatment are not abundant; we must make the best of the few available. The main question is whether the *few* in *a few* is a noun, and therefore to be qualified by an adjective, or an adjective, and therefore to be qualified by an adverb. There is first the familiar *a good few*, still current though colloquial; next, there are *a good many* and *a great many*, extant modifications of the old *a many*; thirdly, we know that *quite a few* and *not a few* are

English, while *a quite few* and *a not few* are impossible. These show sufficiently that while *a few* taken together may be modified by an adverb, a modifying word placed between *a* and *few* can only be an adjective; in fact, the *few* of *a few* is itself a noun meaning small number. That it can be followed by a plural noun without an intervening *of* (*there are a few exceptions*) is nothing against this; it is parallel to *dozen*, and *hundred*: *a dozen eggs, a hundred men*, where, whether *of* is inserted or not, any modifying word is an adjective after, or an adverb before, the *a* (*a round dozen eggs, a good hundred men*, but *roughly* or *fully* or *quite a dozen* etc.). Consequently, if *comparative(ly)* is to be sandwiched it must be *a comparative few*, but if it is to precede the whole, or if it is to qualify *few* without *a*, it must be *comparatively*. On this showing all the examples in the first paragraph of this article are wrong, the last as well as the others.

The objection will probably occur to some readers: What, then, about *a very few*? May we not say *In a very few years all will be changed*? The answer is, first, that *a very few* is no doubt the origin of the mistaken constructions, and secondly that *very* is here not an adverb, but an adjective, as in *She is a very woman* or *devil*, or in *Living on a very minimum of food*; just as we can say *a poor* or *a wretched few*, so we can say *a mere* or *a very few*, with *very* an adjective.

It may be added that *Very few people were there* is better than *A very few people were there*, because *few* means some and not many, while *a few* means some and not none, so that *few* is better fitted than *a few* for combination with words that, like *very*, express degree.

2. *Fewer number(s)* is a solecism, obvious as soon as one thinks, but becoming common; correct to *smaller* in: *Fortunately the number of persons on board was fewer than usual.* / *The bird seems to have reached us in fewer numbers this year.* See ADJECTIVES MISUSED and LESS 3.

**fey.** See FAIRY.

**ff.** In old manuscripts the capital F was sometimes written *ff*. This is the origin of the curious spelling of some English surnames: ffolliot, fforde, ffoulkes, ffrench, and others. The distinction of possessing such a name is naturally prized: readers of *Cranford* will remember Mrs. Forrester's cousin Mr. ffoulkes who always looked down on capital letters and said they belonged to lately invented families; and it was feared he would die a bachelor until he met a Mrs. ffarringdon and married her, 'and it was all owing to her two little *ffs*'.

**fiancé -ée.** See INTENDED.

**fiat.** The legal term is *fī′at* and the motor car *fē′at*.

**fibula.** Pl. *-lae* or *-las*. Pronounce *fī′bula*. See LATIN PLURALS.

**fictitious.** See FACTIOUS.

**fiddle.** Of the verb the OED says 'now only in familiar or contemptuous use'. It is more in demand to convey the meaning of potter or cheat than that of play the violin, and it is a component of derisive words such as *fiddle-de-dee*, *fiddle-faddle*, and *fiddlesticks*. Perhaps the story of Nero's fiddling while Rome was burning started the decline. But the noun has escaped this taint, and is a partner in some most respectable expressions—*fiddle-back chair*, *fit as a fiddle*. So a violinist will still speak of his instrument as a *fiddle*, but not of his playing as *fiddling* or of himself as a *fiddler*.

**fidget,** vb, makes *-eting* etc.; see -T-, -TT-.

**fiducial, fiduciary.** The second is the ordinary form, *fiducial* being used only in some technical terms in surveying, astronomy, etc.

**fidus Achates.** Pronounce *fī′dus ăkā′tēz*. See LATIN PHRASES.

**-fied.** We have not yet decided how to spell our (mostly jocular) verb-compounds in *-fy* (*countrified* or *coun-*

*tryfied* etc.). It seems best to use *-i-* when the noun or adjective does not provide a convenient connecting syllable, but, when it does, not to alter it; so *cockneyfied*, *countryfied*, *dandyfied*, *Frenchified*, *ladyfied*, *townified*, *whisk-(e)yfied*, *yankeefied*. But the present tendency is to spell with *i*. *Yankeefied* is the only one of these words that is never so spelt.

**field,** in the sense of space proper to something (*f. of action, each in his own f.*, etc.). The near-synonyms for this are remarkably numerous. The distinctions and points of agreement between them are fortunately obvious enough not to need elaborate setting forth; but a list, necessarily incomplete, and a characteristic phrase or so for each word may be useful.

Area, branch, compass, department, domain, field, gamut, last, limit, line, locale, point, province, purview, question, radius, range, realm, record, reference, region, register, scale, scene, scope, sphere, subject, tether, theme. *A debate covering a wide* area. *Unsurpassed in his own* branch. *Expenses beyond my* compass. *In every* department *of human activity. Belongs to the* domain *of philosophy. Distinguished in many* fields; *is beyond the* field *of vision. In the whole* gamut *of crime. Stick to your* last. *Unconscious of his* limits. *Casuistry is not in my* line. *A very unsuitable* locale. *Talking beside the* point. *It is not our* province *to inquire. Comes within the* purview *of the Act. Constantly straying from the* question. *Operating within a narrow* radius. *Outside the* range *of practical politics. In the whole* realm *of Medicine. Don't travel outside the* record. *Such evidence is precluded by our* reference. *In the* region *of metaphysics. Any note in the* lower register. *Whatever the* scale *of effort required. A* scene *of confusion. Find* scope *for one's powers; limit the* scope *of the inquiry. Useful in his own* sphere. *Wanders from the* subject. *Get to the end of one's* tether. *Has chosen an ill defined* theme.

**figure, figurant, figurative,** etc. While it is pedantic (in Britain, though not in U.S.) to pronounce *figure* otherwise than as *fi′ger*, it is slovenly to let the natural English laxity go to this extreme with the less familiar *figuration, figurative, figurant, figurine*, etc.; in them the long *u* should be given its full value. See PRONUNCIATION.

**figure of speech** as a term of rhetoric includes any recognized form of abnormal expression adopted for the sake of emphasis, variety, etc., such as APOSIOPESIS, HYPERBOLE, and METAPHOR. In ordinary usage it is applied only to the last two.

**filthy lucre.** See HACKNEYED PHRASES.

**filtrate,** vb, by the side of *filter*, vb, is a SUPERFLUOUS WORD suggesting BACK-FORMATION from *filtration*, but *infiltrate* as a technical term of warfare (hot or cold) has every right to maintain its place against the unnecessary upstart *infilter*.

**finalize.** *In view of the importance of this contract to the future of our forces, why is it taking such a long time to get it finalized?* There can be few occasions on which the neologism *finalized* is an improvement on *completed* or *finished*, and this is certainly not one of them. See NEW VERBS IN -IZE.

**fine,** adj. *Not to put too f. a point upon it* is an apology for a downright expression, and means 'to put it bluntly'. See SUPERIORITY.

**fine,** n. *In fine*, a phrase now seldom used except in writing of a rather formal kind, has entirely lost the sense, which it once had, of at last. It is still sometimes used for finally or lastly, i.e. to introduce the last of a series of parallel considerations; but in the interests of clearness it is better that it should be confined to its predominant modern use, meaning in short or in fact or to sum up, and introducing a single general statement that wraps up in itself several preceding particular ones. Cf. *en fin*.

**finger.** The fingers are now usually

numbered exclusively of the thumb—*first* (or *fore* or *index*), *second* (or *middle*), *third* (or *ring*), and *fourth* (or *little*); but in the marriage service the third is called the fourth. It is also the fourth in the modern method of indicating the fingering of keyboard music, in which the thumb, formerly marked x, is now 1, and the fingers are numbered correspondingly.

**finical, finicking, finicky, finikin.** All that can be said with certainty about the derivation of the words and their mutual relations seems to be that *-al* is recorded 70 years earlier than the others. As to choice between them, the English termination *-cking* is best calculated to express a hearty British contempt for the tenuity naturally symbolized by the three short *i*s; cf. *niggling* and *piffling*; *-cal* is now chiefly in literary and not colloquial use; there *-icky* is commonest and this is also the variant favoured in the U.S. *Finikin* is virtually obsolete.

**fiord, fjord.** The OED gives precedence to *fi-*. The other spelling is apparently used in English only to help the uninformed to pronounce it correctly—*fyord*. As, instead of helping, it only puzzles them, it might well be abandoned.

**fire,** in the sense expel or dismiss (a person), is still an American colloquialism, though making headway among us at the expense of the verb to SACK. 'Often with *out*', says the OED, but this is no longer true; a PHRASAL VERB seems to have had the unusual experience of being shortened to a simple one. *Fire out* in the sense of expel by fire is as old as Shakespeare: *Till my bad angel fire my good one out* (Sonnet 144).

**firearms.** See ARMS.

**first.** **1.** For *first* etc. *floor*, see FLOOR. **2.** *First thing* is equally idiomatic with *the first thing* (*shall do it f. t. when I get there*). **3.** *The first two* etc., *the two* etc. *first*. When the meaning is the first and second, not the possible but uncommon one of 'the two of which each alike is first', modern logic has decided that *the first two* is right and *the two first*, though the older idiom, wrong. Since many find themselves unable to remember which is logical without working it out, and disinclined to do that afresh every time, the simplest way is to suit the treatment of *first two*, *f. three*, *and f. four* (beyond which the doubt hardly arises) to that of larger numbers; no one would say *the 23 first* instead of *the first 23*, and neither should one say *the two first* instead of *the first two*.

4. *First(ly)*, *secondly*, *lastly*. The preference for *first* over *firstly* in formal enumerations is one of the harmless pedantries in which those who like oddities because they are odd are free to indulge, provided that they abstain from censuring those who do not share the liking. It is true that the Prayer Book, in enumerating the causes for which matrimony was ordained, introduces them with *First*, *Secondly*, *Thirdly*; it is true that *firstly* is not in Johnson; it is true that De Quincey labels it 'your ridiculous and most pedantic neologism of *firstly*'; but the boot is on the other leg now. It is the pedant that begins his list with *first*; no one does so by the light of nature; it is an artificialism. Idioms grow old like other things, and the idiom-book of a century hence will probably not even mention *first*, *secondly*.

**firth, frith.** *Firth* is both the older form and the prevailing one.

**fish.** For pl. see COLLECTIVES 3.

**fisher(man).** See ANGLE(R).

**fissionable** is a word that was coined by atomic scientists for their own purposes and met with some criticism. But plenty of our adjectives are made that way—*questionable*, *objectionable*, *impressionable*, etc., and it must be presumed that the old word *fissile* did not give them quite the meaning they wanted.

**fistula.** Pl. *-as*.

**fix**, etc. 1. *Fixedness, fixity*. *Fixedness* is preferable in the sense of intentness, perhaps from the connexion with *fixedly*; in other senses the doubt about its pronunciation (it should have three syllables) has caused it to give place to *fixity*; compare *Looking at her with mild fixedness* with *The unbending fixity of a law of nature*.

2. *Fix*, or *fix up*, is in U.S. 'a serviceable word-of-all-work which saves the trouble of finding the specific term for almost any kind of adjustment or repair' (MAU). Only American examples are given in the OED Supp., but the temptation it offers to the lazy has proved irresistible, and it is now a common colloquialism in Britain. *Fixings* (trimmings) is still an Americanism.

3. *Fixation*. Most literary men know some Latin; that Latin is chiefly of the classical kind, and a little of it is enough to make them aware that *figere*, and not *fixare*, is the classical Latin for fix. Consequently they have always felt an instinctive repugnance to the word *fixation*, and, perhaps unreasonably, preferred to say *fixing* instead of it whenever they could, leaving *fixation* mostly to those who need it in technical contexts. A technical term of chemistry, it has now been borrowed by the psychologists and, in the sense in which they use it, has become a POPULARIZED TECHNICALITY. *The Commissioners on local government seem to have acquired a fixation about county boroughs.*

**fiz(z).** See -Z-, -ZZ

**flaccid.** Pronounce *-ks-*.

**flageolet.** Pronounce *flăjŏlĕ't*.

**flair** means keen scent, capacity for getting on the scent of something desired, a good nose *for* something. The following quotation illustrates the risks taken (see FOREIGN DANGER) by writers who pick up their French at second hand: *And I was eager to burst upon a civilian world with all the flaire [sic] of a newly discovered prima donna.*

**flambeau.** Pl. -*s*, or -*x* (pron. -*z*); see -x.

**flamboyant** is a word borrowed from writers on architecture, who apply it to the French style (contemporary with English perpendicular) characterized by tracery whose wavy lines suggest the shape or motion of tongues of flame. It is now fashionable in transferred senses; but whereas it should be synonymous with flowing or flexible or sinuous or free, it is made to mean florid or showy or vividly coloured or courting publicity. That is unfortunate, for when the primary and the popular meanings of a word are at odds the latter tends to corrupt the former.

**flamingo.** Pl. -*oes*; see -O(E)S 1.

**flammable.** See INFLAME.

**flannel.** Spell *flannelled*, *flannelly*, but *flannelette*. See -LL-, -L-, and, on the suffix -*ette*, SUFFRAGETTE.

**flaunt, flout.** The derivation of both words is obscure, but their meanings are not in doubt. To *flaunt* is to make a parade of, to show off; to *flout* is to display a contemptuous disregard for. They are often confused, especially by the use of *flaunt* for *flout*. *The trouble is that there are some governments which patently resort to the United Nations when it suits them and flaunt it otherwise.* / *We are to a large extent at the mercy of our biological nature and we cannot repeatedly flaunt its demands with impunity.* / *There have been many recent examples where the police have been unable to enforce the provisions of the law, and where the law was flaunted and brought into disrepute.* The true meanings of the words being almost exactly opposite, it is the more curious that one should be made to serve for the other. Perhaps the origin of the confusion is that the same action may sometimes be described by either verb: a flouter may flaunt his flouting. But in the examples given it is clear that *flout* was intended and *flaunt* is a blunder.

**flautist, fluter, flutist.** Of these three names for a flute-player the oldest, *fluter*, is dead, and would be forgotten but for *Phil the Fluter's Ball*. *Flutist* is more than 350 years old; *flautist* (from Italian *flautista*) dates only from the middle of the 19th c., and there seems no good reason why it should have prevailed over the others. But it has.

**fledg(e)ling.** Of the eight quotations in the OED, not one has the -*e*-; but see JUDGEMENT.

**flee.** The verb is now little used except in the form *fled*; *fly* and *flying* have taken the place of *f.* and *fleeing*. But colloquially the word is used with a differentiation. *I must flee* suggests the approach of an unwelcome visitor. *I must fly* suggests the recollection of a forgotten engagement.

**fleshly, fleshy.** The distinction is much the same as between *earthly* and *earthy*. *Fleshy* has the primary senses 'consisting of flesh' (*fleshy tables of the heart*), 'having a large proportion of flesh' (*fleshy hands, fruit,* etc.), and 'like flesh' (*fleshy softness, pink,* etc.); while *fleshly* has the secondary senses of 'proper to the flesh or mortal body, sensual, unspiritual, worldly' (*fleshly lusts, perception, inclinations, affairs,* etc.).

**fleur-de-lis.** Pl. *fleurs-de-lis*; pronunciation, alike in sing. and pl., *fler'dĕlē*. 'The form *flower-de-luce* survives as a poetical archaism and in U.S.'—OED.

**flier, flyer.** The first is better on the principles suggested in the article DRY, but *flyer* is more usual.

**flippant.** For synonymy see JOCOSE.

**flock,** = tuft of wool etc. *Flocks* or *flock* (see COLLECTIVES) is used as the name for the material.

**floor, storey.** In Britain a *single-storey* house is one with a *ground-floor* only; a *two-storey* house has a *ground-floor* with a *first-floor* above it; a *three-storey* house has a *second floor* above the first, and so on. (There may be an *entresol* or *mezzanine* floor between the ground and first floors, but it does not count in the reckoning.) In America, more logically, our *ground floor* is the *first floor*, and our *first* is the *second* and so on; thus in (say) a ten-storey building (U.S. *story*) the top floor would be the tenth in America and the ninth in Britain. See also STOREY.

**flora.** See FAUNA.

**floruit** (*-or'ōŏīt*) is a Latin verb meaning he flourished, used with a date to give the period to which a person's activity may be assigned, especially when the dates of his birth and death are not accurately known; it is also used as a noun—*his f.* etc., i.e. the date at which he was active.

**flotsam and jetsam.** The distinction is between goods found afloat in the sea and goods found on land after being cast ashore. The original sense of *jetsam* was what had been jettisoned or thrown overboard. The words are generally used in combination—almost as SIAMESE TWINS—often figuratively, e.g. of human down-and-outs.

**flour, meal.** Flour is boulted meal, i.e. meal from which the husks have been sifted out after grinding. Meal is the ground product of any cereal or pulse. *Flour* used by itself means wheat-flour; applied to other kinds it is qualified —*corn-flour* (i.e. maize), *rye-flour*, etc. *Meal* when used of wheat has *wheat* prefixed.

**flout.** See FLAUNT.

**flower-de-luce.** See FLEUR-DE-LIS.

**fluid, gas, liquid.** *Fluid* is the wide term including the two others; it denotes a substance that on the slightest pressure changes shape by rearrangement of its particles; water, steam, oil, air, oxygen, electricity, ether, are all fluids. Liquids and gases differ in that the first are virtually incompressible, and the second elastic; water and oil are liquid and fluid, but not gaseous; steam and air and oxygen are gases and fluids, but not liquids.

**fluorine.** For pronunciation see -IN AND -INE.

**flurried, flustered, fluttered.** There is often a doubt which is the most appropriate word; the following distinctions are offered:

A person is flurried who, with several things to attend to, lets each interfere with the others; a person is flustered (possibly *fuddled* with drink) in whom different impulses or emotions contend for expression; a person is fluttered who, being of a timid or apprehensive disposition, is confronted with a sudden emergency.

**flute.** 1. *Fluty*, not *flutey*; see -EY AND -Y. 2. For *fluter* and *flutist* see FLAUTIST.

**fly.** 1. The noun is used as a COLLECTIVE for the insect parasite and the injury it causes to plants and animals (*There is a lot of fly on the roses*).

2. The verb makes *is flown* as well as *has flown*; see INTRANSITIVE P.P.

3. *Fly a kite* means (*a*) raise money by accommodation bill, (*b*) make an announcement or take a step with a view to testing public opinion. Cf. *ballon d'essai*.

4. *Fly-leaf* is a blank leaf in a printed document, especially one between the cover and the title-page of a book, or at the end of a circular or leaflet; it is not another name for a leaflet, which is, however, sometimes called a *fly-sheet*.

5. A *fly-wheel* is one whose sole function is by its inertia and momentum to make the movement of the shaft that works it continuous and regular; hence its metaphorical use.

6. For *flyer* see FLIER.

7. For *fly* and *flee* see FLEE.

**foam, froth.** The natural definition of foam would be the froth of the sea, and that of froth the foam of beer. That is to say, foam suggests the sea, froth suggests beer, and while one word is appropriate to the grand or the beautiful or the violent, the other is appropriate to the homely or the ordinary or the dirty. One demands of foam that it be white; froth may be of what colour it pleases. Froth may *be* scum, but foam, though it may *become* scum, ceases to be foam in the process. It is perhaps also true that froth is thought of mainly as part of a liquid that has sent it to the top, and foam as a separate substance often detached from its source in the act of making. But the difference is much less in the meanings than in the suitable contexts.

**fob.** See FOIST.

**focus. 1.** The noun has pl. *-cuses* or *-ci* (pron. *-sī*); the verb makes *focused*, *-cusing* ('in England commonly, but irregularly, written *focussed*, *-ing*— OED); see -S-, -SS-.
**2.** The verb is liable to loose application, as in: *At one moment it seemed to be quite near, and at the next far away; for the ears, unaided by the eyes, can but imperfectly focus sound or measure its distance*. The f. of a sound being 'the point or space towards which the sound-waves converge' (OED), ears cannot f. sound except by taking the owner to the right point. The eyes do measure distance by focusing, having an apparatus for the purpose; the ears do not.

**foetid.** See FETID.

**foetus, fetus.** British dictionaries favour *foetus*, American the etymologically preferable *fetus*.

**foist.** *The general public is much too easily* foisted off *with the old cry of the shopman that 'there's no demand for that kind of thing'*. The public can be *fobbed* off with something, or the something can be fobbed off on the public; but *foist off* has only the second of these constructions; see ANALOGY and OBJECT-SHUFFLING.

**folio.** Pl. *-os*; see -O(E)S 4. Uses of the word are many and varied. The chief of these are: (Accounting) two opposite pages, or single page, of ledger used for the two sides of account; (Parliamentary and Legal) number of words (72 or 90) as unit of length in document; (Bookbinding) once-folded sheet of printing-paper

giving two leaves or four pages (*in f.*, made of ff.), (also *f. volume* etc.) a book or volume in f.; similarly of smaller sheets and books resulting from various foldings and named after the number of leaves to the sheet: the most usual are *quarto* or *4to*, folded twice into four leaves, and *octavo* or 8*vo*, thrice into 8.

**folk** has almost passed out of the language of the ordinary educated person in England, so far as he talks unaffectedly; *people* has taken its place. Even such well-worn phrases as *menfolk, womenfolk, old folks, little folks, north-country folk*, cannot now be used without suggesting a touch of the archaic or the sentimental: *I do not believe that there is a single beer-drinker who would not have preferred that the Chancellor's concession on beer should have gone to help the old folk. | But where does the lonely 50s. a week old-age pensioner come in his or her desire for the modern amenity of television? Who can better enjoy it, ease their lonely hours, or keep in touch with the world than these folks?* The word is current in the use, imitated from the German, of such compounds as *folk-song, folk-dancing*, and *folklore*, but here too it implies something belonging to an earlier period. It survives more strongly in America, e.g. in the phrase *We're just folks*, i.e. unpretentious, and in the vocative (*Hullo folks; good night folks*) that American entertainers and their English imitators use to their audiences.

**follow.** *As follows. The main regulations of the new Order are* as follow: *First . . . | The principal items of reductions stand* as follow:
In all such contexts, *as follows* should be written. The OED ruling is: 'The construction in *as follows* is impersonal, and the verb should always be used in the singular.' And among its quotations is one from a *Rhetoric* of 1776: 'A few late writers have inconsiderately adopted this last form' [*as follow*] 'through a mistake of the construction.' However, persons who

are pluming themselves on having detected a vulgar error that they can amend are not likely to admit that it is a mare's-nest on the unreasoned *ipse dixit* of an 18th-c. rhetorician, or even of a 20th-c. OED; and some discussion will be necessary. Unfortunately, full demonstration is hardly possible; but several considerations raise separate presumptions in favour of *follows*:

1. It is certain that we all say *as follows* by the light of nature; it is only to the sophisticated intelligence that *as follow* occurs (or would the reformers prefer *occur?*).

2. Similar but more obvious maltreatment of other phrases suggests that the correctors of this may also be mistaken, though it does not prove that they are. Consider for instance the phrases *id est* and *so far as concerns*. *Section* 15 (4), *which deals with persons* (ea sunt, *all present and future members of societies*) *entitled to receive medical attendance*. The author of this (why, by the way, does he stop short of *ii sunt* or *eae sunt?*) would presumably like Byron to have said *Arcades ambo, ea sunt blackguards both*; but *id* does not mean that Arcadian or those Arcadians, it means that phrase. *Many of these stalks were failures, so far as* concern *the objective success*; what the writer means is not so far as the stalks or the failures concern success, but so far as our discussion concerns it. The familiar *as regards* is liable to the same mutilation.

3. The phrase *as follows*, which is very old, no doubt originated in sentences where there was no plural in the neighbourhood to raise awkward questions. The OED quotes (1426) *Was done als her fast folowys* (= as here directly follows), and (1548) *He openly sayde as foloweth. He spoke as follows* may be taken as the type; that is obviously not a piece of normal grammar; what would be the normal way of putting it? *He spoke thus*, which is, at full length, *He spoke so as I shall tell you*, or *He spoke so as it shall be told*, or *He spoke so as the tale follows*, whence, by ellipse, *He spoke as follows*. This pro-

gress is surely natural; but it is equally natural in *His words were so as I shall tell you*, or *His words were so as it shall be told*, or *His words were so as the tale follows*, whence *His words were as follows*. It is true that, when the idiom was being evolved, it was open to its makers to say, instead of *were so as the tale follows*, *were so as words follow*; but they chose otherwise, hundreds of years ago, and the idiom is now fixed. No one would want to change it except under the impression that it was ungrammatical; to show that it is no more ungrammatical than the innovation is enough to condemn the latter.

**following.** used as a preposition. In the article UNATTACHED PARTICIPLES the reader is reminded that 'there is a continual change going on by which certain participles or adjectives acquire the character of prepositions, no longer needing the prop of a noun to cling to. ... The difficulty is to know when this development is complete.' Has *f.* achieved this status? That it is freely so used is notorious ('loosely' says the SOED, 'with slightly formal effect', says the COD). Often it is merely a FORMAL WORD for *after*, as PRIOR TO is for *before*, and as such deserves the same condemnation. Its prepositional use, like that of *prior to*, can be justified only if it implies something more than a merely temporal connexion between two events, something more than *after* but less than *in consequence of*. It can do so in the newspaper report *Following the disturbances in Trafalgar Square last night six men will appear at Bow Street this morning*, but not in the broadcast announcement *Following that old English tune we go to Latin America for the next one*. The second quotation illustrates the absurdity into which a speaker may be led by addiction to *f.*, with its lingering participial sense, as a formal word for *after*. We were not following the old English tune; we were leaving it well behind.

**foot.** vb. *The bill*, or *the cost, foots up to £50* means that £50 is the amount at the f. of the paper on which the addition is done. The origin of *Who will f.*

(i.e. pay) *the bill?* is not so clear; perhaps pay the sum to which it foots up or perhaps undertake responsibility by initialling or signing it at foot. Both phrases are good colloquial English but the first has gone out of fashion; *adds* or *tots* would be a more usual word today.

**footing.** *We have not the smallest doubt that this is what will actually happen, and . . . we may discuss the situation on* the f. *that the respective fates of these two bills will be as predicted.* To give f. the sense of *assumption* or *hypothesis* is a SLIPSHOD EXTENSION; the writer, in fact, on however intimate a f. he may be with lobby prophets, is on a slippery f. with the English vocabulary.

**for,** conj. Two questions of punctuation arise. *F.* is a coordinating conjunction, i.e. one that connects two independent sentences; it is neither, like *therefore* and *nevertheless*, strictly speaking an adverb though serving a conjunctive purpose; nor, like *since* and *because*, a subordinating conjunction that joins a mere clause to a sentence; hence the two points.
  1. Whereas in *Therefore A is equal to B*, and in *Nevertheless he did it*, it is a mere matter of rhetoric, depending on the emphasis desired, whether a comma shall or shall not follow *therefore* and *nevertheless*, it is with *for* a matter of grammatical correctness that there should be no comma; *For, within it is a house of refinement and luxury* is wrong. This naturally does not apply to places where a comma is needed for independent reasons, as in *For, other things being equal, success is a fair test.*
  2. Whereas *since* and *because*, connecting a clause to a preceding sentence, are rightly preceded by a comma only, the presumption with *for*, which connects two sentences, is that a semicolon should be written. This does not rule out the comma, which will often pass when the *for* sentence is a short one; but in such passages as the following the comma is clearly inadequate, and in general the semicolon should be regarded as normal, and the comma as the licence: *This is no party

question, for it touches us not as Liberals or Conservatives, but as citizens.*

**for- and fore-.** The prefix of the words *forbear* (vb), *forbid, forby* (Scot. for besides etc.), *forfend, forgather* (assemble), *forget, forgive, forgo* (relinquish), *forlorn, forsake,* and *forswear,* is unconnected with the English words *for* and *fore,* and means away, out, completely, or implies prohibition or abstention. All these should be spelt with *for-,* not *fore-.* On the other hand the noun *for(e)bears,* and *foregoing* and *foregone* in *the foregoing list, a foregone conclusion,* contain the ordinary *fore,* and should be spelt with the *e. Foreclose* and *forfeit* contain another prefix again (Lat. *foris* outside), though *foreclose* has had its spelling affected by natural confusion with English *fore.* All the words, whether established or made for the occasion, compounded with *fore,* as *forebode, forewarn, foreman, fore ordained,* are spelt with the *e* and should have the *for-* sound distinct

**forbears,** n. See FOREBEARS.

**forbid.** 1. *forbad(e).* The pronunciation is -*ăd,* not -*ād,* and the spelling -*ad* is, to judge by the OED quotations, nearly twice as common as -*ade.*
  2. *Unlike his predecessor, Pope John uses the telephone sparingly . . .; etiquette forbids anyone from calling him.* To f. one *from doing* is an unidiomatic construction on the ANALOGY of *prohibit* or *prevent.*

**forceful, forcible.** The main distinction in sense is that, while *forcible* conveys that force rather than something else is present, *forceful* conveys that much as opposed to little force is used or shown; compare *forcible ejection* with a *forceful personality.* There the words could not be interchanged without altering the meaning, but elsewhere it is often immaterial, so far as sense goes, which word is used. The sense distinction, however, is the less important part of the matter. By usage, *forcible* is the ordinary word, and *forceful* the word reserved for abnormal use, where its special value depends partly on its infrequency and partly on the more picturesque suggestion of its suffix. Unluckily writers have taken to exploiting, and in the process destroying, this special value, by making a VOGUE WORD of *forceful* and always using it in place of *forcible*. If this continues we shall shortly find ourselves with a pair of exact synonyms either of which could well be spared, instead of a pair serving different purposes. Such writers injure the language, which perhaps leaves them cool; but they also injure their own interests; by avoiding the obvious word they lose more in the opinion of the educated than they gain in that of the ignorant. In the first of the following extracts *forceful* is right, but in the other two there is no need whatever to say it instead of the natural *forcible*: *Certainly he was a forceful and impressive personality at a time when the stature of international statesmen was not particularly great. | M. Briand had rightly calculated that he would have the people of France behind him in his forceful endeavour to restore order. | It is his programme to urge upon the Throne peaceful abdication as the only alternative to forceful expulsion.*

**forceps.** Plural the same; but see SINGULAR -S.

**fore.** *To the fore* originally meant at hand, available, surviving, extant. In being borrowed by English from Scottish and Irish writers as a picturesque phrase, it has suffered a change of meaning and is now established in the sense of conspicuous.

**for(e)bears.** As to the form, the prevalent but not sole modern spelling is without the *e*; but the *e* seems better both as separating the noun from the verb *forbear* and as not disguising the derivation (*forebeers*, those who have *been before*); see FOR- AND FORE-.

As to the use of the word by English writers, its only recommendation is that, being Scottish and not English, it appeals to the usually misguided instinct of NOVELTY-HUNTING. *Ancestors, forefathers,* and *progenitors,* supplemented when the tie is not of blood

by *forerunners* and *predecessors,* are the English words, though *ancestors* may sometimes be avoided from modesty, as seeming to claim an *ancestry,* i.e. *forebears* of a superior sort. *By his* forebears *Lord Tankerville is connected with the* ancien régime *of France. His great grandfather, the Duc de Grammont . . .* (read *ancestors*). | *Birmingham is now being afforded an opportunity for offering some kind of posthumous reparation for the great wrong its* forbears *inflicted, close upon 120 years ago, on the illustrious Dr. Priestley* (For *its forbears* read *it.* Birmingham's forebears would be not an earlier generation of Birmingham people, but any villages that may have stood where Birmingham now stands.)

**forecast.** *So far as the operation of the guillotine resolution on the Bill can be* forecasted, *it seems probable that . . .* If etymology is to be our guide, the question whether we are to say *forecast* or *forecasted* in the past tense and participle depends on whether we regard the verb or the noun as the original from which the other is formed. If the verb is original (= to guess beforehand) the past and p.p. will be *cast,* as it is in that verb uncompounded; if the verb is derived (= to make a forecast) they will be *forecasted,* the ordinary inflexion of a verb. The verb is in fact recorded 150 years earlier than the noun, and we may therefore thankfully rid ourselves of the ugly *forecasted*; it may be hoped that we should do so even if history were against us, but this time it is kind. The same is true of *broadcast*; and *broadcasted,* though dubiously recognized in the OED Supp., may be allowed to die.

**forecastle.** Pronounce *fō'ksl*; sometimes so spelt, *fo'c's'le.*

**forego** go before, *forgo* relinquish. See FOR- AND FORE-.

**foregone conclusion.** The phrase is used when an issue supposed to be still open has really been settled beforehand, e.g. when a judge has made up his mind before hearing the evidence;

or again, when an event is so little
doubtful that the doubt is negligible.
But this was not the meaning of the
phrase in its original setting (*Othello*,
III. iii. 434) where Othello says that
Cassio's dream denoted a foregone
conclusion. The precise significance
of *conclusion* may be debatable, but the
purport of the passage is clear enough:
Othello meant that the source of
Cassio's alleged dream must have been
previous actual experience of being
in bed with Desdemona.

**forehead.** See PRONUNCIATION 1.

**foreign danger.** Those who use
words or phrases belonging to languages with which they have little or
no acquaintance do so at their peril.
Even in *e.g.*, *i.e.*, and *et cetera*, there
lurk unsuspected possibilities of exhibiting ignorance. With *toto caelo*,
*bête noire*, *cui bono*, *bona fide*, *qua*, and
*pace*, the risk is greater; and such
words as *alibi* and *phantasmagoria*,
which one hesitates whether to call
English or foreign, require equal caution. See all or any of the words and
phrases mentioned, and FLAIR. Two or
three specimens follow, for those who
do not like cross-references: *I suggest
that a Compulsory Loan be made* pro
ratio *upon all capital* (pro rata). / *Rica-
soli, another of his* bêtes noirs (noires). /
*A man who claimed to be a Glasgow
delegate, but whose* bona fides *were disputed, rose to propose the motion* (was). /
*We are calmly told that Cambridge was
neither worse nor better than the rest of
the world; in fact, it was, we are assured*,
in petto *the reflex of the corrupt world
without* (*in petto* is not in little, but in
one's heart, i.e. secretly). / *THE
TRAMP AS CENSOR* MORES
(morum).

**forename.** Officials who prepare the
forms we have to fill up to get something we want—a premium bond, for
instance, or our name on the register
of electors—have taken to asking us to
give not our *Christian names*, as we
used to do, but our *forenames*. This has
provoked some criticism of them for
being what one critic called 'meticu-

lous-minded'. They deserve rather to
be commended for their good sense in
availing themselves of a ready means
of avoiding the incongruity of asking
for the Christian names of someone
who may not be a member of the
Christian faith. *Forename* is no neologism: a translation of the Latin *prae-
nomen*, it has been an English word for
over 400 years, and goes well with
*surname*. But to ask for 'Christian
names or other forenames', as at least
one form does, is indeed to be
'meticulous-minded'; *forenames* will
do for all.

**forenoon.** *The Church Congress sat in
two sections this forenoon . . . The afternoon programme was divided into three
sections.* Even in contexts that, by the
occurrence as here of *afternoon* in contrast, most suggest the use of *f.*, the
natural English is *morning*. F. is still
current in Scotland and Ireland and
in the Royal Navy, but elsewhere has
fallen out of use as *the* name for the
first half of daylight. Perhaps this is
because the definition of both by reference to noon suggests a sharpness of
division that we no longer observe. We
ordinarily regard the morning as ending at the time of our midday meal and
the afternoon as beginning at the time
when we have finished it—or ought to
have finished it—and call the period
between them lunch-time or dinner-
time, according to our feeding habits.

**for ever.**
*Forever; 'tis a single word!
  Our rude forefathers deem'd it two:
Can you imagine so absurd
        A view?*

     .     .     .     .     .

*And nevermore must printer do
  As men did longago; but run
'For' into 'ever', bidding two
        Be one.*

Calverley's fears have proved groundless. 'Two words' says the *Authors'
and Printers' Dictionary* firmly, a hundred years later.

**foreword, preface.** F. is a word invented in the 19th c. as a SAXONISM
by anti-latinists, and caught up as

a VOGUE WORD by the people who love a new name for an old thing. *P*. has a 500-year history behind it in English, and, far from being antiquated, is still *the* name for the thing. The vogue now seems to be passing, and it looks as though a decent retirement might be found for *f*. by confining it to the particular kind of preface that is supplied by some distinguished person for a book written by someone else who feels the need of a sponsor.

**forgo.** See FOR- AND FORE-.

**forgot,** as a past participle for the current *forgotten*, is now, except in uneducated speech, a deliberate archaism.

**forlorn hope** is not an abstract phrase transferred by metaphor to a storming party, but has that concrete sense in its own right, and only gets the abstract sense of desperate enterprise etc. by misunderstanding. *Hope* is not the English word, but is a mis-spelling of the Dutch *hoop* = English *heap*; the forlorn hope is the devoted or lost band, those who sacrifice themselves in leading the attack. The spelling of *hope* once fixed, the mistake was inevitable; but it is well to keep the original meaning in mind; see TRUE AND FALSE ETYMOLOGY.

**formalism, formality.** It is only from the more abstract sense of *formality*, from *formality* as the name of a quality and not of an action, that *formalism* requires to be distinguished; and there, while *formality* means the observance of forms, *formalism* is the disposition to use them and belief in their importance. Formality is the outward sign of formalism; see -ISM AND -ITY.

**formal words.** There are large numbers of words differing from each other in almost all respects, but having this point in common, that they are not the plain English for what is meant, not the forms that the mind uses in its private debates to convey to itself what it is talking about, but translations of these into language that is held more suitable for public exhibition. We tell

our thoughts, like our children, to put on their hats and coats before they go out; the policeman who has *gone* to the scene of disturbance will tell the magistrate that he *proceeded* there; a Minister of the Crown may *foresee* the advantages of his policy and *outline* it to his colleagues but in presenting it to Parliament he may *visualize* the first and *adumbrate* the second. These outdoor costumes are often needed; not only may decency be outraged sometimes by over-plain speech; dignity may be compromised if the person who thinks in slang writes also in slang. To the detective who has arrested a receiver of stolen property it comes natural to think and speak of the culprit as a *fence* but that is not what will appear on the charge-sheet. What is intended in this article is not to protest against *all* change of the indoor into the outdoor word, but to point out that the less of such change there is the better. A short haphazard selection of what are to be taken as formal words will put the reader in possession of the point; a full list would run into thousands. It must be observed that no general attack is being made on these words as words; the attack is only on the prevalent notion that the commoner synonyms given after each in brackets ought to be translated into them: *accommodation* (rooms); *adumbrate* (outline); *bear* (carry); *cast* (throw); *cease* (stop); *commence* (begin); *complete* (finish); *conceal* (hide); *desist* (stop); *dispatch* (send off); *donate* (give); *endeavour* (try); *evince* (show); *expedite* (hasten); *extend* (give); *felicitate* (congratulate); *locate* (find); *obtain* (get); *proceed* (go); *purchase* (buy); *remove* (take away); *repast* (meal); *seek* (try, look for); *summon* (send for); *sustain* (suffer); *transmit* (send); *valiant* (brave); *veritable* (true); *vessel* (ship); *visualize* (foresee).

There are very few of our notions that cannot be called by different names; but among these names there is usually one that may be regarded as the thing's proper name, its κύριον ὄνομα or dominant name as the Greeks called it, for which another may be substituted to add precision or for many other reasons, but which is present to the mind even behind the substitute. A destroyer is a ship, and, though we never forget its shiphood, the reader is often helped if we call it a destroyer; a vessel also is a ship, but the reader is not usually helped by our calling it a vessel. Though to evince is to show, it does not help him to call showing evincing; what happens is first the translation of *show* into *evince* by the writer, and then the retranslation of *evince* into *show* by the reader. Mind communicates with mind through a veil, and the result is at best dullness, and at worst misunderstanding. The proper name for a notion should not be rejected for another unless the rejector can give some better account to himself of his preference than that he thinks the other will look better in print. If his mental name for a thing is not the proper name, or if, being the proper name, it is also *im*proper, or essentially undignified, let him translate it; but there is nothing to be ashamed of in *buy* or *see* that they should need translating into *purchase* and *observe*; where they give the sense equally well they are fit for any company and need not be shut up at home. Few things contribute more to vigour of style than a practical realization that the κύρια ὀνόματα, the sovereign or dominant or proper or vernacular or current words, are better than the formal words. See also GENTEELISMS and WORKING AND STYLISH WORDS.

**former.** For *the f.* as a pronoun, see LATTER. When the reference is to one of three or more individuals, *the first*, not *the f.*, should be used: *Among the three representatives of neutral States, Dr. Castberg and Dr. Nansen stand for Norway and M. Heringa for Holland; the former is so convinced of* . . .

**formidable.** For pronunciation see RECESSIVE ACCENT.

**formula.** The plural *-las* has become more common than *-lae* except in scientific writings. See LATIN PLURALS.

**forswear.** For *a forsworn lover, witness*, etc., see INTRANSITIVE P.P.

**forte,** person's strong point. For the spelling, which should have been (but is not) *fort*, cf. MORALE. Pronounce *fort*, unlike the musical term *for'tĕ*.

**forth.** *And so f.* is (cf. *and the like*) a convenience to the writer who does not wish to rehearse his list at length, but shrinks from the suggestion, now so firmly attached to *etc.* as to disqualify it for literary use, that he breaks off because it is too much trouble to proceed. The slightly antique turn of the phrase acquits him of unceremoniousness; *and so on* is in this respect midway between *and so forth* and *etc.*

**fortuitous** means accidental, undesigned, etc. That it is sometimes confused with fortunate, perhaps through mere sound, perhaps by the help of lucky, is plain from: *All's well that ends well, and his divorced wives, whom the autobiographer naïvely calls Divorcées Nos. 1, 2, and 3, seem to have borne no kind of ill-will to their more fortunate successor. Reviewing my own Algerian experiences, I must say that I should not have expected so* fortuitous *a termination of a somewhat daring experiment.* | *When first produced, its popularity was limited. Nevertheless it may now sail into a more* fortuitous *harbour on the strength of its author's later reputation.* For such mistakes see MALAPROPISMS.

**fortune.** The verb (*it fortuned that, I fortuned upon*) is an ARCHAISM and a POETICISM.

**forward(s),** adv. The OED says: 'The present distinction in usage between *forward* and *forwards* is that the latter expresses a definite direction viewed in contrast with other directions. In some contexts either form may be used without perceptible difference of meaning; the following are examples in which only one of them can now be used: "The ratchet-wheel can move only *forwards*"; "the right side of the paper has the maker's name reading *forwards*"; "if you move at all it must be *forwards*"; "my companion has gone *forward*"; "to bring a matter *forward*"; "from this time *forward*".' To this it must be added that there is a tendency, not yet exhausted, for *forward* to displace *forwards*, and that since the publication of the OED there has been change. The reader will notice that, while he can heartily accept the banishment of *forwards* from the last three examples, he may well ask himself whether *forward* is not possible in some or all of the first three. But the phrase *backwards and forwards* is still so written. The old pronunciation *fŏr'ăd* survives only at sea but the colloquialism *can't get any forrader* may be heard anywhere.

**foul**, adv. See UNIDIOMATIC -LY.

**foulard.** The OED gives precedence to *fŏōlahr* over *fŏōlar'd* but the *d* is now usually sounded.

**fount(ain).** *Fount* (apart from its use in typography, where it is another word, connected with the metal-casting *found* and pronounced *fŏnt*) is the poetical and rhetorical form of *fountain*; to use it in ordinary contexts, as for the reservoir of a fountain pen, is VULGARIZATION. Nor is it suitably chosen in *A good test of the standing of any force, service, or profession is to ask a man: 'Would you choose it now as a career for your son?' Getting the right answer to that question is a strong but neglected fount of recruits.*

**four.** *On all fours*, apart from its literal application to a person crawling, has now for its chief use the meaning of correspondence between two things at all and not merely some points (*The cases are not o. a. f.*; *The analogy suggested is not o. a. f. with the actual facts*). This seems due to a misunderstanding of the earlier but now less familiar metaphorical use by which a theory, tale, plan, etc., was said to run or be o. a. f. when it was consistent with itself or proof against objections or without weak points—in fact did not limp like a dog on three legs or rock like a table with one leg too short.

The step is easy, though illegitimate, from *The comparison is o. a. f.* (i.e. complete at all points) to *The things compared are o. a. f.* (i.e. alike at all points), and thence to *o. a. f. with*. Whether this is or is not its origin, *o. a. f. with* is now an established idiom.

**fowl.** The collective use of the singular (see COLLECTIVES, *all the fish and f. in the world*) still exists, but is not common except in compounds such as *guinea-fowl*, *wildfowl*.

**fracas.** Pronounce *fră'kah*; pl. spelt *fracas*, and pronounced *fră'kahz*.

**fraction.** The use of *f.* in the sense of a small f., however illogical, is now so common that it can fairly claim to rank as a STURDY INDEFENSIBLE. (*The number of red squirrels is now only a f. of what it used to be.*) Those who use PERCENTAGE in the same way cannot have the same indulgence.

**fragile.** 1. Like many other adjectives ending -ile (e.g. *docile*, *sterile*, *fertile*), *fragile*, after some hesitation, has come down firmly in favour of the long *i* in its second syllable in Britain, though not in America. See -ILE.
2. *fragile*, *frail*. *Frail* is wider both in application and in sense. Whatever is fragile is also frail, but a woman may be frail (i.e. weaker than others in moral strength) who cannot be called fragile (i.e. weaker in physical strength). Where, as in most cases, either word is applicable, there is a certain difference of sense between (*fragile*) liable to snap or break or be broken and so perish and (*frail*) not to be reckoned on to resist breakage or pressure or to last long; that is to say, the root idea of break is more consciously present in *fragile* owing to its unobscured connexion with *fragment* and *fracture*.

**framework.** Few modern clichés have become more pervasive than the phrase *within the framework of*. Like most clichés, it can sometimes be used happily, as it is in *He notes that* The Cloud *describes the contemplative life w. t. f. o. Christian orthodoxy and the*

*Bible, but evidently he believes that for the purposes of psychological study the framework can be disregarded.* More often it is a reach-me-down periphrasis for some more simple way of saying what needs to be said. *Since with the forces at its disposal the maintenance of law and order could not be achieved w. t. f. o. the ordinary law, the Government had to resort to emergency powers* (under the ordinary law). | *These scandals can only be dealt with w. t. f. o. the Trade Union organization* (by the Trade Unions themselves). | *He tells the story of Hannibal's crossing of the Alps w. t. f. o. his own journey* (in the light of his own journey). | *Any negotiations for German bases in foreign countries should be carried out w. t. f. o. NATO* (through NATO). | *His willingness to talk with Molotov and Chou En-Lai, to consider their objectives w. t. f. o. normal diplomatic enterprise, aroused serious doubts in the minds of many Americans* (as part of normal diplomatic enterprise). | *The belief that the penal code as it exists at present w. t. f. o. the traditional concept of criminal responsibility produces the social behaviour we wish our society to possess* (in conformity with the traditional concept). Even those who choose the metaphor deliberately, and not because it is the first thing that comes into their heads, would be wise to refrain for a time; it has become so trite that the very sight of it may nauseate the sensitive reader. It is sometimes varied by using another word instead of *framework*—usually *context*, *setting*, or *climate*—but this rarely avoids the reproach that it introduces a periphrastic vagueness into what ought to be a plain statement. *It is necessary to think of literature as existing not in isolation but as central to the play of historical and political energies. In the context of Russian literature this might almost be regarded as a truism.* (About Russian literature.) | *In the good old days this could have been treated as an ample margin for concessions to the taxpayer, but no such hopes can be entertained in the present context of affairs* (in present conditions). | *This will have to be reviewed within the whole context of* *the future of broadcasting* (as part of the whole future). | Perhaps we should take it as an encouraging sign that the search for an alternative to *framework* is occasionally pursued to fanciful lengths. *In fact the Prime Minister interferes with senior Ministers less than some of his predecessors. He has a keen regard for the traditional landscape of Cabinet government.*

**Frankenstein.** *I tell you this country may have to pay a long price for Carsonism, and if Toryism returned to power tomorrow the Frankenstein of its own creating will dog its steps.* A sentence written by the creatress of the creator of the creature may save some of those whose acquaintance with all three is indirect from betraying the fact: 'Sometimes I endeavoured to gain from Frankenstein the particulars of his creature's formation; but on this point he was impenetrable ' (*Frankenstein or The Modern Prometheus* by Mary Shelley). *Frankenstein* is the creator-victim; the creature-despot and fatal creation is *Frankenstein's monster.* (*Lord Beveridge was sad, even soured, in later years about the inflationary and inconsistent ways in which others had evolved his Frankenstein's monster—the 'Welfare State'.*) The blunder is very common indeed—almost, but surely not quite, sanctioned by custom: *If they went on strengthening this power they would create a F. they could not resist.* | *In a concentration camp at the edge of Elisabethville there were 40,000 Baluba tribesmen, and nothing could be done about them. They were like Frankensteins.*

**frantic.** 1. *Frantically, franticly.* The first is now standard; *-ically* is almost universal as the adverbial form of adjectives in *-ic*, and there is no gain (as there is with *politicly* and *politically*, where two meanings have to be distinguished) in keeping up two forms.
2. Synonyms are *frenzied, furious, mad, passionate, rabid, raging, raving, wild.* Of these: *frantic* and *frenzied* both mean beside oneself or driven into temporary madness by a cause either

specified or apparent from the context (*frantic with pain, excitement*, etc.; *the frenzied populace refused him a hearing*); in mere exaggerations, e.g. when joy is the cause, *frantic* is the word. *Furious* implies no more than anger that has got out of hand—or, of inanimate things, a degree of force comparable to this. *Passionate* applies primarily to persons capable of strong emotions, especially if they are also incapable of controlling them, and secondarily to the sort of action that results. *Rabid* now usually implies the carrying to great excess of some particular belief or doctrine, religious, political, social, medical, or the like (*a rabid dissenter, tory, teetotaller, faddist; rabid virulence*). *Raging* chiefly describes the violence in inanimate things that seems to correspond to madness in man (cf. *furious; a raging storm, pestilence, toothache*). *Raving* is an intensifying epithet for madness or a madman. The uses of *mad* and *wild* hardly need setting forth.

**free.** 1. *Freeman, free man.* The single word has two senses, (*a*) person who has the 'freedom' of a city etc., and (*b*) person who is not a slave or serf, citizen of a free State; in other senses (*at last I am a free man*, i.e. have retired from business, lost my wife, etc.) the words should be separate.

2. *Free will, free-will, freewill.* The hyphened form should be restricted to the attributive use as in *a free-will offering, the free-will theory.* In non-philosophical use *free will* should be written, and the OED prefers it even for the philosophical term. See HYPHENS.

**French words.** Use and Pronunciation. Display of superior knowledge is as great a vulgarity as display of superior wealth—greater indeed, inasmuch as knowledge should tend more definitely than wealth towards discretion and good manners. That is the guiding principle alike in the using and in the pronouncing of French words in English writing and talk. To use French words that your reader or hearer does not know or does not fully understand, to pronounce them as if you were one of the select few to whom French is second nature when he is not one of those few (and it is ten thousand to one that neither you nor he will be), is inconsiderate and rude.

1. USE OF FRENCH WORDS. It would be a satisfaction to have a table divided into permissible words, forbidden words, and words needing caution; but anyone who starts sanguinely on the making of it is likely to come, after much shifting of words from class to class, to the same conclusion as the writer of this article—that of the thousand or so French words and phrases having some sort of currency in English none can be prohibited, and few can be given unconditional licences; it is all a matter of the need, the audience, and the occasion. Only faddists will engage in alien-hunting and insist on finding native substitutes for words that supply a real need: those for instance for which there are no English synonyms with exactly the same shade of meaning, such as *blasé, chic*, and *naïf*; or those that we have adopted as the standard name for the thing, such as *ballet, coupon, debris*; or those that form part of the language of diplomacy, such as *bloc, démarche, pourparlers*, or those that express compendiously what would otherwise need circumlocution, such as *enfant terrible, tête-à-tête, rendezvous*. Only fools will think it commends them to the English reader to decorate incongruously with such bower-birds' treasures as *au pied de la lettre, à merveille, bien entendu, les convenances, coûte que coûte, quand même, dernier ressort, impayable, jeu de mots, par exemple, robe de chambre, sans doute, tracasseries*, and *sauter aux yeux*; yet even these, even the abominations beginning and ending that list, are in place as supplying local colour or for other special reasons on perhaps five per cent. of the occasions on which they actually appear. Every writer who suspects himself of the bower-bird instinct should make and use his own black list, and remember that acquisitiveness and indiscriminate display are pleasing to contemplate only in birds and savages and children.

2. PRONUNCIATION. To say a French word in the middle of an English sentence exactly as it would be said by a Frenchman in a French sentence is a feat demanding an acrobatic mouth; the muscles have to be suddenly adjusted to a performance of a different nature, and then as suddenly recalled to the normal state. It is a feat that should not be attempted. The greater its success as a *tour de force*, the greater its failure as a step in the conversational progress; for your collocutor, aware that he could not have done it himself, has his attention distracted whether he admires or is humiliated. All that is necessary is a polite acknowledgement of indebtedness to the French language indicated by some approach in some part of the word to the foreign sound, and even this only when the difference between the foreign and the corresponding natural English sound is too marked to escape a dull ear. For instance, in *tête-à-tête* no attempt need or should be made to distinguish French ê from English ā, but the calling it *tā′tahtā′t* instead of the natural English *tātātā′t* rightly stamps it as foreign; again, *tour de force* is better with no unEnglish sound at all; neither r need be trilled, and *tour* and *force* should both be exactly like the English words so spelt. On the other hand, there are some French sounds so obviously alien to the English mouth that words containing them (except such as are, like *coupon*, in daily use by all sorts and conditions of men) should either be eschewed by English speakers or have these sounds adumbrated. They are especially the nasalized vowels (*an, en, in, on, un, am,* etc.), the diphthong *eu*, the unaccented *e*, and *u*; to say *bŏng* for *bon* is as insulting to the French language as to pronounce *bulletin* in correct French is insulting to the man in the English street; and *kŏŏldĕsă′k* for *cul-de-sac* is nearly as bad. Anyone in need of particular guidance will find in an appendix to the COD a list of foreign words used in English with both their anglicized and their foreign pronunciations. The former show how a speaker should pronounce each word in English if he would neither exhibit a conscious superiority of education nor be suspected of boorish ignorance; it is at least as important that those who know the foreign language should mitigate their precision as that those who do not should be enlightened. Broadcasters, conscious of the critical millions listening to them, seem specially apt to be carried away by their anxiety to pronounce impeccably. It is, of course, a tricky business, for we have no system. Sometimes when naturalizing a foreign word we take over its foreign pronunciation (*debris, blasé*); sometimes we anglicize it (*baton, calibre*); sometimes what looks like the same word is given a different pronunciation in different associations (*prē-* for *prima donna* but *prī-* for *prima facie*). The broadcaster, determined to be on the safe side, tends to give a foreign pronunciation to every word that has a foreign look, and so, when taken unawares, may find himself slipping into such absurdities as saying *Capulā* for *Capulets, confĕdong* for *confidant, Blongsh* for *Blanche* and *nēsh* for *niche*. See also DIDACTICISM.

**frequentative.** F. verbs are formed with certain suffixes to express repeated or continuous action of the kind denoted by the simple verb. The chief f. suffixes in English are -le, -er, as in *sparkle, chatter, dribble* (drip). Most of the nouns in *-sation, -tation,* come from Latin frequentatives in *-so, -to,* as *conversation* (L *verto* turn, *versor* move about), *hesitation* (L *haereo* stick, *haesito* keep sticking).

**fresco.** Pl. *-os* or *-oes*; the COD puts them in that order. See -O(E)S.

**Freudian English.** See POPULARIZED TECHNICALITIES.

**friable.** Confusion between the common word meaning crumbly and the -*able* adjective from *to fry* is not likely enough to justify the irregular spelling *fryable* for the latter, though oddly enough the OED's first quotation for *friability* illustrates the possibility: *Codfish for . . . friability of substance is commended.*

**friar, monk.** By the word *f.* is meant a member of one of the mendicant orders, i.e. those living entirely on alms, especially 'the four orders' of Franciscans(grey), Dominicans(black), Carmelites (white), and Augustinians. *M.* is used sometimes of all male members of religious orders including friars, but properly excludes the mendicants. The general distinction is that while the monk belongs essentially to his particular monastery, and his primary object is to make a good man of himself, the friar's sphere of work is outside, and his primary object is to do a good work among the people.

**frith.** See FIRTH.

**fritillary.** The OED prefers the accent on the second *i*; and there it is likely to remain, in spite of the M. Arnold line (*I know what white, what purple fritillaries*); the difficulties of articulation presented by an attempt to stress the first syllable are too great; cf. LABORATORY and GLADIOLUS. See RECESSIVE ACCENT.

**frivol.** See BACK-FORMATION.

**friz(z).** See -Z-, -ZZ-.

**frock** was originally a male garment, especially the mantle of a monk or priest (hence to *unfrock*), then the *smock-frock* that was the overall of an agricultural labourer, and finally the *frock-coat* that was for many years the uniform of the man-about-town. Discarded by men, the word has gained increasing favour with women. It was applied in the 19th c. (at first as a nurseryism) to little girls' dresses; they 'went out of frocks' when they 'put their hair up'. The extension of the word to dresses for grown-ups was described by the OED in 1901 as 'recent'; its progress has been rapid, and a *cotton frock* in particular is seldom called anything else, although the higher 'creations' of the dressmaker's art remain *dresses* and *gowns*.

**frontage, frontal, frontier, frontispiece.** It seems best to make the o in all these conform to that in *front* (ŭ, not ŏ). The OED separates *frontier*, in which it prefers ŏ, from the rest. But usage has not conformed; the ordinary pronunciation is now *frŭn'tyĕr* in Britain and *frŭntēr'* in U.S.

**froth.** See FOAM.

**fruition** is not a synonym of *fructification*, though both are derived from the same Latin word fruor, *fruition* from its sense of enjoying and *fructification* from its sense of bearing fruit. *Fruition*, often wrongly supposed to be associated with the English word *fruit*, is the enjoyment that comes from the fructification of hope, especially from possession.

**fryable, fryer.** See FRIABLE and DRY.

**fugacious.** Chiefly in PEDANTIC HUMOUR. Cf. *sequacious*.

**fugue** makes *fugal*, *fuguist*.

**-ful.** The right plural for such nouns as *handful, spoonful, cupful, basketful*, is *handfuls* etc., not *handsful* etc. See PLURAL ANOMALIES.

**fuliginous.** Chiefly in POLYSYLLABIC HUMOUR. *At present it is a f., not to say mysterious, matter.*

**full** for *fully*, meaning quite or completely, is idiomatic in such phrases as *f. twenty miles* (cf. *full fathom five*), *f. grown, f. blown*, and sometimes with the sense of quite sufficiently or rather too, such as *f. early, f. ripe*. In the sense *very*, as in *f. fain, f. many a, f. weary*, where *fully* cannot be substituted, it is poetical.

**ful(l)ness.** Use -*ll*-; see DULLNESS, and -LL-, -L-, 4.

**full stop.** See STOPS; and PERIOD IN ABBREVIATIONS.

**fulsome.** Though the OED recognizes only the pronunciation *fŭl-, fool* is now general.

**function. 1.** That such and such a thing 'is a function of' such another or such others is a POPULARIZED TECHNICALITY: *A man's fortitude under given painful conditions is a f. of two variables.* As not everyone can cope unaided with mathematical technicalities, the following may be useful: 'When one quantity depends upon another or upon a system of others, so that it assumes a definite value when a system of definite values is given to the others, it is called a function of those others.' Knowledge of the mathematical meaning may help writers to avoid using it merely as a showy substitute for *depends on*.

2. *Function* has also the wider sense of the kind of activity proper to a person or thing, the way in which his or its purpose is fulfilled. The f. of a policeman is to preserve law and order; the f. of a clock is to tell the time. As a noun with this meaning *f.* is old; as a verb it is comparatively new and is having its full share of the popularity that novelty brings. We got on quite well without it for a long time and we should be the better if it could have a close season; then we might remember the old words, and say that the clock is *going*, the buses are *running*, the shops are *open*, the heating system is *working*, the medicine is *acting*, the heart is *beating*, the markets are *operating*, and so on.

3. The noun meaning a social or ceremonial occasion (originally a ceremony in the Roman Catholic Church) is suitable only for gatherings of importance conducted with ceremony; for ordinary social occasions *party* is the right word

**fundamental.** See BASAL, BASIC for the choice between these words.

**funebrial, funeral** (adj.), **funerary, funereal.** The continued existence of the first and third words, which no one uses if he can help it, is due to what has happened to the other two. *Funeral*, though originally an adjective, has so far passed into a noun that it can no longer be used as an adjective except in the attributive position, as in *funeral customs, the funeral procession*; *funereal* has become so tied to the meaning of of a funeral, gloomy enough for a funeral, that it can no longer be used to mean simply of or for a funeral. In such a sentence as *The origin of the custom is* ——, it only remains, if an adjective must be found, to choose

between *funebrial* and *funerary*, of which the first is so rare as to be pedantic and the second is generally associated with *urn*.

**fungus.** Pl. *fungi* (pronounced *-jī*) or, less commonly, *funguses*.

**furiously.** Some British journalists find it so amusing that the Frenchman should say *penser furieusement* where we say *think hard*, and *donner furieusement à penser* for *puzzle*, that they bore us intolerably with their discovery. *Ça donne furieusement à penser* is quoted, translated, paraphrased, and alluded to, till we are all heartily sick of it; see GALLICISMS. *That word 'although' caused us f. to think, but when we come to read the leading article in* The Times *we fancy that we get a clue to what may be meant.* / *That sentence of Professor Dicey's makes one think f.* / *The reduction in the majority from 6,000 to 1,400 has given many Coalition members f. to think.*

**furore.** Three syllables (*fūror'ĕ*).

**furry.** See PRONUNCIATION 7.

**further,** adj. and adv. See FARTHER.

**furze, gorse, whin.** The first two would appear to be that very great rarity, a pair of exact synonyms, meaning the same thing and used indifferently in all localities and all contexts. The third differs not in sense, but in being chiefly in use in Scotland, Ireland, and the North of England.

**fuse.** 1. The verb makes *fusible*; see -ABLE 2.

2. The derivations of the *fuse* (electrical) and the *fuse* (explosive) are different. The former is the verb *fuse* (from Lat. *fundere*), to melt by heat, used as a noun. The latter is so named solely from its shape (Lat. *fusus* a spindle); see TRUE AND FALSE ETYMOLOGY.

**fused participle** is a name given to the construction exemplified in its simplest form by 'I like *you pleading* poverty', and in its higher development by 'The collision was owing to *the signalling instructions* laid down by the international regulations for use by ships at anchor in a fog *not having been* *properly followed'*. The name was invented (*The King's English*, 1906) for the purpose of labelling and so making recognizable and avoidable a usage considered by the authors of that book to be rapidly corrupting modern English style.* A comparison of three sentences will show the meaning of the term.

1. Women having the vote share political power with men.
2. Women's having the vote reduces men's political power.
3. Women having the vote reduces men's political power.

In the first, the subject of the sentence is *women*, and *having* (*the vote*) is a true participle attached to women. In the second, the subject is the verbal noun or gerund *having* (*the vote*), and *women's* is a possessive case (i.e. an adjective) attached to that noun. The grammar in these two is normal. In the third, the subject is neither *women* (since *reduces* is singular), nor *having* (for if so, *women* would be left in the air without grammatical construction), but a compound notion formed by fusion of the noun *women* with the participle *having*. Participles so constructed, then, are called fused participles, as opposed to the true participle of No. 1 and the gerund of No. 2.

We are given to ridiculing the cumbrousness of German style, and the particular element in this that attracts most attention is the device by which a long expression is placed between a noun and its article and so, as it were, bracketed and held together. Where we might allow ourselves to say *This never to be forgotten occasion*, the German will not shrink from *The since 1914 owing to the world-war befallen destruction of capital*; only a German, we assure ourselves, could be guilty of such ponderousness. But the fused participle is having exactly the same effect on English as the article-and-noun sandwich on German, the only difference being that the German device is grammatically sound, while the English is indefensible. The examples that follow, in which the two members of each fused participle are in roman

type, all exhibit both the bracketing capacity that makes this construction fatally tempting to the lazy writer, and its repulsiveness to a reader who likes clean sentences. In the last two may be observed a special fault often attending the fused participle—that the reader is trapped into supposing the construction complete when the noun is reached, and afterwards has to go back and get things straight.

*No one is better qualified than Mr. Charles Whibley to write the biography of W. E. Henley; and there is some likelihood of the* life-story *of that influential and strenuous littérateur from his hand* appearing *before the close of the year.* | *The machinery which enables one man to do the work of six results only in the others losing their job, and in* skill *men have spent a lifetime acquiring* becoming *suddenly useless.* | *Regulations for permitting* workmen *who are employed under the same employer, partly in an insured trade and partly not in an insured trade,* being treated . . . *as if they were wholly employed in an insured trade.* | *A dangerous operation, in which everything depends upon* the General Election, *which is an essential part of the operation,* being won. | *We have to account for* the collision *of two great fleets, so equal in material strength that the issue was thought doubtful by many careful statisticians,* ending *in the total destruction of one of them* . . . .

It need hardly be said that writers with any sense of style do not, even if they allow themselves the fused participle, make so bad a use of the bad thing as is shown above to be possible. But the tendency of the construction is towards that sort of cumbrousness, and the rapidity with which it is gaining ground is portentous. A dozen years ago, it was reasonable, and possible without much fear of offending reputable writers, to describe as an 'ignorant vulgarism' the most elementary form of the fused participle, i.e. that in which the noun part is a single word, and that a pronoun or proper name; it was not very easy to collect instances of it. Today, no one who wishes to keep a whole skin will venture on so frank a description. Here are some examples, culled without any difficulty whatever from the columns of a single newspaper, which would be very justly indignant if it were hinted that it had more vulgarisms than its contemporaries. Each, it will be seen, has a different pronoun or name, a sufficient proof in itself of abundant material. *We need fear nothing from* China *developing her resources* (China's)./ *Which will result in* many *having to go into lodgings* (many's). / *It should result in* us *securing the best aeroplane for military purposes* (our). / *It is no longer thought to be the proper scientific attitude to deny the possibility of* anything *happening* (anything's). / *They wish to achieve this result without* it *being necessary to draw up a new naval programme* (its). / *I insisted on* him *at once taking the bill down* (his). / *The reasons which have led to* them *being given appointments in these departments* (their). / *He is prepared to waive this prohibition upon* you *giving him a written undertaking as follows* (your).

It is perhaps beyond hope for a generation that regards *upon you giving* as normal English to recover its hold upon the truth that grammar matters. Yet every just man who will abstain from the fused participle (as most good writers in fact do, though negative evidence is naturally hard to procure) retards the progress of corruption; and it may therefore be worth while to take up again the statement made above, that the construction is grammatically indefensible. At the first blush everyone probably grants this; it is obvious, in any sentence so made as to afford a test (e.g. *Women having the vote reduces men's power*), that the words defy grammatical analysis. But second thoughts bring the comforting notion that the fusion must after all be legitimate; it is only our old friend *occisus Caesar effecit ut*, which means not *Caesar when killed*, but *The killing of Caesar, had such and such results*; why should not *Women having* mean *The possession by women of*, if *occisus Caesar* can mean *The killing of Caesar?* The answer is that the Romans did resort

to sense-fusion, but did not combine it with grammar-confusion; *The deaths of the Caesars had such effects* is *occisi Caesares effecerunt* (not *effecit*); but the fused-participlists say *Women having the vote reduces* (not *reduce*), and *You saying you are sorry alters* (not *alter*) *the case.* The Latin parallel is therefore of no value, and with it goes the only palliation of the bad grammar.

And now, in order that the reader may leave this disquisition sick to death, as he should be, of the fused participle, a few miscellaneous specimens are offered: *We cannot reckon on* the unrest ceasing *with the end of one strike, or on* its not being renewed *in the case of other trades* (Compare *unrest* with *its*). / *It may be that this is part of the meaning and instinctive motive of* fish *such as the perch, going in shoals at all.* / *Developments have occurred in consequence of the action of* one *of the accused, a man 31 years of age, and an ex-student of several colleges,* having turned *approver.* / *The holiday habit is growing upon us, possibly owing partly to the persistent and recurrent habit of* Christmas Day falling *at the week-end.*/ *This habit of* Ministers putting forth *their ideas through newspaper articles sometimes produced curious results.* / *Some similar scheme can be introduced without* the school *doing so* suffering *pecuniary loss.* / *Good criticism combines the subtle pleasure in* a thing being well done *with the simple pleasure in* it being done at all. / *There is a big enough area for the speed men even in the narrow limits of these isles, without* them *making the exquisite little corner of English lakeland the special field for their trials.* / *The truth of the old saw about* being a better thing *to wear out than to rust out.* / *The same objections apply to* the patient telling *the head attendant as to his telling the medical officer* (compare *patient* with *his*).

*Note.* The foregoing article is reproduced just as Fowler wrote it, except that some illustrations have been omitted and others shortened. It provoked some controversy.

Jespersen (SPE Tract XXV) vigorously defended the construction condemned by Fowler. He gave numerous examples of its use by famous authors from Swift to Shaw; he made light of the argument that it defied grammatical analysis, and maintained that it represented 'the last step in a long line of development, the earlier steps of which ... have for centuries been accepted by everybody. Each step, including the last, has tended in the same direction, to provide the English language with a means of subordinating ideas which is often convenient and supple where clauses would be unidiomatic or negligible.' Fowler in his rejoinder (SPE Tract XXVI) admitted that he had underestimated the extent of its use, but was otherwise unshaken. 'I confess', he said, 'to attaching more importance to my instinctive repugnance for *without you being* than to Professor Jespersen's demonstration that it has been said by more respectable authors than I had supposed.'

Thirty years later the dust had still not settled. During the passage through the House of Lords of the Homicide Act, 1957, a noble lord, who must have been a disciple of Fowler, moved to purge the Bill of a fused participle by substituting *other's* for *other* in the clause providing that it should no longer be murder for one party to a suicide pact to be 'a party to the other killing himself'. He was unsuccessful. But Fowler would have been unlikely to accept even the House of Lords as a final court of appeal on such a point.

It is clear that Fowler was right in deprecating the use of the fused participle with a proper name or personal pronoun in a simple sentence: *upon your giving* is undoubtedly more idiomatic than *upon you giving*. But it is by no means so clear that when a more complicated sentence makes a possessive impossible we must deny ourselves the convenience of writing *We have to account for the collision of two great fleets . . . ending in the total destruction of one of them . . . .* Or even when a possessive, though not impossible, is ungainly. 'If this rule were pressed, we should have to say: "His

premature death prevented *anything*'s *coming* of the scheme", which can hardly be called English' (Onions).

# G

**gag.** See CLOSURE for the parliamentary sense.

**gainsay** is a LITERARY WORD, and now little used except in negative contexts such as *There is no gainsaying it*, *Without fear of being gainsaid*, *That can scarcely be gainsaid*.

**gala.** The pronunciation *gah-* is ousting the *gā-* still preferred by most dictionaries.

**gallant.** The ordinary pronunciation is *gă'lănt*. Certain senses, 'politely attentive to women', 'amorous', 'amatory', are traditionally distinguished by the pronunciation *gălă'nt*; but these senses, and still more the special accent, are perhaps moribund. In its ordinary sense too the word is becoming old-fashioned and is rare except in conventional uses such as *The honourable and gallant member* and in citations of acts of bravery. See WORSENED WORDS.

**gallery, galley.** *Que diable allait-il faire dans cette galère?* is a famous line, and so often applicable that it is often applied. It is not possible for anyone who has seen it in its original place to be unaware that *galère* means galley; and therefore to put it, or an allusion to it, into English with *gallery* betrays infallibly the jackdaw with borrowed plumes. To write *galerie* (*Mr. M.*, *who has at least escaped being mixed up in that galerie*) is to say 'Yes, I know the French', and so to add the sin of lying to the peccadillo of pretension. But then, whether one is caught out with *gallery* or *galerie*, one can always explain 'It was the printer; I wrote *galley*, or *galère*'. See GALLICISMS, and FOREIGN DANGER.

**Gallic, Gallican, Gaulish, French.** *Gallican* is a purely ecclesiastical word, corresponding to *Anglican*. *Gaulish* means only 'of the (ancient) Gauls', and is, even in that sense, less usual than *Gallic*. The normal meaning of *Gallic* is the same as that of *Gaulish*, but it is also much used as a synonym in some contexts for *French*. It means not simply 'French', but 'characteristically', 'delightfully', 'distressingly', or 'amusingly', 'French'—'so French, you know', etc.—or again not 'of France', but 'of the typical Frenchman'. We do not, or should not, speak of Gallic wines or trade or law or climate, but we do of Gallic wit, morals, politeness, and shrugs; and the symbolic bird is invariably the Gallic cock. So far as *Gallic* is used for *French* without any implication of the kind suggested, it is merely a bad piece of ELEGANT VARIATION or AVOIDANCE OF THE OBVIOUS.

**Gallicisms.** By *Gallicisms* are here meant borrowings of various kinds from French in which the borrower stops short of using French words without disguise.

1. One form consists in taking a French word and giving it an English termination or dropping an accent or the like, as in *actuality* and *redaction*.

2. Another in giving to an existent English word a sense that belongs to it only in French or to its French form only, as in *assist* (be present at), *impayable* (= priceless for absurdity, impudence, etc.), *arrive* (= attain success etc.), *exposition* (= exhibition), and *actual* (= concerned with the present, topical, *The most actual and instructive article is on broadcasting*).

3. Another in giving vogue to a word that has had little currency in English but is common in French, such as *veritable* and *envisage*.

4. Another in substituting a French form or word that happens to be English also, but in another sense, for the really corresponding English, as when *brave* is used for *honest* or *worthy*, or *ascension* for *ascent*.

5. Another in literally translating a French word or phrase, as in *jump* or *leap to the eyes*, *to the foot of the letter*, *give furiously to think*, *knight of industry*,

*daughter of joy*, *gilded youth*, *the half-world*, *colour of rose*, *do* one's *possible*, *to return to our muttons*, *suspicion* (= soupçon), and *success of esteem*.

To advise the abandonment of all Gallicisms indiscriminately would be absurd. There are thousands of English words and phrases that were once Gallicisms, but, having prospered, are no longer recognizable as such; and of the number now on trial some will doubtless prosper in like manner. What the wise man does is to recognize that the conversational usage of educated people in general, not his predilections or a literary fashion of the moment, is the naturalizing authority, and he will therefore adopt a Gallicism only when he is of opinion that it is a Gallicism no more. To use Gallicisms for the worst of all reasons—that they *are* Gallicisms —to affect them as giving one's writing a literary air, to enliven one's dull stuff with their accidental oddities, above all to choose Gallicisms that presuppose the reader's acquaintance with the French original: these are confessions of weakness or incompetence. If writers knew how 'leap to the eye' does leap to the eye of the reader who, in dread of meeting it, casts a precautionary glance down the column, or how furious is the thinking that 'give furiously to think' stirs in the average Englishman, they would leave such paltry borrowings alone for ever.

Some of the Gallicisms here mentioned, as well as others, are commented upon in their dictionary places. Words and phrases for which the reader is simply referred without comment to this article are to be regarded as undesirable Gallicisms. See also FRENCH WORDS.

**gallop** makes *-oped*, *-oping*; see -P-. -PP-; so does *galop*, the dance, used as a verb.

**gallows,** though originally a plural form, is now singular (*set up a g.* etc.); the plural is usually avoided, but when unavoidable is *gallowses*.

**Gallup poll.** See POLLSTER.

**galore,** an Irish or Gaelic word, and

no part of the Englishman's natural vocabulary, except as a jocular colloquialism (*whisky galore!*), is chiefly resorted to by those who are reduced to relieving dullness of matter by oddity of expression.

**galumph.** See FACETIOUS FORMATIONS.

**gambit** is a type of chess opening in which a player sacrifices a piece or pawn in the hope of greater gain later, and, by reasonable extension, the first move, especially with an implication of cunning, in any contest or negotiation. To use the word merely as a showy synonym of *opportunity* is an example of that SLIPSHOD EXTENSION that almost always goes with POPULARIZED TECHNICALITIES. *German firms are being attracted to Canada by such obvious gambits as the huge growth of Toronto.*

**gamesmanship.** See BRINKMANSHIP.

**gamut.** For synonyms, in the extended sense, see FIELD.

**gang agley** is a BATTERED ORNAMENT.

**gangway.** *Below the g.*, as a parliamentary phrase, used to be applied to members whose customary seat did not imply close association with the official policy of the party on whose side of the House they sat, and to some extent this implication survives: it is still customary for a minister who resigns to take the end seat of the fourth row below the gangway. See also AISLE.

**gaol, gaoler, jail, jailor,** etc. 'In British official use the forms with G are still current; in literary and journalistic use both the G and the J forms are now admitted as correct, but all recent dictionaries give the preference to the latter.'—OED. In the British prison service both terms are now obsolete for the agent noun. Their disappearance is an example of the working of EUPHEMISM. *Gaoler* was superseded by *warder*, and *warder* in turn by *prison officer*. It may be added that the very anomalous pronunciation of g soft before a or o (only in MORTGAGOR and in

the popular pronunciation of MARGARINE?) is a strong argument for writing *jail*.

**garage,** like many other French words in constant necessary use (e.g. *billet-doux, bulletin, cadre, chaperon, commissionaire, cordon, coupon, liqueur, restaurant, valet*), might well be completely anglicized in pronunciation (*gă′rĭj*) and is in fact often so spoken. But the compromise *gă′rahzh* or *gărahj′* is more usual.

**garble.** The original meaning is to sift, to sort into sizes with a view to using the best or rejecting the worst. The modern transferred sense is to subject a set of facts, evidence, a report, a speech, etc., to such a process of sifting as results in presenting all of it that supports the impression one wishes to give of it and deliberately omitting all that makes against or qualifies this. Garbling stops short of falsification and misquotation, but not of misrepresentation; a garbled account is partial in both senses. To use *garbled* in the sense of inaccurate or confused without any element of the tendentious is a SLIPSHOD EXTENSION.

**garden.** For *the G.* in philosophy, see ACADEMY.

**garret, attic.** The two words mean the same thing, but the former is usually chosen when poverty, squalor, etc., are to be suggested.

**gar(r)otte.** The right spelling is *garrotte.*

**gas.** See FLUID. As an abbreviation of *gasoline* (U.S. for petrol) it has gained little foothold in Britain except in the jocular colloquialism *Step on the gas.*

**gaseous.** 1. The pronunciations recognized by the OED are *gă′sĭŭs, gā′sĭŭs,* in that order of preference, but *gā-* is now commoner.
2. *gaseous, gassy.* The first prevails in scientific use; the further the substitution of *gassy* for it can be carried, the better.

**Gaulish.** See GALLIC.

**geezer,** i.e. queer character (usually *old g.*) was originally *guiser* or mummer. It has no connexion with GEYSER.

**gelatin(e).** The form without final -e is only in scientific (or pseudo-scientific) use in Britain, though standard U.S. The OED gives only the pronunciation *-ĭn*, but *-ēn* is now commoner. See -IN AND -INE.

**gemma.** Pl. *-ae.*

**gender,** n., is a grammatical term only. To talk of *persons* or *creatures of the masculine* or *feminine g.*, meaning *of the male* or *female sex*, is either a jocularity (permissible or not according to context) or a blunder.

**genealogy.** See -LOGY.

**general.** For the use of hyphens in such compounds as *Attorney G.*, *Lieutenant G.*, see HYPHENS 2, and for the plurals see PLURAL ANOMALIES.

**generalissimo.** Pl. *-os*; see -O(E)S 7.

**generic names and other allusive commonplaces.**
    When Shylock hailed Portia as *A Daniel come to judgement*, he was using a generic name in the sense here intended; the History of Susanna was in his mind. We do the same when we talk of a *Croesus* or a *Jehu* or a *Hebe* or a *Nimrod* or of *Bruin*, *Chanticleer*, and *Reynard*. When we talk of *a Barmecide feast*, of *Ithuriel's spear*, of *a Naboth's vineyard*, of *being between Scylla and Charybdis*, of *Procrustean regulations*, or *Draconian severity*, or an *Achilles' heel*, we are using allusive commonplaces. Some writers revel in such expressions, some eschew them of set purpose, some are ill provided with them from lack of reading or imagination; some esteem them as decorations, others as aids to brevity. They are in fact an immense addition to the resources of speech, but they ask to be employed with discretion. This article is not intended either to encourage or to deprecate their use; they are often in place, and often out of place; fitness is all. An allusion that strikes a light in one

company will only darken counsel in another; most audiences are acquainted with the qualities of *a Samson*, fewer with those of *a Dominie Sampson* and fewer still with those of *Ithuriel's spear*. Nevertheless, to some audience or other each of these may well be, apart from any decorative value attaching to it, the most succinct and intelligible name for what is meant. It is for the writer to see that he does not try Ithuriel's spear on those whose knowledge stops short at Samson; for if the test reveals them as ignoramuses they will not like it, or him.
    It is perhaps worth while to call attention to a practical difference between the useful and the decorative allusions. When an allusive term is chosen because it best or most briefly conveys the meaning, triteness is no objection to it; intelligibility is the main point. But the choice for decorative purposes is a much more delicate matter; you must still be intelligible, but you must not be trite. The margin in any audience between what it has never heard of and what it is tired of hearing of is rather narrow; it is necessary to hit it between wind and water.
    These few remarks may suffice on the unanswerable question whether allusive terms should be sought or avoided. The purpose of this article is not to answer it, but to point out that, if they are used, it is inexcusable and suicidal to use them incorrectly; the reader who detects his writer in a blunder instantly passes from the respect that beseems him to contempt for this fellow who after all knows no more than himself. It is obvious that the domain of allusion is full of traps, particularly for the decorative allusionist, who is apt to take the unknown for the fine, and to think that what has just impressed him because he knows little about it may be trusted to impress his readers. For an example or two see the articles BENEDICK, FRANKENSTEIN, DEVIL'S ADVOCATE, IRRELEVANT ALLUSION, and MISAPPREHENSIONS.

**-genic** is a suffix used by scientists to form adjectives with the meaning 'of, pertaining to, or relating to, generation or production' (OED Supp.). Few of these are in ordinary use; *pathogenic* (causing disease) is probably the best known. Now that pseudo-scientific jargon is all the rage, however (see POPULARIZED TECHNICALITIES), we may soon have a crop of new -*genic* words, following the lead of the already common use of *photogenic* (a word of long standing in the sense 'productive of light') as an attribute of one who, in more homely but more suitable phrase, 'takes well'. If people must coin these words, we might at least ask that they should be properly made; we should not be invited to swallow such a BAR-BARISM as *crimogenic* by an author who, since he does write *criminology* and not *crimology*, ought to have known better.

**genie.** Pronounce *jē′nĭ* pl. *genii* pron. *jē′nīī*; see LATIN PLURALS. Another form is *jinnee*, pl. *jinn* (often used also as singular).

**genius.** Pl. -*uses*; the form *genii* is now used only as pl. of *genie* (or of *genius* in the sense of *genie*); see LATIN PLURALS. For g. and talent, see TALENT.

**gent** (= gentleman). Apart from commercial jargon (*gents' underwear* etc.) and the colloquial EUPHEMISM *The Gents*, this abbreviation is now only used as a jocular term of praise in the phrase *a perfect gent*.

**genteel** is now used, except by the ignorant, only in mockery. A WOR-SENED WORD.

**genteelism.** By *genteelism* is here to be understood the rejecting of the ordinary natural word that first suggests itself to the mind, and the substitution of a synonym that is thought to be less soiled by the lips of the common herd, less familiar, less plebeian, less vulgar, less improper, less apt to come unhandsomely betwixt the wind and our nobility. The truly genteel do not *ask* but *enquire*, invite one to *step*, not to *come* this way, may detect an *unpleasant odour* but not a *nasty smell*, never *help*, but *assist*

each other to potatoes, of which they may have *sufficient*, but never *enough*, do not *go to bed* but *retire for the night*: and have quite forgotten that they could ever have been guilty of *sitting-room*, *napkin*, and *dirty clothes* where nothing will now do for them but *lounge*, *serviette*, and *soiled linen*.

The reader need hardly be warned that the inclusion of any particular word in the small selection of genteel-isms offered below does not imply that that word should never be used. All or most of these, and of the hundreds that might be classed with them, have their proper uses, in which they are not gen-teel, but natural. *Lounge* is at home in hotels, *step* in the dancing class, *retire* for the superannuated man, and so forth; but out of such contexts, and in the conditions explained above, the taint of gentility is on them. To illustrate a little more in detail, 'He went out without shutting the door' is plain English; with *closing* substituted for *shutting* it becomes genteel; nevertheless, to *close* the door is justified if more is implied than the mere not leaving it open: 'Before beginning his story, he crossed the room and closed the door', i.e. so placed it as to prevent eavesdropping; 'Six people sleeping in a small room with closed windows', i.e. excluding air. Or again, 'The schoolroom roof fell in, and two of the boys (or girls, or children) were badly injured'; *scholars* for boys etc. would be a genteelism, and a much more flagrant one than *closing* in the previous example; yet *scholar* is not an obsolete or archaic word; it is no longer the natural English for a schoolboy or schoolgirl, that is all. The point is that, when the word in the second column of the list that follows is the word of one's thought, one should not consent to displace it by the word in the first column unless an improvement in the meaning would result. The reader will easily increase the list for himself; he may also be disposed to amend it by omission. Genteelisms, if they attain a wide enough currency, soon rub their taint off by use.

# gentle

223

**gentlewoman**

| Genteelisms | Normal words |
|---|---|
| assist | help |
| bosom | breast |
| close | shut |
| couch | sofa |
| enquire | ask |
| dentifrice | tooth-powder |
| desire | want |
| expectorate | spit |
| hard of hearing | deaf |
| help | servant |
| lady dog | bitch |
| lingerie | underclothes |
| lounge | sitting room |
| odour | smell |
| paying guest | lodger |
| perspire | sweat |
| require | want |
| retire | go to bed |
| scholar | schoolboy, -girl |
| serviette | table napkin |
| soiled linen | dirty clothes |
| step, vb. | come, go |
| sufficient | enough |

See also EUPHEMISM, FORMAL WORDS, and WORKING AND STYLISH WORDS.

**gentle. 1.** *The gentle art.* This phrase, long a favourite with anglers as an affectionate description of their pursuit, was cleverly used by Whistler in his title *The Gentle Art of Making Enemies.* The oxymoron was what made it effective; but imitators, aware that Whistler made a hit with *the gentle art,* and failing to see how he did it, have now, by rough handling on inappropriate occasions, reduced it to a BATTERED ORNAMENT (cf. IRRELEVANT ALLUSION). Thus: *We have not the smallest doubt that this is what will actually happen, and without any undue exercise in* the gentle art of intelligent anticipation, *we may discuss the situation.* / *In a Committee* the gentle art *of* procrastinating *may prove very deadly to progress.*
**2.** *Gentle* as what the OED calls a form of polite or conciliatory address (*Have patience, gentle friends*) lingered in writers' apostrophes to their *gentle readers* after it had disappeared from general use. Victorian novelists, especially Thackeray, were much given to it. Authors have now invented other ways of trying to create a sense of intimacy between themselves and their readers, and if the gentle reader is now invoked it will only be by way of a jocular archaism.

**gentleman.** Our use of *g.,* like that of ESQUIRE, is being affected by our progress towards a classless society, but in the opposite way: we are all esquires now, and we are none of us gentlemen any more. The word has remained in the vocative plural for those addressing a male audience or writing a formal letter, as a title of a courtly office (*g. in waiting, g. usher,* etc.), as a distinguishing sign on a public convenience (though here *Men* seems to be displacing it) and, until recently, as an anachronism in the cricket match *Gentlemen* v. *Players;* elsewhere in sport those who used to be called *gg.* to distinguish them from professionals have long been given the more suitable designation *amateur.* But as merely a word for an adult of the male sex g. has become taboo. A girl may say 'He's ever such a nice *man* or *chap* or *fellow* (though the last is becoming dated), but not, as she would once have done, *gentleman,* unless his age warrants the addition of *old.*

**gentlemanly, gentlemanlike.** If the ugly *-like* form were understood to suggest, while the other did not, a warning that all is not gold that glitters, there would be sufficient justification for their coexistence; but the OED quotations do not bear out, nor does the OED emphasize, such a distinction. It seems right, then, that *-like* should be fading away. But since we cannot say *ladyly, ladylike* must be the corresponding adjective for the other sex. See SUPERFLUOUS WORDS.

**gentlewoman, lady.** The first has no sense that does not belong to the second also, but *l.* has half a dozen for which *g.* will not serve—the Virgin, titled woman, wife, an alternative to *madam* as a mark of respect (*Thank you, Lady*), a woman or girl described politely or sometimes as a genteelism or jocularly (*a perfect lady*), in the vocative plural (*Ladies and Gentlemen*) and in numerous compounds—Lady in Waiting, Lady Mayoress, etc. It follows that in the one sense common to both (fem. of *gentleman*, i.e. woman of good birth and breeding, or woman of honourable instincts) *g.* is sometimes preferred as free of ambiguity or as more significant. It is, however, an old-fashioned if not quite archaic word, and as such tends to be degraded by facetious use, and to have associated with it stock epithets, of which some are derisive or patronizing (*ancient, decayed,* and *distressed*) and others are resorted to as protests against such use (*true, Nature's,* etc.). It is therefore to be used with caution.

**genuine.** See AUTHENTIC.

**genus.** Pronounce *jē-*; pl. *genera,* pron. *jĕn-*; see LATIN PLURALS and -US.

**geo-.** A warning is not superfluous against the slovenly pronunciation *jog-* in *geography* and *jom-* in *geometry. Geology* does not seem to offer the same temptation.

**geographic(al).** The short form 'now somewhat rare except in *Geographic latitude*'—OED. See -IC(AL). It is now still rarer.

**geometric(al).** 1. The long form prevails, and there is no difference in meaning; see -IC(AL). 2. *G. progression.* For the misuse of this, see PROGRESSION.

**germ.** See MICRO-ORGANISM.

**German.** High and Low G. High G. is the language known ordinarily as *German; Low G.* is a comprehensive name for English, Dutch, Frisian, Flemish, and some G. dialects. The words *High* and *Low* are merely geographical, referring to the Southern or mountainous, and the Northern or low-lying, regions in which the two varieties developed.

**gerrymander.** Strictly the *g* should be hard; the word derives from the electoral manipulations of Governor Gerry of Massachusetts. But the erroneous pronunciation *j* is now universal, and the word is sometimes even spelt so, perhaps from a false analogy with *jerrybuilding* etc.

**gerund.** 1. G. and gerundive. 2. G. and participle. 3. G. and infinitive. 4. G. and possessive.

1. Gerund, gerundive. The second word is of importance only with regard to the languages that possess the thing, of which English does not happen to be one. But since its occasional use for the other word *gerund,* which *is* of importance in English grammar, may cause confusion, the difference between the Latin gerund and gerundive should be explained. The gerund is a noun supplying a verb's infinitive or noun-form with cases; thus *amare* to love has the gerund *amandi* of loving, *amando* by loving, *amandum* the act of loving; correspondingly the word *loving* as a noun (but not as an adjective) is the gerund in English, though it is of the same form as the participle. From the same stem as *amandi* etc. is formed in Latin an adjective *amandus* lovable, and this in Latin grammar is named the gerundive as being formed from the gerund. The English adjectives formed in *-ble* from verbs, like *lovable,* might well enough be called gerundives from their similarity in sense to the Latin gerundive; but they are not in practice so called, and the word *gerundive* has accordingly no proper function in English grammar.

2. Gerund and participle. The English gerund is identical in form, but only in form, with the active participle; *loving* is a gerund in 'cannot help loving him', but a participle in 'a loving husband'. A grammarian quoted by the OED says 'Gerundives' [by which he means gerunds] 'are participles governed by prepositions; but, there being little or no occasion to distinguish them from other participles, we seldom use this name'. The distinction is, on the contrary, of great importance, and the occasion for making it constantly occurs. In the article FUSED PARTICIPLE an attempt is made to show the fatal effects on style of disregarding it.

**3.** Gerund and infinitive. Among the lapses that are concerned not with particular words but with a whole class of phrases, and that, without being describable as grammatical blunders, reveal a writer's ignorance of idiom, few are more insidious than failure to recognize when the gerund with a preposition is required rather than an infinitive. *I look forward to meet,* or *to meeting, him? I aim to remove,* or *at removing, the cause? The duty is laid on us to do,* or *of doing, our best?*

The variety of cases in which the question arises is so vast, and the rules that should answer it would be so many and need so many exceptions, that it is better not to try to formulate any. Three general remarks may suffice instead, to be followed by some specimens. A. There is very little danger of using the gerund, but much of using the infinitive, where the other would be better. B. Lapses are usually due not to deliberate choice of the worse, but to failure to think of the better. C. The use of the infinitive is often accounted for, but not justified, by the influence of ANALOGY; because *able,* or *sufficient,* or *adequate, to perform* is English, we assume that *equal to perform,* which is to bear the same meaning, must be English too. In the specimens, where analogy seems to have been at work, the analogous word is suggested in the correction bracket.

Specimens after nouns

*But they have been blocked by* the objections *of farmers and landlords* to provide *suitable land, and by the reluctance of local authorities to use their powers of compulsion* (to providing. Hesitation, refusal. Observe that the infinitive after *reluctance* is quite idiomatic). / *I refer to the growing* habit *of a few hooligans* to annoy and assault *those who* . . . (of annoying. Tendency)./ *They have been selected with a* view to illustrate *both the thought and action of the writer's life* (illustrating. So as or in order to). / *Russia assures us that she has no* intention to encroach *upon it* (of encroaching. But idiom after *intention* is less fixed than after most such nouns). /

*You have likened the* resistance *of Ulster Unionists* to be driven *out of the Constitution of Great Britain to the economic opposition of a number of scattered citizens to a reform of the tariff* (to being driven. Refusal, reluctance).

Specimens after adjectives

*A simplicity that seems quite* unequal to treat *the large questions involved* (to treating. Incompetent). / *The navy is not* equal *in numbers or in strength* to perform *the task* (to performing. Sufficient).

Specimens after verbs

*He* confesses to have seen *little of the great poets of his time* (to having. Profess). / *The cab-drivers* object to pay *their proportion of the increase* (to paying. Refuse). / *All the traditions in which she has been brought up* have not succeeded to keep *her back* (in keeping. Availed). / *Mr. Lloyd is committed to* introduce *a new tax on capital gains* (to introducing. Pledged).

**4.** Gerund and possessive. The gerund is variously describable as an *-ing* noun, or a verbal noun, or a verb equipped for noun-work, or the name of an action. Being the name of an action, it involves the notion of an agent just as the verb itself does. *He went* is equipped for noun-work by being changed to *his going,* in which *his* does for *going* the same service as *he* for *goes,* i.e. specifies the agent. With the verb the agent is usually specified, but not always; it is seldom, e.g., used with the imperative (*go,* not *go you* or *you go*) because to specify the agent would be waste of words. With the gerund it is the other way; the agent is usually not specified, but sometimes must be, i.e. a possessive must sometimes be inserted; and failure to distinguish when this is required and when it is superfluous leads to some ugly or unidiomatic writing. Scylla is omission of the possessive when the sense is not clear without it; Charybdis is the insertion of it when it is obvious waste of words; but these are only the extremes, rarely run into. *Jones won by Smith's missing a chance;* if you omit

*Smith's*, and say *Jones won by missing a chance* (as in fact he did, only the missing was not his), Scylla has you. If you say *He suffers somewhat, like the proverbial dog, from his having received a bad name*, you and your *his* are in Charybdis. The second is a real extract; of Scylla it was necessary to invent an illustration; but even Charybdis is rare. What is not rare is something between the two. It will be noticed that the reason why that *his* (with *having received*) was felt to be so intrusive is that the receiver is the same person as the subject of the sentence; compare Smith's missing, where *Smith's* was indispensable just because the misser was *not* the subject of the sentence. Hence has come a subconscious assumption that the possessive will be omitted if, and only if, the agent it would have specified is the same as the agent in the action denoted by the main verb, i.e. either the subject, or, if the verb is passive, an agent following *by* or perhaps not even expressed. The following sentences are bad because they flout this assumption; and, though they escape both Scylla and Charybdis, neither leaving out an essential possessive nor using a superfluous one, they offend against idiom by jumping from one agent to another without giving notice: *By conniving at it, it will take too deep root ever to be eradicated* (*By our conniving* would give the necessary notice. *We shall root it too deeply* would avoid the jump. But better abandon the gerund and write *If connived at*). / *Why should not the punishment for his death be confined to those guilty of it, instead of launching expeditions against three tribes?* (*Why should we not confine*, or *instead of our launching* or *instead of expeditions' being launched*. The first is best). / *By allowing month after month to pass without attempting to defend our trade, von Tirpitz had some excuse for supposing that we recognized it to be indefensible* (*By our allowing*—clumsy—, or—better—*we gave von Tirpitz some excuse*).

'The agent in the action denoted by the verb' was spoken of above, and not simply 'the subject'. This complication was necessary because there is a common type of sentence in which the possessive is regularly omitted, and which would have seemed to contradict the rule if 'the subject' had been allowed to pass as sufficient. That type is seen in *This danger may be avoided by whitewashing the glass*; the agent of the whitewashing is not the same as the subject, i.e. *danger*, but is the same as the agent in avoiding, i.e. the owner of the plants that are not to be scorched; consequently the possessive is not required.

A few wrong forms are added without comment: *Sure as she was of* her never losing *her filial hold of the beloved.* / *I cultivated a cold and passionless exterior, for I discovered that* by assuming *such a character certain persons would talk more readily before me.* / After following *a country Church of England clergyman for a period of half a century, a newly appointed, youthful vicar, totally unacquainted with rural life, comes into the parish.* / *Grateful thanks to the three Musketeers who carried Mrs. Pride home after breaking her leg on Wednesday* (from a 'personal column', quoted by *Punch*, whose comment is 'Least they could do').

**gesticulation, gesture.** The usual relation between the two is that of abstract to concrete: gesticulation is the using of gestures, and a gesture is an act of gesticulation. On the other hand, *gesture* also is sometimes used as an abstract, and then differs from *gesticulation* in implying less of the excited or emotional or theatrical or conspicuous. Similarly, if *a gesticulation* is preferred to *a gesture*, it is in order to imply those characteristics. The use of *gesture* in political and diplomatic contexts, = advance, manifestation of willingness to treat or compromise or make concessions, exhibition of magnanimity or friendliness, etc., is so recent that the OED (1901) has no example of it. It dates from the first world war (the earliest example in the OED Supp. is 1916), and is apparently a GALLICISM, having been substituted for the French *beau geste*.

**get. 1.** *Have got* for *possess* has long been good colloquial English, but its claim to be good literary English is not universally conceded. The OED calls it 'familiar', the COD 'colloquial'. It has, however, the authority of Dr. Johnson ('*He has got a good estate*' does *not always mean that he has acquired, but barely that he possesses it*), and has long been used by many good writers. Philip Ballard in a spirited defence, citing not only Johnson but also Shakespeare, Swift, and Ruskin, concludes 'The only inference we can draw is that it is not a real error but a counterfeit invented by schoolmasters'. Acceptance of this verdict is here recommended. Perhaps the intrusion of *got* into a construction in which *have* alone is enough originated in our habit of eliding *have*. *I have it* and *he has it* are clear statements, but if we elide we must insert *got* to avoid the absurdity of *I've it* and the even greater absurdity of *He's it*, with its ambiguity between *has* and *is*. See also DO 2.

**2.** For *got to* = must (*I've got to go now*) no higher claim can be made than good colloquial.

**3.** *Gotten* still holds its ground in American English. In British English it is in verbal uses (i.e. in composition with *have, am*, etc.) archaic and affected; but as a mere participle or adjective it occurs in poetical diction (*On gotten goods to live contented*) and in mining technicalities (*The hewer is paid only for the large coal gotten*; *There is no current wage rate per ton gotten*) and in the cliché *ill-gotten gains*.

**4.** *Get-at-able.* See COME-AT-ABLE.

**geyser.** Dictionaries differ in their preferences between *gāz-, gēz-,* and *gīz*.- For the thermal spring *gāz-* seems usual in Britain and *gīz-* in U.S. For the waterheating apparatus in Britain *gēz-* is universal.

**-g-, -gg-.** Words ending in g preceded by a single vowel double the g before a suffix beginning with a vowel: *waggery, priggish, froggy, sluggard, sandbagged, zigzagging, periwigged, leapfrogging, humbugged.*

**ghetto.** Pl. *-os*; see -O(E)S 6.

**ghoul.** Pronounce *gōōl.*

**gibber, gibberish.** The first is usually pronounced with soft *g*, and occasionally spelt *ji-*; the second is pronounced with hard *g*, and was sometimes spelt *gui-* or *ghi-* to mark the fact. It is doubtful whether one is derived from the other. For *gibberish, lingo*, etc., see JARGON.

**gibbous.** Pronounce with hard *g*.

**gibe, gybe, jibe.** All three spellings can be found for both the nautical term and the verb meaning to taunt. The standard are *gybe* for the former and *gibe* for the latter. The pronunciation is always with soft *g*.

**gill,** ravine, and gills of fish, have *g* hard. *Gill*, the measure, has *g* soft; so has *Gill*, now usually *Jill*, the stock name for a lass (*Jack shall have Jill; naught shall go ill*), and so ordinarily, though not invariably, has *Gillian*, from which it is derived.

**gillie** has *g* hard; *gillyflower* has *g* soft.

**gimmick.** An early definition of this Americanism in a dictionary of its land of origin is any small device, especially one used secretly or in a tricky manner. That is substantially the same definition as is given by the older dictionaries for *dodge* in its colloquial use. But the new word has far outdistanced the old in popularity. It entered Britain after the second world war and quickly became a VOGUE WORD; it must have passed in record time through the slang and colloquial stages to the dignity of use without inverted commas in leading articles and reviews in *The Times*. Its only modern rival to that distinction is KNOW-HOW; the meteoric rise of these two words perhaps gives a clue to our mid-20th-c. sense of values. *Gimmick* is now used especially of a device 'adopted for the purpose of attracting attention or publicity'—COD. The variety of the following quotations shows how widespread the demand for it is. *The ideas were not Mr. Colin*

*Wilson's own. His original contribution was the 'outsider' gimmick. | The gimmick the White House advisers are most fearful of is the amendment tacked on in the House. | There are great risks involved, and not the least that politics may be turned into a branch of entertainment where the slogan is worth more than the argument, the jester more than the thinker, the gimmick more than the policy. | Over the years the Federal German defence budget has been shamelessly robbed by the other ministries, and has in fact proved a useful gimmick to help balance the federal budget. | Some Labour men have been speculating on the need for a new gimmick to project Labour more positively as an alternative government. They have struck on the idea of making Mr. Bevan party chairman during election year. | The particular gimmick of the new brand is the addition to custard powder, blancmange and cornflour of glucose, which it is claimed makes them less liable to form lumps. | This Eastern love of horror qualifies the argument that horror films are solely the gimmick of an industry in decline. | A lot of gimmicks are being tried out; one of them is the Common Market. | Some had seen in Mr. Macmillan's Moscow talks an inspired gimmick.* | The derivation is obscure; it is usually regarded as a corruption of *gimcrack*, but the *g* is hard.

**gimp, gymp.** Spell *gi-* and pronounce *g* hard.

**gingerly.** This word, which is at least four centuries old, has probably no connexion with *ginger*; see TRUE AND FALSE ETYMOLOGY. Skeat connects it with *gang* (going). It has long been used in the sense in which the translators of the Old Testament used *delicately* to describe the way in which Agag obeyed Samuel's summons to come and be hewed in pieces.

**gipsy.** See GYPSY.

**girl** rhymes with *curl*, *whirl*, and *pearl*, with the first syllable of *early*, not of *fairly*. But a pronunciation *gairl*, not

very easily distinguished from *gal*, was at one time general in upper-class society and, though now dying, is still affected by some persons who aim at peculiar refinement. Novelists who write *gurl* as a representation of coarse speech are presumably of this refined class.

**give.** 1. *Give* one *right*, in the sense justify him or allow that he is in the right, is both French (*donner droit à quelqu'un*) and German (*einem Recht geben*); but it is not English, and the OED appears to quote no example of it. In the first passage below it has been resorted to under the ELEGANT-VARIATION impulse, *justify* having been already used up: *The local Liberals and the Chief Whip who supported them from headquarters are abundantly justified in their belief that a radical candidate had a better chance of winning this particular constituency than a Labour one, and the working-class voters have themselves given them right. | M. Millerand is much praised in France for having resisted Mr. Lloyd George's efforts, and M. Clemenceau apparently gives him right.* It sounds rather less odd with the definite article—*I give him the right of it*—but is still a GALLICISM.

2. *Give to think.* The phrase is commented on in the article GALLICISM as one of the two or three that surpass all other Gallicisms for ineptitude. It has, however, had a lamentable vogue, and a few examples follow; others will be found under FURIOUSLY. *This is a powerful impressionistic sketch, true to life, which gives to think. | In every chapter the author has that to say which arrests attention and gives to think. | But what we are told as to coal and cotton gives furiously to think, as they say in France.* This last gentleman seems to think he has got hold of a striking novelty; he is mistaken.

**glacial, glacier, glacis.** Of the different pronunciations of these words offered us by the dictionaries, those here recommended are *glā'shial*, *glā'sier*, and *glă'sē*.

**glad(den).** See -EN VERBS for the distinction.

**gladiolus.** To give this word the four short syllables that it has in Latin, and pronounce *glă′diŏlŭs* is more than can reasonably be expected from the human tongue: we must choose between *gladī′olus* and *gladiŏ′lus*. The latter is preferred by the OED and all gardeners. Pl. *-li* or *-luses*. See LATIN PLURALS, and RECESSIVE ACCENT.

**gladsome.** See -SOME.

**glamour** makes *glamorous*; see -OUR- AND -OR-.

**glimpse. 1.** As nouns *glance* and *g.* are synonyms only in a very loose sense; the glimpse is what is seen by the glance, and not the glance itself. You *take* or *give* a glance at something, but *get* a glimpse of it; the following sentences are not English: *Was there a member of either House who* gave a glimpse at *this schedule to see for himself whether all these documents deserved to be destroyed?* / A glimpse at *the map will show why Turkey was not receiving munitions from Germany or Austria at that time.*
**2.** *Glimpse* as a transitive verb (catch a glimpse of) was known in the 18th c. but seems to have fallen into disuse until it was usefully reintroduced from America in the 20th c.

**glissade.** Pronounce *glĭsah′d*. See -ADE, -ADO.

**global.** The original meaning, now archaic, was globular. Towards the end of the 19th c. it acquired a new one: 'pertaining to or embracing the totality of a group of items, categories or the like' (OED Supp.). With that meaning it was a useful word, but there seems to be a curious attraction in it (cf. OVERALL) that leads to its misuse for aggregate or total, with which it is properly in antithesis. For instance, the compensation paid to the coal industry on nationalization was a *global* figure representing the estimated value of the industry as a whole, to be apportioned among its constituent units, not an *aggregate* figure arrived at by adding together the estimated values of the several units. *Global*, moreover, seeking wider fields, has now established itself, unnecessarily but firmly, as a synonym for what we used to call *world-wide*. *Mondial* is also available for writers who dislike both words.

**gloss, gloze.** The two nouns *gloss* (a, = comment; b, = lustre) are of different origins, the first Greek, the other Scandinavian; but the meaning of the first, and of its derived verb *gloze*, has no doubt been considerably affected (see TRUE AND FALSE ETYMOLOGY) by ignorance of this fact. Greek γλῶσσα, tongue, had as secondary senses, word or locution, word needing explanation, marginal word serving as explanation, comment. The notions of falsity or misrepresentation or imputation or explaining away by which it (and still more *gloze*) is now so often coloured are not essential to it. But the development of a word meaning explanation into one meaning misrepresentation is not unnatural even without the help of this misunderstanding, and the confusion of the two nouns has meant that in popular as opposed to learned speech the first *gloss* is seldom without the suggestion of something sophistical. The two verbs, *gloss* (or *gloze*), to comment, and *gloss*, to put a lustre on, have been even more closely assimilated into the meaning of extenuate in a specious way, especially the PHRASAL VERB *gloss* or *gloze over*.

**glossary, vocabulary.** Both are partial dictionaries, and to that extent synonymous; but the *g.* is a list to which a reader may go for explanation of vernacular words (e.g. archaic, dialect, or technical) likely to be unfamiliar to him (see GLOSS), while *v.* supplies the reader of a book in a foreign language (e.g. a school edition of a classic) with the English equivalents of the words used in it. The *g.* selects what is obscure; the *v.* assumes that all is obscure. *V.* has also the meaning of the whole stock of words used by a nation, by any set of persons, or by an individual. For *lexicon* etc. see DICTIONARY.

**glycerin(e).** In pharmacy, manuals of chemistry, etc., *-in* was preferred until it was superseded by *glycerol*. In everyday use *-ine* is much commoner, and *-in* something of an affectation; see -IN AND -INE.

**gnaw** has p.p. *gnawed* or *gnawn*. The OED examples from the 17th and later centuries show *-ed* eleven times, and *-n* six; half the six are 19th-c. (Jefferies, Southey, Browning), but the *-n* form may nevertheless be regarded as an ARCHAISM.

**gnomic.** Pronounce *nōm-*. Gnomic literature is writing that consists of or is packed with maxims or general truths pithily expressed. The gnomic aorist in Greek is the use of the aorist —a tense normally referring to the past—to state a fact that is true of all times, e.g. in proverbs. *Men were deceivers ever.*

**go,** v. *Goes without saying* is a GALLICISM, but one of those that are virtually naturalized, ceasing to serve as meretricious ornaments, and tending to present themselves as first and not second thoughts. Still, the English stalwart has 'needless to say', 'need hardly be said', 'of course', and other varieties to choose from.

**gobbledygook,** see JARGON.

**godlily.** See -LILY.

**God's acre,** as a name for churchyard or cemetery, though its beauty may be admitted, has not succeeded in establishing itself in English. It is not a phrase of home growth, but a translation from German; and it is interesting that of four quotations for it in the OED only one shows it used simply, without a reference to its alien nationality. Such a preponderance may be accidental, but remains significant.

**golf.** The OED gives precedence to the natural pronunciation (*gŏlf*), and

remarks: 'The Scotch pronunciation is (gōf); the pronunciation (gŏf), somewhat fashionable in England, is an attempt to imitate this'. This fashion has now passed.

**goloptious, golup-.** See FACETIOUS FORMATIONS.

**goodness knows** has two curiously divergent senses. In *Goodness knows who it can have been* it means God only knows, and I do not; in *Goodness knows it wasn't me* it means God knows and could confirm my statement. Ambiguity is unlikely, but not impossible. (The equivalent Irish euphemism is *The dear knows,* i.e. the dear God.)

**goodwill, good will, good-will.** Except in the attributive use (*as a goodwill token,* that is, as a token of good will), the choice should be between the unhyphened forms, see HYPHENS. *Good will* is required when the notion is virtuous intent etc., and *goodwill* is better when it is benevolence, business asset, etc.

**gormandize, gourmandise.** The first is the English verb, the second the French noun.

**gorse.** See FURZE.

**gotten.** See GET.

**gourd.** The OED gives precedence to the sound *gōrd* over *goord* but the latter is now probably commoner.

**gourmand, gourmet.** The first ranges in sense from greedy feeder to lover and judge of good fare; the second from judge of wine to connoisseur of delicacies. The first usually implies some contempt, the other not.

**governance** has now the dignity of ARCHAISM, its work being done, except in rhetorical or solemn contexts, by *government* and *control.*

**Gr(a)ecism, gr(a)ecize, Gr(a)eco-,** etc. The spelling *grec-* is recommended; see Æ, Œ. See also GRECIAN.

**grammar, syntax,** etc. There was a time when *grammar*, in its broadest sense, might have been loosely defined as the science of language. In these days of scientific specialization that will no longer do. The science of language is *philology*, or, in more recent jargon, *linguistics*. Grammar is a branch of that science, and can be defined as the branch that deals with a language's inflexions (*accidence*), with its phonetic system (*phonology*), and with the arrangement of words in sentences (*syntax*). Other branches of linguistics are: Morphology—How words are made. Orthoepy—How words are said. Orthography—How words are written. Composition—How words are fused into compounds. Semantics—How words are to be understood. Etymology—How words are derived and formed.

Of these, *orthography*, *accidence*, and *syntax*, as the bare essentials for writing and reading, represent for most of us the whole of grammar; and *morphology*, *orthoepy*, *phonology*, and SEMANTICS, are meaningless terms to the average person. The last deserves to be more widely studied as a means of promoting clarity of thought and an antidote to the habit of attributing a numinous value to words.

It has become fashionable to speak disrespectfully of grammar—a natural reaction from the excessive reverence formerly paid to it. The name *Grammar School* remains to remind us that the study of Latin grammar was once thought to be the only path to culture. We took a long time to realize that there is not much sense in trying to apply the rules of a dead synthetic language to a living analytical one; perhaps we have not yet quite abandoned the attempt. 'The vulgar grammar-maker, dazzled by the glory of the ruling language, knew no better than to transfer to English the scheme that belonged to Latin. What chance had our poor mother-tongue in the clutch of this Procrustes?' (J. W. Hales, quoted by the Departmental Committee on the Teaching of English

in England). Fortunately in this matter, as in others, the Englishman has stoutly defended the liberty of the individual. It is, for instance, despite the grammarians, not thanks to them, that over the centuries our language has won ease and grace by getting rid of almost all its case-inflexions; some day perhaps this good work will be completed, and we shall no longer be faced with the sometimes puzzling task of choosing between *who* and *whom*. But it is going too far, if we give the word grammar its proper meaning, to say, as Orwell said, that grammar is of no importance so long as we make our meaning plain. We have developed our own 'noiseless' grammar, as Bradley called it; what are generally recognized for the time being as its conventions must be followed by those who would write clearly and agreeably, and its elements must be taught in the schools, if only as a code of good manners. For some legacies from the Procrustean grammarians see FETISHES and SUPERSTITIONS.

**gram(me).** There seems to be no possible objection to adopting the more convenient shorter form, except that the *-me* records the unimportant fact that the word came to us through French.

**gratis.** Pronounce *ā*, not *ă*, still less *ah*. See LATIN PHRASES.

**grave,** v. (carve etc.). P.p. *graved* or *graven*, the second much the commoner; but the whole verb is archaic except so far as it has been kept alive in particular phrases, especially *graven image* by the Second Commandment.

**gray.** See GREY.

**great.** For the differences between *g.*, *big*, and *large*, see BIG. For *the g. Cham*, *the g. Commoner*, see SOBRIQUETS.

**Grecian, Greek.** The first is now curiously restricted by idiom to architecture, facial outline, the Grecian bend and knot, Grecian slippers, and one or two other special uses. We usually speak of *Greek* history, fire, calends, lyrics, tyrants, Church, dialects, aspirations, but *Grecian* noses and brows, colonnades and pediments, may still be heard of; a boy in the highest form at Christ's Hospital is known as a *Grecian*, and a Greek scholar may be described as *a good Grecian*. See also HELLENE.

**Greek g.** There is something to be said for retaining the hard sound of g even before e, i, and y, in such Greek-derived words as are not in popular but only in learned, technical, or literary use. To those who know some Greek the sound of *-ŏji* in *pedagogy* or *jĕri-* in *geriatrics* or *jīnĭ-* in *gynaecology* either obscures the meaning, which they would catch with the aid of the hard *g*, or, if they happen to be prepared for it and so do not miss the meaning, is still repulsive. To those who do not know Greek, the sound of the words is immaterial, and they might allow the other party the indulgence of a harmless pedantry that affects after all but a few words. A list of deserving cases is given below with the pronunciations reminiscent of the Greek origin. In support of the proposed hard *g* it may be pleaded that the *ch* representing Greek chi is often or usually hard in similar cases (*diptych, trochee, trichinosis, tracheotomy, pachyderm, catechism*, etc.).

Specimen words: *anagoge*; *anthropophagi*; *antiphlogistin*; *demagog(ic)(y)*; *geriatrics*; *gynaecology*; *hegemony*; *(hemi)(para)plegia*; *(laryn)(menin)gitis*, etc.; *misogynist*; *monologist*; *paralogism*; *pedagog(ic)(y)*; *philogynist*.

It should now be added that this advice was given in the nineteen-twenties. Since then the words ending *-gitis* have firmly adopted the pronunciation deprecated; for most of the others the issue remains in the balance, with a general tendency towards the soft *g*.

**grey, gray.** 'In Great Britain the form *grey* is the more frequent in use, notwithstanding the authority of Johnson

and later English lexicographers, who have all given the preference to *gray*.' OED.

**greyhound** is known to be unconnected with *grey*, though the meaning of its first part is doubtful; see TRUE AND FALSE ETYMOLOGY.

**gridiron, griddle, grid.** What the light of nature would suggest as to their relations would be that *grid* was the original word, *griddle* its diminutive, and *gridiron* a compound of it with *iron*. Inquiry seems to reveal, on the contrary, that *grid* is a mere curtailment of *gridiron*, which in turn has nothing to do with the word *iron*, but is a corruption of the earlier form *gredire*, a variant of *gredile* (the source of *griddle*). The particular question is of no practical importance, but is here mentioned as illustrating well the kind of mistake, sometimes dangerous, against which a knowledge of etymology may be a protection; see TRUE AND FALSE ETYMOLOGY.

**grievous.** The mispronunciation *grievious* is surprisingly common, presumably owing to the analogy of *previous* and other words with a long stressed vowel before the adjectival suffix, e.g. *devious* and *abstemious*. HEINOUS is another word that is sometimes given a superfluous *i*, especially when mispronounced *hē-*, and the vulgarism *mischievious* has not wholly disappeared.

**griffin, griffon, gryphon.** *Griffon* is the regular zoological form, i.e. as the name of a kind of vulture; it is also a breed of dog. For the fabulous creature *griffin* is the ordinary, and *gryphon* an ornamental spelling.

**group, bracket.** This is the century of the common man, with the ideal of a classless society before him. It is also a scientific age, and we like to show that we think scientifically and express ourselves accordingly. There is a scientific flavour in the word *group* and a stronger one in *bracket*, suggesting those mathematical formulae in which economists convey their meaning to one another and conceal it from ordinary people. So where our fathers would have said *class* we use one of these words. The *poor* are no longer with us, or the *rich* either, or the *young* or the *old*; we must all be sorted into our proper groups or brackets. With the advance of the social sciences we are in fact being more classified than ever before, but the old word will no longer do; it has become indelicate. *Six of the suicides were of undergraduates who had been to higher-income-group schools, and five from among those who had been to lower-income-group schools. | Will the Chancellor of the Exchequer consider the financial hardships of the small-income groups. | It is some comfort to learn that among the juvenile delinquents the eight to thirteen bracket is the only one that involved more arrests. | Those at the lowest level of the income bracket should have been relieved of income tax altogether. | He would not reveal the price, but said it was well within the million pound bracket. | It is not necessary today to be in the surtax bracket in order to stalk in Scotland.* Those examples show what may happen when EUPHEMISM and POPULARIZED TECHNICALITIES work together to corrupt plain speech.

**grow.** For *a grown man* etc. see INTRANSITIVE P.P.

**groyne.** It appears that the word usually so spelt, and meaning breakwater, is of different origin from *groin* the part of the body or of a vault; the separate spelling is therefore useful.

**gryphon.** See GRIFFIN.

**guarantee, guaranty.** Fears of choosing the wrong one of these two forms are natural, but needless. As things now are, *-ee* is never wrong where either is possible. As a verb, *-y* is called by the OED 'now rare, superseded by *-ee*', and *-ee* should therefore always be used. As a noun, *-y* is correct in some senses, but *-ee* is

established in all. Those who wish to avoid mistakes have in fact only to use *-ee* always.

The contexts in which *-y* may still be reasonably preferred are those in which the sense desired is rather the act or fact of giving security than the security given or its giver; for instance, 'willing to enter into a *-y*', 'contracts of *-y*', 'a league of *-y*', 'an act of *-y*', 'treaties of *-y*', 'be true to one's *-y*', in all of which *-y* is a verbal noun and means guaranteeing.

**guer(r)illa.** The original spelling is with *-rr-*, not *-r-*; and the original meaning is not a person, but a kind of fighting, *guerrillero* being the word for the person. But the *-r-* is four times as common as the *-rr-* in the OED quotations, and we should assert our right to spell foreign words as we choose when we have adopted them (cf. MORALE). And as to the meaning, the phrase *g. warfare* is now so firmly established in place of *g.* itself that the use of *g.* as a personal noun may be considered almost an inevitable BACK-FORMATION from it. The best course is to accept the spelling *guerilla*, and the sense (as old as Wellington's dispatches and still very much alive) 'irregular fighter'. See RESISTANCE.

**guillemot.** Pronounce *gǐ'lǐmŏt*.

**guillotine.** For the parliamentary sense, see CLOSURE. Pronounce *gǐ'lŏtēn*.

**gulf, bay.** Apart from the fact that each has some senses entirely foreign to the other, there are the differences (1) that *g.* implies a deeper recess and narrower width of entrance, while *b.* may be used of the shallowest inward curve of the sea-line and excludes a landlocked expanse approached by a strait; and (2) that *b.* is the ordinary word, while *g.* is chiefly reserved as a name for large or notable instances.

**gumma.** Pl. *-as* or *-ata*. See LATIN PLURALS.

**gunwale, gunnel.** The pronunciation

is always, and the spelling occasionally, that of the second.

**gutta-percha.** Pronounce *-chă*.

**guy.** The noun in Britain means someone of grotesque appearance like the conventional effigies of Guy Fawkes. In America it has no disparaging implication; it is as colourless as our *chap*; indeed, as Chesterton has recorded, to be called *a regular guy* is 'one of the most graceful of compliments'. The meaning of the verb (to make fun of) is the same in both countries.

**gybe.** See GIBE.

**gymnasium.** Pl. *-ums* or *-a*; as the name of a German place of education the plural is *Gymnasien*.

**gymp.** See GIMP.

**gyp.** See SCOUT.

**gypsy, gipsy.** In contrast with the words into which y has been introduced instead of the correct i, apparently from some notion that it has a decorative effect (*sylvan, syphon, syren, tyre, tyro*, etc.), there are a few from which it has been expelled for no better reason than that the display of two *y*s is thought an excessive indulgence in ornament. In *gypsy* and *pygmy* the first y is highly significant, reminding us that *gypsy* means Egyptian, and *pygmy* foot-high (Gk. πυγμή elbow to knuckles). It is a pity that they should be thus cut away from their roots, and the maintenance of the y is desirable. The OED's statement is: 'The prevalent spelling of late years appears to have been *gipsy*. The plural *gipsies* is not uncommon, but the corresponding form in the singular seems to have been generally avoided, probably because of the awkward appearance of the repetition of y'. See Y AND I.

**gyves.** The old pronunciation was with the *g* hard, as indicated by a former spelling *gui-*; but the *g* is now soft.

# H

**habiliments.** See POLYSYLLABIC HUMOUR.

**habitude.** In some of its obsolete senses (relation *to*, intimacy or familiarity) the word was not exchangeable with *habit*. But in the senses that have survived it is difficult to find or frame a sentence in which *habit* would not do as well or better, the only difference being a slight flavour of archaism attaching to *habitude*. The following examples from the OED are chosen as those in which, more than in the rest, *habit* may be thought inferior to *habitude*; *In the new land the fetters of habitude fall off.* | *All the great habitudes of every species of animals have repeatedly been proved to be independent of imitation.* | *They can be learned only by habitude and conversation.* The sense constitution or temperament, though not called obsolete in the OED, is very rare, and *habitude* may fairly be classed as a SUPERFLUOUS WORD.

**hackneyed phrases.** When *Punch* set down a heading that might be, and very likely has been, the title of a whole book, 'Advice to those about to marry', and boiled down the whole contents into a single word, and that a surprise, the thinker of the happy thought deserved congratulations for a week. He hardly deserved immortality, but he has—anonymously, indeed—got it; a large percentage of the great British people cannot think of the dissuasive 'don't' without remembering, and, alas! reminding others, of him. There are thousands for whom the only sound sleep is the *sleep of the just*, the light at dusk must always be *dim, religious*; all beliefs are *cherished*, all confidence is *implicit*, all ignorance *blissful*, all isolation *splendid*, all uncertainty *glorious*, all voids *aching*. It would not matter if these associated reflexes stopped at the mind, but they issue by way of the tongue, which is bad, or of the pen, which is worse. King David must surely writhe as often as he hears it told in Sheol what is the latest insignificance that may not be told in Gath. How exasperating it must be for King Canute to be remembered only by those who have forgotten the purpose of his little comedy on the beach! How many a time must Mahomet have regretted his experiment with the mountain as he has heard his acceptance of its recalcitrance once more applied or misapplied! And the witty gentleman who equipped coincidence with her long arm has doubtless suffered even in this life at seeing that arm so mercilessly overworked.

Hackneyed phrases are counted by the hundred, and those registered below are a mere selection. Each of them comes to each of us at some moment in life with, for him, the freshness of novelty upon it; on that occasion it is a delight, and the wish to pass on that delight is amiable. But we forget that of any hundred persons for whom we attempt this good office, though there may be one to whom our phrase is new and bright, it is a stale offence to the ninety and nine.

The purpose with which these phrases are introduced is for the most part that of giving a fillip to a passage that might be humdrum without them. They do serve this purpose with some readers —the less discerning—though with the other kind they more effectually disserve it. But their true use when they come into the writer's mind is as danger-signals; he should take warning that when they suggest themselves it is because what he is writing is bad stuff, or it would not need such help. Let him see to the substance of his cake instead of decorating with sugarplums. In considering the following selection, the reader will bear in mind that he and all of us have our likes and our dislikes in this kind; he may find pet phrases of his own in the list, or miss his pet abominations; he should not on that account decline to accept a caution against the danger of the hackneyed phrase. Acid test. / Balm in Gilead. / Blessing in disguise. / Blushing honours thick upon him. / Clerk of the weather. / Conspicuous by his ab-

sence. / Consummation devoutly to be wished. / Cups that cheer but not inebriate. / Curate's egg. / Damn with faint praise. / Defects of his qualities. / Dim religious light. / Explore every avenue. / Fair sex. / Feast of reason. / Few and far between. / Filthy lucre. / Free gratis and for nothing. / Guide philosopher and friend. / Hardy annual. / His own worst enemy. / Ill-gotten gains. / In a Pickwickian sense. / Inner man. / Irony of fate. / Last but not least. / Leave no stone unturned. / Leave severely alone. / Method in his madness. / More in sorrow than in anger. / More sinned against than sinning. / Neither fish flesh nor good red herring. / Neither rhyme nor reason. / Not wisely but too well. / Observed of all observers. / Of that ilk. / Of the —— persuasion. / Olive branches. / Powers that be. / Psychological moment. / Shake the dust from off one's feet. / Sleep the sleep of the just. / Speed the parting guest. / Splendid isolation. / Strain every nerve. / Take one's name in vain. / Tender mercies. / There's the rub. / To be or not to be. / Through thick and thin. / Tower of strength. / Weaker vessel. / Wheels within wheels. / Wise in his generation. / Withers are unwrung. See also BATTERED ORNAMENTS, CLICHÉ, IRRELEVANT ALLUSION, WORN-OUT HUMOUR.

**had.** 1. *had, had have.* There are two dangers—that of writing *had . . . have* where *had* is required, and that of writing *had* where *had . . . have* is required. The first has proved fatal in *Had she have done it for the Catholic Church, she would doubtless have been canonized as St. Angela*; and in *Had I have been in England on Monday, I should certainly have been present at the first performance.* This is no better than an illiterate blunder, and easily shown to be absurd. *Had she, had I,* are the inverted equivalents of *if she had, if I had*; no one would defend *If she had have done,* nor *if I had have been,* and it follows that *Had she done, Had I been,* are the only correct inverted conditionals.

The other wrong form is seen in '*The country finds itself faced with arrears of legislation which for its peace and comfort* had far better been *spread over the previous years*'. It ought to be *had far better have been spread*; but the demonstration is not here so simple. At the first blush one says: This *had* is the subjunctive equivalent of the modern *would have*, as in *If the bowl had been stronger My tale* had been *longer*; i.e. *had far better been spread* is equivalent to *would far better have been spread*. Unfortunately for this argument it would involve the consequence that *You had far better done what I told you* must be legitimate, whereas we all know that *You had far better have done* is necessary. The solution of the mystery lies in the peculiar nature of the phrase *had better*. *You had better do it*; *It had better be done*; *You had better have done it*; *It had better have been done*; it will be granted at once that these are correct, and that *have* cannot be omitted in the last two. But why? Because the word *had* in this phrase is not the mere auxiliary of mood or tense, but a true verb meaning find; *You had better do it* = You would find to-do-it better; *You had better have done it* = You would find to-have-done-it better.

To return to the *arrears of legislation* sentence, those arrears would find to-have-been-spread-over-the-previous-years far better, i.e. would have been in a better state if they had been so spread. This reminds us that there is another possible way of arriving at the same sense; *The arrears would have been better if they had been spread* is compressible into *The arrears had been better spread*; *better* then agrees with *arrears*, not with *to-have-been-spread*. But that the writer did not mean to take that way is proved by the impossible order 'had better been' instead of 'had been better' (cf., in *Othello*, *Thou hadst been better have been born a dog*); he has perhaps combined the two possible forms, one idiomatic, and the other at least grammatical, into a third that is neither idiomatic, grammatical, nor

possible. Another example like the 'arrears' one of the wrong omission of *have* is: *The object of his resistance was to force Great Britain to expend men and material in dealing with him which* had better been *utilized elsewhere*.

2. *Had* in parallel inverted clauses. *Had we desired twenty-seven amendments, got seven accepted, and were in anticipation of favourable decisions in the other twenty cases, we should think* . . . To write *Had we desired and were in anticipation* is wrong (see ELLIPSIS 6); to write *Had we desired and were we in anticipation*, though legitimate, is not only heavily formal, but also slightly misleading, because it suggests two separate conditions whereas there is only a single compound one. This common difficulty is best met by avoiding the inversion when there are parallel clauses; write here *If we had desired and were in anticipation*.

**haem-, hæm-, hem-.** See Æ, Œ. The compounds are usually *haem-* in Britain and *hem-* in U.S.

**hail,** vb. *H. fellow well met* is now chiefly used as an adj., and should be, in that use, *hail-fellow-well-met*.

**hair-do.** This now common compound noun has reached the dictionaries, and deserves to supersede the alien *coiffure* and to be written *hairdo*.

**half. 1.** *A foot and a h., One and a h. feet.* In all such mixed statements of integers and fractions ($7\frac{1}{4}$ mill., $3\frac{3}{4}$ doz., $27\frac{1}{2}$ lb., etc.), the older and better form of speech is the first—*a foot and a h., seven millions and a quarter*, etc. In writing and printing, the obvious convenience of the second form, with figures instead of words, and all figures naturally placed together, has made it almost universal. It is a pity that speech should have followed suit; the $1\frac{1}{2}$ ft. of writing should be translated in reading aloud into *a foot and a half*; and when, as in literary contexts, words and not figures are to be used, the old-fashioned *seven millions and a quarter* should not be changed into the seven and a quarter millions that is only due to figure-writing. But perhaps the cause is already lost; we certainly cannot say *a time and a half as large* instead of *one and a half times*. For sing. or pl. after *one and a half*, use pl. noun and sing. vb. *One and a half months is allowed for completion.*

2. The intruding a. *President Eisenhower had a private meeting which lasted* a *half an hour.* | *The industry could have produced* a *half a million tons more.* | *The six o'clock news follows in* a *half a minute.* This vulgarism seems to be getting curiously common.

3. *H. as much again* is a phrase liable to misunderstanding or misuse. *The train fares in France were raised this year* 25%, *and have again been increased by half as much again.* That should mean by a further 37½%, making altogether 62½%; the reader is justified, though possibly mistaken, in suspecting that 12½ (half as much, not half as much again) was meant, making altogether 37½% instead of 62½. The phrase is better avoided in favour of explicit figures when such doubts can arise. See MORE 7 for similar ambiguities.

4. *Half-world* = *demi-monde.* See GALLICISMS.

5. *Better half* = wife. See WORN-OUT HUMOUR.

6. *Half-weekly, -yearly,* etc. For the superiority of these to *bi-weekly, bi-annual,* etc., see BI-.

7. *Halfpennyworth* is best written ha'p'orth and pronounced *hă'păth* but *hă'pniworth* is now often heard.

8. *H. of it is, h. of them are, rotten.* See NUMBER 6 (b).

9. For *half-breed, half-caste,* see MULATTO 1, 4.

**hallelujah, alleluia, alleluya.** 'Now more commonly written as in the A.V. of the O.T. *hallelujah*'—OED. That spelling is preserved in the *H. Chorus,* but it is *alleluia* in *Hymns Ancient and Modern,* and *alleluya* in the *English Hymnal* and *Songs of Praise.* In the *Book of Revelation* A.V. and N.E.B. give *alleluia* and R.V. *hallelujah.*

**halliard.** See HALYARD.

**hallmark.** For synonymy see SIGN.

**halloo** etc. The multiplicity of forms is bewildering; there are a round dozen at the least—*hallo, halloa, halloo, hello, hillo, hilloa, holla, holler, hollo, holloa, hollow, hullo. Holler* may perhaps be put aside as an American verb, *hillo* and *hilloa* as archaic, and *hollow* as confusable with another word. *Hello,* formerly an Americanism, is now nearly as common as *hullo* in Britain, (*Say who you are; do not just say 'hello'* is the warning given in our telephone directories) and the Englishman cannot be expected to give up the right to say *hello* if he likes it better than his native *hullo.* Subject to this, the best selection from the variants to provide for an interjection, a noun, and a verb is perhaps this: *Hullo* for the interjection and for the noun as the name of the interjection; *halloo* for the noun as the name of a shout, and for the verb in dignified contexts; *holla* (with past *holla'd*) for the verb in colloquial contexts. We thus get: *Hullo! is that you?*; *He stopped short with a hullo*; *The minstrel heard the far halloo*; *Do not halloo until you are out of the wood*; *He holla'd out something that I could not catch.* The forms *hallo(a), holler,* and *hollo(a),* would thus be got rid of as well as *hillo(a),* and *hollow.*

**halo.** Pl. *-oes,* see -O(E)S 1; adj. *halo'd,* see -ED AND 'D.

**halyard, halliard.** The first spelling is better, not on etymological grounds, but as established by usage. It is true that the original form is *halier* or *hallyer* = the thing one hales with, and that *-yard* is no better than a popular-etymology corruption; but tilting against established perversions (cf. AMUCK, and see DIDACTICISM) is vanity in more than one sense.

**hamstringed, hamstrung.** See the discussion of FORECAST(ED). With *h.,* no doubt of the right form is possible; in *to hamstring, -string* is not the verb *string*; we do not string the ham, but do something to the tendon called the hamstring; the verb, that is, is made not from the two words *ham* and *string*, but from the noun *hamstring*; it must therefore make *hamstringed*. On *bowstring* vb, where the notion that *-string* is verbal is not quite so obviously wrong, the OED says 'The past tense and p.p. ought to be *bowstringed*, but *bowstrung* is also found'. The case for *hamstringed* is still clearer, but *hamstrung* is a strong rival, if it has not actually won. See also STRING, STRUNG.

**hand.** 1. *Hand and glove, h. in glove.* Both forms are common; the OED describes the second as 'later', and *h. and glove* gives best the original notion, as familiar as a man's h. and glove are, while *h. in glove* suggests, by confusion with *h. in h.* (which is perhaps responsible for the *in*), that the h. and the glove belong to different persons. *H. and glove* is therefore to be preferred, but *h. in glove* seems to be the popular choice.
2. *At close h.* Those who follow the *intricacies of German internal policy at close h. are able to* . . . seems to be an unidiomatic mixture of *from close at h.* and *at close quarters.*
3. *Get the better h. If the Imperial troops get the better h., the foreigners would be in far greater danger* similarly mixes *get the better of* with *get the upper h.*
4. *Handful* makes *-ls*; see -FUL.

**handicap.** The use of *handicapped* as a euphemism descriptive of children not fully equipped mentally or physically is recent. It has been criticized as an unsuitable metaphor on the ground that in the ordinary use of the word in sport the competitor with the greatest handicap will be the one with the greatest natural ability. This is splitting hairs; the usage is established and convenient and the OED dates from 1823 the use of the word in the general sense of 'any encumbrance or disability that weighs upon effort'.

**handsel, hansel.** The OED gives precedence to the first; *h.* makes *-lled* etc., see -LL-, -L-.

**hang.** Past and p.p. *hanged* of the capital punishment and in the imprecation; otherwise *hung*.

**hanging-up.** The indicating of your grammatical subject and leaving it to hang up and await your return from an excursion is not common in modern writing; it belongs rather to the old days of the formal period. When a writer of today does try his hand at it, he is apt, being a novice in the period style, to overdo things; the subject and verb are here italicized for the reader's assistance: 'A stockbroker friend of the Z—s and of the Y—s, and then Lord Z—himself, passed through the box before the *interest* of the audience, which had languished as Lady Z— resumed her place at the Solicitors' table, and "Babs", in her demure grey hat, with the bright cherries, and her deep white fichu, struggled through the crowd from the body of the Court in answer to the call of "Miss Z—X—", *revived.*' Hanging-up may also result, especially in OFFICIALESE, from a writer's failure to look where he is going and just meandering along until he reaches the verb with a bump. 'The *cases* where a change in the circumstances affecting the fire prevention arrangements at the premises is such that, if the number of hours stated in the certificate were recalculated, there would be a reduction (or an increase) in the number of hours of Fireguard duty which the members concerned would be liable to perform for the local authority in whose area they reside, *stand*, however, in an entirely different position.'

**haply.** See WARDOUR STREET.

**happening(s).** It is only in the 20th c. that the word has set up for itself—i.e. has passed from a mere verbal noun that anyone could make for the occasion if he chose, but very few did choose, into a current noun requiring a separate entry in the dictionaries. There is nothing to be said against it on the score of correctness, but it is a child of art and not of nature. It comes to us not from living speech, but from books; the writers have invented it, how far in SAXONISM (*event* is the English for it), and how far in NOVELTY-HUNTING, is uncertain. We cannot help laughing to see that, while the plain Englishman is content that *events* should *happen*, the Saxonist on one side requires that there should be *happenings*, and the anti-Saxonist on the other that things should *eventuate*. The purpose of the quotations appended is to suggest that the use of the word is an unworthy literary or journalistic affectation: *There was, first of all, one little happening which I think began the new life.* / *Mr. William Moore (who has up to now played singularly little part in recent happenings) said ...* / *So clear and vivid are his descriptions that we can almost see the happenings as he relates them.* / *There have been fears expressed of terrible happenings to crowded liners.*

**hara-kiri.** Pronounce *-kǐ'rǐ*. The popular form *hari-kari* is a blunder.

**harbour.** See PORT.

**hardly.** 1. *Hardly, hard.* 2. *Hardly ... than.* 3. *Without h., no — hardly.*
  1. *Hardly, hard.* Except in the sense scarcely, the idomatic adverb of *hard* is *hard*, not *hardly*: 'We worked hard, lodged hard, and fared hard'—DeFoe. It is true that in special cases *hardly* must be substituted, or may be, as in *What is made slowly, hardly, and honestly earned*—Macaulay; if Macaulay had not wanted a match for his two other adverbs in *-ly*, he would doubtless have written *hard*. But there is now a tendency, among those who are not conversant enough with grammar to know whether they may venture to print what they would certainly say, to amend *hard* into *hardly* and make the latter the normal wording. It is even more advisable with *hard* than with other such adverbs to avoid the *-ly* alternative, since, as the following quotations show, a misunderstood *hardly* will reverse the sense: *For attendance on the workhouse he receives £105 a year, which, under the circumstances, is hardly*

earned. / *It must be remembered that Switzerland is not a rich country, and that she is hardly hit by the war.* / *That is the fruit of the hardly contested October battles.* For more examples of the misuse of *hardly* see UNIDIOMATIC -LY.
  2. *Hardly ... than.* This, and *scarcely ... than*, are among the corruptions for which ANALOGY is responsible; *hardly ... when* (or *before*) means the same as *no sooner ... than*, and the *than* that fits *no sooner* ousts the *when* that fits *hardly*. The OED marks the phrases (under *than*) with the ¶ of condemnation; but the mistake is so obvious that it should not need pointing out. It is, however, surprisingly common: *The crocuses had hardly come into bloom in the London parks than they were swooped upon by London children.* / *Scarcely had they arrived at their quarters on Ruhleben racecourse than their relations came to visit them.*
  3. For *without hardly*, see WITHOUT 4. Equally bad is *no — h.*, as in *There is no industry h. which cannot be regarded as a key industry. There is h. any* is the English.

**Harley Street.** One would naturally suppose a *Harley Street physician* to be one who has a consulting room in that street. But those who use the phrase (mostly journalists) clearly do not intend it to have so narrow a connotation. Harley Street is only one of numerous streets lying within that part of the borough of St. Marylebone from which it has long been customary for most London consultants to practise; there is also for instance the no less important Wimpole Street—'that long unlovely street' where Tennyson used to call on Arthur Hallam and Browning on Elizabeth Barrett. Perhaps the part is used as a name for the whole (an example of SYNECDOCHE); if so the attribute signifies no more than that the person to whom it is applied is a consultant practising in London and therefore probably to be found in that district. But it seems to be generally intended to imply the enjoyment of a fashionable and lucrative practice. To give it that meaning is to ascribe to

Harley Street an exclusiveness it never possessed. In any case it is a foolish and misleading phrase that we could well do without.

**harmony, melody, counterpoint.** When the first two words are used not in the general sense, which either can bear, of musical sound, but as the names of distinct elements in music, *h.* means 'the combination of simultaneous notes so as to form chords'—OED, and *m.* 'a series of single notes arranged in musically expressive succession'—OED. *C.* means the combination of a melody with one or more other melodies.

**harness.** *Him that putteth on his h.* is a MISQUOTATION (*girdeth*).

**hart, stag, buck, hind, doe.** The following definitions based on the OED will make the distinctions clear:

Hart—The male of the deer, esp. of the red deer; a stag; *spec.* a male deer after its fifth year.

Stag—The male of a deer, esp. of the red deer; *spec.* a hart or male deer of the fifth year.

Buck—The he-goat, *obs.* . . . The male of the fallow-deer. (In early use perh. the male of any kind of deer.) . . . The male of certain other animals resembling deer or goats, as the reindeer, chamois. In S. Africa (after Dutch *bok*) any animal of the antelope kind. Also, the male of the hare, the rabbit, and the ferret.

Hind—The female of the deer, esp. of the red deer; *spec.* a female deer in and after its third year.

Doe—The female of the fallow deer; applied also to the female of allied animals, as the reindeer. . . . The female of the hare, rabbit, and ferret.

**ha(u)lm, haunch, haunt.** The *-aw-* sound has prevailed over the variant *-ah-*.

**hautboy, oboe.** Pronounce *hō'boi*, *ō'bō*; they are adapted spellings, obsolete and current respectively, of the French *hautbois*.

**have. 1.** *No legislation ever has or ever will affect their conduct.* For this common mistake see ELLIPSIS 2.

**2.** *Some Liberals would have preferred to have wound up the Session before rising.* For this mistake see PERFECT INFINITIVE 2.

**3.** *For if the Turks had reason to believe that they were meditating the forcible seizure of Tripoli, it was not to be expected that facilities for extending Italian influence would readily have been accorded. Would have been,* as often happens, is wrongly substituted for *would be.*

**4.** For *does not have* etc. instead of *has not* etc., see DO 2, and for the wrong use of *do* as a substitute for *have* see DO 3.

**haven.** See PORT.

**haver.** Almost all Englishmen, misled by the picture this word conjures up of simultaneous hovering and wavering, think it means to vacillate, and persist in so using it to the indignation of the Scots, who know that its true meaning is to blather, not to swither.

**hay.** *Look for a needle in a bottle of h.* This is the original form of the saying, *bottle* being a different word from the familiar one, and meaning truss; but having become unintelligible it is usually changed into *bundle of hay* or *haystack.*

**haziness.** By this is meant a writer's failure to make a clear line between different members of a sentence or clause, with the result that they run into one another. If he does not know the exact content of what he has set down or is about to set down, either the word or words that he is now writing will not fit without overlapping, or a gap will be left between them. This sounds so obvious that it may seem hardly worth an article; but even the more flagrant transgressions of the principle are so numerous as to make it plain that a warning is called for. Those more flagrant transgressions are illustrated first

*It is a pity that an account of American activities in aircraft production cannot yet be described* (overlapping; *account* is contained in *described*; omit *an account of*, or change *described* to *given*). | *The need of some effort, a joint effort if possible, is an urgent necessity for all the interests concerned* (*need* and *necessity* overlap). | *It is almost incomprehensible to believe at present that such works as his Five Orchestral Pieces can ever undergo such a total change of character as to . . .* (*to believe* is part of the content of *incomprehensible*). | *The welfare of the poor and needy was a duty that devolved especially on those who had a seat in that House* (gap; it is not the welfare, but the securing of the welfare, that is a duty). | *The mischief aimed at by the Act was to prevent members of local authorities who had occasion to enter into contracts from being exposed to temptation.* (The mischief aimed at was the exposure to temptation, not its prevention). | *The rather heavy expense of founding it could have been more usefully spent in other ways* (spend money; incur expense). | *Hitherto the only way of tackling the evil was by means of prohibiting the exportation from certain places* (*way* and *means* overlap; the only way of tackling was to prohibit; it could only be tackled by means of).

Certain words seem to lend themselves especially to this sort of haziness, as AGO with *since*, BUT with superfluous negative, PREFERABLE with *more*, REASON and CAUSE with BECAUSE or DUE, REMAIN with CONTINUE, SEEM with *appear* and TOO illogical. Examples will be found under the words printed in small capitals. See also LEGERDEMAIN and PLEONASM 4.

**he.** We all claim, by quoting *The Jackdaw of Rheims*, to know the grammar of *he* and *him*, and perhaps have thus paradoxically helped *That's him* to the status of idiomatic spoken English it has now won. Nevertheless a less pardonable *him* occasionally appears; for instance: *It might have been him and not President Wilson who said the other day that . . ..* The tendency to use *he*

where *him* is required is, however, much commoner in print. The mistake occurs when the pronoun is to stand in some out-of-the-way or emphatic position; it looks as if writers, pulled up for a moment by the unusual, hastily muttered to themselves 'Heedless of grammar, they all cried "That's him!"', and thanked God they had remembered to put 'he': *The bell will be always rung by he who has the longest purse and the strongest arm.* | *The distinction between the man who gives with conviction and he who is simply buying a title.* | *One of its most notable achievements was the virtual 'warning off' Newmarket Heath, though not in so many words, of a Prince of Wales*, he *who was afterwards George the Fourth.* | Even Dickens, usually impeccable in points of grammar, was tempted into writing *Clennam had never seen anything like his magnanimous protection by that other Father,* he *of the Marshalsea.* See CASES 3 c, SHE 2, and THEY 4.

**headline language.** It would be unreasonable to criticize headlines for not conforming to literary standards, or even for lacking any grammatical structure. If sometimes there is a touch of vulgarity in them, that is not likely to lessen their appeal to the average newspaper reader. If they should occasionally convey no immediate meaning, they are the more likely to excite curiosity about the articles they profess to describe, and so to serve as a corrective of the habit of relying wholly on newspaper headlines for a knowledge of what is happening in the world. It is not beyond guessing that *JOBLESS JUMPS* means that there has been an increase in unemployment, or that *POLIO JABS FOR ALL* means that vaccine against poliomyelitis is to be available for everyone. But it is not self-evident that *PAIRS SWITCH RIDDLE OF 38 MAJORITY* refers to an allegation that a drop in the government's majority to 38 was due to the failure of certain members of parliament to observe their pairing arrangements, or that *W. H. SMITH*

*OFFER SUCCESS* means that an issue of shares by that company has been over-subscribed, or that *THREE AFRICA PROBE CHAIRS STILL EMPTY* means that three members have still to be appointed to the Commission on Central African Federation, or that *CLUBS RAP SOCCER PROBE* is shorthand for 'The chairmen of the League Clubs regard the Football League's inquiry into the state of English association football as little more than an impertinent intrusion into their authority'. The people who devise the headlines may be presumed to know their job. But when the peculiarities of headline language begin to corrupt literary style it becomes a matter of public interest, and protest is legitimate.

The main peculiarities are two; both come from the need to economize space. One is the constant use of short general-purpose words, mostly worn so smooth with over-use that any precise meaning they may once have had has been obliterated. Thus *bid* stands for any form of human effort, *ban* for any sort of restriction or prohibition, *cut* or *slash* for a reduction in prices or wages or anything else, QUIZ for any kind of interrogation, especially by the police (who are *cops* if constables and *sleuths* if detectives), and *swoop* for the activity that may have preceded it. Any violent accident to a vehicle is a *crash*; to criticize is to *rap*; all agreements and treaties are *pacts*; all ambassadors and others on a mission abroad are *envoys*; couples may be wed but never married, and one may *quit* one's occupation but not abandon or resign it. The other peculiarity is the omission of articles and conjunctions and prepositions, leading, especially the last, to a monstrous abuse of our ancient and valuable right to use nouns as attributive adjectives. *MACMILLAN REFUSES BANK RATE RISE LEAK PROBE* will serve as an illustration. PROBE is the general-purpose headline word for an investigation or inquiry, and the use of *bank-rate*, *rise*, and *leak* as attributive adjectives saves three prepositions.

These tricks, when allowed to affect literary style, destroy both precision and elegance; sentences stumble along painfully and obscurely in synthetic lumps instead of running easily and lucidly with analytical grace. The corruption has gone far, affecting especially political speeches, official writing, and commercialese. Examples, with tentative translations, are: *Where retirement dissatisfaction existed advance activity programming had been insignificant.* (The people who were unhappy after retirement were those who had taken little trouble to plan their activities beforehand.) | *Major vehicle expansion projects must depend on steel availability.* (Major projects for expanding the production of vehicles must depend on how much steel is available.) | See also ABSTRACTITIS, HYPHENS 5, NOUN ADJECTIVES, and PREPOSITION DROPPING.

**heap.** *There* are *heaps more to say*, *but I must not tax your space further. Are*, or *is*? see NUMBER 6.

**heave.** Past and p.p. *heaved* or *hove*. They are to some extent differentiated. *He heaved a sigh as he hove in sight.* | *The boat was hove to and the anchor heaved overboard.*

**Hebrew, Israeli, Israelite, Jew, Semite.** Persons to whom all these words are applicable are thought of by the modern Englishman as *Jews*; if he uses in speech one of the other words instead of *Jew*, it is for some reason, known or possibly unknown to himself. He may be deliberately discarding *Jew* in favour of whichever of the others he first thinks of, and that either at the bidding of ELEGANT VARIATION or NOVELTY-HUNTING or facetiousness, or for the better reason that *Jew* has certain traditional associations, such as usury or anti-Christianity, that are unsuited to the context. Or on the other hand he may be not merely avoiding *Jew*, but choosing one rather than another of the alternatives for its own sake. *Hebrew* suggests the pastoral and patriarchal, or again the possession of a language and a literature; *Israelite* the Chosen People and the theocracy and him in whom was no guile; *Israeli* a citizen of the modern state of Israel and *Zionist* one who was active in the founding of it; *Hebraism* scholarship and *Judaism* strict religious observance; *Semite* the failure of a nation to assimilate its Jews. The fact remains that *Jew* is the current word, and that if we mean to substitute another for it, it is well to know why we do so. A remark or two of the OED bearing on the distinctions may be added: (On *Hebrew*) 'Historically, the term is usually applied to the early Israelites; in modern use it avoids the religious and other associations often attaching to Jew' (it is also the only word for the language, ancient or modern); (on *Jew*) 'Applied comparatively rarely to the ancient nation before the Exile, but the commonest name for contemporary or modern representatives of the race; almost always connoting their religion and other characteristics which distinguish them from the people among whom they live, and thus often opposed to *Christian,* and (esp. in early use) expressing a more or less opprobrious sense'.

**hecatomb.** Pronounce -*ŏm*. The word has no connexion with *tomb*: it comes from the Greek ἑκατόν = 100 and βοῦς = ox, and means the sacrifice of a large number of victims, originally a hundred oxen.

**hectic.** *For a* h. *moment.* / *M. Coué was taken up by some of our* h. *papers, and then dropped because he did not do what he never professed to do.* / *They have got pretty well used to the* h. *undulations of the mark.* The blossoming of h. into a VOGUE WORD, meaning excited, rapturous, intense, impassioned, wild, uncontrolled, and the like, is very singular. The OED (1901) shows hardly a trace of it, and explains its one quotation of the kind ('vehement and h. feeling') as an allusion to the h. flush —no doubt rightly. The 1933 Supp. adds several examples and calls the

usage colloquial; the COD has downgraded it to slang. Now a h. flush is one that is accounted for not, like other flushes, by exceptional and temporary vigour or emotion, but by the *habit* (Gk. ἕξις) *of body* called consumption. The nearest parallel to this queer development seems to be the use of CHRONIC for severe, the only difference being that while that is confined to the uneducated or facetious this has had the luck to capture a wider area.

**hecto-.** See CENTI-.

**hedonist, Cyrenaic, epicurean, utilitarian.** The first (literally, 'adherent of pleasure') is a general name for the follower of any philosophy, or any system of ethics, in which the end or the *summum bonum* or highest good is stated as (in whatever sense) pleasure.

The Cyrenaic (follower of Aristippus of Cyrene) is the hedonist in the word's natural acceptation—the pleasure-seeker who only differs from the ordinary voluptuary by being aware, as a philosopher, that the mental and moral pleasures are pleasanter than those of the body.

The epicurean (follower of Epicurus), bad as his popular reputation is, rises above the Cyrenaic by identifying pleasure, which remains nominally his *summum bonum,* with the practice of virtue.

The utilitarian (follower of Bentham and J. S. Mill), by a still more surprising development, while he remains faithful to pleasure, understands by it not his own, but that of mankind—the greatest happiness of the greatest number.

It will be seen that the hedonist umbrella is a broad one, covering very different persons. Both the epicurean and the utilitarian have suffered some wrong in popular usage; it has been generally ignored that for Epicurus pleasure consisted in the practice of virtue. We now apply *epicure* to one who is given to refined enjoyment of food and drink; and the utilitarian is unjustly supposed (on the foolish ground that what is useful is not beautiful and that beauty is of no use) to rate the bulldozer higher than *Paradise Lost*. It may be worth while to quote the OED's statement of 'the distinctive doctrines of Epicurus:—1. That the highest good is pleasure, which he identified with the practice of virtue. 2. That the gods do not concern themselves at all with men's affairs. 3. That the external world resulted from a fortuitous concourse of atoms'.

**hegemony.** The pronunciation *hēgĕ'mŏni* is recommended; see GREEK G.

**hegira.** Pronounce *hĕ'jĭra* (not *īr'a*).

**heinous** is a word that those who speak it find exceptionally puzzling. *Hā- hē-* and *hī-* may all be heard, sometimes, for greater variety, with a terminal *-ious* instead of *-ous*. *Hā-* is right.

**heir.** *H. apparent, h. presumptive.* These phrases are often used, when there is no occasion for either and *heir* alone would suffice, merely because they sound imposing and seem to imply familiarity with legal terms. And those who use them for such reasons sometimes give themselves away as either supposing them to be equivalent or not knowing which is which. Thus: *By the tragedy of the death of the Crown Prince Rudolph in 1889 the Archduke Ferdinand became the Heir Apparent to the throne.* Rudolph, it is true, was heir apparent; but by his death no one could become h. a. except his child or younger brother (whereas Ferdinand was his cousin), since the Emperor might yet conceivably have a son who would displace anyone else. An h. a. is one whose title is indefeasible by any possible birth; an h. p. is one who will lose his position if an h. a. is born. Mistakes are no doubt due to the double sense of the word *apparent*. Its old sense, retained in *h. a.*, and still possible elsewhere in literary use, but avoided for fear of confusion with the other and prevailing sense, is manifest or unquestionable. But the current

sense is almost the same as that of *seeming*, though with slightly less implication that the appearance and the reality are different; *apparent* in this sense means much the same as *presumptive*, but in the other something very different; hence the error.

**heliotrope.** The prefixes *heli-* and *helio-* are from different Greek words, ἕλιξ = twisted and ἥλιος = sun. The Greek value of the initial vowel is ordinarily retained in the English compounds (e.g. *hēlicopter, hēliograph*); for *heliotrope*, however (the plant that turns to the sun), though the dictionaries recognised the long *e*, the short is more usual, and is the pronunciation approved by the experts who in 1928 drew up recommendations to the BBC about the pronunciation of some doubtful words.

**hellebore.** Pronounce *hĕ'lĭbōr*.

**Hellene, Hellenic** etc. The function of these words in English, beside *Greek* etc., is not easy to define; but the use of them is certainly increasing. They were formerly scholars' words, little used except by historians, and by persons concerned not so much with Greeks in themselves as with the effects of Greek culture on the development of civilization in the world (*Hellenism*). With the modern spread of education, the words have been popularized in such connexions (e.g. *H. cruise, H. games, H. Journal*). At the same time the national aspirations of Greek irredentists called newspaper attention to pan-Hellenism and to the name by which the Greeks and their king call themselves; so that the proportion of people to whom *Greek* means something, and *Hellene* and *Hellenic* nothing, is smaller than it was. Nevertheless, *Greek* remains the English word, into whose place the Greek words should not be thrust without special justification. See also GRECIAN, GREEK.

**hello.** See HALLOO.

**help,** n. For *help* = servant see DOMESTIC.

**help**, v. *Than*, and *as*, one *can help*. *Don't sneeze more than you can help*, *Sneeze as little as you can h.*, are to be classed as STURDY INDEFENSIBLES. Those who refrain from the indefensible, however sturdy it may be, have no difficulty in correcting: *Don't sneeze more than you must*, *Sneeze as little as you can*. Out of *Don't sneeze if you can help it* is illogically developed *Don't sneeze more than you can help*, which would be logical, though not attractive, if *cannot* were written for *can*. And out of *Don't sneeze more than you can help* by a further blunder comes *Sneeze as little as you can help*; a further blunder, because there is not a mere omission of a negative—'you can*not* help' does not mend the matter —but a failure to see that *can* without *help* is exactly what is wanted: the full form would be *Sneeze as little as you can sneeze little*, not *as you* either *can*, or *cannot*, *keep from sneezing*. The OED, which stigmatizes the idiom as 'erroneous', quotes Newman for it: *Your name shall occur again as little as I can help, in the course of these pages* (where *as little as may be* would have done, or, more clumsily, if the *I* is wanted, *as little as I can let it*). A future edition of the OED will be able to add Sir Winston Churchill to the defenders of this sturdy indefensible: *They will not respect more than they can help treaties extracted from them under duress*.

Another indefensible use of *help*, not yet, we may hope, too sturdy for its growth to be checked, is the expression *cannot help but be*, a curious confusion between *cannot help being* and *cannot but be*. | *When we look back over past ages . . . we cannot help but be impressed with the supreme difficulties our ancestors had to overcome . . . I cannot help feeling. . . .* | The speaker's second thoughts were clearly right.

**helpmate, helpmeet.** The OED's remark on the latter is: A compound absurdly formed by taking the two words *help meet* in Gen. ii. 18, 20 ('an help meet for him', i.e. a help suitable for him) as one word.

**hem-.** See HAEM-, HAEMORRHAGE, and Æ, Œ.

**hempen.** See -EN ADJECTIVES.

**hendiadys.** The expressing of a compound notion by giving its two constituents as though they were independent and connecting them with a conjunction instead of subordinating one to the other, as 'pour libation from bowls and from gold' = from bowls of gold. Chiefly a poetic ornament in Greek and Latin, and little used in English; but 'nice and warm', 'try and do better', 'grace and favour', instead of 'nicely warm', 'try to do better', 'gracious favour', are true examples. It should be noticed that such combinations as *brandy and soda, assault and battery, might and main, toil and moil, spick and span, stand and deliver*, since their two parts are on an equal footing and not in sense subordinate one to the other, do not fit the name, and should not be called by it. See SIAMESE TWINS.

**her.** 1. Case. For questions of *her* and *she*, see SHE, and cf. HE.
2. For questions of *her* and *hers* (e.g. *Her and his tasks differ*), see ABSOLUTE POSSESSIVES.
3. For *her* and *she* in irresolute or illegitimate personifications (e.g. *The United States has given another proof of* its *determination to uphold* her *neutrality*. | *Danish sympathy is writ large over all* her *newspapers*), see PERSONIFICATION.

**Herculean.** The normal sound of words in -*ean* is with the -*e*- accented and long; so *Periclē′an, Cytherē′an, Sophoclē′an, Medicē′an, Tacitē′an, pygmē′an*, and scores of others. Of words which, like *Herculean*, vacillate between this sound and that given by shifting the accent back and making the -*e*- equivalent to *ĭ*, most develop a second spelling to suit; so *Caesarean* or *Caesarian, cyclopean* or *-pian, Aristotelean* or *-lian. Herculean*, like *protean*, changes its sound without a change of spelling; and many people in consequence doubt how the words should be said. *Herculḗ'an* seems now to be commoner and there is nothing against it; on the contrary its ponderousness seems to suit the meaning better than *Hercu'lĭan* does. But *Hercu'lĭan* is not a modern blunder to be avoided; it has long-standing authority. In the only three verse quotations given by the OED, -*ē'an* is twice impossible, and once unlikely:

Robust but not Herculean—to the sight
No giant frame sets forth his common height.—*Byron*
Let mine out-woe me; mine 's Herculean woe.—*Marston*
    So rose the Danite strong,
Herculean Samson, from the harlot-lap
Of Philistean Dalilah.—*Milton*

**heredity.** The word is now used, by good writers, only in the biological sense, i.e. the tendency of like to beget like. The extract below, where it has been substituted for *descent* solely because *descendant* is to follow, illustrates well what happens when zeal for ELEGANT VARIATION is not tempered by discretion: *The Aga Khan . . . is unique because of his* heredity—*he is a lineal descendant of the Prophet Mohammed—though he is more noteworthy because of his being the leader of the neo-Moslems.*

**hermetically.** This word is now so constant a partner of *sealed* that one would almost suppose sealing that was not hermetic to be a botched job, just as a *part* seems no longer to be a part unless it is *integral* or a *danger* a danger unless it is *real*. The word is not derived from the Greek god Hermes. He had remarkable talents—before he was a day old he had invented a musical instrument and done some cattle-rustling—but it was his Egyptian counterpart *Thoth*, or *Hermes Trismegistos*, that was the specialist in magic and alchemy whose skill in fusing metals enabled him to make airtight containers.

**herr.** See MYNHEER.

**hers.** See ABSOLUTE POSSESSIVES.

**hesitance, hesitancy, hesitation.** The last has almost driven out the others; -*ce* may be regarded as obsolete, but -*cy* is still occasionally convenient when what is to be expressed is not the act or fact of hesitating, but the tendency to do so. Two examples from the OED will illustrate: *She rejected it without hesitation.* | *That perpetual hesitancy which belongs to people whose intelligence and temperament are at variance.*

**Hibernian** differs from *Irish*(*man*) as GALLIC from *French*, and is of the nature of POLYSYLLABIC HUMOUR.

**hiccup** makes -*uping*, -*uped*; see -P-, -PP-. The spelling -*ough* is a perversion of popular etymology, and 'should be abandoned as a mere error'—OED.

**hide,** vb. P.p. *hidden* or *hid*, the latter still not uncommon.

**highbrow.** See INTELLIGENT, INTELLECTUAL.

**highly. 1.** It should be remembered that *high* is an adverb as well as *highly*, and better in many contexts; e.g. *It is best to pay your men high*; *High-yielding securities*; see UNIDIOMATIC -LY. **2.** Though *highly* in the sense to a high degree is often unobjectionable (*a highly contentious question*; *highly farmed land*), it, like DISTINCTLY, sometimes suggests, when used with adjectives of commendation, a patronizing air (*a highly entertaining performance*), and is best avoided in such connexions by those who wish to give genuine praise.

**hillo(a).** See HALLOO.

**him.** See HE.

**hind,** deer. See HART.

**hindsight,** originally another name for the backsight of a rifle, acquired its figurative use towards the end of the 19th c. and has become deservedly popular for the quality that finds expression in being wise after the event. Write as one word.

**hinge,** vb., makes *hinging*; see MUTE E.

**hippopotamus.** Pl. *-muses* better than *-mi.* See -US.

**his.** 1. *A graceful raising of* one's *hand to* his *hat.* For the question between *his* and *one's* in such positions, see ONE 7.
2. *The member for Morpeth has long been held in the highest respect by all who value sterling character and whole-hearted service in the cause of* his *fellows.* For this type of mistake see PRONOUNS.

**historic(al).** The DIFFERENTIATION between the two forms has reached the stage at which it may fairly be said that the use of one in a sense now generally expressed by the other is a definite backsliding. The ordinary adjective of *history* is *historical*; *historic* means memorable, or assured of a place in history, now in common use as an epithet for buildings worthy of preservation for their beauty or interest; *historical* should not be substituted for it in that sense. The only other function retained by *historic* is in the grammarians' technical terms *historic tenses, moods, sequence, present,* etc., in which it preserves the notion appropriate to narration of the past, especially in the expression *h. present,* a device for imparting vividness to a narrative which is not now so popular with story-tellers as it once was.

Although both adjectives are now always aspirated when not preceded by the indefinite article, the use of *an* with them lingers curiously. See A, AN.

**historicity.** The earliest OED example of this ugly word is dated 1880; but, being effective in imparting a learned air to statements intended to impress the unlearned, it has had a rapid success, and is now common. It has, however, a real use as a single word for the phrase *historical existence,* i.e. the having really existed or taken place in history as opposed to mere legend or literature. To this sense, in which it makes for brevity, it should be confined. *The historicity of St. Paul* should mean the fact that, or the question whether, St. Paul was a real person. The following quotation shows the word in a quite different sense, in which it would not have been worth inventing—why not *accuracy*? If it is given two or more senses liable to be confused it loses the only merit it ever had—that of expressing a definite compound notion unmistakably in a single word: *He is compelled to speak chiefly of what he considers to be exceptions to St. Paul's strict historicity and fairness; and he tells us that he is far from intending to imply that the Apostle is usually unhistorical or unfair.*

**hither,** described by the OED as 'now only literary', is even in literature, outside of verse, almost disused. ('Now usually *here*' says the COD). It is still tolerable, perhaps, in one position, i.e. as the first word in an inverted sentence following a description of the place referred to—*Hither flocked all the* ... Elsewhere, it produces the effect of WARDOUR STREET English, being used mainly by the unpractised writers who bring out their best English when they write to the newspapers. The same is true of *thither*; but, as often happens with stereotyped phrases, *hither and thither* retains the currency that its separate elements have lost. See SIAMESE TWINS.

**hock, hough.** Except in Scotland, the older spelling *hough* is now pronounced like *hock*, which 'has largely superseded' it (OED) in spelling also; it is better to abandon the old spelling.

**hodge-podge.** See HOTCHPOT(CH).

**hoi polloi.** These Greek words for the majority, ordinary people, the man in the street, the common herd, etc., meaning literally 'the many', are equally uncomfortable in English whether *the* (= *hoi*) is prefixed to them or not. The best solution is to eschew the phrase altogether, but it is unlikely to be forgotten as long as *Iolanthe* is played. ''*Twill fill with joy and madness stark the Hoi Polloi* (*a Greek remark*)'.

**hoist** was originally a variant past participle of the now obsolete verb *hoise.*

Both it and *hoised* were current at the turn of the 16th c. '*Tis the sport to have the engineer hoist with his own petard* (i.e. blown up by his own bomb), *Hamlet* III.iv.206./ *When they had hoised up the mainsail to the wind,* Acts xxvii. 40. At the same time *hoist* was already in use as a verb in its own right: *Shall they hoist me up and show me to the shouting varletry of censuring Rome? Ant. and Cl.* v. ii. 55.

**holla, holler, hollo(a), hollow.** See HALLOO.

**Holland.** See NETHERLANDS.

**home,** n., makes *homy*, not *homey*; see -EY AND -Y. The adverbial use of *home* in the sense of at home (*I shall stay home*) is an Americanism; in the sense of to home (*I shall go home*) it is idiomatic in both countries. See PREPOSITION DROPPING.

**homo-** etc. The prefixes *homoeo-* (as in *homoeopath*), *homoio-* (as in *homoiousian*), and, much more common, *homo-* (as in *homosexual*) are all from the Greek word meaning same; it is a vulgar error to suppose that in the last it comes from the Latin *homo* = man. In all these compounds except *homoeopath* etc. the first *o* is short in English, as in Greek.

**homonym, synonym.** Any confusion between the two is due to the fact that *s.* is a word of rather loose meaning. Broadly speaking, homonyms are separate words that happen to be identical in form, and synonyms are separate words that happen to mean the same thing. *Pole*, a shaft or stake, is a native English word; *pole*, the terminal point of an axis, is borrowed from Greek; the words, then, are two and not one, but being identical in form are called homonyms. On the other hand *cat*, the animal, and *cat*, the flogging instrument, though they are identical in form and mean different things, are not separate words, but one word used in two senses; they are therefore not homonyms. True synonyms, i.e. separate words exactly

equivalent in meaning and use, are rare, and the word is applied more frequently to pairs or sets in which the equivalence is partial only; see SYNONYMS.

**homophone.** 'When two or more words different in origin and signification are pronounced alike, whether they are alike or not in their spelling, they are said to be homophonous, or homophones of each other. Such words if spoken are of ambiguous signification.' This definition opens Robert Bridges's essay on English Homophones published in 1919 as Tract II of the newly formed Society for Pure English. He went on to emphasize that true homophones must be *different* words that have, or have acquired, an illogical fortuitous identity of sound; the term should not be applied to words that were originally the same but have acquired different meanings, even though they may be spelt differently. For instance *draft* and *draught*, both meaning something drawn (however different a thing), are not homophones as are *air* and *heir*, *son* and *sun*, *vane* and *vein*.

Of homophones thus defined Bridges compiled lists containing 835 entries involving about 1,775 words, and on this evidence he submitted and argued the following propositions: that homophones are a nuisance; that English is exceptionally burdened with them; that they are self-destructive and tend to become obsolete; that this loss threatens to impoverish the language, and that the 'South English dialect' (now often called the 'Received Pronunciation') is a direct and chief cause of homophones by its smudging of unaccented vowels (e.g. *lesson* and *lessen*), the loss of trilled *r* (e.g. *source* and *sauce*), and the failure to pronounce the *h* in *wh-* (e.g. *whether* and *weather*).

For some later trends in English pronunciation, which on the whole Bridges might have thought faintly encouraging, see the articles PRONUNCIATION, RECEIVED PRONUNCIATION. and RECESSIVE ACCENT.

**hon.** That this prefix may be an abbreviation of either *honourable* or *honorary* is a source of some confusion to foreigners. It stands for *honourable* in reports of debates in the House of Commons, where members may not be referred to by name, except by the Speaker, but must be called *the honourable member for* . . . (*hon. and gallant* if a member of the Armed Forces, *hon. and learned* if a lawyer). It stands for *honorary* when prefixed to the holder of an office (*hon. secretary, hon. treasurer*, etc.) and indicates that he is unpaid. As an abbreviation of *honourable* it is also a courtesy title of the sons and daughters of viscounts and barons and of the younger sons of earls, as well as of the holders of certain high offices, especially Puisne Judges in England and Lords of Session in Scotland. (Privy Councillors and peers below the rank of marquess are *right honourable*; so are Lords Justices of Appeal, the Lord Mayor of London, and the Lord Provost of Edinburgh and a few other civic dignitaries; marquesses are *most honourable* and a duke *his grace*.) *The Hon.*, when used as a courtesy title, requires the person's Christian name or initial, not his surname alone (*the Hon. James* or *J. Brown*, not *the Hon. Brown*), a common mistake is to suppose that the Christian name is unnecessary before a double-barrelled surname. The same applies to the prefixes REVEREND and SIR.

**honeyed, honied.** The first is better.

**honorarium.** Pl. *-iums* or *-ia.* The COD follow the OED in still giving precedence to the pronunciation with a sounded *h*. This, and the pl. *-ia,* seem proper tributes to the word as a distinguished foreigner but there is now no hope that it can retain the *h* sound that all other words beginning *honor-* have dropped.

**hono(u)r.** 1. Keep the -u-; but see -OUR AND -OR. 2. *A custom more honoured in the breach than the observance.* Whoever will look up the passage (*Hamlet* I. iv. 16) will see that it means, beyond a doubt, a custom

that one deserves more honour for breaking than for keeping; but it is almost always quoted in the wrong and very different sense of a dead letter or rule more often broken than kept. So: *The Act forbids entirely the employment of boys* . . . '*by way of trade or for the purpose of gain*'. *Therefore, unless the Act be honoured more in its breach than in its observance, the cherubic choirboy* . . . *is likely* . . . *to be missing from his accustomed place in cathedral and church.* For similar mistakes, see MISAPPREHENSIONS.

**hoof.** Pl. *-fs*, sometimes *-ves*; see -VE(D).

**hope.** In the OED, the examples illustrating the use of the verb are nearly 60 in number; of all these not a single one bears the slightest resemblance or gives any hint of support to any of the sentences here to be quoted. This seems worth mention as showing how modern these uses are; in 1901, the date of H in the OED, they could apparently be ignored. That they were not quite non-existent even then is shown by the fact that one of the offenders quoted below is Emerson, but it may be safely assumed that they were rare; nowadays the newspapers are full of them and have been for years.

First, two examples of the passive use of *hope* which has naturally followed the similar use of *fear* discussed in the article DOUBLE PASSIVES: *No greater thrill can be hoped to be enjoyed by the most persistent playgoer of today than* . . . | *There was a full flavour about the Attorney-General's speech against him in the Assize Court at Launceston which cannot be hoped to be revived in these indifferent times.*

Secondly, ANALOGY has been at work, and as *hope* and *expect* are roughly similar in sense, the construction proper to one (*I expect them to succeed*) is transferred to the other (*I hope them to succeed*, whence *They are hoped to succeed*) with which it is far from proper; so: *I need not say, how wide the same law ranges, and how much it can be hoped to effect.* | *In the form of a bonus intended to cover the rise,* hoped

to be *temporary, in the cost of living.* |
*A luncheon at which the King* is hoped
to be *present.* But the notion that,
because *hope* means hopefully expect,
therefore it can have the construction
that that phrase might have is utterly
at variance with the facts of language.

Thirdly, writers have taken a fancy
to playing tricks with 'it is hoped', and
working it into the sentence as an essential part of its grammar instead of
as a parenthesis; the impersonal *it* is
omitted, and *is* (or *are*) *hoped* is forced
into connexion with the subject of the
sentence, with deplorable results. See
also IT 1. *The final arrangements for*
what is hoped will *prove a 'monster
demonstration'.* | *Who has held two of
the most distinguished positions under
the Crown, and whose self-sacrificing
services for the Empire* may be hoped
*even yet not to be at an end.* In the
first example, *it* should be reinstated; in the second, read *are not
even yet, it may be hoped, at an end.*

**hopeless.** See DESPERATION.

**horrible, horrid** etc. The distinctions
between the two are (1) that *horrid* is
still capable as an archaism in poetical
use of its original sense of bristling or
shaggy; and (2) that while both are
much used in the trivial sense of disagreeable, *horrible* is still quite common
in the graver sense inspiring horror,
which *horrid* tends to lose, being 'especially frequent as a feminine form of
strong aversion'—OED. In this remark there is still some truth, though
not so much as when it was written.
*Horrifying* retains its old force better
than either of them, and not even for
an adjective appropriate to nuclear war
is there any justification for picking
*horrendous* out of the oblivion into
which it has rightly fallen. *By a tacit
understanding not to resume tests the
U.S and Russia could go far to slowing
the absurdly costly and horrendous
spiral.* | *President Kennedy, grappling
with the horrendous problems of the
thermonuclear arms race, must at times
be sorely tempted to see specifically
European anxieties as tiresomely peri-
pheral.*

**horse** makes *horsy,* not *horsey*; see -EY
AND -Y.

**hose** (stockings) is archaic, or a shop
name.

**hospitable.** The stress should be on
*hos-,* not on *-pit-.* For doubtful cases of
such stress see RECESSIVE ACCENT, but
the stress on *hos-* is as old as Shakespeare and Drayton (lines quoted in
OED).

**hospitaller** is better than *-aler*; see
-LL-, -L-.

**hotchpot, hotchpotch, hodgepodge,
hotpot.** The first is nearest to the
original form (Fr. *hochepot* = shake-
pot); 2, 3, and possibly 4, are successive corruptions dictated by desire for
expressiveness or meaning when the
real sense was forgotten. *Hotpot* is
standard in cookery; for other purposes
*hotchpotch* is now the prevailing form,
and it would be best if the two later
ones might perish. *Hotchpot,* being a
technical legal term, would naturally
resist absorption in *hotchpotch,* but
might be restricted to its special use.

**hotel.** The old-fashioned pronunciation with the *h* silent (cf. *herb, hospital,
humble, humour*) is almost dead, though
*an otel* may still be heard, perhaps
because it is less trouble to say than
*a hotel.*

**hough.** See HOCK.

**housewife(ry).** The shortened pronunciation (*hŭ′zĭf* or *hŭ′zwĭf*), which
is almost invariable for the sewing-
case, is now obsolete for the human *h.*
Its displacement by *how′swīf* was in
part brought about in the 16th c.,
when *housewife* and *hussy* were still
realized to be the same word a
distinction between the two was felt
to be the reputable matron's due. The
case for it is even stronger today,
when *housewife* is a highly respectable
statutory occupation. Similarly *how′-
swĭfrĭ* has superseded *hŭ′zĭfrĭ.*

**hovel, hover.** The dictionaries still
give alternative pronunciations (*hŏv-*

and *hŭv-*) for both words. But *hŭv-* (preferred by the OED for *hover* in 1901) is certainly dying. See PRONUNCIATION 5.

**howbeit,** according to the OED, is archaic in one of its senses (nevertheless) and obsolete in the other (although). The archaic has its place in modern writing, the obsolete has not; see ARCHAISM and WARDOUR STREET. Those who, without much knowledge of the kind of literature in which archaism is in place, are tempted to use this word should carefully note the distinction. It is often a delicate matter to draw it aright; but there is little doubt that the OED has done so here.

**however.** Several small points require mention. 1. *however, how ever, how . . . ever.* In everyday talk, *how ever* is common as an emphatic form of the interrogative *how* (*How ever can it have happened?*). Being purely colloquial, it should not appear in print except when dialogue is to be reproduced. This does not apply to cases where *ever* has its full separate sense of at any time or under any circumstances, but it is then parted from *how* by some other word or words. *We believe that before many years have passed employers and employed alike will wonder* however *they got on without it*; this should have been *how they ever got on*; the other order is an illiteracy in itself, and the offence is aggravated by the printing of *however* as one word. See EVER.

2. *But* with *however*. *But it must be remembered, however, that the Government had no guarantee.* / *But these schemes, however, cannot be carried out without money.* And for other examples of this disagreeable but common redundancy see BUT 5. Either *but* or *however* suffices; one should be taken, and the other left; sitting on two stools is little better than falling between them.

3. *However* too late. *These extravagant German counter-attacks in mass on the Cambrai front, however, materially helped the French operations in Champagne.* The excuse for such late placing

of the conjunction—that *these . . . front* is in effect a single word—is sound only against a suggestion that it should be placed after *attacks*; it, or *Nevertheless,* or *All the same,* could have stood at the head of the sentence. The undue deferring of *however* usually comes from the same cause as here, i.e. the difficulty of slipping it in where it interrupts a phrase, and should be recognized as a danger to be avoided.

4. *However* too early. It should be borne in mind that the placing of *however* second in the sentence has the effect, if the first word is one whose meaning is complete (e.g. *He* as compared with *When*), of throwing a strong emphasis on that word. Such emphasis may be intended, or, though unintentional, may be harmless; but again it may be misleading. Emphasis on *he* implies contrast with other people; if no others are in question, the reader is thrown out. *The Action Commission wished to get permission for meetings and had telephonic communication with Wallraff, who declared that he would not negotiate with the workmen.* He, however, *would receive the Socialist members of Parliament.* The only right place for *however* there is after *would*, the contrast being not between him and anyone else, but between *would not* and *would*. The mistake is made with other conjunctions of the kind usually cut off by commas, but is especially common with *however* and *therefore*.

5. For *however* as a cause of ambiguity see FALSE SCENT.

**hue.** For synonymy see TINT.

**hugeous.** Those who use the form perhaps do so chiefly under the impression that they are satirizing the ignorant with a non-existent word, as others of their kind do with *mischevious* or *underconstumble* or *high-strikes* for *mischievous, understand,* and *hysterics*. It is in fact a good old word, and corresponds rather to *vasty* and *stilly* by the side of *vast* and *still*; but it is practically obsolete, and, as its correctness robs it of its facetious capabilities, it might be allowed to rest in peace.

**huguenot.** Pronounce *hŭg'ĕnŏt* in preference to *-nō*.

**hullo.** See HALLOO.

**humanist.** The word is apt to puzzle or mislead, first because it is applied to different things, and uncertainty about which is in question is often possible, and secondly because in two of these senses its relation to its parent word *human* is clear only to those who are acquainted with a long-past chapter of history. The newspaper reader sometimes gets the impression that *humanist* means a great classical scholar. Why? he wonders, and passes on. Another time he gathers that a humanist is a sceptic or an agnostic or a freethinker or something of that sort, you know; again he wonders why, and passes on. Another time he feels sure that a humanist is a Pragmatist or Positivist or Comtist, and here at last, since he knows that Comte founded the Religion of Humanity, there seems to be some reason in the name. And lastly he occasionally realizes that his writer is using the word in the sense in which he might have invented it for himself—one for whom the proper study of mankind is man, the student, and especially the kindly or humane student, of human nature, a humanitarian, in fact, in the popular sense of that word, as in *The Congo is not a unitary State, and it is sad that humanists like Mr. N.-B. . . . seek to force one group into the domination of another.*

The original humanists were those who in the early Middle Ages, when all learning was theology, and nearly all the learned were priests or monks, rediscovered pre-Christian literature, turned their attention to the merely human achievements of Greek and Roman poets and philosophers and historians and orators, and so were named *humanists* as opposed to the divines; hence the meaning classical scholar. But this new-old learning has, or was credited with, a tendency to loosen the hold of the

Church upon men's beliefs; hence the meaning free-thinker. The third meaning—Comtist—was a new departure, unconnected in origin with the first two, though accidentally near one of them in effect, but intelligible enough on the face of it. As to the fourth, it requires no comment.

**humanity.** For *the Humanities*, or *Literae humaniores*, as an old-fashioned name for the study of classical literature, history, and philosophy, see HUMANIST.

**humble-bee.** See BUMBLE-BEE.

**humour,** n., makes *humorous* and *humorist* but *humourless* and *humoursome*; see -OUR- and -OR-. Except by a few old-fashioned people *humour* and its derivatives are now always given the *h* sound.

**humour, wit, satire, sarcasm, invective, irony, cynicism, the sardonic.** So much has been written upon the nature of some of these words, and upon the distinctions between pairs or trios among them (wit and humour, sarcasm and irony and satire), that it would be both presumptuous and unnecessary to attempt a further disquisition. But a sort of tabular statement may be of service against some popular misconceptions. No definition of the words is offered, but for each its motive or aim, its province, its method or means, and its proper audience is specified. The constant confusion between sarcasm, satire, and irony, as well as that now less common between wit and humour, seems to justify this mechanical device of parallel classification; but it will be of use only to those who wish for help in determining which is the word that they really want. See also SATIRE.

**hundred.** See COLLECTIVES 5.

**hussy, huzzy.** In the OED examples, the spelling with -*ss*- occurs nearly five times as often as that with -*zz*- and may now be regarded as established. The traditional pronunciation (*hŭ'zĭ*, cf. HOUSEWIFE) is giving way before

| | MOTIVE or AIM | PROVINCE | METHOD or MEANS | AUDIENCE |
|---|---|---|---|---|
| humour | Discovery | Human nature | Observation | The sympathetic |
| wit | Throwing light | Words and ideas | Surprise | The intelligent |
| satire | Amendment | Morals and manners | Accentuation | The self-satisfied |
| sarcasm | Inflicting pain | Faults and foibles | Inversion | Victim and bystander |
| invective | Discredit | Misconduct | Direct statement | The public |
| irony | Exclusiveness | Statement of facts | Mystification | An inner circle |
| cynicism | Self-justification | Morals | Exposure of nakedness | The respectable |
| The sardonic | Self-relief | Adversity | Pessimism | Self |

*hŭ'sĭ*, which, with the assistance of the spelling, will no doubt prevail.

**hybrids and malformations.** In terms of philology, hybrids are words formed from a stem or word belonging to one language by applying to it a suffix or prefix belonging to another. It will be convenient to class with these the words, abortions rather than hybrids, in which all the elements belong indeed to one language, but are so put together as to outrage that language's principles of word-formation. English contains thousands of hybrid words, of which the vast majority are unobjectionable. All such words as *plainness* or *paganish* or *sympathizer*, in which a Greek or Latin word has become English and has afterwards had an English suffix attached to it, are hybrids technically, but not for practical purposes. The same is true of those like *readable*, *breakage*, *fishery*, *disbelieve*, in which an English word has received one of the foreign elements that have become living prefixes or suffixes; *-able*, *-age*, *-ery*, *dis-*, though of Latin-French origin, are all freely used in making new forms out of English words.

At this point it may be well to clear the ground by collecting a small sample of words that may be accused of being misformed in either of the senses explained above—i.e. as made of heterogeneous elements, or as having their homogeneous elements put together in an alien fashion: *amoral*, *automation*, *backwardation*, *bi-weekly*, *bureaucracy*, *cablegram*, *climactic*, *coastal*, *coloration*, *dandiacal*, *flotation*, *funniment*, *gullible*, *impedance*, *pacifist*, *speedometer*. (Extreme examples, still-

born it may be hoped, are *breathalyser* and *triphibious*.) An ill-favoured list, of which all readers will condemn some, and some all. It will not be possible here to lay down rules for word-formation, which is a complicated business; but a few remarks on some of the above words may perhaps instil caution, and a conviction that word-making, like other manufactures, should be done by those who know how to do it. Others should neither attempt it for themselves, nor assist the deplorable activities of amateurs by giving currency to fresh coinages before there has been time to test them.

A great difficulty is to distinguish, among the classical suffixes and prefixes, between those that are, though originally foreign, now living English, and those that are not. Of the former class *-able* and *dis-* have already been mentioned as examples; to the latter *-ation*, *-ous*, *-ic*, and *a-* (not), may be confidently assigned. But others are not so easy to place; how about *-nce* (*-ance* and *-ence*)? An electrician, in need of a technical term, made the word *impedance*. 'I want a special word' we may fancy him saying 'to mean much the same as *hindrance*, but to be sacred to electricity; I will make it from *impede*; *hinder* makes *hindrance*, so let *impede* make *impedance*.' And why not? *Impede* and *-nce* are both from Latin, so why should it be wrong to combine them? Again, if *-ance* is a living suffix it can be put straight on to a verb that is now English even if it was not so by origin; and *hindrance*, *forbearance*, *furtherance*, and *riddance*, all from English verbs, are enough to prove that *-ance* is a living suffix. So it might plausibly be argued. The answer to the first argument is that if these combinations are made they should be made correctly; the word should have been *impedience* (cf. *expedient*). As to the second argument it is no longer true to say that *-ance* is a living suffix; suffixes, like dogs, have their day, and to find whether *-ance*'s day is today we need only try how we like it with a few English verbs of suitable sense, say *stoppance* (cf. *quittance*), *hurriance* (cf. *dalliance*), *dwellance* (cf. *abidance*), *keepance* (cf. *observance*). By all means let the electricians have their *imped-ance*, but in the interests of both electricity and English let it be confined to the former.

Another suffix that is not a living one, but is sometimes treated as if it were, is *-al*; and it will serve to illustrate a special point. Among regrettable formations are COASTAL, *creedal*, and *tidal*. Now, if *-al* were to be regarded as a living suffix, it would be legitimate to say that *coast* and *creed* are now English words, and could have the suffix added straight to them; but if it is tried with analogous English words (*shore, hill, belief, trust*), the resulting adjectives *shoral, hillal, beliefal*, and *trustal*, show that it is not so. The defence, then, would be different— that *coast* and *creed* are of Latin origin, and so fit for the Latin suffix. But then comes in the other requirement—that if both elements are Latin, they should be properly put together; *coastal(is)* and *creedal(is)* are disqualified at sight for Latin by the *-oa-* and *-ee-*; *costal* and *credal* would have been free from that objection at least. Such words may be described as not hybrids but spurious hybrids; and whether the qualification aggravates or lessens the iniquity is a question too hard for a mere grammarian. All that can be said is that the making of words that proclaim themselves truly or falsely as hybrids by showing a classical suffix tagged on to some purely English vowel combination is a proof of either ignorance or shamelessness. The best examples of such curiosities are perhaps those words in *-ometer* (e.g. *floodo-*

*meter* and *speedometer*), with their impossible English vowels, that imitate the form of the Greek compounds such as *barometer* and *thermometer*. The wordmakers have missed an opportunity with *meter*; we have an English METER that we use in *gas-meter* and *water-meter*; why could they not have given us *flood-meter* and *speed-meter*, as they later gave us *ohmmeter* and *ammeter*, instead of our present monstrosities? The classical connecting vowel *-o-* is quite out of place at the end of an English word; *gasometer* gave the analogy, but *gas*, being a word native in no language, might fairly be treated as common to all, including Greek, whereas *flood* and *speed*, with their double-letter vowels, were stamped as English. See -o-.

It will not be worth while to pursue the matter further, or to explain in detail why each word in the above 'ill-favoured' list is a correct or incorrect formation, since complete rules cannot be given. The object of the article is merely to suggest caution. A reference to it without comment against any word in this book indicates that that word is, in the author's opinion, improperly formed for a reason connected with the making of words from different languages, but not necessarily specified in so slight a sketch as this. And the article is intended only for the serious wordmaker. It would be naïve to tilt at the surprising compounds that manufacturers sometimes invent for their products, as for instance when *-matic* or *-vator* is treated as a living suffix to indicate that the article so described works automatically or cultivates the land. Such conceits are in the same class as facetious misspellings like *eezi, kleen*, and *nu*; sensitive people may find them distasteful, but they are unlikely to corrupt the language. See also BARBARISMS.

**hygiene, hygienic.** *Hī'jiēn, hījiē'nic* are the orthodox pronunciations, but the temptation to simplify the double vowel sound *iē* into *ē* has proved too much for us, and this is now recognized by the dictionaries (cf. *rabies, species,* etc.). See -IES, -EIN. As the form of *hygiene* often puzzles even those who know Greek, it is worth while to mention that it is the French transliteration of Gk. ὑγιεινή (τέχνη), (art) of health.

**hypallage.** An inversion of the natural relation of two terms in a sentence, especially the transfer of epithets, as when Virgil speaks of 'the trumpet's Tuscan blare' instead of 'the Tuscan trumpet's blare', or Spenser of 'Sansfoy's dead dowry', i.e. dead Sansfoy's dowry.

**hyperbaton.** Transposition of words out of normal order, as in Browning's title *Wanting Is—What?*, or in *That whiter skin of hers than snow.*

**hyperbole.** Use of exaggerated terms for the sake not of deception, but of emphasis, as when 'infinite' is used for great or 'a thousand apologies' for an apology.

**hyper-, hypo-.** The ordinary pronunciation of the first syllable of words with one of these prefixes is *hīp-*; for *hyper-* this is invariable. But *hypocri(te)* *(sy)* etc. are always *hĭp-* (the *hypocritical* being thus distinguished from the *hypercritical*), and *hĭp-* is an alternative recognized by the COD for all the other *hypo-* compounds though not for *hypo* itself as a curtailed word.

**hyphens.** No attempt will be made here to describe modern English usage in the matter of hyphens; its infinite variety defies description. No two dictionaries and no two sets of style rules would be found to give consistently the same advice. There is, however, one principle that seems to command at least lip service from all authorities. This is that the hyphen is not an ornament but an aid to being understood, and should be employed only when it is needed for that purpose. (For instance, to take random examples from contemporary writing, the sense is radically altered if the hyphens are omitted in *The Russians would be well content if they could get all-German talks started on something like their terms* and in *There is stodge as well as sublimity hidden away in Bach's 200-odd cantatas.*) 'I am in revolt about your hyphens' wrote Sir Winston Churchill to Sir Edward Marsh. 'One must regard the hyphen as a blemish to be avoided wherever possible.' The purpose of this article is to suggest certain conclusions that seem to flow from acceptance of this dictum. No one should presume to do even this dogmatically. There will remain ample room for differences of opinion whether in particular cases a hyphen is needed as an aid to being understood, and individual judgement will decide.

1. The primary function of the hyphen is to indicate that two or more words are to be read together as a single word with its own meaning. It is always used in such phrasal compounds as *stick-in-the-mud, ne'er-do-well, happy-go-lucky.*

2. Composite nouns consisting of a noun preceded by an attribute can be treated in one of three ways. Some can be left alone without risk of ambiguity, *post office* and *motor car* for instance. Some cannot. In *cross word*, meaning the puzzle, and *black cap*, meaning the bird, the words must somehow be visually linked so as to make the distinction that is made in speech by stressing the first. Usually such composite words are first hyphened and then, if appearances permit, made into a single word, as *crossword* and *blackcap* have been. 'My feeling' says Sir Winston in his protest against hyphening 'is that you may run them together or leave them apart, except when nature revolts.' And if it is right to rid ourselves of hyphens as far as possible the sooner they are run together the better.

Combinations of a noun with an agent-noun take more kindly than most to consolidation, e.g. *bookbinder, doorkeeper, housemaster, pallbearer, platelayer.* In other compounds nature does sometimes revolt. *Seaplane* and *seaside* make pleasant enough words; *seaair* and *seaurchin* do not. But we can often fall back on the alternative: *sea air* and *sea urchin* are not ambiguous. What is an aesthetically tolerable compound is a question that everyone will have his own ideas about. There are some who would reject such compounds as *publichouse, porthole,* and *loophole,* lest they should seem to suggest the pronunciations *-chouse, -thole,* and *-phole;* others may think this over-nice. (The COD would have us write *public house* and *loop-hole;* it admits *porthole* though insisting on *masthead.*) In America they are less squeamish than we are, and do not shrink from such forms as *coattails* and *aftereffects.*

Compounds in which the attribute follows the noun (e.g. *court martial, heir apparent*) can ordinarily do without hyphens, though a precisian will hyphen or consolidate them when they are used as verbs or in the possessive case (*he will be courtmartialled: the heir-apparent's death*). Nor, on the principle suggested in this article, is there any need for the hyphens often used in compound designations of rank or office such as *Attorney General, Lord Lieutenant, Vice Admiral, Under Secretary,* and scores of others. Here both practice and precept are chaotic. Even if we pin our faith to a single current work of reference, it is not easy to discover a uniform principle underlying such decisions as that we must hyphen *Field-Officer* but not *Field Marshal, Quartermaster-General* but not *Attorney General, Commander-in-Chief* but not *Secretary of State, Lieutenant-Governor* but not *Lieutenant Colonel, Lord-Lieutenant* but not *Lord Mayor.* And if we try to get more light by turning to another no less authoritative work of reference it is discouraging to be told that, on the contrary, we must hyphen *Field-Marshal* but not *Field Officer,* and that, in general, the advice there given has little in common with that of the first except its apparent arbitrariness.

**3.** Composite adjectives when used attributively are usually given hyphens, mostly with good reason. They may be adjective + adjective (*red-hot, dark-blue*) or noun + adjective (*pitch-dark,*

*sky-high*), or adjective + participle(*easy-going, nice-mannered*), or noun + participle (*weight-carrying, battle-scarred*) or verb + adverb (*made-up, fly-over*) or phrases such as *door-to-door, up-to-date.* Noun and participle compounds are specially likely to need clarifying hyphens. *The tailor-made dresses,* / *He was surprised to come across a man-eating tiger,* / *Near the hotel is a large moor reserved for shooting-visitors:* these do not give the same sense without the hyphens.

When the first word of the compound is an adverb no hyphen is ordinarily needed, though one may often be found there. It is the business of an adverb to qualify the word next to it; there should be no risk of misunderstanding. To quote Sir Winston again, '*Richly embroidered* seems to me two words, and it is terrible to think of linking every adverb to a verb by a hyphen.' But this will have to be done when the adverb might be mistaken for an adjective. *A little used car* is not necessarily the same as *a little-used car* or *a hard working man* as *a hard-working man* or *extra judicial duties* as *extra-judicial duties.* In *pretty fair guess* on the other hand, *pretty* is unmistakably an adverb and the reader does not need the guidance of a hyphen. So is *over* in the sentence *others may think this over-nice* written earlier in this article, but there the hyphen saves the reader from being put momentarily on a false scent by *think this over.* This possible confusion between adjective and adverb probably accounts for the unnecessary hyphen that often appears with *well* and *ill,* and for the distinction drawn by the textbooks between *well-known* (attribute) and *well known* (predicate).

It is true that combinations of two or more words needing hyphens when used attributively can usually do without them as predicates. *An ill educated man* is ambiguous but *the man is ill educated* is not. We must write *Up-to-date figures, a balance-of-payments crisis, the Africa-for-the-Africans slogan.* But hyphens would be worse than useless in *The figures are up to date, a crisis over the balance of payments, the slogan 'Africa for the Africans'.*

**4.** In composite adjectives, as in the nouns, hyphens can often be got rid of by consolidation, especially of the participial compounds. *Panicstricken, thunderstruck, seaborne, aircooled,* and many others may be so written. Adjectives and nouns formed from phrasal verbs (a *put-up* job, a *knock-out*) must obviously keep their hyphens unless they can be consolidated. Here again the Americans are ahead of us: they make one word of *frameup* for instance. We, if the COD is right, boggle even at *pullover*. *The Times* is to be congratulated on defying the dictionaries and writing of *breakthroughs, holdups, makeup,* and *takeovers.*

**5.** Carelessness, or perhaps a laudable desire to economize in hyphens, sometimes leads to the omission of one where it is manifestly a case of all or none. Neither can be dispensed with in *Two-year-old horses, three-quarter-hour intervals, submarine-cable-laying ships.* Some pretty problems in hyphening are set by the unpleasant modern habit (see HEADLINE LANGUAGE) of forgetting the existence of prepositions and using a long string of words as a sort of adjectival sea serpent (e.g. *a large vehicle fleet operator mileage restriction has now been made imperative*). Those who like writing in this way can be left to solve their problems for themselves. Indeed many of our difficulties with hyphens are of our own making; we can avoid them by remembering prepositions and writing, say, *intervals of three quarters of an hour* instead of *three-quarter-hour intervals.*

**6.** Prefixes follow a similar course. The hyphen used at first is later discarded, and the word is joined up except where nature revolts. Here too the British, unlike the Americans, are curiously hesitant about taking the second step. *Non-stop* should be ready to follow the lead of *nonsense, postmortem* that of *postscript, pre-natal* that of *predecease, off-chance* that of *offshoot, by-product* that of *bystander,* and so on. Awkward juxtapositions, espe-

cially of vowels, may be an impediment, but we are too timid about this. Having swallowed *coeval* and *coefficient,* we are unreasonable to strain at *coeducational.* There is no risk of puzzling the reader if we write *preeminent* or *cooperative,* though consideration for him may make us write the abbreviation of the latter as *co-op.* We already write *coalesce* and *coaxial* without fearing that they may suggest coal or coax; perhaps some day we may write *coworker,* as Americans already do, without feeling nervous about the cow. Some new words with awkwardly placed vowels undoubtedly need hyphens. We cannot yet do without one in *de-ice,* but in time *deice* may look no odder than *deist.*

There is no apparent justification for the hyphen that now accompanies all *self-* compounds except *selfsame,* especially as it leads to the absurdity of having to hyphen the *un* in such words as *un-self-conscious. Ex* on the other hand readily coalesces except when used in the sense of quondam. Here it cannot do without a hyphen, and if what it qualifies consists of more than one word they must be hyphened too, even though they are not naturally so. *Ex-Prime Minister* suggests a minister who is past his prime, as *pre-first war* suggests a war before the first one.

**7.** The practice of using a hyphen in the sense of *to* or *and* or *with* as in the *London-Birmingham motorway,* the *Eisenhower-Dulles partnership,* breaks down when one or more of the things linked consists of two or more words. Thus the natural meaning of *The Chipping Norton-Chipping Camden* road is the road that runs from Chipping to Camden by way of Norton-Chipping. To hyphen all the words is no remedy. Either a suitable conjunction or preposition must be substituted for the hyphen or, if a symbol is used in the middle, it must be a dash, or perhaps a stroke (virgule).

**8.** *Chaplains (whole- or part-time) should be appointed. | Both four- and six-cylinder models are made.* The function of a hyphen is to link a word with its immediate neighbour, and to separate them in this way in order to avoid doubling the linked word is a clumsy device that should be avoided if possible; which is the lesser evil is a matter of taste in each case. The stuttering effect of the separations in the following sentence is a warning against choosing that alternative lightly. *The era is one of the masses then; yet they are either tyrannized by the totalitarians or un- or mis-managed by democrats, shoved from behind by despots, or un- or mis-led by democratic 'leaders'.*

For articles touching on particular uses of the hyphen, see BY BYE BY-, CO-, COMMON SENSE, EVERY ONE, FELLOW, FREE, GOOD WILL, -LESS, -LIKE, MEAN TIME, MIS-, NO 5, RE-, SECOND, -S-, -SS-, -SSS 2, STATE, SUMMER, THREE QUARTER, TODAY ETC., WELL AND WELL-, WILD.

**hypocorisma.** Use of pet names, nursery words, or diminutives, or the like, either simply, as *Molly* for Mary, *comfy* for comfortable, *hanky* for handkerchief, etc., or by way of euphemism, as *fib* for lie, *undies* for underclothes.

**hypothecate. 1.** *H.* makes *-cable*, *-tor*; see -ABLE I, -OR. **2.** *H.* means only to mortgage or pledge. In the following extract—*The Nahua race, which, by tradition, served the Aztecs in much the same way as to origin as the hypothecated Aryans serve ourselves*— it is used as a verb corresponding to *hypothesis*; if an allied verb is really necessary, *hypothesize* is the right form, though it is to be hoped that we may generally content ourselves with *assume*.

**hypothetic(al).** There does not seem to have ever been any distinction in meaning, and the longer is now the ordinary form; see -IC(AL).

**hysteric(al).** The short form has virtually gone out of use as an adjective; see -IC(AL).

**hysteron proteron.** Putting the cart before the horse in speech, as in Dogberry's *Masters, it is proved already that*

*you are little better than false knaves, and it will go near to be thought so shortly.* A less crude example: *The N.C.B. are running the mines as a prosperous and viable industry.*

# I

**I. 1.** *Between you and I* is a piece of false grammar which, though often heard, is not sanctioned, like its opposites *It's me* and *That's him*, even in colloquial usage. But it has distinguished ancestry. Shakespeare wrote *All debts are cleared between you and I*, and Pepys *Wagers lost and won between him and I*. A similar lapse is seen in *It was a tragedy of this kind which brought home to my partner and I the necessity for . . . . | It was these gazelle that set Charles and I talking of hunting.* These solecisms, which would never be committed if the pronoun of the first person stood alone (*brought home to I; set I talking*), are curiously common when it is the second of two words governed by a verb or preposition. It is as though *and I* were treated as a suffix to the preceding word, forming a composite whole not admitting of inflexion. To Dickens it seems to have been natural that Nell's grandfather should say *Leave Nell and I to toil and work*, and more than a hundred years later it seems to have been no less natural for a broadcaster giving a serious talk to say *These things are for the betterment of you and I*. Perhaps this use of *I* is partly due to a vague recollection of hearing *me* corrected when wrongly used in a subjective partnership, as Lydia Bennet should have been when she said *Mrs. Forster and me are such friends.* See also LET and ME.

**2.** *I*, like WE, is liable to be used in successive sentences with different meanings. In the extract below, the first two *I*s mean the average moralist, while the third means the reviewer of Dr. Westermarck's book. It is an insidious trap, but more often baited with *we*, which frequently means in one sentence the editor of his paper, and in the next the country or the Party or any other of the many bodies of which he is a member: *In this respect Dr. Westermarck has given a less adequate account of the moral sentiment than Adam Smith, who declares that our ideas of merit and demerit have a double origin, not only in sympathy with the resentment of the sufferer, but in want of sympathy with the motives of the doer. I condemn theft partly because I dislike thieving and sympathize with the sufferer's claim to keep his property. I cannot help thinking that, though every now and then he does justice to sympathy with the direct motives or impulses from which action arises, Dr. Westermarck overlooks them in favour of retributive sympathy with the recipient.*

**-i.** For plurals with this ending see LATIN PLURALS.

**-iana.** This suffix, applied to the name of an author, originally meant writings or sayings *by* him. It was first used of Dr. Johnson; *Johnsoniana, or a collection of Bon Mots by Dr. Johnson and Others* was the title of a book published in 1776, during his lifetime. The suffix is still sometimes used in this sense (*Something of the Mencken of the final phase would have rounded off the Menckeniana very piquantly*), but nowadays it more often means writings *about* an author: a modern dictionary definition of *Shakespeariana* is 'a collective term for all literary matter on the subject of the poet and his works'. Thus, *The time is overdue for a closed season in Dickensiana. | Miss Hopkins adds a facet to Bronteana and helps considerably towards the final shaping. | He amassed a notable collection of contemporary Mozartiana.* Or it may be used of a period. *Odds and ends of Victoriana in everything from short squibs to full-length books pour from the British Press.*

**ibidem.** Pronounce *ibī'dĕm*.

**-ic(al).** A great many adjectives appear with alternative forms in *-ic* and *-ical*. Often the choice between them on any particular occasion is immaterial, so far as the writer's imme-

diate object is concerned. To those who can afford time to think also of the interests of the English language it may be suggested that there are two desirable tendencies to be assisted.

The first of these is DIFFERENTIATION. There are many pairs in *-ic* and *-ical*, each form well established and in constant use, but with a difference of meaning either complete or incipient. The final stage of differentiation is seen in *politic* and *political*, which are not even content, as usual, to share an adverb in *-ically*, but make *politicly* by the side of *politically*. Between *economic* and *economical* the distinction is nearly as clear, though the seal has not been set upon it by a double provision of adverbs; most writers are now aware that the two words mean different things, and have no difficulty in choosing the one required. This can hardly be said of *comic(al)*, the short form of which is often made to do the other's work. And so the differentiations tail off into mere incipiency. Every well-established differentiation adds to the precision and power of the language; every observance of an incipient one helps it on the way to establishment, and every disregard of it checks it severely. It is therefore clear that writers have a responsibility in the matter.

The second laudable tendency is that of clearing away the unnecessary. When two forms coexist, and there are not two senses for them to be assigned to, it is clear gain that one should be got rid of. The scrapping process goes on slowly by natural selection; sometimes the determining cause is apparent, as when *hysteric, cynic,* and *fanatic,* give way to *hysterical, cynical,* and *fanatical,* because the former have acquired a new function as nouns; sometimes the reasons are obscure, as when *electric* and *dynamic* supersede the longer forms while *hypothetic* and *botanic* are themselves superseded. But that one or other should prevail is a gain; and it is a further gain if the process can be quickened. With this end in view, this dictionary states about many *-ic(al)* words, in their places, which appears to be the winning side, so that writers may be encouraged to espouse it.

Separate entries will be found (omitting -ic, -ical) for botan-, casuist-, com-, cub-, cyn-, diabol-, dynam-, econom-, electr-, fanat-, geograph-, geometr-, hypothet-, hyster-, ident-, lyr-, mag-, period-, philosoph-, sto-, trag-.

**-ics. 1.** *-ics*, *-ic*. Among the names of sciences, arts, or branches of study, are a few words in -ic that rank as real English; the chief are *arithmetic*, *logic*, *magic*, *music*, and *rhetoric*; but the normal form is -ics, as in *acoustics*, *classics*, *dynamics*, *mathematics*, *physics*, *politics*. The substitution of -ic for -ics (*dialectic*, *gymnastic*, *linguistic*, *metaphysic*, etc.) in compliance with French and German usage has the effect, whether it is intended or not, of a display of exotic learning, and repels the possibly insular reader who thinks that 'English is good enough for him'. It should be added, however, that the -ic and -ics forms can sometimes be usefully kept for separate senses; thus, *dialectic* might be reserved for the art of logical disputation and *dialectics* for a particular person's exhibition of skill in it; conversely, a *personal ethic* is an individual's code of ethics and a *tactic* a particular exercise of the art of tactics. But it is not with many words, nor on many occasions, that this need arises, and it is not usually with this end in view that the -ic words are made.

**2.** Grammatical number of *-ics*. This is not so simple a matter as it is sometimes thought. The natural tendency is to start with a fallacy: We say Mathematics *is* (and not *are*) a science; therefore *mathematics* is singular. But there the number of the verb, whether legitimately or not, is at least influenced, if not determined, by that of *a science*. The testing should be done with sentences in which there is not a noun complement to confuse the issue: *Classics are*, or *is*, *now taking a back seat*; *Conics is*, or *are*, *easier than I expected*; *What are*, or *is*, *his mathematics like?*; *Politics are*, or *is*, *most fascinating*; *Your heroics are*, or *is*,

*wasted on me*; *Athletics are*, or *is*, *rampant in the big schools*; *Tactics are*, or *is*, *subordinate to strategy*. The rules that seem to emerge are: (1) Singular for the name of a science strictly so used; *Metaphysics*, or *Acoustics*, deals *with abstractions*, or *sound*. (2) Plural for those same names more loosely used, e.g. for a manifestation of qualities; often recognizable by the presence of *his*, *the*, etc.: *His mathematics* are *weak*; *Such ethics* are *abominable*; *The acoustics of the hall* are *faulty*. (3) Plural for names denoting behaviour or the like: *Heroics* are *out of place*; *Hysterics* leave *me cold*. (4) The presence of a singular noun complement often makes the verb singular: *Mathematics*, or even *Athletics*, is *his strong point*. (For the general rule in such cases see NUMBER 1.)

**idea.** *Humperdinck had the happy idea one day to write a little fairy opera.* *Of writing* is more idiomatic. See GERUND 3.

**identic(al).** The short form has been so far ousted by the long as to be now a mere ARCHAISM except in the language of diplomacy (*identic note* etc.), where it means the same in substance rather than completely identical in wording. See -IC(AL).

**identify.** The extended use of *identify* as a reflexive verb in the sense of to associate oneself closely and inseparably dates from the 18th c., but is only suitably used of some close, constant, and well-known association such as, say, that of Mr. Willett with the campaign for summer time or of Mrs. Pankhurst with that for women's suffrage, or in the intransitive sense in which psychologists use it. To treat *identify with* as a stylish word for associate with or take part in, in a temporary or casual way (*There was a demonstration going on and the accused, on arrival at the scene, identified himself with it*), is a SLIPSHOD EXTENSION.

**id est.** See I.E.

**ideologue, -logist, -logy,** etc. So spelt, not *ideal-*. The words are formed from the Gk. *ἰδέα*, and the Greek combining vowel is *-o-* with rare exceptions for substantives of all declensions. See -O- and -LOGY.

The modern vogue of the word *ideology* is a natural result of the decline of religious faith. We have had to find a word, free from the religious associations of *faith* and *creed*, for belief in those politico-social systems vaguely indicated by such words as *democracy*, *socialism*, *communism*, and *fascism*, which excite in their adherents a quasi-religious enthusiasm. *Ideology* (the science of ideas) lay ready to hand, the more acceptable because it seemed to suggest striving for an ideal. It was therefore pressed into this new service, which has now become its main occupation

**idiom.** This dictionary being much concerned with idiom and the idiomatic, some slight explanation of the terms may perhaps be expected. 'A manifestation of the peculiar' is the closest possible translation of the Greek word. In the realm of speech this may be applied to a whole language as peculiar to a people, to a dialect as peculiar to a district, to a technical vocabulary as peculiar to a profession, and so forth. In this book, 'an idiom' is any form of expression that has established itself as the particular way preferred by Englishmen (and therefore presumably characteristic of them) over other forms in which the principles of abstract grammar, if there is such a thing, would have allowed the idea in question to be clothed. 'Idiom' is the sum total of such forms of expression, and is consequently the same as natural or racy or unaffected English; that is idiomatic which it is natural for a normal Englishman to say or write. To suppose that grammatical English is either all idiomatic or all unidiomatic would be as far from the truth as to suppose that idiomatic English is either all grammatical or all ungrammatical. Grammar and idiom are independent categories; being applicable to the same material, they sometimes agree and sometimes disagree

about particular specimens of it. The most that can be said is that what is idiomatic is far more often grammatical than ungrammatical; but this is worth saying, because grammar and idiom are sometimes treated as incompatibles. The fact is that they are distinct, but usually in alliance. To give a few illustrations: *You would not go for to do it* is neither grammatical nor idiomatic English; *I doubt that they really mean it*, *The distinction leaps to the eyes*, and *A hardly earned income*, are all grammatical, but all for different reasons unidiomatic; *It was not me*, *Who do you take me for?*, *There is heaps of material*, are idiomatic but ungrammatical; *He was promoted captain*, *She all but capsized*, *Were it true*, are both grammatical and idiomatic. For examples of special idioms see CAST-IRON IDIOM.

**idiosyncrasy, -cratic.** The right spelling (-*sy*, not -*cy*) is of some importance, since the wrong one distorts the meaning for all who have a tincture of Greek by suggesting a false connexion with *autocracy* and the many other words in -*cracy*. Those words are from Greek *κράτος* power; this is from Greek *κρᾶσις* mixture. Its meaning is peculiar mixture, and the point of it is best shown in the words that describe Brutus: His life was gentle, and *the elements So mixed in him that* Nature might stand up And say to all the world 'This was a man'. One's idiosyncrasy is the way one's elements are mixed, and the nearest synonyms for it are *individuality*, *personality* or *'make-up'*. But since all these have positive implications not present in *idiosyncrasy*, the continued existence of the latter in its proper sense is very desirable, and it should be kept to that sense. Thus it is reasonable to say that a person has no personality or no individuality, but a person without an idiosyncrasy is inconceivable. Since *idiosyncrasy* means all the ingredients of which a unit is composed, and their proportions and reactions—a valuable compound notion that we may be thankful to find compressed into a single word—it is a pity that it is often used as a polysyllabic substitute for various things that have good simple names of their own; it is both pretentious and absurd to say that so-and-so is one of your idiosyncrasies when you mean one of your habits, ways, fads, whims, fancies, or peculiarities. In each of the following quotations a more suitable word is suggested: *It is an idiosyncrasy of this grumbler that he reads his own thoughts into the minds of others* (characteristic). / *For one reason or another—lack of money, lack of men, sometimes the idiosyncracies* [sic] *of committees— the library has been far less useful than it might have been to the serious student* (fads). / *I do not find him, though he is very quick in observing outward idiosyncrasies, a truthful or an interesting student of the characters, the minds and hearts, the daily actions and reactions, of men and women* (peculiarities). / *Moreover, it* [a liturgy] *is desired as a protection against the idiosyncrasies of the minister, whether in his doctrine or its expression* (vagaries). / *There are several kinds of food freaks; some people have an idiosyncrasy to all fish, particularly shellfish and lobsters* (antipathy or allergy).

*Idiosyncratic* is the adjective of *idiosyncrasy*—unfortunately, because it encourages by an accident the confusion between *-crasy* and *-cracy*. If *idiosyncrasy* is a word that has a real value, but should be much less used than it is, *idiosyncratic*, its hanger-on, should be kept still more severely in its place. The quotations show that there is a danger of its getting more vogue than it deserves; what the reader feels is not that his author has used the word in a wrong sense—he has not— but that he would have done better to circumvent, somehow, the need for it: *We continue to read for much the same reason as incites a Purple Emperor to feed on carrion, a cat on mice, a queen bee on nectar, the South Wind on a bank of violets; we are in pursuit of the idiosyncratic* (of what appeals to *us*). / *What we cannot help learning of their maker, or discoverer—his philosophy, his idiosyncratic view of things—is there, not* *because he wittingly put it there, but because he could not keep it out* (individual. Here, at any rate, the writer could have kept *idiosyncratic* out). / *To be thinking and pondering, roving and exploring between the lines of a book is a less arduous and fussy, a quieter and more idiosyncratic enterprise* (rewarding?). / *He never hesitates at any joke, however idiosyncratic* (however little amusing to anyone but himself?).

**idola fori, idols of the market (place).** This learned phrase (pronounce *idō'lă fō'rī*), in Latin or English, is not seldom used by the unlearned, who guess at its meaning and guess wrong. It is a legitimate enough phrase in writing meant for the educated only, but hardly so in the ordinary newspaper, where it is certain not to be understood by most readers, and where it therefore tends to be given, by SLIPSHOD EXTENSION, the false sense that those who have never been told what it means may be expected to attach to it. That false sense is 'vulgar errors' or 'popular fallacies', one of which expressions should be used instead of it, since it in fact has a much more limited meaning than they, and one not obvious without explanation. See POPULARIZED TECHNICALITIES.

It is the third of Bacon's four classes of fallacies, more often mentioned than the other three because its meaning seems plainer, though it is not so in fact. There are the idols (i.e. the fallacies) of the tribe, the cave, the market, and the theatre, which are picturesque names for (1) the errors men are exposed to by the limitations of the human understanding (as members of the *tribe* of man); (2) those a person is liable to owing to his idiosyncrasy (as enclosed in the *cave* of self); (3) those due to the unstable relation between words and their meanings (which fluctuate as the words are bandied to and fro in the conversational exchange or *word-market*); and (4) those due to false philosophical or logical systems (which hold the *stage* successively like plays). The tribe is the human mind, the cave is idiosyncrasy, the market is talk, and the theatre is philosophy. Who would guess all that unaided? Who, on the contrary, would not guess that an idol of the market-place was just any belief to which the man in the street yields a mistaken deference? The odd thing is that no better instance could be found of an idol of the market than the phrase itself, oscillating between its real meaning and the modern misuse, so that the very person who pours scorn on idola fori is often propagating one in the very act of ridiculing the rest. Well, 'tis sport to have the engineer hoist with his own petard. The mistake is common enough, but is not easily exhibited except in passages of some length, so that one must here suffice; the tendency to exalt the man of action above the man of theory may be ill-advised, but it has nothing to do with shifting acceptations of words, and is not an idolum fori: *With us the active characters, the practical men, the individuals who, whether in public or in private affairs, 'get on with the job', have always held the first place in esteem; the theorists and philosophers a place very secondary by comparison. It is not easy to account for this common estimate. For one thing, as soon as inquiry is made into it, the belief proves to be without foundation—just one of the* idols of the market-place.

**idolatry.** See -LATRY.

**idyl(l).** The form with *-ll* is now usual. The OED recognized only one pronunciation, *ī-*, not *ĭ-*; on this, however, there is room for difference of opinion. (1) It is certain that many people say *ĭ'dĭl*; (2) with *idol* and *idle*, both commoner words, ready to confuse the hearer, a separate pronunciation is all to the good, if there is nothing against it; and (3) it has been pointed out in the article FALSE QUANTITY that the length of the first syllable in the Greek is nothing against its being shortened in the English word. Modern dictionaries recognize both.

**i.e., id est.** 1. To write, or even to say, this in the full instead of in the abbreviated form is now so unusual as to convict one of affectation.

2. *i.e.* means that is to say, and introduces another way (more comprehensible to the hearer, driving home the speaker's point better, or otherwise preferable) of putting what has been already said. It does not introduce an example, and when substituted for *e.g.* in that function, as in the following extract, is a blunder: *Let your principal stops be the full stop and comma, with a judicious use of the semicolon and of the other stops where they are absolutely necessary* (i.e. *you could not dispense with the note of interrogation in asking questions*).

3. It is invariable in form; the changing of it to *ea sunt* etc.—*which deals with persons* (ea sunt, *all present and future members*)—is due to the same misconception (explained under FOLLOW) as the incorrect *as follow*; cf. also INTER ALIA.

4. It is naturally preceded by a stop; it should not be followed by a comma unless the sense requires one, to introduce a parenthesis for instance. *He attacked reactionaries, i.e. those whose opinions differed from his own,* but *He attacked reactionaries, i.e., it would seem, those whose opinions etc.*

**-ies, -ein.** Until recently the dictionaries prescribed a disyllabic pronunciation of *-ies* for words of Latin origin such as *series, species, rabies, caries, scabies.* But in fact few doctors pronounce the second syllable of *rabies* and *caries* differently from that of *herpes,* and almost everyone takes the same liberty with *series* and *species.* It is better to bow to the inevitable than to persist in a vain attempt to preserve what we suppose to have been the way the Romans pronounced these words, and the dictionaries now recognize *ēz* for most of them, if only as alternatives.

Words ending *ein(e)* (*protein, caffeine, codeine*) are rebelling in the same way, even though they are less commonly used; *ēn* is likely to establish itself; modern life is too hurried for these niceties.

**if.** To avoid possible ambiguity it may be prudent to confine *if* to its proper duty of introducing the protasis of a conditional sentence, and not to use it as a substitute for *though* or *whether* or (with *not*) to introduce a possible alternative. *Such experiences are agreeable, if rare.* (Though they are rare or only if they are rare?) *Please inform the secretary if you intend to be present.* (Whether you intend or only if you do intend?) *He enjoyed a large, if not overwhelming, practice in the best class of litigation.* (Though not overwhelming or you might almost call it overwhelming?)

**if and when.** Any writer who uses this formula lays himself open to entirely reasonable suspicions on the part of his readers. There is the suspicion that he is a mere parrot, who cannot say part of what he has often heard without saying the rest also. There is the suspicion that he likes verbiage for its own sake. There is the suspicion that he is a timid swordsman who thinks he will be safer with a second sword in his left hand. There is the suspicion that he has merely been too lazy to make up his mind between *if* and *when*. Only when the reader is sure enough of his author to know that in his writing none of these probabilities can be true does he turn to the extreme improbability that here at last is a sentence in which *if and when* is really better than *if* or *when* by itself.

This absurdity is so common that it seems worth while to quote some examples, bracketing in each either 'if and' or 'and when', and asking whether the omission would in any way change the meaning or diminish the force of the sentence: *The Radicals do not know quite clearly what they will be at (if and) when the fight is renewed.* / *It is to be hoped that the Labour Party, if (and when) they come to power, will be courageous enough to free the ether from the bondage of commercialism.* / *But if (and when) the notices are tendered it will be so arranged that they all terminate on the same day.* / *Mr. Macmillan should try to come to a preliminary*

*understanding with General de Gaulle about how Britain might dispose of its nuclear armoury if (and when) it joins the common market.* / *They must, of course, be certain that they are getting what they are bargaining for, but (if and) when they have made sure of that, they would be wisely advised to pay the price.*

It was admitted above that cases were conceivable in which the *if* and the *when* might be genuinely and separately significant. Such cases arise when one desires to say that the result will or does or did not only follow, but follow without delay; they are not in fact rare, and if a really good writer allows himself an *if and when*, one such must have presented itself. But in practice he hardly ever does it even then, because any strong emphasis on the absence of delay is much better given by other means, by the insertion of *at once* or some equivalent in the result clause. So true is this that, when the devotees of *if and when* have had the luck to strike a real opportunity for their favourite, they cannot refrain from inserting some adverb to do over again the work that was the only true function of their *and when*; in the following quotations these adverbs that make *and when* otiose are in roman type: *The electors knew perfectly well that if and when the Parliament Bill was placed on the Statute-book it would* immediately *be used to pass Irish Home Rule.* / *If and when the Unionist Party win a General Election we are to have* at once *a general tariff on foreign manufactured goods.* / *It is true that if and when an amendment giving women the vote is carried this amendment is* thenceforward *to become part and parcel of the Bill.*

*When or if* is not so purposeless as *if and when*; *or if* does serve to express that the writer, though he expects his condition to be realized, has his doubts: *An official pronouncement as to what particular items of socialist legislation it is proposed to repeal, when, or if, the opportunity arrives.* As *and when* and UNLESS AND UNTIL are open to the same objections as *if and when*, but are much less common.

**ignoramus.** Formerly a law term, meaning *we don't know*, endorsed on a Bill of Indictment by a Grand Jury unwilling to return a True Bill because 'they mislike their evidence as defective or too weak to make good the presentment'. (Blount's Law Dict. 1691, quoted by Skeat.) Plural *-uses*. See LATIN PLURALS.

**ignoratio elenchi.** 'Ignoring of the (required) disproof.' A fallacy that consists in disproving or proving something different from what is strictly in question; called in English *the fallacy of irrelevant conclusion*. If the question is whether the law allows me to pollute water passing through my garden, and I show instead that it *ought to* allow me, since the loss to me by abstaining is a hundred times greater than my neighbour's from the pollution, I am guilty of i. e.

**-ile.** In words with this suffix the modern tendency, says the SOED, is to pronounce *-īl*, 'with some exceptions, in all cases'. This tendency has continued, and many of the words whose conventional pronunciation was once -il have now -īl as the preferred or only one; these include DOCILE, *domicile*, FACILE, FERTILE, FRAGILE, *missile*, *mobile*, *prehensile*, *servile*, STERILE, TACTILE, and VIRILE. (In U.S. -*īl* is more tenacious.) The few that are normally pronounced *-ēl* are also variable; dictionary authority can be found for -il and -īl in *imbecile*, and, though the dictionaries recognize nothing but -ēl for *automobile*, *īl* is often heard. For PROFILE see that article.

**ilex.** Pl. *ilexes*; see -EX, -IX, 3.

**ilk** means same; it does not mean family or kind or set or name. *Of that ilk* is a form constructed for the case in which proprietor and property have the same name; *the Knockwinnocks of that ilk* means the Knockwinnocks of Knockwinnock. The common maltreatments of the phrase, some of which are illustrated below, are partly unconscious and due to ignorance of the meaning of *ilk*, and partly facetious; indulgence in such WORN-OUT HUMOUR is much less forgivable than for an Englishman not to know what a Scottish word means: *The Walkers are a numerous race . . . one of that ilk has suggested that an ancestor probably walked to the Crusades.* | *Lord's was simply the field rented out to the M.C.C. by Thomas of that ilk.* | *Robert Elsmere, the forerunner of so many books 'of that ilk'.* | *This publication was undertaken by John Murray, the first of that ilk.* | *The mighty figures hold the centre of the stage, but the Bunters and their ilk have a good right to frolic at their feet.* | *Part II of the Act, which dealt with multi-occupation, was designed to deal with Rachman and his ilk.* This SLIPSHOD EXTENSION has become so common that the OED Supp. was constrained to add to its definitions 'also by further extension, often in trivial use,—kind, sort'. The COD calls it vulgar.

**illegal, illegitimate, illicit, unlawful.** These words, though largely overlapping, have developed certain differentiations. *Illegal* is the most precise, with its meaning of contrary to the law of the land. The scope of *illegitimate* is wider; it includes not only what is not authorized by law but also what is against propriety or reason, and its possible application ranges from a child born out of wedlock to a deduction not justified by premises. *Illicit* also covers what is impermissible though not necessarily illegal, including logical fallacies; in its sense of contrary to law it is applied especially to activities that the law allows only subject to compliance with certain conditions, e.g. gambling and the manufacture and sale of alcoholic drinks. (This meaning goes naturally with its derivation from Lat. *licere*, permit.) *Unlawful*, with its sweeping implication of comprising what is forbidden not only by the law of the land but also by higher authority such as international law or divine ordinance, is falling into disuse in common currency.

**illegible, unreadable.** The i. is not plain enough to be deciphered; the u. is not interesting enough to be perused.

**illiteracies.** There is a kind of offence against the literary idiom that is not easily named. The usual dictionary label for some specimens of it at least is *vulg.*; but the word *vulgar* is now so imbued on the one hand with social prejudices and on the other with moral condemnation as to be unsuitable. The property common to these lapses seems to be that people accustomed to reading good literature do not commit them and are repelled by them, while those not so accustomed neither refrain from nor condemn them. They may perhaps be more accurately as well as politely called illiteracies than vulgarisms; their chief habitat is in the correspondence columns of the press. A few familiar types may be here collected for comparison, with just enough in the way of illustration to enable each usage to be recognized; actual quotations will be found under many of the words mentioned in their dictionary places. See also INCOMPATIBLES and STURDY INDEFENSIBLES. Some of the examples listed in the former article could equally well be classed as illiteracies, and the reader may think that some of those called sturdy indefensibles deserve the same sterner verdict.

*Likewise* as a conjunction (*Its tendency to wobble . . . l. its limited powers of execution*).

*However, whatever, whoever,* etc., interrogative when written as a single word (*However did you find out?; Whatever can this mean?*).

*Same, such,* and *various,* as pronouns (*Will submit same,* or *the same, for approval; Have no dealings with such; Various have stated*).

Undiscriminating use of split infinitives (*Am ready to at once carry out my promise*).

*Re* in unsuitable contexts (*The author's arguments re predestination*).

*Write* with personal object only (in U.S. a common colloquialism) (*Though she had promised to write him soon*).

*I* for *me* when in company (*Between you and I*).

*Me* etc. for *my* etc. in gerund construction (*Instead of me being dismissed*).

*Between . . . or* for *between . . . and* (*The choice is between glorious death or shameful life*).

*Neither* with a plural verb (*For two reasons neither of which are noticed by Plato*).

**illogicalities.** The spread of education adds to the writer's burdens by multiplying that pestilent fellow the critical reader. No longer can we depend on an audience that will be satisfied with catching the general drift and obvious intention of a sentence and not trouble itself to pick holes in our wording. The words used must nowadays actually yield on scrutiny the desired sense; to plead that anyone could see what you meant, or so to write as to need that plea, is not now permissible; all our pet illogicalities will have to be cleared away by degrees.

Though Milton might be excused or even commended for calling Eve fairest of her daughters, the modern newspaper man must not expect pardon for similar conduct. *Sir Ernest Cassel's Christmas gift to the hospitals of £50,000 is only the latest of many acts of splendid munificence by which he has benefited his fellows before now.* If it is the latest of them, says the pestilent one, it is one of them; if one of them, it was given before now; but it is in fact given now, not before now; which is absurd.

Take, again, the following comment on a quotation the commentator thinks unjustified: *Were ever finer lines perverted to a meaner use?* We know well enough what he is trying to do—to emphasize the meanness of the use— and if that had been all it would have been better to write *Never were lines perverted to a meaner use*; that comment would be made weaker, not stronger, if changed to *never were fine lines* etc. and that again would be further weakened, not strengthened, by a change of *fine* to *finer*; everything that narrows the field of rivals for the distinction of meanest perversion, as *fine* and *finer* do progressively, has an effect contrary to what was intended. True, in this case, it would have been worth while to insert *fine*, since without it *perverted* lacks point; but the change to *finer* weakens the force without adding to the accuracy. Richard III says *Was ever woman in this humour won?*; to have said *Princess*, or *prouder Princess*, instead of *woman* would have made the marvel less and not greater.

Another common, and more conspicuous, illogicality is the unintended anticlimax. *Masters, it is already proved that you are little better than false knaves, and it will go near to be thought so shortly.* Dogberry felt no uneasiness about putting it that way, and some writers seem to agree with him: *A scepticism about the result of military operations which must have had and probably has had a damping effect upon the soldier* (If it must have had, it certainly, not probably, has had).

The abandonment of blind confidence in *much less* is another compliment that will have to be paid to the modern reader's captious logic It is still usual to give no hearing to *much more* before deciding for its more popular rival: *It is a full day's work even to open, much less to acknowledge, all the presents.* See MUCH 2.

A stray variety or two may now bring this subject to an end, though it might be treated at much greater length: (From a notice in a public park): *Any person not putting litter in this basket will be liable to a fine of £5.* Those who have no litter must, it seems, go and find some. | *The schedule we shall have to face will be a much longer one than it would have been if we had undertaken the work this year, and longer still than it would have been if we had been able to do the work last year.* We may deeply sympathize with a writer who has brought himself to the pass of having to choose between saying *still more longer* and being illogical, but we cannot let him off that *more.* | *That would quite easily and fairly redress what he admitted to be the only grievance he could see in Establishment.* The *he* is a

supporter of the Established Church; he would *maintain,* not *admit,* that it is the only grievance, and should have said 'what he admitted to be a grievance, though it was the only one he could see'. | *The gown is not normally worn in the proper manner by ladies in statu pupillari if it is worn with trousers.* What then is the proper manner of wearing a gown with trousers to which these ladies normally neglect to conform?

Other examples or remarks will be found in BUT 3, -ER AND -EST 7, 8, HAZINESS, HELP, REASON, STURDY IN-DEFENSIBLES, THOUGH 4, TOO 2, and YET.

**illume, illuminate, illumine.** The first is a POETICISM and the last obsolescent. *Illuminate* is today the only current word for both literal and figurative senses.

**illusion.** See DELUSION for the differences between the two words.

**illustrate.** The pronunciation *ĭ'lŭstrāt* (as opp. *ĭlŭ'strāt*) has been slowly arrived at, but is now general; see RECESSIVE ACCENT. For *illustrative* the OED gives *ĭlŭ'strătĭv* only; but the fixing of *ĭ'lŭstrāt* has produced the alternative *ĭ'lŭstrătĭv* now recognized by some dictionaries.

**im-.** For spelling of words with variants in *em-*, see EM- AND IM-. The following, not there mentioned, should have *im-*, and not the rarely used or obsolete *em-*: *imbrue, imbrute, impale, imparadise, impark, impawn, imperil.*

**image.** It has been said that people can think only in images. If that is true it must always have been so, and cannot by itself account for the recent vogue of the word. *This could do the Liberals' own i. some harm in blurring the party's distinctive theme.* | *If the Church is to regain any power it will need a complete face-lift of its i.* | *The pay-pause has done much to undermine the Government's i.* | *Asked for a definition of his duties, the vice-chancellor of one of the new universities* *answered without hesitation 'A vice-chancellor is an i.-giver'. | The Australians have invented a literary i. of themselves.* Anyone can find half a dozen similar examples in almost any day's newspapers. The word, used thus, means the 'idea'—the general impression—of some person or institution received by the mind's eye of an outsider, and the image he sees will determine for him whether the person is a good chap or a bad chap, the institution a good show or a bad show. Perhaps it is television that has done it. Politicians and advertisers and other advocates of themselves or their causes can now project their images into our very homes. This at least is clear: that though we may not care very much nowadays about the gift of seeing ourselves as others see us, we put a high value on that of persuading others to see us as we see ourselves. 'Members of the British Insurance Association, it seems, are concerned about the public image of the insurer. . . . It is perfectly understandable that insurers should want to appear, and remain, in the best possible light *vis-à-vis* their clientele. Publishers, barristers, chartered accountants, and architects probably feel the same way; although what may be termed the "ideal public image" may differ a bit as between these various professions. . . . But on one point they would all be agreed. From the neck up the image should be the kind of image that inspires immediate and lasting confidence.'—*The Times*, 29th June 1962. The use of *image* in this sense does no violence to the language. It is one of those VOGUE WORDS (OVERALL and PROPOSITION are others) whose mischief lies not so much in misuse as in over-use. They offer themselves as handy reach-me-downs to people who would do better to think exactly what they want to say. If the writer who said that the Government's image was being 'undermined' had done this, he might have expressed himself differently. See also PERSONA.

**imaginary, imaginative.** The meanings of the two are quite distinct, and never interchangeable. That is imaginary which exists only in someone's imagination; he (or his powers or products) is imaginative who is able or apt to form mental pictures. Any confusion between the two is due to the fact that there are things to which either can be applied, though in different senses, and with some such things the distinction is not always apparent. The difference between an imaginary and an imaginative person is clear enough, but that between imaginary and imaginative distress is elusive; the begging impostor exploits the former; the latter is created and experienced by the tragic or lyric poet (Such a price The Gods exact for song, To become what we sing). *The place is described with such wealth of detail as to lead one to the conclusion that it must have existed; but, of course, on the other hand, it may have been purely imaginative.* Justifiable, or not?

**imago.** Pl. *imagines (-ājinēz)* or *-ágos.* See LATIN PLURALS.

**imbalance.** For several hundred years *un* has been used as the negativing prefix of the verb *balance* and its adjectival p.p., and when more recently need was felt for a noun, *unbalance* naturally followed. In the present century the medical profession, especially the psychiatrists, have rejected this word in favour of the neologism *imbalance.* They may have had good reason for wanting an esoteric word of their own; the layman cannot presume to say. If so they have been disappointed, for their invention has been stolen by the economists, and is on the way to becoming a POPULARIZED TECHNICALITY. *He described as the largest single cause of imbalance the volume of unproductive expenditure which governments felt they had to undertake. | Waste was also causing imbalance in the less developed countries. | We must also remember that the general imbalance which has always characterized the development of the Soviet economy has not yet been fully overcome.* There seems no hope of rescuing *unbalance* from this usurper, but it is really too much if the literary critics are content with neither, and must coin another variant for themselves. *Now Mr. Hughes comes along and begins to redress this disbalance between romance and reason.* And, to make matters worse, this has been stolen too. *If car ownership is doubled in 10 or 12 years, the disbalance between on-peak and off-peak traffic is likely to get worse.* See also IN- AND UN- and DE- AND DIS-.

**imbroglio.** Pronounce *-ōlyō*; pl. *-os*, see -O(E)S 4.

**imbue.** See INFUSE.

**immanent.** The word is something of a stumbling-block; the unlearned hearer or reader is not sure whether it and *imminent* are the same or different; the Latin scholar feels that he does not recall *immaneo* in his Cicero, and wonders whether (*-ant* and *-ent* often playing hide-and-seek with each other) *māno* may be the source instead of *māneo*. (The pronunciation *immān'ent* was prescribed for BBC announcers in 1928 'to avoid confusion with *imminent*', but this has fallen flat.) Under these circumstances it is thought by some that the divines and philosophers who chiefly affect the word should be asked whether they would not gain in intelligibility what they might lose in precision by choosing according to context between *indwelling, pervading, pervasive, permeating, inherent,* and other words that do not mystify us. 'All which though I most powerfully and potently believe, yet I hold it not honesty to have it thus set down', and shall not venture to label *immanent* and *immanence* SUPERFLUOUS WORDS. The OED's note on the use of *immanent* may be useful to those who, not reading philosophic and religious books, find the word an enigma when it makes one of its occasional appearances in the newspaper: 'In recent philosophy applied to the Deity regarded as permanently pervading and sustaining the universe, as distinguished from the notion of an external *transcendent* creator or ruler.'

**immense.** The slang use in the senses excellent or amusing is an instance of NOVELTY-HUNTING which has lost its freshness and grown stale, as such perversions do, and is now rarely heard except in the adverb *immensely* = extremely.

**immortal,** as a compliment to an author or one of his productions or personages, requires to be used with caution. Its real use is to make sure that a reader who may or may not be an ignoramus shall realize that the person or book referred to is well known in the literary world, and that without telling him the fact in too patronizing a manner. But, delicate as the device may originally have been, it is now too well known to escape notice; and whether the reader will be offended or not depends on the exact depth of his ignorance. There are few who will not be angry if they are reckoned to require 'the immortal Shakespeare', or 'Don Quixote', or 'Pickwick Papers'; those who can put up with 'the immortal Panurge', or 'Dobbin', or 'Mrs. Poyser', will be rather more numerous; and so on in many gradations. The author of the following was probably ill inspired in immortalizing Cervantes; but not so ill as if he had done the same—and he might have—for Don Quixote: *Lovers of* Don Quixote *will remember that the immortal Cervantes fought with great courage in this battle.*

**immovable.** Though the differentiation between *immovable* (impossible to move) and *irremovable* (not liable to dismissal) is fully established, blunders sometimes occur; *The President, save for successful impeachment, is immovable by Congress. | By suspending conscription and restoring the immovability of the judges.*

**immunity, impunity.** These words are sometimes ignorantly confused. *Immunity* means exemption from some unpleasant or tiresome thing, as one may become immune from small-pox by vaccination or from jury service by attaining the age of 60. *Impunity* means immunity from one particular unpleasant or tiresome thing, namely punishment.

**impact** (n.) means primarily the striking of one thing against another, a collision and, by extension, its effect on the object struck. Used figuratively in this last sense, it has become a VOGUE WORD. There is no need to multiply examples; some can be found in almost any day's newspapers. It will be enough to quote four that happen to present themselves at the time of writing. *As a senator he had been alarmed at the i. of the first Russian sputnik.* | *A committee is to be set up to investigate the i. of television on children.* | *Perhaps the best yardstick by which to measure the i. of the tax reliefs is. . . .* | *Although the group profit before taxation is a record, the i. of a considerably higher charge for overseas taxation has resulted in a lower net profit.* In the first quotation (where the writer is referring to the reaction of American public opinion to the Russian achievement) the metaphor is not yet 'dead' enough (see METAPHOR I) to be used without incongruity of a moving physical object. In the second the natural word to which i. is preferred is *effect*, and in the third *incidence*. In the fourth i. is used otiosely, as vogue words tend to be (cf. OVERALL); the omission of *the i. of* would leave the sense unchanged and improve the style.

**impassable, impassible.** The two are different in derivation, spelling, pronunciation, meaning, and currency. The first is ultimately from Latin *pando* stretch, the second from Latin *patior* feel; in the first the second syllable is (at least in Southern England) pronounced *pahs*, while in the other it is always *păs*; the first means that cannot be passed, the second that cannot feel. The first is in common use, the second rare, having been superseded by *impassive*.

**impeachment.** For *own the soft i.*, see IRRELEVANT ALLUSION and HACKNEYED PHRASES.

**impenitence, -cy.** There is no perceptible difference of meaning; *-ce* is recommended; see -CE, -CY.

**imperialism.** See WORSENED WORDS.

**impertinent.** See IRRELEVANT.

**impetus.** Pl. *-tuses,* not *-ti*; see -US.

**impinge** makes *-ging*; see MUTE E.

**implement,** n. and v. See NOUN AND VERB ACCENT. The verb, meaning to carry out (a contract etc.), is of Scottish origin. The following quotation is from *Elements of English Composition* by David Irving, 11th ed. 1841: 'To *impleme'nt*, signifying to fulfil, is likewise derived from the barbarous jargon of the Scotish [*sic*] bar.' As recently as 1933 the SOED still called the word chiefly Scottish. Since then it has taken England by storm and become almost a VOGUE WORD with politicians, officials, and the Press. Undertakings, recommendations, promises, and obligations are never now *fulfilled* or *carried out* or *kept* or *observed* or *performed* or *discharged*; *implemented* must always be the word. An occasional change would be refreshing.

**implicit.** *I.* and *explicit*; *i.* and *implied*; *i. faith* etc. The human mind likes a good clear black and white contrast; when two words so definitely promise one of these contrasts as do *explicit* and *implicit*, and then dash our hopes by figuring in phrases where the contrast ceases to be visible—say in 'explicit support' and 'implicit obedience', with *absolute* or *complete* or *full* as a substitute that might replace either or both—we ask with some indignation whether after all black is white, and perhaps decide that *implicit* is a shifty word with which we will have no further dealings. It is in fact noteworthy in more than one respect.

First, it means for the most part the same as *implied*, and, as it is certainly not so instantly intelligible to the average man, it might have been expected to be so good as to die. That it has nevertheless survived by the side of *implied* is perhaps due to two causes. One is that *explicit* and *implicit* make a neater antithesis than even *expressed* and *implied* (*all the conditions whether explicit or implicit*; but *all the implied conditions*; *implied* is much commoner than *implicit* when the antithesis is not given in full). The other is that the adverb, whether of *implicit* or of *implied*, is more often wanted than the adjective, and that *impliedly* is felt to be (see -EDLY) a bad form; *implicitly*, preferred to *impliedly*, helps to keep *implicit* alive.

Secondly, there is the historical accident by which *implicit*, with *faith*, *obedience*, *confidence*, and such words, has come to mean absolute or full, whereas its original sense was undeveloped or potential or in the germ. The starting-point of this usage is the ecclesiastical phrase *implicit faith*, i.e. a person's acceptance of any article of belief, not on its own merits, but as a part of, as 'wrapped up in', his general acceptance of the Church's authority. The steps from this sense to unquestioning, and thence to complete or absolute, are easy; but not everyone who says that implicit obedience is the first duty of the soldier realizes that the obedience he is describing is not properly an absolute one, but one that is based on acceptance of the soldier's status. See POPULARIZED TECHNICALITIES.

**imply.** See INFER for confusion between the two.

**importune, v.** The stress is variable, and the OED allows it on either the second or the third syllable. Of the numerous verse quotations, there are twelve clear for *impor'tune*, and four for *importu'ne*; Shakespeare, Spenser, Chapman, Gray, and Byron, all favour the former. But the latter is now usual; the COD puts it first and some dictionaries give no other.

**impost.** See TAX.

**impost(h)ume.** The *h*, which is not pronounced, and often not written, is better away, though the word is too well established to have its other corruptions removed and its sound altered. It should be, and was, *apostem*, from Greek ἀπόστημα abscess; the *h* comes in by confusion with POSTHUMOUS, in which it is due to a false theory of the etymology.

**impractical, un-.** The second is better; see IN- AND UN-, and PRACTICABLE. The constant confusion between *practicable* and *practical* is a special reason for making use of im- and un- to add to the difference in the negatives: *Its inability to address itself to the questions of the hour produces the impression that the Labour movement is all impracticable agitation* (read *unpractical*).

**impregnate** makes -*natable* (exceptionally, see -ATABLE), since *impregnable* would be inconvenient.

**imprescriptible** is one of the words that are often used without a clear conception of their meaning. A right or property or grant is imprescriptible, if it is 'not subject to prescription'. What then is prescription? If we exclude doctors' prescriptions, most people take it to mean 'uninterrupted use or possession from time immemorial, or for a period fixed by law as giving a title or right; hence, title or right acquired by virtue of such use or possession; sometimes called positive prescription'—OED. But clearly 'not subject to prescription' in this sense does not give us the meaning we want, but something very like the opposite of it. The reading of the riddle requires a piece of legal knowledge that most of us do not possess, viz., that there is another kind of prescription 'now commonly called negative prescription', defined as 'Limitation or restriction of the time within which an action or claim can be raised'—OED. An imprescriptible right, then, is a right not subject to negative prescription, i.e. a right that is not invalidated by any lapse of time. See also INDEFEAS-IBLE, INDEFECTIBLE and PRESCRIPTIVE RIGHT.

**impress,** n. For synonym see SIGN.

**impressible, impressionable.** It is singular that the second form, adapted from the French, should have displaced the first, which might have done the work quite well, although the French verb *impressionner* has failed to produce a current English verb *to impression*. Whatever the reason, *impressionable* is undoubtedly the established form.

**in. 1.** *In years. The play certainly opened to the noisiest first night i. y. | Kew, with 73 hours sunshine, had its dullest March in 13 years. | Cuba, the Congo, and Russia's charges of American 'aggressive action' have been giving the Security Council its busiest time i. y.* These are examples of the conquest of British idiom (*for years*) by American. See AMERICANISMS.

**2.** The combinations *inasmuch as, in order that* or *to, in so far, in that,* and *in toto,* are taken separately in their alphabetical places.

**inadequate.** *Since otherwise the number of troops available might be inadequate to those which might be brought into the field against her.* Though it is true that *adequate* and *inadequate* originally meant respectively made equal and not made equal, and therefore might have been used as in this quotation, modern practice has restricted the words to the notion equal or not equal to a requirement. Vague additions like *to the need, to the occasion, to the task,* are still possible, though felt to be pleonastic; but direct comparisons like that in the above extract, or like *His revenues were found inadequate to his expences* (Gibbon) are abandonments of the DIFFERENTIATION that has taken place between *adequate* and *equal, inadequate* and *unequal. His resources were inadequate,* or *inadequate to the occasion,* but not *inadequate to those of his opponent.*

**inadvertence,-cy.** The first is recommended; see -CE, -CY.

**-in and -ine.** The distinction in chemistry between the two terminations is outside the scope of this dictionary. Although in certain words, e.g. *gelatine, glycerine, margarine,* the *-ine* of popular use violates that distinction, the correct spellings *gelatin* etc. should be left to technical writers or kept for scientific moments, and the *-ine* forms used without hesitation when we are not thinking in terms of chemistry—unless, indeed, the word *pedantry* has no terrors for us.

On the question whether these and similar words should be pronounced *-in* or *-ēn,* or in some cases (e.g. the elements *bromine, chlorine, fluorine,* and *iodine*) *-in,* modern dictionaries vary greatly in their choices. Popular usage, except for a preference for *-īn* in *iodine,* favours *-ēn* (e.g. *margarine*), and this is likely to prevail, as the dictionaries, agreeing for once, recognize that it has in *nicotine.*

**in- and un-.** There is often a teasing *un*certainty—or *in*certitude—whether the negative form of a word should be made with *in-* (including *il-, im-, ir-*), or with *un-.* The general principle that *un-* is English and belongs to English words, and *in-* is Latin and belongs to Latin words, does not take us far. The second part of it, indeed, forbids *in-wholesome* (since *wholesome* has certainly no Latin about it) and thousands of similar offences; but then no one is tempted to go astray in this direction. And the first part, which is asked to solve real problems—whether, for instance, *unsanitary* or *insanitary* is right—seldom gives a clear answer. It forbids *undubitable, uneffable, unevitable,* and other such words of which the positive form does not exist as an English word; but about *sanitary* and the rest it says you may consider them English words and use *un-,* or Latin words and use *in-.* Fortunately the number of words about which doubts exist is not large; for the great majority usage has by this time decided one way or the other. Fashion has varied: 'The practice in the 16th and 17th c.' says the OED 'was to prefer the form with *in-*, e.g. *inaidable, inarguable, inavailable*, but the modern tendency is to restrict *in-* to words obviously answering to Latin types, and to prefer *un-* in other cases, as in *unavailing, uncertain, undevout*'.

Before a list of doubtful pairs, with recommendations, is attempted, some suggestive contrasts may serve to show the conflicting tendencies that are at work. First, markedly Latin endings, as opposed to nondescript ones, produce *in-*: *unjust* but *injustice, unable* but *inability, unquiet* but *inquietude, uncivil* but *incivility*. Second, *-ed* endings have an aversion to *in-*: *undigested* but *indigestible, unanimated* but *inanimate, uncompleted* but *incomplete, undetermined* but *indeterminate, unseparated* but *inseparable, undistinguished* but *indistinguishable, unlettered* but *illiterate, unlimited* but *illimitable, unredeemed* but *irredeemable, unreconciled* but *irreconcilable*. Third, *-ing* endings have a similar aversion to *in-*: *unceasing* but *incessant, undiscriminating* but *indiscriminate*. Fourth, *in-* tends to be restricted to the forms that are closest to the Latin, even in the very open-minded *-ble* group: *unapproachable* but *inaccessible, undestroyable* but *indestructible, undissolvable* but *indissoluble, unbelievable* but *inconceivable, unprovable* but *improbable*. Lastly, *unaccountable* but *insurmountable*, and *unmelodious* but *inharmonious*, are examples of apparent caprice fixed by usage.

The commonest cause of error is thus the existence of a familiar allied word beginning rightly with the prefix that, in the word actually used, is wrong. One other point is perhaps worth stressing. It is a general truth that, while it is legitimate to prefix *un-*, but not *in-*, to any adjective of whatever form, those negative adjectives in *in-* that exist are normally preferred to the corresponding *un-* forms; but when an *in-* (or *il-* or *im-* or *ir-*) adjective has developed a sense that is something more than the negation

of the positive adjective, an *un-* form is sometimes used to discharge the merely negative function without risk of ambiguity; *immoral* having come to mean offending against morality, *unmoral* is called in to mean not moral or outside the sphere of morality. Others are *inept* and *unapt*; UNARTISTIC and *inartistic*; *inhuman* and *unhuman*: UNMATERIAL and *immaterial*; UNRELIGIOUS and *irreligious*; UNSANITARY and *insanitary*; UNSOLVABLE and *insoluble*.

A list is now given of the words about which doubt is most likely, with a statement of the prefix recommended for each; the recommendations are sometimes supported by special reasons, but sometimes merely based on a general impression that one form is more likely than the other to prevail. (But *in-* has been making headway against *un-* in the words starred, and may win.)

| | | |
|---|---|---|
| acceptable | un- | *In-* form labelled rare in OED |
| *advisable | un- | As *acceptable* |
| alterable | un- | |
| appeasable | un- | Delatinized by *-eas-* |
| approachable | un- | |
| communicative | un- | |
| completed | un- | The only indisputable *in——ed* word is *inexperienced* |
| consolable | in- | Established |
| controllable | un- | Much delatinized |
| *decipherable | un- | |
| describable | in- | Established |
| digested | un- | As *completed* |
| discriminating | un- | Words in *-ing* abhor *in-* |
| distinguishable | in- | Established |
| edited | un- | See *completed*; French *inédit* has kept the *in-* form in being |
| effaceable | in- | Established |
| *escapable | un- | Much delatinized |
| *essential | un- | |
| excusable | in- | Established |
| expensive | in- | |
| expressive | un- | Danger of confusion with *inexpressible* |
| frequent | in- | Most *-ent* words so |
| navigable | un- | |
| practical | un- | As *acceptable*; and confusion with *impracticable* |
| recognizable | un- | |
| responsive | un- | Danger of confusion with *irresponsible* |
| retentive | ir- | Most words in *re-* so |
| *substantial | un- | |
| supportable | in- | Established |
| surmountable | in- | Established |
| susceptible | in- | Most *-ible* words so |

**inapt(ness)(itude), unapt(ness), inept(ness), (itude).** In modern usage these overlapping words have sorted themselves out thus: in the sense unfitted, inappropriate, unlikely (to do something), the *-apt-* forms; in sense foolish, the *-ept-* forms. The less suitable noun is chosen in *de Gaulle is said to have had an ineptitude for happiness*.

**inasmuch as** has two meanings: one the original, now rarely met with, i.e. to the same extent as or to whatever degree or so far as (*God is only God inasmuch as he is the Moral Governor of the world*); and the other worn down, with the notion of a correspondence between two scales gone, and nothing left but a four-syllable substitute for since (*I am unable to reply that I am much the better for seeing you, inasmuch as I see nothing of you*); this is the ordinary modern use, and its only recommendation as compared with *since* is its pomposity. On the other hand, the old sense has been supplanted by *so far as* and *in so far as*, and is now unfamiliar enough to be misleading when a literary-minded person reverts to it. *At any rate, Mr. Chamberlain's proposals, inasmuch as they were intended to secure continued loyalty and union amongst the Australian people, were considered altogether unnecessary.* Do we gather that the proposals were in fact rejected, and the reason for this was that their intention was so-and-so? Or that, whether rejected or accepted on other grounds, that intention was not held to justify them? In other words, does *inasmuch as* mean since, or so far as? We cannot tell, without extraneous information. A word that in one sense is pompous, and in another obscure or ambiguous, and in both has satisfactory substitutes, is better left alone.

**incarnation.** *This unfortunately is not the prisoner's first lapse from honesty, for when the Chief Constable said 'he was the very quintessence of cunning and* the incarnation of a book-thief', *he was not speaking without knowledge.* Either the C.C. has been misreported or he

was playfully suggesting that a book-thief is not a human being, but a fiend or possibly a Platonic Idea; for so eminent a person must be aware that incarnation of what is incarnate already is as idle as painting the lily, and much more difficult. Some of us, however, do need to be reminded that while a person may be an incarnation of folly, or Folly clothed in flesh, it is meaningless to call him the incarnation of a fool, because all fools are flesh to start with and cannot be fitted with a new suit of it.

**inceptive, inchoative.** Names given to verbs meaning 'to begin to do something'; in Greek *-σκω* and in Latin *-sco* are the i. terminations, as γιγνώσκω learn (i.e. come to know), *calesco* grow warm. The many English words in *-esce*, *-escent*, as *recrudesce*, *iridescent*, are from Latin verbs of this kind.

**inchoate** means just begun, undeveloped. Those who use it must be on their guard against allowing the analogy of *incoherent* to lead them either into writing *incohate* or into supposing that it has an opposite, *choate*. It comes from the Latin *inchoare*, to begin, which does not consist of a verb *choare* with a negative prefix, but is one in its own right. Pronounce *in'kōāt*.

**incident** (adj.) **incidental.** Two tendencies may be discerned. One is for the shorter form with its less familiar termination to be displaced by the longer; thus we should more usually, though not more correctly, now write *incidental* in such contexts as (shortened from OED examples): *All the powers incident to any government*; *Those in the highest station have their incident cares and troubles*; *The expedition and the incident aggressive steps taken*; *The incident mistakes which he has run into.* The other tendency, cutting across the first, is a differentiation of meaning, based on no real difference between the two forms, but not the less useful on that account. This is that, while *incidental* is applied to side occurrences with stress on their independence of the main action, *incident* implies that, though not essential to it, they not merely happen to arise in connexion with it but may be expected to do so. A consequence of this distinction is that *incident* is mostly used in close combination with whatever word may represent the main action or subject, and especially with *to* as the link; *Youth and its incident perturbations*, or *The enthusiasms incident to youth*. It would be well if the swallowing up of *incident* by *incidental* could be checked, and a continued existence secured to it at least in the special uses indicated. *Half the money has gone in incidental expenses*, and *Our failure brought us an incidental advantage*; but *Office and the incident worries*, and *The dangers incident to motor-racing*.

The survival of *incident* as an adjective is, however, now threatened by its familiar use as a noun in the sense of unfortunate occurrence, especially as a euphemism for *affray*. During the second world war every fall of an enemy bomb was officially called an 'incident', however disastrous its consequences.

**incidentally** is now very common as a writer's apology for an irrelevance. Naturally, those who find it most useful are not the best writers. It is even commoner in speech. It may have point as a convenient way of saying 'It occurs to me to add', in place of the now outmoded 'By the way'. But more often it is just a meaningless piece of padding, like *actually*. It is not easy, for instance, to see what purpose the word was intended to serve in *They are movements representing Majorca and Minorca, in major and minor keys respectively, and blended into a finale incidentally*, or in *You can re-read this acrostic incidentally on page 3 of the 'Radio Times'*. See MEANINGLESS WORDS.

**incline.** 1. See NOUN AND VERB ACCENT. 2. *I incline*—or *am inclined*—*to think* is a formula that may pass if what is intended is to express a provisional opinion that might be changed by a fuller knowledge of the relevant facts. Its much commoner use as a handy cliché for those who shrink from all positive statements deserved the tart rejoinder of Sherlock Holmes to Dr. Watson that he would be better advised to do so; and recalls Sir Winston Churchill's protest that 'The reserve of modern assertions is sometimes pushed to extremes in which the fear of being contradicted leads the writer to strip himself of almost all sense and meaning.'

**include, comprise.** As used in the newspapers, these may be called a pair of WORKING AND STYLISH WORDS. The one used in ordinary life is *include*; the inferior kind of journalist therefore likes to impress his readers with *comprise*. The frequent confusion between *comprise* and *compose* (for which see COMPRISE) is an indication that *include*, which writer and compositor alike know all about, would be in general a safer word. Given the two, however, it would be possible to turn our superfluity to much better purpose than as a chance for the stylish journalist. When two words have roughly the same meaning, examination will generally reveal a distinction; and the distinction in meaning between the present two seems to be that *comprise* is appropriate when what is in question is the content of the whole, and *include* when it is the admission or presence of an item. With *include*, there is no presumption (though it is often the fact) that all or even most of the components are mentioned; with *comprise*, the whole of them are understood to be in the list. The Guards, for instance, include the Coldstreams and the Life Guards, but comprise the Household Cavalry and the Brigade of Guards. *Comprise* is in fact, or would be if this partly recognized distinction were developed and maintained, equivalent to be composed of, whereas *include* is not. The following extracts show *comprise* in contexts where *include* would be the right word: *The German forces . . . exceed twenty-three corps; this number does not* comprise *the corps* *operating in the Masurian Lakes. | The Commission points out that the ample crop of information it has gathered only comprises irrefragably established facts.*

**incognito.** The inflexions are of no great importance; they are now little used, and the abbreviation *incog.* will serve for all. But they should be done right if at all. Of the personal noun *incognito, incognita, incogniti,* are the masculine, feminine, and plural, man, woman, people, of concealed identity. The abstract noun meaning anonymity etc., is *incognito* only, with possible plural *incognitos* (*never dropping their incognitos,* or usually *incognito*). The adverb or predicative adjective (*travelling i.*) is usually *-to* irrespective of gender and number; if declined, it is like the personal noun.

**incommunicado** is a word borrowed from the Spanish *incomunicado* to describe the condition of a prisoner deprived of communication with the outside world. Now that we have anglicized it to the extent of doubling its m, we might well round off the job by pronouncing it *-ādo* instead of the *-ahdo* still given by the dictionaries. See -ADE, -ADO.

**incompatibles.** Under this heading are collected some phrases each consisting of ill-assorted elements. They differ greatly both in degree of badness and in kind; neither point is here discussed, and each phrase is set down in as few words as will enable the usage to be identified. Discussion of those that contain an italicized word will be found on reference to that word; the object of this list is to give the mistakes an extra advertisement. The phrases are: Almost quite; without *scarcely*; finally *scotched*; *decimate* by 50%; rather *unique*; *somewhat* amazing; more *preferable*; *ago* since; *both . . .* as well as; but that *however* is doubtful; *hardly*-earned wages; a line *worth* while pursuing; people *seem*ed to have been bolder in those days; makes *one* forget his manners; I would *like*; those *kind* of; no reason for *undue* alarm. Examples of other similar faults

will be found in UNEQUAL YOKEFELLOWS, and see also ILLITERACIES.

**incompetence, -cy.** The form recommended is *-ce,* cf. COMPETENCE; in legal use, however, *-cy* seems to be preferred.

**incondite.** Pronounce *ĭnkŏ′ndĭt* (see -ITE). The word is of the learned kind, and should be avoided except in what is addressed to a literary audience. It may not be out of place to mention that *condĭtus* composed, not *condītus* seasoned, is the Latin source, and that artless, rude, rough, unpolished, come near the sense.

**incongruous vocabulary.** *Austria-Hungary was no longer in a position, an' she would, to shake off the German yoke. Be in a position to* is a phrase of the most pedestrian modernity; *shake off the yoke,* though a metaphor, is one so well worn that no incongruity is felt between it and the pedestrianism; but what is *an' she would* doing here? Why not the obvious *even if she had the desire?* Or, if *an' she would* is too dear to be let go, why not *Austria now could not, an' she would?* The goldfish *an'* cannot live in this sentence-bowl unless we put some water in with it, and gasps pathetically at us from the mere dry air of *be in a position.* Only a child would expect a goldfish to keep its beauty when out of its right element; and only the writer who is either very inexperienced or singularly proof against experience will let the beauties of a word or phrase tempt him into displaying it where it is conspicuously out of place. Minor lapses from congruity are common enough, and a tendency to them mars the effect of what a man writes more fatally than occasional faults of a more palpable kind, such as grammatical blunders; but they do not lend themselves to exhibition in the short form here necessary. A few of the grosser and more recurrent incongruities, connected with particular words, must suffice by way of illustration. The words out of their element are printed in roman type, and under most of them, in their dictionary places, will be found further examples: *Amongst Smithfield men 'boneless bag meat' has completely ousted the sausage from its erstwhile monopoly of jest and gibe. | Christmas books are put in hand long ere the season comes round. | It is really very difficult to imagine that the reply of the ballot can be* aught *but an answer in the affirmative. | Having in mind the approaching General Election, it appears to me that the result of* same *is likely to be as much a farce as the last. | There are, it may be noted, fewer marquesses than any other section of the peerage* save *dukes. | The Covenanted Civil Service with its old traditions and its hereditary hatred of interlopers, be they merchants, journalists, doctors,* etc. *(be they* is nothing if not stiff, *etc.* nothing if not slack). See also WARDOUR STREET.

**incontinent** = straightway *(Then was there with the angel An host incontinent)* is an archaic adverb. Its connexion with the adjective now in use, which is apt to suggest wrong guesses, is not a close one; the OED explains that it is from Latin *in continenti tempore* (in unbroken time), so that the *in-* of the adverb is the preposition meaning in, whereas that of the adjective is the prefix meaning not.

**incrust, en-.** *Encrust, encrustment,* but *incrustation*; see EM- AND IM-.

**incubus.** Pl. *-bi,* or preferably *-buses.*

**inculcate.** One inculcates something upon or into someone. A curious mistake often occurs, shown in the quotations following: *A passer-by saved him, formed a close friendship with him, and* inculcated *him with his own horrible ideas about murdering women. | An admirable training-place wherein to* inculcate *the young mind with the whys and wherefores of everything which concerns personal safety.* It is possible that the compositor found each time *inoculate* and printed *inculcate,* but a more probable explanation is that *inculcate* is one of the words liable to the maltreatment called OBJECT-SHUFFLING, *i. him with his own ideas* being substituted for *i. his own ideas upon*

*him* on the ANALOGY of *indoctrinate.* Cf. INFUSE and *imbue.*

**incur** makes *-rred, -rring*; see -R-, -RR-. For *incurring* see PRONUNCIATION 7.

**indecorous.** The orthodox pronunciation used to be *ĭndĕkōr′us,* but *ĭndĕk′ŏrŭs* must now be at least as common though most dictionaries still give it second place.

**indefeasible, indefectible.** The distinction between the two, not always very carefully observed, may perhaps best be kept in mind by associating them respectively with *defeat* and *deficient.* That is indefeasible which is not liable to defeat, i.e. to being impaired or annulled by attack from outside; the word is a legal term applied to rights, titles, possessions, and the like. That is indefectible which is not liable to become deficient, i.e. to failing for want of internal power; the word was originally applied to qualities such as holiness, grace, vigour, resolution, affection, or abundance, and later, more loosely, treated as a synonym of perfect, faultless—without defect. Neither word lends itself to the sort of everyday use seen in: *And yet Mr. Barnstaple had the most subtle and indefeasible doubt whether indeed Serpentine was speaking.*

**independence, -cy.** The *-cy* form has only some special senses—Congregationalism, an independent State, and an independent income; and in these, though still preferred to *-ce,* it is now usually displaced by *Congregationalism, sovereign* or *independent State,* and *competency.*

**index.** For pl. see -EX, -IX. 4. and LATIN PLURALS. For synonymy see SIGN. *Index Expurgatorius* is often used loosely for the list of books that the Roman Catholic Church forbids its members to read or permits them to read only in expurgated form. The proper title of that book is *Index Librorum Prohibitorum*; the Index Expurgatorius lists the passages to be expunged.

indifferen(t)(ce)(cy). 1. These were useful words once: *indifferent* (subjective) with *indifference* for the person who feels no preference for either of a pair of things over the other, and *indifferent* (objective) with *indifferency* for a pair of things for neither of which a preference is felt. 'In choice of committees for ripening business for the counsel' wrote Bacon, 'it is better to choose indifferent persons than to make an indifferency by putting in those that are strong on both sides'— a warning as apposite today as three hundred years ago. But the words have since lost most of their virtue. *Indifferent* in its objective sense and *indifferency* are well on their way to archaisms; and, though we still use *indifferent* and *indifference* in the subjective sense, they are becoming rather stilted, in conversation at any rate. *I am indifferent* (or *It is a matter of indifference to me*) *whether you go or stay* is now more naturally expressed by *I couldn't care less*. The fact is that the whole family has been poisoned by the popular use of *indifferent* in the sense of 'rather bad', a euphemistic extension of its meaning neither good nor bad. This has gone to such lengths that clergy sometimes substitute *impartially* in the Prayer for the Church Militant, lest the congregation should be puzzled at their praying that justice may be indifferently ministered.

2. *In doing so it showed an indifference for the interests of its passengers which can only damage its reputation.* Idiom requires *to*, not *for*.

indirect object. The person or thing secondarily affected by the action stated in the verb, if expressed by a noun or pronoun alone (i.e. without *to*, *for*, etc.) is called the i. o.; in Latin and Greek it is recognizable, as it once was in English, by being in the dative, while the direct object is in the accusative. The English dative now having no separate form, the i. o. must be otherwise identified, and this is done by putting it between the verb and the object (Hand *me* that book; Call *me* a taxi), and if it is to follow the object,

must be replaced by a preposition phrase (Hand that book *to me*; Call a taxi *for me*). Variations are (1) when no direct object is expressed, as *You told me yourself*, (2) when the direct object is a mere pronoun and is allowed to precede, as *I told it you before* (but not *I told the story you before*), (3) when the i. o. is after a passive verb, as *It was told me in confidence.*

indirect question is the grammarian's name for a modification of what was originally a question in such a way that it no longer stands by itself as a sentence, but is treated as a noun, serving for instance as subject or object to a verb outside of it. Thus: direct question, *Who are you?*; indirect question, *I asked who he was*, or *Tell me who you are*, or *Who you are is quite irrelevant*. Two points arise, one of grammar, and one of style.

1. It must be remembered that an indirect question is in grammar equivalent to one noun in the singular; the number of its internal subject has no influence on the number of the external verb. To disregard this fact, as when *rest* is written instead of *rests* in the following extract because *terms* happens to be plural, is an elementary blunder—*What terms Bulgaria may be ultimately given rest with the Peace Conference.*

2. The point of style is of much greater interest. How far is it legitimate, in an indirect question, to substitute the order of words that properly belongs to direct questions? The lamentable craze for INVERSION among writers who are fain to make up for dullness of matter by verbal contortions is no doubt responsible for the prevailing disregard of the normal order in indirect questions; for inversion, i.e. the placing of the subject later than its verb, is a mark of the direct, but not of the indirect question. Take these five types:

A. How old are you?
B. Tell me how old you are
   *or* Tell me how old are you?

C. He wondered how old she was
*or* He wondered how old was she?
D. He doesn't know how old I am
*or* He doesn't know how old am I?
E. How old I am is my affair
*or* How old am I is my affair.

A is the direct question; in B, C, D, and E, the first form contains the normal, and the second the abnormal form of the indirect question. It will be seen that the abnormal form becomes progressively disagreeable as we recede from interrogative governing verbs. In B all that is needed to set things right is a comma after *Tell me*, converting the indirect question into a direct one by making *tell me* paratactic. But when we reach D the form might fairly be thought impossible. To contortionists all things may be possible, but readers possessed of the grammatical sense, or of literary taste, will find the following examples of the abnormal order repugnant in the same degree as the types to which the letters B etc. assign them; it is only the encroachments of inversion in general that palliate this special abuse in indirect questions. *I have been asked by the Editor to explain what* are the duties *of the Army towards the civil power, how* is it *constituted, to whom* does it owe *allegiance, by whom* is it *paid, and what* is the source *of its authority* (B. The reason why the first and last clauses here are less distasteful than the others is explained later). | *It shows inferentially how powerless* is that body *to carry out any scheme of its own* (D. Normal order—how powerless that body is). | *Experience has taught in what a restricted region* can the State *as trader or owner act to the general advantage* (D. Normal order—the State can act to the general advantage as trader). | *How bold* is *this attack may be judged from the fact that . . .* (E. Normal order—How bold this attack is). | *Why* should we *be so penalized must ever remain a mystery* (E. Normal order—Why we should).

The further remarks promised on the first example are these: three of the five indirect-question clauses in that are clear cases of abnormal order—

*how is it* instead of *how it is, to whom does it owe* instead of *to whom it owes*, and *by whom is it paid* instead of *by whom it is paid*; but about the other two, which, whether designedly or not, act as advance-guard and rearguard covering those between and almost preventing us from discovering their character, it is not so easy to say whether they are abnormal or not. That is a characteristic of the special type of question consisting of subject, noun complement, and the verb *be*; in the answer to such questions, subject and complement are transposable. Question, *What are the duties?*; answer, *These are the duties*, or *The duties are these*. The indirect question corresponding to the first form is *Explain what are the duties*, and to the second, *Explain what the duties are*; and it can therefore hardly be said that one is more normal than the other. But to questions made up of other elements than subject + *be* + noun complement, e.g. *How is it constituted?*, the two answers (*It is constituted thus*, and *Thus is it constituted*) are by no means transposable; one is plainly normal and the other abnormal. This minor point has been discussed only because sentences like *Explain what are the duties* might be hastily supposed to justify all other uses of direct-question order in indirect-question constructions. See also STOPS (Question Mark).

**indiscreet, indiscrete**, should be distinguished in accent—*ĭndĭskrē't, ĭndĭ'skrēt*; cf. DISCRETE.

**indiscriminate, undiscriminating** are the right forms.

**indisputable.** The stress is on the second syllable. See RECESSIVE ACCENT.

**indissoluble.** *Indissol'uble*, though not yet standard, is likely to prevail over *indiss'oluble*. See RECESSIVE ACCENT.

**individual**, n. The following remarks concern the noun only, not the adjective. '*Individual*, which almost made the fortune of many a Victorian novelist, is one of the modern editor's shibboleths for detecting the unfit'. So it has been said, but editors seem to relax their vigilance occasionally, and the word slips through on its sad old errand of soliciting a smile in vain. Here are a couple of passages in which the choice of it can have been dictated by nothing but WORN-OUT HUMOUR: *It is a most spirited episode, with a supernatural ending according to Tom Causey; this wily individual is the hero of some highly diverting stories.* | *Taking a leaf out of the book of the individual who some years ago put forth his recollections under the title 'Reminiscences of a Young Man'.* Its use contemptuously rather than humorously (cf. Fr. *individu*) is hardly less outmoded. That was the sense in which Mr. Jorrocks understood it when Mr. Moonface's referring to him as one provoked the retort 'You are another indiwidual'; and that was no doubt the significance intended by the M.P. who in more recent times said of a leading article that he thought unsympathetic 'The individual who wrote that leader does not live on £2. 10s a week.'

The test for the right use of the word as opposed to the 'colloquial vulgarism' (OED) is the question whether the writer means to contrast the person he calls an individual with society, the family, or some body of persons. If he does, he may say *individual* with a clear conscience; if not, he must expect us to like his evocation of this ghost of a past jocularity as little as we enjoy the fragrance of the smoking room visited early next morning. A pair of examples will make the difference clear. In the first, the individual member of parliament is directly contrasted with the House of Commons as a body, and is therefore rightly so called. In the second it is true that there is a body of persons in question, but the individual is so far from being contrasted with this body that he is it; the right way to have written the sentence is added in brackets, and the efficiency with which *his* does all the work of *of this longsuffering individual* (19th c. perfume excepted) reveals the writer's style as one not to be imitated: *The House of Commons settled down very quietly to business yesterday afternoon; all trace of the preceding sitting's violent protestation appeared to have been obliterated from the political mind; the only individual who attempted to revive the spirit of animosity was Mr. ——.* | *We are little inclined to consider the urgency of the case made out for the patient agriculturalist; it would seem at first sight as if the needs of this long-suffering individual were such as could be supplied by* . . . (as if his needs could).

**induction, deduction.** The first is reasoning from particular ('cited') cases to general principles; inferring of a law from observed occurrences. If I argue, from the fact that all the MacGregors I have known are Scotch, that MacGregor is a Scotch name, I make an induction. Deduction is reasoning from the general to the particular; basing the truth of a statement upon its being a case of a wider statement known or admitted to be true. If I argue that I shall die because I have been credibly informed that all men do so, and I am a man, I am performing deduction.

Whether the conclusion reached by induction or deduction is true depends on many conditions, which it is the province of Logic to expound; but the broad difference between the two is that induction starts from known instances and arrives at a generalization, or at the power of applying to new instances what it has gathered from the old, while deduction starts from the general principle, whether established by induction or assumed, and arrives at some less general principle, or some individual fact, that may be regarded as being wrapped up in it and therefore as having the same claim to belief as the general principle itself.

**indulge.** *But here and there flashes out a phrase or a sentence that strikes the note of emotion and pride in the achievements of our armies* which *the most reticent of men may* indulge. That passes the limit of what even this very elastic verb can be stretched to. You may i. your emotion, or i. in emotion, or i. yourself in emotion; further, you may i. in, or i. yourself in, a note of emotion; but you cannot i. a note, whether of emotion or of anything else (you can only strike or utter or blow it), and no one who knows any grammar would deny that *which* represents *note*, not *emotion and pride*. The object of *i.* as a transitive verb must be either a person or at least something that can be credited with a capacity for being pleased or gratified; a passion, a fancy, an emotion, may be gratified, but not a note. The mistake is less a misunderstanding of the meaning of *i.* than an example of HAZINESS, *note of emotion* being confused with *emotion*, and the confusion escaping notice under cover of *which*.

**industry.** The accentuation of the second syllable, sometimes heard, is a solecism, perhaps due to analogy with *industrial* and *industrious*.

**-ine.** For *glycerin(e)* etc. see -IN AND -INE.

**ineffective, -fectual, -ficacious, -ficient.** For distinctions see EFFECTIVE.

**inept.** See INAPT.

**inevitable(ness).** To those of us who read reviews of books and picture-shows and acting and music it has been apparent for some time that these words have been added to what may be called the *apparatus criticus*, making up, with other LITERARY CRITICS' WORDS, the reviewing outfit. A search through all the English and French dictionaries within reach when this book was being written showed them all ignorant of the specialized modern use; the OED in particular, dated 1901 for the letter I, has no inkling of it. Even in 1933 the OED Supp. gave only one illustration, and that puts the word in inverted commas: *Illustrations of French wit . . . of the 'inevitable' phrase, that gift to the world past all praise.* A further example or two may therefore be welcome: *And even when a song is introduced, such as Ariel's*

*'Where the bee sucks there suck I', its effect is so great because it seems dramatically inevitable.* | *The mere matters of arrangement, of line therein, show how great was his power, how true his perception; he has the inevitableness of the Japanese.* | *Both themes are well, that is to say inevitably, worked out.* | *Miss —— may not always sing inevitably and spontaneously, simply for the love of beauty.* The COD at first called this use of the word 'critics' slang', but now recognizes it without comment.

What the critic means by *inevitable* is perhaps this: surveying a work of art, we feel sometimes that the whole and all the parts are sufficiently consistent and harmonious to produce on us the effect of truth; we then call it, for short, *convincing*: thus and thus, we mean, it surely may have been or may be; nothing in it inclines us to doubt. To be convincing is but a step short of being inevitable; when the whole and the parts are so admirably integrated that instead of *Thus and thus it may have been* we find ourselves forced to *Thus and thus it must have been* or *was* or *is*, when the change of a jot or tittle would be plain desecration, when we know that we are looking at the Platonic idea itself and no mere copy, then the tale or the picture or the music attains to *inevitableness*. This is an outsider's guess at the meaning; whether the guess is a good one or not, the meaning seems to be one deserving expression in a single word—but only on the condition that that word shall be strictly confined to the works or parts of works that are worthy of it. Now it is, in fact, so often met with that one is compelled to infer the existence of a great deal more inevitability in 20th-c. art of all kinds than one at all suspected; so many things seem inevitable to the critic in which the reader could contemplate extensive alterations without a pang.

**inexactitude** (Terminological). *It* [the employment of indentured Chinese labour on the Rand] *cannot in the opinion of His Majesty's Government be classified as slavery in the extreme* acceptance of the word without some risk of terminological inexactitude. Thus young Mr. Winston Churchill, Parliamentary Under Secretary of State for the Colonies, addressing the House of Commons as spokesman of a government that had just won an overwhelming victory in an election in which denunciation of their predecessors for having sanctioned 'Chinese slavery' had played no small part. This piece of POLYSYLLABIC HUMOUR has worn better than most, thanks to the appeal of its sly whimsicality and the subsequent fame of its author.

**inexpensive.** See DEAR.

**inexpressive, un-.** The second is recommended; see IN- AND UN-.

**infantile, infantine.** It would be convenient if these words had developed a DIFFERENTIATION of the same kind as between CHILDISH and childlike. But this has not happened. Infantine is virtually obsolete, and -ile is used both as a term of contempt for behaviour which, however natural in infancy, does not befit those of riper years (cf. puerile), and also in the sense, without any derogatory implication, of pertaining to infancy, e.g. infantile paralysis, as poliomyelitis used to be called.

**infer** 1. makes -rred etc.: see -R-, -RR-. 2. You clearly infer that your policy was influenced to some extent by your feeling of loyalty to the Labour Government. This misuse of i. for imply is sadly common—so common that some dictionaries give imply as one of the definitions of infer without comment. But each word has its own job to do, one at the giving end and the other at the receiving (What do you imply by that remark? What am I to infer from that remark?) and should be left to do it without interference.

**inferable, -rible, -rrable, -rrible.** The first (with stress on in- not -er-), in the pattern of preferable, referable, transferable, has deservedly prevailed over its rivals, the chief of which

(-rrible) is described by the OED as a 'mongrel' between inferible and inferrable, neither of which has found favour. See also CONFER(R)ABLE.

**inferiority complex.** See COMPLEX.

**infinite(ly).** There are naughty people who will say i. when they only mean great or much or far. Their offence is here dealt with by a triple bench; the first member is a correspondent of a well-known journal; the second is its editor, a meek man, it would seem; the third is he who should have shared the writing of this book, among whose papers was found the cutting with his comment appended.*

1. Sir,—May I appeal to your love of accurate English against the common use in writing, as in speaking, of the word 'infinitely' as equivalent to 'considerably' or 'indefinitely'?—you write that 'oil is infinitely less bulky than coal in proportion to the energy derived from it'. You write that 'the habitual loafer does infinite mischief'. In the first case you intend 'considerably' and in the second case you can only mean that the mischief is indefinite, sometimes great, sometimes no worse than this letter from your obedient servant, AN HABITUAL LOAFER OF NECESSITY.

2. We stand corrected. Our use was a vulgarism. And yet we must not run into a taboo of this noble word. Swinburne uses it finely, accurately, and therefore without vulgarity, in the line 'In the infinite spirit is room for the pulse of an infinite pain'. There the use is exact, because it does not imply mere magnitude.—Ed.

3. Rot. Infinite is no more a vulgarism than any other deliberate exaggeration. And indefinitely is a totally wrong substitute; I have known at least one person habitually use it, with ludicrous effect.

It was naughty of that Editor, though, to say infinite and then take his punishment lying down.

* F. G. Fowler (d. 1918), brother of H. W. Fowler and joint author with him of The King's English.

**infinitive.** 1. For unidiomatic infinitives after nouns that prefer the gerund, as in the extract, see GERUND. *The habit of mapmakers to place lands and not seas in the forefront has obscured the oneness of the Pacific.* The writer probably wrote *to place* because he rightly disliked the repeated *of* in *of mapmakers of placing.* But that is no excuse. 2. See SPLIT INFINITIVE.

**infinitude** does not appear to be now entitled to any higher rank than that of a NEEDLESS VARIANT of infinity. It might well have been differentiated with the sense quality of being infinite, but it is too late for that now. Milton and Sterne, however, will keep it in being for poets to fly to and stylists to play with when *infinity* palls on them. An escape from -ity is sometimes welcome: *It is just this infinitude of possibilities that necessitates unity and continuity of command.*

**infirmity.** 'The last i. of noble minds' is a MISQUOTATION; the last word should be *mind,* and the first *that.*

**inflame** etc. *Inflam(e)able,* formed from the English verb, and used in 16th–17th cc., has been displaced by *inflammable* adapted from French or Latin. *Inflammable* and *inflammatory* must not be confused (see PAIRS AND SNARES) as in *Sir Edward Carson declared before an inflammatory audience that in the event of the Parliament of these realms doing certain things that were distasteful to him he would call out his Volunteers.* It must have been a supposed ambiguity in *inflammable* that led to the coining of the word *flammable.* But that could only make things worse, and *flammable* is now rare, usually in the compound *non-flammable,* a more compact version of *non-inflammable.*

**inflection, -xion.** The second is better; see -XION. In its grammatical sense the word is the general name, including declension, conjugation, and comparison, for changes made in the form of words to show their grammatical relations to their context or to modify their meaning in certain ways.

*Cats, him, greater, sued,* are formed by i. from, or are ii. of, *cat, he, great,* and *sue.*

**inflict,** owing especially to confusion with *afflict,* is peculiarly liable to the misuse explained in the article OBJECT-SHUFFLING. The right constructions are: he inflicted plagues upon them, he afflicted them with plagues, plagues were inflicted upon them, they were afflicted with plagues. Examples of the blunder: *At least the worst evils of the wage system would never have* inflicted *this or any other present-day community.* | *The misconception and discussion in respect of the portraits of Shakespeare with which the world is in such generous measure* inflicted *are largely due to . . .* | *Lively young girls are* inflicted *with stout leather hand-bags.*

**inform, -ation.** *Inform,* a FORMAL WORD for *tell,* is too much used, especially in COMMERCIALESE and OFFICIALESE. There is something about it that makes for verbiage. *I have the honour to inform you, I beg to inform you, I would inform you,* are generally unnecessary preludes to giving the information promised, and the cliché *for your information* is always otiose, if not absurd, unless it means 'for your information only; you are not expected to take any action'. *For your information this Council have two expert rodent operators with vans,* says a local authority's circular letter. If that is all the Council have the operators for it seems a waste of ratepayers' money. Moreover, constant reliance on *inform* leads writers astray into unidiomatic constructions. *Please inform your messenger to wait* will not do. He might be *told* or *asked* or *instructed* to wait, but if *informed* is used idiom requires that *he should wait.*

**infringe.** 1. *I.* makes *infringeable* (or preferably *infringible*), but *infringing*; see MUTE E. 2. *I., i. upon.* Many of those who have occasion for the word must ask themselves before using it what its right construction is. Do you i. (or *i. upon*) a rule? Do you i. (or *i.*

*upon*) a domain? Is the verb, that is, transitive, or intransitive, or sometimes one and sometimes the other? Latin scholars, aware that both *frango* and *infringo* are transitive only, will probably start with a prejudice against *upon*; but Latin is not English, as some of them know. A study of the OED examples leaves no doubt about which construction has predominated from the 16th to the 19th c.; there are 25 quotations for the transitive verb to 4 for *on* or *upon*. But 20th-c. newspaper columns give a very different impression; from them one would infer that *infringe* can no longer stand at all without *upon*: *Is it wise to* i. upon *their rights and susceptibilities?* | *You are* infringing on *our prerogative and trespassing on some of the ground that we intend taking up later.* | *It is suddenly desired to* i. upon *and restrict my Sovereign rights.*

What seems to have been happening is that (1) an imperfect knowledge of Latin has suggested that *infringo* means break in = intrude, whereas it really means break in = damage or violate or weaken; (2) it has therefore been identified in sense with trespass and encroach and assimilated to them in construction, this being further helped by confusion with impinge upon; (3) pretentious writers like to escape from encroach and trespass, familiar words, to *i.*, which will better impress readers with their mastery of the unfamiliar. The advice here tendered is (1) to conceive *i.* as a synonym rather of *violate* and *transgress* than of *encroach* and *trespass*; (2) to abstain altogether from *i. upon* as an erroneous phrase; (3) to use *i.* boldly with *right, rule, law, privilege, patent, sovereignty, boundary, restriction, constitution,* or the like, as object; and (4) when the temptation to insert *on* or *upon* becomes overpowering, as it chiefly does before words like *domain* and *territory,* to be contented with *trespass* or *encroach* rather than say *i. upon.*

**infuse. 1.** *Infusable, infusible. Fusible* being the word for that can be fused, and *infusible* being therefore (see -ABLE

3) the word for that cannot be fused, it is convenient as well as allowable (see -ABLE 2 s.f.) to make from the verb *infuse* not *infusible* but *infusable. Infusable,* then, = that can be infused; *infusible* = that cannot be fused.

**2.** *Infuse, imbue. Infuse* is one of the verbs liable to the OBJECT-SHUFFLING mistake. You can i. courage into a person, or imbue or inspire him with courage, but not infuse him with courage. Examples proving the need of the caution: *The work he did at one school has been repeated at others, until young Australia has been* infused with *the spirit of games.* | *One man, however, it has not affected; say, rather, it has* infused him with *its own rage against itself.* | *He* infused his pupils with *a lively faith in the riches that were within.* This misuse has become so common that the OED Supp. recognizes without comment the meanings *impregnate, pervade, imbue* for *infuse.*

**-ing. 1.** *I would also suggest that,* while admitting *the modernity, the proofs offered by him as to the recent date are not very convincing.* For liberties of this kind taken with the participle, see UNATTACHED PARTICIPLE.

**2.** For the difference between participles in *-ing* and the gerund, see GERUND.

**3.** *On the Press Association's Oldham* representative informing *a leading Liberal of . . ., he replied . . . .* For such mixtures of participle and gerund, see FUSED PARTICIPLE.

**4.** *In all probability he suffers somewhat, like the proverbial dog, from* his having *received a bad name.* For the use of *his* and other possessives in such contexts, see GERUND 4.

**5.** Tender grammatical consciences are apt to vex themselves, sometimes with reason and sometimes without, over the comparative correctness of the *-ing* form of a verb and some other part, especially the infinitive without *to,* in certain constructions. It is well, on the one hand, not to fly in the face of grammar, but rather to eschew what is manifestly indefensible; and, on the other hand, not to give up what one feels is idiomatic in favour of an alternative that is more obviously defensible. Let us examine a few specimens.

(a) *The wearing down phase by phase has been an integral part of the plan, and it has enabled the attack to be kept up* as well as insuring *against hitches.* | As well as closing *the railway, it should make the Danube impracticable for traffic.* We surely all condemn these two examples without a regret. *As well as* is not a preposition, but a conjunction; it therefore cannot govern the gerunds *insuring* and *closing,* as *besides* would have done. If *as well as* is to be kept, *insuring* must become *insured* to match *enabled,* and *closing close* to match *make.* That the latter change is not possible with the sentence in its present order is irrelevant; so much the worse (unless *besides* is written) for the present order. The grammatical conscience was there asleep.

(b) *But America is doing more than furnishing us with loans.* | *We are bound to suspect that Italy is doing something more than raise a diplomatic question.* These are not so simple. The grammatical conscience was certainly awake at one point, for *furnishing* represents second thoughts; *raise* may represent first thoughts, if conscience slept, or third thoughts if conscience let *raising* have its say and then went deliberately back to the idiomatic *raise.* Everyone's first idea in these sentences would be *furnish, raise.* 'But why infinitive?' says Conscience. 'We must write out the sentence at length, clearing away doubts of the exact sense of *do,* the part of speech of *more* etc.; and we get—*America is executing* (doing) *an achievement that is wider* (more) *than* furnish us *is wide;* obviously *furnish* is impossible; write down *furnishing,* which works out.' So far second thoughts. Third thoughts succeed in constructing a defence for *raise* or *furnish,* thus: *I will raise the question; I will do-more-than-raise-the-question;* in this the hyphened group is one verb, and the part of it that takes inflexions is *do: I am-doing-more-than-raise-the-question.* The summing-up is: *raising*

is easily defensible but unidiomatic; *raise* is less easily defensible, but idiomatic; and *raise* has it.

(c) *Dying at their posts rather than surrender(ing).* From this we can extract some confirmation of the defence set up for *raise* in the previous example. There are misguided persons who would actually write *surrendering* there; but they are few, the rest of us feeling that we must either find a justification for *surrender* or else write it without justification. This feeling is strengthened if we happen to remember that we should have no such repugnance to *rather than surrendering* after a participle if the relation to be expressed were a quite different one; compare *acquiring rather than surrendering* with *dying rather than surrendering;* one must have its *-ing,* and the other must not. Well, the justification is the same as with *raise: I am doing more than raise; I will die rather than surrender;* it is true that the form of *surrender* there is decided by *will,* like that of *die,* so that, when *will die* is changed to *dying, surrender* is left depending on air; but meanwhile *die-rather-than-surrender* has become a single verb of which *die* is the conjugable part: *they died rather than surrender; dying rather than surrender.*

**ingeminate.** The phrase *ingeminate peace* means to say Peace, peace! again and again (Latin *geminus* twin); the following sentence looks as if *i.* were in danger of confusion with *germinate* or *generate* or *engender* or some such word: *We have great hopes that the result* [of a discussion on a Royal Commission's report] *will be to i. peace and to avoid the threatened recurrence of hostilities.*

**ingenious, ingenuous.** Both words have deteriorated: *ingenious,* once implying high intellectual ability, now means no more than clever at contrivance; *ingenuous,* once implying noble in birth or character, now means little more than naïve. That *ingenuity* should be the noun for *ingenious,* and not for *ingenuous* as one might expect, is probably due to the frequent misuse of *ingenuous* for *ingenious* by Shake-speare and others in the 17th c. *Ingenuous* has to make do with *ingenuousness*.

**ingratiate** has one sense and one construction only in modern English; it is always reflexive and means only to make (one*self*) agreeable; even in older English, the use shown below is, to judge from the OED, unexampled: *He set himself energetically to the art of ruling his island and ingratiating his new subjects. | Even if it does i. the men, it will only be by alienating the women.*

**inhabit.** The use of this verb in the sense of to house or accommodate or seat may be found in Shakespeare but has long been obsolete and there is no need to resuscitate it. Such modern examples as the following are more likely to be due to ignorance (cf. the similar confusion between *compose* and COMPRISE) than to deliberate use of the artifice of REVIVAL. *This awesome stadium which, with its vast new treble-decker stand can now inhabit 100,000 plus. | The chief aim of the new stand at Lord's is to inhabit at the big matches the overflow from the Pavilion of the 8,000 full members. | Some other grounds which do not inhabit test matches are still older.*

**initiate.** *I.* is liable to the OBJECT-SHUFFLING mistake; you i. persons or minds into things, not things into persons or minds as in: *The Russian Review, a quarterly which is doing so much* to i. into the minds of the British public what is requisite *for them to know about the Russian Empire.* Instil is perhaps the word meant. See also WORKING AND STYLISH WORDS.

**initiative.** The sense of *i.* has been narrowed down by modern usage. Taking 'the first step' as the simple-word equivalent, we might understand that of the first step as opposed to later ones, or of the lead as taken by one person and not another or others. *Initiate* has both meanings, but the latter is the only current sense of *take the i.* It appears in all the special uses; (a) the military,

where the i. is the power of forcing the enemy to conform to *your* first step, so deciding the lines of a campaign or operation; (b) the political, where the i., technically so called, is the right of some minimum number of citizens to originate legislation; (c) the two phrases in which *i.* is chiefly used, 'take the i.', i.e. act before someone else does so, and 'of (or on) one's own i.', i.e. without a lead from someone else.

**-in-law,** describing relationship, was formerly also used in the sense of *step-*. To Sam Weller his father's second wife was always mother-in-law; we are not told what he called his own wife's mother after he married. Today *-in-law* is never so used; my *mother-in-law* becomes so by my marriage, my *stepmother* by hers. The expression *in-law* derives from the Canon Law prescribing the degrees of affinity within which marriage is prohibited.

**innate** and *instinct* (adj.) have complementary uses, e.g. *Courage is innate in the race,* and *A race instinct with courage.* To exchange the words (*The leisurely solidity, the leisurely beauty of the place, so innate with the genius of the Anglo-Saxon*) is the same sort of mistake as OBJECT-SHUFFLING.

**innings.** The pl. *inningses* is colloquial only, *innings* (originally plural) being used for either number—*an innings,* or *several innings.* In U.S. the singular *inning* is used.

**innocence, -cy.** The latter is an archaism, chiefly kept alive by Ps. xxvi. 6. *I will wash my hands in i.*

**innocent of,** in the sense without (*windows innocent of glass*) is a specimen of WORN-OUT HUMOUR. 'She might profitably avoid such distortions as "windows i. of glass" and trays "guiltless of any cloth"'—says a *Times* review.

**innuendo.** Pl. usually *-oes.* As originally used the word meant *viz., to wit* (Lat. *by nodding towards*); Skeat quotes an example from Blount's Gloss. 1674: *He (innuendo the plaintiff) is a thief.*

Hence, by extension, the injurious implication contained in the parenthesis, and, by further extension, any injurious insinuation.

**in order that** is regularly followed by *may* and *might*; *i. o. t. nothing may*, or *might, be forgotten*. The use of the subjunctive without a modal auxiliary (*i. o. t. nothing be forgotten*) is archaic. In some contexts, but not in most, *shall* and *should* may pass instead of *may* and *might* (*i. o. t. nothing should be forgotten*) but certainly the second, and perhaps the first also, of the *shall* examples below is unidiomatic. The other examples, containing *can* and *could*, *will* and *would*, are undoubtedly wrong: *The effort must be organized and continuous* i. o. t. *Palestine shall attract more and more of the race.* | *To influence her in her new adolescence* i. o. t. *we shall once more regain the respect and admiration we enjoyed under the old Russia.* | *It will conclude before lunch-time* i. o. t. *delegates can attend a mass meeting in London.* | *To supplement the work of the doctors on the panel* i. o. t. *every insured person in London* will *be able to obtain the very best medical attention.* | *If the duty had been left on wheat* i. o. t. *the farmer* could *have purchased the offals at a reasonable price . . .* | *A special sign* i. o. t. *the motorist* would *be able to stop in time.* These solecisms are all due to ANA-OGY, *in order that* being followed by *what could properly have followed so that*. Although *in order that* has its uses, as the examples show, *that* or *so that* is less stiff and should be preferred when it will serve.

**in petto.** See FOREIGN DANGER.

**inquire, -ry, en-.** There is a tendency, which deserves encouragement, to differentiate *enquir(e)(y)* and *inquir(e)(y)* by using *en-* as a FORMAL WORD for *ask* and *in-* for an investigation, e.g. *They enquired when the Court of Inquiry was to sit.*

**insinuate.** *Since the outside world looks to us for a moral lead, I hesitate to* give publicity to the corrupt practices that have insinuated into our national life. This use of *i.* as an intransitive verb in the sense of enter subtly is an ARCHAISM. In modern idiom it is only transitive or reflexive; in the sense of hint disparagingly it takes a *that-* clause.

**in so far.** He must have a long spoon that sups with the devil; and the safest way of dealing with *in so far* is to keep clear of it. The dangers range from mere feebleness or wordiness, through pleonasm or confusion of forms, and inaccuracy of meaning, to false grammar. The examples that follow are given in that order; the offence charged against each is stated in a word or two, and the verdict is left undiscussed for the reader to give for himself. If he is sufficiently interested to wish for fuller treatment, he should turn to FAR 4, 5, where different uses of *so far* are considered. The prefixing of *in* is for the most part not dictated by reasons either of grammar or of sense, so that much of what is there said applies to *in so far* also:

*He did not, with such views, do much to advance his object,* save in so far that *his gracious ways everywhere won esteem and affection* (Wordy. Read *though* for *save i. s. f. that*). | *The question . . . is not in any way essentially British,* save i. s. f. as *the position of Great Britain in Egypt makes her primarily responsible* (Wordy. Read *except that* for *save i. s. f. as*). | *The large majority would reply in the affirmative,* i. s. f. as to *admit that there is a God* (Confusion between *so far as to* and *i. s. f. as they would*). | *No such department under present conditions is really requisite,* i. s. f. as *the action of the Commander-in-Chief is thwarted in cases where he should be the best judge* (Wrong sense. Read *since* for *i. s. f. as*). | *The officials have done their utmost to enforce neutrality, and have* i. s. f. *succeeded as the Baltic fleet keeps outside the three-mile limit* (Wrong sense. Read *have so far succeeded that*). | *It has the character of a classic* i. s. f. as *the period it covers* (Ungrammatical. *In so far as* is not a preposition, and cannot govern *period*).

**insouciance, -ant.** The adjective is usually, and the noun often, anglicized in pronunciation to *insoo's-*.

**instance.** The abuse of this word in lazy PERIPHRASIS has gone far, though not so far as that of CASE. Here are two examples: *The taxation of the unimproved values in any area, omitting altogether a tax on improvements, necessarily lightens the burden* in the instance of *improved properties.* | *The stimulation to improve land, owing to the appreciable rating of the same, is more clearly established whenever the outgo is very direct and visible,* such as in the instance of *highly priced city lands.* In the first, *in the instance of* should be simply *on*; and in the second *such as in the instance of* should be *as it is on.* There is some danger that, as writers become aware of the suspicions to which they lay themselves open by perpetually using *case*, they may take refuge with *instance*, not realizing that most instances in which *case* would have damned them are also cases in which *instance* will damn them. The crossing out of one and putting in of the other will not avail; they must rend their hearts and not their garments, and learn to write directly instead of in periphrasis. *Instance* has been called *case's* understudy; in the articles CASE, and ELEGANT VARIATION, will be found many examples of the substitution.

**instant.** See COMMERCIALESE.

**instantly, instantaneously.** *Instantly* is virtually a synonym of at once, directly, and immediately, though perhaps the strongest of the four. *Instantaneously* is applied to something that takes an inappreciable time to occur, like the taking of an instantaneous photograph, especially to two events that occur so nearly simultaneously that the difference is imperceptible.

**instil(l).** The OED gives precedence to *-il*. In either case, *-lled, -lling*; see -LL-, -L-. The word is liable to the OBJECT-SHUFFLING confusion. *Her ef-*

*forts to instil them with culture by reading aloud Rosebery's Life of Pitt are entertainingly described.* You can imbue or inspire children with culture; but you can only instil it into them, not them with it. See ANALOGY.

**instinct, intuition.** See INTUITION.

**instinctual.** The adjective of *instinct* is *instinctive*. Why anyone should have thought it necessary to coin a new one on the analogy of *contractual, habitual*, etc. is not clear; perhaps the psychologists wanted an adjective of their own. But those dictionaries that recognize it do not give it any markedly different meaning from that of *-ive*. The SOED, for instance, defines *-ive* as 'of the nature of instinct, operating by or resulting from innate prompting', and *-ual*, in its Addenda, as 'of or pertaining to, involving or depending on instinct'. It looks as if *-ual* might be a SUPERFLUOUS WORD. The COD ignores it.

**institute, institution.** The two nouns have run awkwardly into and out of one another. The neat arrangement would have been for *-ution* to mean instituting, and *-ute* a thing instituted; but *-ution* has seized, as abstract words will, on so many concrete senses that neatness is past praying for. *Institution* is in fact the natural English word capable of general use, and *-ute* a special title restricted to, and preferred for, certain institutions. An *-ute* is deliberately founded; an *-ution* may be so, or may have established itself or grown. Cricket, five-o'clock tea, the House of Lords, Eton, Guy's Hospital, the National Gallery, marriage, capital punishment, the Law Courts, are all *-utions* and not *-utes*. Whether a particular *-ution* founded for a definite purpose shall have *-ute* or *-ution* in its title is a matter of chance or fashion—*Commonwealth* (formerly *Imperial*) *-ute*, but *the Royal -ution*; *The -ute of Metals* but *the British Standards -ution.* A child is to be got into some *-ution*, and is placed in the National *-ute* for the Blind or the Masonic -ution for Boys. Accountants, architects, and journalists have their -utes, engineers, surveyors, and valuers their -utions. The usual name for new formations is -utes, and they now greatly outnumber the -utions.

**insufficient.** *But Austria also excludes altogether a food-product like meat, of which she produces insufficient.* This noun use (= not enough or too little) is worse than the corresponding use of SUFFICIENT.

**insular** is a mild pejorative, bestowed normally with condescension. It is characteristic of the British that, although quick to discern intellectual and moral virtues in small self-sufficient primitive communities, such as the Tibetans and the Esquimaux, they affect to despise the mentality of subunits of their own population. *Insular*, used of an attitude towards some aspects of international affairs is merely *provincial* or *parochial* writ large. Cf. CONTINENTAL.

**insure.** See ASSURE.

**intaglio.** Pronounce -ă′lyō. Pl. -os, see -O(E)S 4. *Intaglio* is opposed to *relief* as a name for the kind of carving in which the design, instead of projecting from the surface, is sunk below it (*carved in i.*); and to *cameo* as the name for gems carved in i. instead of in relief.

**integra(te)(l).** To *integrate* is to combine components into a single congruous whole. Psychology borrowed it from mathematics and invented the expression *integrated personality*, a reasonable enough piece of jargon to describe someone in whom, as Antony said of Brutus, the elements are rightly mixed by nature. The public have now borrowed the verb from the psychologists with such freedom that it has become a VOGUE WORD, habitually preferred to less stylish but often more suitable words such as *join, combine, unite, amalgamate, merge, fuse, consolidate. Integral*, outside mathematics, is seldom to be found except as the

inseparable companion of *part*. See ADJECTIVES MISUSED.

**intelligent, intellectual.** While an intelligent person is merely one who is not stupid or slow-witted, an intellectual person is one in whom the part played by the mind as distinguished from the emotions and perceptions is greater than in the average man. An intellectual person who was not intelligent would be, though not impossible, a rarity; but an intelligent person who is not intellectual we most of us flatter ourselves we can find in the looking-glass. *Intelligent* is always a commendatory though sometimes a patronizing epithet; *intellectual*, though implying the possession of qualities we should all like to have, is tainted in the communist ideology by its use in disparaging contrast to *workers*; elsewhere too it is seldom untinged by suspicion or dislike—called by a leader-writer in the TLS 'a rather fly-blown word beloved only of sociologists'. The same writer reminds us that Bertrand Russell once wrote to a correspondent 'I have never called myself an intellectual, and nobody has ever dared to call me one in my presence. I think an intellectual may be defined as a person who pretends to have more intellect than he has, and I hope this definition does not fit me.' This echoes the opinion of Bishop Parker in the 17th c.: 'These pure and seraphic intellectualists, forsooth, despise all sensible knowledge as too grosse and materiall for their nice and curious faculties.' That is not unlike the definition given by the OED Supp. of the colloquial equivalent *highbrow* (U.S. *egghead*) as 'a person of superior intellectual attainments or interests: always with derisive implications of conscious superiority to ordinary human standards'.

**intelligentsia** is a word coined in Russia about 1870 and originally applied to intellectuals associated with the revolutionary movement. After the revolution its meaning changed; the official Soviet definition is 'a social stratum consisting of people profes-

intended290intensive

sionally occupied in mental work'. As these include not only those who follow the arts and professions but also white-collar workers generally, the word seems to have become in its country of origin little more than a polite name for what used to be called the upper middle classes. Elsewhere it has never found much favour, and it is now an outmoded word with its leftist colouring washed out of it. Attempts by the Oxford Dictionaries to define it are 'The class of society to which culture, superior intelligence, and advanced political views are attributed' (OED Supp. and OID), 'The class consisting of the educated portion of the population and regarded as capable of forming public opinion (SOED), 'The part of a nation that aspires to independent thinking' (COD).

**intended,** n. It is curious that betrothed people should find it so difficult to hit upon a comfortable word to describe each other by. 'My intended', 'my fiancé(e)', 'my sweetheart', 'my love(r)', 'my young (wo)man', 'my boy (girl) friend', 'my future wife (husband)', 'my wife (husband) to be'—none of these is much to their taste, too emotional, or too French, or too vulgar, or too evasive. The last two objections are in fact one; evasion of plain words is vulgarity, and 'my intended' gives the impression that the poor things are shy of specifying the bond between them, an ill-bred shyness; so too with 'my engaged', and the modern word, 'steady', does not necessarily imply serious intentions. And so in *fiancé(e)* they resort to French instead of vague English for their embarrassing though futile disguise. It is no doubt too late to suggest that another chance should be given to *betrothed*. It means just what it should, i.e. pledged to be married, and is not vulgarized and would be a dignified word for public use. But it is so out of fashion as to sound facetious.

**intensive.** Just as *definitive* and *alternative* are ignorantly confused with *definite* and *alternate*, and apparently

liked the better for their mere length, so *intensive* has become a fashionable word where the meaning wanted is simply *intense*. It must be admitted that there was a time before DIFFERENTIATION had taken place when Burton, e.g., could write *A very intensive pleasure follows the passion*; it there means intense, but the OED labels the use obsolete, and its latest quotation for it is from over two centuries ago. The modern relapse had not come under its notice in 1901, when letter I was issued; nor is it mentioned in the 1933 Supp. *Intensive* perished as a mere variant of *intense*, but remained with a philosophic or scientific meaning, as an antithesis to *extensive*; where *extensive* means with regard to extent, *intensive* means with regard to force or degree: *The record of an intensive as well as extensive development. | Its intensive, like its extensive, magnitude is small.* This is the kind of word that we ordinary mortals do well to leave alone; see POPULARIZED TECHNICALITIES. Unfortunately, a particular technical application of the philosophic use emerged into general notice, and was misinterpreted—intensive method especially of cultivation. To increase the supply of wheat you may sow two acres instead of one—increase the extent—or you may use more fertilizers and care on your one acre—increase the intensity—; the second plan is intensive cultivation, the essence of it being concentration on a limited area. Familiarized by the newspapers with *intensive cultivation*, which most of us took to be a fine name for very hard or intense work by the farmers, we all became eager to show off our new word, and took to saying *intensive* where *intense* used to be good enough for us. The war gave this a great fillip by finding the correspondents another peg to hang *intensive* on—*bombardment*. There is a kind of bombardment that may be accurately called intensive; it is what in earlier wars we called concentrated fire, a phrase that has the advantage of being open to no misunderstanding; the fire converges upon a much narrower front than that from which it is discharged. But as often as not the intensive bombardment of the newspapers was not concentrated, but was intense, as the context would sometimes prove; a bombardment may be intense without being intensive, or intensive without being intense, or it may be both. Not that the confusion is confined to the newspapers; it seems to have affected even those whose duty it is to plan bombardments. 'Why must you write "intensive" here?' wrote Sir Winston Churchill to the Director of Military Intelligence on the 19th March 1944. ' "Intense" is the right word. You should read Fowler's *Modern English Usage* on the use of the two words' (*The Second World War*, v. 615).

**intensive,** gram. Said of words or word-elements that add emphasis; in *vastly obliged, perdurable, vastly* and *per-* are ii. Often in contrast with PRIVATIVE; the in- of *incisive* (and *intensive*) is intensive, and that of *incivility* privative.

**intention. 1.** Ordinary use. **2.** *First, second, ii.* 1. A defining phrase is so often appended to *i.* that the question between gerund and infinitive, treated generally under GERUND 3, is worth raising specially here. Choice between the two is freer for *i.* than for most such nouns, and it can hardly be said with confidence that either construction is ever impossible for it. It will perhaps be agreed, on the evidence of the illustrations below, offered as idiomatic, that when *i.* is used in the singular and without *the, his, an, any,* or other such word, *to do* is better, but otherwise *of doing: Intention to kill is the essential point.* | *You never open your mouth but with i. to give pain.* | *He denied the i. of killing.* | *He concealed his i. of escaping.* | *Some i. than ever there may have been.* | *I have no i. of allowing it.* | *Have you any i. of trying again?* | *I have every i. of returning.* | *He renounced all i. of retaliating.* | *Not without ii. of finding a loophole.*

**2.** *First, second, i.* These phrases have special senses in medicine and in logic, apt to puzzle the layman and to be confused with each other. In medicine, *first i.* denotes (OED) 'the healing of a lesion or fracture by the immediate reunion of the severed parts, without granulation'; and *second i.* 'the healing of a wound by granulation after suppuration'. In logic, *first ii.* are (OED) 'primary conceptions of things, formed by the first or direct application of the mind to the things themselves; e.g. the concepts of *a tree, an oak*'; and *second ii.* 'secondary conceptions formed by the application of thought to first intentions in their relations to each other; e.g. the concepts of *genus, species, variety, property, accident, difference, identity*'.

**inter alia** is Latin for amongst others when 'others' are things. If the others are persons, *alia* must be changed to *alios* (or rarely *alias*); the OED quotes, from 1670, *The Lords produce inter alios John Duke of Lancaster*. But when persons are meant, it is much better nowadays to use English. The writer of the following sentence was either ignorant both of *inter alia* and of Latin, or else pedantic enough to expect us to know that the Latin for *costs* is the masculine *sumptus*: *She will pay twenty thousand million marks within two years (covering,* inter alios, *the costs of the armies of occupation and of food and raw material allowed by the Allies*).

**interdependence, -cy.** No difference in sense; *-ce* is recommended; see -CE, -CY.

**interest,** vb. On *interesting*, the OED, after giving the sound as *i'nteristing*, adds 'formerly, and still dialectically, *interĕ'sting*'. All the longer inflexions— *interestedly, disinterested*, etc. and even the simple verb, are often said by more or less illiterate speakers with the accent on *-ĕst-*. For the maltreatment to which *interest* (n.) is liable see LEGERDEMAIN.

**interior, internal, intrinsic.** See
EXTERIOR.

**intermediary,** n., should be confined
to its concrete sense of a go-between
or middleman or mediator. In its
abstract sense of medium or agency
or means, it is worthy only of the
POLYSYLLABIC HUMOURist; and the
OED's only two quotations for it
(representing, alas! a much larger body
than would be guessed by anyone who
did not make it his business to observe
such things) are clearly in that spirit:
*Mysteriously transmitting them through
the intermediary of glib Jew boys with
curly heads. | We are the only European
people who teach practical geometry
through the recondite intermediary of
Euclid's Elements.*

**intermission** is used in U.S. for what
we call an *interval* (in a musical or
dramatic performance). Under the
influence of LOVE OF THE LONG WORD,
it is beginning to infiltrate here and
should be repelled; our own word does
very well.

**internecine** has suffered an odd fate:
being mainly a literary or educated
man's word, it is yet neither pronounced in the scholarly way nor
allowed its Latin meaning. Strictly it
should be called *ĭnter'nĕsĭn* but it is in
fact called *ĭnternē'sĭn*; see FALSE
QUANTITY. And the sense has had the
Kilkenny-cat notion imported into it
because mutuality is the idea conveyed by *inter-* in English; the Latin
word meant merely of or to extermination (cf. *intereo* perish, *intercido*
slay, *interimo* destroy) without implying the extermination of both parties.
The imported notion, however, is
what gives the word its only value,
since there are plenty of substitutes
for it in the true sense—*destructive,
slaughterous, murderous, bloody, san-
guinary, mortal,* and so forth. The
scholar may therefore use or abstain
from the word as he chooses, but it
will be vain for him to attempt correcting other people's conception of the
meaning, which for some seems to be
little more than *intestine.* That is

presumably how it is used in the following comment on the Queen's having
to choose between Mr. Macmillan and
Mr. Butler as Prime Minister. *We
are running the risk of bringing the
Crown into internecine political warfare.*

**interpellate, interpolate.** The first
word and its noun *interpellation* are
little used now except in the technical
sense proper to parliamentary proceedings, and especially those of the
French Chamber. They are therefore
felt to be half French words, and so
the unnatural pronunciation given to
the verb by the OED (*ĭntĕrpĕ'lāt*) is
perhaps accounted for. It is a pity that
it has not prevailed (modern dictionaries give *ĭnter'pĕlate*); for it would
have the advantage of distinguishing
the sound from that of *interpolate*—a
need illustrated by: *M. Barthou inti-
mated that, on the return of M. Miller-
and from London, he would* interpolate
*him on the question.* The proper meaning of *interpolate* is to make an insertion
in a book or other written matter, used
generally with the implication that the
purpose is to give some false impression. The recent extension by which
the word is sometimes used in the
sense of making a remark that interrupts a conversation is both unnecessary and undesirable, unnecessary because *interject* and *interpose* can do
all that is needed, and undesirable
because it increases the likelihood of
confusion between *interpellate* and
*interpolate.*

**interpretative,** not *interpretive,* is the
right form, *-ive* adjectives being normally formed on the Latin p.p. stem,
i.e. here *interpretat-.* Read *-ative* in:
*They should be at the same time illustra-
tive and interpretive. | The literal and
the interpretive are difficult to reconcile
in a single statement.*

**interregnum.** Pl. *-ums* or *-a*; see
-UM. For the facetious use, = gap, see
PEDANTIC HUMOUR.

**interstice.** Pronounce *ĭnter'stĭs.* See
RECESSIVE ACCENT.

**intestinal.** The Oxford Dictionaries prefer *inte'stinal* to *intesti'nal* (see RECESSIVE ACCENT); the Latin i is long, but on this point see FALSE QUANTITY s.f.

**in that** is a conjunction that has gone a little out of fashion and does not slip from our tongues nowadays. It is still serviceable in writing of a formal cast, but, like other obsolescent idioms, is liable to ill treatment at the hands of persons who choose it not because it is the natural thing for them to say, but because, being unfamiliar, it strikes them as ornamental. So: *This influence was* so far indirect in that *it was greatly furthered by Le Sage.* | *The legislative jury sat to try the indictment against Mr. Justice Grantham* in that *during the Great Yarmouth election petition he displayed political bias.* In the first, two ways of saying the thing are mixed (*was so far indirect that*, and *was indirect in that*); and in the second *in that* is used in a quite suitable context, but wrongly led up to; a man is guilty in that he has done so-and-so, but an indictment against him is not in that anything.

**in the circumstances,** see CIRCUMSTANCE.

**intimidate.** *Similar threats were uttered in the endeavour to* i. Parliament from *disestablishing the Irish Episcopal Church. From* is idiomatic after *deter* and *discourage,* but not after *i.* or *terrify*; see ANALOGY.

**into, in to.** The two words should be written separately when their sense is separate. Correct accordingly: *The Prime Minister took her* into *dinner.*/ *All the outside news came* into *us immediately.*

**in toto** means not on the whole, but wholly, utterly, entirely, absolutely, and that always or nearly always with verbs of negative sense—*condemn, decline, deny, reject, disagree, i. t.* The following is nonsense: *Nor do we produce as much* in toto *as we might if we organized.*

**intransigent** dates in England from about 1880; being now established, it should neither be pronounced as French nor spelt -*eant* any longer.

**intransitive p.p.** This article is less severely practical than most in the book, and is addressed only to those few enthusiasts who find grammatical phenomena interesting apart from any rules of writing that may be drawn from them. As grammatical terminology is far from fixed in English, it must be premised that *p.p.* (past participle) is here taken as the popular name for the single-word participle that does not end in -*ing*, i.e., by the p.p. of *hear* is meant *heard*, not *hearing* nor *having heard* nor *being heard*. All verbs, with negligible exceptions such as *must* and *can*, have this p.p.'s, though in many it is used only as an element in making compound parts like *has climbed* or *will have died.* That function of the p.p. is familiar to everyone and needs no comment. Further, the p.p. of all transitive verbs can be used as an adjective (*a broken jug*). What is not so fully realized is the part played by the adjectival p.p. in many intransitive verbs. It is in the first place much commoner than is supposed. Most of us, perhaps, would say that p.p. adjectives were all passive, i.e. were only made from transitive verbs. A moment's search is enough to correct that notion—*fallen angels, the risen sun, a vanished hand, past times, the newly arrived guest, a grown girl, a gone coon, absconded debtors, escaped prisoners, the deceased lady, the dear departed, a collapsed lorry, we are agreed, a couched lion, an eloped pair, an expired lease.*

Secondly, when a verb is both transitive and intransitive, it is often difficult to say whether in some particular phrase the p.p. is active or passive, and the answer may affect the sense; e.g., *a deserted sailor,* if *deserted* is passive, is one who has been marooned, but, if it is active, is one who has run from his ship; *an angel dropped from heaven* has possibly been passive, but more likely active, in the descent; *a capsized boat* may have capsized or have been capsized; *a failed B.A.* may be one whom the examiners have failed or one who has failed to satisfy them; *my declared enemy* is more often one who has declared enmity than one I have declared an enemy; *a flooded meadow* shows a passive p.p., *a flooded river* perhaps an active one; *a well grown tree* means one thing in the virgin forest, and another in a nursery garden.

Thirdly, recognition of the frequency of the intransitive p.p. will sometimes throw light on expressions whose origin is otherwise not quite obvious: *a determined man* is perhaps one who *has* determined, not been determined; a person is *ill advised* who has advised, i.e. taken thought, badly, not one who has had bad advice given him; he is *well read* who *has* read well; he is *drunk* who *has* drunk; *-spoken* in *soft-spoken* etc. is more intelligible if it is regarded as active, and cf. *well-behaved*; *mistaken clemency* seems to be clemency that *has* erred; *an aged man* may be one who *has* aged, since the verb *age*, = grow old, dates from before 1400; *the dissipated* may be those who *have* wasted their substance, and *the experienced* those *having* experienced things rather than those possessed of experience.

**intrigue,** v. t. The meaning 'puzzle, perplex' is given by the OED, but illustrated by only a single modern quotation, and labelled 'now rare'. Would that were still true! the one quotation (19th-c.) is from a newspaper from which I have before me sixteen 20th-c. cuttings with the word used in that sense. The other chief dictionaries either ignore the sense or treat it contemptuously—English dictionaries, that is, for it is naturally well enough known to the French; but it is one of the GALLICISMS, and LITERARY CRITICS' WORDS, that have no merit whatever except that of unfamiliarity to the English reader, and at the same time the great demerit of being identical with and therefore confusing the sense of a good English word. Besides *puzzle* and *perplex*, there are *fascinate*,

*mystify, interest,* and *pique,* to choose from. Will the reader decide for himself whether the Gallicism is called for in any of the following places?—*A cabal which has intrigued the imagination of the romanticists.* | *The problem, however, if it intrigues him at all, is hardly opened in the present work.* | *Thus it is we read of Viper—that delightful dog—mouthing a hedgehog, much intrigued with his spines.* | *But her personality did not greatly intrigue our interest.* | *When theologian, scientist, and philosopher have intrigued our minds with the subtlety of their argument.*

The reader will not be surprised to learn that, since the foregoing was written in the nineteen-twenties, the resistance of the dictionaries has been stormed. The OED itself, in its 1933 Supp., gives copious examples of the use of *intrigue* in the sense 'to excite the curiosity or interest of; to interest so as to puzzle or fascinate', including one from the writings of a Cambridge Professor of English Literature. The popularity of the word is no doubt due partly to its novelty and partly to the better reason that it can do something more than serve as a synonym for one of the words listed above; it can convey the meaning of two of them at once, *puzzle* and *fascinate* for instance. But it is still true that *intrigue* is often used in place of a simpler and better word, as in some of the examples given above.

**intrinsic.** See EXTERIOR.

**intuition** and *instinct*. The word *intuition* being both in popular use and philosophically important, a slight statement of its meaning, adapted from the OED, may be welcome. The etymological but now obsolete sense is simply inspection (Latin *tueor* look): *A looking-glass becomes spotted and stained from their only intuition* (i.e., if they so much as look in it). With the schoolmen it was The spiritual perception or immediate knowledge ascribed to angelic and spiritual beings, with whom vision and knowledge are identical: *St. Paul's faith did not come by hearing, but by intuition and revelation.*

In modern philosophy it is The immediate apprehension of an object by the mind without the intervention of any reasoning process: *What we feel and what we do, we may be said to know by intuition*; or again (with exclusion of one or other part of the mind) it is Immediate apprehension by the intellect alone, as in *The intuition by which we know what is right and what is wrong*, or Immediate apprehension by sense, as in *All our intuition takes place by means of the senses alone*. Finally, in general use it means Direct or immediate insight: *Rashness if it fails is madness, and if it succeeds is the intuition of genius*.

How closely this last sense borders on *instinct* is plain if we compare *A miraculous intuition of what ought to be done just at the time for action* with *It was by a sort of instinct that he guided this open boat through the channels*. One of the OED's definitions of *instinct*, indeed, is: 'intuition; unconscious dexterity or skill'; and whether one word or the other will be used is often no more than a matter of chance. Three points of difference, however, suggest themselves as worth keeping in mind: (1) an intuition is a judgement issuing in conviction, and an instinct an impulse issuing in action; (2) an intuition is conceived as something primary and uncaused, but an instinct as a quintessence of things experienced in the past whether by the individual or the race; and (3) while both, as faculties, are contrasted with that of reason, intuition is the attribute by which gods and angels, saints and geniuses, are superior to the need of reasoning, and instinct is the gift by which animals are compensated for their inability to reason. OED quotes Addison: 'Our Superiors are guided by Intuition and our Inferiors by Instinct.'

**inure, enure.** Both the connexion between the verb's different senses (*The poor, inured to drudgery and distress; The cessions of land enured to the benefit of Georgia*) and its derivation are so little obvious that many of us,

at any rate when minded to use the less usual sense, feel some apprehension that we may be on the point of blundering. There is also a tendency to spell *in-* and *en-* for the two meanings as if they were different words; *en-* is often preferred for the legal (intr.) sense. The origin is the obsolete noun *ure* (*We will never enact, put in ure, promulge, or execute, any new canons*), which is from French *œuvre*, which is from Latin *opera* work. To inure a person you set him at work or practise him; a thing inures that comes into practice, or operates, in such and such a direction. Variant spellings are therefore unnecessary, and *in-* is preferred by the OED.

**invalid.** The word meaning not valid is pronounced *invă'lĭd*. For the adjective and noun meaning sick (person) the popular verdict, after some vacillation, has been given in favour of *ĭ'nvălēd*, though some dictionaries still admit *-ĭd* as an alternative for the last syllable. The verb is *-ēd* only, and for it the COD, reversing the SOED, would stress the last syllable in preference to the first (see NOUN AND VERB ACCENT), but the word is rarely used except in the passive (*He was invalided out of the service*).

**inveigle.** The OED pronunciation is *invē'gl* without the alternative of *-vā'gl*, but the latter must now be at least as common, and the COD admits it to second place.

**inventory.** Pronounce *ĭ'nvĕntŏrĭ*.

**inversion.** By this is meant the abandonment of the usual order of words in an English sentence and the placing of the subject after the verb as in *Said he*, or after the auxiliary of the verb as in *What did he say?* and *Never shall we see his like again*. Inversion is the regular and almost invariable way of showing that a sentence is a question. It has therefore an essential place in the language, and there are other conditions under which it is usual, desirable, or permissible. But the abuse of it ranks with ELEGANT VARIATION as one of the most repellent vices of modern writing. Inversion and variation of the uncalled-for kinds are like stiletto heels—ugly things resorted to in the false belief that artificiality is more beautiful than nature; but as heels of a practical kind may be useful or indeed indispensable, so too is inversion.

In questions and commands, as contrasted with the commoner form of sentence, the statement, inversion is the rule: *Doth Job fear God for nought?/ Hear thou from heaven thy dwelling-place.* The subject being usually omitted in commands, these do not much concern us; but in questions the subject regularly follows the verb or its auxiliary except when, being itself the interrogative pronoun or adjective, it has to stand where that pronoun almost invariably stands (Browning's *Wanting is—what?* supplies an exception): *Who did it? What caused it?* In the other exceptional sentence-form, the exclamation, inversion is not indeed the rule as in questions, but was once common and is still legitimate: *How dreadful is this place! / What a piece of work is a man! / Few and evil have the days of the years of my life been. / Bitterly did he rue it. / And so say all of us.*

Inversion, then, is the natural though by no means invariable order of words in sentences other than statements. In exclamations particularly, when they do not contain a special exclamatory word such as *how* or *what*, the inversion is what announces their nature; and one form of bad inversion arises from inability to distinguish between an exclamation and a mere statement, so that the latter is allowed the order that marks the former (*Hard is it to decide*, on the pattern of *Hard, very hard, is my fare!*). To these forms of sentence must be added the hypothetical clause in which the work ordinarily done by *if* is done in its absence by inversion: *Were I Brutus. / Had they known in time.*

These inversions—Interrogative, Imperative, Exclamatory, and Hypothetical—form a group in which inversion itself serves a purpose. With statements it is otherwise; there

inversion is not used for its own significance, but because the writer has some other reason for wishing to place at the beginning either the predicate or some word or phrase that belongs to it. The usual reason for putting the whole of the predicate at the beginning is the feeling that it is too insignificant to be noticed at all after the more conspicuous subject, and that it must be given what chance the early position can give it; hence the *There is* idiom; not *No God is*, but *There is no God*. That is Balance Inversion in its shortest form, and at greater length is seen in: *Through a gap came a single level bar of glowing red sunlight peopled with myriads of gnats that gave it a quivering solidity*; if *came through a gap* is experimentally returned to its place at the end of that sentence, it becomes plain why the writer has put it out of its place at the beginning. Another familiar type is *Among the guests were A, B, C . . . Z*

Often, however, the object is not to transfer the predicate bodily to the beginning, but to give some word or words of it first place. This may be meant to give hearer or reader the connexion with what precedes (Link Inversion), to put him early in possession of the theme (Signpost Inversion), or to warn him that the sentence is to be negative (Negative Inversion): *On this depends the whole course of the argument. / By strategy is meant something wider. / Never was a decision more abundantly justified.* Here *on this*, *by strategy*, *never* are the causes of inversion; each belongs to the predicate, not to the subject; and each when placed first tends to drag with it the verb or auxiliary, so that the subject has to wait—tends, but with different degrees of force, that exercised by a negative being the strongest. We can if we like, instead of inverting, write *On this the whole course of the argument depends*, or *By strategy something wider is meant*, but not *Never a decision was more abundantly justified*; similarly *Not a word he said* is a very out-of-the-way version of *Not a word did he say*.

If we now add Metrical Inversion,

our catalogue of the various kinds may perhaps suffice. Where the Bible gives us *As the hart panteth after the water brooks*, and the Prayer Book *Like as the hart desireth the water-brooks*, both without inversion, the hymn-books have *As pants the hart for cooling streams*. That is *metri gratiâ*, and it must not be forgotten that inversion is far more often appropriate in verse than out of it for two reasons—one this of helping the versifier out of metrical difficulties, and the other that inversion off the beaten track is an archaic and therefore poetic habit. A very large class of bad inversions will be seen presently to be those in subordinate clauses beginning with *as*; they arise from failure to realize that inversion is archaic and poetic under such circumstances, and non-inversion normal. It is therefore worth while to stress this contrast between *As pants the hart* and both the prose versions of the same clause.

To summarize these results:

Interrogative Inversion: *What went ye out for to see? | Doth Job fear God for naught?*

Imperative Inversion: *Hear thou from heaven thy dwelling-place.*

Exclamatory Inversion: *How dreadful is this place! | What a piece of work is a man! | And so say all of us! | Few and evil have the days of the years of my life been. | Bitterly did he rue it. | Bang went saxpence!*

Hypothetical Inversion: *Were I Brutus. | Had they known in time.*

Balance Inversion: *There is no God./ Through a gap came* [an elaborately described ray]. *| Among the guests were* [long list].

Link Inversion: *On this depends the whole argument. | Next comes the question of pay.*

Signpost Inversion: *By strategy is meant something wider.*

Negative Inversion: *Never was a decision more abundantly justified. | Not a word did he say.*

Metrical Inversion: *As pants the hart for cooling streams.*

We may now proceed to consider with the aid of grouped specimens

some of the temptations to ill-advised inversion. It may conciliate anyone who suspects that the object of this article is to deprive him altogether of a favourite construction if we admit at once that, though bad inversion is extremely common, non-inversion also can be bad. It is so rare as to call for little attention, but here are two examples: *But in neither case* Mr. Galsworthy tells *very much of the intervening years. | Least of all* it is *to their interest to have a new Sick Man of Europe.* In negative sentences there is the choice whether the negative shall be brought to the beginning or not, but when it is so placed inversion is necessary; read *does Mr. G.*, and *is it*.

### INVERSION AFTER RELATIVES AND COMPARATIVES

The problems offered are interesting, but most difficult to grapple with by way of argument. The line here taken is that the sort of inversion now being dealt with, however devoutly one may believe it to be mistaken, can hardly be proved illegitimate, at any rate without discussion of more tedious length than could be tolerated. On the other hand, it is hardly credible, after a look through the collection shortly to follow, that the writers can have chosen these inversions either as the natural way of expressing themselves or as graceful decoration; so unnatural and so ungraceful are many of them. It follows that the motive must have been a severe sense of duty, a resolve to be correct, according to their lights, at any sacrifice. And from this again it follows that no demonstration that the inversions are incorrect is called for; the task is only to show cause why non-inversion should be permitted, and these idolaters will be free of the superstitions that cramped their native taste.

1. *A frigate could administer roughly half the punishment that could a 74.*

Comment: Compare some everyday sentence: *You earn twice the money that I do*, never *that do I*. The misconception is perhaps that the putting of the object first (here *that*) should draw the verb; but this is not true of relative clauses; *the people that I like*, not *that like I*.

2. *It costs less than* did administration *under the old companies*.

Comment: A simple parallel is *I spend less than you do*, for which no one in talk would substitute *than do you*. Many, however, would write, if not say, *I spend less than do nine out of ten people in my position*. The difference must lie in the length of the subject, and the misconception must be that it is a case for balance inversion, i.e. for saving the verb from going unnoticed. But so little does that matter that if the verb is omitted no harm is done; *did* in the quotation should in fact either be omitted or be put in either of its natural places, after *administration*, or after *companies*.

3. *He looked forward, as* do we *all, with great hope and confidence to Monday's debate.* | *It represents the business interests of Germany as* does no other organization.| . . . *his fondness for the game, which he played as* should an Aberdonian. | *These were persons to be envied, as* might be someone *who was clearly in possession of a sixth sense.* | *The French tanks have had their vicissitudes, as* have had ours.

Comment: *As*, in such sentences, is a relative adverb; it and the unexpressed *so* to which it answers are equivalent to (*in the way*) *in which*, and what was said above of relatives and inversion holds here also. *Try to pronounce it as I do*, not *as do I*; and when the subject is longer, e.g. *the native Frenchman*, though *as does the native Frenchman* becomes defensible, it does not become better than *as the native Frenchman does*, nor as good.

4. *Each has proven ably that the other's kind of Protection would be quite as ruinous as* would be Free Trade. | *We are unable to . . . without getting as excited over the question of funds as* is a cat *on a hot iron.* | *He was as far removed as* are the poles asunder *from the practices which made the other notorious.* | *The lawn-tennis championships will be attracting as much attention as* has the golf championship. | *Thirteen divisions taken from reserve is now as serious as* would have been some fifty divisions *four months ago.*

Comment: The *as* of this batch differs from that of batch 3 in that its fellow *as* of the main sentence belongs to an adjective (*ruinous, excited*, etc.) or adverb (*far, much*). This allows the inversionist a different defence, which he needs, since (to take the first example) balance inversion is clearly not available for *as would be Free Trade* with its short subject. He might appeal here to exclamatory inversion. When the compound sentence is reduced to its elements, they are either (a) *Free Trade would be ruinous*; *Protection would be equally ruinous* (the first clause being a statement); or (b) *Ruinous would Free Trade be! Protection would be equally ruinous* (the first clause being an exclamation). He chooses, how reasonably let the reader judge, the (b) form, and retains its order in the compound sentence. The truth is that in the first two of these sentences the verb should have been omitted, and in the others kept in its ordinary place— *as the poles are asunder, as the golf championship has attracted, as 50 would have been.*

5. *Bad as* has been our record *in the treatment of some of the military inventions of the past, it may be doubted whether the neglect of the obvious has ever been more conspicuously displayed than in . . .* | *And, hopeless as* seem the other divisions *of Belfast, progress is being made in them.*

Comment: The meaning of this *as* idiom is clear; it is Though our record has been so bad, or However bad our record has been; but how it reached its present shape is less apparent. Some light is thrown by the presence in earlier English of another *as*, now dropped; Swift writes *The world, as censorious as it is, hath been so kind . . .*; this points to (*Be our record as*) bad *as our record has been* (*bad*) for the unabbreviated form. Omission of the bracketed words gives the uninverted order, which will only be changed if exclamatory inversion (*Bad has been our record!*) or balance inversion is needlessly applied.

**6.** *It is not all joy to be a War Lord in these days, and gloomy though* is the precedent, *the only thing left for a War Lord to do is to follow the example of Ahab at Ramoth Gilead.*

Comment: *Gloomy is the precedent!* is a not impossible exclamatory inversion; and, if the words were kept together with the effect of a quotation by having *though* before instead of in the middle of them, the exclamatory order might be tolerable, though hardly desirable, even in the subordinated form; but not with *though* where it is. This may be tested by trying a familiar phrase like *Bad is the best. Though bad is the best*, yes; but not *Bad though is the best*; instead of that we must write *Bad though the best is.*

**7.** *The work stands still until* comes the convenient time *for arranging an amicable rupture of the old engagement and contracting of the new.*

Comment: There is no doubt about the motive. It is a balance inversion, and one that would be justified by the great length of the subject if the only place for the uninverted *comes* were at the end of the whole sentence. But what is too often forgotten in such cases is that there is usually a choice of places for the verb; here *comes* would be quite comfortable immediately after *time.*

The conclusion suggested is that, so far as relative clauses are concerned, especially those containing *as*, the writer whose taste disposes him to use the natural uninverted order is at the very least free to indulge it.

## INVERSIONS OF THE LITERARY PARAGRAPHIST

Those who provide newspapers with short accounts of newly published books have an inversion form all to themselves. The principle seems to be to get the title of the book to a place where the reader can find it, and at the same time to avoid the catalogue look that results if the title is printed at the head before the description, and to give a literary air to the paragraph. The title is therefore worked to the end, by the use of odd inversions that

editors would do well to prohibit. But, once broken in to inversion by this special use of it, the minor literary critics learn to love their chains, and it is among them that the false exclamatory inversions dealt with in the next section are most rife. Here, meanwhile, are some specimens:

*Most racily written, with an easy conversational style about it, is Mr. Frank Rutter's 'The Path to Paris'.* | *Diplomatic and military are the letters that comprise the Correspondence of Lord Burghersh, edited by his daughter-in-law.* | *From the point of view of the English reader timely is the appearance of M. Frédéric Masson's historical study* [title]. | *Lively and interesting are the pictures of bygone society in town and country presented in the two volumes, 'The Letter-bag of Lady Elizabeth Spencer-Stanhope'.* | *Mainly concerned with the rural classes, who form something like two-thirds of the whole population, are the sketches and tales collected in 'The Silent India'.* | *Written in his most vivacious vein is Lieut. Colonel Haggard's latest historical study* [title].

## FALSE EXCLAMATORY INVERSION

It has already been pointed out that a statement may be turned into an exclamation by inversion; an adjective or adverb that conveys emotion is put first out of its place, and inversion follows. If Jacob had said *The days of the years of my life have been few and evil*, he would have been stating a bald fact; by beginning *Few and evil have been*, he converts the statement into a groan, and gives it poignancy. Writers who observe the poignancy sometimes given by such inversion, but fail to observe that 'sometimes' means 'when exclamation is appropriate', adopt inversion as an infallible enlivener; they aim at freshness and attain frigidity. In the following examples there is no emotional need of exclamation, and yet exclamatory inversion is the only class to which they can be assigned: *Futile were the endeavour to trace back to Pheidias' varied originals, as we are tempted to do, many of the later statues.* |

*Finely conceived is this poem, and not less admirable in execution. | Facile and musical, sincere and spontaneous, are these lyrics. | Hard would it be to decide which of his many pursuits in literary study he found most absorbing. | Sufficient is it to terminate the brief introduction to this notice by stating . . . | Irresistibly is the reader reminded, though direct analogy is absent, of Sheridan's reference to . . . | Appropriately does the author prelude his recollections with . . .*

## YET, ESPECIALLY, RATHER, ETC.

A curious habit has grown up of allowing these and similar words to dictate a link inversion when the stressing of the link is so little necessary that it gives a noticeable formality or pomposity to the passage. It is a matter not for argument, but for taste; will the reader compare the quoted forms with those suggested in the brackets? *Especially* and *rather* usually change their place when inversion is given up, but *yet* remains first. *His works were burnt by the common hangman; yet was the multitude still true to him* (yet the multitude was). | *Henry Fox, or nobody, could weather the storm which was about to burst; yet was he a person to whom the court, even in that extremity, was unwilling to have recourse* (yet he was). | *The set epistolary pieces, one might say, were discharged before the day of Elia; yet is there certainly no general diminution of sparkle or interest* (yet there is). | *. . . springs of mineralized water, famous from Roman times onward for their curative properties; especially did they come into renown during the nineteenth century* (they came into renown especially). | *Mr. Campbell does not recognize a change of opinion, but admits a change of emphasis; especially is he anxious at the present time to advance the cause of Liberal Evangelism* (he is especially anxious). | *His love of romantic literature was as far as possible from that of a mind which only feeds on romantic excitements; rather was it that of one who was so moulded . . .* (it was rather that). | *There is nothing to show that the Asclepiads took any prominent*

share in the work of founding anatomy, physiology, zoology, and botany; rather do these seem to have sprung from the early philosophers* (these seem rather). | *His book is not a biography in the ordinary sense; rather is it a series of recollections culled from . . .* (it is rather).

## INVERSION IN INDIRECT QUESTIONS

This point will be found fully discussed under INDIRECT QUESTION. Examples of the wrong use are: *How bold is this attack may be judged by . . . | Why should we be so penalized must ever remain a mystery.* The right order would be *How bold this attack is*, and *Why we should be so penalized.*

## SUBORDINATED INVERSIONS

Certain kinds of these have been discussed in the section on relatives and comparatives. A more general point is to be made here—that it is often well, when a sentence that standing by itself would properly be in the inverted form is subordinated as a clause to another, to cancel the inversion as no longer needed. The special effect that inversion is intended to secure is an emphasis of some sort, and naturally emphasis is more often suitable to a simple independent sentence than to a dependent clause. Examples are grouped under A, B, and C, according to the kind of inversion that has been subordinated, and comment on each group follows:

A. Negative Inversion. *The amount involved is no less a sum than £300,000 per annum*, to not a penny of which have the drivers *a shadow of claim. | To give to all the scholars that firm grounding upon which alone can we hope to build an educated nation. | He laid down four principles on which alone could America and Austria go further in exchanging views. | Now that not only are public executions long extinct in this country, but the Press not admitted to the majority of private ones, the hangman has lost his vogue. | But it had only been established that on eighteen of those days did he vote.

Comment: In the first three it will be

admitted that, while *to not a penny of this* etc. (the independent forms) would require the inversion, *to not a penny of which* etc. (the subordinate forms) are at least as good, if not better, without it. The fourth example (executions) will on the other hand be upheld by many who have no inordinate liking for inversion; *not only* is so little used except in main sentences, and therefore so associated with inversion, that *not only public executions are long extinct* has an unfamiliar sound even after *Now that*. It may moreover put the reader on a false scent by suggesting that *not only* qualifies *public* alone. The subordinate inversion in the last example is not quite what it seems, being due to irresolution between an inverted and an uninverted form; the former would be *But only on eighteen of those days had it been established that he voted*; and the latter, *But it had only been established that he voted on eighteen of those days.*

B. Exclamatory Inversion. *Suffice it to say* that in almost one-half of the rural district areas is there *an admitted dearth of cottage homes.* | Though once, at any rate, does that benign mistily golden irony *of his weave itself in.*| While for the first time, he believed, did naval and military history *appear as a distinctive feature.*

Comment: The subordination in two of these only makes more conspicuous the badly chosen pegs on which the inversion is hung. *In almost one-half of the rural district areas,* and *once at any rate,* are not good exclamatory material; *Many a time have I seen him!* shows the sort of phrase that will do. Even if main sentences had been used with these beginnings, they should have been put as statements, i.e. without inversion, and still more when they depend on *Though* and *Suffice it to say that.* In the third example *for the first time* is not incapable of beginning an exclamation; it would pass in a sentence, but becomes frigid in a clause.

C. Link Inversion. When, three years later, came the offer *of a nomination, it was doubtless a welcome solution.*|

Whilst equally necessary is it *to press forward to that unity of thought without which . . .*

Comment: About these there can hardly be a difference of opinion. If the *when* and *whilst* constructions were absent, it would have been very natural to draw *Three years later, Equally necessary,* to the beginning to connect the sentences with what preceded, and inversion might or might not have resulted. But with the interposition of *when* and *whilst* they lose their linking effect, and the natural order should be kept—*When three years later the offer came, Whilst it is equally necessary.*

## INVERSION IN PARALLEL CLAUSES

As with combinations of a negative and a positive statement into one (see NEGATIVE MISHANDLING 2), so with inverted and uninverted members of a sentence care is very necessary.

Not only is it *so necessarily bounded by that moving veil which ever hides the future,* but also is it *unable to penetrate . . . into . . . the past* (*but also is it* is an impossible inversion, brought about by the correct one that precedes). | Not only *in equipment but in the personnel of the Air Battalion* are we *suffering from maladministration* (*Not only in equipment* requires *are we suffering*; *in the personnel* requires *we are suffering.* To mix the two is slovenly; the right form would be *We are suffering not only in* etc.). | *Even* were this tract *of country level plain and* the roads lent themselves *to the manœuvre, it would be so perilous to . . .* (*were this tract* is inverted; *the roads lent themselves* is not, and yet, since there is no *if,* it absolutely requires inversion. Begin *Even if this tract were*; for the only ways to invert the second clause are the fantastic *and lent themselves the roads* and the clumsy *and did the roads lend*). | Had we desired *twenty-seven amendments, got seven accepted, and* were in anticipation *of favourable decisions in the other twenty cases we should think . . .* (Mend like the previous one. To read *and were we* would disguise the fact that the whole is one hypothetical clause and not several). For other examples see ELLIPSIS 6.

## INVERSION IN DIALOGUE MACHINERY

Novelists and others who have to use dialogue as an ingredient in narrative are some of them unduly worried by the machinery problem. Tired of writing down *he said* and *said he* and *she replied* as often as they must, they mistakenly suppose the good old forms to be as tiring to their readers as to themselves, and seek relief in whimsical variations. The fact is that readers care much what is said, but little about the frame into which a remark or a speech is fitted; or rather, the virtue of frames is not that they should be various, but that they should be inconspicuous. It is true that an absolutely unrelieved monotony will itself become conspicuous; but the variety necessary to obviate that should be strictly limited to forms inconspicuous in themselves. Among those that are not inconspicuous, and are therefore bad, are many developments of the blameless and inconspicuous *said he*, especially the substitution of verbs that are only by much stretching qualified for verbs of saying, and again the use of those parts of verbs of saying that include auxiliaries. Most of the following examples exhibit a writer trying not to bore his reader; nothing bores so fatally as an open consciousness that one is in danger of boring, and a sure sign of this is the very tiresome mannerism initiated perhaps by Meredith ('*Ah*' *fluted Fenellan*), and now staled by imitation: '*Hand on heart?*' *she doubted.* | '*Need any help?*' *husked A.* | '*They're our best revenue*' *defended B.* | '*I know his kind*' *fondly remembered C.* | '*Why shouldn't he?*' *scorned D.* | '*Yes*', *moodily consented John*, '*I suppose we must*'. | '*Oh?*' *questioned he.* | '*Oh, what a sigh!*', *marvelled Annunziata.* | '*But then*', *puzzled John*, '*what is it that people mean when they talk about death?*' | '*The sordid sort of existence*', *augmented John.* | '*You misunderstand your instructions*',

*murmured rapidly Mr. Travers.* | '*I couldn't help liking the chap*', *would shout Lingard when telling the story.* | '*I won't plot anything extra against Tom*', *had said Isaac.* | '*At any rate, then*', *may rejoin our critic, 'it is clearly useless...*'|

The ordinary 'said he' etc. (Thou art right, Trim, in both cases, said my uncle Toby) was described above as blameless and inconspicuous. Its place among inversions is in the 'signpost' class. The reader is to be given the theme (i.e., here, the speech) at the earliest possible moment; the speech, being grammatically the object of 'said', yet placed first, draws 'said' to it, and 'he', or my uncle Toby, has to wait. But only such insignificant verbs as *said, replied, continued,* will submit to being dragged about like this; we must treat with greater respect verbs that introduce a more complicated notion, or that are weighted with auxiliaries or adverbs (compare 'went on my uncle Toby' with 'continued my uncle Toby'), or that cannot rightly take a speech as object. These stand on their dignity and insist on their proper place. Or perhaps it would be more accurate to say that they used to do so. The fashion of introducing quotations by inversions set by the more sprightly American periodicals has led to a riot of inversion in popular journalism of the kind parodied by P. G. Wodehouse in 'Where it will all end knows God, as *Time* magazine would say'. See also SAID.

**inverted commas.** See STOPS. For their use by way of apology for slang etc., see SUPERIORITY.

**invite,** n. The OED compares *command* and *request* for the formation, but describes the noun use as colloquial; and it has never, even as a colloquialism, attained to respectability. After more than 350 years of life, it is less recognized as an English word than *bike*. 'Coll. or vulg.', says the COD.

**involve.** This word is overworked as a general-purpose verb that saves the trouble of precise thought. *A collision* took place involving a private motor-car and a lorry (between). | There was no reduction last year in the number of cases involving cruelty to horses (of). | Ground troops and aircraft were ininvolved (used). | This was the first disaster involving a Viscount air-liner (to). | Traffic on the up line was not involved (affected). | Some applicants are still coming forward but the numbers involved are falling (their numbers)./ The proper meaning of *involve* is wrap up, and so entangle, embarrass, as in the common use of its p.p. (*an involved subject, involved in financial difficulties*); in the interests of precision its use in the sense of to produce consequences should be confined to those that are unforeseen, or incidental to something done for another purpose, as in *The abolition of cheque endorsement inevitably involved disuse of receipts on the reverse of cheques.* | *General Grivas's self-imposed task of saving Greece involves strangling the infant republic of Cyprus.*

**inwardness.** *The i., the real i., the true i., of* something has a meaning that it would not occur to us to give it out of our own heads, but that we some time or other discover to be attached to it by other people, especially such as write books. That meaning is, as defined by the OED, 'the inward or intrinsic character or quality of a thing; the inner nature, essence, or meaning'. It is a literary phrase fit for a literary man to use when he is writing for or talking to literary people, but otherwise pretentious. True wisdom is to abstain from it till it seems the really natural phrase; and any inclination to put inverted commas round it is a fair proof that one has not reached, or that one doubts whether one's readers have reached, the stage of so regarding it. There is a certain intrusiveness about the word in these quotations; omission, or a simpler substitute, would have done no harm: *When the First Lord gets to understand* (*the real i. of*) *the present situation, I have every confidence that he will do full justice to the Thames.* | *In this connexion I would warn readers who are unacquainted with* (*the i. of*) *South African affairs not to attach undue importance to a recent declaration.* | *Will you allow me to send a few lines on the true i. of the situation?* (realities). | *We have always contended that the true 'inwardness' of the Land Bill was not the wish to stop evictions, but the wish to stop the scandal of evictions* (motive).

**iodine.** For pronunciation see -IN and -INE.

**-ion and -ment.** Many verbs have associated with them nouns of both forms, as commit, *commission* and *commitment*; require, *requisition* and *requirement*; excite, *excitement* and *excitation*. When both are well established, as in these cases, the two nouns usually coexist because they have come by DIFFERENTIATION to divide the possible meanings between them and so tend to lucidity. How little the essential difference of meaning is in the two terminations may be seen by comparing *emendation* with *amendment* (where the first means rather correction made, and the second rather correcting), and *requisition* with *requirement* (where the first means rather requiring, and the second rather thing required), and then noticing that the two comparisons give more or less contrary results. Further, when there is only one established form, it is not apparent to the layman, though the philologist sometimes knows, why one form exists and the other does not—why for instance we say *infliction* and not *inflictment*, but *punishment* and not *punition*. The conclusion is that usage should be respected, and less usual forms such as *abolishment* and *admonishment*, or rare ones such as *incitation*, and *punition*, should not be resorted to when *abolition*, *admonition*, *incitement*, and *punishment*, are to hand. See also -MENT, and for some similar questions see the next article, and also -ISM AND -ITY, and -TY, -NESS, -ION.

**-ion and -ness.** The question between variants in *-ion* and *-ness* differs from that discussed in the preceding article in several respects. First, *-ness* words can be made from any adjective or participle, whereas the formation of *-ment* words from verbs is by no means unrestricted; by the side of *persuasion* you can make *persuasiveness*, but not *persuadement*. Secondly, there is more possibility of a clear distinction in meaning; *-ion* and *-ment* are both attached to verbs, so that neither has any more claim than the other to represent the verbal idea of action. But between *-ion* and *-ness* that line does exist; though *-ion* and *-ness* are often appended to exactly the same form, as in *abjectness* and *abjection*, one is made from the English adjective *abject*, and the other from the Latin verbal stem *abject-*, with the consequence that *abjectness* necessarily represents a state or quality, and *abjection* naturally a process or action. Thirdly, while both *-ion* and *-ment* pass easily from the idea of a process or action into that of the product—*abstraction* for instance being equivalent either to abstracting or to abstract notion—, subject *-ness* to that treatment is to do it violence; we can call virtue an abstraction, but not an abstractness. In compensation for this disability, the *-ness* words should be guaranteed as far as possible the exclusive right to the meaning of state or quality; e.g. we should avoid talking of the *abstraction* or the *concision* of a writer's style, or of the *consideration* that marks someone's dealings, when we mean abstractness, conciseness, and considerateness. *Concision* means the process of cutting down, and *conciseness* the cut-down state; the ordinary man, who when he means the latter says *conciseness*, shows more literary sense than the literary critic, who says *concision* just because the French (who have not the advantage of possessing *-ness*) have to say it, and he likes gallicizing. It is not always easy to prove that writers do not mean the process rather than the quality, but appearances are often against them. In the following examples, if the words *short-winded* in the first and *pungency* in

the second are taken into account, it is pretty clear that the quality of the style was meant in both, and *conciseness* would have been the right word: *I really think any Muse (when she is neither resting nor flying) ought to tighten her girdle, tuck up her skirts, and step out. It is better than Tennyson's short-winded and artificial concision—but there is such a thing as swift and spontaneous style. | But then as a writer of letters, diaries, and memoranda, Mr. Gladstone did not shine by any habitual concision or pungency of style.* If it were not for this frequent uncertainty about what is really meant, it would be as bad to say *concision* for *conciseness* as to use *correction* (which, like *concision*, could be defended as a Gallicism) for *correctness*, or *indirection* (for which *Hamlet* II. i. 66 might be pleaded) for *indirectness*.

Simple reference of any word in *-ion* to this article may be taken to mean that there is a tendency for it to usurp the functions of the noun in *-ness*. See also **-TY** and **-NESS**.

**I.Q.** (Intelligence Quotient), a measure of mental capacity applied especially to children and mental defectives, is the ratio (expressed as a percentage) that the 'mental age' of the subject of the test (i.e. the age at which the same mental capacity would be found in a normal person) bears to his actual age. Thus the I.Q. of a clever child of 6 with a mental age of 9 would be 150. As the qualities tested are such as ordinarily mature at 15, that figure is taken instead of actual age for all older persons. Thus the I.Q. of an adult mental defective with a mental age of nine would be 60 (900 ÷ 15). Psychologists make the following gradation of mental capacity expressed in terms of I.Q.: *Very superior* 130 plus, *superior* 120/129, *bright average* 110/119, *average* 90/109, *dull average* 80/89, *borderline* 70/79, *feeble-minded* 50/69, *imbecile* 25/49, *idiot* 24 minus. Some lawyers are apt to be intolerant of these niceties, and to subscribe to the dictum of a Lord Chief Justice: 'Nowadays people use expressions like *slightly maladjusted* and *borderline high-grade mental defective* which mean nothing. It is all words.'

**Irene.** A Greek word of three syllables (*īrē'nē*) meaning peace. As a Christian name it has now been largely adopted by those who take it for a disyllable like Doreen, Eileen, etc., and, when they hear others make three syllables of it, account for it to themselves as an optional addition like those in Johnny and Jeanie.

**iridescent.** So spelt, not *irri-*; the origin is Greek *iris* rainbow, not Latin *irrideo* laugh.

**iron.** For *the i. Chancellor, Duke*, see SOBRIQUETS.

**iron curtain.** The use of this term (literally the fire-proof curtain of a theatre) to describe the political division between the communist countries of Europe and the rest is generally ascribed to Sir Winston Churchill. He was not the first to use it; but he popularized it in his Fulton speech, and it is unlikely that he consciously borrowed it. According to an editorial note in the TLS (14 July 1961) the earliest instance of its written use in this sense is in Mrs. (later Viscountess) Snowden's account of her visit to Russia, *Through Bolshevik Russia*, published in 1920.

**iron out.** The use of this PHRASAL VERB in the sense of remove (difficulties etc.) as an iron removes creases and wrinkles is American in origin and has deservedly won a place in our own vocabulary. But as with most new metaphors (cf. BREAKDOWN, CEILING, TARGET) its popularity leads to its being used in contexts so incongruous with its literal meaning as to be absurd, e.g. its head-on collision with another popular new metaphor in *These bottlenecks must be ironed out*.

**irony.** For a tabular comparison of this and other words, see HUMOUR.

Irony is a form of utterance that postulates a double audience, consisting of one party that hearing shall hear and shall not understand, and

another party that, when more is meant than meets the ear, is aware both of that more and of the outsiders' incomprehension.

1. *Socratic irony* was a profession of ignorance. What Socrates represented as an ignorance and a weakness in himself was in fact a non-committal attitude towards any dogma, however accepted or imposing, that had not been carried back to and shown to be based upon first principles. The two parties in his audience were, first, the dogmatists moved by pity or contempt to enlighten this ignorance, and secondly, those who knew their Socrates and set themselves to watch the familiar game in which learning should be turned inside out by simplicity.

2. The double audience is essential also to what is called *dramatic irony*, i.e. the irony of the Greek drama. That drama had the peculiarity of providing the double audience—one party in the secret and the other not—in a special manner. The facts of most Greek plays were not a matter for invention, but were part of every Athenian child's store of legend; all the spectators, that is, were in the secret beforehand of what would happen. But the characters, Pentheus and Oedipus and the rest, were in the dark; one of them might utter words that to him and his companions on the stage were of trifling import, but to those who hearing could understand were pregnant with the coming doom. The surface meaning for the dramatis personae, and the underlying one for the spectators; the dramatist working his effect by irony.

3. And the double audience for *the irony of Fate*? Nature persuades most of us that the course of events is within wide limits foreseeable, that things will follow their usual course and that violent outrage on our sense of the probable or reasonable need not be looked for. These 'most of us' are the uncomprehending outsiders; the elect or inner circle with whom Fate shares her amusement at our consternation are the few to whom it is not an occasional

maxim, but a living conviction, that what happens is the unexpected.

That is an attempt to link intelligibly together three special senses of the word *irony*, which in its more general sense may be defined as the use of words intended to convey one meaning to the uninitiated part of the audience and another to the initiated, the delight of it lying in the secret intimacy set up between the latter and the speaker. It should be added, however, that there are dealers in irony for whom the initiated circle is not of outside hearers, but is an *alter ego* dwelling in their own breasts.

For practical purposes a protest is needed against the application of 'the irony of Fate', or of 'irony' for short, to every trivial oddity: *But the pleasant note changed to something almost bitter as he declared his fear that before them lay a 'fight for everything we hold dear'* —*a sentence that the groundlings by a curious irony were the loudest in cheering* (oddly enough). | *'The irony of the thing' said the dairyman who now owns the business 'lies in the fact that after I began to sell good wholesome butter in place of this adulterated mixture, my sales fell off 75 per cent.'* ('It's a rum thing that . . .' seems almost adequate). *The irony of fate* is, in fact, to be classed now as a HACKNEYED PHRASE.

**irrefragable.** Accent the second (*ĭrĕ'frăgăbl*).

**irrefutable.** For pronunciation see REFUTABLE.

**irrelevance, -cy.** See -CE, -CY.

**irrelevant.** It is stated in the OED, which does not often volunteer such remarks, and which is sure to have documentary evidence, that 'a frequent blunder is *irrevalent*'; that form, however, does not get into print once for a hundred times that it is said. The word is one of those that we all know the meaning of, but seldom trouble to connect with their derivations—a state of mind commoner with Englishmen than with other people because so many of our words are borrowed that we are accustomed to apparently arbi-

trary senses. It is worth remembering that *relevant* and *relieving* are the same word; that, presumably, is irrelevant which does not relieve or assist the problem in hand by throwing any light upon it. There are signs that usage is trying to force *irrelevant* along the path followed by *impertinent*. That, for instance, seems to be its meaning in *To Buchan's old admirers his unique stance is beyond parody, and to mock at it seems irrelevant*. It would be regrettable if this succeeded, for we should then have no word for 'not pertinent.'

**irrelevant allusion.** We all know the people—for they are the majority, and probably include our particular selves —who cannot carry on the ordinary business of everyday talk without the use of phrases containing a part that is appropriate and another that is pointless or worse; the two parts have associated themselves together in their minds as making up what somebody has said, and what others as well as they will find familiar, and they have the sort of pleasure in producing the combination that a child has in airing a newly acquired word. There is indeed a certain charm in the grown-up man's boyish ebullience, not to be restrained by thoughts of relevance from letting the exuberant phrase jet forth. And for that charm we put up with it when one draws our attention to the methodical by telling us there is *method in the madness*, though method and not madness is all there is to see, when another's every winter is *the winter of* his *discontent*, when a third cannot complain of the *light* without calling it *religious* as well as *dim*, when for a fourth nothing can be *rotten* except *in the state of Denmark*, when a fifth, dressed after bathing, tells you that he is *clothed and in his right mind*, or when a sixth, asked whether he does not owe you 4s. 6d. for that cab fare, *owns the soft impeachment*. Other phrases of the kind will be found in the article HACKNEYED PHRASES. A slightly fuller examination of a single example may be useful. The phrase *to leave severely alone* has two reasonable uses: one in the original sense of to leave alone as a method of severe treatment, i.e. to send to Coventry or show contempt for, and the other in contexts where *severely* is to be interpreted by contraries—to leave alone by way not of punishing the object, but of avoiding consequences for the subject. The straightforward meaning and the ironical are both good; anything between them, in which the real meaning is merely to leave alone, and *severely* is no more than an echo, is pointless and vapid and in print intolerable. Examples follow: (1, straightforward) *You must show him, by leaving him severely alone, by putting him into a moral Coventry, your detestation of the crime*; (2, ironical) *Fish of prey do not appear to relish the sharp spines of the stickleback, and usually seem to leave them severely alone*; (3, pointless) *Austria forbids children to smoke in public places; and in German schools and military colleges there are laws upon the subject; France, Spain, Greece, and Portugal, leave the matter severely alone*. It is obvious at once how horrible the faded jocularity of No. 3 is in print; and, though things like it come crowding upon one another in most conversation, they are not very easy to find in newspapers and books of any merit. A small gleaning of them follows: *The moral, as Alice would say, appeared to be that, despite its difference in degree, an obvious essential in the right kind of education had been equally lacking to both these girls* (as Alice, or indeed as you or I, might say). | *Resignation became* a virtue of necessity *for Sweden* (If you do what you must with a good grace, you make a virtue of necessity; without 'make', *a virtue of necessity* is meaningless). | *I strongly advise the single working-man who would become a successful backyard poultry-keeper to ignore* the advice of Punch, *and to secure a useful helpmate.* | Like John Brown's soul, *the cricketing family of Edrich goes marching on.* | *The beloved* lustige Wien [merry Vienna] *of his youth had* suffered a sea-change. *The green glacis . . . was blocked by ranges of grand new buildings* (Ariel must

chuckle at the odd places in which his *sea change* turns up). | *Some may remember that when the disturbances first occurred the first reaction of the Home Office bore a close resemblance to* Pilate's notorious gesture from the Lithostrotos. (Most will remember what the gesture was; some will remember that St. John tells us that Pilate was sitting in a place called the pavement; all are invited to admire the learning of one who knows that the Greek word translated *pavement* is *lithostrotos*.) | *Many of the celebrities who in that most frivolous of watering-places do congregate.* | *When about to quote Sir Oliver Lodge's tribute to the late leader, Mr. Law* drew, not a dial, *but what was obviously a penny memorandum book from his pocket* (You want to mention that Mr. Bonar Law took a notebook out of his pocket; but pockets are humdrum things; how give a literary touch? Call it a *poke*? no, we can better that; who was it drew what from his poke? Why, Touchstone a dial, to be sure! and there you are).

**irrespective(ly)**, adv. When *of* does not follow, the adverb is still *-ly*: *Mercy that places the marks of its favour absolutely and* irrespectively *upon whom it pleases.* When *of* follows, the modern idiom is to use the adjective as a QUASI-ADVERB (cf. *regardless*), as in *All were huddled together*, irrespective *of age and sex*; see UNIDIOMATIC -LY. But good writers perhaps retain the *-ly* in sentences where *irrespective* might be taken for an adjective agreeing with the subject and meaning not taking account, whereas what is desired is an adverb meaning without account taken; so *He values them*, irrespectively of *the practical conveniences which their triumph may obtain for him* (quoted from Matthew Arnold, who would doubtless have refused to drop the *-ly* here). This rather fine (if not imaginary) point of idiom has no practical effect on the meaning of a passage, but does imply an appreciation of the exact meaning and construction of the word *irrespective*—namely, that (unlike *regardless*) it does not mean careless and does not agree with a person.

**is.** 1. *Is* and *are* between variant numbers. 2. *Is* and *are* in multiplication table. 3. *Is* auxiliary and copulative. 4. *Is* after compound subject. 5. *Is,* or *has, nothing to do with.*

1. *Is* and *are* between subject and complement of different numbers. *What* are *wanted* are *not small cottages, but larger houses with modern conveniences.* | *The plausible suggestions to the contrary so frequently put forward is an endeavour to kill two birds with one stone.* | In the first example *are* should be *is* in both places; in the second, *is* should see *are*; for discussion of the first see WHAT 1 and of the second NUMBER 1.

2. *Is* and *are* in the multiplication table. *Five times six is,* or *are, thirty?* The subject of the verb is not *times,* but *six,* the meaning of the subject being 'six reckoned five times'. Before we know whether *is* or *are* is required, then, we must decide whether *six* is a singular noun, the name of a quantity, or a plural adjective agreeing with a suppressed noun; does it mean 'the quantity six', or does it mean 'six things'? That question each of us can answer, perhaps, for himself, but no one for other people; it is therefore equally correct to say *twice two is four* and *twice two are four.* Moreover, as the two are equally correct, so they appear (OED, *s.v.* time) to be about equally old; *four times six* was plural as long ago as 1380, and *ten times two* was singular in 1425.

3. Confusion between auxiliary and copulative uses. *The risk of cards being lost or mislaid under such circumstances is considerable, and great inconvenience ∧ experienced by any workman to whom this accident occurs.* This mistake of leaving the reader to supply an *is* of one kind out of a previous *is* of another kind is discussed under BE 5.

4. *Is* after compound subjects. This is discussed in NUMBER 2.

5. *Is,* or *has, nothing to do with.* A correspondent writes to a newspaper: 'Sir,—Why do I see today, in a celebrated morning contemporary, the following sentence: "The trouble *is* nothing to do with education" (Italics

mine)?' The facts are, first, that *has nothing to do with* requires no defence, secondly, that *is nothing to do with* is said by many to be indefensible, and, thirdly, that *is nothing to do with* is nevertheless very common, perhaps far commoner than the other. When a form of speech that one regards as a corruption gains wide currency, the question whether one should tilt at it is not quite simple. If it is an obvious outrage on grammar, yes; if, on the other hand, its wrongness is of the kind that has to be pointed out before it is noticed, and its hold on the public strong enough to take a good deal of loosening, then perhaps it is better to buttress it up than to tilt at it. Here, then, is an attempt to justify *is.*

Most of us, when we have occasion to repel an impertinent question, and are not in the mood for weighing words in the scales of grammar, feel that *That is nothing to do with you* expresses our feelings better than *That has* etc.; that is to say, the instinctive word is *is,* not *has.* But, says the champion of grammar, instinctive or not, it is a mere wrong mixture of two right ways of saying the thing: *That is nothing to you,* and *That has nothing to do with you.* He is very likely right, but it is not quite so certain as he thinks; and the popular phrase that is on its trial for impropriety should always be given the benefit of the doubt if there is one. Now it does not seem impossible that *It is nothing to do with* may have arisen from sentences in which *to do* has acquired the status of an adjective meaning concerned or connected; such sentences would be: *There is nothing to do with prisons that he cannot tell you.* | '*A Wife's Secret*' (*nothing to do with the old play of that name*). | *Anything to do with spiritualism is interesting.* In the first of those *nothing to do* means not a single thing concerned, *nothing* being a noun; and in the second it means not at all connected, *nothing* being an adverb. No doubt this use of *to do* is elliptical for *having to do*; but the point is that it gives us a different construction for *nothing* (or any corresponding word) which here is not the object of the omitted *having*, as it is of *has* in *It has nothing to do with*, but is either the noun with which the supposed *having* agrees or an adverb negativing it. On this theory, the two forms may be paraphrased thus: *It has nothing to do with you* = It has *no function* to perform with you; and *It is nothing to do with you* = It is *not a matter* concerned with you. The first is simpler to arrive at than the second, but the second is not impossible. The precisian who likes an easily analysable sentence, and the natural man who likes to say the thing that springs to his lips, had better agree to live and let live; and they will do this the more readily if the first can believe that the two ways of putting the thing differ not only in the visible distinction between *has* and *is*, but also in the invisible one between two or more constructions of *nothing*. It may fairly be maintained that there are three right ways of saying the thing: *It is nothing to you*; *It has nothing to do with you*; *It is nothing to do with you*: instead of two right ways and a wrong. And if the speaker is excited, as he well may be, we shall be unlikely to know which of the last two he has chosen; for he will say *it's*. Perhaps that is how the double form arose.

**-ise.** 1. On the general question of the spelling of verbs ending in the sound *īz*, see -IZE, -ISE. 2. For the coining of verbs in *-ize* see NEW VERBS IN -IZE. 3. Terminal *-ise* in words always so spelt is usually pronounced *īz*, but not always; it may be *īs* (*promise, mortise*) or *ēz* (*chemise, expertise*). That is awkward for an announcer who comes unexpectedly on an *-ise* word he has never heard spoken; he may for instance be misled into calling DEMISE *demēz*, partly perhaps by its resemblance to *chemise* and partly by the prevalent superstition that unfamiliar words ought to be given exotic vowel sounds. When in doubt say *īz* is the safest course.

**-ism and -ity.** Many adjectives may have either ending appended and give two nouns of different meaning. Occasionally choice between the two is

doubtful. Roughly, the word in *-ity* usually means the quality of being what the adjective describes, or, concretely, an instance of the quality, or, collectively, all the instances; and the word in *-ism* means the disposition to be what the adjective describes, or, concretely, an act resulting from that disposition, or, collectively, all those who feel it. A few of the more notable pairs follow, to enable the reader to judge how far this rough distinction will serve him in deciding which to use where the difference is less established: BARBARITY and *barbarism*; *catholicity* and *catholicism*; *deity* and DEISM; *fatality* and FATALISM; *formality* and FORMALISM; *humanity* and HUMANISM; *ideality* and *idealism*; *latinity* and LATINISM; *legality* and *legalism*; *liberality* and *liberalism*; *modernity* and *modernism*; *reality* and *realism*; *spirituality* and *spiritualism*; *universality* and *universalism*. See also -TY AND -NESS.

**Israeli(te).** See HEBREW.

**issue,** v. To speak of issuing (sc. from store) an article of equipment to a soldier, or of issuing him the article, is a natural use of the verb. The modern construction, which speaks of issuing him *with* the article, on the analogy of *supply* or *provide*, is not, and has been deservedly criticized for its absurdity. But it has been much popularized by two wars, is recognized without comment by the OED Supp., and has evidently come to stay, whether we like it or not.

**-ist, -alist, -tist, -yist,** etc. The use of the suffix *-ist* in English is so wide and various that any full discussion of it is not here possible. But there are (A) some words whose exact form is still uncertain and should be fixed, and there are (B) others that are both established and badly formed, so that there is danger, unless their faultiness is pointed out, of their being used as precedents for new formations.

A

*agricultur(al)ist*, *constitution(al)ist*, *conversation(al)ist*, *education(al)ist*, and others of the kind. Either form is legitimate; the shorter, besides being less cumbersome, usually corresponds more naturally to the sense. Expert in agriculture (-*turist*), for instance, is simpler than expert in the agricultural (-*turalist*); but in *constitution(al)ist*, perhaps, knowledge of or devotion to what is constitutional, rather than of or to the constitution, is required. Unless there is a definite advantage of this kind in the -*al*- form, the other should be preferred: *agriculturist, horticulturist, constitutionalist, conversationist, educationist*. Popular taste, however, seems to have a curious liking for the longer words, especially for the last two.

*accompan(y)ist*. Neither form is satisfactory; the adding of -*ist* to verbs other than those in -*ize* is unusual (*conformist* is an example), and it is a pity that *accompanier* was not taken; but, of the two, -*nyist* (cf. *copyist*) would have been better than the -*nist* which is now the standard form.

*pacif(ic)ist*. This cannot be classed among those still awaiting decision, the barbarous *pacifist* has taken so strong a hold; 'the shorter form', says the OED Supp., 'is generally preferred to the more correct *pacificist* on the grounds of convenience and euphony'. It has since established itself unshakably. The word is formed on *pacific*, to mean believer in pacific methods; the -*f*- in *pacifist*, with -*ic*- left out, has no meaning, and *pacist* would have been a better word. The omission of an essential syllable by what is called syncope (as in *idolatry*, syncopated from *idololatria*) belongs in English to the primitive stages of the language, and is not now practised; *symbology*, for *symbolology*, is an unprepossessing exception.

*ego(t)ist*. The -*t*- is abnormal; but both forms are established, and a useful differentiation is possible if both are retained; see EGO(T)ISM.

### B

*analyst, ironist, separatist*, and *tobacconist*, are open to objection, though they are all, except perhaps the least

offending of them, *ironist*, firmly established.

*analyst* results from the mistaking of *analyse* for one of those -*ize* verbs from which so many nouns in -*ist* are formed; *analyse*, derived from *analysis*, should itself have been *analysize*, and then *analysist* would have been correct; given *analyse*, *analyser* should have been the noun.

*ironist* supplies the need of a word to match *satirist* and *humorist*. The choice, if it was to end in -*ist*, lay between *ironicalist, ironicist, ironyist*, and *ironist*, of which the last is technically the least justifiable. If regarded as made on the English noun *irony*, the -*y* ought not to be omitted; if Greek is to be called in, the Greek verb and noun ought to have been εἰρωνίζω and εἰρωνία, whereas they are εἰρωνεύομαι and εἰρωνεία; *philanthropist* and *telegraphist* do not obviate the objection, because they are made not on *philanthropy* and *telegraphy*, but on Greek φιλανθρωπία and English *telegraph*.

*separatist*, like all -*ist* words made on other verbs than those in -*ize* (*conformist, computist, controvertist, speculatist*, are the best of the few quoted by the OED), is at once felt to be an uncomfortable and questionable word; but it and (*non*)*conformist*, having attained to real currency, may unfortunately be imitated. *Separationist* would have been the right form.

*tobacconist*, like *egotist*, has no right to the consonant inserted before -*ist*.

**isthmus.** Pl. -*uses*; see -US. The OED gives a choice of three pronunciations of the first syllable, *isth-, ist-*, and *iss-*, in that order; but most of us are likely to find anything but the last too difficult. Cf. *asthma*, where the OED more mercifully narrows the choice to two, *asth-* and *ass-*.

**it.** 1. Omission of anticipatory *it* owing to confused analysis. 2. Other mistakes with anticipatory *it*. 3. Obscure or wrong pronoun reference. 4. *Its, it's*.

The pronoun is so much used in various idiomatic constructions that considerable knowledge, instinctive or acquired, of the ins and outs of syntax is needed to secure one against lapses. The collecting of a few specimens, and comments on them, may put writers on their guard.

1. There is a present tendency to omit the anticipatory *it* in relative clauses, i.e. the *it* that heralds a deferred subject as in *It is useless to complain*. An example is: *The House of Commons is always ready to extend the indulgence which ∧ is a sort of precedent that the mover and seconder of the Address should ask for*. If we build up this sentence from its elements, the necessity of *it* will appear, and the reader can then apply the method to the other examples. *That the mover should ask for indulgence is a precedent*; that, rearranged idiomatically, becomes *It is a precedent that the mover should ask for indulgence*. Observe that *it* there does not mean indulgence, but means 'that the mover should ask for indulgence', *it* being placed before the predicate (*is a precedent*) as a harbinger announcing that the real subject, which it temporarily represents, is coming along later. *It is a precedent that the mover should ask for indulgence*; *the House extends the indulgence*; there are the two elements. To combine them we substitute *which* for *indulgence* in the clause that is to be subordinate, and place this *which* at the beginning instead of at the end of that clause: *the House extends the indulgence which*. . . . Now, if *it* had meant indulgence, i.e. the same as *which* now means, it would have become superfluous; but, as has been mentioned, it means something quite different, and is just as much wanted in the compound sentence as in the simple one. A parallel will make the point clear: *A meeting was held, and it was my duty to attend this*; whether *which* or *and this* is placed at the beginning of the second member instead of the present arrangement, no one would dream of dropping *it* and writing *which was my duty to attend*, or *and this was my duty to attend*. After this rather laboured exposition it will suffice to add to the

more or less similar examples that follow less elaborate indications of the essential construction: *The debate on the Bill produced a tangle of arguments which ∧ required all Mr. Chamberlain's skill to untie. Which* means tangle; the missing *it* means 'to untie which'. Here, however, if an *it* had been inserted after *untie*, *which* would have been subject to *required* instead of object to *untie*, so that the sentence as it stands is perhaps a muddle between two possibilities. / *It has already cost the 100 millions which ∧ was originally estimated would be the whole cost*. The missing *it* means 'that which would be the whole cost' (*that* the conjunction, not the pronoun). / *Faith in drugs has no longer any monetary motive such as ∧ has been asserted was formerly the case*. Without *it*, this implies as one of the elements 'A monetary motive has been asserted was the case'. / *The great bulk of the work done in the world is work that ∧ is vital should be done*. The missing *it* means not *work*, as *that* does, but the doing of it. / *What ∧ was realized might happen has happened*. Elements: 'It was realized that a thing might happen; that thing has happened'.

At the end of the article HOPE the common omission of *it* with *is hoped* is illustrated.

2. Certain points have to be remembered about the anticipatory *it* besides the fact that it may be wrongly omitted: *In connexion with the article by* ——, *it may be worth recalling the naïve explanation given to Dickens by one of his contributors*. Anticipatory *it* heralds a deferred subject; it cannot be used when there is no subject to herald. Where is the subject here? *Explanation* is engaged as object of *recalling*; *recalling* is governed by *worth*; *worth* is complement to *may be*; *it* neither has any meaning of its own nor represents anything else. The author might possibly claim that the construction was a true apposition like that in 'He's a good fellow, that', and that a comma after *recalling* would put all to rights; but anyone who can read aloud can hear that that is not true.

The real way to correct it is to write *worth while* instead of *worth*, which releases *recalling* to serve as the true subject; see WORTH for other such mistakes. | *It is such wild statements as that Mr. Sandlands has made that does harm to the Food Reform cause.* By strict grammatical analysis *does* would be right; but idiom has decided that in the *it . . . that* construction, when *that* is the relative, the number is taken not from its actual antecedent *it*, but from the word represented by it—here *statements*. | *He was a Norfolk man, and it was in a Norfolk village where I first ran across him.* There is no doubt that idiom requires *that* instead of *where*, and the sense of the idiomatic form is plain; *it that I ran across* means *my running across*; *my first running across him was in a village*. The use of *where*, besides being unidiomatic, is also less reasonable; *where* is equivalent to *in which*, and if *in which* or its equivalent is used we require *a Norfolk village* and not *in a Norfolk village*: *and it was a Norfolk village in which I* etc. The use of *in a village* together with *where* is analogous to the pleonasms discussed under HAZINESS. | *It is impossible to enter on the political aspects of Mr. ——'s book, but* ∧ *must suffice to say that he suggests with great skill the warring interests.* The reader of that at once thinks something is wrong, and on reflection asks whether the anticipatory *it*, which means to enter etc., can be 'understood' again before *must suffice* with the quite different meaning of to say etc. It cannot; but some more or less parallel types will show that doubts are natural. Here are (A) two in which the understanding of *it*, though the subjects are different, is clearly permissible: *It is dangerous to guess, but humiliating to confess ignorance.* | *It must please him to succeed and pain him to fail.* And here are (B) two that will not do: *It is dishonest to keep silence, and may save us to speak.* | *It cannot help us to guess, and is better to wait and see.* The distinction that emerges on examination is this: in the 'A' examples *is*, and *must*, are common to both halves; in the 'B' examples it is

otherwise, *is* being answered by *may*, and *cannot* by *is*. *It* may be understood, even if the real subject is changed, when the verb or auxiliary is common to both parts, but not otherwise. If, in the sentence we are criticizing (*It is impossible to enter on . . .*), *and sufficient* were substituted for *but must suffice*, all would be well.

**3.** Examples of *it* and *its* used when the reference of the pronoun is obscure or confused, or its use too previous or incorrect. These faults occur with *it* as with all pronouns, and are discussed generally under PRONOUNS; a few examples are here printed without comment: *Though* it *was not debated, delegates going home will have to give far more thought to the growing dislike of young people for the trade unions than to the H-bombs.* | *Again, unconsciousness in the person himself of what he is about, or of what others think of him, is also a great heightener of the sense of absurdity.* It *makes it come the fuller home to us from his insensibility to* it.| *Where a settlement is effected a memorandum of the same, with a report of its proceedings, is sent by the Board to the Minister of Labour.* | *Both these lines of criticism are taken simultaneously in a message which* its *special correspondent sends from Laggan, in Alberta, to the Daily Mail this morning.*

**4.** The possessive of *it*, like the absolute forms in *-s* of *her*, *their*, *our*, and *your*, has no apostrophe: *its*, *hers*, *theirs*, *ours*, *yours*, not *it's* etc.

**Italian** SOUNDS. A rough notion of how Italian words should be said is sometimes needed. Certain consonant peculiarities are all that require notice; for the vowels it suffices that they have the continental values, not the English. The letters or letter-groups with which mistakes may be made are: c, cc, ch, ci; g, gg, gh, gi, gli, gn, gu; sc, sch, sci; z, zz. If a few words, most of them to be met in English writing, are taken as types, the sounds may easily be remembered:

*cicerone* (*chĭche-*); *c*, and *cc*, before *e* and *i*, = *ch*

*Chianti* (*kĭ*); *ch* always = *k*

*cioccolata* (*chŏk-*); *ci* before *a, o, u,* often = *ch,* the *i* merely showing that *c* is soft
*Gesù* (*jā-*); *g,* and *gg,* before *e* or *i* = *j*
*ghetto* (*gĕ-*); *gh* always = *g*
*Giotto* (*jŏ-*); *gi* before *a, o, u,* often = *j,* cf. *ci* above
*intaglio* (*-ahlyō*); *gli* often = *ly*
*bagni* (*bah'nyē*); *gn* = *ny*
*Guelfo* (*gwĕ-*); *gu* always = *gw*
*fascista* (*-shĭs-*); *sc* before *e* or *i* = *sh*
*scherzo* (*sk-*); *sch* always = *sk*
*sciolto* (*shŏl-*); *sci* before *a, o, u,* often = *sh,* cf. *ci* above
*scherzo* (*-tsō*); *z* = *ts*
*pizzicato* (*pĭtsĭ-*); *zz* usually = *ts*
*mezzo* (*mĕ'dzō*); *zz* sometimes = *dz*

**italics.** Printing a passage in italics, like underlining one in a letter, is a primitive way of soliciting attention. The practised writer is aware that his business is to secure prominence for what he regards as the essence of his communication by so marshalling his sentences that they shall lead up to a climax, or group themselves round a centre, or be worded with different degrees of impressiveness as the need of emphasis varies; he knows too that it is an insult to the reader's intelligence to admonish him periodically by a change of type that he must now be on the alert. The true uses of italics are very different from that of recommending to attention whole sentences whose importance, if they *are* important, ought to be plain without them. And these real uses are definite enough to admit of classification. Some of them may be merely mentioned as needing no remark: a whole piece may be in italics because italics are decorative; text and notes may be distinguished by roman and italic type just as they may by different-sized types; quotations used as chapter-headings, prefaces, dedications, and other material having a special status, are entitled to italics. Apart from such decorative and distinctive functions, too obvious to need illustration, italics have definite work to do when a word or two are so printed in the body of a roman-type passage. They pull up the reader and

tell him not to read heedlessly on, or he will miss some peculiarity in the italicized word. The particular point he is to notice is left to his own discernment; the italics may be saying to him:

(a) 'This word, and not the whole phrase of which it forms part, contains the point': It is not only *little* learning that has been exposed to disparagement.

(b) 'This word is in sharp contrast to the one you may be expecting': It would be an ultimate benefit to the cause of morality to prove that honesty was the *worst* policy.

(c) 'These two words are in sharp contrast': But, if the child never *can* have a dull moment, the man never *need* have one.

(d) 'If the sentence were being spoken, there would be a stress on this word': The wrong man knows that if *he* loses there is no consolation prize of conscious virtue awaiting *him.* / To Sherlock Holmes she is always *the* woman.

(e) 'This word wants thinking over to yield its full content': Child-envy is only a form of the eternal yearning for something better than *this* (i.e., the adult's position with all its disillusionments).

(f) 'This word is not playing its ordinary part, but must be read as the word "——" ': Here *will* is wrongly used instead of *shall.*

(g) 'This is not an English word or phrase': The maxim that deludes us is the *progenies vitiosior* of one to which the Greeks allowed a safer credit.

(h) 'This word is the title of a book or a newspaper, or the name of a fictitious character': The Vienna correspondent of *The Times* reports that . . . / The man in *Job* who maketh collops of fat upon his flanks./A situation demanding *Mark Tapley.*

Such are the true uses of italics. To italicize whole sentences or large parts of them as a guarantee that some portion of what one has written is really worth attending to is a miserable confession that the rest is negligible.

**-ite.** The adjectival suffix *-ite,* from

the Latin p.p. *-itus* is pronounced ĭt in some words and īt in others, with little regard, if any, to whether the Latin verb is of the third conjugation (making *-ĭtus*) or of the fourth (making *-ītus*). The short *i* in *apposite* and *opposite* follows the Latin; that in *definite* does not. The long *i* in *erudite* and *bipartite* follows the Latin; that in *recondite* does not. *Composite* (ĭt in Latin) is still hesitating which way to go. The different suffix *-ite* (originally from Greek *-ĭtēs*), found in such words as *Jacobite*, *anthracite*, *dynamite*, is always *-īt*.

**-ize, -ise, in verbs.** In the vast majority of the verbs that end in *-ize* or *-ise* and are pronounced *-īz*, the ultimate source of the ending is the Greek *-izo*, whether the particular verb was an actual Greek one or a Latin or French or English imitation, and whether such imitation was made by adding the termination to a Greek or another stem. Most English printers, taking their cue from Kent in *King Lear*, 'Thou whoreson zed! Thou unnecessary letter!', follow the French practice of changing *-ize* to *ise*. But the Oxford University Press, the Cambridge University Press, *The Times*, and American usage, in all of which *-ize* is the accepted form, carry authority enough to outweigh superior numbers. The OED's judgement may be quoted: 'In modern French the suffix has become *-iser*, alike in words from Greek, as *baptiser*, *évangéliser*, *organiser*, and those formed after them from Latin, as *civiliser*, *cicatriser*, *humaniser*. Hence, some have used the spelling *-ise* in English, as in French, for all these words, and some prefer *-ise* in words formed in French or English from Latin elements, retaining *-ize* for those of Greek composition. But the suffix itself, whatever the element to which it is added, is in its origin the Greek *-izein*, Latin *-izare*; and, as the pronunciation is also with *z*, there is no reason why in English the special French spelling should be followed, in opposition to that which is at once etymological and phonetic'.

It must be noticed, however, that

a small number of verbs, some of them in very frequent use, like *advertise*, *devise*, and *surprise*, do not get their *-ise* even remotely from the Greek *-izo*, and must be spelt with *-s-*. The difficulty of remembering which these *-ise* verbs are is in fact the only reason for making *-ise* universal, and the sacrifice of significance to ease does not seem justified.

The more important of these exceptions are here given: advertise, advise, apprise, chastise, circumcise, comprise, compromise, demise, despise, devise, disfranchise, enfranchise, enterprise, excise, exercise, improvise, incise, premise, revise, supervise, surmise, surprise, televise. For the creation of verbs in *-ize* see NEW VERBS IN -IZE.

# J

**jacket.** Apart from some special uses (e.g. *Eton, dinner, mess, Norfolk*), *j*. was formerly a tailor's name for what his customer called a *coat*, but seems now to be in general use for all short coats worn by men.

**Jacobin, Jacobite.** These adjectival forms of Jacobus (= James) have been used as sobriquets of several different groups of people (and one of pigeons), but the commonest use of *Jacobins* (the name earlier given in France to Dominican friars) is for the group of extreme revolutionaries formed in Paris in 1789 (who used to meet in what was once a Dominican convent), and the commonest use of *Jacobites* is for adherents of the exiled House of Stuart in the 18th c. The latter is also sometimes used jocularly for devotees of the works of Henry James.

**jail, jailer, jailor.** See GAOL.

**Jansenism** and *Erastianism* are liable to be confused under the general notion of resistance to ecclesiastical authority. It may be said roughly that those who hold that the State should be supreme in ecclesiastical affairs are Erastians, while Jansenists are (for the purpose of this comparison) those who hold that a national branch of the Church is entitled to a certain independence of, or share in, the authority of the Pope. *Jansenism* is now loosely used by Roman Catholics to suggest the more puritanical kinds of Roman Catholicism. Erastus was author of a treatise against the tyrannical use of excommunication by the Calvinistic Churches. Jansen was author of an exposition of St. Augustine's doctrines which was designed to reform the Church of Rome and was condemned by the Pope. He was long prominent in the struggle between Gallicanism and Ultramontanism.

**jargon** is perhaps the most variously applied of a large number of words that are in different senses interchangeable, and so is a suitable heading for an article pointing out the distinctions between them. The words are: *argot, cant, dialect, gibberish, idiom, jargon, lingo, lingua franca, parlance, patois, shop, slang, vernacular.* The etymologies, several of which are indeed unknown, do not throw much light, but may be given for what they are worth: *dialect* and *idiom* are Greek (διαλέγομαι I talk; ἴδιος private or proper or peculiar); *cant* and *vernacular* are Latin (*cantus* song, chant, whine; *verna* homeborn slave); *lingo* is Italian (probably a corruption of *lingua franca*); *argot, jargon, parlance* and *patois* are French; *gibberish* and *shop* and *slang* are English, the first probably an imitation of the sound meant, the second a particular application of the common word, and the third of unknown origin.

*argot* is primarily the vocabulary of thieves and tramps in France, serving to veil their meaning, and is applied secondarily to the special vocabulary of any set of persons. There is in these senses no justification for its application to any English manner of speech instead of whichever English word may be most appropriate.

*cant* in current English means the insincere or parrotlike appeal to principles, religious, moral, political, or scientific, that the speaker does not believe in or act upon, or does not understand. It is best to restrict it to this definite use; but its earlier sense— special vocabulary of the disreputable —survives in the expression *thieves' cant*. As a general term for the special vocabulary of an art, profession, sport, etc. it has been superseded by *jargon, lingo*, and *slang*.

*dialect* is essentially local; *a d.* is the variety of a language that prevails in a district, with local peculiarities of vocabulary, pronunciation, and phrase.

*gibberish* is the name for unintelligible stuff: applied by exaggeration to a language unknown to the hearer (for which, however, *lingo* is more usual), and to anything either too learnedly worded, or on the other hand too rudely expressed, for him to make out its meaning.

*idiom* is the method of expression characteristic of or peculiar to the native speakers of a language; i.e. it is racy or unaffected or natural English (or French etc.), especially so far as that happens not to coincide with the method of expression prevalent in other languages; and *an i.* is a particular example of such speech. An earlier sense, the same as that of *dialect*, still occurs sometimes. See also IDIOM.

*jargon* is talk that is considered both ugly-sounding and hard to understand: applied especially to (1) the sectional vocabulary of a science, art, class, sect, trade, or profession, full of technical terms (cf. *lingo, slang*); (2) hybrid speech of different languages; (3) loosely the use of long words, circumlocution, and other clumsiness. It would be well if *jargon* could be confined to the first sense. There is plenty of work for it there alone, so copiously does jargon of this sort breed nowadays, especially in the newer sciences such as psychology and sociology, and so readily does it escape from its proper sphere to produce POPULARIZED TECHNICALITIES—words that cloud the minds alike of those who use them and those who read them. It is a pity that the edge of *jargon's* meaning has been blunted by its being used in

sense (3); and attempts have been
made to relieve it of the duty by
inventing another word for that style
of writing, as Ivor Brown invented
*barnacular, gargantuan*, and *pudder* in
England and Maury Maverick *gobbledygook* in America. But these have
made little headway; *jargon* remains
for most people the pejorative name
for the style of writing of which the
civil service is rather unfairly supposed to be the chief exponent. For
particular kinds of jargon see COM-
MERCIALESE, LITERARY CRITICS' WORDS,
OFFICIALESE, POPULARIZED TECHNICALI-
TIES, SOCIOLOGESE.

*lingo* is a contemptuous name for any
foreign language (*I can't speak their
beastly lingo*). It is sometimes used,
like *jargon* (1), for a sectional vocabulary.

*lingua franca* is a mixture of languages (Italian, French, Greek, and
Spanish) used by traders in the
Levant; and, by extension, any language, or linguistic mixture, that
serves as a medium of communication
between different peoples.

*parlance*, which means manner of
speaking, has the peculiarity of possessing no significance of its own and
being never used by itself. You can say
That is dialect, That is slang, etc., but
not That is parlance; *parlance* is always accompanied by an adjective or
defining word or phrase, and that adjective, not *parlance*, gives the point:
*in golfing* or *nautical parlance, in the
parlance of the literary critics*, etc.

*patois*, as used in English, means
nothing different from *dialect*, and,
like *argot*, is not used about England.
The French distinguish two stages:
dialects exist until a common literary
language is evolved from them, after
which, if they still linger, they become
patois; but in English we let them
retain their title.

*shop* describes business talk indulged
in out of business hours, or any un-
seasonable technical phraseology, and
is thus distinct, in the special-vocabu-
lary sense, from *jargon* and *slang*.

*slang* is the diction that results from
the favourite game among the young

and lively of playing with words and
renaming things and actions; some
invent new words, or mutilate or
misapply the old, for the pleasure of
novelty, and others catch up such
words for the pleasure of being in the
fashion. Many slang words and phrases
perish, a few establish themselves; in
either case, during probation they are
accounted unfit for literary use. *S.* is
also used in the sense of *jargon* (1), but
with two distinctions: in general it
expresses less dislike and imputation
of ugliness than *jargon*; and it is
naturally commoner about sporting
vocabularies (*golf s.* etc.) than *jargon*,
because many of the terms used in
sports are slang in the main sense also.
*Backslang* is a puerile type, consisting
merely in pronouncing words backwards, e.g. *ynnep* for *penny* and *cool*
for *look*. For RHYMING SLANG see that
article.

*vernacular* describes the words that
have been familiar to us as long as we
can remember, the homely part of the
language, in contrast with the terms
that we have consciously acquired.
*The vernacular* was formerly common,
and is still occasional, for the mother
tongue as opposed to any foreign language; and, by an unessential limitation, it is often applied specially to
rustic speech and confused with *dialect*.

**jaundice, jaunt(y).** The pronunciation *jaw-* has prevailed over the *jah-*
originally preferred by the OED, a fact
recognized by the SOED.

**jazz** is primarily the name given to
a type of dance music of American negro
origin that first became widely known
at the beginning of the 20th c. and is
now popular all over the world. As a
musical term it has resisted all attempts
at exact definition; all that can be said
is that in its most characteristic form it
is in common time with a pizzicato
bass in crotchets to offset a syncopated
solo, often improvised. Or, as Sir
Kenneth Clark has put it more pic-
turesquely, comparing jazz with 'action' painting, 'The trumpeter rises
from his seat as one possessed and

squirts out his melody like a scarlet scrawl against a background of plangent dashes and dots'. Jazz has been classified into *trad*(itional), *mainstream*, and *modern*, but these distinctions are chronological rather than stylistic. It would be outside the scope of this dictionary to trace its stylistic development, marked by a series of engaging designations such as *Ragtime, Swing, Jam, Boogie-Woogie*, and *Bebop*; the curious should consult *The Oxford Companion to Music*.

The derivation of the word is unknown, and has been the subject of almost as much speculation as that of O.K. (q.v.). Early alternative spellings were *jas* and *jass*, and one theory attributes it, improbably, to the French *jaser* meaning babble, another to an early player or conductor called Jas., or Jasbo Brown, another to an old minstrel-show word *jasbo* for a turn that could be relied on to please an audience when all else was failing. According to another, the word originally had an obscene connotation in American slang, and when it was applied by a newspaper as a term of abuse to the first band of the kind that visited Chicago, it was adopted by them as a label likely to assure good attendances.

The word has now strayed outside the confines of music, and is freely used to describe any kind of restless gaiety and garish colouring. It has an adjective, *jazzy*, but the noun used attributively is more common and may be applied to anything from language to stockings. It is also a verb, both intransitive ('to move in a grotesque or fantastic fashion'—OED Supp.) and transitive, usually with *up*, sometimes in its musical sense (*jazz up the classics*), sometimes figuratively of enlivening a person or thing. Thus *jazz*, in both its primary and its transferred senses, has suffered the fate common to all VOGUE WORDS of losing any precision of meaning it may once have had.

**jejune.** Accented on the last syllable by the OED and most other dictiona-

ries, but often now *jĕ'jŏŏn* by RECESSIVE ACCENT.

**jetsam, jettison.** See FLOTSAM. *Jetsam* is what is thrown overboard; *jettison* is, as a noun, the throwing of it.

**jewel(le)ry.** Both forms are established; the longer is the more usual. The pronunciation is always *jŏŏ'ĕlrĭ*.

**jibe.** See GIBE.

**jihad, je-.** The second spelling was formerly usual; but the first is now preferred; it is the form of the word in Arabic.

**jingles,** or the unintended repetition of the same word or similar sounds, are dealt with in the article REPETITION OF WORDS OR SOUNDS. A few examples of the sort of carelessness that, in common courtesy to his readers, a writer should remove before printing may be given here:

The sport of the air is still far *from* free *from* danger. / Mr. Leon Dominian has *amassed* for us a valuable *mass* of statistics. / The situation had *so far* developed *so* little that nothing useful can be said about it, save that *so far* the Commander-in-Chief was satisfied. / We can now look *forward* hopefully to further steps *forward*. / Market stability is a necessary *condition* of industry under modern *conditions*. / The figures I have *obtain*ed put a very different complexion on the subject than that generally *obtain*ing. / The observation *of* the facts *of* the geological succession *of* the forms *of* life. / He served his apprentice*ship* to statesman*ship*. / I aw*aited* a bel*ated* train. / In such a union there is no prob*ability* of st*ability*. / The earliest lists, still so sad*ly* and probab*ly* irretrievab*ly* imperfect (for this commonest form of the jingle, see under -LY). / Hardworking folk should participate in the p*leasures* of *leisure* in goodly m*easure*. / Balzac dwells on the enthusiasm with which they redecorate it with *disparate* adornments in a *desperate* attempt to outshine their neighbours.

**jingo.** Pl. *-oes*, see -O(E)S I.

**jobation, jawb-.** The first is the right

form; from *Job* came the verb *jobe* to reprove, common in the 17th and 18th cc., and from that *jobation*. *Jawb-* is a FACETIOUS FORMATION that had a short life in the late 19th c.

**jocose, jocular,** etc. These and several other words—*arch, facetious, flippant, jesting, merry,* and *waggish*—are difficult to separate from each other; the dictionaries establish no very clear or serviceable distinctions, tending to explain each by a selection of the rest. They are marked off from *funny, droll,* and others, by the fact that in the latter the effect, but in these the intent, is the main point; that is funny etc. which amuses, but that is jocular etc. which is meant (or, if a person, means) to amuse. In the following remarks no definition of the whole meaning of any word is attempted; attention is drawn merely to the points of difference between the one in question and some or all of the others. All of them are usable in contrast with *serious*, but for most a more appropriate opposite may be found for the present purpose, and that word is given in brackets.

*arch* (opp. *staid*) implies the imputation of mischief or roguishness of some sort; the imputation is ironical, or the offence is to be condoned; the meaning is conveyed chiefly by look, tone, or expression. *An arch look, girl, insinuation.*

*facetious* (opp. *solemn*) implies a desire to be amusing; formerly a laudatory word, but now suggesting ill-timed levity or intrusiveness or the wish to shine. *A facetious remark, fellow, interruption.*

*flippant* (opp. *earnest*) implies mockery of what should be taken seriously, and want of consideration for others' feelings. *A flippant suggestion, young man; f. treatment.*

*jesting* (opp. *serious*) differs from the rest in having perhaps no distinctive implication, *A jesting mood, j. Pilate, a j. proposal.*

*jocose* (opp. *grave*) implies something ponderous, as of Adam and Eve's elephant wreathing his lithe proboscis

to make them mirth. *A jocose manner, old boy, description.*

*jocular* (opp. *prosaic*) very commonly implies the evasion of an issue by a joke, or the flying of a kite to test the chances. *A jocular reply, writer, offer.*

*merry* (opp. *melancholy*) implies good spirits and the disposition to take things lightly. *A merry laugh, child, tale.*

*waggish* (opp. *sedate*) implies on the one hand willingness to make a fool of oneself and on the other fondness for making fools of others. *A waggish trick, schoolboy, disposition.*

**jollily, jolly** adv. As a colloquial substitute for very (*a j. good hiding; you know j. well*) the adverb is *jolly*; in other uses (*he smiled j. enough*) it is *jollily*. See -LILY. The former usage was not always colloquial only. The OED quotes from a 17th-c. commentary on the Old and New Testaments *All was jolly quiet at Ephesus before St. John came hither.*

**jonquil.** The OED gives precedence to the older pronunciation *jŭ'ngkwĭl* but fashion has now pronounced in favour of *jŏn-.*

**journal.** Objections have sometimes been made to the extension of this to periodicals other than the daily papers. But 'Our weekly journals o'er the land abound' (Crabbe, 1785) shows that it is much too late to object. Those who do so have presumably just learnt the connexion of *journal* with L. *dies;* for, if it had been long familiar to them, they would surely have been aware also that language is full of such extensions. May a woman not be said to *cry* till she howls? are there no *clerks* but those in Holy Orders? are there no *hypocrites* off the stage? And, to come back to *dies,* is it a blunder to call a sea voyage from London to Tokyo a *journey,* or a pedantry to call it anything else?

**joust, just.** Though *just* is 'the historical English spelling' (OED), *joust* was preferred by Johnson and used by Scott. It is consequently now more intelligible and to be preferred, and the pronunciation *joost* (sometimes *jowst*) has superseded the old *jŭst*.

**judgematical.** See FACETIOUS FORMATIONS.

**judg(e)ment.** See MUTE E for the principle governing the retention and omission of e in derivatives, viz., that it is dropped only before vowels. *Judgement* is now preferred to the once orthodox *judgment*. The latter is the spelling of the A.V. but the R.V. and NEB have *judgement*, and the OED favours the retention of the *e* both here and in similar words such as *abridg(e)-ment*, *acknowledg(e)ment*, *fledg(e)ling*, and *lodg(e)ment*. There are some who would differentiate by using *judgment* for the judicial pronouncement and *judgement* for all other purposes. This may be too subtle for popular taste, but authority for some such distinction can be found in the rules of the OUP, which favours *judgment* in legal works and *judgement* in all others.

**judicial, judicious.** The first has to do with judges and lawcourts and legal judgements, the second with the mental faculty of judgement, that is to say, if the distinction mentioned at the end of the preceding article were to prevail, the first with judgment and the second with judgement. *Judicial murder* is murder perpetrated by means of a legal trial; *judicious murder* is murder that is well calculated to serve the murderer's interests. The distinction is clear enough, except that *judicial* has one use that brings it near *judicious*; this use is impartial or such as might be expected of a judge or a lawcourt, applied to such words as *view, conduct, care, investigation*, to which *judicious* is also applicable in the sense of wise or sagacious or prudent. In the following example, one may suspect, but cannot be sure, that the writer has meant one word and written the other: *The chapter on the relations between Holland and Belgium after the war in connexion with a suggested revision of the treaty of 1839 is* fairly *written in a* judicious *spirit*.

For other such pairs, see PAIRS AND SNARES.

**jugular.** The OED and some other dictionaries want us to say *joog-*; but for ordinary mortals, familiar from childhood with *the jŭgular vein*, it is as much out of the question as to make *kŏ'kăin* out of COCAINE, and the COD has recognized this to the extent of putting *jŭg-* first.

**ju-jutsu, jiu-jitsu.** The first is the preferable spelling, but the art is now more commonly known as *judo*.

**jump.** *J. to the eye(s)* is a bad GALLICISM (5). Examples: *The desperate discomfort of these places as living houses judged by our standards jumps to the eyes.* | *How little there is essentially in common between Virgil and Isaiah jumps to the eye as we read the clever and tasteful paraphrase into Biblical language of the 4th Eclogue.*

**juncture.** *At this juncture* means at this convergence of events; the phrase is emasculated if used, as it often is, merely as a stylish way of saying at this time. It should be reserved for an occasion that is not yet a crisis but may well become one.

**junta, junto.** The first is the Spanish form, which is used in English also. *Junto* is an erroneous form.

**just,** v. and n. See JOUST.

**just,** adv. 1. *Just exactly* is bad tautology. *Mr. Gladstone's dearest friend in political life, who himself passed away* just exactly *half a century ago*.
2. *Just how many* and similar indirect-question forms are Americanisms which have now a firm footing with us, colloquially at least. *Just what makes the best lodgement for oyster spawn has been greatly discussed*.
3. *I don't think the Tories are worse people than we are. I just think they don't belong to this century. They have not just grown up.* Here the speaker in his last sentence has confused *not just* (= not merely, or not in the immediate

past) with *just not* (= simply not or barely). In fact he has not just used the word, he has just not understood how to use it.

4. Pronounce *jŭst*; a warning against the vulgarisms *jest* and *jist* is not superfluous.

**juvenile.** See TEENAGER.

# K

**kadi.** *C*- is the usual spelling.

**kangaroo.** For the parliamentary sense, see CLOSURE.

**kaolin.** Usually pronounced *kā′ŏlĭn*.

**kartell.** See CARTEL.

**Kelt(ic).** See CELT(IC).

**kerb.** See CURB.

**kerosene, paraffin, petrol, petroleum.** The popular use of the words is all that is here in question. *Petroleum* is the crude mineral oil; *petrol*, or *petroleum spirit* (U.S. *gasoline*), is refined petroleum as used in internal combustion engines; *kerosene* and *paraffin* (*oil*) are oils got by distillation from coal or shale, *kerosene* being the usual name in America, and *paraffin* in England.

**ketchup** is the established spelling; formerly also *catchup* and *catsup*, of which the second at least is due to popular etymology. A Chinese or Malay word is said to be the source.

**kilo-, milli-.** In the metric system, *kilo-* means multiplied, and *milli-* divided, by 1000; *kilometre* 1000 metres, *millimetre* 1/1000 of a metre; cf. DECA-, CENTI-.

**kind,** n. The irregular uses—*Those k. of people, k. of startled, a k. of a shock*—are easy to avoid when they are worth avoiding, i.e. in print; and nearly as easy to forgive when they deserve forgiveness, i.e. in hasty talk. *Those k. of* is a sort of inchoate compound, = those-like (cf. *such*, = so-like); *k. of startled* = startled-like. *A k. of a shock*

is both the least criticized and the least excusable of the three. Shakespeare puts the first into the mouth of Cornwall in *King Lear* (*These kind of knaves I know*), and it may perhaps be classed as a STURDY INDEFENSIBLE. The OED's verdict is 'still common colloquially though considered grammatically incorrect'. See also SORT.

**kindly.** *Authors are kindly requested to note that Messrs.* —— *only accept MSS. on the understanding that* . . . *Messrs.* —— *may be kind in making the request, but did they really mean to boast of it?* This misplacement is very common; for the ludicrous effect, compare the confusion between *It is our* PLEASURE and *We have the pleasure*.

**king.** Under *King-of-Arms*, the OED says 'less correctly *King-at-Arms*'. Both phrases are shown by its quotations to have been in use at all periods, but *of* has become established in the designation of those who still hold such offices—*Garter, Clarenceux, Norroy and Ulster, Lord Lyon, Bath,* and the other Orders of Chivalry.

**kith and kin.** This archaism is commonly thought to be one of those phrases like *toil and moil* in which the words have much the same meaning and are doubled for emphasis. (See SIAMESE TWINS.) That is not so. *Kith* means acquaintances and *kin* means kinsfolk. The modern equivalent is friends and relations.

**knee.** The adjective from *knock-knees, broken knees,* etc., is best written with an apostrophe—*knock-knee′d* etc.; see -ED AND 'D.

**kneel.** For *kneeled* and *knelt*, see -T AND -ED.

**knit(ted).** Both forms are still in use for both the past tense and the past participle, but the short form is now unusual in the special sense of making with knitting-needles. *She knit(ted)*, or *had knit(ted), her brows*, but *she knitted* or *had knitted a pair of socks; a well-knit frame*, but *knitted goods* in ordinary use, though *knit wear* survives in the trade. *Knitten* is a pseudo-archaism.

**knoll.** Pronounce *nōl*. The word being chiefly literary, so that most of us have to guess its sound from its spelling, and the sound of final *-oll* being very variable (dŏll, lŏll, Mŏll, Nŏll, Pŏll, against drŏll, rōll, strŏll, tōll, and trōll, among clear cases), it is a pity that the rival spelling *knole* has not prevailed. But *-ll* is now firmly established.

**knot.** The nautical term is a measure of speed meaning one nautical mile (6080 feet) an hour. To suppose it to be a measure of distance and to say that the speed of a ship is 15 knots an hour is to incur the contempt of those who know the true meaning; but the OED has nothing worse to say about it than that it is 'more loose', and gives examples from unimpeachably nautical sources—Anson's *Voyages*, Cook's *Voyages*, and Marryat's *Peter Simple*. The term is derived from the old method of calculating the speed of a ship by running a log line with equally spaced knots in it over the stern and counting the number that slip past in a minute.

**know.** For the interjectory *you know* see MEANINGLESS WORDS.

**know-how.** *There are signs that the Atomic Energy Authority would go a long way in the exchange of information and k.-h.* | *It is not unnatural that some departments are anxious that Britain should not lose all the commercial value of the technical k.-h., of which much derives from the activities of manufacturers.* | *The Greeks themselves are accused by the Turks of those very vices which they are always attributing to the British: superiority, an arrogant assumption of k.-h., unscrupulous pursuit of their own ends and so forth.* This Americanism, said by an American dictionary to be 'standard American usage for the knowledge of how to do something, faculty or skill' has made itself thoroughly at home in England. With us it is mostly applied to manufacturing techniques, especially to the skill needed to overcome some novel difficulty. Its use in 1957 by the President of the Board of Trade during a debate in the House of Commons provoked a protest that he was lowering the dignity of the House by using slang; but this was not supported by the Speaker, who reminded members that 'the slang of yesterday is often the current English of today'. He has proved right. Five years later the word was accorded unassailable status when the House of Lords decided that know-how is 'an ambience pervading a highly specialized production organization' and is properly treated as one of its capital assets (*Rolls Royce Ltd.* v. *Inland Revenue Commissioners*). '*Know-how*' said the Master of the Rolls in a later revenue case 'has a meaning as well recognized as *goodwill*'. The expression evidently fills a need, and its plebeian appearance ought not to count against it. The more elegant *savoir faire* has been put to a different use.

**knowledge.** The first edition of this dictionary quoted from the OED: 'the pronunciation *nōl-*, used by some, is merely a recent analytical pronunciation after *know*', and commented that this was on the same level as *often* with the *t* sounded. If it is true that the speak-as-you-spell movement (see PRONUNCIATION 1) was then attacking both these words, its failure with one has been as marked as its success with the other. *Nōl-* has disappeared even from those pulpits that were its last refuge; the *t* in OFTEN may now be heard from them almost as a matter of course.

**kotow, ko-tow, kowtow.** The dictionaries pronounce this *kŏtow'*, and it is often printed *ko-tow* by way of showing that the first syllable is not to be weakened in the normal way, which would give *kŏtow'*. The real choice lies between both writing and pronouncing *kowtow*, and allowing the weakening to *kŏtow'*; for the word is now fairly common, and cannot possibly maintain under popular wear and tear the full vowel sound in the unaccented syllable. There can be little doubt that the popular verdict has gone in favour of *kowtow*, with equal stress, faintly suggesting *bow down*.

**kudos,** Greek for *glory*, became an English slang word of limited currency at a time when Greek was more widely learnt, and is now, it seems, sometimes mistaken for a plural. See SINGULAR S.

**kyrie eleison.** Of many competing pronunciations the OED prefers *ker'ĭĭ lā'ĭson* (seven syllables). However the vowel sounds may be pronounced, the stress must be on the first syllable of the first word and on the second of the second.

# L

**laboratory.** The pronunciation favoured by the OED and most other dictionaries is *lă'bŏrătŏrĭ*; those who find four successive unaccented syllables trying, liable to result in *lă'brătrĭ*, do better to say *lăbŏ'rătŏrĭ*, which has indeed in practice become the normal pronunciation. But most of us now dodge these difficulties by saying *lab.* See RECESSIVE ACCENT.

**lac, lakh.** In its 15 quotations the OED shows 9 different spellings, but choice now lies between these two; and of the two it treats *lac* as preferable. But *lakh* has long been the official spelling.

**laches** is a singular noun, pronounced *lă'chĕz*, meaning negligence of certain kinds, rarely used with *a* but often with *the* and *no*, and not requiring italics. Its formation is similar to that of *riches* (formerly *lachesse*, *richesse*), but not having become a popular word it has escaped being taken for a plural.

**lachrym-.** The true spelling for all the words would be *lacrim-*, and it would be at least allowable to adopt it; but the *h* and the *y* are still usual.

**lack.** See DEARTH.

**lacuna.** Pl. *-ae* or *-as*; see LATIN PLURALS.

**lade,** apart from the passive use of the p.p., is now almost restricted to the *bill of lading* that the master of a ship gives to the consignor as a receipt for his cargo. Even *laden*, though still in use, tends to be displaced by *loaded* and to sound archaic except in particular phrases and compound words: *heavy-laden buses*, but *loaded* rather than *laden buses*; *sin-laden*, *sorrow-laden*; *a hay-laden* rather than *a hay-loaded lorry*, but *loaded*, rather than *laden*, *with hay*; on the other hand *a soul laden*, rather than *loaded*, *with sin*, because the dignity attaching to slight archaism is in place.

**lady.** 1. *L. Jones*, *L. Mary Jones*, *L. Henry Jones*. The first form is proper only for a peeress below the rank of duchess or a baronet's or knight's wife or widow; the second for one called *L.* because she is the daughter of a duke, marquess, or earl; the third for the wife or widow of the younger son of a duke or marquess.
  2. *L.* by itself in the vocative is often loosely used for *madam* as a term of respect, as *governor* is for *sir*, though this last is passing.
  3. *L.* prefixed to names indicating vocation as a mark of sex (*l. doctor*, *barrister*, etc.) is a cumbrous substitute for a FEMININE DESIGNATION, which should be preferred when there is one in common use. In default of that, the present practice of using *woman* is to be welcomed as better than *l.*, not confusing the essential point with irrelevant suggestions of social position, as in 4.
  4. *L.* prefixed to vocation words to indicate social pretensions (*l. cook*, *nurse*, *companion*, *help*, etc.) is a GENTEELISM that has lost its point now that all women are ladies and none are servants.
  5. For *l.* as undress substitute for *marchioness*, *countess*, *viscountess*, see TITLES. See also GENTLEWOMAN, LADY.

**laid, lain.** See LAY AND LIE.

**lakh.** See LAC.

**lama, llama.** *La-* for the Tibetan priest; *lla-* for the animal.

**lamentable.** Pronounce *lă'm-*. See RECESSIVE ACCENT.

**lamina.** Pl. *-ae*.

**lampoon, libel, pasquinade, satire, skit, squib.** There is often occasion to select the most appropriate of these words, and the essential point of each may be shortly given. A *lampoon* is a virulent or scurrilous published attack on an individual; a *libel* in popular usage is a defamatory statement made publicly or privately (for its meaning in law see LIBEL); a *pasquinade* (now rare) is an attack of unknown or unacknowledged authorship posted in a public place; SATIRE holds up prevailing vices and follies to ridicule; a *skit* is a making game of a person or his doings especially by caricature or parody; a *squib* (now rare) is a casual published attack, short and sharp.

**land, n.** *L. of the leal* means heaven, not Scotland. *L. of cakes* means Scotland, not heaven.

**landslide, landslip** in their literal senses are respectively U.S. and British words for the same thing. In the figurative sense of an overwhelming victory in an election (originating in U.S.) *landslide* is used in both countries, and there are signs that this may be making us forget *landslip*. *The heavy rain of the last twenty-four hours has caused a landslide blocking the main line to Dover.*

**languor, languorous, languid, languish.** The pronunciation is anomalous: *languid* and *languish* have always the *-gw-* sound (*-gwĭ-*); for *languorous* the OED gives only that sound (*-gwŏr-*); but for *languor* it prefers the *-g-* sound (*-gŏr*), though *-gw-* (*-gwŏr-*) is allowed as alternative. Popular choice, however, has now firmly decided that the *gw* sound must be confined to *languid* and *languish*; *languor* and *languorous* can be only *-g-*. On the merits, this seems unreasonable, and is perhaps due to misapprehen-

sion; either *-uor* is confused with the *-our* of *vigour*, *honour*, etc.; or else the *-u-* is mistaken for one of the kind seen in *guest*, *guile*, *guess*, *guild*, where its function is to show that g is not as usual soft before e or i. *Liquor* and *liquid*, *conquer* and *conquest*, show similar inconsistencies.

**lantern, -thorn.** The second, now obsolete, is a corruption due to the use of horn for the sides of old lanterns.

**lapel.** Pronounce *lăpĕ'l*; adj. *lapelled*.

**lapsus.** Pl. *lapsūs*, not *-si*; see *-US*.

**larboard.** See PORT.

**large.** For a comparison of this with *great* and *big*, see BIG.

**large(ly).** After the verbs *bulk* and *loom*, the idiomatic word is not largely, but large, as in the now archaic *writ large*. See UNIDIOMATIC-LY. Examples of the wrong form are: *The Monroe doctrine of late years has loomed so largely in all discussions upon . . . | A phase of the Irish question which has bulked largely in the speeches of the Unionist leaders.*

**largess(e).** Pronounce *lar'jĕs*, and omit the final *-e*. If the word had remained in common use, it would doubtless have come to be spelt, as it often formerly was, *larges*; cf. *riches* and *laches*.

**lasso** is pronounced *lăsoo'* by those who use it, and by most English people too, in spite of a tendency of the dictionaries to favour *lă'sō*. Pl. *-os* or *oes*; COD gives *-os* only. See *-O(E)S* I.

**last.** I. *The l. two* etc., *the two l.* etc. For this see FIRST 3.
2. *Last, lastly.* In enumerations *lastly* is recommended on the same grounds as *firstly*, for which see FIRST 4.
3. *At* (*the*) *long l.* is an idiom labelled 'now rare' by the OED; but it has experienced a revival (without the *the*) due more perhaps to its odd

sound than to any superior significance over *at l.*, and is now often
heard and seen; 'in the end, long as it
has taken or may take to reach it' is
the sense.

**4. Last, latest.** In this now favourite
antithesis (*Dr. Marshall's latest, but
we hope not his last, contribution*) we
are reminded that *latest* means last up
to now only, whereas *last* does not
exclude the future. The distinction is
a convenient one, and the use of *latest*
for *last* is described by the OED as
'now archaic and poetical'. But no
corresponding agreement has yet been
reached for abstaining from *last* when
*latest* would be the more precise word,
and many idioms militate against it
(*last Tuesday; last year; for the last
fortnight; on the last occasion; as I said
in my last*).

**late, erstwhile, ex-, former(ly),
quondam, sometime, whilom.**
With all these words to choose from,
we are yet badly off: *erstwhile* and
*whilom* smack of WARDOUR STREET;
*ex-*, which tends to swallow up the
rest, is ill fitted for use with compound words such as *Lord Mayor*
(see HYPHENS 6), which nevertheless
constantly need the qualification; *late*
is avoided because of the doubt
whether it means that what is over is
the person's life or his tenure of office;
*quondam* and *sometime* have become,
partly owing to the encroachments
of *ex-*, unusual enough to sound
pedantic except in special contexts
(*my quondam friend; sometime rector
of this parish*). The best advice is to
refrain from *ex-* except with single
words (*ex-Mayor*, but not *ex-Lord-
Mayor*, and still less *ex-Lord Mayor*),
and from *l.* except either in the sense
of no longer living or when the person
described is in fact dead, and to give
*former(ly)*, and perhaps *sometime*,
more work to do.

**lather.** The OED gives only *lă-*
(rhyming with *gather*, not *father*); and
an obsolete spelling *ladder* shows the
old vowel sound. Though *lah-* is
often heard, *l.* apparently does not

belong to the class of words in which
*ah* and *ă* are merely southern and
northern variants (*pass* etc.).

**latifundia** is a plural.

**latine,** = in Latin, is a Latin adverb,
pronounced in English *lătīnē*; similar
adverbs (mostly rare and some very
rare) are *anglice, celtice, gallice, graece,
hibernice, scottice, teutonice*. All are
sometimes printed *-è* to show that the
*e* is sounded (*-sē*).

**latinism, latinity.** The first is a
disposition to adopt Latin ways,
especially of speech, or a particular
idiom that imitates a Latin one; the
second is the quality of Latin (classical,
debased, etc.) that characterizes a
person's or a period's style. See -ISM
AND -ITY.

**Latin phrases.** 'The reformed pronunciation of Latin', wrote R. W.
Chapman, 'has caused great confusion
in the pronunciation of Latin words
that have been adopted in English.
The main differences are these:
1. Short vowels, whether stressed or
   not, are unaffected.
2. The long vowels:
   Stressed ā, formerly as in *mate*,
   now as in *ah*.
   Stressed ē, formerly as in *mete*,
   now as in *mate*.
   Stressed ī, formerly as in *mite*,
   now as in *mete*.
3. U, stressed or not, was formerly as
   in *must* (Sulla) or as in *mute*
   (Punica fides). It is now as in *bull*
   or *moot*.
'The safe course is "stare super
antiquas vias". If we impose the new
Latin on English, consistency will
demand that we say *bŏnus* and *mĭnus*
and *Rōmulus* and *Rĕmus* and a thousand other horrors. But in fact many
people, unaware of the trap, do say
*wahdi maycoom, O see sick omnays* and
the like.
'The consonants give little trouble.
But there are some people who say
*Kikero*, and make *Gellius* begin like
*get*.'

This advice is respectfully recommended to the reader in the hope, perhaps over-sanguine, that it may not be too late. The tendency deprecated is powerfully reinforced by the superstition, rife today, that all foreign-looking words ought to be given a foreign-sounding pronunciation (see DIDACTICISM, FRENCH WORDS, and elsewhere), and is leading to the maltreatment even of those Latin phrases that have long been part of our language, such as *a priori, bona fide, pari passu, prima facie, sine die, via.* For the pronunciations recommended for an English-speaking person see the separate entries.

**Latin plurals** (or latinized-Greek). Of most words in fairly common use that have a Latin as well as or instead of an English plural the correct Latin form is given in the word's alphabetical place. A few general remarks may be made here.

1. No rule can be given for preferring or avoiding the Latin form. Some words invariably use it; nobody says *specieses, thesises,* or *basises,* instead of the Latin *species, theses,* and *bases* (*bā'sēz*). Others nearly always have the Latin form, but occasionally the English; *bacilluses, lacunas,* and *genuses,* are used at least by anti-Latin fanatics instead of *bacilli, lacunae,* and *genera.* More often the Latin and English forms are on fairly equal terms, context or individual taste deciding for one or the other; *formulas, indexes, narcissuses, miasmas, nimbuses,* and *vortexes,* are fitter for popular writing, while scientific treatises tend to *formulae, indices, miasmata, narcissi, nimbi,* and *vortices.* Sometimes the two forms are utilized for real differentiation, as when *genii* means spirits, and *geniuses* men. All that can safely be said is that there is a tendency to abandon the Latin plurals, and that, when one is really in doubt which to use, the English form should be given the preference.

2. With a few exceptions too firmly rooted to be dislodged (e.g. *Adelphi*), Latin plurals in *-i* should be pronounced distinctly *-ī,* and not *-ē* or *-ĭ* like the Italian *dilettanti, pococuranti,* etc.; the reformed pronunciation of Latin does not obtain in naturalized Latin words (see LATIN PHRASES), and to say *glă'dĭŏlē* reveals that one is ignorant either that the word is Latin or how Latin words are pronounced. Plurals of words in *-is* (*theses, metamorphoses, neuroses*) should be plainly pronounced *-ēz,* not *-ĭz* like English plurals.

3. In Latin plurals there are naturally some traps for non-Latinists; the termination of the singular is no sure guide to that of the plural. Most Latin words in *-us* have plural in *-i,* but not all; and so zeal not according to knowledge issues in such oddities as *hiati, octopi, omnibi,* and *ignorami.* See -US. Similarly most Latin nouns in *-a* have plural in *-ae,* but not all: *lacuna, -nae;* but *dogma* and *gumma, -mata; Saturnalia,* not singular but plural. *Animalcula* is the plural of *animalculum,* but is sometimes treated as a singular (= *animalcule*), with plural *animalculae.* And, though *-us* and *-a* are much the commonest Anglo-Latin endings, the same danger attends some others (*-ex, -er, -o,* etc.).

4. The treatment of a Latin noun as an English plural because it ends in *-s* is surprising when of modern introduction. The Latin plural of *forceps* is *forcipes,* and the English plural should be *forcepses; a forceps, a set of forcipes* or *forcepses;* and both these were formerly in use. But *shears* and *scissors* and *pincers* and *pliers* have so convinced us that no such word can have a singular that instead of *a forceps* we usually say *a pair of forceps,* and *forceps* has to serve for both singular and plural. See PAIR.

5. For discussion of some of these points in greater detail see -EX, -IX; TRIX; -UM and -US.

**-latry.** For words like *bardolatry* and *babyolatry,* see FACETIOUS FORMATIONS. The archetype of the *-latry* words, *idolatry,* is a shortened form of the medieval *idololatry,* which is a more correct way of combining *idol* with *-latry*, and sounds even more reprehensible.

**latter** survives almost solely in *the l.*, which provides with *the former* a pair of pronouns obviating disagreeable repetition of one or both of a pair of previously mentioned names or nouns. Such avoidance of repetition is often desirable; for the principles, see ELEGANT VARIATION, and REPETITION. But *the l.* is liable to certain special misuses: (1) *The l.* should not be used when more than a pair are in question, as in: *The difficult problems involved in the early association of Thomas Girtin, Rooker, Dayes, and Turner are well illustrated by a set of drawings that . . .; and what was undoubtedly the best period of* the l. *artist is splendidly demonstrated by. . . . (2)* Neither should it be used when fewer than two are in question; the public and its shillings cannot be reasonably regarded as a pair of things on the same footing in: *The mass of the picture-loving public, however, may be assured of good value for the shillings—whatever be the ultimate destination of* the l. (3) The true elegant-variationist, who of course works *the l.* very hard, should observe that a mere pronoun will not do for the antecedent of *the l.*, even though there may be a name in the background; a writer who varies *Gordon* with *the hero of Khartoum* and *his relative* naturally does not shrink from picking up *him* with *the l.*; it is all of a piece, and a bad piece:' *Mr. Hake was a cousin of the late General Gordon, of whom he entertained a most affectionate remembrance. On one occasion, when the hero of Khartoum was dining with* him, the l. *invited his relative to take wine with him, but Gordon imperiously declined.* (4) The true use of it is not to mystify, as in: *The only people to gain will be the Tories and the principal losers will be the working-class voters whose interests the Labour Party is supposed to have at heart. It is a very poor compliment to the intelligence of* the l. [which, in heaven's name?] *to believe, as many Labour members seem to*

*do, that their support of the Labour cause will be all the more ardent if their interests are thus disregarded.*

**laudable** means praiseworthy, generally with a tinge of the patronizing or the ironical. The quotation shows it confused with *laudatory*: see PAIRS AND SNARES: *He speaks in the most laudable terms of the work carried out by Captain Thompson in the Anglo-Egyptian Soudan.*

**laughable.** For the peculiar formation, see -ABLE 4. For *'would be laughable if it were not tragic'* etc., see HACKNEYED PHRASES.

**laughter.** *Homeric l.* is a now common phrase whose meaning must be vague to many readers. It is especially the laugh that runs round a circle of spectators when a ludicrous or otherwise pleasing incident surprises them. In Olympus, when Zeus and Hera have had words, the limping Hephaestus counsels his mother to deal in soft answers. When he, in that former quarrel, had tried to protect her, had he not been flung forth and fallen nine days through air till he landed in Lemnos? And were not nectar and ambrosia in Olympian halls better than such doings? And therewith he hastened round and filled the cups of all the gods; 'and inextinguishable was the laughter of the blessed gods as they watched Hephaestus bustling about the hall'.

And again, when Penelope's suitors set the beggar bully Irus to box with the seeming beggar Odysseus, 'then the twain put up their hands, and Irus struck at the right shoulder, but the other smote him on his neck beneath the ear, and crushed in the bones, and straightway the red blood gushed up through his mouth, and with a moan he fell in the dust, and drave together his teeth as he kicked the ground. But the proud wooers threw up their hands, and died outright for laughter'.

Such is Homeric l.; but whether the frequent use of the phrase is justified by present-day familiarity with Homer is doubtful.

**launch.** The pronunciation *law*, not *lah-*, is now standard.

**laurustinus.** So spelt; *tinus*, a Latin plant-name, not a suffix, was used in apposition to *laurus*; *laures-* is a corruption. Pronounce *lawr-* as in laureate, not *lŏr-* as in laurel.

**lavatory.** See EUPHEMISM.

**lay and lie.** 1. Verbs. Except in certain technical terms of seamanship, the intransitive use of *lay* (= *lie*) is 'now only illiterate' (OED). In modern usage *lay* is transitive only (= put to rest), and makes *laid*; *lie* is intransitive only (= be at or come to rest), and makes *lay*, *lain*, never *laid*. But confusion even between the words *lay* and *lie* themselves is very common in talk. Still commoner, sometimes making its way into print, is the use of *laid* (which belongs to the verb *to lay* only) for *lay* the past tense and *lain* the p.p. of *lie*, as in *We laid out on the grass, and could have laid there all day*. 2. Nouns. *Lie* and *lay* are both used in the senses configuration of ground, direction or position in which something lies. Neither has a long established history behind it. The OED has only one quotation earlier than the 19th c., and that is for *lie* (*the proper lye of the land*, 1692); *lie* is certainly commoner today, and seems also the more reasonable form; *lay* perhaps issued from sailors' and rustic talk, in which the verbs are not kept distinct.

**lay-by.** The criticism of this expression voiced by some purists is misconceived. It was used in waterways and railways long before it was applied to road transport, and may be a survival of the old intransitive use. But anyone who is shocked at this can reassure himself by supposing that the motorist is being invited not to lie in the recess indicated but to lay his vehicle there.

**lay figure** has no connexion with any of the English words *lay*, but is from Dutch *leeman* (*led* joint), and means literally jointed figure.

**leaded,** and *double-leaded*, in printing, mean set with more than the ordinary space between the lines, as is done with matter in the newspapers for which special attention or a special status is desired; the space is made by inserting strips of lead.

**leaden.** See -EN ADJECTIVES. *Leaden*, however, is less disused in the literal sense than most of the words among which it is there placed; *lead roof* or *pipe* is commoner than *leaden*, but *a leaden pipe* is not as unidiomatic as *a golden watch*.

**leaderette.** For the uses of the suffix *-ette* see SUFFRAGETTE.

**leadership.** For the use of *l.* for *leaders* see MEMBERSHIP.

**leading question** is often misused for a poser or a pointed question or one that goes to the heart of the matter (as though *leading* meant principal, or intended, in the slang phrase, to lead the witness up the garden path). Its real meaning is quite different; a l. q. is not hostile, but friendly, and is so phrased as to guide or lead the person questioned to the answer that it is desirable for him to make, but that he might not think of making or be able to make without help. It is used especially of the form of question not permitted to counsel examining one of his own witnesses. To object, as people do when they are challenged to deny or confirm an imputation, 'That is a leading question' is meaningless. See POPULARIZED TECHNICALITIES.

**(‥) leafed, (-) leaved.** See -VE(D).

**leak.** The verb *leak* (*out*) and the noun *leakage* have long been used figuratively of the unauthorized disclosure of secret information. The transitive use of the verb in a similar way, now common in journalism (*he leaked the news*), seems to be a revival of a usage marked by the SOED as obsolete but now recognized by the COD without comment.

**lean.** For *leant* and *leaned*, see -T AND -ED.

**leap.** For *leapt* and *leaped*, see -T AND -ED. Of *l. to the eye(s)*, as wearisome a GALLICISM as exists, some examples must be given to suggest its staleness; others will be found under JUMP. *Bath, it may be admitted, does not l. t. t. eyes as an obvious or inevitable meeting-place for the Congress.* | *This, however, does not l. t. t. eye, and for the moment I am concerned only with the impressions which strike a new-comer.* | *We have not the smallest doubt that there is a perfectly satisfactory explanation of these widely differing totals, but certainly it does not l. t. t. eyes.*

**learn.** 1. For *learnt* and *learned*, see -T AND -ED. The existence of the disyllabic *learned* as an adjective is an additional reason for preferring *-nt* in the verb; and so with *unlearned* and *-nt*. 2. The use of *learn* for teach (*Lead me forth in Thy truth and learn me*) is now only vulgar, jocular, or dialect.

**leasing.** The biblical word, = lying, is pronounced *lē'zĭng*.

**least.** For the common confusion between *much less* and *much more* see MUCH 2. *Least of all* and *most of all* get mixed up in the same way: *If that is the case, what justification exists for the sentences, l. of all for the way in which they were carried out?* | *An active statesman of marked political allegiance, l. of all one burdened with heavy political and governmental responsibility, is not the most suitable for the office of Chancellor.*

**leastwise, -ways.** The OED labels the first 'somewhat rare', and the second 'dialectal and vulgar'. Today the latter description might be applied to both, see -WISE AND -WAYS.

**leather.** 1. For *leather* and *leathern*, see -EN ADJECTIVES. 2. In *l. or prunella* (usually misquoted *l. and prunella*) the meaning is not two equally worthless things, but the contrast between the rough l. apron of a cobbler and the fine gown of a parson. Pope's couplet is *Worth makes the man, and want of it the fellow; The rest is all but l. or prunella*: what makes a man is his worth, not his clothes.

**lectureship, -turership.** The first is of irregular formation, as a parallel for which the OED quotes the now obsolete *clergyship* (though a person can be clergy better than he can be a lecture); but it is long established, and those who use the second instead perhaps make it in momentary forgetfulness that the irregular form exists.

**leeward.** Pronounce *lū'ărd* at sea *lē'ward* on land.

**legal, legitimate,** etc. See ILLEGAL, ILLEGITIMATE, etc.

**legalese.** See OFFICIALESE.

**legalism, legality.** For the distinction, see -ISM AND -ITY.

**legerdemain with two senses,** or the using of a word twice (or of a word and the pronoun that represents it, or of a word that has a double job to do) without observing that the sense required the second time is different from that already in possession. A plain example or two will show the point: *The inhabitants of the independent lands greatly desire our direct* government, which government *has, however, for years refused to take any strong measures.* | *Although he was a very painstaking and industrious pupil, he never indicated any signs of developing into* the great naval genius by which *his name will in future be distinguished.* | *Mark had now got his first* taste *of print, and he liked it, and* it was a taste *that was to show many developments.* In the first of these, government *means successively governance, and governing body*—either of them a possible synonym for it, but not both to be represented by it in the same sentence. In the second, genius *means a singularly able person, but* which, *its deputy, means singular ability.* In the third, whereas the taste *he got was an experience, the* taste *that showed developments was an inclination.* Such shiftings from one sense to another naturally occur sometimes in reasoning, whether used by the disingenuous for the purpose of deceiving others, or by the over-ingenuous with the result of deceiving themselves. Here, however, we are concerned with their formal not with their material aspect; apart from any bad practical effects, they are faults of style. *Interest* is peculiarly liable to maltreatment: *Viscount Grey's promised speech in the House of Lords on Reparations and inter-Allied debts* furnished *all the* interest *naturally* aroused. *Interest* is here virtually, though not actually, used twice—the speech furnished interest, interest was aroused; but what was furnished was interesting matter, and what was aroused was eager curiosity; *interest* can bear either sense, but not both in one sentence. / *For while the Opposition beat their drums as loudly as ever, it was well known that there was very little behind all this fuss, and that* in the very interests *which they so furiously protected they were anxious to meet the Government half-way. Which* stands for *interests*; they furiously protected certain interests, i.e. certain persons or sets of persons or rights or privileges; they were inclined to compromise *in* some people's interests, i.e. in their behalf or favour or name; but *behalf* is not a person or a privilege or the like. The difficulty of expressing the inconsistency, however, explains why the word *interest* is often thus abused.

In the examples that follow, less flagrant than the typical specimens above, the fault is a want of clear thinking on small points, and in this they resemble the contents of the article HAZINESS; other examples will be found under I 2, and WE 2 and 3.

*If* the statements *made are true,* they *constitute a crime against civilization.* Whereas *the statements* means the things alleged, *they* means the things done. / *Even where it includes within its borders no important differences of* nationality, which *has no serious jealousies among* its *people, a completely unitary organization is becoming impossible.* Whereas *nationality* means an abstract property (the belonging to one

or another nation), *which* and *its,* both representing it, mean a concrete nation. / *The vital* differences *of their respective elders make* none *to their bosom friendship.* Whereas *the differences* are quarrels, *none* is (no) alteration. / *Is he, however, correct in ascribing this misnomer to confusion between the English terms 'bend', and 'bar'? Is it not rather due to a mistake in* spelling, which *should be the French form 'barre sinistre'? Spelling* is an art, but when repeated by *which* it means the correct way to spell a particular word. See also SWAPPING HORSES.

**legible, readable.** See ILLEGIBLE.

**legislation, legislature.** By a long-established and useful differentiation, the first is the making of laws, and the second the body that makes them; there should be no going back upon such distinctions, as in: *It is physical science, and experience, that man ought to consult in religion, morals,* legislature, *as well as in knowledge and the arts.*

**legitimate drama.** A phrase denoting what are now more often called 'straight plays' and intended to distinguish these from 'musicals', REVUES, etc.

**legitimate vb., legitimatize, legitimize.** The second and third are mere substitutes for the first without difference of meaning; it has a longer history by two or three centuries, and is neither obsolete nor archaic. We may guess that the others exist only because -*ize,* now so common, saves a moment's thought to those who want a word and forget that there is one ready to hand. See NEW VERBS IN -IZE.

**leisure.** The OED puts the pronunciation *lĕzh*- first, admitting *lēzh*- as an alternative. U S. dictionaries reverse the preference. In England *lē* has virtually disappeared.

**leit-motiv, -f.** The right (German) spelling is with -*v.* Pronounce *lītmōtē'f.* The word means a theme in an opera or other musical composition that recurs in association with some person, situation, or sentiment.

**lengthways, -wise.** See -WISE AND -WAYS.

**lengthy** is an Americanism long established in Britain, sometimes used as a jocular or stylish synonym for *long* but more commonly and more usefully as implying tediousness as well as length.

**lenience, -cy.** An incipient differentiation, which deserves encouragement, would use *-ce* for an action and *-cy* for a disposition. See -CE, -CY.

**lens.** Pl. *lenses*; see SINGULAR -S.

**lèse-majesté.** The English *lese-majesty* is not now a legal term, *treason* having taken its place. The French form is often used of treason in foreign countries, and either is applied jocularly (cf. PEDANTIC HUMOUR) to anything that can be metaphorically considered treason.

**less(er).** 1. *Nothing less.* 2. *Much* and *still less.* 3. *Less, lesser, smaller, lower, fewer.*

1. For the two meanings of *nothing l. than,* a possible source of ambiguity, see NOTHING LESS THAN.

2. The illogical use of *much l.* instead of *much more* is discussed under MUCH 2. Here are two examples of *still l.* for *still more,* interesting in different ways: *Of course social considerations, still l. considerations of mere wealth, must not in any way be allowed to outweigh purely military efficiency.* Here, if *still . . . wealth* had been placed later than *must not,* it would have passed; coming before it, it is wrong; you can understand *must* out of a previous *must not,* but not out of a *must not* that is yet to come. / *Perhaps Charles's most fatal move was the attempted arrest of the five members, undertaken on the Queen's advice, and without the knowledge, and still l. without the consent of, his three new advisers.* The writer of this has curiously chosen, by needlessly inserting that second *without,* to deprive

himself of the usual excuse for using *less* instead of *more,* i.e. the fact that some ellipsis of a word prevents the illogicality from being instantly visible and permits a writer to lose sight of what the full phrase would require while he attends to the broad effect. For similar slips with *least of all* see LEAST.

3. *L., lesser, smaller, lower, fewer,* etc. *The letters and memoirs could have been published, we should imagine, at a less price.* | A lesser prize *will probably be offered which will be confined to British manufacturers.* These extracts suggest ignorance of, or indifference to, modern idiomatic restrictions on the use of *less* and *lesser.* The grammar of both is correct; but, when the context —unemotional statement of everyday facts—is taken into account, *at a less price* ought to be *at a lower price,* and *a lesser prize* ought to be *a smaller prize.* It is true that *less* and *lesser* were once ordinary comparatives of *little* (*lesser* differing from *less* in being used only as an adjective and only before a noun), and that therefore they were roughly equivalent in sense to our *smaller,* also that this piece of archaism, like many others, is permissible in emotional passages or such as demand exceptionally dignified expression. But the extracts have no such qualification.

The modern tendency is so to restrict *less* that it means not *smaller,* but *a smaller amount of;* it is the comparative rather of *a little* than of *little,* and is consequently applied only to things that are measured by amount and not by size or quality or number, to nouns with which *much* and *little,* not *great* and *small,* nor *high* and *low,* nor *many* and *few,* are the appropriate contrasted epithets. *Less butter, courage;* but *a smaller army, table; a lower price, degree; fewer opportunities, people.* Plurals, and singulars with *a* or *an,* will naturally not take *less. Less tonnage,* but *fewer ships; less manpower,* but *fewer men; less opportunity,* but *a worse opportunity,* and *inferior opportunities;* though a few plurals like *clothes* and *troops,* really equivalent to singulars of indefinite amount, are exceptions: *could do with less* (or *fewer*) *troops* or *clothes*. Of *less*'s antipathy to *a*, examples are: *I want to pay less rent*, but *a lower rent is what I want.* | *That is of less value*, but *a lower value attaches to this.* | *Less noise, please*, but *a slighter noise would have waked me.* | *Less size means less weight*, but *I want a smaller size.* Such is the general tendency: to substitute *smaller*, *lower*, *fewer*, or other appropriate word, for *less* (except where it means 'a smaller amount of'), and for *lesser*, and to regard the now slightly archaic *less* in other senses as an affectation. There are no doubt special phrases keeping it alive even in quite natural speech, e.g. *in* or *to a less degree*, where *lower* is not yet as common as *less*; but the general tendency is unmistakable, and since it makes for precision, is one that should be complied with.

**-less.** The original and normal use of this suffix is to append it to nouns, so producing adjectives meaning without the thing, e.g. *headless*, *tuneless*; to this use there are no limits whatever. Words made from verbs, with the sense not able to do or not liable to suffer the action or process, as *tireless* and *fadeless*, are much fewer, are mostly of a poetical cast, and when new-minted strike the reader, at least of prose, as base metal. They have an undeniable advantage in their shortness (compare *resistless* and *fadeless*, with *irresistible* and *unfadable*), but this is outweighed for all except fully established ones by the uneasy feeling that there is something queer about them. Apart from a few so familiar that no thought of their elements and formation occurs to us, such as *dauntless*, *-less* words made from verbs are much better left to the poets. This does not apply to the many in which, as in *numberless*, formation from the noun gives the sense as well, if not as obviously, as formation from the verb (*without number?* or not able to be *numbered?*); *dauntless* itself may perhaps have been made from the noun *daunt*, which in the 15th and 16th cc.

was current in the sense discouragement.

To those who have any regard for the interests of the language as distinguished from its pliability to their immediate purposes, it will seem of some importance that it should not become necessary, with every word in which *-less* is appended to what can be either a noun or a verb, to decide which is this time intended. If the verb-compounds become much more frequent, we shall never know that *pitiless* and *harmless* may not mean 'that cannot be pitied' and 'secure against being harmed' as well as 'without the instinct of pity' and 'without harmfulness'. We ought to be able to assume that, with a few well-known exceptions, *-less* words mean simply without what is signified by the noun they contain; and the way to keep that assumption valid is to abstain from reckless compounding of *-less* with verbs.

A hyphen should not be used unless one is necessary to sort out the three *l*s in one of those rare words in which *-less* has been appended to a word ending *-ll* (the OED gives *bell-less*, *skill-less*, *smell-less*, and *will-less*). There is no need for one to separate two *l*s, as in *heelless*, *soulless*, *tailless*, etc. And it is an abuse of the hyphen to make it serve as an indication that a familiar compound is to be given a meaning different from its usual one, as in *These present cause-less days.*

**lest.** The idiomatic construction after *l.* is *should*, or in exalted style the pure subjunctive (*l. we forget*; *l. he be angry*); good writers rarely use *shall*, *may*, and *might*. The variations in the quotations below are entirely against modern idiom; *will* and *would* after *l.* are merely a special form of the inability to distinguish between SHALL and *will*. *We do not think Mr. Lloyd George need be apprehensive l. the newspaper reader* will *interpret his little homily in Wales yesterday as . . .* | *There must be loyal cooperation, l. the last state of the party* becomes *worse than the first.* | *The German force now* lost no time in retreat, l. they would be cut off and surrounded. Mistakes corresponding to those after l. are still more frequent after IN ORDER THAT.

**let.** **1.** There is little opportunity for mistakes in case in English, since only the personal pronouns have case inflexions, but let used in exhortations is responsible now and then for one of the commonest of them—the use of I for me when linked by and to a noun or another pronoun: And now, my dear, l. you and I say a few words about this unfortunate affair. For other examples and discussion see 1. 2. The use of the imperative l. in the phrase l. alone ( = not to mention) was called by the OED colloquial, but has since won literary status.

**letter forms.** Of the usual forms preceding the signature some are better suited than others to certain correspondents or occasions. 'I am, Sir' etc., or 'Believe me (to be)', or 'I remain', used to precede most of the following forms, but these are now more often omitted except in formal official communications. In those to and from ambassadors these preliminary words are still elaborated in stately formulae: I have the honour to be, with great truth and respect (or with the highest consideration), Sir (or Your Excellency or My Lord), your (or Your Excellency's or Your Lordship's) obedient servant; and an ambassador writing to the Foreign Minister of the country to which he is accredited must not end without giving this assurance: I avail myself of this opportunity to renew to Your Excellency the assurance of my highest consideration.

Your obedient servant: From or to officials in formal communications, and a variant (occasionally Yours obediently) of the more usual Yours faithfully in letters to the editor in newspapers.

Yours respectfully, or (old-fashioned) Your obedient servant, or (old-fashioned) Yours to command: Servant to master, etc.

Yours faithfully: To firms and unknown persons on business and now usual to newspapers, one at least of which prints all its correspondence in this form.

Yours truly: To slight acquaintances and sometimes to unknown persons on business.

Yours very truly: Ceremonious but cordial.

Yours sincerely: In invitations and friendly but not intimate letters. It is now more usual than it was to write letters to strangers Dear Mr —— and yours sincerely instead of Dear Sir—yours faithfully.

Yours ever, or Ever yours, or Yours: Unceremonious between intimates.

Yours affectionately: Between relations etc.

**levee.** The OED gives the pronunciation lĕ'vĭ but popular usage prefers lĕ'vā, which serves to distinguish the word from levy. In the U.S. word meaning embankment the last syllable is pronounced -vē.

**level.** **1.** Do one's l. best, originally American, has lived long enough in England to be no longer slang. **2.** As a noun, level has become almost a VOGUE WORD in distinguishing different social classes and different grades of officials as well as many other things. Examples of this use can be found ranging from the pub-and-street-corner l. to the world l.; though the meaning of the latter is not easy to guess. Its most frequent use is in connexion with international or interdepartmental discussion, especially in the phrases highest (or summit) l. and appropriate l. Thence it has spread widely and, like most vogue words, is often a symptom of a hazy mind that finds it more natural to express itself in clichés than in a straightforward way; it is habitually tacked as an abstract appendage to words that would be better without it (see TAUTOLOGY) and sometimes applied even to dividing lines that would be more suitably regarded as vertical than horizontal. At the legislative and administrative levels we are still largely governed by a traditional governing class. (In legislation and administration.) / This has to await a *decision at Cabinet l.* (by the Cabinet). | *Temperature levels will not differ greatly from today's* (temperatures). | [Of motor bicycles used by the G.P.O.] *There is research into the practicability of achieving a further reduction in their noise l.* (of making them less noisy).

**levy,** n. For synonymy see TAX.

**lexicon.** See DICTIONARY. Pl. *-ns* not *-ca*; see -ON 2.

**Leyden.** Pronounce *lī-*, not *lā-*.

**liable,** possibly because it is a more or less isolated word lacking connexions to keep it steady, constantly has its meaning shifted. For its proper use, see APT, with which there is much excuse for confusing it. *Political and religious bias are also* l. *to operate.* | *The President having to take note of the Nanking Assembly inferentially superseded, but still* l. *to assert itself, can hardly be held as invested with dictatorial power.* | *Walking through England must have been stripped of most of its charms, when every policeman is* l. *to demand the production of a variety of tickets.* | *Duncan has been for several years* l. *to win one of the big prizes of golf.* The first of these quotations illustrates the confusion with *apt*; in the second, *l. to assert* should be capable of asserting; in the third, *is l. to demand* should be may demand or is likely or not unlikely to demand; and in the last the sporting reporter should have stuck to his last and said in the running for instead of *l. to win*.

**liaison.** Pronounce as English (*lĭā′zn*); the military use during war has completed its naturalization. For *liaise* see BACK FORMATIONS.

**libel and slander.** The much quoted saying 'The greater the truth the greater' (or 'worse') 'the libel' makes us all occasionally curious about what a l. is, and how it differs from a slander. In popular usage they are synonymous, meaning a deliberate, untrue, derogatory statement, usually about a person, whether made in writing or orally. In legal usage there

are important differences. Each is an untrue and defamatory imputation made by one person about another which, if 'published' (i.e. communicated to a third person), can be a ground for a civil action for damages. Such an imputation is a *libel* if made in permanent form (writing, pictures, etc.) or by broadcasting. It is a *slander* if made in fugitive form (e.g. by speaking or gestures). A further distinction is that an action for slander cannot ordinarily succeed without proof that actual damage has been caused; in an action for libel this is unnecessary. In both cases proof that the allegation was true is a good defence.

Deliberate defamation of a person in permanent form can also be treated as a crime (*criminal l.*) on the ground that it tends to provoke a breach of the peace. A criminal l. differs from a civil l. in that publication to a third person is not a necessary element in it, and proof of its truth is not in itself a good defence. This explains the seeming paradox in the saying, attributed to Lord Mansfield, quoted above. Since 1843 truth may be pleaded in defence, but only if the accused can show also that he acted in the public interest. The most famous case of criminal l. in modern times was Oscar Wilde's unsuccessful prosecution of Lord Queensberry.

Popular synonyms of *libel* and *slander* (*calumny, defamation, scandal*) are now being forgotten under the influence of the VOGUE WORD *smear*, which gives its user the satisfaction of hitting back by an imputation of baseness or cowardice on the part of the smearer.

**liberal.** In *l. education* the adjective retains a sense that is almost obsolete, and yet is near enough to some extant senses to make misunderstanding possible. A liberal education is neither one in which expense is not spared, nor one in which enlightened methods of teaching prevail, nor even one that instils broadmindedness; or rather it is not so called because it is any of these. It is the education that used to be considered the only fitting one for what used to be called a gentleman (Latin *liber* a free man), and is opposed on the one hand to technical or professional or any special training, and on the other to education that stops short before manhood is reached. The L. *Arts* of the Middle Ages were Grammar, Dialectic, Rhetoric, Music, Arithmetic, Geometry, and Astronomy.

**libertine.** For *chartered l.*, see HACKNEYED PHRASES. The expression comes from the Archbishop's eulogy of King Henry V in the opening scene of that play: *When he speaks, The air, a c. l., is still.*

**libretto.** Pl. *-etti* (pronounce *-ē*) rather than *-os*.

**licence, -se.** The first is right for the noun, the second for the verb. Compare, for this convenient distinction, *advice, -se, device, -se, practice, -se, prophecy, -sy,* in all of which the c marks the noun. (In U.S. *license* and *practice* are preferred for both.)

**lichen.** Pronounce *lī′kn*; Gk. λειχήν is the source. 'The pronunciation *lĭch′ĕn* is now rare in educated use' (SOED), and the COD does not recognize it.

**lich-gate, -house.** So spelt; the OED gives *lych-* only as a variant; see Y AND I. The *l.-g.* is the churchyard gate with a roof over it where the bier may be rested; from ME *lich*, a body, which as *lyke* survives also in *lyke-wake.*

**lickerish, liquorish.** The first is the right form, and the second, being wrongly associated with *liquor*, inevitably alters and narrows the meaning. The word means fond of dainties, sweet-toothed, greedy, lustful, and is connected with the verb *lick* and with *lecher*, not with *liquor*. See TRUE AND FALSE ETYMOLOGY.

**lie.** See LAY AND LIE.

**lien,** n. Pronounce *lē′ĕn.*

**-lier.** For comparative-adverb forms, see -ER AND -EST 3.

**lieutenant.** Pronounce *left-*, but in naval use *lĕt-* and in U.S. *loot-*.

**lifelong, livelong.** *Life-* is the word for lasting a lifetime. *Live-* though used by Shakespeare and Milton and others as an alternative spelling of *life-* is in fact a different word, a compound of *lief* and *long*, serving as an emotional intensive of *long* (esp. *livelong day, livelong night*) now archaic or poetical.

**ligature.** See DIGRAPH.

**light,** n. 1. For *dim religious l.*, see IRRELEVANT ALLUSION. 2. *In l. of* will not do for *in the l. of*, as in *That it should have been so*, in light of *all the facts, will always be a nine-days wonder to the student of history*; see CAST-IRON IDIOM.

**light,** v. Both verbs (kindle, descend) make *lighted* or *lit* for past tense and p.p.; but *lighted* is commoner for the p.p. of the first verb used attributively: *Is the fire lighted* or *lit*, but *Holding a lighted torch.*

**like,** adj. For *and the l.*, see FORTH.

**like** in questionable constructions. 1. It will be best to dispose first of what is, if it is a misuse at all, the most flagrant and easily recognizable misuse of *l.* A sentence from Darwin quoted in the OED contains it in a short and unmistakable form: *Unfortunately few have observed l. you have done.* Most people use this construction daily in conversation. It is the established way of putting the thing among all who have not been taught to avoid it; the substitution of *as* for *l.* in their sentences would seem to them artificial. But in good writing this particular *l.* is rare, and even those writers with whom sound English is a matter of care and study rather than of right instinct, and to whom *l.* was once the natural word, usually weed it out. The OED's judgement is as follows: 'Used as conjunction, = "like as", as. Now generally condemned as vulgar or slovenly, though examples may be found in many recent writers of standing'. Besides the Darwin quoted above, the

OED gives indisputable examples from Shakespeare, Southey, Newman, Morris, and other 'writers of standing'. The reader who has no instinctive objection to the construction can now decide for himself whether he will consent to use it in talk, in print, in both, or in neither. He knows that he will be able to defend himself if he is condemned for it, but also that, until he has done so, he will be condemned. It remains to give a few newspaper examples so that there may be no mistake about what the 'vulgar or slovenly' use in its simplest form is: *Or can these tickets be kept* (1. the sugar cards were) *by the retailer?* / *The retail price can never reach a prohibitive figure* l. petrol has done. / *The waves of China's revolution have risen* l. the waters of the rivers did last year. / *They studied the rules of a game* l. a lawyer would study *an imperfectly-drawn-up will.* / *Our great patron saint 'St. George' was a Greek,* l. a good many of the saints are. / *The idea that you can learn the technique of an art* l. you can learn *the multiplication table or the use of logarithms.*

In U.S. the colloquial use of *l.* as a conjunction has been carried a step further by treating it as equivalent not only to *as* but also to *as if,* a practice that still grates on English ears. Examples from the OED Supp. are *The old fellow drank of the brandy* l. he was used to it. / *None of them act* l. they belonged to the hotel.

2. The rest of this article is intended for those who decide against the conjunctional use that has been already discussed, and wish to avoid also some similar questionable uses of a less easily recognizable kind. All the examples in 1 were of the undisguised conjunctional use, and contained a subordinate clause with its verb; most of those now to come have no subordinate verb, and in all of them *l.* may be regarded as an adjective or adverb having the additional power (cf. *worth*) of directly governing nouns as if it were a preposition.

The first type is perhaps not really different from that discussed in 1.

Examples are: *These lovely shores of Ullswater will be just sterile shores* like you see at Thirlmere. / *But in an industrialized county* l. so great a part of Lancashire is, *the architecture can hardly fail to . . .* / *He even looks exactly* l. the member for Stratford-upon-Avon should. The peculiarity of these is that in each there is a previous noun or pronoun, *shores, county, he,* with which *l.* may agree as an adjective, and an ellipsis of 'those' or 'what' may be supposed. Such a defence is neither plausible nor satisfactory, and the sentences are no better than others containing a verb.

Of sentences in which *l.* is not followed by a verb, certain forms are unexceptionable, but are liable to extensions that are not so. The unquestioned forms are *He talks l. an expert* and *You are treating me l. a fool,* in which *l.* is equivalent to a prepositional adverb = similarly to; and *You, l. me, are disappointed,* in which *l.* is equivalent either to an adverb as before, or perhaps rather to a prepositional adjective = resembling in this respect. The second, third, and fourth types are faulty and represent neglect of various limitations observed in the correct forms.

Second type: *The Committee was today,* l. yesterday, *composed of the following gentlemen.* / *The Turks would appreciate the change, as,* unlike Koweit, *their political title is* here *beyond dispute.* / *It is certain that* now, unlike the closing years *of last century, quotation from his poetry is singularly rare.* / *We may have 110,* like last year, *when Paignton . . . and Jersey all enjoyed a sun-bath of nearly 200 hours.* The limitation here disregarded is that the word governed by *l.* must be a noun, not an adverb or an adverbial phrase. *Yesterday* and *last year* are not nouns, but an adverb and an adverbial phrase; and *Koweit* and *the closing years,* meaning *at* Koweit and *in* the closing years, have also only a deceitful appearance of being nouns.

Third type: *People get alarmed on each occasion* on which (l. the present case) *dying children suddenly appear.*

*He has completed a new work in which,* l. its author's recent books, *no failing in sparkle or vigour will be traceable.* | *And then came the war*; l. many another English village, *it filtered slowly, very slowly, through* to his. The limitation (suggested with diffidence) that has here been disregarded is that the preceding noun to which *l.* is attached must be not one governed by a preposition, but subject or object of the main verb. The preceding nouns are *which* (i.e. occasion), *which* (i.e. work), and *his* (i.e. village), governed by *on, in,* and *to*; instead of *l.*, read *as in the present case, as in its author's recent books,* and *as to many another.*

Fourth type: L. his Roman predecessor, *his private* life *was profligate; l. Antony, he was an insatiate gambler.* | *Although,* l. his colleague, *his* conduct *had not been above reproach.* The limitation is that the word governed by *l.* must be *in pari materia* with the one to which it is compared. The *predecessor* and *his colleague* are not so related to *his life* and *his conduct*; but *Antony* is to *he,* and that sentence alone will pass muster. This mistake, however, of comparing unlike things, though the commonest type of misuse of *l.* by educated writers, is not peculiar to *l.*, but is a slovenly parsimony of words that may occur in many other constructions. It is perhaps even commoner with *unlike* than *l.*: *Unlike Great Britain, the upper house in the United States is an elected body.*

A warning should, however, be added against going too far in anxiety to avoid all questionable uses of *l.* A fashion seems to be growing, even among some good writers, to prefer *as* to *l.* not only, rightly, as a conjunction, but also, ill-advisedly, as a prepositional adjective. *In Paris he was dissuaded from a literary career and, as his friend and rival Bernard, entered on the study of medicine.* | *Lord A, as his fellow crusaders, B and C, is no pedant.* | *As Macaulay before him, Professor Trevor Roper generally mentions the name of the book he is reviewing.* On another page of the newspaper from which the last quotation is taken we may read *Like*

*Shaw, Gilbert had the pleasure of becoming comfortably rich.* If these two are compared it is clear that *l. Shaw* is preferable to *as Macaulay,* both as more idiomatic and as an insurance against ambiguity. Shakespeare did not make Antony say 'I am no orator l. Brutus is', but he felt no qualms about 'It is excellent to have a giant's strength, but it is tyrannous to use it l. a giant'.

Again, the fact that there is a verb *to like,* and that one cannot say *I very l.,* or *I too l.,* seems to create in the minds of some would-be purists the idea that even when *l.* is not a verb but a prepositional adjective *very* or *too* must be helped out by *much. The new measure proved to be very much l. the original one.* | *It was too much l. flogging a dead horse.* Polonius did not say 'Very much l. a whale', and there is no more reason for writing *too much l. flogging* than there would be for saying 'You are driving too much near the car in front of you'.

**-like.** 'In formations intended as nonce-words, or not generally current, the hyphen is ordinarily used'—OED. For instance the OED prints cowl-like, eel-like, flail-like, jail-like, owl-like, and pearl-like. But established *-like* compounds are commonly written as one word—*ladylike, lifelike, statesmanlike,* etc. See HYPHENS.

**like,** v. **1.** *I would l.* Even on those who use *should* and *would* according to the Englishman's idiom under all ordinary temptations the verb *l.* seems to exercise an irresistible influence; a couple of examples follow *pro formâ,* but anyone can find as many as he pleases with very little search: *We would l. to ask one or two questions on our own account.* | *There is one paragraph in it that I would l. to refer to.* There is indeed no mystery about why people do this; it is because, if the thing had to be said without the use of the verb *like, would* and not *should* is the form to use: *We would ask; that I would refer to.* But that has nothing to do with what is right when the verb *l.* is used. If anything it makes the matter worse;

*would* is properly used without *l.* because it contains the idea of volition; to use it with *l.* is equivalent to saying *I should l. to l. I would l.* is no better than any of the *wills* and *woulds* that are well recognized as Scottish, Irish, American, and other kinds of English, but not English English. If the SHALL and WILL idiom is worth preserving at all (though reasons are given in that article for fearing that it is crumbling irretrievably), *I should like* must be treated as its proper form.

2. *L. to see* and *l. seeing* are equally idiomatic, but *dislike* prefers *seeing*.

**likely,** adv. *Yet it was not easy to divine the thought behind that intentness of gaze*; likely *it was far from the actual scene apparently holding its attention.* In educated speech and writing in England the adverb is never used without *very, most,* or *more,* except by way of poetic archaism or of stylistic NOVELTY-HUNTING. But in Scotland and Ireland it is common in speech and in U.S. may be found in print: *The climate in America is so severe in winter that stocks will l. die out.* | *It will l. be financed largely by capital raised in the United States.* For *likelily,* see -LILY; for *likelier* adv., -ER AND -EST 3. See also APT.

**likewise.** The use as a conjunction (*Its tendency to wobble and its uniformity of tone colour, l. its restricted powers of execution*) is, like the similar use of ALSO, an ILLITERACY; the OED quotes no example.

**-lily.** Avoidance of the adverbs in *-lily,* i.e. adverbs made regularly from adjectives in *-ly,* is merely a matter of taste, but is very general and increasing. Neither the difficulty of saying the words nor the sound of them when said is a serious objection so long as there are not more than three syllables; *holily* and *statelily* and *lovelily* are not hard to say or harsh to hear; but with *heavenlily* and *ruffianlily* hesitation is natural; and the result has been that adverbs in *-lily,* however short, are now with a few special exceptions seldom heard and seldomer seen. Methods of avoidance are various:

1. It is always possible to use a periphrasis or a synonym, and to say *in a masterly manner, at a timely moment,* and the like, instead of *masterlily, timelily,* or to be content with *decorously* etc. instead of *mannerlily.*

2. A large number of adjectives in *-ly* are established as adverbs also. So *early,* (*most* or *very*) LIKELY, and the adjectives of periodical recurrence like *daily* and *hourly.* A single quotation will show the danger of making one's own adverbs of this kind: *External evidence, however, is rare; and its rarity gives value to such work as Mr. —— here* masterly *does.*

3. Before adjectives and adverbs the *-ly* adjective often stands instead of the *-lily* adverb, making a kind of informal compound. Though we should say *horribly pale* and not *horrible pale,* we allow ourselves *ghastly pale* rather than use *ghastlily;* so *heavenly bright, beastly cold, jolly soon,* etc.—all without the hyphen that would mark regular compounds.

4. In sentences where it is just possible, though not natural, for a predicative adjective to stand instead of an adverb, that way is sometimes taken with an adjective in *-ly* though it would not be taken with another: *it happened timely enough,* though not *opportune enough;* she nodded friendly, though not *she nodded amiable.*

5. Perhaps any adjective formed by appending *-ly* either to an adjective (*kind, kindly; dead, deadly*) or to a noun of the kind that is easily used in apposition like an adjectival epithet (*cowardly,* cf. *the coward king; soldierly,* cf. *a soldier colonist; scholarly,* cf. *the scholar gipsy*) is sometimes, though always consciously and noticeably, allowed to pass as an adverb: *it was ruffianly done, soldierly conducted, scholarly expounded.*

On the other hand, avoidance is not always called for; some *-lily* words are current, though not many. Those that might suggest themselves (*he laughed jollily; sillily complacent; live holily; dodged it wilily*) seem to be all from adjectives in which the *l* is part of the word-stem, not of *ly* as an adjectival ending; and though we are most of us not conscious of that fact nowadays, it may have had its effect in separating these from the others.

**limb.** When we first come across an eclipse in the newspapers and read of *the sun's lower l.*, we suspect the writer of making jokes or waxing poetical, so odd is the association of limbs with that globular form. It is a relief to learn that *l.* cannot be used for *edge* without the help of a metaphor; the *l.* in astronomy etc. is from Latin *limbus* hem, and the *l.* of ordinary speech is a separate and native word.

**lime** the mineral makes *limy*, the fruit *limey*; see -EY AND -Y. *Limey* as U.S. slang for an Englishman, especially a sailor, originated in the compulsory use of lime juice as an anti-scorbutic in the British Navy.

**limit, delimit.** These verbs should not be used as synonyms. To *limit* is to confine within bounds; to *delimit* (or delimitate) is to determine precisely what those bounds are, e.g. of a territorial frontier.

**limited** is a victim of SLIPSHOD EX-TENSION. Legitimate uses such as *The time for discussion of this motion is l.* | *Only a l. amount of money can be devoted to this project* have led to a lazy habit of treating *l.* as a convenient synonym for many more suitable and more exact words. *These houses are intended for people of l. means* (small). | *He has l. interests outside his work* (few). | *Information about his early life is l.* (meagre). | *The occasions on which such drastic measures are likely to be needed will be l.* (rare).

**linage** (*līn*- meaning number of lines) should be so spelt. *Lineage*, though often seen, is, owing to the existence of *lineage* (*linĕ*- meaning descent), still less desirable than other spellings with intrusive MUTE E.

**line, n.** For some synonyms in sense *department* etc. (*What's my line?*) see FIELD.

**lingo, lingua franca.** See JARGON. Pl. of the first -*os*; see -O(E)S 6.

**liqueur.** The anglicized *likūr'* is now standard. Cf. AMATEUR.

**liquid.** See FLUID for fluid, gas, and l.

**liquid(ate)(ize).** The euphemistic sense of putting political opponents to death, individually or collectively, dates from the Russian revolution. For a time, it became a VOGUE WORD (often facetious) for getting rid of or doing away with anything by any means, especially obstacles to the liquidator's ambitions. But the freshness that made it attractive seems now to be wearing off. That *liquidate* is always used figuratively is no doubt the reason for coining *liquidize* for the literal sense. But seeing that *liquefy* has been available to us for 500 years we ought to be able to do without the *pis aller* of an *-ize* verb. See NEW VERBS IN -IZE.

**liquorish.** See LICKERISH.

**lira,** Italian monetary unit, has pl. *lire* (pronounce *lēr'ā*) or anglicized *liras*. To use *lira* as pl. (*We had not enough l. to stay longer in Italy*) is absurd.

**lissom(e).** The standard form is -*om*.

**list, please.** The third sing. pres. is *list* or *listeth*, the past tense *list* or *listed*. The verb being in any form archaic, it is of no importance whether the more obviously archaic impersonal construction (*as him list* etc.) or the now commoner personal one (*as he list* etc.) is used.

**litany, liturgy.** The two words have come so close to each other in use that it is a surprise when one first finds that the initial syllables are not the same in origin, nor even connected. For those who know the Greek words, a litany is a series of prayers, a liturgy is a canon of public service; the latter in practice includes prayer, but does not say so.

**literally.** We have come to such a pass with this emphasizer that where the truth would require us to insert with a strong expression 'not l., of course, but in a manner of speaking' we do not hesitate to insert the very word that we ought to be at pains to repudiate; cf. VERITABLE. Such false coin makes honest traffic in words impossible. *If the Home Rule Bill is passed, the 300,000 Unionists of the South and West of Ireland will be* l. thrown to the wolves. / *The strong tête-de-pont fortifications were rushed by our troops, and a battalion crossed the bridge* l. on the enemy's shoulders./ (At election time) *My telephone wires have been kept* l. red-hot. / *H. B. Stallard in the half-mile* l. 'flew' *round the track.* / *She* l. lifted her horse *over the last jump.* / *He* [a climber] *came through safely, but he had* l. to cling on with his eyebrows. / *Our eyes were* l. pinned to the curtain. / *The Prime Minister sat throughout the debate* l. glued *to the Treasury bench.* / *Marie Corelli, when she settled in Shakespeare's native town,* l. took the bard to her bosom. / *Sir Stanley Spencer was a brilliant talker who could* l. take you to the stars.

**literary critics' words.** The literary critics here meant are not the writers of books or treatises or essays of which the substance is criticism; readers of that form of literature are a class apart, and if a special lingo exists between them and its writers, the rest of us are not concerned to take exception to it. Anything said in this book about literary critics is aimed only at the newspaper reviewers of books and other works of art. Those reviewers, as anyone knows who examines them critically in their turn, give us work that ranges from the very highest literary skill (if the power of original creation is set aside as here irrelevant) to the merest hack-work. The point is that, whether they are highly accomplished writers, or tiros employed on the theory that anyone is good enough to pass an opinion on a book, their audience is not the special class that buys critical works because its tastes are literary, but the general public, which buys its criticism as part of its newspaper, and does not know the critics' lingo. It follows that,

the better the critic, the fewer literary critics' words he uses. The good critic is aware that his public wants to understand, and he has no need to convince it that he knows what he is talking about by parading words that it does not understand. With the inferior critic the establishment of his status is the first consideration, and he effects it by so using, let us say, *actuality, engaged,* and *inevitable,* that the reader shall become aware of a mysterious difference between the sense attaching to the words in ordinary life and the sense now presented to him. He has taken *actuality* to mean actualness or reality; the critic perplexes him by giving it another sense, which it has a right to in French, where *actuel* means present, but not in English—the sense of up-to-dateness, or resemblance not to truth in general but to present-day conditions; and he does this without mentioning that he is gallicizing. And so with the other words; the reader is to have it borne in upon him that a more instructed person than himself is talking to him even if it means coining a new word; *cretinocratic,* for instance, is the term by which one reviewer, evidently a very superior person, expresses his opinion of television programmes. One mark of the good literary critic is that he is able to explain his meaning without resort to these lingo words and under no necessity to use them as advertisements.

Specimens of literary critics' words, under some of which (printed in capitals) further remarks will be found, are: Actuality, AMBIENCE, AMBIVALENT, awareness, COMMITTED, compelling, CREATIVE, DEDICATED, DICHOTOMY, DISTINCTION, engaged, evocative, immediacy, INEVITABLE, perceptive, seminal, SIGNIFICANT, SYMPATHETIC.

**literary words.** A l. w., when the description is used in this book, is one that cannot be called archaic, inasmuch as it is perfectly comprehensible still to all who hear it, but that has dropped out of use and had its place taken by some other word except in writing of a poetical or a definitely literary cast. To use literary words instead of the current substitutes in an unsuitable context challenges attention and gives the impression that the writer is a foreigner who has learnt the language only from books. See also what is said of FORMAL WORDS. *Chill* for *chilly*, *eve* for *evening*, *gainsay* for *deny*, etc., *loathly* for *loathsome*, *visage* for *face*, etc., may be instanced; but literary words are reckoned by thousands.

**literature.** The meaning of this word has been extended (OED's first example is dated 1895) to include written matter of any sort, especially that issued by commercial or industrial firms to commend or explain their goods and services. *Please send me any l. you have about your Autumn Pleasure Cruises.* This usage is still only colloquial but is unlikely to remain so, however much we may regret that so reputable a word should be put to so menial a duty, and that we should thus be left without one for the special kind of written matter for which *l.* used to be reserved (see BELLES-LETTRES).

**lithesome** is, between *lithe* and *lissom*, a SUPERFLUOUS WORD.

**litotes.** The same as, or a variety of, MEIOSIS. Sometimes confined to the particular kind of rhetorical understatement in which for the positive notion required is substituted its opposite with a negative. In 1 *Cor*. xi. 17 and 22, *I praise you not* has the effect of an emphatic I blame; *not a few* means a great number; *Not bad, eh?*, after an anecdote, means excellent. But often used, like *meiosis*, of other understatements meant to impress by moderation. In the Greek word (λιτότης) the i is long and the o short, and that is the pronunciation (lī'tŏtēz) given by the OED. Some modern dictionaries prefer a long o, but surely this scholars' word can claim a place in that 'small province' described in FALSE QUANTITY 'in which the false-

quantity principle may fairly reign', and change should be resisted.

**little.** See SMALL. Comparison *less(er)* (for limitations of sense see LESS 3), *least*, or more usually *smaller*, *-est*.

**littoral,** n., has a technical sense in which it is doubtless of value; marine life being distributed into abyssal, pelagic, and littoral, the last (sc. zone or region) is the shallow waters near the shore. But that is not the sense in which most of us know it; it meets us as a name for the land region bordering and including the shore. In that sense it may be important in treaties and the like to have a word that does not mean strictly the mere line of coast or shore; but in ordinary contexts it should never be preferred to *coast*, and its present popularity is due to pretentiousness. Why not *coast* in *The towns along the Mediterranean l.*, *The Russian settlements on the Eastern Caspian l.*? See FORMAL WORDS.

**liturgy.** See LITANY.

**-lived.** In *long-l.* etc. the correct pronunciation is *līvd*, the words being from *life* (cf. *-leaved* from *leaf* etc.) and not from *live*; but *līvd* is almost always heard.

**livelong.** See LIFELONG.

**liven.** See -EN VERBS.

**llama.** See LAMA.

**-ll-, -l-.** Final *l* is treated differently from most final consonants in British, but not American, usage. The rule is to double it, if single, in inflexions and in some derivatives, irrespective of the position of the accent.

1. When verbs in *-l* (except those in which a long vowel sound, made up either of two vowels or a vowel and a consonant, such as *ai*, *ea*, *ee*, *oi*, *ow*, *ur*, precedes the *-l*) make inflected or derived words in *-able*, *-ed*, *-en*, *-er*, or *-ing*, *-ll-* is written—*controllable*,

*carolled, befallen, traveller, equalling*; but *failed, boiling, curled*, etc., and before *-ment l* is not doubled; see also PARALLEL, WOOL.

2. When nouns or adjectives in *-l* (with exceptions as in the preceding paragraph) make other words by addition of *-ed, -er*, or *-y*, the *l* is doubled: *flannelled, jeweller, gravelly*; but parallel is an exception. Before *-ish* and *-ism* and *-ist, l* is not doubled: *devilish, liberalism, naturalist.* Irregular superlatives vary, most using one *l*, but words in *-ful* always two: *brutalest loyalest, civil(l)est, joyfullest.*

3. The simple form of a good many verbs vacillates between *-l* and *-ll*, and no rule is possible that will secure the best form for all words and not conflict with the prevailing usage for some. APPAL, for instance, seems to have come down in favour of one *l*; but as a general rule it is perhaps safe to say that where vacillation exists *-ll* is better if *a* precedes (*befall, enthrall, install*), and *-l* if another vowel, especially *i* (*distil, instil, enrol, annul*); verbs in *-ll*, however, take single *l* before *-ment* (*enthralment, instalment*).

4. Derivatives and compounds of words in *-ll* sometimes drop one *l*; so *almighty, almost, already, altogether, always* (but not *alright*, see ALL RIGHT), *chilblain, fulfil, skilful, thraldom, wilful.* This is perhaps helped by some apparent but not real examples such as *belfry, bulwark*, and *walnut*, which are not from *bell, bull*, and *wall. Dul(l)ness* and *ful(l)ness* are debatable; the older spelling, though (according to the OED) the one less 'in accordance with general analogies', has only one *l*, but the spelling with two seems to be gaining ground and is recommended. See DULLNESS.

**Lloyd's,** the underwriters' office. So written, not *-ds* or *ds'*.

**load, lode.** In the compounds with *stone* and *star* it is usual to spell *loadstone*, but *lodestar*. The first element is the same, and is the ordinary *load*, of which the original sense was way, connected with the verb *lead*; the spelling distinction is accidental, and

both *lodestone* and *loadstar* are sometimes used.

**loan.** The verb, formerly current, was expelled from idiomatic English by *lend.* But it survived in U.S., and has now returned to provide us with a NEEDLESS VARIANT.

**lo(a)th.** *Loth* was once the standard form, but the OED gives preference to *loath*; and that spelling avoids obscuring the connexion with the verb *loathe.* The verb is always *loathe*, and *loathly* and *loathsome* have always the *a.*

**lobby** has long been used as a verb meaning to frequent the l. of a legislative assembly for the purpose of influencing the members' votes. It originated in America. As a collective noun meaning a body of lobbyists, it was, according to the SOED, still an Americanism in 1933. But a few modern quotations will show that it is now used freely in Britain in this sense, and indeed for any kind of what is alternatively called a pressure group. *This is heartening to the Opposition, and particularly to the trade union group, who had organized a strong l. in the Bill's favour.* | *The Minister of Agriculture is running into trouble with the pig and bacon l. among the farming M.P.s.* | *Their counsel is likely to be that the Group should never become a militant pressure l. but should remain essentially a research society.* | *The scheme has aroused the opposition of many who object to the details of its planning, as well the permanent Christ Church l., which opposes any Meadow road.*

**local(e).** 1. The 'erroneous form' (OED) *locale* is recommended for the noun meaning scene of operations; cf. MORALE. 2. Pronounce *lŏkah'l*. 3. The word's right to exist depends on the question whether the two indispensable words *locality* and *scene* give all the shades of meaning required, or whether something intermediate is useful. The defence of *l.* would be on these lines: A locality is a place, with features of some sort, existing independently of anything that may happen there. If something happens in a locality, the locality becomes that something's locale, or place of happening. If the something that happens is seen or imagined or described in connexion with its locale, the locale becomes its scene or visible environment.

**locate.** The earliest example given by the OED of the use of this verb in the sense 'discover the exact place or locality of' is dated 1882. If giving the word this meaning is to be of any value to us, it should not be treated (as it often is) merely as a dignified synonym of *find*; there is a differentiation that should be respected. A successful search ends in *locating* what is sought if the primary purpose is to discover the place where a person or object is, in *finding* it if the purpose is to discover the person or object, wherever he or it may be. One may try to *locate* the enemy's guns or a fault in an electrical circuit, but one will try to *find* a lost child or a suitable parking-place.

**loch.** See LOUGH.

**locus.** Everyone says *lōcum tēnens* and almost everyone *in lōco parentis* and *lōcus standi*; but scholars generally give the *o* its Latin value in *lŏcus classicus*. Pl. *-ci (-sī)*.

**locution** is a potentially convenient word as equivalent to word or phrase; not more than potentially, because it so far smacks of pedantry that most people prefer to say *word* or *phrase* on the rare occasions when *expression* is not precise enough for the purpose, and *l.* gets left to the pedants. *His style is comparatively free from locutions calculated to baffle the English reader*; does anyone really like that better than *expressions*?

**lode.** See LOAD.

**lodg(e)ment.** Retention of the *-e-* is recommended; see JUDGEMENT.

**logan.** Pronounce *lŏ'găn(berry)* but *lŏ'găn(stone)*.

**logistics.** See STRATEGY.

**-logy.** This suffix denoting *science of* normally has *o* as its combining vowel; hence the jocular coinage *ologies* (cf. *ismx*). The principal exception is *genealogy*; *mineralogy*, an apparent exception, is a telescoping of *mineralology*. See also -O-.

**Lombard(y).** For pronunciation *lŏm-* or *lŭm-* see PRONUNCIATION 5.

**lonelily.** See -LILY.

**longhand.** The coming of the typewriter has upset the meaning of more than one word. *Manuscript* (generally spoken of as *the ĕmĕss*) is no longer confined to its etymological sense of written by hand but is applied to any script, whether written or typed (opposite *print*); *longhand*, though the dictionaries do not yet recognize any meaning other than ordinary writing (opposite *shorthand*) is often used for a script that is written (opposite *typescript*). The confusion could easily be set right if the correct opposites were always used: *longhand* opp. *shorthand*, *manuscript* opp. *typescript*, *script* opp. *print*. But the corruption of *manuscript* has probably gone too far to be mended.

**long-lived.** See -LIVED.

**long variants.** 'The better the writer, the shorter his words' would be a statement needing many exceptions for individual persons and particular subjects; but for all that it would be broadly true, especially about English writers. Those who run to long words are mainly the unskilful and tasteless; they confuse pomposity with dignity, flaccidity with ease, and bulk with force; see LOVE OF THE LONG WORD. A special form of long word is now to be illustrated. When a word for the notion wanted exists, some people (1) forget or do not know that word, and make up another from the same stem with an extra suffix or two; or (2) are not satisfied with a mere current word, and resolve to decorate it, also with an extra suffix;

or (3) have heard a longer form that resembles it, and are not aware that this other form is appropriated to another sense. Cases of (1) and (2) are often indistinguishable; the motive differs, but the result is the same; and they will here be mixed together, those of (3) being kept apart.

(1) and (2). Needless lengthenings of established words due to oversight or caprice: administrate (administer); assertative (assertive); contumacity (contumacy); cultivatable (cultivable); dampen (damp, v.); denunciate (denounce); dubiety (doubt); epistolatory (epistolary); experimentalize (experiment, v); extemporaneously (ex tempore); filtrate (filter, v); preventative (preventive); quieten (quiet, v); transportation (transport).

### Examples

*The capability of the Germans to* administrate *districts with a mixed population.* | *Still speaking in a very loud* assertative *voice, he declared that . . .* | *Mlle St Pierre's affected interference provoked* contumacity. | *If you add to the* cultivatable *lands of the immediate Rhine valley those of . . .* | *His extreme sensitiveness to all the suggestions which* dampen *enthusiasm . . .* | *Lord Lansdowne has done the Liberal Party a good turn by putting Tariff Reform to the front; about this there can be no* dubiety. | *Cowper's Letters . . . the best example of the* epistolatory *art our language possesses.* | *A few old masters that have been* experimentalized *on.* | *M. Delcassé, speaking* extemporaneously *but with notes, said . . .* | *A Christianity* filtrated *of all its sectional dogmas.* | *Jamaica ginger, which is a very good* preventative *of seasickness.* | *Whether that can be attributed to genuine American support or to a* quietening *down of the speculative position is a matter of some doubt.*

**3.** Wrong use of longer forms due to confusion: advancement (advance); alternative (alternate); correctitude (correctness); creditable (credible); definitive (definite); distinctive (distinct); estimation (estimate); evaluate (value); excepting (except); intensive (intense); partially (partly); prudential (prudent); reverential (reverent); transcendental (transcendent). The differences of meaning between the longer and shorter words are not here discussed, but will be found, unless too familiar to need mention, under the words in their dictionary places.

### Examples

*It was only by* advancement *of money to the tenant farmers that the calamity could be ended.* | *When the army is not fully organized, when it is in process of* alternative *disintegration and rally, the problems are insoluble.* | *Baron —— believes himself to be the oldest living Alsatian; and there is small reason to doubt the* correctitude *of his belief.* | *It is* creditably *stated that the length of line dug and wired in the time is near a record.* | *But warning and suggestion are more in evidence than* definitive *guidance.* | *Trade relations of an ordinary kind are quite* distinctive *from those having annexation as their aim.* | *Since November 11 the Allies have been able to form a precise* estimation *of Germany's real intentions.* | *The sojourn of belligerent ships in French waters has never been limited* excepting *by certain clearly defined rules.* | *The covered flowers being less* intensively *coloured than the others.* | *The two feet, branching out into ten toes, are* partially *of iron and* partially *of clay.* | *It is often a very easy thing to act* prudentially, *but alas! too often only after we have toiled to our prudence through a forest of delusions.* | *Their behaviour in church was anything but* reverential. | *The matter is of* transcendental *importance, especially in the present disastrous state of the world.*

It only remains to say that nothing in this article must be taken as countenancing the shortening of such words as *quantitative* and *authoritative*; and see INTERPRE(TA)TIVE. *It is as if the* quantitive *theory of naval strategy held the field.* | *Her finely finished* authoritive *performance was of great value.*

**longways, -wise.** See -WISE, -WAYS.

**loom,** v. For *l. large(-ly)* see LARGE(LY).

**loose, loosen,** vv. For the distinction, see -EN VERBS.

**lord.** Younger sons of dukes and marquesses are spoken of by the title of *L.* followed by Christian and family name, as *L. Arthur Smith*. Omission of the Christian name is wrong; the permissible shortening is not *L. Smith*, but *L. Arthur*. Dickens and Conan Doyle are among the writers of fiction who have tripped over this. A man with the title *L. Decimus Tite Barnacle* could not be an 'overpowering peer' with a seat in the House of Lords, nor could anyone be called with equal propriety *L. Verisopht* and *L. Frederick Verisopht*, or *L. St. Simon* and *L. Robert St. Simon*. For *l.* as an undress substitute for *marquess, earl, viscount*, see TITLES.

**Lord Bacon** is a mixture, given undeserved currency by being the title under which Macaulay's Essay on him was first published. The possible correct styles are *Bacon, Francis Bacon, Sir Francis Bacon, Lord Verulam, Lord* or *Viscount St. Albans*, of which the best to use is the first or second, both having been his names throughout his life.

**lose.** *L. no time in* is a notoriously ambiguous phrase: *No time should be lost in exploring the question.*

**lot.** 1. *A l. of people* say *so, Lots of paper* is *wanted*, etc. See NUMBER 6 (b). 2. *L.* in the sense of a large or excessive number is still called colloquial by the COD, but the following quotations show that modern writers do not hesitate to use it in serious prose. (The first is from Sir Winston Churchill's account of the Battle of Jutland, the second from a well-known writer's book on style). *The chance of an annihilating victory had been perhaps offered at the moment of deployment, had been offered again an hour later when Scheer made his great miscalculation, and for the third time when a little before midnight the Commander in chief decided to reject the evidence of the Admiralty message. Three times is a l. | A l. of writing is too confined and obscure; a l. is too wordy; a l. is too peevish or pompous or pretentious; a l. is too lifeless; a l. is too lazy.*

**loth.** See LOATH.

**lotus.** Pl. *-uses.*

**louden.** See -EN VERBS.

**lough.** The Irish *l.* and the Scottish *loch* are pronounced alike, i.e. with the breathed guttural, though by the English often anglicized into *lŏk*.

**lour, lower.** The meaning is frown. Spell *lour* and pronounce *lowr*. The word is not connected with *low* and the other verb *lower* (*lō'er*) and it is a pity that it should be confused with that verb by the second spelling (the oldest forms are *lour* and *lure*) and so have its meaning narrowed and its pronunciation altered. The confusion is due chiefly to the word's being often applied to clouds.

**lovelily.** See -LILY.

**love of the long word.** It need hardly be said that shortness is a merit in words. There are often reasons why shortness is not possible; much less often there are occasions when length, not shortness, is desirable. But it is a general truth that the short words are not only handier to use, but more powerful in effect; extra syllables reduce, not increase, vigour. This is particularly so in English, where the native words are short, and the long words are foreign. I open *Paradise Lost* and *The Idylls of the King*, and at each first opening there face me: '*Know ye not, then' said Satan, fill'd with scorn; 'Know ye not me? ye knew me once no mate For you, there sitting where ye durst not soar.' | And in those days she made a little song And call'd her song 'The Song of Love and Death', And sang it; sweetly could she make and sing.* Fifty-six words, of which fifty-two are monosyllables. Slightly selected passages, indeed, but such as occur on nearly every page; and these are not exercises in one-syllable words for teaching children to read; they are the natural as well as the best ways of saying what was to be said. Nor is it in verse only that good English runs to monosyllables; I open a new religious book, and find at once this passage about the Kingdom of Heaven: *His effort was, not to tell mankind about it, but to show it to them; and He said that those who saw it would be convinced, not by Him, but by it. 'To this end was I born, and for this cause came I into the world, that I should bear witness unto the truth. Every one that is of the truth heareth my voice.' There for once he spoke in general and abstract terms. Those who are of the truth, those who seek truth for its own sake, will listen to Him and know that what he says is true.* Twelve words that are not monosyllables in 101 words; and there is no taint whatever of affected simplicity in it. Good English does consist in the main of short words. There are many good reasons, however, against any attempt to avoid a polysyllable if it is the word that will give our meaning best; moreover the occasional polysyllable will have added effect from being set among short words. What is here deprecated is the tendency among the ignorant to choose, because it is a polysyllable, the word that gives their meaning no better or even worse. Mr. Pecksniff, we are told, was in the frequent habit of using any word that occurred to him as having a good sound, and rounding a sentence well, without much care for its meaning. He still has his followers. In the article LONG VARIANTS, examples are given of long forms chosen in place of shorter ones of the same word or stem. Attention is here confined to certain words frequently used where unrelated shorter ones would be better. They are doubtless chosen primarily not for their length, but because they are in vogue; but their vogue is in turn due to the pompous effect conferred by length. They are: *alternative, mentality, meticulous, overall, percentage, proportion, proposition, protagonist,* all of which will be found in their dictionary places. There are

many similar words, under which bare references to this article may be made; but these will serve as types. A quotation or two is given under each, and a fitter word offered.

Alternative: *The men dismissed will all be offered alternative work* (other work). / *The trouble between landlord and tenant in the country is due to the shortage of alternative accommodation* (shortage of houses).

Mentality: *A twenty-foot putt by Herreshoff at the twenty-fourth hole did not help Hilton's golfing mentality* (nerve). / *No one has so wide a knowledge of Afghan politics and of the mentality of the Pathan* (mind).

Meticulous: *These meticulous calculations of votes which have not yet been given rather disgust us* (exact). / *Owing to a meticulous regard for the spirit of the party truce, their views have not been adequately voiced by their leaders* (strict). / *Most of the British and American proposals have been too vague and sentimental on the one hand and too elaborate and meticulous on the other* (detailed).

Overall: *The overall production of coal is likely to be two million tons more this year than last* (total). / *I can quite understand that the Conservative Party are unwilling to look at the overall picture* (whole).

Percentage: *This drug has proved successful in a percentage of cases* (some).

Proportion: *The greater proportion of these old hands have by this time already dropped out* (part).

Proposition: *Dexter decided that his seamers were not a proposition* (unlikely to succeed). / *The agriculturist asks that 'corn-growing shall become a paying proposition'* (made to pay).

Protagonist: *The two great Western Powers who have acted as protagonists among the Allies in this war* (leaders). / *But most of the protagonists of this demand have since shifted their ground* (champions)

A few lines of the long-word style we know so well are added: *Vigorous condemnation is passed on the foreign policy of the Prime Minister, 'whose temperamental inaptitude for diplomacy and preoccupation with domestic issues have rendered his participation in external negotiations gravely detrimental to the public welfare'.* Vigorous indeed; a charging hippopotamus hardly more so. That is what comes of preferring abstract words to concrete. See ABSTRACTITIS.

**Low Countries.** See NETHERLANDS.

**lower, lour.** See LOUR.

**lu** (pronunciation). See PRONUNCIATION 6.

**lunatic fringe.** See ULTRA.

**lunch, luncheon.** *Lunch,* once a vulgarism for *luncheon,* has become the ordinary word for the meal, and *luncheon* is a FORMAL WORD.

**lung(e)ing.** See MUTE E; omit the *e*.

**lustrum.** Pl. *-tra,* sometimes *-trums*; see -UM.

**luxuriant, luxurious.** *Luxurious* is the adjective that belongs in sense to *luxury* and conveys the ideas of comfort or delight or indulgence; *luxuriant* has nothing to do with these, implying only rich growth, vigorous shooting forth, teeming, prolific; as *luxurious* to *luxury,* so *luxuriant* to *exuberance. Luxurious houses, habits, life, people, climate, idleness, times, food, cushions, dreams, abandonment, desires; luxuriant vegetation, crops, hair, imagination, invention, style.* The points at which they touch and become liable to confusion are, first, that abundance, essential to luxuriance or exuberance, also subserves luxury, though not essential to it; and, secondly, their common property in the verb *luxuriate,* which means both to enjoy luxury and to show luxuriance. A luxurious fancy is one that dwells on luxury; a luxuriant fancy one that runs riot on any subject, agreeable or other. The writer of the following has used the wrong word in the first place and the right one in the second: *Mr. H. was a man of somewhat striking outward appearance: he wore a somewhat* luxurious

*beard; ...his taste for a beard apparently resulted in a somewhat bushy* luxuriant *growth of hair all round his face.*

**-ly.** 1. For the tendency among writers and speakers who are more conscientious than literary to suppose that all adverbs must end in *-ly,* and therefore to use *hardly, largely, strongly,* etc., where idiom requires *hard, large, strong,* etc., see UNIDIOMATIC -LY.

2. For participial adverbs like *determinedly,* see -EDLY.

3. It was said in the article JINGLES that the commonest form of ugly repetition was that of the *-ly* adverbs. It is indeed extraordinary, when one remembers the feats of avoidance performed by the elegant-variationist, the don't-split-your-infinitivist, and the anti-preposition-at-ender, to find how many people have no ears to hear this most obvious of all outrages on euphony. Not indeed on euphony pure and simple, but on euphony and sense in combination; for as many *-ly* adverbs as one chooses may be piled on each other if one condition of sense is fulfilled—that all these adverbs have the same relation to the same word or to parallel words. *We are utterly, hopelessly, irretrievably, ruined*; *It is theoretically certain, but practically doubtful*; *He may probably or possibly be in time.* These are all irreproachable. In the first, each of the three adverbs expresses degree about *ruined*; in the second, each limits the sense of an adjective, the two adjectives being contrasted; in the third, the two give degrees of likelihood about the same thing, that is to say, in all three cases the *-ly* adverbs are strictly parallel. Euphony has nothing to say against repetition of *-ly* if there is point in it, which there is if the adverbs are parallel. But, when parallelism is not there to comfort her, Euphony at once cries out in pain, though too often to deaf ears.

*Russian industry is at present* practically completely *crippled. Practically* is not marching alongside of *completely,* but riding on its back; read *almost.*/

He found himself sharply, and apparently completely, *checked*. *Sharply and completely*, by all means; but not *apparently completely*; read *as it seemed*. | *Maeterlinck* probably and wisely *shrank from comparison with* '*Hérodias*'. Though *probably* and *wisely* both apply directly to the same word *shrank*, their relation to it is not the same, *probably* telling us how far the statement is reliable, and *wisely* how far the course was justified; read *It is probable that Maeterlinck wisely shrank*.

**Lyceum.** Pl. *-ms*; see -UM. For the meaning in Greek Philosophy see ACADEMY.

**lych-gate** etc. See LICH-GATE.

**lyric(al).** *Lyric* is now the established adjective for most uses; we speak of *lyric poets, poetry, verse, drama, muse, elements*, and not *lyrical*. *Lyrical* is in some sort a parasite upon *lyric*, meaning suggestive of lyric verse. *Lyric* classifies definitely, while *lyrical* describes vaguely. With some words either can be used, but with different effect; a lyric rhapsody is one actually composed in lyric verse; a lyrical rhapsody is talk full of expressions, or revealing a mood, fit for lyric poetry. *Lyrical emotion, praise, sorrow*, etc.; or again, a person may *grow lyrical*. See also -IC(AL).

**lyrics.** The OED definition (as regards modern usage) is: 'Short poems (whether or not intended to be sung), usually divided into stanzas or strophes, and directly expressing the poet's own thoughts and sentiments'. The short pieces between the narrative parts of Tennyson's *Princess* (*Home they brought her warrior dead* etc.), are typical examples. Wordsworth's *Daffodils*, Shelley's *Skylark*, Keats's *Grecian Urn*, Milton's *Penseroso*, Burns's *Field Mouse*, Herrick's *Rosebuds*, Lovelace's *Lucasta*, Shakespeare's *It was a lover*, may serve to illustrate; but attempts to distinguish lyric poetry clearly from other kinds (epic, dramatic, elegiac, didactic, etc.) have not been successful, the classes

not being mutually exclusive. The term is now applied to the songs, whatever their subject, in what are called 'musicals'; and that is the sense in which most people today would understand *lyrics*; the word has suffered VULGARIZATION.

# M

**macabre.** The OED gives only *-ahbr* for the pronunciation, but some more recent dictionaries allow *-ahber* as an alternative, and it may win. Cf. CALIBRE.

**Machiavel(li(an(ism.** The formerly current shortening *Machiavel* is now obsolete not only as the personal but even as the generic name; *a very Machiavel*, once common, is not now used. The adjective is accordingly now spelt *Machiavellian*, not *-elian*. For the *-ism* noun, choice lies between *Machiavellianism* and *Machiavellism*; in spite of greater length, the first is the better; the clipping of the word to which *-ist* and *-ism* are to be added is always disagreeable, and yet *Machiavelliism* is clearly impossible; see on *accompan(y)ist* in -IST A.

**machicolate.** Pronounce *măchǐ'kǒlāt*, not *măk-*.

**machination.** Pronounce *măk-*.

**macula.** Pronounce *mă'cūla*. Pl. *-lae*.

**mad,** v. For this and *madden*, and *the madding crowd*, see -EN VERBS.

**madam(e).** In the English word, whether as appellation (*I will inquire, Madam*; *Dear Madam*; *What does Madam think about it?*), as common noun (*the City madams*), or as prefix (*Madam Fortune, Madam Venus*), there should be no *-e*. As a prefix to a lady's name instead of *Mrs.*, *Madame* is right, with plural *Mesdames*. *Madam*, the appellation, suffers from having no plural, *Ladies* being the substitute, for which *Mesdames* is sometimes jocularly used. The shop-assistant's odd pronunciation (*mŏdm*) is perhaps due to a notion that French

*Madame* is more in keeping with haunts of fashion than English *Madam*.

**madness.** For *method in m.*, see IRRELEVANT ALLUSION.

**maestro.** The dictionaries give the pronunciation *mǎĕs* or *mahs-*, but popular usage makes it *mī-*. Plural *-tri*, pron. *ē*.

**Magdalen(e).** The spellings and pronunciations are:

1. In the names of the Oxford (*-en*) and Cambridge (*-ene*) colleges, pronounce *mau'dlin*.
2. In the use as a noun meaning reformed harlot etc., use *mag'dălĕn*.
3. When used with *the* instead of the name Mary M., *the Magdalene* (*-ēn*) and *the Magdalen* (*-ĕn*) are equally correct.
4. In the full name *Mary Magdalene* the four-syllable pronunciation (*măgdălē'nĕ*) is the best, though if it were *Mary the Magdalene -lĕn* would be right, as it is in *the Magdalene*, i.e. the famous person of Magdala. *Magdalene* may be regarded either as an English word = of Magdala, like *Lampsacene, Cyzicene, Tyrrhene,* etc., in which case *the* could not be omitted, or as the actual Greek feminine of *Magdalenos* become part of her name, in which case the final *-e* cannot be silent. *Mary Magdalen*, however, is also possible.

**magic(al),** adjectives. See -IC(AL). *Magic* tends to lose those adjective uses that cannot be viewed as mere attributive uses of the noun. First, it is very seldom used predicatively; *the effect was magical* (never *magic*); *the ring must be magical* (not *magic*, though *must be a magic one* is better than a *magical one*). Secondly, the chief nonpredicative use is in assigning a thing to the domain of magic (*a magic ring, carpet, spell, crystal*; *the magic art*), or in distinguishing it from others and so helping its identification (*magic lantern, square*), rather than in giving its characteristics descriptively (*with magical speed*; *what a magical transformation*).

**Magna C(h)arta.** Until recently authority seemed to be for spelling *charta* and pronouncing *kar'ta*, which was hard on the plain man. In a Bill introduced in 1946 authorizing the Trustees of the British Museum to lend a copy to the Library of Congress, *Charta* was the spelling used. But when the Bill reached committee stage in the House of Lords, the Lord Chancellor (Lord Jowitt) moved to substitute *Carta* and produced conclusive evidence that that was traditionally the correct spelling. The amendment was carried without a division; so *Carta* has now unimpeachable authority.

**mahlstick.** See MAULSTICK.

**Mahomet, Mohammedan,** etc. A middle-aged lady, on being asked whom she understood by the Prophet of Allah, hesitated, suspecting some snare, but being adjured to reply said quite plainly that he was *Mahomet* and further that his followers were called *Mahometans*—thus fulfilling expectations. The popular forms are *Mahomet(an)*; the prevailing printed forms are *Mohammed(an)*.

The worst of letting the learned gentry bully us out of our traditional *Mahometan* and *Mahomet* (who ever heard of *Mohammed and the mountain*?) is this: no sooner have we tried to be good and learnt to say, or at least write, *Mohammed* than they are fired with zeal to get us a step or two further on the path of truth, which at present seems likely to end in *Muhammad* with a dot under the h; see DIDACTICISM. The literary, as distinguished from the learned, surely do good service when they side with tradition and the people against science and the dons. *Muhammad* should be left to the pedants, *Mohammed* to historians and the like, while ordinary mortals should go on saying, and writing in newspapers and novels and poems and such general reader's matter, what their fathers said before them.

The fact is that we owe no thanks to those who discover, and cannot keep silence on the discovery, that *Mahomet*

is further than *Mohammed*, and *Mohammed* further than *Muhammad*,
from what his own people called him.
The Romans had a hero whom they
spoke of as *Aeneas*; we call him that
too, but for the French he has become
*Énée*; are the French any worse off
than we on that account? It is a
matter of like indifference in itself
whether the English for the Prophet's
name is *Mahomet* or *Mohammed*—in
itself, yes; but whereas the words
*Aeneas* and *Énée* have the Channel
between them to keep the peace,
*Mahomet* and *Mohammed* are for ever
at loggerheads; we want one name for
the one man; and the one should have
been that around which the ancient
associations cling. It is too late to
recover unity; the learned, and their
too docile disciples, have destroyed
that, and given us nothing worth having in exchange.

**maieutic.** Pronounce *māū'tĭk*. The
word means performing midwife's
service (to thought or ideas). Socrates
figured himself as a midwife (μαῖα)
bringing others' thoughts to birth
with his questionings. *Educative* contains the same notion, but much overlaid with different ones, and the
literary critic and the pedagogue consequently find *m.* useful enough to
pass in spite of its touch of pedantry.

**major** means greater, and those who
like pomposities are within their
rights, and remain intelligible, if they
call the greater part the m. portion;
they can moreover plead that *m. part
and portion* have been used by good
writers in the times when pomposity
was less noticeable than it now is.
Those who do not like pomposities
will call it *the greater part* and deserve
our gratitude, or at least escape our
dislike. *I, who had described myself as
'sick of patriotism' . . . found myself
unable to read anything but a volume*
the m. portion of which *consisted of
patriotic verse.*
M. is a convenient word to describe
something of more than ordinary importance or likely to have unusually
serious consequences, e.g. *m. road, m.*

*railway accident, m. war.* That no
doubt explains, but does not justify,
its having become a VOGUE WORD of
the sort that attains popularity because of the ease with which it can
be used to save the trouble of thinking
of some other word, e.g. *chief, main,
principal*, etc.

**major general.** See HYPHENS and
PLURAL ANOMALIES.

**majority.** 1. Distinctions of meaning. 2. Number after *m.* 3. *Great*
etc. *m.*

1. Three allied senses, one abstract
and two concrete, need to be distinguished if illogicalities are to be
avoided: (A) *Majority* meaning a
superiority in number, or, to revive an
obsolete unambiguous word, a plurity
(*. . . was passed by a bare, small, great,
m.; the m. was scanty but sufficient*).
(B) *Majority* meaning the one of two
or more sets that has a plurity, or the
more numerous party (*The m. was*, or
*were, determined to press its*, or *their,
victory*). (C) *Majority* meaning most
of a set of persons, or the greater part
numerically (*The m. were fatally
wounded*; *A m. of my friends advise it*).
But it should not be used as a dignified
substitute for the greater part of a
whole that is not numerical, as in *It is
a book with sociological merits in the m.
of it.*
2. Number. After *m.* in sense (A) the
verb will always be singular. After *m.*
in sense (B), as after other nouns of
multitude, either a singular or a plural
verb is possible, according as the body
is, or its members are, chiefly in the
speaker's thoughts. See NUMBER 6.
After *m.* in sense (C), in which the
thought is not of contrasted bodies at
all, but merely of the numbers required to make up more than a half,
the verb is almost necessarily plural,
the sense being more people than not,
out of those concerned. Correct *was*
to *were* in *The vast m. of Conservatives
was willing to vote for going in to Suez
one day and for coming out a few days
afterwards.*
3. *Great* etc. *m.* With *m.* in sense (A),
*great, greater, greatest*, etc., are freely

used, and cause no difficulty. With
*m.* in sense (B) they are not often
used, except to give the special sense of
party having a great, greater, plurity
as compared with that enjoyed by
some other (*This great m. is helpless*;
*having the greatest m. of modern times
devoted to him*). With *m.* in sense (C),
*great* is possible and common, *the
great m.* meaning most by far, much
more than half; but the use of *greater*
and *greatest* with it, as if *m.* meant
merely part or number, is, though
frequent, an illiterate blunder; examples of it are: *By far* the greatest m.
*of American rails, apart from gambling
counters, have gone across the Atlantic.* /
*The club is representative of several
hundreds,* the greater m. *of whom are
repatriated Britishers from Russia.*

**make.** *M. him repeat it,* not *to repeat*;
*He must be made to repeat it,* not *made
repeat.*

**make-believe** is the true form of the
noun as well as the verb, and *make-
belief* a false correction; to *make
believe* has meant to pretend from
the 14th c.

**malapropisms.** When Mrs. Mala-
prop, in Sheridan's *Rivals,* is said to
'deck her dull chat with hard words
which she don't understand', she
protests 'Sure, if I *reprehend* anything
in this world, it is the use of my *ora-
cular* tongue, and a nice *derangement*
of *epitaphs*'—having vague memories
of *apprehend, vernacular, arrangement,*
and *epithets.* She is now the matron
saint of all those who go wordfowling
with a blunderbuss. Achievements so
heroic as her own do not here concern
us; they pass the bounds of ordinary
experience and of the credible. Her
votaries are a feebler folk; with them
malaprops come single spies, not in
battalions, one in an article, perhaps,
instead of four in a sentence, and not
marked by her bold originality, but
monotonously following well beaten
tracks. In the article PAIRS AND SNARES
a number of words are given with
which other words of not very different sound are commonly confused,

and under most of the separate words
contained in that list illustrations will
be found; *predict* and *predicate, reversal* and *reversion, masterful* and
*masterly,* will suffice here as examples.
But it is perhaps hardly decent to
leave the subject without a single concrete illustration. Here are one or two
less staled by frequent occurrence than
those mentioned above: *He thought it
desirous that the House of Lords should
determine the tests to be applied.* /
*Mr. —— has circulated what* portends
*to be a reply to a letter which I had
previously addressed to you.* / *His capacity for continuous work is* incredulous. /
*It* abrogates *too many functions to itself.* / *Mr. —— said that the air
raids had been so* destructible *that
many roads had been roped off.* / *He has
skilfully piloted the company through
practically* unchartered *seas.* / *I hope
my inexperience will not* mitigate *against
my chances.* / *The sole* benefactors *of
this 'revindication' appear to be victims
imprisoned or executed.* / *That* in-
sinuendo *is quite unwarranted.*

**male.** *M., masculine.* The distinction drawn between *female* and
*feminine* is equally true for *m.* and
*masculine*; the reader will perhaps be
good enough to look through the
article, FEMALE, FEMININE, and make
the necessary substitutions. The only
modification needed is in the statement about the original part of speech
of *female. Male* was not, like that, a
noun before it was an adjective; but
this difference does not affect present
usage.

**malignancy, -nity.** These nouns
almost reverse the relation between
the adjectives to which they belong.
The general distinction between
*malignant* and *malign* is that the first
refers rather to intention and the
second rather to effect (see BENIGN);
it would therefore be expected that
*malignancy* would be the word for
spitefulness, and *malignity* for harmfulness. But the medical use of
*malignant* (see BENIGN 3) has so
strongly affected *malignancy* that

*malignity* has had to take over the sense of spite, and almost lost that of harm.

**mall.** Originally the pronunciation of The M. was *maul* (from its association with *m.*, a shady walk, so pronounced) and that of Pall M. *pĕlmĕ'l* (probably from its supposed association with the adverb *pell-mell*). Popular pronunciation today is tending to a uniform -*ăl*, but this has made more headway against the old pronunciation of The M. than it has against that of Pall M.

**mandatary, -tory.** The -*ary* form is noun only, = one to whom a mandate is given; the -*ory* form is primarily adjective, = of the nature of a mandate, and secondarily a noun, = mandatary. A distinction in spelling between the personal noun and the adjective is obviously convenient, and the form *mandatary* might suitably have been used instead of *mandatory* in the Covenant of the League of Nations for the trustees of what were then called Mandated Territories. Similar personal nouns, some of them with associated forms in -*ory* of more or less different sense, are ACCESSARY, *adversary*, *commissary*, DEPOSITARY, *emissary*, *notary*, *registrary* (Cambridge form of *registrar*), *secretary* (cf. the adjective *secretory*), *tributary*.

**manes,** spirit of dead person. Pronounce *mā'nēz*; a plural noun, with plural construction though sometimes used as singular in sense.

**mangel, mangold.** The first is 'in English the now prevailing form' (OED), and, as it is not less significant to the Englishman, and nearer the pronunciation, than the original German *mangold*, it might have been expected to prevail. The dictionaries do indeed put it first, but the farmers and merchants who are the chief users of the word have an odd preference for -*old*.

**manifold.** Pronounce *măn-*, not *mĕn-*. Owing to this difference in pronunciation between *m.* and *many*, the word is no longer felt to be a member of the series twofold, threefold, thirtyfold, a hundredfold, and attempts to treat it as such result in unidiomatic English. It is better to coin *many-fold* for the occasion (cf. BUSINESS, BUSYNESS) than to imitate the writers of the quotations below. Both the uses illustrated in them are called obsolete by the OED, and the revival of them after centuries of dormancy is perhaps accounted for by the adaptation of the 'now literary' word to commercial and engineering uses in *m. writing*, *m. pipes*, etc., and its consequent popularization. *Such elimination would recoup that expense*, m., *by the saving which it would effect of food valuable to the nation—namely, salmon.* | *This organization in capable hands should repay in* m. *the actual funds raised on its behalf.*

**Manil(l)a.** 'The form *Manila* is correct, but rare except in geographical use'—OED. The established -*lla* is recommended.

**manœuvre,** v., makes -*vred*, -*vring*; see MUTE E. For the n. and v., see -RE and -ER.

**-manship.** For such compounds see BRINKMANSHIP.

**mantle,** v. The common use in which the subject is *face, cheek, brow, flush, blush, colour, blood,* etc., appears to come not directly from the original sense to clothe as with a mantle, but from the special application of that to liquids that cover themselves with foam etc.; otherwise the natural construction would be the less usual. *A blush mantled her cheek* etc. and not the more usual *A blush* or *The blood mantled in her cheek* or *Her cheek mantled with a blush.*

**many.** *While there have been* m. a *good-humoured smile about* . . . Like *more than one* (see MORE B), and no less illogically, *many* a requires always a singular verb.

**Maquis.** See RESISTANCE.

**marathon.** This now familiar word for a long-distance race and, by extension, for any other long-drawn-out test of endurance, was introduced in the first revived Olympic Games at Athens in 1896. The battle of Marathon, in which the Greeks defeated the invading Persians in 490 B.C., is said to have been marked by two notable long-distance runs. One, recorded by Herodotus, was that of the professional runner Phidippides, who ran from Athens to Sparta (150 miles) to ask for help in the impending battle. The other, not mentioned by Herodotus, was that of a soldier who ran 'in full armour, hot from the battle' (Plutarch) to Athens (22 miles) to announce the victory, and fell dead from exhaustion as he did so. Plutarch says that it was uncertain whether the name of this soldier was Thersippus or Eucles, but Lucian later attributed the exploit to Phidippides himself, a version popularized by Browning in his poem with that title.

**margarin(e).** The pronunciation *marj-* instead of *marg-* is clearly wrong, and is not even mentioned in the OED as an alternative. But that was before two wars had made everyone familiar with the substance, and the dictionaries now admit the popular variant *marj-*. We may suppose that *Margaret* and *Margery* fought for the analogy and *Margery* won. It does not seem likely that what is now universally called *marge* for short will regain the hard *g* in its full name, but as this is how it is pronounced by those who advertise it on television the battle is not lost. Perhaps the only English words in which g is soft before a or o or u are *gaol* (with its derivatives) and *mortgagor*. See -IN AND -INE for the termination.

**marginal, minimal.** These words are favourites with writers who find *small* too drab for their taste. They have their special meanings, and, though it would be unreasonable to expect them to be strictly confined, they should not be allowed to stray far. *Marginal* is properly applied to what is

so close to the dividing line between two opposing states that there is no saying which side it will go (cf. *border-line*), as a *m. seat* is one held by so small a majority as to be in peril at the next election, *m. land* is land that might or might not be profitable to cultivate, and, by a slight extension, the same meaning of 'doubtfully worth while' is seen in *Talking about critics without close reference to the authors they discuss is bound to be a m. activity. Minimal* is properly applied to something that is the smallest possible. An example of its misuse for effect is: *Ireland gave a magnificent performance in holding the all-conquering visitors to such a minimal margin.* A minimal margin in rugby football is one point. In this game the score was 8–3.

**marital.** The OED gives *mă'rĭtăl*, without even permitting *mărī'tal* but the COD allows the latter as an alternative. The short *i* is no doubt a shock to those who know the sound of *marītus* in Latin better than that of *m.* in their own language; see, however, FALSE QUANTITY for a batteryful of such shocks. See also RECESSIVE ACCENT.

**mark.** For synonymy see SIGN.

**marquetry, -eterie.** Spell -*try*, and pronounce *mar'kĭtrĭ*.

**marquis, -ess.** *Marquess* is now the usual spelling. To judge from *Who's Who*, there is an overwhelming preference for it among mm. themselves; only 2 of the 39 peers of that rank and only 1 of the 9 bearers of that courtesy title call themselves *marquis*. In books of reference that spell the word uniformly it is now always *marquess*. *Burke* and *Debrett* changed from *marquis* early in the 19th c. and *Whitaker's Almanack* towards the end of it. For Marquess Smith and Lord Smith see TITLES.

**marten, -in.** The beast has -*en*, the bird -*in*.

**masculine.** See MALE.

**massage, -eur, -euse.** *Măsah'zh, măser', măser'z* are the dictionary pronunciations, but the stress, especially in the first, is shifting to the first syllable as the words become naturalized.

**massive** in its figurative sense is a useful and expressive word that deserves to be treated with respect and discrimination. The virtue is being taken out of it now that it has become a VOGUE WORD, ousting more ordinary and often more suitable adjectives. Almost every day's newspapers will provide evidence of its popularity; in the following small selection, which might be multiplied indefinitely, other words are suggested that could have been used to give *massive* a rest. *A far greater effort should be made to produce accounts which do not require m. adjustment in future years* (sweeping). / *A call for local authorities to embark on a m. research scheme to determine what Britain's towns should be like in future* (comprehensive). / *The next country to break through into a m. economic advance may well be Spain* (vigorous). / *M. security precautions were in force today at General de Gaulle's review* (extraordinary). / *The objectors are probably right in opposing any increase in the m. capital grants which are made available in Northern Ireland* (lavish). / *It hardly seems that it can profit China to drive India into such m. anger* (intense). / *It will be difficult to represent the result of the referendum as the m. victory General de Gaulle demanded as the price of going on* (overwhelming). / *As I presided over the meeting I was able to note the intense enthusiasm of the m. audience when Mr. Wilson made his statement* (huge).

**masterful, masterly.** Some centuries ago both were used without distinction in either of two very different senses: (A) imperious or commanding or strong-willed, and (B) skilful or expert or practised. The DIFFERENTIATION is now complete, *-ful* having the (A) and *-ly* the (B) meanings, and is nicely observed in *The presentation in each case was masterly* (perhaps in a few rare instances a trifle too masterful) *and always the playing was crystal clear.* Disregard of it is so obviously inconvenient that it can only be put down to ignorance. *Masterly* is less liable to misuse, but *masterful* often appears instead of *masterly.* A few examples follow, in all of which *masterly* should have been the word: *When he began to outplay the Englishman and picked up hole after hole the crowd was carried away by* his masterful work *and driven to applauding.* / *The influence of the engineering and mechanical triumphs of the staff of the canal zone has been dealt with by* masterful writers. / *The judge told the jury that the prosecution had been put to them in a* masterful and restrained *fashion.* For the adverbs of adjectives in *-ly* see -LILY.

**mat,** lustreless. Correctly so spelt; it is a French adjective. But *matt*, no doubt due to an instinct of differentiation (cf. SET(T)) is said by the OED Supp. to be now the usual form.

**mate, checkmate.** The full form is now chiefly in metaphorical use, while the shortened one is preferred in chess.

**material,** adj. There are at least four current antitheses in aid of any of which *m.* may be called in when an adjective is required. There is matter and form (*m.* and *formal*); there is matter and spirit (*m.* and *spiritual*); there is MATÉRIEL and personnel(*m.* and *personal*); and there is what matters and what does not matter (*m.* and *trifling*). Before using *m.*, therefore, with reference to any one of these, the writer should make sure that there is no risk of confusion with another. *Agriculture, though the most m. of all our pursuits, is teaching us truths beyond its own direct province.* / *The old bonds of relationship, and community of* m. interests. / *A comparison between the French peasant-proprietor and the English small-holder as he might conceivably become under a freehold system, a comparison, be it said, to the m. advantage of the former.* The curious dislike of the preposition *of* that seems so widespread today has increased the confusion. As if the adjective had not enough different senses already, the noun is used adjectivally in yet another. *M. allocations will be made* seems to be more satisfying to those who deal with such things than either *allocations of m. will be made* or *m. will be allocated.* See NOUN ADJECTIVES.

**materialize.** The word has plenty of uses of its own, e.g. *Those who would m. spirit. A soul materialized by gluttony. Virgil having materialized a scheme of abstracted notions. Ghosts or promises of ghosts which fail to m.* It should not be forced to do the work of *happen* or be fulfilled or *form*, e.g. *There would seem to be some ground for hope that the strike will not m. after all.* / *Year after year passed and these promises failed to m.* / *Out of the mist of notes and protocols a policy seems gradually to be materializing.* In these latter senses *m.* is on the level of *transpire* (happen), *proposition* (job), *eventuate* (happen), *unique* (notable), *envisage* (foresee), *individual* (man), and such abominations.

**matériel.** In antitheses with *personnel*, expressed or implied, the French spelling and pronunciation should be kept, and not replaced by those of the English *material.* In practice the Services compromise by writing *materiel* (without the accent) and saying *material.*

**mathematics.** For the grammatical number, see -ICS 2.

**matins, matt-.** The OED treats *matins* as the standard form, but *mattins* is common. Possibly the double *t* was introduced at the time of the Reformation (the Act of Uniformity has *mattens*) to distinguish the use of the word for the Morning Prayer of the Anglican Church from its use in the Roman Catholic Church for the first of the canonical hours of the breviary. For *m.* and *morning prayer*, see MORNING.

**matrix.** For pl. see -EX, -IX, 4, and -TRIX.

**matter.** *The distribution shows that, as exceptional bravery is confined to no rank in the Army, so recognition is given to it* by no m. whom *it is displayed.* If elliptical phrases like *no m. who* are to be treated freely as units, care must be taken that the ellipsis can be filled in correctly. *By it is no m. whom it is displayed* is wrong, and *it is no m. by whom it is displayed* is right; accordingly the order should be *no m. by whom.* The principle is—by all means save your reader the trouble of reading more words than he need, but do not save yourself the trouble of rehearsing the full form by way of test. The real cause of the mistake here is the superstition against prepositions at the end; *no m. whom it is displayed by* would have been correct; but the writer was frightened at his final preposition, made a grab at it, and plumped it down in a wrong place; see SUPERSTITIONS, and OUT OF THE FRYING-PAN. The offence is aggravated by the inevitable impulse to connect *by* with *is given.*

**matutinal.** Chiefly in POLYSYLLABIC HUMOUR. *Here they were found by a m. gardener.*

**maugre.** See WARDOUR STREET.

**maulstick,** not *mahl-*, is the standard form.

**maunder, meander.** Though the etymology of *maunder* is uncertain, it is clear that it is not a corruption of *meander*, its earlier sense being to complain, growl, grouse. But it is also clear from the way some people use *meander* that they take the two words to be merely variant pronunciations. *Meander* means to follow a winding course, was originally used of rivers, is still often so used, describes frequent but not violent change of direction rather than aimlessness, and is applied more often to actual locomotion than to vagaries of the tongue. *Maunder* is best confined to speech, and suggests futility rather than digression, dull discontent rather than quiet enjoyment, and failure to reach an end rather than loitering on the way to it.

**maxim.** See APHORISM.

**maximum.** Pl. *-ma*, rarely *-mums*.

**maybe** (= perhaps) was long ago normal English, as natural as *perhaps*, if not more. In America it has remained the ordinary word. But in Britain it became a novelistic property, the recognized rustic or provincial substitute for *perhaps*. Having acquired, during this rustication, a certain unfamiliarity, it emerged stylishly archaic, so that *perhaps* and *m.* were for a time a pair of WORKING AND STYLISH WORDS, the only suitable function of *m.* being to replace *perhaps* in a context whose tone demanded a touch of primitive dignity; so *Our Lord speaking quite simply to simple Syrian people, a child or two m. at his knees*. Now, under American influence, we are bringing it back into use as a natural alternative to *perhaps*. But maybe we are not yet quite at home with it; for why else should we feel the need of the colloquial *could-be*?

**me.** The use of *me* in colloquialisms such as *It's me* and *It wasn't me* is perhaps the only successful attack made by *me* on *I*. There is a greater temptation to use *I* for *me*, especially when *and me* is required after another noun or pronoun that has taken responsibility for the grammar and has not a separate objective case; *between you and I, let you and I try*, are not uncommon. For discussion see I.

**meal, flour.** See FLOUR.

**mealies.** The singular (chiefly in combinations as *m.-field*, *m. porridge*) is *mealie*, not *-ly*, the etymological connexion being not with *meal* and *mealy*, but with *millet*.

**mean.** *He is no m. cricketer.* This use of *no m.*, echoing St. Paul's description of himself as a citizen of no m. city, was the subject of correspondence between A. A. Milne and Fowler in the TLS shortly after the publication of *Modern English Usage*. Grammatically the expression is admittedly above reproach, provided that it is without the indefinite article (see NO 2). But Milne took Fowler to task for not having denounced it as an overworked archaism, 'a penny-in-the-slot adjective which leaves nothing in the writer's mind as he puts it down but a hopeful feeling that he is being more amusing, more like Shakespeare and the Bible, than if he had said *good*'. Fowler disagreed. 'For Mr. Milne', he said, '*no mean* is the perfect cliché: for me it is not indeed a favourite substitute for *considerable* or *meritorious* or *first-class*, but a blameless one. . . . There is life yet in *no mean*, and I trust that as long as it lives it will bear in mind, as between Mr. Milne and me, that Codlin's the friend, not Short.'

Today the verdict would probably go to Milne. It is fashionable to reprobate the use of clichés so sweepingly as to induce in conscientious but timid writers a morbid dread of using any expression that might be so described. See CLICHÉ.

**meander.** See MAUNDER.

**meaningless words.** Words and phrases are often used in conversation, especially by the young, not as significant terms but rather, so far as they have any purpose at all, as aids of the same kind as are given in writing by punctuation, inverted commas, and underlining. It is a phenomenon perhaps more suitable for the psychologist than for the philologist. Words and phrases so employed change frequently, for they are soon worn out by overwork. Between the wars the most popular were DEFINITELY and *sort of thing*. One may suppose that they originated in a subconscious feeling that there was a need in the one case to emphasize a right word and in the other to apologize for a possibly wrong one. But any meaning they ever had was soon rubbed off them, and they became noises automatically produced. Their immediate successors have been *actually* (pronounced *akshally*) and *you know*. *Actually*, says the OED, may be added to vouch for statements which seem surprising, incredible, or exaggerated. That is how Mrs. Nickleby and Mr. Pyke used it. '"I had a cold once," said Mrs. Nickleby, "I think it was in the year 1817 ... that I thought I should never get rid of; actually and seriously that I thought I should never get rid of".... "And I'll tell you what", said Mr. Pyke, "if you'll send round to the public house for a pot of mild half-and-half, positively and actually I'll drink it."' Many people today seem to find it impossible to trust any assertion, however commonplace, to be believed without this warranty. We have all had experiences like that recorded by Ivor Brown: 'I met some young people recently who used *actually* in almost every sentence. "Are you living in London?" "Actually I am." "Actually we must be going now."'' 'Actually', he adds, 'I did not mind if they did.' The now ubiquitous *you know* (cf. the obsolete *dontcherknow*) seems to be a compendious way of saying 'I know I am expressing myself badly, but I am sure you are intelligent enough to grasp my meaning'.

WELL is a permanent member of the class of words thus used, and INCIDENTALLY is having a long innings. The 'interrogative expletive' *what*, as the OED calls it, quoting *Goodbye Miss Thornton, awfully jolly evening, what?*, was once fashionable, but has had its day.

**means,** n. In the sense income etc.,*m.* always takes a plural verb: *My m. were* (never *was*) *much reduced.* In the sense way to an end etc.: *a m.* takes singular verb; *m.,* and *the m.,* can be treated as either singular or plural. *All m.* (pl.) and *every m.* (sing.) are equally correct; *the m. do not,* or *does not, justify the end*; *the end is good, but the m. are,* or *is, bad*; such *m. are* (not *is*) *repugnant to me,* because *such* without *a* is necessarily plural; cf. *such a m.*

*is not to be discovered*; and similarly with other adjectives, as *secret m. were found,* but *a secret m. was found.*

**mean time.** For the solar time, two words unhyphened. For the adverb, one word.

**measure.** *Lord Curzon's policy has been overthrown by the present announcement, which* to a great m. *restores Bengal to her former greatness.* To a great extent, but *in great m.*; see CAST-IRON IDIOM.

**measure up to.** See PHRASAL VERBS and RISE.

**meatus.** For plural (*-ūs* or *-uses*) see -US 2.

**medi(a)eval.** The shorter spelling is recommended; see Æ, Œ.

**mediatize.** To m. a ruler is to reduce him to dependence on another State, but without changing his titular dignity. The word originated in the Holy Roman Empire, and meant that the prince now owed *mediate* (i.e. indirect) allegiance instead of *immediate* to the Emperor.

**medicine.** The conventional pronunciation is two syllables (*měd'sn*), but under the influence of the speak-as-you-spell movement (see PRONUNCIATION 1) the word is often given three (*měd'isin*) now recognized by the COD as an alternative. The adjective is *mědĭ'sĭnăl.*

**mediocr(e)(ity).** Pronounce *mē'dĭōkĕr, mědĭŏk'rĭty,* but the stressing of the third syllable of the first and a long e in the first syllable of the second are not without dictionary recognition.

**medium.** In the spiritualistic sense, the plural is always -ums. In all other senses, -a and -ums are both in use, and -a seems to be the commoner. See -UM.

**meet.** For *we are met together* etc., see INTRANSITIVE P.P. For *meet up with* see PHRASAL VERBS.

**mein Herr.** See MYNHEER.

**meiosis.** Pl. *-oses* (*-ēz*). The use of understatement not to deceive, but to enhance the impression on the hearer. Often applied to the negative opposite illustrated under LITOTES but taking many other forms, and contrasted with HYPERBOLE. Very common in colloquial and slang English; the (now outmoded) emphatic RATHER, with its stress on the second syllable, the use of *some* that we have borrowed from U.S. (*Some chicken, some neck*: see SOME I), the schoolboy *decent* (= very nice), the retort *I'll see you further* (i.e. in hell) *first*, and the strangely inverted hyperbole *didn't half swear* (= swore horribly), are familiar instances.

**melodrama.** The term was first applied to plays in which music (μέλος) accompanied the spoken word, but there was no singing. Later m. approximated to opera, but with the difference that the dialogue was entirely spoken and the songs interspersed, as in *The Beggar's Opera*. Today the term is used of plays of a certain kind without regard to their musical content, if any. It is generally used with some contempt, because the appeal of such plays is especially to the unsophisticated, whose acquaintance with human nature is superficial, but whose admiration for goodness and detestation of wickedness are ready and powerful. The melodramatist's task is to get his characters labelled good and wicked in his audience's minds, and to provide striking situations that shall provoke and relieve anxieties on behalf of poetic justice. Whether a play is or is not to be called a melodrama is therefore often a doubtful question, upon which different critics will hold different opinions. The typical characters of a m. in its crudest form have been described as 'a diabolically clever villain customarily engaged in the pursuit of a pure and lovely heroine, constantly foiled and finally defeated by a manly and honest hero, who is frequently aided by a comic personage'—*Enc. Brit.* All that the m. now so called inherits from the early form is the appeal to emotion; the

emotional effect of musical accompaniment is obvious, and it is on emotional sympathy that m. still relies.

**melody, harmony.** See HARMONY.

**melted, molten.** *Molten*, apart from its use as a poetic variant of *melted*, is now confined to what needs great heat to melt it. *Molten iron, melted butter*.

**membership, leadership.** The use of *m.* in the sense number of members (of a club etc.) is, though not a very desirable one, established (*The necessity of adding to the m. of the House*; *A large m. is necessary*). Much less desirable is the extension from number of members to members, a practice now rife and corrupting other words, especially by the use of *leadership* for leaders. *He cannot, even if the entire m. of a union marched up to his office, investigate alleged irregularities into the conduct of a ballot.* | *There is a growing restlessness among a section of its m.* | *I hope our m. will listen to the advice of their elected representatives.* | *Leadership* is used in this way so constantly that we seem to be in danger of forgetting that there is such a word as *leaders*. Examples like the following could be multiplied indefinitely. *They have refrained from making declarations that the union's policy is not in the best interest of the m. or that the leadership has failed to implement the policy.* | *It was decided to proceed against the leadership of the E.T.U. under Rule 13.* | *The new Soviet leadership now launched its propaganda campaign for peace.* | *The leadership of the Parliamentary Party behaves as though it were a Shadow Administration.*

Needless substitution of the abstract for the concrete is one of the surest roads to flabby style (see ABSTRACTITIS). In the following quotation, where the correct use of the second *leadership* in its abstract sense might have been expected to put the writer on his guard, he seems to have been so bemused by the lure of the abstract that he could not bring himself, by writing *leaders* for the first, to clothe in flesh and blood those whom he was urging to act wholeheartedly and in good conscience. *If the present leadership will wholeheartedly and in good conscience give the country that leadership, they will not lack loyal and enthusiastic support.* Even the book-reviewers are becoming infected. *It is hard to see what British readership there can be much longer for books about such topics in American society.* It is no less hard to see why anyone should think *readership* a more suitable word than *readers*.

**memorandum.** Pl. *-da* or, less usually, *-ums*. The commercial abbreviation *memo*, often pronounced *mĕmo*, is best left unspoken.

**mendacity, mendicity.** The first is the conduct of a liar, the second that of a beggar. See PAIRS AND SNARES.

**-ment.** For differences between this and *-ion*, see -ION AND -MENT. The stems to which *-ment* is normally appended are those of verbs; freaks like *oddment* and *funniment* should not be made a precedent of; they are themselves due to misconception of *merriment*, which is not from the adjective, but from an obsolete verb *merry* to rejoice.

**mentality.** This word has been going through a strange experience. Originally it meant that which is of the nature of mind, or mental action. In the 19th c. it was used in a narrower sense— intellectual power. *An insect's very limited m.* / *Pope is too intellectual and has an excess of m.* It was only in the present century that it acquired its meaning of 'mental character or disposition' (OED Supp.). '*First I would give you an insight into his m.*', said Holmes. '*It is a very unusual one—so much so that I think his destination is more likely to be Broadmoor than the scaffold.*' / *It is useless to pretend that there will be anything but hostility between the partners in industry so long as this m. persists.* It is not easy to say why *m.* should have ousted for this purpose other serviceable words such as *disposition, attitude, character, mind,*

etc.; perhaps NOVELTY HUNTING accounts for it. But we may surmise that its progress to a VOGUE WORD has been helped by the disparaging flavour it has taken on: it affords a convenient means of being rude politely. *The m. of the politician is a constant source of amazement to the engineer.* / *It is difficult to comprehend a m. which is bounded entirely by finance and expediency.* / *When I read that rather pathetically hopeful suggestion, with a fairly long experience of Treasury m., I could not help being reminded of the old music-hall song 'You don't know Nelly like I do'.* Possibly this pejorative use derives from that association of *mental* with *patient, defective, hospital,* and the like which has made it a slang word for insane. It would be a striking example of the waywardness of words if one that was formerly used for excess of intellectuality ended by meaning an insufficiency of it. Perhaps it is destined to be superseded by *psychology*, already used in the same sense, as the slang *mental* is being superseded by *psycho*. See POPULARIZED TECHNICALITIES.

**Mephistopheles.** The adjective is *Mephistophĕle'an* or *Mephistophē'lĭan*; the latter perhaps more likely to last; see HERCULEAN.

**mercy.** For *the tender mercies of*, see HACKNEYED PHRASES.

**mesembryanthemum** is now so spelt by botanists, not *mesembrianthemum*. Both forms were used in the early eighteenth century: that with *y* by Dillenius who said it contained Greek μέσος middle, ἔμβρυον embryo, and ἄνθος flower, and that with *i* by Breyne because some species opened their flowers only for a short time at midday (Greek μεσημβρία noon). Night-flowering species were discovered and in 1753 Linnaeus settled the issue for botanical scholars by choosing the Dillenian spelling. The form with *i* is reflected in the rendering of the name in English as *midday flower* and in German as *Mittagsblume*.

**metal, mettle,** are the same word, whose difference of spelling in its figurative sense reflects a difference of meaning: *metal* the stuff of which a man is made, *mettle* the stuff of which a particular kind of man (or horse) is made. Ophelia was of *metal* more attractive than Gertrude; Lady Macbeth was of undaunted *mettle* that should compose nothing but males.

**metamorphosis.** Generally accented on the middle syllable (*-mor-*); but the more regular accent on *-pho-* is often heard. As *m.* seems to be the only word in *-osis* irregularly accented, and as it retains the classical plural (*-oses*, pronounced with *-ēz*), and as the *-osis* ending is now familiar in *tuberculosis* and other medical terms, it may be expected to revert to *mĕtămorfō'sis*; cf. *metempsychosis*, which is stated by the OED to have formerly had the accent on the *-sy-*, and has now recovered.

**metaphor.** *Our task when we are returned to power will be to restore to agriculture the twin pillars of efficiency and security*. 'How infinite', wrote Sir Winston Churchill long before these words were spoken, 'is the debt owed to metaphors by politicians who want to speak strongly but are not sure what they are going to say.' Hardly less, as no one knew better than Sir Winston, is the debt owed to metaphors by those who, knowing what they are going to say, wish to illumine and vivify it. Moreover, our vocabulary is largely built on metaphors; we use them, though perhaps not consciously, whenever we speak or write. The purpose of this article is to give some advice about the handling of this indispensable but ticklish instrument. See also CLICHÉ.

1. Live and dead m. 2. Some pitfalls. 3. Self-consciousness and mixed m. 4. For m. and simile, see SIMILE AND METAPHOR.

1. Live and dead m. In all discussion of m. it must be borne in mind that some metaphors are living, i.e. are offered and accepted with a conscious-

ness of their nature as substitutes for their literal equivalents, while others are dead, i.e. have been so often used that speaker and hearer have ceased to be aware that the words used are not literal. But the line of distinction between the live and the dead is a shifting one, the dead being sometimes liable, under the stimulus of an affinity or a repulsion, to galvanic stirrings indistinguishable from life. Thus, in *The men were sifting meal* we have a literal use of *sift*; in *Satan hath desired to have you, that he may sift you as wheat, sift* is a live m.; in *the sifting of evidence*, the m. is so familiar that it is about equal chances whether *sifting* or *examination* will be used, and that a sieve is not present to the thought—unless indeed someone conjures it up by saying *All the evidence must first be sifted with acid tests*, or *with the microscope*. Under such a stimulus our m. turns out to have been not dead but dormant. The other word, *examine*, will do well enough as an example of the real stone-dead m.; the Latin *examino*, being from *examen* the tongue of a balance, meant originally to weigh; but, though weighing is not done with acid tests or microscopes any more than sifting, *examine* gives no convulsive twitches, like *sift*, at finding itself in their company. *Examine*, then, is dead m., and *sift* only half dead, or three-quarters.

2. Some pitfalls: A. Unsustained m.; B. Overdone m.; C. Spoilt m.; D. Battles of the dead. E. Mixed m.

2. A. Unsustained m. *He was still in the middle of* those 20 years of neglect *which only began to* lift *in 1868*. The plunge into m. at *lift*, which presupposes a mist, is too sudden after the literal 20 years of neglect; years, even gloomy years, do not lift. / *The means of education at the disposal of the Protestants and Presbyterians of the North were* stunted and sterilized. *The means at disposal* indicates something too little vegetable or animal to consort with the metaphorical verbs. Education (personified) may be stunted, but means may not. / *The* measure *of Mr. A's shame does not* consist *in the*

*mere* fact *that he has announced his intention to* . . . Metaphorical measuring, like literal, requires a more accommodating instrument than a stubborn fact.

2. B. Overdone m. The days are perhaps past when a figure was deliberately chosen that could be worked out with line upon line of relentless detail, and the following well-known specimen is from Richardson: *Tost to and fro by the high winds of passionate control, I behold the desired port, the single state, into which I would fain steer; but am kept off by the foaming billows of a brother's and sister's envy, and by the raging winds of a supposed invaded authority; while I see in Lovelace, the rocks on one hand, and in Solmes, the sands on the other; and tremble, lest I should split upon the former or strike upon the latter*.

The present fashion is rather to develop a m. only by way of burlesque. All that need be asked of those who tend to this form of satire is to remember that, while some metaphors do seem to deserve such treatment, the number of times that the same joke can safely be made, even with variations, is limited. The limit has surely been exceeded, for instance, with 'the long arm of coincidence'; what proportion may this triplet of quotations bear to the number of times the thing has been done?—*The long arm of coincidence throws the Slifers into Mercedes's Cornish garden a little too heavily. | The author does not strain the muscles of coincidence's arm to bring them into relation. | Then the long arm of coincidence rolled up its sleeves and set to work with a rapidity and vigour which defy description*.

Modern overdoing, apart from burlesque, is chiefly accidental, and results not from too much care, but from too little: *The most irreconcilable of Irish landlords are beginning to recognize that we are on the eve of the dawn of a new day in Ireland. On the eve of* is a dead m. for about to experience, and to complete it with *the dawn of a day* is as bad as to say *It cost one pound sterling, ten*, for *one pound ten*.

2. C. Spoilt m. The essential merit of real or live m. being to add vividness to what is to be conveyed, it need hardly be said that accuracy of detail is even more needed in metaphorical than in literal expressions. The habit of m., however, and the habit of accuracy do not always go together: *Yet Jaurès was the* Samson *who* upheld the pillars *of the Bloc. | Yet what more distinguished names does the Anglican Church of the last reign boast than those of F. D. Maurice, Kingsley, Stanley, Robertson of Brighton, and even, if we will* draw *our* net *a little* wider, *the great Arnold? | He was the very essence of cunning, and the* incarnation *of a* book-thief. Samson's way with pillars was not to uphold them; we draw nets closer, but cast them wider; and what is the incarnation of a thief? Too, too solid flesh indeed. Similarly a m. may be spoilt if so used that the picture it is intended to evoke becomes incongruous or ridiculous. *We must not allow ourselves to be stampeded into stagnation. | This is a virgin field pregnant with possibilities. | She drew the teeth of Miss Reynolds's forehand with cross-court volleys*. For other examples SEE BOTTLENECK, BREAKDOWN, CEILING, and TARGET.

2. D. Battles of dead metaphors. In *The Covenanters took up arms* there is no m.; in *The Covenanters flew to arms* there is one only—*flew to* for quickly took up; in *She flew to arms in defence of her darling* there are two, the arms being now metaphorical as well as the flying. Moreover, the two metaphors are separate ones; but, being dead ones, and also not inconsistent with each other, they lie together quietly enough. But dead metaphors will not lie quietly together if there was repugnance between them in life; e'en in their ashes live their wonted fires, and they get up and fight: *It is impossible to* crush *the Government's* aim *to restore the means of living and working freely*. Crush for baffle, aim for purpose, are both dead metaphors so long as they are kept apart; but the juxtaposition forces on us the thought that you cannot crush an aim. |

*National military training is the* bed-rock *on which alone we can hope to* carry through *the great struggles which the future may have in store for us. Bedrock* and *carry through* are both moribund or dormant, but not stone-dead. | *The vogue of the motor car seems destined to help forward the provision of good road communication, a* feature *which is sadly* in arrear. Good road communication may be a feature, and it may be in arrear, and yet a feature cannot be in arrear; things that are equal to the same thing may be equal to each other in geometry, but language is not geometry. | *They are* cyphers *living* under the shadow of *a great man*.

2. E. Mixed metaphors. For the examples given in D, tasteless word-selection is a fitter description than mixed m., since each of the words that conflict with others is not intended as a m. at all. *Mixed m.* is more appropriate when one or both of the terms can only be consciously meta-phorical. Little warning is needed against it; it is so conspicuous as seldom to get into speech or print undetected. *This is not the time to throw up the sponge, when the enemy, already weakened and divided, are on the run to a new defensive position.* A mixture of prize-ring and battle-field. | *The Rt. Hon. Gentleman is leading the people over the precipice with his head in the sand.* A strange con-fusion between the behaviour of Gadarene swine and that of ostriches. | *There is every indication that Nigeria will be a tower of strength and will forge ahead.* A mixture of a fortress and a ship. | *The Avon and Dorset River Board should not act like King Canute, bury its head in the sands, and ride rough-shod over the interests of those who live by the land and enjoy their fishing.* A picture that staggers the imagination, and a libel on a great king.

In the following extract from a speech it is difficult to be sure how many times metaphors are mixed; readers versed in the mysteries of oscillation may be able to decide: *No society, no community, can place its*

house *in such a condition that it is* always *on a rock, oscillating* between *solvency and insolvency. What I have to do is to see that our house is* built upon a solid foundation, *never allow-ing the possibility of the Society's* life-blood being sapped. *Just in proportion as you are careful in looking after the condition of your income, just in pro-portion as you deal with them carefully, will the solidarity of the Society's financial condition remain intact. Im-mediately you begin to* play fast and loose *with your income* the first blow *at your financial stability* will have been struck.

3. Self-consciousness and mixed m. The gentlemen of the Press regularly devote a small percentage of their time to accusing each other of mixing metaphors or announcing that they are themselves about to do so (*What a mixture of metaphors! If we may mix our metaphors*, or *change the m.*), the offence apparently being not to mix them, but to be unaware that you have done it. The odd thing is that, whether he is on the offensive or the defensive, the writer who ventures to talk of mixing metaphors often shows that he does not know what mixed m. is. Two typical examples of the offen-sive follow: *The Scotsman says: 'The crowded benches of the Ministerialists contain the germs of disintegration. A more ill-assorted majority could hardly be conceived, and presently the Opposi-tion must realize of what small account is the manœuvring of the Free-Fooders or of any other section of the party. If the sling be only properly handled, the new Parliamentary Goliath will be over-thrown easily enough. The stone for the sling must, however, be found on the Ministerial side of the House, and not on the Opposition side'. Apparently the stone for the sling will be a germ. But doubtless mixed feelings lead to mixed metaphors. | 'When the Chairman of Committees—a politician of their own hue—allowed Mr. Maddison to move his amendment in favour of secular education, a decision which was not quite in accordance with precedent, the floodgates of sectarian controversy were*

opened, and the apple of discord—the endowment of the gospel of Cowper-Temple—was thrown into the midst of the House of Commons.' What a mixture of metaphor! One pictures this gospel-apple battling with the stream released by the opened floodgates.

In the first passage, we are well rid of the germs before we hear of the sling, and the mixture of metaphors is quite imaginary. Since literal benches often contain literal germs, but crowded benches and germs of disintegration are here separate metaphors for a numerous party and tendencies to disunion, our critic had ready to his hand in the first sentence, if he had but known it, something much more like a mixture of metaphors than what he mistakes for one. In the second passage, the floodgates and the apple are successive metaphors, unmixed; the mixing of them is done by the critic himself, not by the criticized; and as to gospel-apple, by which it is hinted that the mixture is triple, the original writer had merely mentioned in the gospel phrase the thing compared side by side with what it is compared to, as when one explains the Venice of the North by adding Stockholm.

Writers who are on the defensive apologize for change and mixture of metaphors as though one was as bad as the other. The two things are in fact entirely different. A man may change his metaphors as often as he likes; it is for him to judge whether the result will or will not be unpleasantly florid. But he should not ask our leave to do it; if the result is bad, his apology will not mend matters, and if it is not bad no apology was called for. On the other hand, to mix metaphors, if the mixture is real, is an offence that should not have been apologized for, but avoided. In either case the motive is the same—mortal fear of being accused of mixed m.:—

. . . showed that Free Trade could provide the jam without recourse being had to Protective food-taxes; next came a period in which (to mix our metaphors) the jam was a nice slice of tariff pie for everybody; but then came the Edinburgh Compromise, by which the jam for the towns was that there were to be . . . When jam is used in three successive sentences in its hackneyed sense of consolation, it need hardly be considered in the middle one of them a live m. at all. However, the as-good-as-dead m. of jam is capable of being stimulated into life if anyone is so foolish as to bring into contact with it another half-dead m. of its own (i.e. the foodstuff) kind; and it was after all mixing metaphors to say the jam was a slice of pie. But then the way of escape was to withdraw either the jam or the pie, instead of forcing them together down our throats with a ramrod of apology. / Time sifts the richest granary, and posterity is a dainty feeder. But Lyall's words, at any rate—to mix the metaphor—will escape the blue pencil even of such drastic editors as they. Since all three metaphors are live ones, and they are the sifter and the feeder, the working of these into grammatical connexion with the blue pencil does undoubtedly mix metaphors. But then our author gives us to understand that he knows he is doing it, and surely that is enough. Even so some liars reckon that a lie is no disgrace provided that they wink at a bystander as they tell it; even so those who are addicted to the phrase 'to use a vulgarism' expect to achieve the feat of being at once vulgar and superior to vulgarity.(See SUPERIORITY.) Certainly we cannot detect the suggested lack of warmth in the speech as it is printed, for in his speech, as in the Prime Minister's, it seems to us that (if we may change the metaphor) exactly the right note was struck. / It is essential, then, that the Labour Party should go into its election campaign with the engine running. And how better to get the engine running than to harry and snap (I say, I am mixing my metaphors today: I hope you don't mind) at the Government? Certainly, gentlemen, you may change or mix your metaphors, if it seems good to you; but you may also be pretty sure that, if you feel the necessity of proclaiming the fact, you had better have abstained from it. / Two of the trump cards played against *the Bill are* (1) *that 'it makes every woman who pays a tax-collector in her own house', and* (2) *that 'it will destroy happy domestic relations in hundreds of thousands of homes'; if we may at once change our metaphor, these are the notes which are most consistently struck in the stream of letters, now printed day by day for our edification in the* Mail. This writer need not have asked our leave to change from cards to music; he is within his rights, anyhow, and the odds are, indeed, that if he had not reminded us of the cards we should have forgotten them in the intervening lines. But how did a person so sensitive to change of m. fail to reflect that it is ill playing the piano in the water? A *stream of letters*, it is true, is only a picturesque way of saying many letters, and ordinarily a dead m.; but once put your seemingly dead yet picturesque m. close to a piano that is being played, and its notes wake the dead—at any rate for readers who have just had the word *m.* called to their memories.

**metaphysics** and *metaphysical* are so often used as quasi-learned and vaguely depreciatory substitutes for various other terms, for theory and theoretical, subtle(ty), (the) supernatural, occult(ism), obscure and obscurity, philosophy and philosophic, academic(s), and so forth, that it is pardonable to forget that they have a real meaning of their own, especially as the usual resource of those who suddenly realize that their notion. of a word's meaning is hazy—an appeal to its etymology—will not serve. It is agreed that *Metaphysics* owes its name to the accident that the part of Aristotle's works that treated of metaphysical questions stood after (μετά) the part concerned with physics (τὰ φυσικά), and that the word's etymology is therefore devoid of significance. It is indeed actually misleading if it suggests the inference, as it has to some, that m. is 'the science of things transcending what is physical or natural'. Even Saintsbury, for instance, though admitting some

justice in the criticism of the label 'metaphysical' invented by Dryden and adopted by Johnson for Donne, Cowley, and their school, maintained that it was 'not inappropriately used for the habit, common to this school of poets, of always seeking to express something after, something behind, the obvious first sense and suggestion of a subject'.

Metaphysics is the branch of philosophy that deals with the ultimate nature of things, or considers the questions, What is the world of things we know? (*ontology*) and, How do we know it? (*epistemology*), though some philosophers would confine the term to the first. Such being the subject of Metaphysics, it is not wonderful, in view of the infinity of theories and subtlety of arguments evoked, that it should have come by some or all of the wrong acceptations mentioned above.

**meter, metre.** The spelling of the measuring instrument (probably from *mete* to measure out) is always *meter*. RECESSIVE ACCENT has shortened the e in its compounds such as *chrono'meter*, *speedo'meter*, etc. For a discussion of the formation of some of these compounds see HYBRIDS AND MALFORMATIONS. For the unit of linear measurement and the term of prosody *meter* and *metre* are alternative spellings of the English equivalent of the Greek noun μέτρον, measure. *Meter* is the older and is preferred in U.S.; *metre*, which comes to us through French, is preferred in Britain. But both countries use the older form for the prosody compounds (*hexameter*, *pentameter*, etc.) as well as for *diameter* and *perimeter*; for the linear compounds we in Britain ordinarily write -*metre* (*centimetre*, *millimetre*, etc.), but are inclined to make an exception of *kilo-*; we probably write *kilometer* as often as *kilometre*, though not often enough for the recessive accent to have established the pronunciation *kilo'meter*, sometimes heard.

**method.** For *m. in madness*, see IRRELEVANT ALLUSION.

**meticulous.** What is the strange charm that at one time made this wicked word irresistible to the British journalist? Did he like its length? Did he pity its isolation (for it has no kindred in England)? Could a Latin scholar like him not get *meticulosus* out of his head? Could so accomplished a Frenchman never be sure whether *méticuleux* or *m.* was the word he knew so well? Or what was it? It is clear, first, that the word is not a piece of latinity that cannot be forgotten. 'Ante- and post-classical' say Lewis and Short; that is, you may read your Cicero and Virgil and Horace and Livy through and never meet it, and when it is unearthed *i* ⁿlautus or somewhere it means ι what the journalists made it mean, but just frightened. It is the word for the timid hare, or the man who is gibbering with fear (*Nullust hoc meticulosus aeque . . . Perii, pruriunt dentes*—Was ever man in such a funk? . . . Lord, how my teeth chatter!). Some centuries ago m. had that meaning, comprehensible enough through the Latin *metus* (fear) to all who have learnt any Latin, but not to others, since *metus* by some odd chance has given no common words to English. But the word died out, and when it was resuscitated in the 19th c., it was given a new sense for which it was not in the least needed, and freely used as an unwanted synonym for careful, exact, punctilious, scrupulous, precise, etc. It would be idle to try to put it back into an etymological strait-jacket and to apply it only to the care that has its origin in terror of being caught breaking rules or misstating facts, but if it is to escape the reproach of being a SUPERFLUOUS WORD it should at least be confined to a degree of care, not necessarily excessive or fussy—we have *pernickety* for that—but greater than what is implied by *punctilious* or *scrupulous*. The first of the two examples that follow illustrates the legitimate use; the second is ludicrous in that it excludes not merely the idea of great care but even that of any care at all. *Gone is the wealth of m. detail*

with which he loved to elaborate his finely finished pictures. | Mr. ——, who has succumbed to the wounds inflicted upon him ten days previously by a pet lion, had his fate foretold with m. accuracy more than 2000 years ago by the greatest Greek dramatist.

**metonymy.** Substitution of an attributive or other suggestive word for the name of the person or thing meant, as when *the Crown, Homer, wealth,* stand for the sovereign, Homer's poems, and rich people. See PERSONIFICATION.

**metope.** The word meaning part of a frieze and that meaning the front part of the face are derived from different Greek words; pronounce respectively *mĕt'ŏpĕ* and *mĕt'ōp.*

**mews,** originally a plural, but now used freely as singular with *a,* should be content to serve both purposes. We have enough HOMOPHONES already without adding such an ill-assorted pair as *mewses* and *Muses.*

**miaow, miaul.** It is better to be content with *mew* and *caterwaul* than to multiply phonetic approximations.

**miasma.** See LATIN PLURALS.

**mickle and muckle** are merely variants of the same word, meaning a large amount. The word for a small amount is *pickle,* and the not uncommon version *Many a mickle makes a muckle* is a blunder; the right forms are *Many a little* (or *Mony a pickle*) *makes a mickle* (or *muckle*), with other slight variations.

**micro-organisms** etc. M. is loosely used as a generic term for any animal or vegetable organism so small as to be invisible without magnification. *Microbe* and *germ* are also popularly used for any of the micro-organisms that are associated with disease and decomposition; so are *bacteria* and *bacilli,* but strictly these last are distinct types of the class of uni-cellular organisms known as *schizomycetes. Virus* is the name given to a specially minute organism that causes diseases in man and other animals (e.g. poliomyelitis and foot-and-mouth disease) and in plants; *rickettsia* is in many ways similar to *virus* and some species cause disease in man and other animals, e.g. typhus.

**middle (article).** Newspaper article of a kind so called from having stood between the leading articles and the book reviews, and being a short essay usually of some literary pretensions on some subject of permanent and general rather than topical or political interest. In the TLS, which preserves the term, it is in fact in the middle of the paper.

**middling(ly).** The *-ly* is unusual and undesirable: *a middling good crop*; *did middling well*; *it went only middling*. See UNIDIOMATIC -LY.

**midwifery.** *Mǐ'dwǐfrǐ* is perhaps usual, but both *mǐdwǐf'ěrǐ* and *mǐd'wǐfrǐ* are also heard; cf. HOUSEWIFE. See RECESSIVE ACCENT.

**mighty.** In the colloquialism *m. fine* etc., *mightily* should not be substituted. See UNIDIOMATIC -LY.

**milage.** See MUTE E. The rule there suggested would make this the right spelling, but *mileage* is probably more usual; some dictionaries give it only.

**millenarian,** pertaining to, or a believer in, the millennium. The apparent inconsistency in spelling (*-n-*, *-nn-*) results from the fact that the *millenarian*, like *millenary*, does not contain the stem of the Latin *annus*, which is present in *millennium*; if it were formed from *millennium*, the form would be *millenniarian*. *Millenarian* strictly means thousander, not thousand-yearer. Cf. CENTENARY.

**millenary.** The standard pronunciation is *mǐ'lǐnǎrǐ*; but see CENTENARY.

**milli-.** See KILO-.

**milliard** means a thousand millions; it is chiefly a French term, though perhaps advancing in general currency. In France it is the equivalent

in ordinary use of the mathematical French BILLION, which, like the American, differs from the English in being a thousand million, not a million million.

**million. 1.** *A m. and a quarter, two millions and a half*, rather than *one and a quarter million(s)* and *two and a half millions*; see HALF.
**2.** *Amongst the eight million are a few hundred to whom this does not apply.* Here *million* and *hundred* are better than *millions* and *hundreds*; but *He died worth three millions* rather than *million*; this because 'a million' is an established noun (as distinguished from a mere numeral) in the sense £1,000,000, but not in the sense a million people.
**3.** *Forty-five million people* rather than *forty-five millions of people* (on the analogy of *dozen, score, hundred, and thousand*); but, with *a few* and *many, millions of* is perhaps the more usual form.

**minacious, minatory.** Both words smack of pedantry; but while the first is serviceable only for POLYSYLLABIC HUMOUR, the second is not out of place in a formally rhetorical context.

**mine.** For (*my* or) *mine and your future depends upon it* etc., see ABSOLUTE POSSESSIVES.

**mineralogy** is a syncopated form (the syncopation done in French) for *mineralology*, and should not be quoted in defence of proposed wrong forms in -alogy. See -LOGY.

**minify, minimize, diminish.** *Minify* is a badly formed and little used word. It owes its existence to the desire for a neat opposite to the correctly formed *magnify*, but is now chiefly used by people who, rightly enough offended by the extension of *minimize* to improper meanings, are too ready to catch at the first alternative. A slight further search would bring them through *minish* (to which the only—but fatal—objection is that it is archaic) to *diminish*.

*Minimize* is both a rightly formed and a current word, but unfortunately current in more senses than it has any right to. It should have been kept strictly to the limits imposed by its derivation from *minimus* (not less or little, but least), and therefore should always have meant either to reduce to the least possible amount (*We must minimize the friction*) or to put at the lowest possible estimate (*It is your interest to minimize his guilt*). The meaning 'lessen' given to *minimize* in the following quotations flagrantly ignores the essential superlative element by qualifying it adverbially: *The utility of our convoy would have been* considerably minimized *had it not included one of these.* | *An open window or door would* greatly minimize *risk.*

*Minify* should be given up as a SUPER-FLUOUS WORD; *minimize* should be kept as near as may be to its proper senses; *magnify* should have as its opposite, in one of its senses *diminish* (*the diminishing end of the telescope*), and in another *underestimate* (*neither magnify nor underestimate the difficulties*). See also BELITTLE.

**minimal.** See MARGINAL, MINIMAL.

**minimum.** Pl. -*ma*. See -UM.

**minister.** The tendency to apply the word, in the sense m. of religion, to non-Anglicans and to avoid applying it to Anglicans is noteworthy, seeing that *m.* is common in the Prayer-Book rubrics. It is explained by historical circumstances; m. was adopted as an acceptable name 'at first chiefly by those who objected to the terms *priest* and *clergyman* as implying erroneous views of the nature of the sacred office'—OED.

**minor** (in logic). See SYLLOGISM.

**minority** is like MAJORITY, only more so, in that odd tricks can be played with its meanings. Corresponding to the A, B, and C, of *majority*, m. has, A, inferiority of number or fewerness or pauciority, B, a party having that quality, and C, less than half of any set of people. 'More so', because, if one presses one's rights, one may say

that a small m. (sense B) is in a considerable m. (sense A) or is the vast m. (sense C), both of which statements happen to sound absurd. Again, in a Board of 51 *a m. of one* may be either 25 persons (A) or one person (B). The point need not be laboured, but should be appreciated. There is a tacit convention, in the interests of lucidity, that adjectives naturally appropriate to magnitude shall not be used with *m.* to emphasize smallness of number, and another that *a m. of one* shall always mean one person. But the first is not always kept to: *With* a considerable m. *of the votes polled, the Tory Party have obtained a clear and substantial majority over all other parties in the House.* Oddly enough, the newspaper whose own words those are has this paragraph about a fellow offender: *Says a motoring writer in a Sunday paper; 'It is time that the interests of the public at large were considered by attacking the real evil—the dangerous and inconsiderate driver. Fortunately, he constitutes the* vast minority *of motor-car owners and drivers'. We know what is meant, but 'the vast minority' is a very unfortunate way of saying it.* In the first passage *m.* is used in sense A, and in the second in sense C; but the convention is applicable to both or neither.

**Minotaur.** Generally pronounced *mĭn-*, though the *i* is long in Greek and Latin; but see FALSE QUANTITY.

**miocene.** A typical example of the monstrosities with which scientific men in want of a label for something, and indifferent to all beyond their own province, defile the language. The elements of the word are Greek, but not the way they are put together, nor the meaning demanded of the compound. See HYBRIDS AND MALFORMATIONS.

**mis-.** The OED hyphens *mis-* when prefixed to a word beginning *s*, e.g. *mis-spelt.* But most modern dictionaries write these as one word, and this is recommended. There is no more need for a hyphen in a *mis*-compound than in a *dis*- one, and these are always written as one word (*dissatisfied*, *disservice*, etc.). It might perhaps be argued that since *miss*, unlike *diss*, is a familiar word in its own right, a hyphen is needed to avoid a FALSE SCENT; but the half-dozen or so *mis-s* words in current use are too familiar to call for this precaution. See HYPHENS. See also -S-, -SS-, -SSS- 2.

**misalliance**, though formed after the French *mésalliance*, is so natural an English word that it is free of the taint of gallicism, and should always be preferred to the French spelling.

**misapprehensions** of which many writers need to disabuse themselves. Discussions will be found under the words printed in small capitals.

That a DEVIL'S advocate, or *advocatus diaboli*, is a tempter of the good, or whitewasher of the bad, or the like.

That a PERCENTAGE is a small part.

That a LEADING QUESTION is a searching one.

That CUI BONO? means What is the good or use?

That *One touch of* NATURE *makes the whole world kin* means much the same as *A fellow-feeling makes one wondrous kind*.

That POLITY is a scholarly word for *policy*.

That *more* HONOURE*d in the breach than the observance* means more often broken than kept.

That King Canute thought he could stop the tide from flowing.

That *madding* in Gray's *Far from the m. crowd* means 'distracting'. (EN VERBS s.v. *mad*.)

That LEATHER and *prunella* are both shoddy material for clothes.

That the natural manifestations of an inferiority COMPLEX are shyness and diffidence.

That many a MICKLE *makes a muckle*.

That a *dead* RECKONING means a reckoning that is dead right.

That ILK means clan or the like.

That *arithmetical*, and *geometrical*, PROGRESSION necessarily mean fast, and very fast, progress.

That *the* COMITY *of nations* means the members of a sort of league.

That PROPORTION is a sonorous improvement on *part*.

That SUBSTITUTE is an improvement on *replace* in the sense take the place of.

That PROTAGONIST is an improvement on *champion* and *leader*.

That an EXCEPTION strengthens a rule.

That FRANKENSTEIN was a monster.

That a PRESCRIPTIVE RIGHT is an indefeasible right.

That to beg a question is to avoid giving a straight answer to one (PETITIO PRINCIPII).

That any sort of a defence can properly be called an ALIBI.

That FRUITION means becoming fruitful.

That INFER means imply.

That COMPRISE means compose.

**miscellany.** Pronounce *mĭ'sĕlanĭ* or *mĭsĕ'lănĭ*; the OED puts the former first, and RECESSIVE ACCENT is in its favour.

**miserere, misericord,** hinged seat in choir-stall. The first is labelled 'an incorrect form' in the OED.

**misogynist.** *-jinist* is the usual pronunciation. See GREEK G.

**misprints to be guarded against.** *adverse* and *averse*, *casual* and *causal*, *casualty* and *causality*, *deprecate* and *depreciate*, *inculcate* and *inoculate*, *interpellate* and *interpolate*, *justiciable* and *justifiable*, *personality* and *personalty*, *principal* and *principle*, *recourse* and *resource* and *resort*, *reality* and *realty*, *risible* and *visible*, *-tion* and *-tive* (e.g. *a corrective* and *a correction*), *uninformed* and *uniformed* are common confusions worth providing against by care in writing and vigilance in proof-correcting. *Concensus* (non-existent) often appears instead of the real word *consensus*, and to *signal out* (non-existent in the sense meant) instead of to *single out*. See also PAIRS AND SNARES.

**misquotation.** The correct words of a few familiar sayings that are more often wrongly than rightly quoted may be useful. The misquoting of phrases that have survived on their own merits out of little-read authors (e.g. of *Fine by degrees* etc. from Prior, usually changed to *Small* etc.) is a very venial offence; and indeed it is almost a pedantry to use the true form instead of so established a wrong one; it would be absurd to demand that no one should ever use a trite quotation without testing its verbal accuracy. Again, the established change made in the *Leave-not-a-rack-behind* quotation by shifting *the baseless fabric of this vision* from some lines earlier into the place of another phrase that does not suit general use so well, and the common telescoping of *Pride goeth before destruction and an haughty spirit before a fall*, though most people no doubt make them without knowing what they are doing, might reasonably enough be made knowingly, and are no offence. But when a quotation comes from such a source as a well-known play of Shakespeare, or *Lycidas*, or the Bible or Prayer Book, to give it wrongly at least requires excuse, and any great prevalence of such misquotation would prove us discreditably ignorant of our own literature. Nevertheless, such words as *A poor thing, but my own*, are often so much more used than the true form that their accuracy is sure to be taken for granted unless occasional attempts like the present are made to draw attention to them.

Water water everywhere *nor any* drop to drink (not *and not a*).

And *whispering I will* ne'er consent consented (not *vowing she would*).

They kept the *noiseless* tenor of their way (not *even*).

A little *learning* is a dangerous thing (not *knowledge*).

In the sweat of thy *face* shalt thou eat bread (not *brow*).

To gild refined gold, to *paint* the lily (not *gild the lily*).

Screw your courage to the sticking-*place* (not *point*).

I will a *round* unvarnished tale *deliver* (not *plain . . . relate*).

An *ill-favoured* thing, sir; but mine own (not *poor*).

Let not him that *girdeth* on his harness boast himself as he that putteth it off (not *putteth*).

*That* last infirmity of noble *mind* (not *the . . . minds*).

Make assurance *double* sure (not *doubly*).

Tomorrow to fresh *woods* and pastures new (not *fields*).

The devil can *cite* Scripture for his purpose (not *quote*).

A goodly apple rotten at the *heart* (not *core*).

A fellow-feeling makes *one* wondrous kind (not *us*).

Chewing the *food* of sweet and bitter fancy (not *cud*).

I am escaped *with* the skin of my teeth (not *by*).

Passing rich *with* forty pounds a year (not *on*).

*He that complies* against his will Is of *his own* opinion still (not *who consents . . . the same*).

*Fine* by degrees and beautifully less (not *small*).

When *Greeks joined Greeks*, then *was* the tug of war (not *Greek meets Greek . . . comes*).

**Miss.** *The Misses Jones* etc. is the old-fashioned plural, still used when formality is required, e.g. in printed lists of guests present etc.; elsewhere *the Miss Joneses* is now usual.

**missile.** Usually pronounced -*ĭl* in Britain; *ĭl* in U.S. See -ILE.

**mitigate.** *M. against* for *militate against* is a curiously common MALAPROPISM.

**mixed metaphor.** See METAPHOR.

**-m-, -mm-.** Monosyllables ending in m double it before suffixes beginning with a vowel if the *m* is preceded by a short vowel, but not if it is preceded by a long one or a vowel and r: *hammy, gemmed, dimmest, drummer*; but *claimant, gloomy, worming*. Words of more than one syllable follow the rule for monosyllables if their last syllable is a word in composition, as *bedimmed, overcramming,* but otherwise they do not double the m: *bedlamite, emblematic, pilgrimage, victimize, venomous, unbosomed, blossoming, bottomed,* except that words in -gram do double it (compare *epigrammatic, diagrammatic,* with *systematic*).

**mocha,** coffee or precious stone. Pronounce *mō′ka.*

**model** makes *-lled, -lling,* etc.; see -LL-, -L-. The popular use of this verb in the sense of to serve as what used to be called a mannequin seems to date from mid-20th c.

**modus vivendi** (literally way of living) is any temporary arrangement that enables parties to carry on pending settlement of a dispute that would otherwise paralyse their activities. Pronounce *mōdus vīvĕndī.* See LATIN PHRASES.

**Mohammed(an).** See MAHOMET.

**moiety,** apart from uses as a legal term and as a FORMAL WORD, exists merely for the delight of the ELEGANT-VARIATIONist in such triumphs as: *The Unionist candidate was returned by exactly half the number of votes polled, the other m. being divided between a Labour and an Independent opponemt.*

**moire, moiré.** *Moire,* or *moire antique,* is the name of the watered silk material; *moiré* is first an adjective meaning watered like moire (often of metal surfaces), and secondly a noun meaning watered surface or effect. *A moire dress; velvets and moire antiques; a moiré surface; the moiré has been improved by using the blowpipe.*

**molecule.** See ATOM.

**molten.** See MELTED.

**momentarily, momently.** The first means for a moment (*he was momentarily abashed*), the second from moment to moment or every moment (*am momently expecting a call from him*). The differentiation is well worth more faithful observance than

it gets; and the substitution of either, which sometimes occurs, for INSTANTLY or *immediately* or *at once* is foolish NOVELTY-HUNTING.

**monachal, monastic, monkish.** Each has its own abstract noun— *monachism, monasticism, monkery.* Of the three sets *monastic(ism)* is the one that suits all contexts; it is useful that *monkish* and *monkery* should also exist, as serving the purpose of those who wish to adopt a certain tone. *Monachal* and *monachism,* though they would have passed well enough if *monastic(ism)* did not exist and were not much better known, seem as it is to have no recommendation unless it is a good thing that scholars writing for scholars should have other names for things than those generally current, even though the meaning is the same. If that is, however, a bad thing, *monachal* and *monachism* should be allowed to die.

**monadism, monism.** Both terms owe their existence to the META-PHYSICAL problem of the relation between mind and matter. The view that regards mind and matter as two independent constituents of which the universe is composed is called *dualism.* In contrast with dualism, any view that makes the universe consist of mind with matter as a form of mind, or of matter with mind as a form of matter, or of a substance that in every part of it is neither mind nor matter but both, is called *monism. Monadism* (Leibniz called in *monadology*) is the name given to a particular form of monism, corresponding to the molecular or atomic theory of matter (see ATOM), and holding that the universal substance (according to the third variety of monism described in the preceding sentence) consists of units called *monads.*

**monarchical, -chic, -chal, -chial.** The first is the current form; *-chic* and *-chal* are occasionally used, the first for antithetic purposes (*the monarchic, the aristocratic, and the democratic branches of our constitution*); the second with a slight rhetorical difference, where *kingly* might serve (*the royal harangue has a certain monarchal tone*); -*ial* seems superfluous. In all *ch* is pronounced *k*.

**monastic(ism).** See MONACHAL.

**mondial.** See GLOBAL.

**-monger.** See WARMONGER.

**monism.** See MONADISM.

**monk.** For *m.* and *friar*, see FRIAR.

**monocle.** That this, a HYBRID, a GALLICISM, and a word with no obvious meaning to the Englishman who hears it for the first time, should have ousted the entirely satisfactory *eyeglass* is a melancholy illustration of the popular taste in language. See also QUIZ.

**monologue.** This and *soliloquy* are precisely parallel terms of Greek and Latin origin respectively; but usage tends to restrict *soliloquy* to talking to oneself or thinking aloud without consciousness of an audience whether one is in fact overheard or not, while monologue, though not restricted to a single person's discourse that *is* meant to be heard, has that sense much more often than not, and is especially used of a talker who monopolizes conversation, or of a dramatic performance or recitation in which there is one actor only.

**monosyllabic.** *His m. answer was 'Nonsense'.* This must have been a remarkable feat of articulation. *M.* bears its precise meaning too plainly on its face to be a suitable subject for an experiment in SLIPSHOD EXTENSION.

**monotonic, -nous.** The secondary sense of *monotonous* (same or tedious) has so nearly swallowed up its primary (of one pitch or tone) that it is well worth while to remember the existence of *monotonic*, which has the primary sense only. In -*ic* the accent is on the third syllable, in -*ous* on the second.

**mood.** It may save misconceptions to mention that the grammar word has nothing to do with the native word meaning frame of mind etc.; it is

merely a variant of *mode*, i.e. any one of the groups of forms in the conjugation of a verb that serve to show the mode or manner by which the action denoted by the verb is represented— *indicative, imperative,* and *subjunctive.*

**moral,** adj. **1.** For distinctions between *m.* and *ethical*, *morals* and *ethics*, see ETHICAL. **2.** *M. victory, m. certainty.* The first is often applied to an event that is from another point of view a defeat; the second is always applied to what is in fact an uncertainty. It is so easy to see why *m. victory* should mean what it does, and so hard to see why *m. certainty* should, that anyone considering the point by the mere light of nature is tempted to guess that *m. certainty* is the illegitimate offspring of *m. victory*, and perhaps to abstain from using it as a solecism. The OED quotations show that, on the contrary, it is much the older of the two phrases; and though this peculiar sense of PRACTICAL or *virtual* in combination with *certainty* is hard to account for, it is established as idiomatic.

**moral(e),** n. During the first world war, when the quality denoted by this word was naturally much talked about, there was no little confusion about how the word should be spelt. The following comments were written at that time, and the spelling *morale*, advocated in them, has now prevailed.

The case for the spelling *moral* is that (1) the French use the word *moral* for what we used to call morale, and therefore we ought to do the same; and that (2) the French use *morale* to mean something different from what we mean by it.

The case against *moral* is (1) that it is a new word, less comprehensible to ordinary people, even after its wartime currency, than the old *morale*; (2) that it must always be dressed in italics owing to the occasional danger of confusion with the English word *moral*, and that such artificial precautions are never kept up; (3) that half of us do not know whether to call it *mŏ′răl, mŏră′l,* or *mŏrah′l,* and that it is a recognized English custom to resolve such doubts by the addition of -e or some other change of spelling.

The view here taken is that the case for *moral* is extraordinarily weak, and the case against it decidedly strong, and in fact that the question is simply one between true pedantry and true English. A few remarks may be made on the points already summarized.

Here are two extracts from bookreviews in *The Times*: *He persistently spells* moral (*state of mind of the troops, not their morality*) *with a final* e, *a sign of ignorance of French.* | *The purist in language might quarrel with Mr.* ——*'s title for this book on the psychology of war, for he means by* morale *not 'ethics' or 'moral philosophy', but 'the temper of a people expressing itself in action'. But no doubt there is authority for the perversion of the French word.* Is it either ignorance of French or a perversion of the French word? Or would a truer account of the matter be that we have never had anything to do with the French word *morale* (ethics, morality, a moral, etc.), but that we found the French word *moral* (state of discipline and spirit in armies and the like) suited to our needs, and put an -e on to it to keep its sound distinct from that of our own word *moral*, just as we have done with the French word *local* (English *locale*) and the German *Choral* (English *chorale*), and as, using contrary means for the same end of fixing a sound, we have turned French *diplomate* into English *diplomat*? Our English *forte* (*geniality is not his forte*, etc.) is altered from the French *fort* without even the advantage of either keeping the French sound or distinguishing the spoken word from our *fort*; but who proposes to sacrifice the reader's convenience by correcting its 'ignorant' spelling?

The right course is to make the English word *morale*, use ordinary type, call it *morah'l*, and ignore or abstain from the French word *morale*, of which we have no need. See for other examples of pedantry with French words, À L'OUTRANCE and DOUBLE ENTENDRE; cf. also GUERILLA.

**more. 1.** For limitations on the use of *the more*, see THE 6, 7. **2.** For the common confusion between *much m.* and *much less*, see MUCH 2. **3.** *M. than one*, though its sense is necessarily plural, is treated as a sort of compound of *one*, and follows its construction; it agrees with a singular noun and takes a singular verb: *m. t. o. workman was killed, m. t. o. was killed*, not *workmen* or *were*. **4.** For *m. and m. than* see THAN 7. **5.** For *m. in sorrow than in anger*, see HACKNEYED PHRASES. **6.** *The new dock scheme affects the whole of the northern bank of the Thames in a more or less degree.* This is wrong because, though *a less degree* is English, *a m. degree* is not; and the reason for that again is that while LESS still preserves to a certain extent its true adjectival use (= smaller) as well as its quasi-adjectival use (= a smaller amount of), the former use of *m.* (= larger) has long been obsolete, and it retains only the latter sense, namely a larger amount of. *Less butter, less courage, a less degree*, and even *a less price*, are possible; but not *a m. degree* or *a m. price*, only *m. butter* or *courage. The m. part*, and *More's the pity*, are mentioned by the OED as survivals of the otherwise obsolete sense and the former seems since to have become extinct. **7.** Writers trying to tell their readers by how much something is m., or less, than something else occasionally puzzle them by departing from ordinary arithmetical terminology. *Our sales during the first six months of this year have been three times m. than in the corresponding period last year.* If the writer intended the literal meaning of his words—that the sales had quadrupled—he chose a clumsy way of putting it. If he meant, as he probably did, that they had trebled, he should have said *three times*, or *three times as much as*, those in the previous period. Or again, *Domesday gross incomes must be multiplied fifty times to give the 1938 equivalent*, and *The degree of radiation would be thirty times less than that fixed by security measures.* Do these mean that the incomes must be multiplied by fifty and that the radiation is one-thirtieth of the permissible? If so, why not say so, and make all plain? If we are to accept this new idiom, what are we to call less than a thirtieth? More than thirty times less or less than thirty times less?

*Cinema attendances were m. than halved this year* may perhaps pass; it is unambiguous but not the happiest way of saying that the attendances were less than half. The same indulgence cannot be given to the advertisement *Architect will, in spare time, survey, plan, and execute all work for m. than moderate fees.* But the advertiser may have been arming himself with a defence against clients aggrieved by his charges.

For other misuses of mathematical terms see N (*to the nth*), HALF 3, and PROGRESSION.

**morning.** *M. Service, M. Prayer, Matins.* The first is perhaps the usual unofficial term; the other two are official, and the last is especially in High-Church and musical use. Similarly *Evening Service, Evening Prayer, Evensong.* See also MATINS.

**morphia, morphine.** The meaning is the same, the second being the scientific term, but the first surviving in ordinary use.

**mortal.** For *all that was m. of*, and *the m. remains of*, see HACKNEYED PHRASES, AND STOCK PATHOS

**mortgagee, -ger, -gor. 1.** As the word *mortgager* is one that could be formed at will from the verb *mortgage* even if it were not recorded (as it is), the maintenance of the form -*gor*, pronounced -*jor*, seems anomalous; the only other English words in which *g* is soft before *a* or *o* or *u* are perhaps GAOL and its derivatives, and the debatable MARGARINE. But lawyers always like, if they can, to call the person doing something -*or* and the person he does it to -*ee*. **2.** The mortgagee is the person who lends money on the security of an estate, the mortgagor

the person who pledges his property in order to get the loan. But, as the owner of a mortgaged estate is often himself described as 'mortgaged up to the eyes' etc., and as -*ee* suggests the passive, and -*or* the active party, those who are not familiar with the terms are apt to have the meanings reversed in their minds.

**mortician** as a euphemism for *undertaker* is a U.S. GENTEELISM. Our equivalent is *funeral furnisher* or *director.*

**mortise, -ice.** The first is better. In *m. and tenon*, the m. is the receiving cavity.

**moslem, muslim.** The OED treats the first as the ordinary English form, but *muslim* has since gained on it. *M.* can be used as adjective or as noun, and the plural of the noun is preferably -*ms*, but sometimes the same as the singular; the use of the plural *moslemin* or *muslimin* is sheer DIDACTICISM. The proper pronunciation is *mŭslim*, like *muslin*; the *mooslim* we often hear is as absurd as the similar mispronunciation of TRUFFLE.

**mosquito.** Pl. -*oes* is more usual; see -O(E)S 1.

**most(ly). 1.** *The internecine conflict has largely killed sentiment for any of the factions, and* the Powers mostly concerned *have simply looked on with a determination to localize the fighting.* The only idiomatic sense of *mostly* is for the most part (*The goods are mostly sent abroad.* / *Twenty-seven millions, mostly fools*). But it is often wrongly used for *most*, as in the quotation; see UNIDIOMATIC -LY.

**2.** As a curtailed form of *almost*, *most* is a common American colloquialism, especially for qualifying comprehensive adverbs and pronouns such as *always, everyone, anything*, but is rarely heard in Britain outside rustic speech.

**mot.** *The mot juste* is an expression which readers would like to buy of writers who use it, as one buys one's neighbour's bantam cock for the sake of hearing its voice no more. It took a long time to get into the dictionaries, French or English (even in 1933 the SOED had nothing to say about it), and those who wanted to know more had to be content to associate it vaguely with Flaubert. The COD has since defined it as 'the expression that conveys a desired shade of meaning with more precision than any other'. If that is so the *m. j.* seems a trifle long in: *The epitaph which she wrote for herself at an early age contains the* mot juste: *'Here lies Sylvia Scarlett, who was always running away. If she has to live all over again and be the same girl, she accepts no responsibility for anything that may occur'.*

**moth.** The collective use of *m.* in the senses of moths or the m. or the ravages of moths (*furs harbour m.*; *m. is the most destructive of these*; *proof against m.*; *to prevent m.*) is neither defined nor illustrated in the OED, but is now common. The well-known Bible passage, however, on which this use is perhaps based (*where moth and rust doth corrupt* etc.), cannot in fact be quoted in defence of it, since in it one may suppose the rhetorical omission of the article that is common enough in paired or contrasted phrases (*eye hath not seen, nor ear heard*), which has no resemblance to the examples of *m.* as a collective given above.

**mother.** For *the M. of Parliaments,* see SOBRIQUETS.

**mother-in-law.** See -IN-LAW.

**mother-of-pearl, -o'-pearl.** The dictionaries favour the *of* form; the other gives what used to be the ordinary pronunciation but is disappearing with the general tendency to make pronunciation conform to spelling.

**motivat(e)(ion).** These importations from U.S. are not without their uses, but under the influence of LOVE OF THE LONG WORD they have become too popular. *What motivated your action?* is not a good

way of saying '*What made you do it?*' The latest duty assigned to *motivation* is to serve, not very suitably, as a slogan word for an industrial technique that aims at instilling into workers a consciousness of the identity of their interests with those of their employers.

**motive.** *The victorious party has every m. in claiming that it is acting not against the Constitution, but in its defence.* We may say *an* or *every interest* in *doing*, but it must be *a* or *every m.* for *doing.* See ANALOGY, and CAST-IRON IDIOM.

**motorcade.** See CAVALCADE.

**motto.** For synonymy, see SIGN. Pl. *-oes*, see -O(E)S I; adj. *motto'd,* see -ED AND 'D.

**moujik, muzhik.** Pronounce *mōō'-zhĭk.* The first is the established form, and correction to the second does no one any good and perplexes those who have just come to know what the old word means; see DIDACTICISM.

**mould.** The three common words so spelt (shape n. and v.; earth; fungous growth) are probably all unconnected; but the identity of form has no doubt caused the second to be tinged with the meaning of the third, and the original notion of powdery earth has had associated with it the extraneous one of rottenness. See TRUE AND FALSE ETYMOLOGY.

**movies.** Americans go to the *m.* English people, after a half-hearted experiment with *flicks*, now go either to the *pictures* or to the *cinema.*

**mow.** 1. The noun meaning heap of hay or corn, and the noun or verb meaning grimace (as in *mop and mow*). Despite the dictionaries, which prefer the pronunciation rhyming with *how*, most people who use these words probably pronounce them *mō*, like the verb meaning to cut grass. 2. The p.p. of this verb, when used as an adjective, should be *mown* (*the mown,* not *mowed, grass*; *new-mown* etc.); when it is verbal, both forms are current (*the lawn was mown*, or *mowed, yesterday*).

**M.P.** Four forms are wanted: ordinary singular, ordinary plural, possessive singular, and possessive plural. They are easily supplied : M.P. (*He is a(n)* M.P.; M.P.s (*M.P.s now travel free*); M.P.'s (*What is your M.P.'s name?*); M.P.s' (*What about* an increase in *M.P.s' salaries?*). The following newspaper extract contains two of the parts, but represents them both by the same form, and that one belonging to another:

M.P.'S *PIGEON RACE*

*A pigeon race, organized by* M.P.'s, *took place on Saturday.* Read M.P.S' for the first and M.P.s for the second.

**Mr, Mrs.** See PERIOD IN ABBREVIATIONS for the question whether *Mr* and *Mrs* or *Mr.* and *Mrs.* are better.

**much. 1.** For the use of *m.* rather than *very* with participles (*m. pleased* etc.), see VERY.

**2.** *M. more* and *m. less.* The adverbs *more*, and *less*, are used in combination with *m.* or *still* to convey that a statement that is being or has been made about something already mentioned applies more forcibly yet to the thing still to be mentioned: *The abbreviating, m. more the garbling, of documents does great harm.* | *Garbling was not permitted, m. less encouraged.* The choice between *more* and *less* is under some circumstances a matter of difficulty even for those who are willing to be at the pains of avoiding illogicality, and a trap for those who are not.

With sentences that are affirmative both in effect and in expression it is plain sailing; *m. more* is invariable. With sentences that are negative in expression as well as in effect there is as little doubt; *m. less* is invariable: *I did not even see him, m. less shake hands with him.* It is when the effect is negative, but the expression affirmative, even if only technically affirmative, that doubts arise. The meaning of

'technically', and the distinction between 'effect' and 'expression' must be made clear. *It will be a year before it is done*; the effect of that is negative, since it means that the thing will *not* be finished in less than twelve months; but its expression is affirmative, there being no negative word in it. *It is not possible to do it under a year*; the effect and the expression of that are obviously both negative. *It is impossible to do it under a year*; the effect of that is negative, but the expression is technically affirmative. Though there is no difference in meaning between the last two, the difference of expression decides between *more* and *less*: *It is not possible to do it under a year*, m. less *in six months*; *It is impossible to do it under a year*, m. more *in six months*. What governs the decision is the words required to fill up the ellipsis: *It is not possible to do it under a year*, much ——? (*is it possible to do it*) *in six months*; *It is impossible to do it under a year*, much ——? (*is it impossible*) *to do it in six months*.

Careless writers make the mistake of letting the general effect run away with them instead of considering the expression. In the example that has just been worked out the fault is a slight one, because the wrong filling up of the ellipsis (*is it possible* instead of *is it impossible*) is so easy as to seem no less natural to the reader than to the writer. In less simple examples the fault is much more glaring. In all the following quotations *more* should have been written instead of *less*.

*It is a full day's work even to open*, m. less *to acknowledge, all the presents, the letters, and the telegrams, which arrive on these occasions.* The (concealed) negative effect is: *You could not open them under a day*; but the expression is actually, not merely technically, affirmative, and the words to be supplied are *is it a full day's work.* | *The machine must be crushed before any real reforms can be initiated*, m. less *carried.* Negative effect: *You cannot initiate till the machine is crushed.* Expression, fully affirmative. | *But of real inven-*

*tion and spontaneity*, m. less *anything approaching what might be classed as inspiration, there is little enough.* Expression technically affirmative. | *It would be impossible for any ruler in these circumstances*, m. less *a ruler who was convinced of his own infallibility, to guide the destinies of an empire.* Supply *would it be impossible for* before choosing between *more* and *less*. | *I confess myself altogether unable to formulate such a principle*, m. less *to prove it*. Supply *unable*.

*M. less*, where m. *more* is required, is in fact so common that it must be classed among the STURDY INDEFENSIBLES. For similar confusion between *still less* and *still more* see LESS 2.

**muchly.** See UNIDIOMATIC -LY.

**mucous, -cus.** The first is the adjective (*m. membrane*), the second the noun.

**Muhammad(an).** See MAHOMET.

**mulatto** (pl. *-os*) and other words of race mixture.

1. *M., half-breed, half-caste, Eurasian*, all denote individuals of mixed race, but each has a more special application from or to which it has been widened or narrowed. These are: *m.*, white and negro; *half-breed*, American-Indian and white or negro; *half-caste*, European and Indian; *Eurasian*, European and Asian.

2. *M., quadroon, octoroon.* The first is the offspring of a white and a negro; the second that of a white and a mulatto, having a quarter negro blood; the third that of a white and a quadroon, having an eighth negro (etc.) blood.

3. *Creole* does not imply mixture of race, but denotes a person either of European or (now rarely) of negro descent born and naturalized in certain West-Indian and American countries.

4. *Half-caste, Eurasian, Anglo-Indian*, are all sometimes used of persons whose descent is partly British or other European and partly Indian. That is

the proper sense of *half-caste* and *Eurasian*, the latter being a polite substitute for the former. *Anglo-Indian*, again, would properly mean a half-caste, and is now sometimes preferred in that sense to *Eurasian* as a further step in politeness; but its traditional meaning is an Englishman who has spent most of his life in India.

*Coloured persons* is the term applied in South Africa to those of mixed white and native blood.

**multitude.** For number with nouns of multitude see NUMBER 6.

**mumps.** Usually treated as singular; see PLURAL ANOMALIES.

**must** as a noun. Apart from its occasional appearance in obvious applications (the OED quotes from Dekker, *M. is for kings and obedience for underlings*), the use of *m.* as a noun in the sense of compelling obligation, necessity, *sine qua non*, is modern. It has become a common colloquialism (*Presentation at Court used to be a m. for debutantes.* | *Another m. for viewers is tonight's play*) and may be found in serious writing. The OED quotes *The absolute m. of duty and right* (1885); more recent examples, which show that the usage may still be felt to need probationary inverted commas, are: *Professional activity, then, is characterized by necessity, be it the 'must' of nature and society or the 'must' and 'ought' of ambition and conscience.* | *The Englishman cannot find time to read the hundreds of his own countrymen whose writings are a 'must'.* | *The Chairman of the National Coal Board said that it was a 'must' for the coal mining industry to get into the black this year.* There are indeed signs that *must* in this sense is an incipient VOGUE WORD.

**must, need.** The following questions with their positive and negative answers illustrate a point of idiom— *Must it be so? Yes, it must; No, it need not.* | *Need I do it? No, you need not; Yes, you must.* For *needs must* see NEED.

**mustachio.** Pl. *-os*, see -O(E)S 4. *M.* is now archaic for *moustache*. The adjective derived from it, *mustachio'd* (see -ED and 'D) survived for a time, but its facetious flavour grew stale; now it too is rarely used, and we say more simply and more sensibly *moustached* or *with a moustache*.

**muster.** *Dental treatment was also kept very prominently before their consideration, so that, at the time of the Armistice, the general condition of these women's mouths would pass a very fair m. M.* in the phrase *pass m.* means an inspection; and to pass an inspection very fairly is quite a different thing from passing a very fair inspection. *Pass m.* is one of the many idioms that must be taken as they are or left alone.

**mute e.** Needless uncertainty prevails about the spelling of inflexions and derivatives formed from words ending in mute e. Is this -e to be retained, or omitted? It is a question that arises in thousands of words, and especially in many that are not separately recorded in the dictionaries, so that the timid speller cannot get it answered in a hurry. It is also one to which different answers are possible; every dictionary-maker probably thinks that if he were recording all words with an internal-mute-e problem he would answer the question with paternal but arbitrary wisdom for each word; but he also knows that it would be absurd for him to attempt to give even all those that are likely to be wanted. The need is not for such a gigantic undertaking, but for a rule of the simplest kind and with the fewest exceptions, to deliver us from the present chaos: the dictionary-makers are not agreed about the spelling even of such common words as *blam(e)able*, *lik(e)able*, *sal(e)able*, *siz(e)able*, and *tam(e)able*, and in the rulings even of a single authority may be found such seemingly arbitrary distinctions as *lateish* but *whitish*, *ageing* but *icing*.

To get an idea of the number of words concerned, the reader should consider the following questions, and realize that some of the items stand for thousands, some for hundreds, and some for dozens, of similar cases. Does pale make *paleish* or *palish*; love, *loveing* or *loving*, *loveable* or *lovable*; strive, *striveing* or *striving*; excite, *exciteable* or *excitable*; move, *moveable* or *movable*; like, *likely* or *likly*; dote, *doteard* or *dotard*; judge, *judgement* or *judgment*; hinge, *hingeing* or *hinging*; singe, *singeing* or *singing*; gauge, *gaugeable* or *gaugable*; notice, *noticeable* or *noticable*; mouse, *mousey* or *mousy*; change, *changeing* or *changing*, *changeling* or *changling*; hie, *hieing* or *hiing*; glue, *gluey* or *gluy*; due, *duely* or *duly*; blue, *blueish* or *bluish*; whole, *wholely* or *wholly*?

The only satisfactory rule, exceptions to which need be very few, would be this: If the suffix begins with a consonant, the mute e should be retained; if the suffix begins with a vowel, the mute e should be dropped. Applying this to the list above, we get (with the wrong results in italics as a basis for exceptions); palish; loving; lovable; striving; excitable; movable; likely; dotard; judgement; hinging; *singing*; *gaugable*; *noticable*; mousy; changing; changeling; *hiing*; *gluy*; *duely*; bluish; *wholely*.

The chief exception (*gaugeable*, *noticeable*, *singeing*) is that e remains even before a vowel when the soft sound of c or g is to be indicated in the spelling (as before -*able*, where it would normally be hard), or where the e must be retained to distinguish the word from another (e.g. *singeing* from *singing*, *dyeing* from *dying*, *holey* from *holy* and *routeing* from *routing*). There need be no other general exceptions: *duly*, *truly*, and *wholly*, are individual ones merely; so is *acreage* (see SPELLING POINTS 4); *hieing* is specially so spelt to avoid consecutive *i*s, much as *clayey* has an e actually inserted to separate two *y*s; and *gluey*, *bluey*, are due to fear that *gluy*, *bluy*, may be pronounced like *guy* and *buy*.

For practical purposes, then, a single rule, with a single exception, would suffice—stated again below. The only sacrifice involved would be that of the power (most arbitrarily and inconsistently exercised at present) of indicating the sound of an earlier vowel by insertion or omission of the e (*mileage* for fear that *milage* may be pronounced *mil-*). The history of *dispiteous* is perhaps the best comment; from *despite* came *despitous* (*dĭspi'tus*); when the spelling changed to *despiteous* the pronunciation changed to *dispi'tius*, and out of this came a false association with *piteous*, cutting the word off from its etymology and attaching it to *pity* instead of to *spite*.

## RULE

When a suffix is added to a word ending in mute e, the mute e should be dropped before a vowel, but not before a consonant.

## EXCEPTION

The e should be kept even before a vowel if it is needed to indicate the soft sound of a preceding g or c or to distinguish a word from another with the same spelling.

## EXAMPLES

change, *changeling, changing, changeable*; singe, *singeing*; hinge, *hinging*; trace, *traceable*; move, *movable*; horse, *horsy*; strive, *striving*; pale, *palish*; judge, *judgement*.

It is not suggested that this rule could now be applied universally; there are some words in which an unnecessary e is too firmly entrenched to be dislodged. For instance the Scottish word *timeous* is ordinarily so spelt; *mileage* has established a strong position against *milage*; *rateable* has been adopted as the official spelling (though *The Times* with commendable boldness prefers *ratable*), and *ageing* is usually so written, with the authority of the OED, though it is not clear why we should so spell it when we are content to write *raging*. But the rule would at least settle the question for the very large number of words whose spelling is still in the balance, and would give much needed guidance to those who

wish to use adjectives in -*able* not given separately in the dictionaries.

For similar problems with adjectives ending -*y* and verbs ending -*ye* see -EY AND -Y IN ADJECTIVES and VERBS IN -IE, -Y, AND -YE 7.

**mutual** is a well-known trap. The essence of its meaning is that it involves the relation *x* is or does to *y* as *y* to *x*, and not the relation *x* is or does to *z* as *y* to *z*. From this it follows that *our mutual friend Jones* (meaning Jones who is your friend as well as mine), and all similar phrases, are misuses of *m*. An example of the mistake, which is very common, is: *On the other hand, if we* [i.e. the Western Powers] *merely sat with our arms folded there would be a peaceful penetration of Russia by the country* [i.e. Germany] *which was the mutual enemy* [i.e. of both Russia and the Western Powers]. In such places *common* is the right word, and the use of *m*. betrays ignorance of its meaning. It should be added, however, that *m.* was formerly used much more loosely than it now is, and that the OED, giving examples of such looseness, goes no further in condemnation than 'Commonly censured as incorrect, but still often used in the collocations *m. friend*, *m. acquaintance*, on account of the ambiguity of *common*'. The Dickens title has no doubt much to do with the currency of *m. friend*. Perhaps it should now be regarded as qualifying for admission to the STURDY INDEFENSIBLES.

Another fault is of a different kind, betraying not ignorance, but lack of the taste or care that should prevent one from saying twice over what it suffices to say once. This happens when *m.* is combined with some part of *each other*, as in: *It is this fraternity of Parliament-men serving a common cause,* mutually *comprehending each other's problems and difficulties, and respecting each other's rights and liberties, which is the foundation of the structure*. It may fairly be said that the sole function of *mutual*(*ly*) is to give the sense of some part of *each other* when it happens to be hard to get *each other* into one's sentence. If *each other* is got in, *m.* is superfluous; in the quotation it adds nothing whatever, and is the merest tautology.

A few bad specimens follow: *The ring was mutually chosen by the Duke and Lady Elisabeth last Wednesday.* | *They have affinities beyond a m. admiration for Mazzini.* | *It involves . . . m. semi-bankruptcy of employers and employed.*

For the distinction between *m.* and *reciprocal*, see RECIPROCAL.

**muzhik.** See MOUJIK.

**my.** For *my and your work* etc. (not *mine*), see ABSOLUTE POSSESSIVES.

**mynheer, mein Herr, Herr.** The first is Dutch and can mean sir or Mr; the second is German for sir; the third is German for gentleman and Mr.

**myriad** is generally used of a great but indefinite number; but it is well to remember that its original sense, still occasionally effective, is ten thousand.

**myself.** For misuses of *myself* see SELF.

**mystic** has been much slower than *mysterious* in becoming a popular word and thereby losing its definitely spiritual or occult or theological implications. Everything that puzzles one has long been called mysterious (who committed the latest murder, for instance), but not mystic. It is very desirable that *mystic* should be kept as long as possible from such VULGARIZATION. Unfortunately the NOVELTY-HUNTERS, tired of *mysterious*, have now got hold of it: *But I don't want to be* mystic, *and you shall hear the facts and judge me afterwards.*

**myth,** from Gr. μῦθος, legend, is a word introduced into English little more than a century ago as a name for a form of story characteristic of primitive peoples and thus defined by the OED: 'A purely fictitious narrative usually involving supernatural persons, actions, or events, and embodying some popular idea concerning natural or historical phenomena.' By those who wished to mark their adherence to this original sense the word was sometimes pronounced *mīth*. But to-day the meaning popularly attached to the word is little more than a tale devoid of truth or a non-existent person or thing or event; always in these senses, and usually even in the original one, the pronunciation is now *mĭth*.

# N

**n.** *To the nth.* As a mathematical symbol, *n* means an unspecified number; it is a dummy occupying a place until its unknown principal comes along, or a masquerader who on throwing off the mask may turn out to be anything. It does not mean an infinite number, nor the greatest possible number, nor necessarily even a large number, but simply the particular number that we may find ourselves concerned with when we come to details; it is short for 'one or two or three or whatever the number may be'. It follows that the common use of *to the nth* for to the utmost possible extent (*The Neapolitan is an Italian to the nth degree.* | *Minerva was starched to the nth*) is wrong. It is true that sentences can be constructed in which the popular and the mathematical senses are reconciled (*Though the force were increased to the nth, it would not avail*), and here, no doubt, from the idea of something indefinitely large, the origin of the misuse is to be found. Those who talk in mathematical language without knowing mathematics go out of their way to exhibit ignorance.

Similar misuses, though more pardonable because they do not profess to employ the precise language of mathematics, are *to a* DEGREE for to the last degree and to *have a temperature* for to be FEVERISH. These have won the status of STURDY INDEFENSIBLES. See MORE 7 and PROGRESSION for other misuses of mathematical terms.

**naïf, naïve, naïveté, naïvety.** If

we were now adopting the French adjective for the first time, and were proposing not to distinguish between masculine and feminine, but to choose either -*f* or -*ve* for all uses, something might be said for the masculine form (in spite of *pensive, effective*, etc.) as being the French word before inflexion. But both forms have been with us for centuries representing both genders, and it is undeniable that *naïve* is now the prevalent spelling, and the use of *naïf* (either in all contexts or whenever the gender is not conspicuously feminine) is a conscious correction of other people's supposed errors. Such corrections are pedantic when they are needless; on the needlessness of correcting established misspellings of foreign words, see MORALE.

The slowness with which the naturalization of the words has proceeded is curious and regrettable. For it will hardly be denied that they deserve a warm welcome as supplying a shade of meaning not provided by the nearest single English words. The OED definition, for instance, 'Natural, unaffected, simple, artless', clearly omits elements—the actor's unconsciousness and the observer's amusement—that are essential to the ordinary man's idea of naïveté. Unconsciously and amusingly simple; *naïve* means not less than that, and is therefore a valuable word. But, as long as the majority of Englishmen are kept shy of it by what is to them queer spelling and pronunciation, its value will not be exploited. The difficulty is greater with the noun than with the adjective; many by this time write *naive* without the diaeresis, and many call it *nāv*, a pronunciation recognized by most dictionaries, though *nä'ēv* is given first place. But *naivety*, though it was used by Hume and other 18th-c. writers, has not yet made much headway against *naïveté*; till it wins, these potentially useful words will be very much wasted.

**nail.** *Hit the* (*right*) *n. on the head.* It is clear from the OED quotations that *right*, which blunts the point by dividing it into two, is a modern insertion; all the quotations up to 1700 are without it, and all after 1700 have it; it is better omitted.

**names and appellations.** 'How now, daughter and cousin', was the greeting given to Celia and Rosalind by Duke Frederick, who afterwards, with more precision, admonished the latter 'You, niece, provide yourself'; 'Thou bringst me peace and happiness, son John' said Henry IV; 'Prepare her, wife, against this wedding-day' said Capulet. Except for *father, mother, grandfather*, and *grandmother*, this custom of using a relationship-term by way of direct address disappeared gradually during the 19th c.; *husband* and *wife* seem to have been the first to go and *uncle, aunt,* and *sister* (often abbreviated to *sis*) to have lasted longest. So far as disparity of age makes it seemly for us still to use these terms in speaking to our relations we now ordinarily add the christian name—*Uncle John, Aunt(ie) Mary, Cousin Tom*—though not, as was often done until comparatively recently, the surname—*Uncle Smith* and *Aunt Jones*.

Though *father* and *mother* are still in use, the more affectionate alternatives *papa* and *mamma*, long current, have been superseded by *daddy* and *mummy*, and are now only used as deliberate revivals. The Victorian schoolboy's words for his parents, *pater* (or *governor*, abbreviated to *gov.*) and *mater* are obsolete.

Modern trend has been away from formality and towards familiarity. Though service with the Forces has increased the use of *sir* by young people to their elders (at the same time as it has been falling into disuse as a mark of difference in social status), the duty of saying *sir* to one's father, once generally recognized, has long ago been dropped. It is indeed far from unknown for young people to call by their christian names not only uncles and aunts but also fathers and mothers, and even sometimes grandfathers and grandmothers. Greater freedom in the use of christian names is in fact one of the most striking of the recent changes in this field. In Victorian days as soon as a girl put her hair up and wore long skirts and a boy went into tails they became *Miss Jones* and *Mr. Smith*; and quite a long apprenticeship, perhaps even formal permission, was needed before they were *Mary* and *John* to each other. Today young people would think it ridiculous not to be on christian-name terms from the start; it may take almost as long for them to become familiar with one another's surnames as it used to take to qualify for dropping them. A natural consequence of this is the discontinuance of the old practice of calling the eldest of the Jones sisters *Miss Jones* and the others *Miss Mary*, *Miss Jane*, and so on.

In Jane Austen's day it seems to have been not uncommon for husbands and wives to address each other in the same way as a stranger would; Mrs. Bennet said *Mr. Bennet* to her husband and Lady Bertram *Sir Thomas* to hers. By the time that Dickens was writing this must have become less usual, for we are told as a circumstance worth recording that Mr. Dombey had been *Mr. Dombey* to his first wife when she first saw him, and he was *Mr. Dombey* when she died. Today it would be something of a solecism, among social equals, for Mr. Dombey to call his wife *Mrs. Dombey* (or vice versa), even when referring to her in conversation with someone else; good manners permit nothing more formal than *my wife* or *my husband* and would favour *Fanny* or *Paul*, unless the person addressed is so slight an acquaintance that he might not know who is meant. For him, however, unless he is on christian-name terms with the spouse referred to, the alternatives would be *your wife* (*husband*) or *Mrs.* (*Mr.*) *Dombey*; what is friendliness on one side would be presumption on the other and what would be stiffness on one side is courtesy on the other.

In modes of direct address between friends and acquaintances the B.B.C. has introduced an innovation. In talking *about* people we habitually refer to them by christian name and surname, as for instance *William Jones* and *Mary Smith*, but we do not call them that in speaking *to* them; they will be *Mr. Jones*, *Jones*, or *William*, *Mrs.* (or *Miss*) *Smith* or *Mary*, according to the degree of our acquaintanceship. In broadcast conversation, however, people not only habitually call one another *William Jones* and *Mary Smith* but repeat the names much more often than would be natural in ordinary talk. The reason for this convention is presumably that it both conduces to an atmosphere of intimacy and also helps an audience who have only their ears to guide them in identifying the speakers. Perhaps some day it will spread into normal usage, but not without resistance, especially to the beginning of letters in this way (*Dear A.B.*). This excites curiously strong feelings. 'An unspeakable usage' says one of the contributors to the symposium *Noblesse Oblige* (see U AND NON-U). 'An odious practice' says a member of Parliament, voicing his indignation in *The Times*, perpetrated mostly by 'party officials who have been instructed to place me on the mat . . . television tycoons and others who could be called ignorant'. Professor Ross, however, who started all the U-and-non-U business, attributes the practice mainly to 'intellectuals of any class'. If the intellectuals and the ignorant are indeed in league over this, the chances of successful resistance do not look bright.

Of the problem how to address a parent-in-law no generally satisfactory solution seems yet to have been found. *Father-in-law* and *mother-in-law* are barely tolerable even facetiously. Present practice apparently varies from calling them *Mr.* and *Mrs. Smith*, as though they were strangers, to using their christian names as though they were contemporaries—as indeed they sometimes are.

More on this subject may be found in R. W. Chapman's essay *Names, Designations and Appellations* (SPE Tract XLVII) to which acknowledgement is made.

**napkin** should be preferred to *serviette*. Perhaps the association of the word with *nappies* accounts for the GENTEELISM.

**narcissus.** Pl. *-ssuses* or *-ssī*. See LATIN PLURALS.

**natter,** said to be a variety of *gnatter*, perhaps from Icelandic *knetta* (grumble), was classed as 'dialect' by the OED, and its present popularity as a colloquialism is recent enough for it to have entered the 1951 COD only through the Addenda. Perhaps its vogue is partly due to its looking like a PORTMANTEAU WORD combining nagging and chattering, for that is exactly what it means.

**nature. 1.** Periphrasis. The word is a favourite with the lazy writers who prefer glibness and length to conciseness and vigour. *The accident was caused through the dangerous nature of the spot, the hidden character of the by-road, and the utter absence of any warning or danger-signal.* The other way of putting this would be 'The accident happened because the spot was dangerous and the by-road hidden, and there was no warning or danger signal of any kind'. *There will be showers this evening in the western part of the region and they may be of a thundery nature.* Why not just *thundery*? It is true that *nature* slips readily off the tongue or pen in such contexts, but the temptation should be resisted; see PERIPHRASIS.
   **2.** *One touch of nature makes the whole world kin.* What Shakespeare meant was: There is a certain tendency natural to us all, viz. that specified in the following lines (*Troilus and Cressida,* III. iii. 176–9), which is, so far as one word may express it, fickleness. What is meant by those who quote him is: A thing that appeals to simple emotions evokes a wonderfully wide response. This is both true and important; but to choose for the expression of it words by which Shakespeare meant nothing of the kind is unfair both to him and to it. That the first words of a cynicism appropriately put into the mouth of the Shakespearian Ulysses should be the stock quotation for the power of sympathy is an odd reversal.

**naught, nought.** The variation of spelling is not a modern accident, but descends from Old English. The distinction, however, now usually observed between the senses borne by each form is a matter of convenience only, and by no means universally recognized. This distinction is that *nought* is simply the name of the cipher o, while the archaic, poetic, and rhetorical uses in which the word is substituted for *nothing* in any other than the arithmetical sense now prefer *naught*: *one, nought, nought, one; noughts and crosses; bring* or *come to, set at, naught; availeth naught; give all for naught.*

**nautilus.** The OED puts the plural *-ī* before *-uses*, but the COD reverses the order. See LATIN PLURALS.

**navy.** For *n. and army*, see ARMY.

**near(ly).** The use of *near* in the sense of *nearly* (*Not near so often; near dead with fright; near a century ago*) has been so far affected by the vague impression that adverbs must end in *ly* as to be obsolescent; see UNIDIOMATIC -LY for other words in which the process has not gone so far. Those who still say *near* for *nearly,* when provincialism and ignorance are both out of the question, are suspected of pedantry; it is a matter in which it is wise to bow to the majority.
   This use of *near* in the sense of 'approximating in kind or degree' has, however, returned to some extent, especially qualifying nouns, e.g. *near-beer, near-wool, near-communist.* The OED Supp. calls this revival 'chiefly U.S.', but it is now common in Britain too. It should not be confused with the use of the adjective *near* in such a phrase as *a near miss.* A near miss (unhyphened) is a miss that was nearly a hit; near-beer (hyphened) is a beverage that professes to be nearly beer.

**near by** has been long established as
an adverb, and there is no good
reason for those who draft police
notices to prefer *in the vicinity*.
The Americanism *near-by* or *nearby*,
used as an adjective, seems to be over-
coming the resistance it first met in
Britain; its convenience is likely to
win it literary status here, if it has not
already done so.

**nebula.** Pl. *-lae*.

**necessary.** For *essential*, *n.*, and
*requisite*, see ESSENTIAL. On the pronunciation of the adverb (*nĕc'ĕssărĭlȳ*)
see RECESSIVE ACCENT.

**necess(it)arian.** The existence of
two forms of a word, unless they are
utilized for differentiation, is inconvenient, putting those who are not
thoroughly familiar with the matter
to the needless pains of finding out
whether the two do in fact stand for
different things or for the same. It
would therefore be well if one of this
pair could be allowed to lapse. There
is no valid objection to the formation
of either; but *necessitarian* is the better
word, (1) as having a less unEnglish or
a somehow more acceptable sound,
(2) because its obvious connexion with
*necessity* rather than with *necessary*
makes the meaning plainer, and (3) as
being already the more usual word.
*Necessarian* should be regarded as a
NEEDLESS VARIANT and is rightly
obsolescent.

**necessities, necessaries.** These
words are distinguishable to the extent
that *necessities* alone can be used in the
abstract sense proper to its singular to
mean the needs of a necessitous person. As concrete words for necessary things attempts are sometimes
made to differentiate them: it is suggested for instance that nature decides
what are necessities and human judgement what are necessaries. But the
fact is that they have long been treated
as synonyms by popular usage and are
past disentangling.

**nectar** has kept the word-makers busy

in search of its adjective; *nectareal,
nectarean, nectared, nectareous, necta-
rian, nectariferous, nectarine, nectarious,*
and *nectarous,* have all been given
a chance. Milton, with *nectared, nectarine,* and *nectarous,* keeps clear of the
four-syllabled forms in which the accent is drawn away from the significant
part; and we might do worse than let
him decide for us.

**need.** 1. *He seems to think that the
Peronne bridge-head was abandoned
earlier than need have been.* | *It was
assumed that Marshal Foch's reserves
and army of manœuvre had been used
up and need no longer to be taken into
account as a uniform, effective body.*
These extracts suffice to show that
lapses in grammar or idiom may occur
with *need*. The first looks like some
confusion between the verb and the
noun *need*; at any rate the two right
ways of putting it would be (*a*) *earlier
than it* (i.e. the bridge-head) *need have
been* (sc. abandoned), where *need* is the
verb, and (*b*) *earlier than there was
need* (sc. to abandon it), where *need*
is the noun.

With uncertainties whether *need* is
a noun or a verb, whether *needs* is
a verb or a plural noun or an adverb,
and what relation is borne to the
verbal *needs* and *needed* by the abnormal *need* often substituted for them,
there are certain difficulties. The
writer of the second extract above has
missed the point of idiom that, while
*needs* and *needed* are ordinary verbs
followed by infinitive with *to*, the
abnormal *need* is treated as a mere
auxiliary, like *must*, requiring no *to*;
the reserves *needed no longer to be
taken,* or *did not need any longer to be
taken,* but *need no longer be taken, into
account.* The rules for the use of *need*
instead of *needs* and *needed* are: It is
used only in interrogative and negative
sentences; in such sentences it is more
idiomatic than the normal forms, which
are, however, permissible; if *need* is
preferred, it is followed by infinitive
without *to*, but *needs* and *needed* require
*to* before their infinitive. Idiomatic
form, *They need not be counted*; normal

form, *They did not need to be counted,* or *They needed not to be counted*; wrong forms, *They need not to be counted, They needed not* (or *did not need*) *be counted.*

2. In the phrase *needs must* or *must needs* (*He must needs go that the devil drives*) *needs* is an archaic adverb (= of necessity) reinforcing *must.* To-day the phrase is mostly used ironically, expressing contempt for some gratuitously foolish or annoying action. *He had been warned the dog was dangerous and he m. n. go and pat it. | All plans had been made with great care and he m. n. interfere and upset them.*

3. *Need* and *want. Need* implies an objective judgement, *want* a subjective, a distinction often blurred by the loose use of *want.* A child may need punishment but is unlikely to want it. *Never was legislation more needed; never was it less wanted* said Lloyd George of his National Health Insurance Act.

**needle.** For *A n. in a bottle of hay* see HAY.

**needless variants.** Though it savours of presumption for any individual to label words needless, it is certain that words deserving the label exist; the question is which they are, and who is the censor that shall disfranchise them. Every dictionary-maker would be grateful to an Academy that should draw up an index expurgatorius and relieve him of the task of recording rubbish. There is no such body, and the dictionary-maker must content himself with recognizing, many many years after the event for fear he should be precipitate, that a word here and there is dead, aware the while that he is helping hundreds of others to linger on useless by advertising them once more. Natural selection does operate in the worlds of talk and literature; but the dictionaries inevitably lag behind. It is perhaps, then, rather a duty than a piece of presumption for those who have had experience in word-judging to take any opportunity, when they are not engaged in actual dictionary-making, of helping things on by irresponsible expressions of opinion. In this book, therefore, reference is made to the present article or to that called SUPERFLUOUS WORDS regarding many words that either are or ought to be dead, but have not yet been buried.

This article is concerned only with those that can be considered by-forms differing merely in suffix or in some such minor point from other words of the same stem and meaning. Sometimes the mere reference has been thought sufficient; more often short remarks are added qualifying or explaining the particular condemnation; some of these references are listed below to enable the reader to examine details. Here the general principle may profitably be laid down that having two names for the same thing is a source not of strength but of weakness, because the reasonable assumption is that two words mean two things, and confusion results when they do not. On the other hand, it may be much too hastily assumed that two words do mean the same thing; they may, for instance, denote the same object without meaning the same thing if they imply that the aspect from which it is regarded is different, or are appropriate in different mouths, or differ in rhythmic value or in some other respect that may escape a cursory examination. To take an example or two: it is hard to see why *hydrocephalic* and *hydrocephalous* should coexist and puzzle us to no purpose by coexisting; but *correctitude* by the side of *correctness* has a real value, since it was expressly made to suggest by its sound *conscious rectitude* and so present correctness in an invidious light. Again, it would be rash to decide that *dissimulate* was a needless variant for *dissemble* on the grounds that it means the same and is less used and less clearly English, without thinking long enough over it to remember that *simulate* and *dissimulation* have a right to be heard on the question.

Some of the words under which reference to this article is made (not always concerning the title-word itself) are acquaintanceship, blithesome, bumble bee, burden, chivalry, cithern, competence, complacence, concernment, concomitance, corpulence, direful, disgustful, dissemble, infinitude, necessitarian, quieten.

**ne'er-do-weel, ne'er-do-well.** The OED's remark is: 'The word being of northern and Sc. origin, the form -*weel* is freq. employed even by southern writers.' But that is no longer true. South of the Border *ne'er-do-well* has won.

**negative.** 'The answer is in the negative' is Parliament language, but deserves much severer condemnation (as a pompous PERIPHRASIS for No, sir) than most of the expressions described as unparliamentary language. The formula no doubt originated in a desire to give an answer less terse, and less likely to provoke supplementaries, than a blunt *No, sir.* It is seldom used now; it seems to be one of the few phrases (cf. *exploring every avenue* and *leaving no stone unturned*) against which ridicule has had some success. (*The answer is in the plural* was the reply once given by a jocular Attorney-General.) But this circumlocutory way of saying *no* (or *yes*) seems to have escaped from the House of Commons and crossed the Atlantic. *I take an affirmative attitude on that* said a Hollywood bystander asked by a television interviewer whether he thought American women were getting too influential.

**negative mishandling. 1.** Generally speaking, English grammar today regards two negatives as cancelling each other and producing an affirmative; the proper meaning of *Nobody didn't go* is that everyone went. But it is by no means true, as is sometimes supposed, that this is axiomatic. In some other languages (Greek for example) negative symbols are multiplied to reinforce one another; a speaker 'spreads as it were a thin layer of negative colouring over the

whole of the sentence instead of confining it to one single place' (Jespersen). We used to do the same (*Nor what he said, though it lacked form a little, was not like madness*), and multiplication for emphasis survives in vulgar speech with little risk of ambiguity. Everyone knew what Mr. Dombey's butler meant when he said that he hoped that he might 'never hear of no foreigner never boning nothing out of no travelling chariot', and everyone knows what is meant when, in a television interview, the question 'Are you still a trades-union official' is answered 'I'm not nothing no more'. In educated speech the idiom survives in such sentences as *I shouldn't wonder if it didn't rain.* See NOT 4.

2. Of actual blunders, as distinguished from lapses of taste and style, perhaps the commonest, and those that afflict their author when he is detected with the least sense of proper shame, are various mishandlings of negatives. Writers who appear educated enough to know whether a sentence is right or wrong will put down the opposite of what they mean, or something different from what they mean, or what means nothing at all, apparently quite satisfied so long as the reader can be trusted to make a shrewd guess at what they ought to have said instead of taking them at their word; to his possible grammatical sensibilities they pay no heed whatever, having none themselves. It is parallel clauses that especially provide opportunities for going wrong, the problem being to secure that if both are negative the negative force shall not be dammed up in one alone, and conversely that if one only is to be negative the negative force shall not be free to spill over into the other. Some classified specimens of failure to secure these essentials may put writers on their guard; the corrections appended are designed rather as proofs of the error than as necessarily the best emendations.

(*a*) If you start with a negative subject you may forget on reaching the second clause to indicate that the subject is not negative there also: No lots *will therefore be put on one side for another attempt to reach a better price*, but must *be sold on the day appointed* (but all must be sold). / [During an air-raid] Very few people *even got out of bed*, and went *through their ordeal by fire as an inescapable fate* (and most went). / None of the ministers *foresaw that the legations would be attacked* and had taken *no precautions to lay in arms and ammunition* (and they had taken). / Neither editor nor contributors *are paid*, but are moved *to give their services by an appreciation of the good work* (but all are). For other examples see NO. 3.

(*b*) You may use negative inversion in the first clause, and forget that the second clause will then require to be given a subject of its own because the inversion has imprisoned the original subject: Nor does he refer *to Hubrecht's or Gaskell's theories*, and dismisses *the paleontological evidence in rather a cavalier fashion* (and he dismisses). / *Not only* was Lord Curzon's Partition detested *by the people concerned*, but was *administratively bad* (it was). / *In neither case* is this due *to the Labour Party*, but to *local Socialist aspirations* (but in both).

(*c*) Intending two negative clauses, you may enclose your negative between an auxiliary and its verb in the first and forget that it cannot then act outside its enclosure in the second: *There is scarcely a big hotel, a brewery company, or a large manufactory, which* has not sunk *a well deep into the London chalk* and is drawing *its own supply of water from the vast store* (and succeeded in drawing; if *has* continues, *not* does so with it). / *No scheme run by Civil Servants sitting in a London office is likely to succeed if these gentlemen* have not themselves lived *on the land*, and by experience are able *to appreciate actual conditions of agriculture* (and learnt to appreciate).

(*d*) Conversely, intending a negative and an affirmative clause, you may so fuse your negative with a construction common to both clauses that it carries on to the second clause when not

wanted: *These statements* do not seem *well weighed*, and to savour *of the catchword* (and savour—cutting the connexion with *do not seem*). / *If the Colonial Secretary* is not going *to use his reserve powers when trial by jury breaks down*, and to acquiesce *in the view that no consequences need follow when a settler shoots a native for stealing a sheep, he may as well give up the business of governing altogether* (and acquiesces—cutting the connexion with *is not going*).

(*e*) You may negative in your first clause a word that when supplied without the negative in the second fails to do the work you expect of it: *To raise the standard of life of the many* it is not sufficient *to divide the riches of the few* but also to produce *in greater quantities the goods required by all* (it is also necessary to produce).

(*f*) You may so misplace the negative that it applies to what is common to both clauses instead of, as was intended, to what is peculiar to one: *It* is not expected *that tomorrow's speech will deal with peace*, but will be confined *to a general survey of* . . . (It is expected . . . will not deal).

(*g*) You may treat a double negative expression as though it were formally as well as virtually a positive one: *It would* not be difficult *to quarrel with Mr. Rowley's views about art*, but not with *Charles Rowley himself* (It would be easy). / He has cast about for and neglected no *device chemical or mechanical that might add to his ability* (and tried every device).

3. Other examples of negative mishandling will be found in BUT 2, NO 3, NOR, and WITHOUT 5. A few miscellaneous specimens are here collected without comment: *Were it not for its liking for game eggs, the badger* could not but be considered other *than a harmless animal*. / *They are unlikely to attempt any very serious work, but rather* to bring *themselves to racing pitch with sharp pieces of rowing and paddling*. / *No rival is* too small to be overlooked, *no device is* too infamous not to be practised, *if it will* . . . / Not a whit undeterred *by*

the disaster which overtook them last week. | Is it quite inconceivable that if the smitten had always turned the other cheek the smiters would not long since have become so ashamed that. . .? | I do not think it is possible that the traditions and doctrines of these two institutions should not fail to create rival schools. | Only in Southern England was there too much cloud to prevent the eclipse being seen. | He slipped off quietly, but not so quietly that his going failed to escape the watchful eye of Inspector Ferris. | It was want of imagination that failed them. | No age can see itself in a proper perspective, and is therefore incapable of giving its virtues and vices their relative places.

**negligible, -geable.** The first spelling is now established; cf. incorrigible, dirigible. The prevalence at one time of -geable is perhaps explained by the word's having been familiarized chiefly in the translated or untranslated French phrase quantité négligeable. Cf. intransigent and intransigeant.

**negotiate.** The use of n. in the sense of 'tackle successfully' (a fence or other obstacle or difficulty) originated in the hunting-field; those who hunt the fox like also to hunt jocular verbal novelties. This usage, though quoted from 1862, was still felt in 1909 to need the apology of inverted commas: the OED Supp. quotes from a writer in the Quarterly Review in that year Some rival has 'negotiated'—we believe this to be the sporting phrase—the 150 miles in 47 hours. Today the use of n. in this sense, though it has not completely shed the taint of its sporting-jocular origin, is recognized by the COD without comment. Perhaps it might be pleaded in defence of this invasion of the domain of words such as clear, get past or round or over, dispose of, surmount, overcome, etc., that n. implies a special need for skill and care.

**neighbourhood.** In the n. of for about (Cowdrey, who had taken so many blows—somewhere in the n. of 20

at least . . .) is a repulsive combination of POLYSYLLABIC HUMOUR and PERIPHRASIS.

**neither.** 1. Pronunciation. 2. Meaning. 3. Number of the pronoun and adjective. 4. Number and person of verb after neither . . . nor. 5. Position of neither . . . nor. 6. Neither . . . or. 7. Neither as conjunction. 8. Neither pleonastic.

1. The pronunciation recommended is nī, not nē; see EITHER.

2. The proper sense of the pronoun (or adjective) is 'not the one nor the other of the two'. Like either, it sometimes refers loosely to numbers greater than two (Heat, light, electricity, magnetism, are all correlatives; neither can be said to be the essential cause of the others); but none or no should be preferred; cf. EITHER 3. This restriction to two does not hold for the adverb (Neither fish nor flesh nor fowl).

3. The number of the adjective and pronoun is properly singular, and disregard of this fact is a recognized grammatical mistake, though, with the pronoun at least, very common: The conception is faulty for two reasons, neither of which are noticed by Plato. | What at present I believe neither of us know. Grammar requires is noticed, and knows. The same mistake with the adjective is so obviously wrong as to be almost impossible; not quite, however: Both Sir Harry Verney and Mr. Gladstone were very brief, neither speeches exceeding fifteen minutes. An almost equally incredible freak with the pronoun is: Lord Hothfield and Lord Reay were born the one in Paris and the other at The Hague, neither being British subjects at the time of his birth (as indeed neither could be unless he were twins).

4. Number and person after neither . . . nor. If both subjects are singular and in the third person, the only need is to remember that the verb must be singular and not plural. This is often forgotten; the OED quotes, from Johnson, Neither search nor labour are necessary, and, from Ruskin, Neither painting nor fighting feed men, where

*is* and *feeds* are undoubtedly required; and we may still read in journals of the highest standing, or hear from the B.B.C., such sentences as *Official quarters in London are confident that neither President Eisenhower nor Mr. Dulles want an armed clash with China.* | *Comment in Paris is guarded as neither General de Gaulle nor the Foreign Minister are immediately available.* | *Neither Mr. Gaitskell nor Mr. Grimond have tried to minimize the Russian provocation in Cuba.* The right course is not to indulge in bad grammar ourselves and then plead that better men like Johnson and Ruskin have done it before us, but to follow what is now the accepted as well as the logical rule.

Complications occur when, owing to a difference in number or person between the subject of the *neither* member and that of the *nor* member, the same verb-form or pronoun or possessive adjective does not fit both: Neither you nor I (was?, were?) chosen; Neither you nor I (is?, am?, are?) the right person; Neither eyes nor nose (does its?, do their?) work; Neither employer nor employees will say what (they want?, he wants?). The wise man, in writing, evades these problems by rejecting all the alternatives—any of which may set up friction between him and his reader —and putting the thing in some other shape; and in speaking, which does not allow time for paraphrase, he takes risks with equanimity and says what instinct dictates. But, as instinct is directed largely by habit, it is well to eschew habitually the clearly wrong forms (such as *Neither chapter nor verse* are *given*) and the clearly provocative ones (such as *Neither husband nor wife is competent to act without* his *consort*). About the following, which are actual newspaper extracts, neither grammarians nor laymen will be unanimous in approving or disapproving the preference of *is* to *are* or of *has* to *have*; but there will be a good majority for the opinion that both writers are grammatically more valorous than discreet: *Neither apprenticeship* systems *nor technical education* is *likely to*

*influence these occupations* (why not have omitted *systems*?). | *Neither Captain C. nor I* has *ever thought it necessary to . . .* (Neither to Captain C. nor to me has it ever seemed . . .).

5. Position of *neither . . . nor. Which neither suits one purpose nor the other.* Read *which suits neither one purpose nor the other. Suits* being common to both members should not be inserted in the middle of the *neither* member. Such displacement has been discussed and illustrated under EITHER 5, and need only be mentioned here as a mistake to be avoided.

6. *Neither . . . or.* When a negative has preceded, a question often arises between *nor* and *or* as the right continuation, and the answer to the question sometimes requires care; see NOR and OR. But when the preceding negative is *neither* (adv.), the matter is simple, *or* being always wrong. Examples of the mistake: *Diderot presented a bouquet which was* neither *well* or *ill received.* | *Like the Persian noble of old, I ask 'that I may* neither *command* or *obey'.* Here again, to say that Morley and Emerson have sinned before us is a plea not worth entering.

7. *Neither* alone as conjunction. This use, in which *neither* means 'nor yet', or ' and moreover . . . not', and connects sentences instead of the ordinary *and not* or *nor* (*I have not asked for help*, neither *do I desire it*; *Defendant had agreed not to interfere*, neither *did he*) is much less common than it was, and is best reserved for contexts of formal tone. But it is still fully idiomatic where it links the speaker with some other person: *Defendant did not interfere; neither did I.*

8. *Neither* with the negative force pleonastic, as in *I don't know that neither* (instead of *either*), was formerly idiomatic though colloquial, but is now a vulgarism. Nor should a pleonastic *one* be attached to *neither*, as in *neither one showed much courage.*

**nepenthe(s).** Three syllables, whether with or without the -s. The -s is part of the Greek word, and should have been retained in English; but it has very commonly been dropped, probably from being mistaken for the plural sign, cf. *pea* for *pease* etc. The prevailing form (except in botany, where the classical word is naturally used) is now *-the*. But in ordinary use the word has been superseded by *tranquilliser*.

**nephew.** The conventional pronunciation is *nev-*, but some modern dictionaries recognize *nef-* as an alternative, a result no doubt of the speak-as-you-spell movement. See PRONUNCIATION I.

**nervous diseases.** A *nerve* is a fibre or bundle of fibres arising from the brain, spinal cord, or other ganglionic organ, capable of stimulation by various means and serving to convey impulses (especially of sensation and motion) between the brain etc. and some other part of the body (OED), and the *nervous system* is the name given to the whole complex of nerves and nerve centres. By no means all nervous diseases impair mental faculties, but the expression is popularly regarded as a euphemism for mental disorder, a misconception that the medical profession try to counter by increasing use of the adjective *neurological* instead of *nervous*. But the *neuro-* compounds are themselves tainted by the special meaning the psychologists have given to *neurosis*, and will no doubt fall under the same suspicion; it is never much use trying to change popular ideas by changing names. See EUPHEMISM.

**-ness.** For the distinction between *conciseness* and *concision*, and similar pairs, see -ION AND -NESS and -TY AND -NESS.

**net.** In the commercial sense (free from deduction etc.) the spelling should, as elsewhere, be *net*, not *nett*. See SET(T).

**nether.** For *n. garments*, *n. man*, etc., see PANTALOONS, PEDANTIC HUMOUR, and EUPHEMISM.

**Netherlands, Low Countries, Holland, Dutch.** *The Netherlands* is now the official name for the kingdom of Holland (*Queen of The N.*, *The N. Ambassador*, etc.), though formerly Holland was only part of *The N.*, or *The Low Countries*, which, like the association of today known as *Benelux*, included also Belgium and Luxemburg. The name of the country is sometimes used adjectivally (e.g. for *holland* the linen fabric, *Hollands gin*, and *sauce hollandaise*), but *Dutch* (formerly also of much wider application) is the ordinary adjective for the language and the people and their habits and characteristics, real or fancied, exemplified in such phrases as the *d. auction* that works in reverse, the heavy-handed *d. uncle*, the artificially induced *d. courage*, the *d. bargain* made over drinks, the cacophonous *d. concert*, and the contributory *d. treat*. The word for a costermonger's wife, immortalized in Albert Chevalier's *My Old Dutch*, is different: it is an abbreviation of *duchess*.

**neurasthenia.** The usual pronunciation is *-thĕ-* not *-thē-*. See FALSE QUANTITY.

**never** need not have a temporal significance; it can be used as an emphatic negative: *I never expected you would be here today*. From this it is a short step to use it in a transferred temporal sense; *I never remember meeting him* (I do not remember ever meeting him). This, however illogical, is idiomatic, at least colloquially.

**never so, ever so,** in conditional clauses (*refuseth to hear the voice of the charmer, charm he never so wisely*). The original phrases, going back to Old English, are *never so*, and *never such*. The change to *ever*, 'substituted from a notion of logical propriety' (OED), seems to date from the later 17th c. only, and *never so* is very common in the Bible and Shakespeare. *Ever so*, however, is the normal modern form, not *never so*, and it is in vain that attempts are occasionally made to put the clock back and restore *never* in ordinary speech. In poetry, and under circumstances that justify archaism, *never so* is unimpeachable; but in everyday style the purism that insists on it is futile. As to that 'notion of logical propriety', it was perhaps that there was nothing negative in the sense; but that is not true, if 'charm he never so wisely' is a compressed form of 'charm he so wisely as never else'; we can at least see how the *never* idiom may have arisen. To account for *ever* (except as a mistaken correction of *never*) is a much harder problem. But the modern phrase, explicable or not, and logical or not, is *ever so*, though it is now rarely used in quite this way. *Ever so* has become a colloquial adverb of emphasis: *Thank you ever so much*, or, in progressive vulgarization, *He's ever so nice.* / *I've enjoyed myself ever so.*

**nevertheless, nonetheless** have the same meaning; the choice between them is a matter of taste. *Never-* used to be the favourite, but *none-* has recently been gaining on it.

**news.** The number varied (*the n. is bad, are bad*) for more than two centuries, but has now settled down as singular.

**new verbs in -ize.** A feature of the second Elizabethan age, as of the first, is that new words proliferate. One way of making them is to add the suffix *-ize* to a noun or adjective, and so increase our stock of verbs. The device itself is far from new; our vocabulary already contains some hundreds of verbs so formed. The Greeks used the suffix ἴζειν to make verbs with, and we have followed them; sometimes we take over what is, except for its termination, an actual Greek word (*apologize, dogmatize, ostracize*, etc.); sometimes we add *-ize* to a word or stem, usually of Greek or Latin origin (*colonize, immunize, summarize*, etc.) but occasionally of later date (*bastardize, jeopardize, standardize*, etc.) especially to names

of places or people (*americanize, bowdlerize, galvanize, pasteurize*, etc.). Within reason it is a useful and unexceptionable device, but it is now being employed with a freedom beyond reason. The purpose is usually to enable us to say in one word what would otherwise need more than one. Whether that justifies the creation of a new *-ize* word is a question on which opinions will differ, and it does not admit of a categorical answer.

Most verbs in *-ize* are inelegant. Sir Alan Herbert has compared them to lavatory fittings, useful in their proper place but not to be multiplied beyond what is necessary for practical purposes. (Perhaps it was this stricture that led to the use of the word *Inthroning* at the coronation of Queen Elizabeth II instead of the word *Inthronization* used at previous coronations.) A different, and often preferable, way of providing ourselves with a new verb is boldly to use a noun as one. We say to *doctor*, not to *doctorize*, to *partition*, not to *partitionize*; if it was necessary for us to have verbs for sending people to hospital and for dividing sleeping accommodation into cubicles, we should have done better to say to *hospital* and to *cubicle* than to coin those ungainly words *hospitalize* and *cubiclize*. *Pressurize, containerize,* and *institutionalize* are useful but to treat *pressure, container,* and *institution* as verbs would have been less clumsy. The lure of the suffix is so powerful that we sometimes find it added absurdly to nouns whose use as verbs is already recognized, such as SERVICE and *fragment*. CONTACT used as a verb excited some criticism at first, but even its critics would admit that this was better than coining a verb *contactize*. And having sensibly invented the verb *delouse*, we ought to have nothing to do with the *deratize* that is now showing its ugly head.

There are obvious limits to the use of this device for avoiding recourse to *-ize* when we want a new verb. Adjectives take less kindly than nouns to being made to serve as verbs and not all nouns are suitable. It could for instance be argued on behalf of *itiner-ize* that, the natural verb *itinerate* having been put to a different use, it is a better one than to *itinerary* would be, with its almost unpronounceable present participle, though this does not answer the criticism that since we already have the verb to *route* we do not need either of them. A similar plea might be made for *diarize* on the ground of its inflexional advantages over to *diary*. If we must have words for employing casual labour instead of regular, or a civilian staff instead of a military, or for spoiling the country-side by extending a town, or for substituting diesel locomotives for steam, we cannot reasonably quarrel with *casualize, civilianize, sub-urbanize*, and *dieselize*. A recent experiment of this kind, seriously made in a serious journal, was to refer to the *comprehensivizing* by a local authority of its educational system, a neologism that certainly deserves high marks if the test of its merit is the number of words it saves; so would *deprole-tarianize* used by a speaker on the Third Programme, if we could be sure we knew what it meant. Lexicographers must be allowed their *alphabetize* (they have indeed had it for about a hundred years), and road engineers their *reflectorize*, and PUBLICIZE can put up a good defence against a charge of being merely an unwanted synonym for *publish*. Whether *randomize* can be similarly justified on the ground that no other word, such as *shuffle* or *jumble*, would do, and whether *finalize* (which seems to have come to stay in spite of the protest that greeted it) can claim to have a meaning not quite the same as that of any established word are arguable questions. But few will deny that *trialize* and *reliablize* are unwanted monstrosities and when simple verbs like *proof* and *perfume* are available we might have been spared *impermeablize* and *fragrantize*.

One reason for the popularity of *-ize* words (and also no doubt for the reaction of the minority who regard all new ones indiscriminately as accursed things) seems to be that those engaged in the advertisement and entertainment industries think, perhaps rightly, that the look and sound of them make a meretricious appeal. We may be expected to respond more readily to an invitation to *slenderize* than to *slim*, to visit a *picturized* film than a *pictured* one; if we want to 'repel the signs of age' we shall be more likely to buy a preparation that *moisturizes* the skin than one that merely *moistens* it, and we shall find *tenderized* prunes more alluring than those that are merely *tender*. Some missionaries of moral uplift have the same idea. Of a book by one of them a TLS reviewer has written: 'Of course it is disconcerting to be told that a certain method is "rather unique", and few people brought up in an older religious convention will find the formula "Prayer-ize, Picturize, Actualize" of much help to them in the practice of meditation. St. Ignatius might, however, have said something similar, though he would have expressed himself with greater regard to the niceties of language.'

**next.** 1. *The n. three* etc. 2. *N. Friday, June,* etc. 3. *N. important* etc.

1. For the question between *the next three* etc. and *the three* etc. *next*, see FIRST 3.

2. *Next June, n. Friday,* etc., can be used as adverbs without a preposition (*Shall begin it next June*); but if *next* is put after the noun, idiom requires a preposition (*may be expected in June next, on Monday next*). See PREPOSITION DROPPING.

3. *The 'No Surrender' party had the rank and file at their back because they fought to the last ditch to save the grandest institution in the country; do they expect support now in wrecking the two* next important *institutions? The two next important institutions* is clearly used in the sense 'the two institutions next in importance'. The OED quotes no example of such a use, but it is perhaps not uncommon colloquially, and must be a conscious or unconscious experiment in extending

the convenient *next best* idiom. That idiom requires a superlative, and such words as *oldest, worst, narrowest, weightiest* suit it well; but it is ugly with adjectives having no superlative other than that with *most,* and there is a temptation to try whether, for instance, *next important* will not pass for *next most important.* It should be resisted; the natural sense of *the two next important institutions* is 'the two next institutions that are of importance', which need by no means be the two that are next in importance.

**nexus.** The English plural *nexuses* is intolerably sibilant, and the Latin, *nexus (-ūs),* not *nexi* (see -US), sounds pedantic; the plural is consequently very rare; if one must be used *-uses* is recommended. See LATIN PLURALS.

**nice.** 1. *N.* makes *nicish;* see MUTE E. 2. *Nice and* as a sort of adverb = satisfactorily (*I hope it will be n. a. fine; Aren't we going n. a. fast?*) is a HENDIADYS that is an established colloquialism, but should be confined, in print, to dialogue. 3. Meaning. 'I am sure', cried Catherine, 'I did not mean to say anything wrong; but it *is* a nice book, and why should I not call it so?' 'Very true,' said Henry, 'and this is a very nice day, and we we are taking a very nice walk; and you are two very nice young ladies. Oh! it is a very nice word indeed! it does for everything' (*Northanger Abbey*). *N.* has been spoilt, like CLEVER, by its *bonnes fortunes;* it has been too great a favourite with the ladies, who have charmed out of it all its individuality and converted it into a mere diffuser of vague and mild agreeableness. That was not how the Duke of Wellington used it when he described the battle of Waterloo as 'a damned nice thing—the nearest run thing you ever saw in your life'. Everyone who gives it its more proper senses which fill most of the space given to it in any dictionary, and avoids the one that tends to oust them all, does a real if small service to the language.

**nigger.** To be called a *n.* is now regarded as an insult by an American negro, unless the word is used affectionately by one negro of another; a murder provoked by the words 'You dirty nigger' is a leading case in U.S. criminal law (Fisher *v.* U.S.). *N.* has been described as 'the term that carries with it all the obloquy and contempt and rejection which whites have inflicted on blacks'. But it survives in the phrases *Work like a n., The n. in the woodpile, n. minstrels.*

**-n-, -nn-.** Monosyllables ending in *n* double it before suffixes beginning with vowels if the sound preceding the *n* is a short vowel but not if it is a long one or a vowel and *r: mannish,* but *darning; fenny,* but *keener; winning,* but *reined; conned,* but *coined; runner,* but *turned.* Words of more than one syllable follow the rule for monosyllables if their last syllable is accented, but otherwise do not double the *n: japanned* and *beginner,* but *dragooned, womanish, turbaned, awakening, muslined.*

**no.** 1. Parts of speech. 2. Confusion of adjective and adverb. 3. *No* in negative confusions. 4. Negative parentheses. 5. Writing of compounds. 6. Plural.
1. *No* is (A) an adjective meaning in the singular not a (or not any), and in the plural not any; it is a shortened form of none, which is still used as its pronoun form: *No German applied; No Germans applied; None of the applicants was,* or *were, German. No* is (B) an adverb meaning by no amount and used only with comparatives: *I am glad it is no worse. No* is (C) an adverb meaning not and used only after *or,* and chiefly in the phrase *whether or no: Pleasant or no, it is true; He must do it whether he will or no. No* is (D) a particle representing a negative sentence of which the contents are clear from a preceding question or from the context: *Is he there?—No* (i.e. he is not there). *No, it is too bad* (i.e. I shall not submit; it is too bad). *No* is (E) a noun meaning the word *no,* a denial or refusal, a negative vote or voter:

*Don't say no; She will not take a no;
The Noes have it.*

**2.** Confusion of adjective and adverb. If the tabulation in (1) is correct, it is clear how the worse than superfluous *a, the,* and *her,* made their way into the following extracts. The writer of each thought his *no* was a (B) or a (C) adverb (against which the absence of the invariable accompaniments should have warned him) and did not see that it was the adjective, which contains *a* in itself and is therefore incompatible with another *a,* or *the,* or *her. We can hardly give the book higher praise than to say of it that it is* a no unworthy companion *of Moberly's 'Atonement'* (Omit *a,* or write *not* for *no*). | *The value of gas taken from the ground there and sold amounted to* the no insignificant value of $54,000,000 (the not). | *Paintings by Maud Earl, who owes* her no small reputation *as an artist to the successes which* . . . (her reputation, no small one). | A fourth example is more excusable because the conditions are obscured by the accidental presence of a comparative: *We could ask for* no more cheerful a by-product *of our discontent than a second volume of this most patriotic of Christmas books.* Such a sentence as *The second volume will be no more cheerful a by-product than the first* would be right, *no* being there actually the adverb. But the phrase in its present setting means no by-product that shall be more cheerful, and *no* is the adjective and contains *a* and refuses to have another thrust upon it. For *no mean city* etc. see MEAN.

**3.** *No,* used in the first of two parallel clauses, ensnares many a brave unwary writer; the modifications necessary for the second clause are forgotten, and bad grammar or bad sense results. See NEGATIVE MISHANDLING; some specimens are: *He sees in England* no attempt to *mould history according to academic plans,* but to *direct it from case to case according to necessity* (it is rather directed). | *Although* no party *has been able to carry its own scheme out,* it *has been strong enough to prevent any other scheme being carried* (each has been). | No place *of any importance,*

and a good many *of none at all, are now without their bowling greens* (All places of importance . . . now have).

**4.** Negative parentheses. The rule here to be insisted on concerns negative expressions in general, and is stated under *no* only because that word happens to be present in violations of it oftener perhaps than any other. The rule is that adverbial qualifications containing a negative must not be comma'd off from the words they belong to as though they were mere parentheses. The rule only needs stating to be accepted; but the habit of providing adverbial phrases with commas often gets the better of common sense. It is clear, however, that there is the same essential absurdity in writing *He will, under no circumstances, consent* as in writing *He will, never, consent,* or *He will, not, consent.* It is worth while to add, for the reader's consideration while he glances at the examples, that it would often be better in these negative adverbial phrases to resolve *no* into *not* . . . *any* etc. *We are assured that the Prime Minister will, in no circumstances and on no consideration whatever, consent* (will not in any . . . or on any . . . Or omit the commas, at the least). | *And Paley and Butler, no more than Voltaire, could give Bagehot one thousandth part of the confidence that he drew from* . . . (could not, any more than . . . Or could no more than Voltaire give). | *Proposals which, under no possible circumstances, would lead to any substantial, or indeed perceptible, protection for a home industry* (which would not lead under any . . . Or which would under no possible circumstances lead).

**5.** Writing of compounds. About *no ball* (noun) and *no-ball* (verb), *nobody,* and *nohow,* doubts are needless; the forms given are the right ones. For *no one* see EVERY ONE, 1, where that form is recommended. The adverbs *noways* and *nowise* are best so written; but *in nowise,* which is often used instead of the correct *in no wise,* is as absurd as *by nomeans* or *on no-account* would be; cf. ANY 1.

**6.** Pl. *noes*; see -O(E)S 2.

**nomad.** The pronunciation *nŏm-* is gaining on *nōm-* especially in the noun, though the COD still puts *nŏm-* first.

**nom de guerre, nom de plume, pen name, pseudonym.** *Nom de guerre* is current French, but, owing to the English currency of *nom de plume*, is far from universally intelligible to Englishmen, most of whom assume that, whatever else it may mean, it can surely not mean nom de plume. *Nom de plume* is open to the criticism that it is ridiculous for English writers to use a French phrase that does not come from France; not perhaps as ridiculous as the critics think (see MORALE), but fear of them will at any rate deter some of us. Nobody perhaps uses *pen-name* without feeling either 'What a good boy am I to abstain from showing off my French and translate *nom de plume* into honest English!', or else 'I am not as those publicans who suppose there is such a phrase as *nom de plume*'. For everyone is instinctively aware that *pen-name*, however native or naturalized its elements, is no English-bred word, but a translation of *nom de plume*. *Pseudonym*, lastly, is a queer out-of-the-way term for an everyday thing. But it is perhaps the best of the bunch except for those who take the commonsense view of *nom de plume*— that it is the established word for the thing, and its antecedents do not concern us. It is now sometimes used for pseudonyms in general, not merely those adopted by writers; and it would be unreasonable to take exception to its being granted a liberty already won by *nom de guerre*.

**nomenclature.** Pronunciation *nŏ'- menklăture* or *nōmen'klăture*; most dictionaries prefer the former.

Dictionaries that give a list of synonyms with each word may be very helpful to a writer if they cast their nets wide and are used with discrimination (see SYNONYMS). Otherwise they may do him a very doubtful service. One can hardly believe but that the authors of the extracts below have looked up *name* in search of some longer and more imposing word, some

(shall we say?) adequately grandiose vocable. That *nomenclature* does not mean a name, but a system of naming or of names, is to such writers what they would perhaps call a mere meticulosity; see LOVE OF THE LONG WORD. *The forerunner of the present luxurious establishment was the well-known Gloucester Coffee House, the nomenclature of which was derived from that Duke of Gloucester who . . | A small committee of City men has just launched a society, under the nomenclature of the 'League of Interpreters', with the object of . . . | The most important race of the season for three-year-old fillies; the nomenclature was obtained from Lord Derby's seat, 'The Oaks', in the little hamlet of . . .*

**nominal.** For this as the adjective of *noun*, see NOUN.

**nominative.** The grammatical word is always pronounced *nŏ'mĭnătĭv*, often slurred into *nom'nătiv* (cf. *rej'ment* for REGIMENT); the adjective connected in sense with *nominate* and *nomination* (e.g. in *partly elective and partly n.*) is often, and perhaps more conveniently, *nominā'tiv.*

**nominativus pendens.** A form of ANACOLUTHON in which a sentence is begun with what appears to be the subject, but before the verb is reached something else is substituted in word or in thought, and the supposed subject is left in the air. The most familiar and violent instance is *which* used in Sarah Gamp's manner (*which fiddle-strings is weakness to expredge my nerves this night*); but the irregularity is not uncommon even in writing, and is always apt to occur in speech. Cf., in Shakespeare, '*They* who brought me in my master's hate, I live to look upon their tragedy' (*Rich. III*, III. ii. 57).

**non-.** This negative prefix, says the COD, 'is now freely used'. Its use is indeed becoming too free. It has given us many useful words of the type of *non-attendance, non-combatant, non-conformist, non-skid,* and *non-stop,* and some useful differentiations by providing a colourless antonym where the ordinary one has acquired a positive implication, e.g. *non-effective* and *ineffective*, *non-human* and *inhuman*, *non-moral* and *immoral*, *non-natural* and *unnatural*, *non-professional* and *unprofessional*. But the ease with which any word can be negatived by this device tempts the indolent to use it unnecessarily and to write for instance *non-concur* for *dissent*, *non-essential* for *unessential*, *non-sick* for *healthy*, *non-success* for *failure*. Cf. the similar abuse of DE- and DIS-.

**nonce** survives only in the phrases *for the nonce* (and that with a WARDOUR STREET flavour) and *nonce word*, meaning a word coined for a single occasion.

**nonchalant, -ance.** Pronounce *nŏn'-shălănt*, *-ăns* (i.e. as English words, but with *-sh-*).

**none. 1.** It is a mistake to suppose that the pronoun is singular only and must at all costs be followed by singular verbs etc.; the OED explicitly states that plural construction is commoner.
    **2.** The forms *none so*, *none too*, are idiomatic (*It is none so pleasant to learn that you have only six months to live*; *The look he gave me was none too amiable*), but are perhaps seldom used (especially the former) without a certain sense of condescending to the vernacular as an aid to heartiness of manner or to emphasis; and condescension is always repellent.

**nonentity,** in the now rare abstract sense of non-existence, should have the *non* pronounced clearly *nŏn*, and perhaps be written with a hyphen (*non-entity*). In the current concrete sense of a person of no account, it is written *nonentity* and said with the *o* obscured.

**nonesuch, nonsuch.** The first is the original form, but the second the now usual one. Pronounce *nun-*.

**nonetheless.** See NEVERTHELESS.

**non sequitur.** The fallacy of assuming an unproved cause. Thus: It will be a hard winter, for holly-berries (which are meant as provision for birds in hard weather) are abundant. The reasoning called POST HOC, PROPTER HOC is a form of n. s.

**no-one, no one.** The second is recommended; see EVERY ONE I.

**nonsense.** *The Government cannot give way to the railwaymen without making a complete nonsense of its restraint policy.* The idiom is *make nonsense* (or *make sense*) of something, without the indefinite article.

**nor** is a word that should come into our minds as we repeat the General Confession. Most of us in our time have left undone those things which we ought to have done (i.e. failed to put in *nor* when it was wanted) and done those things which we ought not to have done (i.e. thrust it in when there was no room for it). The negative forms of *He moves and speaks*, *He both moves and speaks*, are *He moves not nor speaks*, *He neither moves nor speaks*; or, with the verb resolved as is usual in modern negative sentences, *He does not move or speak*, *He does not either move or speak*. The tendency to go wrong is probably due to confusion between the simple verbs (*moves* etc.) and the resolved ones (*does move* etc.); if the verb is resolved, there is often an auxiliary that serves both clauses, and, as the negative is attached to the auxiliary, its force is carried on together with that of the auxiliary and no fresh negative is wanted. Two cautions are necessary on this carrying on of the negative force and consequent preference of *or* to *nor*. The first is that it will not do to repeat the auxiliary and yet use *or* under the impression that the previous negative suffices; that is what has been done in: *He was naturally and properly at pains to prove that his company had not acted negligently or carelessly or had been unduly influenced by reasons of economy* (There was a choice here between *or been* and *and had not been*; *or had been* makes nonsense).
    The other caution, much more often required, is that if the negative is

attached not to an auxiliary (or other word common to two clauses) that will carry it forward, but to some other part of the first clause, the negative force is cut off and has to be started afresh by *nor*. The following examples illustrate the danger; in each *or* must be corrected to *nor* if the rest of the sentence is to remain as it is, though some slight change of arrangement such as is indicated would make *or* possible: *In this kind of work there was often little oral preparation of material, little systematic collection of facts and views, well assimilated and digested, or much discussion of balance and proportion.* The writer has forgotten that he began *there was often little* and has ended as though he had said *there was seldom much. Or* must be corrected to *nor was there.* | *In its six months of power it has offered not one constructive measure or done a single thing to relieve suffering* (it has not offered one). | *He did* nothing *without consulting Lovel*, or *failed in anything without expecting and fearing his admonishing* (he did not do anything . . . or fail).

The above are the ordinary types of mistake with *nor*. Others that should hardly require mention are *either . . . nor* and the poetical omission of the first negative. *Either . . . nor* is as bad as NEITHER . . . *or: There was not,* either *in 1796 in Italy,* nor *on the Mediterranean coast of Spain in 1808, any British force at work which . . .* | *As we have not got the world's tonnage production for April,* nor *yet* either *the British* nor *the world's losses for the same month, it is only possible to . . .*

The insertion of a clumsy *and* where *nor* should stand by itself is shown in: *Mr. Burton never underestimates Othello, and nor in consequence do we.* | *The secret encouragement to Kruger from the Germans in 1899 is no part of school history, and nor is Kruger's obstinate withholding of civic rights from the English burghers.*

*Do nor undo* is legitimate in poetry, but not in prose of so ordinary a kind as: *For her fingers had been so numbed that she* could do nor undo *anything.*

**normalcy** (= *normality*) is a word of the 'spurious hybrid' class (see HYBRIDS AND MALFORMATIONS), and seems to have nothing to recommend it. It is said by the OED to be originally U.S., but there is no ground for the charge made against President Harding of having coined it; others had used it long before he did.

**northerly.** For the special uses and meanings of this set of words, see EASTERLY.

**nostalgi(a)(c)** is formed by compounding two Greek words so as to give the meaning of suffering caused by an unfulfilled wish to return home; it was invented as a medical name for homesickness so severe as to amount to a disease. It would be unreasonable to condemn, on etymological grounds, its now common use to describe the wistful melancholy that comes from thinking not of the home that cannot be revisited but of the years that cannot be relived. But its popularity seems to be putting it in danger of less venial extensions which would deprive it of its essential ingredient of pain or suffering. *The point where Sussex, Kent, and Surrey meet—one of the most poetically isolated and nostalgic— in England.* Here the writer seems to have intended no more than *attractive* —a place that one wants to return to.

**nostrum.** Pl. *-ums*, not *-a*; see -UM.

**not.** 1. *Not all* and *all* . . . not. 2. *Not* in meiosis and periphrasis. 3. *Not* in exclamations. 4. *Not* pleonastic. 5. *Not . . . but.* 6. *Not only.* 7. *Not because.* 8. *Not that* etc.

1. *Not all* and *all . . . not. All is not gold that glisters; Every land does not produce everything.* Precisians would rewrite these sentences as *Not all is gold that glisters* (or *Not all that glisters is gold*) and *Not every land produces everything.* The negative belongs logically to *all* and *every*, not to the verbs, and the strict sense of the first proverb would be that glistering proves a substance to be not gold. A valued correspondent has written—'Do not you think that the use of *all . . . not* ought

to be restricted to propositions of the type *All A is not-B*, and where *Not all A is B* is meant, that should be the order? Of course that never has been a rule, from "All of you have not the knowledge of God" onwards, but it would save a great deal of ambiguity if it could be made one. I notice that Somerville and Byrne, in their German Grammar, with *Nicht alle Menschen sprechen Deutsch* before them, translate it "All men do not speak German", neglecting the plain guidance of their original'. This gentleman has logic on his side, logic has time on its side, and probably the only thing needed for his gratification is that he should live long enough. The older a language grows, and the more consciously expert its users become, the shorter shrift it and they may be expected to grant to illogicalities and ambiguities. *All . . . not* for *Not all*, like *the two* FIRST for *the first two*, the displacements of BOTH and NEITHER and ONLY, the omission of *not* in *than you can* HELP, and the use of *much* LESS for *much more*, is already denounced by those who have time to spend on niceties; but it is still, like many other illogicalities, the natural and idiomatic English. It may pass away in time, for *magna est veritas et praevalebit*; in the meantime it is worth anyone's while to get on speaking terms with the new exactitudes (i.e. to write *Not all* himself), but worth nobody's while to fall foul of those who do not choose to abandon the comfortable old slovenries.

2. *Not* in MEIOSIS and PERIPHRASIS. 'We say well and elegantly, not ungrateful, for very grateful'—OED quotation dated 1671. It was a favourite figure of Milton's: Eve was 'not unamazed' at finding that a snake could speak, and Comus's well-placed words were baited with 'reasons not unplausible'. It is by this time a faded or jaded elegance, this replacing of a term by the negation of its opposite; jaded by general over-use; faded by the blight of WORN-OUT HUMOUR with its *not a hundred miles from*, *not unconnected with*, *not unmindful of*, and other

once fresh young phrases. ('One can cure oneself of the *not un-* formation', said Orwell, 'by memorizing this sentence: *A not unblack dog was chasing a not unsmall rabbit across a not ungreen field.*') But the very popularity of the idiom in English is proof enough that there is something in it congenial to the English temperament, and it is pleasant to believe that it owes its success with us to a stubborn national dislike of putting things too strongly. It is clear too that there are contexts to which, for example, *not inconsiderable* is more suitable than *considerable*; by using it we seem to anticipate and put aside, instead of not foreseeing or ignoring, the possible suggestion that so-and-so is inconsiderable. The right principle is to acknowledge that the idiom is allowable, and then to avoid it except when it is more than allowable. Examples occur in every day's newspapers, in which their authors would hardly claim that elegance or point was gained by the double negative, and would admit that they used it only because they saw no reason why they should not; such are: *The style of argument suitable for the election contest is, no doubt,* not infrequently *different from the style of argument suitable for use at Westminster* (often). *One may imagine that Mr. —— will* not be altogether unrelieved *when his brother actor returns tomorrow* (will be much relieved).

3. *Not* in exclamations. *But if you look at the story of that quadrilateral of land, what a complex of change and diversity do you not discover!* A jumble of question and exclamation. The right exclamation would be: *What a complex you discover!* The possible question would be: *What complexity do you not discover?* What a *complex*, and the stop, are essentially exclamatory: *not* is essentially interrogative; *do* is characteristically interrogative, but not impossible in exclamations. The forms in a simpler sentence are: Exclamation: *What I have suffered!*; Question: *What have I not suffered?*; Exclamation with inversion: *What have I suffered!*; Confusion: *What*

*have I not suffered!* See STOPS (question and exclamation marks).

**4.** *Not* pleonastic. The point discussed in (3) was the intrusion of a *not*, unnecessary indeed but explicable, into exclamations that are confused with rhetorical questions. Much less excusable, as needing no analysis to show that it is wrong and often destructive of the sense, is the *not* that is evoked in a subordinate clause as a mere unmeaning echo of an actual or virtual negative in the main sentence, as in *The Home Secretary said he had found nothing to make him doubt that H. was not rightly convicted.* We all know people who habitually say *I shouldn't wonder if it* didn't *turn to snow soon* when they mean *if it turned.* But the same mistake in print is almost as common as it is absurd— so common indeed with wonder and SURPRISE as to rank as a STURDY INDEFENSIBLE. *Nobody can predict with confidence how much time may* not *be employed on the concluding stages of the Bill.* | *I do not of course deny that in this, as in all moral principles, there may* not *be found, here and there, exceptional cases.* | *He is unable to say how much of the portraiture of Christ may* not *be due to the idealization of His life and character.* | *It would not be at all surprising if, by attempting too much, Mr. —— has* not *to some extent defeated his own object.*

**5.** *Not . . . but.* Mrs. Fraser's book, however, is not *confined to filling up the gaps in Livingstone's life . . .* but it *deals most interestingly with her father's own early adventures. . . .* See BUT 3 for more flagrant mishandlings of *not* followed by *but.* The difference between right and wrong often depends on the writer's seeing that the subject, for instance, of the *not* sentence must not be repeated (or taken up by a pronoun) in the *but* sentence; it must be allowed to carry on silently. The above double sentence, which is not idiomatic English as it stands, is at once cured by the omission of *it.*

**6.** *Not only* out of its place is like a tintack loose on the floor; it might

have been most serviceable somewhere else, and is capable of giving acute and undeserved pain where it is. To read the following extracts one after another, all of them requiring only a preference for order over chaos to have tidied them up, must surely call a blush to the Englishman's cheek for his fellow countrymen's slovenly ways: *Ireland, unlike the other Western nations,* preserved n. o. *its pre-Christian literature, but when Christianity came, not direct from Rome but from Britain and Gaul,* that literature received *a fresh impulse from the new faith* (N. o. did Ireland . . . preserve). | *We must remember that n. o. are we concerned in the present situation because South Africa is a member of the Commonwealth but because of our Protectorates* (We are concerned in the present situation n. o. because . . .). | *N. o. had she now a right to speak, but to speak with authority* (She had now a right n. o. to speak). | *N. o. does the proportion of suicides vary with the season of the year, but with different races* (The proportion of suicides varies n. o. with). | *N. o. would this scheme help the poorer districts over their financial difficulties,* but would remove from London the disgrace that in some parts of London the streets are . . . (This scheme would n. o. help). | *N. o. was the audience drawn from central London; those privileged to hear the speech came from all parts* (The audience was not drawn from central London only. The blunder is here double, and before this tintack can be harmless it must be not merely picked up, but smashed up).

**7.** For *not because* as a cause of ambiguity (*Mr. Dayal said he was not returning to New York because of the outbreak of fighting with the Congolese*) see BECAUSE 2.

**8.** *Not that* etc. *Not that, not but what, not so* beginning a sentence (sometimes VERBLESS) to introduce a modification, clarification, or contrast, is established idiom. *The outstanding fact about Mrs. Gaskell is her femininity. Not that Charlotte Brontë and George Eliot are unfeminine.* | *Not* *but what the accident may have been of Herbert's making. | Most men novelists cannot resist it. Not so Trollope.*

**nothing less than.** The OED remarks: 'The combination *nothing less than* has two quite contrary senses', and gives as the first 'quite equal to, the same thing as', with, for illustration, *But yet methinks my father's execution Was nothing less than bloody tyranny*; and as the second 'far from being, any thing rather than', with, for illustration, *Who, trusting to the laws, expected n. l. t. an attack.* To the second sense it adds the description 'Now rare'. As a matter of grammar, either sense is legitimate, *less* being different parts of speech in the two, as appears in the light of paraphrases: my father's death was *no smaller thing than* tyranny (i.e. *less* is an adjective); they expected *nothing in a lower degree than they expected* an attack (i.e. *less* is an adverb); grammar, then, leaves the matter open. But the risks of ambiguity are very great. If the sense of *they expected n. l. t. an attack* did not happen to be fixed by *trusting to the laws*, who would dare decide whether they expected it very much or very little? The sense called by the OED 'now rare' should, in the interests of plain expression, be made rarer by total abandonment; a speaker can show what he means by *n. l. t.* by the way he says it, but a writer cannot. It is unfortunately less rare than the label would lead one to suppose; passages like the two that follow are not uncommon, and are very puzzling to the reader: *It recognizes also both the necessity of reform and liberation from dead dogmas and rubrics, and the impossibility of reform coming from a House of Commons* desiring n. l. t. *to occupy its debates with discussions of the validity of the thirty-nine articles. | Now we are introduced to inspired 'crowd-men' or heroes who have a passion for making order out of the human chaos and finding expression for the real soul of the people; these heroes or crowd-men* resemble n. l. t. *the demagogue as popularly conceived.*

**nought.** See NAUGHT.

**noun** has two adjectives—*nominal* and *nounal*, but is comfortable with neither. The objection to the first is that it is a word much used in other senses. This has induced grammarians to form the word from which they of all people should have shrunk—*nounal*. It is what is described in the article HYBRIDS AND MALFORMATIONS as a *spurious hybrid*; see that article for a discussion of similar words. The grammarian's right course is to work with the word *noun* as far as possible, and, when an adjectival form or an adverb is indispensable, use *nominal(ly).* See also ADJECTIVAL.

**noun-adjectives.** 'Too many *of*s have dropped out of the language', said Lord Dunsany in 1943, 'and the dark of the floor is littered with this useful word.' Some twenty years earlier this phenomenon had provoked the following comment in the first edition of the present dictionary: 'It will be a surprise, and to some an agreeable one, if at this late stage in our change from an inflexional to an analytic language we revert to a free use of the case that we formerly tended more and more to restrict. It begins to seem likely that *drink's victims* will before long be the natural and no longer the affected or rhetorical version of *the victims of drink*. The devotees of inflexion may do well to rejoice; the change may improve rather than injure the language; and if that is so let due praise be bestowed on the newspaper press, which is bringing it about. But to the present (or perhaps already past) generation, which has been instinctively aware of differences between *drink's victims* and *the victims of drink*, and now finds them scornfully disregarded, there will be an unhappy interim. It is the headline that is doing it.'

The last sentence was prophetic. The headline has gone on doing it. The principal destroyer of *of*s is no longer the possessive case, though it still claims its victims: recent examples are *They assume the rumour's truth, | The* *scene of the story's most thrilling complications, | In Johnson's despite, | The two shoes' disparity.* But as chief culprit it has been supplanted by the noun-adjective. There is of course nothing new in putting a noun to this use when no convenient adjective is available; examples abound in everyday speech—*government department, nursery school, television set, test match,* and innumerable others. But the noun-adjective, useful in its proper place, is now running riot and corrupting the language in two ways. It is throwing serviceable adjectives onto the scrap heap; why, for instance, should we speak of an *enemy attack,* a *luxury hotel,* a *novelty number,* an *England eleven,* when we have the adjectives *hostile, luxurious, novel,* and *English?* And, what is worse, it is making us forget that to link two words together with *of* may be both clearer and more graceful than to put the second before the first as an attribute: to forget for instance that, though *nursery school* is a legitimate use of the noun-adjective, *nursery school provision* is an ugly and obscure way of saying *provision of nursery schools;* that if *a large vehicle fleet* were translated into either *a large fleet of vehicles* or *a fleet of large vehicles* an ambiguity would be removed, and that a girl who *could not allay her guilt feelings* would probably find her *guilty feelings* (or her *feelings of guilt*) no less persistent. It is significant that we have dropped the old phrase *state of the world;* today it must always be the *world situation.* For grosser examples of this corruptive influence see HEADLINE LANGUAGE and HYPHENS 5.

**noun and adjective accent.** When a word of more than one syllable is in use both as a noun and as an adjective, there is a certain tendency to differentiate the sound by accenting the last syllable in the adjective, but not in the noun. Thus *There is a co'mpact between us,* but *His style is compa'ct.* The tendency is much less marked than the corresponding one with nouns and verbs (see next article) and seems to be

weakening. A few examples are given, of which the first two are, like *compact,* undisputed, and the rest questionable; from these the reader will be able to form an opinion for application to similar cases: insti'nct a., i'nstinct n.; minu'te a., mi'nute n.; conte'nt a., co'ntent(s) n. (sometimes); adu'lt a., a'dult n. (formerly undisputed but now tending towards a uniform a'dult); expe'rt a., e'xpert n. (same as adult); supi'ne a. (sometimes; and cf. the adv. supi'nely), su'pine n.; upri'ght a. (sometimes when predicative; and cf. the n. uprigh'tness), u'pright n. (= post etc.); u'pstairs adj., upstai'rs adv.

**noun and verb accent, pronunciation, and spelling.** When there is both noun and verb work to be done by a word, and the plan of forming a noun from the verb, or a verb from the noun, by adding a formative suffix (as in *stealth* from *steal*) is not followed, the one word may be called on to double the parts. In that case there is a strong tendency to differentiate by pronunciation, as in *use* (n. *ūs*, v. *ūz*); such a distinction is sometimes, as in *use*, unrecorded in spelling, but sometimes recorded as in *calf* and *calve*. It is not possible to draw up a complete list of the words affected, because the impulse is still active, and the list would need constant additions, especially of words whose pronunciation can be modified without change of spelling. But, as this means that the pronunciation of many words is for a time uncertain, a slight analysis of a fair number of examples may help those who are in doubt. It can be laid down, to start with, that DIFFERENTIATION is in itself an aid to lucidity, and therefore that, when one does not suspect oneself of being the innovator, and the only question is between accepting and rejecting a distinction initiated by others, acceptance is wisdom.

1. The largest class is that of words whose accent is shifted from the first syllable in the noun to the last in the verb. A specimen list follows in which the words marked with an asterisk are

doubtful, either because a tentative differentiation is not yet established, or because an established one is weakening: accent; ally*; commune*; compound; compress; concert; confine(s); conflict; consort; construe*; contest; contract; contrast; converse; convert; convict; convoy*; costume*; decrease; descant; detail(s)*; dictate; digest; discard; discount; discourse; entail*; escort; essay; exploit; extract; ferment; import; imprint; incline*; increase; indent; inlay*; insult; interchange*; invalid (sick person)*; perfume; premise; present; produce; record; reject; suspect. There are also a few words in which some speakers shift the accent, and others go half way by giving the last syllable of the verb with a clear instead of an obscure vowel; so complement, compliment, experiment, implement, etc.

2. Other words, especially monosyllables, are differentiated not by accent but by a modification in noun or verb of the consonantal sound at the end, which is hard in the noun and soft in the verb. This difference is often for the ear only and does not affect spelling; so abuse, betroth (unlike *troth*), close (hard *s* in *cathedral close*), excuse, grease, house, misuse, mouse, mouth. In this class, as in the classes in the preceding paragraph, are words about which usage varies. More often the change of sound is recorded in the spelling, and about such words no doubts arise; but examples are worth giving to confirm the fact that the distinguishing of the parts of speech by change of sound is very common, and that its extension is natural to words whose spelling fails to show it. If the use of *leaf* for *leave* (furlough) by the soldiers of the first world war was an instinctive application of this principle, it provided a rare embryo specimen (though not, it seems, destined to live) to set beside the fully developed ones of which this class mainly consists. Examples are: advice and advise, bath and bathe, belief and believe, brass and braze, breath and breathe, calf and calve, cloth and clothe, device and devise, glass and glaze, grass and graze, grief and grieve, half and halve, life and live, loss and lose, proof and prove, relief and relieve, safe and save, sheath and sheathe, shelf and shelve, strife and strive, thief and thieve, teeth and teethe, wife and wive, wreath and wreathe.

**novelty hunting.** It is a confession of weakness to cast about for words of which one can feel not that they give one's meaning more intelligibly or exactly than those the man in the street would have used in expressing the same thing, but that they are not the ones that would have occurred to him. Anyone can say *surroundings* and *combination* and *total*; I will say *ambience* and *synthesis* and *overall*. Anyone can say *mixed feelings* and *shock* and *workable*; I will say *ambivalence* and *trauma* and *viable*. Everyone is talking about *angry young men*; I will call them *professional iracunds*. Why? Obviously because, there being nothing new in what I have to say, I must make up for its staleness by something new in the way I say it. And if that were all, if each novelty-hunter struck out a line for himself, we could be content to register novelty-hunting as a useful outward sign of inward dullness, and leave such writers carefully alone. Unluckily they hunt in packs, and when one of them has a find they are all in full cry after it, till it becomes a VOGUE WORD, to the great detriment of the language. See that article.

**nth.** For *to the nth*, see N.

**nucleus.** Pl. -*lei* (-*lĕī*).

**nugae.** Pronounce *nū′gē* rather than *nū′jē*.

**number.** Several kinds of mistake are common, and various doubts arise, involving the question of number. With some of them pure grammar is competent to deal; in others accommodations between grammar and sense are necessary or usual or debatable; occasionally a supposed concession to sense issues in nonsense. The following numbered sections are arranged accordingly, the purely grammatical points coming first. 1. Subject and complement of different numbers. 2. Compound subjects. 3. Alternative subjects. 4. Red herrings. 5. Harking back with relatives. 6. Nouns of multitude etc. 7. Singular verb preceding plural subject, and vice versa. 8. *As follow(s)* etc. 9. *Other(s)*. 10. *What*. 11. Pronouns and possessives after *each*, *every*, etc. 12. Nonsense.

1. If subject and complement are of different numbers, how is the number of the verb to be decided? That is, to come to particulars in the simplest form, shall we use *are* in *Clouds are vaporized water*, and *was* in *The last crop was potatoes*, because the subject *clouds* is plural and the subject *crop* singular, or shall we prefer *is* and *were* to suit the number of the complements *water* and *potatoes*? The natural man, faced with these examples has no doubt: 'Of course, *Clouds are*, *The crop was*, whatever may be going to follow.' The sophisticated man, who thinks of *The wages of sin is death*, hesitates, but probably admits that that is an exception accounted for by the really singular sense of *wages* (= guerdon). Both are right; it may in fact be fairly assumed that, when the subject is a straightforward singular (not a noun of multitude, such as *party* etc.), or a straightforward plural (not used in a virtually singular sense like *wages*) and does not consist of separate items (as in *he and she*), the verb follows the number of the subject, whatever that of the complement may be. That it is not as needless as it might be thought to set this down will be plain from the following extracts, some of the simplest form, all violating the rule: *Our only guide* were *the stars*. / *Its strongest point* are *the diagrams*. / *The plausible suggestions to the contrary so frequently put forward* is *an endeavour to kill two birds with one stone*. / *Mr. Coulton contests the idea that the pre-Reformation days* was *an age of religious instruction*. The only comment necessary on these is that when, as in the

first two examples, it makes no difference to the meaning which of two words (*stars* or *guide*, *point* or *diagrams*, is made the subject and which the complement, the one that is placed first must (except in questions) be regarded as subject and have the verb suited to its number: *Our only guide* was *the stars*, or *The stars* were *our only guide*. Such apparent exceptions as *Six months was the time allowed for completion* / *The few days Mrs. Kennedy will spend in London is in the nature of a rest for her*, are not true ones, for here the complement makes it clear that the subject, though plural in form, is singular in sense (*a period of* —). See also THE 5 and THIS 1.

2. Compound subjects. In *Mother and children were killed* we have a compound subject; in *Mother or children are to die* we have not one compound subject, but two alternative subjects; for the latter see 3 below. The compound subject is necessarily plural, whether its components are both plural, of different numbers, or both singular. It is a not uncommon mistake to make the verb singular, as in *They are prepared to retire if and when proper pledges and security is given* / *Their lives, their liberties, and their religion is in danger*. The wrong singulars here, unless they are due to carelessness of the 'red herring' variety described in 4 below, seem to point to a mistaken theory that, when the parts of a compound subject differ in number, the verb follows the nearest. True, grammar may sometimes be overridden if there is a better justification than carelessness or ignorance. For instance, a singular verb is natural where the group forms a compound word like *bread and butter*, and sometimes legitimate when it conveys a single notion, as in *The traditional feeling that killing and violence was against the moral law*, where the two words amount to a HENDIADYS meaning violent killing. But the exception is a narrow one; it is stretched too far if it is pleaded in defence of the earlier quotations; still less can it

justify *Sunshine and showers is again forecast for today.* For sentences of this type in which the verb precedes the compound subject see 7 below.

3. Alternative subjects. *The reading aloud of poetry or prose, or speaking them by heart, compel the student to study their meaning closely.* / *United Nations troops in the Congo have been ordered to shoot if life or property are in danger.* When both alternatives are singular in grammar and in sense the verb can only be singular. These quotations, one from an educational pamphlet and the other from a news agency, show how easy it is to blunder into a plural. (For the same mistake with *neither . . . nor* see NEITHER 4.) But when the alternatives differ in number, as in *Mother or children are to die, Is the child or the parents to be blamed?*, the methods in order of merit are: (*a*) Evade by finding a verb of common number: *Mother or children must die,* Shall *the child or the parents be blamed?*; (*b*) Invoke ellipsis by changing the order: *The mother is to die, or the children, Is the child to be blamed, or the parents?*; (*c*) Give the verb the number of the alternative nearest it: *Mother or children are to die, Is the child or the parents to be blamed?*

4. Red herrings. Some writers are as easily drawn off the scent as young hounds. They start with a singular subject; before they reach the verb, a plural noun attached to an *of* or some other similar distraction happens to cross, and off they go in the plural; or vice versa. This is a matter of carelessness or inexperience only, and needs no discussion; but it is so common as to call for a few illustrations: *This argument for Mr. Macmillan's attending the Assembly does not alter the irksome fact that he, in common with President Eisenhower, are dancing to Mr. Khrushchev's tune.* / *The results of the recognition of this truth is . . .* / *The foundation of politics are in the letter only.* / *Offering opinions are as far as they are permitted to go.* / *The partition which the two ministers made of the powers of government were singularly happy.*

5. Harking back with relatives. *Who, which,* and *that,* can in themselves be singular or plural, and there is a particular form of sentence in which this produces constant blunders. *He is one of the best men that have ever lived* (with which compare *He is one that has lived honestly*). In the first sentence there are two words capable of serving as antecedent to *that,* viz. *one* (as in the bracketed sentence) and *men.* A moment's thought shows that *men* is the antecedent necessary to the sense: Of the best men that have ever lived (or of the best past and present men) he is one. But with *one* and *men* (or their equivalents) to attach the relative to, writers will hark back to *one* in spite of the nonsense it gives, and make their verbs singular: *He is another of the numerous people who is quite competent in the art of turning what he has to say into rhyme and metre.* / *Vaughan Williams is one of those contemporary composers who does not feel the need for a new medium.* / *Detective-inspector J. H., one of the officers who has been helping in investigating the great train robbery.*

An example or two offering peculiarities may be added: *Describing him as one of those busy men who in some remarkable way* find *time for adding to* his *work;* to have got safely as far as *find,* and then break away with *his,* is an odd freak. *Houdin was a wonderful conjurer, and is often reckoned the greatest of his craft who* have *ever lived;* this reverses the usual mistake: Is the greatest who *has,* Is one of the greatest who *have.* *She was wearing one of those loose, light, almost childish raincoats which was faintly reminiscent of an academic gown.* This writer has left us guessing whether he means raincoats, which *was* or raincoats that *are.* See THAT (REL.).

6. Nouns of multitude etc.

(*a*) Such words as *army, fleet, Government, company, party, pack, crowd, mess, number, majority,* may stand either for a single entity or for the individuals who compose it, and are called nouns of multitude. They are treated as singular or plural at discretion—and sometimes, naturally, without discretion. *The Cabinet is divided* is better, because in the order of thought a whole must precede division; and *The Cabinet are agreed* is better, because it takes two or more to agree. That is a delicate distinction, and few will be at the pains to make it. Broader ones that few will fail to know are that between *The army* is *on a voluntary basis* and *The army* are *above the average civilian height*, and that between *The party lost* their *hats* and *The party lost* its *way*. In general it may be said that while there are always a better and a worse in the matter, there are seldom a right and a wrong, and any attempt to elaborate rules would be waste labour.

But if the decision whether a noun of multitude is to be treated as a singular or as a plural is often a difficult business, and when ill made results at worst in a venial blemish, failure to abide by the choice when made, and plunging about between *it* and *they*, *have* and *has*, *its* and *their*, and the like, can only be called insults to the reader. A waiter might as well serve one with a dirty plate as a journalist offer one such untidy stuff as: *The University of London Press hopes to have ready the following additions to* their *series of* . . . */ The Times also* gives *some interesting comments by* their *special correspondent.* / *During* their *six years of office the Government* has *done great harm.* / *The village is at work now and ready to do* their *bit.* / *When the generation which participated, and which* was *represented at yesterday afternoon's affair,* have *gone* . . . / *With* their *children's programme the B.B.C. never* puts *a foot wrong.*

(b) *There* are *heaps more to say, but I must not tax your space further.* The plurals *heaps* and *lots* used colloquially for a great amount now always take a singular verb unless a plural noun with *of* is added: *There* is *heaps of ammunition,* but *There* are *heaps of cups; There* is *lots to do,* but *Lots of people* think *so.* Compare the use of *half* in *Half of it* is *rotten,* but *Half of them* are *rotten.*

(c) When the word *number* is itself the subject it is a safe rule to treat it as singular when it has a definite article and as plural when it has an indefinite. *The number of people present was large,* but *A large number of people were present.* In *Before the conclave begins in a fortnight's time a number of details has to be settled* the singular is clearly wrong; it is the details that have to be settled not the number; *a number of details* is a composite subject equivalent to *numerous details.* This use of *a number of* in the sense of more than one but not a great many is idiomatic, but the almost absurd vagueness of the expression if interpreted literally makes careful writers prefer an adjective such as *some, several, many, numerous*; this has the advantage too of leaving no doubt that the verb must be plural.

7. Singular verb preceding plural subject and vice versa. The excuse for this in speaking—often a sufficient one—is that one has started one's sentence before fixing the precise form of the subject, though its meaning may have been realized clearly enough. But the writer both can and ought to do what the speaker cannot, correct his first words before the wrong version has reached his audience. If he does not, he too, like the waiter with the dirty plate (see 6 (a)), is indecently and insultingly careless. Examples: *For the first time there* is *introduced into the Shipyard Agreement clauses which hold the balance equally.* / *A book entitled 'America's Day', by Ignatius Phayre, in which* is *discussed the pressing problems of home and foreign policy that* ... / *On these questions there* is *likely to be acute differences among the political groups and parties.* / *Where only three years ago* was *pasture land now stands vast engineering shops, miles of railway sidings, and the constant hum of machinery.*

The converse mistake is seldom made; in the following, the red herring of *these* no doubt accounts for *are*: *The Thames has certain natural disadvantages as a shipbuilding centre; to these* are *added an artificial disadvantage.*

When the verb precedes a subject compounded of singular and plural, some questions of more interest than importance may arise. *There were a table and some chairs in there*; *were* is better because the compound subject is compact. *There were a plain deal table in there and some wicker armchairs which Jorgenson had produced from somewhere in the depths of the ship.* The alteration of *were* to *was* would now be an improvement; but why, if *were* was best in the bare framework given first? How has the author elaborated it? First and least, he has made table and chairs less homogeneous, less the equivalent of 'some articles of furniture', by describing one as plain deal and the others as wicker; secondly, he has attached to *chairs* and not to *table* a long relative clause; third and most important, in order to cut off the relative clause from *table* he has had to shift *in there* to an earlier place. The result is that the verbal phrase (*there were . . . in there*) is so arranged that it encloses one item of the compound subject (table) and leaves the other (chairs) out in the cold. The author would have done better to write *was* and let the second part be elliptical with *there were in there* to be understood out of *there was in there*.

**8.** *As follow(s), concern(s), regard(s),* etc. *For higher incomes than £1,000 the new rates will be as follow. As follow* is not English; *as follows* is; for discussion of the point see FOLLOW.

**9.** *Other(s). The wrecking policy is, like* other *of their adventures in recent times, a dangerous gamble.* Read *other adventures of theirs*; for discussion see OTHER 4.

**10.** *What. What* provoke *men's curiosity are mysteries.* See WHAT for the question whether it can be plural.

**11.** Pronouns and possessives after *each, every, anyone, no one, one,* etc. *Everyone without further delay gave* themselves *up to rejoicing.* | *But, as* anybody *can see for* themselves, *the quotation of the actually relevant portion of the argument in our columns would have destroyed . . . Each* and the rest are all singular; that is undisputed; in a perfect language there would exist pronouns and possessives that were of as doubtful gender as they and yet were, like them, singular; i.e., it would have words meaning him-or-her, himself-or-herself, his-or-her. But, just as French lacks our power of distinguishing (without additional words) between his, her, and its, so we lack the French power of saying in one word his-or-her. There are three makeshifts: first, *as anybody can see for himself or herself*; second, *as anybody can see for themselves*; and third, *as anybody can see for himself.* No one who can help it chooses the first; it is correct, and is sometimes necessary, but it is so clumsy as to be ridiculous except when explicitness is urgent, and it usually sounds like a bit of pedantic humour. The second is the popular solution; it sets the literary man's teeth on edge, and he exerts himself to give the same meaning in some entirely different way if he is not prepared to risk the third, which is here recommended. It involves the convention (statutory in the interpretation of documents) that where the matter of sex is not conspicuous or important the masculine form shall be allowed to represent a person instead of a man, or say a man (homo) instead of a man (vir.) Whether that convention, with *himself or herself* in the background for especial exactitudes, and paraphrase always possible in dubious cases, is an arrogant demand on the part of male England, everyone must decide for himself (or for himself or herself, or for themselves). Have the patrons of *they* etc. made up their minds yet between *Everyone* was *blowing their noses* (or *nose*) and *Everyone* were *blowing their noses?* For a further discussion of this question see THEY.

**12.** Nonsense. *He comes for the first time into the Navy at an age when naval officers—unless* they *are so meritorious or so fortunate as to be* one of *the three Admirals of the Fleet—are compelled by law to leave it.* Naval officers cannot be one admiral; and what is wrong with *unless they are Admirals of the Fleet?*

**numeracy** is a word coined by the Committee on Education presided over by Sir Geoffrey Crowther in 1959 as a term for that complement which is desirable in the sixth-form education of arts specialists in the same way as literacy is in that of science specialists. It is defined as 'not only the ability to reason quantitatively but also some understanding of scientific method and some acquaintance with the achievements of science'. Clearly there is need for such a word; whether this one has come to stay remains to be seen.

**numerous** is not a pronoun, as the following extract makes it: *These men have introduced no fewer than 107 amendments, which they know perfectly well cannot pass, and* numerous of which *are not meant to pass.* See VARIOUS, which is more often misused in the same way.

**nurs(e)ling.** The form recommended, though rather less common hitherto than the other, is *nurseling*; see MUTE E for the criterion.

# O

**-o-** is a connecting vowel of Greek origin, its extended modern function being so to shape the end of a Greek or Latin word that when a suffix or another word is applied to it the two will coalesce recognizably into a single derivative or compound. Three points should be noticed: 1. The part ending in *-o-* is not a word, even though it might be used by itself as a curtailed one (*hydro*, *photo*, etc.) In the compound it is essentially the beginning only of a word; *We owe it to the genius of Hertz that we are now able to measure directly the velocity of* electro and *magnetic waves*; *electro* is there used as an adjective instead of *electric*, and is indefensible. 2. The words fit for the *-o-* treatment are, if not necessarily authentic ancient Greek or Latin, at

least such as may pass for Greek or Latin. If the ancient Romans did not call the Russians *Russi* or talk of *America* and *Americani*, we can suppose that was only because they had not the chance, and we are therefore entitled to make *Russo-*, *Americo-*, and *Americano-*. But the Greeks and Romans did know what speed was, and no one supposes they called it *speed*, whence it follows that *speedo-* and *speedometer* are 'barbaric' formations. 3. It is not enough that the word to be treated should be actual or possible Latin or Greek; the shaping must be done in the right way. *We must take account of* religio-*philosophic speculations with regard to the nature of Eternal Life*; Latin, it is true, has both *religio* and *religiosus*, but only the second admits of the *-o-* treatment, and it gives *religioso-philosophic*. See also HYBRIDS AND MALFORMATIONS and -LOGY.

**oaf.** Plural *oafs* (*the muddied oafs at the goals*); but some dictionaries still gives *oaves* as an alternative.

**o and oh.** Usage has changed, *Oh* having formerly been prevalent in many contexts now requiring *O*, and is still by no means fixed. The present tendency is to restrict *Oh* to places where it has a certain independence, and to prefer *O* where it is proclitic or leans forward upon what follows; which means for practical purposes that as the sign of the vocative (*O God our help*; *O mighty-mouthed inventor of harmonies*) *O* is invariable, and as an exclamation the word is *O* when no stop immediately follows it, but otherwise *Oh* (*Oh, what a lie!*; *Oh! how do you know that?*; *O for the wings of a dove!*; *O who will o'er the downs with me?*; *O worship the King!*). This distinction is observed in *Hymns Ancient and Modern*, but *The English Hymnal* and *Songs of Praise* have a uniform *O*.

**oath.** Pl. pron. *ōdhz*; see -TH AND -DH.

**obdurate,** adj. The OED quotations show Shakespeare, Milton, and Barham, for *ŏbdūr'ăt* and Shelley for *ob′dŭrăt*. The former is still sometimes heard, but is old-fashioned. See RE-CESSIVE ACCENT.

**obedient.** For *yours obediently* etc. see LETTER FORMS.

**object**, vb., **objection.** The infinitive construction is deprecated and the gerund recommended. *The cab-drivers object to pay their proportion of the increase* (read *paying*). | *They have been blocked by the objections of farmers and landlords to provide suitable land* (read *providing*).

**objective, object,** nn. I. LOVE OF THE LONG WORD has led to the popular use of *objective* for *object* in its sense of purpose, aim, or end. *Objective* (*point*) is a military term meaning the specific aim or topographical target of a military operation. It should be confined to contexts that do not strain the metaphor. 2. *The object of the exercise*, meaning the purpose for which something was done (even though the doing of it may have been in no sense an 'exercise') is a CLICHÉ born in operational training during the second world war. Sometimes it is used jocularly (it has the same kind of attraction as POLYSYLLABIC HUMOUR) and sometimes seriously, intruding, as all clichés will, in unsuitable places, e.g. in this comment on a deportation order: *What one has here as the object of the exercise is not only the intention to place him on a ship leaving the country but the express intention of delivering him to the U.S. government.*

**objective and subjective** (adjj.) are terms of philosophy and physiology distinguishing concepts and sensations that have an external cause from those that arise only in the mind. (They are also now the usual grammatical terms for what used to be called the accusative and nominative cases.) They have become POPULARIZED TECHNICALITIES, treated as stylish substitutes for commoner and more precise words such as (un)biased, (dis)interested, (im)partial, (un)prejudiced. *The answer, as every objective Australian will admit, was hysterical batting.* | *Television in*

many *of its aspects is a battlefield of subjective judgements.* In the first quotation the writer meant fair-minded, but that was too commonplace a word for him. In the second, if the writer had paused to ask himself what he meant, he might have had to answer that all he was saying, rather tritely, was that people disagree about the merits of television programmes because tastes differ.

**objective genitive.** The genitive that stands to a verbal noun or noun of action in the same relation as the object to a verb. In *fear God, God* is the object of the verb, and, in *put the fear of God into them, God* is in the same relation to the noun *fear,* and *of God* is called the objective genitive. In English the 'of' genitive is usual, but the inflected genitive or the possessive adjective also occurs, as in *the President's murder* and *the deep damnation of his taking-off.*

**object-shuffling.** The conferring of a name on a type of mistake, making it recognizable and avoidable, is worth while if the mistake is common. *Object-shuffling* describes what unwary writers are apt to do with some of the many verbs that require, besides a direct object, another noun bearing to them a somewhat similar relation, but attached to them by a preposition. You can inspire courage in a person, or inspire a person with courage; the change of construction is object-shuffling, which, with the verb *inspire,* is legitimate and does not offend against idiom. But with *instil* the object-shuffling would be wrong; you can instil courage into a person; to instil a person with courage is contrary to idiom. Wherever reference is made under any word to this article, the meaning is that with that word object-shuffling is not permissible. Most of the verbs liable to this maltreatment are derived from Latin verbs compounded with prepositions and therefore beginning with *in-, sub-, pre-,* etc. The Latin scholar, aware that the verbal parts of *substituo* and *instillo* and *praefigo* mean to put and to pour and to fasten, instinctively chooses for their direct objects the stopgap, the influence, and the appendage, not the thing displaced, the person influenced, and the main body; and in writing of the more educated kind his example is followed. But the non-Latinist, if he is also unobservant, gives *substitute* the construction of *replace*, *instil* that of *fill*, and *prefix* that of *preface*. It is seldom that the mistake is made with non-Latin words; an example will be found under FOIST. Two specimens may be here given; the reader who wishes for more will find them under the words ENFORCE, INCULCATE, IN- FLICT, INFUSE, INSTIL, PREFIX, SUBSTI- TUTE, and others. *A quarterly which is doing so much to* initiate into the minds *of the British Public what is requisite for them to know* (to initiate the B.P. in what is requisite). / *The ecclesiastical principle was* substituted by *the national* (the national principle was substituted for the ecclesiastical).

**obligated** as a synonym for *obliged* (having received a favour etc.) is now a mere solecism; but in the full sense of bound by law or duty *to* do some- thing it is still used, especially in legal language.

**obligatory.** The pronunciation recom- mended is *ŏblĭ′gătŏrĭ* in preference to either *ŏb′lĭgătŏrĭ* or *ŏblĭgāt′ŏrĭ* given as alternatives by some dictionaries. See RECESSIVE ACCENT.

**oblige.** The derivatives of *o.* and *obligate* (see OBLIGATED above) are troublesome; there are two possible adjectives in *-able* (see -ABLE 1), viz. *obligable* from *obligate* (= that can be legally bound; pronounce *ŏ′blĭgăbl*), and *obligeable* from *oblige* (= that can have a favour conferred; pron. *ŏblī′- jăbl*). *Obligee* and *obligor* belong in sense to *obligate*, and have curious meanings: *obligor*, not one who confers an obligation, but one who binds or obligates himself to do something; *obligee*, not one who is obliged, but one to whom a service is due (towards whom a duty has been undertaken). The dictionaries are not agreed

whether we should say *-ĭjor* or *ĭgor*, *-ĭjē* or *ĭjē*; *igor* and *ĭ′jē* are recom- mended.

**obliqueness, obliquity.** There is some tendency to confine the latter to the secondary or figurative senses; *obliquity* of mind or judgement or out- look, but *obliqueness* of the line or ground; cf. OPACITY, and see -TY AND -NESS. It is perhaps well to encourage such DIFFERENTIATION.

**oblivious.** A word misused in two ways. (1) Its original sense is no longer aware or no longer mindful, not simply unaware or unconscious or insensible. But it has strayed beyond recall from the area of forgetfulness into the wider one of heedlessness. *A contempt to which the average Englishman in his happy self-sufficiency is generally o.* / *He may have driven off quite o. of the fact that any harm had been done.* / *Singly or in groups, o. to the traffic in the streets, they pursued their eager quest.* (2) Even when the word might bear its true sense of forgetful (as opposed to unaware), it is often followed by the wrong preposi- tion (*to*). This is a natural result of the misuse explained in (1); it will be noticed that two of the three quota- tions there given show *to* instead of *of*, perhaps on the analogy of *insensible to*. But in the following examples *to* has been used even where the meaning might otherwise be the correct one of forgetful: *Each of them oblivious to the presence of anybody else, and intent on conversation.* / *A principle to which the romances of the eighteenth century were curiously oblivious.* / *Mr Humphreys is always oblivious to the fact that the minority in one part of the kingdom is represented by the majority in another part.*

The making of these mistakes is part of the price paid by those who reject the homely word, avoid the obvious, and look about for the imposing. *Forgetful, unaware, unconscious, un- mindful,* and *heedless,* while they usually give the meaning more pre- cisely, lay no traps.

**obnoxious** has two very different senses, one of which (exposed or open or liable *to* attack or injury) requires notice because its currency is now so restricted that it is puzzling to the uninstructed. It is the word's rightful or *de jure* meaning, and we may hope that scholarly writers will keep it alive, as they have hitherto succeeded in doing. Meanwhile the rest of us need not scruple to recognize the usurping or *de facto* sense of offensive or objectionable. This has perhaps no right to exist ('apparently affected by association with noxious' says the OED), but it does and will, and, unlike the other, it is comprehensible to everyone.

**oboe.** See HAUTBOY. The player is an *oboist*, not *oboeist*.

**observance, observation.** The useful DIFFERENTIATION in virtue of which neither word can be substituted for the other, and each is appropriated to certain senses of *observe*, should not be neglected. *Observance* is the attending to and carrying out of a duty or rule or custom; it has none of the senses of *observation* (watching, noticing, etc.), and *observation* in turn does not mean performing or complying. Though the distinction is modern, its prevalence in good writing may be judged from the OED's having only one 19th-c. and no 20th-c. example of *observance*, as against many of *observation*, in the sense consciously seeing or taking notice. It has been strengthened by the use of *observation* in such expressions as *o. car*, *o. balloon*, *o. post*, and *o. ward*. Unfortunately a perverted taste for out-of-the-way forms has been undoing this useful achievement, and such uses as the following, almost unknown for two or three centuries, have again become common: *To reinforce observance with imagination.* / *Emerson does not check his assumptions; he scorns observance.* / *From him Mr. Torr inherits both his gift for exact observance and lively humour.* / *His early poetry, the product of exalted sensation rather than of careful observance.* / *Whose powers of observance and memory have combined to make as varied a raconteur as . . . In all these the word should be *observation*; one quotation is added in which *observation* is wrongly used for *observance*: *The British Government has failed to secure the observation of law* and has lost the confidence of all classes.*

**obstetric(al).** See -IC(AL). The short form is much commoner, and is recommended; its formation is in fact faulty (a midwife is *obstetrix*, *-icis*, so that *obstetricic* would be the true adjective), while that of *obstetrical* would pass; but only pedantry would take exception to *obstetric* at this stage of its career.

**obtain.** See FORMAL WORDS. *Customer* —Can you get me some? *Shopman*— We can o. it for you, madam.

**octavo.** See FOLIO. Pl. *-os*; see -O(E)S 6. Pronounce *ā* not *ah*.

**octingentenary, octocentenary, octocentennial.** See CENTENARY.

**octopus.** Pl. *-uses*; *-pi* is wrong and *-podes* pedantic.

**octoroon, -taroon.** Both are bad forms, the *-r-* being imported from *quadroon*, which has a right to it. But the second is worse than the first, since *octa-* is not (like *quadr-*) Latin, but Greek. For meaning, see MULATTO 2.

**oddment.** Though the word itself is established and useful, its formation is anomalous (see -MENT) and should not be imitated.

**ode.** The OED definition of the word in its prevailing modern sense may be given: 'A rhymed (rarely unrhymed) lyric, often in the form of an address; generally dignified or exalted in subject, feeling, and style, but sometimes (in earlier use) simple and familiar (though less so than a *song*).' But what with confusion between this very comprehensive modern sense and the more definite Greek sense (as in *choric ode* and *Pindaric ode*), what with the obvious vagueness of the modern sense itself, and what with the fact that 'elaborate' and 'irregular' are both

epithets commonly applied to ode metres, the only possible conception of the ode seems to be that of a Shape
If shape it might be called that shape had none
Distinguishable in member, joint, or limb.

**oe, œ, e.** See Æ, Œ. The following spellings of words beginning with *oe* or its substitutes are recommended: ecology etc.; ecumenical; oedema etc.; Oedipus; oesophagus; oestrum. The pronunciation in all is simply *ē*.

**-o(e)s.** The Englishman has a legitimate grievance against the words in *-o*. No one who is not prepared to flout usage and say that for him every word in *-o* shall make *-oes*, or all shall make *-os*, can possibly escape doubts; one kind of whole-hogger will have to write *heros* and *cargos* and *potatos* and *gos* and *negros*, while the other kind must face *embryoes, photoes, cameoes, fiascoes,* and *generalissimoes*. In this book, many words in *-o* have been entered in their dictionary places with the plurals that seem advisable; here, one or two guiding principles may be tentatively suggested. Although there are several hundred nouns in *-o*, the ending is one that is generally felt to be exotic, and only a small minority are allowed the plural in *-oes*, which, since it is invariable in many very familiar words (*no, go, cargo, potato, hero, negro,* etc.) might be considered the normal form. It must be understood that the following rules are not more than generally true, and that sometimes they come to blows with each other over a word, and that such battles may remain long undecided.

1. Words used as freely in the plural as in the singular usually have *-oes*, though there are very few with which it is invariable; names of animals and plants fall naturally into this class. So *banjoes; bravoes; cargoes; dingoes; dominoes; flamingoes; heroes; potatoes.*

2. Monosyllables take *-oes*; so *goes, noes.*

3. Words of the kind whose plural is seldom wanted or is restricted to

special uses have *-os*; so *dos* (the musical note); *crescendos; dittos; guanos; infernos; lumbagos.*

4. When a vowel precedes the *-o*, *-os* is usual, perhaps because of the bizarre look of *-ioes* etc.; so *arpeggios; bagnios; cameos; embryos; folios; punctilios.*

5. The curtailed words made by dropping the second element of a compound or the later syllables have *-os*; so *chromos; dynamos; magnetos; photos; pros; stylos.*

6. Alien-looking or otherwise queer words have *-os*; so *albinos; alto-relievos; centos; commandos; duodecimos; fiascos; ghettos; lingos; medicos; negrillos.*

7. Long words tend to *-os*; so *archipelagos* (or *-oes*); *armadillos; generalissimos; manifestos.*

8. Proper names have *-os*; so *Gallios; Lotharios; Neros; Romeos.*

**of.** This preposition shares with another word of the same length, AS, the evil glory of being accessary to more crimes against grammar than any other. But, in contrast with the syntax of *as*, which is so difficult that blunders are very excusable, that of *of* is so simple that only gross carelessness can lead anyone astray with it. Nevertheless, straying is perpetual, and the impression of amateurishness produced on an educated reader of the newspapers is discreditable to the English Press. Fortunately, the commonest type of blunder with *of* is very definite and recognizable, so that the setting of it forth with sufficient illustration has a real chance of working some improvement. That type is treated in the first of the following sections, the list of which is: (1) Wrong patching; (2) Patching the unpatchable; (3) Side-slip; (4) Irresolution; (5) Needless repetition; (6) Misleading omission; (7) Some freaks of idiom.

1. Wrong patching. In the examples that follow the same thing has happened every time. The writer composes a sentence in which some other preposition than *of* occurs once but governs two nouns, one close after it

and the other at some distance. Looking over his sentence, he feels that the second noun is out in the cold, and that he would make things clearer by expressing the preposition for the second time instead of leaving it to be understood. So far, so good; care even when uncalled for is meritorious. But his stock of it runs short, and instead of ascertaining what the preposition really was he hurriedly assumes that it was the last in sight, which happens to be an *of* that he has had occasion to insert for some other purpose; that *of* he now substitutes for the other preposition whose insertion or omission was a matter of indifference, and so ruins the whole structure. In the examples, the three prepositions concerned are in roman type; in each case the reader will notice that to correct the sentence it is necessary either to omit the later of the two *of*s or to alter it to the earlier preposition: *An eloquent testimony* to *the limits* of *this kind of war, and* of *the efficiency of right defensive measures.* | *He will be in the best possible position* for *getting the most out* of *the land and* of *using it to the best possible advantage.* | *The definite repudiation of militarism as the governing factor* in *the relation of States and* of *the future moulding of the European world.* | *A candidate who ventured to hint* at *the possible persistence* of *the laws of economics, and even* of *the revival of the normal common-sense instincts of trade.* | *The Ministry aims not merely* at *an equitable division* of *existing stocks, but* of *building up reserves against the lean months.*

2. Patching the unpatchable. These resemble the previous set so far as the writers are concerned. They have done the same thing as before; but for the reader who wishes to correct them there is the difference that only one course is open; *of* must be simply omitted; the other preposition (*between* or *without*) cannot be repeated. We can say *for you and for me* instead of *for you and me* if we choose, but not *between you and between me* instead of *between you and me*; *with cries and with tears* means the same as *with*

cries and tears, but *without cries or without tears* does not mean the same as *without cries or tears*; on this point, see OVERZEAL. *It could be done without unduly raising the price of coal,* or of *jeopardizing new trade.* | *He will distinguish between the American habit of concentrating upon the absolute essentials, of 'getting there' by the shortest path, and* of *the elaboration in detail and the love of refinements in workmanship which mark the Latin mind.* | *Without going into the vexed question of the precise geographical limitations,* or of *pronouncing any opinion upon the conflicting claims of Italy and* of *the Yugo-Slavs, what may be said is that . . .*

3. Side-slip. Besides the types given in the previous sections, so beautifully systematic in irregularity as almost to appear regular, there are more casual aberrations of which no more need be said than that the sentence is diverted from its track into an *of* construction by the presence somewhere of an *of*. Analogous mistakes are illustrated in the article SIDE-SLIP. *The primary object was not the destruction of the mole forts,* or of *the aeroplane shed,* or of *whatever military equipment was there,* or even of *killing or capturing its garrison.* | *Its whole policy was, and is, simply to obstruct the improvement of the workingman's tavern, and* of *turning every house of refreshment and entertainment in the land into that sort of coffee tavern which . . .*

4. Irresolution. *Here again we have illustrated Germany's utter contempt for her pledged word and* of *her respect for nothing but brute force.* | *His view would be more appropriate in reference to Hume's standpoint than* of *the best thought of our own day.* These are the results of having in mind two ways of putting a thing and deciding first for one and then for the other: *we have illustrated,* and *we have an illustration of*; *to Hume's standpoint* (*than to the thought*), and *to the standpoint of Hume* (*than of the thought*).

5. Needless repetition of *of*. *There is a classical tag about the pleasure of being on shore and* of *watching other folk in a big sea.* A matter not of

grammar, but of style and lucidity; in style the second *of* is heavy, and in sense it obscures the fact that the pleasure lies not in two separate things but in their combination.

6. Omission of *of*. *The prohibition of meetings and the printing and distribution of flysheets stopped the Radicals' agitation*. Unless an *of* is inserted before *the printing*, the instinct of symmetry compels us to start by assuming that *the printing* etc. *of fly-sheets* is parallel to *the prohibition of meetings* instead of, as it must be, to *meetings* alone. For the modern tendency to displace *of* and other prepositions by using nouns attributively see NOUN-ADJECTIVES and HEADLINE LANGUAGE.

7. Some freaks of idiom. *You are the man* of *all others that I did not suspect. He is the worst liar* of *any man I know. A child* of *ten years* old. *That long nose* of *his*. The modern tendency is to rid speech of patent ILLOGICALITIES; and all the above either are, or seem to persons ignorant of any justification that might be found in the history of the constructions, plainly illogical: the man of all *men*; the worst liar of *all liars*; a child *of ten years*, or a child *ten years old*; a *friend* of mine, i.e. among my friends, but surely not that *nose* of his, i.e. among his noses: so the logic-chopper is fain to correct or damn; but even he is likely in unguarded moments to let the forbidden phrases slip out. Some will perhaps be disused in time; meanwhile they are recognized idioms—STURDY INDEFENSIBLES possibly, though not without their defenders. Jespersen, for instance, has shown that the use of *of* in such constructions as *A child of ten years old* and *That long nose of his* is as old as Caxton, and has argued that *of* is here not partitive but appositional—merely a grammatical device to make it possible to join words which for some reason or another it would otherwise be impossible to join. The latter construction may be found in the opening lines of *Antony and Cleopatra*: 'Nay but this dotage of our general's O'erflows the measure'.

officialese is a pejorative term for a style of writing marked by peculiarities supposed to be characteristic of officials. If a single word were needed to describe those peculiarities, that chosen by Dickens, *circumlocution*, is still the most suitable. They may be ascribed to a combination of causes: a feeling that plain words sort ill with the dignity of office, a politeness that shrinks from blunt statement, and, above all, the knowledge that for those engaged in the perilous game of politics, and their servants, vagueness is safer than precision. The natural result is a stilted and verbose style, not readily intelligible—a habit of mind for instance that automatically rejects the adjective *unsightly* in favour of the PERIPHRASIS *detrimental to the visual amenities of the locality*.

This reputation, though not altogether undeserved, is unfairly exaggerated by a confusion in the public mind between *officialese* and what may be termed *legalese*. For instance a correspondent writes to *The Times* to show up what he calls this 'flower of circumlocution' from the National Insurance Act 1959; it ought not, he says, to be allowed to waste its sweetness on the desert air. *For the purpose of this Part of this Schedule a person over pensionable age, not being an insured person, shall be treated as an insured person if he would be an insured person were he under pensionable age and would be an employed person were he an insured person.*

This is certainly not pretty or luminous writing. But it is not officialese, nor is it circumlocution. It is legalese, and the reason why it is difficult to grasp is not that it wanders verbosely round the point but that it goes straight there with a baffling economy of words. It has the compactness of a mathematical formula. Legalese cannot be judged by literary standards. In it everything must be subordinated to one paramount purpose: that of ensuring that if words have to be interpreted by a Court they will be given the meaning the draftsman intended. Elegance cannot be expected from anyone so circumscribed. Indeed it is hardly an exaggeration to say that the more readily a legal document appears to yield its meaning the less likely it is to prove unambiguous. It is fair to assume that if the paragraph quoted were to be worked out, as one would work out an equation, it would be found to express the draftsman's meaning with perfect precision.

If an official were to use those words in explaining the law to a 'person over pensionable age not being an insured person' he would indeed deserve to be pilloried. The popular belief that officials use an esoteric language no doubt derives partly from the reluctance they used to feel to explain the law in their own words lest they should be accused of putting a gloss on it. But, now that the daily lives of all of us are affected by innumerable laws, officials have had to overcome this inhibition and act as interpreters; they could not get through their work if they had not learned to express themselves to ordinary people in a way that ordinary people can understand. Circumlocution is rife in present-day writing and speaking, but officials are no longer markedly worse than other people; they are probably better than most. But the following examples show that they still sometimes fall into the old bad habit of giving explanations in terms only fit for an Act of Parliament, if that:

*'Appropriate weekly rate' means, in relation to any benefit, the weekly rate of personal benefit by way of benefit of that description which is appropriate in the case of the person in relation to whom the provision containing that expression is to be applied. | Unemployment benefit is not payable in respect of 13.2.56 to 17.2.56 which cannot be treated as days of unemployment on the ground that the claimant notwithstanding that this employment has terminated received, by way of compensation for the loss of the remuneration which he would have received for that day each of those days if the unemployment had not been terminated, payment of an amount which exceeds the amount arrived at by deduct-*

*ing the standard daily rate of unemployment benefit from two thirds of the remuneration lost in respect of that day each of those days.*

That circumlocution may occasionally be found even in the utterances of those official 'spokesmen' who ought to know better, may be illustrated by this extract from a London evening newspaper:

*Discussing Anglo-American talks on the Barnes Wallis folding-wing plane, a Ministry spokesman said:*

*'The object of this visit is a pooling of knowledge to explore further the possibility of a joint research effort to discover the practicability of making use of this principle to meet a possible future NATO requirement, and should be viewed in the general context of interdependence.'*

*Or, (our version):*

*'This visit is to find out whether we can, together, develop the folding wing for NATO.'*

**officious** has a meaning in diplomacy oddly different from its ordinary one A diplomatist means by *an o. communication* much what a lawyer means by one without prejudice; it is to bind no one, and, unless acted upon by common consent, is to be as if it had not been. It is the antithesis of *official*, and the notion of meddlesomeness attaching to it in ordinary use is entirely absent. But the risk of misunderstanding is obvious, and the word is now rarely used.

**offing, offish,** etc. The vowel sound *awf* in *off* and its compounds that formerly prevailed in Southern or 'standard' English (see RECEIVED PRONUNCIATION) is tending to disappear, and *ŏff* is now usual.

**often.** The pronunciation *aw'fn* is becoming old-fashioned and *ŏf'n* or *ŏf'ten* is now usual. According to the OED the sounding of the *t* was not then recognized by the dictionaries. But that was before the speak-as-you-spell movement got under way, and as long ago as 1933 the SOED recorded that the sounding of the *t* was then frequent in the south of England. That would now be an understatement of its currency. The long-drawn-out joke in *The Pirates of Penzance*—'When you said *orphan* did you mean a person who has lost his parents or *often*, frequently'—will soon be unintelligible to the audience. See PRONUNCIATION I.

**oh.** For *oh* and *O* see o and OH.

**O.K.** The most weighty consideration ever given to the origin and meaning of *O.K.* must have been that of the Judicial Committee of the Privy Council in 1935, when hearing the case of *Nippon Menkwa Kabushiki Kaesha* v. *Dawson's Bank*. The issue turned on the responsibility that a rice merchant must be held to have assumed by writing *O.K.* and his initials on certain invoices. 'Without some assistance in the way of evidence', said Lord Russell of Killowen, 'their Lordships might have found themselves in a difficulty, and all the more so since the origin of this commercial barbarism (which according to the OED was in use as far back as 1847) is variously assigned in different works of authority. The general view seems to be that the letters hail from the U.S.A. and represent a spelling, humorous or uneducated, of the words *All Correct*.' The case was accordingly decided as it would have been if those words had been written on the invoices.

After this authoritative pronouncement, the search for other origins and meanings, in which many people have exercised much ingenuity, may be regarded as of academic interest only. But it is perhaps worth while briefly to record other suggestions, leaving the reader to distinguish for himself between the serious and the flippant. The strongest rival of *All Correct* used to be that the letters represented the Choctaw *oke* (= it is), but this is rejected by the OED Supp. as not according with the evidence. Others ascribe the origin to the O.K. Club of New York, so called after 'Old Kinderhook', the nickname of Martin van Buren (1782–1862); to the initials of

Otto Kaiser, a German-born American industrialist; to the Southern French dialect word *oc* (= oui); to *aux quais* stencilled on the casks of Puerto Rico rum specially selected for export; to *Aux Cayes*, a place in Haiti noted for the excellence of its rum; to the letters indicating rank appended to the signature of a German Oberkommandant; to a misreading of O.R. (= order recorded); to the Scots *Och aye*; to the Finnish *oikea*; and to the Latin *omnia correcta*.

Recent research is said to have traced the earliest known use of *O.K.* to the *Boston Morning Post* of 23 March 1839. It was not until nearly a hundred years later that, greatly helped by radio and television, it won its present popularity in England. It is made to serve as an adjective, often predicative (*That's O.K.*) and occasionally attributive (*Advertising is in these days a socially okay profession*); it supersedes the old formulas of assent *Very well*, *All right*, and *Right oh*, and the questions *Do you understand?* and *Are you ready?*; as a verb it means *sanction* or *approve*, and as a noun is similarly used (*Put an O.K. on* or *Give the O.K. to*). It has bred a jocular variant in *Okidoke(y)*. Apart from its use in COMMERCIALESE, it is still a colloquialism and some day will no doubt share the fate of all popular colloquialisms and have to yield to a newcomer.

**old. 1.** For the distinction between *older*, *oldest*, and *elder*, *eldest*, see ELDER. **2.** For the phrase *a boy* etc. *of ten* etc. *years old*, see OF 7. **3.** For *the o. lady of Threadneedle Street* see SOBRIQUETS.

**olden. 1.** The adjective, which is of a strange formation and not to be reckoned among the numerous -EN ADJECTIVES, is also peculiar in use; *the olden time(s)* or *days* is common, but outside that phrase the word is usually as ridiculous as *Ye* substituted for *the* in the sham-archaic advertisements of shopwindows. The combination of *olden* with *régime* in the following example is what one might expect the author to call very tasty; see INCONGRUOUS VOCABULARY. *They form part of the olden railway régime, when every Great Western main-line train was deliberately halted for ten minutes at Swindon for refreshment.* 2. For the verb, = make or grow older, see -EN VERBS.

**olfactory.** For *o. organ*, see POLYSYLLABIC HUMOUR.

**olive-branches.** See HACKNEYED PHRASES, and SOBRIQUETS.

**-ology.** See -LOGY.

**Olympian, Olympic, Olympiad.** The distinction between the first two, not as old as Shakespeare and Milton, but now usually observed, is useful. *Olympian* means of Olympus, of or as of the Greek gods whose abode was on that mountain: *Olympian* Zeus, splendour, indifference. *Olympic* means of Olympia, of the athletic contests there held: *Olympic* games, victors. *Olympiad*, sometimes misused for the period during which the Olympic games are celebrated, means the intervals of four years between celebrations, used by the ancient Greeks in dating events.

**omelet(te).** The shorter spelling is preferable.

**omen.** For synonymy see SIGN.

**ominous.** Pronounce ŏm-. The alternative ōm-, as in *omen*, is no longer recognized by the dictionaries.

**omnibus book.** This term was coined for something which, like the vehicle, was for everyone's use—'a book or volume (usually containing several works) published at a price intended to place it within the reach of all' (OED Supp.). It is now generally applied, irrespective of price, to a book whose contents, like the passengers in an omnibus, are many and varied, though probably linked either by common authorship or a common topic.

**omniscience.** The dictionaries are almost unanimous for the pronunciation -sh(i)ens but -siens is a strong rival and likely to prevail.

**on.** For *on all fours*, see FOUR. For *onto*, *on to*, and *on*, see ONTO. For *on* and *upon* see UPON.

**-on.** Of words derived from Greek and having in English the termination -on: 1. Some usually (or invariably) form the plural in -a; so *asyndeton, criterion, hyperbaton, noumenon, organon, oxymoron, phenomenon.* 2. Others seldom or never use that form (though it would not be incorrect) but prefer the ordinary English -s; so *electron, lexicon, neutron, proton, skeleton.* 3. In others again, the substitution of -a for -on to form the plural would be a blunder (the relevant Greek plurals being of some quite different form), and -s is the only plural used; such are *anion, archon, canon, cation, cotyledon, cyclotron, demon, mastodon, nylon, pylon, siphon, tenon.*

**once.** 1. The use as a conjunction (i.e. = *if once* or *when once*, as in *Once you consent you are trapped*) is sound English enough, but writers who use it should remember both that there is a vigorous abruptness about it that makes it more suitable for some contexts than others, and also that, unless preceded by *if* or *when*, it may lay a FALSE SCENT, as in *But their aloofness might have quite the opposite result of that which they desire; for once the crisis had arrived, home affairs would indeed be swamped.*

2. *Once and away, once in a way.* The two phrases seem properly to have distinct meanings, the first 'once and no more' (*It is not enough to harrow once and away*—1759 in OED), and the second 'not often'. Later, no doubt because the phrases are almost indistinguishable when spoken, both were used in the second sense. But eventually *in a way* or *in a while* became the ordinary phrase for not often; *once and away* is obsolescent and the phrase for once and no more is *once (and) for all.*

**one.** 1. Pronunciation. 2. Writing of *anyone, no one,* etc. 3. *One and a half years* etc. 4. *One of the, if not the,* best book(s). 5. *One of the men who does things.* 6. Kind of pronoun—numeral, indefinite or impersonal, or first-personal? 7. Possessive of the numeral and the impersonal—*his? one's? their?* 8. Mixtures of *one, you, we,* etc. 9. *One* as a prop-word.

1. It is much to be desired that those who teach theological students, broadcast announcers, and others whose voices we have to listen to, should urge their pupils to pronounce the word briskly as *wŭn* instead of the drawled *wahn* that is now so common.

2. The forms recommended are *anyone, everyone, no one, someone.* For discussion see EVERY ONE 1.

3. *One and a half years* or *a year and a half.* The second is recommended, when words and not figures are used; for discussion, see MILLION 1, and HALF 1.

4. *One of the, if not the, best book(s).* Grammar is a poor despised branch of learning; if it were less despised, we should not have such frequent occasion to weep or laugh at the pitiful wrigglings of those who feel themselves in the toils of this phrase. That the victims know their plight is clear from the way they dart in different directions to find an outlet. Here are half a dozen attempts, all failures, but each distinguishable in some point of arrangement from the rest: (*a*) *Given in the Costume Hall— one of, if not the most, spacious of salons for dresses and costumes—the dancing has been* . . . (*b*) *One of the finest, if not the finest, poem of an equal length produced of recent years.* (*c*) *I think the stage is one of, if not the best of all, professions open to women.* (*d*) *Fur was one of the greatest—perhaps the greatest —export articles of Norway.* (*e*) *The Japanese were one of the most, if not the most, enterprising nations in the East.* (*f*) *One of, if not, the oldest Voortrekkers of South Africa has just passed away.*

The nature of the problem is this: we have two expressions of the type 'one of the best books' and 'the best book'; but we have been taught to avoid repetition of words, and therefore desire that part of one of these nearly similar expressions should be understood instead of said or written. Let us then enclose the partially expressed one inside the other, as a parenthesis. Can this be done? It will be seen that (*a*), (*b*), and (*c*), though they differ in minor points, are alike in failing to pass the most obvious test— does the enclosing expression read right if the parenthesis is left out? *One of spacious of salons, One of the finest poem, One of professions open to women:* the first and second nonsense, the third the wrong sense. In (*d*), (*e*), (*f*), the enclosing expression taken alone does give sense; the further test they have to pass is—if the words understood in the parenthesis are written in, does the whole read as sound, though perhaps inelegant, English? *One of the greatest (perhaps the greatest export article) export articles; One of the most (if not the most enterprising nation) enterprising nations; One of (if not the oldest Voortrekker) the oldest Voortrekkers.* Not sound English, but nonsense; compare it with the expanding of a correctly compressed sentence: *He was, if not a perfect, a great orator,* which being filled up gives *if not a perfect orator, a great orator; that* is not nonsense, but sound English. The rule that has been broken in the supposed compressions (*d*), (*e*), (*f*), and not broken in the real one, is that you cannot understand out of a word that is yet to come another word (as *article* out of a coming *articles, nation* out of a coming *nations, Voortrekker* out of a coming *Voortrekkers*), but only the same word, as *orator* out of *orator.* When, as always happens in this idiom, there is a change of number, the only thing is to see that the place from which the understood word is omitted is after, not before, the word from which it is to be supplied; for from a word that has already been expressed the taking of the other number is not forbidden. Accordingly, the right form for the words that concern us in the examples (*a*)–(*f*) will be: *One of the most spacious, if not the most spacious, of salons*; *One of the finest poems of an equal length produced of recent years, if not the finest*; *One of the best professions open to women, if not the best of all*; *One of the greatest export articles of Norway, perhaps the greatest*; *One of the most enterprising nations in the East, if not the most*; *One of the oldest Voortrekkers, if not the oldest.*

It may be thought that for (*a*) the best has not been done, and that *One of the, if not the, most spacious of salons* would have been less clumsy, and yet legitimate. It is an improvement on the original, and by inserting a *the* and correcting the stops makes a plausible attempt at compromise. But it is not legitimate, because *most spacious* has to be taken as at the same time singular and plural. English disguises that fact by its lack of inflexions, but does not annul it; and, though most people are not quite sure what is the matter, they can feel that something is.

**5.** *One of the men who does things. Does* should be *do.* This blunder, easier to deal with than that in 3, but not less frequent, will be found discussed in NUMBER 5.

**6.** Kind of pronoun. To avoid confusion in this and the later sections between certain uses of the pronoun *one* that tend to run into each other, it will be necessary to ask the reader to accept, *pro hac vice*, certain names. *One* is a pronoun of some sort whenever it stands not in agreement with a noun, but as a substitute for a noun preceded by *a* or *one*: in 'I took one apple' *one* is not a pronoun, but an adjective; in 'I want an apple; may I take one?' *one* stands for *an apple* or *one apple*, and is a pronoun. But for the purpose of this article it is more important to notice that *one* is not always the same kind of pronoun; it is of three different kinds in these three examples: *One of them escaped*; *One is often forced to confess failure*; *One knew better than to swallow that.* In the first, *one* may be called a numeral pronoun, which description will cover also *I will take one, One is enough*, and so on. In

the second, *one* has a special sense; it stands for *a person*, i.e., the average person, or the sort of person we happen to be concerned with, or anyone of the class that includes the speaker; it does not mean a particular person. It might be called an indefinite, or an impersonal, pronoun; for the sake of contrast with the third use, *impersonal pronoun* will here be the name. In the third, *one* is neither more nor less than a substitute for *I*, and the name that best describes it is the false first-personal pronoun. The distinction between the numeral and the impersonal, which is plain enough, is important because on it depend such differences as that between *One hates his enemies* and *One hates one's enemies*; those differences will be treated in section 7. The distinction between the impersonal and the false first-personal, a rather fine one in practice, is still more important because it separates an established and legitimate use from one that ought not to exist at all. The false first-personal pronoun *one* is an invention of the self-conscious writer or speaker, and its suppression before it can develop further is very desirable. Outside this section, the rest of which will be devoted to illustrating the attempts to popularize this usage, it will be assumed that it does not exist except as a mere misuse of the impersonal *one*.

Let us take a fictitious example and pull it about, in order to make the point clear: *He asked* me *to save his life, and* I *did not refuse*; the true first-personal pronoun, twice. *He asked* me *to save his life; could* one *refuse?*; true first-personal pronoun, followed by impersonal pronoun. *He asked* me *to save his life, and* one *did not refuse*; true first-personal pronoun, followed by false first-personal pronoun. The *one* of *could one refuse?* means I or anyone else of my kind or in my position, and is normal English; the *one* of *one did not refuse* cannot possibly mean anything different from *I* by itself, and is a fraud. But the self-conscious writer sees in this fraud a chance of eating his cake and having it; it will enable him to be impersonal and personal at once. He has repined at abstention from *I*, or has blushed over not abstaining; here is what he has longed for, the cloak of generality that will make egotism respectable. The sad results of this discovery are shown in the following extracts. In none of them is there any real doubt that *one* and *one's* mean I and my simply; but in some more than in others the connexion with the legitimate impersonal use is traceable. The writer should make up his mind that he will, or that he will not, talk in the first person, and go on the sound assumption that *one* and *one's* do not mean I and me or my and mine.

### The false first-personal ONE

*But* one *must conclude* one's *survey at the risk,* I *am afraid, of tedious reiteration) by insisting that . . . | I have known in the small circle of* one's *personal friends quite a number of Jews who . . . | This is not,* I *think, ecclesiastical prejudice, for* one *has tried to be perfectly fair. | His later poems have their great limitation, as* one *will presently suggest, but they are extraordinarily powerful. | The book is bound in red and gold, and has the novelist's autograph in gold upon the front;* one *mentions gold twice over, because . . .*

7. Possessive, and other belongings, of *one*. By other belongings are meant the reflexive, and the form to be used when the pronoun *one* has already been used and is wanted again either *in propria persona* or by deputy; when Caesar has been named, he can be afterwards called either *Caesar* or *he,* so, when *one* has been used, does it matter whether it is repeated itself or represented by *he* etc.?

In the first place, there is no doubt about the numeral pronoun *one*; its possessive, reflexive, and deputy pronoun, are never *one's, oneself,* and *one,* but always the corresponding parts of *he, she,* or *it. I saw one drop his stick; Certainly, if one offers herself as candidate; One would not go off even when hammered it.*

Secondly, the impersonal *one* always can, and now usually does, provide its own possessive etc.—*one's, oneself,* and *one*; thus *One does not like to have* one's *word doubted; If one fell,* one *would hurt* oneself *badly.*

But thirdly, in American, in older English, and in a small minority of modern British writers, the above sentences would run *One does not like to have* his *word doubted; If one fell,* he *would hurt* himself *badly.*

The modern fashion (*one's, oneself,* etc.) gives a useful differentiation between the numeral and the impersonal, and it makes recourse to the horrible *their* etc. needless (*One does not like to have* their *word doubted;* see THEY I). The following examples will suffice to show that not all writers yet accept the modern idiom, though it is certainly in the interests of the language that they should: *There are many passages which* one *is rather inclined to like than sure he would be right in liking* (19th-c. American). | *Assuredly, there is no form of 'social service' comparable to that which* one *can render by doing* his *job to the very best of* his *ability. | Let us, in fact, substitute a 'graceful raising of* one's *hand to* his *hat, with a nod'. | As* one *goes through the rooms, he is struck by the youth of most. | If seeing sixteenth-century Europe implied spending the nights in sixteenth-century inns,* one *imagines he would rather have stayed at home.*

The difference between *One hates* his *enemies* and *One hates* one's *enemies* is at once apparent if to each is added a natural continuation: *One hates* his *enemies and another forgives them*; *One hates* one's *enemies and loves* one's *friends.* The first *one* is numeral, the second impersonal, and to make *his* and *one's* exchange places, or to write either in both places, would be plain folly.

Let it be added, for anyone who may regard *one's* and *one(self )* in the impersonal use as fussy modernism, that they are after all not so modern: *I hope, cousin, one may speak to* one's *own relations*—Goldsmith. It is perhaps a feeling that the repetition of *one* is awkward, or even slightly pompous, that has produced a tendency among recent writers to avoid the impersonal *one* and to say instead *a man* or *people* or, informally, *you*. The writer of the following seems to have had this feeling so strongly that he could not bring himself to insert the *one* after *oneself* that is needed to give the sentence a grammatical construction. *One becomes ashamed: what in oneself has identified with the hero is seen to be shabby and selfish.* A case of CANNIBALISM.

8. Mixtures of *one* with *we, you, my,* etc. These are all bad, though the degrees of badness differ; for instance, it is merely slipshod to pass from *one* in an earlier sentence to *you* in the next, but more heinous to bring two varieties into syntactical relations in a single sentence. *As* one *goes through the rooms,* he *is struck by the youth of most of those who toil; the girls marry,* you *are told.* Here *he* belongs to section 7, in which the sentence has been quoted; *you* illustrates the more venial form of mixture. / *As* one *who vainly warned* my *countrymen that Germany was preparing to attack her neighbours for many a long day before the declaration of war,* I *say that* ...; *My* should be *his, one* being the numeral pronoun; but this kind of attraction in relative clauses (*my* taking the person of *I* instead of that of *one* and *who*) is very common. / *To listen to his strong likes and dislikes* one *sometimes thought that* you *were in the presence of a Quaker of the eighteenth century;* a bad case; *you were* should be *one was.* / *Perhaps there are too many of them;* we *might have enjoyed making their acquaintance still more had* one *been given pause;* either *we* should be *one,* or *one* should be *we.* / *No* one *likes to see a woman who has shared* one's *home in distress; no one* contains the numeral, not the impersonal, *one,* and *one's* should be *his.*

9. *One* as a 'prop-word' is a name given by grammarians to the use of *one* (or *ones*) to support an adjective or other qualifying word or words that would be awkward or ambiguous standing alone. *The second resolution was the U.S. one.* / *The satellite was a small test one.* This is established idiom, but should not be employed unnecessarily. It could perhaps be argued that, even when not needed to remove awkwardness or ambiguity, *one* or *ones* may be justified as contributing a subtle emphasis: that *His life was a sedentary and lonely one* gives a sharper picture than *His life was sedentary and lonely,* and that in *We are apt to find that the very men who block a scheme are the ones who clamour loudest* . . . the use of *ones* emphasizes the inconsistency of behaviour (they are the *very ones*). But in such a sentence as the following *ones* is a clumsy and unwanted intruder. *Handbooks which are too popular, or ones too exclusive, and technical, or both.*

**one another.** See EACH 2.

**one word or two or more.** For ALL RIGHT and *alright,* ALREADY and *all ready,* ALTOGETHER and *all together,* ANY *way* and *anyway,* at ANY *rate* and *at anyrate,* COMMON SENSE and *commonsense,* EVERY ONE and *everyone,* INTO and *in to,* ONTO and *on to,* see the words in small capitals. For *blackcap* and *black cap,* see HYPHENS; for *no-one* and *no one, someone* and *some one,* EVERY ONE, and for *in no wise* and *in nowise,* NO 5.

**only,** adv.: its placing and misplacing. *I read the other day of a man who 'only died a week ago', as if he could have done anything else more striking or final. what was meant by the writer was tha he 'died only a week ago'.* There speak one of those friends from whom th English language may well pray t be saved, one of the modern pre cisians who have more zeal than dis cretion, and wish to restrain libert as such, regardless of whether it i harmfully or harmlessly exercised. I is pointed out in several parts of thi book that illogicalities and inaccuracie of expression tend to be eliminated a a language grows older and its user attain to a more conscious mastery c their materials. But this tendency has its bad as well as its good effects; the pedants who try to forward it when the illogicality is only apparent or the inaccuracy of no importance are turning English into an exact science or an automatic machine. If they are not quite botanizing upon their mother's grave, they are at least clapping a strait waistcoat upon their mother tongue, when wiser physicians would refuse to certify the patient.

The design is to force us all, whenever we use the adverb *only*, to spend time in considering which is the precise part of the sentence strictly qualified by it, and then put it there—this irrespective of whether there is any danger of the meaning's being false or ambiguous because *only* is so placed as to belong grammatically to a whole expression instead of to a part of it, or as to be separated from the part it specially qualifies.

It may at once be admitted that there is an orthodox placing for *only*, but it does not follow that there are not often good reasons for departing from orthodoxy. For *He only died a week ago* no better defence is perhaps possible than that it is the order that most people have always used and still use, and that, the risk of misunderstanding being chimerical, it is not worth while to depart from the natural. Remember that in speech there is not even the possibility of misunderstanding, because the intonation of *died* is entirely different if it, and not *a week ago*, is qualified by *only*; and it is fair that a reader should be supposed capable of supplying the decisive intonation where there is no temptation to go wrong about it. But take next an example in which, ambiguity being practically possible, the case against heterodox placing is much stronger: *Mackenzie only seems to go wrong when he lets in yellow; and yellow seems to be still the standing difficulty of the colour printer.* The orthodox place for *only* is immediately before *when*, and the antithesis between seeming to go and really going, which, though not intended, is apt to suggest itself, makes

the displacement here ill advised. Its motive, however, is plain—to announce the limited nature of the wrong before the wrong itself, and so mitigate the censure: a quite sound rhetorical instinct, and, if *goes* had been used instead of *seems to go*, a sufficient defence of the heterodoxy. But there are many sentences in which, owing to greater length, it is much more urgent to get this announcement of purport made by an advanced *only*. For instance, the orthodox *It would be safe to prophesy success to this heroic enterprise only if reward and merit always corresponded* positively cries out to have its *only* put early after *would*, and unless that is done the hearer or reader is led astray; yet the precisian is bound to insist on orthodoxy here as much as in *He died only a week ago*.

The advice offered is this: there is an orthodox position for the adverb, easily determined in case of need; to choose another position that may spoil or obscure the meaning is bad; but a change of position that has no such effect except technically is not only justified by historical and colloquial usage but often demanded by rhetorical needs.

See also POSITION OF ADVERBS.

A specimen or two of different kinds are added for the reader's unaided consideration: *The address to be written on this side only.* | *Europe only has a truce before it, but a truce that can be profited by.* | *Some of the Metropolitan crossings can only now be negotiated with considerable risk.* | *If only the foundry trades had been concerned, probably the employers would not have greatly objected to conceding an advance.* | *I only know nothing shall induce me to go again.* | *We can only form a sound and trustworthy opinion if we first consider a large variety of instances.* | *Butter only served in this establishment* (i.e. no margarine).

For *not only* see NOT 6.

**only too.** In this combination, idiomatic when properly used, *too* (says the OED) means 'more than is desirable or more than might be expected', and *only* 'emphasizes the exclusion of any different quality or state of things such as might be desired or expected'. Recent examples, the first illustrating 'more than is desirable' and the second 'more than might be expected' are: *It seems only too probable that the Geneva Conference will open with long and perhaps bitter wrangles about membership and procedure.* | *Proceedings were interrupted during a debt case at Bristol County Court yesterday when a solicitor discovered that his gown had been set alight by an electric fire. When he apologized for creating a disturbance the Registrar said he was only too glad to have some interest introduced into the case.* Colloquially the most frequent use of *only too* is with *glad* or *pleased* or some other adjective of gratification, as the Bristol Registrar used it, and is equivalent to *Don't mention it* or *You're welcome.* But as long ago as 1933 the SOED noted that in recent use *only too* was often a mere intensive, equivalent to *extremely.* This loose usage has since become deplorably common and threatens to destroy the point of a convenient idiom. See also TOO.

**onomatopoeia.** Formation of names or words from sounds that resemble those associated with the object or action to be named, or that seem suggestive of its qualities; *babble, cuckoo, croak, ping-pong, quack, sizzle,* are probable examples. The word is also used of a sentence whose sound suggests what it describes, as in Tennyson's.

Myriads of rivulets hurrying thro' the lawn,
The moan of doves in immemorial elms,
And murmuring of innumerable bees.

**onto, on to, on.** *The logic of this electioneering leads straight to the abolition of the contributions and the placing of the whole burden* on to *the State.* | *The Pan-Germans are strong enough to depose a Foreign Secretary and force their own man* on to *the Government in his place.* Writers and printers should make up their minds whether there is

such a preposition as *onto* or not. If there is not, they should omit the *to* in such contexts as the above, which are good English without it. If there is, and they like it better than the simple *on* or *to* (an odd taste, except under very rare conditions), they should make one word of it. Abstain from the preposition if you like; use it and own up if you like; but do not use it and pretend there is no such word; those should be the regulations. The use of *on to* as separate words is, however, correct when *on* is a full adverb; and doubts may occasionally arise whether this is so or not. Is *on* an adverb, or is *onto* a preposition, for instance, in *He played the ball on to his wicket*? As *He played on* could stand by itself, being a PHRASAL VERB, it is hard to deny *on* its independent status. Occasions for *on to* are: *We must walk on to Keswick*; *Each passed it on to his neighbour*; *Struggling on to victory.* Occasions for *on* or *to* or *onto*, but on no account *on to* are: *Climbed up on(to) the roof*; *Was invited (on)to the platform*; *It struggles (on)to its legs again*; *They fell* 300 *ft. on(to) a glacier.*

**onward(s).** The shorter form is much commoner in all senses, except possibly in phrases of the type *from the tenth century onwards*: *onward* is both adjective and adverb, *onwards* adverb only.

**opacity, opaqueness.** The figurative senses are avoided with the second, but the literal senses are not confined to it, though there is perhaps a tendency to complete DIFFERENTIATION: *The opacity of his understanding*; *Owing to the opaqueness, or opacity, of the glass.*

**operate** makes *-rable, -tor*; see -ABLE 1, and -OR. Used as a transitive verb (*o.* a machine, *o.* a business) it is an Americanism formerly frowned on here but now established.

**opine** is a stilted and obsolescent word for express an opinion, seldom used now except either jocularly or to suggest that the opinion expressed is a dogmatic one based on inadequate data. *After a brief look at the current trend of costs they o. that there should be a good opportunity in 1958 to stop the rise in prices that has troubled the country for 20 years.* (Cf. *opinionated.*)

**opinion.** For *Climate of o.* see CLIMATE.

**opinionated, -ative.** Both have existed long enough in English to justify anyone in using either. But for those who do prefer a sound to a faulty formation it may be said that the first and commoner is unobjectionable, and the second not. A Latin *opinionatus* might have been correctly made from the noun *opinio*; cf. *dentatus* from *dens*, and many others; and the English representative of *opinionatus* would be either *opinionate* or *opinionated*. But Latin *-ativus* belongs to verb-derivatives only, and *opinativus* from the verb *opinor*, giving English *opinative* (which once existed), would have been the true source for a word in *-ative*.

**opportunity.** *He rapidly rose by the display of rare organizing ability to be superintendent over the affairs of the company in the Far East, with practically a free hand—a fact of which he took every opportunity.* You *take the o.*, or *an o.*, or *every o.*, of doing something. You *take advantage*, or *all possible advantage, of* a fact or event or state of affairs. The two sets of phrases must not be mixed; see CAST-IRON IDIOM, and ANALOGY.

**opposite** tempts careless writers to the slovenly telescoping seen in: *He can thwart him by applying it to the opposite purpose for which it was intended.* Insert *to* (or *from*) that after *purpose*; and for similar temptations cf. AS 3 (*the question as to whom it belongs* etc.), and CANNIBALISM.

**optic.** For the noun use, = eye, see PEDANTIC HUMOUR; 'Formerly the learned and elegant term'—OED.

**optimism, -ist(ic).** *Besides optimism, which affirms the definitive ascendancy* of good, and pessimism, which affirms the definitive ascendency of evil, a third hypothesis is possible. | *The optimistic or sentimental hypothesis that wickedness always fares ill in the world. | He is reported to have been optimistic that the work of the Monckton Commission will help to steer the London Conference to a successful conclusion. | There is cause for optimism that terms can be arranged for a return to work tomorrow.* The first two quotations show the words in their proper sense, the last two in their modern popular triviality. They have become VOGUE WORDS, on much the same level as INDIVIDUAL and REALISTIC. They belong in time between those two; they are not yet discredited like the former, but have lost the charm of novelty that still lingers about the latter. Like both those, they owe their vogue to the delight of the ignorant in catching up a word that has puzzled them when they first heard it, and exhibiting their acquaintance with it as often as possible; and, like both those, they displace familiar words that would exactly express the intended meaning by others that do not. In the third and fourth quotations, *hopeful* and *hoping* would have given the sense not less but more exactly than *optimistic* and *optimism*. Pessimism and *pessimistic* are similarly misused. *The Admiral said he was very pessimistic about the possibility of finding anyone alive* (had very little hope that anyone would be found alive). See also SANGUINE.

**optimum** is suffering a debasement of the same kind as OPTIMISM. It has a precise meaning to which it should be confined—the conditions in which an organism will thrive best or a machine work most satisfactorily. For the same reason as *optimism* is preferred to *hope*, *optimum* is preferred to *best*, with the same lamentable results.

**opus.** For a musical composition or in the phrase *magnum opus*, usually pronounced *ōpus*. Pl., seldom used, *opera*; see LATIN PLURALS.

**or. 1.** *Or, nor.* **2.** Number, pronouns, etc., after *or*. **3.** *Or* in enumerations. **4.** Wrong repetition after *or*.

**1.** *Or, nor.* There are sentences in which it affects neither meaning nor correctness whether *or* or *nor* is used. *I can neither read nor write* requires *nor*. *I cannot either read or write* requires *or*. But in *I cannot read nor* (or *or*) *write* we may use either. The alternatives in the last are differently arrived at, but are practically equivalent: *I cannot read, nor* (can I) *write*; *I cannot read(-)or(-)write*, where the supposed hyphens mean that *write* may be substituted for *read* if desired. The use of *nor* in such cases was formerly in fashion, and that of *or* is now in fashion; that is all. But the modern preference for *or* where it is equally legitimate with *nor* has led to its being preferred also where it is illegitimate; so: *It is of great importance that they should face them in* no academic spirit, or *trust too much to conclusions drawn from maps.* / *No Government Department or any other Authority has assisted.* The test of legitimacy has been explained in NOR; and it suffices here to say that what rules out *or* in the first extract is the position of *no* (alter to *they should not face them in any*), and in the second it is the presence of *any* (precluding the carrying on of *no*).

**2.** Number, pronouns, etc., after *or*. When the subject is a set of alternatives each in the singular the verb must be singular however many the alternatives, and however long the sentence; in the extract below, *account* should be *accounts*; for discussion see NUMBER 3: *Either the call of patriotism and the opportunity of seeing new lands, or conscription, or the fact that tramping was discouraged even by old patrons when the call for men became urgent,* account *for it.* If the alternatives differ in number or person, the nearest prevails (*Were you or he, was he or you, there?*; *either he or you were, either you or he was*), but some forms (e.g. *Was I or you on duty?*) are avoided by inserting a second verb (*Was I on duty or were you?*). Forms in

which difference of gender causes difficulty with pronouns (*A landlord or landlady expects their, his or her, his, rent*) are usually avoided, *their rent* or *the rent due to them* being ungrammatical (see THEY 1), *his or her rent* or *the rent due to him or her* clumsy, and *his rent* or *the rent due to him* puzzling; some evasion, as *expects rent,* or *the rent,* is always possible.

**3.** *Or* in enumerations. *I never heard a sermon that was simpler, sounder, or dealt with more practical matters.* In the very numerous sentences made on this bad pattern there is a confusion between two correct ways of saying the thing, viz. (*a*) *that was simpler, sounder, or more practical,* (*b*) *that was simpler* or *sounder or dealt with more practical matters.* See ENUMERATION, and for full discussion AND 2. The abundant illustration of the latter makes similar quotations here needless.

**4.** Wrong repetition after *or*. A misguided determination to be very explicit and leave no opening for doubt results in a type of mistake illustrated in the article OVERZEAL. It is peculiarly common with *or*, and to put writers on their guard a number of examples follow. False analogy from *and* explains it; with *and*, it does not matter whether we say *without falsehood and deceit* or *without falsehood and without deceit,* except that the latter conveys a certain sledge-hammer emphasis; but with *or* there is much difference between *without falsehood or deceit* (which implies that neither is present) and *without falsehood or without deceit* (which implies only that one of the two is not present). In all the examples except the last, either *or* must be changed to *and*, or the word or words repeated after *or* must be cut out; in the last example, if *or* is to be retained, it will be necessary, besides omitting the second *no*, to change *one* to *person. No great economy or no high efficiency can be secured.* / *There would be nothing very surprising* or *nothing necessarily fraudulent in an unconscious conspiracy to borrow from each other.* / . . . *prevents the labourer from being a free agent* or *from having a free market for his* *labour.* | *Every arrangement ends in a compromise, and* no one or no party *may ever be expected to carry its own views out in their entirety.*

Another type of wrong repetition after *or*, not accidental but deliberate, has of late been much favoured by writers of popular tales, especially in describing the cogitations of their characters. This consists of a rhetorical question followed by another question posed as an alternative to the first but in fact a repetition of it. *Hadn't he, in her, the truth he had been seeking? Or hadn't he?* | *Was this an important clue that should be followed up at once? Or was it?* Perhaps those who employ this gimmick want to titillate the reader by surprising him with *hadn't he* when he was expecting *had he* and with *was it* when he was expecting *wasn't it*, or perhaps they want to give a realistic touch to a picture of a fuddled mind. Whatever the purpose, it is the kind of conceit that is unlikely to outlast its freshness.

**-or** is the Latin agent-noun ending corresponding to the English *-er*; compare *doer* and *perpetrator*. English verbs derived from the supine stem of Latin ones—i.e. especially most verbs in *-ate*, but also many others such as *oppress, protect, act, credit, possess, invent, prosecute*— usually prefer this Latin form to the English one in *-er*. Some other verbs, e.g. *govern, conquer*, and *purvey*, not corresponding to the above description have agent nouns in *-or* owing to their passage through French or other circumstances that need not here be set forth. A few odd differences may be of interest: *decanter* and *castor*; *digester* and *collector*; *corrupter* and *corrector*; *deserter* and *abductor*; *eraser* and *ejector*; *promoter* and *abettor*. Some verbs have alternative forms; generally preferring *-er* for the personal and *-or* for the mechanical agent (e.g. *adapt, convey, distribute, resist*), or *-or* for the lawyers and *-er* for ordinary use (e.g. *settle, pay, devise, vend*).

**oral.** See VERBAL.

**orate.** A BACK-FORMATION from *oration*, and marked by the slangy jocularity of its class.

**oratio obliqua, recta.** Latin names, the second for the actual words used by a speaker, without modification, and the first for the form taken by his words when they are reported and fitted into the reporter's framework. Thus *How are you? I am delighted to see you* (recta) becomes in obliqua *He asked how I was and said he was delighted to see me*; or, if the framework is invisible, *How was I? He was delighted to see me.* Most newspaper reports of speeches, and all third-person letters, are in oratio obliqua or reported speech.

**orchis, -chid.** The first form is applied chiefly to the English wild kinds and is accordingly the poetic and the country word; pl. *-ises*, but *-ids* is often used for both.

**ordeal.** All the verse quotations in the OED (Chaucer, Spenser, Cowley, Butler, Tennyson) show the accent on the first syllable. But fashion has changed, and *ordē'al* or *ordē'l* is now the ordinary pronunciation. The second is etymologically correct; the *ea* has the same value as in *meal, steal*, etc., not as in adjectives such as *lineal, corporeal*, etc.

**order.** For wrong constructions after *in order that* (*i. o. t. the complaint that colliery proprietors are diverting domestic coal for industrial purposes* can *be considered*), see IN ORDER THAT. For the periphrasis *of the order of* (= *about*) see COMPOUND PREPOSITIONS.

**orderly.** See -LILY.

**organ(on)(um).** Pronounce *org'-ăn(on)(um)*.

**orient,** v., **orientate.** The second, a LONG VARIANT of the first, seems likely to prevail in the common figurative use.

**orison.** Pronounce *ŏ'rĭzn*.

**ornithology** etc. In Greek the *i* is long, but in the English words the

short *i* is now general. See FALSE
QUANTITY.

**orotund.** A PORTMANTEAU WORD from
*ore rotundo*. The odd thing about the
word is that its only currency, at least
in its non-technical sense, is among
those who should most abhor it, the
people of sufficient education to realize
its bad formation; it is at once a monstrosity in its form and a pedantry
in its use. If the elocutionists and
experts in voice-production like it as
a technical term, they are welcome to
it; the rest of us should certainly leave
it to them, and not regard it as a good
substitute for *grandiloquent, highflown,
pompous,* and the like.

**osculatory** has its serious uses in
biology and mathematics, but to
most of us is known only as a POLY-
SYLLABIC-HUMOUR word: *The two
ladies went through the o. ceremony. |
At the end of one letter were a number
of dots which he (counsel) presumed
were meant to represent an o. performance.*

**ostensibly, ostentatiously.** Both
mean 'by way of making a show', but
the purpose in the first case is to
conceal a truth; in the second it is
merely display—'showing off'. What
the writer of the quotation that follows
intended *ostensibly* to mean is obscure;
perhaps it is just a pleonastic reinforcement of *seem* of the kind described in
the article HAZINESS. *The Royal Colleges ostensibly seem smugly satisfied
with their role in directing the nation's
medical future.*

**other.** 1. *Each o., one another.*
2. *On the other hand.* 3. *Of all others.*
4. *Other, others* or *another.* 5. *Other
than.*

1. *Each o., one another.* For the
syntax of these, and for the distinction
sometimes made between them, see
EACH 2.

2. *On the o. hand.* For the difference
between this and *on the contrary,* see
CONTRARY 2.

3. *Of all others. You are the man of
all others I wanted to see.* A mixture
of *You are the man of all men* etc. and

*You are the man I wanted to see beyond
all others.* A still popular ILLOGICALITY,
used by good writers and perhaps to
be counted among the STURDY IN-
DEFENSIBLES that are likely to survive
their critics.

4. *Other, others* or *another.* The
writers of the following sentences may
be supposed to have hesitated between
*other* and *others.* If they had decided
for *others,* they would have been more
in tune with modern usage; to say they
would have chosen more correctly is
hardly possible: *The Unionist Party
will do well to remember that the wrecking policy is, like* other *of their adventures in recent times, a dangerous
gamble. | We find here, as in* other *of
his novels, that he has no genius for . . . |
Mrs. —— will, we hope, incite* other
*of her countrymen and countrywomen to
similar studies. | We were quite prepared
for the most rigid prohibition of trade
with Germany; so was France and* other
*of our Allies. | A Privy Councillorship,
an honour which has but rarely been won
by* other *than those who were British
subjects from the moment of their birth.*

In four of these we have what the
OED calls the absolute use of the
adjective, the noun represented by
*other* being present elsewhere in the
sentence, but not expressed with *other*
(*like* other *adventures of the adventures*
etc. would be the fully expressed
forms); in the fifth we have the full
pronoun use, *other* meaning *other
persons,* and *persons* not being expressed
either with *other* or elsewhere. But in
both uses the OED describes the
plural *other* as archaic, and the plural
*others* as the regular modern form. In
older English, however, *other* was
normal in such contexts, so that those
who like the archaic can justify themselves. And since the OED was published *others* itself has been falling out
of fashion for the absolute use; a
modern writer will probably feel that
neither *other* nor *others* is idiomatic:
he will avoid the absolute use of the
adjective and fall back on the usage referred to in OF 7 as a STURDY IN-
DEFENSIBLE—*like* other *adventures of
theirs, in* other *novels of his,* other

*countrymen and countrywomen of hers, other allies of ours.*

5. Abuses of *other than*. The existence of an adverbial use of *other* is recognized by the OED, but supported by very few quotations, and those from no authors whose names carry weight; its recent development may be heartily condemned as both ungrammatical and needless. In each of the following quotations it will be seen that the phrase on the other side of *than* is adverbial like *otherwise*, and not adjectival like *other*. In the article OTHERWISE the converse mistake is shown to be at least equally common. Both mistakes are as stupid as they are common; and, though the substitution of *otherwise* for *other* or vice versa removes the blunder, it would usually be better to use neither *other than* nor *otherwise than*, but some different expression.

### Other for *otherwise*

*A subordinate sprite will no more obey a conjuration addressed to him by a magician o. t. in the name of his proper superior than* . . . (in any other name than that). | *The Court is not at liberty to construe the words other than strictly* (otherwise). | *Yet how many of the disputants would know where to look for them—o. t. by a tiresome search through the files of the daily Press—if they desired to consult them?* (short of). | *Although the world at large and for long refused to treat it o. t. humorously* (otherwise).|*I should think the 'Times Literary Supplement' would want to avoid the services of a reviewer who . . . cannot cope with allegedly bad work o. t. by recommending its suppression* (except).

But simple confusion between *other* and *otherwise* does not account for every bad *other than*. A notion seems to prevail that one exhibits refinement or verbal resource or some such accomplishment if one can contrive an *other-than* variant for what would naturally be expressed by some other negative form of speech: *could not leave o. t. restless* is thought superior in literary tone to could not but leave restless; *be other than flattered* to help

being flattered; *o. t. when Parliament was sitting* to when Parliament was not sitting; so: *Four years of war could not leave a people* o. t. restless. | *Mr. Collier has some faults to find, but no Englishman can be* o. t. flattered *by the picture which he paints of British activities.* | *The Premier sent telegrams to the various States suggesting that they should concur in the Governor-General residing in New South Wales* o. t. when Parliament was sitting. One or two of these are justifiable, while one is certainly not, from the grammarian's point of view; regarded as ornaments, they are clearly of no great value. On the whole, *other than* should be registered as a phrase to be avoided except where it is both the most natural way of putting the thing and grammatically defensible.

**otherwise** has recently been having very curious experiences—emphatically *has recently*, because, while the OED originally showed no trace whatever of the two uses to be illustrated below, both of them have become so common that probably no one ever reads his newspaper through without meeting them and the OED Supp. had to add to its definitions 'after a noun, adjective or adverb followed by *or*: equivalent to a noun, adjective or adverb having an opposite or different meaning'. Being itself an adverb, it can of course properly be so used with another adverb (*He decided, sensibly or o., to take a chance*). What we are here concerned with is the popular extension of this usage to adjectives and nouns. Whether this popularity is a sign of lately developed indispensability, or merely a new example of the speed with which a trick of bad English can be spread by fashion, it is hard to say with confidence; but, as one use is a definite outrage on grammatical principles, and the other not very easy to reconcile with them, we may perhaps hope that they are freaks of fashion only.

The first is the ungrammatical use of the adverb *otherwise* when the adjective or pronoun *other* would be correct;

cf. OTHER for the converse mistake. Comment will be better reserved till the reader has seen some examples: *This reduction in total expenditure has been made concurrently with certain increases*—automatic and otherwise—*in particular items.* / *There are large tracts of the country,* agricultural and otherwise, *in which the Labour writ does not run.* / *No further threats,* economic or otherwise, *have been made.* / *This is a common incident in all warfare,* industrial or otherwise. / *No organizations,* religious or otherwise, *had troubled to take the matter up.* / *An advisory council composed entirely of those persons,* musicians or otherwise, *who have had a genuine acquaintance with operatic tradition.*

An apology may fairly be expected for presenting so long a string of monotonous examples. The apology is that, before asking writers and speakers to give up a favourite habit, one should convince them that it is their habit. That the habit is a bad one needs no demonstration; but it is worth while to consider how it has come into existence, and whether abstention from it is really a serious inconvenience. In all the above quotations, the structure is the same—an adjective (or descriptive noun) deferred till after its noun and followed by an *and* or an *or* joining to it the adverb *otherwise*. Now, what should possess anyone, under those circumstances, to match the adjective with *otherwise* instead of *other*? Is it not (far-fetched as the explanation may seem) that the old saying 'Some men are wise and some are otherwise' once struck the popular consciousness as witty, and has incidentally inspired a belief that *otherwise,* and not *other,* is the natural parallel to an adjective? The justification of the proverb's own wording is simple—that it is a pun, and that puns treat grammar as love treats locksmiths, with derision. A pun, however, and still more the faded memory of a pun, is a bad basis for a general idiom. Nor is there any difficulty whatever in abstaining from this bit of bad grammar. It is true that things have now

reached the stage when many people feel that to change the popular *otherwise* to the correct *other* is sometimes pedantic; but it is only sometimes, and there are other resources. The above examples would none of them be less natural if the offending expression were rewritten thus: *certain automatic and other increases*—some agricultural and some not—*no further economic or other threats*—industrial or non-industrial—*religious or non-religious*. In correcting, the simple change of *otherwise* to *other* has been avoided, though in fact the critic who would cry 'pedantry' at it must be a little crazy on the subject. And when *other* would be not an adjective but a pronoun, as in the last example, the change is natural and obvious—*those persons, musicians and others*.

In the second use now to be deprecated, the word to which *otherwise* supplies the parallel is not an adjective but a noun. Take the three forms: *What concerns us is his solvency*; *What concerns us is his solvency or insolvency*; *What concerns us is his solvency or otherwise*. Most of the sentences in which *or otherwise* answers to a noun are of this type, and it makes no appreciable difference to the meaning and effect which of the three forms is chosen. The first and the second are as much and as little different in most contexts as *Are you ready?* is different from *Are you ready or not?*, though there is a possibility that the expression of the alternative, which if not expressed would be implied, conveys a special emphasis. The third differs from the second (if grammar is put aside) only as a piece of ELEGANT VARIATION differs from the same meaning given without the variation; *otherwise* is used to escape repeating, in *insolvency,* the previous *solvency*. Few readers who will compare without prejudice the three forms will refuse to admit that the best of the three, wherever it is possible, is the first and shortest—*What concerns us is his solvency*—, the additions *or insolvency* and *or otherwise* being mere waste of words. If writers in general put the question to themselves, made the admission, and acted upon it, not one *or otherwise* in the list that follows would have been written. *Or otherwise* after a noun is (*a*) nearly always superfluous, (*b*) when it is not superfluous, an inferior substitute for *or* with the negative form of the preceding noun or an equivalent, and (*c*) grammatically questionable. Examples of the ordinary foolish use now follow, and the reader is invited to agree that each would be improved by the simple omission of *or otherwise*: *The electorate may be consulted on the merits, o. o., of a single specific measure. | Whatever an up-to-date survey may reveal as to the necessity o. o. for inner relief roads within the city. | I wanted to learn Mr. Khrushchev's views on the possibility, o. o., of the conflict leading to war between the two blocs.* Finally, to point the contrast between the foolish way and the sensible way, the opening sentences of two 'letters to the editor' appearing in the same column: *The variegated assortment of opinions and arguments given in your columns as to the suitability o. o. of Lord Home for the premiership. . . . | In view of recent discussions as to the advantages or disadvantages of a tote monopoly. . . .*

It must be conceded that *o. o.* in this construction is not invariably superfluous. It is not superfluous in *With the view of showing the applicability (o. o.) to the practical affairs of government of the principles which . . .* Without *o. o.* this would mean that the purpose was to demonstrate that the principles were applicable, not to examine whether they were or not. But *o. o.* has become such a plague that even when defensible it is better avoided, as it easily can be here by writing either *examining* for *showing* or *how far these principles are applicable* for *the applicability o. o. of these principles*.

**ought,** v., **1.** is peculiarly liable to be carelessly combined with auxiliary verbs that differ from it in taking the plain infinitive without *to*. *Can and ought to go* is right, but *Ought and can*

*go* is wrong. *We should be sorry to see English critics suggesting that they ought or could have acted otherwise*; insert *to* after *ought*, or write *that they could or ought to have acted.* See ELLIPSIS 2.

**2.** *You didn't ought to have done that* is a not uncommon colloquial vulgarism. *Ought,* the past tense of *owe* (now used as present also) is the only surviving form of that verb in its sense of be under a duty to, or be expected to. An auxiliary cannot therefore be used with *ought* as though it were an infinitive, and it must be negatived with a bare *not* in the old-fashioned way; *you ought not to have done that.* Cf. the somewhat similar restrictions on *used.* See USE.

**our.** **1.** *Our, ours.* **2.** *Our* editorial and ordinary. **3.** *Our, his.*

**1.** *Our, ours. Ours and the Italian troops are now across the Piave.* The right alternatives are: *The Italian troops and ours, The Italian and our troops, Our and the Italian troops*; the wrong one is that in the quotation; see ABSOLUTE POSSESSIVES.

**2.** The editorial *our*, like *we* and *us* of that kind, should not be allowed to appear in close proximity to any non-editorial use of *we* etc. In the following, *our* and the second *we* are editorial, while *us* and the first *we* are national: *For chaos it is now proposed to substitute law, law by which* we *must gain as neutrals, and which, in* our *view, inflicts no material sacrifice on* us *as belligerents.* We *do not propose to argue that question again from the beginning, but . . .*

**3.** *Our, his. Which of us would wish to be ill in* our *kitchen, especially when it is also the family living-room?* If a possessive adjective were necessary, *his* and not *our* would be the right one, or, at greater length, *his or her.* People of weak grammatical digestions, unable to stomach *his,* should find means of doing without the possessive; why not simply *the* kitchen, here? But many of them, who prefer even the repulsive *their* to the right forms (see THEY), are naturally delighted when *of* *us* gives them a chance of the less repulsive though slovenly *our*. It is undeniable that *which of us* is a phrase denoting a singular, and that the possessive required by it is one that refers to a singular.

**-our and -or.** The American abolition of *-our* in such words as *honour* and *favour* has probably retarded rather than quickened English progress in the same direction. Our first notification that the book we are reading is not English but American is often, nowadays, the sight of an *-or*. 'Yankee' we say, and congratulate ourselves on spelling like gentlemen; we wisely decline to regard it as a matter for argument. The English way cannot but be better than the American way; that is enough. Most of us, therefore, do not come to the question with an open mind. Those who are willing to put national prejudice aside and examine the facts soon realize, first, that the British *-our* words are much fewer in proportion to the *-or* words than they supposed, and, secondly, that there seems to be no discoverable line between the two sets so based on principle as to serve any useful purpose. By the side of *favour* there is *horror*, beside *ardour pallor*, beside *odour tremor*, and so forth. Of agent-nouns *saviour* (with its echo *paviour*, itself now tending towards *pavior*) is perhaps the only one that retains *-our*, *governor* being the latest to shed its *-u-*. What is likely to happen is either that, when some general reform of spelling is consented to, reduction of *-our* to *-or* will be one of the least disputed items, or that, failing general reform, we shall see word after word in *-our* go the way of *governour*. It is not worth while either to resist such a gradual change or to fly in the face of national sentiment by trying to hurry it; it would need a very open mind indeed in an Englishman to accept *armor* and *succor* with equanimity. Those who wish to satisfy themselves that it is right to deny any value to the *-our* spelling should go to the article *-or* in the OED for fuller information

than there is room for here. See also SPELLING POINTS.

**-our- and -or-.** Even those nouns that in our usage still end in *-our* (see -OUR and -OR), as opposed to the American *-or*, e.g. *clamour*, *clangour*, *humour*, *odour*, *rigour*, *valour*, *vapour*, *vigour*, have adjectives ending in *-orous*, not *-ourous*—*humorous*, *vaporous*, etc.

Derivatives in *-ist*, *-ite*, and *-able*, should perhaps be regarded as formed directly from the English words, and therefore more ready to retain the *u*; so *colourist*, *labourite* (cf. *favourite*, of different formation), *colourable* and *honourable*; while derivations in *-ation* and *-ize* are more suitably treated, like those in *-ous*, as formed first in Latin, and therefore spelt without the *-u-*; so *coloration*, *invigoration*, *vaporize*, and *deodorize*.

**ours, our.** See OUR 1.

**outcome** is one of the words specially liable to the slovenly use described in the article HAZINESS; so: *The outcome of such nationalization would undoubtedly lead to the loss of incentive and initiative in that trade*. The o. of nationalization would *be* loss; *nationalization would lead to loss*.

**out-herod.** In view of the phrase's continuing popularity and many adaptations (*Mr. Acheson's is a programme that out-Adenauers Adenauer*), two cautions are perhaps called for. The noun after *out-Herod* should be *Herod* and nothing else. The OED quotes 'out-Heroding the French cavaliers in compliment'. Other examples are *Ecclesiastical functionaries who out-Heroded the Daughters of the Horseleech. | The stench emanating from the barrels containing the fruit before boiling would never be forgotten. It even out-Heroded smog.* Similarly after adaptations of the out-herod phrase, the name should be repeated (*out-McCarthy McCarthy*, not *out McCarthy the communist-hunters*). Secondly, the name used should be one at least that passes universally as typifying something; to out-Kautsch Kautsch (*The similar German compilation edited by*

*Kautsch was good; but Charles easily out-Kautsches Kautsch*) is very frigid.

**out of the frying-pan.** A very large proportion of the mistakes that are made in writing result neither from simple ignorance nor from carelessness, but from the attempt to avoid what are rightly or wrongly taken to be faults of grammar or style. The writer who produces an ungrammatical, an ugly, or even a noticeably awkward phrase, and lets us see that he has done it in trying to get rid of something else that he was afraid of, gives a worse impression of himself than if he had risked our catching him in his original misdemeanour; he is out of the frying-pan into the fire. A few typical examples will be here collected, with references to other articles in which the tendency to mistaken correction is set forth more at large.

*Recognition is given to it by no matter whom it is displayed.* The frying-pan was 'no matter whom it is displayed by', which the writer did not dare keep, with its preposition at end; but in his hurry he jumped into nonsense; see MATTER, and PREPOSITION AT END. / *When the record of this campaign comes dispassionately to be written, and in just perspective, it will be found that* . . . The writer took 'to be dispassionately written' for a SPLIT INFINITIVE, and by his correction convinces us that he does not know a split infinitive when he sees it. / *In the hymn and its setting there is something which, to use a word of Coleridge, 'finds' men.* 'A word of Coleridge's' is an idiom whose genesis may be doubtful, but it has the advantage over the correction of being English; *a word of Coleridge* is no more English than *a friend of me*. See OF 7. / *But the badly cut-up enemy troops were continually reinforced and substituted by fresh units.* The frying-pan was RE-PLACE in the sense 'take the place of'; the fire is the revelation that the writer has no idea what the verb SUBSTITUTE means. / *Sir Starr Jameson has had one of the most varied and picturesque careers of any Colonial statesmen.* 'Of *any* statesman', idiomatic but appa-

rently illogical, has been corrected to what is neither logical (*of all* would have been nearer to sense) nor English. / *The claim yesterday was for the difference between the old rate, which was a rate by agreement, and between the new.* The writer feared, with some contempt for his readers' intelligence, that they would not be equal to carrying on the construction of *between*; he has not mended matters by turning sense into nonsense; see OVERZEAL. / *The reception was held at the bride's aunt.* The reporter was right in disliking *bride's aunt's*, but should have found time to think of 'at the house of'.

The impression must not be left, however, that it is fatal to read over and correct what one has written. The moral is that correction requires as much care as the original writing, or more; the slapdash corrector, who should not be in such a hurry, and the uneducated corrector, who should not be writing at all, are apt to make things worse than they found them.

**outstanding,** with its two meanings of standing out, conspicuous, and standing over, unsettled, needs careful handling to avoid ambiguity. The announcer of election results who began *The other o. result is* . . . may have misled listeners into thinking that he was about to tell them of a result of special interest; he would have made his meaning clear if he had said *The other result still o.*

**outworn.** *There is, however, a little more in Mr. Bonar Law's speech than these husks of a controversy outworn.* Allusions like this, shown to be such by the position of *outworn*, to *A pagan suckled in a creed outworn* betray mortal dread of being commonplace, and draw attention to the weakness they are meant to cloak. See IRRELEVANT ALLUSION.

**overall** in its established senses is used as a noun for the garment, as an adverb (= all over) in the naval phrase *a ship dressed o.*, and as an adjective to describe a measurement taken between the extreme points of the thing measured, especially the length of a ship. In recent years it has blossomed in its adjectival use into one of the most pervasive of VOGUE WORDS; few can be playing greater havoc with the vocabulary. Among its victims are *absolute, aggregate, average, complete, comprehensive, general, inclusive, net, overriding, supreme, total,* and *whole.* Its allure is indeed so strong that it is often used in contexts where it contributes nothing to the sense. *It was not until the third ballot that Mr. Michael Foot secured an absolute o. majority.* | *In more than 3,000 miles of running there was an o. net gain of 14 minutes on schedule.* | *The o. growth of London should be restrained.* Perhaps Ivor Brown is right in attributing the popularity of the word to 'the presence of sonority, almost of poetry, in its composition'.

**overlay, -lie.** It has been mentioned (see LAY AND LIE) that the two simple verbs are sometimes confused even in print especially in the p.p. form, for which *-lay* makes *-laid* and *-lie* makes *-lain.* It is still more common for *overlay* and *underlay* to be used where *-lie* is wanted, because the *-lie* verbs too are transitive, though in different meanings from those in *-lay. The talk about things in general which overlays the story is quite dull.* This should be *overlies*; similarly newborn pigs with a clumsy mother are liable to be *overlain,* not *overlaid.*

**overlook, oversee.** There is much avoidable confusion in the use of these words and their derivatives. *Overlook* though sometimes still used in the sense of inspect or supervise (*Lord Hailsham, as Lord Privy Seal, is to overlook scientific development*), and often in that of to command a view of, has as its chief current use the meaning of to look beyond and therefore fail to see, ignore, condone. Thus there is now an unfortunate ambiguity in the notice *Gentlemen are requested not to overlook the Ladies' Bathing Place. Oversee* and *overseer* retain their meanings of supervise and supervisor, but *oversight* has followed *overlook* instead

of its parent verb, and is now little used except in the sense of a failure to observe. Confusion would be removed if the following differentiation, representing the main current usages, were strictly observed:

*Look over*: 1. inspect casually. 2. look beyond.

*Overlook*: 1. command a view of, 2. fail to notice, ignore, condone.

*Oversee*: supervise.

*Oversight*: failure to notice, inadvertence.

**oversea(s).** According to the OED *-sea* is the adjective and *-seas* (rarely *-sea*) the adverb. The plural form has now become usual for both—*British Overseas Airways Corporation, Overseas Services of the B.B.C., British Overseas Trade*—and the singular is disappearing, though *Oversea Settlement* remains as an item in the Civil Estimates. It is right that *overseas* should prevail; it is more fitting for the sense of *far and wide* in which it is generally used; to an island people *oversea* is no different from *abroad.*

**overthrowal.** *The drama lies in the development of a soul towards the knowledge of itself and of the significance of life, and the tragedy lies in the overthrowal of that soul.* This NEEDLESS VARIANT of *overthrow* was unknown to the original edition of the OED, but seems to have had some ill-advised patronage since; the 1933 Supp. inserts it with examples. See -AL nouns.

**overtone and undertone.** There is a difference worth preserving in the figurative uses of these words. *Overtones* are the higher notes produced by the vibration of a string or column of air or piece of metal, seldom separately distinguishable by the ear from the main note. *Overtone* is therefore an apt metaphor for suggesting that a word has implications over and above its plain meaning. '*Artificial*' *cannot be used without an overtone of disparagement.* | '*Split in the Party!*' '*Split*' *is one of those convenient words that not only fit easily within the limits of a headline but are also rich in dramatic* *overtones.* *Undertone* is not a technical term of music, though sometimes misused as one. It means a low or subdued utterance or sound (OED) and is suitably used figuratively for a communal state of mind inferred from evidence rather than expressed outright. *There was an undertone of optimism in Geneva yesterday.* / *There was little business in the gilt-edged market, but the undertone was firm.*

**overzeal.** Readers should be credited with the ability to make their way from end to end of an ordinary sentence without being pulled and pushed and admonished into the right direction; but some of their guides are so determined to prevent straying that they plant great signposts in the middle of the road, often with the unfortunate result of making it no thoroughfare. In the examples the signpost word, always needless, often unsightly, and sometimes misleading, is enclosed in square brackets:

*But it does not at all follow that because Mr. Long is 65* [*that*] *he will not be equal to* . . . See THAT, CONJ. 4, for more. / *We agree that the Second Chamber would be differently constituted according as we went forward to other schemes of devolution and federation,* [*and according as we*] *decided to make Home Rule for Ireland our one and only experiment.* Read *or decided*; see ACCORDING AS for more. / *The working man has to keep his family on what would be considered a princely wage in England, but* [*which*], *in point of fact, is barely enough to keep body and soul together.* See WHAT 4 for more. / *The object for which troops were sent was* [*for*] *the protection of British property.* The object was not *for* protection; it *was* protection. / *But what no undergraduate or* [*no*] *professor in the art of writing verse could achieve is* . . . See OR 4 for more. / *There are others who talk of moving and debating a hostile amendment, and then* [*of*] *withdrawing it.* Moving, debating, and then withdrawing make up a single suggested course; but the superfluous *of* implies that the talkers vacillate between two

courses. / *Had Bannockburn never been fought, or* [*had*] *seen another issue, Scotland would have become a second Ireland.* The motive is to exclude *never* from the second clause; but either that ambiguity must be risked and *had* omitted, or *had it* must be inserted instead of *had.*

**owing to** is often used in combination with *the fact that* as a clumsy periphrasis for a simple conjunction such as *because, since, for, as*; and for this purpose we could do with less of it. But for another we should like to see more of it; its rights are now perpetually infringed by DUE TO (q.v.) which has not yet won, as *o. t.* has, unquestioned recognition as a preposition.

**Oxbridge.** See PORTMANTEAU WORDS.

**oxymoron.** Pl. *-s* or *-ra*; see -ON. The combining in one expression of two terms that are ordinarily contradictory, and whose exceptional coincidence is therefore arresting. *A cheerful pessimist; Harmonious discord; His honour rooted in dishonour stood, And faith unfaithful kept him falsely true.* The stock example is Horace's *Splendide mendax.* The figure needs discreet handling or its effect may be absurd rather than impressive. We are too much accustomed to saying that we are *terribly pleased* to notice that absurdity (see TERRIBLY), but the demonstrator who says *As you can see, the discrepancy is immensely slight,* does not make a happy choice of adverb.

# P

**pace** (ablative of *pax*, pronounced *pāsĕ*). This latinism (*p. tuâ* by your leave, or if you will allow me to say so; *p. Veneris* if Venus will not be offended by my saying so) is one that we could very well do without in English. Not only is it often unintelligible to many readers even when rightly used; it is also by many writers wrongly used. In the two following pieces, which unfortunately have to be long if the point is to be clear, the meaning is 'according to Mr. Begbie' or 'according to the Jungborn enthusiasts'; it ought to be just the opposite—'though Mr. B. (or the enthusiasts) will doubtless not agree': *After the beauty of rural life in the South his picture of Belfast is a vision of horror. On the details of that picture we need not dwell; but the moral which Mr. Begbie appears to draw from his contrast is that a Conservative Irish Parliament will do little to better the conditions of town life, and that the industrial classes would find relief from those conditions more quickly under the rule of the English Parliament, which*, pace *Mr. Begbie, is advancing rapidly towards some form of Socialism. | For more than ten thousand years these things have been recognized in some part of the world; during that lapse of time, at least, some men and women have been living according to their own lights rather than according to the light of nature. Now*, pace *the Jungborn enthusiasts, the time has come to change all this. If man would survive as a species, we learn in effect, he must begin the return journey to the place whence he came.*

Minor objections are that the construction is awkward in English (it needs a genitive but *pace Mr. Smith* is the nearest we can get to *p. Caesaris*), and that the Latinless naturally, but distastefully to those who know Latin, extend the meaning or application as they do those of VIDE, RE, and E.G. So: *But in the House of Lords there is no hilarity*—pace *Lord Salisbury's speech last night. Pace* does not mean notwithstanding a fact or instance, but despite someone's opinion, as in *The great poet need not be* (pace *Coleridge) a philosopher.| Modern history*, pace *Ford, is not wholly bunk.*

**pachydermatous.** A favourite with the POLYSYLLABIC HUMOURists.

**pacif(ic)ist.** The longer form is the better etymologically, but euphony favours the shorter and probably accounts for its having prevailed. See -IST A.

**package,** long established as a noun, has been given fresh duties both as a verb and as an adjective. The appearance of a new verb, to *package*, when we already have to *pack*, puts on those who use it the onus of showing that it is not merely a LONG VARIANT. They have a good case: to *package* means—or should mean—something more than to *pack*. Anyone can pack by cramming things anyhow into a suitcase; to package is to fit things neatly and securely into parcels; it is a skill that has to be taught. Moreover, *packaging* is the name suitably given to a busy and growing industry engaged in devising and manufacturing new types of wrappers and containers. As an adjective, *package* is used attributively of terms of agreement offered by one negotiating party to another, or of the deal that results, meaning that what is offered all hangs together; it may be accepted or rejected as a whole, but the package must not be undone and some of its contents taken and others left. *Has the untying of the Western package plan for German reunification and European security already begun, and will a separate agreement on Berlin now be discussed?* In this sense it can claim to be a useful metaphor, saying in one word what would otherwise need several.

**pair.** In such expressions as *p. of scissors, p. of spectacles, p. of trousers*, etc.— 'articles compounded of two corresponding parts which are not used separately' (OED)—the word denoting the thing of which there are two parts (*scissors* etc.) is obviously plural. But when we drop the *pair(s) of*, as we usually do, and just say *scissors* etc., we may mean either one pair or more than one; and even when we mean only one we always treat the word as grammatically plural. It follows that, although its use for more than one pair may be unambiguous (*There are plenty of scissors in the house*), its use for a single pair (*These scissors need sharpening*) must always be grammatically ambiguous, however clear the intention may be. If we want to indicate unambiguously that we mean a single article, or a specified number of single articles, we cannot do without *pair(s) of* (*Get me a pair*, or *two pairs, of scissors*). The less familiar words are, however, not so insistent on being treated as plurals as the more familiar. No one would say *Hand me that trousers*, but some might say *Hand me that shears*. For other difficulties with words plural in form but singular in meaning see PLURAL ANOMALIES and SINGULAR S.

**pact.** See HEADLINE LANGUAGE.

**pairs and snares.** Of the large number of words that are sometimes confused with others a small selection is here given. It will be noticed that nearly all are of Latin origin; the confusion arises largely from the Englishman's natural failure, if he has not learnt Latin, to realize instinctively the force of suffixes that are not native. Those who have any doubts of their infallibility may find it worth while to go through the list and make sure that these pairs have no terrors for them; under one or other of most of the pairs in its dictionary place they will find remarks upon the difference, and usually proofs that the confusion does occur. While the Englishman's vagueness about Latin suffixes or prefixes is the most frequent cause of mistakes, it is not the only one. Often the two words might legitimately have been equivalents, or actually were in older usage, and the ignorance is not of Latin elements but of English idiom and the changes that DIFFERENTIATION has brought about. And again there are pairs in which the connexion between the two words is only a seeming one. To exemplify briefly, *contemptuous* and *contemptible* are a pair in which suffixes may well be confused; *masterful* and *masterly* one in which differentiation may well be overlooked; and *deprecate* and *depreciate* one of the wholly false pairs.

The list is as follows: acceptance and acceptation; advance and advancement; affect and effect; alternate and alternative; antitype and prototype; apologue and apology; ascendancy and ascendant; autarchy and autarky; ceremonial and ceremonious; comity and company; complacent (-ency) and complaisant (-ance); compose and comprise; consequent and consequential; contemptible and contemptuous; definite and definitive; deprecate and depreciate; derisive and derisory; discreet and discrete; disinterested and uninterested; e.g. and i.e.; euphemism and euphuism; fatal and fateful; forceful and forcible; fruition and fructification; immovable and irremovable; infer and imply; impassable and impassible; inflammable and inflammatory; ingenious and ingenuous; judicial and judicious; laudable and laudatory; luxuriant and luxurious; masterful and masterly; mendacity and mendicity; militate and mitigate; observance and observation; perspicacity (-acious) and perspicuity (-uous); policy and polity; precipitate and precipitous; predicate and predict; prescribe and proscribe; proportion and portion; purport and purpose; regretful and regrettable; resource, recourse, and resort; reverend and reverent; reversal and reversion; seasonal and seasonable; sensual and sensuous; titillate and titivate; transcendent and transcendental; triumphal and triumphant; unexceptionable and unexceptional. See also MISPRINTS.

**pajamas** is the American word for the garment known to us as *pyjamas*.

**palace.** Educated usage is exceptionally divided between the two pronunciations *pǎ'lăs* and *pǎ'lǐs*. In spite of Milne's well-known lines *They're changing guard at Buckingham Palace; Christopher Robin went down with Alice*, the speak-as-you-spell movement will probably secure the victory of the first.

**palaeo-, palæo-, paleo-.** The first is recommended; see Æ, Œ.

**Pall Mall.** See MALL.

**palpable.** *The work that has yet to be done is palpable from the crowded paper of amendments with which the House is faced.* A good illustration of the need for caution in handling dead metaphors.

*Palpable* means literally touchable, or perceptible by touch; that meaning is freely extended to perceptible by any of the senses, and even to appreciable by the intelligence. The final extension is necessary here, and would pass but for the *from* phrase that is attached. *From the paper* implies sensuous perception; when intellectual inference is intended the dead metaphor is stimulated into angry life by the inconsistency; see METAPHOR. *P.* is one of the words that are liable to clumsy treatment of this sort because they have never become vernacular English, and yet are occasionally borrowed by those who have no scholarly knowledge of them.

**pander,** n. and v. Though *-ar* is the older and better form, it is waste of labour to try to restore it.

**pandit.** See PUNDIT.

**panegyric, -rize, -rist.** The pronunciations recommended are: *pănĕjĭ'rĭk, păn'ĕjĭrīz, pănĕjĭ'rist.*

**pantaloons, pants.** The words we use for what were once called generically *nether garments* have undergone some curious variations, partly due to changes of fashion in the way we like to cover or expose our legs, and partly under the influence of EUPHEMISM, which demands constant change in the names we give to anything supposed to have indelicate associations, in this case even going so far at one time as to throw the cloak of *unmentionables* or *inexpressibles* over the whole shocking business. *Pantaloons,* a word imported from France for the trousers that superseded *breeches* or *small-clothes,* is now obsolete except as the formal word for the *trews* worn by certain Scottish regiments; so is the female equivalent *pantalettes,* the long frilly drawers worn by little girls in the early Victorian age. But the diminutives survive: in U.S. *pants* for trousers and in England *pants* (for women *panties*), or, when particularly exiguous, *briefs,* or *scanties,* as undergarments. *Breeches* has returned and, with the help of *jodhpurs, plus-fours, shorts,* and *trunks,*

has ousted *knickerbockers* which, in its curtailed form *knickers,* has been appropriated by women. *Bloomers* was short-lived for cyclists, but survives for schoolgirls; *drawers* is falling into genteel obsolescence, and even the highly respectable *trousers* is menaced by *slacks* and *jeans.*

**papyrus.** The horse that won the Derby in 1923 was known to many as *Păp'ĭrus,* but the right pronunciation is *Păpī'rus.* Pl. *-ri.*

**para-.** Two prefixes of different origin are used in forming English words: the Greek preposition meaning alongside of or beyond, and (through Italian) the Latin imperative meaning guard against. The first gives us a large number of words such as *paradox, paragraph, parallel, paraphrase* and *paramilitary,* the second such words as *parapet, parasol,* and *parachute.*

**parable.** For p. and allegory, see SIMILE AND METAPHOR.

**paradise** rivals NECTAR in the number of experiments that the desire for a satisfactory adjective has occasioned. But, whereas *nectar* is in the end well enough provided, no one uses any adjective from *paradise* without feeling that surely some other would have been less inadequate. The variants are *paradisaic*(al*), paradisal, paradisean, paradisiac(al), paradisial*, paradisian*, paradisic(al),* of which the asterisked ones are badly formed. *Paradisal* seems the least intolerable, perhaps because it retains the sound of the last syllable of *paradise;* but the wise man takes refuge with *heavenly* or other substitute.

**paraffin.** See KEROSENE.

**paragraph.** The purpose of paragraphing is to give the reader a rest. The writer is saying to him: 'Have you got that? If so, I'll go on to the next point.' There can be no general rule about the most suitable length for a paragraph; a succession of very short ones is as irritating as very long ones are wearisome. The paragraph is essentially a unit of thought, not of length:

it must be homogeneous in subject-matter and sequential in treatment. If a single sequence of treatment of a single subject means an unreasonably long paragraph, it may be divided into more than one. But passages that have not this unity must not be combined into one, even though each by itself may seem to make an unduly short paragraph.

Paragraphing is also a matter of the eye. A reader will address himself more readily to his task if he sees from the start that he will have breathing-spaces from time to time than if what is before him looks like a marathon course.

**parallel.** 1. Exceptionally among verbs in -l (see -LL-, -L-), *p.* does not double the l: *paralleled* etc.; the anomaly is due to the -ll- of the previous syllable. 2. *We have already had occasion to comment on the re-markable parallelity between . and ...* The noun used, where *p.* itself will not serve, is *parallelism*, not *parallelity*; the latter is not even recorded in the OED.

**parallel-sentence dangers.** 1. Negative and affirmative. 2. Inverted and uninverted. 3. Dependent and inde-pendent.

1. Negative and affirmative. A single example may be given here to show the kind of difficulty that occurs: *There is not a single town in the crowded district along the Rhine which is* not open *to these attacks, and* must be prepared *for defence with guns, troops, and aeroplanes.* For discussion and illustration of this and many other varieties, see NEGATIVE MISHANDLINGS.

2. Inverted and uninverted. *And not merely in schools and colleges, but as organizers of physical training,* are women readily finding *interesting and important employment.* The *not merely* part requires the inverted *are women finding;* the *but* part requires the un-inverted *women are finding.* The right solution is to start the sentence with *And women are finding employment not merely* etc. In INVERSION the section

headed *Inversion in Parallel Clauses* is devoted to this and similar types.

3. Dependent and independent. *The municipality charged itself with the pur-chase of these articles in wholesale quan-tities, and it was to the Town Hall that poor people applied for them, and were served by municipal employees.* The parallel clauses in question were, in their simple form, (*a*) *The poor people applied for them to the Town Hall,* and (*b*) *The poor people were served by municipal employees.* The writer has decided, for the sake of emphasizing *Town Hall,* to rewrite (*a*) in the *it was . . . that* form; but he has forgotten that he cannot make (*a*) dependent and leave (*b*) independent unless he sup-plies the latter with a subject (*and* they *were served*). The correct possibilities are: (i, both independent) *The people applied to the Town Hall for them, and were served by municipal employees;* (ii, both dependent) *It was to the Town Hall that the people applied, and by municipal employees that they were served;* (iii, dependent and indepen-dent) *It was to the Town Hall that the people applied, and they were served by municipal employees.*

**parallelepiped.** Pronounce *pără-lĕlĕ'pīpĕd* or more conveniently *pără-lĕlĕpī'pĕd.*

**paraplegia** is usually pronounced with a soft *g.* See GREEK G.

**paraselene.** Five syllables (-*ē'nē*). Pl. *ae.*

**parasitic(al).** The longer form has no special function, and might well be discarded. See -IC(AL).

**parenthesis** (pl. -*theses*). 1. Length. 2. Relevance. 3. Identification. 4. Stops after parentheses.

1. Length. A parenthesis is a con-venient device, but a writer indulges his own convenience at the expense of his readers' if his parenthesis is so long that a reader, when he comes to the end of it, has little chance of remem-bering where he was when it began.

*Mr. B. makes full use of both these advantages; after a rather slow start—Bevin was himself a slow starter, of minor importance, save for the much-publicized appearance as the Dockers' K.C., until the early twenties were well begun, and Mr. B. finds little fresh to say about the early period—the narrative is full of interest and documentation.*
2. Relevance. A parenthesis may or may not have a grammatical relation to the sentence in which it is inserted. In *This is, as far as I know, the whole truth* there is such a relation, and in *This is, I swear, the whole truth* there is not; but one is as legitimate as the other. It is not equally immaterial whether the parenthesis is relevant or not to its sentence; parentheses like the following cannot possibly be justified: *In writing this straightforward and workmanlike biography of his grandfather (the book was finished before the war, and delayed in publication) Mr. Walter Jerrold has aimed at doing justice to Douglas Jerrold as dramatist, as social reformer and as good-natured man.* The time of writing and the delay in publication have no conceivable bearing on the straightforwardness, workmanlikeness, biographicality, grandfatherliness, justice, drama, reform, or good nature, with which the sentence is concerned. If it had been called a long-expected instead of a straightforward biography, it would have been quite another matter; but, as it is, the parenthesis is as disconcerting as a pebble that jars one's teeth in a mouthful of plum pudding. The very worst way of introducing an additional fact is to thrust it as a parenthesis into the middle of a sentence with which it has nothing to do. A similar example is: *Napoleon's conversations with Bertrand and Moncholon (it is unfortunate that there are several misprints in the book) are a skilful blending of record and pastiche.*
3. Identification. Still more fatal than readiness to resort to parenthesis where it is irrelevant is inability to tell a parenthesis from a main sentence. *A remarkable change had come over the Government,* he suggested, since the Bill had left the Committee, and expressed doubts *as to whether the Minister altogether approved of the new turn of affairs.* In this, *he suggested* is as much a parenthesis as if it had been enclosed in brackets; if it were not parenthetic, the sentence would run *He suggested that a change had come.* But the writer, not knowing a parenthesis when he sees one (or even when he makes one), has treated it as parallel with *expressed,* and so fully parallel that its *he* may be expected to do duty with *expressed* as well as with *suggested.* Either the first part should be rewritten as above with *suggested* for its governing verb, or another *he* must be inserted before *expressed.*
4. For stops after parentheses and for the choice between square brackets, round brackets, dashes, and commas see STOPS s.f.

**parenthetic(al).** In most uses the longer form is obsolescent; but it has still a special sense worth preserving, i.e. full of or addicted to parentheses (*a horribly -ical style*). See -IC(AL).

**pariah.** About the four possible pronunciations (*păr′ĭa, păr′ĭa, pahr′ĭa,* and *pări′a*) there is little agreement among the dictionaries, except that the modern ones have dropped the last, and, on the whole, *pā-* and *pă-* are more favoured than *pah-.*

**pari passu.** Pronounce *pār′ī pă′sū.* See LATIN PHRASES.

**park,** v. To *park,* in the sense of to deposit temporarily, especially a motor car, is modern; but the seeds of it can be found in Shakespeare's *How are we parked and bounded in a pale.* True, it there implies inability to get out, but that experience is not unknown to the motorist either. The usage started in America; its acceptance in Britain and its extension, often jocular, to things other than cars have been wholehearted. Hats and coats are now *parked* at least as often as they are *left* or *deposited,* and the OED Supp. gives examples of its use for deposits varying from children to chewing-gum.

**Parkinson.** Though there may be little risk of confusion between the two men who have won eponymous fame, it is perhaps worth while to distinguish them. *P.'s disease* (paralysis agitans, or shaking palsy) is so called because it was first fully described by James Parkinson in 1817. *P.'s law* ('Work expands so as to fill the time available for its completion'), based on a statistical study of the staffing of British government departments, was propounded by Professor C. Northcote Parkinson in 1957 in a whimsical exposition of a fundamental truth.

**parlance.** See JARGON.

**parliament.** Dictionaries differ about the pronunciation *par'lăment* or *par'-liment*. The former is recommended.

**parlous** is a word that wise men leave alone. It is the same by origin as *perilous*, but centuries ago it suffered the same fate that has befallen *awful* and *chronic* in more recent times; it became a VOGUE WORD applied to many things very remote from its proper sense. It consequently lost all significance, 'died of its own too much', and was for a long time (for most of the 18th c.) hardly heard of. In the 19th c. it was exhumed by ARCHAISM and PEDANTIC HUMOUR, and the adepts in those arts should be allowed exclusive property in it.

**parody.** See BURLESQUE for synonyms.

**paronomasia.** Puns, plays on words, making jocular or fanciful use of similarity between different words or of a word's different senses. The best known of all (though concealed in English) is perhaps that of *Matt.* xvi. 18: *Thou art Peter* (Greek *Petros*), *and upon this rock* (Greek *petra*) *I will build my church.* Now that we regard puns merely as exercises in jocularity, and a pretty debased form even of that (see PUN), we are apt to be jarred by the readiness of Shakespeare's characters to make them at what seem to us most unsuitable moments, as Macbeth does just before the murder of Duncan, and

Lady Macbeth just after it: *If the assassination could catch . . . with his surcease success.* / *If he do bleed, I'll gild the faces of the grooms withal; for it must seem their guilt.*

**parricide, patricide.** The first is the orthodox form. *Patricide* has no doubt been substituted by some deliberately, in order to narrow the meaning to murder(er) of a father, as *matricide* and *fratricide* are limited, and by others in ignorance of the right word. *Parricide* includes not only the murder of either parent or any near relative or anyone whose person is sacred, but also treason against one's country; and the making of *patricide* to correspond to *matricide* is therefore natural enough.

**Parthian shot.** It seems to be a coincidence that the popular corruption *parting shot*, which no doubt owes its origin to the similarity of sound, has a meaning akin to that of the parent phrase. *Parthian shot* refers to the tactics of the Parthian mounted archers, who would discharge a volley into the enemy while moving smartly out of range of retaliation; *parting shot* is ordinarily used to describe a 'last word' fired by one of the parties to an argument at the other before breaking off the verbal engagement. Although the Parthian tactics were undoubtedly formidable, it is a MISAPPREHENSION to use *Parthian shot* to mean merely an attack that strikes home; the essence of it is that the attack is made at the moment of retreat.

**partially** is often used where *partly* would be better. This is, no doubt, because it is formed normally, by way of the adjective *partial*, while *partly* formed direct from the noun *part* is abnormal. There is much the same difference between the two words as between *wholly* (opp. *partly*) and *completely* (opp. *partially*); in other words, *partly* is better in the sense 'as regards a part and not the whole', and *partially* in the sense 'to a limited degree': *It is partly wood; This was partly due to cowardice; A partially drunken sailor; His partially re-established health.*

**participles** 438 **participles**

Often either will give the required sense equally well; *partly* is then recommended, since it is *partially* that tends to be over-used; see LONG VARIANTS for other such pairs. An example or two of the wrong *partially* are: *The two feet are partially of iron and partially of clay.* | *Whether 'The Case is Altered' may be wholly or partially or not at all assignable to the hand of Jonson.*

**participles.** 1. Unattached p. 2. Absolute construction. 3. Fused p. 4. Sentry p. etc. 5. Accent and pronunciation in p.p. (or participal adjective or noun) and verb.

1. Unattached p. For this danger, as insidious as notorious, see UNATTACHED PARTICIPLE.

2. Absolute construction. *The Municipal Council, having refused their assistant clerks' demand for a rise in salary, those in the Food Supply offices today declared a strike.* This false stopping (there should be no comma after *Council*) is an example of what is perhaps both the worst and the commonest of all mistakes in punctuation. See ABSOLUTE CONSTRUCTION.

3. Fused p. The construction so called is fully discussed in the article FUSED PARTICIPLE.

4. Sentry participle etc. If newspaper editors, in the interest of their readers, maintain any discipline over the gentlemen who provide inch-long paragraphs, they should take measures against a particular form that, by a survival of the unfittest, bids fair to swallow up all others. In these paragraphs, before we are allowed to enter, we are challenged by the sentry in the guise of a participle or some equivalent posted in advance to secure that our interview with the C.O. (or subject of the sentence) shall not take place without due ceremony. The fussiness of this is probably entertaining while it is quite fresh; one cannot tell, because it is no longer fresh to anyone. It is likely to result in jamming together two unrelated ideas in one sentence. Examples: *Described as 'disciples of Tolstoi', two Frenchmen sentenced at Cheltenham to two months' imprisonment for false statements to the registration officer are not to be recommended for deportation.* | *Winner of many rowing trophies, Mr. Robert George Dugdale, aged seventy-five, died at Eton.* | *Believed to be the youngest organist in the country, Master Herbert Woolverton, who officiates at Hutton Church, Essex, has passed the examination as Associate of . . .* | *Thirty-four years in the choir of the Chapel Royal, Hampton Court Palace, Mr. Francis P. Hill, of Milner Road, Kingston, has retired.* | *Found standing in play astride the live rail of the electric line at Willesden and in danger of instant death, Walter Spentaford, twelve, was fined 12s. for trespass.* The device of the sentry participle is now worked so hard that the participle is liable to become UNATTACHED, though it cannot often be left in the air quite so egregiously as are the two participles in this description of the departure of Grivas from Cyprus. *Seen off at Nicosia airport by the head of the security forces which had failed to track him, boastfully claiming that he had spent his time 'with the British', seldom has an Imperial Power looked so ridiculous.*

5. Accent and pronunciation in p.p. (or participial adjective or nouns) and verb. Beside many of the verbs formed from Latin supine stems (*animate, dilute, extract*, etc.) there are passive participles of the same spelling, now used as adjectives or nouns. They are often distinguished from the verbs by a difference in sound. This may be (A) a shifting of the accent, as in *attri'bute* v., *a'ttribute* n.; *co'nsummate* v., *consu'mmate* a.; *convi'ct* v., *co'nvict* n.; *dicta'te* v., *di'ctates* n.; *dige'st* v., *di'gest* n.; *extra'ct* v., *e'xtract* n.; *refu'se* v., *re'fuse* n.; (B) the obscuring of the vowel of *-ate* into an indeterminate light sound, as *moderāt* v. becomes *moderĕt* a., and the same phenomenon can be seen in *advocate, animate, articulate, degenerate, delegate, deliberate, designate, desolate, elaborate, estimate, legitimate, regenerate, reprobate, separate, subordinate.* (This differentiation is weakening with the progress of the speak-as-you-spell

movement—see PRONUNCIATION 1; delegates to the United Nations are now as likely to be called -*ates* as -*ets* by the reader of a news bulletin.) Or, (C) there may be a change of consonant sound, as in *diffuse* (-*z*) v., *diffuse* (-*s*) a.; *refuse* (-*z*) v., *refuse* (-*s*) n. See also NOUN AND VERB ACCENT.

**particular.** This is a strong adjective, emphasizing that there is a reason why the noun to which it is attached should be singled out and distinguished from others. It is emasculated when it is used as an unnecessary reinforcement of *this* etc. *It is time that consideration was given to this particular problem.* | *Availabilities of this particular material are extremely limited.* | *That performance brings 'Music at Night' to a close for this particular Thursday.* This trick is mostly to be found in political speeches; perhaps it comes from a vague feeling that the monosyllabic demonstrative is not impressive enough by itself.

**parting.** 1. *The United Nations Organization is at the p. of the ways.* Organizations and men are now so familiar with that position that, when told they are there once more, they are not disquieted; their only impulse is to feel in their breeches pockets for the penny with which they may toss up. See HACKNEYED PHRASES. 2. For *parting shot* see PARTHIAN.

**partisan.** Unlike the obsolete weapon, the word meaning adherent of a cause or guerilla fighter (see RESISTANCE) is usually accented on the last syllable.

**pasquinade.** See LAMPOON.

**pass.** 1. The verb makes *passed* for its past tense (*You passed me by*), and for its p.p. used verbally (*It has passed out of use*); but when the p.p. has become mere adjective it is spelt *past* (*In past times*). The distinction between p.p. and adjective is rather fine in *Those times have passed away* (p.p.), *Those times are passed away* (INTRANSITIVE P.), *Those times are past* (adjective).

2. To *make a pass at* in the sense of to make amorous advances to is a naturalized American colloquialism.

**passable, passible.** See IMPASSABLE.

**passive disturbances.** 1. The double passive. 2. Passive of *avail oneself of*. 3. *Do* after passive. 4. The misplaced *as*. 5. The impersonal passive.
The conversion of an active-verb sentence into a passive-verb one of the same meaning—e.g. of *You killed him* into *He was killed by you*—is a familiar process. But it sometimes leads to bad grammar, false idiom, or clumsiness.

1. The double passive. *People believed him to have been murdered* can be changed to *He was believed to have been murdered*; but *They attempted to carry out the order* cannot be changed to *The order was attempted to be carried out* without clumsiness or worse. For full discussion see DOUBLE PASSIVES.

2. Passive of *avail oneself of*. *We understand that the credit will be availed of by three months' bills, renewable three times, drawn by the Belgian group on the British syndicate.* A passive is not possible for *avail oneself of*; see AVAIL.

3. Active of *do* after passive verb. *Inferior defences could then, as now, be tackled, as Vernon did at Porto Bello, Exmouth at Algiers, and Seymour at Alexandria.* The active form would be [*An admiral*] *could then, as now, tackle inferior defences*; if *defences could be tackled* is substituted, it is better to change the voice of *did* too—*as was done*, or *as they were, by Vernon* etc. But the use of the active is common; see DO 3 c.

4. The misplaced *as*. *The great successes of the Co-operators hitherto have been won as middlemen.* Active form, sound enough—*The Co-operators have won their successes as middlemen.* Conversion to the passive has had the effect of so tying up *the co-operators* with *of* that it is not available, as in the active form, for *as middlemen* to be attached to. A common and perhaps venial lapse.

5. The impersonal passive—*it is felt, it is thought, it is believed*, etc.—is a

construction dear to those who write official and business letters. It is reasonable enough in statements made at large—*It is believed that a large green car was in the vicinity at the time of the accident.* | *It is understood that the wanted man is wearing a raincoat and a cloth cap.* But when one person is addressing another it often amounts to a pusillanimous shrinking from responsibility (*It is felt that your complaint arises from a misunderstanding.* | *It is thought that ample provision has been made against this contingency*). The person addressed has a right to know who it is that entertains a feeling he may not share or a thought he may consider mistaken, and is justly resentful of the suggestion that it exists in the void. On the other hand, the impersonal passive should have been used in *For these reasons the effects of the American recession upon Britain will be both smaller and shorter than were originally feared. Were* should be *was* (i.e. than it was originally feared they would be).

**past.** See PASS.

**past master.** The OED derives the use of this expression in the sense of expert from the phrase to pass master, i.e. to graduate as master in some faculty.

**pastiche.** The Italian form *pasticcio* is usual for the musical medley, the French for the literary. Pronounce *pastēsh', pastǐ'cho*.

**pastorale.** Pronounce *-ahlĕ*; pl. *-li* (*-ē*).

**pasty,** n. The pronunciations *pā-* and *pǎ-* seem to be encroaching on the once orthodox *pah-*. The adjective is always *pā-*.

**patent.** *Pā-*, or *pǎ-*? *Pā-* predominates in Britain, *pǎ-* in America. But even in Britain many retain *pǎ-* for the sense connected with *letters p.*, i.e. for the technical uses (e.g. *Patent Office*) as opposed to the general or etymological sense open and plain. This distinction is based on the fact that *p.* in

the general senses comes direct from Latin, and in the technical sense through French. The one pronunciation *pā-*, however, is recommended for British use in all senses. It should be remembered that the Latin quantity (*pǎ-*) is of no more importance than it is in the opposite word *latent*, now universally *lā-*; see FALSE QUANTITY.

**paterfamilias.** In Roman history, or references to it, the plural should be *patresfamilias*; but as an adopted English word it makes *paterfamiliases*. See LATIN PLURALS. Pronounce *pāt-* in spite of the FALSE QUANTITY.

**pathetic fallacy** is a phrase made by Ruskin; the OED quotes from 'Modern Painters': *All violent feelings ... produce ... a falseness in ... impressions of external things, which I would generally characterize as the 'Pathetic fallacy'.* In ordinary modern use *pathos* and *pathetic* are limited to the idea of painful emotion; but in this phrase, now common, the original wider sense of emotion in general is reverted to, and *the p.f.* means the tendency to credit nature with human emotions. *Sphinxlike, siren-sweet, sly, benign, impassive, vindictive, callously indifferent the sea may seem to a consciousness addicted to pathetic fallacies.* Burns was under the influence of the p.f. when he wrote:

*Ye banks and braes o' bonnie Doon,*
  *How can ye bloom sae fresh and fair!*
*How can ye chant, ye little birds,*
  *And I sae weary fu' o' care!*

**pathos.** For this and *bathos* the OED recognizes only the pronunciations *pā-*, *bā-*, and this is now established. But in compounds (e.g. *pathology, bathometer*) the *a* is short.

**patois.** For *p.*, *dialect*, etc., see JARGON.

**patriot(ic).** The sounds usually heard are perhaps *ā* in the noun and *ǎ* in the adjective though neither is fixed. There is no objection to the difference, and the FALSE QUANTITY *ā* is of no importance.

**patron, -age, -ess, -ize.** 1. The

dictionaries agree that *patron* should have a long *a* and *patronize* a short. About the other two words they differ. *Pătronage* and *pătroness* are here recommended.

2. The pejorative use of *patronize* (treat with condescension) has almost driven out its older meaning, protect and encourage, be a patron of (*Chatterton, angry at being, as he felt, merely patronized where he had sought patronage, grew insolent*). The expression to patronize a shop, once common, is now rare, and this need not be regretted; one's relationship to a shop is that of a customer, not a patron.

**pave** makes the exceptional agent-noun *paviour*.

**pawky.** The Englishman is tempted to use the word merely as a synonym in certain contexts for *Scotch*; any jest uttered by a Scot is pawky, and pawky humour is understood to be unattainable except by Scots. The underlying notions are those of craftiness, concealment of intention, apparent gravity, ironical detachment (cf. U.S. *deadpan*). The pawky person says his say, and, if the hearers choose to find more point in the words than a plain interpretation necessitates, that is their business; for him more than other people his Jest's prosperity lies in the ear Of him that hears.

**pay.** *Suddenly a girl was heard screaming for help. The crowd paid no notice.* We *pay* attention and heed but we *take* notice. See CAST-IRON IDIOM.

**pay off.** See PHRASAL VERBS.

**peccadillo.** Pl. preferably -*os*; see -O(E)S 7.

**pedantic humour.** No essential distinction is intended between this and POLYSYLLABIC HUMOUR; one or the other name is more appropriate to particular specimens, and the two headings are therefore useful for reference. But they are manifestations of the same impulse, and the few remarks needed may be made here for both. A warning is necessary, because we have all of us, except the abnormally

stupid, been pedantic humorists in our time. We spend much of our childhood picking up a vocabulary; we like to air our latest finds; we discover that our elders are tickled when we come out with a new name that they thought beyond us; we devote some pains to tickling them further; and there we are, pedants and polysyllabists all. The impulse is healthy for children, and nearly universal—which is just why warning is necessary; for among so many there will always be some who fail to realize that the clever habit applauded at home will make them insufferable abroad. Most of those who are capable of writing well enough to find readers do learn sooner or later that playful use of long or learned words is a one-sided game boring the reader more than it pleases the writer, that the impulse to it is a danger-signal—for there must be something wrong with what they are saying if it needs recommending by such puerilities—and that yielding to the impulse is a confession of failure. But now and then even an able writer will go on believing that the incongruity between simple things to be said and out-of-the-way words to say them in has a perennial charm. Perhaps it has for the reader who never outgrows hobbledehoyhood; but for the rest of us it is dreary indeed. It is possible that acquaintance with such labels as *pedantic* and *polysyllabic humour* may help to shorten the time that it takes to cure a weakness incident to youth.

An elementary example or two should be given. The words *homoeopathic* (small or minute), *sartorial* (of clothes), *interregnum* (gap), are familiar ones, or were so in their day; the popularity of such conceits is short. *To introduce 'Lords of Parliament' in such homoeopathic doses as to leave a preponderating power in the hands of those who enjoy a merely hereditary title.* / *While we were motoring out to the station I took stock of his sartorial aspect, which had changed somewhat since we parted.* / *In his vehement action his breeches fall down and his waistcoat runs up, so that there is a great interregnum.*

These words are, like most that are much used in humour of either kind, both pedantic and polysyllabic. A few specimens that cannot be described as polysyllabic are added here, and for the larger class of long words the article POLYSYLLABIC HUMOUR should be consulted: adipose tissue; aforesaid; beverage; bivalve (the succulent); digit; eke (adv.); ergo; erstwhile; Jupiter Pluvius; optic (eye); parlous; sanctum sanctorum; save the mark; this thusness; tonsorial artist.

**pedantry and purism.** *Pedantry* may be defined, for the purpose of this book, as the saying of things in language so learned or so demonstratively accurate as to imply a slur upon the generality, who are not capable or not desirous of such displays. The term, then, is obviously a relative one; my pedantry is your scholarship, his reasonable accuracy, her irreducible minimum of education, and someone else's ignorance. It is therefore not very profitable to dogmatize here on the subject; an essay would establish not what p. is, but only the place in the scale occupied by the author. There are certainly many accuracies that are not pedantries, as well as some that are; there are certainly some pedantries that are not accuracies, as well as many that are; and no book that attempts, as this one does, to give hundreds of decisions on the matter will find many readers who will accept them all.

*Purism* is like pedantry, except that it does not necessarily imply a parade of superior learning. Now and then a person may be heard to 'confess', in the pride that apes humility, to being 'a bit of a purist'; but *purist* and *purism* are for the most part missile words, which we all of us fling at anyone who insults us by finding not good enough for him some manner of speech that is good enough for us. It is in that disparaging sense that the words are used in this book; by *purism* is to be understood a needless and irritating insistence on purity or correctness of speech. Pure English, however, even

apart from the great number of elements (vocabulary, grammar, idiom, pronunciation, and so forth) that go to make it up, is so relative a term that almost every man is potentially a purist and a sloven at once to persons looking at him from a lower and a higher position in the scale than his own. The words have therefore not been very freely used; that they should be renounced altogether would be too much to expect, considering the subject of the book.

It follows that readers who find a usage stigmatized as pedantry or purism have a right to know the stigmatizer's place in the scale, if his stigma is not to be valueless. Accordingly, under headings of various matters in which these propensities may colour judgement, a few articles are now mentioned by referring to which the reader who has views of his own will be able to place the book in the scale, and judge what may be expected of it.

*Carelessness*: CANNIBALISM: JINGLES: HAZINESS.

*Choice of Words*: SAXONISM: FORMAL WORDS: WARDOUR ST.

*Idiom*: ANALOGY 3: CAST-IRON IDIOM: INCOMPATIBLES.

*Grammar*: CASES: ELLIPSIS: NUMBER.

*Misuse of Words*: POPULARIZED TECHNICALITIES: SLIPSHOD EXTENSION: VOGUE WORDS.

*Pretentiousness*: DIDACTICISM: LOVE OF THE LONG WORD: SUPERIORITY.

*Pronunciation*: FALSE QUANTITY: RECEIVED PRONUNCIATION: RECESSIVE ACCENT.

*Punctuation etc.*: HYPHENS: ITALICS: STOPS.

*Spelling*: SPELLING POINTS.

*Style*: ELEGANT VARIATION: ONLY: PREPOSITION AT END: SPLIT-INFINITIVE.

*Verbosity*: COMPOUND PREPOSITIONS: PERIPHRASIS: TAUTOLOGY.

*Word Formation*: ANALOGY 2: BARBARISMS: HYBRIDS AND MALFORMATIONS.

**pediment** is a corruption of *periment* (itself probably a corruption of *pyramid*) and means a triangular structure over a portico. The natural assumption that it was derived from Latin *pes*, foot, led as early as 1726 to its being used in the sense of foundation. The use of the same word to mean both foundation and superstructure is not to be encouraged, and the choice of *p.* is unfortunate in *Efficient farming is one p. of national prosperity*.

**peewit.** See PEWIT.

**pellucid.** See TRANSPARENT.

**pendant, pendent, pennant, pennon.** There is much confusion between these; the reasonable distribution of meanings to forms would be as follows: *pendent*, the adjective, hanging; *pendant*, a noun, a hanging ornament or appurtenance; *pennant*, a noun in nautical use for certain pieces of rigging and certain flags; *pennon*, a noun in chivalric and military use for a lance-streamer or the like. *Pendent* should not be used as a noun; *pendant* should be neither an adjective nor the nautical noun; *pennon* should not be the nautical noun.

**pendente lite.** Pronounce *pĕndĕ'ntĕ lī'tĕ*. See LATIN PHRASES.

**pendulum.** Pl. *-ms*; see -UM.

**peninsula(r).** Uses of the noun (*-la*) instead of the adjective (*-lar*), as *the Peninsula War*, or vice versa, as *the Spanish Peninsular*, are not uncommon. The latter is clearly wrong, but the former can be justified on the ground that the noun is used attributively.

**penman** should be used with reference to handwriting only, not to the writing of books or articles; in the sense writer or author it is an affectation—not indeed a new invention, but a REVIVAL.

**pen-name.** See NOM DE GUERRE.

**pennant, pennon.** See PENDANT.

**per.** It is affected to use Latin when English will serve as well; so much *a year* is better than *per annum* and much better than *per year*, and there is no point in saying *per* passenger train when we can say *by*. But when a

preposition is needed, and no English one is available, *per* may be useful even with an English noun. *Output a man-shift a day* is an awkward way of expressing a formula which is clear if *per* is used. Generally, however, it is best to confine *per* to its own language in established phrases such as *per cent.*, PER CAPITA, *per contra*, and *per stirpes*. For *as per* see COMMERCIALESE and for PER PRO see that heading.

**peradventure.** See ARCHAISM.

**per capita.** *The consumption of tobacco and alcohol has increased during the year as follows: spirits, 1·112 gallons per capita, compared with 1·030 in 1911.* | *The entire production of opium in India is two grammes per capita yearly.* This use is a modern blunder, encouraged in some recent dictionaries. '(So much) a head', or 'per man', which is the meaning here, would not be *per capita* (any more than it would be 'per men'), but *per caput* if Latin had to be used. *Per capita* describes the method of sharing property in which persons, and not families, are the units, and its opposite is *per stirpes* (*Patrimonial estates are divided per capita; purchased estates, per stirpes*); it is out of place, and something of a barbarism, however lately popular, except in such a context. Even more out of place is the invention of an adjectival inflexion: *Not a day passes but what the Moscow press summons the Soviet people to catch up with and surpass America in the per capital production of meat, milk, and butter.*

**percentage.** See LOVE OF THE LONG WORD. The notion has gone abroad that a percentage is a small part. Far from that, while a part is always less than the whole, a percentage may be the whole or more than the whole; there is little comfort to be had today from reflecting that our cost of living can be expressed as a percentage of 1939's. The uneducated public prefers a word that sounds scientific, even if it gives the sense less well, to another that is commonplace; see POPULARIZED TECHNICALITIES. In all the following examples but the last the word *percentage* has no meaning at all without the addition of *small* or of something else to define it; and in the last the greater part would be the English for *the large percentage. But in London there is no civic consciousness; the London-born provides only a percentage of its inhabitants. | This drug has proved of value in a percentage of cases. | It is none the less true that the trade unions only represent a percentage of the whole body of railway workers. | A mere percentage of listeners tune in to the Third Programme. | The largest percentage of heat generated is utilizable, but the rest escapes and is lost.* For an exact parallel see PROPORTION, and for a contrast see FRACTION.

**perchance** is a POETICISM very much out of place in pedestrian prose, as, for instance, in *There is nothing, perchance, which so readily links the ages together as a small store of jewels and trinkets.*

**peremptory.** Pronounce *pĕ′rĕmtŏrĭ* or *pĕrĕ′mtŏrĭ*. The former is recommended. See RECESSIVE ACCENT.

**perfect infinitive**, i.e. *to have done* etc. These are forms that often push their way in where they are not wanted, and sometimes, but less often, are themselves wrongly displaced by present infinitives.

1. After past tenses of *hope, fear, expect*, and the like, the perfect infinitive is used, incorrectly perhaps, but so often and with so useful an implication that it has become idiomatic. That implication is that the thing hoped etc. did not in fact come to pass, and the economy of conveying this without a separate sentence justifies the usage. So: *Philosophy began to congratulate herself upon such a proselyte from the world of business, and hoped* to have *extended her power under the auspices of such a leader. | It was the duty of that publisher* to have *rebutted a statement which he knew to be a calumny. | I was going* to have *asked, when . . .*

2. After past conditionals such as *should have liked, would have been possible, would have been the first to,* the present infinitive is (almost invariably) the right form, but the perfect often intrudes, and this time without the compensation noted in 1; for the implication of non-fulfilment is inherent in the governing verb itself. So: *If my point had not been this, I should not have endeavoured* to have *shown the connexion. | Jim Scudamore would have been the first man* to have *acknowledged the anomaly. | Peggy would have liked* to have shown *her turban and bird of paradise at the ball. | The Labour members opened their eyes wide, and except for a capital levy it is doubtful whether they would have dared* to have gone *further.* Sometimes a writer, dimly aware that 'would have liked to have done' is usually wrong, is yet so fascinated by the perfect infinitive that he clings to that at all costs, and alters instead the part of his sentence that was right: *On the point of church James was obdurate; he* would like to have insisted *on the other grudging items* (would have liked to insist).

3. With *seem, appear,* and the like, people get puzzled over the combinations of the present and past of *seem* etc. with the present and perfect of the infinitive. The possible combinations are: He seems to know, He seems to have known, He seemed to know, He seemed to have known. The first admits of no confusion, and may be left aside; the last is very rarely wanted in fact, but is constantly resorted to as an *en-tout-cas* by those who cannot decide whether the umbrella of *He seems to have known* or the parasol of *He seemed to know* is more likely to suit the weather. Thus: *I warned him when he spoke to me that I could not speak to him at all if I was to be quoted as an authority; he seemed* to have taken *this as applying only to the first question he asked me* (seems to have). | *It was no infrequent occurrence for people going to the theatre in the dark to fall into the marshes after crossing the bridge; people* seemed to have been *much more willing to run risks in those days* (seemed to be if the writer was a contemporary; seem to have been if a historian).

**perfect,** adj., in its sense 'free from imperfection, faultless' (OED), ought logically to be treated as one of those absolutes (cf. UNIQUE) that will not tolerate adverbs of degree or comparison such as *rather, very, more*; one thing may be more nearly p. than another, says logic, but it cannot be more p. But logic is an unsure guide to usage (see IDIOM), and those who choose to regard this restriction as pedantry can cite the OED: 'Often said of a near approach to such a state and hence capable of comparison' and G. M. Trevelyan on Lady Jane Grey: *As learned as any of the Tudor sovereigns, this gentle Grecian had a more perfect character than the best of them.*

**perhaps.** Of the pronunciations, that with the r and the h both sounded can only be managed by a Scot; that with the r slurred and the h sounded is orthodox; that in two syllables with r sounded but h silent is rare among the educated; that in one syllable (*prăps*) is used by many more than would plead guilty, and does not deserve the scorn heaped on it by those who parody mispronunciations in print.

**period.** For synonyms see TIME. For the full stop, see STOPS. As a term of rhetoric, strictly, any complete sentence; but applied usually to one consisting of a number of clauses in dependence on a principal sentence, and so, in the plural, to a style marked by elaborate arrangement.

**period (full stop) in abbreviations.** Abbreviations are chiefly made in two ways: one by giving the beginning of the word in one or more letters and then stopping, the other by dropping out some portion of the middle. Those of the first kind are rightly ended with a period, but the common practice of doing the same to the second is ill advised. Abbreviations are puzzling, but to puzzle is not their purpose, and everything that helps the reader to guess their meaning is a gain. One such help is to let him know when the first and last letters of the abbreviation are also those of the full word, which can be done by not using the period, but

writing *wt* (not *wt.*) for weight, *Bp* (not *Bp.*) for bishop, *Mr* (not *Mr.*) for Mister, *Bart* or *Bt* (not *Bart.* or *Bt.*) for baronet, *bot.* for botany but *bot* for bought, *Capt.* for captain but *Cpl* for corporal, *doz.* for dozen but *cwt* for hundredweight, *Feb.* for February but *fcp* for foolscap, *Frl.* for Fraulein but *Mlle* for Mademoiselle, *in.* for inches but *ft* for feet, *Geo.* for George but *Thos* for Thomas, *Lat.* for Latin but *Gk* for Greek, *h.w.*, but *ht wt*, for hit wicket. As to abbreviations formed by combining the initial letters of two or more words, practice is not uniform, but the tendency is to omit the periods —OED, BBC, UNO, NATO, etc. See also CURTAILED WORDS.

**periodic(al).** The -ic form is not used of publications (*periodical literature, periodicals*); the -ical form is not used of literary composition (*Johnson's periodic style*) or in certain scientific terms (*periodic table, motion*, etc.); otherwise the two words do not differ in meaning, but the longer tends to oust the shorter.

**peripeteia.** A sudden change of fortune in a drama or tale, e.g. in *The Merchant of Venice*, the downfall of Shylock, with Gratiano repeating to him his own words 'O learned judge'.

**periphrasis** is the putting of things in a round-about way. *The cost may be upwards of a figure rather below £10m.* is a periphrasis for The cost may be nearly £10m. *In Paris there reigns a complete absence of really reliable news* is a p. for There is no reliable news in Paris. *Rarely does the 'Little Summer' linger until November, but at times its stay has been prolonged until quite late in the year's penultimate month* contains a p. for November, and another for lingers. *The answer is in the negative* is a p. for No. *Was made the recipient of* is a p. for Was presented with. The periphrastic style is hardly possible on any considerable scale without much use of abstract nouns such as *basis, case, character, connexion, dearth, description, duration, framework, lack, nature, reference, regard, respect.* The existence of abstract nouns is a proof that abstract thought has occurred; abstract thought is a mark of civilized man; and so it has come about that periphrasis and civilization are by many held to be inseparable. These good people feel that there is an almost indecent nakedness, a reversion to barbarism, in saying No news is good news instead of *The absence of intelligence is an indication of satisfactory developments*. Nevertheless, *The year's penultimate month* is not in truth a good way of saying November.

Strings of nouns depending on one another and the use of compound prepositions are the most conspicuous symptoms of the periphrastic malady, and writers should be on the watch for these in their own composition. For examples of the first see ABSTRACTITIS, HEADLINE LANGUAGE, NOUN ADJECTIVES, and -TION, and for examples of the latter see COMPOUND PREPOSITIONS and most of the words cited above as periphrasis-makers.

**periwig** is the same word as *peruke* and much older than *wig*, which is a shortened form of it.

**permanence, -cy.** One of the pairs (see -CE, -CY) in which the distinction is neither broad and generally recognized, nor yet quite non-existent or negligible. Writers whose feeling for distinctions is delicate will prefer -*ce* for the fact of abiding, and -*cy* for the quality or an embodiment of it: *The essential quality of a monument is permanence.* | *His new post is not a permanency.*

**perorate** is not in fact one of the modern BACK-FORMATIONS like *revolute*, *enthuse*, and *burgle*, but it suffers from being taken for one, and few perhaps use it without some fear that they are indulging in a bold bad word.

**per proc., per pro., p.p.,** are abbreviations of *per procurationem*, by the agency of. There are differences of opinion about the proper placing of the words. One would suppose the natural sequence to be *AB* (*principal*) *per pro CD* (*agent*); but it is not the usual one. It is said to be common in Scotland and occasionally used by English solicitors, and has been called the 'proper form' in a judicial obiter dictum (Scrutton L.J. in *Slingsby* v. *District Bank*, 1 K.B. 1933). But the matter is one of custom, not of law, and the sequence customary in English banking practice is *per pro CD* (*agent*) *AB* (*principal*). For other uses of *per* see PER and COMMERCIALESE.

**persiflage,** 'whistle-talk'. Irresponsible talk, of which the hearer is to make what he can without the right to suppose that the speaker means what he seems to say; the treating of serious things as trifles and of trifles as serious. 'Talking with one's tongue in one's cheek' may serve as a parallel. Hannah More, quoted in the OED, describes French p. as 'the cold compound of irony, irreligion, selfishness, and sneer'. Frivolity and levity, combined with gentle 'leg-pulling', are perhaps rather the ingredients of the compound as now conceived, with *airy* as its stock adjective. Yeats said of it that it was 'the only speech of educated men that expresses a deliberate enjoyment of words.... Such as it is, all our comedies are made out of it.'

**persistence, -cy.** The distinction is the same as with *permanence, -cy*, but is more generally appreciated: *the persistence of poverty* or *of matter*; *courage and persistency are high gifts.* See -CE, -CY.

**person.** 1. Verb forms. 2. Person of pronoun.

1. When a compound subject consists of two or more alternative parts differing in person, there is sometimes a doubt about the right verb form to use (*Are you or I next?* etc.). See NEITHER 4, OR 2, for discussion.

2. Person of pronoun. Two questions arise which are exemplified in (*a*) *To me, who has* [or *have?*] *also a copy of it, it seems a somewhat trivial fragment*, for which see WHO AND WHOM 5 and (*b*) *Most of us lost our* [or *their?*] *heads,* for which see US 2.

**persona.** *One can only conclude that Mr. N. is correct to wonder whether he was ever really cut out for politics, an enterprise in which a man's persona is at least as important as his private person.* Persona (originally the mask worn by an actor) is a term of Jungian psychology which Jung himself defined as 'The individual's system of adaptation to, or the manner he assumes in dealing with, the world. . . . One could say, with a little exaggeration, that the persona is that which in reality one is not, but which oneself as well as others think one is'. It would seem therefore, so far as the ordinary man can understand these things, that a *persona* is much the same as an IMAGE; if it becomes a POPULARIZED TECHNICALITY it may relieve that greatly overworked word of some of the burden now put upon it.

**personage, personality.** Both words are used for exceptional kinds of persons, but with a difference. A *personage* is one who owes his importance to birth or high office; he is now known more familiarly as a VIP (= very important person), a term adopted during the second world war by pilots and others to whose care such persons might be temporarily entrusted. A *personality* (long established in the sense of a 'man of parts') is now popularly applied, generally with an attributive noun (*film p.*, *television p.*), to a person who has won his fame by his talents in the world of entertainment. In terms of monetary reward a personality is rated much higher than a personage. Sometimes the two merge. *The party is rather startled to find that it has its biggest electoral asset in Mr. G. He has already broken through as a TV personality.* And it was clearly right to choose the word of greater appeal when coining a phrase (*personality cult*) for that factitious veneration of a personage that commonly follows a successful revolution.

**persona grata.** Pronounce *grāta*, say the dictionaries; see LATIN PHRASES.

But *grahta* is usual when the words are used not as a naturalized English phrase but in their original sense as part of the international language of diplomacy.

**personal equation** is a phrase of definite meaning; it is the correction, quantitatively expressed, that an individual's observation of astronomical or other phenomena is known by experiment to require; minutely accurate assessment is essential to the notion. The learned sound of *equation*, however, has commended it to those who want some expression or other with *personal* in it, and are all the better pleased if such commonplace words as *view* or *opinion* or *taste* or *judgement* can be replaced by something more imposing. So: *M. Poincaré likes Mr. Bonar Law better than he liked Mr. Lloyd George; let us hope that the improved p. e. will count for something.* | *If Lady Astor's entrance upon the parliamentary scene is worthy of commemoration, the cost of it . . . should have been under the control of the House, which naturally resents the treatment of this matter as a family affair; in general there is too much p. e. about Astorian politics.* See POPULARIZED TECHNICALITIES.

**personal(i)ty.** Personal property in the legal sense is *-alty*; the other noun work of *personal* is done by *-ality*; cf. *real(i)ty.* See PERSONAGE.

**personally** is apt to be used redundantly with the personal pronoun, like PARTICULAR with the demonstrative adjective. *He was undeniably a man with the power and the courage to record the life of the Antwerp campine exactly as he p. saw it.* | *He spoke p. to each of his supporters as they left the meeting.* He could not have seen or spoken otherwise than p.; *he* by itself is enough. The legitimate uses of *p.* are (*a*) to signify that something was done by or to someone in person and not through an agent or deputy (*The writ was served on the defendant p. at his residence.* | *The appointment was* *made by the Secretary of State p.*) and (*b*) to exclude considerations other than personal (*I welcome the decision although I am not p. interested*).

**personification, nouns of multitude, metonymy.** When a country is spoken of as *She*, we have personification; when we doubt whether to write *The Board refuse* or *The Board refuses*, we are pulled up by a noun of multitude; when we call Queen Elizabeth *the Crown*, we use metonymy. Some mistakes incident to these forms of speech run into one another, and are therefore grouped together here, under the headings: 1. Ill-advised personification. 2. Vacillation. 3. Unattached possessives.

1. Ill-advised personification. To figure 'the world' as a female, a certain 'quarter' or certain 'circles' as sentient, or 'Irish womanhood' as a woman, is to be frigid—the epithet proper to those who make futile attempts at decoration. Such personifications are implied in *Just now* the world *wants all that America can give* her *in shipping* (read *it* for *her*), in *But on application to* the quarter most likely to know *I was assured that the paper in question was not written by Dickens* (The quarter is no doubt a person or persons, and capable of knowledge; but it will surely never do to let that secret out), in *According to Foreign Offices circles reports on him by the British Embassy in Moscow were ignored by the Admiralty* (Perhaps the writer thought that 'circles' had a more intimate air than the usual 'spokesman', who can only hand out what he has been given, cf. SOURCE), and in *The womanhood of Ireland stands for individualism as against co-operation, and presents the practical domestic arguments in* her *support* (Whether *her* implies the personification of *womanhood* or of *individualism* does not much matter; it must be one or the other, and neither is suited for the treatment). It is in places like these, where a writer hardly intends personification, but slips unconsciously or half-heartedly into implying it, that he reveals his want of literary instinct. On the other hand a personification may be a more convenient object of attack than persons; it cannot answer back: *He accused imperialism and colonialism of being behind the plot to murder him*.

2. Vacillation. *Britain, Paris,* and the like, are words naturally admitting of personification, and can be referred to in their literal sense by *it* and *its*, or in their personified sense by *she* and *her*. So much everyone knows; what will perhaps surprise the reader is to find how many writers are capable of absurdly mixing the two methods in a single phrase. In the following examples the words in which the vacillation is exhibited are in roman type: *Germany deserted and rearmed. So did Japan,* which *defied efforts to stop* her *invading Manchuria.* | *When Poplar no longer maintains* its *own paupers* she *must no longer determine the standard on which they are to be maintained.* | *The United States has given another proof of* its *determination to uphold* her *neutrality.* For other examples see WHICH, THAT, WHO 8, and for similar vacillation between singular and plural verbs with nouns of multitude see NUMBER 6.

3. Unattached possessives. *Danish sympathy with Finland is writ large over all* her *newspapers, literature, and public speeches, as the most casual visitor to Copenhagen can see. Her* means 'of (the personified) Denmark'; we can all see that; but we most of us also resent, nevertheless, a personification that is done not on the stage, but 'off'; a Denmark personified and not presented is a sort of shadow of a shade. | *This is a timely tribute from a man who has spent a large part of his life in Friendly Society work, and who would be the last to sanction anything that imperilled* their *interests. Their* means 'of the Friendly Societies'; but where are they? The adjective *Friendly Society* is as unavailing here as *Danish* in the previous example. | *The true doctrine is that every public act of the Crown is an act for which* her *advisers are responsible.* In some contexts it does not matter whether one says *the Queen, Her Majesty,* or *the Crown*; but while

the Queen has *her* advisers, the Crown can only have *its*.

**personnel.** Pronounce *persŏnĕ'l*. The word is widely used for the people composing one or other of the Services of the Crown, military and civilian, or any large-scale organization. Although it is far from new, its increasing use has incurred some criticism as an unnecessary and undesirable innovation. It can claim in justification that, when all Services include women, the old expression *men and material* will no longer do, and some other word must be used. But it has no doubt become too popular and is often made to serve as jargon for such words as *staff, employees, men, women, people*, etc.

**perspic-.** *Perspicacious, -acity*, mean having or showing insight; *perspicuous, -uity*, mean being easy to get a clear idea of; see PAIRS AND SNARES. *Shrewd* and *shrewdness, clear* and *clearness*, or other short words, are used in preference by those who are neither learned nor pretentious. The learned can safely venture on the *perspic-* pairs, but when the unlearned pretender claims acquaintance with them, they are apt to punish the familiarity by showing that he is in fact a stranger to them. The usual mistake is to write *-uity* for *-acity*, as in: *He claims for it superiority to other alternatives, the defects of which he sees with that perspicuity which the advocates of each ideal system invariably display towards rival systems. | The high-class West End and provincial tailors are displaying considerable perspicuity in buying checks.*

**persuade** makes *-dable* as well as *persuasible*; the latter is now more usual (see -ABLE 2).

**persuasion.** Parodies of the phrase 'of the Roman, Protestant, etc. p.', e.g. *Hats of the cartwheel p.*, are to be classed with WORN-OUT HUMOUR.

**pertinence, -cy.** There is no useful distinction; the first will probably prevail. See -CE, -CY.

**peruke.** See PERIWIG.

**pessimism.** See OPTIMISM for comments on the popular use.

**pestle** is perhaps the only *-stle-* word in which the *t* is sometimes sounded (see PRONUNCIATION 2); but this is not recommended.

**petitio principii** or 'begging the question'. The fallacy of founding a conclusion on a basis that as much needs to be proved as the conclusion itself. ARGUING IN A CIRCLE is a common variety of p. p.; other (not circular) examples are that capital punishment is necessary because without it murders would increase, and that democracy must be the best form of government because the majority are always right.

**petrol(eum).** For synonyms see KEROSENE.

**pewit, pee-.** The OED puts first the form *pewit*, but the pronunciation is *pē'wit*, an imitation of the bird's cry. In *Will Waterproof* Tennyson rhymes the word with *cruet*, but the pronunciation *pū'it*, given in some dictionaries as an alternative, is dialect.

**phalanx.** Ordinary pl. *-xes*, but in anatomy and botany *phalanges (fală'-njēz)*; singular usually *phalange*. See LATIN PLURALS.

**phantasmagoria** is sing., not (as in the following) pl.: *We shall then be able to reach some conclusion as to the meaning and effect of these bewildering phantasmagoria.* The word was designed to mean 'crowd of phantasms'.

**phantasm, phantom.** The two are by origin merely spelling variants, differentiated, but so that the differences are elusive; the following tendencies are discernible, but sometimes conflict. 1. *Phantom* is the more popular form, *-asm* being chiefly in literary use. 2. Both meaning roughly an illusive apparition, *phantom* stresses the fact that the thing is illusive, and *-asm* the fact that it does appear, so that they give respectively the negative and the positive aspect. 3. A phantom presents itself to the eye bodily or mental, a phantasm to any sense or to the intellect. 4. *Phantasm* has an adjective (*phantasmal*) of its own; *phantom* has not, but is used attributively (*phantom hopes* etc.) with much freedom, and where a true adjective is necessary borrows *phantasmal*; the two nouns are no doubt kept from diverging more definitely than they do by this common property in *phantasmal*.

**Pharisee.** The adj. *Pharisaic* is preferable to *Pharisaical*; see -IC(AL). The *-ism* noun is *Pharisaism*, not *-seeism*.

**pharmacopoeia.** Pronounce -*pē′ă*. For the spelling see Æ, Œ.

**phenomenal** means 'of the kind apprehended by (any of) the senses': everything that is reported to the mind by sight, hearing, taste, smell, or touch—whether the report answers to reality or not—is phenomenal. If the report is correct, the thing reported is also real; if not, it is 'merely phenomenal'. The question of real existence and its relation to perception and thought is the concern of META-PHYSICS, and *p*. is a metaphysical word, contrasted variously with real, absolute, and noumenal. But the object here is not to expound the metaphysical meaning of these terms; it is only to point out that *p*. is a metaphysical term with a use of its own. To divert it from this proper use to a job for which it is not needed, by making it do duty for remarkable, extraordinary, or prodigious, was a sin against the English language, but the consequences seem now to be irremediable; this meaning is recognized without comment by most dictionaries.

**phenomen(al)ism.** The longer form is recommended; see -IST, -ALIST.

**phenomenon.** Pl. *-ena*; see -ON. *P*. in the sense 'notable occurrence' or 'prodigy' is open essentially to the same objections as PHENOMENAL used correspondingly. It also has dictionary recognition, but seems to be less freely used; perhaps it has been tainted by the absurdity of Mr. Crummles's 'infant phenomenon'.

**philander(er).** 1. It is odd that these words (compounded of φιλο- love and

ἀνδρ- man), once used by the Greeks as a term of praise for a faithful wife, should now be a term of disparagement for a less than faithful husband. 2. That certain marsupial animals are so called is due to their having been first described by a naturalist with this curious christian name.

**philately, -ist.** The derivation is, through French, from Greek ἀτέλεια, exemption from tax; the word thus means fondness for the symbols that vouch for no charge being payable, namely stamps. It is a pity that for one of the most popular scientific pursuits one of the least popularly intelligible names should have been found. The best remedy now is to avoid the official titles whenever *stamp-collecting* and *-collector* will do.

**-phil(e).** The *-e* originally taken on from French is now often dropped, with the good result of bringing back the pronunciation from the queer *-fīl* to *-fĭl*.

**philharmonic, philhellenic,** etc. On the question whether the *h* should be sounded see PRONUNCIATION 3.

**Philistine.** The special modern meaning is thus given by the OED: 'A person deficient in liberal culture and enlightenment, whose interests are chiefly bounded by material and commonplace things (But often applied contemptuously by connoisseurs of any particular art or department of learning to one who has no knowledge or appreciation of it; sometimes a mere term of dislike for those whom the speaker considers bourgeois).' The Philistine retorts by calling the speaker a *highbrow*, or, if American, an *egghead*. See INTELLIGENT.

**philosophic(al).** Except where *-ical* is stereotyped by forming part of a title (*Philosophical Society*, *Transactions*, etc.), the *-ic* form is now commoner in all the more specific senses; *-ical* still prevails in the very general sense 'resembling' or 'befitting a philosopher', i.e. wise or unperturbed or well balanced; and this gives a basis for DIFFERENTIATION; see -IC(AL).

**phobia.** See POPULARIZED TECHNICALITIES.

**phon(e)y.** This slang word for 'sham' first became widely known when it was applied by Americans in the autumn of 1939 to a war of which they were then spectators only. The COD makes no guess at its derivation, but Eric Partridge, rejecting the suggestions that it may come from *funny* or from *telephoney* (simulation being easier on the telephone than face to face), or from a seller of imitation jewellery named Forney, traces the word back to 1781 when 'the ring-dropping game, one of the old ever-lastings for fooling the credulous, was known as *the fawney rig*, the fawney trick, *fawney* being an English attempt at the Irish *fainné*, a finger ring'.

**photogenic.** See -GENIC.

**phrasal verbs.** A name given by Henry Bradley to those fixed combinations of verb and adverbial particle from which (to quote Pearsall Smith) 'we derive thousands of vivid colloquialisms and idiomatic phrases by means of which we describe the greatest variety of human actions and relations'—the combinations for instance of verbs such as *get, put, take, set,* with adverbs such as *in, out, to, from, up, down,* to give differences of meaning which in languages of a more synthetic structure are represented by compound verbs. The use of phrasal verbs as nouns, a prominent feature of contemporary English (e.g. *set-to, take-over, hold-up, show-down, wash-out*), has also proved an invaluable method of enriching our vocabulary vigorously from native material instead of relying on foreign borrowing.

This useful resource is now being abused. Owing perhaps to a craving for prolixity we have got into the habit of using phrasal verbs in senses no different from that of the parent verb alone, though some of them, when coined, may have been intended to have a different nuance. This preference is largely of U.S. origin, but we have proved ready pupils. It may be that new combinations are invented, as *meet up with* is used for *meet, visit with* for *visit, lose out on* for *lose, match up* for *match, miss out on* for *miss, man up* for *man,* and *win out* for *win.* Or it may be that an established combination is used not in its proper sense but as a synonym of the simple verb. The differences in meaning between *check* and *check up on, close* and *close down,* *face* and *face up to, start* and *start up, stop* and *stop off, try* and *try out,* though real, are subtle—in some the particle is merely an intensification—and it is perhaps natural, however regrettable, that they should be disappearing owing to this curious dislike of the verb standing alone. A less excusable example of the ousting of a simple verb by a phrasal one is that of *pay,* in the sense of prove profitable, by *pay off,* which already had its special and very different meanings of finally discharging an obligation or a crew and of letting a sailing-boat fall away from the wind. This novel use of *pay off* has become so common that *pay* as an intransitive verb seems likely to disappear. *Nothing paid off more quickly on the railways than the substitution of diesel locomotives for steam. | His view that if Mr. Cousins wanted to get tough he could not carry the bulk of the trade-union leaders with him has paid off fairly well. | He recognizes that Pope's conscious artistry, his endless patient polishing, his wish always to write at his best, pays off. | The Prime Minister's remaking of his Cabinet has paid off handsomely.*

A few examples of other phrasal verbs used for the simple verb are: *You may be interested to listen to an item at ten past one if you have missed out on this week's* Radio Times. *| The duke has gone to Inverary to rest up for a few days. | It is not an issue of whether we meet up with public disapproval. | Mr. Herter has not yet been sounded out as to his willingness to succeed Mr. Dulles. | I have not seen the colonel for many years and am very anxious to* visit with him. | He told me that the merger could be successfully consummated only if I agreed to join and head it up. | All we can do in this way will be of no avail if we are going to lose out to the communists in other ways. | There are at least twenty shades of cream, so matching up needs some care. | It is expected that by that time the usual afternoon temperatures of about 90° will have started to drop off. | Which do you think will win out, Mr. ——, Mr. Lumumba, or Mr. Tshombe? | The rumble of the great guns drowned out all other noises. | The countries that go up to make the Commonwealth. In the last example the writer perhaps intended to say go to make up, using the phrasal verb in its proper sense of compose; but the up has lost its way.

**phthisis.** The old pronunciation dropped the ph-; this might have recovered its sound now that everyone can read it if the word were still in general use, but T.B. (the initial letters of tubercle bacillus, used for tuberculosis) has the advantage of presenting no difficulties of pronunciation. The Greek word had short i, but thī- is now the accepted pronunciation; see FALSE QUANTITY.

**physician, doctor, surgeon.** In the United Kingdom every medical practitioner is required to have a qualification as Physician and also as Surgeon (i.e. to have at least a Bachelor's degree in both medicine and surgery). Many general practitioners still formally style themselves Physician and Surgeon, and the survival of the name surgery for the general practitioner's consulting-room is evidence that the two branches of the profession were once less separate than they are now. In ordinary parlance physician is more often used to distinguish the specialist or consultant from the general practitioner than both of them from the surgeon; for that purpose the ordinary distinction is that the former are called Dr even though they may not have a doctor's degree and the latter Mr even though they may have one.

**physics, physiology.** The two words had once the same wide meaning of natural science or natural philosophy. They have now been narrowed and differentiated, physics retaining only the properties of matter and energy in inorganic nature, and physiology only the normal functions and phenomena of living organisms. Each has been copiously subdivided, but they have met again in the science of biophysics. Physic, as the old term for the science of medicine, survives in the title of the chair of medicine at Cambridge—the Regius Professorship of Physic. For the adjective of physiology, -ical is so much the commoner that it might be treated as the only form. See -IC(AL).

**physiognomy.** Phiz as slang for face, one of the abbreviations that annoyed Swift, had a long life but is now dead; so is the conceit of using the word in full for the same purpose by way of POLYSYLLABIC HUMOUR. Pronounce preferably with silent g.

**piano,** as an adjective or adverb, the Italian for soft, has retained its Italian pronunciation (pyah'no), whether used as a musical direction or in its figurative sense of listless or depressed. The noun, a curtailed word from pianoforte, has been anglicized into piă'no; pyah'no, sometimes heard, is an affectation.

**piazza.** The OED gives piă'ză; but the Italian consonant sound -ătsă (or -ahtsă) is now more usual. The proper meaning of the Italian word is a public square or market-place, but in America the word is used, like porch, for what we call a veranda and in England sometimes for a colonnade.

**pibroch.** Pronounce as Scotch (pē-, and ch as in loch). The word means a type of music played on the bagpipe, not, as is often supposed, the instrument itself.

**picaresque.** The p. novel is defined in the Enc. Brit. as: 'The prose autobiography of a real or fictitious personage who describes his experiences as a social parasite, and who satirizes the society which he has exploited.' The type is Spanish, but the most widely known example is the French *Gil Blas*. *Pícaro* is a Spanish word meaning vagabond. The type has its modern analogue. *It* [the 1950s] *was the decade of the p. novel, and the new, shambling, oafish anti-heroes, flotsam and jetsam of the Welfare State.* See ANTI-.

**picket, picquet, piquet.** The second form serves no purpose at all; the third (a different word from the other two) should be reserved for the card-game, and *picket* be used for all other senses, including that of the military outpost often spelt with -*qu*- or -*cqu*-.

**picture.** See PRONUNCIATION 1.

**pidgin, pigeon.** 'Business-English' was the name given by the Chinese to the Anglo-Chinese lingua franca; but they pronounced *business* pidgin, and we have confused the meaningless *pidgin* with the significant *pigeon*. The same explanation accounts for the expression *That's not my p.*

**piebald, skewbald.** *P.* is properly of white and black, *s.* of white and some other colour.

**pigeon, dove.** Used absolutely, the words are coextensive in application, every d. being a p., and vice versa; but *p.* is the ordinary word, and *d.* is now the rarity, suited for poetical contexts, symbolism, etc., and surviving in *dovecote* as the home of the domesticated pigeon. *D.* is also still used without special significance of particular kinds of p. especially the turtle and other natives, but not of exotics; and much more often the kind is specified, as in *stock, ring, turtle, -d.*

**pigmy.** For spelling see GYPSY.

**pilule.** Preferably so spelt, not -*ll*-; the Latin is *pilula*.

**pindarics.** The form of English verse in which a poem consists of several stanzas, often of unequal length, with the rhymes within the stanza irregularly disposed, and the number of feet in the lines arbitrarily varied. In

Pindar's own odes the structure is an elaborate one of strophe, antistrophe, and epode, far removed from irregularity; but the English imitators noted the variety of metre within his strophes and neglected the precise correspondence between them. Horace too must have found him difficult to scan, for he says of him *Numeris fertur lege solutis*. *P.* came consequently to be the name for verse in which regularity of metre was scorned under the supposed impulse of high emotion.

**piscina.** Dictionaries differ in their preferences between *pĭsē'na* and *pĭsī'na*; the former is probably now more usual. Plural -*ae* or -*as*.

**pistachio.** Pl. -*os*; see -O(E)S 4. The pronunciation put first in the OED is *pĭstā'shĭō*.

**piteous, pitiable, pitiful.** There are three broadly different senses for the words: 1. Feeling pity; 2. Exciting pity; 3. Exciting contempt. It would have been easy, then, if the problem had been posed beforehand, to assign a word to a sense, *piteous* to no. 1, *pitiable* to no. 2, and *pitiful* to no. 3. But language-making is no such simple affair as that, and spontaneous development has worked badly here. *Piteous* has senses 1 and 2 (though now archaic in the former), *pitiable* senses 2 and 3, and *pitiful* senses 1, 2, and 3—a very wasteful confusion, but too inveterate to be got into order now. See also PLENTEOUS.

**pituitary.** Pronounce *pĭtū'ĭtărĭ*. Latinists grieved by the accent and the short second *i* may find consolation in FALSE QUANTITY.

**pity,** n. *In the meantime, we can only muse upon the pity of it.* For *the p. of it*, and *p. 'tis 'tis true*, see STOCK PATHOS, and HACKNEYED PHRASES.

**placable.** The OED gives *plā*- precedence, but custom has disregarded this guidance; *plă*- is now usual.

**placard** is one of the words that have resisted the tendency described in

NOUN AND VERB ACCENT. Though some dictionaries give the pronunciation *plakar'd* as an alternative for the verb, it is ordinarily *pla'kard*, like the noun.

**placate.** The pronunciation recommended is *plăkā't*; the variants *plā'kāt* and *plă'kāt* are dying. The word is quoted from the 17th c. Beside the adjective *placable*, *placatable* can be made for the gerundive use; see -ABLE 1.

**place.** For *going places* see PREPOSITION DROPPING.

**plague** makes *-guable*, *-guing*, *-guy*, *-guily*; see MUTE E, -EY AND -Y.

**plaid.** Pronounced *plād* in Scotland, but *plăd* in England. Confusion between *plaid* and *tartan* is common outside Scotland. *Plaid* is the shawl-like outer garment of the Highlander's national dress. *Tartan* is the woollen cloth woven in coloured stripes with distinctive patterns for the different Highland clans; it can be made into plaids or kilts or any other sort of garment. The English outfitter's advertisement *Traditional Plaid Kilts* is nonsense to the Scotsman. Dickens made the same mistake in describing Lord George Gordon as dressed in trousers and waistcoat of the Gordon plaid.

**plain** makes *plainness*. *P. sailing* is ('probably'—OED) a popular use of the nautical term *plane sailing*, which means navigation by a plane chart, 'a simple and easy method, approximately correct for short distances'. The corruption, if it is one, is so little misleading, since *plain sailing* is as intelligible in itself as *clear going* or any such phrase, that any attempt to correct it is needless as well as vain.

**plateau** is so far naturalized that the sound *plă'tō*, and the pl. *-s*, are now usual; see also -X and RECESSIVE ACCENT.

**platform.** The political sense of party programme is still rather American than English, but is increasingly common with us. It is a revival rather than an imported novelty; as early as the Shakespearian era *p.* was used in the sense of design or plan, especially for a scheme of Church government.

**platitude, -dinous.** For the differences between *p.*, *commonplace*, and *truism*, see COMMONPLACE.

**Platonic love.** For the origin of the expression, see Plato's *Symposium*. For its meaning, the definition, and one or two quotations, from the OED here follow: (Definition) Applied to love or affection for one of the opposite sex, of a purely spiritual character, and free from sensual desire. (Quotations): (Howell) It is a love that consists in contemplation and idaeas of the mind, not in any carnall fruition. (Norris) Platonic Love is the Love of Beauty abstracted from all sensual Applications, and desire of Corporal Contact. (Lewes) Love is the longing of the Soul for Beauty; the inextinguishable desire which like feels for like, which the divinity within us feels for the divinity revealed to us in Beauty. This is the celebrated Platonic Love, which, from having originally meant a communion of two souls, and that in a rigidly dialectical sense, has been degraded to the expression of maudlin sentiment between the sexes.

**platypus.** Pl. *-puses*, not *-pi*; see -US.

**plausible** has moved a long way from its original meaning 'deserving of applause'. Applied to a person it is always pejorative; a *p.* man is one who obtains a credence he does not deserve. Applied to an argument the word has not travelled so far on the downward path: it may still be used of one that commends itself, though speculative.

**plead** in Scots law and U.S. usage has past tense, *pled*, elsewhere *pleaded*. For *special pleading* see SPECIAL.

**pleased.** For *very p.*, see VERY.

**pleasure.** *I have the pleasure of* (or *I have pleasure in*) *doing so-and-so* means I do it and am glad to do it, (*I have already had the pleasure of meeting him*; *I have pleasure in declar-*

*ing this exhibition open). It is a pleasure to do* means the same (*It is a pleasure for me to be here*). On the other hand *It is* my *pleasure to do so-and-so*, or *that so-and-so should be done*, means I choose to, and therefore of course shall, do it or have it done—an imperious statement of intention. The second idiom is based on the definite special sense of *p.* with possessives (*my, his, the king's*, etc.), viz. one's will, desire, choice (*The accused was found guilty but insane and was ordered to be detained during Her Majesty's pleasure*). Insensibility to idiom often causes *It is my* or *our p.* to be substituted for *it is a p.* or for *I* or *We have the p.* Examples of the mistake are: *Once again* it is our p. *to notice the annual issue of 'The Home Messenger'.* | *In the experiment which* it was my p. to *witness, M. Bachelet used only two traction coils.* | *When* it was my p. to *address a public meeting of more than 2,000 at the Royal Theatre the organized opposition numbered less than seven score.*

**plebiscite, -tary.** Pronounce *plĕ'bĭsĭt, plĭbĭ'sĭtărĭ.*

**plectrum.** Pl. *-tra.*

**Pleiad.** Pronounce *plī'ad.* Pl. *-ds* or *-des* (*-ēz*). In the singular the word has been applied to more than one brilliant cluster of persons or things (usually seven), notably to the Pléiade of poets of the French Renaissance.

**pleistocene, pliocene, miocene,** are regrettable BARBARISMS. It is worth while to mention this, not because the words themselves can now be either ended or mended, but in the hope that men of science may remember their duties to the language—duties much less simple than they are apt to suppose.

**plenteous, -iful.** As with other pairs in *-eous* and *-iful* (e.g. from *bounty, beauty, duty, pity*), the meaning of the two is the same, but the *-eous* word is the less common and therefore the better suited to the needs of poetry and exalted prose; for these it should be reserved.

**plenty.** *Excuses are plenty* (i.e. plentiful), *There is p. wood* (i.e. p. of), *That is p. hot enough* (i.e. quite), are irregularities of which the first is established in literature, the second is still considered a solecism (though the omission of *of* is easily paralleled, as in *a little brandy, a dozen apples, more courage, enough food*), and the third is recognized colloquial, but not literary, English. For *aplenty* see WARDOUR STREET.

**pleonasm** is the using of more words than are required to give the sense intended (see also TAUTOLOGY).
1. It is often resorted to deliberately for rhetorical effect (*Lest at any time they should see* with their eyes *and hear* with their ears). The writer who uses p. in that way must be judged by whether he does produce his effect and whether the occasion is worthy of it.
2. There are many phrases originally put together for the sake of such emphasis, but repeated with less and less effect until they end by boring instead of impressing the hearer. Such are the pairs of synonyms *if and when, unless and until, save and except, in any shape or form, of any sort or kind.* These and many others have long worn out their force, and what those who would write vigorously have to do with them is merely to unlearn them; see IF AND WHEN, the apparently least pleonastic of these stock phrases, for fuller discussion. Those who use this form of p. can hardly be unconscious that they are saying a thing twice over, the *and* or *or* being there as a reminder. See also SIAMESE TWINS.
3. In other phrases the offender is evidently unconscious, and expresses the same notion twice over in the belief that he is saying it once. Such are EQUALLY AS, *more* PREFERABLE, *more especially*, and *continue to* REMAIN, which mean neither more nor less than *equally* (or *as*), *preferable, especially*, and *remain*, by themselves, but which can be defended, by those who care to defend them, as not worse than uselessly pleonastic. With these may be classed the queer use of *both*, repugnant

to sense but not to grammar, where it takes the place of *they* or *the two* though the emphasis necessarily attaching to it is absurd; so: *Both men had something in common.* | *Archer Bey telephoned to General Morris and both conferred at the Residency.* See BOTH 2 for more varieties of this very common ineptitude.

4. A further downward step brings us below the defensible level, and we come to the overlappings described in the article HAZINESS: *The resolution was unanimously passed by the whole meeting.* | *He always preserved his equanimity of mind.* | *He will not prejudge the verdict* in advance. | *The Suez Canal's international control must be perpetuated* for all time. | *The decline in attendances at matches may have caused some apathy towards the* outcome of the *results.* | *The* future of the *Federation would be short-lived.* See also AGO and BECAUSE.

5. Lastly, there are the pleonasms in which, by wrongly repeating a negative or a conjunction, the writer produces a piece of manifest nonsense or impossible grammar. So: *We can only say that if the business men who read* The Times *are really of opinion that this is a sensible procedure, and* that, *if they find any satisfaction whatever in the writing down of a huge sum, which everybody knows can never be recovered, they will have only themselves to thank if* ... See also NEGATIVE MISHANDLING, THAT CONJ., and OVERZEAL.

**plethora.** Pronounce *plĕ'thŏra* in spite of the FALSE QUANTITY this gives to both first and second syllables.

**pleura.** Pl. *-rae.*

**plexus.** Pl. *-uses* or rarely *plexus (-ūs)*, not *-xi*; see *-US.*

**plumb-.** The *b* is silent in *plumber, plumbery, plumbing,* and *plumbless,* but sounded in *plumbago, plumbeous, plumbic, plumbiferous,* and *plumbism.*

**plural anomalies.** 1. See *-ICS* 2 for the question whether words in *-ics* are singular or plural.

2. Plurals of words ending *-s.* Names of diseases, such as *mumps, measles,*

*glanders,* can be treated as singular or plural; *chickenpox* and *smallpox,* originally plural, are now reckoned singular. *Innings, corps,* and some other words in *-s,* are singular or plural without change of spelling, but, while *corps* has *-s* silent in singular and sounded in plural, *an innings* and *several innings* show no distinction, whence arises the colloquial double plural *inningses.* For proper names ending *-s* the correct plural form is *-es—the Joneses, the Rogerses,* etc. Cf. FORCEPS and GALLOWS. The growing tendency to indicate the plural of proper names, especially if of more than one syllable, by writing *the Rogers', the Evans'* is an abuse of the apostrophe on which see POSSESSIVE PUZZLES 7.

3. Plurals of compound words. These ordinarily form their plurals logically, by attaching the *-s* to the noun element in them, or, if there is more than one noun, to the significant one. *Listeners-in, sons-in-law, heirs presumptive, master mariners, tugs of war, deeds poll.* But many familiar compounds now make their plurals as if they were single words (*char-a-bancs, will-o'-the-wisps, four-in-hands*), especially if they contain no noun (*ne'er-do-wells, forget-me-nots*), or are *-ful* compounds (*handfuls, mouthfuls, spoonfuls*), or are PHRASAL VERBS used as nouns (*take-offs, knock-outs, call-ups*). Compounds containing *man* or *woman* make both elements plural (*menservants, women clerks*), and so usually, though not always, do two words linked by *and* (*ins-and-outs, pros and cons, rights-and-lefts, ups-and-downs*). Our compounded drinks we usually think of as a single entity, and are likely to ask for two *whisky-and-sodas,* two *gin-and-tonics,* etc.; if we make an exception in asking for two *gins-and-French,* it will only be because of the awkwardness of *Frenches.*

*Lord Chancellors, Lord Mayors,* and *Lord Provosts,* being special kinds of chancellors, mayors, and provosts, logically put the *s* on the second word as the significant one; and for the same reason *Lords of Appeal, Lords of*

*Session, Lords in Waiting,* etc., being special kinds of lords, put it on the first. But the *Lords Commissioners of the Treasury* and the *Lords Justices* of the Court of Appeal traditionally add the *s* to both words. For *Lord Lieutenant* the official plural is *Lords Lieutenant,* but the OED gives as alternatives both *Lords Lieutenants* and *Lord Lieutenants. Lieutenant General* and *Major General,* being now regarded as special kinds of general, and not, as they once were, special kinds of lieutenant and major, logically have the *s* at the end, and *Adjutant Generals* and *Quartermaster Generals* have followed suit by analogy. The officials called *General* in civil life, e.g. *Attorney G., Solicitor G., Governor G., Postmaster G., Paymaster G.,* being special kinds of attorney, solicitor, etc., should be Attorneys General and so on. But Attorney Generals and Solicitor Generals are now the usual plurals for those officers (said by the OED to be 'better'); and, although *Whitaker's Almanack* still gives *Governors General* and the OED would have us say *Postmasters General,* titles such as these, so far as they have need of a plural, will no doubt eventually fall into line, following the popular tendency to disregard these niceties that has already made *court martials* and *poet laureates* sound at least as natural to us as the more correct *courts martial* and *poets laureate.*

**4.** Plurals of words ending *-y.* There are few exceptions to the rule that, when the *y* is preceded by a vowel, the plural is formed by adding *-s* in the ordinary way, and when it is preceded by a consonant the plural is *-ies (days, donkeys, spies, jetties). Soliloquies* is not a true exception, for *qu* in English is in effect a single consonant; *monies* as an alternative plural to *moneys* is obsolescent, and so is the vehicle that used to make its plural *flys* to distinguish them from insects. But proper names do not conform. *There are now two Germanys.* / *The three Marys at the Crucifixion.* In applying proper names to other purposes we are inconsistent.

Private soldiers used to be *tommies,* and *johnnies* are so spelt; but the hats called after du Maurier's heroine are *trilbys.*

**5.** For the number of *mews* and *news,* see those words, and for the plurals of initial-letter abbreviations (*M.P., A.D.C., N.C.O.,* etc.) see M.P. Other articles on plural formations are -AE, -AS (of words ending in a); -EX, -IX (of words so ending); LATIN PLURALS; -O(E)S (of words ending in *o*); -ON; -UM; -US; -TRIX (of words so ending); VE(D), -VES (of words ending in *f*); -X (or *s* of words ending in *-eau*).

**plurality.** With three-cornered contests as common as they now are, we may have occasion to find a convenient single word for what we used to call an *absolute majority,* but now, under the baneful influence of OVERALL, have rechristened an *overall majority,* i.e. a majority comprising more than half the votes cast. In America the word *majority* itself has that meaning, while a poll greater than that of any other candidate, but less than half the votes cast, is called a *plurality.* It might be useful to borrow this distinction, but to better it by changing *plurality* to *plurity.* The correct meaning of *plurality* is not moreness (which is the notion wanted, but which would be *plurity*), but pluralness or severalness or more-than-oneness. *Plurity* is an obsolete English word exactly suited to the need; cf. REVIVALS. See also MAJORITY.

**pn-.** For the pronunciation of *pneumatic* and *pneumonia* the OED gave *nu-* only, but preferred *pnu-* for less familiar words from the same stems, such as *pneumatology, pneumonometer.* But all such compounds, now more numerous, have taken the easier way, and *nu-* is now invariable. Cf. PS- and PT-.

**pochard.** The OED puts *pōch-* first, and it is still the most common pronunciation; but *pŏch-, pŏk-,* and *pŏk-,* are also recognized.

**pocketful.** Pl. *-ls*; see -FUL.

**podium.** Pronounce *pō-*; pl. *-ia.*

**poetess.** See FEMININE DESIGNATIONS.

**poetic(al).** See -IC(AL). The two forms are more or less peculiar in being both in constant use without as yet any clear division of functions between them. Certain tendencies, not always operative, there are: *poetical* labels, while *-ic* admires (*The -ical works of* —; *Conceived in a truly -ic spirit*); *-ical* is the form for 'written in verse', and *-ic* for 'instinct with poetry' (*Poetical composition*; *The -ic impulse, -ic justice, -ic licence*); *-ical* is the commonplace, and *-ic* the rhetorical form (*In a poetical mood*; but *In -ic mood*); *-ical* is sometimes used at the end of a sentence when in another position *-ic* would be more natural (*An idea more true than -ical*, cf. *A no less true than poetic idea*); and *-ic* is sometimes jocularly substituted for *-ical* (*The -ic effusions of an advertising soapboiler*).

**poeticisms.** By these are meant modes of expression that are thought (or were once thought) to contribute to the emotional appeal of poetry but are unsuitable for plain prose: 'To most people nowadays, I imagine,' says T. S. Eliot, '*poetic diction* means an idiom and a choice of words which are out of date and which were never very good at their best.' Poeticisms are not favoured even by poets any more. The revolt against them advocated by Wordsworth in his preface to *Lyrical Ballads* has gone to lengths that would have surprised him. Nevertheless injudicious writers of prose are still occasionally tempted to use them as tinsel ornaments. See INCONGRUOUS VOCABULARY and WARDOUR STREET. Simple reference of any word to this article is intended as a warning.

**pogrom.** *Pŏgrŏ'm* is the orthodox pronunciation, but with the wider currency of the word that has followed greater recourse to such practices it is often pronounced with what is to us a more natural stress on the first syllable.

**point.** I. For synonyms in the sense province etc., see FIELD. 2. *P. of view* is the native phrase; *standpoint* is a translation of the German *Standpunkt*, and appears in the form *standpunct* in one of the earliest OED quotations. What is against *p. o. v.* is the awkwardness of following it, as is constantly necessary, with another *of* (*from the p. o. v. of philosophy*). There is no valid objection to *standpoint* but *p. o. v.* holds its own where the *of* difficulty does not present itself. *Viewpoint*, an earlier product of the repugnance to *standpoint*, has the disadvantage of calling to mind what *standpoint* allows to be forgotten, that the idiomatic English is undoubtedly *p. of view*. The perplexed stylist is at present inclined to cut loose and experiment with *angle*. What is here recommended is to use *p.o.v.* as the normal expression, but not be afraid of *standpoint* on occasion. 3. But to say this is not to condone the too common use of *p. o. v.* merely as a clumsy PERIPHRASIS: *The amount offered seems not unreasonable* from the p. o. v. of *a living wage* (as). | *Bare boards are inconvenient* from the p. o. v. of *cleaning* (for). *Trees not worth planting* from either the use or beauty p. o. v. (either for use or for beauty). Another misuse of *p. o. v.* and *viewpoint* is where the right word would be *view* or *views*. Examples: *He had to buy a newspaper to secure expression of his p. o. v.* | *To many Americans this p. o. v. appears cynical or even immoral.* | *A chief constable supports my p. o. v. that traffic control duties should be separated from a policeman's other duties.* | *Mr. Haxell's viewpoint was that the affairs of the E.T.U. were in the control of its members.*

**polemic(al).** It would be convenient, and not be counter to any existing distinctions, if *-ic* were kept to the noun use and *-ical* to the adjectival; see -IC(AL).

**polity** is a word that has emerged from its retirement in the writings of philosophic historians or political philosophers, become a newspaper word, and suffered the maltreatment usual in such cases. It has been seized upon as a less familiar and therefore more impressive spelling of *policy* (with which it is indeed identical in origin), and the differences that have long existed between the two have been very vaguely grasped or else neglected. A useful indication that the two words are of widely different meanings is that *policy* is as often as not without *a* or *the* in the singular, whereas *polity* in its right senses is very rarely so. *Polity* is not (like *policy* or *principle*) a line of action, nor (like *politics*) a branch of activity, nor (like *statesmanship*) an art or quality. But in the following newspaper extracts it will be seen that one of those senses is required, and that one of those words, or at any rate some other word, would be the right one instead of *polity*: *This Newspaper Trust has during the last two years increasingly assumed the right and the power to upset ministries, to nominate new ministers and discharge others, and to dictate and veto public polity.* | *The main obstacles to advancement have always been social superstitions, political oppression, rash and misguided ambitions, and gross mistakes in polity.* | *Habits of living from hand to mouth engendered by centuries of crude polity will not die out in a month.* | *And now that by their feats in arms peace has been brought within sight, the work in the field has admittedly to be rounded off, completed, and made lastingly effective for the common good by a work of Polity.*

The true meanings of *polity* are: 1 (now rare) a condition, viz. the being organized as a State or system of States; 2 (most frequent) some particular form of such organization, e.g. a republic, monarchy, empire, confederation, etc.; 3 (not uncommon) a people organized as a State. The first three of the following examples are newspaper extracts showing the correct and usual sense 2; the fourth and fifth are OED quotations from Gladstone and Huxley illustrating senses 1 and 3: *Dr. Hezeltine's lecture is an interesting account of the influence of English political and legal ideas upon the American polity.* | *If the terms are accepted the future polity of Europe must be more than ever based*

*on force.* | *Keynes points out that the commercial and industrial system of Europe has grown up with the pre-war polity as its basis.* | (Gladstone) *At a period antecedent to the formation of anything like polity in Greece.* | (Huxley) *Those who should be kept, as certain to be serviceable members of the polity.*

**polloi.** See HOI POLLOI.

**pollster,** a word slow to gain entry to the dictionaries, is now commonly used for one who conducts a 'poll' of public opinion by getting samples of it through questioning individuals. This now popular method of finding out what people think about current issues, especially for forecasting the results of elections and for obtaining data for sociological conclusions, is generally associated with the name of Dr George Gallup, who founded the American Institute of Public Opinion in 1936. He did much to develop and improve the device, but did not originate it. Its first recorded use was as a journalistic enterprise in Delaware and North Carolina for forecasting the result of the presidential election of 1824. See also CROSS-SECTION. Complementary to the activities of the pollster are those of the *psephologist* (a recent coinage from Greek ψῆφος, voting-pebble) who analyses the results of elections.

**polypus.** Pl. *-pi* (*-ī*) or *-puses*, see -US. The inconsistency between this and OCTOPUS is due to its having come to us through classical Latin, in which it was declined like the ordinary Latin nouns in *-us*.

**polysyllabic humour.** See PEDANTIC HUMOUR for a slight account of the impulse that suggests long or abstruse words as a means of entertaining the reader or hearer. Of the long, as distinguished from the abstruse, *terminological inexactitude* for lie or falsehood is a favourable example (see INEXACTITUDE), but much less amusing at the hundredth than at the first time of hearing. *Oblivious to their pristine nudity* (forgetting they were stark naked) is a less familiar specimen.

Nothing need here be added to what was said in the other article beyond a short specimen list of long words or phrases that have been used for this purpose. Polysyllabic humour was a favourite device of 19th-c. humorous writers, Dickens, Surtees, and Gilbert for instance. It is out of fashion now, though still occasionally ventured; modern examples are: *The bath in that day and age was still a mark of social distinction. Regular and lengthy unavailability by immersion a matter of social prestige.* | *I went into the Memorial Library.... There was no visible librarian in its vasty desuetude.* | *East was surveying his prospects pleasurably, though he knew that something piscatorial was afoot.* We do not think it funny any more. There is plenty of polysyllabic writing today, especially by officials and scientists, but it is done in all seriousness. The doctor who comments on the scantiness of a patient's eyebrows by saying *The supraorbital ridge hair is partially depleted* has no thought of raising a laugh; nor has the official who explains to an applicant for rehousing the difficulty of finding *alternative accommodation.* Examples of the old-fashioned type of humour are: *Solution of continuity, femoral habiliments, refrain from lacteal addition,* and *olfactory organ,* for gap, breeches, take no milk, and nose. *Osculatory, pachydermatous, matutinal, fuliginous, fugacious, esurient, culinary,* and *minacious,* for kissing, thick-skinned, morning, sooty, fleeting, hungry, kitchen, and threatening. *Frontispiece* or *physiognomy, cachinnation, epidermis,* and *natatory art* for face, laughter, skin, and swimming. *Paterfamilias, perambulate,* and *peregrinate* for father, walk, and travel.

**pomade.** The OED gives -*ād* as the English pronunciation, and -*ahd* as a foreign one; the latter, however, is now more common. See -ADE, -ADO.

**pomegranate.** Of the four possible pronunciations of *pome-* (*pŏmĕ-, pŏm-, pŭmĕ-,* and *pŭm-*) the second is now probably the commonest. See PRONUNCIATION 5.

**pommel, pu-.** The first spelling is usual for the noun, the second for the verb, though the verb is merely a use of the noun, and not of different origin. Both are pronounced *pŭm-,* and make -*lled* (see -LL-, -L-).

**poncho.** Pronounce with -*ch-.* Pl. -*os*; see -O(E)S 6.

**pontificate, pontify.** The first is now the more usual word for play the pontiff.

**poor.** For *poorness* and *poverty* see the latter. For 'a p. thing but mine own', see MISQUOTATION. For pronunciation see RECEIVED PRONUNCIATION.

**popularized technicalities.** The term of this sort most in vogue when this article was first written in the 1920s was undoubtedly *acid test (The measure, as our correspondent says, provides an acid test for every Free Trader),* which became familiar through a conspicuous use of it by President Wilson during the first world war. It still shows remarkable vigour. In contrast with this comparatively recent acquisition may be set *intoxicated,* so long popular as to be not now recognizable for a medical term at all; it is just a ponderous GENTEELISM for *drunk.* Have we to fear something of the kind with *allergic*? Its popularity raises apprehension. *The audience showed itself highly allergic* (i.e. very hostile) *to the exhortations of the speaker* is the kind of thing one may read without surprise in any newspaper today. A few examples of these popularized technicalities may be gathered together; they will be only as one in a score or a hundred of those that exist, but will serve as specimens. Upon many of them some remarks will be found in their dictionary places. Two general warnings will suffice: first, that the popular use more often than not misrepresents, and sometimes very badly, the original meaning; and secondly, that free indulgence in this sort of term results in a tawdry style. It does not follow that none of them should ever be used; many are valuable in their proper places.

From Philosophy—*optimism* and *pessimism, category, concept, dualistic*.

From Mathematics—*factor, progression*(*arithmetical* and *geometrical*), *to the* n*th, to be a function of, percentage* and *proportion* (= part), *curve* (= tendency), *brackets*(= groups), *differential*.

From Religion—*devil's advocate, immanent, incarnation*.

From Law—*special pleading, leading question, party* (= person), *aforesaid* and *such* and *same, re, exception that proves the rule, prescriptive right*.

From War—*decimate, echelon, internecine, objective* (n.).

From Logic—*dilemma, idols of the market, beg the question, dichotomy*.

From Commerce—*asset, liquidate*.

From Architecture—*flamboyant, baroque, rococo*.

From Agriculture etc.—*intensive, hardy annual, aftermath*.

From Astronomy—*ascendant, personal equation*.

From Chemistry and Physics—*eliminate, acid test, reaction, end product, potential, ultimate analysis*.

From Literature—*protagonist, euphuism, Homeric laughter, myth, pathetic fallacy*.

From Chess—*checkmate, gambit, stalemate*.

From Seamanship — *by - and - large, dead reckoning, plain sailing, under way*.

From Medicine and Psychology. This section calls for special treatment. Our interest in our bodies has always made us prone to popularize medical terms, generally to their detriment, as we have spoiled *hectic* and *chronic*, and are now spoiling *allergic*. Freud and his successors have awakened in us a similar interest in our minds, and the cult of psychoanalysis, especially in America, has produced a new jargon, sometimes called *Freudian English*, that employs with greater freedom than accuracy terms such as *ambivalent, antitype, complex, ego, egocentric, euphoric, extrovert, father figure, fixation, id, imbalance, inhibition, introvert, libido, manic, masochistic, moron, narcissism, persona, phobia, psyche, psychopath*

(abbreviated to *psycho*. and replacing the old *mental*), *repression, schizophrenic* (abbreviated to *schizo*.), *subconscious*, and *trauma*. Gossip columnists report this jargon as current in what might be thought improbable places. *Mrs. R., wife of a Texas businessman, a tall handsome woman, refused a glass of champagne. 'It makes me sneeze', she said. 'And I don't feel masochistic enough to drink champagne right now.'* / *Raven-haired actress Miss C. gave me this quote from New York last night: 'I am introverted to the extent that I do not care to discuss my most personal affairs in public.'* And, on a higher level of journalism, *The American bases have superimposed an encirclement complex on the older interventionist trauma*. As has been well said, these quasi-scientific clichés 'have many advantages: their use evokes and even releases emotion, they have the knowing look of key concepts, and no one is quite sure what they mean'. But any gratification they give to their users is at the cost of the harm done to the language by wearing down the points of words which, one suspects, may not always have been very sharp, even when confined to esoteric use.

**porcelain** is china, and china is p.; there is no recondite difference between the two things, which indeed are not two, but one; and the difference between the two words is merely that *china* is the homely term, while *porcelain* is exotic and literary. See WORKING AND STYLISH WORDS. A tendency to differentiate by using *china* as a generic term and confining *porcelain* to the finest semi-transparent sorts has been rudely checked by the modern use of *porcelain* for sanitary ware.

**Porch.** For *the P*. in Philosophy see ACADEMY.

**port, harbour, haven.** The broad distinction is that a haven is thought of as a place where a ship may find shelter from a storm, a harbour as one offering accommodation (used or not) in which ships may remain in safety for any purpose, and a port as a town whose harbour is frequented by naval or merchant ships.

**port, larboard.** The two words mean the same, but *p.* has been substituted for *l.* (the earlier opposite of *starboard*) because of the confusion resulting, when orders were shouted, from the similarity between *l.* and *starboard*.

**portmanteau (word).** Pl. *-s* (or *-x*; see -x). For *p. word* the OED quotes from *Through the Looking-glass*: 'Well, "slithy" means "lithe and slimy" . . . You see it's like a portmanteau—there are two meanings packed up into one word.' Some examples will be found under FACETIOUS FORMATIONS. But *p.* words are not always facetious; we have of late made free use of this device, as we have of CURTAILED WORDS and acronyms, to provide us with new words for new things. There are, for instance, *transistor*, telescoping *transfer* and *resistor*, *motel* for a motorist's hotel, *moped* for a motor-assisted pedal bicycle, *trafficator* for a contrivance by which a motorist can indicate his intentions to other traffic, and SUBTOPIA for a utopia consisting entirely of suburbs. *Oxbridge*, as the name of a fictitious university for a fictitious character, is a p.w. of long standing; it was there that Arthur Pendennis had the misfortune to end his undergraduate days by being 'plucked'. In its current sense of 'Oxford or Cambridge or both' it is in constant use; its obvious convenience has almost, though perhaps not quite, won it literary status. Its rival *Camford* (where Sherlock Holmes went to find out what Professor Presbury was up to) has the fatal handicap of reversing the natural order of the component words. *Pakistan* is a mixture of a portmanteau word and an acronym: it is said to be compounded of elements from *Punjab, Afghan Frontier, Kashmir, Sind,* and *Baluchistan.*

**Portuguese,** n., is both singular and plural. 'In modern times a sing. *Portug(u)ee* has arisen in vulgar use' —OED. Cf. *Chinee*.

**pose.** The verb meaning nonplus (with

its noun *poser* unanswerable question) is a different word from that meaning to lay down or place, being shortened from *appose*.

**position.** 1. For the overworking of abstract nouns such as *position* and *situation* see -TION. 2. The use of *position* as a verb has met with some criticism. But there are instances, going back many years, of its use in the senses both of to place in position and of to ascertain the position of; and if it can claim a useful role on the ground that neither *place* nor *post* nor *pose* will always give quite the same meaning (which is at least arguable), it need not be rejected merely because it is primarily a noun. Cf. *petition, partition, condition.*

**position of adverbs.** The word *adverb* is here to be taken as including adverbial phrases (e.g. *for a time*) and adverbial clauses (e.g. *if possible*), adjectives used predicatively (e.g. *alone*), and adverbial conjunctions (e.g. *then*), as well as simple adverbs such as *soon* and *undoubtedly*. To lay down and illustrate exhaustive rules would not be possible in reasonable compass. Nor is there any need to do so; the mistakes that occur are almost always due to certain false principles, and these may be isolated for treatment. Many readers may justly feel that they do not require advice on so simple a matter as where their adverbs should go. To save them the trouble of reading this long article, here is a string of sentences exhibiting all the types of misplacement to be discussed. Those who perceive that the adverb in each is wrongly placed, and why, can safely neglect the rest; the bracketed number after each refers to the section in which its type is discussed: *The people are now returning and trying to again get together a home* (1). | *He came to study personally the situation* (2). | *He exercised an influence that is still potent and has yet adequately to be measured on the education of our younger artists* (3). | *It deals with matters as to which most persons* long ago *have made up their minds* (4). | *We still are of opinion that* *the only way of getting rid of 'abuses' is a root-and-branch alteration of the thing itself* (5). | *The Food Ministry must either take action or defend* effectively *their inactivity* (6). | *To decry the infantry arm for the sake* unduly *of piling up artillery and what not, is the notion of persons who* . . . (7). | *As 'the Monroe doctrine'* of late years *has loomed so largely in all discussions upon the international policy of the United States, an attempt to trace its growth and development as a popular 'cry' might prove of some service* (8).

There are certain verb groups about which the question is conceivable— Should they be allowed to be interrupted by adverbs? Such are the infinitive, e.g. *to try* (may we say *to earnestly try*?), the compound verb, e.g. *have thought* (may we say *I have never thought so*?), the copula and complement, e.g. *was a riddle* (may we say *He was in some ways a riddle*?), the verb and its object, e.g. *passed the time* (may we say *It passed pleasantly the time*?), the gerund and its governing preposition, e.g. *by going* (may we say *by often going*?). The first of these questions is a very familiar one; almost all who aspire to write English have had the split infinitive forced on their attention, and the avoidance of it has become a fetish. The other questions are not familiar, but the points here to be made are that they also require consideration, that a universal yes or a universal no is not the right answer either to the split-infinitive question or to any of the others, that the various answers sometimes come into conflict, and that to concentrate on the split-infinitive question and let the others take care of themselves is absurd.

The misplacements to be considered will be taken under the heads: 1. Split infinitive. 2. Fear of split infinitive. 3. Imaginary split infinitive passive. 4. Splitting of the compound verb. 5. Separation of copulative verb and complement. 6. Separation of transitive verb and object. 7. Separation of preposition and gerund. 8. Heedless misplacings.

1. Split infinitive. The heinousness

of this offence is estimated in the article SPLIT INFINITIVE. Here the general result of that estimate is merely assumed, viz.: (A) that *to love* is a definitely enough recognized verb-form to make the clinging together of its parts the natural and normal thing, (B) that there is, however, no sacro-sanctity about that arrangement, (C) that adverbs should be kept outside if there is neither anything gained by putting them inside nor any difficulty in finding them another place, but (D) that such gain or difficulty will often justify the confessedly abnormal splitting. One pair of examples will throw light on C and D: *The people are now returning and trying* to again get together *a home*. | *With us outside the Treaty, we must expect the Commission* to at least neglect *our interests*. In the first, it is easy to write *to get a home together again*, and, as *again* does not belong to the single word *get*, but to *get a home together*, nothing is gained by its abnormal placing. In the second, *at least* cannot be put before *to* because it would then go with *Commission* (i.e. the Commission, even if not other people), nor after *neglect* because it would then be doubtful whether it referred back to *neglect* or forward to *interests*, nor after *interests* because it would then belong either to *interests* or to *neglect our interests*, neither of which is what is meant; where it stands, it secures our realizing that the writer has in mind some other verb such as *injure* or *oppose* with which the weaker *neglect* is to be contrasted. In a split infinitive, however, we have not so much a misplacing of the adverb as a violence done to the verb. It is by repulsion, not by attraction, that the infinitive acts in effecting the many misplacings, to be shown below, for which it is responsible.

2. Fear of split infinitive. The order of words in the following examples is bizarre enough to offend the least cultivated ear; the reason why the writers, whose ears were perhaps no worse than their neighbours', were not struck by it is that they were obsessed by fear of infinitive-splitting. It will

be seen that the natural (not necessarily the best) place for the adverb in each is where it would split an infinitive. *Such gentlemen are powerless to analyse* correctly *agricultural conditions.* / *A body of Unionist employers which still has power to influence* greatly *opinion among those who work for them.* / *Might I* kindly *ask you to forward?* The place into which each adverb has been shifted is such that one or other of the faults explained in later sections is committed, and the writer is OUT OF THE FRYING-PAN into the fire; see especially 6.

But the terrorism exercised by the split infinitive is most conspicuous where there is in fact (see next section) no danger.

**3.** Imaginary split infinitive passive. In the following examples it is again clear that the natural place for the adverb is not where it now stands, but invariably after the words *to be.* To insert an adverb between *to* and *be* would be splitting an infinitive; to insert one between *to be* and *concerned, pained,* etc., is nothing of the kind, but is a particular case of the construction explained in 5. The position after *to be* is not only the natural one in these sentences, but the best. The mistake— and that it is a definite mistake there is no doubt whatever—is illustrated in the following extracts: *Every citizen worth the name ought* vitally *to be concerned in today's election.* / *All of us who believe in parliamentary institutions cannot fail* deeply *to be pained at reading the story.* / *The nuisance of allowing visitors to cross the footlights had begun* so much *to be felt by the London theatrical managers that they . . .* / *We think the public will not fail* unfavourably *to be impressed by the shifting nature of the arguments.* / *An Act has been passed enabling agricultural land* compulsorily *to be acquired at a fair market price.* / *At a time when NATO needs* drastically *to be reformed in its structure and purpose.*

**4.** Splitting of the compound verb. By *compound verb* is meant a verb made up of an auxiliary (or more than one) and an infinitive (without *to*) or

participle. When an adverb is to be used with such a verb, its normal place is between the auxiliary (or sometimes the first auxiliary if there are two or more) and the rest. Not only is there no objection to thus splitting a compound verb, but any other position for the adverb requires special justification: *I have never seen her*, not *I never have seen her*, is the ordinary idiom, though the rejected order becomes the right one if emphasis is to be put on *have* (*I may have had chances of seeing her but I never have*). But it is plain from the string of examples now to come that a prejudice has grown up against dividing compound verbs. It is probably a supposed corollary of the accepted split-infinitive prohibition; at any rate, it is entirely unfounded. In each of the first four extracts there is one auxiliary; and after that, instead of before it, the adverb should have been put; the other four have two auxiliaries each (which raises a further question to be touched upon in the following paragraph): Single auxiliary: *If his counsel* still is *followed,* '*the conflict*' is indeed *inevitable.* / *Its very brief span of insect-eating activity* hardly can *redeem its general evil habit as a grain-devourer.* / *Politicians of all sorts in the United States* already are *girding up their loins for the next election.* / *Yet one of the latest Customs rulings by the United States Board of Appraisers* assuredly, *to use the phrase its members* best would *understand, is 'the limit'.* / Double auxiliary: *Oxford* must heartily be *congratulated on their victory.* / *If the desired end is ever attained it* earnestly may be *hoped that especial care will be taken with the translation.* / *The importance which* quite rightly has been *given to reports of their meetings.* / *The Maharaja made arrangements for her education, which* never since has been *permitted to languish.* Write *must be* heartily *congratulated, it* may be earnestly *hoped, which has* quite rightly been *given, which has* never since been *permitted.*

This minor point of whether the adverb is to follow the first auxiliary or the whole auxiliary depends on the answer to a not very simple riddle—Is it in intimate connexion with the verbal notion itself independently of the temporal or other limitations imposed by the auxiliaries? Fortunately this riddle can be translated into simpler terms—Do the adverb and verb naturally suggest an adjective and noun? If so, let them stand next each other, and if not, not. *Heartily congratulated, earnestly hoped,* suggest hearty congratulations and earnest hope; but *rightly given* does not suggest right gift or right giving, and still less does *never since permitted* suggest no subsequent permission. That means that the notions of giving and permitting are qualified by *rightly* and *never since* not absolutely, but under the particular limitations of the auxiliaries, and that the adverb is better placed between the auxiliaries than next to *given* and *permitted.* This, however, is a minor point, as was said above; the main object of this section is to stress the certain fact that there is no objection whatever to dividing a compound verb by adverbs.

5. Separation of copulative verb and complement. This is on the same footing as the splitting of the compound verb discussed in 4; that is, it is a delusion to suppose that the insertion of an adverb between the two parts is a solecism, or even, like the splitting of the infinitive, a practice to be regarded as abnormal. On the contrary, it is the natural arrangement, and in the following examples *fundamentally, also,* and *often,* have been mistakenly shifted from their right place owing to a superstition: *It would be a different thing if the scheme had been found* fundamentally to be *faulty, but that is not the case.* | *It is not always in these times that the First Lord of the Treasury* also *is Prime Minister.* | *The immense improvement which they have wrought in the condition of the people, and which* often *is quite irrespective of the number of actual converts.*

6. Separation of transitive verb and its object. The mistakes discussed in sections 2 to 5 have this in common, that they spring from a desire, instinc-

tive or inculcated, to keep the parts of a verb group together and allow no adverb to intrude into it. But there is one kind of group that is only too often broken up by adverbs that ought to have been placed before or after the whole. That is the group consisting of a transitive verb and its object. *I had to second by all the means in my power diplomatic action. To second diplomatic action* is the verb and object, separated by a seven-word adverb. It is a crying case; everyone will agree to deferring the adverb, and the writer had either no literary ear or some grammatical or stylistic fad. The longer the adverb in proportion to the object, the more marked is the offence of interpolating it. But the same mistake is seen, though less glaringly, in the following examples; the roman-type adverb in each should be removed, sometimes to a place before the verb, sometimes to one after the object: *Are they quite sure that they have interpreted* rightly *the situation?* | *A lull of the breeze kept for a time the small boat in the neighbourhood of the brig.* | *He spoke in a firm voice, marking* strongly *the syllables, but in tones rather harsh.* | *The only conceivable exception is some great question affecting* vitally *human liberty and human conscience.* | *Continuation with the university courses would most certainly elevate* further *the people.*

There are conditions that justify the separation, the most obvious being when a lengthy object would keep an adverb that is not suitable for the early position too remote from the verb. One of the extracts below may be adapted to illustrate; if it had run 'would expose to ridicule an authority that, as it is, is not very imposing' instead of 'expose to the ridicule of all the restless elements in East Europe their authority which' etc., the shortness of 'to ridicule' compared with the length of the object would have made that order the best and almost necessary one. But anyone who applies this principle must be careful not to reckon as part of the object words that either do not belong to it at all or are unessential to it; otherwise he will offend the discerning reader's ear as cruelly as the authors now to be quoted: *They are now busy issuing blue prints and instructions, and otherwise helping* in all sorts of ways *our firms to get an efficient grip of the business of tractor-making in a hurry.* The object is *our firms* alone, not that and the rest of the sentence; put it next to *helping.* | *Who are risking* every day with intelligence and with shrewdness *fortunes on what they believe. Fortunes* alone is the object; put it after *risking.* | *His make-up, which approached* too nearly *sheer caricature to be reckoned quite happy.* A very odd piece of tit for tat; *too nearly* divides *approached* from *caricature*, and in revenge *caricature* divides *to be reckoned* from *too nearly*; put *sheer caricature* next to *approached.* | *Failure of the Powers to enforce their will as to the Albanian frontier would expose* to the ridicule of all the restless elements in East Europe *their authority, which, as it is, is not very imposing.* There are two differences from the adaptation made above—first that the adverb has eleven words instead of two, and secondly that the relative clause is not an essential part of the object; *their ... imposing* should be put directly after *expose*.

7. Separation of preposition and gerund. This hardly needs serious treatment. But here is amusingly apparent somebody's terror of separating *of* and *piling* by an adverb—which is no more than an exaggeration of the superstitions dealt with in 3, 4, and 5. *To decry the infantry arm for the sake unduly of piling up artillery and what not, is the notion of persons who ...*

8. Heedless misplacings. It would appear from the analysis attempted above that when adverbs are found in wrong positions it is usually due to mistaken ideas of correctness. But now and then it is otherwise, and an example or two of merely careless placing may be given: *The terms upon which the British 'governing classes' have obtained their influence are those upon which it alone may be retained* (upon which alone it may). | *But a work of art that is all form and no emotion*

*(and we doubt whether, in all deference to M. Saint-Saens, such an anomaly did ever or could ever exist) would seem to belong more properly to the sphere of mathematics* (the putting of the *deference* adverb after instead of before *whether* makes nonsense). | *Should, too, not our author be considered?* (*too* might go after *not*, or *author*, or *considered*, according to the meaning wanted; but no meaning can justify its present position). See also ONLY and EVEN 1.

**possessive puzzles.** 1. *Septimus's, Achilles'.* 2. *Whose, of which.* 3. *Mr. Smith* (*now Lord London*)*'s.* 4. '*The Times'* '*s opinion.* 5. *Somebody's else.* 6. *Five years' imprisonment.* 7. *The non-possessive* '*s.*

1. *Septimus's, Achilles'.* It was formerly customary, when a word ended in -*s*, to write its possessive with an apostrophe but no additional *s*, e.g. *Mars' hill, Venus' Bath, Achilles' thews.* In verse, and in poetic or reverential contexts, this custom is retained, and the number of syllables is the same as in the subjective case, e.g. *Achilles'* has three, not four syllables, *Jesus'* two, not three. But elsewhere we now usually add the *s* and the syllable—always when the word is monosyllabic, and preferably when it is longer, *Charles's Wain, St. James's Street, Jones's children, the Rev. Septimus's surplice, Pythagoras's doctrines.* Plurals of proper names ending *s* form their possessives in the same way as ordinary plurals (*the Joneses' home, the Rogerses' party*). For *goodness' sake, conscience' sake,* etc., see SAKE.

2. *Whose, of which.* See WHOSE for the question whether the use of *whose* as the possessive of *which*, and not only of *who*, is permissible.

3. (A) *Mr. Smith* (*now Lord London*)*'s intervention was decisive?* or (B) *Mr. Smith* (*now Lord London*) *intervention?* or (C) *Mr. Smith's* (*now Lord London's*) *intervention?* or (D) *The intervention of Mr. Smith* (*now Lord London*)? C is clearly wrong because the intervention was not Lord London's; B is intolerable because we cannot be happy without the *'s* close

before *intervention*, just as we cannot endure *someone's else umbrella* though we can with an effort allow the umbrella to be *someone's else*. A is the reasonable solution, but has no chance against the British horror of fussy correctness; and, failing it, the only thing is to run away, i.e. to use D. The same difficulty arises in other forms. *This was this most gallant and consistent horse* (? horse's) *and his veteran rider's second win in this event.* The only satisfactory way out is to write *This was the second win in this event of* . . . | *The people in the house opposite's geraniums* and *I never knew that the woman who laced too tightly's name was Matheson* will pass as dialogue in a novel; that is the way people do talk. But it will not do in serious prose. Even in a thriller the reader ought not to be asked to solve the puzzle set by *Photographs of the Piccadilly Palace lounge, the smart waitress, the doorkeeper, and the wife of one of the charwomen's brother appeared in every journal.*

**4.** *In 'The Times' 's opinion.* This also has to be run away from. To write *in 'The Times's' opinion* is not running away, but merely blundering; if the newspaper title is to have inverted commas and the possessive is to be used, the form at the top with two independent apostrophes jostling each other is the only correct possibility. But there are two ways of escape; one is to write the title in italics instead of inverted commas, but the possessive *s* in roman type (*The Times's*)—illogical, because the possessive *s* is an inflexion, and therefore part of the word—and the other is to fly to *of* (*in the opinion of 'The Times'*).

**5.** For *somebody else's* or *somebody's else* see ELSE.

**6.** *Five years' imprisonment, Three weeks' holiday*, etc. *Years* and *weeks* may be treated as possessives and given an apostrophe or as adjectival nouns without one. The former is perhaps better, so as to conform to what is inevitable in the singular—*a year's imprisonment, a fortnight's holiday*.

**7.** The non-possessive *'s*. The ordinary purpose of inserting an apostrophe before a final *s* is to show that the *s* is possessive, not plural; it originally indicated the omission of the *e* from the possessive inflexion *es*. It may occasionally be used before a plural *s* as a device for avoiding confusion, but this should not be extended beyond what is necessary for that purpose. We may reasonably write *dot your i's and cross your t's*, but there is no need for an apostrophe in *but me no buts* or *one million whys*, or for the one we sometimes see in such plurals as M.P.s, A.D.C.s, N.C.O.s, the 1920s, etc. To insert an apostrophe in the plural of an ordinary noun is a fatuous vulgarism which, according to a correspondent of *The Times*, is infecting display writing. *'TEA'S* outside the wayside cottage is bad enough, but I have seen *SHIRT'S* and *VEST'S* in a large Oxford St. shop, and at one of London's terminal stations a beautifully written board calls attention to *ALTERATION'S AND ADDITION'S.*' A notice at the edge of a wood *KEEP OUT, POLICE DOG'S WORKING* cannot escape suspicion of using an apostrophe in the same way.

**possible.** 1. *Do* one's *p*. 2. Construction. 3. *P., probable*.

**1.** *Do* one's *possible* is a GALLICISM; and, with *do what* one *can* in established existence, it is superfluous.

**2.** Construction. *But no such questions are* possible, *as it seems to me*, to arise *between your nation and ours.* | *No breath of honest fresh air is suffered to enter, wherever it is* possible *to be excluded.* These are wrong. Unlike *able*, which ordinarily requires to be completed by an infinitive (*able to be done, to exist*, etc.), *p*. is complete in itself and means *able to be done* or *occur*. The English for *are p. to arise* and *is p. to be excluded* is *can arise, can be excluded*. The mistakes are perhaps due to the frequency of such forms as *It is p. to find an explanation*, in which *it* is not an ordinary pronoun, but merely anticipatory; that is, the sentence in its

simpler form would not be *An explanation is p. to find,* but *To find an
explanation is p.* When it is felt that *p.*
does require to be amplified, it is done
by *of* with a verbal noun—*Limits that
are p. of exact ascertainment*; but *admit
or are susceptible* or some other word
is usually better.

3. *P., probable.* It would be too
much to demand that *p.* should always
be kept to its strict sense and never so
far weakened that *impossible* (or *possible*
in a negative context) means no more
than very unlikely; but, when *probable*
and *p.* are in explicit contrast, the
demand may fairly be made. *The
Prohibition Amendment can only be
revoked by the same methods as secured
its adoption. I met no one in America
who deemed this probable, few who
thought it even possible.* As all sensible
people knew that, whatever its improbability, it was possible, the picture
of American intelligence was uncomplimentary; but this literal absurdity
is common enough, and ranks with the
abuse of LITERALLY and UNTHINKABLE.

**post hoc (ergo) propter hoc.** The
fallacy of confusing consequence with
sequence. On Sunday we prayed for
rain; on Monday it rained; therefore
the prayers caused the rain.

**posthumous.** The *-h-* is silent, and
also, though never omitted, etymologically incorrect. *Postumus,* = *last,*
superlative of Latin *post,* was applied
in English usage especially to the
last-born of a family, and so to one
born after his father's death. The *h*
was inserted because the word was
wrongly supposed to be derived from
*post humum*—i.e. after the father had
been laid in earth—and so it eventually
became confined to that meaning.
Another example of an intrusive *h* is
*lachrymose.*

**postprandial.** Chiefly in PEDANTIC
HUMOUR.

**potboiler, potwalloper.** These words
have the same primary meaning (see
WALLOP) but in colloquial use have
diverged widely. A *potboiler* is a work

of literature or art executed merely to
make a living. A *potwalloper* (or *pot-
waller*) was an elector who before 1832
derived his franchise as a householder
from the possession of a separate fireplace to cook his food on.

**poteen, -th-.** The OED treats *-teen*
as the established spelling. The stress
is on the second syllable.

**potency, -nce.** In general senses *-cy*
is universal, and *-ce* is confined to its
technical senses in engineering, watch-
making, etc. See -CE, -CY.

**potential** has no longer the meaning
of *potent,* which should have been the
word in: *The Labour Party . . . was
exercising most potential influence on
some social problems.* The substantival
use in the sense of available resources
(*the army's war potential*) is a modern
POPULARIZED TECHNICALITY from physics.

**pother** is now, except in dialects, a
LITERARY WORD. Its old form was
*pudder,* and the more corrrect, but
now less usual, pronunciation is
*pŭ'ther* rhyming with *other.* There
is no proof of connexion with either
*bother* or *powder,* though it is thought
that *bother* may be an Irish corruption of *pother.* Between *pother*
and *bother* there is the difference in
meaning that *p.* denotes ado or bustle
or confusion in itself, while *b.* emphasizes the annoyance or trouble caused.

**pot(t).** The paper size is so named
from the pot that it formerly bore as
a watermark; the right spelling is *pot,*
the *-tt* being merely like that in *matt,*
NETT, and SET(T).

**poverty, poorness.** The dominant
sense of *poor* is having little money or
property. The noun corresponding to
this dominant sense is *poverty,* and
*poorness* is never so used in modern
English. The further the dominant
sense is departed from, the more does
*poverty* give way to *poorness*—*Poverty
is no excuse for theft*; *The poverty* (or
*poorness*) *of the soil*; *The poorness* (or
*poverty*) *of the harvest*; *The poorness
of his performance.* See -TY AND -NESS.

**-p-, -pp-.** Monosyllables ending in -*p* double it before suffixes beginning with vowels if the sound preceding it is a short vowel, but not if it is a long one or a vowel and *r*: *trapped, scrappy, uppish, popping, sleepy, carping, leaper.* Words of more than one syllable follow the rule for monosyllables if their last syllable is accented (*entrapped*, but *escarped*); they also double the *p* if, like *handicap* and *kidnap*, they have a clear vowel sound as opposed to the obscure sound in *wallop* and *gallop*, or if, like *horsewhip* and *sideslip*, they are compounded with a monosyllable; but otherwise they do not double it except *worship*. Thus: *chirruped, enveloping, galoping, galloper, gossipy, filliped, equipped, trans-shipping, hiccuped, handicapper, kidnapped, walloping, horsewhipping, worshipper, sideslipped.* In U.S. the final *p*, like the final *l*, is doubled less freely.

**practicable, practical. 1.** The negative forms are *impracticable*, but *unpractical*; *impractical* is often wrongly written (*The most impractical of all persons—the man who works by rule of thumb*); see IN- AND UN-.
**2.** Meanings. Each word has senses in which there is no fear that the other will be substituted for it; but in other senses they come very near each other, and confusion is both natural and common. Safety lies in remembering that *practicable* means capable of being effected or accomplished, and *practical* means adapted to actual conditions. It is true that the practicable is often practical, and that the practical is nearly always practicable; but a very practical plan may prove impracticable owing to change of circumstances, and a practicable policy may be thoroughly unpractical. In the extracts that follow each word is used where the other was wanted: *In the case of a club, if rules are passed obnoxious to a large section of the members, the latter can resign; in our national relationships, secession is not practical nowadays.* The last sentence is in clear antithesis to *the latter can resign*, and means You cannot secede, or in other

words Secession is not *practicable.* / *But to plunge into the military question without settling the Government question would not be good sense or* practicable *policy;* and *no wise man would expect to get serviceable recruits in this way.* The policy was certainly practicable, for it was carried out; and the writer, though he had not the proof that we have of its practicability, probably did not mean to deny that, but only to say that it was not suited to the conditions, i.e. practical. / *We live in a low-pressure belt where cyclone follows cyclone; but the prediction of their arrival is at present not* practical.

**practically.** It is unfortunate that *practically* should have escaped from its true meaning into something like the opposite. It is easy to see how this came about. The word is reasonably used in such sentences as *It is practically worn out, He is practically insane,* meaning that the thing, though it still works after a fashion, cannot be relied on as effective, or that the person, though not certifiable, is as incompetent to deal with the practical affairs of life as if he were. From this it is a short step to treating *practically* as synonymous with almost, and to absurdities such as saying of the losing horse in a photographic finish that it practically won, which is exactly what it did not do. This straining of *practically* is the less excusable because more suitable adverbs are plentiful: e.g. *almost, nearly, all-but, virtually, substantially.*

**practice, -se.** Noun -*ce*, verb -*se* but in U.S. commonly -*ce*; see LICENCE.

**practitioner.** See PHYSICIAN.

**pragmatic(al).** In the diplomatic, historical, and philosophical senses, the -*ic* form is usual. In the general sense of officious or opinionated, -*ical* is commoner. In the interests of DIFFERENTIATION these tendencies should be encouraged; see -IC(AL).

**pray.** *Pray in aid.* One of the picturesque phrases that people catch up and use without understanding: *We* are disturbed *to find that this principle of praying in aid the domestic circumstances of the woman appears to have been sanctioned officially by the Committee on Production.* This writer, and most of those who use the words, suppose that *in aid* is an adverb, and that *pray* is therefore free to take an object—here *circumstances*. The fact is that the object of *pray* is *aid*, and *in* is not a preposition but an adverbial particle, *to pray in aid* being word for word to call in help; if the helper or helping thing is to be specified, it must have an *of* before it, as in the following OED quotations: *A city or corporation, holding a fee-farm of the King, may pray in Aid of him, if anything be demanded of them relating thereto.* / *An incumbent may pray in aid of the patron and ordinary.*

**pre-.** On the question whether this prefix should be hyphened see HYPHENS, where the advice is given that a hyphen should not be used unless it is needed to prevent confusion, and the practice of hyphening such words as *preeminent* is deprecated.

**precedence, precedent.** The pronunciation is tricky. The OED gives for the first *prĕcē'dence* only and for the second *prĕcē'dent* only in adjectival use, but *prĕ'cĕdent* only in noun use. This, which is a very disputable account of present usage, is not likely to remain true; the COD now puts *prĕ'cĕdence* before *prĕcē'dence* and admits *prĕ'cĕdent* as an alternative for the adjective. It looks as if *prĕ'cĕ-* might prevail for all except perhaps the adjective. This has been superseded in ordinary use by *preceding* and is now rarely used except in the expression *condition precedent*, which lawyers usually pronounce *prĕ-* (or *prē-*)*cĕ'dent*.

**precedent.** *The House of Commons is always ready to extend the indulgence which* [*it*] *is a sort of p. that the mover and seconder of the Address should ask for.* A bad piece of SLIPSHOD EXTENSION; a p. is not a custom or a tradition (though it may start one; cf. HAZINESS), but a previous case.

**preciosity** and *preciousness* illustrate well the DIFFERENTIATION that should be encouraged whenever there is an opening for it between the two terminations; see -TY AND -NESS. Since -*ty* is now never used except in the special sense of excessive fastidiousness in diction, pronunciation, and the like, -*ness* might well have been confined to the more general senses instead of being treated, as it now is, as a synonym of -*ty*. See also PRECISENESS, PRECISION.

**precipitance, -ancy, -ation.** The most economical way of dealing with the words would have been to let -*ancy* perish, and make -*ance* mean rashness of action or suddenness of occurrence or speed of motion, and -*ation* the bringing or coming to pass with especial rashness or speed. But what is happening is that all three exist side by side, -*ance* and -*ancy* slowly giving way to -*ation* just as their parent *precipitant* has given way to *precipitate*. This is the more regrettable because *precipitation* has also to do duty for the technical senses of the verb *precipitate* in chemistry, physics, and meteorology. In the adjectives, on the other hand, a useful differentiation has taken place. See next article.

**precipitate, precipitous.** *It appears that Mr. Campbell has prevailed upon his executive not to take any precipitous action at this stage.* / *The step seems a trifle rash and precipitous when one remembers the number of banking and commercial failures that* ... Those who write thus either are ignorant of the established difference between *precipitous* (steep) and *precipitate* (rash), or must not be surprised if they are taken to be so. Formerly, -*ous* was freely used where we now always say -*ate*; but that time has long passed away. See PAIRS AND SNARES.

**preciseness, precision.** For the natural difference (-*ness* representing a state or quality and -*ion* a process or action) see -ION AND -NESS. So far as the words are used with overlapping meanings, *preciseness* is differentiated by implying that the importance of precision is exaggerated. *Preciseness* rather than *precision* is the attribute of a *precisian*.

**predacious, predatory.** *Predacious* is applied only to animals and organisms that prey on others. *Predatory*, an older word, besides having the meaning of predacious, is applied also to human beings who prey on others. *Predacious* is usually pronounced with a long *e* and *predatory* with a short.

**predicate.** 1. The OED pronounces *p.*, and its derivatives *predicable* and *predication*, with *prĕd-*, not *prēd-*, and this is still usual. The verb is said with -*āt*, the noun with -*ĕt*; see PARTICIPLES 5 B. 2. *P.*, *predict*. The Latinless have great difficulty in realizing that the words are not interchangeable variants. *P.* is from Latin *praedicare* to cry forth or proclaim, but *predict* from Latin *praedīcere* to say beforehand or foretell; the Latin simple verbs are different, and *prae* has not the same meaning in the two compounds. *P.* makes *predicable* and *predication*, *predict* makes *predictable* and *prediction*. It is naturally *predicate* and its derivatives that are misused; examples of the misuse are: *The case for establishing compulsory and voluntary systems side by side in the same country is not only not proven, but involves a change in strategic theory that predicates nothing but disaster* (threatens? foreshadows? presages? just possibly predicts; certainly not predicates). / *A profound change in the balance of the Constitution predicable by anyone who had searched the political heavens during the last four years and observed the eccentric behaviour of certain bodies and their satellites is now upon us* (predictable). / *What she would say to him, how he would take it, even the vaguest predication of their discourse, was beyond him to guess* (anticipation? outline? prevision? just possibly prediction; certainly not predication).

*Predicate* and its derivatives mean to assert, and especially to assert the existence of some quality as an attribute of the person or thing that is spoken of (*Goodness or badness cannot

with any propriety be predicated of motives.* / *To predicate mortality of Socrates*, i.e., to state that Socrates is mortal). The words (apart from *predicate* n., the grammatical term) are mainly used in logic, and are best left alone by those who have no acquaintance with either logic or Latin. See PAIRS AND SNARES.

**preface.** 1. For *p.* and FOREWORD see the latter. 2. For *p.* and *prefix*, vv., see PREFIX.

**prefect.** The adjective *prefectorial* is better than -*toral*.

**prefer(able).** 1. -*r(r)*-. 2. *More preferable*. 3. *To, rather than, than*.

1. *Prefer* makes -*rring*, -*rred* (see -R-, -RR-), but *preferable* (*prĕ'fĕrabl*); the latter formation is anomalous but established; see CONFER(R)ABLE for similar words.

2. *More preferable* is an inexcusable PLEONASM. *The cure for that is clearly the alternative vote or the second ballot, the former alternative being, in our view, on every ground the more preferable.*

3. *To, rather than, than.* If the rejected alternative is to be expressed, the normal construction for it is *to*: *I p. pears to apples, riding to walking.* The OED, defining the construction, gives nothing besides *to* except *before* and *above*, both of which it obelizes as archaic or disused. A difficulty arises, however, with *to*: the object of *prefer* is often an infinitive, but the sound of *I p. to die to to pay blackmail*, or even of *I p. to die to paying*, is intolerable. It is easy sometimes to change the verb to a noun (*to die* to *death*), but by no means always. When the infinitive is unavoidable, the way out is to use *rather than* instead of *to*: *I p. to die rather than pay blackmail.*

To use simple *than* instead of *rather than* (*I p. to die than pay*) is clean against established idiom, as bad as saying *superior than* or *prior than* instead of *superior to* or *prior to*. But this solecism, of which there is hardly a trace in the OED article (1908), has become common; the array of quotations given below is in amusing contrast with the solitary specimen (dated 1778) that the OED could show. Even the *rather than* mentioned above is not much to be recommended. If the writer is bent on using *prefer*, it will pass, but a better plan is to change the verb *prefer* to *choose rather* or *would rather* (*He chose to die rather than pay*; *I would rather die than pay*). The main point is that *prefer than* without *rather* is not English: *The majority of them, we rather think, would p. to bear the ills they know than to fly to the untried remedy of the State regulation of wages* (Shakespeare preferred *rather bear . . . than* to *prefer to bear . . . than*; the other *rather* has caused him to be corrected, but not improved)./ *They have always preferred to speculate on the chance of winning a General Election than to settle with their opponents* (rather than settle). / *Surely the public would prefer to arrive half an hour later than run the ghastly risks* (would choose . . . rather than run)./ *The nine deportees would p. to go home than to undergo sentence after trial by Court-martial* (would sooner go . . . than undergo). / *He is persuasive rather than dogmatic, and prefers to suggest than to conclude* (suggesting to concluding).

**prefix. 1.** The noun is accented on the first syllable, the verb on the second, see NOUN AND VERB ACCENT. **2.** The meaning of the noun is an affix attached to the beginning of a word or stem to make a compound word, as *re-, ex-, be-, a-,* in *reform, ex-officer, belabour, arise.* **3.** For derivative nouns it is better to rub along with *prefix* and *prefixing* than to resort to *prefixion* and *prefixture.* **4.** *Prefix*, v., *preface*, v. *P.* is one of the verbs liable to the OBJECT-SHUFFLING abuse. You can prefix a title to your name, but not prefix your name with a title. Several examples of the confusion follow; in each the construction must be turned inside out if *p.* is to be kept, but in most of them the change of *prefix*(*ed*) to *preface*(*d*) would put things right: *The speeches in the present volume are* prefixed by *a clear and connected account of the administration*

*of India. | Many others are Austrian Barons of modern creation, these titles being very numerous, because every son is allowed to* prefix *his name with the title. | A 'Collection of Poems and Essays by Mary Queen of Scots',* prefixed by *an essay on the character and writings of Mary Stuart. | Two notes dealing with recent cases on the subject of company directors are* prefixed by *the catchwords in very prominent type: 'Retirement and Remorse'. | The story is* prefixed by *an introductory sketch of Pope Alexander VI's Spanish ancestry.| Every paragraph is* prefixed with *a kind of title to it.*

The poor old word *preface*, with FOREWORD assailing it on one front and *prefix* on another, is going through troubled times.

**pregnant construction.** 'But Philip was found *at* Azotus' is in the Greek 'But Philip was found *to* Azotus'; i.e. the expressed sentence contains an implied one—Philip was conveyed to and Philip was found at Azotus. Though we cannot (except in the dialect of Devon etc.) say He was found to Azotus, we do habitually say Put it in your pocket, meaning Put it in(to and keep it in) your pocket.

**prejudice,** n. *The Committee's Report adds that without doubt a marked* prejudice *to the eating of eels exists in Scotland.* The prepositions after *p.* when it means a preconceived opinion are *against* and *in favour of*; this *to* is on the analogy of *objection* or perhaps of *without p. to* which has its special meaning of without abandoning (a right or claim); see ANALOGY.

**preliminary,** adv. See QUASI-AD-VERBS.

**prelude.** The noun is *prĕ'lūd*; the verb used to be *prĕlū'd* ('All the verse quotations and the dictionaries down to *c* 1830'—OED), but is now pronounced like the noun—a remarkable exception to the tendency mentioned in NOUN AND VERB ACCENT. See also PRONUNCIATION 6.

**premature.** The pronunciation *prĕ'-mătūr* is recommended, but the *e* is sometimes long and the stress is sometimes on the last syllable; in any case, the last syllable is fully pronounced and not weakened to -*cher*.

**premier** as an adjective is now suggestive of tawdry ornament, though it was formerly not avoided by good writers and has shown signs of coming back into favour in the wake of the now popular *première*. The ELEGANT-VARIATIONist finds it useful (*There was a time when the School of Literae Humaniores stood first in point of number, but of late the History School has taken* p. *place*), but would do better to find some other way out. It is wise to confine it now to such traditional phrases as the *Duke of Norfolk is* p. *duke and earl of the U.K. Premier* (n.) for Prime Minister dates from 1726. The usual pronunciation is *prĕmier* but the dictionaries admit *prēmier* as an alternative.

**premise(s), -ss(es). 1.** The noun is *prĕ'mĭs* and the verb *prĕmī'z*, see NOUN AND VERB ACCENT. **2.** The verb is spelt *premise*, not -*ize*; see -IZE, -ISE. **3.** The two noun spellings (-*ises* and *isses* in the plural) may perhaps be thought useful; but ambiguity cannot often arise between the parts of a SYLLOGISM (-*isses*) and those of a parcel of real property (-*ises*); and there is no reason for the variation. The two words are one, the parts of a syllogism being 'the previously stated', and the parts of the parcel of real property being 'the aforesaid' (buildings, land, etc., previously set out in the deed). The uniform spelling *premise* (pl. *premises*) is recommended.

**premium.** Pl. -*ms* only; see -UM.

**preparatory.** For the use in *They were weighing it* p. *to sending it to town*, see QUASI-ADVERBS.

**prepare.** The common use of *prepared*, especially in OFFICIALESE, in such contexts as *he was not prepared to disclose the source of his information | I am prepared to overlook the mistake* for *he refused* and *I am willing* is to be deprecated as wantonly blurring the meaning of p. *Prepared to* should be reserved for cases in which there is some element of preparation, e.g. *I have read the papers and am now prepared to hear you state your case.*

**preposition at end.** It was once a cherished superstition that prepositions must be kept true to their name and placed before the word they govern in spite of the incurable English instinct for putting them late ('They are the fittest timber to make great politics *of*' said Bacon; and 'What are you hitting me *for*?' says the modern schoolboy). 'A sentence ending in a preposition is an inelegant sentence' represents what used to be a very general belief, and it is not yet dead. One of its chief supports is the fact that Dryden, an acknowledged master of English prose, went through all his prefaces contriving away the final prepositions that he had been guilty of in his first editions. It is interesting to find Ruskin almost reversing this procedure. In the text of the *Seven Lamps* there is a solitary final preposition to be found, and no more; but in the later footnotes they are not avoided (*Any more wasted words . . . I never heard of. | Men whose occupation for the next fifty years would be the knocking down every beautiful building they could lay their hands on*). Dryden's earlier practice shows him following the English instinct; his later shows him sophisticated with deliberate latinism: 'I am often put to a stand in considering whether what I write be the idiom of the tongue, . . . and have no other way to clear my doubts but by translating my English into Latin'. The natural inference from this would be: you cannot put a preposition (roughly speaking) later than its word in Latin, and therefore you must not do so in English. Gibbon improved upon the doctrine, and, observing that prepositions and adverbs are not always easily distinguished, kept on the safe side by not ending sentences with *on, over, under,* or the like, even when they would have been adverbs.

The fact is that the remarkable freedom enjoyed by English in putting its prepositions late and omitting its relatives is an important element in the flexibility of the language. The power of saying *A state of dejection such as they are absolute strangers to* (Cowper) instead of *A state of dejection of an intensity to which they are absolute strangers*, or *People worth talking to* instead of *People with whom it is worth while to talk*, is not one to be lightly surrendered. But the Dryden–Gibbon tradition has remained in being, and even now immense pains are sometimes expended in changing spontaneous into artificial English. *That depends on what they are cut with* is not improved by conversion into *That depends on with what they are cut*; and too often the lust of sophistication, once blooded, becomes uncontrollable, and ends with, *That depends on the answer to the question as to with what they are cut.* Those who lay down the universal principle that final prepositions are 'inelegant' are unconsciously trying to deprive the English language of a valuable idiomatic resource, which has been used freely by all our greatest writers except those whose instinct for English idiom has been overpowered by notions of correctness derived from Latin standards. The legitimacy of the prepositional ending in literary English must be uncompromisingly maintained; in respect of elegance or inelegance, every example must be judged not by any arbitrary rule, but on its own merits, according to the impression it makes on the feeling of educated English readers.

In avoiding the forbidden order, unskilful handlers of words often fall into real blunders (see OUT OF THE FRYING-PAN). A few examples of bad grammar obviously due to this cause may fairly be offered without any suggestion that a rule is responsible for all blunders made in attempting to keep it. The words in brackets indicate the avoided form, which is not necessarily the best, but is at least better than that substi-

tuted for it: *The War Office does not care, the Disposal Board is indifferent, and there is no one* on whom *to fix the blame* or to hang (no one to fix the blame on or to hang). / *The day begins with a ride with the wife and as many others as want to ride and* for whom *there is horseflesh available* (and as there are horses for). / *This was a memorable expedition in every way, greatly appreciated by the Japanese, the Sinhalese, the Siamese, and with whomever else B.O.A.C. briefly deposited their valuable cargo* (and whomever else B.O.A.C. briefly deposited their valuable cargo with). / *It is like the art* of which *Huysmans dreamed but never executed* (the art that Huysmans dreamed of). / *That promised land for which he was to prepare, but scarcely to enter* (that he was to prepare for).

It was said above that almost all our great writers have allowed themselves to end a sentence or a clause with a preposition. A score or so of specimens follow ranging over six centuries, to which may be added the Bacon, Cowper, and Ruskin examples already given: (Chaucer) But yit to this thing ther is yit another thing y-ijoigned, more to ben wondred upon. (Spenser) Yet childe ne kinsman living had he none To leave them to. (Shakespeare) Such bitter business as the day Would quake to look on. (Jonson) Prepositions follow sometimes the nouns they are coupled with. (Bible) I will not leave thee, until I have done that which I have spoken to thee of. (Milton) What a fine conformity would it starch us all into. (Burton) Fit for Calphurnius and Democritus to laugh at. (Pepys) There is good ground for what he goes about. (Congreve) And where those qualities are, 'tis pity they should want objects to shine upon. (Swift) The present argument is the most abstracted that ever I engaged in. (Defoe) Avenge the injuries . . . by giving them up to the confusions their madness leads them to. (Burke) The less convincing on account of the party it came from. (Lamb) Enforcing his negation with all the might . . . he is master of. (De Quincey) The average, the prevailing tendency, is what we look at. (Landor) The vigorous mind has mountains to climb, and valleys to repose in. (Hazlitt) It does for something to talk about. (Peacock) Which they would not otherwise have dreamed of. (Mill) We have done the best that the existing state of human reason admits of. (Kinglake) More formidable than any . . . that Ibrahim Pasha had to contend with. (M. Arnold) Let us see what it amounts to. (Lowell) Make them show what they are made of. (Thackeray) So little do we know what we really are after. (Kipling) Too horrible to be trifled with.

If it were not presumptuous, after that, to offer advice, the advice would be this: Follow no arbitrary rule, but remember that there are often two or more possible arrangements between which a choice should be consciously made. If the final preposition that has naturally presented itself sounds comfortable, keep it; if it does not sound comfortable, still keep it if it has compensating vigour, or when among awkward possibilities it is the least awkward. If the 'preposition' is in fact the adverbial particle of a PHRASAL VERB, no choice is open to us; it cannot be wrested from its partner. Not even Dryden could have altered *which I will not put up with* to *up with which I will not put*.

**preposition dropping. 1.** For the disappearance of *of*s and other prepositions due to the habit of using nouns attributively see NOUN ADJECTIVES and HEADLINE LANGUAGE. **2.** *The Puritan way of eating fish is to eat it Saturday instead of Friday.* | *Can you dine with us Thursday at eight?* Of this construction the OED says: 'The adverbial use of the names of the days of the week is now chiefly U.S. except in collocations like *next Saturday, last Sunday*.' This is still substantially true, though the usage is less uncommon here than it was. It is true also of the similar use of the plural in such sentences as *He sees patients mornings.* | *We always go there summers.* **3.** The adverbial use of *place* in such phrases as *living some p. near*,

a common American colloquialism, is beginning to infiltrate into Britain, but is not yet enough at home to have lost its air of self-conscious jocularity. Another Americanism, *going places*, seems to appeal to some of our political correspondents as a more sprightly idiom than their native equivalent 'getting somewhere'. *As soon as the Liberals succeeded in peeling off one Lloyd George to the Tories and another to the Socialists they began to do better. A party without a Lloyd George can hope to go places.* | *Mr. Gaitskell used to speak of 'keeping the ship on an even keel' . . . but a ship has sometimes to be handled a little roughly if the captain wants to go places.*

**prescience.** The OED gives *prēshyĕns* only; but *prĕ-* has since become at least as common, and is likely to prevail. The sounding of *-sc-* as *-s-* (as in *science*) instead of *-sh-* is now often heard, and the speak-as-you-spell movement will probably establish it.

**prescribe, proscribe.** These words are often confused, especially by the use of *pro-* for *pre-*. *Pro-* means to put outside the protection of the law, to denounce as dangerous; *pre-* means to lay down as a rule or direction to be followed. *If I look at the list of proscribed authors in our various universities, I notice with pleasure that since 1940 no year has passed without Jane Austen appearing in the syllabus of at least one.* The speaker clearly did not mean, as one might infer from the word he used (or perhaps the printer substituted), that Jane Austen's works were on the INDEX.

**prescriptive.** That a *p. right* has some special sanctity is a common MIS-APPREHENSION, perhaps due to confusion with IMPRESCRIPTIBLE. A *p. right* is one acquired by prescription, i.e. by uninterrupted use or possession. So far from having any special sanctity, it is more likely to be open to challenge than titles derived in other ways.

**present**, adj. *The p. writer* is a periphrasis for *I* and *me* that is not entirely avoidable under existing journalistic conditions (cf. *your reviewer*) and is at any rate preferable to the false first-personal *one* (see ONE 6) that is sometimes tried as a substitute. But it is very irritating to the reader; personality, however veiled, should be introduced into impersonal articles only when the necessity is quite indisputable. The worst absurdity occurs when a contributor or correspondent whose name appears above or below his article or letter puts on this *Coa vestis* of a veil; but they often do it. See also WE 3.

**presently.** **1.** *P.* = by - and - by. When Sir Andrew exclaimed *For the love of God a surgeon! Send one p. to Sir Toby* he was using the most urgent adverb he could think of. The barren fig-tree that p. withered away did so with a promptitude that astonished the disciples. But today no one who was told *The doctor will see you p.* would expect to be shown immediately into the consulting-room. That a word which should mean, and once meant, *instantly* has come to mean *by-and-by* must reflect a stubborn resistance of hope to experience. Cf. the cliché *I won't keep you a moment*, generally the prelude to a longish wait. **2.** *P.* = at present. *Five per cent. of British steel capacity is p. idle.* | *Actor Mr. C. S., p. playing Professor Higgins in 'My Fair Lady'.* | *Europe p. can only be defended by American nuclear weapons.* This sense of p. was said by the OED to be obsolete in literary English since the 17th c., though in regular use in most English dialects and common in Scottish writers. It is now enjoying a vigorous revival, though whether for any better reason than NOVELTY HUNTING may be doubted, seeing that we have available for the same purpose not only *now* but also for those who dislike monosyllables *at present* and *currently*.

**pressure group.** See LOBBY.

**prestidigitator, -tion.** Now chiefly in POLYSYLLABIC HUMOUR.

**prestige.** It is surprising that the pronunciation has not been anglicized, like that of *vestige*, which also came to us from Latin through French. *Prĕst'ĭj* is indeed given as an alternative pronunciation by the OED and some other more recent dictionaries, but it is never heard. *Prestige* is one of the few words that has had an experience opposite to that described in WORSENED WORDS. It formerly meant illusion or imposture.

**presumptive.** For *heir p.*, see HEIR 2.

**pretty** is in good usage as an ironical adjective: *He has made a p. mess of the job.* | *Things have come to a p. pass.* It can also be used as an adverb meaning *fairly, moderately* (*The performance was p. good,* | *He did p. much what he liked*), but only when qualifying another adverb or an adjective. Otherwise the adverb is *prettily*; in the colloquialism *sit pretty* and the notice to motorists *PLEASE PARK PRETTY* it can be argued that the word is a predicative adjective. See UNIDIOMATIC, -LY.

**prevent.** The idiom is *p. me from going* or *p. my going*; *p. me going*, though common colloquially, is better avoided in writing; see FUSED PARTICIPLE. *Preventable* is recommended rather than *-ible*; see -ABLE 2.

**prevent(at)ive.** The short form is better; see LONG VARIANTS.

**previous.** **1.** For the construction in *will consult you previous to acting*, see QUASI-ADVERBS. **2.** *Too p.*, originally amusing both because the sense of *p.* was a specially made one, and because *too* was with that sense deliberately redundant, has passed into the realm of WORN-OUT HUMOUR. **3.** *The p. question* is a phrase that does not explain itself. We all know that in the House of Commons moving the p. q. is somehow a way of attempting to shelve the matter under debate, but the light of nature would suggest only, and wrongly, that the proposal was to go back to what the House had been engaged upon before this present matter. The p. q. is in fact a proposal that the matter under debate should not now (formerly, should now) be divided upon. Those who wish to shelve the matter move this p. q., to which they now vote ay (formerly no). If the motion is negatived the original question must be put forthwith.

**pre-war.** The only justification for saying *p.* instead of *before* the war is that *before the war* makes a very unhandy adjective, and we are now constantly in need of a handy one; *before-the-war conditions, politics, prices*, as phrases for everyday use, will never do, and the only justification is also sufficient. But it fails to cover the use of *pre-war* as an adverb, a practice that began after the first world war but is still called vulg. by the COD. There is nothing unhandy in that use of *before the war*, which might well be restored in all contexts of the kind here shown—*The suggestion is utterly untrue, as a comparison of present prices with those* prevailing pre-war *will show.* | *The season-ticket holder, too, is to pay about 75 per cent. more than he did pre-war.* | *The number of houses demolished annually pre-war is again not accurately known.*

**pride.** For *P. goeth before a fall*, see MISQUOTATION.

**prideful.** This Scotticism has been taken up by English NOVELTY HUNTERS, but the meretricious attraction of novelty seems to be the only advantage it can claim over *proud* or *arrogant*.

**pride of knowledge** is a very unamiable characteristic, and the display of it should be sedulously avoided. Some of the ways in which it is displayed, often by people who do not realize how disagreeable they are making themselves, are illustrated in the following among many articles: DIDACTICISM, FRENCH WORDS, GALLICISMS, IRRELEVANT ALLUSIONS, LITERARY CRITICS' WORDS, NOVELTY-HUNTING, POPULARIZED TECHNICALITIES, QUOTATION, SUPERIORITY, WORD-PATRONAGE.

**prig** is a word of variable and indefinite meaning; the following, from an anonymous volume of essays, may be useful: 'The best thing I can do, perhaps, is to give you the various descriptions that would come into my head at different times if I were asked for one suddenly. A prig is a believer in red tape; that is, he exalts the method above the work done. A prig, like the Pharisee, says: "God, I thank thee that I am not as other men are"—except that he often substitutes *Self* for *God*. A prig is one who works out his paltry accounts to the last farthing, while his millionaire neighbour lets accounts take care of themselves. A prig expects others to square themselves to his very inadequate measuring-rod, and condemns them with confidence if they do not. A p. is wise beyond his years in all the things that do not matter. A p. cracks nuts with a steam hammer: that is, calls in the first principles of morality to decide whether he may, or must, do something of as little importance as drinking a glass of beer. On the whole, one may, perhaps, say that all his different characteristics come from the combination, in varying proportions, of three things—the desire to do his duty, the belief that he knows better than other people, and blindness to the difference in value between different things.'

**prima donna.** Pronounce *prē-*. Pl. *prima donnas*.

**prima facie.** Pron. *prī′mă fā′shĭē*. See LATIN PHRASES.

**primary colours.** As the phrase is used in different senses, the OED definition is here given: 'Formerly, the seven colours of the spectrum, viz. red, orange, yellow, green, blue, indigo, violet; now, the three colours red, green, and violet (or, with painters, red, yellow, and blue), out of different combinations of which all the others are produced.' OED adds: 'In speaking of the colours of objects, *black* and *white*, in which the rays of light are respectively wholly absorbed and wholly reflected, are included.'

**primates.** Pronounce *prīmā′tēz* when used as a term for the highest order of mammals. Archbishops, though they belong to that order, must be content with a disyllabic plural *prī′mātz*.

**primer.** The traditional pronunciation is *prī′mĕr*, and the word was very commonly spelt with *-mm-*. This pronunciation was retained in the obsolescent names of types (*long primer* is now *10 point* and *great primer 18 point*); in the names of school manuals and other uses of *p*. *prī′mer* is universal.

**principal, principle.** Misprints or even mistakes of one for the other are very frequent, and should be guarded against.

**prior.** For the adverbial and prepositional use (*p. to* = before) see QUASI-ADVERBS. But the phrase is incongruous, and ranks merely with FORMAL WORDS, except in contexts involving a connexion between the two events more essential than the simple time relation, as in *Candidates must deposit security prior to the ballot.* The use deprecated is seen in: *Prior to going to Wiltshire, Mr. —— very successfully hunted the —— Hounds.* Cf. FOLLOWING.

**Priscian.** To *break Priscian's head* is to violate the rules of grammar. Priscian was a 6th-c. Roman grammarian greatly respected in the Middle Ages.

**prise.** This spelling is often used, as is *pry* in U.S., to differentiate the verb meaning to force up by leverage from the other verb or verbs spelt *prize*; it is also the old spelling of the nautical verb meaning to capture.

**privacy.** The OED recognizes only *prīv-*, but *prĭv-* must now be at least as common (cf. *privet*, *privilege*, and *privy*) and is recognized as an alternative by the COD.

**privative.** Prefixes that deny the presence of the quality denoted by the simple word are called *p*. or *negative*. The a- of *aseptic* and the in- of *innocent* are privative, whereas the a- of *arise* and the in- of *insist* are not.

**privilege, v.** 1. *He was generally believed to be an exceptionally taciturn man, but those who were* privileged with *his friendship say that this was a habit assumed against the inquisitive.* An unidiomatic use, on the ANALOGY of *honoured with*. 2. A privileged person is one who enjoys some special right or immunity. Those who—like most of us—possess no special right or immunity might reasonably be described as *unprivileged*. But who are the people in between, whom we often hear referred to as the *underprivileged classes*? Those who find an emotive value in that cliché must be presumed to have some notion of the answer; the rest of us are left guessing.

**probable.** Two temptations call for notice. The first is that of attaching an infinitive to *p.*; cf. POSSIBLE; a thing may be *likely to happen*, but not *p. to happen*; ANALOGY is the corrupter *Military cooperation against Russia is scarcely* probable to be *more than a dream.* The second is the wrong use of the future after *p*. *The result will probably be* is right; but *The probable result will be* is a mixture between that and *The probable result is*; correct accordingly to *is* in: *It is believed that Said Pasha will be forced to resign, and that his most probable successor will be Kiamil Pasha.*

**probe.** *We may well have to wait for the close approach of a space-probe to Jupiter before this problem can be settled* A *p.*, says the OED, is a surgical instrument for exploring the direction and depth of wounds and sinuses. So perhaps to apply the word to an instrument that explores direction and depth in space is a reasonable extension. But *p.* has reached this suitable and respectable niche only accidentally, after doing much damage on the way; for of all the monosyllabic VOGUE WORD produced by HEADLINE LANGUAGE it is, with the exception of *bid*, the most popular and therefore the most mischievous.

If an *-able* adjective is required, as seems likely, it must be *probeable* for fear of confusion with the ordinary *probable*—one of the extremely rare necessary exceptions to the rule given under MUTE E.

**problematic(al).** The longer form is slightly more common; there is no clear difference in usage. See -IC(AL).

**proboscis.** The pl. recommended is *-scises*; the Latin form is *-scides* (*-ēz*), and *probosces* is wrong. For *p.* = nose, see POLYSYLLABIC HUMOUR. Pronounce *-bŏ'sis*.

**procatalepsis.** A figure by which an opponent's objections are anticipated and answered in advance.

**procedural.** The earliest example given by the OED Supp. is dated 1919. The word is now much used of international conferences, almost always with some inauspicious noun, e.g. *p. difficulties*, *p. wrangles*.

**process, n.** The OED gives *prŏ'sĕs* as the better pronunciation; but *prō'sĕs* is winning.

**process, v.**, meaning to institute legal *p.* or to treat material or food, is pronounced like the noun; meaning to go in procession it is a BACK-FORMATION; pronounce *prŏsĕ's*.

**proem, proemial.** Pronounce *prō'm*, *prōē'mĭăl*. But the words, not having made their way like *poem* and *poetic* into common use, remain puzzling to the unlearned and are better avoided in general writing.

**proffer** makes *-ering*, *-ered*; -R-, -RR-.

**profile.** Popular use in the sense of character-sketch perhaps explains

why the anglicized *prō'fīl* is supplanting *-fēl*. See -ILE.

**program(me). 1.** Spelling. It appears from the OED quotations that *-am* was the regular spelling until the 19th c., and the OED's judgement is: 'The earlier *program* was retained by Scott, Carlyle, Hamilton, and others, and is preferable, as conforming to the usual English representation of Greek *gramma*, in *anagram*, *cryptogram*, *diagram*, *telegram*, etc.' But the British preference for *-amme* seems to be as firmly established as the American for *-am*.

**2.** *P.* as a verb. 'He [the Minister of Education] has done some large things well, but he did a small thing badly when he informed Miss Bacon that 250,000 more secondary places were "programmed" to start building in the next 12 months. That is a sad verb to receive a Ministerial *imprimatur*.' It is true that the use of *p.* as a verb is new. It was unknown to the SOED in 1933; in 1951 the COD admitted it without comment. Its newness is not necessarily against it; to use a noun as a verb is a recognized way of adding to our vocabulary. Whether we are justified in doing so in any particular case depends on whether we are supplying a need or merely inventing an unwanted synonym for an existing word. To the argument that *p.* does not pass this test because *plan* will do just as well, the Minister might reply that his department works on a building programme and that he chose a convenient way of saying that these places were included in it. The verb is established in the vocabulary of those who use electronic computers.

**progress.** The OED gives *prō-* as preferable to *prŏ-* and it has maintained its lead. Noun *pro'grĕs*, verb *progrĕ's*; see NOUN AND VERB ACCENT. But *prō'grĕs* is usual for the transitive verb, now much used in the manufacturing and building industries in the sense of pushing a job forward by regular stages. Cf. PROCESS.

**progression.** *Arithmetical p. and geometrical p.* These are in constant demand to express a rapid rate of increase, which is not involved in either of them, and is not necessarily even suggested by a. p. Those who use the expressions should bear in mind (1) that you cannot determine the nature of the progression from two terms whose relative place in the series is unknown, (2) that every rate of increase that could be named is slower than some rates of a. p. and of g. p., and faster than some others, and consequently (3) that the phrases 'better than a. p., than g. p.', 'almost in a. p., g. p.', are wholly meaningless.

*In 1903 there were ten thousand 'paying guests', last year* [1906] *fifty thousand. The rate of increase, is* better, *it will be observed,* than *a. p.* Better, certainly, than a. p. with increment 1, of which the fourth annual term would have been 10,003; but as certainly worse than a. p. with increment a million, of which the fourth term would have been 3,010,000; and neither better nor worse than, but a case of, a. p. with increment 13,333⅓. The writer meant a. p. with annual increment 10,000; but as soon as we see what he meant to say we see also that it was not worth saying, since it tells us no more than that, as we knew before, fifty thousand is greater than forty thousand.

Even g. p. may be so slow that to raise 10,000 in three years to as little as the 10,003 mentioned above is merely a matter of fixing the increment ratio low enough. Neither a. p. nor g. p. necessarily implies rapid progress. The point of contrast between them is that one involves growth or decline at a constant pace, and the other at an increasing pace. Hence the famous sentence in Malthus about population and subsistence, the first increasing in a geometrical and the second in an arithmetical ratio, which perhaps started the phrases on their career as POPULARIZED TECHNICALITIES. Of the following extracts, the first is a copy of Malthus, the second a possibly

legitimate use, according to what it is meant to convey, and the third the usual absurdity: *The healthy portion of the population is increasing by a. p., and the feeble-minded by g. p.* | *Scientific discovery is likely to proceed by g. p.* | *As the crude prejudice against the soldier's uniform vanished, and as ex-Regular officers joined the Volunteers, and Volunteers passed on to the Army, the idea that every man owes willing service to his country began to spread in an almost geometrical ratio.*

**prohibit.** The modern construction, apart from that with an object noun as in *an Act prohibiting export*, is *from doing*, not *to* do; the OED marks the latter as archaic, but it is less archaism than ignorance of idiom and the analogy of *forbid* that accounts for it in such contexts as: *Marshal Oyama prohibited his troops to take quarter within the walls.* | *The German Government has decided to issue a decree prohibiting all Government officials to strike.*

**prolate, -lative.** Many verbs have meanings that are not self-sufficient, but need to be carried forward by another verb in the infinitive; such are the auxiliaries, and other verbs meaning be able or willing or wont or desirous, begin, cease, seem, be said, etc. This infinitive is called prolate or prolative.

**prolepsis.** Anticipatory use of an epithet, i.e. the applying of it as if already true to a thing of which it only becomes true by or after the action now being stated. A strong example is
So the two brothers and their *murder'd* man
Rode past fair Florence
i.e., the man who was afterwards their victim. More ordinary examples are
He struck him dead, Fill full the cup, etc.

**prolific 1.** The adjective is in common use, but to make a satisfactory noun from it has passed the wit of man. *Prolificacy, prolificalness, prolificity,* and *prolificness,* have been tried and found wanting; substitutes such as *fertility, productiveness, fruitfulness,* are the best solution. 2. *P.* can only be properly applied to what produces (*a p. writer, orchard,* etc.), not to what is produced, as in *His works, which are p., include many which have been translated into English.*

**promenade** is still ordinarily *-ahd* though modern dictionaries allow *-ād* as an alternative. See -ADE, -ADO.

**Promethean.** Pronounce *Prométh'-ĕăn,* as Othello did. See also HERCULEAN.

**promiscuous.** The colloquial use for random, chance, casual, etc., springs from POLYSYLLABIC HUMOUR and is out of place in serious writing.

**promise.** The noun *promisor* is confined to legal use, and *-er* is the ordinary word. *P., v.,* is liable to the abuse discussed in DOUBLE PASSIVES: *If it had been taken down, even though* promised *to be re-erected, it might have shared the fate of Temple Bar.*

**promissory.** So spelt, not *-isory.* The stress is on the first syllable. See RECESSIVE ACCENT.

**prone.** See APT and SUPINE.

**pronounce** makes *-ceable;* see -ABLE I. *Pronouncedly* has four syllables; see -EDLY. *Pronouncement* is kept in being by the side of *pronunciation* owing to complete differentiation; it means only declaration or decision, which the other never does.

**pronouns** and pronominal adjectives are tricky rather than difficult. Those who go wrong over them do so from heedlessness, and will mostly plead guilty when they are charged. It is enough to state the dangers very shortly, and prove their existence by sufficient citations. 1. There must be a principal in existence for the pronoun or proxy to act for. 2. The principal should not be very far off. 3. There should not be two parties justifying even a momentary doubt about which the pronoun represents.

4. One pronoun should not represent two principals on one occasion. 5. The pronoun should seldom precede its principal.

1. No pronoun without a principal in being. *Viscount Wolverhampton, acting under medical advice, has resigned the office of Lord President, and His Majesty the King has been pleased to accept it* (*it* is resignation; but as that word has not been used we can only suppose H.M. to have accepted the office). | *Now, the public interest is that coal should be cheap and abundant, and that it should be got without the dangerous friction which has attended the disputes between masters and men in this trade. And, if nationalization is to be the policy,* it *looks to an assured peace in the coal-trade as its main advantage. For this it will pay a fair price and be willing that a considerable experiment should be made, but without the sure prospect of such a peace it will see no benefit to itself and a very doubtful benefit to the miners in the change from private to State ownership* (Each of the *its* in roman type means the public, not the public interest). | *The number of these abstainers is certainly greater than can be attributed to merely local or personal causes, and those who have watched the election agree that a portion of* them *are due to doubts and uncertainties about the Act* (A portion, that is, of the abstentions, not of the abstainers). | *An American Navy League Branch has even been established in London, and is influentially supported by* their *countrymen in this city* (Whose countrymen?).

2. The principal should not be very far off. We have to go further back than the beginning of the following extracts to learn who *he* and *she* are: *And yet, as we read the pages of the book, we feel that a work written when the story is only as yet half told, amid the turmoil of the events which* he *is describing, can only be taken as a provisional impression.* | *It is always a shock to find that there are still writers who regard the war from the standpoint of the sentimentalist. It is true that this story comes from America and bears the*

*traces of its distance from the field of action. But even distance cannot wholly excuse such an exterior view as* she *permits herself.*

**3.** There should not be two parties justifying even a moment's doubt about which the pronoun represents. *There is no doubt that the Home Secretary is justified in law in issuing a deportation order and therewith goes the right, absolutely within his discretion, to place him in a ship or an aircraft bound for a destination of* his *selection* (The Home Secretary's selection or the deportee's?). *In the December previous to* his *raid on the Tower he was chief of a gang who, overpowering* his *attendants, seized the Duke of Ormonde in St. James Street when returning from a dinner-party* (His refers not to the preceding *he*, but to the following *Duke*; see 5, and FALSE SCENT). | *Professor Geddes's fine example of sociology applied to Civics,* his *plea for a comprehensive and exact survey of* his *own city as a branch of natural history required for the culture of every instructed citizen* (The professor's own city? Ah, no; here comes, perhaps better late than never, the instructed citizen, the true principal). | *As it is, the short-sighted obstinacy of the bureaucracy has given* its *overwhelming strength to the revolution* (Not bureaucracy's, but revolution's, strength; see also 5). | *Coriolanus is the embodiment of a great noble; and the reiterated taunts which* he *hurls in play after play at the rabble only echo the general temper of the Renascence* (Not Coriolanus, but Shakespeare, is the hurler; the interloping of Coriolanus between Shakespeare and his proxy makes things difficult for the reader).

**4.** One pronoun, one job . . . *which opens up the bewildering question as to how far the Duma really represents the nation. The answer to this is far from solving the Russian riddle, but without answering it it is idle even to discuss it* (It represents, first, the bewildering question, secondly, the discussion of that riddle, and last, the riddle itself— which is not the same as the question).|

*This local option in the amount of outdoor relief given under the Poor Law has always operated inequitably and been one of the greatest blots on the system; to extend it to the first great benefit under the Insurance Act will greatly lessen* its *usefulness* (It is the blot, but its is the Act's). | *Again, unconsciousness in the person himself of what he is about, or of what others think of him, is also a great heightener of the sense of absurdity; it makes it come the fuller home to us from his insensibility to it* (It is first the unconsciousness, secondly the sense of absurdity, and thirdly absurdity).

**5.** The pronoun should seldom precede its principal. *For Plato, being then about twenty-eight years old, had listened to the 'Apology' of Socrates; had heard from* them *all that others had heard or seen of* his *last hours* (had heard from others all that they had heard etc.). | *The old Liberal idea of cutting expenditure down to the bone, so that* his *money might fructify in the pocket of the taxpayer, had given place to the idea of* . . . (the taxpayer's money might fructify in his pocket). | *Both these lines of criticism are taken simultaneously in a message which* its *special correspondent sends from Laggan, in Alberta, to the* Daily Mail *this morning* (which the D.M. prints this morning from its correspondent etc.).

**pronunciam(i)ento.** The Spanish spelling is with the *i*, but we have dropped it in anglicizing the word. Pl. *-os*; see -O(E)S 6.

**pronunciation.** In the article RECEIVED PRONUNCIATION (R.P.) some account is given of the system of pronunciation that is regarded as correct for an educated Englishman. The article RECESSIVE ACCENT deals with one feature of the R.P. that calls for special notice. The present article is concerned with certain trends now discernible in our pronunciation and with a few particular points.

### I. GENERAL

The R.P. has always been to some extent conventional; the spelling of a

word is not necessarily a safe guide to
its sound. If, for instance, the conventional pronunciation of *forehead*
is *fŏ'red*, and of *knowledge nŏ'lĭj*,
precisians who try to restore the
supposed true sounds of those words
expose themselves to an imputation of either ignorance or pedantry;
the right rule is to speak as our neighbours do, not better. But pronunciation is never static, and there are
forces at work today that make it more
than usually subject to change.

One of the forces is a movement,
not springing from the vagaries of
individual precisians, but more
broadly based, towards speaking
words as they are spelt. This is
a natural result of the growth of
popular education, of words being
visible and not merely audible symbols
to far more people than they used to
be, of the teaching of careful articulation in schools as an antidote to slipshod speech and local accent, and of
the precise enunciation cultivated by
some professional broadcasters.

Modern dictionaries afford plenty of
evidence that the speak-as-you-spell
movement is encroaching on many
conventional pronunciations; not only
the older and more esoteric, such as
that *hotel* and *humour* must not be
aspirated and that *girl* must be gairl,
*Ralph Rāf*, and *golf gŏf*, but also
many others in general use. The
experiences of OFTEN (q.v.) are typical;
and those who say *for'tūn* instead of
*for'chŭn*, *clothes* instead of *clōz*,
*pik'tūr* instead of *pik'cher*, and
make six syllables of *extraordinary* can
claim the COD's recognition of these
either as the now prevalent pronunciations or as admissible alternatives: *Regiment* and *medicine* are given
their three syllables quite as often as
their conventional pronunciations *rej-
ment* and *medsin*. The same thing is
happening to proper names. *Marylebone, St. Mary Axe, Cirencester,
Daventry,* and *Pontefract* are now
rarely given their old telescopic pronunciations *Maribn, Simeryax, Sisiter,
Dāntri,* and *Pumfret* (see 10 below).
To a future generation *forchŭn* and *clōz*

may sound as odd as the once correct
*awspitl* and *obleeged* do to us.

The other force that can hardly fail
to have its effect is American pronunciation, now an easy second to
the R.P. over the air in Britain, and
markedly different from it, especially in some of its vowels, such as
sounding *o* as *ah* and *ū* as *oo*, and its
more even distribution of stress.
American influence no doubt accounts for the fact that Englishmen
seem to be taking to saying *rĕ'search*
and *dĕ'fect* instead of putting all the
emphasis on the second syllable with
a hardly perceptible vowel sound in
the first, and also perhaps for the
increasing number of words in which
we pronounce *lu* as *loo*. See 6 below.

A third trend, deprecated elsewhere
in this book, is our greater tendency
to give foreign values to the vowels in
any word, however well acclimatized
in England, that has any foreign look,
however slight. Even our Marias and
Theresas are being turned into *Marē'as*
and *Therā'sas*. For this see FRENCH
WORDS and LATIN PHRASES.

## 2. SILENT *T*

With the possible exception of *pestle*,
the speak-as-you-spell movement has
not yet made any impression on the
large class of words ending in -*sten* and
-*stle* in which the *t* is not sounded
(*fasten, listen, castle, bustle*, etc.). But
it is encouraging the sounding of *t* in
words in which -*st*- is followed by a
consonant. The COD now puts *wăs(t)-
kŏt* before the once correct *wĕskŏt* and
anyone using the R.P. is likely to give
full value to the *t* in such words as
*postpone, coastguard,* and *dustbin*. On
the other hand, *soften* shows no disposition to follow the lead of *often*, or
not enough to have reached the dictionaries.

But the convention that condones the
omission of the *t*-sound in many words
gives no excuse for the slovenly dropping of either that or any other consonant from a word in which convention requires it to be sounded, such as
(to take examples that may sometimes
be heard over the air) saying *fax* for

*facts, instinx* for *instincts, seketary* for *secretary,* and *Febyuary* for *February.* It was all very well for Miss Pinkerton to tell her sister to put Becky's *dixonary* back in the closet, but the headmistress of today is expected to articulate more carefully. *Artic* for *arctic* and *strenth* for *strength* give the same impression of slovenliness to the hearer, though in fact they differ by being old alternative pronunciations that have gone out of fashion.

### 3. SILENT *H*

In the first edition of this dictionary it was said that in many compounds whose second element begins with *h* the *h* is silent unless the accent falls on the syllable that it begins; thus *phil-hellenic* and *philharmonic* should not sound the *h*; in *nihilism* also it should be silent. Here too the speak-as-you-spell movement has been at work, and though the COD does not favour the pronunciation of the *h* in these words, it is in fact often heard, and some modern dictionaries give it. See also A, AN I, HONORARIUM, HOTEL, and WH-.

### 4. *A* AND *O*

The variations *ah* and *ă* for *a*, and *aw* and *ŏ* for *o*, widely prevalent in large classes of words (*pass, telegraph, ask, gone, soft, loss*) are largely local distinctions, *ah* and *aw* roughly southern and *a* and *o* northern, with *o* tending to displace *aw*.

### 5. SHORT *O*

Short *o* in a stressed position, not content with vacillating between *aw* and *ŏ*, is sometimes given the sound of a short *ŭ*. This is one of the many arbitrary features of our pronunciation that defy all rule. It is not apparent, for instance, to the ordinary man, though there may be an explanation known to the etymologist, why we should give the vowel different sounds in *brother* and *brothel, company* and *compact, colour* and *column, dozen* and *lozenge, honey* and *honest, mongrel* and *mongoose*; or why, when the *o* is followed by a *v*, we should show unusual consistency in pronouncing it *ŭ* with only rare excep-

tions such as *novel* and *sovereign*. All that can be said is that the speak-as-you-spell movement is likely to bring down on the side of *ŏ* those words that are still hesitating and probably also some of those that now seem to have settled for a *ŭ*. Among the words classed as hesitants by the COD are: (preferring *o*) *Covent Garden, Coventry, hovel, hover, pomegranate, sojourn,* and (preferring *u*) *combat, comrade, conduit, Lombard.*

### 6. LONG *U*

Long *u* may be pronounced either *yoo* or *oo*, and what determines the choice is not obvious. It seems that we in Britain, unlike the Americans, prefer *yoo* unless difficulties of articulation deter us. It is not easy to say *yoo* after *ch, j, r,* or *sh,* and so we fall back on *oo* after those consonant sounds, e.g. *chute, jute, rude,* and *sugar*. No other consonant inhibits *yoo* quite as much, and with most of the others it is standard: *abuse, acute, duty, funeral, argument, huge, kudos, music, nude, putrid, tune. L* is, however, exceptional. It is not so difficult to say *yoo* after *l* as after *ch* etc., but it is more difficult than after most other consonants, and there is clearly a movement going on, shared to a small extent by *su-*, in the pronunciation of *lu-*. It was formerly *de rigueur* to put in the *y* sound; a *lute*, and even a *flute*, had to be called *lyoot* and *flyoot*, not *loot* and *floot*, or the speaker was damned in polite circles, just as *Syoosan* was once the only genteel way of pronouncing that name, and the dictionaries still give that sound for *suit*. Some people seem to count the victorious progress of *loo* one of the vulgarities of modern speech. Among these was the OED which went so far as to prefer *glyoo* to *gloo* for the pronunciation of *glue*, though it reversed the order for *blue* (*bloo, blyoo*). For most of us, as the dictionaries now concede, anything but *bloo* and *gloo* is surely impossible, however refined we like to be where the trials of articulation are not so severe. Indeed, it seems clear that *loo* is slowly but surely displacing *lyoo*, helped no doubt by the general preference for the *oo* sound in U.S. What governs its progress is not easy to say: there is no apparent reason why *dilute* and *prelude* should cling firmly to *yoo* while *delude* and *allude* are still hesitating and *conclude* and *recluse* are complete converts to *oo*, or why *aluminium* should be *yoo* and *aluminous oo*.

### 7. -ER- OR -UR-

What should be the vowel sounds in participles and other inflexions and derivatives containing -err- or -urr-? Is *erring*, for instance, to have the same vowel sound as *err*, or does it change to that of *error*? Is the vowel sound in *furry* and *currish* to be that of *err* or that of *hurry*? The OED is nearly but not quite consistent; in the words *concurring, currish, demurring, deterring, erring, furry, purring, slurring*, and *spurring* the *err* sound is prescribed; *recurring, recurrence*, and *occurrence*, however, are to be said like *hurry*. It may be taken that the *err* sound is the orthodox one. *Demurrer* has two pronunciations: as *hurry* for the legal term and as *err* for the person who demurs.

### 8. *AL* FOLLOWED BY A CONSONANT

*Al* followed by a consonant is pronounced in six different ways: *ăl*, as in *calculate*; *awl* (or *ŏl*) as in *altar*; *aw*, as in *walk*; *ah*, as in *almond*; *ă*, as in *salmon*; and *ā* as in *Ralph* and *halfpenny*. Where such a combination occurs in a word that is a syncopated compound of *all* (*albeit, also, altogether*, etc.) the pronunciation is always *awl*. Elsewhere it varies capriciously.

With B it is *ăl* (*Albania, albatross*) or, rarely, *awl* (*Albany, Talbot*). With hard C and K it is *aw* in words ending *k* (*talk, walk*, etc.) and, ordinarily, in *falcon*; elsewhere it is *ăl* (*talc, balcony, Dalkeith, alkali*) or rarely *awl* (*Balkan* and, sometimes, *falcon*). With D it is *awl* (*bald, alder*) or, rarely *ăl* (*aldehyde*). With F and PH it is *ah* in *calf* and *half* and their compounds; in others *ăl* (*Alfred*,

*alphabet*) or rarely *awl* (*palfrey*) and, exceptionally, *ā* in the syncopated pronunciation of *halfpenny* and the old pronunciation of *Ralph*. With G it is *ăl* (*algebra, hidalgo*). With L it is *awl* in words ending -*all* (*call, fall*, etc.) and their compounds and inflexions but elsewhere *ăl* (*ballot, callous*). (The long *a* in *Balliol* comes from the old spelling *Baliol*.) With M it is *ah* (*alms, palm*) with variants *awl* (*almanac*), *ăl* (*halma*) and *ă* (*salmon*). With N it is rare and unruly: *awl* in *walnut*, *ah* in *Calne*, and *ăl* in the *balne*-compounds. With P it is *ăl* (*alpaca, scalp*) but *awl* in *Walpole*. In the few words in which it occurs with R it is *awl* (*Alresford, walrus*). With S and soft C it is ordinarily *awl* (*false, palsy*) but sometimes *ăl* (*halcyon, Alsatia, salsify*). With T it is *awl* (or *ŏl*) (*salt, falter*) with some *ăl* exceptions (see below). With V it is *ah* in inflexions of the words ending -*alf* (*halve, calve*) otherwise *ăl* (*valve, salvo*) but *awl* in *Malvern*.

The commonest mistake is the mispronunciation of *alt* by speakers who forget that in some words it is *ălt* instead of the normal *awlt*. These are words derived from the Latin adjective *altus* such as *alto* and *altitude*, the *alter* of *alter ego* and *altruism* and its kindred.

### 9. *OUGH*

This combination of letters has deservedly become the classic example of the notorious inconsequence of English spelling. There are nine different ways of saying it: *ō* as in *though*, *oo* as in *through*, *ow* as in *bough*, *aw* as in *ought*, *off* (or *awf*) as in *cough*, *uff* as in *rough*, *ŏk* as in *hough*, *ŏch* as in *lough* and an indeterminate *er* sound as in *borough*. A tenth might be added (*up*) if it were not that *hiccough* is merely a misspelling of HICCUP. Most of the words containing *ough* are familiar, and their pronunciation established, however unreasonably. A few exceptions will be found in their dictionary places. See for instance HOCK (for *hough*), LOUGH, SLOUGH, SOUGH, and TROUGH.

## 10. PROPER NAMES

Many proper names have a traditional pronunciation not easily inferred from their spelling. The list that follows gives some examples; more will be found in an appendix to the COD. In titles and surnames these curiosities are naturally more persistent than in place-names, where the speak-as-you-spell movement is eroding them (see section I above). But even among the former this is beginning to tell. The Pepys family are now not *Peeps* but *Peppiss*; the Auchinlecks no longer call themselves *Affleck* as they did in Boswell's day, and the Dalziels and Menzieses who pronounce their names as they are spelt must outnumber those who are still the traditional *Dĕĕl'* and *Mingis*.

| Name | Pronounced |
|---|---|
| Abergavenny (the title) | Abergenny |
| Althorp | Awltrup |
| Beauchamp | Bēcham |
| Beauclerk | Bōclār |
| Beaulieu | Būlĭ |
| Belvoir (the castle) | Bĕver |
| Bethune | Bēten |
| Blount | Blunt |
| Blyth | Blī |
| Bohun | Bōōn |
| Broke | Brŏŏk |
| Caius (the college) | Kēz |
| Cherwell | Charwell |
| Cholmondeley | Chumlĭ |
| Cockburn | Cōburn |
| Coke | Cŏŏk |
| Colquhoun | Cohōōn' |
| Cowper | Cōōper |
| Crespigny | Crĕ'pĭny |
| Devereux | Deverōōks |
| Fiennes | Fīnz |
| Glamis | Glahmz |
| Harewood | Harwood |
| Home | Hūm |
| Knollys | Nōlz |
| Ker | Kar |
| Keynes | Kānz |
| Legh | Lē |
| Leveson Gower | Lōōson Gore |
| Magdalen(e) (the colleges) | Maudlin |
| Marjoribanks | Marchbanks |
| Pole Carew | Pōōl Kārĭ |
| Poulett | Pawlet |
| Ruthven | Riven |
| St. John (the title) | Sin'jun |
| St. Leger (the surname) | Sent'lejer |
| Sandys | Săndz |
| Tyrwhitt | Tĭrĭt |
| Waldegrave | Wawlgrave |
| Wavertree | Wawtry |
| Wemyss | Wēmz |
| Woburn | Wōōbŭn |

**propaganda** is not unnaturally mistaken for a Latin neuter plural = things to be propagated; it is in fact an ablative singular from the title *Congregatio de Propagandâ Fide* = Board [of Cardinals] for Propagating the Faith.

**propensity.** *That propensity of lifting every problem from the plane of the understandable by means of some sort of mystic expression is very Russian.* Propensity *to* do or *for* doing, not *of* doing; the ANALOGY of *practice, habit,* etc., is responsible.

**prophecy, -sy.** The noun *prophecy̆*, the verb *prophesy̆*; see LICENCE.

**prophetic(al).** The *-al* form perhaps lingers only in such phrases as *the -al books*, in which the meaning is definitely 'of the Prophets'. See -IC(AL).

**proportion.** It has been recorded as a common MISAPPREHENSION that *p.* is a sonorous improvement upon *part.* What was meant will be plain from the following examples, in all of which the word has been wrongly used because the writers, or others whom they admire and imitate, cannot resist the imposing trisyllable; the greater part, most, etc., should be substituted. *The greater proportion of these old hands have by this time already dropped out; it is estimated that only 25,000 of them remain now* (Most of). / *A few years ago the largest proportion of the meat coming through Smithfield had its origin in the United States* (the greater part). / *There was a large and fashionable audience, and, as might be expected, the greater proportion of them were natives of India* (most of them). / *By far the* *largest proportion of applications for using the machinery of the Act came from the employees* (the most applications). | *The larger proportion of the children received are those of unmarried mothers* (Most of).

'The word has been wrongly used.' It is not merely that here are two words, each of which would give the sense equally well, and that the writer has unwisely allowed LOVE OF THE LONG WORD to decide the choice for him. *Proportion* does not give the sense so well as *part*. Where *p.* does so far agree in sense with *part* that the question of an exchange between them is possible, i.e. where it means not a ratio but a quota or amount, there is nevertheless a clear difference between them. A proportion is indeed a part, but a part viewed in a special light, viz. as having a quantitative relation to its whole comparable with the same relation between some analogous whole and part. Thus a man who out of an income of £1000 spends £400 upon house-rent is rightly said to spend a large p. of his income on rent, if it is known that most people's rent is about 1/5 of their income; *p.* is there a more precise and better word than *part*, just because other ratios exist for comparison. But to say 'A large p. [instead of *a large part*] of these statements is unverified', where there is no standard of what ratio the verified facts bear to the unverified in most statements, is to use a worse long word instead of a better short one.

The case is much stronger against *p.* when it is accompanied, as in the extracts given above, by a comparative or superlative (*greater, largest*, etc.) showing that the comparison implied by *p.* is not between two ratios (e.g. between the ratio of the part in question to the whole in question and that of some other part to some other whole) but simply between the two parts into which one whole is divided. Of these two parts of course one is greater or less than or equal to the other, but that relation is adequately given by *greater* etc. *part*, and only confused by the dragging in of the comparison of ratios

expressed by *p.* It is a clumsy blunder to use words like *greater* and *largest* with *p.* when the comparison is between the parts of one whole and not between the ratios borne by parts of different wholes to their respective wholes. To give contrasted examples of the wrong and the right: *We passed the greater proportion of our candidates* is wrong; read *part*; *We hope to pass a greater proportion of our candidates next year* is right.

For a parallel, see PERCENTAGE.

**proportionable, -nal, -nate.** All three adjectives have existed since the 14th c.; so it may be presumptuous to advise the superannuation of any of them. But the fact is that, so far from needing three words, we can hardly provide two with separate functions. If any of the three is to go, it should be *-able*, for which the latest OED quotation is dated 1832. *Proportional* is better suited to the most general sense of all, 'concerned with proportion', and *-ate* to the particular sense 'analogous in quantity to', *-al representation* but *punishment -ate to the offence*. But both are so fully in possession of the most usual sense 'in proportion' or 'in due proportion' that it is useless to think of confining that sense to either.

**proposal.** See PROPOSITION.

**proposition.** The use as a VOGUE WORD is American in origin. The OED Supp. quotes from Owen Wister: *'Proposition' in the West does in fact mean whatever you at the moment please.* This remark, made in 1902, seems now to have become true in Britain also. Those who will look through the examples collected below may perhaps be surprised to see the injury that this single word is doing to the language, and resolve to eschew it. It won its popularity partly because it combined the charms of novelty and length, and partly because it ministers to laziness; there is less trouble in using it than in choosing a more suitable word from the dozen or so whose places it is apt to usurp.

It may be granted that there is nothing unsound in principle about the development of sense. *Proposition* does or did mean propounding, and, like other *-tion* words, may naturally develop from that the sense of thing propounded, from which again is readily evolved the sense thing to deal with, and that sufficiently accounts for all or nearly all the uses to be quoted. The mischief, as with all vogue words, is that they drive more precise words out of business and make for loose thinking. *P.* ought to have been kept to its former well-defined functions in logic and mathematics, instead of being given this new status as Jack-of-all-trades.

Used for proposal: *What's your proposition* (i.e. how much will you offer?)/ *'Let us pull down everything' seems to be his proposition.* | *Newman said to Mr. Hastings 'You must share my room and bed'. This (says Mr. Hastings) was to me a curious proposition, but one I had to accept.* | *This is a 50:50 proposition* (i.e. we go halves).

Used for task, job, problem, objective: *England has now to meet France, which is a different proposition.* | *There are a good many stages at which a disciplinary proposition may present itself.*

Used for undertaking, occupation, trade: *He has got a foothold mainly because the English maker has been occupied with propositions that give a larger proportion of profit.* | *Establishing floating supply depots at frequent intervals across the ocean, a proposition which only a multi-millionaire could have undertaken.* | *Railways are carrying the burden of a capital structure related to the cost of their assets, many of which cannot be made paying propositions.*

Used for opponent: *Australia's four seam-bowlers are sure to be a nasty proposition.* | *The former is a very tough proposition as an opponent in singles.* | *This Sixth Army now standing opposite us was not a very fearsome proposition.*

Used for possibility, prospect: *Petrol at 6½d. or 7½d. a gallon was hardly a commercial proposition.* | *The only way to increase the recruiting standard of the Territorial Force is to make the service*

*a more attractive proposition to the man and the employer.*

Used for enterprise worth undertaking: *Middlesex had everything to gain by going for the runs, but Warr decided that it was not a proposition.*

Used for area, field: *The mining district, according to the best information obtainable, is a placer proposition, and placer mining ruins the land.* | *Lancashire is vitally interested to secure a sufficient supply of cotton on the Gezira plains in the Sudan, this locality being what one speaker described as 'the very finest cotton-growing proposition in the whole world'.*

Used for method, experiment: *The territories will certainly require many novel propositions for their development.*

The crowning outrage on this long-suffering word is its use as a verb in the sense of make amatory advances. This originated as U.S. slang, but now seems to have won a foothold in serious writing on both sides of the Atlantic. *The central idea is of a donnish household in which the stingy economist husband is suddenly propositioned in the middle of the night by his wife's college girl friend.*

**proscribe.** See PRESCRIBE

**prosecutrix.** For plural see -TRIX.

**prosody.** That part of the study of a language which deals with the forms of metrical composition. The adjective recommended is *prosodic*, and the agent noun *prosodist*.

**prospect,** v., makes *-tor*; see -OR. The OED accents *pro'spect*, not *prospe'ct*, in the only current verb senses; but the analogy of similar NOUN AND VERB ACCENTS has since prevailed.

**prospectus.** Pl. *-tuses*, not *-ti*. See LATIN PLURALS.

**prostrate.** The adjective *pro'strate*, the verb *prostra'te*; see PARTICIPLES 5 A. For meaning see SUPINE

**protagonist.** The word that has become a prime favourite, and is more often than not made to mean champion or advocate or defender, has no right whatever to any of those meanings, and almost certainly owes them to the mistaking of the first syllable (representing Greek πρῶτος first) for *pro* on behalf of—a mistake made easy by the accidental resemblance to *antagonist*. 'Accidental', since the Greek ἀγωνιστής, the second part of both compounds, has different meanings in the two words, in *antagonist* combatant, but in *protagonist* play-actor. The Greek πρωταγωνιστής means the actor who takes the chief part in a play—a sense readily admitting of figurative application to the most conspicuous personage in any affair. The deuteragonist and tritagonist take parts of second and third importance, and to talk of several protagonists, or of a chief p. or the like, is an absurdity as great, to anyone who knows Greek, as to call a man the p. of a cause or of a person, instead of the p. of a drama or of an affair. It is now a rarity to meet *p.* in a legitimate sense; but two examples of it are put first in the following collection. All the others are (for Greek scholars, who perhaps do not matter) outrages on this learned-sounding word, because some of them distinguish between chief pp. and others who are not chief, some state or imply that there are more pp. than one in an affair, and the rest use *p.* as a mere synonym for *advocate*.

Legitimate uses: *In* Jeppë *the subsidiary personages do little more than give the p. his cues.* | *Marco Landi, the p. and narrator of a story which is skilfully contrived and excellently told, is a fairly familiar type of soldier of fortune.*

Pro- and ant- in contrast: *Protagonists and antagonists make a point of ignoring evils which militate against their ideals.*

Absurd uses with *chief* etc.: *The chief p. is a young Nonconformist minister.* | *It presents a spiritual conflict, centred about its two chief pp.,* ('co-starring' them, no doubt) *but shared in by all its characters.*

Absurd plural uses: *By a tragic but rapid process of elimination most of the pp. have now been removed.* | *As on a stage where all the pp. of a drama assemble at the end of the last act.* | *The pp. in the drama, which has the motion and structure of a Greek tragedy, are ... (*Fie! fie! a Greek tragedy and pp.?*).

Confusions with *advocate* etc.: *Enthusiastic p. of militant Protestantism.* | *It was a happy thought that placed in the hands of the son of one of the great pp. of Evolution the materials for the biography of another.* | *But most of the pp. of this demand have since shifted their ground.*

It was admitted above that we need perhaps not consider the Greek scholar's feelings; he has many advantages over the rest of us, and cannot expect that in addition he shall be allowed to forbid us a word that we find useful. Is it useful? Or is it merely a pretentious blundering substitute for words that are useful? *Pro-* in *protagonist* is not the opposite of *anti-*; -*agonist* is not the same as in *antagonist*; *advocate* and *champion* and *defender* and *combatant* are better words for the wrong senses given to *p.*, and *p.* in its right sense of *the* (not *a*) chief actor in an affair has still work to do if it could only be allowed to mind its own business.

But in the time that has passed since the above was written there has been little sign of its being allowed to do so. Forty years later we may come across such examples as these almost any day: *The Bishop of Bath was the Crown's chief p. against the Church courts.* | *Do the Channel bridge pp. suppose that navigational interests will allow the littering of the waterway with obstacles at every 250 yards?* The temptation to regard *protagonist* as the antonym of *antagonist* seems irresistible, and its use in this sense may soon have to be classed as a STURDY INDEFENSIBLE.

**protean.** The dictionaries that recognize *prŏtē'ăn* as well as *prō'tĭan* still give it second place, but it is a strong rival and likely to win, see HERCULEAN.

**protest.** Verb *prŏtĕ'st*, noun *prō'tĕst*; see NOUN AND VERB ACCENT. As a transitive verb *protest*, in British idiom, means either (*a*) to state formally or solemnly (something about which a doubt is stated or implied), e.g. *he* *protested his innocence,* or (*b*) to refuse to accept (a bill of exchange). In the sense of to protest against (*the students will probably protest the decision*) it is recognized by the OED Supp. without comment, but is still an Americanism; to a British reader such a sentence as the following cries out for the insertion of *against*: *It has been an impressive piece of passive resistance, protesting the rigid pass laws, police shooting, and the arrests of African leaders.*

**protestant,** when used as adjective or noun without reference to the specialized sense in religion, is often pronounced *protĕ'stănt* for distinction.

**protocol** has travelled a long way from its original meaning of the first leaf glued on to a manuscript. As a term of diplomacy it is now used in two very different senses. One is for an agreement that supplements, amends, or qualifies an existing treaty, or deals with some temporary aspect of it. The other is for the ceremonial etiquette observed in diplomacy. The second is the sense in which it is popularly understood; the word seldom appears in the newspapers except in connexion with some delicate procedural question in diplomatic relations. *A long sigh of relief floated down the corridors of the Foreign Office yesterday when advices arrived from Gravesend that Nina was safely on board ship. So long as she remained incarcerated in the Soviet Embassy there was a danger of agonizing problems of p. and extraterritoriality.* | *Meanwhile Mr. Khrushchev was conforming to his usual p. on these occasions by making a threatening speech attacking the country which his guest represented.*

**prototype.** See -TYPE.

**protrude** makes -*dent* and -*sive* (protruding), -*sible* (able to be protruded) and -*sile* (able to be protruded and withdrawn).

**prove** makes -*vable*; see MUTE E. *Proved*, not *proven*, is the regular p.p., the latter being properly from the verb *preve* used in Scotland after it had given way to *prove* in England; cf.

*weave woven, cleave cloven.* Except in the phrase *not proven* as a quotation from Scots law, *proven* is better left alone.

**provenance, provenience.** The word is, and will doubtless continue to be, in literary and artistic use only. It is therefore needless to take exception to the first much better known form on the ground that it is French and try to convert users to the second, even if it is better in itself.

**provided (that)** is better than *providing* as an introduction to a proviso. The following examples show that care is needed in substituting it for *if*: *Ganganelli would never have been poisoned provided he had had nephews about to take care of his life.* | *The kicks and blows which my husband Launcelot was in the habit of giving me every night, provided I came home with less than five shillings.* | *She and I agreed to stand by each other, and be true to old Church of England, and to give our governors warning, provided they tried to make us renegades.* | *The chances are that the direction to proceed to Vladivostok at all costs, provided such instruction were ever given, may have been reconsidered.*

It will be agreed that *if* should have been written in all, and the object-lesson is perhaps enough. Those who wish for an abstract statement in addition may find that the following test, applied to each of the examples, will compel their rejection: A clause introduced by *provided* must express a stipulation (i.e. a demand for the prior fulfilment of a condition) made by the person who in the main sentence gives a conditional undertaking or vouches conditionally for a fact.

**province.** For synonyms, see FIELD.

**provost.** Provosts may be found in the Church, in education, in local government, and in the armed services. In the more modern dioceses of the Church of England, where the cathedral is also a parish church, the provost corresponds to the dean in the older; he is head of the chapter, and has the same style and status as a dean, and is also a parish priest. As an academic term provost (instead of the more usual *master*) is the title of the head of three colleges at Oxford (Oriel, Queen's, and Worcester) and one at Cambridge (King's), and also of the resident head of the governing body of Eton College (*The Provost and Fellows*). In local government a provost is the Scottish equivalent of an English *mayor*. In the armed services it is the designation of certain officers with police duties, e.g. *p.-marshal*. A provost of the last kind is called *prŏvō*; the others are pronounced as spelt.

**proxime accessit.** Pl., used in naming more than one, *proxime accesserunt* (*ăksĕsēr'ŭnt*).

**prox(imo).** See COMMERCIALESE.

**prudent(ial).** While *-ent* means having or showing prudence, *-ial* means pertaining to, or considered from the point of view of, or dictated by, prudence. To call an act *-ent* is normally to commend it; to call it *-ial* may be to disparage it: the prudential act may prove to have been dictated by a mistaken idea of what was prudent; e.g. a prisoner's decision not to go into the witness-box. But the difference is often neglected, and *-ial* preferred merely as a LONG VARIANT.

**prunella.** For the meaning of *leather or* (usually misquoted *and*) *p.*, see LEATHER 2.

**prurience, -cy.** There is no differentiation; *-ence* is recommended; see -CE, -CY.

**pry.** See PRISE.

**ps-.** With the advance of literacy and of the speak-as-you-spell movement (see PRONUNCIATION 1) it might be expected that the pronunciation of the *p* in words beginning thus would be restored except in *psalm* and its family, e.g. in the compounds of *pseud(o)-* and such important words as *psychical* and *psychology*. The OED describes the dropping of the *p* sound as 'an unscholarly practice often leading to

ambiguity or to a disguising of the composition of the word'. It may be significant that the COD now gives the *ps-* pronunciation first, unlike the SOED (1933) which preferred *s-*. But *ps-* is a long way off victory; the normal pronunciation is still *s-*. Cf. PN- and PT-.

**psephologist.** See POLLSTER.

**pseudonym.** See NOM-DE-GUERRE.

**psychic(al).** Both forms have been and are in common use in all senses, but *-al* is tending to prevail, partly perhaps as corresponding in form to the frequent antithesis *physical*, and partly because *-ic* is now ordinarily used to describe a person who possesses extra-sensory perception.

**psychological moment.** The original German phrase, misinterpreted by the French and imported together with its false sense into English, meant the psychic factor, the mental effect, the influence exerted by a state of mind, and not a point of time at all, *das Moment* in German corresponding to our *momentum*, not our *moment*. Mistake and all, however, it did for a time express a useful notion, that of the moment at which a person is in a favourable state of mind (such as a skilled psychologist could choose) for one's dealings with him to produce the effect one desires. But, like other POPULARIZED TECHNICALITIES, it has lost its special sense and been widened till it means nothing more definite than the critical moment or the nick of time, to which as an expression of the same notion it is plainly inferior. It should be avoided in the extended sense as a HACKNEYED PHRASE, and at least restricted to contexts in which *psychological* is appropriate; see also IRRELEVANT ALLUSION. Three examples follow, going from bad to worse: *It is difficult to believe that grievances which have been spread over many years have suddenly reached the breaking-point at the precise p. m. when the Franco-German settlement was reaching its conclusion.* | *There is a feeling that the* *p. m. has come to fight with some hope of success against la vie chère. | Everything goes right, no sleeping calf or loudcrowing cock grouse is disturbed at the p. m., the wind holds fair.*

**psychopathic, psychotic, neurotic.**
No brief definition of these terms would be likely to command universal assent; the disorders of the human mind are too multifarious, and still too mysterious, to admit of sharp classification. This is specially true of *psychopathic*, which, it has been said, 'has been used for years as a convenient psychiatric waste-paper-basket for cases difficult to classify'. (The Mental Health Act 1959 tries to remove this reproach by defining the condition as a persistent disorder of personality which results in abnormally aggressive or seriously irresponsible conduct.) But there seems to be fairly general agreement that *psychotic* alone of the three can properly be used of mental derangement amounting to insanity. In the language of analogy it has been said that 'in the psychotic we suppose that there has been some radical breakdown in the machinery; in the neurotic we suppose that it is working badly, though perhaps only temporarily; in the psychopath we suppose that the machinery was built to an unusual pattern or is faulty'.

**pt-.** In *ptarmigan* and *ptomaine*, and in *Ptolemy* and its derivatives, the *p* is always silent. In other *pt-* words the OED favours its being sounded, and the COD puts this first in *ptosis* and the *ptero-* compounds, though whether in fact this feat is ordinarily attempted in these rarely spoken words may well be doubted. Cf. PN- and PS-.

**ptomaine.** Pronounce *tōmā'n*. The OED allows that the *p* is not sounded, but stigmatizes the two-syllable pronunciation as illiterate. But, as with *cocaine*, it is impracticable to maintain the three-syllable *tōmā'in*. (Cf. -IES, EIN.)

**publicize.** We have long had a word

*publish* and we once had a word *publicate* but have forgotten it. *Publicize* is new; the first example quoted by the OED Supp. is dated 1928. *P.* can justify its existence only if it is used in a sense different from that of *publish*. To *publish* is to make available to the public; to *publicize* should mean to advertise what has thus been made available. Unfortunately these NEW VERBS IN -IZE seem to have an attraction that makes for their use merely as synonyms for older words, and there are signs that *publicize* is no exception to the operation of this Gresham's Law.

**puisne** from *puis né*, born later and so inferior, is the same word as *puny* (undersized) and is so pronounced. A *puisne judge* is any judge of the High Court of Justice other than (by statute) the Lord Chancellor, the Lord Chief Justice, and the Master of the Rolls, and (by custom) the President of the Probate, Divorce, and Admiralty Division.

**puissant.** The disyllabic *pwĭ'sănt*, the older pronunciation, is recommended, the word itself being archaic.

**pummel.** See POMMEL.

**pun.** The assumption that puns are *per se* contemptible, betrayed by the habit of describing every pun not as *a pun*, but as *a bad pun* or *a feeble pun*, is a sign at once of sheepish docility and desire to seem superior. Puns are good, bad, and indifferent, and only those who lack the wit to make them are unaware of the fact. See PARONOMASIA.

**punctuation.** See STOPS.

**pundit.** In serious use as the title of a learned Hindu, the original spelling, *pandit*, is preferred. *Pundit* and *punditry* have passed into the language as facetious and slightly contemptuous terms for learned specialists and the characteristics attributed to them by more ignorant people.

**pupa.** Pl. *-ae.*

**pupil.** For the derivatives *pupil(l)age*, *pupil(l)ary*, etc., the double *l* is recommended; see -LL-, -L-.

**purchase.** As a substitute for *buy* (goods for money), *p.* is to be classed among stylish words (see WORKING AND STYLISH WORDS), but in figurative use (*p. victory by sacrifice* etc.) it is not open to the same objection.

**puritanic(al).** The long form is commoner, and there is no perceptible difference in meaning. The existence of a third adjective *puritan*, which suffices for the mere labelling function (= of the puritans), makes the -*ic* form even less useful than it might otherwise be, and it seems to have been squeezed out; see -IC(AL).

**purlieu** is a WORSENED WORD.

**purport.** 1. Noun *pur'port*, verb *purpor't*; see NOUN AND VERB ACCENT.
2. Meaning. The word is one that, whether as noun or as verb, requires cautious handling. The noun may be said to mean 'what appears to be the significance' (of a document, an action, etc.); its special value is that it is noncommittal, and abstains from either endorsing or denying the truth of the appearance, but lightly questions it. When such an implication is not intended, the word is out of place, and *tenor, substance, pith, gist, drift*, or other synonym, should be preferred. But NOVELTY-HUNTING discovers *p.* sometimes in place of *scope* or *purview*, and even of *purpose*. Read *purview* or *scope* in: *In 'A Note on Robert Fergusson' he touches a theme outside the general purport of the book.*

As to the verb, there are certain well-defined idiomatic limitations on its use, one of which, in an ugly development, is sometimes neglected. This development is the use of the passive, as in: *Professor Henslow compiles from published works the information as to the other world, Christian life and doctrines, the nature of man, etc., purported to be conveyed in communications from 'the other side'.* | *Many extracts from speeches* purported to have been made *by Mr. Redmond are pure

fabrications.* | *An alternative, briefer, and much more probable account of the Controversial Parts of the Dialogue* Purported to be Recorded *in the Republic of Plato.* Though the verb is an old one, there is in the OED quotations only one passive use, and that dated 1894. The above extracts are doubtless due to the corrupting influence of the DOUBLE PASSIVE; that construction is especially gratuitous with *p.*, the sense of which fits it to serve, in the active, as a passive to *suppose, represent*, etc. In all the extracts *supposed* would stand; pretentiousness has suggested *purport* as a less familiar and therefore more imposing verb, and ignorance has chosen the wrong part of it (*purported*) instead of the right (*purporting*).

The first idiomatic limitation, then, is that the verb, though not strictly intransitive only, should never be used in the passive. The second is that the subject, which is seldom a person at all, should at any rate not be a person as such—only a person viewed as a phenomenon of which the nature is indicated by speech, actions, etc., as the nature of a document is indicated by its wording. Normal subject: *The story purports to be an autobiography.* Legitimate personal subject: *The Gibeonites sent men to Joshua purporting to be ambassadors from a far country.* Illegitimate personal subject: *She purports to find a close parallel between the Aeschylean Trilogy and* The Ring, *but she does it by leaving out* Siegfried *altogether.* | *Sir Henry is purported to have said 'The F.A. are responsible for everything inside the Stadium'.* | *From observation I would say that the strength of the demand for a Republic is not what the Nationalists purport it to be.*

**purpose,** n. *It serves very little p. to ask the Chancellor of the Exchequer to give a little more in this direction or in that.* There are three idioms: Be *to the, to* (very) *little, to no, p.*; Do something *to some, to much, to no, to* (very) *little, p.*; Serve *the, my,* etc., *no, p.* These should not (see CAST-IRON IDIOM) be confused. *Serve very little p.* is a mixture of the third with one of the others.

**purposive** ('an anomalous form'—
OED) is a word of the kind described
as 'spurious' in the article HYBRIDS
AND MALFORMATIONS; the Latin suffix
-*ive* is unsuited to the delatinized and
anglogallicized *pur*-, which represents
but conceals the Latin *pro*. *Purposeful*
in some contexts, and *purposed* in
others, should meet most needs, and
there are *deliberate*, *designed*, *calculated*, and many more synonyms. But
the psychologists have now adopted
*purposive* for a colourless word to
mean the opposite of *aimless*, without
the element of resolution or determination that most people feel to be contained in *purposeful*. *Pre-Freudian
psychologists supposed that only some
behaviour has some purposive explanation. All behaviour, Freud said, has
some purposive explanation*. That might
be a useful differentiation if it were not
that other writers, always eager to
seize on technical terms, especially of
psychology, now use -*ive* in contexts
where -*ful* would be better.

**pur sang.** *The men who direct it are*
pur-sang *mandarins, trained in all the
traditions of a bureaucracy which lives
not for, but on, the people.* If one is
brave enough to use the French words,
one should be brave enough to place
them as such—*are mandarins pur sang.*

**pursuant(ly).** See QUASI-ADVERBS.

**pursuivant.** Pronounce *per'swivănt.*

**purview.** For synonyms see FIELD.

**put(t).** According to the OED the
pronunciation *pŭt*, with or without the
additional -*t*, and with verbal forms
*putted* instead of *put*, universal in golf,
in weight-putting is confined to Scotland.

**pygmean, -aean.** The first is recommended; see Æ, Œ. Pronounce *pĭg-
mē'ăn.*

**pygmy, pi-.** For the reason why *py-*
is the better, see GYPSY.

**pyramidal.** Pronounce *pĭră'mĭdl*, not
*pĭrămĭ'dl.*

**pyrites.** Pronounce *pĭrī'tēz.*

**pyrrhic.** The *p. dance* was the Spartan
war-dance, said to have originated in
Crete; the origin of the name is uncertain. A *p. foot* is a term of prosody,
meaning the metrical foot used in the
dance.(ᵁᵁ). A *p. victory* is a victory won
at too great a cost, so called from the
remark attributed to Pyrrhus king of
Epirus after his victory over the
Romans at Asculum in 279 B.C. 'One
more such victory and we are undone.'

# Q

**qua** is sometimes misused like other
Latin words; see E.G., I.E., PACE, RE, VIA,
VIDE. The real occasion for the use of
*q.* occurs when a person or thing
spoken of can be regarded from more
than one point of view or as the holder
of various coexistent functions, and a
statement about him (or it) is to be
limited to him in one of these aspects,
*Qua lover he must be condemned for
doing what qua citizen he would be condemned for not doing*; the lover aspect
is distinguished from another aspect in
which *he* may be regarded. The two
nouns (or pronouns) must be present,
one denoting the person or thing in all
aspects (*he*), and the other singling out
one of his or its aspects (*lover*, or *citizen*). *Qua* is wrongly used in the following two extracts: *The root of this
conviction,* qua *Great Britain, is the
preposterous fiction of the military value
of the Ulster volunteers; and the root
of this conviction,* qua *Ireland, is the
shameful and cruel bamboozling of a
section of my unfortunate fellow-Provin-
cials into the delusion that few soldiers
and no artillery will be available against
them.* | *The familiar gentleman burglar
who, having played wolf to his fellows*
qua *financier, journalist, and barrister,
undertakes to raise burglary from being
a trade at least to the lupine level of
those professions.* In the first of these,
a gross misuse, Great Britain, and
Ireland, are not aspects of the conviction, but things as different from
a conviction as an hour from a walking-
stick. Perhaps the writer was confusing

*qua* with *quoad* (so far as . . . is concerned). In the second, a much less definite offence, financier etc. do not give aspects of the man to be distinguished from other coexistent aspects, but merely successive occupations; the fault is that the occasion does not justify the substitution of the very precise *qua* for the here quite sufficient *as*.

**quad** in all compound words of which it forms a part is pronounced as it is when used as a CURTAILED WORD (*kwod*), though the dictionaries still admit *kă-* as an alternative for *quadrille*. Cf. -QUAT-.

**quadroon.** See MULATTO 2.

**quagmire.** Pronounce *kwăg-*. The COD does not admit *kwŏg-* even as an alternative, though some dictionaries allow it a second place.

**quality. 1.** For 'has the defects of his qq.' see HACKNEYED PHRASES. **2.** The adj. is *-itative*, not *-itive*; see QUANTITATIVE.

**qualm.** Pronounce *kwahm*. It seems clear that its rival *kwawm* has lost the battle.

**quandary.** Now that the word has come into common use, the pronunciation *kwŏndār'i* ('the original stressing'—OED) is less often heard than *kwŏ'ndări*. See RECESSIVE ACCENT, and cf. *boundary* and many other words in which *-ary* is unstressed.

**quantit(at)ive.** The long form is right. But the use of *quantitive* for quantitative is more frequent, and perhaps less of a mere inadvertence, than that of *qualitive* for qualitative, and *authoritive* for authoritative. *And what is true of railway traffic is true, so far as this* quantitive *economy of labour is concerned, of all industry in which mechanical power and labour-saving appliances are employed*; and see LONG VARIANTS s.f. In the light of the Latin words *tempestivus, primitivus*, and *adoptivus*, anomalous in different ways, it would be rash to say that *quantitive, qualitive*,

and *authoritive* were not defensible forms; but at any rate good English usage is against them.

**quantity. 1.** *A negligible q.*, a POPULARIZED TECHNICALITY and a GALLICISM, is often used where *negligible* by itself gives all that is wanted. **2.** As a term of prosody *quantity* is applied chiefly to Greek and Latin verse, the metres of which are based on 'quantity', i.e. on the length or shortness of sounds or syllables measured by the time taken to pronounce them; the long indicated by a macron (–) and the short by a breve (◡). The difference between the long and short vowels in English, though indicated in the same way (*ā* and *ă*; *ē* and *ĕ* etc.), is qualitative and has no true analogy with the difference between the long and short quantities of classical prosody.

**quarter, n. 1.** Constructions. *For a q. of the price, for q. of the price, for a q. the price, for q. the price*, are all blameless English. *After three and a q. centuries*, or *three centuries and a q.*? See HALF and THREE-QUARTERS. **2.** For the use of the word in such a sentence as *It is learned in official qq.* see PERSONIFICATION.

**quartet(te)** etc. The *-et* forms should be used.

**quarto.** See FOLIO. Pl. *-os*; *-o(e)s* 6.

**quasi.** Pronounce *kwāsī* not *kwahzē*; see LATIN PHRASES.

**quasi-adverbs.** *The Leninist maxim of taking a man by the hand* preparatory *to taking him by the throat*. From a narrowly grammatical point of view, the word should be *preparatorily*; but it never is, except in the mouths of those who know just enough grammar to be timid about it. The adjective is loosely attached to the action described in 'taking a man by the hand'. Most of those who would correct, or be tempted to correct, *preparatory* to *preparatorily* feel no temptation to write *accordingly*, instead of *according, as* or to, because the latter is so familiar as not to draw their attention. See also UNIDIOMATIC -LY, in which words of a slightly different kind are considered. It should be observed that it is only certain adjectives with which the use is idiomatic; for instance, *He did it contrary to my wishes*, but neither *opposite to* nor *different from* them. A few of the adjectives concerned are: *according* and *pursuant*; *contrary*; *doubtless*; *preliminary* and *preparatory*; *irrespective* and *regardless*. Another pair of adjectives exhibiting the same arbitrary distinction of idiom as that between *contrary* and *opposite* is OWING and DUE. PREVIOUS *to* and PRIOR *to* are grammatically blameless, but that does not justify their use as substitutes for *before* because they are thought to be grander or more genteel.

**quassia.** Of the three pronunciations *kwăs'ya*, *kwă'sha*, and *kwŏ'sha*, the first seems likely to prevail.

**-quat-.** Pronunciation varies capriciously in different words, and in some is not yet settled. Reference to a selection of modern dictionaries suggests the following conclusions. *Kwŏt* is established in *quatrain* and *squat* and preferred (to *kwăt*) in *quaternary* and *quaternion*. *Kwăt* is established in *aquatic* and preferred (to *kwŏt*) in *quatercentenary*. *Kăt* is established in *quatorzain* and preferred (to *kāt* or *kwăt*) in *quatrefoil*. *Kwaht* is preferred (to *kwăt*) in *quattrocento*.

**queer.** It has become dangerous to apply this apparently innocent adjective to a person, since, as a noun, it is now a slang euphemism for a homosexual.

**question.** 1. For LEADING QUESTION, see that article. 2. For *previous q.*, see PREVIOUS 3. 3. For order of words in indirect questions (*He asked what was he to do* etc.), see INDIRECT QUESTION. 4. For the question-mark wrongly and rightly used, see STOPS. 5. For *beg the question*, see PETITIO PRINCIPII 6. *Question as to*. This ugly and needless but now common formula is discussed

and illustrated under AS 3; but it is worth while to repeat here that it is at its worst when *question* has *the*, as in: *When the nation repudiated Papal authority*, the q. *naturally arose* as to *who were to have the endowments.* | *From time to time there appears in the weekly Revenue Statement an item on the expenditure side of 'War Loans and Exchequer Bonds', and* the q. *has cropped up* as to *its meaning.* The reason why the formula is worst with *the* is that you do not say *the* instead of *a* q. unless either it is already known what q. is meant or you are about to supply that information at once. To fulfil that expectation—that is to explain what the q. *is*, not what it concerns—you must use not the *as to* phrase but an interrogative clause in simple apposition with q. (*the q. who was to have*), or, if a noun is to be used instead of such a clause, attach that noun to q. by *of* (*the q. of its meaning*); *of* is the preposition that expresses identity, as in *the city of Exeter, the crime of murder*.

**questionnaire** is too recent an importation to be in the original OED (1904). In the 1933 Supp. the first example of its use is dated 1901. It is a pity that we could not be content with our native *questionary* (called 'rare' by the OED but quoted from the 16th and 17th cc.); *commentary, glossary, dictionary*, and *vocabulary*, with many less common words, would keep it well in company. It is still used occasionally, but since *-aire* is preferred by government departments and POLLSTERS and others with a thirst for information, *-ary*'s chances against it are poor. The anglicized pronunciation *kwěschŏnăr'* has not yet ousted *kěstiŏnăr'* but will no doubt do so eventually.

**question mark.** See STOPS.

**quiet,** n., *quietness, quietude*. The first is much more used than the others; it is possible to distinguish roughly the senses to which each is more appropriate, but often there is a legitimate choice between two points of view.

*Quiet* is a state of things or an atmosphere: *A period of quiet followed*; *Seeking quiet and rest*. *Quietness* is a quality exhibited by somebody or something: *The quietness of his manner, of rubber tires*. *Quietude* is a habit or practice: *Quietude is out of fashion in these days*. An example of each follows in which (if what has been said above is true) one of the others would have been preferable: *How becomingly that self-respecting quiet sat upon their high-bred figures* (quietude); *Enjoying the fruit of his victory, peace and quietness* (quiet); *The quietude of the meadows made them his favourite resorts* (quietness or quiet).

**quieten.** See -EN VERBS FROM ADJECTIVES.

**quire, choir.** See CHOIR.

**quite. 1.** Excessive use of *q.* often amounts to a mannerism, and many writers would do well to convict and cure themselves of it by looking over a few pages or columns of their work. **2.** The colloquial formula 'quite all right' is an apparent PLEONASM, *quite* and *all* being identical in sense; 'quite right' is all right, and 'all right' is quite right, but 'quite all right' is all quite wrong, unless indeed *all right* is here used in its sense of adequate but no more, and *quite* is added for reassurance. **3.** *Quite* (*so*). Many people are in the habit of conveying their assent to a statement that has just been made to them in talk by the single word *quite*, instead of the usual *quite so*; perhaps *quite* sounds to them neater, conciser, than *quite so*. What they do not realize is that choice between the two is sometimes open to them, but by no means always; used in wrong places, *quite* is an example of SLIPSHOD EXTENSION. Three specimen exchanges will make the matter clear: (a) He seems to be mad.—Quite. (b) To demand that Englishmen should act on logic is absurd.—Quite (so), but . . . (c) Well, anyhow, he did it.—Quite so, but the question is . . . In (a), *quite so* would be out of place, because what is to be qualified by *quite* is simply the word *mad*, understood directly from

what precedes. In (b), choice is open; *quite* will amount to *quite absurd* (as in (a)); *quite so* will amount to *it is quite as you say*; and the general effect of each is the same. In (c), *quite* would be wrong, because the other speaker's words do not supply anything, as in (a) and (b), for *quite* to qualify; the sense is clearly not *he quite did it*, but, as in the second alternative of (b), *it is quite as you* say. The bad use, in actual life, is well shown in this scrap of lawcourt examination: *There was no power in anyone to bring the child back?—Quite*. The vogue of *quite* in this sense is, however, waning. Other substitutes for *yes* have been challenging its supremacy—DEFINITELY, for instance, and *That's correct*. **4.** It is interesting that we now use *quite* colloquially, and generally with a special intonation, to mean not quite. *Did you enjoy the play? Yes, quite*. No one would interpret the answer as unqualified praise. **5.** For *quite a few* see FEW.

**quiz** as a verb meaning to make sport of was popular in the 18th c., and as a noun was applied alike to the quizzer, the victim, and the process; a monocle was called a *quizzing-glass* because for a man, like a lorgnette for a woman, it was a powerful aid to silent quizzing of the more malicious sort. In this sense the word went out of fashion, but was revived in America towards the end of the 19th c. with the less colourful meaning of any kind of oral interrogation or viva-voce examination, and has now become on both sides of the Atlantic one of the most used of the monosyllabic products of HEADLINE LANGUAGE. As a term for the general-knowledge competitions that have become a popular form of entertainment it is useful, indispensable indeed. But the aura of flippancy that still hangs about it makes it unsuitable for the more serious purposes to which it is increasingly put. Some may still feel that there is a touch of incongruity, if not indecency, in giving the headline 10-HOURS QUIZ to the news that a man suspected of murder

has been subjected to prolonged interrogation by the police.

**quondam.** See LATE etc.

**quorum.** Pl. *-ums*, not *-a*; see -UM.

**quotation.** Didactic and polemical
writers quote passages from others to
support themselves by authority or to
provide themselves with something
to controvert; critics quote from the
books they review in illustration of their
estimates. These are matters of business
on which no general advice need be
offered. But the literary or decorative
quotation is another thing. A writer
expresses himself in words that have
been used before because they give his
meaning better than he can give it
himself, or because they are beautiful
or witty, or because he expects them
to touch a chord of association in his
reader, or because he wishes to show
that he is learned or well read. Quotations due to the last motive are invariably ill advised. The discerning
reader detects it and is contemptuous;
the undiscerning is perhaps impressed,
but even then is at the same time
repelled, pretentious quotations being
the surest road to tedium. The less
experienced a writer is, and therefore
on the whole the less well read he is
also, the more is he tempted to this
error. The experienced knows he had
better avoid it; and the well-read, aware
that he could quote if he would, is not
afraid that readers will think he cannot.
Quoting for association's sake has
more chance of success, or at any rate
less certainty of failure; but it needs
a homogeneous audience. If a jest's
prosperity lies in the ear of him that
hears it, so too does a quotation's; to
each reader those quotations are agreeable that neither strike him as hackneyed nor rebuke his ignorance by
their complete novelty, but rouse dormant memories. Quotation, then,
should be adapted to the probable
reader's degree of cultivation, which
presents a very pretty problem to those
who have a mixed audience to face;
the less mixed the audience, the safer
is it to quote for association. Lastly,
the sayings wise or witty or beautiful

with which it may occur to us to adorn
our own inferior matter, not for business, not for benefit of clergy, not for
charm of association, but as carvings
on a cathedral façade, or pictures on
the wall, or shells in a bower-bird's
run, have we the skill to choose and
place them? Are we architects, or bric-
à-brac dealers, or what?

Enough has perhaps been said to
indicate generally the dangers of quoting. A few examples follow of oddities
that may serve as particular warnings;
see also IRRELEVANT ALLUSION and
MISQUOTATION.

### PRETENTIOUSNESS

*In the summer of 1867 England re-
ceived with strange welcome a strange
visitor.* 'Quis novus his nostris suc-
cessit sedibus hospes?' *Looking forward into the future we may indeed
apply yet other words of Dido, and say
of the new comer to these shores* 'Quibus
ille jactatus fatis!' *It was the Sultan of
Turkey who came to visit England.*

### MANGLINGS

*It may seem somewhat unfair to quote
the saying of the old Latin poet,* 'Montes
parturiunt, ridiculus mus est', *in relation to the Government's achievements in
matters of domestic legislation.* (Something seems to have happened to the
old Latin poet's metre and tense.) / *His
treatment of the old, old story of the
Belgian franc-tireur is typical.* 'L'animal est très méchant, il se défend quand
on l'attaque.' (Something has happened
to the French poet's rhyme, as well as
his metre.) / *Here again, however, there
was a fly in the amber—the incoming of
the Italians.* (A fly in amber, or a fly in
the ointment—what can it matter?) /
*The Chancellor of the Exchequer finds
himself on the horns of a quandary.* (A
quandary is no doubt an awkward
place to get out of but not because
of a fear of impaling oneself on any
horns.)

### QUOTATION SANDWICH

*Yet if we take stock of our situation
today, even those of us who are 'fearful
saints' can afford 'fresh courage' to*

'take'. | The 'pigmy body' seemed 'fretted to decay' by the 'fiery soul' within it. (Original: A fiery soul which, working out its way, Fretted the pigmy body to decay.)

### FOREIGN OIL AND ENGLISH WATER

Who will be pleased to send details to all who are interested in strengthening l'entente cordiale. (Read the entente cordiale.) | Even if a change were desirable with Kitchener duce et auspice. | Salmasius alone was not unworthy sublimi flagello. | The feeling that one is an antecedentem scelestum. | The clergy in rochet, alb, and other best pontificalibus.

### CLUMSY ADAPTATION

But the problem of inducing a refractory camel to squeeze himself through the eye of an inconvenient needle is and remains insoluble. | Modern fashions do not presuppose an uncorseted figure; that way would modish disaster lie. | Gossip on a subject which is still on the knees of the future.

**quotation marks.** See STOPS.

**quote. 1.** Q. makes -table; see MUTE E. **2.** The devil can q. etc.; q. should be cite; see MISQUOTATION. **3.** Quote, n., a CURTAILED WORD for quotation, is a colloquialism much used by gossip columnists for statements extracted by them from interviewees; in their jargon So-and-so gave me this quote means So-and-so said I might quote this as his opinion. Quotes, n. pl., for quotation marks may be left to those whose occupation makes such a shortening indispensable.

**quoth, quotha.** See ARCHAISM.

**quotient.** For intelligence q. see I.Q.

**q.v.** See VIDE.

# R

**rabbit** makes -iting, -ity; see -T-, -TT-. For Welsh r. see TRUE AND FALSE ETYMOLOGY.

**rabies.** Pronounce rā'bēz. See -IES, -EIN.

**rack and ruin.** The OED, though it calls rack a variant of wrack, recognizes this spelling; it is no doubt helped by the visible alliteration. The normal uses of the words are rack for driven clouds and wrack for seaweed cast up on shore. See SIAMESE TWINS.

**racket.** The implement for striking a ball and the game of rackets are properly so spelt, not racquet(s). The word meaning disturbance, uproar, etc., is of different origin. Its use in the sense of a lucrative enterprise of questionable honesty was known in England at the beginning of the 19th c.; it seems to have then died, but was born again in New York a hundred years later, fostered in Chicago, and re-exported to Britain, where it has become thoroughly at home. See also -EER.

**raddle.** See RUDDLE.

**radiance, -cy.** The second is rare, but kept in being as metrically useful or rhetorically effective; -CE, -CY.

**radius.** Pl. -ii (-iī). For synonyms in the sense scope etc., see FIELD.

**radix.** Plural -ices (pronounce rā'-dīsēz), see -EX, -IX, and, for the quantity, -TRIX.

**railroad** as a noun is U.S. for railway and, as a verb, for to send by rail or, by extension, to drive something (e.g. a legislative measure) ruthlessly forward.

**rain or shine**, as a phrase for 'whatever the weather', is mentioned in few dictionaries, and has an Irish sound. It is quoted from Dryden—Be it fair or foul, or rain or shine—in the Century Dictionary.

**raise.** Raising and removing one's hat are FORMAL WORDS for taking it off. Raise (v.) in the sense of breeding animals or bringing up children is an Americanism. Raise (n.) was originally U.S., but the OED Supp. recognizes its now general use for an increase of pay (though rise is still more usual in

Britain) and for an increase in a bid at bridge or a stake at poker.

**raison d'être.** See FRENCH WORDS. How not to use it can hardly be better shown than in: *It has been proposed by the Liberal Nonconformist M.P.s that it shall be sufficient for the Sovereign to affirm a belief in the Protestant Faith without pledging himself to be a member of the Church of England; the raison d'être is obvious, but . . .*

**Ralph** is a contraction of *Radulf*. The pronunciation *Rāf* is a survival from the 17th c. spelling *Rafe*. It is now disappearing in favour of *Rălf*, though still, according to the exponents of U AND NON-U principles, one of the shibboleths by which the U-speaker may be recognized.

**range,** n. For synonyms in the sense *scope*, see FIELD.

**range,** v. 1. R. makes *-ging*, *-geable*; see MUTE E. 2. *Gratuities ranging from 10 lire for each of the singers in the Sistine Chapel choir up to much larger sums for higher officials.* If one has not provided oneself with figures for both extremes, one should not raise expectations by using *r. from . . . to.* It is as bad as saying 'Among those present were A, B, *and others'.*
3. *Range oneself* for settle down (*He had no intention of marrying and ranging himself just yet*) is a bad GALLICISM.

**ranunculus.** Pl. *-luses* or *-li (-ī)*; see LATIN PLURALS 1.

**rapport,** formerly common enough to be regarded and pronounced as English (*răpor't*), may now perhaps be called again a FRENCH WORD; it is seldom heard except in the phrase *en r.,* where it retains its French pronunciation.

**rapt,** meaning originally carried off, raped, snatched away, but now usually absorbed or intensely concentrated, has perhaps been affected by the identical sound of *wrapped* or *wrapt*, though the similar extension of the meaning of *ravish* is enough to show that such an explanation is not necessary. The best

known passage (*Thy rapt soul sitting in thine eyes*) has doubtless helped. A concordance to Milton supplies also: *Wrapped in a pleasing fit of melancholy | Thus wrapped in mist of midnight | Rapt in a balmy cloud.*

**rara avis** is seldom an improvement on *rarity*; see IRRELEVANT ALLUSION.

**rarebit.** See TRUE AND FALSE ETYMOLOGY.

**rarefaction, rarefy.** So spelt (in contrast with *rarity*), but pronounced *rārĕ-* (in contrast with *rarely*). *Rarefaction*, not *-fication*, is the correct as well as the usual form, the Latin verb being *rarefacere* (not *-ficare*).

**rase.** See RAZE.

**raspberry.** See RHYMING SLANG.

**ratcatcher.** A *r.* was originally a self-employed man, usually a rustic, who earned his living by what would now be called LIQUIDATING, or at any rate reducing, the rat population in or around houses, farm buildings, etc. Later the word was used to denote informal attire in the hunting-field, e.g. a tweed hacking-jacket rather than the black covert coat worn by gentlemen who, though regular subscribers to the Hunt, do not aspire to pink. In its original sense the word is now largely obsolete. Most British rats now meet their doom at the hands of a local authority official with the designation of *pest control officer* or *rodent operator*.

**rate,** n. For *r.* in the sense *impost*, see TAX. For *ratable* or *rateable* see MUTE E.

**rather.** 1. *R. is it* etc. 2. *R. superb* etc. 3. *I had r.* 4. *Dying r. than surrender.* 5. *Rather'.*
1. *R. is it* etc. Towards the end of the long article INVERSION will be found a section headed *Yet, especially, rather*, deprecating the use of inversion after a linking *rather*, e.g. in *The responsible leaders of the Opposition have abandoned the view that another General Election would 'probably but stereotype the last verdict'*; rather is it

*felt that* . . . (r. it is felt, *or* it is felt, r..).
It should be remembered, however,
with *r.*, that care is needed, in mending
or avoiding the inversion, not to put *r.*
where it might be interpreted as *somewhat*; to write *it is r. felt* in this
example would be worse than the inversion itself.

2. *R. superb* etc. *There is something*
rather delicious *in the way in which
some of these inventors ignore previous
achievements.* | *This was* r. a revelation. |
*While exercising generosity and kindli-
ness more than most the doctor* r. loved
*a quarrel.* What is the use of fine warm
words like *delicious* and *revelation* and
*love* if the cold water of *r.* is to be
thrown over them? 'R. agreeable' if
you will; 'r. surprising' by all means;
'r. enjoyed' certainly; but away with
*r. delicious, r. a revelation,* and *r. loved*!
Cf. *somewhat* (SOME 4).

3. *I had* r. is as idiomatic as *I would*
*r.*; *had* is the old subjunctive, = I
should hold or find, and is used with *r.*
on the analogy of *I had liefer* = I
should hold it dearer; see HAD I.

4. *Dying* r. *than surrender, He resigned* r. *than stifle his conscience,* etc.
The use of the infinitive after *r. than* in
such contexts is discussed in -ING 5 (c).

5. *Rather'* with the stress on the
second syllable, as a colloquial way of
expressing emphatic assent, is a typically British MEIOSIS, cf. *not 'alf, I don't
think.* It dates from the middle of the
19th c. and now sounds old-fashioned,
even in the field of juvenile slang
where it mostly flourished.

**ratiocinate** and its derivatives, as
exclusively learned words, may fairly
be pronounced *rătĭ-* rather than *răshĭ-*;
and that is now usual although the
OED gives only *răshĭ-*.

**ration.** Pronounce *ră-*. *Rā-* was formerly common and is still standard
U.S. But in the British armed services
it has always been *ră-*, and when the
civilian population were obliged to
accept the system they accepted the
pronunciation too.

**rationale** is the neuter of the Latin
adjective *rationalis,* and should therefore be pronounced *răshŏnā'lĕ*. But
confusion with such French words as
*morale* and *locale,* together with our
present fondness for giving a foreign
sound to the vowels in any word that
has a foreign appearance, has produced
the mispronunciation *-ah'l,* which
seems likely to have an undeserved
victory.

**rationalize** in the sense of to reorganize an industry so as to eliminate
wasteful production is recent; the first
example in the OED Supp. is dated
1927. It was greatly in demand in the
between-war period but has since been
elbowed out by the VOGUE WORDS
INTEGRATE, *streamline,* and *tailor.* An
even more recent use, still with us, is as
a term of psychology meaning to put
a veneer of reason over thought and
actions that are in fact emotional.

**ratlin, ratline, ratling.** The derivation is uncertain; but the last syllable
probably contains neither the word
*line* nor the participle termination *-ing.*
The spelling *ratlin* is perhaps, therefore, the best.

**rat-race.** *There is a tendency that the
independent schools might become an
academic* r.-r. 'The word has come into
great favour' comments *The Times.*
'Those who look round in anger constantly see this curious contest wherever they believe ill deeds to be done.
. . . There are rat-races, so the indignant say, in the world of entertainment
and the salesmanship of books and
papers, and now even in education.'
At a time when the word has not yet
reached the dictionaries, its meaning
can only be guessed; but this definition, offered to *The Times* by a correspondent, cannot be far out: 'Any
occupation or profession in which the
participant can find little purpose or
inspiration but by pressure of environment is bound to show signs of activity
and "get ahead".'
The origin of the usage is less easy to
guess. One suggestion is that it comes
from those research laboratories in
which a rat is put into a sort of tread-
mill and compelled to keep on walking

or running at the whim of the experimenter. That must certainly be a frustrating experience, but it seems to lack the element of rivalry that we should expect to find in a race. Another theory, perhaps more probable, is that we need look no further than the migratory habits of the Scandinavian lemmings, or the story of the Pied Piper of Hamelin: that millions of rodents scuttling and tumbling blindly on until in the end they plunge into the water to die are a perfect description of our city life today under the lure of the piper's tune *Material Prosperity*.

However that may be, few will question *The Times*'s conclusion 'If there were Pest Control Officers in the vocabulary as there are in the rodent world . . . they might attend to a usage that is indeed becoming tiresome.'

**ravel** makes *-lled, -lling*, etc.; see -LL-, -L-. The verb is curiously applied both to the tangling and the disentangling process though the disentangling (now commonly *unravel*) is generally *r. out*. *Must I r. out my weaved up follies?* The verbs that can mean either to deprive of, or to provide with, what is expressed by the noun of the same spelling (compare *will but skin and film the ulcerous place* with *skin 'em alive*) are not parallel, because with them the noun is the starting-point.

**raze, rase.** *Rase* is the older spelling, but *raze* now prevails. There is some tendency to use *rase* still for senses, such as erase or scrape off, that are now archaic; but the distinction corresponds to no difference of etymology, and *raze* should be the only form.

**re.** For the use of this telltale little word see COMMERCIALESE and IL-LITERACIES. A quotation or two follow: *Dear Sir,—I am glad to see that you have taken a strong line re the Irish railway situation. | Why not agree to submit the decision of the Conference re the proposed readjustment to the people so that they alone can decide? | Sir,— I have had sent me a cutting from your issue of the 14th inst., from which I* gather that reference had been made in a former issue to some alleged statements of mine re the use of the military during the recent railway dispute. | Sir,— There is another fact re above. Twice with, and twice without, italics. But the word has a respectable legal use. *In re John Doe deceased.*

**re-** (prefix). 1. Hyphening. In *re-*compounds the hyphen is usual before *e*, though seldom necessary, (*re-entrant, re-examine*, etc.) but not before other vowels (*rearmament, re-iterate, reorganize, reunion*). It is sometimes used when the compound follows the simple verb (*make and re-make, discussion and re-discussion*) and it is necessary when a compound such as *re-cover* = to put a new cover upon, *re-count* = count again, *re-form* = form afresh, *re-sign* = sign again, is to be distinguished from a better-known and differently pronounced old word (*recover* = get back, *recount* = narrate, *reform* = improve, and *resign* = relinquish). See HYPHENS 6.

2. Pronunciation. In Latin the *e* is short. That is not necessarily a guide to its pronunciation in English (see FALSE QUANTITY) but in this case we do in fact ordinarily conform. The main exceptions are (a) where the verb to which *re-* is prefixed begins with a vowel (for examples see above) and (b) to emphasize that the prefix is having a significant effect on a common verb, not only, as in the extreme examples given above, to reinforce the hyphen as a protection against misunderstanding, but often also where there is no risk of ambiguity (*recapture, recharge, regroup, retouch*, etc.). It seems, however, that there is a modern tendency to give the $\bar{e}$ sound to words that were formerly pronounced $\breve{e}$. Among words for which the SOED gave only $\breve{e}$ but at least one modern dictionary gives only $\bar{e}$ are *reclaim, regenerate, repair, repeal, replace, replenish*. This may be partly due to U.S. influence, especially to the American pronunciation of disyllabic *re-* words such as *research* and *recess*. See PRONUNCIATION 1.

**reaction.** *Darwin's observations upon the breeds of pigeons have had a* reaction *on the structure of European Society.* | *Any apparent divisions in this country, even the threat of a vote of censure, might have had its* reaction *on public opinion in Italy.* Owing to its use in chemistry, *r.* has become a POPULAR-IZED TECHNICALITY liable like other such terms to be used by SLIPSHOD EXTENSION where it is not wanted, e.g. where nothing more is meant than effect or influence or the simple action. This misuse is betrayed in the above quotations by the word *on*, which suits *action* etc., but does not suit *r.* except in senses in which it means more than any of those three. The senses of *r.* may be distinguished thus: 1. The process of reversing what has been done or going back to the status quo ante: *progress and r.; the forces of r.* 2. The recoil from unusual activity or inactivity, producing an equally unusual degree of the reverse: *extremes and r.; the r. from passion, despair, a cold bath.* 3. The second half of interaction, B's retaliation upon the first agent A, making up with action the vicissitudes of a struggle etc.: *after all this action and r.* 4. The reflex effect upon A of his own actions: *the r. of cruelty upon the cruel.* 5. The action called forth from B by A's treatment: *stimulus and r.; the r. of copper to sulphuric acid.* This last is the sense that covers the chemical use, and the one also that is often interchangeable with effect etc.; but *on* or *upon* is out of place with it; not *the r. of sulphuric acid on copper*, but either *the r. of copper to sulphuric acid* or *the action of sulphuric acid on copper.* Similarly, not the *r.*, but *the action* or *effect* or *influence*, of Darwinism *on* Europe and of English votes of censure *on* Italian public opinion. 6. A further extension of this too popular word is its use in the sense of opinion. *May I have your r. to this proposal?* This is excusable if what is wanted is an opinion given immediately without reflection. But if a considered opinion is asked for it is an example of slipshod extension at its worst. *It is understood that the Governor wishes to have further consultations with the authorities in London before formulating his final r. to the proposal.* We do not formulate our reactions; they formulate themselves unbidden. 7. *Chain reaction.* A dictionary definition is 'Chemical reaction forming intermediate products which react with the original substance and are repeatedly renewed.' This is much in favour as a more suitable metaphor in these days than the old-fashioned 'snowball'. *If arbitration awarded, say, a four or five per cent increase, the Government could not possibly accept this without triggering off a c.-r. of comparable increases throughout the economy.*

**reactionary.** 'Except for its technical scientific senses, to which it would be a mercy if it were confined, *reactionary* is a word so emotionally charged as to be little more than a term of abuse' (Evans). That is no less true of Britain than of America. The word derives its pejorative sense from the conviction, once firmly held but now badly shaken, that all progress is necessarily good.

**readable, legible.** See ILLEGIBLE.

**real** is an indispensable adjective in its proper place. But when used, as it so often is, with such words as *danger, risk, facts, truth,* not to qualify a noun but merely to give it a support it ought not to need, the word serves only as an illustration of the truth that the adjective is the enemy of the noun. See ADJECTIVES MISUSED. The adverbial use of *real* in the sense of *very (That's real good of you)* is an American colloquialism that has made little headway in Britain, except perhaps as a stage property of entertainers.

**realistic.** Besides serving as an adjective for *realism* in its application to the arts, *realistic* (with *unrealistic*) is usefully extended to distinguish those who have their feet on the ground from those who have their heads in the clouds. But both have now become

VOGUE WORDS, with the usual consequences. *Sensible, practical, practicable, workable, wise, reasonable, politic* —all these words and their antonyms are forgotten; every action or project today is pronounced *realistic* by those who approve of it and *unrealistic* by those who do not. There is no need to multiply examples; plenty can be found every day in the Press, especially in the reports of political speeches. A single issue of *The Times* will provide us with three. *The corporation was morally bound to take up underwriting so long as the terms were realistic at the time of issue.* (From the evidence of a witness before the Bank Rate Leak Inquiry.) | *Lord Woolton made a most realistic observation when he suggested that the House of Lords should meet later in the afternoon in order to allow younger members to attend outside of work hours.* (From a letter to the Editor.) | *It would be unrealistic not to recognize the special difficulties the T.U.C. may have in giving evidence.* (From a leading article.) The witness meant *reasonable* and the correspondent *sensible*. Only the leader-writer could fairly claim that he chose the right word; and it shows up the unruliness of vogue words to reflect that *unimaginative*, a word fundamentally opposite in meaning to *unrealistic*, would have served him no less well.

**realize.** 1. *R.* makes *-zable*; see MUTE E. 2. *What was realized might happen has happened.* The insertion of *it* between *what* and *was* is, however ugly, indispensable unless the sentence is to be recast. For discussion and parallels see IT 1.

**realm.** For synonymy see FIELD.

**-re** AND **-er.** Many words spelt *-re* are pronounced as if the spelling were *-er*; so *centre, fibre, acre, manœuvre.* In American usage the spelling of these is now *-er*, except when, as in *acre* and *lucre*, a preceding *c* would have its sound changed from *k* to *s*.

In English usage the *-re* is preferred in the words in which it has not completely disappeared (as it has in *diameter, number,* and many others). The American usage is thus more consistent; but it does not follow (cf. -OUR AND -OR) that we should do well to adopt it. The prophecy may be hazarded that we shall conform in time, one word in *-re* after another changing to *-er*; for *kilometre*, for instance, the most used of the *-metre* words, the variant *kilometer* is now recognized by the dictionaries (see METER, METRE).

**reason.** 1. *Have r.* = be in the right, and *give one r.* = admit that he is in the right, are GALLICISMS.
2. *It stands to r.* is a formula that gives its user the unfair advantage of at once invoking r. and refusing to listen to it; or rather he expects it to do that for him, but is disappointed, few of us being ignorant nowadays that it is the prelude to an arbitrary judgement that we are not permitted to question.
3. *The r. is because* etc. *The R. why so few Marriages are Happy is because Young Ladies spend their time in making Nets not in making Cages.* | *The r. Adam was walking along the lanes at this time was because his work for the rest of the day lay at a country house about three miles off.* Swift and George Eliot could be called for the defence by the modern journalist who writes *The r. was because they had joined societies which became bankrupt*, or *The only r. for making a change in the law is because the prostitute has an annoyance value.* But there is obviously a tautological overlap between r. and *because*: the *young ladies' so spending their time, the work's lying in that direction, their joining those societies,* and *the prostitute's having an annoyance value* are the reasons, and they can be paraphrased into the noun clauses *that the young ladies spend* etc., but not into the adverbial clauses *because the young ladies spend* etc. And so, although *the r. is because* often occurs in print and

oftener in speech, *the r. is that* is more correct and no more trouble. For analogous overlaps see HAZINESS. Forms nearly as common as this are *the r. is due to*, and *the r. is on account of*, as in: *My only r. for asking your permission to comment upon his remarks* is due to *the fact that many of your readers will not have seen my previous replies* | *The* r. why *I put such a poem as 'Marooned' so very high* is on account of *its tremendous imaginative power*.

**rebate.** This spelling should be reserved for the financial deduction; the carpentry term pronounced *rábbĕt* should be so spelt.

**rebus.** Pl. *-uses*; see -US 5.

**recall.** See RECOLLECT.

**receipt, recipe.** In the sense 'formula for the making of a food or medicine', with its transferred applications 'remedy', 'cure', 'expedient', 'device', etc., either word is as good as the other. But in practice *prescription* has displaced both of them as a name for the doctor's formula, and *recipe* is the usual word for a formula for preparing food, *receipt* being left to its use as an acknowledgement of payment etc. These facts are worth mention because it is sometimes debated, idly for the most part, which of the two is the right word. *Recipe* is pronounced *rĕ'sĭpĕ*, being a Latin imperative = take, originally the first word of prescriptions written in Latin.

**received pronunciation** is the name given to a system of pronunciation used by only a minority of the people of Britain but heard, since the invention of broadcasting, constantly by almost all of them. It is readily recognizable but not easy to define; nor are its boundaries sharply marked. It is a name coined by Professor Wyld

for what was called 'standard English' by Sweet as long ago as 1908 and described by him as 'a class dialect rather than any local dialect—the language of the educated all over Britain'. (Where educated northern speech differs from educated southern, as in the different value given to *a* in such words as *past* and *bath*, the R.P. adopts the southern. Robert Bridges's name for it was 'Southern English'.) Wyld's definition was that it is 'the pronunciation of the great public schools, the universities and the learned professions, without local restriction'. But it is of course not confined to them. Teachers who use the R.P. now cover a far wider field than 'the great public schools' and millions of people listen daily to broadcasts in it, so that the scope of that pronunciation, or something like it, is expanding and that of local variations correspondingly contracting.

Opinions about the merits of the R.P. differ sharply and are apt to be expressed with curious acrimony. On the one hand, a Professor of English at Oxford (Wyld) has called it 'the best kind of English, not only because it is spoken by those often very properly called the best people, but also because it has two great advantages that make it intrinsically superior to every other type of English speech—the extent to which it is current throughout the country and the marked distinctiveness and clarity of its sounds'. On the other hand, a Lecturer in English at Cambridge (Rossiter) has said of it that 'it is not the accent of a class but the accent of the class-conscious . . . the dialect of an effete social clique, half aware of its own etiolation, capitalizing linguistic affectations to convert them to caste-marks. . . . Its taint of bogus superiority, its implicit snobbery make it resented. Its frequent slovenliness and smudge condemn it on purely auditory grounds.'

In America there seems to be a similar conflict of opinion about the way the 'best people' among us speak. On the one hand, the lexicographer Vizetelly attacked the R.P. in 1931 as a 'debased, effete and inaudible form of speech'. 'The best people of England today', he wrote, 'talk with the cockney voice that, leaving the purlieus of Limehouse, has reached the purlieus of Mayfair. This is the aftermath of the war, during which the spirit of democracy prevailed, and the pronunciation of the common people left its impress indelibly on the so-called best people, with a few languid drawls, terminal *aws*, clipped *g*'s and feeble *h*'s thrown in for good measure, which later acquired the name of the Oxford voice [and] steadily debased the coinage of English speech with emasculated voices and exaggerated idiosyncrasies.' On the other hand, Professor Clark of Minnesota wrote in 1951: 'Educated Southern British pronunciation certainly has unique prestige throughout the English - speaking world. . . . In America the attitude toward [it] is in the strict sense ambivalent. It is hardly an exaggeration to say that *all* Americans envy the Southern Englishman his pronunciation. The envy expresses itself in varying degrees and in very various ways, sometimes by the appearance of loathing, but it is envy still.'

Whatever other inferences may be drawn from these discordant voices, there is one that seems pretty safe: that even men of erudition may find difficulty in appraising the R.P. dispassionately; class prejudice creeps in and colours the verdict one way or the other. It strains our credulity to be told either by Vizetelly that the R.P. erupted from Limehouse during the nineteen-twenties or by Rossiter that it is a vestigial survival of the mannerisms of the Victorian fop, which, if we are to believe those who guyed him in *Punch* and elsewhere, consisted mostly of pronouncing *s* as *th* and *r* as *w*, dropping the terminal *g* from present participles and interspersing his speech with *haws* and *whats*. Nor is it any easier to believe, with Wyld, that the R.P.'s superiority to every other type of English speech is to be found in the distinctiveness and clarity of its sounds. As R. W. Chap-

man has said: 'Of all the charges levelled against Standard English by Dr. Vizetelly and his like—that it was born in Limehouse, Mayfair or Oxford, that it is inaudible, emasculated, exaggerated, stilted, and chaotic, only one is gravely embarrassing to its defenders. The charge of careless articulation is serious indeed.'

The charge of foppishness brought against the R.P. may be in part the result of its giving no consonantal value to the letter *r* except before a vowel. In *hear* and *poor* for example the value given to the *r* is that of the indeterminate vowel sound (*er*) which in lexicography is represented by an inverted e; *hear* is *hĕ'ĕr* and *poor pŏo'ĕr*. Drawling speech may allow these to degenerate into sounds like *hĕ'ah* and *paw'ah* and so excite justifiable derision, especially from those who, like the Americans and the Scots, give some consonantal value to their terminal *rs*. But such noises are a distortion of the R.P., not a manifestation of it. The charge of careless articulation arises mainly from our habit of concentrating on one syllable of a word, or in longer words sometimes on two, and leaving the rest to take care of themselves (see next article). The vowel sounds of the unaccented syllables are thus obscured, and tend, so far as they are audible at all, to be given the indeterminate *er* sound. Bridges used to fulminate against what he called the derderderderdiness of our speech (*symperthy*, *melerncherly*, *melerdy* and so on) maintaining that our pronunciation was 'on the road to ruin' owing to the smudging of unaccented syllables. On the other hand it can be argued that uneven stressing is a central feature of spoken English, and that a pronunciation which conforms strictly to a spelling fixed probably two centuries or more ago is not necessarily preferable to a conventional one. Meanwhile the R.P. goes its own way, regardless of admonition. Its present trend, for better or worse, is undoubtedly towards more precise enunciation of unaccented syllables; see PRONUNCIATION I.

recessive accent. The title of this article has been criticized on the ground that the word 'recessive' implies the existence of a tendency for the stress in English words to drift backwards, syllable by syllable. It should not be so understood; it is merely a convenient label for the phenomenon described in the article.

The accentuation of English words is finally settled by the action of three forces on the material presented to them in each word. First, the habit, mentioned in the preceding article, of concentrating on one syllable, or in long words sometimes on two, and letting the others take care of themselves. English words of three or four syllables are common in which there is only one clear vowel sound, *corruption* and *enlightenment* for instance. This habit is in marked contrast to the French equality of syllables, and is responsible for that obscuring of the English vowel-sounds which is the main count in the indictment that has been brought against the RECEIVED PRONUNCIATION.

The second force is R.A., the 'recessive accent', or the tendency to place this usually single stress early in the word. The most obvious illustration is what happens to the French words we borrow; *château, plateau, tableau, garage, menu, charlatan, souvenir, nonchalant*, and hundreds of others, come to us with their last syllables at least as clear and fully stressed as any, but the ancient Germanic tendency to stress the beginning of uncompounded words soon turns them into *shă'to* and *shăr'lătăn* and the like. (The movement of stress in longer words is more complicated. In a word like *formidable* for instance the French, from whom we took it, would put a strong stress on the *a* of *able* and a secondary one on *for*; English first puts the main stress where the French secondary one was and a secondary stress where the French main one was, and later gives up the secondary one). Again, owing to our traditional preference for stressing first syllables, words that were long pronounced in English with stress on the middle syllable now have it on the first: *aggra'ndize, compen'sate, contrar'y, demon'strate, eque'rry* become *a'grandiz, com'pensate, con'trări, dem'onstrate, e'kwĕri*.

These first and second forces work well enough together, and, as they are always extending their influence and gradually assuming control of new words, account for many of the variant pronunciations so much more numerous in English than in most languages. In deciding which of two renderings should be preferred, it should be remembered that when R.A. has once opened an attack it will probably effect the capture, and that it is well to be on the winning side.

But working against these forces is a natural repugnance to a rapid succession of light syllables hardly differing from one another. Hence come reactions about many polysyllabic words whose surrender to R.A. is on record in the dictionaries. These tell us that *laboratory* and *disciplinary* should be, and *Deuteronomy* may be, pronounced with the stress on the first syllable and *recriminatory* with it on the second. These are not easy to say, and the attempt is apt to result in the omission of syllables—*la'brătrĭ, di'splinrĭ, Dŭ'trŏnmĭ*, and *recrim'inătrĭ*. Such dangers are dodged by the use of two stresses (*dĭ'siplī'năry, Dŭ'terŏ'nomy, rĕcrim'ină'tory*) or by shifting the stress forward again (*labŏ'ratory*).

For this and other reasons the R.A., though it has made great progress, has by no means won its battle. Differently stressed pronunciations even of words about which the dictionaries have surrendered may still be heard on the platform and over the air. These cannot be wholly due to the difficulties of articulation that the R.A. may present in words of four or more syllables; for they include stressed penults in many three-syllable words such as *communal* and *industry*. Whatever the reason, it seems clear that there is still considerable popular resistance to the R.A., showing itself not only in refusing to follow a shift of stress to the first syllable of a word but also in shifting forwards an established first-syllable stress, and saying, for instance, *formid'able* and *controv'ersy*. *Commendable*, once accented on the first syllable ("'Tis sweet and commendable in thy nature, Hamlet'), is now *commend'able* by the unanimous verdict of the dictionaries; and, though they agree in awarding victory to the R.A. in the other four-syllable adjectives ending -ble, there are many people who prefer to say *hospit'able*, *disput'able*, etc. Apart from admittedly unorthodox pronunciations such as these, the swaying fortunes of the R.A.'s attack are shown by the number of words for which the dictionaries themselves recognize alternative pronunciations. A few examples from the 1964 COD are (preferring stress on the first syllable) *aspirant, contemplative, decorous, dissoluble, doctrinal, marital, obdurate, peremptory, recondite, secretive, vertigo*, and (preferring stress on the second) *abdomen, albumen, centenary, coronal, environs, interstice, remonstrate, sonorous, trachea*.

No one can say with confidence how the battle will go; the opinion expressed earlier in this article that the R.A.'s was likely to be the winning side may still be sound, but it was formed before the speak-as-you-spell movement (see PRONUNCIATION 1) was as strong as it has since become, and this makes for more even stressing.

**recidivist.** Pronounce *rĕsĭ'dĭvĭst*. A FORMAL WORD for an *old lag*.

**recipe.** See RECEIPT.

**recipient.** 1. *The Serjeant-at-Arms and Lady Horatia Erskine were yesterday the recipients of presentations from members of the Press Gallery.* / *Mr. Albert Visetti, who has just been the r. of a pleasant presentation from his pupils.* / *Mr. John D. Clancy, K.C., M.P., who enjoys the unique distinction of having represented continuously an Irish constituency for a quarter of a century, has just been made the r. of a presentation to mark the event.* Can any man say that sort of thing and retain a shred of self-respect? Perhaps it is a good sign

that the only illustrations available are obviously dated. 2. As an adjective *recipient* may usefully be differentiated from *receptive*: *recipient minds* are those that in fact receive; *receptive minds* are those that are quick to receive.

**reciprocal, mutual.** To the difficulties presented by MUTUAL itself must be added that of the difference between it and *r*. *M*. regards the relation from both sides at once: *the m. hatred of A and B*; never from one side only: not *B's m. hatred of A*. Where *m*. is correct, *r*. would be so too: *the r. hatred of A and B*; but *m*. is usually preferred when it is possible. *R*. can also be applied to the second party's share alone: *B's r. hatred of A*; *r*. is therefore often useful to supply the deficiencies of *m*. A, having served B, can say 'Now may I ask for a r. [but not for a m.] service?' Two parties can take m. or r. action, and the meaning is the same; one party can take r., but not m., action. *Mr. Wilson said*: '*I trust your Government saw in the warmth of the greetings accorded to his Royal Highness the manifestation of friendly goodwill which the people of the United States hold for those of Britain. Believing in the reciprocal friendship of the British people it will be my aim in the future to . . .'.* In this passage, *m*. could not be substituted for the correct *r*.; if, however, the words had been not 'of the British people', but 'of the two peoples', *m*. would have been as good as *r*., or indeed better. But it must be added that, since it takes two to make a friendship, which is essentially a m. or r. relation, to use either adjective is waste.

**reckon.** 1. *R*. in the sense of consider is not in origin an Americanism, like *guess* and CALCULATE, though it became one by adoption in the Southern States in the 19th c., corresponding to the Yankee *guess*. It was a literary word in England more than 400 years ago, used by the translators of the Bible: *I r. that the sufferings of the present time are not worthy to be compared with the glory which shall be revealed in us*. There is good authority for its use in this way in more modern times too; OED quotes Jowett: *I r., said Socrates, that no one could accuse me of idle talking.* But it is now little used outside rustic talk. 2. Dead reckoning. *The German staff calculated the BEF would be mobilized by the tenth day, gather at the embarkation ports on the eleventh, begin embarkation on the twelfth, and complete the transfer to France on the fourteenth. This proved to be almost dead reckoning.* It is a MISAPPREHENSION to suppose that, in the phrase *dead reckoning, dead* is used in the sense of exact, as it is in such phrases as *d. right, d. on the target.* On the contrary, this method of calculating a ship's position (without observation of the heavenly bodies) can only give a not wholly reliable approximation.

**recognizance.** The old pronunciation *rĕkŏn'-* is now giving place to *rĕkŏgn'-*, a change to be commended since the word has a close kinship with *recognize* and none at all with *reconnaissance.*

**re-collect, recollect, remember, recall.** To re-collect is to collect or rally what has been dissipated (*but he soon re-collected his courage* or *himself*); the distinction between this and the ordinary sense of *recollect* is usually though not always kept up in pronunciation, and should be marked by the hyphen, see RE-. Between *recollect* and *remember* there is a natural distinction often obscured by the use of *recollect* as a FORMAL WORD for the 'dominant' term *remember. Recollect* is preceded by *I can't* as naturally as *remember* is by *I don't*; i.e. *recollect* means not so much remember as succeed in remembering, and implies a search in the memory. *Peter remembered* (not *recollected*) *the word of Jesus, which said unto him, Before the cock crow, thou shalt deny me thrice.* Another word, *recall*, has the peculiarity of suggesting memory stimulated by association. But in practice these distinctions are rarely consciously observed.

**reconcil(ement)(iation).** The first is comparatively little used, but has the special function of representing the act of reconciling rather than the act or state of being reconciled, which means in practice that it is more fitly followed than is *reconciliation* by an objective genitive, as in *The reconcilement of duty with pleasure is no easy problem.*

**recondite.** The old pronunciation *rĕkŏ'ndĭt* (or *-īt*) has not been completely conquered by the RECESSIVE ACCENT, but modern dictionaries put *rĕ'kŏndīt* first.

**recondition.** See CONDITION.

**reconnaissance.** Pronounce as an English word—*rĕkŏ'nĭsăns.*

**recount(al).** For *re-count* see RE-. For the noun *recountal* (*When the very interesting stories of crime have been unfolded, we can follow the recountal of detection without any bewilderment*), see -AL NOUNS; 'Frequent in recent journalistic use' says the OED, perhaps not designing to commend it. Its popularity seems to have been short; it rarely appears now.

**recourse.** See RESORT.

**recriminate.** *Idle people who pass their time in recriminating France.* For this transitive use, 'now rare', the OED has only a single quotation later than the 18th c.; the COD does not recognize it.

**recriminatory.** For the rival pronunciations (*-crĭm'ĭnătŏrĭ, -crĭminā'tŏrĭ*) see RECESSIVE ACCENT.

**recrudescence.** *Hong Kong, Friday.* —*There is an alarming r. of piracy in the West River.* | *A literary tour de force, a r., two or three generations later, of the very respectable William Lamb* (*afterwards Lord Melbourne*), *his unhappy wife, Lady Caroline Lamb, and Lord Byron.* | *First, we have the unfortunate circumstances which caused England to be weakly represented in the second test match; secondly, we have the r. of Mr. Laver.* To recrudesce is to become raw again or renew morbid activity, as a wound or ulcer may, or, metaphorically, a pestilence or vice or other noxious manifestation. That being so, the first example above is proper enough; but what have Mr. Laver and Lord Melbourne done that their reappearance should be a r.? Nothing, except fall into the hands of journalists who like POPULARIZED TECHNICALITIES and SLIPSHOD EXTENSION. This disgusting use is apparently of the 20th c. only; the recrudescences in the OED quotations are of 'abuses', 'calumny and malignity', 'Paganism', 'the epidemic', 'the wound', 'a varicose ulcer', and that is all.

**recruital.** See -AL NOUNS.

**rectilinear, -neal.** There is no objection to either in itself; but -ar is so much commoner that, as there is no difference of meaning, -al should be abandoned as a NEEDLESS VARIANT.

**rector.** See VICAR.

**recur.** For the pronunciation of *recurring, recurrence,* see PRONUNCIATION 7.

**recusancy, -ce.** The second is much less common, and should be dropped as a NEEDLESS VARIANT. The stress is on the first syllable.

**reddle.** See RUDDLE.

**reduce** makes -cible; see -ABLE 2. After r. to and be reduced to the gerund, not the infinitive, is idiomatic: He was reduced to retracting (not to retract) his statement; see GERUND 3.

**reductio ad absurdum.** The method of disproving a thesis by producing something that is both obviously deducible from it and obviously contrary to admitted truth, or of proving one by showing that its contrary involves a consequence similarly absurd. A r. a. a. of the theory that the less one eats the healthier one is would be 'Consequently, to eat nothing at all gives one the best possible health'. The proof, as opposed to the disproof, by r. a. a. is a form often used by Euclid, e.g. in I, vi, where the contrary of the thing to be proved is assumed, and shown to lead to absurdity.

**redundant.** This word affords a striking example of the way in which VOGUE WORDS blunt the language. Until well into the 20th c. it was a scholarly word, not in common use. When it was first applied, not unaptly, to employees dismissed because there was no longer work for them to do, it had all the attractions of novelty, and quickly ousted the more commonplace superfluous. Then it was widened into a synonym of unnecessary, unsuitable, disused, without any regard to the idea of too much which is strictly inseparable from it. *The thieves were disappointed because the strongroom they broke into had become r.* says a cathedral verger telling the story to pressmen. It is a depressing thought that vogue words can so bemuse the minds of simple people as to make them choose this ridiculous way of saying that the strongroom was not being used any more. Academic writers too are not immune from the contagion of this cliché. *Once you do question this assumption* [that the moral code of a society is the product of its penal code] *the problem of criminal responsibility becomes r.* (Here what the writer was trying to say seems to be that the problem loses its importance.) Or again, this time from an official source: *The Authority are now reluctant to proceed with the provision of services for a 10,000 population, in case their work becomes r. due to the subsequent need for catering for a larger population.* This amounts to saying that a water-supply designed for a population of a certain size might prove insufficient if the population afterwards increased.

**reduplicated words,** under which heading are included both plain reduplications such as *teeny-weeny* and also those with a change of vowel sound ('apophony') such as *tip-top.* Most of us find an engaging quality in these words. Perhaps that is because it is through them that we enter the world of speech, progressing from the first tentative *mum-mum* and *dad-dad* to the more ambitious *teeny-weeny tootsy-wootsy,* and *piggy-wiggy.* And i we have the misfortune eventually to fall into a second childhood there will be another—*ga-ga*—for our friends to see us out with.

In most of the r. words used in adult life the second half is merely a booster of the first, so that if we know the meaning of the first we know that of the whole. With many not even that is needed; they are self-explanatory. We do not have to consult a dictionary to know what Hotspur meant when he complained of being made to listen to such a deal of skimble-skamble stuff from Glendower, or King Claudius when he doubted the wisdom of having had Polonius interred in hugger-mugger, or Fluellen when he opined that there was no tiddle-taddle nor pibble-pabble in the camp of Pompey, or Parolles when he told Bertram that he would do better to go to the wars than to hug his kicky-wicky here at home. (If the last had been speaking today he would no doubt have said *popsy-wopsy* or *hotsy-totsy*.)

Many—perhaps most—of these words have a disparaging or contemptuous flavour. A few examples of those that are, on the contrary, colourless, though in some cases onomatopoeic, are *boogie-woogie, chiff-chaff, crinkum-crankum, criss-cross, flip-flap, hokey-pokey* (in one of its senses), *hurdy-gurdy, ping-pong, roly-poly, rub-a-dub, pitter-patter, see-saw, tick-tack, walkie-talkie,* and *zig-zag*. One or two are actually commendatory: *ding-dong* implies perseverance and *ram-stam* and *willy-nilly* determination, *lovey-dovey* affection, and *tip-top* and *super-duper* high praise. For greetings and farewells we have *chin-chin, bye-bye,* and *ta-ta*.

Of the many disparaging ones some examples follow grouped according to the objects they are naturally used of.

Of our fellow men, their qualities, appearance, and behaviour: *clever-clever, creepy-crawly, dilly-dally, fuddy-duddy, goody-goody, hanky-panky, hocus-pocus, jiggery-pokery, lardy-dardy, namby-pamby, niddle-noddle, niminy-piminy, rag-tag, riff-raff, shilly-shally, whipper-snapper.*

Of trifling or confused talk: *argle-bargle, chit-chat, clap-trap, flim-flam, hob-nob, hubble-bubble, mumbo-jumbo, pow-wow, tittle-tattle.*

Of disorderliness: *harum-scarum, helter-skelter, higgledy-piggledy, hodge-podge, hulla-baloo, hurly-burly, jim-jams, mish-mash, pell-mell, pully-hauly, razzle-dazzle, rowdy-dowdy, topsy-turvy.*

Of trifling or inferior things: *fiddle-faddle, gew-gaws, knick-knacks, wishy-washy.*

For comprehensive use: *fiddle-de-dee, hoity-toity, pooh-pooh, tut-tut.*

**reeve** makes *rove* or *reeved* both in past and in p.p.

**refection** (meal). A FORMAL WORD.

**refectory.** The pronunciation *rĕ'fĕktŏrĭ* is still in monastic use, but the popular pronunciation is *rĕfĕk'tŏrĭ*.

**referable.** Pronounce *rĕ'fĕrabl*. For the irregular form (cf. -R-, -RR-), see CONFER(R)ABLE.

**reference.** 1. For synonyms in the sense scope or purview, see FIELD. 2. Perhaps because *referee* has been largely appropriated to other purposes, *reference* is now commonly used to mean a person to whom reference is permitted as a witness to character, and, by further extension, the written testimonial that he gives.

**referendum,** properly meaning a question to be referred (to the people), has been appropriated as a name for the system of so referring questions and for any particular occasion of its exercise. The normal form would have been *reference*, but *referendum* has the advantage of not bearing several other senses. Plural preferably -*dums*; -*da* is confusing as suggestive of the original sense—questions to be referred—for which we now use *terms of reference*.

**reflection, reflexion.** Though the second is 'the etymological spelling' (OED), the first is in general senses (thought, remark, censure, etc.) almost invariable, and even in the physical senses (casting back of light etc.) at least as common as -*xion*. A clear differentiation being out of the question, and the variation of form being without essential significance, the best thing to do is to use the commoner spelling, *reflection*, in all senses. For the change from the older *reflexion* see -XION, -XIVE.

**reflective, reflexive.** The case is simpler with these than with *reflection* and *reflexion*. *Reflexive* has now lost all its senses except the grammatical one, and *reflective* has resigned that and kept the rest; the differentiation wanting in the nouns has been accomplished for the adjectives. But *reflective*, though it can at need have any of the adjective senses corresponding to *reflection*, is current chiefly as synonymous with meditative, and *reflecting* or *reflected* is substituted for it as often as possible in referring to the reflection of light etc.—*reflecting surface, reflected colour*, rather than *reflective*.

In grammar *reflexive* verbs are those of which the object and the subject are the same person or thing. *Pride*, v., is reflexive, since one prides oneself, not someone else; and many verbs that are not solely r. can be used reflexively, e.g. *kill* one*self*. R. pronouns are those serving as object to r. verbs, *myself* etc.; the personal pronouns *me, you*, etc., are still occasionally used as reflexives, e.g. in *He sat him down, I bethought me*.

**reform, re-form.** See RE-.

**refract** makes *-tor*. For *refractable* and *refrangible*, of which the first is recommended, see -ABLE 2.

**refrigerate.** *Refrigeratory* is pronounced by the OED *rĭfrĭ′jĕrătŏrĭ*, not *-ātŏrĭ*—a hard nut for some jaws; perhaps that is why we have dropped the *y* from the machine now in popular use, and, not content with that, have shortened its name to *fridge*. See RECESSIVE ACCENT.

**refutable, irrefutable.** The OED prefers the accent on *-fu-* in both, but allows the other also. The RECESSIVE-ACCENT force, and the analogy of words so familiar as *(dis)reputable* and *(in)disputable*, seem likely to result in the pronunciation here advised—*(ĭ)rĕ′fūtăbl*.

**refutal.** For this SUPERFLUOUS WORD see -AL NOUNS.

**refute.** *He sharply refuted the suggestion and said he could produce ample evidence that it was wholly without foundation.* He could r. the suggestion only by producing the evidence; till then he could only deny it.

**regalia.** The word meaning royal emblems etc. is a plural; it is slipshod writing to treat it as singular, and strictly it is slipshod extension to use it of insignia other than royal, e.g. of those of an Order of Chivalry, but this is often done.

**regard.** 1. *R*. in periphrasis. 2. *Take r*. 3. *R., consider*.

1. *R*. in periphrasis. The noun is much used in COMPOUND PREPOSITIONS; see that article for excesses of the kind. The two examples that follow, in which *about* would have served for *with r. to*, and *in* for *in r. to*, are mere everyday specimens of a practice that is not strikingly bad on each occasion, but cumulatively spoils a writer's style and injures the language: *It is well said, in every sense, that a man's religion is the chief fact with r. to him.* | *In r. to three other seats there will be a divided Unionist vote.*

The verb is also much over-used periphrastically in *as regards*: *Turkish rule cannot be tolerated in future over any country the population of which is Christian as regards the majority of its inhabitants*. This should run—any country (a) whose population is chiefly Christian, or (b) with a predominantly Christian population, or (c) in which the majority of the population is Christian, or (d) in which the majority are Christians, or (e) where Christians are in a majority. See AS 3 for the disfigurements to which the very similar *as to* leads those who indulge in such phrases.

2. *Take r. The vast majority, it would be safe to say, have patients over a field which* takes no r. *to borough or other boundaries*. A mixture of the two

phrases *take account of* and *have r. to*; see CAST-IRON IDIOM.

3. *R., consider.* (See also CONSIDER.) *I consider it monstrous* or *a shame* is English; *I r. it monstrous* or *a shame* is not, but requires *as*: *I r. it as monstrous, as a shame.* In the following quotations *as* should be inserted, or *consider* or *think* used. *He regards it beyond question that Moses wrote practically the whole of the Pentateuch.* | *But the Generals present regarded the remedy worse than the evil.*

Sometimes the omission of *as* is perhaps caused, though not excused, by the proximity of another *as*, or by an abnormal order of words. *Consider* is the remedy in all. *We in Ireland r. no insult so supreme as the insult that we are intolerant.* | *The man who regards the postal system as stable as the solar system.* | *We r. this attempt to create enthusiasm for the Union Jack by statutory enactment as ill-advised as the policy of 'Say Suzerain'.* | *No leader, however inviolable he regards himself, can hope to escape criticism.* | *Both these mansions were designed by the same architect, the late Louis Vulliamy, whose masterpieces they are generally regarded.* To block a side-issue, let it be said that two or three of the pieces might be technically defended on the ground that since, for instance, *sound as a bell* means the same as *as sound as a bell*, the *as* before *stable* in the second and *ill-advised* in the third may be the one that belongs to the phrase *regard as*, and not correlative to the later *as*. It is obviously not so; moreover anyone who takes that line and omits those examples has still the others on his hands.

*R. to be* (or do) is as unidiomatic as *r. it monstrous* etc., though less common, and can be corrected in the same way. Examples: *Some County Associations r. it to be their first duty to accumulate large invested funds.* | *He regards Spiritism as practised today to be full of the gravest dangers.* | *But for a long time it seemed to be regarded that the heads of important trade departments could be relegated to any gentleman of influence*

**regardless.** See QUASI-ADVERBS.

**regiment.** The pronunciation *re'ji-ment* for the noun is now heard at least as often as the *rej'ment* formerly usual; in the derivatives (*regimental* etc.) the *i* is always sounded.

**region.** For some synonyms in the sense sphere etc., see FIELD. *In the r. of* for about or nearly is a bad COMPOUND PREPOSITION.

**register,** v., makes *-trable*. The agent-noun is *-trar*; the old form *-trary* is retained only in Cambridge University. The official term for the office of the kind of registrar the public is most familiar with (that of births, deaths, and marriages) is *Register Office*. To call it a *registry office* is said by the OED to be a colloquialism; but it is a very common one, encouraged by the fact that the Registrar of Friendly Societies and the Registrar of Lands, unlike the Registrar General, keep their records in registries, not registers. Registry is also the word for the domestic employment agency and for the branch of a government department that records and files papers.

**regretful** means feeling or manifesting regret, not causing it; the latter sense belongs to *regrettable*. In the extracts below the wrong word has been chosen; see PAIRS AND SNARES. *The possession of those churches was unfortunately the reason of the r. racial struggles in Macedonia.* | *He was obviously a man of education and culture, which made it all the more r. that he had elected to lead a life of crime.* | *It was not surprising, however r., that Scotland had lagged behind.*

**rehabilitation** is an old word put to a new duty. In the 16th c. it meant the 'restoration by formal act or declaration (of a person degraded or attainted) to former privileges, rank, and possessions' (OED). Today the sense in which the word is most commonly used, not unsuitably, has been defined as 'a continuous process by which disabled persons should be translated from the state of being incapable under full medical care to the state of being producers and earners' (Beveridge).

**reinforce, re-enforce.** The ordinary form (*rein-*) has been so far divorced from the simple verb (formerly *inforce* or *enforce*, now always the latter) that it seldom or never means to enforce again, as when a lapsed regulation is revived. For that sense *re-enforce* should be used; see RE-. Both make *-ceable*; see -ABLE I.

**reiterant.** *But the booing and r. cries of 'No' grew louder, and at length he sat down.* This AVOIDANCE OF THE OBVIOUS, as often, has resulted in a blunder; *r.* means repeating, not repeated. In any case, what are *booing* and *r.* doing in one sentence?

**rejoin, re-join.** See RE-. The hyphened form should be restricted to actual reuniting. (*The parts will re-join if laid close end to end*, or *should be re-joined with care*).

**relation, relationship, relative.** 1. These words, as terms of kindred, have seen some changes. *Relative* started as an adjective meaning what we call related, but, being used as short for related person, became a noun denoting a person. *Relation* started as an abstract noun meaning our relationship (in its only right sense; see next article); but, being transferred from the abstract to the concrete, came also to denote a person. We have had to take to *related* and *relationship* because the others in their original senses have failed us, and we now find ourselves with *relation* and *relative* as two names for the same thing, only so far different as *-ive* is something of a FORMAL WORD and *-ion* the dominant term.

2. The word for one kind of relative —*cousin*—has a curious history. The Latin word *consobrinus* from which it is derived means, etymologically, the child of a mother's sister, but was also used in the wider sense of any first cousin, or, in the legal phrase, cousin german. Later the meaning was enlarged to include any collateral relative more distant than brother or sister,

and the English word, sometimes affectionately abbreviated to *coz*, was used in the same way as in Shakespeare's day ('Me uncle?' 'You cousin'). We have now restricted the meaning again and classified the relations properly so called: my first cousins are the children of my parents' brothers and sisters; my second cousins are the children of my parents' first cousins; my third cousins are the children of my parents' second cousins, and so on. My first cousin once (or twice) removed is the child (or grandchild) of my first cousin, and so on.

It was formerly customary for the Sovereign to call a foreign monarch cousin (*Our cousin of France*) and to do the same to noblemen of his own country. Royal writs and commissions are still so phrased if addressed to a peer above the rank of baron, with archaic epithets nicely graded. A duke is *Our right trusty and right entirely beloved c.*, a marquess or an earl *Our right trusty and entirely beloved c.*, a viscount *Our right trusty and well-beloved c.* See also TRUSTY.

The verb *cozen* is said by Skeat to be 'merely a verb evolved out of *cousin*', i.e. to claim kindred for one's own advantage; 'the change of meaning from *sponge* to *beguile* or *cheat* was easy'. The OED queries this explanation, but does not offer any other.

**relation(ship).** The word *relation* has many senses, most of which are abstract. It approaches the concrete in the rather rare sense of a story or narrative, and it is fully concrete in the very common sense a related person, i.e. a son or mother or cousin or aunt or the like. Now *sonship, cousinship*, etc., being words for which there is a use, it is entirely natural that -ship should be affixed also to the word that summarizes them; *sonship* the being a son, *relationship* the being a relation —with the extension (due to the generalizing sense of *relation*) into 'the being this, that, or the other relation', or 'degree of relatedness'. To that use of *relationship*, then, there is no objection. But to affix -ship to *relation* in any of its other, or abstract, senses is against all analogy; the use of -ship is to provide concretes (*friend, horseman, clerk, lord*) with corresponding abstracts; but *relation*, except when it means related person, is already abstract, and one might as well make *connexionship, correspondenceship,* or *associationship*, as *relationship* from *relation* in abstract senses. Of the following extracts the first, in which *relationship* should have been used, shows how that word, when it is justifiable, may lend precision to the meaning; the second suggests, by the writer's shifting from one to the other, that *relationship* in the improper sense has no superiority whatever to *relation* or *relations*; and the others show how needlessly the LONG VARIANT is often resorted to: *The king was therefore not necessarily of royal blood, though usually he was the son of the previous Pharaoh; the relation of Tut-ankh-Amen to his predecessor is not known. | Why not leave* the relations *of landlord and farmers, as well as those of farmers and labourers, to the beneficial effects of the policy? Why is a tribunal necessary in the one case and not in the other if mutual frankness will adjust* all relationships? | *A state of things may be created which is altogether inconsistent with the* relationship *which should properly exist between police and public. | A step which must have great effect on the commercial* relationship *between America and Europe. | She declared that she and her husband had no business* relationships.

**relative.** For the use in *I wrote to him r. to renewing the lease,* see COMPOUND PREPOSITIONS. For the slipshod use of *relatively* and *comparatively* see COMPARATIVELY.

**relative pronouns.** See the separate words—*who, which, what, that, such as, as.*

**relax.** In the middle of the 20th c. the exhortation invariably addressed to anyone showing signs of strain is *Relax!* In the 19th c. it was *Buck up!* There is matter for an essay here, but this is not the place for it.

**relegate** makes -*gable*; see ABLE 1. *The large terrace, usually a dining-room, has also been* relegated *to the King's use, and will be adorned with groups of Alpine plants.* Devoted? *Relegated* is not very polite to His Majesty; has the writer looked up *assign* in a synonym dictionary and decided that *r.* is the least familiar of the list? Familiar to him it does not seem to be; see NOVELTY HUNTING.

**relevance, -cy.** The OED treats -*cy* as the standard form; see -CE, -CY. In practice they are probably equally common.

**reliable.** For a discussion of the objection that used to be taken to *r.* and some other adjectives in -*able* see -ABLE 4.

**relict.** The word is now hardly used except as a legal term for *widow*. Pronounce rĕ'lĭkt.

**relievo.** Pl. -*os*; see -O(E)S 6. But the form might well be dropped as a needless mixture between the Italian *rilievo* (rēlyā'vō) and the English *relief*; cf. BAS-RELIEF.

**religious.** For *dim r. light* see IR-RELEVANT ALLUSION.

**remain.** 1. *There remain(s).* 2. *Continue to r.* 3. *I r.*

1. *There* remains *to be said* a few words *on the excellence of M. Vallery Radot's book.* The use of a singular verb before a plural subject is discussed in NUMBER 7. The present example is perhaps due to confusion between *It remains to say* and *There remain to be said.*

2. *Continue to r  R.* (in the sense that concerns us) means in itself 'continue to be'; *to continue to continue to be* is, except in some hardly imaginable context, a ridiculous tautology, and would not call for mention if it were not surprisingly common; see HAZINESS and PLEONASM 3. *And yet through it all I c. t. r. cheerful. | It is expected that very soon order will be restored, although the people c. t. r. restive.*

3. *I r.* For this see LETTER FORMS.

**remark,** v., has as one of its senses 'to say by way of comment' or 'say incidentally'. It would be absurd pedantry to insist that it should never be used for *say* except when 'by way of comment' is clearly justified, and it is often very difficult to decide whether it is justified or not. Nevertheless, it is well to remember the qualification, and be thereby saved from two bad uses of *r.*, (1) as a mere FORMAL WORD, and (2) as a word relied on to give by its incongruity a mildly facetious touch —one of the forms of WORN-OUT HUMOUR: *You may drive out Nature with a pitchfork but she will always return*, as Horace remarked *in a language no longer quoted in the House of Commons*.

**remedy.** *Remediable* and *remedial* are pronounced *rĕmē-*, but in the unlikely event of anyone's wanting to say *remediless* he should call it *rĕ'mĕdĭless*, and not use what the OED calls 'the original stressing' *rĕmĕ'dĭless*.

**remember.** See RECOLLECT for the distinction.

**reminisce,** a BACK FORMATION from *reminiscence*, though quoted from 1829, is still only colloquial.

**remise** as a law term pronounce *rĕmīz*; in other senses (fencing etc.) *rĕmēz*.

**remit** makes *-tted, -tting*, etc.; see -T-, -TT-; but *remissible*; see -ABLE 2. Of the nouns *remittance* is right for the sending of money, *remittal* for referring a case from one court to another, and *remission* for all other senses.

**remonstrate** is ordinarily pronounced (in contrast with *demonstrate*) *rĕmŏ'n-strāt*, perhaps because the current noun is *remonstrance*. But *rĕ'monstrate* is also heard and gaining ground. (See RECESSIVE ACCENT.) The other noun, *-ation*, is now rare, and should not be preferred to *-ance*: *Although every attempt is made at this office to save people from being misled, our remonstrations have not hitherto met with success.*

**remote** is not one of the adjectives that can be used as QUASI-ADVERBS; it must have a noun with which it can be more reasonably conceived to agree than it can with *knowledge* in the following extract: *Even somewhat remote from the main tourist routes the knowledge of English in shops is remarkable* (Even some distance from . . .).

**remove** makes *-vable*; see MUTE E. For *r.* one's *hat, the cloth*, etc., see FORMAL WORDS.

**remunerate** makes *-rable*: see -ABLE 1. *R.*, *-ation*, and *-ative*, are, as compared with *pay(ing)*, FORMAL WORDS, and should not be preferred, as they often are, without good reason.

**renaissance.** *R.* was so far established as the English word for the thing before it was latinized or anglicized into *renascence* that it is still the more intelligible of the two, and may well be left in possession, at any rate for the word in its specialized meaning. Pronounce as English—*rĕnā'sns.*

**rendezvous.** Pronounce *rŏ'ndĕvōō* (or *-dā-*), but in the plural of the noun *-ōōz*. The verb makes *-voused* (pron. *-vōōd*), *-vousing* (pron. *-vōōing*).

**rendition** in the sense of surrender (of armed forces or fugitives) is now rare. In the sense of rendering (a translation or a musical or dramatic performance) it is standard U.S. but in Britain a NEEDLESS VARIANT with no other credential than its more imposing appearance. The same is true of its occasional adoption in the Services for *rendering* in the sense of sending in returns, accounts, etc., a usage not recognized by the dictionaries of either country.

**renounce** makes *-ceable*; see -ABLE 1. Between *renouncement* and *renunciation* there is no such differentiation as that which preserves the two nouns of *pronounce*, and *renouncement* is accordingly passing out of use.

**rep.** 1. The OED treats this, not *repp* or *reps*, as the right form of the textile name. 2. As a CURTAILED WORD *rep* has been put to several uses: as schoolboy slang for *repetition*, as slang now obsolete for *reputation* and *reprobate*, and as a colloquialism for REPERTORY.

**repa(i)rable.** *Reparable* (*rĕ′pă-*) is used almost only of abstracts such as *loss, injury, mistake,* which are to be made up for or to have their effects neutralized; *repairable* sometimes in that way also, but chiefly of material things that need mending. The negatives are *irreparable,* but *unrepairable;* see -ABLE 3.

**repellent, repulsive.** That is repellent which keeps one at arm's length; that is repulsive from which one recoils; that is, the second is a much stronger word.

**repercussion** in its modern figurative sense is useful, but, having become something of a VOGUE WORD, is apt to be overdone. It is wasted if used merely as an imposing synonym of *consequence* or *result.* It should be confined to consequences that are unexpected and unpleasant, and multiply themselves; the idea underlying the word is of something hitting back —consequences such as Macbeth was thinking of when he spoke of 'bloody instructions which, being taught, return to plague the inventor'.

**repertoire, repertory** are essentially the same word, being the French and English equivalents of the Latin *repertorium,* but are to some extent differentiated. *Repertoire* is confined to the collection of pieces that a musician or dramatic company is prepared to perform; *repertory,* besides having this meaning, is also applied to collections of other kinds, especially of information. In their application to the stage *repertory* is ordinarily used as an attribute of a company or theatre and *repertoire* for its stock-in-trade.

**repetition of words or sounds.** The first thing to be said is that a dozen sentences are spoilt by ill-advised avoidance of repetition for every one that is spoilt by ill-advised repetition. Faulty repetition results from want of care; faulty avoidance results from

incapacity to tell good from bad, or servile submission to a rule of thumb —far graver defects than carelessness. This article is accordingly of slight importance compared with that in which the other side of the matter is presented; see ELEGANT VARIATION, where the rule of thumb against repetition is shown to have the most disastrous consequences. It will therefore be fitting to begin with an example (from Henry James) of deliberate and effective repetition: *I daresay I fancied myself a remarkable young woman . . . but I needed to be remarkable to offer a front to the remarkable things that presently gave their first sign.*

The fact remains, however, that repetition of certain kinds is bad; and, though the bad repetitions are almost always unintentional, and due to nothing worse than carelessness, and such as their authors would not for a moment defend, yet it is well that writers should realize how common this form of carelessness is. The moral of the examples that will be given is the extremely simple one—read what you have written before printing it. The examples are divided into batches under headings, and little comment need be added.

DEPENDENT SEQUENCES, i.e. several *of* phrases, or two or more *which* clauses or *that* clauses or *-ly* words, each of which is not parallel or opposed, but has a dependent relation, to the one before or after it. For examples of the last see -LY 3.

*The founders* of *the study* of *the origin* of *human nature.* | *The atmosphere of mutuality must be created* which *will make it possible to discuss proposals* which *would have seemed impracticable.*

TWO ACCIDENTALLY SIMILAR BUT NOT PARALLEL USES OF A WORD. Some other examples may be found in JINGLES.

*Space forbids us to give a translation of the entire article,* which *would run to* several *columns; but there are* several *points* which, *if quoted from the rest of the article, would give the impression that . . .* | *In these days American revolutionary upsets appear small enough beside the other afflictions of the world; yet the situation is interesting* enough. | *It was entitled 'Le Comité de Lecture', and it* resented, *in language which our feminists would strongly* resent, *the presence of ladies on that committee.* Doubtful specimens of this kind sometimes occur in which the repetition may have been intended, but the parallel or contrast is so little significant or so untidily expressed that it was probably accidental, as in *The Japanese democracy are* affronted *at what they regard as an* affront *to their national dignity.*

HAPHAZARD REPETITION, IN A DIFFER-ENT SENSE, OF A WORD (or such use of one of its inflexions or derivatives or other belongings).

*The cure for that is clearly the* alternative *vote or the second ballot, the former* alternative *being the more preferable.* | *This may have been* due *to* undue *power placed in his hands by the Constitution.* | *To this* last unsuccessful *attempt* succeeded *the boredom of the trenches.* | *We cannot believe that the Bill will be shipwrecked on this point, for that would be not only disastrous to* itself, *but disastrous to the reputation of the House of Lords* itself. | *Such a misfortune would give the impression that the English do not treat their religion* seriously—*an impression which would have a* serious *effect politically as well as morally.* | *Sir William White has now received the crowning distinction of the Presidency of the Royal* Association; *his* association *with the Navy may be said to date almost from his birth.* | *They are kept in vigour for a time by the automatic* generation *of enthusiasm, but after a while the ebb begins; a movement generally grows and dies with a* generation. Here again it is sometimes possible to suspect a writer of what is worse than carelessness, a pointless but intended repetition meant to have the effect of a play on words or the mildest of puns: *The triple* bill *of* Bills *which are down for the autumn sitting, the Mines Bill, the Shops Bill, and the Insurance Bill.* | *Of the octogenarians twenty-three* died *in the first, and thirty-three in the second half of the century; while if we add the nonagenarians twenty-five* an-

cients *died in the more* ancient, *and thirty-eight in the modern time.* | *Anonymity seems to be a peculiar delight to writers on naval matters, though perhaps necessity has something to do with the* matter.

ASSONANCE, RHYME, etc.

*'Worser and worser' grows the plight of the Globe* over *the* oversea *trade figures.* | *If no such Council existed, the Secretary of State would have to* form *an* informal *one if not a* formal *one.* | *The features which the* present *Government in this country* presents *in common with* representative *and responsible government are few and formal.* | *No* factual *report had by then been* actually *received.*

**repetitional, repetitionary, repetitious, repetitive.** With all these on record, *repetition* would seem to have a good stock of adjectives at need; but few writers have the hardihood to use either of the first two. Between the other two a DIFFERENTIATION seems to be developing: that *repetitious* is preferred in the pejorative sense, to describe speaking or writing that is tediously iterative, and *repetitive* is a more neutral word, used especially of repetition other than verbal, e.g. manual work on an assembly line.

**replace** makes *-ceable*; see -ABLE I. There is the literal sense of put (thing or person) back in the same place as before; and there are, broadly different from this, various uses in which substitution is the idea—return an equivalent for, fill or take the place of, find a substitute for, supersede, and so forth. All the dictionaries, or certainly most of them, give the substitute uses without comment, and they are established in the language; but some wise men of Gotham have discovered that, if one is perversely ingenious enough, one can so use *r.* that it shall not be clear whether literal putting back or substitution is meant. This is true; here is an example in which a little thought is required: *We do not regard the situation as a simple one; a large proportion of the men on strike have been replaced, and as complete* rein-statement *is one of the demands of the* *union, there are obvious difficulties to be overcome.* To use *r.* there was foolish; 'have had their places filled' was the way to put it. But the wise men of Gotham were so proud of a discovery that ordinary people had made about hundreds of other words that they issued a decree against using *r.* at all in any sense except that of put back. The consequences in over-use and misuse of the verb SUBSTITUTE and the noun *substitution* have been lamentable, but need not be set forth here. It is enough to state that the objections to the secondary senses of *replace* and *replacement* are idle, and that only the same kind of care is required that is taken not to use *trip* in the special sense stumble, or *mistress* in the special sense female paramour, where the context makes confusion likely with the unspecialized senses.

**replenishment, repletion.** The first is the process of filling something up or the amount of matter that effects the process; the second is the filled-up condition. See -ION AND -MENT.

**replete.** *No teacher's bookcase is replete without it.* Everyone at once rightly corrects to *complete*; but why not *r.?* You can say 'a bookcase r. with works of genius'. Because quite full (*r.*) is not the same as adequately filled (*complete*).

**replica.** *The 'Devil' over the gateway,* a copy *of the grotesque on Lincoln cathedral, which gave rise to the proverb 'As sure as the Devil looks over Lincoln'. The present 'Devil' is a mere modern replica of the original imp erected by the founder.* 'Properly one made by the original artist' says the OED, after defining *r.* as a copy or duplicate of a work of art. *Properly,* therefore, there is no such thing as a modern r. of an ancient original; and it is this *proper* sense that alone makes the foreign word *r.* worth maintaining in English by the side of the abundant English words for copies or duplicates.

**reportage,** marked obsolete by the OED in its former sense of rumour or gossip, has since been revived with the

meaning typical style of reporting news, or the action of doing so.

**repp.** See REP.

**reprimand.** Some dictionaries give the pronunciation *rĕprĭmah'nd* for the verb (see NOUN AND VERB ACCENT), but it is usually the same as the noun, *rĕ'prĭmahnd.*

**repulsive.** See REPELLENT.

**request.** *The German Commission requested the Allied Commission for information as to whether an extension of the Armistice could be relied upon.* R. information from the A. C., r. to be informed by the A. C., r. that the A. C. would inform; any of these will do, but 'requested . . . for' is unidiomatic, and due to the ANALOGY of *ask* or of *r.* used as a noun.

**require.** 1. For the shop assistant's *Do you require anything more?* see GENTEELISMS. 2. *The Australian innings was wound up just before lunch, and England did not require to start on their long trail until afterwards.* Though there is some dictionary authority for this intransitive use of *r.,* the reporter, like the shop-assistant, expressed himself in a stilted and unnatural way.

**requirement, requisite** n. The two are so far synonyms that in some contexts either will do: *The requirements,* or *The requisites, are courage and callousness.* But *requirement* means properly a need, and *requisite* a needed thing: *That sum will meet my requirements,* never *my requisites;* but, just as the abstract *need* is often used for the concrete *needed thing,* so *requirement* may perhaps always be substituted for *requisite: Sponge, toothbrush, and other requirements* will pass, though *requisites* is better and more usual. For *requisite* (adj.) see ESSENTIAL.

**reredos.** Two syllables (*rēr'dŏs*).

**rescind** has *rescission. Recision* (cutting back), which was formerly used in the same sense, is now virtually obsolete.

**resentment.** *May I, as one in complete sympathy with the general policy of the Government, give expression to the strong r. I feel to the proposed Bill.* R. *at* or *against*, not *to*. Repugnance? See ANALOGY 3 and CAST-IRON IDIOM.

**residue, -uum, -ual, -uary.** There are two special uses, to each of which one noun and one adjective are appropriated—the legal sense concerned with what remains of an estate after payment of charges, debts, and bequests; and the mathematical, chemical, and physical sense of what remains after subtraction, combustion, evaporation, etc. The legal noun and adjective are *residue* and *residuary*, the mathematical etc. are *residuum* and *residual*, though the differentiation is occasionally infringed in both directions. In more general use, *residuum* implies depreciation, differing from *residue* as do *leavings* or *sweepings* from *remainder*. *Residuum* has plural *-dua*.

**resilience, -cy.** There is a very slight difference of sense: that *-ce* can and *-cy* cannot mean an *act* of rebounding. But since there is no chance of *-ce*'s being confined to that special sense this does not make the existence of the two anything better than an inconvenience; it is therefore best to use *-ce* always; see -CE, -CY.

**resin.** See ROSIN.

**resistance. 1.** *You have likened the r. of Ulster Unionists to be driven out of the Constitution . . . to the opposition . . .* Read *to being driven*; see GERUND 3.
**2.** *R.* was given a new specific meaning during the second world war, when it was used attributively with *movement*, or by itself, sometimes preceded by *underground*, for irregular operations carried on by the inhabitants of occupied territory against the invaders. As a generic term it embraces the activities of the bodies known variously as GUERILLA, *partisan*, *maquis*, and *chetniks*. Guerilla war, defined by the OED as 'an irregular war carried on by small bodies of men acting independently', was first applied to the r. movement in Spain during the Napoleonic occupation. It is a diminutive of the Spanish *guerra*, war. Partisan, an older word, originally meant 'a member of light irregular troops employed on special enterprises', but as early as 1706 was used for a commander who would later have been called a guerilla chief. In the Russian civil war it was the name given to isolated bodies of revolutionary forces, and in the second world war to those who took part in resistance movements in occupied Russia and Yugoslavia. Maquis was the French equivalent. It is a Corsican word meaning scrubland; a dictionary definition is 'Le maquis nourrit le betail, abrite le gibier et parfois les bandits'. The kind of country that provided a natural refuge and base of operations for bandits in Corsica served the same purpose for the members of the r. movement in France. Chetniks (from *cheta*, a band) was the name given to the guerilla forces raised by General Mikhailovitch in Yugoslavia.

**resoluble, resolvable.** Both are in use without distinction of meaning, the first being more a literary, and the other more a colloquial word. The negatives should be *irresoluble*, but *unresolvable*; see -ABLE 3; in *The number of irresolvable difficulties is relatively small*, correct either the prefix or the suffix. Accent *resoluble* on the first syllable and *resolvable* on the second. See also DISSOLUBLE and SOLUBLE.

**resolution, motion.** As names for a proposition that is passed or to be passed by the votes of an assembly, the two differ in that the passing of a motion results in action, and a m. is that something be done; while a resolution is not necessarily more than an expression of the opinion that something is true or desirable. Since, however, opinion often becomes operative, and since also resolutions as well as motions are moved, i.e. are at least in one sense motions, the distinction is elusive. It is nevertheless of some value if not too rigidly applied.

**resolve, solve.** Both words can be

used in the sense to clear up, explain, settle, a problem, difficulty, puzzle, etc. Popular choice is now showing a misguided preference for *resolve*, perhaps for the bad reason that it seems more dignified. Misguided, because *resolve* has a variety of other meanings; *solve* has only the one.

**resort, resource, recourse.** Confusion between these three is very frequent, and, since in some senses each is really synonymous with each, the confusion is, if not excusable, at least natural. The usual mistake is to say *resource* when one of the others is required. Of the following examples, the first four are unquestionably wrong; in the other two, the most idiomatic expression has not been chosen: *Such ships of the German Navy as remain in the Southern Seas must now have resource to the many sparsely-inhabited islands* (recourse). | *She will not be able to do so without resource to the sword* (recourse, resorting, resort). | *Surely he was better employed in plying the trades of tinker and smith than in having resource to vice* (recourse). | *. . . should an autonomous régime for Macedonia have been agreed to by Turkey without resource to war* (recourse, resort). | *. . . binding all Powers to apply an economic boycott, or, in the last resource, international force, against any Power which . . .* (resort). | *The question of having to send troops is only considered as a last resort* (resource).

The words are chiefly used in certain established phrases, given below; when alternatives appear in brackets, they are to be taken as less idiomatic. *To resort to; to have recourse (resort) to; without recourse (resort, resorting) to. Without resources; at the end of his resources; had no other resource left; the only resource (resort); as a last resource; in the last resort. His usual resource was lying; his usual recourse (resort) was to lying; his usual resort was Brighton. A man of great or no resource; a man of many or no resources. Golf is a great resource; Hoylake is a great resort.*

*Without resource* in the sense 'irreparably', though it has been used by good

writers, is rather a GALLICISM than an English idiom.

**respect.** The COMPOUND PREPOSITION *with r. to, in r. of*, should be used not as often, but as seldom, as possible. *Rules for making provision with r. to any matter with r. to which the Council thinks that provision should be made.* Why was *about* not good enough? See also REGARD and PERIPHRASIS.

**respective(ly).** Delight in these words is a widespread but depraved taste. Like soldiers and policemen, they have work to do, but, when the work is not there, the less we see of them the better; of ten sentences in which they occur, nine would be improved by their removal. The evil is considerable enough to justify an examination at some length. Examples may be sorted into six groups: A, in which the words give information needed by sensible readers; B, in which they give information that may be needed by fools; C, in which they say again what is said elsewhere; D, in which they say nothing intelligible; E, in which they are used wrongly for some other word; and F, in which they give a positively wrong sense.

A. RIGHT USES

*There are two other chapters in which Strauss and Debussy take respectively a higher and a lower place than popular opinion accords them.* But for *r.*, the reader might suppose that both composers were rated higher on some points and lower in others; *r.* shows that *higher* goes with Strauss, and *lower* with Debussy. | *That training colleges for men and women respectively be provided on sites at Hammersmith and St. Pancras.* But for *r.* he might take both colleges to be for both sexes; *r.* shows that one is for men and the other for women. | *This makes it quite possible for the apparently contradictory messages received from Sofia and Constantinople respectively to be equally true. R.* shows that the contradiction is not, e.g., between earlier and later news from the Near East, but between news from one and news from the other town.

### B. FOOLPROOF USES

The particular fool for whose benefit each r. is inserted will be defined in brackets. *Final statements are expected to be made today by Mr. Bonar Law and M. Millerand in the House of Commons and the Chamber of Deputies respectively* (r. takes care of the reader who does not know which gentleman or which Parliament is British, or who may imagine both gentlemen talking in both Parliaments). | *The Socialist aim in forcing a debate was to compel the different groups to define their r. attitudes* (the reader who may expect a group to define another group's attitude). | *The Chiefs of Staff will remain the heads of their r. services* (the reader who might think that the First Sea Lord was to take over the Army). | *Each of the Rugby first three pairs won their r. matches against opposition not to be despised* (the reader who might think that one of the Rugby pairs had won a match between two of the others).

### C. TAUTOLOGICAL USES

After each is given in brackets the expression or the fact that makes r. superfluous. *Having collected the total amount, the collector disburses to each proper authority its r. quota* (each . . . its). | *He wants the Secretary for War to tell the House in what countries they are at present stationed, and the numbers in each country respectively* (each). | *Madame Sarah Bernhardt and Mrs. Bernard Beere respectively made enormous hits in 'As in a Looking Glass'* (*hits*, plural). | *The October number of the Rassegna is chiefly remarkable for the r. articles of the Marchese Crispolto Crispolti on Pope Benedict V and the War and by the Marchese Colonna di Cesarò on Zionism and the Entente* (the mention of each article immediately after its author's name). | *I publish the banns of marriage between A and B; also between C and D. If any of you know cause or just impediment why these persons respectively should not be joined*

*together* etc. (the separation into pairs by *also between*).

### D. UNINTELLIGIBLE r.

*The writing-room, silence-room, and recreation-room, have respectively blue and red arm-chairs.* | *A certain estate is for sale; its grounds border three main roads, namely, Queen's, Belmont, and King's respectively.*

### E. r. FOR ANOTHER WORD

The writers of these mean no more than *both* (to be placed in the second after *Fellow*). *The two nurses' associations respectively organized in Scotland make no secret of their membership.* | *He was a Fellow of Balliol College, Oxford, and of the University of London respectively.*

### F. REVERSAL OF SENSE

*It is recognized that far too little is known by Englishmen and Americans about their r. countries; in this country there is only one lectureship on American history, and that is at King's College, Strand.* This can only mean that Englishmen know too little of England, and Americans know too little of America—which is no doubt true, but is not the truth that the writer wished to convey; 'about each other's countries' would have served both writer and reader.

The simple fact is that *respective(ly)* are words seldom needed, but that pretentious or meticulous writers drag them in at every opportunity for the air of thoroughness and precision they are supposed to give to a sentence, a fault to which lawyers and officials are specially prone.

**respirable.** *Rĕspĭr'able* was the pronunciation preferred by the OED, but the COD puts *rĕs'pĭrable* first. Perhaps our greater familiarity with respirators, together with the pull of the RECESSIVE ACCENT, accounts for the change.

**respite.** For the verb the OED gives *rĕ'spĭt* only, and prefers it to *rĕ'spīt* for the noun. But the latter seems now to be gaining ground for both, though not yet with much support from the dictionaries

resplendence, -cy. The first is recommended; see -CE, -CY.

restive is often used wrongly as a synonym of *restless*. *Restive* implies resistance. A horse may be restless when loose in a field, but can only be restive if it is resisting control. A child can be restless from boredom, but can only be restive if someone is trying to make him do what he does not want.

resurrect. See BACK FORMATION.

retina. Pl. *-as* or *-ae*; see LATIN PLURALS I.

retiral, little used outside Scotland, is a NEEDLESS VARIANT of *retirement*. See -AL NOUNS.

retire. For *retired admiral* etc., see INTRANSITIVE P.P., and for *r.* = go to bed see GENTEELISM.

retract makes *-tor*; see -OR. Of the two nouns *retrac(ta)tion*, the shorter is used in all senses, the longer only in the secondary or non-literal ones, i.e. where the meaning is not 'pulling backwards', but 'apologizing for' or 'cancelling' or 'revoking'. *Protrusion and retraction of the tongue*; *Offer and retrac(ta)tion of terms*; *Publication and retrac(ta)tion of a libel*.

retrieve makes *-vable*; see MUTE E. Of the nouns *retrieve* and *retrieval*, the first is used in particular phrases (*beyond*, *past*, *retrieve*), and the other elsewhere (*for the retrieval of his fortunes* etc.).

retro-. Pronunciation varies capriciously between *rĕtro-* and *rētro-*. In the two commonest compounds, *retrospect* and *retrograde*, it is usually *rĕtro-*.

retrograd-, retrogress(-). There are two series: (1) adj. and v. *retrograde*, n. *retrogradation*; (2) v. *retrogress*, n. *retrogression*, adj. *retrogressive*. But, as most of us have a preference for *retrograde* as the adj. and *retrogression* as the noun, and no great liking for either verb, there is unfortunately little prospect that one

series will oust the other, though *retrograde* v. and *retrogradation* are now virtually confined to astronomy.

return. For *the returned exile* etc. see INTRANSITIVE P.P.

rev. See REVEREND.

reveille. Usually pronounced *rĕvĕ'li* by officers and *rĕvă'li* by other ranks.

Revelation(s). Though the Bible title is *The Revelation of St. John the Divine*, the plural *Revelations* is quite established in ordinary speech, and to take exception to it is PEDANTRY. But *The Revelations* is a confusion of the correct *The Revelation* with the popular *Revelations*.

revenge. For *r.* v. and *avenge*, *r.* n. and *vengeance*, see AVENGE.

reverend, rev., reverent(ial). *Reverend* means deserving reverence, and *reverent* feeling or showing it.

Archbishops are *most reverend*, bishops *right r.*, and deans *very r.*; archdeacons are not *r.* but *venerable*. *Reverend* is abbreviated *Revd* or now usually *Rev.* To describe a clergyman as *Rev. Smith* instead of *Rev. J. Smith* or the *Rev. Mr* or *Dr Smith* is a common vulgarism, so common indeed (especially in Scotland and Ireland) that it may soon cease to deserve that description. Reporters giving lists of clergy have difficulties with the plural of the abbreviation; but, since *reverend* is an adjective (and not, like *parson* in the now disused 'Parsons Jones and Smith', a noun), there is neither occasion for nor correctness in such forms as *Revs.* and *Revds*; if *the Rev. J. Smith*, *W. H. Jones*, and *P. Brown* seems to do less than justice to the status of anyone but the first, each must be given a *Rev.* of his own. In other contexts, however, *reverend* can be treated as a noun, though rarely in the singular, except as a vulgarism. SOED quotes (1894) *We are not so quarrelsome as you reverends are*.

Between *reverent* and *reverential* the difference is much the same as that between PRUDENT and *prudential*, *reverential* being as applicable to what simulates reverence as to what is truly instinct with it, while *reverent* has only the laudatory sense. But *reverential* is often wrongly chosen merely as a LONG VARIANT; when *reverent* would not be out of place, *reverential* is a substitute as much weaker as it is longer.

**reverse,** n. Such phrases as 'remarks the r. of complimentary', meaning uncomplimentary remarks, are cumbrous specimens of WORN-OUT HUMOUR.

**reverse,** v. For the adjective, *-sible* is the prevalent form, negative *irreversible*.

**reversion** has various senses, chiefly legal or biological, to be found in any dictionary, and not needing to be set forth here. It suffices to say that they all correspond to the verb *revert*, and not to the verb *reverse*, whose noun is *reversal*. In the following extracts it has been wrongly given the meaning of *reversal*: *The reversion of our Free Trade policy would, we are convinced, be a great reverse for the working class.* / *But to undertake a complete reversion of the Bolshevik policy is beyond their powers.*

**reviewal.** See -AL NOUNS, and use *review* n.

**revisal.** See -AL NOUNS. *The Union demands a 'thorough revisal of the whole tariff'*; why not the established *revision*?

**revivals.** When towards the close of the 19th c. a method of curtailing debate in the House of Commons was found necessary, there was much talk of the French *clôture*, and it seemed for some years as if the French name would have to be taken over with the French thing; the English equivalent CLOSURE, naturalized in the 14th c., had become so unfamiliar that it did not suggest itself readily, and when proposed was not cordially received. 'Moving the closure' is now familiar enough; but, though the word had not become strictly obsolete, it was so rare as to look like either a new formation or a revival, and it is at once a good

specimen of the kind of revival that justifies the reviver and a good proof of how effectually a more or less disused word may come to life again. To anyone alive today it would hardly occur that *closure* was on a different footing from *budget* or *motion* or *dissolution* or *division* or any other parliamentary term. As to 'the kind of revival', the occasion may be defined as one on which a name has to be found for a new thing, and a question arises between a foreign word and a disused English one that might well have served if the thing and the word had been alive together. Another acceptable revival is the substitution of *Forenames* for *Christian names* as the heading of the appropriate column in certain returns and application forms required by government departments and local authorities. (See FORENAME.) Another is the re-emergence of DISASSEMBLE to meet the needs of the machine age — a word originally marked obs. in the OED but restored to current status in the 1933 Supp. For another see ASIAN. Another is *carrel* (a monk's study in a cloister) pleasantly revived for a small apartment or recess set apart for private use in a public library.

It is by no means uncommon for very ordinary words to remain latent for long periods. To take only some notable cases in the letter B, the OED records such disappearances of *balsam* (600 years), *bloom*, the iron-foundry word (600 years), *bosk* (500 years), *braze*, to make of brass (550 years); but the reappearance of these, except perhaps of *bosk*, was not so much a deliberate revival as a re-emergence out of the obscurity of talk into the light of literature. It is only with deliberate revivals, however, that it is worth while to concern ourselves here —words like *carven* (carved), *childly*, *dispiteous*, *dole* (grief), and *loan* (v.), or uses of words in obsolete senses such as *egregious* meaning excellent or *enormity* meaning hugeness. *Carven* seems to have been disused for 300 years, *childly* for 250; *dispiteous* (formerly *despite/ous* full of despite, now *dĭs/*

*piteous* unpitying) for 200; *loan* (v.) and *dole* for a long time in England at least. Revivals like these, and those of obsolete senses, not to fill gaps in a deficient vocabulary as *closure* and *disassemble* did, and as it has been suggested (see MAJORITY) that *plurity* might, but to impart the charm of quaintness to matter that perhaps needs adornment, are of doubtful benefit either to the language or to those who experiment in them. Is it absurdly optimistic to suppose that what the stream of language leaves stranded as it flows along consists mainly of what can well be done without, and that going back to rake among the debris, except for very special needs, is unprofitable? At any rate, the simple referring of any word to this article is intended to dissuade the reader from using it.

**revue** is a word imported from France early in the 20th c. for 'a theatrical entertainment purporting to give a review (often satirical) of current fashions, events, plays, etc.' (OED). This form of entertainment and that called a *musical* have since approximated to one another, but it is still a characteristic of the *r.* to have even less continuity of structure than the *m.*

**rhapsodic(al).** The short form is now usually limited to the original sense 'of the Greek rhapsodes', while *-ical* has usually (and might well have only) the secondary sense of ecstatically expressed or highflown; see -IC(AL).

**Rhenish.** Pronounce *rĕn-*.

**rhetorical question.** A question is often put not to elicit information, but as a more striking substitute for a statement. The assumption is that only one answer is possible, and that if the hearer is compelled to make it mentally himself it will impress him more than the speaker's statement. So *Who does not know . . .?* for *Everyone knows, Was ever such nonsense written?* for *Never was* etc.

**rhino** = rhinoceros. Pl. *-os*, see -O(E)S 5, or (see COLLECTIVES 3) *-o*. For *rhino* = money see BEAN.

**rhombus.** Pl. *-buses* or *-bi*; see LATIN PLURALS.

**rhyme.** **1.** As a technical term of English prosody, *r.* has a more limited meaning than its popular one. It consists of identity of sound in the terminal syllables of words, from the last fully accented vowel to the end, but only if the preceding consonantal sound is different. Thus *greet* rhymes with *deceit* but *seat* does not; *relation* rhymes with *station* but *crustacean* does not; *risible* rhymes with *visible*, but *invisible* does not. Words that, to judge from spelling, might have been rhymes, but have not in fact the required identity of sound, as *phase* and *race*, *love* and *move* and *cove*, are often treated as rhyming, as indeed in some cases they once did, but are called imperfect rhymes. One-syllable rhymes are called *male* or *masculine* or *single*, two-syllable *female* or *feminine* or *double*, three-syllable and four-syllable *triple* and *quadruple*.

**2.** *rhyme, rime.* Nothing seems to be gained, except indeed a poor chance of the best of three reputes (learning, pedantry, and error), by changing the established spelling. The OED states that *rhyme* 'finally established itself as the standard form', and that the revival of *rime* 'was to some extent due to the belief that the word was of native origin and represented OE *rím*' (= number). *Rhyme* is in fact the same word as *rhythm*, and ultimately from Greek *ῥυθμός*, though it came into English from French in the altered form *rime*, and was only later restored, like many other words, to a spelling more suggestive of its origin. It is highly convenient to have for the thing meant a name differently spelt from *rhythm*, but that convenience *rhyme* gives us as fully as *rime*, while it has the other advantage of being familiar to everyone.

**rhyming slang** originated in the cockney underworld in the early part of the 19th c., and later spread to Australia, the United States, and Ireland. The slang term consists of two or more words of which the last is a rhyme or assonance of the word to be represented. A few examples are: *Anna Maria* (fire), *apples and pears* (stairs), *bull and cow* (row), *Cain and Abel* (table), *elephant's trunk* (drunk), *France and Spain* (rain), *hot potato* (waiter), *plates of meat* (feet), *round the houses* (trousers), *trouble and strife* (wife), *Uncle Ned* (bed). The slang term is often abbreviated: thus the slang for *feet* is *plates* and for *trousers* is *round me*. Rhyming slang sometimes passes into ordinary colloquialism; it is for instance the origin of *brass tacks* (facts), *dicky* (unsound), and *raspberry* (expression of disapproval).

**rhythm.** Rhythmless speech or writing is like the flow of liquid from a pipe or tap; it runs with smooth monotony from when it is turned on to when it is turned off, provided it is clear stuff; if it is turbid, the smooth flow is queerly and abruptly checked from time to time, and then resumed. Rhythmic speech or writing is like waves of the sea, moving onward with alternating rise and fall, connected yet separate, like but different, suggestive of some law, too complex for analysis or statement, controlling the relations between wave and wave, waves and sea, phrase and phrase, phrases and speech. In other words, live speech, said or written, is rhythmic, and rhythmless speech is at the best dead. The rhythm of verse is outside the scope of this book, and that of prose cannot be considered in its endless detail; but a few words upon it may commend the subject as worth attention to some of those who are stirred by the mere name to ribald laughter at faddists and aesthetes.

A sentence or a passage is rhythmical if, when said aloud, it falls naturally into groups of words each well fitted by its length and intonation for its place in the whole and its relation to its neighbours. Rhythm is not a matter of counting syllables and measuring the distance between accents; to that misconception is due the ridicule sometimes cast upon it by sensible people conscious of producing satisfactory English but wrongly thinking they do it without the aid of rhythm. They will tell you that they see to it, of course, that their sentences sound right, and that is enough for them; but, if their seeing to it is successfully done, it is because they are, though they do not realize it, masters of rhythm. For, while rhythm does not mean counting syllables and measuring accent-intervals, it does mean so arranging the parts of your whole that each shall enhance, or at the least not detract from, the general effect upon the ear; and what is that but seeing to it that your sentences sound right? Metre is measurement; rhythm is flow, a flow with pulsations as infinitely various as the shape and size and speed of the waves; and infinite variety is not amenable to tabulation such as can be applied to metre. So it is that the prose writer's best guide to rhythm is not his own experiments in, or other people's rules for, particular cadences and stress-schemes, but an instinct for the difference between what sounds right and what sounds wrong. It is an instinct cultivable by those on whom nature has not bestowed it, but on one condition only—that they will make a practice of reading aloud. That test soon divides matter, even for a far from sensitive ear, into what reads well and what reads tamely, haltingly, jerkily, lopsidedly, topheavily, or otherwise badly; the first is the rhythmical, the other the rhythmless. By the time the reader aloud has discovered that in a really good writer every sentence is rhythmical, while bad writers perpetually offend or puzzle his ear—a discovery, it is true, not very quickly made—he is capable of passing judgement on each of his own sentences if he will be at the pains to read them, too, aloud. In all this, reading aloud need not be taken quite literally; there is an art of tacit reading aloud ('My own voice pleased me, and still more the mind's Internal echo of the imperfect sound'), reading with the eye and not the mouth, that is, but being as fully aware of the unuttered sound as of the sense.

Here, to conclude, are a few examples of unrhythmical prose, followed by a single masterpiece of rhythm. If these are read through several times, it will perhaps be found that the splendour of the last, and the meanness of the others, become more conspicuous at each repetition: *Mr. Davies does not let his learning cause him to treat the paintings as material only to be studied by the Egyptologist with a critical and scientific eye.* Never a chance of pausing, or an upward or downward slope, in the four lines. | *But, so far as I could see, nobody carried away burning candles to rekindle with holy fire the lamp in front of the ikon at home, which should burn throughout the year except for the short time it is extinguished in order to receive anew the light that is relit every year throughout the Christian world by Christ's victory over death.* Inordinate length of the last and subordinate member beginning at *except*, which throws the whole sentence off its balance. | *But some two or three months ago I asked the hospitality and assistance of your columns to draw public and civic attention to the above position of affairs, and to the fact that the use of the Embankment, as a thoroughfare, was limited, and, in fact, almost prohibited, by the very bad and deterrent condition of the roadway at both ends of the portion from Chelsea to Westminster, the rest of the road being fairly good, of fine proportions, and easily capable of being made into a most splendid boulevard, for all ordinary traffic, as a motor road, in which respect it was dangerously impossible at parts, and as a typical drive or walk.* This writer has produced a single sentence seventeen lines long without a single slip in grammar. That so expert a syntactician should be rhythm-deaf is amazing. | *Some simple eloquence distinctly heard, though only uttered in her eyes, unconscious that he read them, as, 'By the death-beds I have tended, by the childhood I have suffered, by our meeting in this dreary house at midnight, by the cry wrung from me in the anguish of my heart, O father, turn to me and seek a refuge in my love before it is too late!' may have arrested them.* Of what use to

talk of simple eloquence in a sentence contorted and disproportioned like that? | *Let anyone ask some respectable casuist whether Lavengro was not far better employed, when in the country, at tinkering and smithery than he would have been in running after all the milkmaids in Cheshire|, though tinkering is in general considered a very ungenteel employment|, and smithery little better|, notwithstanding that an Orcadian poet, who wrote in Norse about 800 years ago, reckons the latter among nine noble arts which he possessed|, naming it along with playing at chess, on the harp, and ravelling runes|, or as the original has it, 'treading runes'|—that is, compressing them into small compass by mingling one letter with another|, even as the Turkish caligraphists ravel the Arabic letters|, more especially those who write talismans.* One of the decapitable sentences from which if piece after piece is chopped off at the end the remainder after each chop is one degree less ill balanced than before.

*And the king was much moved, and went up to the chamber over the gate and wept: and as he went, thus he said: O my son Absalom, my son, my son Absalom! would God I had died for thee, O Absalom, my son, my son!*

**rhythmic(al).** Both forms are too common to justify any expectation of either's disappearance; yet there is no marked differentiation. What there is perhaps amounts to this, that *-al* is the more ordinary pedestrian term, and therefore better suited for the merely classifying use (*and other rhythmical devices*: cf. *so rhythmic a style*). See -IC(AL).

**ribbon, riband.** The second is 'now archaic'—OED. But the word is still so spelt sometimes in the expression *blue r.*, especially the *Blue riband of the Atlantic*.

**riches.** *But the promoters will certainly not need to go back to ancient history for it; they will have* an embarrassment of riches *from the immediate past*. See GALLICISMS and, for the number of *riches*, SINGULAR S.

**rick** (twist, sprain). See CRICK.

**rickettsia.** See MICRO-ORGANISMS.

**ricochet.** The spelling, accent, and pronunciation recommended are: *ricochet* (rĭ′kŏshā); *ricocheted* (rĭ′kŏshād); *ricocheting* (rĭkŏshā′ing). Cf. CROCHET.

**rid.** There is no clear line between *rid* and *ridded* in past inflexions, but the prevailing usage favours *rid*, almost always for past tense (*when he rid—* rarely *ridded—the world of his presence*) and for p.p. as active (*we have rid—* rarely *ridded—the land of robbers*) and always for p.p. as passive (*I thought myself well rid—* never *ridded—of him*).

**right. 1.** Except in the phrase *right away*, the American use of *right* as an intensive adverb (even as an apparently indispensable reinforcement of *now*) is still not quite comfortable in England, though often used by entertainers imitating American technique. It is one of those usages that were current when the American colonies were founded and were preserved by the emigrants but forgotten by those they left behind. When the entertainer says 'I'm right glad to be here' or 'Good-bye folks; I'll see you again right soon', he is using the word in just the same way as did the translators of the Psalmist's 'Then Israel should be right glad' and 'O hear me and that right soon', as the Queen still does when she issues a commission to one of her 'right trusty' cousins. **2.** *Right*, *righten*, vv. See -EN VERBS. **3.** *Right(ly)*, advv. The adverb *right* in the senses 'properly', 'correctly', is being squeezed out by the tendency to UNIDIOMATIC -LY. It is well, before using *rightly* in these senses, to consider whether *right* is not better, though usage is much less decided than with many alternative adverbs of the kind. In all the following types *rightly* is possible, but *right* is better: *He guessed* or *answered right* (but *He rightly guessed that it was safe* or *answered twenty-seven*); *You did right in apologizing* or *to apologize* (but *You rightly apologized*); *If I remember right* (but *I cannot rightly recollect*); *I*

*hope we are going right*; *If it was tied right, it will hold*; *Teach him to hold his pen right*.

**righteous.** Pronounced rī′chŭs, but under the influence of the speak-as-you-spell movement (see PRONUNCIATION) rī′tyŭs is encroaching on this and may displace it.

**rigour,** meaning strictness, is so spelt in Britain, but *rigorous*; see -OUR AND -OR, and -OUR- AND -OR-. The pathological term is *rigor*.

**rise. 1.** For *the risen sun* etc. see INTRANSITIVE P.P. **2.** *It is hoped that the Joint Committee will r. equal to the occasion,* . . . Either *r. to* or *be equal to*. With those to choose between there is no need for the popular PHRASAL VERB *measure up to*, which has only novelty to commend it. **3.** For *rise* and *raise* see RAISE.

**risible. 1.** Pronounce rĭz′ĭbl. **2.** *R.* is very liable to MISPRINTING as *visible*. **3.** *Were I to send my library of sixty specimens to auction I really expect some r. bid of, say, ten or fifteen pounds would be offered.* Originally meaning 'having the faculty or power of laughing, inclined to laugh', *r.* nearly perished except in the special sense 'of laughter' (muscles etc.), but has now enjoyed a revival in the sense 'provoking laughter', especially, as in this quotation, derisive laughter.

**road, street,** etc. Of the many different names we give to thoroughfares for vehicles, *road* is the most comprehensive. Any prepared surface along which vehicles may pass can be referred to in ordinary parlance as a road, though its title may be 'street' or even 'lane'. As an appellation, *road* is also the most important of them—or was until the revival of *way*; it is the natural word for the titles of our long-distance highways such as the Great North Road and the Great West Road, and for what we called the Dover Road, the Portsmouth Road, and the Bath Road, until we rechristened them, less romantically, A2, A3, and A4.

*Street* means, etymologically, a

paved way—*via strata*; hence its use, strange to our ears, for the great roads of the Roman occupation—Ermine Street, Stane Street, and Watling Street. Its current meaning is 'road in town or village comparatively wide as opposed to lane or alley, running between two lines of houses or shops' —OED. What was at first styled a 'road' will therefore become a 'street' when houses or shops are built beside it. Sometimes the name is changed when this happens (London's Oxford Street was once Oxford Road). More often it is not, and so our expanding cities and towns abound with streets that are still called 'roads'. *Road* seems to go naturally with the name of the place it leads to, and *street* with the name of a person associated with it—Hampstead Road and Old Kent Road, but Bond Street and Regent Street. *High Road* is a main thoroughfare in open country, *High Street* the principal street of a town.

These are the two commonest designations. But a street may also be called a *grove*, a *place*, a *row*, a *lane*, or an *avenue*, not to mention other names more or less self-explanatory, such as *circus*, *crescent*, *gardens*, *gate*, *hill*, *rise*, *square*, *terrace*, and *vale*. There was presumably some good reason at the time of christening for the selection of one of these fancy names, but time will usually have obliterated the features attributed to it in its dictionary definition. Westbourne Grove will disappoint anyone who hopes to find there 'a small wood or a group of trees affording shade or walks'. A distinguishing feature of a *place*, according to the OED, is that it is 'not properly a street'. That is still true of some of London's 'places', but most of them (Devonshire, Grosvenor, Portland, for instance) might now be called 'streets' with perfect propriety. To an even greater extent than streets and squares 'places' seem to be called after some noble family, no doubt because they were originally associated with a family mansion. A *row* is a 'street, especially a narrow one, formed by two continuous lines of

buildings'. But several of London's 'rows' are now broadish streets and bordered by lines of buildings that are by no means continuous. Even less does Rotten Row conform to the definition; but there *row* may be a different word; the origin of the name is unknown. One or two of London's 'lanes' have remained typically narrow and tortuous, especially in the City, but most have broadened out of recognition. *Avenues*, on the other hand, are still mostly 'broad roadways with trees (or other objects) at regular intervals'. But so are a great many other roads, and the reason why 'avenue' was preferred is not always apparent. Some (Northumberland for instance) were once true 'avenues' in that they marked the chief approach to a great house. It was an imposing word, and may have commended itself for that reason. In any case, no one could question the fitness of Eastern Avenue and Western Avenue for the names given in our time to the main approaches to London from east and west. (That an avenue may be expected to lead to some definite and desirable destination no doubt accounts for the use of that word, rather than *road* or *street*, in the well-worn cliché *explore every avenue*).

*Way*, as the name for a road is as old as Fosse Way and Icknield Way. It has been preserved in *highway* and *railway*, and agreeably revived for modern arterial roads, both generically (*motorway*) and for bypass sections of some of them (Lewisham Way, Watford Way, etc.).

Certain of these words have become stereotyped components of certain phrases, some intelligibly, others rather less so. No through *Road*, but One Way *Street*; Cross-*roads*, but *Street*-corners; Rule of the *Road*, but High*way* Code; Traffic *lane*, but Dual carriage*way*; Major *Road* ahead, but Clear*way*.

**roast. 1.** The use of the p.p. *roast* is very narrowly limited: *roast beef* or *lamb*, but *roasted coffee beans*; *a roast joint*, but *a well roasted joint*; *is*

**robustious** 530 **rostrum**

*better roast(ed) than boiled*, but *should certainly be roasted*. 2. For *rule the r.*, see RULE.

**robustious, rumbustious.** *Robustious* was in common use in the 17th c. (*O! it offends me to the soul to hear a r. periwig-pated fellow tear a passion to tatters*); in the 18th it became rare (Johnson said that it was now only used in low language and in a sense of contempt); in the 19th it had a revival, especially by archaizing writers. It is now less known than the colloquial *rumbustious*, probably a corruption of it.

**rocketry.** See SUMMITRY.

**rococo.** See BAROQUE.

**rodent operator.** See RATCATCHER.

**rodomontade,** not *rho-*.

**role, rôle.** Though the word is etymologically the same as *roll*, meaning the roll of MS. that contained an actor's part, the DIFFERENTIATION is too useful to be sacrificed by spelling always *roll*. But, there being no other word *role* from which it has to be kept distinct, both the italics and the accent might well be abandoned. As to the sanctity of the French form, see MORALE.

**Roman Catholic.** See CATHOLIC.

**Romanes, Romany** (gypsy language). Pronounce *rŏ'mănĕz, -nĭ*. But the name of the biologist who founded a lecture at Oxford is pronounced *Rōmah'nĕz*.

**Romansh, Roumansh, Rumans(c)h** (the Romance dialects of S.E. Switzerland). The OED treats the first as the standard form.

**rondo.** Pl. *-os*; see -O(E)S 6.

**roof.** Pl. *-fs*; see -VE(D).

**roost.** For *rule the r.*, see RULE.

**root** (philol.). Roots are the ultimate elements of words, not admitting of analysis. In the word *unhistorically*, un-, -ly, -al, -ic, -tor, can all be set aside as successive affixes modifying in recognized ways the meaning of what each was added to. There remains HIS, which would be called the root

if *unhistorically* were an isolated word; investigation shows that the same element, with phonetic variations that are not arbitrary, is present in many other words, e.g. in English *wit*, in the Latin-derived *vision*, and in the Greek-derived *idea*; and that the Indo-European or Aryan root is VID, with the sense sight or knowledge. Cf. STEM.

**root, rout** (poke about). The second form is called by the OED an 'irregular variant' of the first. The two, with the other verb *root* directly connected with the noun, naturally cause some difficulty. It would be a convenient differentiation if the spelling *root* could be confined to contexts in which the notion of *roots* is essential, and *rout* were adopted where search or bringing to light is the point. So we should get rooting up trees, rooting out weeds or sedition, but routing about in a lumber-room or among papers, routing out secrets, routing a person out of bed, routing up a recluse or a reference. Pigs, being equally intent on roots and search, may root or rout (or rootle) indifferently.

**rosary, -ery** (rose-garden). The first is the old word (from 15th c. in OED), direct from Latin *rosarium*. The second is a 19th-c. formation made presumably, from *rose* and *-ery*, by someone not aware that *rosary* has this sense. *Rose-garden* or *-bed* is recommended for ordinary use, and *rosary* for verse.

**rosin** is by origin merely a form of *resin* changed in sound and spelling; but the two are now so far differentiated that *resin* is usual for the liquid in or taken from the tree, and as the general chemical term for substances having certain qualities, while *rosin* denotes the distilled solid.

**roster.** The dictionaries, some of which still give *rŏster*, have been slow in catching up with the practice of the Services, the chief users of the word, who say *rōster*, and with the dictum of Skeat, 'The *o* is properly long: pron. *roaster*'.

**rostrum.** Plural usually *-ra*, rarely *-ums*. See -UM.

**rota(to)ry.** 1. Pronounce *rō'tătory*; see RECESSIVE ACCENT. 2. *Rotary* is not, like *authoritive, deteriate,* and *pacifist,* a shortening of a more correct form, but is a separate word: *rota* wheel gives *rotarius* (English *rotary*) wheel-like; *roto* revolve gives *rotatorius* (English *rotatory*) revolving etc. On the other hand there is no important difference in meaning either essential or customary, and that *-atory* has survived as well as *-ary* in mechanics may be because the most familiar use of *-ary* is now as the name of the clubs. But the risk of confusion is so small that *rotatory* may well be regarded as a SUPERFLUOUS WORD.

**rotten.** For *something r. in the state of Denmark,* see IRRELEVANT ALLUSION.

**rough(en),** vv. See -EN VERBS; but the relation between this pair demands some further treatment. 1. The intransitive verb in its literal sense of become rough is always *roughen,* except that the addition of *up* occasionally enables *rough* to serve, e.g. of the sea. 2. In the simple transitive senses also (= make rough), *roughen* is usual, but if *up* is added *rough* is preferred, and *rough* by itself is the word for arming horseshoes against slipping. 3. In the other transitive senses of to treat roughly or shape roughly (the latter usually with adverbs, *in, off, out*), the verb is *rough*: *rough a horse,* break it in; *rough a calf,* harden it by exposure; *rough* (up) *a person,* abuse or maltreat him (*Briton freed after being roughed up*); *rough in the outlines*; *rough off timber*; *rough out a scheme*; *rough a lens,* shape without polishing it. 4. To take things in the rough is to *rough it.*

**rout** (poke about). See ROOT.

**route.** The ordinary pronunciation *root* has now largely displaced the *rowt* that formerly prevailed in military phrases such as *r. march, column of r.*

**rowan.** The OED gives *rō'ăn* as the English and *row'ăn* as the Scottish pronunciation, but the latter is now common in England also.

**rowlock.** Pronounce *rŭ'lŏk.*

**-r-, -rr-.** Monosyllables ending in -r double it before suffixes beginning with vowels if the vowel sound preceding it is short, but not if it is long: *barring* but *nearing, stirred* but *chaired, currish* but *boorish.* Words of more than one syllable follow the rule for monosyllables if their last syllable is accented (with the exception noted below), but otherwise do not double the r; *preferred* but *proffered, interring* but *entering, abhorrent* but *motoring.* Exception: *infer, prefer, refer,* and *transfer,* though accented on the last syllable, give adjectives in *-erable,* and shift the accent to the first syllable: *prĕ'ferable* etc. For *confer* see CON-FER(R)ABLE.

**ruddle** (red ochre, and, as verb, colour with this) has the two variants *raddle* and *reddle,* of which *raddle* is the form usually preferred as a contemptuous synonym for rouge and rouging, and *reddle* is an occasional variant of *ruddle. Ruddle* itself is applied chiefly to sheep-marking.

**ruff** (bird) has fem. *reeve.* 'A very remarkable form, which has not been explained' (Skeat).

**ruination** is not, like *flirtation, flotation,* and *botheration,* a HYBRID, being regularly formed from *ruinate*; but it now has the effect of a slangy emphatic lengthening of the noun *ruin.* This is only because the parent verb *ruinate,* which was common in serious use 1550–1700, is no longer heard; but the result is that *ruination* is better avoided except in facetious contexts.

**rule.** 1. For 'The exception proves the *r.*' see EXCEPTION. 2. *Rule the roast* (roost). The OED gives no countenance to *roost,* and does not even recognize that the phrase ever takes that form. But most people say *roost* and not *roast*; they have never heard of *rule the roast,* and think that the reference is to a cock keeping his hens in order. Against this tempting piece of popular etymology the OED offers us nothing more succulent than 'None of the early examples throw any light on the precise origin of the expression'. In seven out of the eight pre-18th-c. examples quoted the spelling is not *roast* but *rost* or *roste*; but the OED philologists would doubtless tell us that *rost(e)* is more likely to represent Old-French *rost* (roast) than Old-English *hróst* (roost). Writers should take warning, at any rate, that *rule the roast* is the orthodox spelling, and that when they have written it the compositor must be watched. But *rule the roost* is so much commoner, and to most of us seems so much more intelligible, that the day may come when we shall not be able to write *roast* without being suspected of DIDACTICISM.

**rumbustious.** See ROBUSTIOUS.

**run.** For *fresh-run salmon* etc., see INTRANSITIVE P.P.

**ruridecanal.** The pronunciation -*dĕ-cā'nal* is preferable to -*dĕ'cănăl*.

# S

**'s.** 1. For *for conscience' sake* etc. see SAKE.
2. For *Achilles'*, *Jones's*, etc., and for questionable uses of *'s*, see POSSESSIVE PUZZLES.
3. For such errors as *to use a word of Coleridge* instead of *Coleridge's*, see OUT OF THE FRYING-PAN.

**Sabbatic(al).** Both forms are now rare, except *sabbatical* in such phrases as *s. term*, *s. year*. See -IC(AL).

**sabotage.** The use in English of this French word (n. and v.) in the sense of malicious damage by workmen to their employers' property dates from the early 20th c. The origin of the usage has been attributed to the practice of throwing sabots into machinery (cf. our 'throw a spanner into the works'), or, alternatively, to 'the cutting of the shoes (*sabots*) holding the railway lines' during the French rail-

way strike of 1912. Neither is convincing; there is no need to look further for an explanation than the figurative use of *sabot*, at least a hundred years old, for any scamped or botched piece of work; hence by the end of the 19th c. *sabotage* had come to mean anything done maliciously by workmen, especially by way of bad workmanship, to injure their employers' interests. Once established in the English language, it quickly became a VOGUE WORD, especially for the supporters of any project to apply to people who successfully opposed it—a synonym for such words as *obstruct, frustrate, wreck, destroy*. It is properly used only in contexts appropriate to its essential implication of malice and disloyalty; perhaps one of the reasons for its popularity is that it enables its users to imply a charge of malice without actually making one. Even within its proper implication of disloyalty it is given plenty of work to do. In the *General Law Amendment Act* passed by the South African Parliament in 1962 the word is defined as including any action that endangers law and order, health, water, or electrical services, medical, sanitary, or fire services, food, the free movement of traffic and postal communications, as well as damage to property in certain cases, and the illegal possession of weapons and explosives. That, as others have said for other reasons, is really too much.

**saccharin(e).** See -IN AND -INE; there is, however, some convenience in using *saccharin* for the noun and *saccharine* (-*ēn*) for the adjective.

**sacerdotage.** See FACETIOUS FORMATIONS.

**sack,** dismiss(al), having been on record for well over a hundred years, has earned promotion from the slang to the colloquial class. See also FIRE.

**sac, sack, sacque.** *Sac* is a medical and biological word, not a dressmaker's or tailor's. For the garment, *sack* is the right form. The other spellings are pseudo-French, wrong in different degrees: there is no French word *sacque*; there is a French word *sac*, but it is not, as the English *sack* is, the name for a particular garment.

**sacrilegious** is often misspelt from confusion with *religious* and the analogy of the ordinary pronunciation of the noun (*sacrilij*). The pronunciation *-ĭjus* is indeed so common that the dictionaries now recognize it as an alternative to *-ējus*.

**saga.** Any of the narrative compositions in prose that were written in Iceland or Norway during the Middle Ages; in English use often applied to any tale of high adventure, and also, following a fashion set by Galsworthy, to a series of novels of contemporary life in which the same characters reappear. But there is an epic quality in the word that should be respected, and save it from the VULGARIZATION of being applied to a strip cartoon. *One suspects she would have figured in one of Flook's sagas, if Flook had only met her.*

**sage.** For *the s. of Chelsea* see SOBRIQUET.

**said.** 1. *S.* = aforesaid. 2. *S. he* etc.
  1. *(The) said.* In legal documents, phrases like 'the s. Robinson', 's. dwelling-house', are traditional precautions against any possible ambiguity that may lurk in *he* and *it*. Jocose imitation of this use (*regaling themselves on half-pints at the s. village hostelries*), once common, but now indulged in only by writers desperately anxious to relieve conscious dullness, is to be classed with WORN-OUT HUMOUR
  2. *Said he* etc. *Said he, said So-and-so,* placed after the words spoken, is entirely unobjectionable; the ingenuity displayed by some writers in avoiding what they needlessly fear will bore their readers is superfluous ('One of the only attempts at a literary heightening of effect', says a famous critic about a popular mid-20th c. novel, 'is the substitution for the simple "said" of other more pretentious verbs—so that the characters are always shrilling,

barking, speculating, parrying, wailing, wheedling, or grunting whatever they have to say'). But to put *said* before the words said (*Said a Minister: 'American interests are not large enough in Morocco to induce us to . . .'*) is not equally irreproachable. The sprightliness of it was indeed denounced as intolerable in the first edition of this dictionary. It can no longer be called that; we have to bear it whether we like it or not, for it has become a commonplace of popular journalism. An extension even more distasteful to old-fashioned people is the use of *says so-and-so* to introduce quotations in serious writing. (*Says Dr. Johnson: 'A writer of dictionaries is a harmless drudge.'*) For further discussion and illustration of the points mentioned in this paragraph see the last section of INVERSION.

**sail.** For *plain sailing* see PLAIN. By the side of the usual but etymologically abnormal *sailor*, the normal agentnoun *sailer* exists for use in such contexts as *She* (ship) *is a slow sailer*.

**Saint.** *St* or *S.* is better than *St.* for the abbreviation (see PERIOD IN ABBREVIATIONS); Pl. *Sts* or *SS*.

**St Stephen's.** See SOBRIQUET.

**sake.** For *God's s., for mercy's s., for Jones's s., for Phyllis's s.*; but when the enclosed word is a common noun with a sibilant ending, whose possessive is a syllable longer than its subjective, the s of the possessive is not used; an apostrophe is sometimes, but not always, written; *for conscience s., for goodness' s., for their office s., for peace' s.*

**salad days** (one's raw youth) is one of the phrases whose existence depends on single passages (see *Ant. and Cleop.* I. v. 73 *My s. d. when I was green in judgement*). Whether the point is that youth, like salad, is green and raw, or that salad is highly flavoured and youth loves high flavours, or that innocent herbs are youth's food as milk is babes' and meat is men's, few of those who use the phrase could perhaps tell us; if so, it is fitter for parrots' than for human speech.

**sal(e)able.** See MUTE E.

**saline.** Pronounce *sā'līn*, and see -IN AND -INE and FALSE QUANTITY.

**salivary.** Pronounce *să'lĭvărĭ*, and see FALSE QUANTITY.

**salve.** The noun and verb meaning remedy ought strictly to be pronounced *sahv*. The verb meaning save or rescue is an entirely separate one, a BACK-FORMATION from *salvage*, pronounced *sălv*. It is no doubt because the first is now usually mispronounced *salv* that *salvage* has assumed the role of a verb and is displacing the second. The Latin word meaning Hail!, and used chiefly as the name of a R.-C. antiphon, is pronounced *să'lvē*.

**sal volatile.** Pronounce *săl vŏlă'tĭlĭ*.

**same.** *S.* or *the s.*, in the sense the aforesaid thing(s) or person(s), as a substitute for a pronoun (*it, him, her, them, they*) was once good English, abundant in the Bible and the Prayer Book, but is now an ARCHAISM, surviving mainly in legal documents and COMMERCIALESE; a modern example of the latter is *This charge was an error and we have struck same from our books. We enclose a revised account and trust you will now be able to pass same.* But it is by no means confined to law and commerce. It has the peculiarity that it occurs chiefly in writing, not often in speech, and yet is avoided by all who have any skill in writing. In all the extracts below, as well as in the letter quoted above, the writers would have shown themselves much more at their ease if they had been content with *it, them,* or other pronoun. *Shops filled to the doors with all kinds of merchandise and people eager to acquire* t. s. | *If not directly, at least through the official presence of their representatives, or by a chosen delegation of* t. s. | *Sir,— Having in mind the approaching General Election, it appears to me that the result of* s. *is likely to be as much a farce as the last.* | *I again withdraw the statements, and express my regret for having made* t. s. | *I consider this question as already*

settled, *and consequently any further discussion on* s. *is pure waste of time.* | *In view of the dissatisfaction caused by the management in dealing with the wage application and the antiquated system of labour relations we feel that immediate steps should be taken to remedy* t. s. Did the writer of the last quotation fall back on *the same* because the choice between *it* and *them* would have faced him with the problem of making up his mind what the pronoun's antecedent was?

**samurai.** Pronounce *să'moŏrī*. Pl. same.

**sanat-, sanit-.** The chief words, as they should be spelt, are: *sanatorium* a healing-place; *sanative* and *sanatory* curative; *sanitary* conducive to public health; *sanitation* measures to secure public health; *sanitarian* one who favours sanitary reform. *Sanitarium* is the usual U.S. word for *sanatorium*; *sanitorium, sanatarium,* and *sanitory,* do not exist.

**sanction,** n. The popular sense (permission, authorization, countenance, consent) has made such inroads into the more original senses still current especially in Law and Ethics that it is worth while to draw attention to these. They have had a popular revival, especially in the phrase *economic sanctions,* as a possible method of enforcing decisions of the League of Nations and its successor the United Nations. The s. of a rule or a system is the consideration that operates to enforce or induce compliance with it; judicial punishment is the *s.* of the law against breaches of it. The OED quotes from T. Fowler: 'Physical ss. are the pleasures and pains which follow naturally on the observance or violation of physical laws, the ss. employed by society are praise and blame, the moral ss. . . . are . . . the approval and disapproval of conscience; lastly, the religious ss. are either the fear of future punishment, and the hope of future reward, or, to the higher religious sense, simply the love of God, and the dread of displeasing Him.'

**sand-blind** is neither (like, say, *purblind*) a current word, nor (like, say, *bat-blind*) intelligible at sight. Its modern existence depends on one passage (*M. of V.* II. ii. 35–80), and it can rank only as an ARCHAISM. The latest quotation in the OED is dated 1905, *But there is a sort of sand-blindness endemic in the Liberal party just now. Sand* is now generally taken to be a corruption of OE *sam* = half. If so, the appearance of the word misled both Shakespeare (*This is my true-begotten father, who, being more than sand-blind, high gravel blind, knows me not*) and Dr. Johnson, whose definition is *Having a defect in the eyes by which small particles appear to fly before them.*

**sanguine** has been virtually put out of business by the very inferior OPTIMISTIC. Candour, however, compels the admission that *optimism* and *optimist* have the advantage in mechanical convenience over *sanguineness* and *sanguine person.*

**Sanhedrim, -in.** 'The incorrect form *sanhedrim* . . . has always been in England (from the 17th c.) the only form in popular use'—OED.

**sans.** As an English word, pronounce *sănz*; but it is at best WARDOUR STREET English: *The poet whom he met sans hat and coat one four-o'clock-in-the-morning.*

**Santa Claus** is from a Dutch-dialect form of *Saint* (*Ni*)*cholas* and there can be no good reason for preferring it to our *Father Christmas.*

**sapid,** unlike its negative *insipid*, is a merely LITERARY WORD.

**sapient.** Except in the expression *homo sapiens* s. is chiefly a LITERARY WORD, and usually ironical, e.g. M. Arnold's doctor who 'shakes his s. head and gives the ill he cannot cure a name', and T. S. Eliot's 'sapient sutlers of the Lord'.

**saponaceous,** apart from its use in chemistry, is a favourite POLYSYLLABIC HUMOUR word.

**sapor,** apart from its use in medicine, is a merely LITERARY WORD; for the spelling *-or*, see -OUR AND -OR.

**sarcasm** does not necessarily involve irony, and irony has often no touch of sarcasm. But irony, or the use of expressions conveying different things according as they are interpreted, is so often made the vehicle of s., or the utterance of things designed to hurt the feelings, that in popular use the two are much confused. The essence of s. is the intention of giving pain by (ironical or other) bitter words. See also IRONY, and HUMOUR.

**sardine** (stone; *Rev.* iv. 3). Pronounce *sar'dīn.* It is probably an error for *sardius*, a variety of cornelian, which is substituted for it in the R.V. and the N.E.B.

**sardonic.** See HUMOUR etc. for some rough distinction between this, *cynical*, *sarcastic*, etc.

**sartorial.** See PEDANTIC HUMOUR.

**satellite.** The word is now much to the fore in more than one sense—*s. states, earth s., s. towns.* Primarily it means (from Lat. *satelles*) a member of a body-guard of an important personage, often with an implication of subservience; it is therefore applied with singular aptness to the s. states of the Soviet Union. Its use for a secondary body revolving round a planet is nearly as old. Its adjectival use for towns built to absorb some of the population and industries of overcrowded cities is recent; it reflects the idea of 'hangers-on' implicit in the original meaning and is a legitimate extension.

**satiety.** Pronounce *sătī'ĕtĭ.*

**satire.** For rough distinction from some near-synonyms, see HUMOUR etc. Here it may be added that s. has recently been suffering VULGARIZATION. A word that suggests the powers of an Aristophanes, a Juvenal, or a Swift, and an impulse of *saeva indignatio* is prostituted when it is applied to mere snook-cockers of whom it has been said by one critic, jealous for the integrity of the word, that their only concern is to 'find someone who is doing something—no matter who, no matter what —and fling a few insults at him', and by another that their conception of satire is 'a cannibal dance round the idea of authority'.

**satiric(al).** The senses addicted to, intending, good at, marked by, satire are peculiar to the long form (*a -al rogue*; *you are pleased to be -al*; *with -al comments*; *a -al glance*). In the merely classifying sense of or belonging to satire (*the —— poems of Pope*; *the Latin —— writers*), either form may be used, but *-ic* is commoner. This DIFFERENTIATION might well be hastened by deliberate support; but the line of demarcation between the two groups is not always clear. See -IC(AL).

**satiric, satyric.** The two spellings represent two different and unconnected words; *satyric*, which is in learned or literary use only, means of satyrs, and especially, in *s. drama*, a form of Greek play having a satyr chorus.

**satisfy.** There is ample authority, going back several centuries, for the use of *s.* in the sense of 'to furnish with sufficient proof or information; to set free from doubt or uncertainty' (OED). But this meaning sometimes clashes with that of to please or content; the fact of which one is satisfied in one sense may be far from satisfactory in the other. For this reason it is unfortunate that, especially in official pronouncements, the word *convinced* seems to be forgotten, and *satisfied* is becoming the standard way of announcing factual conclusions. Perhaps its extra syllable seems to give it a more authoritative air. An announcement of the type in which *convinced* would have been more suitable is *The rescue party, on returning to the surface, said they were satisfied that there was no possibility of any more of the men being found alive.*

**Saturnalia.** See LATIN PLURALS 3.

The word is originally plural, but, being the name of a festival, comes to be construed, both in literal and metaphorical use, more often as singular (*the S. was*, or *were*, *at hand*; *now follows a s. of crime*). When a real plural is required (*the sack of Magdeburg, the French Revolution, and other such s. of slaughter*), the form is *-ia* not *-ias*.

**satyr.** See FAUN for distinctions.

**save** (except). **1.** For *s. and except*, see PLEONASM 2. **2.** Trench (*English Synonyms*, 4th ed., 1858), writing on 'except, excepting, but, save', has no more to say of the last than that ' "Save" is almost exclusively limited to poetry'. He would have a surprise if he were to see a modern newspaper; we can still say that it ought to be almost limited to poetry, but no longer that it is. Though nearly everyone uses *except* or *but*, not *s.*, in speaking, and perhaps everyone in thinking, and though the natural or 'dominant' word *except* is neither undignified nor inferior in clearness, some people seem to have made up their minds that it is not good enough for print, and very mistakenly prefer to translate it, irrespective of context, into *s*, making *s.* a FORMAL WORD, like the policeman's *proceed* for *go*. No doubt it is natural to fall back on *s.* to avoid a jingle such as *Permission will not be granted except in very exceptional circumstances.* But does anyone not a writer—and does any good writer—think that the substitution of the formal *s.* for the natural *except* in the following sentences (or *but* in the first of them) has improved them? *The handful of ship's officers could do nothing s. summon the aid of a detachment of the Civic Guard.* | *The spur proved to be so admirably adapted to its purpose that it has existed unaltered, s. in detail, to the present day.* | *So completely surrounded by other buildings as to be absolutely invisible—s. from a balloon or an aeroplane.* | *The baby takes no special harm, s. that it is allowed to do as it likes, and begins to walk too soon.* | *The increased rates will take effect on* *the Underground lines, s. on one stretch between Bow and Barking.*

**save,** v. *S. the mark* (with variants *God s., bless, God bless, the mark*) is a stylistic toy, of which no one can be said to know the original meaning, though different people make different guesses at it. The OED's description of it, as it now survives, is: 'In modern literary use (after some of the examples in Shakespeare), an expression of impatient scorn appended to a quoted expression or to a statement of fact.'

**saw** has p.p. *sawn*, rarely *sawed*.

**Saxonism and anti-Saxonism.** Saxonism is a name for the attempt to raise the proportion borne by the originally and etymologically English words in our speech to those that come from alien sources. The Saxonist forms new derivatives from English words to displace established words of similar meaning but Latin descent; revives obsolete or archaic English words for the same purpose; allows the genealogy of words to decide for him which is the better of two synonyms. Examples of the first kind are FOREWORD (earliest OED quotation, 1842) for *preface*, and *birdlore* (1830) for *ornithology*, and BODEFUL (1813) for *ominous*; of the second, BETTERMENT for *improvement*, HAPPENINGS for *events*, *english* for *translate* (*into English*), FOLK for *people*, and FOREBEAR for *ancestor*; of the third, BELITTLE for *depreciate*, *burgess* or *burgher* for *citizen*. The wisdom of this nationalism in language—at least in so thoroughly composite a language as English—is very questionable. We may well doubt whether it benefits the language; that it does not benefit the style of the individual, who may or may not be prepared to sacrifice himself for the public good, is pretty clear. Here is the opinion of the *Dictionary of National Biography* on Freeman's English: 'His desire to use so far as possible only words which are purely English limited his vocabulary and was some drawback to his sentences.'

The truth is perhaps that conscious deliberate Saxonism is folly, that the choice or rejection of particular words should depend not on their descent but on considerations of expressiveness, intelligibility, brevity, euphony, or ease of handling; at the same time any writer who becomes aware that the Saxon or native English element in what he writes is small will do well to take the fact as a danger-signal. But the way to act on that signal is not to translate his Romance words into Saxon ones; it is to avoid abstract and roundabout and bookish phrasing whenever the nature of the thing to be said does not require it.

Anti-Saxonism is not, like Saxonism, a creed. There are indeed, properly speaking, no anti-Saxonists. The term is here used as a name for the frame of mind that turns away not so much from the etymologically English vocabulary as from the homely or the simple or the clear. It is a practice and a propensity that go far to account for the follies of Saxonism. *Happenings* and *birdlore* and *bodeful* and the like are the products of a healthy revulsion from the turgid taste that finds satisfaction in such words as *adumbrate*, *ameliorate*, and *eventuate*, and in the many other ways that are described in, for instance, ABSTRACTITIS, AVOIDANCE OF THE OBVIOUS, LOVE OF THE LONG WORD, PEDANTIC HUMOUR, PERIPHRASIS, and -TION endings. That the meaning of many of the words and phrases favoured by the anti-Saxonist is vague is a recommendation to one kind of writer as saving him the trouble of choosing between words of more precise meaning, and to one kind of reader as a guarantee that clear thought is not going to be required of him.

**say. 1.** Except as a poeticism, the noun survives only in such phrases as *to have a s.* (to have the right to be consulted) and *to have said one's s.* (to have finished expressing one's opinion). **2.** The use of the verb's imperative to introduce an hypothesis or an approximation (*Let us meet soon—say next Monday*; *You will need some cash—say*

£5) is established idiom. 3. The ordinary pronunciation of *says* (v.) is *sez*, and it is odd that many writers of fiction, including Kipling, should spell it *sez* in dialogue by way of indicating Irish brogue. 4. For *said* or *says So-and-so* see SAID.

**saying.** 'As the s. is', or 'goes', is often used by simple people, speaking or writing, who would fain assure us that the phrase they have allowed to proceed from their lips or pen is by no means typical of their taste in language; it only happens to be 'so expressive' that one may surely condescend to it for once. Well, *qui s'excuse s'accuse*; if the rest of their behaviour does not secure them from insulting suspicions, certainly the apology will not. See SUPERIORITY.

**scabies.** Now usually two syllables: *skā'bēz*. See -IES, -EIN.

**scallop, sco-.** The spelling is usually with *-a-*, but the pronunciation with *-ŏ-*. The verb makes *-oping, -oped*; see -P-, -PP-.

**scandalum magnatum.** The second word is the genitive plural of Latin *magnas* a magnate, not a p.p. agreeing with *scandalum*. The phrase means the offence of uttering a malicious report against some high official, and the use of it in such senses as 'a crying scandal' is a blunder.

**scant**, adj., is a LITERARY WORD, preferred in ordinary contexts to *scanty*, *small, few, short*, etc., only by those who have no sense of incongruity (*The attendance was so scant as to suggest that many members must have anticipated the holiday*). It survives as a current word, however, in some isolated phrases, as *s. courtesy, s. attention, s. regard*, and, echoing Hamlet's mother, *s. of breath*.

**scarce**, adv., used instead of *scarcely*, is a LITERARY WORD. It is true that the OED says: 'Before adverbs in *-ly* the form *scarce* is often adopted instead of *scarcely*, to avoid the iteration of the suffix.' On that iteration, see -LY; but

such avoidance is a case of OUT OF THE FRYING-PAN.

**scarcely.** 1. *S. . . . than*. 2. *Not* etc. *. . . s*.

1. *S. . . . than*. *S. was the nice new drain finished than several of the children sickened with diphtheria.* For this construction, condemned in OED (s.v. *than*) as erroneous, see HARDLY 2. The ANALOGY of *no sooner . . . than* is no doubt responsible. *Before* or *when* is what should be used with *scarcely*.

2. *Not* etc. *. . . s*. We most of us feel safe against even saying 'I don't s. know', with *not* and *s.* in hand-to-hand conflict; but, if a little space intervenes, and the negative is disguised, the same absurdity is not very rare in print: *The services of the men who have worked the railway revolution* without *the travelling public being* scarcely *aware that we are at war should not be forgotten.* | *It has been* impossible *to tell the public* s. *anything about American naval cooperation with the British.* The English for *without s. realizing* is either *s. realizing*, or *without quite realizing*, or *not fully realizing*.

**scarf.** The dictionaries have not yet decided whether to prefer the plural *-fs* or *-ves*. In practice, *-ves* is probably more usual for the article of clothing, and *-fs* is invariable for the carpentry term.

**scavenge(r)**, vv. *Scavenger*, n., is the origin, in English, from which *to scavenge* is a BACK FORMATION, the normal verb being *to scavenger*; cf. to soldier, to filibuster, to buccaneer, to privateer, to mountaineer, to volunteer, to solder, to bicycle, and hundreds of other verbs that are in fact verbal uses of nouns. *Scavenge*, however, is now much commoner than *scavenger* as the verb.

**scena** (mus.). Pronounce *shā'nah*. But the cognate and commoner *scenario* is now anglicized into *sĕnãrio*, at least in the world of films

**scene.** For synonyms in the sense locale, see FIELD.

# sceptic(al)

## science and art

**sceptic(al), scepsis,** etc. The established pronunciation is *sk-*, whatever the spelling; and with the frequent modern use of *septic* and *sepsis* (the latter a 19th-c. word only), it is well that it should be so for fear of confusion. But to spell *sc-* and pronounce *sk-* is to put a needless difficulty in the way of the unlearned, for *sce-* is ordinarily pronounced *se* even in words where the *c* represents a Greek *k*, e.g. *scene* and its compounds and *ascetic*. America spells *sk-*; we might pocket our pride and copy.

**schedule.** Pronounce *shĕ'dūl* in Britain; *skĕ'dūl* in U.S.

**schismatic(al).** See -IC(AL). The short form is now ordinarily used for the noun, and the long one for the adjective.

**schist.** Pronounce *sh-*. The apparent inconsistencies of English treatment of Greek words are well illustrated by *schism* (*sĭ-*), *schist* (*shĭ-*), and the *schizo-* compounds (*skĭ-* or *skī-*), all being from the same Greek word. The explanation is that only the last were consciously taken direct from Greek; the other two came to us through French (*scisme* and *schiste*).

**schizomycetes.** See MICRO-ORGANISMS.

**scholar.** Though there is no apparent reason why *s.* and *ss.* should not mean pupil(s) at a school, schoolboy, schoolgirl, schoolchildren, etc., it is something of a solecism to use them in those senses. A scholar at a school or university is a pupil who holds a scholarship, or, more loosely, one who is of scholarship standard intellectually, and the use of the word in the other sense implies that the user is unacquainted with school idiom. *It is the sincere hope of the council that its endeavour to promote the 'sport' in the schools will be recognized by the masters, and that they will bring the proposed championships to the notice of their scholars.*

**scholarly.** For adverb see -LILY.

**scholiast,** an ancient commentator on a classical text, is liable to be confused by the unlearned with *sciolist*, a superficial pretender to knowledge.

**school** (of fish etc.), **shoal.** The two words are etymologically one, and equally unconnected with the ordinary word *school*; both are also current, and without difference of sense except that *school* is more usual for the cetacea; a *school* of porpoises but a *shoal* of herring.

**sciagraphy** etc., **ski-.** The regular representative in English of Greek *sk-* (here σκιά shadow) is *sc-*; but it is legitimate to pronounce *c* as *k*, cf. SCEPTIC. This particular set of words has been taken into English twice—in the 16th c. as terms in perspective, usually with the spelling *sc-*, and in the 19th as equivalent to radiography etc., usually with the spelling *sk-*. To maintain both the *sc-* and the *sk-* forms would have been very unsatisfactory, and, with *radiography* in existence, also needless. The X-ray sense has now been properly abandoned to the *radio-* words and the *scia-* words are restricted to their older use in perspective, spelt only *sc-* and pronounced *sk-*.

**science and art.** S. knows, a. does; a s. is a body of connected facts, an a. is a set of directions; the facts of s. (errors not being such) are the same for all people, circumstances, and occasions; the directions of a. vary with the artist and the task. But, as there is much traffic between s. and a., and, especially, a. is often based on s., the distinction is not always clear; the a. of self-defence, and the boxer's s.— are they the same or different? The OED, on s. 'contradistinguished from art', says: 'The distinction as commonly apprehended is that a s. is concerned with theoretic truth, and an a. with methods for effecting certain results. Sometimes, however, the term s. is extended to denote a department of practical work which depends on the knowledge and conscious application of principles; an a., on the other hand, being understood to require merely knowledge of traditional rules and skill acquired by habit.'

**scilicet,** usually shortened to *scil.* or *sc.*, is Latin (*scire licet* you may know) for 'to wit'. It is not so often misused as *e.g.* and *i.e.*, not having been popularized to the same extent. Its function is to introduce: (a) a more intelligible or definite substitute, sometimes the English, for an expression already used: *The policy of the I.W.W* (sc. *Independent Workers of the World*); *The Holy Ghost as Paraclete* (scil. *advocate*); (b) a word or phrase that was omitted in the original as unnecessary, but is thought to require specifying for the present audience: *Eye hath not seen, nor ear heard* (sc. *the intent of God*). See also VIZ.

**scintilla** is rarely used except in the singular (*a s. of doubt*; *not a s. of evidence*); if a plural should be needed it is sufficiently at home to make *-as* rather than *-ae*.

**scleroma, sclerosis.** Pl. *-ō′mata, -ō′ses (-ēz)*; see LATIN PLURALS 2. The derivation is from Gk. σκληρός, hard. Laymen should be on their guard against calling the disease *scelerosis* as though it had something to do with Lat. *scelerosus*, wicked.

**scon(e).** The spelling *scone*, and the pronunciation *skōn*, are given preference by the OED; but in Scotland, its land of origin, the pronunciation is *skŏn*, and English people who know this so pronounce it. The place is pronounced *Skoon*.

**scope.** For synonyms see FIELD.

**score,** *n.* (= 20). See COLLECTIVES 5.

**scoria** is a singular noun, pl. *-iae*; but, as the meaning of the singular and of the plural is much the same (cf. *ash* and *ashes, clinker* and *clinkers*), it is no wonder that the singular is sometimes wrongly followed by a plural verb (*The scoria were still hot* etc.), or that a false singular *scorium* is on record.

**scotch.** This verb owes its currency entirely to the sentence in *Macbeth*— 'We have scotch'd the snake, not kill'd it' (Theobald's famous emendation of *scorch'd*). The contrast between scotch-

ing (or disabling) and killing is expressly drawn in five quotations given in the OED for the correct use, and is understood to be implied even when it is not expressed. *S.*, then, can say in six letters and in one syllable 'put temporarily out of action but not destroy' —a treasure, surely, that will be jealously guarded by the custodians of the language, viz. those who write. But no; too many of them are so delighted at finding in *s.* an uncommon substitute for such poor common words as *kill* or *destroy* to remember that, if they have their way, the value of a precious word will be not merely scotched, but killed and destroyed, or, as they would put it, 'finally scotched'. *Finally* or *entirely* with *s.* should be, in view of the history of *s.*, an impossibility. But it is now to be met with often in the newspapers; and, after all, a writer who, like the author of the first extract below, does not know the difference between a rumour and the contradiction of a rumour, can hardly be expected to recognize so super-subtle a distinction as that between wounding and killing: *The contradiction of a rumour affecting any particular company, although it may have a certain effect upon the price of shares at the time, is seldom entirely scotched by directorial statements.* / *Nine months have gone by since the Crown Estate Commissioners finally scotched rumours that wholesale demolition of the Nash Terraces in Regents Park was contemplated.* / *The idea is so preposterous that by the time this is in print it may be definitely scotched.* These writers might perhaps plead that they were not using Macbeth's word but drew their metaphor from the *scotch*, thought to be of different origin, that is placed under a wheel to prevent it from moving. But the plea is not convincing, and they cannot escape the reproach of having in fact blunted the point of a useful word.

**Scotch, Scots, Scottish.** 1. (as adjj.). The third represents most closely the original form, the first and second being the contractions of it adopted in England and Scotland respectively. *Scottish* is still both good English and good Scotch. The English form *Scotch* had (OED, 1914) 'before the end of the 18th c. been adopted into the northern vernacular; it is used regularly by Burns, and subsequently by Scott. . . . Within the last half century there has been in Scotland a growing tendency to discard this form altogether, *Scottish*, or less frequently *Scots*, being substituted.' Regimental titles vary: the King's Own Scottish Borderers and the Scottish Rifles, but the Scots Guards, the Royal Scots, the Scots Fusiliers, and the Scots Greys. Scots law and pound Scots are the only forms. Out of deference to the Scotsman's supposed dislike of *Scotch*, that word has been falling into disuse in England also, but it remains in both countries in numerous associations, e.g. *whisky, broth, tweed, mist, terrier, fir.*

2 (as nn.). For the name of the Scotch dialect, the noun *Scottish* is little used; *Scotch* is the English noun, and *Scots* the usual Scotch one.

**Scot, Scots(wo)man, Scotch(wo)man.** Englishmen use the third forms by habit, the first sometimes for brevity or for poetical or rhetorical or jocular effect, and the second occasionally in compliment to a Scottish hearer, *Scots*- being (OED) 'the prevalent form now used by Scotch people'.

**scot(t)icism.** The usual spelling is *-tt-*.

**scout, gyp.** College servants at Oxford and Cambridge respectively.

**scream, screech, shriek.** The first is the 'dominant' word for a cry uttered, under emotion, at a higher pitch than that which is normal with the utterer. Those who wish to intensify the pitch and the emotion substitute *shriek*; those who wish either to add the notion of uncanny effect, or to make fun of the matter, substitute *screech*.

**screw** *your courage to the sticking-place* (not *point*); MISQUOTATION.

**scrimmage, scru-.** The form with *-u-*, abbreviated to *scrum*, is preferred in rugby football, that with *-i-* in more general uses.

**scrip, script.** The commonest uses of these words today are: *scrip* for the provisional acknowledgement of a subscription to the capital of a company, to be replaced eventually by a *certificate*; *script* for print that imitates handwriting and handwriting that imitates that kind of print, for the text of a talk or play broadcast on radio or television, and, by examiners, for candidates' written answers. Both are derived from Latin *scribere* and have no connexion with the obsolete *scrip* meaning a pilgrim's wallet. See also LONGHAND.

**scrumptious.** See FACETIOUS FORMATIONS.

**scull, skull.** The one-handed oar has *sc-*, the cranium *sk-*. The notion that the words are ultimately the same is discountenanced by the OED, though Skeat derived both from Icelandic *skal*, a hollow, the *scull* being so called on account of the shape of its blade.

**sea.** *S. change. Suffer a s. c.* is one of the most importunate and intrusive of IRRELEVANT ALLUSIONS, and HACKNEYED PHRASES. *We hope that the Prime Minister will on this occasion stick to his guns, and see that his policy does not for the third or fourth time suffer a s. c. when its execution falls into the hands of his colleagues.* On the hyphening of *sea-* compounds see HYPHENS 2, 4.

**seal.** For some synonyms, see SIGN.

**seamstress, semps-.** The OED treats the first as the word, and the second as the variant.

**sear, sere.** *Sear* for the nouns (part of gunlock, mark of burn), and for the verb (burn); *sere* for the adjective (withered). *Sere and yellow* is an example of PEDANTIC HUMOUR.

**seasonable, seasonal.** These adjectives are fully differentiated: *-able* means appropriate to the time of year, especially seasonable weather, or, more generally, opportune; *-al* means depending on or varying with the seasons, e.g. seasonal employment. OED Supp. quotes this example of the misuse of *seasonable* for *seasonal*: *Persons engaged in seasonable trades in which the duration of seasonable employment is too short to enable them to qualify for benefit.*

**second.** 1. *S. chamber.* 2. *S. floor.* 3. *S. (-)hand* etc. 4. *S. intention.* 5. *S. sight.* 6. *S.*, v. (mil.).

1. *S. chamber*, in a parliament, is the upper house, as concerned chiefly with rejection, confirmation, and revision.

2. For *s. floor* and *s. storey*, see FLOOR.

3. *S. (-)hand* etc. The second-hand of a watch is so written. The adjective meaning not new or original is best written as one word, but if with two they should be hyphened (*secondhand clothing* or *information*). The adverb should be two words unhyphened (*always buys s. hand*; *heard only at s. hand*). So too with other *s.* compounds: *s.-class carriage* but travel *s. class*; *s.-best bed* but *come off s. best*. See HYPHENS.

4. For *s. intention*, see INTENTION.

5. *S. sight.* Two words unhyphened; see HYPHENS.

6. The verb *s.* in its technical military sense is pronounced *sĕko'nd.*

**secretive** (pronunciation). The OED gives only *sĕkrē'tĭv*; but *sē'krĕtĭv* has since gained ground, and the COD puts it first. Probably those who conceive the meaning as fond of secrets say *sē'krĕtĭv*, and those who conceive it as given to secreting say *sĕkrē'tĭv*. See RECESSIVE ACCENT.

**sect** is a word whose sense is to some extent affected by its user's notion of its etymology. The OED favours Latin *sequor* (follow) as the origin, so that *s.* would mean a following, i.e. a company of followers. The popular etymo-logy that derives the word from the Latin *seco* (cut)—called by Skeat 'baseless and unworthy of serious mention' —is naturally interpreted as giving 'a part cut away' from a Church etc., and so a company of schismatics. According to the first, and probably correct, derivation, the Church of England, or the Roman Catholic Church, may be called a s. without offence to its members, according to the second it will not.

**secundum quid.** See SIMPLICITER.

**sedilia.** Pronounce *sĕdĭ'lyă.* A plural noun, rarely used in singular (*sedile*, pr. *sĕdĭ'lĕ*).

**see, bishopric, diocese.** A bishopric is the office belonging to a bishop; a diocese is the district administered by a bishop; a see is (the chair that symbolizes) a bishop's authority over a particular diocese. A b. is conferred on, a d. is committed to, a s. is filled by, such and such a man. *My predecessors in the see*; *All the clergy of the diocese*; *Scheming for a bishopric.*

**seek.** For two abuses to which the word is liable, see FORMAL WORDS, and DOUBLE PASSIVES.

**seem.** 1. Pleonasms with *s.* 2. *Seem-(ed) to (have) be(en).* 3. *To my* etc. *seeming.* 4. *As seem(s) to be the case.*

1. Pleonasms. *These conclusions*, it seems to me, appear *to be reached naturally.* Such absurdities are not uncommon with *s.*; see PLEONASM 4, and HAZINESS.

2. For the very common confusion between *seem(s) to have been* and *seemed to be* see PERFECT INFINITIVE 3.

3. *To my* etc. *seeming.* An example of the unsuitable use of this idiom is *From wherever he may start, he is sure to bring us out very presently into the road along which*, to his seeming, *our primitive ancestors must have travelled.* The phrase *to my* etc. *seeming* has been good English in its time; its modern representative is *to my* etc. *thinking*, and *to his seeming* will pass only in archaic writing. That the author of the extract is an archaizer is plain independently, from the phrase

'very presently' (see PRESENTLY); but he has no business to be archaizing in a sentence made unsuitable for it by the essentially unarchaic 'primitive ancestors'.

**4. As seem(s) to be the case.** An example of the wrong use of the plural is *How can the Labour Ministry acquire proper authority if it has powers so limited as seem to be the case?* To write *as seem to be the case* is always wrong, because the relative pronoun *as*, for which see AS 5, never represents an expressed plural noun (such as *powers* here), but always a singular notion like fact or state of affairs, and that not expressed, but extracted out of other words. *As seems to be the case* is, then, the only right form of the phrase; but even that will not do here, because it involves the doubling of two parts by *as*, that of the relative adverb, indispensable after the preceding *so*, and that of the relative pronoun required by the otherwise subjectless verb *seems*. What has happened is this. The writer wanted to say *if it has powers so limited as its powers seem to be*. He shied at the repetition of *powers*, and felt about for *as seems to be the case* as a substitute, though he forgot to alter *seem* to *seems*. But, since *so* makes the relative pronoun *as* impossible, the true solution was to let the *as* be a relative adverb, writing *if its powers are so limited as they seem to be*.

**seemly.** For the adverb, see -LILY.

**seigneur** etc. Spellings recognized in the OED are seigneur, seignior; seigneuress; seigneury, seigneurie, seign(i)orage, seign(i)ory; seigneurial, seign(i)or(i)al. The pronunciation in all begins with *sān* followed by the *y* sound. Differences in meaning or use between alternative forms (as *seigneur* and *seignior*, *seigneury* and *seigniorage*) cannot be detailed here, but exist and are sometimes of importance.

**seise, seisin.** Pronounce *sēz, sē'zĭn*. The words are sometimes (but less often) spelt *-ze, -zin*, and belong etymologically to the ordinary verb *seize*;

but in the legal phrases *to s. a person of*, i.e. put him in possession of, and *to be -ed of*, i.e. to possess, the *-s-* spelling is usual. Mistakes are sometimes made over the preposition. *With* should have been *of* in *It must not be thought that the cooperative movement was satisfied that the Labour Party was fully seised either with their deeds or needs*.

**self.** *As both* self and wife *were fond of seeing life, we decided that . . . | He ruined* himself and family *by his continued experiments.* Correct the first to *both I and my wife*, and the second to *himself and his family.* Such uses of *s.* are said by the OED to be 'jocular or colloquial' extensions of a 'commercial' idiom (e.g. *your good self*, see COMMERCIALESE); and, unless the jocular intent is unmistakable, they are best avoided. So is the common use of *myself* for *I* or *me* when in association with someone else—*My partner and myself will be glad to see you.* The use of *myself* without *I* in the subjective case (*Myself when young did eagerly frequent*) is now poetical only. The word is best confined to its uses as a reflexive or for emphasis. *I have hurt myself. | I did it myself. | I can manage by myself. | I am quite myself again.* But when a personal pronoun has been preceded by the same word used as a possessive adjective the addition of *self* may make a sentence run more smoothly, e.g. *The difficult relations between her father, her mother, and her(self).*

**self-.** *Self-* compounds (sometimes invented) are often used when the *self-* adds nothing to the meaning. *Agricultural depression and the rural exodus had made village life* self-despondent *and anaemic. | Hence it is* self-evident *that economic changes in the agricultural system must greatly affect the general well-being.* There could hardly be any difference of meaning between *despondent* and *self-despondent*. *Self-evident*, on the other hand, has its own valuable sense of evident without proof (*res ipsa loquitur*, as the lawyers say) or intuitively certain; it should not be used in the sense of no more than evident, without any implication that proof is needless, as *hence* shows that it has been in the above extract. Other words resembling *self-despondent* in being never preferable to the simple form without *self* are *self-collected* (calm etc.), *self-conceit(ed)*, *self-consistent*, *self-diffidence*, *self-opinionated*. And others resembling *self-evident* in having a real sense of their own but being often used when that sense is not in place are *self-assurance, self-complacent, self-confidence, self-consequence*. But these are samples only; there are scores of possible compounds that a writer should not use without first asking himself whether the *self-* is pulling its weight. It is not to be supposed that the otiose use of *self-* is a modern trick; on the contrary, the modern tendency is to abandon many such compounds formerly prevalent, and the object of this article is merely to help on that sensible tendency. On the other hand, the practice of affixing *self-* to a noun or an adjectival participle to mean automatically is common and useful; *Self-starter, self-closing, self-raising, self-sealing*. But it needs keeping in its place. *This apparatus self empties* (to quote from an advertisement) is no more English than *This man self supports*. On the hyphening of *self-* compounds see HYPHENS 6.

**semantics** is the branch of the science of linguistics that is concerned with the meaning of words. The name is modern; it seems to have come into English from the publication in 1900 of Mrs H. Cust's translation of the *Essai du Semantique* by the French philologist Michel Breal (1832–1915). Unlike the many scientific terms that are invented merely to give a new look to an old concept, this one signified a new approach to what had previously been called *semasiology*. Its progress was at first slow; all that the SOED said about semantics in 1933 was that it was another term for the older word. But, largely

through the work of I. A. Richards and C. K. Ogden (notably their book *The Meaning of Meaning*), it has since been greatly popularized, and, in spite of excessive claims made for it by some enthusiasts, has established itself not only in philology but also in philosophy and psychology. Postulating that words 'mean' nothing by themselves, and that 'the kind of simplification typified by the once universal theory of direct meaning relation between words and things is the cause of almost all the difficulties which thought encounters', it aims at removing the confusion between 'verbal' problems and 'real' ones and at furthering the use of words as instruments of precision for the conveyance of thought from one mind to another.

**semi-.** Compounds are innumerable, and restrictions little called for: but the claims of *half-*, which is often better, should at least be considered: *This would be an immense gain over the existing fashion of a multitude of churches ill-manned and semi-filled*. Pronounce *sĕ-* regardless of FALSE QUANTITY. See also DEMI.

**semicolon.** See STOPS.

**Semite.** See HEBREW.

**sempstress.** See SEAMSTRESS.

**senior.** For *the s. service*, see SOBRIQUET.

**sensational** means pertaining to or perceptible by the senses. About the middle of the 19th c. it acquired its special meaning of calculated to produce a violent impression, and in that capacity it has been greatly in demand, to the disgust of the purists. Its length makes it unsuitable for modern popular journalese, and useless for the headlines that now form so important a part of the medium between the Press and its readers. But *sensational* is finding a new outlet: it is challenging FABULOUS and *fantastic* for top place among the epithets by which advertisers hope to entice buyers.

**sense,** n. *S. of humour* is properly the power of finding entertainment in people's doings, especially in such of them as are not designed to entertain. But the phrase has received an extension, or perhaps rather a limitation, that bids fair to supersede the original meaning. When we say nowadays that a person 'has no s. o. h.', or 'lacks humour', we often mean less that he is not alive to the entertainment provided by others' doings than that he is unaware of elements in his own conduct or character likely to stir the s. o. h. in others—has not, in fact, the gift of seeing himself as others see him even in the degree in which it is possessed by the average man.

**sense,** v. *We s. the tragedy of Anna Wolsky as she steps light-heartedly into Sylvia Bailey's life.* / *The water rail ... is somewhat unwieldy in flight, and senses so much, for it seems to prefer to run.* The verb has been used for some three centuries in philosophic writing as a comprehensive form of 'see or/and hear or/and smell or/and taste or/and feel by touch', i.e. of 'have sense-perception of'. From that the use illustrated above is distinct, meaning according to the OED definition 'to perceive, become aware of, "feel" (something present, a fact, state of things, etc.) not by direct perception but more or less vaguely or instinctively'. The OED's earliest example is dated 1872, and the meaning has only recently become part of ordinary English. It has the advantage of brevity as compared with become conscious of, get an inkling of, and other possibilities; and that is no doubt the reason why it has largely overcome the irritation and suspicion of preciosity which most readers at first felt—and some still feel—when confronted with it.

**sensibility.** Just as *ingenuity* is not ingenuousness, but ingeniousness (see INGENIOUS), so *sensibility* is not sensibleness, but sensitiveness; to the familiar contrasted pair *sense and sensibility* correspond the adjectives *sensible and sensitive*—an absurd arrangement, and doubtless puzzling to foreigners, but beyond mending; -TY AND -NESS.

**sensible, sensitive, susceptible.** In certain uses, in which the point is the effect produced or producible on the person or thing qualified, the three words are near, though not identical, in meaning. *I am sensible of your kindness, sensitive to ridicule, susceptible to beauty.* Formerly *sensible* could be used in all three types of sentence; but its popular meaning as the opposite of *foolish* has become so predominant that we are no longer intelligible if we say *a sensible person* as the equivalent of *a sensitive* or *a susceptible person*, and even *sensible of* is counted among LITERARY WORDS though surviving as a cliché for the opening of after-dinner speeches: *I am deeply sensible of the honour you have done me.* . . . The difference between *sensible of, sensitive to,* and *susceptible to,* is roughly that *sensible of* expresses emotional consciousness, *sensitive to* acute feeling, and *susceptible to* quick reaction to stimulus: *profoundly, gratefully, painfully, regretfully,* sensible of; *acutely, delicately, excessively, absurdly,* sensitive to; *readily, often, scarcely,* susceptible to. With *of* the meaning of *susceptible* is different; it is equivalent to admitting or capable. *A passage susceptible of more than one interpretation*; *an assertion not susceptible of proof.*

**sensitize** is a word made for the needs of photography, and made badly. It should have been *sensitivize*; one might as well omit the adjective ending of *immortal, signal, fertile, human,* and *liberal,* and say *immortize, signize, fertize, humize,* and *liberize,* as leave out the *-ive.* The photographers, however, have made their bed, and must lie in it; the longer the rest of us can keep clear, the better. But the OED quotes: *Education, while it sensitizes a man's fibre, is incapable of turning weakness into strength.* Just as, failing *pacificist, pacist* would have been better than *pacifist* (see -IST), so, failing *sensitivize, sensize* would have been better than *sensitize.*

**sensuous** is thought to have been expressly formed by Milton to convey what had originally been conveyed by the older *sensual* (connexion with the senses as opposed to the intellect) but had become associated in that word with the notion of undue indulgence in the grosser pleasures of sense. At any rate Milton's own phrase 'simple, sensuous, and passionate' in describing great poetry as compared with logic and rhetoric has had much to do with ensuring that *sensuous* should remain free from the condemnation now inseparable from *sensual*.

**sentence** is defined in every grammar book and every dictionary, but it would not be easy to find two that gave the same definition, and some of them make heavy weather of it. Here are some examples from standard works:

1. A word or set of words followed by a pause and revealing an intelligible purpose.
2. A group of words which makes sense.
3. A combination of words which is complete as expressing a thought.
4. A collection of words of such kind and arranged in such a manner as to make some complete sense.
5. A meaningful group of words that is grammatically independent.
6. A complete and independent unit of communication, the completeness and independence being shown by its capability of standing alone, i.e. of being uttered by itself.
7. A group of words, or in some cases a single word, which makes a statement, or a command (or expression of wish), or a question or an exclamation.
8. A number of words making a complete grammatical structure.
9. A combination of words that contains at least one subject and one predicate.
10. A set of words complete in itself, having either expressed or understood in it a subject and a predicate and conveying a statement or question or command or exclamation; if its subject or predicate or verb (or more) is understood, it is an elliptical sentence.

These definitions show a difference of approach, depending on whether 'sentence' is given its popular meaning of

'such portion of a composition or utterance as extends from one full stop to another' (OED) or the meaning grammarians give it a combination of words in an analysable grammatical structure. Failure to distinguish between the two has been the cause of much sterile argument about what the word really means. The first seven of these ten definitions take the 'popular' approach, the eighth and ninth the 'grammarians'. The tenth tries to reconcile the two by giving a grammarians' definition with a procrustean device for fitting into it apparently unconformable sentences of the 'popular' kind. It is more realistic to admit that the two may be irreconcilable; that what may suitably be placed between one full stop and another may lack even an elliptical grammatical construction.

Modern writers show greater freedom than was once customary in what they place in that position. *And what of the will to power? | Finally on one small point. | So far so good. | So then. | Now for his other arguments.* These, taken from scholarly writings by contemporary men of letters, cannot be denied the right to be called sentences, but it would be straining language to say that they are elliptical in the sense that 'a subject or predicate or verb (or more)' must be 'understood'. Grammarians are free to maintain that no sequence of words can be called a s. unless it has a grammatical structure, but they should recognize that, except as a term of their art, the word has broken the bounds they have set for it. For more on this subject see VERBLESS SENTENCES, and on the danger of letting sentences get out of hand see HANGING UP and TRAILERS.

**sentinel, sentry.** The first is the wider and literary word, and the fitter for metaphorical use; the second is the modern military term.

**septcentenary.** See CENTENARY.

**sequelae.** A plural word with rare singular *sequela*.

**sequence of tenses. 1.** A certain assimilation normally takes place in many forms of sentence, by which the tense or mood of their verbs is changed to the past or conditional when they are made into clauses dependent on another sentence whose verb is past or conditional, even though no such notion needs to be introduced into the clause. Thus, *Two will do* is a sentence; turn it into a clause depending on *I think*, and the tense remains unaltered: *I think that two will do*. Next, into one depending on *I thought* or *I should think*; it becomes *two would do*; after *I thought* there is a change in the clause to past time, and therefore *would do* is not only normal, but invariable; after the conditional *I should think* the conditional *would do* is also normal. But this is not invariable, s. of tense being sometimes neglected and s. of mood often. *Two will do; I think that two will do; I thought that two would do; I should think that two* (normal sequence) *would do*, or (vivid sequence) *will do*. (In these examples, the usually omitted *that* has been inserted merely to make it clear that a real clause is meant, and not a quotation such as *I thought 'two will do'*.) The point to be noticed is that the change of tense or mood is normal **s.**, and the keeping of it unchanged (called *vivid s.* above) is, though common and often preferable, abnormal. Further examples are: *He ·explained what relativity* (normal) *meant,* or (vivid) *means; I should not wonder if he* (normal) *came,* or (vivid) *comes*.

**2.** Sequence out of place. *One would imagine that these prices* (normal) *were,* or (vivid) *are, beyond the reach of the poor; These prices, one would imagine, are beyond* etc. The base is *These prices are;* if made dependent on *One would imagine, are* may be changed, or may not, to *were;* but if *one would imagine* is a parenthesis instead of being the main verb, the change is impossible. Nevertheless it happens: *The shops have never had such a display of Christmas presents, but here again the prices, one would imagine, were beyond the reach of any but the richest persons.* The mistake, a common one, results from

not knowing a parenthesis when one sees it; see PARENTHESIS 3.

**seq., seqq., et seq(q).,** are short for Latin *et sequentes* (*versus*) 'and the subsequent lines', or *et sequentia* 'and the words etc. following'. The abbreviation differs from ETC. in two ways: *et seq.* is literary and *etc.* is not; and *et seq.* refers to words elsewhere specified; *etc.* may do so but is more likely to leave the reader to think of them for himself.

**seraglio.** Pronounce *sĕrah'lyō*. Pl. *-os;* see -O(E)S 4.

**sere.** See SEAR.

**sergeant, -j-.** For the military and police rank, *-g-;* in legal titles (*Common S.* etc.), *-j-*. In *S. at arms,* the official spelling is *-j-*.

**serial.** For the musical term see ATONAL.

**seri(ci)culture.** The full form is the right one etymologically but is not now used; cf. *pacif(ic)ist* in -IST.

**serpent.** See SNAKE.

**service** as a verb was unknown to the original OED; its insertion in the 1933 Supp. is supported by a quotation from Stevenson's *Catriona* in which its meaning seems to differ little, if at all, from that of *serve*. When used merely as an imposing synonym of that verb it must be condemned as a NEEDLESS VARIANT. But it has since established itself usefully in the special sense of giving periodical attention to a machine. That being so, the attempt to bring into currency a new verb *servicize* is worse than useless.

**sestet(te), sex-.** The second is a late variant of the first. The tendency today is to use *ses-* in prosody and *sex-* in music. Spell *-et*.

**set(t).** The extra *t* is an arbitrary addition in various technical senses, from a lawn-tennis to a granite set. Each class of persons interested in some special sense has doubtless added a *t* to distinguish that sense from all others; but so many are the special senses that the distinction is now no more distinctive than an ESQ. after a man's name, and all would do well to discard it. Cf. the less futile *matt* for MAT and *nett* for NET.

**severely.** For *leave s. alone*, see IRRELEVANT ALLUSION. There are degrees of badness; in the first of the two following extracts, for instance, *s.* is less pointless than in the other: *That immortal classic which almost all other pianists are content to l. s. a. on the topmost shelf.* | *If our imports and exports balance, exchanges will be normal, whatever the price, and I am glad that Mr. Mason agrees that exchanges should be left s. a.*

**sew.** P.p. *sewed* or *sewn*. The first is, perhaps contrary to general belief, the older form and (to judge by the OED 19th–20th-c. examples) was then slightly the commoner. But *sewn* has since gained on it.

**sew(er)age.** It is best to use *sewage* for the refuse, and *sewerage* for the sewers or the sewer system. *Sewage* is defensible as a derivative of the formerly recognized but now dialectal verb *sew* to ooze out.

**sexcentenary.** See CENTENARY.

**sextet.** See SESTET.

**shade,** n. For colour synonym see TINT.

**shade, shadow,** nn. It seems that the difference in form may be fairly called an accidental one, the first representing the nominative and the second the oblique cases of the same word. The meanings are as closely parallel or intertwined as might be expected from this original identity, the wonder being that, with a differentiation so vague, each form should have maintained its existence by the side of the other. The OED's main heads of meaning are three for each, one set hardly distinguishable from the other. For *shade*: I. Comparative darkness; II. A dark figure 'cast' upon a surface by a body intercepting light, a shadow; III. Protection from glare and heat. For *shadow*: I. Comparative darkness; II. Image cast by a body intercepting light; III. Shelter from light and heat. The most significant point is that, in II of *shade*, *shadow* is offered as a definition of *shade*, without reciprocity in II of *shadow*, the inference from which is that in division II *shadow* is the normal word, and *shade* exceptional. This almost identity of meaning, however, branches out into a considerable diversity of idiom, one word or the other being more appropriate, or sometimes the only possibility, in certain contexts. The details of this diversity are too many to be catalogued here, but it is a sort of clue to remember that shadow is a piece of shade, related to it as, e.g., pool to water. So it is that shade is a state— viz. partial absence of light—and not thought of as having a shape, nor usually as an appendage of some opaque object, both which notions do attach themselves to shadow. So too we say *light and shade* but *lights and shadows, in the shade* but *under a shadow*; and so too *shady* means full of shade, but *shadowy* like a shadow. The use of *shady* in the sense 'of a nature or character unable to bear the light, disreputable' dates from the mid 19th c.

**Shakspere, Shakespear(e), -erian, -earian, -ean,** etc. The forms preferred by the OED are *Shakspere, Shaksperian*. It is a matter on which unanimity is desirable, and it is unfortunate that the OED's verdict has not been accepted as authoritative. But the preference today is undoubtedly for Shakespeare. It is no use trying to withstand a strong popular current in such a matter; even the SOED has had to conform. *Shakespeare, Shakespearian* are therefore recommended.

**shall** and **will, should** and **would.** 1. Plain future. 2. Plain conditional. 3. *I would like* etc. 4. Indefinite future and relative. 5. Elegant variation. 6. *That*-clauses.

'To use *will* in these cases is now a mark of Scottish, Irish, provincial, or extra-British idiom'—Henry Bradley in the OED. 'These cases' are of the type most fully illustrated in 1 and 2 below, and the words of so high an authority are here quoted because there is an inclination, among those who are not to the manner born, to question the existence, besides denying the need, of distinctions between *sh.* and *w.* The distinctions are elaborate; they are fully set forth in the OED; no formal grammar can be held to have done its duty if it has not stated them, and their essence is briefly summarized in WILL (v.) which should be read with this article. It will therefore be assumed here that the reader is aware of the normal usage, and the object will be to make the dry bones live by exhibiting some sentences containing common types of violation of it. The 'Scottish, Irish, provincial, or extra-British' writer will thus have before him a conspectus of the pitfalls that are most to be feared if he wishes to observe the English idiom as described by the OED. But it is necessary to add that the position is no longer as it was when Bradley wrote. The 'Scottish, Irish, etc. idiom', especially as followed on the American continent, has made formidable inroads; and insistence on the rules laid down in the OED and illustrated in this article may before long have to be classed as insular pedantry. This is regrettable. The English idiom affords a convenient means of distinguishing delicate shades of meaning; and that is a valuable element in a language.

1. Plain future. In the first person '*shall* has, from the early ME period, been the normal auxiliary for expressing mere futurity without any adventitious notion'—OED. *Will* conveys an implication of intention, volition, or choice. In the following examples, in which there is clearly no such implication, the use of *will* is contrary to English idiom: *This is pleasant reading; but we* won't *get our £2,000 this year.* |

*Perhaps we* will *soon be surfeited by the unending stream of 'new' literature, and will* turn with relief to . . . | *If we compare these two statements, we* will *see that so far as this point goes they agree.| But if the re-shuffling of the world goes on producing new 'issues', I* will, *I fear, catch the fever again. | We never know when we take up the morning paper, some of us, which side we* will *be on next.*

2. Plain conditional. Similarly the right auxiliary for a colourless conditional in the first person is *should*, and in the following examples the use of *would* is contrary to the English idiom: *If we traced it back far enough we* would *find the origin was* . . . | *I* would *not be doing right if I were to anticipate that communication. | I think I* would *be a knave if I announced my intention of handing over my salary.* Two other examples will provide for a common exception to the rule that *shall* or *should* is the colourless auxiliary for the first person. In sentences that are, actually or virtually, reported, a verb that as reported is in the first person but was originally in the second or third often keeps *will* or *would*: *People have underrated us, some even going so far as to say that we* would *not win a single test match* (the people said *You will not*, which justifies, though it by no means necessitates, *we would not* in the report). | *He need not fear that we* will *be 'sated' by narratives like his* (his fear was *You will be sated*, which makes *we will* not indeed advisable, but defensible).

3. The verbs *like, prefer, care, be glad, be inclined*, etc., are very common in first-person conditional statements (*I should like to know* etc.). In these *should*, not *would*, is the correct form in the English idiom. But here *would* has long been encroaching. 'I would be glad to receive some instruction from my fellow-partner' says Pompey in *Measure for Measure*. The OED (1928), quoting this and other early examples, goes no further than to say that *should* is 'regarded as more correct', but that *would* is 'still frequent'. It is now even more frequent—as common as *should* if not commoner. Its

use with *like* is illustrated in LIKE (v.)
Examples with other verbs are: *We
cannot go into details, and* would prefer
*to postpone criticism until* . . . | *Nor has
he furnished me with one thing with
which I* would care *to sit down in my
little room and think.* | *If we should take
a wider view, I* would be inclined *to
say that.*

It has been suggested (see LIKE)
that the common use of *would* in the
first person with such verbs may be
partly due to a confusion between the
modern *I should like to* and the archaic
*I would* with the same meaning.
Another more general factor favouring
*will* and *would* is that, when the auxiliary is not emphasized, elided forms
such as *I'll, We'll, I'd, We'd* are
habitually used in speech and increasingly in print (see ELISION), and these
are naturally resolved into *I will* etc.
Possibly too the great popularity of
that curious Americanism *I wouldn't
know* has had something to do with it
by making *I* and *would* seem natural
partners. Whatever the reason, it seems
clear that attempts to repel this particular invasion from the other side of
the Atlantic have now about as much
chance of success as Mrs. Partington
had with the Atlantic itself.

4. In clauses of indefinite future time,
and indefinite relative clauses in future
time, *will* is now entirely unidiomatic;
either *shall* is used, chiefly in formal
contexts and legal documents or, much
more often futurity is allowed to be inferred from the context and the present
tense is used. This mistake is now rare;
but it may be worth while to give one
or two examples: *He has now had to go
clean out of the county to find employment,
leaving his wife with her mother until he
will be able to make another home for her*
(until he is able to make). | *So long as this
will not be made clear, the discussion will
go on bearing lateral issues* (is not made). |
*When this will be perceived by public
opinion the solution will immediately
become obvious* (is perceived). Here
also a borderline example may be
of service: *We have strong faith that
a rally to the defence of the Act will be
a feature of next year's politics, if the*
*Tory Party* will *have the courage to
come into the open and declare war upon
it.* An assurance from the writer that
by *will have the courage* he meant
*chooses to have the courage* would be
received perhaps with incredulity, but
would secure him a grudging acquittal;
*has the courage* is what he should have
written.

5. *Shall* and *will, should* and *would,*
are sometimes regarded as good raw
material for elegant variation. 'I wrote
*would* in the last clause; we will have
*should* in the next for a change.'
*If we found the instances invariably in
mutual support we* would be content with
*but a few, but if we found even one in
contradiction we* should require a large
*body of evidence.* | *We* should have been
*exposed to the full power of his guns, and,
while adding to our own losses,* would
have forgone the advantage of inflaming
*his.* | *You* shall not find two leaves of
*a tree exactly alike, nor* will *you be able
to examine two hands that are exactly
similar.* But the follies to which ELE-
GANT VARIATION gives rise are without
number.

6. *That*-clauses after *intend* or *intention, desire, demand, be anxious,* etc.,
have *shall* or (more usually) *should* for
all three persons. Among the etc. are
not included *hope, expect,* and the like;
but the drawing of the line is not easy.
Roughly, *shall* and *should* are used
when the word on which the *that*-
clause depends expresses an influence
that affects the result, as a demand
does, but a hope or a fear does not.
In 'England expects that every man
will do his duty', the substitution of
*shall* for *will* would convert an expression of confidence into an exhortation. Examples of the wrong *will* are:
*I am anxious that, when permanently
erected, the right site* will *be selected.* |
*And it is intended that this* will *be
extended to every division and important
branch.* On the other hand, *will* in the
next quotation is justifiable, since here
*desire* cannot be regarded as affecting
the result; it means merely *hope.*
*The strong desire that the relations of
the English-speaking peoples* will *be so
consolidated that they may act as one*

*people.* But the American practice of omitting the auxiliary in such sentences is becoming increasingly common in Britain. *I am anxious that . . . the right site be selected* etc. See SUB-JUNCTIVE (S.V. ALIVES).

**shambles.** *The Colonial Secretary denied a statement by Mr. B. that the conference on the future of Malta had been a shambles.* So sensational an event could hardly have remained secret. *Shambles*, originally a board on which meat was exposed for sale, and then applied to the slaughter-house from which it came, is legitimately used by extension for any scene of blood and carnage. But to describe as a *s.* a condition of mere muddle and disorder that is wholly bloodless is a SLIPSHOD EXTENSION emasculating the word. *The Monte Carlo rally has become a s. Snowbound roads are littered with abandoned cars. | We have the decorators in the house and the place is an absolute s.*

**shamefaced, -fast.** It is true that the second is the original form, that *-faced* is due to a mistake, and that the notion attached to the word is necessarily affected in some slight degree by the change. But those who, in the flush of this discovery, would revert to *-fast* in ordinary use are rightly rewarded with the name of pedants. To use *shamefast* as an acknowledged archaism in verse is another matter.

**shanty,** sailors' song. See CHANTY.

**shape.** For *in any s. or form* see PLEONASM 2 and SIAMESE TWINS; *Lord A— states that 'he is absolutely unconnected i. a. s. o. f. with the matter'.* The p.p. is *-ed* and *-en* is archaic.

**shapely.** For the adv., see -LILY.

**shard.** In the sense fragment of pottery, the OED treats *shard* as the normal form and *sherd* as the variant; on the other hand, the greater familiarity of *potsherd* tends to keep *sherd* in being. In the well-known phrase 'the shard-borne beetle' (*Macbeth* III. i. 42), the interpretation 'borne through the air on shards' (i.e. the

wing-cases), which has so far prevailed as to set up *shard* as an entomological term for wing-case, appears to be an error; the real meaning was 'born in shard', there being another word *shard*, now obsolete except in dialects, meaning cowdung.

**sharp,** not *sharply*, is the right adverb in matters of time, direction, and pitch. *Pull up s., turn s. left, look s., at eight o'clock s., you are singing s.* See UNIDIOMATIC -LY.

**she.** 1. For *she* and *her* in ill-advised personifications (e.g. *The world wants all that America can give* her), see PERSONIFICATION I.
2. Case. A few violations of ordinary grammar rules may be given; cf. HE. *I want no angel, only she* (read *her*). | *When such as her die* (read *she*). | *She found everyone's attention directed to Mary, and she herself entirely overlooked* (omit *she*). | *But to behold her mother—she to whom she owed her being* (read *her*). | *I saw a young girl whom I guessed to be she whom I had come to meet* (read *her*). | *Nothing must remain that will remind us of that hated siren, the visible world, she who by her allurements is always tempting the artist away* (read *her* or preferably omit *she*).

**sheaf.** The noun has pl. *-ves*. For the verb, *-ve* or *-f*, see -VE(D).

**shear,** v., has past *sheared* in ordinary current senses (*We sheared our sheep yesterday*; *A machine sheared the bar into foot-lengths*; *This pressure sheared the rivets*). It has past *shore* in archaic and poetical use (*shore through the cuirass, his plume away*, etc.). For the p.p., *shorn* remains commoner in most senses than *sheared*, but is not used in the technical sense of distorted by mechanical shear, nor usually in that of divided with metal-cutting shears.

**shear-hulk, shearlegs, sheer-.** The spelling *sheer* is due to and perpetuates a mistake. Shears or shearlegs are two (or more) poles with tops joined and feet straddled (and so resembling shear-blades) carrying tackle for hoisting weights. A shear-hulk is an old ship used for hoisting and provided with shearlegs. The spelling *sheer hulk* results from confusion with the adjective *sheer* (i.e. mere), and the omission of the hyphen and shifting of the accent from *shear* to *hulk* naturally follow, assisted by the rhythm of the line in *Tom Bowling*. It would be well to restore *shear-hulk* and make *shearlegs* (already often so spelt) invariable.

**sheath(e).** The verb is *sheathe* (*-dh*); the noun is *sheath* (*-th*) singular and *sheaths* (*-dhs*) plural.

**sheer(ly).** *They would say the money has, to the present, been sheerly wasted. | A collection of brief pieces in which the sheerly poetical quality is seldom looked for and seldom occurs. | The economic condition of the people is sheerly desperate.* Perhaps owing to the adverbial use of *sheer* (*falls sheer down* etc.), the adverb *sheerly* is usually avoided, and always gives the reader a shock. Though the OED quotes it from Burns, Scott, and Stevenson, it may fairly be called unidiomatic. Possibly it is current in Scotland; at any rate the OED quotations include no well-known English writer.

**sheikh** is the correct spelling, and *shāk* the better pronunciation.

**shelf.** There are two separate nouns, one meaning ledge, board, etc., and the other sand-bank etc. Each has pl. *-ves*, verb *-ve*, adjectives *-ved*, *-fy*, and *-vy*; see *-VE*(D). *Shelf-ful* (of books etc.), n., is ordinarily written with hyphen; pl. *-ls* (unless the two words *shelves full* are suitable and preferred).

**shelty, -ie.** The word meaning Shetland pony is usually *-ie*. That meaning a hut (which the OED perhaps makes out to be rarer than it is, and condemns as 'prob. some error') is usually *-y*.

**shereef, sherif, sheriff.** The Mohammedan and the English titles are not etymologically connected. For the former both the spelling *-eef* is preferable to *-if* both as indicating the accent (*shĕrē´f*) of an unfamiliar word, and as avoiding assimilation to the English *-iff*.

**sheriffalty, sheriffdom, shrievalty, sheriffship.** The first three are four or more centuries old, and are still current. The last is a newcomer. *Sheriffdom* is faintly suggestive of *bumbledom*; *shrievalty*, though still the standard word, has the disadvantage of not instantly announcing its connexion with *sheriff*; and *sheriffalty* is unlikely to hold its own against the competition of the more commonplace formation *sheriffship*.

**shew, show.** 'The spelling *shew*, prevalent in the 18th c. and not uncommon in the first half of the 19th c., is now obs. exc. in legal documents'—OED. In *shewbread* the old spelling naturally persists.

**shibboleth** is a WORSENED WORD. Ability to pronounce it properly was the means by which Jephthah distinguished his own Gileadites from the refugee Ephraimites among them; the true meaning of the word would today be expressed by the cliché ACID TEST. It is now rarely used except in the sense of a catchword adopted by a party or sect, especially one that is old-fashioned and repeated as a parrot-cry, appealing to emotion rather than reason. (*Mr. C.'s own 'programme of radical reform', however, makes more than one concession to the ancient ss.*) Sometimes it seems to be thought of merely as an ornamental synonym of maxim or cliché. *The s. of 'No duty towards trespassers' cannot be applied generally. | We are offended by the needless repetition of ss.—'subliminal uprush' for instance.*

**ship.** The saying *Lose* (or *spoil*) *the ship for a halfpennyworth of tar* is puzzling; it is not easy to picture so trifling a parsimony having so dire a result. The fact is that *ship* is a rustic pronunciation of *sheep* ('Mutton's mutton' said Sir Pitt Crawley. 'What *ship* was it Horrocks and when did you kill?') The original saying was *Lose the sheep for a halfpennyworth of tar*; that a sheep might die from neglect to dress a sore is a much less improbable contingency. But to attempt to restore the original form would of course be gross DIDACTICISM.

**-ship.** For the ordinary significance of this suffix see RELATION(SHIP); for *-manship* compounds see BRINKMAN-SHIP and for the use of *membership*, *leadership*, etc., for *members*, *leaders*, etc., see MEMBERSHIP.

**shire.** The ruling principle for Great Britain is that *-shire* is added when the county and its eponymous town (not always the 'county' town) would otherwise be identically named. (Sometimes the relation is obscured by change of form. Shropshire reflects A.S. Chronicle 'Scrobbesbyriscir', i.e. Shrewsburyshire. Hampshire goes back to Hamtun, now Southampton.) But there need not be any eponymous town. Berkshire is an old name of which the first element was originally a Celtic word meaning 'hilltop'. The towns corresponding to Dorset, Somerset, and Wiltshire are Dorchester, Somerton, and Wilton. Where the relation between town and county is obscure, *-shire* is optional, as in Dorset(-shire), Somerset(-shire), but Duke of Somerset. Berkshire is invariable but Devon (with no town name) alternates with Devonshire, the shorter form prevailing in North and South Devon and in picturesque phrases like Sunny Devon. (There is a Duke of Devonshire and an Earl of Devon.)

Durham is an exception, the county being a relatively modern institution, and never takes *-shire*. Cornwall, Cumberland, Essex, Kent, Norfolk, Suffolk, Surrey, Sussex, and Westmorland, never take *-shire*, but good writers have committed Rutlandshire.

Some abbreviated forms are current in writing, though little used in speech —Berks., Bucks., Hants, Herts., Wilts., Notts., Yorks. Salop serves for both Shropshire and Shrewsbury, and a member of Shrewsbury School is a Salopian.

*-shire* is never applied to an Irish county, though there is a Marquess of Downshire.

In Scotland counties came later and *-shire* is often absent. Fifeshire is

unnecessary and hardly in good use. Buteshire is found on maps but rarely said—Bute, Isle of Bute, being preferred. Edinburgh can mean the county, and Edinburghshire, like the Loch Ness monster, is believed in only by those who have seen it. Midlothian is the name of the county containing Edinburgh, and East and West Lothian are much commoner than Haddington-shire and Linlithgowshire.

'The Shires' can mean all the counties ending in *-shire* or, historically, all the counties as distinct from the boroughs. But usually it means the midland counties famous for fox-hunting.

**shoal.** See SCHOOL.

**shoot, chute, shute.** The last is 'app. in part a dial. form of *shoot* sb. and partly a variant spelling of *chute*'—OED. Between the English *shoot* and the French *chute* (lit. fall) there has been much confusion, and there seems to be no good reason why *shoot* should not have been made the only spelling and allowed to retain such senses as it has annexed from *chute*. But the opportunity for this is past; *chutes*, so spelt, have become too common and too useful.

**shop.** For the talk called *s.* see JARGON.

**short supply.** *In s. s.* is an expression coined about the time of the second world war: it was unknown to the SOED in 1933 but appeared in the 1951 COD. It is a harmless enough phrase, scarcely deserving the rude things that have been said about it by some purists or the apologetic inverted commas in which it is sometimes dressed. We can say without offence that a commodity is *in great demand*; why should we not be allowed to say that one is *in s. s.*? But it was overworked as soon as it appeared: so many things were then *in s. s.*, and there is a rotundity about it that to the official mind no doubt made it seem preferable to *scarce*. With the passing of the period of chronic shortages the vogue waned. But it lasted long enough to tempt even good writers sometimes to use the phrase unsuitably as a PERIPHRASIS for *scarce*. *So Indonesia has had to start life woefully understaffed. Administrators, engineers, doctors, teachers—all are in s. s.*

**should.** For *s.* and *would*, see SHALL.

**show.** For spelling see SHEW. For the p.p. *shown* has ousted the variant *showed*.

**shred,** v. In the p.p. *shredded* and *shred* are both old and both extant; but the shorter only as an archaism.

**shriek.** See SCREAM.

**shrievalty.** See SHERIFFALTY.

**shrink** has past *shrank* (arch. *shrunk*), p.p. usu. *shrunk* as verb or pred. adj., and *shrunken* as attrib. adj.: *has shrunk, is shrunk* or *shrunken, her shrunken* or *shrunk cheeks.*

**shy.** For the spelling of inflexions and derivatives see DRY, and VERBS IN -IE etc.

**Siamese twins.** This seems a suitable term for the many words which, linked in pairs by *and* or *or*, are used to convey a single meaning. As with the human variety, some verbal twins can be divided and each partner live separately; on others this cannot be attempted without fatal results. Their abundance in English is perhaps partly attributable to legal language, where the multiplication of near-synonyms is a normal precaution against too narrow an interpretation, and also contributes a pompous sonority to ceremonial occasions. A Royal patent creating a peer, with a splendid prodigality of words, will *advance, create, and prefer* him to the *state, degree, style, dignity, title, and honour* of his rank, and guarantee that he shall *enjoy and use all the rights, principles, pre-eminences, immunities, and advantages* appertaining to it. The phraseology of the Prayer Book, seldom content with one word if two can be used, may also have something to do with it: in the first few minutes the congregation at Morning Prayer will *hear acknowledge and confess, sins and*

*wickedness, dissemble nor cloke, assemble and meet together, requisite and necessary.*

In ordinary writing Siamese twins are a fruitful source of clichés, and it may be worth while to examine a sample of the commoner of them. Many are merely tautological: a synonym or near-synonym is added for the sake of emphasis: *Alas and alack, betwixt and between, bits and pieces, gall and wormwood, heart and soul, jot or tittle, leaps and bounds, lo and behold, nerve and fibre, rags and tatters, shape or form, sort or kind, toil and moil.* These are separable twins, but others, superficially similar, are indivisible, either because one of the components is used in an archaic sense and would not now be understood by itself, or because the combination has acquired a meaning different from that of either component alone. Such are *chop and change, fair and square, hue and cry, kith and kin, might and main, rack and ruin, odds and ends, part and parcel, spick and span, use and wont.* Others again consist not of synonyms but of associated ideas (e.g. *bill and coo, bow and scrape, flotsam and jetsam, frills and furbelows, hot and strong, hum and ha, thick and fast, ways and means*) or of opposites or alternatives (e.g. *cut and thrust, fast and loose, hither and thither, by hook or by crook, thick and thin, to and fro*). Some are from the law—*act and deed, aid and abet, each and every, let or hindrance, null and void*—and some quotations or literary allusions—*fear and trembling, hip and thigh, prunes and prisms, rhyme nor reason, sackcloth and ashes, sere and yellow, whips and scorpions.* A few can be classed as HENDIADYS (*grace and favour = gracious favour, rough and ready = roughly ready*).

Whenever a Siamese twin suggests itself to a writer he should be on his guard; it may be just the phrase he wants, but it is more likely to be one of those clichés that are always lying in wait to fill a vacuum in the brain.

**sibling** is a very old word that passed out of use for centuries but was revived at the end of the 19th c. as a technical term of anthropology and allied sciences. *Ss.* are brothers and sisters, other than twins. *Sib* is an archaic word meaning related, and is the second element in *gossip* (*godsib*), which originally meant godparent.

**sibyl(line).** The spelling (not *sybi-*) should be noted; see Y AND I. The accent is on the first syllable, not, as might be expected, the second. Cf. *Apennine*. But the misspelling *Sybil*, as in the title of Disraeli's novel, is common in the modern use as a Christian name.

**(sic),** Latin for *so*, is inserted after a quoted word or phrase to confirm its accuracy as a quotation, or occasionally after the writer's own word to emphasize it as giving his deliberate meaning; it amounts to Yes, he did say that, or Yes, I do mean that, in spite of your natural doubts. It should be used only when doubt *is* natural; but reviewers and controversialists are tempted to pretend that it is, because *sic* provides them with a neat and compendious form of sneer. *The industrialist organ is inclined to regret that the league did not fix some definite date such as the year 1910* (sic) *or the year 1912.* The *sic* is inserted because the reader might naturally wonder whether 1910 was meant and not rather 1911; a right use. / *The* Boersen Courier *maintains that 'nothing remains for M. Delcassé but to cry Pater peccavi to Germany and to retrieve as quickly as possible his diplomatic mistake* (sic)'. *Mistake* is the natural term for the quoted newspaper to have used; the quoting one sneeringly repudiates it with (*sic*). / *An Irish peer has issued a circular to members in the House, with an appeal for funds to carry on the work of enlightening* (sic) *the people of this country as to the condition of Ireland.* What impudence! says (*sic*); but, as no one would doubt the authenticity of *enlightening*, the proper appeal to attention was not (*sic*), but inverted commas (see STOPS). / '*A junior subaltern, with pronounced military and political views, with no false modesty in expressing them, and who* (sic) *possesses*

*the ear of the public . . .*' The quoter means 'Observe by the way this fellow's ignorance of grammar—*and who* without a preceding who!' As the sentence is one of those in which the *and-who* rule of thumb is a blind guide (see WHICH WITH AND OR BUT), and is in fact blameless, the (*sic*) recoils, as often, and convicts its user of error.

**sice, size, syce.** *Sice* is correct for the six at dice etc., *size* for the glutinous substance, and *syce* for the Indian groom.

**sick, ill.** The original and more general use of *sick* was suffering from any bodily disorder. English love of EUPHEMISM, shrinking from the blunt word *vomit*, has appropriated *sick* to that use, especially predicatively (*be, feel, s.*), transferring its more general sense to *ill*. But we still so use it attributively (*s. people*, a *s. child, s. bed, s. pay*, etc.) and predicatively in the army phrase for declaring oneself ill, *go s.*, which has now entered civil life. Instead of either *iller* or *sicker*, *more ill* or *more s., worse* is the comparative wherever it would not be ambiguous.

**sickly.** For adverb see -LILY.

**side-slip.** The grammatical accident to which this name is here given is most often brought about by the word *of*, and in OF 3 its nature has been so fully explained that nothing more is now required than some examples of the same accident not caused by *of*. In the earlier quotations other prepositions play the part of *of*; in the later ones the mistakes, though also due to the disturbing influence of what has been said on what is to be said, are not of quite the same pattern, and will need slightly more comment.

### Prepositional side-slips

*. . . possessing full initiative after its success, and able at will to expend a minimum force* in *defending itself* against one half *of the defeated body, and a* maximum effort against *destroying the other half* (*in*, for the second *against*). / *But there is one that deserves special*

mention because it lies at the root of the nation's confidence in the Navy and in the Navy's own cohesion as a loyal and united service (read *of* for the second *in*). / He has little in common with *those* union leaders who seem to be little more than faithful retainers to the ex-public school socialists, nor for *those* brawny strong-arm 'We'll show 'em' leaders who now seem equally out of date (read *with* for *for*). / In a plea for *the setting aside* of *this accord*, or at least for *certain* parts of the accord, by the Conference, the 'Temps' intimates that . . . (omit the second *for*). / The Independents would then be in the position in which the pledged Liberals now are of *being unable* to *appear on a platform or* helping *any* Liberal movements in any of the 330 Tory constituencies (read *to help* for *helping*).

### Miscellaneous

*Today we can but be thankful* that *the nerve of Fisher proved cool at the crisis, and* that *to him we mainly owe it that we have not to record a disaster of almost historical importance in the history of the railway.* Who is Fisher, that we should prefer him as saviour to other signalmen? The second *that* is there only because the first has sent the writer off at a tangent. To mend, either (a) omit the second *that*; or, (b) omit 'to him we mainly owe it that'. / *If it can be done, and* only *if it can be done, shall we be in the position to re-establish civilization.* The intervention of the parenthesis with its *only* is allowed to upset the order of words, viz. *we shall be*, required by the start of the sentence; this variety of side-slip is further illustrated in the section of INVERSION dealing with inversion in parallel clauses. / *Whether* the cessation of *rioting, looting, and burning which has been secured largely by the declaration of martial law and rigorous shooting of leaders of the rabble is merely temporary or* has been put an end to for good *remains to be seen.* If the cessation of rioting *has been put an end to for good*, a lively time is coming. To mend, read *permanent* instead of the words just italicized; and for this variety see

HAZINESS. / *He therefore came round to the view that simple Bible-teaching were better abolished altogether and that the open door for all religions were established in its place.* If the writer had been content with *would be* in place of the first *were*, he would certainly not have been trapped into thinking that the same auxiliary gave the right sense where the second *were* stands; but venturing on dangerous ground, which the subjunctive always is except to skilled performers, he side-slips. See SUBJUNCTIVES.

**sidle** is a BACK FORMATION from *sideling*, an obsolete form of *sidelong*.

**sien(n)a, Sien(n)a, Sien(n)ese.** The place is now usually spelt *Siena*, but the paint and the school of painting retain the old-established *-nn-*.

**sign** (indication) and some synonyms. The synonyms are so many that it seems worth while to collect some of them and add sentences showing each of them in a context to which it is better suited than any, or than most, of the others. The selected words are: badge, cachet, character, characteristic, cognizance, criterion, device, differentia, emblem, hall-mark, impress, index, indication, mark, motto, note, omen, portent, prognostic, seal, shibboleth, sign, slogan, stamp, symbol, symptom, test, token, touch, trace, trait, type, watchword.

*Sufferance is the* badge *of all our tribe. All his works have a grand* cachet. *These attributes of structure, size, shape, and colour, are what are called its* 'specific characters'. *Superstition is not the* characteristic *of this age. Geoffrey assumed as his* cognizance *the Sprig of Broom. Success is no* criterion *of ability. Shields painted with such* devices *as they pleased. The chief* differentiae *between man and the brute creation. The spindle was the* emblem *of woman. Lacking the* hall-mark *of a university degree. Lucerne bears most strongly the* impress *of the Middle Ages. The proverbs of a nation furnish the best* index *to its spirit. There is no* indication *that they had any knowledge of agriculture.*

*Suspiciousness is a* mark *of ignorance.* 'Strike while the iron's hot' *was his* motto. *Catholicity is a* note *of the true* Church. *Birds of evil* omen *fly to and fro. A* sky *dark* with *portent of* rain. *From* sure *prognostics* learn *to know the* skies. *Has the* seal *of death in his* face. *Emancipation from the fetters of party* shibboleths. *An outward and visible* sign *of an inward and spiritual* grace. *Our* slogan *is Small Profits and Quick Returns. Bears the* stamp *of* genius. *The Cross is the* symbol *of Christianity. Is already showing* symptoms *of* decay. *Calamity is the true* test *of friendship. By what* token *could it* manifest *its presence? One* touch *of nature makes the whole world* kin. Traces *of Italian influence may be* detected. *They have no national* trait *about them but their language. The paschal lamb is a* type *of Christ. The old Liberal* watchword *of Peace, Retrenchment, and Reform.*

**signal, single,** vv. *But there is intense resentment that Japan should be* signalled out *for special legislation.* | *There was one figure more sinister than the rest, whom Lloyd George* signalled out *for his wrath in true revivalist style.* | *The German Emperor has been spared an inglorious end in obscurity; but why has he been* signalled out *for the dignity of a special trial?* Three specimens of a very common MISPRINT or blunder; *singled* should be the word. Unfortunately, there is just nearness enough in meaning between the verb *single* on the one hand and, on the other, the adjective *signal* and the verb *signalize* to make it easy for the uncharitable to suspect writer rather than printer; and therefore especial care is called for, as with *deprecate* and *depreciate*.

**signif(y)(icant).** The dictionaries give *important* as one of the definitions of *significant*, but to use it merely as a synonym for that word is to waste it. The primary sense of s. is conveying a meaning or suggesting an inference. A division in the House of Commons may be important without being significant; the failure of some members to vote in it may be significant without being important. *There is no important change in the patient's condition* means that he is neither markedly better nor markedly worse. *There is no significant change in the patient's condition* means that there is no change which either confirms or throws doubt on the previous prognosis.

*It doesn't signify,* meaning it doesn't matter, once fashionable, has had its day.

**Signor(a), -rina,** Italian titles. Pronounce *sēn'yor, sēnyor'a, sēnyŏrē'nă*.

**silk(en).** See -EN ADJECTIVES.

**sillabub, syl-.** The OED attributes the *-y-* to 'the influence of *syllable*'. See Y AND I for intrusions of *y*.

**sillily.** One of the few current *-lily* adverbs; see -LILY s.f.

**silo.** The noun has pl. *-os*; see -O(E)S 6. The verb makes *-o'd* or *-oed*; see -ED AND 'D.

**silvan, sylvan.** There is no doubt that *si-* is the true spelling etymologically (Latin *silva*, a wood, changed in MSS. to *syl-* under the influence of Greek *ὕλη*); there is as little doubt that *sy-* now preponderates, and the OED does the word under that spelling, giving *silvan* as a variant. *Silvan* is here recommended, just as in Y AND I restoration of the right letter is recommended in other words. Though the false form does prevail nowadays, it is by no means universal; and it is worth notice that, out of seven Scott quotations in the OED, four show *sy-* and three *si-*. It is often too late to mend misspellings, but hardly so in this case; recent dictionaries agree in putting *si-* first.

**simian.** Pronounce *sĭ-*; the Latin noun is *sī*, but see FALSE QUANTITY.

**similar** is apt to bring disaster to certain writers—those to whom it is a FORMAL WORD to be substituted in writing for the like or the same with which they have constructed a sentence in thought. *It is claimed that the* *machine can be made to turn on its own centres*, similar *to the motor-boats which the inventor demonstrated at Richmond.*/ *Nevertheless, although adjoining New York all along its northern border and in its farming, manufacturing, and general industrial development swayed by similar business considerations that govern the Empire State, its people went as strongly for Roosevelt as their neighbours in New York went against him.* In the first quotation, *like* would stand, being both adjective and adverb, but *similar*, being adjective only, must be changed to *similarly*. In the second, *the same considerations that* would have been English, but *similar considerations that* must be corrected to *s. c. to those that*.

**simile.** To let this specialized and literary word thrust itself, as in the following quotation, into the place of the *comparison* or *parallel* that we all expect and understand is to come under suspicion of blindly using a synonym dictionary: *The advent of Kossovo Day cannot but suggest* a s. between *the conflict then raging and that in which we are engaged today.* A s. is always a comparison; but a comparison is by no means always a simile, and still less often deserves to be called one.

**simile and metaphor,** allegory and parable, apologue and fable. *Allegory* (uttering things otherwise) and *parable* (putting side by side) are almost exchangeable terms. The object of each is, at least ostensibly, to enlighten the hearer by submitting to him a case in which he has apparently no direct concern, and upon which therefore a disinterested judgement may be elicited from him, as Nathan submitted to David the story of the poor man's ewe lamb. Such judgement given, the question will remain for the hearer whether Thou art the man: whether the conclusion to which the dry light of disinterestedness has helped him holds also for his own concerns. Every parable is an allegory, and every allegory a parable. Usage, however, has decided that *parable* is the fitter name

for the illustrative story designed to answer a single question or suggest a single principle, and offering a definite moral, while *allegory* is to be preferred when the application is less restricted, the purpose less exclusively didactic, and the story of greater length. *The Faerie Queen* and *The Pilgrim's Progress* are allegories. The object of a parable is to persuade or convince; that of an allegory is often rather to please. But the difference is not inherent in the words themselves; it is a result of their history, the most important factor being the use of *parable* to denote the allegorical stories told by Christ.

It is of *allegory* that the OED gives as one of the definitions 'an extended or continued metaphor'. But the comment may be hazarded that there is some analogy between the relation of allegory to parable and that of simile to metaphor, and that the OED definition would, if that is true, have been still better suited to *parable* than to *allegory.* For between simile and metaphor the differences are (1) that a simile is a comparison proclaimed as such, whereas a metaphor is a tacit comparison made by the substitution of the compared notion for the one to be illustrated (*the ungodly flourishing 'like' a green bay-tree* is a confessed comparison or simile; *if ye had not plowed with my heifer,* meaning dealt with my wife, is a tacit comparison or metaphor); (2) that the simile is usually worked out at some length and often includes many points of resemblance, whereas a metaphor is as often as not expressed in a single word; and (3) that in nine out of ten metaphors the purpose is the practical one of presenting the notion in the most intelligible or convincing or arresting way, but nine out of ten similes are to be classed not as a means of explanation or persuasion, but as ends in themselves, things of real or supposed beauty for which a suitable place is to be found.

It cannot be said (as it was of allegory and parable) that every simile is a metaphor, and vice versa; it is rather that every metaphor presupposes a simile, and every simile is compressible or convertible into a metaphor. There is a formal line of demarcation, implied in (1) above; the simile is known by its *as* or *like* or other announcement of conscious comparison. There is no such line between allegory and parable, but in view of distinctions (2) and (3) it may fairly be said that parable is extended metaphor and allegory extended simile. To which may be added the following contrast. Having read a tale, and concluded that under its surface meaning another is discernible as the true intent, we say This is an allegory. Having a lesson to teach, and finding direct exposition ineffective, we say Let us try a parable. To reverse the terms is possible, but not idiomatic. See also METAPHOR.

The only difference between *apologue* and *fable* is that the second word is in common use and the first is not. Their distinctive feature is that, in the stories they tell, the participants are animals or even inanimate things. 'A fable or apologue seems to be, in its genuine state, a narrative in which beings irrational, and sometimes inanimate, are, for the purpose of moral instruction, feigned to act and speak with human interests and passions' (Johnson).

**simony.** The pronunciation *sī'mony* is probably commoner than *sĭm-* and conforms to the ordinary pronunciation of the name (Simon Magus) of the sorcerer who seems to have been the first to attempt this malpractice (*Acts* viii. 18).

**simpliciter, secundum quid.** These convey, the first that the statement etc. referred to need not, the second that it must, be restricted to certain cases or conditions.

**simulacrum.** Pl. -*cra*.

**sin.** 1. 'To sin one's mercies', which puzzles everyone to whom it has not been familiar from childhood, is paraphrased by the OED, but without explanation, as 'to be ungrateful for one's blessings or good fortune'. 2. 'More

sinned against than sinning' (*King Lear* III. ii. 60) has become a HACK-NEYED PHRASE; descent from the height of Lear to the latest triviality of 'tempted and fell' lands us, naturally, in bathos, and STOCK PATHOS.

**since.** For the very common mistake of using *s.* after *ago*, see AGO. For 'P.S. Since writing this your issue of today has come to hand', see UN-ATTACHED PARTICIPLES.

**sincerely.** For 'yours s.' etc. see LETTER FORMS.

**sine-.** For *sinecure* (= Lat. *sine cura*) the OED gives the pronunciation *sīnĕkūr*, adding that 'in Scotland and America the first vowel is freq. pronounced short'. It is now often pronounced short in England too, but the dictionaries still incline to *sĭn-*, and this conforms to the standard pronunciation of phrases such as *sine die* and *sine qua non*. The trigonometrical term is *sĭn*.

**sing.** For the past tense *sang* has prevailed over *sung*, formerly usual.

**Singalese.** See SINHALESE.

**sing(e)ing.** See MUTE E, and use the -*e*- in the participle of *singe*.

**singular -s** (or sibilant ending). The feeling that the z sound at the end of a noun proves it plural has played many tricks in the past; *pea*, *caper* (the herb), and jocularly *Chinee*, have been docked under its influence of their endings, *riches* is now always treated as a plural, and many other examples might be collected. The process continues today. '*Kudos*', wrote Edmund Wilson of American usage in 1963, 'seems now to be well established as the plural of a word meaning honourable mention or prize or something of the sort. A correspondent has sent me a clipping of a headline *KUDOS ARE IN ORDER* from a newspaper article on "Business Trends".' On the other hand it may be worth while to notice that the glasses of spectacles are *lenses* and not *lens*, that the plural of a *forceps* should certainly be what it unfortunately is not at present, *for-cepses*. Cf. GALLOWS and INNINGS.

**Sinhalese, Sing(h)alese, Cingalese.** The first and the last are recognized as the standard forms in the OED and the first has since made the greater progress.

**sinister** in heraldry means left (and *dexter* right), but with the contrary sense to what would naturally suggest itself, the left (and right) being that of the person bearing the shield, not of an observer facing it. For *bar, baton, bend, s.*, see BAR.

**sink, v. 1.** Past tense *sank* or *sunk*, the former now prevailing, especially in intransitive senses. **2.** *Sunk(en)*. The longer form is no longer used as part of a compound passive verb: *the ship would have been, will be, was, sunk*, not *sunken*. But *sunken* has not a corresponding monopoly of the adjectival uses: *sunken cheeks* or *eyes*; *a sunken (or sunk) rock*; *a sunk (or sunken) ship*; *a sunk (or sunken) fence* or *road*; *sunk carving*; *a sunk panel, shelf, storey*. Roughly, *sunken* is used of what has sunk or is (without reference to the agency) in the position that results from sinking, i.e. it is an INTRANSITIVE P.P.; and *sunk* is used of what has been sunk especially by human agency.

**sinus.** Pl. *-uses*, see -US 2.

**Sioux.** Pronounce *sōō*. Plural spelt *Sioux* and pronounced like singular, or with final *z* sound.

**siphon,** not *sy-*. See Y AND I.

**sir** (as prefix). To say *Sir Jones* is a mistake peculiar to foreigners. But writers often forget, as with HON., that a double-barrelled surname will not do instead of Christian name and surname: Sir Douglas-Home cannot be written for Sir Alec Douglas-Home. The same is true of the corresponding feminine prefix *dame*.

**siren,** not *sy-*. See Y AND I. Pronounce *sī'rĕn. Sīrĕn'*, perhaps on the analogy of the vulgar pronunciation of IRENE, was often heard during the second world war, and indeed is etymologically correct, for in the Greek word the *e* is long. But on this see FALSE QUANTITY.

**Sirius.** Pronounce *sĭ-*, not *sī*; for neglect of classical quantities see FALSE QUANTITY.

**sirloin.** The knighting of the loin attributed to various kings seems to have been suggested by, and not to have suggested, the compound word; it has, however, so far affected the spelling (which should have shown French *sur* = upper) that *sir-* may now be taken as fixed.

**sister,** in hospital use, is applied properly to one in charge of a ward or in authority over other nurses—matron, sisters, staff nurses, nurses, and probationers, being the hierarchy. But *s.* is often substituted, especially by soldiers in hospital, as a courtesy title for *nurse*.

**sisterly.** For the adv., see -LILY.

**situate(d).** The short form is still common in house-agents' advertisements and legal documents, but elsewhere out of favour.

**situation.** For excessive use of *s.* and *position* see -TION.

**sizable** is the recommended spelling. See MUTE E.

**skeptic(al), skepsis,** etc. See SCEPTIC(AL) etc.

**skew,** adj., though still current technically, chiefly in architecture, mathematics, and mechanics, and in a few compounds such as *s. bald* and, colloquially, *s.-eyed* and *s.-whiff*, has so far gone out of general use as to seem, in other applications, either archaic or provincial. The current word (adv. and pred. adj.) is *askew*.

**skewbald.** See PIEBALD.

**ski.** The anglicized pronunciation *skē* is now more usual than *shē*, which is how it is pronounced in Norway, its place of origin. Plural *ski* or *skis*. For the verb *ski'd* is preferable to *skied*; see -ED AND 'D.

**skiagraphy** etc. See SCIAGRAPHY.

**skier, skyer.** The user of ski is a *skier* obviously. The skied cricket-ball is spelt sometimes with *y* and sometimes with *i*; the OED prefers *skyer*, which has also the advantage of saving confusion; and, as it is more reasonable to derive it from *sky* n. than from sky v., there is no need to make it conform to *crier* and *pliers*; in any case there is little consistency in the spelling of derivatives of monosyllables ending -y; see DRY.

**skilled.** The *skilled* and the *unskilled* are sheep and goats, distinguished by having had or not having had the requisite training or practice; the two words exist chiefly as each other's opposites, or terms of a dichotomy. The point of the limitation is best seen by comparison with *skilful*. *Skilled* classifies, whereas *skilful* describes. You are skilled or not in virtue of your past history, you may be classed as semi-skilled, but will not (in idiomatic speech at least) be, as an individual, very or most or fairly skilled. You are skilful according to your present capacity, and in various degrees.

**skin.** *With the s. of my teeth*; see MISQUOTATION.

**skull.** See SCULL.

**slack(en), vv.** In the article -EN VERBS it is implied that the relation between the adjective and verb *slack* and the verb *slacken* is not simple enough to be there treated with the rest. One's first impression after a look through the OED articles on the two verbs is that whatever either means the other can mean too—an experience familiar to the synonym-fancier. The following distinctions are therefore offered with the caution that quotations contravening them may be found in the OED and elsewhere. **1.** *Slacken* is the ordinary word for to *become* slack, and for to make (or let become) slacker:

the tide, breeze, pace, demand, rope, one's energy, slackens; we slacken our efforts, grip, speed, opposition, the girth, the regulations. **2.** To slack, if it is to have such senses, is reinforced by *off*, *out*, *up*, etc.: *the train slacked up*; *had better slack off*; *slack out the rope*. **3.** *Slack*, not *slacken*, trespasses on the territory of *slake*: *slack* one's *thirst*, *lime, the fire* (see next article). **4.** *Slack*, not *slacken*, means to *be* slack or idle: *accused me of slacking*. **5.** *Slack* (trans.), not *slacken*, means to come short of or neglect (one's duty etc.), now archaic.

**slake, slack, vv.** Both are derived from the adjective *slack*, and *slake* had formerly such senses as loosen and lessen, which have now passed to the newer verb *slack* owing to their more obvious sense-connexion with it. *Slake* tends more and more to be restricted to the senses assuage, satisfy, moisten, (thirst, desire, vengeance, lips, lime).

**slander.** See LIBEL AND SLANDER.

**slang.** See JARGON and RHYMING SLANG.

**slaver, slobber, slubber, vv.** The three words, as well as *slabber*, which is virtually obsolete, may be assumed to be of the same ultimate origin, and, though they may have reached us by different routes and had more or less separate histories, they have so far acted and reacted upon one another that for people not deep in historical philology they are now variants of one word, partly but not completely differentiated. The basic meaning is (1) to run at the mouth, and its developments are (2) kissing, (3) licking, (4) fulsome flattery, (5) emotional gush, and (6) superficial smoothing over or mere tinkering. All three have sometimes any of the first four senses, though *slubber*, which is now chiefly in archaic literary use, tends to be confined to sense 6; and in that sense *slobber* is exceptional and *slaver* not used. The difference between *slaver* and *slobber* is partly of status, the former being the more literary and dignified and the latter colloquial and vivid, and partly of extent, *slaver* not going beyond sense 4, while *slobber* covers sense 5 and even 6. The now much commoner word for to run at the mouth—*dribble* —has had somewhat similar experiences with *drivel* (a word of different origin but assimilated to it) and *drool* (a contraction of *drivel*), which have retained their basic meaning besides acquiring a new one—to talk nonsense.

**slay**, though poetic or rhetorical in Britain, is still in use in America, for violent killing. It is a convenient word for HEADLINE LANGUAGE; *slayer* takes less space than *murderer*. For this reason it is likely to come back to journalistic use in Britain. It still makes *slew*, *slain*, not *slayed*.

**sled(ge), sleigh.** Though all three are interchangeable, they tend to be distinguished in use as follows: *sled*, drag for transporting loads; *sledge* English, *sleigh* U.S. and Canadian, for carriage on runners. 'Chiefly U.S. and Canada' is the OED label on *sleigh*; but the use of sledges in Great Britain is comparatively so rare that the Canadian idiom has naturally prevailed; *sleigh* is always the world at winter sports.

**sleep.** For *the s. of the just* see HACK-NEYED PHRASES.

**sleight.** Pronounce *slīt*; it is related to *sly* as *height* to *high*.

**sling, slink.** Past tenses and p.p. *slung, slunk*; the OED records but does not countenance the pasts *slang, slank*.

**slipshod extension.** To this heading, which hardly requires explanation, reference has been made in the articles on many individual words. Slipshod extension is especially likely to occur when some accident gives currency among the uneducated to words of learned origin, and the more so if those words are isolated or have few relatives in the vernacular; examples are *alibi, protagonist, recrudescence, optimism, meticulous, feasible, dilemma*.

The last two of these offer good typical illustrations. The original meaning of *feasible* is simply doable (Lat. *facere* do);

but to the unlearned it is a mere token, of which he has to infer the value from the contexts in which he hears it used, because such relatives as it has in English—*feat, feature, faction, fashion, malfeasance, beneficence*, etc.—either fail to show the obvious family likeness to which he is accustomed among families of indigenous words, or are (like *malfeasance*) outside his range. He arrives at its meaning by observing what is the word known to him with which it seems to be exchangeable—*possible*, and his next step is to show off his new acquisition by using it instead of *possible* as often as he can, without at all suspecting that the two are very imperfect synonyms; for examples see FEASIBLE. He perhaps notices now and then that people look at him quizzically as if he were not quite intelligible, but this does not happen often enough to prevent him from putting it comfortably down to their ignorance of the best modern idiom.

The case of *dilemma* as a word liable to slipshod extension differs in some points from that of *feasible*, though a dilemma is confused with a difficulty just as *feasible* is with *possible*. A person who has taken a taxi and finds on alighting that he has left his money at home is in a difficulty; he is not in a dilemma, but he will very likely say afterwards that he found himself in one. The differences are first that the mere Englishman has still less chance than with *feasible* of inferring the true meaning from related words, *dilemma* being an almost isolated importation from Greek; and second that the user need hardly be suspected of pretension, since *dilemma* is in too familiar use for him to doubt that he knows what it means. Nevertheless, he is injuring the language, however unconsciously, both by helping to break down a serviceable distinction, and by giving currency to a mere token word in the place of one that is alive. He is in fact participating in what has been called the crime of verbicide.

Slipshod extension, however, though naturally more common with words of learned antecedents, is not confined to them, and in the following selection will be found several that would seem too thoroughly part of the vernacular to be in danger of misuse. In many of the articles referred to, further illustration of slipshod extension is given: *Alibi, anticipate, chronic, claim, complex, crisis, crucial,* dead letter, *decimate, desiderate, dilemma, echelon, factor, feasible, identify, ilk, involve, level, limited, liquidate, literally, meticulous, mutual, nostalgic, optimism, practically, proposition, protagonist, reaction, recrudescence, redundant, sabotage, scotch, shambles, shibboleth, significant, transpire, unthinkable,* as well as many of the words listed as POPULARIZED TECHNICALITIES and as VOGUE WORDS.

A stray example may be added of a word with which such abuse is exceptional and apparently unaccountable. This will serve to illustrate the truth that slipshod extension is not the sort of blunder against which one is safe if one attends to a limited list of dangerous words; what is required is the habit of paying all words the compliment of respecting their peculiarities. *An excellent arrangement, for there are thus none of those smells which so often* disfigure *the otherwise sweet atmosphere of an English home.* What has no figure or shape cannot be disfigured. Not that the limitation need be closely pressed; we need not confine the word to a face or a landscape; an action, a person's diction, or a man's career (to take things of which the OED quotes examples), can also be disfigured, because each of them can be conceived, with the aid of metaphor, as a shapely whole. But a shapely atmosphere?

**sloe-worm.** See SLOW-WORM.

**slogan.** Though the great vogue of the word as a substitute for the older *motto, watchword,* etc., is of the 20th c. only, and we old fogies regard it with patriotic dislike as a Gaelic interloper, it was occasionally so used earlier; the OED has a quotation from Macaulay. And we have turned it to good account by appropriating it to those

catchwords with which in the modern world politicians, ideologists, and advertisers try to excite our emotions and atrophy our minds. For some synonyms, see SIGN.

**slosh.** See SLUSH.

**slough.** The noun meaning bog is pronounced *-ow*; the meaning cast skin etc. and the verb meaning cast or drop off are pronounced *-ŭf*.

**slovenly.** For the adv., see -LILY.

**slow(ly),** advv. In spite of the encroachments of *-ly* (see UNIDIOMATIC -LY), *slow* maintains itself as at least an idiomatic possibility under some conditions even in the positive (*how slow he climbs!, please read very slow*), while in the comparative and superlative *slower* and *slowest* are usually preferable to *more* and *most slowly*; see -ER AND -EST 3. Of the 'conditions', the chief is that the adverb, and not the verb etc., should contain the real point; compare 'We forged slowly ahead', where the slowness is an unessential item, with 'Drive as slow as you can', where the slowness is all that matters. In the phrase *go slow* (e.g. of a clock or of workmen) *slow* alone is idiomatic.

**slow-worm, sloe-.** Though the derivation of the first part is obscure (Skeat attributes it to the supposed poisonous properties of the creature—the 'worm that slays'), it is certainly not connected with either the noun *sloe* or the adj. *slow*; *slow-* is now the established form, and the OED calls *sloe-* obsolete.

**slubber.** See SLAVER.

**sludge.** See SLUSH.

**slumber.** Apart from mere substitutions of *s.* for *sleep* dictated by desire for poetic diction or dislike of the words that common mortals use, *slumber* is equivalent to the noun *sleep* with some adjective or the verb *sleep* with some adverb. Slumber (often used in the plural without difference of sense) is easy or light or half or broken or daylight sleep, or again mental or stolen or virtual or lazy sleep. The implied epithet or adverb, that is, may be almost anything; but the choice of *slumber* instead of *sleep*, if not due to mere stylishness (see WORKING AND STYLISH WORDS), is meant to prevent the reader from passing lightly by without remembering that there is sleep and sleep. For *slumberwear* see GENTEELISMS.

**slumb(e)rous.** The shorter form is recommended; cf. DEXT(E)ROUS. But analogies for either are plentiful: *cumbrous, wondrous, monstrous, leprous, idolatrous*; but *thunderous, slanderous, murderous*.

**slush, sludge, slosh.** The differences are not very clear. There is the natural one, resulting from the stickier sound, that *sludge* is usually applied to something less liquid than *slush* or *slosh*, e.g. to slimy deposits or clinging mud, especially to precipitates in processes such as coal-washing and sewage-disposal, whereas thawing snow is typical slush. *Slush* and *slosh* are both used to describe what is metaphorically watery stuff—twaddle or sentimentality, and *slosh* alone is the slang word for a violent blow or, as a verb, giving one.

**sly** makes *slyer, slyest, slyly, slyness, slyish*; for comparison with other such words, see DRY.

**small.** Relations with *little* are complicated, and the task of disentangling them might excusably be shirked, if not as difficult, then as unprofitable. But examination of the differences between seeming equivalents does give an insight into the nature of idiom. Under BIG some attempt has been made at delimiting the territories of *great, large,* and *big. Small* and *little* have to divide between them the opposition to those three as well as to *much*, and the distribution is by no means so simple and definite as the pedantic analyst might desire.

Of the possible pairs of opposites let some be called patent pairs, as being openly and comfortably used with both members expressed, and the rest latent pairs. The patent pairs

start with three that are pretty clearly distinguishable in meaning. Contrasts of size or extent are given by *large and small*; 'large and small rooms', 'of large or small size', 'large or small writing', 'large and small appetites', 'large and small dealings, dealers'. Contrasts of quantity or amount are given by *much and little*; 'much or little butter, faith, exercise, damage, hesitation, study'. Contrasts of importance or quality are given by *great and small*; 'the Great and the Small Powers', 'great and small occasions', 'a great or a small undertaking', 'great and small authors'. Besides these, the main divisions, there are two minor patent pairs sometimes substituted for one or other of them—*great and little* and *big and little. Great and little* as a patent pair is preferred to *large and small* in distinctive names ('the Great and the Little Bear', 'Great and Little Malvern'; Great and Little Cumbrae; it is also common (see below) as a latent pair in two senses. *Big and little* is a patent pair often substituted for either *large and small* ('big and little farms, motor cars') or *great and small* ('big and little wars, ships') or *great and little* ('the big and the little toe').

The patent pairs are sets of opposites so far felt to correspond that one does not hesitate to put them together as in all the examples quoted. Or again either member can be used when the other is not expressed but only implied; e.g., 'the Great Powers' is more often used alone, but 'the Great and the Small Powers' is also an ordinary expression; the 'Big Five' depends for its meaning on the existence of an unspecified number of smaller banks and the 'little-go' on that of an examination no longer called the 'great-go'.

By latent pairs are meant sets of opposites in which one member has the meaning opposite to that of the other but could not be expressly contrasted with it without an evident violation of idiom. For instance, no one would put *large and little* together; 'large and little lakes' sounds absurd; but one speaks of 'a (or the) little lake' without hesitation, though 'large lakes' (not 'great lakes', which ranks with the distinctive names above referred to) is the implied opposite. Another latent pair is *much and small*; though 'much or small hope' is impossible, and 'much or little hope' is felt to be required instead, yet 'small hope', 'small thanks', 'small credit', 'small wonder', are all idiomatic when the irregular opposite *much* is not expressed. Similarly with *big and small*; we never contrast them openly, but in 'the big battalions', 'big game', 'a big investment or undertaking', the opposite in reserve is *small*. *Great and little* was said above to rank both as a patent and as a latent pair. In the latter capacity it allows us to talk of 'great damage', 'great doubt', 'great hesitation', and again of 'little damage' etc., but forbids us to put the pair together; it is 'much or little (not 'great or little') doubt'.

**smear.** See LIBEL AND SLANDER.

**smell,** v. 1. For *smelt* and *smelled* see -T AND -ED. 2. The intransitive sense to emit an odour of a specified kind is idiomatically completed by an adjective, not an adverb; a thing smells sweet, sour, rank, foul, good, bad, etc., not sweetly, badly, etc. But the tendency referred to in UNIDIO-MATIC -LY sometimes misleads the unwary into using the adverbs. The mistake is the easier because (a) when the character of the smell is given by 'of so-and-so' instead of by a single word, an adverb is often added; compare *smells strong* or *delicious* (i.e. has a strong or delicious smell) with *smells strongly* or *suspiciously of whisky* or *deliciously of violets*; and (b) when to *smell* is used, as it may be, for to *stink*, an adverb is the right addition—*this water smells outrageously*. *Smells disgusting* and *smells disgustingly* are both idiomatic, but are arrived at in slightly different ways, the first meaning 'has a disgusting smell', and the second 'stinks so as to disgust one'.

**smite** is an archaic word, now little

used except jocularly, especially with the addition *hip and thigh* in IRRELE-VANT ALLUSION to what Samson did to the Philistines, and in the p.p. *smitten*, meaning *épris*.

**smog,** though rarely allowed as yet to do without probationary inverted commas, is older than would be supposed by the many people who first heard it at the time of the great London fog of December 1952. The OED Supp. quotes from *The Globe* of 27 July 1905 'The other day at a meeting of the Public Health Congress Dr. Des Voeux did a public service in coining a new word for the London fog, which was referred to as *smog*, a compound of *smoke* and *fog*.' It has now entered the field of metaphor: *For much of the way the author's meaning is all but buried in a dense smog of sub-epigrams and superfluous images.* | *Talk of technical superiority at this stage . . . is simply a smoke screen which should be dispelled without further ado—a rather bad case of technical smog.* See PORTMANTEAU.

**smouch.** It may be by coincidence, or it may be by the emergence of a dialect usage long latent, that this obsolete word for *kiss* was adopted by young people in the middle of the twentieth century as the slang term for indulging in those amatory exercises that in less sophisticated times were called *kissing and cuddling*, and in contemporary U.S. slang *necking*.

**smudge, smutch.** The earlier noun is *smutch*, the earlier verb *smudge*; but this has had no apparent effect on usage; -*dge* now prevails in ordinary literal use, -*tch* being preferred in metaphor; a painting is smudged but a reputation smutched. *Smutch*, however, is now less used than *smirch* or *besmirch*.

**snake, serpent.** *Snake* is the native and *serpent* the alien word; it is also true, though not a necessary consequence, that *snake* is the word ordinarily used, and *serpent* the exceptional. The OED's remark on *serpent* is 'now, in ordinary use, applied chiefly to the larger and more venomous species; otherwise only *rhetorical* . . . or with reference to serpent-worship'. We perhaps conceive serpents as terrible and powerful and beautiful things, snakes as insidious and cold and contemptible; hence *sea-serpent*, but *water-snake*. The serpent shines in the night sky; the snake lurks in the grass.

**snapshot,** vb. The OED recognizes no verb to *snapshoot*, though it gives *snapshooter* and *snapshooting* (chiefly in the original sense, i.e. with gun, not camera); but *snapshot* is now established both as noun and as verb, usually abbreviated to *snap*. In short *snapshoot* has been divided and its two parts set to different tasks. The holiday-maker on the beach *snaps*; the camera-man in the film studio *shoots*.

**so.** 1. Phrases treated elsewhere. 2. *So long, and so to* —, *do so*. 3. Appealing *so*. 4. Didactic *so*. 5. Repeated *so*. 6. *So* with p.p. 7. Explanatory *so*. 8. *So* with superlatives and absolutes. 9. *So* introducing a clause of purpose.

1. For *so far from, so far as, so far that*, see FAR; for *and so on, and so forth*, see FORTH; for *quite* (*so*) see QUITE; for *so to speak* see SUPERIORITY; for *ever, never, so* see NEVER SO.

2. *So long, and so to* —, *do so*. *So long* used colloquially for *goodbye* or *au revoir*. It perhaps matters little for practical purposes, but the OED gives no countenance to the derivation from *salaam*, and treats the phrase as a mere special combination of *so* and *long*; those who are inclined to avoid it as some sort of slang may be mollified by its naturalness as a short equivalent for Good luck till we meet again.

*And so to a division, and so to dinner*, etc. This formula for winding up the account of a debate or incident, borrowed directly or indirectly from Pepys, is apt to take such a hold upon those who once begin upon it that, like confirmed cigarette-smokers, they lose all count

of their indulgences; it is wise to abstain from it altogether.

*Do so*. For absurdities such as the following, which are too common, see DO 3. *It is a study of an elderly widower who, on approaching sixty, finds that he knows hardly anything of his three daughters, and sets out to do so*.

3. The appealing *so*. The type is *Cricket is so uncertain*. The speaker has a conviction borne in upon him, and, in stating it, appeals, with his *so*, to general experience to confirm him; it means *as you*, or *as we all, know*. A natural use, but more suitable for conversation, where the responsive nod of confirmation can be awaited, than for most kinds of writing. In print, outside dialogue, it has a certain air of silliness, even when the context is favourable, i.e. when the sentence is of the shortest and simplest kind (for this use of *so* is really exclamatory), and the experience appealed to is really general. Readers will probably agree that in all the following extracts the context is not favourable.—*In the case of Ophthalmology in the tropics a work of authority is so sadly overdue. | But he does combine them ingeniously, though in instancing this very real power we feel that it might have been so much more satisfactorily expended. | He was always kind, considerate, and courteous to his witnesses, this being so contrary to what we are led to expect from his successors. | Slade would seem to have some of the philosophy of his kind, as well as the technique, which chiefly is the reason why one so hopes he will not be rushed on too rapidly*.

4. The didactic *so*. This is a special form of the appealing *so*, much affected by Walter Pater: *In the midst of that aesthetically so brilliant world of Greater Greece* is an example. The *so* is deliberately inserted before a descriptive adjective, and is a way of saying, at once urbanely and concisely, Has it ever occurred to you how brilliant etc. it was? That is to say it differs from the *sos* in 3 in being not careless and natural, but didactic and highly artificial. Effective enough on

occasion, it is among the idioms that should never be allowed to remind the reader, by being repeated, that he has already met them in the last hundred pages or so. *Here an Englishman has set himself to follow in outline the very distinctive genius of Russia through the centuries of its difficult but always so attractive development. | And still no one came to open that huge, contemptuous door with its so menacing, so hostile air.*

**5.** *So* in repetition. A change from the artificial to the entirely artless. *So* is a much used word, but not indispensable enough to justify such repetitions of it as the following: *The pity is that for so many men who can so hardly keep pace with rising prices it should become so difficult to follow the sport. | For ironically enough the very complexity of modern political life, which today makes it so necessary for the Government to improve their lines of communication with the people, has also done much to weaken the principal bridge that previously helped so much towards this end—the House of Commons. | The situation was well in hand, but it had so far developed so little that nothing useful can be said about it, save that so far the Commander-in-Chief was satisfied.*

**6.** *So* with p.p. The distinction usually recognized with VERY between a truly verbal and an adjectival p.p. is not applicable to *so*; but it is well worth while, before writing plain *so*, to decide between it and *so much, so well,* etc. The insertion of *much* in the first and *well* in the second quotation after *so* would certainly be an improvement: *Admiral Faravelli reports that Tripoli batteries have been so damaged that Turkish soldiers have been forced to retire into town. | Ireland being mainly an agricultural country, and England industrial, the Bill is not so suited to Ireland as to his country.*

**7.** The explanatory *so.* Type: *He could not move; he was so cold.* The second member is equivalent to a sentence beginning with *for,* and the idiom is mainly, but not solely, collo-

quial. What requires notice is that, when it is used in formal writing, it is spoilt if *for,* whose work is being done for it by *so,* is allowed to remain as a supernumerary. Two examples follow, the first right, the second wrong: *The dangers of the situation seem to us very real and menacing; both sides, in maintaining a firm attitude, may so easily find themselves bluffing over the edge into the precipice. | It would seem particularly fitting that an American professor of literature should discuss the subject of Convention and Revolt,* for *in that country the two tendencies are at present so curiously and incongruously mingled.*

**8.** *So* with superlatives and absolutes. *So,* when it qualifies adjectives and adverbs, means to such a degree or extent; it is therefore not to be applied to a superlative, as in *The difficult and anxious negotiations in which he has taken so foremost a part in Paris,* or to an absolute, as in *It is indeed a privilege to be present on so unique an occasion.* See also SUCH 6.

**9.** *So* introducing a clause of purpose or result. To introduce such a clause with *so that* (or *so as to*) is standard usage. But British idiom does not countenance the use of *so* alone, as in *The Nigerian authorities asked for him to be returned under the Fugitive Offenders Act so he could stand trial on charges of treasonable felony. | K. gave up a staff job to become a freelance journalist so he could fit in training.* See also IN ORDER THAT.

**sobriquet.** I. *sob-* is much longer established in English than *soub-* besides being the only modern French form. Pronounce *sō'brǐkā*. 2. Under this heading, for want of a better, are here collected eighty or so out of the thousands of nicknames or secondary names that have become so specially attached to particular persons, places, or things, as to be intelligible when used instead of the real or primary names, each of which is thus provided with a deputy or a private pronoun. The deputy use is seen in 'It was carried to the ears of that famous hero and warrior, the Philosopher of Sans Souci', where 't. P. o. S. S.' acts for Frederick the Great; and the private-pronoun use in 'He employed his creative faculty for about twenty years, which is as much, I suppose, as Shakespeare did; the Bard of Avon is another example ...', where 't. B. o. A.' means Shakespeare or the latter. Some names have a large retinue of sobriquets; Rome, e.g., may be the Eternal City, the City of the Seven Hills, the Papal City, the Empress of the Ancient World, the Western Babylon and her list of sobriquets is not half told; Napoleon may be Boney, the Little Corporal, or the Man of Destiny.

Now the sobriquet habit is not a thing to be acquired, but a thing to be avoided; and the selection that follows is compiled for the purpose not of assisting but of discouraging it. The writers most subject to temptation are sportsmen writing about the sports in which they have excelled. Games and contests are exciting to take part in, interesting or even exciting also to watch, but essentially (i.e. as bare facts) dull to read about. Such a writer, or the ghost he employs, conscious that his matter and his audience are both dull enough to require enlivening, thinks that the needful fillip may be given if he calls fishing the gentle craft, a ball the leather, a captain the skipper, or a saddle the pigskin, and so makes his description a series of momentary puzzles that shall pleasantly titillate inactive minds. Here is a *Times* reviewer, who sighs over 'One sad fault, which runs through this, and, alas! a good many other excellent books—the habit of seldom calling a spade a spade. Does it really help, or is it really humorous, to call the fox "Charles James", a hare "Madam", a nose a "proboscis", and Wales "Taffyland"? Of course, a sporting book will tend to use sporting expressions; but a good deal of this irritating circumlocution is unnecessary, and might well be left for colloquial use'.

It is by no means true, however, that the use of sobriquets is confined to this, or to any, class of writers; the Philosopher of Sans Souci and the Bard of Avon quoted above are from Thackeray and Conan Doyle, though they are unfavourable specimens of those authors' styles. Moreover, the sobriquet deputy has its true uses. Just as Bacon knows of 'things graceful in a friend's mouth, which are blushing in a man's own', so the sobriquet may often in a particular context be more effective than the proper name; though 'the Papal City' means Rome, its substitution may be a serviceable reminder, when that is appropriate, that Rome in one of its aspects only is intended. Again, some sobriquets have succeeded, like mayors of the palace, in usurping their principals' functions; the Young Pretender is actually more intelligible, and therefore rightly more used, than Charles Edward, and to insist on 'came over with William I' in preference to 'with the Conqueror' would be absurd.

No universal condemnation of sobriquets is therefore possible; but even the better sort of journalist, seldom guilty of such excesses as the sporting writer, is much tempted to use them without considering whether they tend to illuminate or to obscure; 'the philosopher of Ferney', he feels, at once exhibits his own easy familiarity with Voltaire the man (*Voltaire* the word, by the way, is itself one of the mayor-of-the-palace sobriquets) and gratifies such of his readers as know who is meant. As for those who may not know, it will be good for them to realize that their newspaper is more cultured than they. The sobriquet style, developed on these lines, is very distasteful to all readers of discretion. Those who may become aware, in glancing through the following alphabetical selection of sobriquets other than those already mentioned, that these and similar substitutes are apt to occur frequently in their own writing should regard it as a very serious symptom of perverted taste for cheap ornament. In most of the expressions an initial *the* is to be supplied: Alma Mater (university); Auld Reekie (Edinburgh); Beefeater (Yeoman of the Guard); Black Country (industrial west midlands); Black Maria (prison van); Black Prince (eldest son of Edward III); Bluecoat school (Christ's Hospital); Blue ribbon of the turf (the Derby); Cœur de Lion (Richard I); Cousin Jacky (Cornishman); Digger (Australian); Emerald Isle (Ireland); Ettrick Shepherd (James Hogg); Farmer George (George III); Father of History (Herodotus); Father of Lies (Satan); First Gentleman of Europe (George IV); Garden of England (Kent); Gilded Chamber (House of Lords); Gloriana (Elizabeth I); G.O.M. (Gladstone); Granite City (Aberdeen); Great Cham (Samuel Johnson); Great Commoner (the elder Pitt); Herring pond (North Atlantic Ocean); House (Chamber of the House of Commons, Stock Exchange, Christ Church Oxford); Iron Chancellor (Bismarck); Iron Duke (Wellington); Jack Tar (common sailor R.N.); John Bull (Englishman); Jollies (Royal Marines); Kingmaker (Warwick); King of beasts (lion); King of Terrors (death); Knight of the Rueful Countenance (Don Quixote); Lion of the North (Gustavus Adolphus); Maid of Orleans (Joan of Arc); Merry Monarch (Charles II); Mother of Parliaments (British Parliament); Ocean greyhound (liner); Old Contemptibles (British Expeditionary Force 1914); Old Nick (devil); Old Lady of Threadneedle Street (Bank of England); Old Pretender (James, son of James II); Paddy (Irishman); Pommy (British immigrant to Australia or New Zealand); Porch (Stoic school of philosophy); Queen of the Adriatic (Venice); Rag (Army and Navy Club); Rupert of debate (14th Earl of Derby); Sage of Chelsea (Carlyle); Sailor King (William IV); St Stephen's (Houses of Parliament); Seagreen incorruptible (Robespierre); Senior (or Silent) service (navy); Soapy Sam (Bishop Wilberforce); Sport of kings (horse-racing); staff of life (bread); Stagirite (Aristotle); Stars and stripes (U.S. flag); Swan of Avon (Shakespeare);

Strawberry leaves (ducal rank); Tommies (British soldiers); Uncle Sam (U.S.A.); Virgin Queen (Elizabeth I); Union Jack (British flag); Warrior Queen (Boadicea); Great Wen (London); Wizard of the North (Scott); Young Chevalier (Charles Edward Stuart).

**soccer, -cker.** *Soccer* did not deserve its victory in the competition between these alternative spellings. *Accept, success, eccentricity, accident, flaccid, coccyx,* show the almost invariable sound of -cc- before e, i, y; perhaps the only exceptions are *baccy* and *recce,* which the hard c sound in the full words makes more excusable then *soccer.*

**sociable, social.** For confusion between pairs of adjectives in -*able* and -*al*, see EXCEPTIONABLE, PRACTICABLE. No such patent misuses occur with the present pair as with those; there is merely a tendency to use *social* not where it is indefensible, but where the other would be more appropriate. Roughly, *social* means of or in or for or used to or shown in or affording society; and *sociable* seeking or loving or marked by the pleasures of company. *Social* is rather a classifying, and *sociable* rather a descriptive adjective: man is a social being; Jones is a sociable person; people are invited to a social evening, and say afterwards (or do not say) that they had a very sociable evening. Obviously, overlapping is likely. The OED, under a definition of *social* that includes 'sociable' as an equivalent, gives two quotations in which *sociable* should have been preferred (*His own friendly and social disposition*—Jane Austen / *He was very happy and social*—Miss Braddon), as well as one that is just on the right side of the border (*Charles came forth from that school with social habits, with polite and engaging manners* —Macaulay).

**sociolegese.** We live in a scientific age, and like to show, by the words we use, that we think in a scientific way. In more than one article of this dictionary, especially in POPULARIZED TECHNICALITIES, reference is made to the harm that is being done to the language by this well-meant ambition (see also GROUP, BRACKET). Sociologese, like COMMERCIALESE and OFFICIALESE, deserves an article to itself. Sociology is a new science concerning itself not with esoteric matters outside the comprehension of the layman, as the older sciences do, but with the ordinary affairs of ordinary people. This seems to engender in those who write about it a feeling that the lack of any abstruseness in their subject demands a compensatory abstruseness in their language. Thus, in the field of industrial relations, what the ordinary man would call an informal talk may be described as *a relatively unstructured conversational interaction*, and its purpose may be said to be *to build, so to speak, within the mass of demand and need, a framework of limitation recognized by both worker and client*. This seems to mean that the client must be persuaded that, beyond a certain point, he can only rely on what used to be called self-help; but that would not sound a bit scientific. Or again, still in the field of industrial relations, results may be summarized in language like this: *The technique here reported resulted from the authors' continuing interest in human variables associated with organizational effectiveness. Specifically, this technique was developed to identify and analyse several types of interpersonal activities and relations, and to provide a method for expressing the degree of congruence between two or more of these activities and relations in indices which might be associated with available criteria of organizational effectiveness.*

There are of course writers on sociological subjects who express themselves clearly and simply; that makes it the more deplorable that such books are often written in a jargon which one is almost tempted to believe is deliberately employed for the purpose of making what is simple appear complicated, exhibiting in an extreme form the common vice (see ABSTRACT-

ITIS) of preferring pretentious abstract words to simple concrete ones. It would be easy but tedious to multiply examples; two will be enough.

1. (On the reason why the 'middle class' speak differently from the 'lower working class'.) *The typical, dominant speech-mode of the middle class is one where speech becomes an object of perceptual activity, and a 'theoretical attitude' is developed towards the structural possibilities of sentence organization. This speech-mode facilitates the verbal elaboration of subjective intent, sensitivity to the implications of separateness and difference, and points to the possibilities inherent in a complex conceptual hierarchy for the organization of experience.* [*The lower working class*] *are limited to a form of language use which, though allowing for a vast range of possibilities, provides a speech form which discourages the speaker from verbally elaborating subjective intent, and progressively orients the user to descriptive rather than abstract concepts.*

2. (On family life.) *The home then is the specific zone of functional potency that grows about a live parenthood; a zone at the periphery of which is an active interfacial membrane or surface furthering exchange—from within outwards and from without inwards—a mutualising membrane between the family and the society in which it lives.*

**soft.** 1. For 's. impeachment' see IRRELEVANT ALLUSION. 2. For *play, sleep, fall*, etc., *s.*, see UNIDIOMATIC -LY.

**soi-disant.** See FRENCH WORDS. English, with *self-styled, ostensible, would-be, professed, professing, supposed*, and other words, is well provided for all needs.

**sojourn.** OED gives *sŭ-, sŏ-, sō-*, in that order. The battle is still undecided, but *sŏ-* has gained on *sŭ* and both on *sō-*. See PRONUNCIATION 5.

**solder.** *Sŏ'der* was formerly the established pronunciation but the speak-as-you-spell movement (see

PRONUNCIATION I) has favoured *sōlder* or *sŏlder*; the sounding of the *l* is now preferred by most dictionaries.

**soldierly.** For adv. see -LILY.

**solecism** ('offence against grammar, blunder in the manner of speaking or writing') is a Greek word (σολοικισμός), said to come from the corruption of the Attic dialect among the Athenian colonists of Soloi in Cilicia. The grammarians used to distinguish between *barbarism*, incorrectness in the use of words, and *solecism*, incorrectness in the construction of sentences.

**solemnness.** See SPELLING POINTS 2, s.f.

**soliloquy.** See MONOLOGUE.

**solo.** Pl. *-os*, see -O(E)S 6, or more formally in music *soli* (*ē-*).

**so long,** = goodbye. See SO 2.

**soluble, solvable,** make *insoluble, unsolvable*; see IN- and -UN-. Substances are soluble (or dissolvable), not solvable; problems are soluble or solvable. See also DISSOLUBLE and RESOLUBLE.

**solve.** See RESOLVE.

**some.** I. *S.* in meiosis. 2. *Some one, someone*. 3. *Sometime, some time*, etc. 4. *Somewhat*. 5. *Somewhen*.

I. Meiosis. 'This is some war', with strong emphasis on *some*, is modern colloquial for 'This is a vast war', 'This is indeed a war, if ever there was one'. It is still felt as slang, and it comes to us from America; but it results from that love of MEIOSIS which is shared with the Americans by us. We say a place is some distance off, meaning a long way; we say 'It needs some faith to believe that', meaning a hardly possible credulity. So far the effect is exactly parallel to the emphatic use of *rather* in answer to a question—'Do you like it?' 'Rather!', meaning not somewhat, but exceedingly (see RATHER 5). The irregular development comes in when *some*, meiosis and all, is transferred

from its proper region of quantity or number to that of quality; some faith is a wonderful amount of faith; but some war is a wonderful kind or specimen of war, and some pumpkins (more than 100 years old, and said to be the original American phrase) were not a great number of pumpkins, but pumpkins of so superior a quality as to be the only fitting description of the speaker's girl friend (*She was some pumpkins*). Phrases of this kind are apt to perish when they become so trite as no longer to sound humorous, but Sir Winston Churchill's *Some chicken! Some neck!* should have given immortality to this one. Compare with it our own equivalent, which lacks the piquant irregularity, 'something like a war'.

2. For *someone, some one*, see EVERY ONE.

3. *Some time, sometime*, etc., advv. *Some time* is often used elliptically for at some time or other. There is no essential objection to writing it *some-time* or *sometime*, but it is convenient to keep it in two separate words for distinction from the *sometime* that appears in such descriptions as 'sometime Fellow of . . .', 'sometime Rector of this Parish', meaning formerly. *Someplace* for somewhere is U.S. only. See PREPOSITION DROPPING.

4. *Somewhat* has for the inferior journalist what he would be likely to describe as 'a somewhat amazing fascination'. Thus: *The evidence furnished in the somewhat extraordinary report of the Federation as to its waste of huge sums of money on . . .* | *His election experiences were somewhat unique.* | *The flocks of wild geese, to which the flamingo is somewhat more or less closely allied.* | *The Labour motion introduced the proviso, somewhat for the first time, that the process should be gradual.* These are examples selected for their patent absurdity, and their authors are doubtless so addicted to the word that they are no longer conscious of using it. What first moves people to experiment in the somewhat style is partly timidity

—they are frightened by the coming strong word and would fain take precautions against shock—and partly the notion that an air of studious understatement is superior and impressive; and so in our newspapers 'the intemperate orgy of moderation is renewed every morning'. Cf. the similar use of COMPARATIVELY and *relatively* as shock-absorbers.

5. *Somewhen* should be regarded as the progeny of *somewhere* and *somehow*, and allowed to appear in public under the wing of either or both of its parents, but not by itself.

**-some.** This suffix has been so much used to make fanciful words, now archaic or poetical (*blithesome, brightsome, gladsome, darksome, lightsome, lovesome,* and the like), that we are disinclined to treat seriously the *-some* words that we do use (e.g. *awesome, cuddlesome, fearsome, gamesome, winsome*) unless they are so familiar that their compound origin has been forgotten, e.g. *fulsome, gruesome, handsome, loathsome, noisome, quarrelsome, tiresome, wearisome.*

**somersault,** *summersault, somerset, summerset.* The italicized alternatives are obsolete, except perhaps *somerset* in rustic talk.

**son-in-law.** See -IN-LAW.

**sonnet,** once used loosely of any short poem, is now applied only to those rhymed poems of fourteen decasyllabic lines of which there are in English three recognized varieties, the Petrarchian, the Shakespearian, and the Miltonic.

**sonorous.** 'Properly *sono′rous*, it will probably sooner or later become *so′nŏrous*' said Skeat in 1884. It has not done so yet; the COD still puts *sono′rous* first. Perhaps it is our familiarity with Milton's *Sonorous metal blowing martial sounds* that has made us resist the RECESSIVE ACCENT.

**sophistic(al).** *Sophistical* is now the usual form. It would be well if, in accordance with what is said in the article -IC(AL), *sophistic* could be con-

fined to the merely defining sense 'of the (Greek) Sophists', *sophistical* being left to the quibbler and *sophisticated* to the worldly-wise and other modern meanings of that rather wayward participle.

**soprano.** Pl. *-os*, see -O(E)S 6, or -ni (*-ē*).

**sore.** *And the people . . . lifted up their voices and wept s.* The use of *s.* as an adverb is WARDOUR STREET, if not archaic, and *sorely* cannot be classed among the adverbs in -ly whose use is deprecated in UNIDIOMATIC -LY.

**sorites** (pronounce *sorī′tēz*), meaning 'heap', is a term applied to two entirely different things.

1. A process by which a predicate is brought into the desired relation to a subject by a series of propositions in which the predicate of one becomes the subject of the next, and the conclusion has the first subject and the last predicate. Thus: Schoolmasters are teachers; Teachers are benefactors; Benefactors are praiseworthy; Therefore schoolmasters are praiseworthy. A sorites may be a short way of exhibiting truth, or, as in the above example, may conceal fallacies at each or any step.

2. A logical trick named from the difficulty of defining a heap. If grains of corn are accumulated one by one, at what point will the addition of a single grain convert into a heap what was not a heap before?

**sorrow.** For 'more in s. than in anger', see HACKNEYED PHRASES.

**sorry, sorrow.** The two words do not belong to each other, as one might suppose; *sorry* is the adjective of the noun *sore*. Sore and sorrow, however, are so near in sense (especially in earlier and wider meanings of *sore*) that the mistake has no ill effects. Still, the knowledge has its practical value; connexion between *sore* and *sorry* helps to account for the use of *sorry* in the sense of paltry, shabby, wretched, worthless (e.g. *s. business, excuse, plight*), or of nasty, as when

Macbeth, looking at his hands, exclaims 'This is a sorry sight'.

**sort,** in the irregular but idiomatic uses touched upon under KIND, is equally common, and subject to the same limitations. *Sort of* and *kind of* preceding a verb (*I s. o. expected it*) differ from the others in being more generally confined in practice to the colloquial, like the *sort of thing* with which persons conscious of their limited powers of expression used to punctuate everything they said, until it was replaced by the now inevitable *you know*. It is worth mention that the OED, always chary in condemnation, records these idioms without seriously questioning their legitimacy. The same is true of the common depreciatory use of *of sorts* (*Yet in principle the Government have a case of sorts*), a convenient idiom that should perhaps now be granted literary status; the quotation is from a leading article in *The Times* in 1962. For *of any sort or kind* (*We can only repeat that there is no inconsistency of any sort or kind in our attitude*) see PLEONASM and SIAMESE TWINS. For *those sort* (or *kind*) *of things* see KIND.

**sough.** The pronunciation alternatives in the OED are *sŭf, sow* and *soo,* the last followed by the breathed guttural. English people, uncertain how to pronounce the word, are shy of using it; when they do they probably give it the first; a Scot, who has no such inhibitions, will certainly give it the last.

**sound,** adv. For *sleep sound(ly)*, see UNIDIOMATIC -LY.

**soupçon.** See GALLICISMS.

**source.** *Mr. M. said that he had met by accident the original s. of the information. He had not asked whether the s. was willing for his name to be disclosed.* / *One of my ss. has given me the wording of a very important valentine. Sources* (of information) is an established and unexceptionable phrase, especially for indicating, without specifying, where a piece of news came from. (*There is no confirmation as yet from official ss.*) Cf. *circles* and *quarters.* But the PERSONIFICATION of *source* in the above quotations, though perhaps a natural result of the imprisonment of two journalists for refusing to reveal their 'sources', is both needless and absurd. *Informant* is the word.

**south-.** Compounds (*s.-east* etc.) are pronounced with th. Of the derivatives, *southerly, southern, southernwood, southron,* have *sŭdh-; souther* and *southing* have sowth-; *southward(s)* is sowthward(z) or (at sea) sŭdhard(z).

**southerly.** For the special uses and meanings of this set of words, see EASTERLY.

**southpaw.** *Eagle-eyed viewers may have noticed a left-handed violinist fiddling the opposite way to everyone else. . . . This fiddler was James Barton, the only southpaw in the business at the moment.* This seems to be an example of NOVELTY HUNTING for the purpose of ELEGANT VARIATION. *Southpaw* is an Americanism originally applied to a left-handed pitcher at baseball, and later extended more widely (especially in sport) to those who do with the left hand what is usually done with the right, and vice versa, e.g. a boxer who leads with the right arm and leg instead of the orthodox left.

**Soviet.** The OED Supp. puts *Sŏv-* before *Sōv-,* and this seems to be the usual pronunciation of English-speaking people who have visited the country, though *Sōv-* is commoner.

**sow,** vb. The p.p. *sown* is four times as frequent, in the OED 19th–20th-c. quotations, as *sowed.*

**spark off** is a PHRASAL VERB of modern coinage, ordinarily used of the immediate cause of some explosive event (e.g. a strike, riot, or war) whose more remote causes have been gradually accumulating. It is an apt metaphor. *These incidents revealed serious deterioration in the security situation; the view generally held in the Southern Province was that any incident might* *spark off immediate violence. Trigger off* is similarly used.

**special.** 1. *Special, especial.* 2. *S. pleading.*

1. For *special(ly)* as distinguished from *especial(ly)*, see ESPECIAL. The two following quotations show each adverb used where the other would have been better: *Ample supplies of food and clothing for the prisoners are now available there, having been shipped from America especially for this purpose. | The neighbourhood is not specially well provided with places where soldiers can get amusement and refreshments.*

2. *S. pleading* is a POPULARIZED TECHNICALITY. When we say that a person's argument is s.p., we mean that he has tried to convice us by calling our attention to whatever makes for the conclusion he desires, and diverting it from whatever makes against it. But this is, not indeed the highest, but at any rate the almost universal, argumentative procedure. That is, it is advocacy or (in the untechnical sense) pleading, and the word *special* adds nothing to the meaning; why then call it special? Pleadings, in law, are a series of formal written statements by the parties to a suit designed to establish clearly, before the case is tried, what is the issue or question to be decided. S. p. is adaptation of the typical outline formulae to the circumstances of a particular case. As one consequence of modern legal reforms, pleadings are now very commonly dispensed with; but formerly the s. p. had to be done with extreme accuracy if cases were not to be lost on points of form that were of no real importance. S. p. accordingly became identified with legal quibbling, and suffered the same fate as casuistry, passing into a byword for dishonest evasion of real issues. This vague and inaccurate sense the name has retained now that the thing itself is no longer familiar outside the legal profession.

**speciality, -alty.** The two words, like many pairs in -IC(AL), while they

seem to cry out for DIFFERENTIATION, have made little progress in that direction. Anyone who thinks he knows which of the chief senses belong to which, and tests his notions by looking through the OED quotations, is likely to have a surprise; he will perhaps conclude that writers use either form for any of the senses according as they prefer its sound in general or find it suits the rhythm of a sentence. Where usage is so undecided, it would be presumptuous to offer a profitable differentiation, or to recommend either of two fully established forms for extinction. The most that can be ventured is that *speciality* is in most senses the commoner, and that *specialty* prevails in the sense of a special subject of study or research and in the legal sense of a contract under seal.

**specie(s).** Plural the same; see LATIN PLURALS. For pronunciation see -IES, -EIN. See also next article.

**specific(ally).** These words, like RESPECTIVE(LY), though their real value need not be questioned, are often resorted to by those who have no clear idea of their meaning but hold them to diffuse an air of educated precision. A short table of the senses of *specific* follows, showing the relation of each to the central notion of species; it is in the last rather loose sense that it may be wise to avoid the word and choose one of the more generally understood synonyms.

1. Characterizing a kind or species. *S. gravity* is that belonging to some substance (e.g. gold or beer) as a kind or as such.

2. Constituting kind or species. *S. difference* is that which entitles audacity, man, etc., to be called by those names rather than by more general ones such as courage, mammal.

3. Indicating species in classification, i.e. the class next below *genus*. In *Pinus sylvestris* (Scotch fir) and *Passer domesticus* (house sparrow) *Pinus* and *passer* are generic, *sylvestris* and *domesticus* specific.

4. Applicable to a kind only. *S. remedy* (or *specific*, used as a noun) is

one used for a particular disease or organ.

**5.** Of a disease due to some identifiable micro-organism or lesion.

**6.** *S. performance* (of a contract) ordered by a Court in cases where damages for breach would not adequately compensate the other party.

**7.** Not universal but limited, not general but particular, not vague but definite. *S. directions, accusation, cause.*

**specious** is a WORSENED WORD. Originally it meant fair or pleasing to the eye or sight; resplendent with beauty (OED). But it has long been used only of people, things, or arguments, whose attractiveness is deceptive. Cf. PLAUSIBLE.

**spectrum.** Pl. usually *-tra*; see -UM, and LATIN PLURALS.

**speculum.** Pl. usually *-la*; see -UM and LATIN PLURALS.

**speed.** Past and p.p *sped*; but *s. up,* = increase the s. of, makes *speeded* (*must be speeded up* etc.), and that is also the natural past of *speed* in the sense of drive (a car) at an excessive speed.

**spell,** vb. 1. For *spelt, spelled,* see -T and -ED. 2. The sense amount to, mean, involve as inevitable result, seen in *Democracy spells corruption,* and esp. in So-and-so *spells ruin* ('common in recent use'—OED), had its merit, no doubt, when new, but has now a rather faded look.

**spelling points.** 1. Spelling reform. 2. Double and single letters for consonantal sounds. 3. Cross references. 4. Miscellaneous.

**1.** Spelling reform. We can no longer do as Swift did, and airily dismiss the subject as 'the foolish opinion advanced of late years that we ought to spell exactly as we speak'. In our age of compulsory education, growing impatience with the notorious difficulty of English spelling, recently showing itself in more than one attempt to effect reform by legislation,

creates an obligation to declare one's general attitude towards reform before touching any details. The line here followed is, then: that the substitution for our present chaos of a phonetically consistent method that did not sacrifice the many merits of the old spelling would be of incalculable value; that a phonetically consistent method is in English peculiarly hard to reconcile with the keeping together of word-families, owing to the havoc played on syllable sounds by variations of stress; that attempts at so radical a reform are likely to meet insuperable prejudice, and so perhaps to delay less ambitious but desirable changes; that most reformers are so much more awake to the obvious advantages of change than to its less obvious evils that we cannot trust them with the disposal of so vastly important a matter; and, finally, that English had better be treated in the English way, and its spelling not be revolutionized but amended in detail, here a little and there a little as absurdities become intolerable, till a result is attained that shall neither overburden schoolboys nor stultify intelligence nor outrage the scholar. 'Those who reverence [the uncompromising tyranny of our spelling system]', said Robert Bridges, 'have to learn that it has no divine right, and if they obstinately uphold its usurpation they are playing into the hands of the revolutionists, who would cast it off altogether and substitute the worse tyranny of a questionable phonetic system.' In this book some modest attempts are made at cleaning up the more obtrusive untidinesses; certain inconsistencies have been regarded as no longer required of us in the present diffusion of literacy. The well-known type theoretic-radical cum practical-conservative covers perhaps a majority of our population, and its influence is as sound and sane in the sphere of spelling as elsewhere.

**2.** Double and single letters for consonantal sounds. If a list were made of the many thousands of words whose spelling cannot be safely inferred from their sound, the doubtful point in perhaps nine-tenths of them would be whether some single consonantal sound was given by a single letter, as m or t or c, or a double letter, as mm or tt, or two or more, as sc or cq or sch. Acquiesce and a*q*ueduct, bivoua*c* and bivoua*ck*ing, Bri*t*ain and Bri*tt*any, co*mm*ittee and co*m*ity, *c*rystal and *ch*rysalis, i*n*oculate and i*nn*ocuous, insta*ll* and insti*l*, ha*r*ass and emba*rr*ass, leve*ll*ed and unparalle*l*ed, perso*n*ify and perso*nn*el, *sch*edule and *sh*ed, *sc*ience and *s*ilence, ti*c* and ti*ck*, are examples enough. The use of double letters (tt etc.) or two letters (ck etc.) to give a single sound is due sometimes to the composition of a word, as when *in-* not and *nocens* harmful are combined to make *innocent*, sometimes to the convention by which the sound of a preceding vowel tends to be of one kind (ā ē ī ō ū) before a single letter and of another (ă ĕ ĭ ŏ ŭ) before two, and sometimes to factors in wordformation, perhaps philologically explicable, but less obvious than in compounds like *innocent*. Of these causes the only one that has a meaning for anyone who knows no language but English is the convention of vowel sounds. He is aware that much more often than not a distinction exists analogous to that between *holy* and *holly*; but the interference of the other causes is so incalculable and so frequent that he soon finds it hopeless to rely upon the principle in doubtful cases. Hence a large proportion of the tears shed over spelling. Little relief can be given; the words in which sound is no guide to whether a consonantal sound is given by one letter or two are not a score or so of which a list could be made and learnt, but thousands. Nothing short of a complete spelling-book will serve the turn of a really weak speller, though it is true that a short list can be made of words in which mistakes are especially common, and that some classes of mistake can be guarded against by rules. Such a list is best made by each person who finds himself in need of it, out of his own experience and to suit his own requirements; a few words

that will usually be included are *abbreviate, accommodate, appal, banister, battalion, bilious, Britannia, Brittany, bulrush, bunion, camellia, canonical, committee, desiccated, disappear, disappoint, embarrass, exaggerate, harass, innocuous, inoculate, install, instil, moccasin, saddler, skilful, tonsillitis, unparalleled*. It is worth remark that words presenting two opportunities for mistake like *disappoint* (dissap-, disapp-, dissapp-, disap-), or three like unpa*rallel*ed, are more than two or three times as dangerous as others, temptations to assimilate or dissimilate the two or more treatments being added to the doubled or trebled opportunity.

Among the rules referred to above are those that govern the doubling or not of a word's final consonant when suffixes are added in inflexion or wordformation. Directions are given for the various consonants under the articles -B-, -BB-, and so on, to be found in their alphabetical places; but it may be useful to state the main principle here:—Words ending in a singleletter consonant preceded by a short vowel sound, when they have added to them a suffix beginning with a vowel (e.g. *-ed* of the past, *-er* of the agent or of comparison, *-able* or *-y* of adjectives), double the final letter if they either are monosyllables or bear their accent on the last syllable; they keep it single if they have their last syllable unaccented. But a final l is doubled irrespective of accent, and with a final s usage varies. Thus the addition of *-ed* to the verbs *pot, regret, limit, travel*, and *bias*, gives *potted* (monosyllable), *regretted* (accented final), *limited* (unaccented final), *travelled* (final l), and *biassed* or preferably *biased* (final s); similarly the verbs *tar, demur, simper, level, focus*, give *tarring, demurring, simpering, levelling*, and *focussing* or preferably *focusing*; the adjectives *thin, common, cruel*, give *thinnest, commonest*, and *cruellest*; the nouns *gas, syrup* give *gassy, syrupy*.

Two more questions of single and double letters are of importance to weak spellers. In forming adverbs in -ly from adjectives in -l or -ll, neither a single nor a triple l is ever right; *full*, *purposeful*, *especial*, and *dull*, have adverbs *fully*, *purposefully*, *especially*, and *dully*. And in forming nouns in -ness from adjectives in -n both *n*s are retained—*commonness*, *rottenness*, *plainness*, etc.; even *solemn*, with its mute n, need hardly be excepted, but the OED gives the orthodox *solemnness* only as a variant of *solemness*.

**3.** Cross references. Various points are discussed in short special articles throughout the book; and many words whose spelling is disputed will be found spelt with or without discussion in their alphabetical places. The following collection of references may serve as a conspectus of likely mistakes and desirable minor reforms.

For such words as *lik(e)able*, *mil(e)age*, *gaugeable*, *pal(e)ish*, *judg(e)ment*, *wholly*, see MUTE E.

For plural of words in -o see -O(E)S; many individual words are also given.

For plural of words in -y see PLURAL ANOMALIES 4.

For *tire tyre*, *tiro tyro*, *silvan sylvan*, *siphon*, *cipher*, *siren*, *sillabub*, *sibyl*, *gypsy*, *pygmy*, etc., see Y AND I, and the words.

For *Aeschylus Æschylus*, *Oedipus Œdipus*, *oecumenical œc- ec-*, *diarrhoea -œa*, *Caesar Cæs-*, *diaeresis -œr-*, etc., see Æ, Œ.

For *dyeing*, *flier*, *triable*, *paid*, *tying*, etc., see VERBS IN -IE, -Y, -YE.

For *one-ideaed -ea'd*, *umbrellaed -a'd*, *mustachioed -o'd*, *feeed fee'd*, etc., see -ED AND 'D.

For the question between -ize and -ise as the normal verb ending, and for a list of verbs in which -ise only is correct see -IZE, -ISE, IN VERBS.

For plural of *handful*, *spoonful*, etc., see -FUL. Choice is not between *handfuls* and *handsful*, but between *handfuls* and *hands full*, either of which is sometimes the right expression.

For adjectives ending -ble see -ABLE, -IBLE.

For choice between hyphening, separation, and consolidation, see HYPHENS.

For inflexions of verbs in *c* like *picnic* and *bivouac* see -C-, -CK-.

For alternatives like *enquiry* and *inquiry*, *undiscriminating* and *indiscriminating*, see EM- AND IM-, and IN- AND UN-.

For 'pet names' like *doggie*, *nannie*, see -EY, -IE, -Y.

For adjectives like *hors(e)y*, *mat(e)y*, *clayey*, *hol(e)y*, see -EY AND -Y.

For *for(e)bears*, *for(e)gather*, *for(e)go*, etc., see FOR- AND FORE-.

For *cooperate co-op- coop-*, *preeminent* etc., recover and *re-cover*, re-*enforce* and *reinforce*, etc., see CO-, and PRE-, and RE-.

For *formulae -las*, *hippopotamuses -mi*, etc., see LATIN PLURALS.

For *burnt -ned*, *leapt -ped*, etc., see -T AND -ED.

For *by and by*, *by the bye*, *by-election*, etc., see BY, BYE, BY-.

For derivatives of *day* and other monosyllabic words in -y, see DRY.

For *no one no-one*, *someone*, etc., see EVERY ONE.

For *countryfied*, *Frenchified*, etc., see -FIED.

For *glycerin(e)*, *gelatin(e)*, etc., see -IN AND -INE.

For *into in to*, *onto on to*, see INTO, and ONTO.

For *prophecy -sy*, *device -se*, etc., see LICENCE.

For *net(t)*, *mat(t)*, *pot(t)*, etc., see SET(T).

For *deserter*, *corrector*, etc., see -OR.

For *governo(u)r*, *labo(u)r*, etc., see -OUR AND -OR.

For *humo(u)rous*, *colo(u)ration*, etc., see -OUR- AND -OR-.

For *cwt. cwt*, *Mlle. Mlle*, *Dr. Dr*, etc., see PERIOD IN ABBREVIATIONS.

For *Jones's Jones'*, *Venus' Venus's*, see POSSESSIVE PUZZLES.

For *referable*, *inferrible*, etc., see CONFER(R)ABLE.

**4.** Miscellaneous. The rule 'i before e except after c' is very useful; it applies only to syllables with the vowel sound $\bar{e}$; words in which that sound is not invariable, as *either*, *neither*, *heinous*, *inveigle*, do not come under it; *seize* is an important exception; and it is useless with proper names (*Leigh, Reith*, etc.). The c exception covers the many derivatives of Latin *capio*, which are in such common use that a simple rule of thumb is useful (*receive, deceit, inconceivable*; but *relieve, belief, irretrievable*).

The writing of the very common *anti-* against instead of the rarer *ante-* before (e.g. *antichamber, antidated*) is to be carefully avoided.

Verbs in *-cede, -ceed*, are so many and so much used, and the causes of the difference are so far from obvious, that mistakes are frequent and a list will be helpful: *cede, accede, antecede, concede, intercede, precede, recede, retrocede, secede*, to which may be added *supersede*; but *exceed, proceed, succeed*. The curious thing is that a division so little reasonable should be so religiously observed; there is no disagreement among good spellers, and the only mistake into which they occasionally slip is *preceeding* for *preceding*.

Adjectives and nouns in *-ble, -cle, -tle*, etc., make their adverbs and adjectives not by adding -ly or -y, but by changing -le to -ly: *humbly, subtly, singly, supply* (not *suppley*), *treacly, tangly*.

Adjectives in *-ale, -ile, -ole*, add -ly for their adverbs: *stalely, vilely, docilely, solely*; but *whole* makes *wholly*.

For verbs ending in *-bre, -tre*, etc., the forms *sabring, accoutring, centring, mitring, manœuvring*, are recommended in preference to *sabreing, manœuvering*, etc. Similarly *ochrous* and *ogrish* seem better than *ochreous* or *ocherous* and *ogreish* or *ogerish*; but impious hands can hardly be laid upon *acreage*.

Of adjectives in *-(e)rous* some never use the e, as *cumbrous, disastrous, idolatrous, leprous, lustrous, monstrous, wondrous*; some have it always, as *boisterous, murderous, obstreperous, slanderous, thunderous*; *dextrous* and *slumbrous* are better than *dexterous* and *slumberous*.

**sphere.** For synonyms in the sense province, see FIELD.

**sphinx.** Although the OED gives the plural *sphinges* it would be pedantry to use it. *Sphinxes* (like *minxes*) is the only tolerable form.

**spif(f)licate.** OED spells -*ifl*-; see FACETIOUS FORMATIONS. This is an old one, dating back to 1785, and is now outmoded. In these grimmer days we LIQUIDATE instead.

**spill.** For *spilt -lled*, see -T AND -ED.

**spilth.** See REVIVALS. There is a gap of 200 years between Shakespeare (who uses it once only) and the earliest modern OED quotation. Its revival has been feeble, and it may fairly be classed as an archaism.

**spin.** For the past tense the OED 19th-c. quotations give *span* and *spun* in exactly equal numbers; *spun* has since made the greater progress, and is likely to prevail.

**spinach, -nage.** The first is the recognized spelling; the other, corresponding to the popular pronunciation, is obsolete.

**spindleage,** not *spindlage*, is the OED spelling of this little-used word, coined on the analogy of *acreage* to mean the total number of cotton spindles in use at a given time in any specified area. But see MUTE E.

**spindrift, spoon-.** The first is the usual modern word. The original *spoondrift* is from an obsolete nautical verb *spoon* or *spoom* meaning (of ship or foam) to scud; there is no profit in trying to restore the correct but now puzzling form.

**spinet.** The OED prefers the accent on the first syllable; among its verse quotations is one in favour of each. But -*et'*, perhaps from the analogy of *duet, motet*, etc., is now more usual.

**spiritism, spiritualism.** *Spiritism* and *spiritistic* mean the same as *spiritualism* in its most frequent and *spiritualistic* in its only acceptation. 'Preferred by those specially interested in the subject, as being more distinctive than *spiritualism*' is the OED comment on *spiritism*. To ordinary people the old noun with a new meaning comes much more natural than the recent invention. What first occurs to the mind of anyone who nowadays hears the word *spiritualism* is not the general sense, i.e. 'tendency towards a spiritual view or estimate of things'; it is the special sense of 'belief that the spirits of the dead can hold communication with the living'. So true is this that the addition of 'modern', at first thought necessary to distinguish the special from the general sense, is no longer made. And in fact the OED's comment is now out of date. Spiritualists no longer prefer the word *spiritism*; on the contrary they resent the idea underlying it that the two meanings of *spiritualism* can properly be divorced.

**spiritual, -ous.** The DIFFERENTIATION (-*al* of soul, -*ous* of liquor) is now complete, and neglect of it is more likely to be due to inadvertence than to ignorance.

**spirituel(le).** The word's meaning is not quite clear to everyone, and is therefore here given in the OED terms: 'Of a highly refined character or nature, esp. in conjunction with liveliness or quickness of mind.' And on the spelling the OED remarks: 'The distinction between the masc. and fem. forms has not been always observed in English.' That is undoubtedly so, and the spelling problem presented is an awkward one. On the one hand, the notion of m. and f. forms for adjectives is entirely alien to English, and if a French adjective is to make itself at home with us it must choose first whether it will go in male or female attire and discard its other garments; on this point cf. NAIF and *naive*. On the other hand, the choice with this particular word is a dilemma; if we decide for -*el* we are sacrificing the much more familiar of the two forms—more familiar because the word has been chiefly applied to women and in this application purposely made feminine by those who recognize both genders. But, if we decide for -*elle*, few of us can rid ourselves of the feeling that the word is

feminine and suitable only to what, for the English, is alone feminine, viz. woman; so that we find ourselves debarred from describing qualities, faces, talk, and above all men, as spirituelle, and cannot give the word its proper extension.

The lesser evil is to spell always *spirituel*; the objection to it is not, like that to -*elle*, one that will endure for ever, but one that, when the form is settled, will no longer be felt.

**spirt, spurt.** The spelling is now very much a matter of personal fancy, and whether more than one word is concerned is doubtful. There are, however, two distinguishable main senses—that of gush, jet, or flow (v. and n.), and that of sprint, burst, hustle (v. and n.); and for the second sense the form *spurt* is far the commoner. It would plainly be convenient if the differentiation thus indicated were made absolute; *a spirt of blood*; *works by spurts*; *oil spirts up*; *Jones spurted past*. See also SPRINT.

**spiv.** In inventing this word the English have emulated the American genius for coining monosyllabic words (cf. *stunt, blurb*) whose sound is curiously suited to their meaning. The origin of *spiv* is obscure, but it is presumably connected with the slang word *spiff* (dandyishness) once familiar in the juvenile slang *spiffing*.

**splendiferous.** See FACETIOUS FORMATIONS.

**split infinitive.** The English-speaking world may be divided into (1) those who neither know nor care what a split infinitive is; (2) those who do not know, but care very much; (3) those who know and condemn; (4) those who know and approve; and (5) those who know and distinguish.

1. Those who neither know nor care are the vast majority, and are a happy folk, to be envied by most of the minority classes. 'To really understand' comes readier to their lips and pens than 'really to understand'; they see no reason why they should not say

it (small blame to them, seeing that reasons are not their critics' strong point), and they do say it, to the discomfort of some among us, but not to their own.

2. To the second class, those who do not know but do care, who would as soon be caught putting their knives in their mouths as splitting an infinitive but have only hazy notions of what constitutes that deplorable breach of etiquette, this article is chiefly addressed. These people betray by their practice that their aversion to the split infinitive springs not from instinctive good taste, but from tame acceptance of the misinterpreted opinion of others; for they will subject their sentences to the queerest distortions, all to escape imaginary split infinitives. 'To really understand' is a s. i.; 'to really be understood' is a s. i.; 'to be really understood' is not one; the havoc that is played with much well-intentioned writing by failure to grasp that distinction is incredible. Those upon whom the fear of infinitive-splitting sits heavy should remember that to give conclusive evidence, by distortions, of misconceiving the nature of the s. i. is far more damaging to their literary pretensions than an actual lapse could be; for it exhibits them as deaf to the normal rhythm of English sentences. No sensitive ear can fail to be shocked, if the following examples are read aloud, by the strangeness of the indicated adverbs. Why on earth, the reader wonders, is that word out of its place? He will find, on looking through again, that each has been turned out of a similar position, viz. between the word *be* and a passive participle. Reflection will assure him that the cause of dislocation is always the same—all these writers have sacrificed the run of their sentences to the delusion that 'to be really understood' is a split infinitive. It is not; and the straitest non-splitter of us all can with a clear conscience restore each of the adverbs to its rightful place: He was proposed at the last moment as a candidate likely *generally* to be accepted. / When the record of this cam-

paign comes *dispassionately* to be written, and in just perspective, it will be found that . . . / New principles will have *boldly* to be adopted if the Scottish case is to be met. / This is a very serious matter, which clearly ought *further* to be inquired into. / The Headmaster of a public school possesses very great powers, which ought *most carefully and considerately* to be exercised. / The time to get this revaluation put through is when the amount paid by the State to the localities is *very largely* to be increased.

3. The above writers are bogy-haunted creatures who for fear of splitting an infinitive abstain from doing something quite different, i.e. dividing *be* from its complement by an adverb; see further under POSITION OF ADVERBS. Those who presumably do know what split infinitives are, and condemn them, are not so easily identified, since they include all who neither commit the sin nor flounder about in saving themselves from it—all who combine a reasonable dexterity with acceptance of conventional rules. But when the dexterity is lacking, disaster follows. It does not add to a writer's readableness if readers are pulled up now and again to wonder— Why this distortion? Ah, to be sure, a non-split die-hard! That is the mental dialogue occasioned by each of the adverbs in the examples below. It is of no avail merely to fling oneself desperately out of temptation; one must so do it that no traces of the struggle remain. Sentences must if necessary be thoroughly remodelled instead of having a word lifted from its original place and dumped elsewhere: What alternative can be found which the Pope has not condemned, and which will make it possible *to organize legally* public worship? / It will, when better understood, tend *firmly to establish* relations between Capital and Labour. / Both Germany and England have done ill in not combining *to forbid flatly* hostilities. / Every effort must be made *to increase adequately* professional knowledge and attainments. / We have had *to shorten somewhat* Lord

D——'s letter. / The kind of sincerity which enables an author to *move powerfully* the heart would . . . / Safeguards should be provided *to prevent effectually* cosmopolitan financiers from manipulating these reserves.

**4.** Just as those who know and condemn the s. i. include many who are not recognizable, since only the clumsier performers give positive proof of resistance to temptation, so too those who know and approve are not distinguishable with certainty. When a man splits an infinitive, he may be doing it unconsciously as a member of our class 1, or he may be deliberately rejecting the trammels of convention and announcing that he means to do as he will with his own infinitives. But, as the following examples are from newspapers of high repute, and high newspaper tradition is strong against splitting, it is perhaps fair to assume that each specimen is a manifesto of independence: It will be found possible *to considerably improve* the present wages of the miners without jeopardizing the interests of capital. / Always providing that the Imperialists do not feel strong enough *to decisively assert* their power in the revolted provinces. / But even so, he seems *to still be allowed* to speak at Unionist demonstrations. / It is the intention of the Minister of Transport *to substantially increase* all present rates by means of a general percentage. / The men in many of the largest districts are declared *to strongly favour* a strike if the minimum wage is not conceded.

It should be noticed that in these the separating adverb could have been placed outside the infinitive with little or in most cases no damage to the sentence-rhythm (*considerably* after *miners*, *decisively* after *power*, *still* with clear gain after *be*, *substantially* after *rates*, and *strongly* at some loss after *strike*), so that protest seems a safe diagnosis.

**5.** The attitude of those who know and distinguish is something like this: We admit that separation of *to* from its infinitive is not in itself desirable, and we shall not gratuit-

ously say either 'to mortally wound' or 'to mortally be wounded'; but we are not foolish enough to confuse the latter with 'to be mortally wounded', which is blameless English, nor 'to just have heard' with 'to have just heard', which is also blameless. We maintain, however, that a real s. i., though not desirable in itself, is preferable to either of two things, to real ambiguity, and to patent artificiality. For the first, we will rather write 'Our object is to further cement trade relations' than, by correcting into 'Our object is further to cement . . .', leave it doubtful whether an additional object or additional cementing is the point. And for the second, we take it that such reminders of a tyrannous convention as 'in not combining to forbid flatly hostilities' are far more abnormal than the abnormality they evade. We will split infinitives sooner than be ambiguous or artificial; more than that, we will freely admit that sufficient recasting will get rid of any s. i. without involving either of those faults, and yet reserve to ourselves the right of deciding in each case whether recasting is worth while. Let us take an example: 'In these circumstances, the Commission, judging from the evidence taken in London, has been feeling its way to modifications intended to better equip successful candidates for careers in India and at the same time to meet reasonable Indian demands.' To better equip? We refuse 'better to equip' as a shouted reminder of the tyranny; we refuse 'to equip better' as ambiguous (*better* an adjective?); we regard 'to equip successful candidates better' as lacking compactness, as possibly tolerable from an anti-splitter, but not good enough for us. What then of recasting? 'intended to make successful candidates fitter for' is the best we can do if the exact sense is to be kept; it takes some thought to arrive at the correction; was the game worth the candle?

After this inconclusive discussion, in which, however, the author's opinion has perhaps been allowed to appear with indecent plainness, readers may like to settle the following question for themselves. 'The greatest difficulty about assessing the economic achievements of the Soviet Union is that its spokesmen try absurdly to exaggerate them; in consequence the visitor may tend badly to underrate them.' Has dread of the s. i. led the writer to attach his adverbs to the wrong verbs, and would he not have done better to boldly split both infinitives, since he cannot put the adverbs after them without spoiling his rhythm? Or are we to give him the benefit of the doubt, and suppose that he really meant *absurdly* to qualify *try* and *badly* to qualify *tend*?

It is perhaps hardly fair that this article should have quoted no split infinitives except such as, being reasonably supposed (as in 4) to be deliberate, are likely to be favourable specimens. Let it therefore conclude with one borrowed from a reviewer, to whose description of it no exception need be taken: 'A book . . . of which the purpose is thus—with a deafening split infinitive—stated by its author: "Its main idea is *to* historically, even while events are maturing, and divinely—from the Divine point of view—*impeach* the European system of Church and States".'

**splodge, splotch.** The second is two centuries older; the first perhaps now more usual and felt to be more descriptive; cf. SLUSH, and SMUDGE.

**splutter, sputter.** Without any clear or constant difference of meaning, it may be said that in *sputter* the notion of spitting is more insistent, and that it tends on that account to be avoided when that notion is not essential.

**spoil.** For *spoiled*, *-lt*, see -T AND ED. For confusion between it and *despoil* see DESPOIL.

**-spoken.** For the curious use in *fair*, *free*, *soft*, *out*, etc., *-s* (where *fair-speeched* etc. might have been expected), see INTRANSITIVE P.P. It should be remembered that in these compounds *fair-* etc. are adverbial as much as *out-*, and that what is remarkable is not the adverbial use of the adjective, but the active use of the participle.

**sponge.** 1. S. makes *spongeable*; but *sponging* and *spongy*, see MUTE E. 2. *Jones on the wing and Meredith in the pack showed no inclination to throw in the sponge.* The idiom is to throw *up* the sponge or throw *in* the towel, alternative ways by which a second may indicate that his man gives up.

**spontaneity, -ousness.** See -TY AND NESS. Pronounce *-eity* to rhyme with *deity*, not *laity*.

**spoondrift.** See SPINDRIFT.

**spouse.** For the use in ordinary writing in preference to *wife*, see FORMAL WORDS; but *s.* is serviceable as short for husband-or-wife in some styles, e.g. in dictionaries or legal documents.

**sprain, strain.** It is natural to wish for a clear line of distinction between two words that, as applied to bodily injuries, are so near in sense and both so well established; but they are often treated as equivalent. *Sprain*, perhaps, describes the result rather of a momentary wrench or twist, especially of an ankle or wrist, and *strain* that of an exertion of muscle too strong or too long for its capacity.

**spring.** The past *sprang* is now established, both in trans. and intrans. senses; *sprung* remains for the p.p.

**sprint, spurt.** The words are to a considerable extent interchangeable; *sprint* is, at least apart from dialect use, a 19th-c. word only, *spurt* going further back, but the newer word is displacing the older. A short race, or a run at high speed, is now a sprint, while for a quickening of pace, or a spasmodic effort bodily or mental, *spurt* is still the more usual term; the differentiation is useful. See also SPIRT.

**spry** makes *spryer*, *spryest*, *spryly*, *spryness*, *spryish*.

**spurt.** For *s*. and *spirt*, see SPIRT; for *s*. and *sprint*, see SPRINT.

**sputter.** See SPLUTTER.

**squandermania(c).** A FACETIOUS FORMATION.

**square.** Of persons, says the SOED, 'honourable, upright'. That is not how the word is used in the teenagers' slang of the mid-20th c.; and it is not clear why they should have given it a meaning ('old-fashioned'—COD) not very different from that of *straitlaced*, which suggests a shape far from square. Perhaps this usage has its origin in footwear rather than figure: *squaretoed* is an epithet for the prim and old-fashioned that goes back to the 18th c. *Stuffy* was square's immediate predecessor; those who used to call their parents stuffy are now squares to their own children.

**squib.** For synonymy see LAMPOON.

**squir(e)archy.** Though 'the spelling with *e* has been by far the more usual' (OED), the spelling without it is preferable (see MUTE E), and Sydney Smith and FitzGerald appear among its patrons in the OED quotations.

**-s-, -ss-, -sss-. 1.** The general rules for the doubling or not doubling of final consonants before suffixes are summarized in SPELLING POINTS 2. So few monosyllables or words accented on the last syllable end in a single -s that rules need not be here stated; it will suffice to say that: (*a*) The plural of *bus* is *buses*; this irregularity is explained by the fact that *buses* is an abbreviation of the regular *omnibuses* and the spelling has become too firmly fixed to be changed to *busses*, even though we now write *bus* and not *'bus*. (*b*) *Biases* and *focuses*, nn. or vv., *biased* and *focusing*, are said by the OED to be 'more regular' than the -ss- forms that are nevertheless sometimes still used in England for the verb inflexions; similarly *canvas* (the fabric) gives *-ases* (pl. n.), *-ased*, and so too *orchises*, *nimbuses*, *portcullised*, *trellised*, *boluses*,

*bonuses*, *incubuses*, *atlases*, etc. (*c*) *Nonplus* makes *nonplussed*.
**2.** For the question whether such words as *mis-shapen* and *mis-spelt* should be hyphened see MIS-, where it is recommended that they should be written as one word. But three *s*'s are felt to be too many to sort themselves out without help; *mistress-ship* and *Inverness-shire* are always so written.

**St.** For the question between *St Peter* and *St. Peter* etc., see PERIOD IN ABBREVIATIONS.

**stadium.** Plural *-dia*, but popularization into *-diums* is likely to follow the popularization of the word.

**staff, stave. 1.** In all modern senses the plural of *staff* is *staffs*; but from the archaic plural *staves* has come the back-formation *stave*, which has taken the place of *staff* in music and cooperage. **2.** For *s. of life* see SOBRIQUETS.

**stag.** See HART.

**Stagirite.** The *S.*; see SOBRIQUETS. Pronounce *Stǎg'ĭrīt* although Aristotle's birthplace was *Stǎgī'ra*. See FALSE QUANTITY.

**stalactite, stalagmite.** Stress on the first, not the second, syllables is recommended; see RECESSIVE ACCENT.

**stall** in the sense of coming to an involuntary stop, though now mainly used of machines, is not a product of the machine age; it was inherited from horse transport. OED Supp. quotes *The last time he passed his horses stalled, that is they were for some time unable to drag the wagon through the worst places* (1807).

**stamen.** Plural *-s*; the Latin plural *stamina* has been put to other work.

**stamp, n.** For synonymy, see SIGN.

**stanch, staunch.** The adjective is *staunch*, the verb *stanch*, and the usual pronunciations *-aw-* and *-ah-* respectively.

**stand.** For *stands to reason*, see REASON 2. For *standpoint*, *point of view*, and *point*, see POINT.

**standard.** The phrase *by any standards* has been described by Edmund Wilson as 'one of the sloppiest of current clichés'. Like many phrases that have become clichés, it is beyond reproach when judiciously used: one might say, for instance, that *suffer a sea change* was a cliché by any standards, meaning that, although opinions differ about what phrases can properly be called clichés, everyone would agree that this was one. It is not judiciously used by a writer who can hardly be supposed to mean literally *any* standards, but does not tell us (perhaps does not even know himself) what standards he does mean; as for instance (to quote Wilson) 'when a new novel is described as "of major significance" or "a superb achievement" "by any standards", without one's knowing whether the standards invoked are supposed to include Tolstoy and Flaubert or whether it merely means that the reviewer does not happen at the moment to remember that he has ever read anything that seemed to excite him more'.

**star,** v. See FEATURE.

**starlight, -lit, -litten,** adjj. The first (in adj. use, e.g. *a starlight night*) may or may not be historically the noun used attributively, but is certainly now to be so regarded. Attributive uses of nouns, like adverbial uses of apparent adjectives (see UNIDIOMATIC -LY), sometimes strike people whose zeal for grammar is greater than their knowledge of it as incorrect; and perhaps *starlit* is often substituted for *starlight* owing to this notion. No harm is done, *starlit* being a blameless word, and indeed better in some contexts; if 'a starlight night' and 'a starlit sea' have their epithets exchanged, both suffer to the extent at least of sounding unnatural. The further step to *starlitten* is not so innocent, *litten* being not archaic but pseudoarchaic; the writer who uses *starlitten* is on a level with the tradesman who relies on such attractions as Ye Olde Curyosytie Shoppe.

**state,** n. It is a convenient distinction to write *State* for the political unit, at any rate when the full noun use is required (not the attributive, as in *state trading*), and *state* in other senses. (See CAPITALS.) The following compound forms are recommended (see HYPHENS): statecraft, stateroom, State socialism, State prisoner, State trial, State paper.

**state,** v. *I may state 'Irish Nationality' was recommended to me by* . . . 'State' is one of the verbs that insist on proper ceremony and resent the omission of THAT, conj.

**stately.** For the adv., see -LILY.

**static(al).** See -IC(AL); there is no marked differentiation, but the *-ic* form has prevailed in the adjective and *-ically* in the adverb.

**stationary, -ery.** The adj. (not moving), *-ary*; the noun (paper etc.), *-ery*. The second is from *stationer*, one who has a station in a market for the sale of books, as distinguished from an itinerant vendor.

**statist, statistician,** etc. The pronunciation of the first (*stā'tĭst*) is very much against it, inevitably suggesting state, and not statistics; and in fact its old sense was statesman, though now, as if it were a back-formation from *statistics*, it means only statistician. Either it should be abandoned and *statistician* always used, or it should be cut off from *state* by being pronounced *stă'tĭst*. Both these seem to be coming about; the word is becoming obsolete, preserved only in the journal of that name, which is now ordinarily pronounced *stăt-*. Thus *statistician* is left in sole possession. Of the alternative adjectives *-ic* and *-ical* the short form is now virtually obsolete; the word is used only as a noun.

**status. 1.** Pronounce *stātus*, in spite of the FALSE QUANTITY. The plural (*statūs*) is not used; if a plural is unavoidable there is no escape from *statuses*.

**2.** 'The *status quo*' is the position in which things (1) are now or (2) have been till now or (3) were then or (4) had been till then; in senses 2 and 4 *ante* (*t. s. q. ante*) is sometimes, but need not be, added. With *in* the phrase becomes *in statu quo* (*ante*), without *the*, and with *ante* similarly optional.

**3.** *Status* is a word now much used in the jargon of sociology (perhaps because it has become indelicate to speak of *class*), especially in the phrase *status-symbol*, which has become so important in the Affluent Society.

**statutable, -tory.** The two words are hardly distinguishable in meaning; *-table* is considerably older but now rare; *-tory* has captured the field. If *-table* were ever revived it might be to provide an adjective meaning capable of being dealt with by legislation.

**stave, v.** The past and p.p. *stove* (instead of *staved*) is comparatively modern and (OED) 'chiefly *Naut.*' For the noun see STAFF.

**stead, n.** *The atmosphere of the home life was favourable to the growth of qualities which were presently to stand him in inestimable stead.* The obsolescent phrase to *stand in stead*, meaning to advantage, has been so narrowed by usage that to stand in good or better stead is the limit within which it can now be used without affectation; words like *inestimable* should not be substituted; see CAST-IRON IDIOM.

**steer, n.** See CATTLE.

**stem, v.** The recent popularity in Britain of the Americanism *stem from* is presumably due to NOVELTY HUNTING, for it has no advantage, either in convenience or in suitability of metaphor, over our *spring from*.

**stem, n.** In grammar a word's stem is the part from which its inflexions may be supposed to have been formed by the addition of affixes; in the inflexions it may be found unchanged, or may have been affected by phonetic tendencies; thus the stem of *man* is *man*, giving *man's*, *men*, and *men's*. Cf. ROOT; of the English verb *wit* the root

is VID, but the stem, giving *wit*, *wot*, *wist*, *wottest*, etc., is *wit*. Different parts of a 'word' may be formed from different stems; there are for instance several stems in what is called the verb *be*.

**step.** For *s. this way*, *s. in*, etc., see FORMAL WORDS.

**sterile.** The older spellings (usually *-il*, *-ill*) suggest that the pronunciation *-īl* is modern, but it is now usual in Britain though not in U.S. (See -ILE.)

**sternum.** Pl. *-na* or *-nums*; see LATIN PLURALS.

**stevedore.** Three syllables (*stē'vē-dōr*).

**sticking-place, -point.** In the *Macbeth* passage, *-place* is the word; see MISQUOTATION.

**stickleback, tittlebat.** The first is the orthodox and etymological form, the other being (OED) 'a variant, of childish origin'.

**stigma.** In the ecclesiastical, botanical, medical, etc. senses the plural is *stig'măta*; *stigmā'ta*, occasionally heard, is a solecism and *stigmas* is used only in the figurative sense of imputation or disgrace, in which a plural is rare. See LATIN PLURALS.

**stigmatize.** The mistake fully dealt with under REGARD 3 occurs sometimes with *s.*: . . . *bravely suffering forfeiture and imprisonment rather than accept what in this same connexion Lord Morley stigmatized the 'bar sinister'*; things are not stigmatized monstrous, but stigmatized *as* monstrous.

**stile, style.** *Stile* is the spelling for the means of passage, and for the carpentry term (*stiles and rails*); *style* for all other senses. This division is not historically correct, being due to the confusing of Latin *stilus* (writing-tool) with Greek στῦλος (column); but it is so generally accepted, and attempts to improve upon it are so conflicting, that it is better to refrain, and leave the y in all the classically derived senses; see also Y AND I.

**stiletto.** Pl. preferably *-os*; see -O(E)S 6.

**stimulus.** Pl. *-lī*; LATIN PLURALS.

**stink.** Past *stank* or less commonly *stunk*.

**stock pathos.** Some words and phrases have become so associated with melancholy occasions that it seems hardly decent to let such an occasion pass unattended by any of them. It is true that such trappings and suits of woe save much trouble; it is true that to mock at them lays one open to suspicion of hardheartedness; it is also true that the use of them suggests, if not quite insincerity, yet a factitious sort of emotion, and those are well advised who abstain from them. A small selection, which might be greatly enlarged, is: In her great sorrow; The land he loved so well; The supreme sacrifice; The pity of it!; The mortal remains of; All that was mortal of; The departed; One more unfortunate; More sinned against than sinning; A lump in one's throat; Tug at one's heartstrings; Stricken; Loved and lost; But it was not to be.

**stockpile** (n. and v.), an importation from America, deserves the welcome we have given it; for its formation is unexceptionable, and the sense it conveys of accumulating reserve stocks, especially of raw materials, against a possible scarcity cannot be as readily conveyed in any other way. But to let it usurp the place of the traditional woodpile as a lurking-place for niggers is to overdo our welcome. *Empirical philosophy, anthropocentric humanism, pseudo-scientific dogma and a wholesale rejection of absolute values—these are rapidly becoming, as it might be, the niggers in the stockpile.*

**stoic(al).** See -IC(AL). Both forms are used as adjectives, *-ic* being indeed the commoner; but points of difference are discernible. In the predicative use *stoic* is rare: *his acceptance of the news was stoical, he was stoical in temper*, rather than *stoic*. In the attributive use, *stoic* naturally pre-serves the original sense more definitely, while *stoical* forgets it. When we say *stoic indifference*, we mean such indifference as the Stoics taught or practised; when we say *stoical indifference* we think of it merely as resolute or composed. The *stoic virtues* are those actually taught by the Stoics, the *stoical virtues* simply those of the sterner kind. Lastly, while either epithet is applicable to abstracts, *stoical* is the word for persons: *with stoic* or *stoical composure*; *stoic* or *stoical life* or *tone* or *temper* or *views*; *he is a stoical fellow*; *these stoical explorers*; *a stoical sufferer*; *my stoical young friend*.

**stokehold, -hole.** The earliest OED quotation for the first is dated 1887; the *-hole* form goes back to 1660. That is no doubt because it was only when steampower was used at sea that a word was needed for that part of the hold of a ship in which stoking was done. But the subsequent encroachment of *hold* on *hole* for stoking-places on land suggests that *hole*, though the true form, is now thought undignified. Though the OED defines the two differently, the impression produced by its quotations is not that there are two names for two different things, but rather that *stokehole* has had in its time, and perhaps still has, more than one meaning. To maintain a distinction between words at once so similar in form and, to the general public, so vague in sense, is clearly impossible. The form *stokehole* is recommended, at least for all such places ashore.

**stomacher,** article of dress. The old pronunciation was with *-cher*, not *-ker*, and it should be kept to as long as the word is historical only.

**stop,** v. Those who use *stop* when others would use *stay* (*Where are you stopping?* etc.) are many, and are frequently rebuked. The OED deals very gently with them: 'Cf. *stay*, which is often preferred as more correct'; and it is not a case for denunciation, but rather for waiting to see which word will win. Meanwhile, careful speakers do prefer *stay*; and it is in its favour that its noun, and not *stop*, is certainly the right one in the corresponding sense (*during our stay*, not *our stop*) and that the verb itself is in undisputed possession of the colloquialism *stay put*. It may also be suggested that, if *stop* is a solecism, there are degrees of enormity in the offence: *I shall stop for the night somewhere on my way, Won't you stop to dinner?, I shall stop in town till I hear, We have been stopping at the Deanery*, of which the last is the worst, point to a limitation—that *stop* is suitable only when interruption of a journey or postponement of departure rather than place of sojourn is in question; in the former case the Americanism *stop off* or *over* is now as likely to be used as the plain verb. See PHRASAL VERBS.

**stops**, etc. (comma, semicolon, colon, full stop, exclamation, question, inverted commas, apostrophe, hyphen, italics, brackets, dashes). There is not room in this book for a treatise on punctuation, nor for discussion of principles even where the question is one between opposed views of correctness, and not between acknowledged correctness and careless or ignorant error. But, if it is assumed (1) that the reader need be warned only against mistakes that experience shows to be prevalent, and (2) that the views here taken on disputed points are sound, an article consisting mostly of ill stopped sentences with corrections may be of use.

## COMMA

A. Separating inseparables, e.g. a verb from its subject or object or complement, a defining relative from its antecedent, or an essential modification from what cannot stand without it. *The charm in Nelson's history, is, the unselfish greatness* (read *history is the*). One comma parts verb from subject, the other complement from verb. / *He has been called the Portuguese Froissart, but he combines with Froissart's picturesqueness, moral philosophy, enthusiasm, and high principles* (read *picturesqueness moral*). The comma

parts the object (*moral . . . principles*) from its verb *combines*. / *A literature of Scotch Gaelic poetry and prose exists, though too little notice has been taken of it, even within the Scotch borders, for the Scot, who ignores such literature, does not deserve his name, which proves him to be a Gael* (read *Scot who ignores such literature does*). The *who* starts a defining relative clause; see THAT (REL.) 1, WHICH 7, WHICH, THAT, WHO 9, and WHO AND WHOM 3. / . . . *whether some disease other than tuberculosis may not account for the symptoms and signs observed. Only, if we do not succeed in our investigations, are we entitled to admit the diagnosis of tuberculosis* (read *Only if we do not succeed in our investigations are*). Without the clause from which the comma parts it, *only* is mere nonsense. / *Situated, as we are, with our vast and varied overseas possessions, our gigantic foreign trade, and our unapproachable mercantile marine, we at any rate can gain nothing by war* (read *Situated as*). We should write not 'How, are we situated?', but 'How are we situated?'; the *as* clause is exactly parallel to, and as essential as, *how*. / *We are assured that the Prime Minister will, in no circumstances and on no consideration whatever, consent to . . .* (read *will in no circumstances . . . whatever consent*). The words that negative *will* must not be cut off from it. See NO 4.

In the foregoing examples the commas are manifestly wrong. A more difficult question is whether it is legitimate to break the rule about not separating inseparables in order to indicate the end of a long or complicated subject. In enumerations, for instance, should there be a comma after *Spanish* in *French, German, Italian, and Spanish, are taught*? (The question whether there should be one after *Italian* is discussed in B below.) The answer here suggested is no; not even when the intrusion of an adverbial phrase between subject and verb tempts a writer to use a comma to prevent ambiguity, as he might write *French, German, Italian, and Spanish, in particular are taught*, to show that *in particular* relates to all four languages and not to Spanish only. The sentence should be recast, and this can easily be done by moving *in particular* from the tail of the procession to the head of it. (Similarly if the writer wanted to make it quite clear that *in particular* qualified *Spanish* only this could be done by inserting those words immediately before *Spanish*.) *Nothing had been allowed to be published except books, pamphlets, and papers, which had secured the approval of the Communist party.* Here the comma after *papers* removes any possibility that that word alone may be taken as the antecedent of *which*, but only at the cost of separating inseparables—the defining relative from its antecedent. Here again reconstruction is easy, e.g. *No books, pamphlets, or papers had been allowed to be published except those that had* etc.

It is not only enumerations that tempt a writer to put a comma between subject and verb. Any long and involved subject will do so. He may for instance feel the need of a comma after *subject* in *The question whether it is ever legitimate to use a comma to mark the end of a long and complicated subject is an arguable one*. Even good writers sometimes do this. But it is surely better to recast an ungainly sentence than to try to mend it by the crude device of an intrusive comma.

B. Within enumerations. The more usual way of punctuating such an enumeration as was used as an example in the preceding section is *French, German, Italian and Spanish*: the commas between *French* and *German* and *German* and *Italian* take the place of *ands*; there is no comma after *Italian* because, with *and*, it would be otiose. There are, however, some who favour putting one there, arguing that, since it may sometimes be needed to avoid ambiguity, it may as well be used always for the sake of uniformity. Examples of sentences calling for a comma before the *and* are: *Tenders were submitted by John Brown, Cammell Laird, Vickers, and Harland*

*and Wolff*. Without the comma after *Vickers* we do not know whether the tendering firms were four or five, or, if they were four, whether *Harland* partners *Vickers* or *Wolff*. / *The smooth grey of the beech stem, the silky texture of the birch, and the rugged pine*. If there is no comma after *birch*, the pine is given a silky texture. The use of a comma before the *and* is here recommended.

C. In the absolute construction. For the cause, and effect, of this common mistake, see ABSOLUTE CONSTRUCTION. *But these objections were overruled, and the accused, having pleaded not guilty, the hearing of evidence commenced* (read *and, the accused having*).

D. In confluences, i.e. when alternatives etc. finish their course together, the necessary comma after the second is apt to be forgotten. *As regards the form of the festival, many, if not most of the customs popularly associated with it may, perhaps, be traced to* . . . (read *most, of*). / *His craftsmanship, again, was superb—more refined, more intellectual than that of Frith* (read *intellectual, than*).

E. In compound appendages to names. *Mr F. Haverfield has collected and edited a volume of 'Essays by Henry Francis Pelham, Late President of Trinity College, Oxford and Camden Professor of History'* (read *Oxford, and*).

F. In ambiguous appositions. Insertion or omission of commas is seldom a sufficient remedy, and indeed is usually impossible. The thing is to remember that arrangements in which apposition commas and enumeration commas are mixed up are dangerous and should be avoided.

*To the expanded 'Life of Shakespeare', first published in 1915, and to be issued shortly in a third edition by Mr. Murray, the author, Sir Sidney Lee, besides bringing the text up to date, has contributed a new preface*. Which is the author? / *Some high officials of the Headquarter Staff, including the officer who is primus inter pares, the Director of Military Operations, and the Director of Staff duties.* . . . How many were there going to St Ives? / *Lord*

*Curzon, Sir Edmond Elles, the present Military Member, and the Civilian Members of Council traverse the most material of Lord Kitchener's statements.* Was Sir Edmond the Military Member? Such ambiguities can often be avoided by the use of a pair of brackets instead of a pair of commas: *Sir Edmond Elles (the present military member)*. . . .

G. Omitted between connected but independent sentences, or used instead of semicolon between unconnected sentences.

*When the Motor Cars Act was before the House it was suggested that these authorities should be given the right to make representations to the central authorities and that right was conceded* (read *authorities, and*). | *The winter was exceptionally cold, once again misery and unemployment were created* (read *cold; once*). | *'Will the mighty Times aid us in this historic struggle?' Dear to the heart of an editor must be such an appeal, we wish someone would seek for our aid in so flattering a formula* (read *appeal; we*).

H. On the use of commas before and after *for* (conj.) see FOR; on the effect of commas round adverbial conjunctions such as *however* and *therefore* see THEREFORE, and round adverbial phrases see NO. 4.

## SEMICOLON

The use of semicolons to separate parallel expressions that would normally be separated by commas is not in itself illegitimate; but it must not be done when the expressions so separated form a group that is itself separated by nothing more than a comma, if that, from another part of the sentence. To do this is to make the less include the greater, which is absurd.

*And therein lies a guarantee of peace and ultimate security, such, perhaps, as none of the States of South America; such as not even Mexico herself can boast* (read *America, such as not even Mexico herself, can*). | *If you say with the enemy pinned upon the West, suffering passively blow upon blow, and never*

*able to restore himself after each blow, or to recover what he has lost; with his territory blockaded; his youngest boys drawn into the struggle, that your victory is impossible; if you say* . . . (read *lost, with his territory blockaded, his*). | *If, as Mr. Gibson Bowles contends, the Law of Nations is all plain sailing; if it is a thing of certainties and plain definitions, it would be strange that a conference of jurists should have* . . . (read *sailing, if*).

## COLON

As long as the Prayer-Book version of the Psalms continues to be read, the colon is not likely to pass quite out of use as a stop, chiefly as one preferred to the semicolon by individuals, or in impressive contexts, or in gnomic contrasts (Man proposes: God disposes); but the time when it was second member of the hierarchy, full stop, colon, semicolon, comma, is past. Some contemporary writers deliberately—almost ostentatiously—so employ it, but in general usage it is not now a stop of a certain power available in any situation demanding such a power, but has acquired a special function: that of delivering the goods that have been invoiced in the preceding words. In this capacity it is a substitute for such verbal harbingers as *viz., scil., that is to say, i.e.,* etc.

## FULL STOP

In abbreviations. For the use as a symbol of abbreviation, as in i.e. for *id est, Capt.* for *Captain,* and less reasonably in *Mr.* for *Mister, cwt.* for *hundredweight,* see PERIOD IN ABBREVIATIONS.

In the spot plague. The essence of the style that has been so labelled is that the matter should be divided into as short lengths as possible separated by full stops, with few commas and no semicolons or conjunctions. This is tiring to the reader, on whom it imposes the task of supplying the connexion, and corrupting to the writer, whose craving for brevity persuades him that anything will pass for a sentence: *It was now clear. The light was that of late evening. The air hardly more than cool. | The letter may be long and garbled. Some of it may have little to do with the Government. Other departments may be concerned. Local authorities may be involved. For even more plaguy examples see* VERBLESS SENTENCES.

## EXCLAMATION

Not to use a mark of exclamation is sometimes wrong: *How they laughed.*, instead of *How they laughed!*, is not English. Excessive use of exclamation marks is, like that of ITALICS, one of the things that betray the uneducated or unpractised writer: *You surprise me, How dare you?, Don't tell such lies*, are mere statement, question, and command, not converted into exclamations by the fact that those who say them are excited, nor to be decorated into *You surprise me!, How dare you!, Don't tell such lies!.* It is, indeed, stated in a well-known grammar that 'A note of exclamation is used after words or sentences which express emotion', with, as example, *How are the mighty fallen in the midst of the battle! I am distressed for thee, my brother Jonathan!.* The second half of this quotation clearly violates the rule laid down above, being, however full of emotion, a simple statement, and yet having an exclamation mark. But anyone who will refer to 2 *Sam.* i. 26 will find that mark to be not the Bible's, but the grammarian's; the earlier one of verse 25 is right. So far, the inference seems obvious and simple—to confine the exclamation to what grammar recognizes as exclamations, and refuse it to statements, questions, and commands. Exclamations in grammar are (1) interjections, as *oh!*; (2) words or phrases used as interjections, as *Heavens!, hell!, by Jove!, my God!, Golly!*; (3) sentences containing the exclamatory *what* or *how*, as *What a difference it makes!, What I suffered!, How I love you!, How pretty she is!*; (4) wishes proper, as *Confound you!, May we live to see it!, God forbid!*; (5) Ellipses and inversions due to emotion, as *Not*

another word!, If only I could!, That it should have come to this!, Much care you!, Pop goes the weasel!, A fine friend you have been!*; (6) apostrophes, as *You miserable coward!, You little dear!.* It is true that the exclamation mark should be given to all expressions answering to the above types, and also that it should not be given to ordinary fully expressed statements, questions, or commands; but the matter is not quite so simple as that. Though a sentence is not to be exclamation-marked to show that it has the excited tone that its contents imply, it may and sometimes must be so marked to convey that the tone is not merely what would be natural to the words themselves, but is that suitable to scornful quotation, to the unexpected, the amusing, the disgusting, or otherwise to imply that the words, if spoken, would have a special intonation. So: *You thought it didn't matter!, He learnt at last that the enemy was—himself!, Each is as bad as the other, only more so!, He puts his knife in his mouth!.* But not: *That is a lie!, My heart was in my mouth!, Who cares!, I wish you would be quiet!, Beggars must not be choosers!*; in all these the words themselves suffice to show the tone, and the exclamation mark shows only that the writer does not know his business

## QUESTION MARK

The chief danger is that of forgetting that whether a set of words is a question or not, and consequently requires or repudiates the question mark, is decided not by its practical effect or sense, but by its grammatical form and relations. Those who scorn grammar are apt to take *Ask him who said so* for a question, and *Will you please stand back* for a request, and to wrongly give the first the question mark that they wrongly fail to give the second. But the first is in fact a command containing an INDIRECT QUESTION, and the question mark belongs to direct questions only, while the second is in fact a direct question, though it happens to be equivalent in sense to a request. Even practised writers may fall into the error of putting a question mark after an indirect question; *The question that comes to mind most forcefully in reading this autobiography is why Mr. Harding should enjoy such immense renown?* Or, from Trollope, *But let me ask of her enemies whether it is not as good a method as any other known to be extant?*. Conversely the question mark is often wrongly omitted, especially when the natural confusion caused by the conveying, for instance, of what is in sense a statement in the grammatical form of a question is aggravated by the sentence's being of considerable length— e.g. when *Will it be believed that* is followed by several lines setting forth the incredible fact. Still more fatal is a type of sentence that may be put either as an exclamation or as a question, but must have its stop adapted to the exclamatory or interrogative nature of the *what* or *how* whose double possibilities cause the difficulty. *How seldom does it happen* can only be an exclamation, and must have *happen!*; but *How often does it happen* may be either a question (answer, Once a month etc.) requiring *happen?*, or an exclamation (meaning, Its frequency is surprising) requiring *happen!*. *In that interval what had I not lost!* (either *lost!* should be changed to *lost?*, or *not* should be omitted).

The archness of the question mark interpolated in brackets infallibly betrays the amateur writer: *Sir,—The following instance of the doubtful advantages (?) of the Labour Exchanges as media . . . seems to deserve some recognition.*

## INVERTED COMMAS

There is no universally accepted distinction between the single form '. . .') and the double (". . ."). The more sensible practice is to regard the single as the normal, and to resort to the double only when, as fairly often happens, an interior quotation is necessary in the middle of a passage that is itself quoted. To reverse this is clearly less reasonable; but, as quotation within quotation is much less common than the simple kind, and conspicuousness is desired, the heavy double mark is the favourite. We can only hope that *The man who says* 'I *shall write to "The Times" tonight'* will ultimately prevail over *The man who says "I shall write to 'The Times' tonight"*.

Questions of order between inverted commas and stops are much debated and a writer's personal preference often conflicts with the style rules of editors and publishers. There are two schools of thought, which might be called the conventional and the logical. The conventional prefers to put stops within the inverted commas, if it can be done without ambiguity, on the ground that this has a more pleasing appearance. The logical punctuates according to sense, and puts them outside except when they actually form part of the quotation. Thus:

*Conventional*: Oxford has been called a 'Home of lost causes.' *Logical*: . . . 'Home of lost causes'.

*Conventional*: 'That Home of lost causes,' as Oxford has been called. *Logical*: 'Home of lost causes', as . . .

*Conventional*: Oxford has been called a 'Home of lost causes;' that cannot be said of it now. *Logical*: 'Home of lost causes'; that cannot . . .

*Conventional* and *logical*: 'Oxford,' said Arnold 'Home of lost causes and impossible beliefs'. Here the logical puts the comma in the same place as the conventional, since it forms part of the quotation. (But it should be noted that commas are not needed merely to mark an interruption in a quotation such as *said so-and-so*; the closing and opening of the inverted commas do that. For instance there should not be commas after *Oxford* and *Arnold* in 'Oxford' said Arnold 'is the Home of lost causes.' They are often so used, but are clearly otiose.)

In the treatment of question and exclamation marks the systems tend to merge, perhaps because those symbols show up so glaringly the illogicality of the conventional one. In the following examples the punctuation is standard under either system:

Did you say 'I am not my brother's keeper'?

I said 'Am I my brother's keeper?'

Did you say 'Am I my brother's keeper'?

They cried out 'We are lost!'

How heartrending was their cry 'We are lost'!

Where, as in the third and fifth examples, strict logic would require two symbols, one each side of the inverted comma, logic must respect appearances and be content with one, as it has to with full stops when a quotation and the sentence containing it end together.

The conventional system is more favoured by editors' and publishers' rules. But there are important exceptions, and it is to be hoped that these will make their influence felt. The conventional system flouts common sense, and it is not easy for the plain man to see what merit it is supposed to have to outweigh that defect; even the more pleasing appearance claimed for it is not likely to go unquestioned.

### APOSTROPHE

For difficulties with this as sign of the possessive case and other uses and abuses of -'s, see POSSESSIVE PUZZLES. For its use in avoiding certain bizarre word-forms, see -ED AND 'D.

### HYPHENS, ITALICS

See those articles.

### PARENTHESIS SYMBOLS

On the uses and misuses of parenthesis see PARENTHESIS. Here only two things remain to be said.

1. Parentheses may be indicated in any one of four ways: by square brackets, by round brackets, by dashes, and by commas. Square brackets are the most disconnective; their main use is for an explanatory interpolation in a quotation. Of the other three, commas are suitable for the parenthesis that least interrupts the run of the sentence, and dashes and round brackets for those that do so progressively more.

2. After the second bracket or dash any stop that would have been used if the brackets or dashes and their contents had not been there should still be used. After the second bracket this is sometimes forgotten; after the second dash it is seldom remembered, or rather, perhaps, is deliberately neglected as fussy. But, if it is fussy to put a stop after a dash, it is messy to pile two jobs at once upon the dash, and those to whom fussiness is repugnant should eschew the double-dash form of parenthesis except where no stop can be needed. *So far as it is true—and how far it is true does not count for much—it is an unexpected bit of truth* (read *much—, it*). | *If he abandons a pursuit it is not because he is conscious of having shot his last bolt—that is never shot—it is because* . . . (read *never shot—; it is*).

**storey, story.** 1. Whether these names for the floor and the tale are etymologically the same word or not—on which the doctors differ—, there is an obvious convenience in the two spellings. It is, for instance, well to know the difference between *storied windows* (illustrating biblical or other stories) and *storeyed windows* (divided by transoms into storeys). The DIFFERENTIATION, however, is still a probationer, and indeed lacks the support of the OED. That is sadly against it, especially when the 19th-c. quotations are found to show -*ry* and -*ries* four times as often as -*rey* and -*reys*; *clerestory* refuses to conform and -*ry* is standard in U.S. On the other hand it is encouraging to find that recent British dictionaries (including the COD) put -*rey* first. 2. For the curious difference in sense between *storey* and *floor* see FLOOR.

**stouten.** See -EN VERBS.

**stove,** = *staved.* See STAVE.

**straightaway, straightway.** The first has proved more to modern taste; *straightway* is now WARDOUR STREET, if not archaic.

**straight(ly).** *Certain members of the*

*Labour Party have spoken very honestly
and straightly about the growth of this
idea. | For once, he did not mince his
words on a labour question; would that
he had spoken as straightly on previous
occasions!* These two examples, of
which the first shows a perhaps defensible *straightly*, and the second a certainly indefensible one, throw some
light on the regrettable but progressive extinction of our old monosyllabic
adverbs; it is the company of *honestly*
that partly excuses the first *straightly*;
see UNIDIOMATIC -LY.

**straight plays.** See LEGITIMATE
DRAMA.

**strain, sprain.** For the distinction,
see SPRAIN.

**strait(en).** The chief phrases in
which these, and not *straight(en)*, must
be used are: *the strait gate, the straitest
sect, strait jacket, strait waistcoat,
straitlaced, straitened circumstances.*

**strategic(al),** pronunciation. In the
penult of adjectives and nouns
in *-ic* (and the antepenult of *-ical*
words), if *-ic* is preceded by a
single consonant, there is an overwhelming preponderance for the
short sound of the previous vowel
(except *u*); so *errătic, barbăric, mechănic, trăgic, poĕtic, acadĕmic, ĕthic, angĕlic, arthrĭtic, prolĭfic, chrŏnic, exŏtic,
microscŏpic, histŏric, spasmŏdic, lўric,
paralўtic,* and hundreds more; but
with *u* we have *scorbūtic, mūsic, cūbic.*
Nevertheless, *strateˊgic* has prevailed
over *strateˊgic*; the most notable of
other exceptions are *scenic* and *basic,*
in which the long vowels are the natural
result of familiarity with *scene* and *base.*

**strategy, tactics.** Etymologically,
strategy is generalship, and tactics is
array, and the modern antithesis retains as closely as could fairly be expected the original difference. The
OED definition of strategy and note
on the distinction follow, with three
quotations, of which the first two are
from the OED. 'Strategy. The art of
a commander-in-chief; the art of projecting and directing the larger military movements and operations of a
campaign. Usually distinguished from
*tactics,* which is the art of handling
forces in battle or in the immediate
presence of the enemy.' [This difference has been preserved in the Air
Arm, where there is a distinction between *strategic* and *tactical* bombing.]
(Quotations) *Strategy differs materially
from tactic; the latter belonging only to
the mechanical movement of bodies set
in motion by the former. | Before hostile
armies or fleets are brought into contact
(a word which perhaps better than any
other indicates the dividing line between
tactics and strategy). | The study of
strategy, which is the art of bringing
forces into contact with the enemy, and
of tactics, which is the art of using those
forces when they are in contact with the
enemy.* A third branch of the military
art—that of the transport, quartering
and supply of troops—has in comparatively modern times been given
the name of *logistics.*

Readers of these quotations, in which
*contact* and *tactics* are juxtaposed,
should perhaps be warned against
supposing that the *tact* in the two
words is etymologically the same. The
likeness is accidental: *contact* (Latin)
is touch, *tactics* (Greek) is array.

**stratum, stratus.** *Strātum* (layer
etc.) pl. *-ta*; see -UM. *Strātus* (cloud)
pl. *-ti.* In the *strati-* compounds
(*stratify, stratiform,* etc.) the *a* is
shortened, but it remains long in the
*strato-* compounds (*s. cirrhus, s. cumulus,* etc.) *Stratosphere* however, being
in more general use than the other *-o*
compounds, is now usually given a
short *a* on the analogy of other well-known *strat-* words such as *strategy.*

**strayed,** adj. See INTRANSITIVE P.P.

**street.** See ROAD.

**stress, strain.** To most of us *stresses*
and *strains* are a pair of SIAMESE TWINS
suitably describing our worries and
the effect they have on us. Those
figurative uses of the words are no
great extension of their literal meanings as terms of mechanics. Here too
strain is the result of stress; stress
being mutual action exerted by contiguous bodies or parts and strain the

alteration of form or dimensions produced by it.

**strew.** P.p. indifferently *-n* and *-ed*.

**stricken.** This archaic p.p. of *strike* survives chiefly in particular phrases, and especially in senses divorced from those now usual with the verb— *stricken in years*, *the stricken deer*, *poverty-stricken*, *panic-stricken*. The use of the word by itself as an adjective = afflicted, in distress, is sometimes justified (e.g. *the stricken population* after an earthquake), but more often comes under the description of STOCK PATHOS.

**stride.** Past *-ode*, p.p. (rare) *-idden*.

**stringed, strung.** The first is formed from the noun and the second from the verb. Strictly therefore, a bow is *stringed* or *unstringed* according as it is provided with a string or not, and *strung* or *unstrung* according as it is bent to the string or not; a tennis racket is *stringed* when the gut is inserted in the frame but *strung* only when the strings are made taut. But in practice this distinction has virtually disappeared and *strung* holds the field, *stringed* remaining only in occasional adjectival use. A piano is a *stringed* instrument, but if the strings need renewing it will be *restrung*. See HAM-STRINGED for discussion of that word and of *bowstring*, v.

**strive.** Past *strove*, p.p. *striven*. The OED adds that 'many examples of *strived*' for both 'occur in writers of every period from the 14th to the 19th c.', but it is rarely used now and not recognized by the COD.

**stroma.** Pl. *-ata*.

**strung.** See STRINGED.

**strychnia, -nine.** See MORPHIA; but *strychnia* has not, like that word, maintained itself in popular use.

**sturdy indefensibles.** Many idioms are seen, if they are tested by grammar or logic, not to say what they are nevertheless well understood to mean. Fastidious people point out the sin, and easy-going people, who are more numerous, take little notice and go on committing it. Then the fastidious people, if they are foolish, get excited and talk of ignorance and solecisms, and are laughed at as pedants; or, if they are wise, say no more about it and wait. The indefensibles, however sturdy, may prove to be not immortal, and anyway there are much more profitable ways of spending time than baiting them. It is well, however, to realize that there are such things as foolish idioms; an abundance of them in a language can be no credit to it or its users, and drawing attention to them may help to keep down their numbers. In the article ILLITERACIES some examples are given of indefensible usages that are common but not yet so sturdy as to have qualified as acceptable idiom. In the present article some examples are given of indefensibles that may fairly claim admission to that status, colloquially at least. The line is not easy to draw, and most readers will probably be disposed to move some examples from one list to the other. See also ILLOGICALITIES.

It 's ME. That 's him (HE).

Don't be longer *than you can* HELP.

*So far from* hating him, I like him (FAR 2).

The man *of all others* for the job (OTHER 3).

The worst liar *of any man* I know (OF 7).

A child *of ten years old* (OF 7).

That long nose *of his* (OF 7).

It is no USE complaining.

Better known than popular (-ER AND -EST 8).

It is a day's work even to open, *much less* to acknowledge, all the letters (MUCH 2).

*All* men do *not* speak German (NOT 1).

He ONLY died a week ago.

It should not be taken TOO literally.

I should not be surprised if it didn't rain (NOT 4).

Receipts are only a FRACTION of what they used to be.

Has he got a temperature? (FEVERISH).

Our MUTUAL friend.

It is only a very APPROXIMATE estimate.

WHO is it for?
I wish I could play LIKE you do.
He is insensitive to a DEGREE.
His hitting was simply PHENOMENAL.
There was no play today DUE to rain.
The REASON is because . . .
Those KIND of things.

**sty,** nn. Pl. *sties.* The separate spelling *stye* (pl. *styes*), sometimes used for the abscess on the eyelid, has not the support of the OED, and the danger of confusion is too slight for artificial differentiation.

**style, stile.** See STILE.

**stymie** is the established spelling. Though abolished in golf, the word is likely to continue in figurative use, especially the verb.

**suave.** Pronounce *-āv* rather than *-ahv.*

**subject.** For synonyms in sense *theme* etc., see FIELD.

**subjective.** See OBJECTIVE, SUBJECTIVE.

**subjective genitive.** See OBJECTIVE GENITIVE for the principle. If from the sentence *God created man* two nouns are taken, *God's creation* contains a subjective genitive, and *man's creation* (or usually *the creation of man*) an objective genitive.

**subjunctives.** The word is very variously used in grammar; so it will be well to explain at the outset that in this article it is taken to mean the use of a verb-form different from that of the indicative mood in order to 'denote an action or a state as conceived (and not as a fact), and [expressing] a wish, command, exhortation, or a contingent, hypothetical, or prospective event'—OED. About the subjunctive, so delimited, the important general facts are: (1) that it is moribund except in a few easily specified uses; (2) that, owing to the capricious influence of the much analysed classical moods upon the less studied native, it probably never would have been possible to draw up a satisfactory table of the English subjunctive uses; (3) that assuredly no one will ever find it either possible or worth while to do so now that the subjunctive is dying; and (4) that subjunctives met with today, outside the few truly living uses, are either deliberate revivals, especially by poets, for legitimate enough archaic effect, or antiquated survivals giving a pretentious flavour to their context, or new arrivals possible only in an age to which the grammar of the subjunctive is not natural but artificial.

We may accordingly divide the uses of the subjunctive into four classes, which we will call Alives, Revivals, Survivals, and Arrivals, and no concealment need be made of the purpose in hand, which is to discourage the last two classes.

ALIVES, i.e. uses that are still our natural form of speech.

Those uses are alive which it occurs to no one to suspect of pedantry or artificiality, and which come as natural in speech as other ways of saying the thing, or more so. Specimens are:

*Go away* (and all 2nd-person imperatives).

*Manners be hanged!* (and such 3rd-person imprecations).

*Come what may, Be that as it may, Far be it from me to . . .,*(and other such stereotyped formulae).

*I shall be 70 come Tuesday.*

*If he were here now* (and all *if . . . were* clauses expressing a hypothesis that is not a fact).

*I wish it were over.*

*Though all care be exercised* (the difference is still a practical one between *Though . . . is,* = In spite of the fact that, and *Though . . . be,* = Even on the supposition that).

*I move that Mr. Smith be appointed Chairman.* This use of the subjunctive in a formal motion is established idiom, and its scope has been widened under American influence; it is now used after any words of command or desire. *Public opinion demands that an inquiry be held | He insists that steps be* taken to meet this danger | It is suggested that a ring road be built to relieve the congestion | He asks that the patent rights be given back to him | He is anxious that the truth be known. British idiom used to require should be; but this use of the subjunctive seems now to have become so well established with us that we can read in a leading article in 'The Times', No one would suggest that a unique, and in the main supremely valuable, work be halted, and in a recent work by an English historian of high repute She had used her stay in Holland to insist that the Dutch release a ship carrying arms for the King.

REVIVALS, i.e. antiquated uses revived for poetic effect or some other special purpose.

What care I how fair she be?
Lose who may, I still can say . . .
If ladies be but young and fair.

But illustration is superfluous; there are no uses of the subjunctive to which poets, and poetic writers, may not resort if it suits them. The point to be made is merely that it is no defence for the ordinary writer who uses an antiquated subjunctive to plead that he can parallel it in a good poet.

SURVIVALS, i.e. uses formerly natural but now falling into disuse.

In the examples that will be given there is nothing incorrect. The objection to the subjunctives in them is that they are used in constructions where the indicative is now so much more usual that they diffuse an atmosphere of formalism over the writing in which they occur; the motive underlying them, and the effect they produce, are the same that attend the choosing of FORMAL WORDS, a reference to which article may save some repetition.

If it have [has] a flaw, that flaw takes the shape of a slight incoherence. | It is quite obvious to what grave results such instances as the above may lead, be they [if they are] only sufficiently numerous. | If Mr. Hobhouse's analysis of the vices of popular government be [is] correct, much more would seem to be needed. | It were [would be] futile to attempt to deprive it of its real meaning. | Do not ring unless an answer be [is] required. | That will depend a good deal on whether he be [is] shocked by the cynicism. | Whether these tales be [are] true or exaggerated it is certain that few of his successors could perform any such remarkable feats.

ARRIVALS, i.e. incorrect uses due to growing unfamiliarity with the idiomatic uses of the mood.

The best proof that the subjunctive is, except in isolated uses, no longer alive, and one good reason for abstaining from it even where, as in the Survival examples, it is permissible, are provided by a collection, such as anyone can gather for himself from any newspaper, of subjunctives that are wrong. A collection follows, roughly grouped.

MIXED MOODS: That two verbs whose relation to their surroundings is precisely the same should be one subjunctive and one indicative is an absurdity that was unlikely to happen until the distinction had lost its reality; but now it happens every day: If that appeal be made and results in the return of the Government to power, then . . . | There are those who, if there be common security and they are all right, not only care nothing for, but would even oppose, the . . . | If the verdict goes against him his home may be sold up, or if an injunction be obtained against him and he defies it he may be imprisoned. | These bes are not themselves wrong; they are Survival subjunctives; but the fact that the verbs associated with them, which have subjunctives ready for use just as much as to be, are allowed to remain indicative shows that the use of be is mechanical and meaningless.

WERE IN CONDITIONALS: The correct type, a common enough 'Survival', is Were that true there were no more to say. The first were, of the protasis, is right only in combination with the other were, of the apodosis, or with its modern equivalent, would be. Neither of them is applicable to past time any more than would be itself; their reference is to present or to undefined time, or more truly not to time at all (and especially not to a particular past time) but to utopia, the realm of non-fact. If it is a hard saying that *were* (singular) in conditionals does not refer to past time, consider some other verb of past form in like case. Such a verb may belong to past time, or it may belong to utopia: *If he heard, he gave no sign* (*heard* and *gave*, past time); *If he heard, how angry he would be!* (*heard* and *would be*, not past time, but utopia, the realm of non-fact or the imaginary); the first *heard* is indicative, the second is subjunctive, though the form happens to be the same. In the verb *be*, conveniently enough, there happens to be still a distinguishable form for the subjunctive, and if that verb were used in sentences similar in form to the two sentences containing *heard* they would be *If it was* (never *were*) *so it did not appear*, *If it were* (or nowadays alternatively *was*) *so how angry we should be! Were* (sing.) is, then, a recognizable subjunctive, and applicable not to past facts, but to present or future non-facts; it is entirely out of place in an *if*-clause concerned with past actualities and not answered by a *were* or *would be* in the apodosis.

It has been necessary to labour this explanation because for the many readers who are not at home with grammatical technicalities the matter is puzzling. Examples: *It is stated that, during the early part of the War of Independence* (1821), *the Greeks massacred Mussulmans; if this were so, it was only in self-defence.* | *If rent were cheap, clothes were dearer than today.* | *If the attitude of the French Government were known to our own Government last week it explains the appeal.* | *We must not look for any particulars as to that lost work* (*if it were ever written*), *"The Life and Adventures of Joseph Sell"*. These four contain *if . . . were* (sing.) in the protasis—an 'Alive' form if the apodosis is *would be* or *were*, i.e. if the conditional is of the utopian kind, but wrong if *were* refers to a particular past time. Read *was* in each. Examples in apodosis: *It were just and

fitting that on such an occasion a Prince of the Royal House and Heir-apparent to the Throne should himself have plied the fires of the record warship with coal.* The newspaper is patting the Prince on the back for what he actually did, viz. stoke; it means not that it would be right on an imaginary occasion, but that it was on that past occasion right for him to stoke; read *was*. | *The dull winter prospect appeared so quiet and peaceful, it were difficult to imagine the Boches over there—on sentry, in their dugouts, eating, drinking, sleeping, just like the men about me.* Paraphrasing so as to get rid of the glamour of the word *were*, we get not 'I should find it difficult', but 'I found it difficult'; read *it was difficult.*

SEQUENCE: To those who have had to do with Latin and Greek Grammar, there will be a familiar sound in *Sequence of tenses and Sequence of moods*; what is implied in the terms is that it may be necessary to use a tense or a mood not to convey the meaning peculiar to it as such, but for the sake of harmony with the tense or mood of another verb on which it depends. The principle has its place, though little is heard of it, in English grammar also (see SEQUENCE OF TENSES); it is mentioned here because the most likely explanation of the subjunctives now to be quoted, some clearly wrong, some at the best uncalled-for, seems to be a hazy memory of sequence of moods. After each example the possible false reasoning is suggested: *But if, during the intercourse occasioned by trade, he finds that a neighbour in possession of desirable property be weaker than himself* (*if he finds* is a conditional; therefore the clause dependent on it must be in the subjunctive). | *By all means let us follow after those things which make for peace, so far as be possible* (*let us follow* is an exhortation; therefore the clause dependent on it must etc.). | *We should be glad to know that every chairman of a Local Education Authority or Education Committee were likely to read this short biography* (*should* is subjunctive, therefore etc.; or, perhaps more probably, *should be glad to know* is in one word *wish*, and *wish . . . were* is beyond cavil). | *And if exceptional action were needed to prove love, what would after all be proved, except that love* were *not the rule?* (*would* is a subjunctive, therefore etc.). | *If I made a political pronouncement I should feel that I* were *outraging the hospitality of the Brotherhood movement* (*should* is a subjunctive, therefore etc.). It may be admitted that some of these are less bad than others, and that, while the group is characteristic of a time that is not at ease with its subjunctives, anyone who wished to parallel its details in earlier writers who used the mood far more frequently than we, as well as more naturally, could doubtless do so; nevertheless they are best classed with Arrivals.

INDIRECT QUESTION: Latin grammar is perhaps also responsible for the notion that indirect question requires the subjunctive. There is no such requirement in English; *Ask him who he be* is enough to show that. *Sir A. N. asked Sir R. R. if he* were *aware that one of the miners' secretaries in Scotland had been . . .* Read *was*; but again such subjunctives may be found in older writers.

MISCELLANEOUS: *He therefore came round to the view that simple Bible teaching* were *better abolished altogether and that the open door for all religions* were *established in its place. Were better abolished* is a correct Survival; but indulging in a phraseology that is now unnatural has tempted the writer into an impossible continuation. | *Be the ventilation of a gaseous mine as efficient as it can be made, nothing will prevent . . .* An unidiomatic extension of the 'Alive' *Be that as it may*, made absurd by its length. | *He replied gently, but firmly, that if his department* were *to be successful, he must accommodate himself to the people who employed him.* His words were not 'If my department be to succeed', but 'is to'. The sequence change of *is* should be to *was*, and to use *were* instead ruins the sense; 'were to be successful' means 'succeeded' or

'should succeed', not 'was to have a chance of succeeding'.

The conclusion is that writers who deal in Survival subjunctives run the risks, first, of making their matter needlessly formal, second, of being tempted into blunders themselves, third, of injuring the language by encouraging others more ignorant than themselves to blunder habitually, and lastly, of having the proper dignity of style at which they aim mistaken by captious readers for pretentiousness.

**submerge.** Though the verb has superseded *submerse*, the adjective remains *submersible*.

**submissible, -ittable.** The second form is unexceptionable; but on the principle explained in -ABLE 2, *submissible* would have been expected to establish itself on the analogy of *ad*, *o*, and *per*, *-missible*. It is in fact, to judge from the OED, hardly existent, but may nevertheless be recommended as preferable if an adjective should be needed.

**subpoena.** Best so written, see Æ, Œ; p.p. *subpoena'd*, see -ED AND 'D. Plural -*as*.

**subsidence.** In view of the frequent discussions of this topic in Parliament and elsewhere, it is surprising that we are not yet of one mind whether to call it *subsī'dence* or *sub'sĭdence*. The OED gives preference to the long *i*. On the other hand *residence*, *confidence*, *providence*, and *coincidence*, all associated with verbs in -*ī'de*, and all disregarding that fact and conforming to the RECESSIVE ACCENT tendency, are a very strong argument on the other side, and the COD puts *sub'sĭdence* first. But if ever a final choice is made it is popular opinion that will make it, and popular opinion seems on balance to be on the side of the OED, perhaps because of a subconscious inclination to make clear that the word the speaker is using is connected with *subside* and not with *subsidy*.

**subsist, exist.** The essential difference is that *subsist* imports the idea of continuing to exist, as in the OED example *Which charter subsists to this day, and is called Magna Carta.* But in present-day usage *subsist* commonly implies (as in the phrase *subsistence level*) existence on the bare necessities of life.

**substantial, substantive.** Both words mean 'of substance', but they have become differentiated to the extent that *-ial* is now the word in general use for real, important, sizable, solid, well-to-do, virtual, etc., and *-ive*, apart from its meaning in grammar, is chiefly used in special senses: in parliamentary procedure a *substantive motion* is one that deals expressly with a subject in due form; in law *substantive law* (that which is to be enforced) is so called to distinguish it from *adjective law* (the procedure for enforcing it); in the Services *substantive* is used to distinguish rank or office that is permanent from one that is acting or temporary.

**substitute, v., substitution.** A very rapid change—according to the view here taken, a corruption—has been taking place in the meaning and use of these words; so rapid, indeed, that what the OED stigmatized in 1915 as 'Now regarded as incorrect' may, if nothing can be done to stop it, become normal usage and oust what is here held to be the words' only true sense. The definition to which the OED adds the above note is still recorded by the 1964 COD, with the comment 'vulg.' It is (for the verb) 'To take the place of, replace'. The true meaning is something entirely different, viz. to put (a person or thing) in the place of another, and the use of the noun follows it. We can set down for comparison a sentence or two that are right and one or two that are wrong, choosing as nouns that will make the points clear *butter* and *margarine, Englishman* and *alien.*

### CORRECT

A. We had to substitute margarine (for butter).

B. Aliens are being substituted (for Englishmen).

C. [Aliens are replacing Englishmen.]

D. The substitution of margarine (for butter) is having bad effects.

E. Let there be no more substitution of aliens (for Englishmen).

F. Its substitution (for butter) is lamentable.

### INCORRECT

A. We had to substitute butter (by margarine).

B. Englishmen are being substituted (by aliens).

C. Aliens are substituting Englishmen.

D. The substitution of butter (by margarine) is having bad effects.

E. Let there be no more substitution of Englishmen (by aliens).

F. Its substitution (by margarine) is lamentable.

One can hardly read those parallels, with the risks of ambiguity that they suggest, without realizing that either the old or the new must go. We surely cannot keep such a treacherously double-edged knife as *substitute* has become; either its original edge or the one into which its back has been converted must be ground off; which is it to be? Another reflection, which may not occur unsuggested to all, is that in the incorrect set the words *replace* or *replacement* would have done, whereas in the correct set to use them instead of the *sub-* words would either have been impossible or have changed the meaning. And here, probably, is what accounts for the whole perversion of our words; *substitute* and *substitution* have been seized upon by people who failed to apprehend with precision the true meaning and fancied they had found equivalents in sense for the words *replace(ment)*, which they had been ignorantly taught to regard as solecisms in the required senses (see REPLACE); so they determined (in their lingo) to substitute *replace* by *substitute*, whereas they ought to have refused (in English) to substitute *substitute* for *replace* or to replace *replace* by *substitute*.

To sum up: The new popular use is

wrong and confusing, and is based upon a superstition; but it shows disturbing signs of growing unchecked, and therefore it will be necessary to give a convincing array of quotations, to satisfy readers that this article is not an attack on the negligible. It is indeed high time that *replace* was reinstated and *substitute* reduced to its proper function. In going through the sentences, those who are new to the question may observe that nearly all can be mended in two ways, shown for the verb in the first example and for the noun in the second. One is the change to *replace(ment)*, and the other the turning of the sentence upside down and changing of *by* to *for*. One or two exceptional types are placed at the end with special corrections.

*The ecclesiastical principle was substituted by the national, the Empire and the Papacy by the Communes* (Either *was replaced*; or *The national principle was substituted for the ecclesiastical, the Communes for the Empire and the Papacy*). / *Chief among these innovations is the substitution of the large and unwieldy geographical unit by a small and compact local administrative unit* (Either *is the replacement of*; or *is the substitution of a compact local unit for the unwieldy geographical unit*). / *Although only a temporary, and liable to be substituted by an ex-service man at any time, because I was physically unfit for the army, I am glad to . . .* / *If it proves successful it will be extended all along the border; if it fails it will be substituted by an arbitrary line along the lakes and rivers.* / *If a good raw hide gear is substituted by a set of laminated gears, they will be found quite as silent.* / *Even the suppression of the provinces, and their substitution by larger spheres of Government, is being considered.* / *The substitution of a voluntary censorship by a compulsory Government one would result in a more onerous authority.* / *The Chancellor of the Exchequer looked forward to the abolition of the excess profits duty and its substitution by a tax on war fortunes.* / *And the very slow diminution is due to the substitution of these barbaric methods by others rational and decent.* /

*The Jordan army formed an impressively smart guard of honour, more British in appearance than ever, as a result of an order substituting the traditional red and white kafiyehs with peaked khaki caps for officers and berets for other ranks.* / *The key to the whole issue is a Stalinist appreciation of the economic consequences of the Second World War, which brought about the disintegration of a simple all-embracing world market and its substitution by two world markets . . .* / *If potatoes substitute bread, what is going to substitute potatoes?* is a question every German will have to ask himself (In the comparatively rare active use, the upside-down method is not quite applicable. Either read *replace*, or *If we substitute potatoes for bread, what are we going to substitute for potatoes?*). / *Money and talent, often substituted by their counterfeits, speculation and trickery, have here broken down all barriers* (*often substituted by* means simply *or often*).

**subtle, subtil(e)**, etc. The modern forms are *subtle*, *subtlety*, *subtler*, *subtlest*, *subtly*, but *subtilize*; b is silent in all. Spellings with the i retained are (except in *subtilize*) usually left to archaists of various kinds; and, as Milton was content with *suttle*, there seems little reason for going back beyond *subtle* to *subtil*.

**subtopia** is a puzzling word. Clearly it is a HYBRID, compounded of Latin and Greek elements, but it does not bear its meaning on its face, even for those who know those languages. At first sight it seems to be a compound of *sub* (under) with τόπος (place), and so should mean somewhere underground. But the contexts it is used in show that that cannot be right. May it then be a compound of *sub* with an aphetic form of *utopia*, meaning a place that just falls short of being ideal? That cannot be right either; the word is certainly not a term of praise. Only by tracing the word to its source can the truth be found. It is a PORTMANTEAU word, a telescoping of *suburb* and *utopia*, invented by *The Architectural Review* (June 1955) as a name for an imaginary country—the Britain of the year 2000 if 'development' is allowed to continue at its present rate, a Britain consisting of 'isolated oases of preserved monuments in a desert of wire and concrete roads, cosy plots and bungalows' a Britain in which 'there will be no real distinction between town and country [but] both will consist of a limbo of shacks, bogus rusticities, wire, and aerodromes set in some fir-poled fields. . . .' The inventor of the word expressed the hope that it would stick. It has.

**subtract(ion), substr-.** The forbidden -s- is recognized by the OED, though called 'now illiterate'; and in the long array of writers who have used it are Bentham, the Duke of Wellington, and Carlyle.

**succedaneum.** Pl. *-ea*. An examination of quotations is so far from suggesting any difference of meaning between this pedantic term and its synonym *substitute* that it may surely be relegated to the SUPERFLUOUS WORDS, especially as it seems no longer to have its former currency in medicine for an inferior drug substituted for another.

**succeed.** *All the traditions in which she has been brought up have not succeeded to keep her back.* Read *in keeping*, and see GERUND 3.

**success.** For *s. of esteem*, see GALLICISMS 5.

**succuba, -bus.** Pl. *-ae, -ī*; the words mean the same, and are not respectively feminine and masculine. The masculine counterpart is *incubus*.

**such.** 1. *S. which, s. who, s. that, s. where*, etc. 2. *S. that* rel. and *s. that* conj. 3. *S.* exclamatory. 4. Illiterate *s.* = it etc. 5. Defining *s.* 6. *S* = *so*. 7. *S. as* for *as*. 8. *Suchlike*. 9. *As s.*

1. *S. which, s. who, s. that* (rel. pron.), *s. where* (rel. adv.). *Such* is a demonstrative adjective and demonstrative pronoun, to which it was formerly common to make other relatives besides *as* correspond, especially *which, who, that*, and *where*. Modern

idiom rejects all these, and confines itself to *as*; the OED's remark on the use of *such . . . which* etc. is 'Now rare and regarded as incorrect'. It is not in fact so very rare; but most modern examples of it are due either to a writer's entire ignorance of idiom or to his finding himself in a difficulty and not seeing how to get out of it. In the following extracts, when a mere change of *which* etc. to *as* is not possible, a way out is suggested: *The third year should be reserved for* such additional or special subjects (*elocution, for instance*) which *need not be regarded as essential.* | *It was proposed to grant to* such casual employees of the Council who *had been continuously employed for three months, and* whose employment *was likely to extend over twelve months, the privilege of additional leave* (read *those*, or *any*, for *such*). | *It is subject, of course, to* such possible changes of plan that *any unexpected turn of events may bring about* (read *the* or *those* for *such*). | *I noticed two cars approaching in* such a manner that *seemed to indicate they would both arrive at the junction together* (omit *such*) | *There was an item 'Hit Europeans or cut throat' but there was also an item that if a person is arrested because of Congress the case must be taken to 'senior Europeans', which presumably means the High Court or* such of them whose *throats had not been cut* (read as had not had their throats cut).

2. *Such that* rel. and *such that* conj. Now and then a *s. that* for *s. as* is perhaps due to the writer's hesitating between two ways of putting a thing, one with the relative *as* and the other with the conjunction *that*, and finally achieving neither, but stumbling into the relative *that*. *They will never learn the truth from this system of military inquiries, because they will only see the results if those are* such that *the Government would like them to see* (such as the Government would like them to see? or such that the Government would like them to be seen?). | *I cannot think that there is* such a different level *of intelligence among Englishmen and Germans* that *would prevent similar*

*papers from being a profitable property in Great Britain* (such . . . as would prevent? *or* such . . . that it would prevent?).

**3.** *Such* exclamatory or appealing. *The Earl of Derby was the titular King of Man—a piece of constitutional antiquarianism of which Scott made* such *splendid use in 'Peveril of the Peak'.* *Such* is liable to the same over-use of this kind as *so*; reference to SO 3 will make further illustration unnecessary here. Use and over-use of an idiom are different things, and there is no need to avoid this *such* altogether. In the above quotation it may be noticed that if the writer had said *the* piece of antiquarianism instead of *a* piece the *such* would have passed well enough.

**4.** The illiterate *such* (used instead of *those, it, them* or other pronoun). The significance of the epithet will be found explained in ILLITERACIES, and a few examples with corrections will suffice: *His seven propositions for non-partisan legislation must appeal to the common-sense of every man and woman in the realm; is it too much to hope that such will* combine *to* render *them realities?* (that all will). / *We have seen during the war how those persons in humble circumstances who came suddenly into possession of moneys* spent such—*i.e., in* . . . (spent them). / *But when it comes to us following his life and example, in all its intricate details, all will, I think, agree* that such *is* impossible (that that is). / *An appeal to philanthropy is hardly necessary, the grounds* for such being *so self-evident* (for it being).

**5.** The defining *such*. A useful device in drafting legal documents, where precision is all-important, is to use *such* in the sense of as defined above, so as to avoid ambiguity without having to repeat the defining words, as in *The particulars required by this section may be furnished by or on behalf of any person who is a party to the agreement* . . . *and where such particulars are duly furnished by or on behalf of any such person the provisions of this section shall be deemed to be duly complied with on the part of all such persons.* Sometimes this device may be legitimate for the

ordinary writer, but more often his *such* is merely a starchy substitute for *that* or for using a pronoun instead of repeating the word that *such* qualifies, as it is in the following examples: *That there is a void in a millionaire's life is not disproved by anyone showing that a number of millionaires do not* recognize such void (recognize it, *or* the *or* that void.) / *If I am refused the Sacrament I do not believe that I shall have less chance of entering the Kingdom of God than if I received* such *Sacrament* (received it).

**6.** *Such* = *so*. Most people have no hesitation in saying *such a small matter, such big apples, with such little justice, such conflicting evidence*; others object and say that it should be *so small a matter, apples so big, with so little justice, evidence so conflicting*. It must first be admitted that (with allowances for phrases of special meaning) the objectors are entitled to claim the support of grammar. In 'such a small matter' it is usually *small*, not *matter* or *small matter*, that is to be modified by *such* or *so*, and, *small* being an adjective, the adverb *so* is obviously the grammatical word to do the job. (At the same time, *such a small matter*, though it usually means so small a matter, may also mean a small matter of the kind that has been described; but, speaking generally, the objectors have grammar on their side.) Shall we then be meek and mend our ways at their bidding? Why no, not wholesale. We will try to say *so* wherever idiom does not protest or stiffness ensue. For instance, we will give up 'with such little justice' without a murmur; but they cannot expect of us 'I never saw apples so big' instead of 'such big apples'. And they must please to remark that the *such* idiom has so established itself that the other is often impossible without a change of order that suggests formality or rhetoric; *so big apples? so conflicting evidence?* No; the adjective has to be deferred (*apples so big*) in a clearly artificial way; but we grant that 'so small a matter' does strictly deserve preference over 'such a small matter', and, if so partial a concession is worth their acceptance, let it be made. But there is no excuse for mixing the two idioms, as in *I think it may be quite a number of years before such obscure a musical is presented again.*

7. *Such as* for *as. Even the effects of unfavourable weather can be partially counteracted by artificial treatment such as by the use of phosphates.* The repetition of *by* results in a *such as* not introducing a noun (*use*), as it should, but a preposition (*by*)—a plain but not uncommon blunder. Omit either *such* or *by. | Some are able to help in one way, such as for instance in speaking; some in another, such as organization.* The second part is right; the first should be either *in one way such as for instance speaking*, or *in one way as for instance in speaking*; *such as* requires a noun (*speaking*) to complete it, not an adverbial phrase (*in speaking*). See AS 5.

8. *Suchlike.* That the word is a sort of pleonasm in itself, being ultimately = *solike-like*, is nothing to its discredit, such pleonasms being numerous (cf. *poulterer* = *pullet* + -*er* + -*er*) and *suchlike* can be found in the A.V., Shakespeare, and Lamb. But now, although perhaps admissible colloquially as an adjective (*barley, oats, and suchlike cereals*), its use as a pronoun (*schoolmasters, plumbers, and suchlike*) is better left to the uneducated, *the like* being used instead.

9. *As such* is liable to be used in curious ways, so curious sometimes that the writer's meaning can only be guessed. *The statistics as such add little to our information. | There is no objection to the sale of houses as such.* In the first example *as such* probably means by themselves, *per se*. In the second the context suggests that the writer meant there was no objection in principle to the sale of houses; if so he chose an absurd way of saying it.

**sufficient(ly)** and **enough.** The words are discussed under ENOUGH; for *sufficient* wrongly preferred in the following extracts, see the first paragraph of that article: *So far as the building trade is concerned the complaint we have made to the Government is that sufficient has not been done to get materials organized. | And there should be sufficient of a historic conscience left in the Midland capital to evoke a large subscription.* See also GENTEELISMS.

**suffragette.** A FACETIOUS FORMATION, now of only historical interest; perhaps the reason why it won immediate acceptance was that it seemed to be a happy fusion of the notions of female suffrage and female advocates of it. But the essential significance of the French suffixes *et* and *ette* is diminutive (*clarionet* = a little clarion, *cigarette* = a little cigar); and many of the militant suffragettes were by no means diminutive. We use the suffix in its proper sense in such words as *leaderette*, and, by a not unreasonable extension, to imply not a diminutive but a synthetic quality, e.g. *flannelette, leatherette.* The belief engendered by *suffragette* that *ette* is a feminine termination is true only to the extent that -*ette*, and not -*et*, is the natural diminutive of French words of the feminine gender. *Midinette* cannot be quoted in support of it if Larousse is right in saying that *midinettes* 'sont celles qui se contentent d'une dînette à midi'. See FEMININE DESIGNATIONS s.f.

**suggestio falsi.** Pronounce -*tĭŏ fǎ'lsī* (see LATIN PHRASES), 'suggestion of the untrue'. The making of a statement from which, though it is not actually false, the natural and intended inference is a false one. For instance, if A, asked whether B is honest, replies, though he in fact knows no harm of B, that his principle is to live and let live and he is not going to give away his old friend, the questioner infers that A knows B to be dishonest. Cf. SUPPRESSIO VERI.

**suit, suite,** nn. *Suite* is pronounced *swēt*; for the pronunciation of *suit* see PRONUNCIATION 6. The two words are the same, and the differences of usage accidental and variable. But where, the sense being a set, either form would seem admissible, we do say at present *a suit of clothes, a suit of armour, a suit of sails, the four suits at* *cards, follow suit*; and on the other hand a *suite (of attendants* etc.), *a suite of rooms* or *apartments, a suite of furniture* or *chairs, a musical suite.*

**sulphureous, sulphuric, sulphurous.** The last has differentiated pronunciations *sŭ'lfūrŭs* and *sŭlfūr'ŭs*, so that there are *four* adjectives to divide the work. *Sulphuric* and *sulphurous* (*-ūr'ŭs*) can for general purposes be ignored as technical terms in Chemistry like other *-ic* and *-ous* pairs. *Sulphureous and sulphurous* (*sŭ'l-*), which remain, have never been effectively differentiated, and the OED refers the reader for most senses of one to definitions given under the other. Differentiation may be expected to come, and perhaps the likeliest course for it to take and therefore the best to fall in with is that *sulphurous*, now the more popular word, should take to itself the secondary or extended senses, and *sulphureous* be restricted to the primary material ones meaning 'of or containing sulphur' without the specific limitations of *sulphuric* and *sulphurous* (*-ūr'ŭs*). This would give—though naturally the borderline is not quite sharp—*sulphureous gases, springs, smells, drugs, substances,* but *sulphurous yellow, light, torments, language, preachers.*

**sumach, -ac.** The OED gives precedence to the first spelling, and pronounces *sū'măk* or *shōō'măk. Sū-* is now more usual.

**summer.** 1. *St. Luke's, St. Martin's, S.* Each of these is often used when the other would be the right one; St. Luke's day is in October (18th), St. Martin's in November (11th). Each of them, or any fine warm spell after the end of September, may be called by the generic term *Indian summer.*

2. *Summer time, summer-time, summertime.* The first is the daylight-saving term; in other senses either of the others should be used; see HYPHENS.

**summitry.** This new word for the pursuit of world peace through meetings of the Heads of States is an example of the use of the suffix *-ery*

(shortened to *-ry*) to describe 'that which is characteristic of, all that is connected with [the word it is attached to], in most cases with contemptuous implication' (OED). Other modern coinages of the same kind are *rocketry* —an activity that summitry aims at keeping within bounds—and *gimmickry.*

**summon(s).** 1. For *summon* and *send for,* see FORMAL WORDS. 2. *Summons,* n., has pl. *summonses.* 3. *Summon* is the verb in ordinary use; *summons* should not be used as a verb except in the special sense to serve with a legal summons or issue a summons against, and even in that sense *summon* is equally good.

**sunk(en).** For idiomatic use of the two forms, see SINK.

**super.** This prefix has been put to many varied uses. By itself, as a CURTAILED WORD, it may stand for *superfine, superficial measure,* or *supernumerary,* in the last capacity distinguishing the actor who has not a speaking part from those who have; in most detective stories it serves as the friendly designation (cf. *sarge*) of the superintendent of police working on the case, and it has for a time shared with *wizard* and *smashing* the duty of expressing the acme of juvenile approval (see FABULOUS). Its use as a prefix not in its primary sense of 'above', 'transcending' (*superhuman, supersonic, superstructure,* etc.) but meaning 'of a superior kind', as in *superman, supermarket, superministry, super-priority,* and scores or hundreds of other words, is so evidently convenient that it is vain to protest when others indulge in it, and so evidently catachrestic that it is worth while to circumvent it oneself when one can do so without becoming unintelligible. Super-cinema, meaning merely a cinema of exceptional size or splendour, and not something that transcends and thereby ceases to be a cinema, and the super-DIFFERENT by which the advertiser tries to make us believe that what he is offering is of unparalleled excellence, may serve as specimens of the worse applications.

**superficies.** Five syllables (-*fĭ′shĭēz*); pl. the same.

**superfluous words.** That there are such things in the language is likely to be admitted, and perhaps it might be safe even to hazard the generality that they ought to be put in a black list and cast out; but woe to the miscreant who dare post up the first list of proscriptions! Brevity and timidity will therefore be the marks of our specification; the victims will be mainly such as have no friends, with just one or two of other kinds slipped in to redeem the experiment from utterly negligible insignificance. Indeed, it is more necessary to account for the tameness of the list than to defend its boldness; but then it must be remembered that most of the words naturally thought of as conspicuously suitable for expulsion are those SLIPSHOD EXTENSIONS and VOGUE WORDS which, abominable as they are in some of their modern senses, are not superfluous, because each of them has somewhere in the background a sense or senses at least worth preserving, and often of importance. The use of them needs to be mended, but not ended, and they are dealt with elsewhere. The list follows; reasons for the condemnation should be looked for under the word concerned, unless a special article is indicated: *dampen* (-EN VERBS); *escalate*; *escapee*; *instinctual*; *faience*; *filtrate*; *gentlemanlike*; *habitude*; *legitimatize* and *legitimize*; *lithesome*; *minify*; *olden*, v. (-EN VERBS); *preventative*; *quieten* (-EN VERBS); *refutal*; *righten* (-EN VERBS); *rotatory*; *smoothen* (-EN VERBS); *succedaneum*; *tactual*; *un-come-at-able*; *vice-regent*; *viceroyal*. It is, however, quite likely that some of these words will prevail over their rivals, and so transfer to them the stigma of superfluity, as for instance *escalate* over *escalade*, and *quieten* over *quiet*.

**superior.** 1. Used of people (*a most superior person*) in a patronizing or ironical way, implying that the person one calls *s*. is nevertheless one's inferior, it resembles the corresponding use of *worthy* and *well-meaning* in producing on the hearer an unfavourable impression of the speaker. 2. Used of things it is often merely an affected way of saying 'of good quality'. It is so used mainly by those who make or sell the things they so describe, and should be left to them. The reviewer who concludes 'This is altogether a superior book' is unlikely to give us confidence in his judgement. 3. *S. to*, not *s. than*, is required by idiom; but such is the power of ANALOGY that even people who obviously cannot be described as uneducated are sometimes capable of treating *s.* as we all treat *better* or *greater* (cf. PREFER, with which the same mistake is much more frequent). The quotations are purposely given at sufficient length to show that the writers are not mere blunderers: *Mr. Ernle, on the other hand, as we gather from his preface, desired first to translate Homer, and in looking about for a metre decided on the hexameter as the most appropriate and* superior *for this style of the heroic* than *the blank or rhymed verse of the great English masters* (read *better . . . than*, or *s. . . . to*). | *Whatever the conditions in the provinces—the present inquiry has dealt only with the Metropolis—able and public-spirited men have refused to accept the dictation of the B.M.A., and are giving far* superior *attention to the insured persons* than *was possible under the cheap conditions of the old club practice* (read *greater . . . than*, or *s. . . . to what*).

**superiority.** Much misplaced ingenuity in finding forms of apology is shown by writers with a sense of their own superiority who wish to safeguard their dignity and yet be vivacious, to combine comfort with elegance, to touch pitch and not be defiled. Among them are: *To use an expressive colloquialism—in the vernacular phrase—if the word may be permitted—so to speak —in homely phrase—not to put too fine a point upon it—if the word be not too vulgar—as they say—to call a spade a* *spade*—*not to mince matters*—*in the jargon of today*—or the use of deprecatory inverted commas. Such writers should make up their minds whether their reputation or their style is such as to allow of their dismounting from the high horse now and again without compromising themselves. If they can do that at all, they can dispense with apologies; if the apology is needed, the thing apologized for would be better away. *A grievance once redressed ceases to be an electoral asset (if we may use a piece of terminology which we confess we dislike). | Turgenev had so quick an eye; he is the master of the vignette*—*a tiresome word, but it still has to serve. | About one thing there is complete unanimity; 'Coalition' must go; it is not a Party name, and in any case it will not do at the next election'; to put it vulgarly, that cock won't fight. | When the madness motif was being treated on the stage, Shakespeare (as was the custom of his theatre) treated it 'for all it was worth'. | With its primary postulate, 'steep' as it is, we will not quarrel. | It is a play that hits you, as the children say, 'bang in the eye'. | The annual conflict between the income-tax demand note and the January sales has ended, it seems, in the more or less complete triumph of what the Upper Fifth would call the former. | He seized my hand in what the lover of a cliché would call an 'iron grip'. | To make use of an overworked phrase, the wall painting requires a more severe application of 'fundamental brainwork'. | England had been compelled, in homely phrase, to 'knuckle down' to America. | Its work was, if we may use a somewhat homely expression, 'done to time'. | Palmerston is to all appearance what would be vulgarly called 'out of the swim'.*

For another form of superiority, that of the famous 'of course', as often exposed and as irrepressible as the three-card trick, see COURSE.

In short, some writers use a slang phrase because it suits them, and box the ears of people in general because it is slang; a refinement on the institution of whipping-boys, by which they

not only have the boy, but do the whipping.

**superlatives.** For some misuses of superlatives see -ER and -EST, 5, 6, 7, and 9.

**superstitions.** 'It is wrong to start a sentence with "But". I know Macaulay does it, but it is bad English. The word should either be dropped entirely or the sentence altered to contain the word "however".' That ungrammatical piece of nonsense was written by the editor of a scientific periodical to a contributor who had found his English polished up for him in proof, and protested. Both parties being men of determination, the article got no further than proof. It is wrong to start a sentence with 'but'! It is wrong to start a sentence with 'and'! It is wrong to end a sentence with a preposition! It is wrong to split an infinitive! See the article FETISHES for these and other such rules of thumb and for references to articles in which it is shown how misleading their sweet simplicity is; see also the article SUBSTITUTE for an illustration of the havoc that is wrought by unintelligent applications of an unintelligent dogma. The best known of such prohibitions is that of the SPLIT INFINITIVE, and the hold of that upon the journalistic mind is well shown in the following, which may be matched almost daily. The writer is reporting a theatre decree for hat-removal: '. . . the Management relies on the cooperation of the public to strictly enforce this rule'. *Even a split infinitive* (he comments) *may be forgiven in so well-intentioned a notice.* Theatre-managers are not stylists; the split this manager has perpetrated, is it not a little one? and to put him, irrelevantly, in the pillory for it betrays the journalist's obsession.

Well, beginners may sometimes find that it is as much as their jobs are worth to resist their editors' edicts, as the champion of 'But' did. On the other hand, to let oneself be so far possessed by blindly accepted conventions as to take a hand in enforcing them on other people is to lose the independence of judgement that would enable one to solve the numerous problems for which there are no rules of thumb.

**supine.** 1. The pronunciation of the term of Latin grammar has always been *sū'pīn*. The English adjective was formerly *sūpī'n*, but RECESSIVE ACCENT has been at work (as in CANINE etc.) and the prevailing pronunciation is now also *sū'pīn*. 2. *Supine* means lying face upwards; the words for lying face downwards are *prostrate* and *prone*, but these are also used loosely for lying flat in any position.

**supple.** *The fine mass of the head, solidly yet* supplely *modelled, is set in a particularly beautiful convention of the hair.* The adverb is *supply*, not *supplely*; cf. SUBTLE. It is true that the OED found more instances in print of *-plely* than of *-ply*, and therefore on its historical principles made *supplely* the standard form. But the pronunciation is undoubtedly *sŭ'plĭ*, not *sŭ'pŭl-lĭ*, and the long spelling has been due to the wish to distinguish it to the eye from *supply* (*suplī'*) n. and v. Such devices are not legitimate except in the last necessity, as with *singeing* and *singing*; and it is to be observed that, whereas the -e- in *singeing* selects the right of two possible pronunciations, the -le- in *supplely* suggests a wrong one. It is unfortunate that adjectives in -bble, -ckle, -ddle, -ffle, -ggle, -pple, -ttle are few and not provided with adverbs common enough to settle the question; *subtly* is in fact the best analogue, and its spelling, though *subtlely* has been occasionally used, is now established. The COD now gives *supply* only.

**supplement.** Noun *sŭ'plĕmĕnt*; verb usually *sŭplĕme'nt*; see NOUN AND VERB ACCENT.

**suppositious, supposititious.** *The* supposititious *elector who imagined that the Parliament Bill was a weapon for show and not for use is, we venture to say, a mythical being.* It is often assumed that the first form is no more

than an ignorant and wrong variant of the other. Ignorant it often is, no doubt, the user not knowing how to spell or pronounce *supposititious*; but there is no reason to call it wrong. *Suppositious* and *supposititious* may as well coexist, if there is work for two words, as FACTIOUS and *factitious*; and, if the support of analogy for the shorter form is demanded, there are *ambitious, expeditious, seditious, nutritious, cautious,* and *oblivious* to supply it. There are moreover two fairly distinct senses to be shared, viz. spurious, and hypothetical. *Supposititious* is directly from the Latin p.p. *suppositus* = substituted or put in another's place, and therefore has properly the meanings foisted, counterfeit, spurious, pretended, ostensible. *Suppositious* is from the English *supposition* = hypothesis, and therefore may properly mean supposed, hypothetical, assumed, postulated, imaginary. It does not follow that *suppositious* is wanted; *suppositional* is now a less unusual form, and probably the work either might do is better done by the more familiar synonyms above given. But it does follow that *supposititious* should not be given, as in the quotation at the head, senses proper to the synonyms of *suppositious*; it should be confined to those implying intent to deceive.

**suppressio veri.** Pronounce *vēr'ī*. Intentional withholding of a material fact with a view to affecting a decision etc.; cf. SUGGESTIO FALSI.

**surcease,** n. and v., is a good example of the archaic words that dull writers at uneasily conscious moments will revive in totally unsuitable contexts; see INCONGRUOUS VOCABULARY. The fact is that in ordinary English the word is dead, though the pun in *Macbeth* (*and catch, with his surcease, success*) is a tomb-stone that keeps its memory alive; there are contexts and styles in which the ghosts of dead words may be effectively evoked, but in newspaper articles and pedestrian writing ghosts are as little in their element as in Fleet Street at midday. The follow-

surgeon 608 swapping horses

ing quotations are borrowed from the
OED, all from 19th- or 20th-c. writers:
*It was carried on in all weathers . . .
with no surcease of keenness.* | *Private
schools for boys give four days' surcease
from lessons.* | *There is no surcease in the
torrent of Princes . . . who continue to
pour into the capital.* | *I . . . thereupon
surceased from my labors.* | *They could
never surcease to feel the liveliest inter-
est in those wonderful meteoric changes.* |
*Intrigues and practices . . . would of
necessity surcease.*

**surgeon.** See PHYSICIAN.

**surly.** Adv. *surlily*; see -LILY s.f. The
adjective was originally *sirly* (sir-like
= arrogant); the change of spelling
disguises the fact that *-ly* in *surly* is
the ordinary suffix, and perhaps accounts for *surlily* on the analogy of
*jollily, sillily, holily*.

**surprise.** '*I should not be surprised if
the Chancellor of the Exchequer* does
not agree *with me.*' For this pleonastic
use of *not* see NOT 4.

**surveillance.** Pronounce *servā'lans*.

**susceptible.** See SENSIBLE.

**suspenders.** To use the word for
*braces* in England, as it is in U.S.,
would be to throw away the advantage
of having two names for two things; in
England *suspenders* keep in place not
men's trousers but women's stockings,
or, as *sock-suspenders*, men's socks.

**suspense, suspension.** In the verbal
sense, = suspending, the second is
right. *Suspense*, though it still retains
that force in suspense of judgement,
suspense account, and some legal uses,
has become so identified with a state
of mind that to revive its earlier use
may be confusing. In the following
quotation it is clear that *suspense* compels one to read the sentence twice,
whereas *suspension* or *suspending* would
have been understood at first sight:
*The state of war is inevitably the
suspense of Liberalism, and in all the
nations at war there are some men who
greatly hope that it may also be the
death of Liberalism.*

**suspicion.** 1. For *s.* = soupçon, see
GALLICISMS. 2. The OED gives examples of *suspicion* as a verb, but it is
charitable to regard its use in this way
today as facetious or slang or dialect.

**sustain.** *Mr. ——, Master of the ——
Hounds, has sustained a broken rib and
other injuries through his horse falling.*
The very common idiom here illustrated is described by the OED as 'in
modern journalistic use'; but with
such abstract objects as *defeat, loss,
hardship, damage*, etc., instead of
*broken rib* it is as old as the 15th c.,
and the extension is not a violent one.
Nevertheless, *sustain* as a synonym
for suffer or receive or get belongs to the
class of FORMAL WORDS, and is better
avoided both for that reason and for
the stronger one that, if it is not made
to do the work of those more suitable
words, it calls up more clearly the
other meaning in which it is valuable,
viz. to bear up against or stand or
endure without yielding or perishing,
as in 'capable of sustaining a siege'.
Cf. *undergo* an operation.

**swap, swop.** The OED prefers *-ap*,
but *-op* is probably now commoner
and is preferred by the COD.

**swapping horses** while crossing the
stream, a notoriously hazardous operation, is paralleled in speech by changing a word's sense in the middle of a
sentence, by vacillating between two
constructions either of which might
follow a word legitimately enough, by
starting off with a subject that fits one
verb but must have something tacitly
substituted for it to fit another, and by
other such performances. These lapses
are difficult to classify and to exemplify,
and any exposition of their nature
naturally incurs the charge of PE-
DANTRY. Nevertheless, the air of
slovenliness given by them is so fatal
to effective writing that attention must
be called to them whenever an opportunity can be made, as by this claptrap
heading.
1. Changing a word's sense. For
this see LEGERDEMAIN WITH TWO
SENSES.

**2.** Shifting from one construction to another. *But supposing nothing changed and this Pope, who is made incompetent by the weight at once of his virtues and his ignorances, enjoys a long life, we should look for a great decline in.* . . . *Supposing* is followed first by an object (*nothing*) and adjectival complement (*changed*), and secondly by a substantival clause (*this Pope enjoys*). Either is right by itself, but to swap one for the other means disaster.
**3.** Tacit modification of the subject etc. *This barbarism could be stopped in a very short time, if it were made a punishable offence to throw rubbish into the street, and would have the added value of reducing the army of scavengers.* It is not the barbarism, but the stoppage of it, that would have the added value. | *Fifty per cent. of the weight could be knocked off practically every new petrol vehicle produced and yet be able to carry exactly the same load.* What would carry the same load is not the 50 per cent. knocked off, but the vehicle without it. | *A. C. Benson recalls a pleasant fiction, supposed to have happened to Matthew Arnold.* A fiction neither happens nor is supposed to happen to anyone. A fiction can be recalled, but before it can be supposed to have happened it must be tacitly developed into a fictitious experience; for it is itself a statement or narrative and not an event. See HAZINESS and SIDE-SLIP for other specimens of similar confusion.

**swat, swot.** *Swat*, as used in *Swat that fly*, is a variant of *squat*, the original meaning of which, as a transitive verb, was to hit with a smart blow. *Swot*, the slang term for to work hard, is a variant of *sweat*.

**swath(e).** The OED gives both spellings for the agricultural noun, and the pronunciations *swawth* or *swŏth* or *swãdh*; see -TH AND -DH. The noun and verb meaning wrap are both *swathe* (*swãdh*). The possible differentiation is easy to see and has made some progress; for the agricultural noun the COD now gives *swath* as the only spelling and *swawth* as the only pronunciation.

**swell.** *Swollen* is now the usual form of the p.p., but *swelled* is occasionally preferred, perhaps as a more colourless word for augmented, without the suggestion of augmented to excess conveyed by *swollen*. On the other hand there can have been no such motive to account for the choice of *swelled* in what is perhaps its chief use today— the phrase *swelled head*.

**swim.** The past *swam* and p.p. *swum* are now invariable, though the OED has a Carlyle quotation for *swam* p.p., and a Tennyson for *swum* past.

**swine.** Sing. and pl. the same. Except as a term of abuse the word is now used only as a COLLECTIVE, and even so is a FORMAL WORD. The animals susceptible to swine fever are called pigs.

**swing.** Past *swung*, though OED quotes for *swang* Wordsworth, Tennyson, Gosse, and Belloc.

**swing(e)ing.** *At the bottom was tripe, in a swinging tureen*—Goldsmith. A capacious one? Or one hung on pivots? See MUTE E, and use the -e in the participle of *swinge*.

**swoop.** See HEADLINE LANGUAGE.

**sybil.** See SIBYL.

**syllabize** etc. A verb and a noun are clearly sometimes needed for the notion of dividing words into syllables. The possible pairs seem to be the following (the number after each word means—1, that it is in fairly common use; 2, that it is on record; 3, that it is not given in OED):

| | |
|---|---|
| syllabate 3 | syllabation 2 |
| syllabicate 2 | syllabication 1 |
| syllabify 2 | syllabification 1 |
| syllabize 1 | syllabization 3 |

One first-class verb, two first-class nouns, but neither of those nouns belonging to that verb. It is absurd enough, and any of several ways out would do; that indeed is why none of them is taken. The best thing would be to accept the most recognized verb *syllabize*, give it the now non-existent noun *syllabization*, and relegate all the rest to the SUPERFLUOUS WORDS; but there is no authority both willing and able to issue such decrees.

**syllabub.** See SILLABUB.

**syllabus.** The plural *-buses* is now more used than *-bi*.

**syllepsis** (taking together) **and zeugma** (yoking) are two figures of speech distinguished by scholars, but sometimes confused in use, the second and more familiar word being applied to both. Examples of syllepsis are: *Miss Bolo went home* in *a flood of tears and a sedan chair.* | *He* lost *his hat and his temper.* | *She was seen washing clothes* with *happiness and Pears' soap.* Examples of zeugma are: Kill *the boys and the luggage!* | *With* weeping *eyes and hearts.* | *See Pan with flocks, with fruits Pomona* crowned.

What is common to both figures is that a single word (that in roman type in each example) seems to be in the same relation to two others but in fact is not. The distinction between the two figures is that syllepsis is grammatically correct, but requires the single word to be understood in a different sense with each of its pair (e.g., in the last *with* expresses first accompaniment, but secondly instrument), whereas in zeugma the single word actually fails to give sense with one of its pair, and from it the appropriate word has to be supplied—*destroy* or *plunder* the luggage, *bleeding* hearts, Pan *surrounded.* Intentional use of these figures has been so much overdone as to be now a peculiarly exasperating form of WORN-OUT HUMOUR. A few specimens follow, of which the first is perhaps not of the intentional kind meant to amuse, and is, as an established formula, hardly realized to be a syllepsis. *The newly elected member for Central Leeds* took *the* oath *and his* seat. | *Mr. —— played the Duke quite ably; and the* flood *of* flowers *and* enthusiasm *was terrific.* | *Half-clad stokers toiled in an atmosphere consisting of one part air to ten parts mixed* perspiration, coal-dust, *and* profanity. | *Such frying, such barbecueing, and everyone dripping in a* flood *of* sin *and* gravy. | *Impassively*

*malignant Chinamen scramble after each other* in *hot* haste, *and three-line* paragraphs.

**syllogism.** Deduction, from two propositions containing three terms of which one appears in both, of a conclusion that is necessarily true if they are true; a s. of the simplest form is:
All men are mortal;
All Germans are men;
Therefore all Germans are mortal.
The predicate of the conclusion (here *mortal*) is called the *major term*, and the preliminary proposition containing it the *major premise*; the subject of the conclusion (here *Germans*) is called the *minor term*, and the preliminary proposition containing it the *minor premise*. The term common to both premises (here *men*) is called the *middle term*.

**sylvan.** See SILVAN.

**symbol.** For synonyms see SIGN.

**symbolic(al).** The short form is more usual; there is no difference in meaning.

**sympathetic.** *The play, in spite of sublime scenes and poetry, is an illustration and a warning to artists who deny, or forget, that no powers of execution and no subordinate achievement can compensate for a central figure who is 'unsympathetic', and that it is better for a 'hero' to provoke active fear or hate than indifference or half-contemptuous pity.* | *Macbeth is not made great by the mere loan of a poet's imagery, and he is not made sympathetic, however adequately his crime may be explained and palliated, by being the victim of a hallucination.* | *Let me first say that Elsie Lindtner is by no means sympathetic to the writer of this paper; if she were, the tragedy of the book would be more than one could bear.* This use of s. to describe not a person who feels sympathy but one who excites it is comparatively recent in English, though of longer standing in French and other languages. It was borrowed by our book reviewers and dramatic critics, presumably because they felt that there was no English word that conveys the same meaning; neither *attractive* nor *engaging* nor *congenial* nor *appealing* will quite do. That is a respectable reason for introducing a GALLICISM, and *s*. in this sense is now fully naturalized, at any rate when used of characters in books and plays. But we overdo our welcome when we extend it to the visual arts, and apply it to inanimate objects. It has no advantage over *pleasing* in *The Venetian glass frequently had a slight brownish or greyish tint, very sympathetic to the eye*. It is true that we may find 'a sympathetic twilight' in Wordsworth, but that is an example of the PATHETIC FALLACY.

**sympathy.** The exception sometimes taken to following *s*. with *for* instead of *with* is groundless. *With* is the usual preposition for *s*. in the sense of sharing an emotion, but under the sense of compassion the OED puts *for* before *with* as the normal construction. *For* would have been better in *The Queen Mother has long had a compassionate sympathy with sufferers from this disease*.

**symposium.** Pron. *-ō'zĭum*; pl. *-ia*. But Plato's *Symposium* is usually pronounced with the short *o* of the word in its Greek form.

**symptom.** For synonyms see SIGN.

**synchronize** is not a word that we need regret the existence of, since there is useful work that it can do better than any other (e.g. synchronized clocks, gears, television records); but it is a word that we may fairly desire to see as seldom as we may, one of the learned terms that make a passage in which they are not the best possible words stodgy and repellent; it may be compared with the lists in POPULARIZED TECHNICALITIES. It might well be reserved for planned concurrences, and not applied to those that are accidental. The extracts below, for instance, would surely have been better without it: *The lock-out mania, therefore, has synchronized* [coincided?] *with an increased willingness for sacrifice on the part of the men.* | *A movement of Russian troops to the Caucasus was ordered. . . . This movement synchro-* nized *with* [There were at the same time] *reports of an extensive movement of Turkish troops near the Persian frontier.* | *The winter solstice, which north of the Equator synchronizes with* [determines] *the first day of the winter quarter, occurs at six minutes to eleven tonight.*

**syncope, syncopation.** *Syncope* as a grammatical term means the shortening of a word by omission of a syllable or other part in the middle. *Symbology* and *pacifist* and *idolatry* for *symbolology*, *pacificist*, and *idolalatry*, are examples. *Syncopation* as a musical term means the suspension or alteration of rhythm by pushing the accent to a part of the bar not usually accented. It is not, as many people think, an invention of the composers of modern dance music; what they have done is to popularize a device freely used in classical music.

**syndrome** (a medical word for a set of symptoms) is, like *syncope*, a trisyllabic Greek word, and the dictionaries at first said that it should be so pronounced. But in practice the analogy of *aerodrome*, *palindrome*, and *hippodrome* has proved irresistible, and we say *sin'drōm*.

**synecdoche.** The mention of a part when the whole is to be understood, as in *A fleet of fifty sail* (i.e. ships), or vice versa as in *England* (i.e. the English cricket XI) *won*. The journalists who wrote the following extracts were using *s*.: *This newspaper—and probably the country—will wait its time and see how the new* faces *perform before judging them.* | *Her Royal Highness will shake hands with many of the big* names *in Variety*.

**synonyms,** in the narrowest sense, are separate words whose meaning, both denotation and connotation, is identical, so that one can always be substituted for the other without change in the effect of the sentence in which it is done. Whether any such perfect synonyms exist is doubtful, except perhaps when more than one name is given to the same physical object or condition, e.g. *gorse* and *furze*, *undernourishment* and *malnutrition*. But if it is a fact that one is much more often used than the other, or prevails in a different geographical or social region, then exchange between them does alter the effect on competent hearers, and the synonymity is not perfect. At any rate, perfect synonyms are extremely rare.

Synonyms in the widest sense are words of which either, in one or other of its acceptations, can sometimes be substituted for the other without affecting the meaning of a sentence. Thus it does not matter (to take the nearest possible example) whether I say a word has 'two senses' or 'two meanings', and *sense* and *meaning* are therefore loose synonyms; but if 'He is a man of sense' is rewritten as 'He is a man of meaning', it becomes plain that *sense* and *meaning* are far from perfect synonyms; see FIELD, and SIGN, for sets of this kind.

It is perhaps worth while to record as curiosities a few apparent antonyms that are sometimes used as near-synonyms. Such are *best* and *worst* (as transitive verbs), *flammable* and *inflammable*, *loose* and *unloose*, *passive* and *impassive*, *ravel* and *unravel*, *valuable* and *invaluable*. In some of their senses *bend* and *unbend* and *certain* and *uncertain* almost qualify for admission to the list.

Synonyms, or words alike in sense but unlike in look or sound, have as their converse HOMONYMS and HOMOPHONES, or words alike in look or sound but unlike in sense. The *pole* of a tent or coach or punt, and the *pole* of the earth or the sky or a magnet, are in spite of their identical spelling separate words and are therefore homonyms. *Gauge* and *gage*, not spelt alike, but so sounded, are homophones.

Misapprehension of the extent to which words are synonymous is responsible for much bad writing of the less educated kind. From the notion that ALIBI is a synonym of *defence*, as it is when the defence is of a particular kind, come the absurdities, illustrated under the word, of its use for any sort

of excuse; so with DECIMATE (and *ravage*), DIFFERENTIAL (and *difference*), DILEMMA (and *difficulty*), FEASIBLE (and *possible*), LOCATE (and *find*), OPTIMISTIC (and *hopeful*), PERCENTAGE (and *part*), PRACTICABLE (and *practical*), PROPORTION (and *portion*), PROTAGONIST (and *champion*), SABOTAGE (and *destroy*), SHAMBLES (and *disorder*), SUBSTITUTE (and *replace*), as well as numberless others. To appreciate the differences between partial synonyms is therefore of the utmost importance. There are unluckily two obstacles to setting them out in this book. One is that nearly all words are partial synonyms of some other words, and the treatment of them all from this point of view alone would fill not one but many volumes. The other is that synonym books in which differences are analysed, engrossing as they may have been to the active party, the analyst, offer to the passive party, the reader, nothing but boredom. Everyone must, for the most part, be his own analyst; and no one is likely to write well who does not expend, whether expressly and systematically or as a subconscious accompaniment of his reading and writing, a good deal of care upon points of synonymy. A writer's concern with synonyms is twofold. He requires first the power of calling up the various names under which the idea he has to express can go. Everyone has this in some degree; everyone can develop his gift by exercise. But copiousness in this direction varies, and to those who are deficient in it ready-made lists of synonyms are a blessed refuge, even if the ease they bring may have as doubtful an effect on their style as the old *Gradus ad Parnassum* on the schoolboy's elegiacs. Such lists, to be of much use, must be voluminous, and those who need them should try Roget's *Thesaurus* or some other work devoted to that side of synonymy. Secondly, he requires the power of choosing rightly out of the group at his command, which depends on his realizing the differences between its items. As has already been said, such differences cannot be expounded for a language in anything less than a vast dictionary devoted to them alone, and no attempt at it has been made in this book except in cases where experience shows warnings to be necessary. Still, a book concerned like the present with English idiom in general cannot but come into frequent touch with synonymy; and those who wish to pursue that particular branch of idiom will find the following list of articles (in addition to those previously referred to) useful as a guide: FORMAL WORDS, GENTEELISMS, INCONGRUOUS VOCABULARY, LITERARY CRITICS' WORDS, LITERARY WORDS, LONG VARIANTS, NEEDLESS VARIANTS, POPULARIZED TECHNICALITIES, SLIPSHOD EXTENSION, VOGUE WORDS, WORKING AND STYLISH WORDS.

**synonymity, synonymy.** There is work for both words, the first meaning synonymousness, and the second the subject and supply of synonyms.

**synopsis.** Pl. *-psēs*.

**syntax.** See GRAMMAR.

**synthesis.** Pl. *-thesēs*. The scientific sound of the word, aided no doubt by the common use of *synthetic* in the sense of artificial, often tempts the pretentious to use it instead of more appropriate words such as *combination, alliance,* or *union,* as in: *A flickering gleam on the subject may be found in a pamphlet called 'The Case against Home Rule', by Mr. Amery, which also propounds the new idea of a synthesis between the tariff and the opposition to Home Rule.* Cf. the similar abuse of INTEGRATE.

**synthetize,** not *synthesize,* is etymologically the right formation, but *-size* is more used.

**Syriac, Syrian.** There is the same difference in application as between *Arabic* and ARAB(IAN).

**syringe** (n. and vb.). Pron. *sĭ'rĭnj,* not *sĭrĭ'nj.*

**systemic,** as compared with the regular *systematic,* is excused by its usefulness in distinguishing a sense required in physiology etc. 'of the system or body as a whole'; other wrong formations, *systemist, systemize,* etc., have no such excuse, and *systematist* etc. should be invariable.

**systole.** Pronounce *sĭ'stŏlĕ.*

# T

**tableau.** For plural see -X.

**taboo.** Accent on last syllable. Though this accent is English only, it is established English, and to correct it is PEDANTRY; to spell *tabu* (except in ethnological dissertations) is no better. Past and p.p. usually *tabooed,* sometimes (see -ED AND 'D) *taboo'd.*

**tabula(rasa).** Pl. *lae* (*-sae*). This term is now applied by psychologists to the human mind at birth, viewed as having no innate ideas; *clean slate* serves well enough for ordinary use.

**tactile, tactual.** Why two words? And, there being two, is any useful differentiation either established or possible? The existence of *tactile* is sufficiently explained by the desire for a form corresponding to a large class of adjectives that mean having the power or quality of doing or suffering some action—contractile, ductile, erectile, fictile, fissile, flexile, pensile, prehensile, protrusile, retractile, sessile, and tensile, not to mention more familiar words such as agile, docile, fragile, textile and volatile. And the existence of *tactual* is sufficiently explained by a natural preference for *tactually* over *tactilely.* The *-ual* words belong to Latin abstract nouns in *-us,* and the *-ile* words to Latin verbs, and on the whole their meanings are true to that difference, however little we may know or remember it. But, *tactile* and *tactual* (unlike other pairs such as *agile* and *actual, textile* and *textual*) are used almost indiscriminately, with a tendency for *-ile* to prevail for all purposes, especially now that *tactile values* is established in the vocabulary of art criticism. Differentiation is no longer possible, and *tactual* might be allowed to die.

**talent, genius.** Henry Bradley, in the OED, sums up the familiar contrast thus: 'It was by the German writers of the 18th c. that the distinction between "genius" and "talent", which had some foundation in French usage, was sharpened into the strong antithesis which is now universally current, so that the one term is hardly ever defined without reference to the other. The difference between *genius* and *talent* has been formulated very variously by different writers, but there is general agreement in regarding the former as the higher of the two, as "creative" and "original", and as achieving its results by instinctive perception and spontaneous activity, rather than by processes which admit of being distinctly analysed.' Carlyle's definition of *genius* as meaning 'transcendent capacity of taking trouble, first of all' provoked from Samuel Butler the comment that 'it might be more fitly described as a supreme capacity for getting its possessors into trouble of all kinds and keeping them therein so long as the genius remains'.

**talus.** Pl. of the word meaning ankle etc., *talī*; pl. of the word meaning slope etc., *taluses*. The first comes to us direct from Latin, the second through French.

**-t and -ed.** Some verbs that once had an alternative -t form for past tense and participle have now lost it, either wholly or to an extent that makes it archaic, e.g. curst, dropt, husht, kist, stopt, tost, and whipt. For many others -t is now the only formation, e.g. crept, dealt, felt, kept, left, meant, slept, swept. Typical of the words that have preserved both alternatives are bereave, burn, dream, kneel, lean, leap, learn, smell, spell, spill, and spoil.

In the first edition of this dictionary an attempt was made to assess the comparative popularity of the two endings when used in print by counting the number of times they occur in all OED quotations of the 19th and 20th cc. The first figure after each word is the number for -ed, the second for -t.

1. Preferring -ed.

> dream—5, 3. (See also DREAM.)
> kneel—3, 2.
> lean—12, 2.
> leap—7, 5.
> learn—5, 0. (See also LEARN.)
> spoil (mar)—9, 5.

2. Equal.

> bereave—3, 3. (See also BE-
> REAVED.)
> spell—4, 4.

3. Preferring -t.

> burn—7, 16. (See also BURNT.)
> smell—2, 8.
> spill—8, 17.

So far as we can accept these figures as a guide, it would seem that the -ed forms then still prevailed in print; if the past tense were distinguished from the past participle, the preponderance of -ed for it would have been slightly greater. But there has since been a movement (advocated in the original edition) towards -t. *Lean* is the only word in the first group for which the COD still puts -ed before -t; for *kneel* it admits no alternative to *knelt*. See also BLESSED.

**tapis.** See CARPET.

**target.** After the second world war this word was much used to express the quantitative result hoped for from some enterprise such as the output of a manufacturing concern or the amount subscribed for some public or charitable purpose. To an exceptional degree it has shared the experience of most popular metaphors of being 'spoilt' (see METAPHOR 2 C) by use in a way flagrantly incongruous with its literal meaning. Targets, it is said, must be 'pursued vigorously'; to be 'within sight' of one and to 'keep fully abreast' of it are, it seems, positions that practically guarantee success, and when a target is 'doubled' the implication that it will be twice as difficult to hit goes unquestioned. But then, as Lord Conesford has remarked, 'those who thus describe their ambitions never seem to entertain the faintest hope of actually hitting their targets, even when these are overall or even global ones; in their most optimistic moods they speak of "reaching" or "attaining" the target, an achievement which, since the bow and arrow went out of use, has never been rated very high'.

**tart, pie.** The current distinction is roughly that a tart always contains fruit or sweet stuff, and a pie usually meat or savoury stuff; but the earlier distinction was that a pie was closed in with pastry above and a tart was not; and as relics of the old use we retain *mince pie* as the only possible form, and *apple pie* as an attribute of *order* or *bed*, and *cherry pie* as the name of a flower. There is a similar distinction between *pasty*, which is enclosed and usually contains meat, and *flan*, which is open and usually contains fruit.

**tartan.** See PLAID.

**tasty** has been displaced, except in uneducated or facetious use, largely in its primary sense by *savoury* and wholly in its secondary by *tasteful*.

**tattler.** Now so spelt; formerly, and especially in the name of the 18th-c. periodical, *tatler*, a spelling preserved in the modern periodical of that name.

**tattoo.** See -ED AND 'D.

**tautology** (lit. 'saying the same thing', i.e. as one has already said) is a term used in various senses.

1. To repeat the words or the substance of a preceding sentence or passage may be impressive and a stroke of rhetoric, or wearisome and a sign of incompetence, mainly according as it is done deliberately or unconsciously. In either case it may be called tautology (though the word is in fact seldom used except in reproach), but it is with neither of these kinds that we are here concerned. Another sense is the allowing of a word or phrase to recur without point while its previous occurrence is still unforgotten. This kind of *t.* will be found

fully discussed in the articles JINGLES, REPETITION and ELEGANT VARIATION; it is of great importance as an element in style, but need not here be treated again. Another form of *t.* is that dealt with in PLEONASM 2, in which synonyms, either capable of serving the purpose by itself, are conjoined, as in *save and except*. Again, the word is sometimes applied to identical propositions such as 'I don't like my tea too hot', or 'There is no need for undue alarm'; for such statements see the *truism* section of COMMONPLACE.

2. Another form is the addition of what have been called 'abstract appendages' to words that have no need of them, such as *weather conditions* for *weather*, *temperature values* for *temperatures*, *height levels* for *heights*. Sometimes these appendages are merely otiose; sometimes they are misused by people who do not know that when properly used they give the compound a meaning different from that of the single word, as when an amateur psychologist, wanting to show off, refers to someone's *behaviour pattern* when he means no more than *behaviour*. A curious recent example is the use of *time-scale* for *time*. An example of the right use of *time-scale* is *Most of these discoveries have thrown light on the background of the Old Testament. With the New Testament archaeology has been less helpful. The time-scale is so much shorter—a century compared with two millennia or more*. An example of its wrong use is *Within the time-scale of the next eight to ten years the weapons system that will be carried by British aircraft will remain valid*. This seems to be no more than a cumbrous way of saying that the type of weapons carried in British aircraft is not likely to change much in the next ten years. *Within the time-scale of* is a striking example of a COMPOUND PREPOSITION; the word needed was *in*. In *He did not think it possible to build such an aeroplane in the same time-scale*, *t.-s.* seems to mean no more than *time*.

3. What remains to be illustrated here is the way in which writers who are careless of form and desirous of

emphasis often fail to notice that they are wasting words by expressing twice over in a sentence some part of it that is indeed essential but needs only to be expressed once. It is true that words are cheap, and, if the cost of them as such to the writer were the end of the matter, it would not be worth considering. The intelligent reader, however, is wont to reason, perhaps unjustly, that if his author writes loosely he probably thinks loosely also, and is therefore not worth attention. A few examples follow, and under BOTH 2 and EQUALLY AS 2 will be found collections of the same kind of *t.*: *The motion on constitutional reforms aims at placing women on* the same equality *with men in the exercise of the franchise* (As no other equality has been in question, *same* and *equality* are tautological; *in the same position as*, or *on an equality with*). | *May I be permitted to state that the activities of the Club are not limited only to aeronautics?* (*Limited* and *only* are tautological; *limited to*, or *directed only to*). | *It is sheer pretence to suppose that speed* and speed alone *is the* only *thing which counts* (Omit either *and speed alone*, or *only*). *He said that only one* additional *train had been* added *from Cannon Street during the rush hour to the restricted war-time service* (Omit *additional*).

**tax, duty** and some synonyms—*cess, contribution, customs, due, excise, impost, levy, rate, toll, tribute*. With such sets of words it is often convenient to have a conspectus of the distinctions and be saved the labour of turning them up for comparison in separate dictionary articles. Such convenience is all that is here aimed at, a rough definition of each word being given after some general remarks on the words *tax* and *duty*.

Historically a *tax* was a direct charge on a taxpayer which bore some relation to his ability to pay and was imposed for revenue only; a *duty* was an indirect charge, levied on transactions or commodities and sometimes imposed primarily for political or economic reasons. But there is no

longer any clear distinction between the two in Britain. Few taxes are so called specifically; the most notable are *income tax* (including *surtax*), *profits tax* and *purchase tax*. The rest are mostly *duties*, e.g. *customs, excise, estate, stamp* and the various *licence duties*. Some, though their statutory designation may be *duty* (e.g. *entertainments, petrol*), are often popularly referred to as *tax*. What were virtually the same taxes were officially called *Excess Profits Duty* in the first world war and *Excess Profits Tax* in the second, and later *Excess Profits Levy*. As a generic term *tax* is the widest, embracing not only the imposts specifically called *duties*, but also, in the phrase *local taxation*, the *rates* imposed by local authorities. At the same time *duties* as a generic term is sometimes used to include those specifically called *taxes*: the statutory expression *Inland Revenue Duties* includes income tax as well as stamp duty and estate duty. The tangle is past unravelling.

*Cess*, another word for local taxation, obsolete in England but still current in Scotland and Ireland.

*Contribution* was the word originally used for the *Profits Tax* (first called *National Defence Contribution*) imposed during the second world war, and for the single-year tax on capital (*Special Contribution*) exacted after it. The word was no doubt chosen as a euphemism (cf. the old *benevolence*) in the hope that payment of it might be thought of as a privilege rather than an obligation; even when the *National Defence Contribution* was rechristened *Profits Tax* by statute after the war, Parliament considerately added a proviso that it would not be unlawful to continue to refer to the tax by its old name.

*Customs*, payment levied upon imports from foreign countries.

*Due*, any obligatory payment, the nature being usually specified by an attributive noun, as *harbour, market, dues*.

*Excise*, duty charged on certain home products, especially alcoholic liquors, before they can be sold.

*Impost*, a generic term for any compulsory payment exacted under statutory authority.

*Levy*, exaction from every person concerned of an equal amount or an amount proportional to his property. Generally of a single non-recurrent impost but now used more loosely; see note on *tax* and *duty* above.

*Rate*, amount of assessment on property for local purposes; see also the note on *tax* and *duty* above.

*Toll*, fixed charge for passage over bridge, ferry, etc., or for permission to sell in a market.

*Tribute*, periodical payment made by one State to another in token of submission (now called, less bluntly, reparations) or as price of protection.

**teasel, teazle.** The first is the standard form.

**techy.** See TETCHY.

**teem.** The word meaning to be prolific and that meaning to pour out are of different origins. There is some natural confusion between them, but it may be presumed that when we say *the river is simply teeming with fish* we are using the first word, and the second when we say *it is simply teeming with rain*.

**teenage(r).** We have given a warm welcome to these Americanisms, no doubt because we felt the need for a suitable and colourless word for what the pseudo-scientific jargon fashionable today calls members of the 13–19 age-bracket. *Juvenile* is tainted by its association with *delinquent* and *court*. *Adolescent* is a starchy word, also faintly tendentious. *Youths* are of one sex only. *Young persons*, the statutory expression, is prim, and unsuitable now that we are in no danger of feeling, as Mr. Podsnap did, that the question about everything was whether it would bring a blush to the cheek of the young person. *Teenage(r)* seemed to fill a gap. Unfortunately it also seems to be acquiring an overtone of disparagement; the fact is that teenagers as a class are rarely in the news except when they misbehave. So it is pleasant to be occasionally reminded that they have their virtues too. *Teenagers tend to be more generous in giving presents than their parents, and are spending as much on gifts as on their own clothing, says an interim report of the Trustee Savings Banks Association.*

**teetotaller,** but *teetotalism*; see -LL-, -L-.

**tele-.** Inevitably, in these days of annihilation of distance, this Greek prefix has been used with a freedom that makes it a prolific source of what are called in this book BARBARISMS. It began respectably enough with *telescope, telegraph, telegram* (though some purists would have had us say *telegrapheme*), *telepathy* and *telephone*, but is now promiscuous in its attachments, e.g. *television, teleprinter, telecommunications, telecontrol*, and others constantly multiplying. Indeed, it is time to recognize that *tele-* (like *anti-, post-,* and *pre*) has gatecrashed into our vocabulary, and, being now naturalized, is free to associate without offence with any other member. *Televise* (a BACK-FORMATION from *television*) has *s* not *z*. See -IZE, -ISE.

**temerarious.** 'Now only literary'—OED; see LITERARY WORDS.

**temperature.** See FEVERISH.

**templet, -plate.** The *-ate* form, due to false association with *plate*, as in *wall-plate* etc., has won an undeserved victory over the better form *-et*.

**temporal, temporary.** The meaning of *temporal* was originally the same as that of *temporary* but was later restricted to the sense 'of or pertaining to time as the sphere of human life, terrestrial as opposed to heavenly'. —OED. Hence the distinction between the temporal (civil) and the spiritual (ecclesiastical) authorities, and the division of the House of Lords into Lords Temporal (lay peers) and Lords Spiritual (bishops).

**ten-.** That the value given to an English vowel is etymologically a FALSE QUANTITY is not necessarily anything against it. But since most of the words derived from Lat. *teneo* (*tenant, tenon,* etc.) preserve the short *e* of that verb there seems no good reason for the long one we sometimes hear in *tenable* and *tenet.*

**tend** (= *attend*). *Dr. Hutton has written an interesting account of the Eskimos of Labrador, among whom he has lived for some years past* tending *to* their *needs in his hospital.* Since this verb *tend* (unlike the one connected with *tendency*) is said to be merely an aphetic form of *attend,* it is remarkable that its construction and that of *attend* should differ. But they certainly do; *tend* one's *needs,* but *attend to* one's *needs*; see CAST-IRON IDIOM. See also TREND.

**tendentious,** a new and useful word (OED dates it from 1900), hesitated at first between *-cious* and *-tious*; the latter is now established.

**tenor.** The form *tenour* is called obsolete by the OED for all senses, though it appears in some of its 19th-c. quotations, especially in the sense course or procedure or purport; see -OUR AND -OR. 'They kept the *even* tenor of their way' is a MISQUOTATION (*noiseless*).

**tenses etc.** Certain points requiring care will be found under SEQUENCE OF TENSES, SUBJUNCTIVE, PERFECT INFINITIVE, AS 4, HAD, LEST, SHALL AND WILL.

**tercentenary.** See CENTENARY.

**teredo.** (*tĕrē-*) English pl., *teredos,* see -O(E)S 6; Latin pl. *terē'dĭnĕs,* see LATIN PLURALS.

**term.** For *major, minor, middle, t.* in logic, see SYLLOGISM.

**terminate.** See WORKING AND STYLISH WORDS.

**terminological.** For *t. inexactitude* see INEXACTITUDE.

**terminus.** Even in the commonest sense of railway *t.,* the plural *terminī*

has long been used; the SOED gave no other. The COD, however, now puts *-uses* first. See LATIN PLURALS. In air travel *terminal* is now the usual word.

**terrain.** The justification of the word is that it expresses a complex notion briefly. When it is used as a substitute for *ground, tract, region,* or *district*—good ordinary words—, it lacks the justification that an out-of-the-way word requires, and becomes pretentious. It means a piece of ground with all the peculiarities that fit or unfit it for military or other purposes; and to speak of 'the peculiarities of the *t.*', 'the nature of the *t.*', etc., instead of simply 'the *t.*', is a pleonasm, though the readers' assumed ignorance may excuse it.

**terribly.** *The other day (said I) I read a love scene in a story that went like this: 'Am I beautiful?' she asked him. 'Terribly' he said. And then he asked her 'Do you love me?' 'Horribly' she said. Why (I was then asked) don't you go home and write something humorous. Don't you want to? 'Frightfully' I replied.* (James Thurber).
It is strange that a people with such a fondness for understatement as the British should have felt the need to keep changing the adverbs by which they hope to convince listeners of the intensity of their feelings, until, by a process of exhaustion, they have arrived at such absurdities as these, to which might be added *dreadfully* and *fearfully.* The early ones of the series, such as *consumedly, excessively, mightily, prodigiously,* and *vastly,* however hyperbolical, were reasonable enough to use of pleasurable emotions. The downward path began with *awfully,* a word now so worn with use as to be reduced to the level of *very.* The Greeks, as the OED reminds us, used their word for it (δεινῶς) in the same way.

**tertium quid.** 'A third something.' Originally something which results from the combination of two things but is itself different from both, with properties not so well ascertained as those of its elements. In this sense an alloy, or a chemical compound, or a chord ('not a fourth sound, but a star') might be called *t. q.* It was later used in the changed sense (the notion of unknown qualities being lost) of another alternative, a middle course; as the Liberal Party might be called a *t. q.* between the Conservatives and the Socialists, or as Browning used it for the heading of Book IV of *The Ring and the Book*, in which views are set forth about the culpability of the murderer Guido which are neither those of the 'Half Rome' who have defended him in Book II nor those of the 'Other Half Rome' who have condemned him in Book III. A new twist was given to the expression when Kipling wrote a story beginning *Once upon a time there was a Man and his Wife and a Tertium Quid*; and the meaning the third party in 'the eternal triangle' is now likely to be suggested by it to the popular mind.

**tessera.** Pl. *-rae.*

**test.** For synonyms see SIGN.

**testatrix.** For pl. see -TRIX.

**test match.** The expression is strictly applicable only to cricket. It was originally used of representative matches between England and Australia, but is now treated as including representative cricket matches between any two countries played under the rules of the Imperial Cricket Conference.

**te(t)chy, touchy.** In the sense irritable, over-sensitive, the OED suggests that *touchy* is perhaps an alteration of *techy*; *techy* (or *teachy*) is the oldest recorded form, but *tetchy* is the usual modern spelling of those who do not prefer *touchy*. As the origin of *te(t)chy* is uncertain, and the much commoner *touchy* gives the same meaning without being a puzzle, any attempt to keep *te(t)chy* alive seems due to a liking for curiosities, especially as *testy* is also available.

**tether.** For synonyms in the fig. sense see FIELD.

**thalamus.** Pl. *-mī.*

**than.** 1. *T.* after *prefer(able).* 2. *T* and inversion. 3. Infinitive or gerund after *rather t.* 4. *Hardly* and *scarcely t.* 5. *T.* after *the more* etc. 6. *T.* as strong conjunction, as weak conjunction, and as preposition. 7. Double standard of comparison. 8. *T.* after non-comparatives. 9. *T.* and ellipsis. 10. Flounderings.

1. For *t.* after *prefer* and *preferable* without *rather*, a common solecism, see PREFER(ABLE) 3.

2. *T.* and inversion. *The evidence could not now be given in the same sense, any more* than could Mr. Chamberlain's speeches *of 1903 be now delivered.* |*The visit will be much more direct in its effect upon the war* than could be any indiscriminate bombing *of open towns.* Such inversions are deprecated; see INVERSION (after relatives and comparatives).

3. Infinitive, or gerund etc., after *rather t.* *They were all in favour of 'dying in the last ditch'* rather than sign *their own death-warrant.* The justification of *sign* instead of *signing* is discussed in -ING 5 (c).

4. *Hardly t., scarcely t. But hardly had I landed at Liverpool than the Mikado's death recalled me to Japan.* Read *no sooner* for *hardly*, or *when* for *than*; and see under HARDLY 2, SCARCELY 1.

5. *T.* after *the more, the less,* etc. *If we simply take the attitude of accepting her theory of naval policy, we make it so much the less probable that she will change her law* than *if we enter into violent contention.* See THE 6 for the wrongness of this construction.

6. *T.* as strong conjunction, weak conjunction, and preposition. In *You treat her worse than I treat her*, *t.* is a strong or subordinating conjunction, attaching an adverbial clause to its owner *worse.* *You treat her worse than I* may also be so described with the explanation that there is an ellipsis of *treat her*; or, alternatively, *t.* may be called a weak or coordinating

conjunction linking the two similarly constructed nouns *you* and *I*. In *You treat her worse than me*, the same two names for *t*. are possible, but the ellipsis is of *you treat*, and, in the alternative, the linked nouns are *her* and *me*; those are the possibilities if the sentence is said with the only sense that an educated person gives it. But an uneducated person may mean by it You treat her worse than I treat her; and, if it is to be so taken, *t*. is not a conjunction of either kind, but a preposition governing *me*. Doubts whether a word is a preposition or a conjunction or both are not unknown; see, for example, BUT I with regard to such phrases as *all but he* (conj.) and *all but him* (prep.); usage, moreover, changes in such matters with time. It is obvious, however, that recognition of *t*. as a preposition makes some sentences ambiguous that could otherwise have only one meaning (*I would rather you shot the poor dog than me*), and is to that extent undesirable. The OED statement on the preposition use is that, with the special exception of *t*. *whom*, which is preferred to *t*. *who* unless both can be avoided, 'it is now considered incorrect'. That so-called incorrectness occurs in the following examples, where, by the strict application of the OED ruling, *us*, *him*, and *them* should be *we*, *he*, and *they*. But the prepositional use of *than* is now so common colloquially (*He is older than me*; *they travelled much faster than us*) that the bare subjective pronoun in such a position strikes the reader as pedantic, and it is better either to give it a more natural appearance by supplying it with a verb or to dodge the difficulty by not using an inflective pronoun at all. Examples: *He could do worse than consider why West Germany can put up 210,000 more houses a year than us* (than we can). / *The butcher of the last few months has been a good deal more obliging than him of the war period* (than the one). / *Do not let us split up our energy by having more than one society; the idea is more than them all* (than all of them).

On the other hand, the subjective *he*s that follow had better have been *him* on the weak-conjunction principle, especially since the ellipsis required for the strong-conjunction explanation is awkward in the first two: *If ever Captain O'Connor gives us a second volume, we beg him to engage no other artist* than he *who illustrated the first*. / *The Entente had no better friend* than he *on the other side of the Atlantic*. / *'I've foiled better men* than he' *the Count answered*.

7. Double standard of comparison; *more and more t.* A ludicrous example of conflicting *than*s, which almost any reader would detect, is: 'I have *less* confidence *than* Mr. Orr in the valuers being obliged to adopt his method of valuation *than* that we all shall be compelled to adopt theirs'. *Less* is clearly unequal to its two jobs; it can put Mr. Orr in his place with regard to me, or the valuers with regard to us, but not both. Such a freak sentence would not be worth quoting but for the light it throws on a less flagrant but more frequent absurdity of the same kind, the following of *more and more* with *than*: *My eyes are more and more averse to light than ever*. / *The order has gradually found more and more room for educational and learned work than was possible in the early centuries*. Both sentences would be right if *and more* were omitted; but the introduction of it implies the tacit introduction of other *than*s which conflict with those that are expressed. *More and more* means more yesterday than the day before, and more today than yesterday; to combine that shifting date with the unshifting dates *ever* and *in early centuries* is impossible. *T*. should never be used after *more and more*.

8. *T*. after non-comparatives. *Else*, *other*, and their compounds, are the only words outside true comparatives whose right to be followed by *t*. is unquestioned; and 'true comparatives' is to be taken as excluding such Latin words as *superior* and *inferior*, *senior* and *junior*, all of which, as well as *prefer(able)*, require not *t*., but *to*. The use of *t*., on the analogy of *other t*., after *different*, *diverse*, *opposite*, etc., is 'now mostly avoided' (OED). Some

examples follow of irregularities that should not be allowed to appear in print: *What then remains if this measure of disagreement still continues than to dispose of the Bill by fair discussion?* (For *what . . . than* read *what . . . but* or *what else . . . than*). | *There is obviously a vastly increased number of people who can and do follow reasoned arguments . . . than there was before educational methods were so efficient.* (For *increased* read *greater*.) | *It would be doubly difficult to influence the South African Government if that country walks out of the Commonwealth than if it stays in even as a republic.* (Read *much more difficult . . . than* or *twice as difficult . . . as*.) | *A mystery virus has filled children's hospitals in Birmingham. Four times as many babies than usual have been affected.* (For *than* read *as*.) | *One would have been readier to consider this if he had given chapter and verse for his astonishing claim that Delius is performed twice as much nowadays than he was ten years ago.* (For *than* read *as*.) But *different than* is sometimes preferred by good writers to the cumbersome *different from that which* etc., as it was by Richardson when he wrote *A very different Pamela than I used to leave all company and pleasure for.* Modern examples are: *He is using the word in quite a different sense than he did yesterday.* | *The changes in the Commonwealth since the war ended . . . have left Great Britain in a very different position than she was in 1938.* | *The air of the suburb has a quite different smell and feel at eleven o'clock in the morning than it has at the hours when the daily toiler is accustomed to take a few hurried sniffs at it.*

**9.** *T.* with ellipsis or brachylogy. Some kinds of ellipsis are so customary in the member of a sentence beginning with *t.* that to write out the whole sense would be much more noticeable than the ellipsis (cf. the use of *different than* illustrated in para. 8). But hasty writers are encouraged by this to think that any slovenliness will pass muster: *The power of this great combine is now greater than any of the nationalized industries* (than that of

any). For other examples and discussion see ELLIPSIS 5.

**10.** Flounderings. There is often a difficulty in getting the things to be compared into such grammatical conformity as will enable them to stand on either side of a *than*, but writers who take so little trouble about it as the authors of the following sentences must not be surprised if their readers are indignant: *In countries where a Referendum is a recognized part of the constitutional machinery, the House of Representatives* is much more ready to pass, *provisionally, constitutional reforms, and submit them to the electorate,* than are Bills passed *by the Houses of Parliament in a country like ours.* | *'The Awkward Age', which was just published, was being received with a little more intelligence and sympathetic comprehension* than had been the habit of greeting *his productions.*

**-th and -dh.** Monosyllabic nouns ending in *-th* after a vowel sound (including *-ar-* etc.) differ in the pronunciation of the plural. Those only need be considered whose plural is in regular use, which excludes *sloth, broth, ruth,* and many others. The common words *lath, mouth, oath, path, sheath, truth,* and *youth,* all sound the plural as -dhz, not -ths; but the equally common words, *berth, birth, breath, death, fourth, girth, growth, smith,* and *myth,* have -ths in sound as well as in spelling. Others again are doubtful; such are *bath, cloth,* and *wreath* (-dhz recommended), and (with -ths usual) *heath, hearth, moth,* and *wraith.* Cf. the article -VE(D). It may be added that the verbs or verbal nouns connected with *bath, breath, cloth, mouth, sheath, teeth,* and *wreath,* have the dh sound (*bathe, breathe, clothe, mouthing, sheathe, teething, wreathe*); cf. also *smithy, worthy, northern,* and *southern* all with -dh-.

**thank you, thanks,** etc. *I thank you* is now reserved for formal occasions or tongues; *thank you* is the ordinary phrase, but tends more and more to be lengthened with or without occasion into *thank you so much,* or *thank you* *very much*, often with the addition of *indeed* for good measure. *Thanks* is a shade less cordial than *thank you*, and *many* and *best* and *a thousand thanks* and *thanks awfully* are frequent elaborations of it; *much thanks* is archaic, surviving through our familiarity with Francisco's *For this relief much thanks*, and now only used jocularly. The colloquial variant *Thanks a lot* is becoming popular. If an acknowledgement of thanks is felt to be needed it will be *Don't mention it*, or *Not at all* or, in U.S., *You're welcome*.

**that**, adj. and adv. 1. *That* = such a, so great a, to such an extent. 2. *That* with noun and participle. 3. *At that*.

1. The adjectival use (*He has* that *confidence in his theory that he would act on it tomorrow*) was formerly normal English, and survives colloquially, but in literary use *such a*, *so great a*, etc., are substituted. The adverbial use ('to that extent or degree, so much, so, esp. with an adv. or adj. of quantity, *that far*, *that much*, *that high*') was said by the SOED to be dialect and U.S. That would now be too round a statement. When qualifying another adverb (or adverbial phrase), the adverbial *that* (or in negative and interrogative sentences *all that*) has become a common colloquialism now well on its way to literary status. *Whatever the Court of Appeal or the House of Lords, assuming the case goes that far, may decide.* | *A side effect of this practice is that the parent flycatcher has to work that much harder.* | *The figures show that even Lazards do not sell £2m. all that frequently.* | *Sometimes the law is not regarded with all that favour by those unaccustomed to it.* When qualifying an adjective, the use of *that* merely as an intensive adverb (*I was that angry*) is still a vulgarism. But it is sometimes so used to indicate a comparison, though in an affirmative sentence *so* or *as* is more idiomatic. *Only half a dozen close advisers, if that many, were privy to his inmost thoughts.* | *But once a girl has chosen her career, is it that easy for her to pursue it?* | *Even the wildfowl are*

*not all that wild.* | *Today the situation is very different. It is American stubbornness that keeps China out of the U.N. But is it all that different?* In the other colloquial use, meaning etcetera, and generally implying a touch of contempt (*1066 and all that*), *that* is not an adverb but a pronoun. See also THIS 2.

2. *T.* with a noun and a participle or other equivalent of a defining relative clause. The type meant is shown in *that part affected*, *that land lying fallow*, *that theory now in question*, and the contention is that it is a bad type. In the OED there is a solitary example, and that justifiable for special reasons; but in modern newspaper use it is very common. Four specimens are: *It was essential that both these phases of his art should be adequately represented in* that branch of the National Gallery devoted *to native talent.* | That part relating *to the freedom of the seas was given fairly fully in the 'Times'.* | *Aphorisms and maxims are treated with* that respect usually reserved *for religious dogma.* | *Shorter hours in all departments of labour prevent* that expeditious handling of cargoes needed.

The use of *that* (demonstrative adjective) with the sole function of pointing forward to a defining relative clause is established English, and 'that part which concerns us' is as common as 'the part that concerns us'; but when a participle or phrase is substituted for the relative clause it is an innovation to keep the *that*; it may safely be said that most good writers take the trouble to clear away the now needless *that*, and write *the* instead. The full form should have been *that branch which is devoted* (or *the branch that is devoted*), and the short form *the branch devoted*; and similarly for the rest.

It should be observed that sentences occur similar at first sight to those condemned, but with the difference that another purpose is served by *that* instead of or as well as that of heralding the participle etc. One such is the OED quotation already referred to: *On that peninsulated rock called La Spilla*; here *that* is justified as meaning 'the well-known'. Compare also: *The world needs peace. You will always find us at your side to preserve that* peace bought *by so much blood.* Here the justification of *that* is its referring back to the *peace* of the previous sentence. So too we might say of the similar use of *that* in conveyancing (*all that land situated* etc.) that it not only heralds the participle but also refers back to the preceding negotiations.

The misuse here objected to is still commoner in the plural; see THOSE 2.

**3.** *At that. The only seats available were at the back, and very uncomfortable at that.* The convenience of this idiom, said by the OED to have been originally U.S. colloquial or slang, may be judged by the elaborate definition there given of its significance: 'even when that has been taken into consideration; estimated at that rate, at that standard, even in that capacity, in respect of that, too; "into the bargain"'.

**that,** conj. 1. Kinds of clause attached by *t.* conj. 2. Substantival clauses without *that.* 3. *T.* and *whether* with *doubt(ful).* 4. Interim *t.* 5. *T.* and *as* after (*in) so far.* 6. Non-parallel *t.*-clauses in combination.

**1.** Kinds of clause attached by *t.* conj. In adjectival or relative clauses that begin with *t.* it is a relative pronoun, not a conjunction; see for these the next article. *T.* conj. attaches a substantival clause to the verb, noun, etc., to which it is object (*I hear that he is dead*), subject (*T. pain exists is certain*), in apposition (*The fact t. pain exists*), etc.; or else an adverbial clause to the word etc. modified (*The heat is such that it will boil water*). The only point needing to be insisted on is that in either case, whether the *t.*-clause is substantival or adverbial, the sentence out of which it is made by prefixing *t.* must be in statement form, not a question, command, or exclamation. Sentences of those other kinds can be subordinated or turned into clauses, but not by prefixing *t.* The mistake is not made by good writers, but occurs often enough to need mention.

One way of avoiding it is to rearrange the sentence so that there is unsubordinated quotation of the question etc., and the other is, before subordinating, to convert the question etc. into a statement giving the same meaning. Two examples will be enough, the first of an impossible adverbial clause and the second of an equally impossible substantival. *Crises, international or national, arise so rapidly in these days* that who can say *what a few years may bring forth?* | *One can only comment that if such a refuge was open to the Romans,* how much more available is it *to our own people.* Such sentences can be remedied in either of the ways mentioned above. Either *Crises . . . arise so rapidly in these days; who can say* etc., or *Crises . . . arise so rapidly in these days that no one can say* etc. Either *One can only comment: if such a refuge was open . . . how much more available is it* etc., or *One can only comment that if such a refuge was open . . . it is much more available* etc.

**2.** Substantival clauses without *that* (already touched on in ELLIPSIS 4). *I know that my Redeemer liveth: I know I can trust you.* These are equally good English; if *that* were shifted from the first to the second, both would still be grammatically correct, but each less idiomatic than as it is: the use or omission of the *t.* of a substantival clause depends partly on whether the tone is elevated or colloquial. But a glance at the following examples of obviously wrong omission will show that there is not free choice after all verbs or in all constructions: *The Italian Olympic Committee announced all officials, regardless of nationality, will wear the same uniform.* | *Looking back, I observe the pattern of the track left behind wanders sometimes uncertainly.* | *That the suggestion has been made by two resident undergraduates indicates little opposition may be expected.* | *We learn he used his cousin as a bait.* | *He urged an international conference be held.* | *The enormous rents which would be asked for new houses would naturally render owners of existing* *properties restless and envious,* with the result they would *continually strive to raise their own rents to a similar level.* / *He said a* major reason was *the Western peoples brought with them a laxity in moral standards.* It at once occurs to the reader that *announce, observe, indicate, learn, urge* are words that stand on their dignity and will not dispense with the attendance of *t.* The lesson of the last two examples is that omission is unadvisable when the substantival clause is in apposition to a noun, as here to *result,* and *reason.*

It may be useful to give tentative lists, to which everyone can make additions for his own use, of verbs that (1) prefer *t.* expressed, (2) prefer *t.* omitted, and (3) vary according to the tone of the context. (1) *T.* is usual with *agree, announce, argue, assume, aver, calculate, conceive, contend, hold, indicate, learn, maintain, observe, reckon, remark, state, suggest;* (2) *T.* is unusual with *believe, dare say, presume, suppose, think;* (3) *T.* is used or omitted with *be told, confess, consider, declare, grant, hear, know, perceive, propose, say, see, understand.* The verbs with which the question may arise are many more than these few, which may however be enough to assist observation. It should be added that the tendency is to omit *that,* and some of the words in the first list may be thought to have become eligible for transfer to the third. Perhaps this is due to U.S. influence, where *that* is omitted much more freely than it is here. It seems clear, from some of the examples given above, that this is having an effect on British journalism.

3. *T.* and *whether* with *doubt(ful).* *It gave him cause for wonder that no serviceable* [petroleum] *'pool' had been revealed in England; that any existed, however, seemed doubtful, for clearly* . . . The choice allowed by idiom is between *Whether any existed seemed doubtful,* and *That any existed seemed unlikely,* according to the shade of meaning required. See DOUBT(FUL).

4. Interim *t.* A writer often embarks on a substantival *t.*-clause only to find that it is carrying him further

than he reckoned, and to feel that the reader and he will be lost in a chartless sea unless they can get back to port and make a fresh start. His way of effecting this is to repeat his initial *t.* This relieves his own feeling of being lost. Whether it helps the inattentive reader is doubtful; but it certainly exasperates the attentive reader, who, from the moment he saw *t.* has been on the watch for the verb it tells him to expect, and realizes suddenly, when another *t.* appears, that his chart is incorrect. These interim *that*s are definite grammatical blunders, which can often be mended by leaving out the offending *t.* with or without other superfluous words; in the examples below the omittenda are bracketed. The first two show the most venial form of the mistake, the resumptive *t.* being inserted at the point from which progress to the expected verb is not to be again interrupted by subordinate clauses; the others are worse: *Sir F. S. described the resolution as disastrous for the party, and suggested that, as it had been already made known publicly,* [that] *it should not be proceeded with.* / *It would be unwise to assume that, merely because Mr. Khrushchev has announced his intention of doing so,* [that] *Mr. Macmillan will follow suit.* / *We can only say that if the business men who read the* Times *are really of opinion that this is a sensible procedure, and* [that,] *if they find any satisfaction whatever in the writing down of a huge sum which everybody knows can never be recovered, they will have only themselves to thank if the politicians continue to make game of them.* / *It should be borne in mind that, whilst many things have increased in cost, and* [that] *therefore the value of the £1 has decreased, there are many items of expenditure which have not increased in anything like the same proportion.* / *It has been shown that if that inheritance be widening, as it is, and* [that if] *the means of increasing it exist, as they do, then growth of numbers must add to the power.*

Another sentence is appended as showing not indeed an interim *t.*, but mistakes curable by the same method

of excision. *The Minister added that
there was no need to say that the Government knew nothing about these statements, still less* [*that it*] *had authorized
them, or* [*that it*] *knew what amount of
truth there might be in them.* If the
writer wishes to keep his *that*s, he must
correct *had authorized* into *had not
authorized*, and *knew* into *did not know*;
the repetition of *t.* has lulled him into
the state in which *yes* and *no* mean the
same thing.

See OVERZEAL for other examples of
needless 'signpost' words.

**5.** *That* and *as* after (*in*) *so far*. For
the rather elusive distinction, and its
importance, see FAR 4, 5, and IN SO
FAR.

**6.** Non-parallel *t.*-clauses in combination. Parallel *t.*-clauses can be
strung together *ad libitum*, and may be
rhetorically effective. It is otherwise
with interdependent or dissimilar *t.*-
clauses; for the principle see REPETI-
TION. The unpleasantness of the
construction deprecated is sufficiently
shown in: *It is thoroughly in accordance
with this recognition that the people
have rights superior to those of any individual that Mr. Roosevelt is seeking
legislation that will perpetuate the
Government's title to the coal and oil
lands in the public domain.*

**that,** rel. pron. 1. Relation between
*that* and *which*. 2. *That*-ism. 3. *That*
as a relative adverb. 4. Ellipsis of *that*.
5. *That*-clause not close up. 6. One
*that* in two cases. 7. Double government.

1, Relation between *that* and *which*.
What grammarians say should be has
perhaps less influence on what shall be
than even the more modest of them
realize; usage evolves itself little disturbed by their likes and dislikes. And
yet the temptation to show how better
use might have been made of the
material to hand is sometimes irresistible. The English relatives, particularly as used by English rather than
American writers, offer such a temptation. The relations between *that*, *who*,
and *which* have come to us from our
forefathers as an odd jumble, and

plainly show that the language has not
been neatly constructed by a master
builder who could create each part to
do the exact work required of it, neither
overlapped nor overlapping; far from
that, its parts have had to grow as they
could. It might seem orderly that, as
*who* is appropriated to persons, so *that*
should have been appropriated to
things, or again that, as the relative
*that* is substantival only, so the relative
*which* should have been adjectival only.
But we find in fact that the antecedent
of *that* is often personal, and *which*
more often represents than agrees
with a noun. We find again that
while *who* has two possessives (*whose*
and *of whom*), and *which* one (*of
which*), *that* has none of its own,
though it often needs it, and has to
borrow *of which* or *whose*. Such
peculiarities are explicable, but not
now curable; they are inherent in the
relative apparatus that we have
received and are bound to work with.
It does not follow that the use we are
now making of it is the best it is
capable of; and perhaps the line of
improvement lies in clearer differentiation between *that* and *which*, and
restoration of *that* to the place from
which, in print, it tends to be ousted.

A supposed, and misleading, distinction is that *that* is the colloquial and
*which* the literary relative. That is a
false inference from an actual but misinterpreted fact. It is a fact that the
proportion of *that*s to *which*s is far
higher in speech than in writing; but
the reason is not that the spoken *that*s
are properly converted into written
*which*s. It is that the kind of clause
properly begun with *which* is rare in
speech with its short detached sentences, but very common in the more
complex and continuous structure of
writing, while the kind properly begun
with *that* is equally necessary in both.
This false inference, however, tends to
verify itself by persuading the writers
who follow rules of thumb actually to
change the original *that* of their
thoughts into a *which* for presentation
in print.

The two kinds of relative clause, to

one of which *that* and to the other of which *which* is appropriate, are the defining and the non-defining; and if writers would agree to regard *that* as the defining relative pronoun, and *which* as the non-defining, there would be much gain both in lucidity and in ease. Some there are who follow this principle now; but it would be idle to pretend that it is the practice either of most or of the best writers.

A defining relative clause is one that identifies the person or thing meant by limiting the denotation of the antecedent: *Each made a list of books that had influenced him*; not books generally, but books as defined by the *that*-clause. Contrast with this: *I always buy his books, which have influenced me greatly*; the clause does not limit *his books*, which needs no limitation; it gives a reason (= for they have), or adds a new fact (= and they have). There is no great difficulty, though often more than in this chosen pair, about deciding whether a relative clause is defining or not; and the practice of using *that* if it is, and *which* if it is not, would also be easy but for certain peculiarities of *that*. The most important of these is its insistence on being the first word of its clause; it cannot, like *whom* and *which*, endure that a preposition governing it should, by coming before it, part it from the antecedent or the main sentence; such a preposition has to go, instead, at the end of the clause. (*The book about which I spoke to you* must become *The book that I spoke to you about* if *that* is used). That is quite in harmony with the closer connexion between a defining (or *that*-) clause and the antecedent than between a non-defining (or *which*-) clause and the antecedent; but it forces the writer to choose between ending his sentence or clause with a preposition, and giving up *that* for *which*. In the article PREPOSITION AT END it is explained that to shrink with horror from ending with a preposition is no more than foolish SUPERSTITION; nevertheless there are often particular reasons for not choosing that alternative, and then

the other must be taken, and the fact accepted that the preposition-governed case of *that* is borrowed from *which*, and its possessive from *who*. Another peculiarity of *that* is that in the defining clauses to which it is proper it may, if it is not the subject, be omitted and yet operative—see 4 below. Another important point is that non-defining clauses need commas and defining clauses do not: *Jones, whom I saw yesterday, told me*, but *The man that I saw yesterday told me*.

The following roman type sentences (or parts of sentences) are re-writings, in conformity with the account already given of the difference between *that* and *which*, of verbatim extracts from newspapers; the original is inserted in italics below each, and the reader is invited to compare the two versions and to say whether, even apart from the grammatical theory here maintained, the re-writings do not offer him a more natural and easy English than the originals. Where the reason for the change is not at once obvious, a note is added. But it will save repetition to state shortly here what is explained more fully under WHICH WITH AND OR BUT, namely that a defining and a non-defining clause, whether *that* is used in both or *which* in both, or *that* in one and *which* in the other, ought not to be coupled by *and* or *but* as if they were parallel things.

a. The Bishop of Salisbury is the third bishop that his family has given to the world.

*The Bishop of Salisbury is the third bishop* which *his family has given to the world*.

b. Visualize the wonderful things the airman sees and all the feelings he has.

*Visualize the wonderful things the airman sees and all the feelings* which *he has*. Two *that*s, one *that*, or no expressed relative (= a suppressed *that*) will do equally well.

c. It seems that the Derna, which arrived safely, was sent in the ordinary way.

*It seems that the Derna* that *arrived*

safely *was sent in the ordinary way*. The defining *that*-clause would be right only if there were several Dernas, of which only one arrived safely.

d. Among the distinguished visitors the Crawfords had at Rome was Longfellow.

*Among other distinguished visitors which the Crawfords had at Rome was Longfellow.*

e. Even in the cathedral organ-loft there are grievances that flourish and reforms that call for attention.

*Even in the cathedral organ-loft there are grievances which flourish and reforms that call for attention.* The change from *which* to *that* is mere ELEGANT VARIATION; even two *which*s would be preferable.

f. A hatred of the rule that not only is unable to give them protection, but strikes at them blindly and without discrimination.

*A hatred of the rule that is not only unable to give them protection, but which strikes at them blindly and without discrimination.* What has caused the change from *that* to *which* here is the writer's realizing that *but that* is somehow undesirable. It is so, because of the repugnance of *that*, mentioned above, to being parted from its antecedent; but the way out is to let the previous *that* carry on for both clauses, a task it is quite equal to.

g. She cannot easily regain control of the threads of culture that she has let drop, which now lie in muddled tangles at her feet.

*She cannot easily regain control of the threads of culture which she has let drop, and now lie in muddled tangles at her feet.* The first clause is defining, and should have *that*; the second may be defining or non-defining, being unessential to the identification and yet capable of being regarded as helping it. Against allowing the *that* to carry on, as in f, there is the objection, disregarded indeed by the writer, that the two relatives are in different cases; it is therefore best to make the second clause non-defining, and use *which*, without *and*.

h. The life-work that Acton collected

innumerable materials for, but never wrote, was a History of Liberty.

*The life-work for which Acton collected innumerable materials but never wrote was a History of Liberty.* Restoration of the defining *that* often solves the difficulty seen here and in the next piece, that of a relative under double government, first by a preposition and then by a verb; the postponing of the preposition, abnormal though possible with *which*, is with *that* not only normal but necessary.

j. You give currency to a subtle fallacy that one often comes across, but does not like to see in one's favourite paper.

*You give currency to a subtle fallacy across which one often comes, but does not like to see in one's favourite paper.*

k. After a search for several days, he found a firm that had a large quantity of them for which they had no use.

*After a search for several days he found a firm which had a large quantity of them and which they had no use for.* Both clauses are defining, and *that* is required; but the relatives have not the same antecedent, and the *and* is therefore wrong (see WHICH WITH AND OR BUT). But there is a legitimate choice between *that . . . for* and *for which*, and the latter gives an escape from one *that*-clause depending on another.

l. There will be a split in the Lutheran Church comparable to the quarrel that has broken out in the Catholic Church on the question of modernism, but seems to have run its course.

*There will be a split in the Lutheran Church comparable to the quarrel that has broken out in the Catholic Church on the question of modernism, but which seems to have run its course.* The second clause may be either defining or non-defining; if defining, *that* (or preferably nothing, cf. f) is required instead of *which*; if non-defining, *but* must be omitted, and *which* kept.

m. The class that I belong to, which has made great sacrifices, will not be sufferers under the new plan.

*The class to which I belong and*

which *has made great sacrifices will not be sufferers under the new plan.* Defining and non-defining wrongly coupled; omit *and*, and prefer (*that*) *I belong to* to the equally legitimate *to which I belong* as better both in clearness and in sound.

n. All honour to these men for the courage and wisdom they have shown, which are of infinitely greater value to the country than . . . .

*All honour to these men for the courage and wisdom they have shown,* and *which are of infinitely greater value to the country than. . . .* The second clause is clearly non-defining; the *and* should go.

2. *That*-ism. As has been explained, the tendency in modern writing is for *which* to supersede *that* even in the functions for which *t.* is better fitted. On the other hand some writers seem deliberately to choose *that*, where most other people would use *which*, under the impression that its archaic sound adds the grace of unusualness to their style. A few examples will show *that* in non-defining clauses to be certainly noticeable, and the reader will perhaps conclude that its noticeability is not a grace: *But her fate, that has lately been halting in its pursuit of her, overtakes her at last.* / *This is clearly recognized by Mr. Macfall in his eloquent and well illustrated monograph, that is more than a mere record of the fortunes of its titular subject.* / *Neither . . ., nor . . ., nor . . ., will save the country if the town, that has all the power in its hands, is content to let it die.* / *His arguments on these points were heard by the great audience of business men in almost unbroken silence, that gave place to an outburst of applause when he . . .*

3. *That* as relative adverb. The familiar yet remarkable fact that a preposition governing *that* does not precede it but follows it at a distance has already been mentioned. The idiom now to be noticed may be traceable to that fact. In the four following examples *that* serves as a sort of relative adverb, equivalent to *which* with a preposition: *We very much question*

*whether the eventual historian will regard it as a period of Rationalism in the sense that we have apparently agreed to regard the eighteenth century as a period of Rationalism* (= in which). / *She found herself after Trafalgar in the same position that Rome found herself after the destruction of the Carthaginian fleet* (= in which). / *He took him for his model for the very reason that he ought to have shunned his example* (= for which). / *Others, watching the fluctuating rates of exchange with all the anxiety that a mariner consults his barometer in a storm-menaced sea, are buying securities that can . . .* (= with which).

This is a freedom that should no more be allowed to lapse than the right of putting a preposition last or of omitting an objective *that* (see 4 below). But idiom requires that *which* should not be so treated; it has been tried, with obviously bad results, in: *It touched them in a way which no book in the world could touch them.* / *The man who cleaned the slate in the way which Sir E. Satow has done both in Morocco and Japan.* And further, *that* itself should not be so treated unless the preposition to be supplied in the clause has been actually expressed with the antecedent; in the following, *at which* must be substituted for *that*: *One of the greatest dangers in London is the pace that the corners in the main streets are turned.*

4. Ellipsis of *that*. Both the relative adverb *that* and the relative pronoun are sometimes omitted, at least in speech. As to the adverb, *At the speed he was going he could not stop* and *It happened on the day we first met* are good colloquial English. As to the pronoun, this is very frequently omitted when it is the *object* of a defining clause. 'In the spoken language' (says Onions) 'the tendency is to omit the relative as much as possible, and to prefer (e.g.) *The book I am reading* to *The book that I am reading.*' 'In the written language', he adds, 'its omission is often felt to be undignified'; but this feeling is probably not so strong now as it may

once have been. On the other hand *which*, in the non-defining clauses to which it is proper, must be expressed. This fact, *which you admit, condemns you* cannot be changed to *This fact, you admit, condemns you* without altering the sense. The omission of the relative pronoun where it is the *subject* of a clause was formerly a not uncommon poeticism, e.g. Shakespeare's *I have a brother is condemned to die* and Tennyson's *What words are these have fallen from me*, and may still be found in dialogue written in Irish vernacular, e.g. *Herself will be safe this night with a man killed his father holding danger from the door.* Otherwise the omission occurs only (as a colloquialism) after *there is*, *he is* etc. (*There's a man wants to speak to you*; *It isn't everyone could do it*; *He is not the man he was*), or before *there is*, e.g. *I know the difference there is between you.*

**5.** *That*-clause not close up. The clinging of the defining *that* to its antecedent has been noticed in 1. It is the gap between it and the antecedent that occasions a certain discomfort in reading the correct sentences below. Each *that*-clause is, or may be meant as, defining; but between each and the actual noun of the antecedent (*thoroughfare, fight, formulae, country*) intervenes a clause or phrase that would suffice by itself for identification. In such circumstances a *that*-clause, though correct, is often felt to be queer, and it is usually possible to regard it as non-defining and change *that* to *which*. The reader will probably agree that the change would be desirable in some of the four, and in others for special reasons undesirable: 'Petty France' *was the name anciently borne by the thoroughfare now known as York-street*, that *runs from the Broadway, Westminster, to Buckingham Gate.* | *Dingwall, which has taken a very active part in the electoral fight for the Wick Burghs*, that *has resulted in so striking a Liberal triumph, has other claims upon . . .* | *The foolish formulae for which the Coalition was responsible, and* that *the Conservatives*

*have taken over, are not good enough.* | *When Mr. Raleigh writes, as he does, as if America was a country of bounding megalomaniacs*, that *measured everything by size and wealth, he is talking nonsense.*

**6.** One *that* in two cases. It is quite in order to let a relative *which* or *that* carry on and serve a second clause as well, but only if three conditions are satisfied: the antecedent of the two must be the same; both must be defining, or both non-defining; and the case of the relative must be the same. This last condition is violated with *that* in the examples now to be given. If there is a change of case, *that* or *which* must be repeated; or, more often, the repetition should be saved by some change of structure, as suggested in the brackets: *The whole thing is a piece of hypocrisy of a kind* that *few associations would care to avow even in committee, but is here exhibited unblushingly in the light of day* (committee; but here it is exhibited). | *The art of war includes a technique* that *it is indispensable to acquire and can only be acquired by prolonged effort* (that must be acquired, but can). See CASES 3D.

**7.** Double government. *A book that I heard of and bought* is a familiar and satisfactory form of speech; *that* is governed first by *of* and again by *bought*; but it is not good enough for those who consider that spoken *that* should become written *which*, and that a preposition should not end a clause; they change it to *A book of which I heard and bought*, forgetting that if they do not repeat 'which I' this commits them to 'A book of which I bought'. Examples have already been given in h and j of the first section; but the efficacy of *that* in making the mistake impossible is so little appreciated as to deserve special treatment. First an example that shows the right form for such needs, with *that*: '*Command*', *by William McFee, is one of those fine roomy books* that *one lives in with pleasure for a considerable time and leaves at the last page with regret.* Next some examples that illustrate the frequency of the mistake, which is naturally not made by those who recognize that in writing as well as in speech *that* is the true defining relative, and the place for a preposition governing it is later in the clause: *A great international conference* to which *America is to be invited, or is to be asked to* convene *at Washington.* / *We must not be faced by a peace* of which *we may disapprove and yet must* accept. / *An ammunition dump* on which *he dropped his remaining bombs* and left *blazing merrily.* / *It is incarcerated in prison-like places*, to which *it objects, and does all in its power to* avoid.

**the.** 1. *The* with titles. 2. *The Times correspondent* etc. 3. *By the hundred* etc. 4. *The good and* (*the?*) *bad.* 5. *The* with two nouns and singular verb. 6. Single adverbial *the* with comparatives. 7. Double adverbial *the* with comparatives.

1. *The* with titles. It is curious that we use *the* when speaking of ancient writings but not of the more recent. We say *The Agamemnon, The Iliad*, but not *The Hamlet, The Paradise Lost.* We of course use *the* if it is part of a title (*The Winter's Tale, The Doctors' Dilemma*) and we usually do so also when referring to a work of which *a* is part of the title; we should probably say *That is a quotation from The* (*not A*) *Midsummer Night's Dream, Tale of Two Cities, Shropshire Lad*, etc.; *Look it up in The* (*not A*) *New Oxford Dictionary*; no doubt the reason why we substitute the definite article for the indefinite is that one or the other is clearly needed and the indefinite does not seem to indicate the work with enough precision.

2. *The Times correspondent* etc. *It is agreed that* The Hague Conference *is to be a meeting of technical experts.* The capital T of *The* raises a question that, however trivial, is for ever presenting itself: in 'the Conference at The Hague', or 'the correspondent of The Times', we know where to use a capital and where a small letter; but when one *the* is cut out by using (*The*) *Hague* and (*The*) *Times* attributively instead of as nouns, is the remaining

*the* that which belonged to *Hague*, or that which belonged to *Conference*, that which belonged to *Times* or that which belonged to *correspondent*? and is it consequently to be *The*, or *the*? Though compositors or writers often choose the wrong alternative and print *The*, a moment's thought shows that it is *Conference* or *correspondent* that must have its *the*, while *Hague* and *Times* can do without it. We say 'a Times correspondent', and 'the last Hague Conference', stripping *Hague* and *Times* of their *The* without scruple; it follows that the indispensable *the* belongs to the other word, and should not be *The* unless after a full stop.

3. *By the hundred* etc. *The mild revelations of a gentle domestic existence which some royal personages have given us command readers by the hundreds of thousands.* The idiomatic English is *by the hundred thousand*; *by hundreds of thousands* will also pass, but with the plural *the* is not used. So also with *dozen, score*, etc.

4. *The good and* (*the?*) *bad. Primitively splendid dresses, which appealed after the manner of barbaric magnificence to* the most complex and elementary *aesthetic instincts.* Is the omission of another *the most* or another *the* between *and* and *elementary* tolerable? The purist will condemn it on principle, and probably most of us will, for this particular case, endorse his condemnation. But he will add that neither must we say 'The French, German, and Russian figures are not yet to hand', unless we are talking of their combined total; the Germans and the Russians, he will say, must have their separating *the*; and in these rigours sensible people will not follow him. What may fairly be expected of us is to realize that among expressions of several adjectives or nouns introduced by *the* some obviously cannot have *the* repeated with each item without changing the sense (*the black and white penguins*), and some can logically claim the repetition (*the red and the yellow tomatoes*). A careful writer will

have the distinction in mind, but he will not necessarily be a slave to logic; 'the red and yellow tomatoes' may be preferred for better reasons than ignorance or indolence. For other attempts to impose a needless rigidity see ONLY, and NOT 1.

5. *The* with two nouns and singular verb. *It is* the single-handed courage and intrepidity *of these men which appeal to the imagination, and are even more marvellous than their adventures.* Two nouns of closely allied meaning are often felt to make no more than a single notion; *courage and intrepidity* is almost a HENDIADYS for *intrepid courage.* That feeling is here strengthened by the writer's choosing to use only one *the* instead of two; and to change *appeal* and *are* to *appeals* and *is* would be not only legitimate, but an idiomatic improvement. See NUMBER 2.

6. Single adverbial *the* with comparatives. In 'the more the merrier' we have double *the*; in 'They are none the better' we have single *the*, and that is the type to be first discussed. In both types *the* is not the ordinary adjective or 'article', as in 'the table' etc., but an adverb (or, in the double type, two adverbs); the original meanings were in the double type *by what* (i.e. *by how much*) and *by that* (i.e. *by so much*), and in the single type *by that* (i.e. *thereby* or *on that account*, or sometimes *by so much* or *by that amount*). These facts are familiar to all students of grammar, and are simple enough; but the modern idiom based on them is less easy to be sure of. It will appear from the extracts presently to be quoted that the usage here ascribed to the best writers is not universal; it is indeed often violated. What is here maintained is that good writers do not, and bad writers do, prefix *the* to comparatives when it conveys nothing at all; and again that good writers do not, and bad writers do, allow themselves a *than* after a comparative that has *the* before it. The second and more limited question may be taken first, and disposed of by saying roundly that the *the* should never be used with a comparative if *than* follows. In

the following extracts for instance the *the* form should be got rid of in the first two by the omission of *the* without any consequential changes and in the last two by changing *none the* to *no. I do not believe that the New Royalty productions would have pleased people* any the more than at present *by having money lavished upon scenery* (any the more = any more than if money had not been lavished). | *A sentence in the courts of summary jurisdiction has not* any the less effect *upon the status and prospects of a prisoner* than *a sentence in the superior courts* (any the less effect = any less effect than if it were not in courts of summary jurisdiction). | *But does that make Sophocles more Greek than Aeschylus or Euripides? Each of the latter may be more akin to other poets; but he is* none the less Greek than *Sophocles* (none the less Greek = no less Greek than if he were not more akin to other poets)./*Meanwhile the intellectual release had been* none the less marked than *the physical.*

As to the more general question of when a single adverbial *the* is appropriate and when it is out of place before a comparative, without the complication of a following *than*, a fashion seems to have grown up of inserting an indefensible *the* in the false belief that it is impressive or literary. Such fashions are deplorable; it is wisdom either to abstain altogether from the adverbial *the* or to clear one's ideas about what one means by it. The function of this *the* is to remind or acquaint the reader that by looking about he may find indicated the cause (or sometimes the amount, if *the* means *by so much* rather than *thereby*) of the excess stated by the comparative. If no such indication is to be found earlier or later in the passage, *the* has no justification, and merely sets readers searching for what they will not find. Correct examples are: *I am the more interested in his exploit because he is my cousin,* where *the* anticipates *because* etc.; *Though he is my cousin I am not the more likely to agree with him,* where *the* refers back to *though* etc.; *As the hour approached*

*I grew the more nervous*, where *the* means *by so much* and refers back to *as* etc. In the examples that follow it will be found impossible to point to such a cause or measure of excess anticipated or recalled by *the*, and moreover it will probably be admitted at once that removal of *the* does not weaken the sense, but improves it.

First will come some quotations each meant to convey something of this sort: 'A says so-and-so;(that really does not much concern us;) what concerns us more is so-and-so else'; but in each a *the* has been gratuitously inserted, with nothing for it to anticipate or recall; the bracketed sentence above is not usually expressed, but it or an equivalent is a necessary part of the sense: *But whilst the origin of words is a very fascinating study, we are at the moment* the more interested *in some of the language used at yesterday's demonstrations.* / *That was the principle asserted in the resolution, but what* the more interests *us is the reasons given for this advertised resistance.* / *It would not be difficult to preach a very effective sermon out of the fact that Professor Dicey uses the word 'England' when he clearly means, so far as we can see,* the *United Kingdom, but we are* the more concerned *to examine the Professor's thesis.*

These are simple affairs; the reader is mystified for a moment by *the*, but soon sees that all he has to do is to neglect it. The next examples are not quite so simple, because each contains some expression, of a kind commonly associated with this *the*, that nevertheless is not to be associated with it here and, if it is so taken, will spoil the sense: *It is socially inexpedient that the diseased should languish unattended because of inability to provide skilled assistance, and it is* not the less inexpedient *that the prisoner should stand unaided before justice because his means cannot secure legal representation.* The *because* clause does not explain *the*, as one might guess, but belongs to *stand unaided.* / *It is gratifying to receive such clear testimony to a widespread interest in an intelligent study of the* Bible; *and it is* not the less gratifying that *many recent books deal with the subject from a special point of view.* The *that* clause looks like the explanation of *the*, but is in fact the subject of 'is not less gratifying'. In these examples the use of *the* goes beyond mere ineptitude, and amounts to the serious offence of laying FALSE SCENT.

It may even be thought that in the vogue of this *the more* etc., where *the* is an adverb, is to be found the explanation of the wrong adjectival *the* in: *It is curiously entertaining to see how, in all essential things, the actor-playwright is invariably* the better craftsman than *the literary man who commences dramatist.* Read *a better craftsman.* Choice in such sentences lies between *A is a better man than B* and *Of the two A is the better man*; the wrong form *A is the better man than B* either confuses those two or apes the adverbial use.

**7.** Double adverbial *the* with comparatives. It has been stated in 6 that in this construction one *the* means *by how much* and the other *by so much*, like the Latin *quo . . . eo* and *quanto . . . tanto.* The most familiar example, 'the more the merrier', is the short for 'by how much we are more, by so much we shall be merrier'. Here again it is confusing that a *the* may be either adverbial or (as the definite article) adjectival. Writers may be thus tempted to try to make one *the* serve two masters. This is what has happened in *The better education a girl can have and the more time which can be spent on her training, the better.* The reader takes the first two *thes* as adjectival—articles attached to *education* and *time*—and nonsense results. The sentence must be corrected by giving those nouns an adjectival *the* apiece—*The better* the *education . . . the more* the *time* etc.—and so leaving no doubt that the others are adverbial.

The idiom may be described as a sliding scale stating that one process of increase or decrease varies with the variation in another, and the two parts are the measure and the thing measured. Even when constructed correctly, with an unmistakable adverbial *the* in each part, it is suitable chiefly for short, emphatic, pointed sentences, elliptical in their simplest form (*The more the merrier*; *The sooner the better*), without relative clauses or parentheses that upset the balance, and with inversion permissible only in the second part. Examples of its suitable use are *Dr. H. falls into the error of believing that the more Bennett knew of psychology, the more he knew of people*, and, with a legitimate though unnecessary inversion in the second part, *The older I grow, the more inclined am I to ask of a book no more than that it should be readable*. The unnatural effect of an inversion in the first part is illustrated in *The less distinct was the message he felt impelled to deliver, the more beautiful is often the speech in which he proclaims it*; if *was the message* were changed to *the message was* this would be a suitable use of the idiom in spite of its length, because of the detailed correspondence between the two parts. But the idiom is unequal to so elaborate a task as has been set it in *The economic welfare of a community is likely to be greater* (1) *the larger is the average volume of the national dividend,* (2) *the larger is the average share of the national dividend that accrues to the poor, and* (3) *the less variable are the annual volume of the national dividend and the annual share that accrues to the poor.*

**their,** as the possessive of *they*, is liable, like *they*, to misuse as a common-sex singular, for discussion of which see NUMBER 11 and THEY. Two specimens will here suffice without further comment: *But each knew the situation of their own bosom, and could not but guess at that of the other.* / *No one can be easy in* their *minds about the present conditions of examination.*

**theirs.** See ABSOLUTE POSSESSIVES.

**theism.** See DEISM for the difference.

**them.** For misuses common to *them* and *they*, see NUMBER 11 and THEY. The reflexive use of *them = themselves* is archaic, and as such usually to be avoided; but the following quotation is enough to show that with an archaic verb the archaic reflexive may be more appropriate: *Together the two— employee and director—hied themselves to the loco. superintendent's office.* Read *hied them to*.

**there.** In the well-known special use of *there* before *be, exist,* and such verbs, two things call for notice. First, the use is anticipatory, i.e. *there* accompanies and announces inversion of verb and subject, standing in the place usually occupied before the verb by the subject; consequently, when there is no inversion, this *there* is out of place, and should be struck out, e.g., in: *Bombay is without a doubt the headquarters of* whatever cricket there exists *in India today*. An exception must however be made for the verb *be* itself; 'whatever cricket there is' is English, though 'whatever cricket there exists' is not. The reason is easy to see. In inversions, *there* has become so regular an attendant on *is, are, was* etc. when used as a substantive verb (i.e. = exist) that even when there is no inversion the need is felt of inserting it as a sign that the verb must be taken in its substantive sense, not as auxiliary or copulative. But with other verbs, about which there is no such doubt, no such sign is wanted, and *there* is used only with inversion.

The second thing that calls for notice is that, since in the *there* idiom verb precedes subject, there is a danger of the verb's being hastily put into the wrong number; for examples see NUMBER 7.

**thereafter, thereat, therein, thereof, thereto, therewith,** etc. The remark at the end of the article THEREFORE applies also to these compounds.

**thereby.** 1. The use of *t.* after a number etc. (*half a dozen or t.*) is Scottish, the English idiom being *or thereabouts* or *or so*. 2. *A special tribunal will be* *constituted to try the accused,* thereby assuring *him the guarantees essential to the right of defence.* For this use of *t.* with an UNATTACHED PARTICIPLE (*assuring*'s noun is not *tribunal,* but an inferred *constitution*), see that article and THUS, which is more frequently resorted to in similar difficulties. In the following example it is clear that *thereby* means by the salary etc.; but whether *affording* agrees with salary etc., so that the salary affords encouragement by the salary, or whether it agrees with 'firm' looming in the distance, the writer probably knows as little as we: *The latter is usually the recipient of a liberal salary and expenses, with periodical increments, holidays, and security,* thereby affording *every encouragement to promote the interests of his firm.*

**therefor, therefore.** The two are now distinct in accent and meaning as well as in spelling. *Therefor* is accented on the second syllable, *therefore* on the first; and *therefor* is to be used only where *for that, for it, for them,* etc., could stand equally well. In grammatical terms, *therefore* is an adverbial conjunction, and *therefor* an adverbial or adjectival phrase (adverbial in *He was punished therefor,* and adjectival in *The penalty therefor is death*). The essential function of *therefore* is to make clear the relation of its sentence to what has gone before; that of *therefor* is the same as that of *thereafter, thereat, therein, thereof, thereto, therewith,* etc., usual in legal documents but elsewhere serving only to give a touch of formality or archaism to the sentence in which it is substituted for the *for it* etc. of natural speech.

**therefore.** Apart from the danger of meaning *therefor* and writing *therefore* or vice versa, the only caution needed is that discretion is necessary in the use of commas before and after words of the class to which *t.* belongs. Like *then, accordingly, nevertheless, consequently,* and many others, it is an adverb often used (itself, indeed, almost always) as a conjunction; and it is a matter of taste whether such adver-

bial conjunctions shall or shall not be comma'd off from the rest of the sentence in which they stand. Light punctuators usually omit the commas (or comma, if *t.* stands first), heavy punctuators usually give them, and both are within their rights. But it must be remembered that the putting of a comma before *t.* inevitably has the effect of throwing a strong accent on the preceding word, and that some preceding words are equal to that burden, and some are not. From the three following examples it will be at once apparent that the *although* can bear the commas, and the *and*s cannot: *Although, therefore, the element of surprise could not come into play on this occasion, the Germans were forced to withdraw.* | *It would be impossible for the State to pay such prices, and, therefore, we must content ourselves with . . .* | *Malaria was the cause of a very large proportion of the sickness, and, therefore, the disease deserves especial study by . . . .*

Again, the word *it* is one that can seldom be emphasized and consequently abhors a comma'd *therefore* such as follows it in: *It, therefore, comes rather as a shock to find simultaneously in many papers this morning articles declaring . . .* | *It, therefore, behoves those who have made the passage of the Bill possible to attend once more.* But where emphasis can reasonably be laid on *it,* and it can mean 'it' more than others' or the like, the commas become at least tolerable; so: *It is a concrete and definite idea, the embodiment of which in practicable shape is by far the most urgent constructive problem of international statesmanship;* and it, therefore, *calls for the most careful examination.*

Many words, however, are neither naturally emphatic like *although* nor naturally unemphatic like *and* and *it,* and after them care should be taken not to use the commas with *therefore* except when emphasis is intended. The personal pronouns are good examples; in the following, we ought to be able to conclude from the commas that 'we' are being deliberately contrasted with others who believe otherwise: *We, therefore, find great comfort in believing that Canadian loyalty depends not on . . ., nor on . . ., but on . . . .* Probably that is the case, and the commas are justified; but if the light punctuation were generally accepted as the rule with these adverbial conjunctions, and commas used only when emphasis on the preceding word was desired, one of the numberless small points that make for lucidity would be gained.

A curious specimen may be added: *We therefore are brought again to the study of symptoms.* Here it is obvious that *We* is unemphatic; but the writer, though he has rightly abstained from commas, has been perverse enough to throw an accent on *We* by other means, viz. by putting *therefore* before instead of after *are*.

**they, them, their.** 1. *One, anyone, everybody, nobody,* etc., followed by *their* etc. 2. Confusions with nouns of multitude and personifications. 3. Unsatisfactory pronoun reference. 4. Case.

1. *One* etc. followed by *their* etc. The grammar of the recently issued *appeal to the Unionists of Ireland, signed by Sir Edward Carson, the Duke of Abercorn, Lord Londonderry, and others, is as shaky as its arguments. The concluding sentence runs: 'And we trust that everybody interested will send a contribution, however small, to this object, thereby demonstrating their* (sic) *personal interest in the anti-Home Rule campaign'. Archbishop Whately used to say that women were more liable than men to fall into this error, as they objected to identifying 'everybody' with 'him'. But no such excuse is available in this case.* Undoubtedly grammar rebels against *their*; and the reason for using it is clearly reluctance to recognize that, though the reference may be to both sexes, the right shortening of the cumbersome *he or she, his or her,* etc., is *he* or *him* or *his,* as *his* and *him* are used with a boldness surprising in a government department in *There must be opportunity for the individual boy or girl to go as far as his keenness and ability will take him.* Whether that

reluctance is less felt by the male is doubtful; at any rate the OED quotes examples from Fielding (*Everyone in the house were in their beds*), Goldsmith, Sydney Smith, Thackeray (*A person can't help their birth*), Bagehot (*Nobody in their senses*), and Bernard Shaw. It also says nothing more severe of the use than that it is 'Not favoured by grammarians'. In colloquial usage the inconvenience of having no common-sex personal pronoun in the singular has proved stronger than respect for the grammarians, and the one that is available in the plural is made to serve for the singular too. But in prose their disfavour is not treated so lightly; few good modern writers would flout them so conspicuously as Fielding and Thackeray did in the sentences quoted, or as Ruskin in *I am never angry with anyone unless they deserve it.* The question is discussed in NUMBER 11: examples of the wrong *their,* in addition to those that follow, will be found under THEIR; and the article ONE 5, 6, 7, may be useful. *The lecturer said that everybody loved their ideals.* | *Nobody in their senses would give sixpence on the strength of a promissory note of that kind.* | *Elsie Lindtner belongs to the kind of person who suddenly discovers the beauty of the stars when they themselves are dull and have no one to talk with.* The last is amusing for the number of the emendations that hurry to the rescue: E. L. is one of the people who discover . . .; . . . kind of people who discover . . .; . . . when he himself is . . .; . . . when she herself is . . .; . . . the kind of woman who discovers . . . when she herself is. . . . As to ' . . . when she herself is . . .' without further change, it is needless to remark that *each, one, person,* etc., may be answered by *her* instead of *him* and *his* when the reference, though formally to both sexes, is especially, as here, female.

2. Confusions with nouns of multitude and personifications. What is meant appears from the following quotation, in which a noun of multitude (section) is treated in the same sentence first as singular (*acknowledges*)

and then as plural (*they*). The British Section of the International Council for Bird Preservation gratefully acknowledges the assistance they have received from readers of 'The Times'. Discussion and other examples will be found in NUMBER 6, PERSONIFICATION 2, and WHICH, THAT, WHO 8.

3. Unsatisfactory reference. For the many possibilities in this kind, see PRONOUNS. One flagrant example will here suffice: *The Germans will argue that, whatever* they *may undertake to keep the French at bay,* they *will still have no guarantee that* they *will evacuate their territory or even refrain from further occupations when* they *prove unable to meet the enormous demands still hanging over* them.

4. Case. Like *him* and HE (which see for comment as well as CASES 3 c.) *them* and *they* occasionally go wrong, as in: *The whole foundation of our constitution depends upon the King being faithfully served by his advisers, and* they *taking complete responsibility for every act which he does.* | *A society in which poverty marked half the population, whereas now it does not now mark one-twentieth, and* they *mostly the old.* Observe that responsibility for the first of these two blunders rests with the FUSED PARTICIPLE; read, *upon the King's being . . . and their taking.*

**thine.** See ABSOLUTE POSSESSIVES.

**thingumajig, thingumabob, thingummy,** are the chief survivors of a large number of variants.

**think.** 1. After *t.*, *that* is usually omitted; see THAT, conj. 2.

2. *Think to* is used idiomatically in the sense expect (*I did not think to see you here*), and also, at least colloquially, in the sense remember (*Did you think to ask him for his address?*). *Think up*, meaning invent or devise, usually a pretext or excuse, is a common colloquialism of U.S. origin.

3. The use of *think* as a noun, also a common colloquialism today, (*You'd better have another think*), is over a hundred years old according to the OED, which calls it 'dial or colloq.'

4. *No thinking man*. One of the bluffing formulae, like *It stands to reason* (see REASON 2), that put the reader's back up and incline him to reject the view that is being forced on him. For *incline to think* see INCLINE.

**thinkable** is a word of the same unfortunate ambiguity as its much more popular opposite UNTHINKABLE. *Protection is only a thinkable expedient on the assumption that competition in the home market is to be made unprofitable.*

**third person.** For badges of anonymity such as *the present writer* and *your reviewer* see WE 3.

**this.** 1. *This three weeks, this five years*, etc., are as good English as *these* etc., the numeral and the plural noun being taken as the singular name of a period; but the modern grammatical conscience is sometimes needlessly uneasy about it. See also NUMBER 1.

2. Like *that* (see THAT adj. and adv. 1), *this* is used in the sense 'to this extent', but only colloquially, except perhaps when helped out by *much. I didn't expect to be this late.* | *It wasn't this hot yesterday.* | *I didn't expect to meet this much opposition.*

**thither.** See HITHER. An OED quotation shows how the word is still available, though rarely indeed, when real ambiguity would result from *there*; it is from a guidebook: *The road thither leaves the main road at right angles.* But *to it* is now the normal English.

**-th nouns.** The remarks made in the article -AL NOUNS apply also to the invention of new or revival of obsolete nouns in -th. There are large numbers of well established words such as *truth, depth, growth*; but the suffix is no longer a living one, and the use of new or revived -*th* nouns is chiefly a poetasters' trick. Some specimens are: *greenth, gloomth*, and *blueth*, all made by Horace Walpole; *blowth* (blossom etc.), more or less obsolete; *spilth*, a revival; and *illth*, made by Ruskin as antithesis to *wealth* in its older and wider sense.

**thoroughbred, pedigree.** In Britain the first is used of horses, the second of cattle and other animals.

**those.** 1. For *those kind of, those sort of*, see KIND.
2. *Those* (adj.)+noun+adjective. (*The winner will be selected from*) *those persons named; persons* is the noun, and *named* the adjective. This arrangement is now very common but is so little warranted by good literary usage that the OED, which illustrates the constructions of which this is a hybrid product, does not quote a single example of it; cf. what is said of the same construction under THAT, adj. 2. The word *adjective* in the formula above is to be taken as including participles active or passive, and adjectival phrases, as well as simple adjectives—whatever, in fact, is equivalent to a defining relative clause (*those persons following, those persons named, those persons in the list below, those persons present*—all equivalent to *the persons that* etc.). *Those named* is a proper substitute for (*the*) *persons named*, the pronoun (not adjective) *those* taking the place of the noun *persons* with or without *the*, and (*the*) *persons named* is itself a shortening of *the persons that are named*. But *those persons named* is a mixture of the long form (*the*) *persons that are named* and the short form *those named*, in which mixture what was gained by using the pronoun *those* instead of *the persons* is thrown away by reinserting the noun and making *those* an adjective. It is true that there is another legitimate form in which *those* does appear as an adjective, viz. *those persons who are named*; but that is a form in which what is aimed at is not lightness and brevity, but on the contrary formality and precision; it is therefore not one that should be abbreviated.

All this is offered not as a proof that *those persons named* is impossible grammar, but as a reasonable explanation of what is believed to be the fact, that good writers do not say it, but say either (*the*) *persons named* or *those named*. The following quotation is useful as containing samples both of

the right and of the wrong usage: *It depends upon the extent to which* those in authority *understand their responsibility, and are able so to make their influence felt as to enlist the active support of* those boys with most influence *in the school. Those in authority* is right, whereas *those persons in authority* would have been wrong; and *those boys with most influence* is wrong, and should be *the boys with* etc.

The following use of *those* is quite another matter, and of no importance, but worth giving as a curiosity: *It is impossible for the Ambassador to issue invitations to* those other than *Americans.*

**though.** 1. *Though, although.* 2. *As though.* 3. (*Al*)*though* with participle or adjective. 4. Illogical use.
1. *Though, although.* The definite differences between the two hardly need stating. They are: first that *though* can and *although* cannot be used as an adverb, placed last (*He said he would come; he didn't, though*); and secondly that *though* is alone possible in the *as though* idiom. In the use common to both forms, i.e. as a complete conjunction, no definite line can be drawn between them, and either is always admissible. But it is safe to say, in the first place, that *though* is much commoner, and, secondly, that the conditions in which *although* is likely to occur are (a) in the more formal style of writing, (b) in a clause that does not follow but precedes the main sentence, and perhaps (c) in stating an established fact rather than a mere hypothesis: *He wouldn't take an umbrella though it should rain cats and dogs*; *Although he attained the highest office, he was of mediocre ability.*
2. *As though. It is not as though there* has *been cruelty and injustice. Had*, in place of *has*, is the only right English; see AS 4 for discussion and examples.
3. (*Al*)*though* with participle or adjective. Like other conjunctions (*if, when, while*, etc.), (*al*)*though* is often used with the significant word of its

clause alone, the subject and the auxiliary or copulative verb being readily supplied; so *Though annoyed, I consented.* The convenience of this is obvious, but care is needed, as appears from the two quotations that follow: *Though* new *to mastership herself, a lady master is not new to the pack, for she follows Mrs. Garvey in the position.* / *Though* sympathizing *as I do with Poland, I cannot resist the impression that it would be doing Poland an ill service to.* . . . The point shown by the first is that the omission must not be made when it leaves the participle or adjective apparently attached to a wrong noun. *New* in fact belongs to *she*, but seems to belong to *a lady master*; if *she is* had not been omitted after *though*, all would have been in order. In the Poland sentence, the correction really required is to omit *though*, 'sympathizing as I do' being self-sufficient. But, even if we suppose *as I do* omitted, there is a wrong sound about *though sympathizing* itself that suggests a limitation on this idiom. *Though*, and other conjunctions, must not be constructed with a participle unless that participle would have been used in the unabridged clause; and that would not have been *though I am sympathizing*, but *though I sympathize*. Contrast with this the perfectly satisfactory *Though living he is no longer conscious*, where the full form would be not *Though he lives*, but *Though he is living.* See also IF.

4. **Illogical use.** The danger of using adversative conjunctions where two propositions are not strictly opposed, but in harmony, is explained and illustrated in BUT 3. In the following example, *though* would be right if the words 'is the only country in Europe that' were not there; as it stands, the sentence is nonsense: *Though it is only in recent times that in England the Jewish civil disabilities were repealed, Turkey is the only country in Europe that has throughout been free of any anti-Jewish propaganda.*

**thrash, thresh.** One word, with two pronunciations and spellings differen-

tiated. To separate grain is almost always *-esh*; to flog is always *-ash*; and that is the usual spelling in figurative and transferred use, e.g. thrash out a problem.

**Threadneedle Street, Old Lady of.** See SOBRIQUETS.

**threaten.** *The Mass Vestments, now threatened to be authoritatively revived, have to be decided upon.* See DOUBLE PASSIVES.

**three-quarter(s).** The noun expressing a fraction has the -s, and, though usually hyphened, is better written as two separate words; see HYPHENS. This noun is often used attributively with another noun, e.g. with *back* at rugby football, or with *length* or *face* in portraiture; in those conditions a hyphen is required to show that the adjective+noun has become one word. But further, it is usual, when a plural noun is used attributively or compounded, to take its singular for the purpose, even if that singular does not otherwise exist (*billiard room*, not *billiards room*; *scissor-shaped*, not *scissors-shaped*; *racket-court*, not *rackets-court*). Accordingly, *three-quarter* back and *three-quarter* face are the normal forms. But the nouns *back, length*, etc., are often dropped when context allows, and the attributive compound is allowed to represent them as well as itself; a *three-quarter* is now the usual football term.

**threnody.** See ELEGY.

**thrive.** The OED gives *throve, thriven*, as the past and p.p., but allows *thrived* for either.

**through.** The Americanism *be through with* in the sense of be finished with is recognized by the OED. Only American examples are given but it is now common in Britain also, though still a colloquialism in both countries. The convenient American use of *through* in such a phrase as *from Monday through Friday*, meaning from Monday to Friday inclusive, is still strange to us.

**thus.** There is a particular use of *thus* that should be carefully avoided. In this use *thus* is placed before a present participle (*thus enabling* etc.), and its function, when it is not purely otiose, seems to be that of apologizing for the writer's not being quite sure what noun the participle belongs to, or whether there is any noun to which it can properly be attached (cf. UNATTACHED PARTICIPLES); the exact content of *thus* itself is often as difficult to ascertain as the allegiance of the participle. To each quotation is appended (1) a guess at the noun to which the participle belongs, and (2) a guess at the content of *thus*; the guesses are honestly aimed at making the best of a bad job, but readers may prefer other guesses of their own: *Our object can only be successfully attained by the substantial contributions of wealthy sympathizers,* thus enabling *us to inaugurate an active policy* (contributions? by being substantial?)./*But now a fresh anxiety has arisen owing to the rising of the Seine,* thus making *the river navigation more difficult and slow* (rising? by occurring?)./ *Production rose quickly from 5m. tons in 1958 to over 13m. tons last year,* thus enabling *Frondizi to claim that the battle had been won* (production? By rising quickly?). It should be noticed that the resolution of the participle into a relative clause, and the omission of *thus*, gets rid of the difficulty every time (which would enable; which makes; which enabled).

**thyme.** Pronounce *tīm*; before the 17th c. the usual spelling was *tyme* or *time*. But *thymus* and *thymol* are pronounced *th-*.

**tibia.** *Tĭb*—in English, though *tīb.* in Latin; see FALSE QUANTITY. Pl. -ae; see LATIN PLURALS.

**tic douloureux.** *The private foundations develop tic douloureux trying to do their manifold duty to creativeness.* There are indications that the phrase *develop t. d.*, signifying conscientious endeavour carried to extremes, may be on its way to replacing the outmoded clichés *strain every nerve and sinew,* *explore every avenue, leave no stone unturned* and their successor *lean over backwards.* Its aptness may seem questionable to doctors, but it certainly suggests even more devoted effort than the old phrases. One may strain nerves and sinews without doing oneself any permanent harm, and turn stones and explore avenues without any ill effects, and lean over backwards without overbalancing. But t. d. (trigeminal neuralgia) is agonizingly painful, and only the most dedicated enthusiasts would risk 'developing' it, however good the cause. The best pronunciation (pace OED, which says 'often mispronounced') is *tĭk dŏlŏrŏō'.*

**tilde** (tĭ'ldĕ). The mark put over n (ñ) in Spanish when it is to be followed by a y sound, as in *señor* (senyor'). Used in the *COD* and *OID* as a symbol for the word being defined.

**tile(r), tyle(r).** The words used in freemasonry are usually spelt with y, but are not of different origin. See Y AND I.

**till, until.** The first is the usual form; for what difference of usage exists, see UNTIL.

**tilth.** A word not open to the remarks made in -TH NOUNS, being very far indeed from a recent formation. It differs, however, from the really common nouns in -th, such as *truth* and *wealth* and *filth*: though still in use for the depth of soil prepared by cultivation, it has become archaic in its general meaning of tillage or tilled land; and, being therefore a favourite with those who affect poetic diction, it has unfortunately begotten a progeny that has not its parent's claims to respect.

**time.** Under this, as the most general term, may be collected some synonyms. Of the six following words each is given a single definition with a view merely to suggesting the natural relation between them. Though each is often used in senses here assigned not to it but to another (or not mentioned at all), the words *aeon, date, epoch, era, period, cycle,* form a series when they are strictly interpreted, and to keep that series in mind is helpful in choosing the right word.

An *aeon* is an infinitely long period of time.

A *date* is the identifiable or intelligibly stated point of time at which something occurs.

An EPOCH is the date of an occurrence that starts things going under new conditions.

An *era* is the time during which the conditions started at an epoch continue.

A *period* is an era regarded as destined to run its course and be succeeded by another.

A *cycle* is a succession of periods itself succeeded by a similar succession.

A *time*, and an *age*, are words often exchangeable with all or most of the above, but less precise in meaning. Cf. also the words *term*, *span*, *spell*, *season*, *duration*, JUNCTURE, *moment*, *occasion*.

**time-scale.** See TAUTOLOGY.

**timous, timeous.** It would be better to omit the *e*; see MUTE E. Whereas its sole function is to indicate the ī sound, the OED states that it actually results in the erroneous pronunciations *tī′mĭus* and *tī′mĭus*. But the Scots, now the only users of the word, cling stubbornly to the *e*.

**timpano.** see TYMPANUM.

**tinker,** v. *It was an undesirable thing to be always tinkering with this particular trade.* The idiomatic preposition is *at*, not *with*; the latter, now at least as common, is probably due to confusion with *tamper with*, and illustrates what was said in CAST IRON IDIOM about the battle between analogy and idiom: that analogy perpetually wins.

**tint,** *shade, hue.* All are available as substitutes for the dominant word *colour*. Different *hues* are, so far as meaning goes, simply different colours, so called because for good or bad reasons the everyday word is held to be unworthy of the context. Different *tints* and *shades* are properly speaking not different colours but varieties

of any particular colour, *tints* produced by its modification with various amounts of white, and *shades* by various admixtures of black. These distinctions, however, are no longer observed; *shade* (less commonly *tint*) is now the usual word for the slightly different varieties of what remains broadly the same colour produced by an admixture not merely of black or white but of any other colour, e.g. a darker, or lighter, or greener shade of blue. *Hue* and *tint* retain certain special associations, e.g. *sunset hues, autumn tints,* and for hairdressers and their customers *tinting* has become a euphemism for dyeing.

**-tion** and other *-ion* endings. Turgid flabby English is full of abstract nouns; the commonest ending of abstract nouns is *-tion*, and to count the *-ion* words in what one has written, or, better, to cultivate an ear that without special orders challenges them as they come, is one of the simplest and most effective means of making oneself less unreadable. It is as an unfailing sign of a nouny abstract style that a cluster of *-ion* words is chiefly to be dreaded. But some nouny writers are so far from being awake to that aspect of it that they fall into a still more obvious danger, and so stud their sentences with *-ions* that the mere sound becomes an offence, as it does in *Speculation* on the subject of the *constitution* of the British *representation* at the Washington *inauguration* of the League of *Nations* will, presumably, be satisfied when Parliament meets. *Position* and *situation*, often in combination, are special offenders. *The situation in the industry has reached a tragic position* (The industry is in a tragic state)./ *They based this opinion largely on the position of the company's financial situation* (on the state of the company's finances). / *The Trades Union Congress should call a halt to the situation* (should stop this)./ *We ought to be told the present position on this matter* (how this matter now stands)./*The position in regard to unemployment has deteriorated* (more people are unemployed). / *At the moment the political situation in Malta* *'s in a strange position* (is strange). Writers given to overworking these words would be wise to try doing without them altogether; they would seldom find any great difficulty in it, and they would have a salutary exercise in clear thinking. See also AB-STRACTITIS.

**tipstaff.** The OED prefers the plural -*staffs* to -*staves*, and there seems no good reason for applying the archaic plural of STAFF to officials who are still very much alive.

**tiptoe**, v., is, like *hoe* and *shoe*, an exception to the MUTE E rule, and makes *tiptoeing*.

**tirade.** The OED prefers *tǐ*- to *tī*-; so does the COD, but it is questionable whether this reflects the prevailing practice.

**tire, tyre.** For other words in which the same spelling question has arisen, see Y AND I. The OED regards the word as a shortening of *attire*—the wheel's attire, clothing, or accoutrement; and it states the spelling facts thus: 'From 15th to 17th c. spelt *tire* and *tyre* indifferently. Before 1700 *tyre* became generally obsolete, and *tire* remained as the regular form, as it till does in America; but in Great Britain *tyre* has been recently revived as the popular term for the rubber rim of. . . .' From this it appears that there is nothing to be said for *tyre*, which has no claim to be etymologically preferable and is needlessly divergent from our own older and present American usage. Some diehards, including *The Times*, made a long stand against it, but its victory seems now assured.

**tiro, ty-.** Spell *ti*-, and see Y AND I; pl. -*os*, see -O(E)S 6.

**tissue.** The OED gives precedence to *tǐ'shū* over *tǐ'sū*; but the latter is now regarded as the better pronunciation. It is clear, however, that the sh sound prevailed in the 16th c., since h, which can only be accounted for as marking sound, occurs in quotations

from 1501; and this may be the reason for the OED's choice.

**titbit, tid-.** The older spelling is *tid-*; but it is now so much less usual in Britain (though not in U.S.), and the significance of *tid* is so doubtful, that there is no case for reverting to it. To make the two parts of such words rhyme or jingle is a natural impulse that need not be resisted unless it involves real loss of meaning.

**titillate** (tickle the fancy) and *titivate* (smarten up) are often confused. See PAIRS AND SNARES.

**titles.** A curious and regrettable change has come about in the present century. Whereas we used, except on formal occasions, to talk and write of Lord Salisbury, Lord Derby, Lord Palmerston, and to be very sparing of the prefixes Marquess, Earl, and Viscount, the newspapers now prefer to tell us of the doings of Marquess this, Earl that, Viscount the other, and similarly Marchioness this and Countess that and Viscountess the other that have replaced the Lady that used to be good enough for ordinary wear. This change of fashion has not affected the lowest rank of the peerage; they are still in common parlance Lord, not Baron, though women of that rank in their own right are often called Baroness. This distinction between the sexes is natural: the title Lord can be borne only by a peer or the son of a peer (or by a Scottish judge), but the title Lady is not necessarily indicative of the birth or merit of the lady herself.

**tmesis.** Separation of the parts of a compound word by another word inserted between them, as when 'toward us' is written *to usward*, or 'whatsoever things' *what things soever*. The classic example is Ennius's *Saxo cere comminuit brum* (for *comminuit cerebrum*). Tmesis is a popular figure of modern slang, e.g. *hoo-bloody-ray*.

**to.** 1. Substitution for other prepositions. 2. Unidiomatic infinitive.

    1. *After three years' experience of*

*the official machine I am of opinion that the causes are to be found in the rottenness of the present system, to the absence of any system at all so far as Cabinet control is concerned, and to the system of bestowing honours on the recommendations of Ministers.* The *to*s result from indecision between *are to be found* and some loosely equivalent phrase such as *may be traced*, perhaps assisted by the writer's glancing back to recover his construction and having his eye caught by *to*. This sort of mistake occurs much more often with OF, under which it will be found fully illustrated.

2. Unidiomatic infinitive. *The impossibility to assert himself in any manner galled his very soul.* | *The two factors are the obvious necessity to put an end once and for all to the Turkish misrule over alien races, and the . . . To assert* and *to put* should clearly be *of asserting* and *of putting*. Discussion will be found under GERUND 3; but it may be added here that it is not difficult to account for this very common lapse, sequences apparently similar being familiar enough. There is, for instance, nothing against saying *It was an impossibility to assert himself*, or *It is an obvious necessity to put an end*; the difference is that *to assert* etc. and *to put* etc. are not there, as in the examples, adjectival appendages of *impossibility* and *necessity*, but the real subjects of the sentences, which might have run *To assert himself was an impossibility*, and *To put an end to so-and-so is a necessity*.

**to and fro puzzles.** It is not fair to a reader to ask him to go backwards and forwards over the line that divides the positive from the negative so many times that he is not sure which side he has ended on. *The Opposition refused leave for the withdrawal of a motion to annul an Order revoking the embargo on the importation of cut glass.* | (Heading of an official circular) *Suspension of Cancellation of Suspension of Withdrawal of Licences.* May cut glass now be imported or not? Can licences now be obtained or not?

**today, tomorrow, tonight.** The lingering of the hyphen in these words (*to-day* etc.), still recognized by the dictionaries as alternative spellings and often appearing in print, even in some national newspapers, is a very singular piece of conservatism. It helps no one to pronounce; it distinguishes between no words that without it might be confused; and, as the *to* retains no vestige of its original meaning, a reminder that the words are compounds is useless. Moreover, it is probably true that few people in writing ever dream of inserting the hyphen; when it appears it has probably been inserted by those who profess the mystery of printing, though by doing so they flout the precepts of the APD and OUP and probably most other style rules.

**together.** See ALTOGETHER.

**toilet, -ette.** The word has become completely anglicized in spelling and sound (*toi'let*)—always for the popular EUPHEMISM for water-closet and ordinarily for the earlier sense of attire etc. But for the latter *toilette* (*twahlet'*) has not wholly disappeared and may perhaps revive for obvious reasons.

**token.** For synonyms see SIGN. *By the same t.*, *more by t.*, are archaic phrases which, when current, came to mean little more than *that reminds me*, or *incidentally*. Today they must be classed as WARDOUR STREET.

**toll.** See TAX.

**tomato.** Pl. *-oes*; see -O(E)S 1. *-ahto* in Britain; *-āto* in U.S.

**tomorrow, to-m-.** See TODAY.

**tonight, to-n-.** See TODAY.

**tonsil** makes *tonsillitis* etc.; -LL-, -L-.

**tonsorial.** A word used almost only in PEDANTIC HUMOUR.

**too.** 1. With passive participle. 2. Illogical uses.

1. With passive participles *t.* is subject to the same limitations as VERY,

though the point has been less noticed. The line, however, between the adjectival and the verbal p.p. is often hard to draw; in the following two quotations the addition of *with* etc. and *in* etc. to the participles may be thought to turn the scale and make *too much* preferable to *too*: *Belfast is* too occupied with *its own affairs, too confident of itself, to be readily stirred to any movement which would endanger its prosperity.* / *But he was* too engrossed in *Northern Europe to realize his failure.*

**2.** Illogical uses. These are very common, so common as to deserve a place among the STURDY INDEFENSIBLES and to be almost idiomatic. They result from confusing two logical ways of making a statement, one with and the other without *too*. *Praise which perhaps was scarcely meant to be taken* too *literally* (a, which may easily be taken too literally; b, which was not meant to be taken literally). / *We need not attach* too *much importance to the differences between Liberal and Labour* (a, We may easily attach too much; b, We need not attach much)./*It is yet far too early to generalize* too *widely as to origins and influences* (a, If we generalize too early we may generalize too widely; b, It is too early to generalize widely). Another illogical use is the common colloquial LITOTES such as *He wasn't too pleased*, meaning He was very cross indeed. See also ONLY TOO.

**top, ace, crack** (adj.) are used variously to denote human excellence. *Crack*, the oldest, is mainly in military usage: *crack regiment, crack swordsman, crack shot*. *Ace* is primarily an Air Force term; it dates from the first world war, and was originally applied to an airman who had shot down at least three enemy machines. Both words, though showing some signs of obsolescence, are occasionally used more widely, especially in sport, e.g. a *crack batsman*, an *ace jockey* or *racing driver*; we have *ace* reporters too, and writers of detective stories sometimes give the title to their supremely successful sleuth. *Top* is the most recent. It is reminiscent of the 18th-c. use of

*topping* (*They both have the honour of being footmen to very topping people*, wrote Fanny Burney), a word that later enjoyed an ephemeral vogue as juvenile slang. Unlike the other two, which are individual, *top* is usually collective, though not always; the President of the Federal Bureau of Investigation is known familiarly as *America's Top Cop*; the description *Top Pop Singer* is applied to any notable exponent of a form of entertainment now much in demand, and *He's the tops* is an expression of high commendation. *Top* appears mostly in journalism and advertising; it is convenient for headlines and slogans. *Top names in international industry talk frankly to the 'Daily Mail'.* / *Top people take 'The Times'* / *Bandleader aims to wed top Deb.* / *Khrushchev's bid to stay Top Red.* The second world war gave us also *Top Secret* and *Top Level*, but the latter no longer represents the highest level at which conferences may be held and decisions taken; it has been downgraded by *Summit Level*. See LEVEL.

**torso.** Pl. *-os*; see -O(E)S 6.

**tortoise.** Pronounce *tor'tŭs*. The pronunciation *-oiz* or *-ois*, given as an alternative in some modern dictionaries, is one of the less agreeable results of the speak-as-you-spell movement. See PRONUNCIATION[1].

**total.** The adjective makes *-ally*, *-alize(r)*, *-alizator*, *-ality*; and the verb *-alled, alling*. See -LL-, -L-. Both adjective and adverb seem to have an attraction that makes for their indiscriminate use. There is a latent sense in them of things being added up: *total* is the right word in *total war* but in the sense of *absolutely, completely, entirely, quite, utterly, wholly* (*He was totally at a loss*) one of those near-synonyms is generally better than *t.* And in the phrase *sum total t.* is tautological.

**totalitarian** dates from the nineteen-twenties, and is defined by the OED Supp. as 'Of or pertaining to a polity which permits no rival loyalties or parties'.

**tote** has been made to serve as a colloquial abbreviation of more than one word: first of *total*, next of *total abstainer*, and lastly of *totalizator*. This last is the only sense in which it is now used in Britain; in America there is also a verb *tote*, a colloquialism meaning to bear or carry.

**tother**, now only colloquial, was formerly in good literary use, and was then more often written *tother* than *t'other*; it is a telescoping not of *t(he) other* but of *(tha)t other*, and there is therefore no need for the apostrophe.

**toto caelo.** Literally, 'by the whole sky', i.e. by the greatest possible distance, 'poles apart'. Properly used only with *differ*, *different*, and words of similar meaning; the writer of the following extract has guessed that it is a high-class variant of entirely: . . . *had the effect of habitually repealing its own canon in part, during the life-time of parties* . . ., *and of repealing it,* toto caelo, *after the death of either of them.* See FOREIGN DANGER.

**touchy.** See TETCHY.

**toupee, toupet.** The first is the form common in England in the 18th c., written without an accent and pronounced *tōōpē'*; the second is the French word, pronounced *too'pā*. But modern English practice is to spell *-ee* and pronounce *-ā*.

**tourniquet.** The OED prefers *-kĕt* but *-kā* is probably now more common, perhaps because of the present tendency to give foreign values to the vowels of any word with a foreign look.

**tow- and towing-.** There is perhaps an impression that in the compounds (e.g. -boat, -line, -net, -path, -post, -rope) *towing-* is the correct form, and *tow-* a slovenly modern abbreviation. But it appears from the OED that *tow-boat* and *tow-line* are the only forms recorded for boat and line (the latter 1719), and *tow-rope* is about a century older than *towing-rope*; *towing-path* however, is as much older than *tow-path*. There is in fact no reason for avoiding either form. Cf. *wash(ing) basin*.

**toward, towards, towardly.** The adjectives *toward* (including the predicative use as in *a storm is toward* i.e. coming) and *towardly* are pronounced *tō'ărd(lĭ)*. The preposition were formerly pronounced *tōrd(z)* but in recent use the influence of spelling is forcing *tōōwor'd(z)* into common use. The adjectives in all senses are obsolescent, or at any rate archaic, but *untoward (untō'ard)* is still current. Of the prepositions the *-s* form is the prevailing one, and the other tends to become literary on the one hand and provincial on the other.

**trace, n.** For synonyms see SIGN.

**trachea.** The pronunciation *trā'kĭ* seems to be superseding the once or thodox *trăkē'a*, perhaps because of the influence of the compound (*tracheotomy* etc.) in which the stress is usually on the first syllable and the *a* usually long.

**trade union.** Plural trade unions, but Trades Union Congress.

**tradition(al)ism, -ist.** For the general question between such variants, see -IST. In this case the longer forms are usual, probably because the words are often opposed to *rationalism -ist*, the form of which is fixed by *ration*'s not having the necessary meaning.

**tragédienne.** See COMEDIAN.

**tragic(al).** See -IC(AL). It may almost be said that the longer form is, in serious use, dead; though the OED quotes it once or twice from modern writers in senses that it does not mark obsolete, in each of them *tragic* would have been the natural word. It survives, however, in playful use, often with a memory of the 'very tragicall mirth' of Pyramus and Thisbe in *Midsummer Night's Dream*. For *tragic* (or *dramatic*) *irony*, see IRONY, 2

**trailers.** Under this name a few specimens are collected of the sort of sentence that tires the reader out by again and again disappointing his hope of coming to an end. It is noticeable that writers who produce trailers produce little else, and that where one fine example occurs there are sure to be more in the neighbourhood. The explanation probably is that these gentlemen have on the one hand a copious pen, and on the other a dislike (most natural, their readers must agree) to reading over what it may have set down. Whatever its cause, the trailer style is perhaps of all styles the most exasperating. Anyone who was conscious of this weakness might do much to cure himself by taking a pledge to use no relative pronouns for a year; but perhaps most of its victims are unconscious. *This type of wicket is always trappy, one ball coming first on to the bat, with another hanging fire, which so frequently causes a catch to be given by the batsman playing too quickly, as Hallows appeared to do when caught and bowled by Macaulay, when he promised a good innings, in spite of being missed at fine leg from a ball which certainly should have been caught, since the ball was played and not hit off the legs.* / *It is true that part of the traffic here is heavy, but at least the surface might be conditioned by modern methods, even if the form of paving cannot well be altered, though I think it ought to be—e.g., if Sydney Smith's suggestion as to the wood pavement problem perplexing an old vestry—'Gentlemen, put your heads together, and the thing's done' — is impracticable, there are now improved means open to a modern City Council, both in surface dressing, in hard woods, and even in macadam, by the use of slag—locally called dross—from the iron furnaces in Yorkshire, which makes the hardest and smoothest surface.* / *It may be that the modification of our Free Trade principles to a sufficient form of Fair Trade will be all that is necessary to prevent the final decline, which probably the pinch of the last few years has prevented from setting in from a previous run of prosperity, which, by causing the* *easy realization of fine old businesses under the seductive lines of Limited Liability, has resulted in the 'Super man' being eliminated in favour of a joint control in which the divergence of opinion among Directors with little personal interest has prevented a uniformity and continuity of policy absolutely essential in the management of any business with widespread interests.* / *Against this portrait of an ogress-devourer of talent one must set the unselfish devotion she gave to Mahler and Werfel and her lifelong friendships with Hauptmann and the Bergs (Madame Berg was an illegitimate daughter of the Emperor Franz Joseph): and so one must give her the benefit of the doubt and assume that, despite her man-eating propensities, she possessed that Austrian charm of the most rare kind which blends inherited good breeding with habitual acceptance of the moods of genius and makes a stimulating listener and comforting friend out of the tweed-capped and finely booted hostess of Semmering or Venice or Vienna or even of Beverley Hills; one of those warm stoves round whom expatriates rally.*

**traipse.** See TRAPES.

**trait.** The final t is sounded in America, but still usually silent in England. For synonyms, see SIGN.

**tranquil** makes -*illity*, -*illize*, -*illy*; see -LL-, -L-, 2. Mis-spellings are very common, esp. *tranquility*.

**transcendence, -cy.** See -CE, -CY.

**transcendent(al).** These words, with their many specialized applications in philosophy, are for the most part beyond the scope of this book; but there are popular uses in which the right form should be chosen. 1. The word that means surpassing, of supreme excellence or greatness, etc., is *transcendent*, and the following is wrong—*The matter is of transcendental importance, especially in the present disastrous state of the world.* See LONG VARIANTS for similar pairs. 2. The word applied to God in contrast with IMMANENT is *transcendent*. 3. The word that means visionary, idealistic, outside of experience, etc., is *transcendental*. 4. The word applied to Emerson and his 'religio-philosophical teaching' is *transcendental*.

**transfer.** Noun *tra'nsfer*, verb *transfer'*, see NOUN AND VERB; *transferred*, *-erring*, *-errer*, see -R-, -RR-; but *transferable*, see CONFER(R)ABLE; and *transference*, *transferee*, and *transferor*. Of *transferrer* and *transferor*, the first is the general agent-noun, a person or mechanism that passes something on, and the second a legal term for the person who conveys his property to another, the *transferee*.

**tranship, transship, trans-ship.** To all who do not happen to have been reconciled by familiarity to the short form it presents itself as an odd sort of monster, which they start by pronouncing *trǎ'nshǐp* (cf. *transom*), and do not at once connect with shipping. And they have at any rate the justification, however little they may be aware of it, that there are no other live words in which *trans* is curtailed to *tran* when it is prefixed to a word of English and not Latin origin like *ship*. But the OED accepts *tranship*, saying only '*less commonly* trans-ship'. Generations of clerks have saved themselves trouble and nearly made away with the s and the hyphen; of 28 OED quotations, including those for *tran(s-)shipment*, nine only show s-s or ss and nineteen *s*, and the progress of the last has since been uninterrupted.

**transient, transitory.** The primary meanings (brief, fleeting) are the same, but *transient* is used with special senses in music and philosophy, and *transitory* in law.

**translucence, -cy.** See -CE, -CY.

**translucent.** See TRANSPARENT.

**transmit** makes *-itted*, *-itter*, *-itting*, see -T-, -TT-; and *-issible* or *-ittable*, see -ABLE 2. A FORMAL WORD.

**transmogrify.** See FACETIOUS FORMATIONS.

**transparence, -ency.** The second is the usual form. The first is marked *rare* in the OED; and indeed, in its only two *-ence* quotations that are as late as 1800, euphony plainly accounts for the avoidance of *-cy*: *Motive may be detected through the transparence of tendency.* | *Adamantine solidity, transparence*, and brilliancy. But *-ence* seems to have been coming back recently in order to distinguish the quality of transparence from the transparent picture (*-ency*).

**transparent,** and the synonyms *diaphanous, pellucid, translucent*. *Transparent* is the general word for describing what is penetrable by sight (lit. or fig.) or by light, and it can be substituted for any of the others unless there is some point of precision or of rhetoric to be gained. All three synonyms have the rhetorical value of being less common than *transparent*, and therefore appear more often in poetical writing. As regards precision, the following definitions of the words' narrower senses are offered, and to each are appended some specially appropriate nouns, and the adjective or participle that seems most directly opposed.

That is *diaphanous* which does not preclude sight of what is behind it; *material, film*; opp. *shrouding*.

That is *transparent* which does not even obscure sight of what is behind it and transmits light without diffusion; *glass, candour, pretence*; opp. *obscuring*.

That is *pellucid* which does not distort images seen through it; *water, literary style*; opp. *turbid*.

That is *translucent* which does not bar the passage of light but diffuses it; *alabaster, tortoise-shell*; opp. *opaque*.

The OED quotes from a newspaper: 'The windows of this classroom were once transparent; they are now translucent, and if not cleaned very soon will be opaque'.

**transpire.** The notorious misuse of this word consists in making it mean happen or turn out or go on; and the legitimate meaning that has been misinterpreted into this is to emerge from secrecy into knowledge, to leak out, to become known by degrees. It is needless to do more than give a single example of the right use, followed by several of the wrong: *The conditions of the contract were not allowed to t.* (right). | *Byron and Claire Claremont met that night at seven o'clock. Precisely what transpired we shall never know.* | *British sources naturally decline to say what transpired at the dinner last night at which Mr. Gromyko was host to Mr. Selwyn Lloyd.* | *The secrecy that was preserved about the time and place of Dr. Beeching's meeting with the union leaders suggests that we are unlikely to hear much about what transpired.* | *Both men opened in a subdued mood in what transpired to be the last game of this grand fight.* The last of these adds to the wrong meaning of *t.* an unidiomatic construction after it in the infinitive *to be*. That construction will not do even when *t.* has its true sense; that sense is complete in itself, and *transpired to be* is as little English as *came to light to be*: here is the right sense followed by the wrong construction: *They must have been aware of the possibility that the facts might be as they ultimately transpired to be.*

**trapes, traipse.** The first seems to be regarded as the orthodox spelling; but the word in this form has so puzzling a look that it would surely be better to use the second, which is allowed by the OED as an alternative, is quoted from Swift and Pope, and can be pronounced only one way. It is indeed more favoured in present usage except in the participle *trapesing*.

**trauma.** Although this word is fast becoming a POPULARIZED TECHNICALITY the plural—*mata* is still more usual than—*mas*.

**travail, travel.** Formerly only slightly distinguished in pronunciation as *tră'vĭl* and *tră'vĕl*, but it is now usual to differentiate more markedly by pronouncing the first *tră'văl*. For *travelled* adj. see INTRANSITIVE P.P.

**traverse.** The verb is ordinarily pronounced, like the noun, with the accent on the first syllable, but may eventually follow the tendency described in NOUN AND VERB ACCENT; it is sometimes heard with the accent on the second.

**travesty.** See BURLESQUE; and, for verb inflexion, VERBS IN -IE etc., 6.

**treachery, treason.** Although the second is sometimes used in the wider sense proper to the first, the distinction, which is worth observing, is that anyone who has trusted anyone else may be the victim of treachery but only a sovereign State relying on the allegiance of its citizens can be the victim of treason.

**treasonable, treasonous.** The meanings are not distinguishable; *treasonous* is now comparatively rare, and more likely to be met in verse.

**treble.** See TRIPLE.

**trecento, -tist.** Pronounce *trāchĕ'ntō, -tist*. This and *quattrocento, -ist, cinquecento, -ist,* are words constantly used by writers on Italian art. Though their true meaning is 300, 400, 500, they are used as abbreviations for the centuries 1300–1399 (1301–1400 is with us the 14th c.), 1400–1499 (our 15th c.), and 1500–1599 (our 16th). There is therefore a double puzzle, Italian 300 for Italian 1300, and Italian 13th c. for English 14th c. The words in *-ist* mean painters etc. of the century.

**trefoil.** OED gives preference to *trē-* over *trĕ-*.

**trek.** *He spent the whole day on the road trekking around from one client to another.* The verb *go* and its innumerable near-synonyms should be enough to satisfy any reasonable person that there is no need for him to maltreat *trek* by using it in a way wholly inconsistent with the associations it still retains of mass migration slowly and painfully accomplished. This SLIPSHOD EXTENSION, though still colloquial only, is increasingly common.

**trend.** A word that, whether as noun or as verb, should be used by no one who is not sure of both its meaning and its idiomatic habits. *There has unquestionably been a trend of German policy to.* . . . | *His chapter on* . . ., *although it has little to do with the rest of his volume, and trends very closely upon the forbidden theme of history, is interesting.* 'There is a t. of German policy to do' is not English, though 'The t. of German policy is to do' would be. *Trends very closely upon* is perhaps a confusion with *trenches* etc.; the essential idea in *t.* is direction, not encroachment. As a verb *trend* is now not often used; *tend* is supplanting it. But there is a difference, worth preserving, similar to that between the nouns *trend* and *tendency*: a trend is a tendency that is continuous and consistent.

**trepan, trephine,** nn. and vv. The first, the older term for the instrument and for operating with it, is probably still the prevailing one in lay use; but in surgical books etc. *trephine*, which as a noun is properly the name of an improvement on the trepan, is now the regular term. The dictionaries prefer the pronunciation *-ēn* to *-īn*.

**tribunal.** Pronounce *trĭbū'năl* or *trībū'năl*; the i is short in Latin, but OED puts *trī-* first, and see FALSE QUANTITY.

**tribute.** 1. For meaning see TAX. 2. A SLIPSHOD EXTENSION of the less excusable kind—since the meaning of *t.* is surely no mystery—is that which nowadays sets 'a t. to' to do the work of a proof (or illustration etc.) of, as in: *The debate on the whole was a tribute to the good taste and good form of the House of Commons.* | *All these and many other prominent English works have been fairly and critically analysed, and it is a tribute to the modesty of the American editors that the European works receive first place.*

**tricksy, tricky.** DIFFERENTIATION is proceeding, in the direction of restricting *tricksy*, now much less used, to contexts in which the quality is regarded not with condemnation or dislike or apprehension (= dishonest, cunning, difficult, etc.) but with amusement or interest (= playful, frolicsome, etc.). It had formerly, to judge from the OED record, all the meanings to itself, being more than two centuries older than *tricky*. At the same time the gap is being widened by the increasing use of *tricky* to describe a task needing adroitness.

**trilogy.** In ancient Athens there were dramatic competitions at which each dramatist presented three plays, originally giving successive parts of the same legend; the extant *Agamemnon, Choephoroe,* and *Eumenides,* of Aeschylus formed a trilogy, and, with the addition of the lost *Proteus,* a *tetralogy.* Later trilogies were connected not necessarily by a common subject, but by being works of the same author, presented on the same occasion. In modern use the word is applied to a work such as Shakespeare's *Henry VI,* comprising three separate plays, or to a novel etc. with two sequels.

**tripe.** 'Tripe is not a joke; it is a valuable foodstuff' was the protest made by an ex-president of the National Association of Tripe Dressers against the contemptuous use of the word by a royal speaker. (Tripe is in fact 'the first or second stomach of a ruminant, especially the ox, prepared as food' —OED). The indignation of the ex-president is understandable, but the custom that provoked it will not be easily dislodged. The use of *t.* as an opprobrious term for a person dates from the 16th c.; the OED quotes *Sayest thou me so thou Tripe, thou hated scorn,* which suggests that even then it must have been felt to be pretty powerful invective. Its application to literature, art, conversation etc. of inferior quality came later; the OED's earliest quotation is dated 1892: *This book . . . very vulgar . . . it is a dish of literary and artistic 'tripe and onions'.* Later still is its now very common application to challenges of a kind that present so little difficulty that they can be treated with contempt—examination papers, bowling at cricket, service at tennis, etc. The Tripe Dressers may find some consolation in the thought that the dictionaries still call these usages 'slang', and that slang, if it does not succeed in winning a respected place in the vocabulary, and so ceasing to be slang, usually disappears when it has lost its freshness.

**triphibious.** See AMPHIBIOUS.

**triple, treble.** If the musical sense of *treble* is put aside, there are perhaps no senses in which one is possible and the other impossible; but they do tend to diverge. First, though either can be adjective, verb, or noun, *treble* is the more usual verb and noun, and *triple* the more usual adjective. Secondly, in the adjectival use *treble* now refers rather to amount (three times as great etc.), and *triple* rather to plurality (of three kinds or parts). A few phrases, in each of which the word used is clearly preferable to the other, will illustrate: *Newspaper has trebled its circulation.| Treble the money would not buy it now.| This is quite treble what I expected.| Going at treble the pace.|He offered me treble wages. | The fight was resumed with treble fury. | Treble difficulty* (= three times the difficulty); *a triple difficulty* (= a difficulty of three kinds)./*Surrounded with a triple wall.|Triple-expansion engines.| The classification is, triple.| Triple alliance, contest, birth.*

**tripod.** OED pronounces *trī′pŏd*, with no alternative (but *trĭ′podal*, also without alternative), and this is still the usual pronunciation (cf. *tripos*) though the COD admits the alternative *trĭpod*.

**triptych.** Pronounce -k.

**triumphal, -phant.** The meanings are quite distinct, but to use the first for the second is usually a worse mistake than the converse, because the idea it ought to convey is nar-

rower and more definite. *Triumphal* means only of or in the celebration of a victory, and belongs to the original 'triumph' or victorious general's procession; *triumphant* belongs to triumph in any of its senses, especially those of brilliant success or exultation. In the following quotations each word is used where the other was required. The 'progress' of the first was not almost, but quite, triumphant; and the 'career' of the second, if it lasted 66 years and was troubled, may have been triumphant, but hardly triumphal. . . . *through the streets of which he had almost a triumphant progress, with women clinging about his car, manifesting in every possible way their delight at his presence. | ... the story he told us of the sixty-six previous years of his troubled, triumphal career.* See also PAIRS AND SNARES.

**triumvir.** Pl. -*rs* or (less usual except of the Roman prototypes) -*rī*; see LATIN PLURALS. But *triumvirate* (set of triumvirs) is now more usual than either plural. The accent is on the second syllable.

**-trix.** Any Latin agent-noun in -*tor* could form a feminine in -*trix*, and some of these when taken into English continue to do so, especially such as are, like *testator* and *prosecutor*, in legal use. It is a serious inconvenience that the Latin plural is -*ices* (-*i′sēz*); if the Latin quantity is preserved, the accent has to be shifted in the plural, which makes the word hardly recognizable. The result is that it is sometimes given up as a bad job; OED gives e.g. *prosecu′trices* and *ra′dices* (radix is like the -trix words, with Latin pl. *radī′ces*), and allows *matrix* a popular *ma′trices* (or even *matrixes*) by the side of a correct *matrī′ces*; but for *cicatrix* it allows only -*trī′ces*, and for *directrix* and *executrix* states only that the pl. is -*ices* and leaves us to deal with quantity and accent as we please.

One way of curing this sort of confusion would be to sink the words' latinity and give them all the ordinary English plural—*testa'trixes* etc. instead of *testatrī'ces* or *testa'trĭces*. For some of them the further anglicizing of *-trix* into *-tress* would also be possible. Another way of escape would be to use the masculine form and drop the feminine. This has distinguished advocacy. 'There is no such word as *executrixes*' said a judge to counsel who had found difficulty in articulating the anglicized plural. 'The word is *executrices*, and even that is bad enough. There is no feminine of *executor*. My father, if he found such a word in a document, would strike it out in wrath and consign the document to the waste paper basket. "Executors" he would say "are always masculine in form though often feminine in fact".' This provoked a letter to *The Times* from the Senior Registrar of Probate. 'It has long been the practice' he said 'of the devotees of the somewhat esoteric cult of probate freely to employ the word *executrixes* . . . and where a testator or testatrix, urged possibly by humility or pride respectively, has appointed persons falling within the category, signature and seal have for many years approved its use.' In view of the present tendency to dispense with FEMININE DESIGNATIONS, the judge's dictum is here respectfully recommended, not indeed as a statement of what the usage is but of what it might well be.

The chief words concerned are: administratrix, cicatrix, directrix, executrix, matrix, prosecutrix, radix, testatrix.

**troche** (the medicinal lozenge). A word that it requires some ingenuity to pronounce wrong, *trōsh*, *trōch*, *and trōk*, being all recognized. The OED draws the line at *trō'kĭ*, which is, it appears, 'commercial and vulgar', but the COD is more tolerant, and admits a disyllabic pronunciation (*trō'kē*) as a fourth alternative. But this would make it indistinguishable from the metrical foot (*trochee*), and we have too many homophones in the language already.

**troop.** *Trooping the colour* is the

orthodox modern phrase; but in the older quotations in the OED it is *colours*. *Troop* is the same word as the French *troupe*, which we have anglicized for the military body but not for the theatrical.

**troublous.** 'Now only literary or archaic' says the OED; and one of its quotations shows well the bad effect of diversifying commonplace contexts with words of that sort; the ordinary *troublesome* was the word wanted: *Mr. Walpole took on himself the management of the Home Office, little knowing what a troublous business he had brought upon his shoulders.*

**trough.** The COD gives pronunciations *trŏf*, *trawf*, and *trŭf* in that order. It might have added *trō*, usual in the Irish countryside. To an Englishman *trŏf* is the established pronunciation.

**trousers.** See PANTALOONS.

**trousseau.** For plural, see -x.

**trout.** Pl. usually the same, see COLLECTIVES 2. But *old trouts* for the affectionately disrespectful slang term.

**trow,** when still in ordinary use, was pronounced *trō*. Now WARDOUR ST.

**truculence, -cy.** See -CE, -CY; and for pronunciation, foll.

**truculent.** OED gave preference to *trŏŏ'kŭ-* over *trŭ'kŭ-*; but the latter has since won, chiefly, no doubt, owing to the much greater ease given by the wider dissimilation of the two vowels; cf. the substitution of *lŏŏ* for *lŭ* in the still more difficult *lugubrious* and *lucubration* (see PRONUNCIATION 6).

**true and false etymology.** English being of all languages the one that has gathered its material from the most varied sources, the study of its etymology is naturally of exceptional interest. It is a study, however, worth undertaking for that interest, and as an end in itself, rather than as a means to acquiring either a sound style or even a correct vocabulary. What concerns a writer is much less a word's history than its present meaning and idiomatic habits. The etymologist is aware, and the person who has paid no attention to the subject is probably unaware, that a *belfry* is not named from its *bell*; that a child's *cot* and a sheep-*cot* come from different languages; that *Welsh rabbit* is amusing and right, and *Welsh rarebit* stupid and wrong; that *isle* and *island* have nothing in common etymologically; and that *pygmy* is a more significant spelling than *pigmy*. But to know when it is well to call an island an isle and when it is not is worth more than to know all these etymological facts. Still, etymology has its uses, even for those whose sole concern with it is as an aid to writing and a preventive of blunders; some knowledge of it may save us from treating *protagonist* as the opposite of *antagonist*, or from supposing a *watershed* to be a river-basin, or from materializing the *comity of nations* into either a committee or a company of them, or from thinking that to *demean* oneself is necessarily to lower oneself or do a *mean* thing or that an *alibi* can be used for any sort of excuse. But the etymology providing such stray scraps of useful knowledge relates much more to the French and Latin elements in our language than to its native or Teutonic substratum.

After this warning that etymological knowledge is of less importance to writers than might be supposed, a selection of words is offered exemplifying the small surprises that reward or disappoint the etymologist. They are arranged alphabetically, and are a very low percentage of what might have been collected; with each word the barest indication only is given of the point. To many readers it will be already known, and by others it may be easily verified in any good dictionary; the object of the list is not to give etymologies, but to provide anyone who is curious about the value of such knowledge with the means of assessing it. The words in small capitals are the

few that happen to have been treated in their places in any way that at all bears upon the present subject.

AMUCK, not E *muck*

andiron and GRIDIRON, not iron

apparel, not L *paro* prepare

arbour, not L *arbor* tree

barberry, not E *berry*

belfry, not E *bell*

blindfold, not E *fold*

bliss, not E *bless*

boon, a prayer, not its granting

bound (homeward etc.), not E *bind*

bridal, not an adjective in *-al*

BRIER (pipe), not E *brier*

bum (buttocks), not a contraction of bottom.

buttonhole (vb.), not *hole* but *hold*

catgut, not made from the intestines of a cat.

CHEVAUX DE FRISE, = Frisian cavalry

cinders, not L *cineres*

cockroach, not *cock* or *roach*

COCOA, COCONUT, unconnected

COMITY, not L *comes* companion

convey, not L *veho* carry

cookie (bun etc.), not E *cook*

COT(E), separate words

court card, a corruption

CRAYFISH, not E *fish*

curtail, not E *tail*

cutlet, not E *cut*

DEMEAN (conduct oneself), not E *mean*

dispatch, not F *dépêcher*

egg on, not *egg* but *edge*

EQUERRY, not L *equus* horse

errand, not L *erro* wander

FAROUCHE, not L *ferox* fierce

FORBEARS, = fore-beërs

FORLORN HOPE not *forlorn* nor *hope*

FUSE (explosive), from L *fusus* spindle

GINGERLY, not E *ginger*

GREYHOUND, not E *grey*

humble pie, a pie made from the *umbles* (intestines of a deer).

incentive, not L *incendo* to fire

ingenuity, stolen by INGENIOUS from *ingenuous*

island, mis-spelt from confusion with *isle*

Jerusalem artichoke, not *Jerusalem* but *girasole* (sunflower)

LITANY, LITURGY, first syllables unconnected

MOOD (gram.), = *mode*, not *mood* (temper)

old dutch, not Dutch but duchess (NETHERLANDS)

pen, pencil, unconnected

PIDGIN, not pigeon.

PROTAGONIST, Gk *prōtos* first, not *prŏ* for

recover, not E *cover*

river, not L *rivus* river

run the gauntlet, not *gauntlet* (glove) but *gantlope* (passage between two files of men)

SANDBLIND, not from *sand*

scarify, not E *scare*

scissors, not L *scindo sciss-* cleave

SLOW WORM, not slow.

SORRY, SORROW, unconnected

vile, villain, unconnected

walnut unconnected with *wall*

WATERSHED, neither a store of water nor a place that sheds water

Welsh rabbit, not *rare bit*.

**truffle.** Pronounce *trŭ'fl*, which is the natural English; association with French cookery leads many people to partly assimilate the sound to that of the differently spelt French word, and absurdly to say *trōō'fl*.

**truism.** The word's two meanings have been compared both with each other and with some synonyms under COMMONPLACE. *It is not permissible to be too sanguine of the outcome of the Conference.* / *The present amount of fall-out does not warrant undue concern.* / *There is no need to be* unnecessarily *anxious about the outbreak.* These are examples of the sort of *t.* that writers should not allow themselves. As to the use of the word itself, the temptation to say that a thing is a truism when no more is meant than that it is true, because it has a smarter sound, should be resisted; so: *It probably owes much to the dialect in which it is played; but that is a truism of almost every Irish or Scotch play.*

**truly.** See LETTER FORMS.

**trusty.** It is still customary for the Sovereign, when issuing a commission to one of her subjects, to address him as her *trusty and well-beloved*, her *right trusty and right well-beloved* if he is a privy councillor. (See also RELATION 2). Otherwise the word is no longer in use except as a deliberate archaism. That is a pity, for it is a good word, much better than its drab cousin *trustworthy*, which enjoyed unmerited popularity in the days when pedants could not bring themselves to use *reliable* (see -ABLE, 4). The nouns particularly likely to attract the adjective *trusty* were *blades* and *steeds* and *henchmen*; it was a product of the age of chivalry and they died together.

**try.** The idiom *t. and do* something is described as colloquial for *t. to do*. Its use is probably commonest in exhortations and promises: *Do t. and stop coughing*; *I will t. and have it ready for you*. And it is hardly applicable to past time or to negative sentences, *He tried and made the best of it* is not English in the sense required, nor is *It is no use to t. and make the best of it*; but *He did t. and make the best of it* will pass, especially if the *did* is emphatic. It is, therefore, colloquial, if that means specially appropriate to actual speech; but not if *colloquial* means below the proper standard of literary dignity. Though *t. to do* can always be substituted for *t. and do*, the latter has a shade of meaning that justifies its existence; in exhortations it implies encouragement—the effort will succeed—; in promises it implies assurance—the effort shall succeed. It is an idiom that should be not discountenanced, but used when it comes natural.

**tryst.** The OED (1926) gives the pronunciation *trīst* only; but the SOED (1933) prefers *trĭst*; the COD gives *trĭst* only, and Thomas Hardy rhymed it with *exist*. But the pronunciation matters little now that the word has become archaic. The parties to a tryst now call it a DATE.

**-t-, -tt-.** Words ending in -t are very numerous, and there seems to be some hesitation about making them conform to the rules that prevail for most consonants: forms like *rivetter*, *carrotty*, *docketted*, are often seen, though good usage is against them. Monosyllables ending in -t double it before suffixes beginning with vowels if the vowel sound preceding the suffix is short, but not if it is long, or followed by r: *pettish*, *potted*, *cutter*, but *flouting*, *sooty*, *skirting*. Words of more than one syllable follow the rule for monosyllables if their last syllable is accented (*coquettish*, but *repeater*); but otherwise they do not double the t: *discomfited*, *combatant*, *wainscoting*, *snippety*, *pilotage*, *balloted*.

**tubercul(ar)(ous)**. These adjectives are often treated as synonymous, but should be differentiated. *Tubercular* means of the nature of, or pertaining to, a tubercle; *Tuberculous* means affected with, or of the nature of, tuberculosis.

**tumultuary, tumultuous**. The distinction between the two is not very definite, and sentences may easily be made in which either might be used and give the same sense. But it may be said, first, that -*tuous* is now the much commoner word, which should be chosen unless there is good reason to prefer the other; and, secondly, what is emphasized by -*tuous* is rather the violence and impetus and force, while -*tuary* emphasizes the irregularity and indiscipline of the thing described: *tumultuous* applause, seas, attack, joy, crowd; *tumultuary* forces (hastily levied), risings (unorganized).

**tumulus**. Pl. -*lī* only.

**turbidity, turbidness**. See -TY AND -NESS and TURGID.

**turbine**. OED recognizes only the pronunciation with -*ĭn*; but that with -*īn*, now often heard, though due only to misguided reverence for spelling, is likely to prevail; the COD gives no other.

**Turc-**. See TURK.

**tureen**. Strictly the right pronunciation is *tĕrē'n*, in accordance with the derivation (*terra* earth) and the older English spelling (*terrene* etc.); but it is now always *tūrē'n*.

**turf**. Pl.; -*fs* and -*ves* appear an equal number of times in the post-18th-c. quotations of the OED, which itself uses -*fs*. But turf in its Irish sense of peat is cut into *turves* for burning. See -VE(D).

**turgid** (swollen) is sometimes confused with *turbid* (muddy). Not only do the words look alike, but their figurative meanings tend to coalesce: writing that is turgid in style is often turbid in sense. That may explain the confusion but does not excuse it; there is the more need for care in preserving the difference.

**Turk**. In spite of the spellings of *Turkey* and *Turkish*, *Turk* makes *Turco*—in such compounds, as *Turcophil* and -*phobe*. The inhabitants of the area formerly known as Turkestan are now usually spelt *Turkoman* (pl. -*mans*).

**turn, v.** In the age idiom two constructions are recognized: *I have turned 20* and *I am turned 20*. The old form *I am turned of 20* is no longer idiomatic. The meaning is not 'over 20' but 'past the 20th birthday'.

**turquoise**. Pronunciation debatable. With Ben Jonson, Shakespeare, Milton, and Tennyson, all for *ter'kĭz* (or something like it), it is a pity that we cannot return to that; but the adoption of the later French spelling has corrupted us, and the OED labels *ter'kĭz* archaic On the other hand it refuses to recognize the *kw* sound for the -*qu*- and complete the triumph of spelling. But popular judgement has gone against it, and *turkwoiz* or *turkwahz*, or something between the two, is now customary.

**tushery**. See ARCHAISM.

**twelve-tone**. See ATONAL.

**twilit.** The earliest OED quotation for the word is 1869, so that, whatever its merits may be, it is not venerable. Its formation implies a verb *to twilight* made from the noun; and that verb, though unknown to most of us, is recorded as having been used. It also implies that *to twilight* has p.p. *twilit* rather than *twilighted*, which is not impossible. But, though *twilit* can therefore not be absolutely ruled out, it is better to use *twilight* attributively where that does the work equally well, as it usually does, and elsewhere to do without. In the two following quotations, *twilight* would have served at least as well: *He found himself free of a fanciful world where things happened as he preferred—a twilit world in which substance melted into shadow. | The years of the war were a clear and brilliantly lit passage between two periods of twilit entanglement.*

**-ty and -ness.** Though any adjective may be formed into a noun on occasion by the addition of -ness, the nouns of that pattern actually current are much fewer than those made from Latin adjectives with -ty, -ety, or -ity as their ending. Thus from *one* and *loyal* and *various* we can make for special purposes *oneness, loyalness,* and *variousness*; but ordinarily we prefer *unity, loyalty,* and *variety*. Of the -ty words that exist, a very large majority are for all purposes commoner and better than the corresponding -ness words, usage and not anti-latinism being the right arbiter. Scores of words could be named, such as *ability, honesty, notoriety, prosperity, sanity, stupidity,* for which it is hard to imagine any good reason for substituting *ableness, notoriousness,* etc. On the other hand words in -ness that are better than existent forms in -ty are rare; though *acuteness* and *conspicuousness* have the advantage of *acuity* and *conspicuity*; and if *perspicuousness* could be established in place of *perspicuity* it might help to obviate the common confusion with *perspicacity*. But in general a -ty word that exists is to be preferred to its rival in -ness,

unless total or partial differentiation has been established, or is designed for the occasion. Total differentiation has taken place between *ingenuity* and *ingenuousness, casualty* and *casualness, sensibility* and *sensibleness, enormity* and *enormousness*; the use of either form instead of the other changes or destroys the meaning. Partial differentiation results from the more frequent use made of the -ty words. Both terminations have, to start with, the abstract sense of the quality for which the adjective stands; but while most of the -ness words, being little used, remain abstract and still denote quality only, many of the -ty words acquire by much use various concrete meanings in addition; e.g., *humanity, curiosity, variety,* beside the senses 'being human, curious, various', acquire those of 'all human beings', 'a curious object', and 'a sub-species'. Or again they may be habitually applied in a limited way so that the full sense of the adjective is no longer naturally suggested by them; *preciosity* is limited to literary or artistic style, *maturity* suggests the moment of reaching rather than the state of matureness, *purity* and *frailty* take a sexual tinge that *pureness* and *frailness* are without, *poverty* is more nearly confined to lack of money than *poorness*. It is when lucidity requires the excluding of some such meaning or implication attached only to the -ty form that a -ness word may reasonably be substituted.

Articles under which special remarks will be found are BARBARISM, BARBARITY, ENORMOUS, INGENIOUS, OBLIQUENESS, OPACITY, POVERTY, PRECIOSITY, SENSIBILITY. For similar distinctions between other nearly equivalent terminations, see -CE, -CY, -IC(AL), -ION AND -NESS, -ION AND -MENT, -ISM AND -ITY.

A few specimens may be added and classified that have not been cited above, but are notable in some way, (A). Some words in -ty for which there is no companion in -ness; the Latin adjective not having been taken into English: celerity, cupidity, debility, fidelity, integrity, lenity, utility. (B).

Some more in which the -ty word has a concrete or other limited sense not shared by the other: ambiguity, capacity, commodity, fatality, festivity, monstrosity, nicety, novelty, speciality, subtlety. (C). Some of the few in -ness that are as much used as those in -ty, or more, though the -ty words exist: clearness (clarity), crudeness, falseness, jocoseness, morbidness, ponderousness, positiveness, tenseness, unctuousness. (D). Some -ness words that have no corresponding form in -ty in common use, though the adjective is of Latin origin and might have been expected to produce one: crispness, facetiousness, firmness, largeness, massiveness, naturalness, obsequiousness, pensiveness, proneness, robustness, rudeness, seriousness, tardiness, tediousness, tenderness, vastness, vileness. (E). If there is also a -tion word, derived from the verb, this naturally signifies the process, and the -ty word, derived from the adjective, the result, e.g. liberty and liberation, multiplicity and multiplication, profanity and profanation, satiety and satiation, variety and variation. But sometimes these pairs develop by usage a sharper differentiation, e.g. inanity and inanition, integrity and integration, sanity and sanitation.

**tycoon** (Great prince) was the title applied by foreigners to the military ruler (Shogun) of Japan in the times (before 1867) when the Mikado's temporal power was usurped. Its adoption in U.S. as a colloquial term for a business magnate dates from the early 20th. c., and it has now taken firm root in Britain also.

**tyle(r).** See TILE(R).

**tympan(um)(o).** Pronounce *tĭm'pă-*. *Tympanum*, the eardrum, has plural *-a*; *tympano*, or more usually *tímpano*, the kettledrum, has plural *-i*, pronounced *ē*.

**-type.** There is much confusion and other misuse of the words *antetype*, *antitype*, *prototype*, and *archetype*, as

in the following extracts. *People may wonder whether he always knows the meaning of the words he uses when they find him calling a wooden copy of the Queen Elizabeth put up to deceive the Germans her* 'prototype' (*antitype*, if any type, but better *counterfeit*). |*The fees of the most successful barristers in France do not amount to more than a fraction of those earned by their* prototypes *in England* (should be *counterparts*, or, colloquially, *opposite numbers.*| *The type of mind which prompted that policy finds its* modern prototype *in Unionist Ulster* (should be *equivalent*). | '*I presume you bring this war figure into dramatic contrast with his* anti-type.'—'*Yes; and with the other types of the* . . .' (should be *opposite*).

The true meanings of the words are as follows:

*ANTETYPE* ('a preceding type, an earlier example'—OED) is a rare word that should hardly ever be used, first because its similarity in sound and opposition in sense to the established *antitype* is inconvenient, secondly as being liable to confusion with *prototype* also from their closeness in meaning, and thirdly because *forerunner* is ready to take its place when it really does not mean prototype.

*ANTITYPE*, unlike the others, is not a type. *Type* and *antitype* are a complementary pair or correlatives: the type is the symbol or emblem or pattern or model, and the antitype ('lit. responding as an impression to the die'—OED) is the person or object or fact or event in the sphere of reality that answers to its specification.

*PROTOTYPE*, on the other hand, serves, with limitations, as a synonym for *type*. In particular it may be preferred (i) to emphasize (like *archetype*, see below) the priority in time of a certain type over its antitype, or (ii) when *type*, which has other senses, might be ambiguous, or (iii) when typification itself is of no great consequence, and the sense wanted is no more than 'the earliest form' of something. With this last meaning it has come into general use for the first of a new type of aircraft or other machine constructed experimentally before being put into production.

*ARCHETYPE* is sometimes used as a dignified synonym of *prototype*: '*The House of Commons, the a. of all representative assemblies* (Macaulay). | *Houdini is the archetypal escapologist, as Sherlock Holmes is the archetypal detective.* Recently it has been given a more specialized meaning as the name applied by Jung to what he first termed the 'primordial images' underlying those unconscious mental processes common to all mankind that he called 'the collective unconscious'. In this sense, or something like it, *archetype* and *archetypal* have had the misfortune to be seized on as POPULARIZED TECHNICALITIES, and have suffered the usual consequences. 'A year ago,' writes a puzzled reviewer, 'Fr. Gerald Vann O.P. published his book on *Trees of Life*, with archetypal bearings on Eden and the Crucifixion. . . . Just over a year ago Mr. F. J. Stopp studied the archetypal structure of the novels of Mr. Evelyn Waugh. Last March Mr. D. Streatfeld equated the underworlds to which were abducted the Persephone of Greek myth and the heroine of *No Orchids for Miss Blandish*, discovering archetypal significance in barmaids and cloakroom attendants, more especially if they were blondes.' As Jung himself sadly admitted, 'the concept of the archetype has given rise to the greatest misunderstandings . . . and must be presumed to be very difficult to understand'.

**typescript.** See LONGHAND.

**typographic(al).** Both forms are in use, and no shade of difference seems discernible in the OED quotations, except that those for -al are more numerous. This trend has continued; -al is now the ordinary adjective. See -IC(AL).

**tyrannic(al).** *Tyrannic* is now not at home outside verse. See -IC(AL).

**tyrannize.** *This attempt to coerce and*

tyrannize us will produce results which the Government will have good reason to regret. | They were 'the strong, rugged, God-fearing people' who were to be tyrannized and oppressed by a wicked Liberal Government.* Most readers of good modern writing will have the familiar slight shock incident to meeting a solecism and want to insert 'over'. But the OED's comment on the transitive use is merely 'now rare', and it produces abundant examples from older writers. Still, the present idiom is to *tyrannize over*, not to *tyrannize*, one's subjects.

**tyrant.** The original Greek sense of the word is so far alive still that readers must be prepared for it. Neither cruel nor despotic conduct was essential to the Greek notion of a tyrant, who was merely an absolute ruler owing his office to usurpation. The word connoted the manner in which power had been gained, not the manner in which it was exercised; despotic or 'tyrannical' use of the usurped position was natural and common, but incidental only. *Dictator*, originally the name of a Roman official appointed in time of grave emergency, has been similarly besmirched by the behaviour of later dictators.

**tyre, tyro.** See TIRE, TIRO.

# U

**u.** For the pronunciation of long *u* (*yoo* or $\overline{oo}$) see PRONUNCIATION 6.

**u and non-u.** These descriptive labels have won wide popularity. They originated in an essay by Professor Alan Ross (1956), on *Linguistic Class-indicators in present-day English*, published in a Finnish philological journal. The English class system, he said, was essentially tripartite—upper, middle, and lower—and it was only by its language that the upper class (U) was clearly marked off from the others (Non-U). 'The line of demarcation', he added, 'is often a line between, on the one hand, gentlemen, and, on the other, persons who, though not gentle-men, might at first sight appear, or would wish to appear, as such.' But he did not define *gentleman* otherwise than by describing the usages and taboos, almost wholly linguistic, that he held to be the distinguishing marks of that class. The authoritative work on the subject is *Noblesse Oblige: an Enquiry into the identifiable Character-istics of the English Aristocracy*, edited by Nancy Mitford: Hamish Hamilton, 1956. It contains Ross's essay and some diverting variations on his theme by other writers.

The shibboleths whose observance is said to distinguish the U from the Non-U, so far as they are matters of pronunciation, are, though more esoteric, similar to those that mark the division between the minority who speak with the RECEIVED PRONUNCIA-TION and the majority who do not. The element common to both—the notion of a select class recognizable by the way it speaks—no doubt accounts for the spleen of some of the attacks made on the Received Pronunciation by critics touchy about such things. See that article. Other shibboleths, no less important, depend on the choice of words. Articles in this dictionary with a bearing on what, by 'U' stan-dards are (to borrow Ross's adjectives) 'correct, proper, legitimate or appro-priate' include those on ESQUIRE, EUPHEMISMS, GENTEELISMS, GIRL, GOLF, HON., LADY, LETTER FORMS, LORD, NAMES and APPELLATIONS, PRONUN-CIATION, RALPH, RECESSIVE ACCENT, REVEREND, SIR, and TITLES.

**-ular.** Adjectives ending thus are something of a trap to those who like words to mean what they seem to say. They are made from diminutive nouns, but cannot be relied on to convey a diminutive sense; a *glandule* is necessarily a small gland; but *glandu-lar* means 'of glands' not 'of small glands'. *Cellular* means consisting of cells, not necessarily of the minute cells known as cellules. The ending -ular has become a favourite with ad-jective-makers, and such an adjective is often preferred to one that is, or might be, made directly from the simple noun instead of from the dimi-nutive. So also *globular* for *globose*, *granular* for *graneous* or *granose*, *tubular* for *tubal*, *valvular* for *valvar*.

**ulna.** Pl. *-nae*

**ultimatum.** Pl. *-ta*, *-tums*; see LATIN PLURALS and -UM. Considering that *-tums* is more than 200 years old (Swift is quoted in OED), it is strange that anglicization has been so long delayed, and that *-ta* was in a large enough majority to justify OED in presenting it alone as the plural; *-tums* (now put first by the COD) is here recommended.

**ultimo, ult.** See COMMERCIALESE.

**ultra,** originally a Latin preposition and adverb meaning beyond, was adopted as an English noun in the early 19th c. (pl. *-as*) meaning a person who goes beyond others in opinion or action of the kind in question. This was no doubt a development of its use as a prefix in such adjectives (and nouns) as *ultra-fashionable(s)*, *ultra-revolutionary* (*-ries*). Such compounds were curtailed into *ultra* adj. and n.; but this was no longer felt to be, like *sub* when used for *subaltern* or *sub-scription*, a CURTAILED WORD. It won independence of any second element, its own meaning being sufficient, and became a synonym for the now more usual *extremist*. (*Even some of the former ultras are making the discovery that the Common Market issue remains a balance of advantages and disad-vantages.*) Fanatical ultras in any cause are now known by the collective term *lunatic fringe*, which has been described as 'a splendidly prejudicial British phrase, with its suggestion of hair dragged villainously low over the forehead or edging the circumference of the face in the way that magistrates so disapprove of'. But the origin of the phrase does not seem to be British; it is attributed to President Theodore Roosevelt.

**ultramontane.** With the full or exact meaning of ultramontanism as now understood we need not concern ourselves, beyond defining it roughly as the policy of raising the authority of the Pope in all matters to the highest possible level. But to those who are not content to accept words as arbitrary tokens, and do not see why a papal zealot should be an 'over-the-hills' man, an explanation may be welcome. The mountains are the Alps, and beyond the mountains means, to an Italian, outside Italy, and, to others, in Italy. So, when there were differences in the Church about the right relation between the Italian bishops and the extra-Italian, each party could describe the other as the Ultramontanes, which makes the historical use of the word confusing. In modern use it is applied, chiefly by opponents, to the party of Italian predominance, whose principle is the absolute supremacy of the Pope, and the denial of independence to national Churches.

**ultra vires.** Pronounce -*ī'ēz*. See LATIN PHRASES.

**ululate, -ation.** OED gives precedence to *ŭlū-* over *ūlū-*, and it may have known that it was stating the prevalent usage. But the pronunciation of words seldom heard is hard to be sure of; and, unless there are reasons against it, it seems plain that the imitative effect got by repeating the same sound should not be sacrificed; *ūlūl-* suggests howling much more vividly than *ŭlūl-*, and is the pronunciation given by at least one modern dictionary.

**-um.** For general remarks on the plural of Latin nouns adopted in English, see LATIN PLURALS. Those in -um are numerous and demand special treatment. The Latin plural being -a, and the English -ums, three selections follow of nouns (1) that now always use -ums, either as having completed their naturalization (as it is to be hoped the rest may do in time), or for special reasons; (2) that show no signs at present of conversion, but always use -a; (3) that vacillate, sometimes with a

differentiation of meaning, sometimes in harmony with the style of writing, and sometimes unaccountably. In deciding between the two forms for words in the third list, it should be borne in mind that, while anglicization is to be desired, violent attempts to hurry the process actually retard it by provoking ridicule.

1. Plural in -ums only (those marked * are not Latin nouns, and the -a plural for them would violate grammar as well as usage): albums; antirrhinums; asylums; conundrums*; decorums; delphiniums; Elysiums; factotums*; forums; harmoniums; laburnums; lyceums; museums; nasturtiums; nostrums; panjandrums*; pendulums; petroleums; pomatums; premiums; quorums*; targums*; vellums*.

2. Plurals in -a only: agenda; bacteria (and many scientific terms); crania; curricula; desiderata; dicta; errata; maxima; minima; momenta; quanta; scholia (and other such learned words); strata; succedanea vela.

3. Words with either plural; the remarks in brackets are inserted as suggestions only: aquarium (usu. -ms); compendium; emporium; encomium (usu. -ms); exordium; honorarium (usu. -a); interregnum (usu. -ms); lustrum (usu. -a); medium (-ms in spiritualism); memorandum (usu. -a); millennium; rostrum (usu. -a); spectrum (usu. -a); speculum (usu. -a); stadium (-ms gaining); trapezium (usu. -a); ultimatum (-ms better); vacuum.

**umbilicus, -ical.** The OED recognizes only *ŭmbĭ'lĭkal* for the adjective, but for the noun gives precedence to *ŭmbĭlī'kus* over *ŭmbĭ'lĭkus* The former is now invariable and is affecting the pronunciation of the adjective, which will probably conform. See FALSE QUANTITY for the question involved.

**umbo.** Pl. *-os* or *-ō'nēs*; see LATIN PLURALS.

**umbra.** Pl. *-rae*.

**un, 'un,** = one, as in *that un, young*

*un, old un, game un*, and such phrases, needs no apostrophe or hyphen.

**un-.** 1. Danger of ellipsis after un-words. 2. Not un-. 3. Un-, in-.

1. Danger of ellipsis after un-. *Untouched* means not touched, but with the difference that it is one word and not two, a difference that in some circumstances is important. In *I was not touched, and you were* the word *touched* is understood to be repeated, and not to carry the *not* with it; but *I was untouched, and you were* cannot be substituted with the same effect; if it means anything, it means that both were untouched, the *un-* having to be understood as well as the *touched*. Needless as such a statement may sound in a simple case like the above, where there is nothing to distract attention from the wording, blunders essentially similar are frequent. A couple of examples follow, and the state of mind that produces them is fully illustrated in NEGATIVE MISHANDLING: *Dr. Rashdall's scholarship is unquestioned; most of his writings and opinions on ecclesiastical matters are.* What is meant is that most of them are questioned, not unquestioned. / *When I sat in the square of Oudenarde, opposite the old Hôtel de Ville, which happily has come through the war untouched by Vandal hands, methought, if it had been, who in Belgium could have built the like of it?* That is, had been touched, not untouched; correct *untouched* into *without being touched*.

2. For the meiosis *not un-* see NOT 2.

3. For the choice between *un-* and *in-* as a negative prefix see IN- AND UN-.

**unaccountable.** *Occurrences that are for the time being, and to the spiritualist, unaccountable by natural causes. U.* itself belongs to the class of words, including *reliable*, whose legitimacy is upheld in -ABLE 4; but to use *by* after it, compelling the reader to resolve it into its elements (*not to be accounted for*), and so discover that *for* is missing, is very indiscreet.

**unanimous, nem. con.** A motion is said to be carried unanimously when everyone present votes for it; nem. con. (nemine contradicente) when, though no one votes against it, one or more persons abstain from voting.

**unapt, inapt, inept.** See INAPT (NESS) etc.

**unartistic, in-.** The second is the usual word; but since it has acquired a sort of positive sense, 'outraging the canons of art' etc., the other has been introduced for contexts in which such condemnation is not desired; the *unartistic* are those who are not concerned with art. See IN- AND UN-.

**unattached participles** and adjectives (or wrongly attached). A firm sent in its bill with the following letter: *Dear Sir, We beg to enclose herewith statement of your account for goods supplied, and* being desirous *of clearing our Books to end May will* you *kindly favour us with cheque in settlement per return, and much oblige.* The reply ran: *Sirs, You have been misinformed. I have no wish to clear your books.* It may be hoped that the desire on which they based their demand was ultimately (though not per return) satisfied, but they had certainly imputed it to the wrong person by seeming to attach *being desirous* not to the pronoun it belonged to (*we*), but to another (*you*). The duty of so arranging one's sentences that they will stand grammatical analysis is much more generally recognized than it formerly was, and it is now not a sufficient defence for looseness of this kind to produce parallels, as can very easily be done, even from great writers of past generations; on this see ILLOGICALITIES. On the other hand it is to be remembered that there is a continual change going on by which certain participles or adjectives acquire the character of prepositions or adverbs, no longer needing the prop of a noun to cling to; we can say *Considering the circumstances you were justified*, or *Roughly speaking they are identical*, and need not correct into *I think you were justified* and *I should call them identical* in order to regularize the participles.

The difficulty is to know when this development is complete; may I write *Referring to your letter, you do not state . . .*, or must it be *I find you do not state . . .?* i.e., is *referring* still undeveloped? In all doubtful cases it is best to put off recognition. DUE to (*There was no play today due to rain*) is an example of such a development that has made great headway but not yet won full recognition. *Based upon* (*Based upon your figures of membership, you suggest that the Middle Class Union has failed*) is another. See also FOLLOWING and PRIOR TO.

The conscious or unconscious assumption that a participle or adjective has acquired the powers of preposition or adverb when it has in fact not done so perhaps accounts for most of the unattached and wrongly attached; but there are many for which no such excuse is possible. Before proceeding to them, let us make a few sentences containing undoubtedly converted participles, sentences in which the seeming participle is not felt to need an associated noun or pronoun: *Talking of test matches, who won the last?*; *Coming to details, the spoilt ballot-papers were 17*; *Barring accidents, it ought to work*; *Seeing that it is fine now, a start may as well be made*; *They are illiterate* (*using the word in its widest sense*); *Granting his honesty, he may be mistaken*; *Failing you, there is no chance left*; *Twelve were saved, not counting the dog*; *Allowing for exceptions, the rule may stand.* It is natural, and perhaps right, to explain this common type as originally not a participle at all, but a shortening of the gerund preceded by the old preposition *a*; *talking of = a-talking of*, i.e. in talking or while there is talk of. However that may be, it is only fanatical purists who will condemn such sentences, and a clear acknowledgement of their legitimacy should strengthen rather than weaken the necessary protest against the slovenly uses now to be illustrated. After each extract will be given in brackets first the word or phrase, whether present or

not, to which the participle or adjective ought to be attached, and secondly the one, if any, to which careless grammar has in fact attached it: *A belief that a Committee of Inquiry is merely an evasion, and that, if* accepted, *the men will be caught out* (Committee; men). | *A girl fell on a pen, which pierced her eye, and,* causing *meningitis, she died* (which; she). | *Having accepted a rearmed Germany into NATO, it is only a matter of time before she will be equipped with nuclear weapons* (the other western powers; she). | *Twice burnt and rebuilt, the present house was designed in 1850* (previous houses; the present house). *Leaning over the bridge was a water-rat* (the narrator; the water-rat). | *Having decided to send a space-ship into orbit with a man inside, the question will be . . .* (the explorers of space; the question). | *While not dissenting from the broad approach of the Government to this matter, would the Prime Minister not consider it desirable to follow up Mr. Khrushchev's suggestion?* (the questioner; the Prime Minister). In fairness it should be added that the elliptical form of the last example has become standard practice in asking supplementary questions, and no doubt saves valuable parliamentary time. But the same plea cannot be made for the correspondent who begins *While grateful for the main points made in your second leader last Sunday, one of secondary liturgical interest arises from it.* (See also WHILE.)

These examples illustrate the most flagrant type of unattached participles. Others, more excusable, but nevertheless frowned on by grammarians, are (a) those in which a possessive adjective serves as the peg to hang the participle on (*Having been defeated by the Larkey Boy, his visage was in a state of such dilapidation as to be hardly presentable.* | *Handing me my whisky, his face broke into an awkward smile*); and (b) those in which the participle is meant to be linked not with the subject of a verb but with its object, direct or indirect, as it might be with *us* in the quotation that opens this article; or is with *me* in Malvolio's ex-postulation to Olivia: *And, acting this in an obedient hope, Why have you suffered me to be imprisoned?* and with *him* in *Anything much before Dryden was to Johnson's mind 'archaic', and, having little historic sense and no adequate conception at all of a theory of development, 'archaic' stood to him for 'imbecile' 'childish'*.

For participles that are unattached because used in an ABSOLUTE CON-STRUCTION without the noun or pronoun that grammarians hold to be necessary to it (*Being August, the oleanders were in flower*) see that article, and for the tendency of 'sentry' participles to be left in the air see PARTICIPLES 4.

**unaware(s).** *Unaware* (with *of*) is the adjective, *unawares* the adverb. *I was unaware of the danger*; *The danger took me unawares*.

**unbeknown(st).** Both forms are now rare except in dialect or un-educated speech or in imitations of these. The *-st* form is more exclu-sively adverbial; cf. *unawares* as the adv. of *unaware*. It is also the one more likely to be chosen in un-educated or facetious speech; *unbe-known* (*to*) is still occasionally used seriously.

**unbending,** as participle of *to un-bend*, means throwing off stiffness, but as a participial adjective compounded of *un-* and *bending* it means never throwing off stiffness. Contrasts, not usually so diametrical as this, often result from the prefixing of *un-* at different stages; e.g., in 'lessons learnt and unlearnt', *unlearnt* may mean either of two very different things as may *undone* in *Our plans are undone*. *Undeceived* might have provided an-other example, but in fact it is used only in the sense of disabused.

**unbias(s)ed.** The spelling varies; see *-s-, -ss-*. The single *s* is now in favour.

**un-come-at-able.** The word had

doubtless, two or three centuries ago, a jolly daredevil hang-the-grammarians air about it. That has long evaporated; it serves no purpose that *inaccessible* or *unattainable* does not; it requires a writer to choose between five forms (*uncomatable* is the other extreme); and it surely deserves a place among SUPERFLUOUS WORDS. Its more modern equivalent *un-get-at-able* is equally un-wanted.

**uncommon.** The old slang use as an adverb = remarkably (*an u. fine girl* etc.) has nearly died out, and is no longer in place outside the dialogue of period novels.

**unconditional.** To emphasize that an offer, loan, or gift is unconditional, the colloquial Americanisms *with no strings attached*, and, less frequently, *with no tabs on* have been welcomed. The former is a development of an old metaphor (cf. *to have on a string, to pull the strings*); the latter comes from the American use of *tab* for a check kept by a creditor of what is owing to him.

**unco-operative.** See CO-, where the advice is given that *cooperate* etc. should be written without hyphen or diaeresis. The ambiguity of *unco-operative* to a Scotsman is the more reason for omitting the hyphen in that word.

**undependable.** For the legitimacy of this and similar words, see -ABLE 4.

**under,** prep. See BELOW, BENEATH, and UNDERNEATH for distinctions.

**under consideration.** See CON-SIDERATENESS, CONSIDERATION.

**underlay, -lie,** vbs. The confusion noticed in LAY and LIE is worse con-founded for the compounds; see the remarks on OVERLAY.

**underneath** (prep.), compared with BELOW and *under*, is not, like BENEATH, a word that tends to become archaic; on the contrary, it is still in full colloquial as well as literary use. Its range is much narrower than those of the other words, being almost confined to the physical relation of material things (cf. 'underneath the arches' with 'below par', 'under the influence', 'beneath contempt'), but within that range it is often preferred as expressing more emphatically the notion of being covered over, and carrying a step further the difference pointed out between *below* and *under*.

**underprivileged.** See PRIVILEGE.

**understatement.** See MEIOSIS.

**under the circumstances.** See CIRCUMSTANCES.

**undertones.** See OVERTONE AND UNDERTONE.

**undigested, undisciplined, undiscriminating.** All better than the *in-* forms; see IN- AND UN-. The *in-*, which is at variance with the prevalent modern usage, owes its escape to the protection afforded by *indigestion, indigestible, indiscipline,* and *indiscriminate.*

**undistributed middle.** A FALLACY. The u. m. is the logical name for a middle term that is not made universal; see SYLLOGISM, where the middle term *men* is made universal by the word *all*, or 'distributed'. Such distribution is necessary to the validity of the conclusion, and *the fallacy of the u. m.* consists in allowing a middle term that is not universalized to pass as universally true. Thus we know or believe that wet feet result in colds; we catch cold, and say 'I must have got my feet wet'; i.e., in syllogistic form:
Colds are wet-feet products.
My trouble is a cold.
Therefore my trouble is a wet-feet product.
Which would be sound if colds meant all colds, but not if it merely means some colds.

**undue, -duly.** *There is no need for undue alarm.* Well, no; that seems likely. See TRUISM; in the making of

truisms *undue* is one of the favourite ingredients.

**uneatable, inedible.** The former is usually applied to what cannot be eaten because of its condition and the latter to what cannot be eaten because of its nature.

**uneconomic(al).** For the distinction see ECONOMIC(AL).

**unedited.** Better than the *in-* form, which some of those who are literary by profession seem to prefer, perhaps under the influence of the French *inédit*; see IN- AND UN-.

**unequal.** *She has been compelled to undertake an offensive for which, as events have proved, she was wholly unequal.* | *A simplicity that seems quite unequal to treat the large questions involved.* The preposition after *u.* is *to*, not *for*; but if a verbal phrase with *to* is used it must be *to* with the gerund, not with the infinitive; see GERUND 3.

**unequal yokefellows.** The phrase is here used in a comprehensive sense enabling a number of faults, most of them treated at length in other articles, to be exhibited side by side as varieties of one species. They are all such as do not obstruct seriously the understanding of the passage in which they occur, but they do inflict a passing discomfort on fastidious readers. For a writer who is not fastidious it is an irksome task to keep in mind the readers who are, and he inclines to treat symmetry as troublesome or even obtrusive formalism; he too could be mechanically regular if he would, but he is not going to be at the pains of revising his first draft into conformity with niceties that are surely of no consequence. It is true that such revising is an ungrateful task; but there must be something wrong with a writer who, by the time he is through his apprenticeship, is not free of the need for this sort of revision; to shape one's sentences aright as one puts them down, instinctively avoiding lopsidedness and checking all details of the framework, is not the final crown of an accomplished writer, but part of the rudiments of his trade. If one has neglected to acquire that habit in early days, one has no right to grumble at the choice that later confronts one between slovenliness and revision.

Conspicuous among the slights commonly inflicted upon the minor symmetries are those illustrated below:

*Scarcely* (temporal) demands *when* or *before*: *Scarcely was the drain finished* than *several sickened with diphtheria.* See SCARCELY.

*Each* demands a singular verb: *The opportunities which each* are *capable of turning to account.* See EACH.

Only the expressed part of a verb used with one auxiliary can be 'understood' with another: *Examples of false and exaggerated reports have always and will always disturb us.* See ELLIPSIS.

A subjunctive in one of two parallel clauses demands a subjunctive in the other: *If the appeal* be *made and* results *in. . . .* See SUBJUNCTIVES (arrivals).

Sealing up of a subject within its verb demands repetition of the subject if it is to serve again: *Does he dislike its methods and* will *only mention . . .?* See PARALLEL-SENTENCE DANGERS.

One or two other types may be added without references: *Either he did not know or was lying* (read *He either*); *The old one was as good if not better than this* (read *as good as this if not better*); *One of the worst kings that has ever reigned* (read *have*); *It is all and more than* I expected (read *all I expected, and more*); *He was young, rich, handsome, and enjoyed life* (read *and handsome.* See also ILLITERACIES and INCOMPATIBLES.

**unexceptionable.** See EXCEPTIONABLE.

**un-get-at-able** see UN-COME-AT-ABLE.

**unhuman.** For the use of this by the side of *inhuman*, see IN- AND UN-.

**unidiomatic -ly.** As the lapses from idiom here to be illustrated probably owe their origin to the modern wider extension of grammatical knowledge, it may be prudent to start by conciliating the sticklers for grammar and admitting that a -ly is sometimes missing where it is wanted. So: *The Carholme course, shaped very similar to the Doncaster Town Moor, is one of the best in England.* | *Proceedings instituted by the local Education Committee against the mother for neglecting to send her girl to school* regular. | *I hope that most teachers in the present day have learnt to read the Old Testament (thanks to the higher critics)* different *from the way I was taught to read it in my youth.* | *Surely no peace-loving man or woman will deny that it would be advisable to prevent strikes and lock-outs* consistent *with the principles of liberty as set forth by John Stuart Mill.*

But, if grammar is inexorable against *consistent* and *different* and the rest, it would in the following sentences allow *contrary* and *irrespective* without a frown, while idiom for its part would welcome them: *The provision is quite inadequate and very grudgingly granted, and often,* contrarily *to the spirit of the Act, totally denied.* | *Loyal obedience is due to the 'powers that be', as such,* irrespectively *of their historical origin.* | *His method is to whitewash them all vigorously with the same brush,* irrespectively *of differences in the careers and characters of his heroes.* Contrary and *irrespective* are among the adjectives that have, with others mentioned in UNATTACHED PARTICIPLES and in QUASI-ADVERBS, developed adverbial force; to ignore that development is bad literary judgement, but, among the mistakes made with -ly, one of the least.

A degree worse is the use of a -ly adverb where idiom requires not an adverb at all, but a predicative adjective. See LARGE(LY) for the phrases *bulk* and *loom large*, and substitute adjectives for adverbs in the following quotations: *In neither direction can we fix our hopes very* highly. | *This country was brought much more* closely *to disaster at sea than ever the Allies were on land* (much closer). |

*It is a gigantic labour before which the labours of Westphalia, of Utrecht, of Vienna, pale* insignificantly. Sometimes indeed it would not be easy to say whether a predicative adjective or an adverb is more idiomatic. It does not matter which we call *hard, bright,* and *wide* in It froze *hard*, the fire burned *bright*, the doors were flung open *wide*; the adverbial or adjectival form can be used with equal propriety in He played the melody *loud(ly)*, it was dangling *loose(ly)*, they sat *idl(e)(y)* by the fire, when trenching dig *deep(ly)*. Pedantic criticism is often heard of harmless injunctions such as *Drive slow, Hold tight, Turn sharp left*: there is an old joke about a drowning lady who when adjured by her rescuer to hold tight 'murmured *say tightly* as she went down for the third time'. If such phrases really need to be justified grammatically, this tendency for the adverb and adjective to merge affords a colourable plea.

Yet a little worse is the officious bringing up to date of such time-honoured phrases as *mighty kind, sure enough* and *safe and sound*: Still, it is *mightily kind of the* Morning Post *to be so anxious to shield the Labour Party from the wrath to come.* / We begin to remember the story of the detective who died murmuring to himself 'More clues!' and towards the end of the book, surely enough, more clues there are. / I hope your daughter will get home safely and soundly, Mrs. F.

But much more to be deprecated than all the particular departures from idiom already mentioned is the growing notion that every common adjective, if an adverb is to be made of it, must have a -ly clapped on to it to proclaim the fact. Of very many that is not true; see for instance DEAR, DIRECT, MOST, PRETTY, RIGHT, and STRAIGHT. Two such words may here be taken for special treatment, *much(ly)* as the least, and *hard(ly)* as the most, important of all. We all know that *much* can be an adverb, and probably most of us would guess that *muchly* was a modern facetious formation, perhaps meant to burlesque the

ultra-grammatical, and at any rate always used jocosely. We should be wrong; it is over 300 years old. Its earliest use was serious, and even now it may occasionally be met in contexts where the point of the joke is not apparent: *Many players who were in the habit of relying muchly upon the advice of their caddies found themselves completely at sea.* Nevertheless, as it seems from the OED to have lain dormant for over 200 years, our guess is not so far out, and its revival in the 19th c. illustrates the belief that adverbs must end in -ly. *Muchly* does not often make its way into print, except in dialogue as a recognized symbol of the mildly jocose talker, and has been worth attention only in contrast with *hardly*. That, as will appear, is substituted in print for the idiomatic *hard* neither seldom nor with any burlesque intention, but seemingly in ignorance. Ignorance that *hard* can be an adverb seems incredible when one thinks of *Hit him hard, Work hard, Try hard,* and so forth; the ignorance must be of idiom rather than of grammar. Neglect of idiom is, in this case, aggravated by the danger that *hardly*, written as meaning hard, may be read as meaning scarcely; for some proofs that that danger is real, see the article HARDLY. The examples that here follow are free from such ambiguity, but in each of them idiom demands expulsion of the -ly: *How hardly put to it the Tories are for argument is shown by* . . . / *Another sign of how hardly the great families are pressed in these times.* / *The invasion of Henley by the fashionable world bears very hardly on those who go only for the sport.* / *They have been as hardly hit as any class in the community by the present state of trade.* / *If there is a man more hardly hit by existing conditions than the average holder of a season ticket he is hard to find* (harder hit).

Other such adverbs are WIDE, *late, deuced*, HIGH, each spoilt in the appended extracts by an unidiomatic -ly: *And then he'd know that betting and insurance were widely apart.* / *Several drawings in the new volume are dated as lately as August and September, 1922. | I bite it—it is deucedly big—I light it and inhale. | M. Millerand has played highly, but he has lost his stake.* MIDDLING, *soft*, FAIR, and SHARP, are specimens of the many others that might be named.

**unilateral.** See BILATERAL.

**uninterested.** See DISINTERESTED.

**unique.** A watertight definition or paraphrase of the word, securing it against confusion with all synonyms that might be suggested, is difficult to frame. In the first place, it is applicable only to what is in some respect the sole existing specimen, the precise like of which may be sought in vain. That gives a clean line of division between it and the many adjectives for which it is often ignorantly substituted—*remarkable, exceptional, fabulous, rare, marvellous,* and the like. In the qualities represented by those epithets there are degrees; but uniqueness is a matter of yes or no only; no unique thing is more or less unique than another unique thing, as a rare thing may be rarer or less rare than another rare thing. The adverbs that *u.* can tolerate are e.g. *quite, almost, nearly, really, surely, perhaps, absolutely,* or *in some respects*; and it is nonsense to call anything *more, most, very, somewhat, rather,* or *comparatively u.* Such nonsense, however, is often written: *What made Laker's achievement* all the *more unique was.... | I am now at one of* the most unique *writers' colonies imaginable. | I have just come across the production of a boy aged seven which is, in my experience,* somewhat unique. | *Sir, I venture to send you a copy of a* rather unique *inscription on a tombstone. | A* very unique *child, thought I.*

But, secondly, there is another set of synonyms—*sole, single, peculiar to,* etc.—from which *u.* is divided not by a clear difference of meaning, but by an idiomatic limitation (in English though not in French) of the contexts to which it is suited. It will be admitted that we improve the two following sentences if we change *u.* in the first into *sole,* and

in the second into *peculiar: In the always delicate and difficult domain of diplomatic relations the Foreign Minister must be* the unique medium *of communication with foreign Powers. | He relates Christianity to other religions, and notes what is* unique to the former *and what is common to all of them.* Unique *so used is a* GALLICISM.

**unity.** *The unities,* or *dramatic unities,* are the u. of time, the u. of place, and the u. of action. The first has been observed if all that happens in a play can be conceived as sufficiently continuous to fill only something like the same time (stretched by generous reckoning to a day) as the performance. The second is observed when changes of scene, if any, are slight enough to spare an audience the sensation of being transported from one place to another. The third is observed when nothing is introduced that has no bearing upon the central action of the play. The notion that these were the essentials of good drama came from a literal interpretation of Aristotle's *Poetics.* Only the last is universally so recognized.

**unlearned, -nt.** See LEARN and cf. UNBENDING.

**unless and until.** See PLEONASM 2, for other such duplications. One of the conjunctions is almost always superfluous, as in the still commoner IF AND WHEN, the discussion in which article may serve for this pair also; but a few quotations will allow the reader to judge whether 'unless and' might not in each be left out with advantage: Unless and until *it is made possible for a builder or householder to obtain an economic rent, so long will building be at a standstill. | Speaking for himself he said that* unless and until *the Second Chamber was reformed and the constituencies were given some constitutional means of expressing their opinion, he treated every measure that proceeded from the House of Commons as at present constituted as coming from a tainted source. | Provided further that any Bill shall not be presented to his Majesty* *nor receive the Royal Assent under the provisions of this section unless and until it has been submitted to and approved by the electors. | There should be no commitment to enter the Common Market unless and until clear and unequivocal safeguards for Commonwealth producers are obtained.*

**unlike,** in its less simple uses, i.e. when we get beyond 'unlike things', 'the two cases are u.', and 'this is u. that', to 'unlike you, I feel the cold' and further developments, is subject to the complications set out in LIKE, though occasions for its misuse are much fewer except in what is there called the fourth type. Here *unlike* leads writers astray perhaps even more often than *like*. A typical sentence is *Unlike Great Britain, the Upper House in the United States is an elected body.* (The Upper House in the United States unlike that in Great Britain is . . . .) In addition to what is said under LIKE two special warnings may be given. *I counted eighty-nine rows of men standing, and* unlike in London, *only occasionally could women be distinguished.* U. is there treated as though it had developed the adverbial power described in the article UN-ATTACHED PARTICIPLES as acquired by *owing (to)* etc. It has not, and something adverbial (*in contrast with what we see in London?*) must be substituted. *M. Berger, however, does not appear to have—unlike his Russian masters—the gift of presenting female characters.* As with many negatives, the placing of *u.* is important; standing where it does, in this quotation, it must be changed to *like*; *unlike* would be right if the parenthesis were shifted to before 'does not appear'.

**unmaterial,** if chosen instead of the ordinary *im-*, confines the meaning to 'not consisting of matter', and excludes the other common meaning of *immaterial*, viz. 'that does not matter', 'not important or essential'; see IN- AND UN-.

**unmentionables.** See PANTALOONS.

**unmoral.** For this and *im-*, see IN- AND UN-.

**unpractical.** See PRACTICABLE.

**unrealistic.** See REALISTIC.

**unreligious,** chosen instead of the usual *ir-*, excludes the latter's implications of sin etc., and means outside the sphere of religion; see IN- AND UN-.

**unsanitary, in-.** *In-* is the established form; but it would not be used, as *un-* might, of a place etc. that neither had nor needed provisions for sanitation: *a primitive and unsanitary but entirely healthy life* or *village*; *insanitary* implies danger to health. See IN- AND UN-.

**unsolvable** differs from *insoluble* in having its reference limited to the sense of the English verb *solve*, and not covering, as *insoluble* does, various senses (dissolve as well as solve) of the Latin verb *solvere*; it is therefore sometimes useful in avoiding ambiguity; see IN- AND UN-.

**unstable.** Better than *in-*, in spite of *instability*; see IN- AND UN-.

**unstringed, unstrung.** See STRINGED.

**unthinkable** is now a sort of expletive. When we say *damn*, it relieves us because it is a strong word and yet means nothing; we do not intend the person or thing or event that we damn to be burnt in hell fire; far from it; but the faint aroma of brimstone that hangs for ever about the word is savoury in wrathful nostrils. So it is with *unthinkable*, 'that cannot be thought'. That a thing at once exists and does not exist, or 'the things which God hath prepared for them that love him', are unthinkable, i.e. the constitution of the human mind bars us from conceiving or apprehending them. But we do not mean all that with our VOGUE-WORD *unthinkable* at present; anything is now unthinkable from what reason declares impossible or what imagination is helpless to conceive (as that our civilization could survive a nuclear war) down to what seems against the odds (as that Britain should win the Davis Cup), or what is slightly distasteful to the speaker (as that corporal punishment should be reintroduced). The word is so attractive because it seems to combine the most forcible sound with the haziest meaning.

Clearly the word cannot now be restored to its severely philosophical sense (as shown in the first of the quotations that follow). Its most frequent use today is as a convenient way of saying that a particular eventuality is by common consent too silly or too wicked or too horrible to contemplate (as in quotations 2, 3, and 4). That is a reasonable extension. But to call some project unthinkable merely by way of expressing one's own emphatic disagreement with it (as in quotations 5 and 6), or to describe as unthinkable a state of affairs actually existing at the time (as in quotations 7 and 8) is to use the word in a sense so plainly incongruous with its natural meaning as to make it pointless; it becomes merely an expletive. 1. *'Ultimate' scientific ideas may be unthinkable without prejudice to the 'thinkableness' of 'proximate' scientific ideas.* 2. *First and foremost we must continue to make it evident that the cost of major aggression is unthinkably high.* 3. *It is unthinkable that we should continue a policy under which a given locality may be allowed to commit a crime against a friendly nation.* 4. *To the list of conductors who offend . . . your music critic by their treatment of Mozart's feminine endings must be added Toscanini, Sir Thomas Beecham, and Fritz Busch, in whose case such criticisms are unthinkable.* 5. *With all respect to the advocates of a third reading amendment, such a course appears to us to be simply unthinkable* (out of the question?). 6. *Mr. F. was loudly cheered yesterday when he said that any road link with the Continent would be unthinkable* (disastrous?). 7. *It is unthinkable that hundreds. upon hundreds of people should be*

*getting their freedom on the ground of adultery, whilst thousands of innocent sufferers under desertion, drink, cruelty, and insanity, are left outside any relief* (monstrous?). 8. *He said we were apt to forget the lessons of the war; some people he met said 'I want to forget'; that was, to his mind, a wrong and unthinkable attitude to adopt* (unthinking?).

Slipshod use of *unthinkable* by some writers has provoked others to paradoxes exposing its absurdities, and this may help to check the vagaries of the word. *It seems unthinkable that a new commander for South Africa should have been appointed without prior consultation with the Commander in Chief, and nothing shows the government's fear of Wolseley better than that they not only thought of the unthinkable but actually did it.* | *The Kennedy administration has dared to think about the unthinkable and has persuaded itself that nuclear war can be used as an instrument of foreign policy.* | *The history of multi-racial progress since the war is essentially the story of the reluctant accepting the unthinkable.*

**until.** 1. *Until, till.* 2. *U.* or *till* for *before* or *when.* 3. *Unless and u.* 4. *Until such time as.*

1. *Until* has very little of the archaic effect as compared with *till* that distinguishes *unto* from *to*, and substitution of it for *till* would seldom be noticeable, except in any such stereotyped phrase as *true till death.* When the clause or phrase precedes the main sentence, *until* is perhaps actually the commoner (*until his accession he had been unpopular*).

2. Neither *until* nor *till* is idiomatic in sentences of a certain type, which require *when* or *before*: *In one of the city parks he was seated at one end of a bench, and had not been there long until a sparrow alighted at the other end.* The reason is that *till* and *until*, strictly defined, mean (if there is no negative) 'throughout the interval between the starting-point (i.e., here, his sitting down) and the goal (here, the sparrow's arrival)'; or (if there is a negative) 'at any point in that interval'; and to say

that it was not long at any point in that interval is meaningless. The OED calls the misuse dial. and U.S.

**3.** For *unless and until*, see UNLESS. The writer of the following has evidently a praiseworthy antipathy to *u. and u.*, which would have given, however verbosely, his meaning; but in struggling to escape he has made nonsense, which is worse than verbosity: *He will still be able to supply his front and to be in touch with Jerusalem by two avenues of supply, the road and the railway*, until, or if, *the critical point of Nablous is lost to him.*

**4.** *Until such time as* has its uses, with an implication of uncertainty whether the event contemplated will ever happen. But more often it is mere verbosity for *until*.

**untoward.** Pronounec *ŭntŏ′ărd*. See TOWARD(S).

**up. 1.** The phrase *up to date* is three words unhyphened, except when it is used as an attributive adjective; then, it is hyphened: *An up-to-date bungalow*; but *You are not up to date, Bring the ledger up to date.* See HYPHENS.

**2.** *Up against* (faced or confronted with), and *up to* (incumbent upon), are good examples of the rapidity with which in modern English new slang phrases make their way through the newspapers into literary respectability.

**3.** Perhaps because *up* is a useful component of so many PHRASAL VERBS (*buck, clean, hurry, tidy, wash,* etc.), its unnecessary addition to simple verbs seems to present a special temptation to phrasal-verb addicts. For examples see that article. Two more may be given here; in the first the *up* is unnecessary and in the second clearly wrong. *They quickened up people's imagination.* / (Of a famous cricketer) *Even at his best he could not have picked up the ball off his legs so quickly.*

**up and down.** As geographical terms these words are ordinarily used in their natural senses. We speak of going *down* to the south and *up* to the north, *down* to the sea and *up* country

from the coast. But two special uses are worth noting.

**1.** In relation to London. The use of *up* for a journey to London and *down* for one from it preceded the adoption of those expressions by the railways. Perhaps the idea of accomplishment latent in *up* made it seem the right word for reaching the more important place. Geographical bearing was immaterial; going north from London was no less *down* than going south. 'At Christmas I went down into Scotland', wrote Lord Chancellor Campbell in 1846, 'and, crossing the Cheviots, was nearly lost in a snowstorm.' 'You don't mean to say', said Miss La Creevy, 'that you are really going all the way down into Yorkshire this cold winter's weather, Mr. Nickleby.'

The railways conformed. They gave the name *up-line* to that on which their trains arrived at their London termini and *down-line* to that on which they left. When a railway was built without any direct connexion with London the up-line was that on which trains ran to the more important terminus. But with the multiplication of railways, and their linking into a network by junctions, the terms have ceased to have any clear and uniform significance, except in the case of main lines to and from London. For instance, in one of the routes connecting Edinburgh and Glasgow, the line on which a passenger travels from Glasgow is at first the up-line, for it leads to the more important terminus, Edinburgh, but at a junction where there is a connexion southwards it becomes the down-line, for on it run trains from London to Edinburgh.

Today, when we all use maps, a Londoner feels the incongruity of saying he is going *down* to a place that is higher on the map. He finds it natural enough to say *down* to Brighton or Portsmouth, and he may not jib at *down* even for a more northerly place if its bearing deviates enough from the vertical, Birmingham or Newmarket for instance. But he will almost certainly say *up* to Edinburgh or Aber-

deen, or even Leeds, though he will
be likely to speak of the return journey
also as *up*; the vigour of the phrase *up to
London* still resists these influences.
2. In relation to Oxford and Cambridge. To a member of these universities *up* means in residence. An undergraduate goes *up* at the beginning of
term and remains *up* until he goes
*down* at the end of it, unless he has
the misfortune to be sent down earlier.
But this special use relates only to the
universities, not to the cities. An
undergraduate will go *up* (not *down*)
to London for the day, and his parents
visiting him from London will go
*down* (not *up*) to see him.

**upon, on.** According to the OED
'The use of one form or the other is
usually a matter of individual choice
(on grounds of rhythm, emphasis,
etc.) or of simple accident, although in
certain contexts and phrases there may
be a general tendency to prefer one to
the other.' The choice seems almost
wholly arbitrary, and there is no saying why one has taken root in some
phrases and the other in others: *Upon
my word* but *on my account*; *Kingston
upon Thames* and *Burton upon Trent*,
but *Henley on Thames* and *Newark on
Trent*. But they should not be treated
as a convenience for ELEGANT VARIA-
TION, as in *I put it to Mr. Grimond that
Mr. Gaitskell depended on the votes of
the Labour M.P.s and not upon the vote
of the conference*.

**upstairs.** See NOUN AND ADJECTIVE
ACCENT.

**urinal.** The natural pronunciation
(see RECESSIVE ACCENT) is *ūr'ĭnăl*;
*ŭrī'năl* accords better with the imaginary sanctity of Latin quantities;
but how little that comes to is shown
in FALSE QUANTITY. The first is, however, the choice of most dictionaries,
and is recommended.

**us.** 1. Case mistakes. 2. *His*, *our*, etc.,
after *of us*.
 1. Case. *Us* and *we*, where in roman
type in the following examples, are ungrammatical: *They are as competent as

us as regards manufacture, and so why
not serve them the same as they serve us?* |
*The Germans are involved like our-
selves in a blind struggle of forces, and
no more than* us *to be blamed or praised.* |
*Age and experience bestow the skill to
recognize in a book only what we require
that we not only read and mark, but
inwardly digest; it becomes* us. | *Let us be
content*—we Liberals, *at any rate*—*to go
on in the possession of our old principles.*
In the first two, after *as* and *than*, there
can be no objection to letting grammar
have its rights, with the correct *we*;
if it is thought prudent to avoid even
the faintest suspicion of pedantry the
verb can be repeated: *as competent as
we are* | *no more to be blamed or praised
than we are*. (See THAN 6.) In the
third, if *becomes we* is thought pedantic,
*becomes a part of us* is an easy way out;
and in the last, if it is obtrusively
formal to keep the required case in
mind for the duration of a dash and
repeat it on the other side, *Let us
Liberals at any rate be content* would
not have been unbearably ordinary.
 2. *Our*, or *his* etc., after *of us*. *Types,
it must be admitted, under which each
of us can classify a good many of his
acquaintances*. | *Most of us lost their
heads.* Those are perhaps the logical
choice of pronoun, but much more
commonly *our acquaintances, our heads*,
are used owing to the attraction of *us*.

**-us.** The plurals of nouns in *-us* are
troublesome. 1. Most are from Latin
second-declension words, whose Latin
plural is *-i* (pronounced ī); but when that
should be used, and when the English
plural *-uses* is better, has to be decided
for each separately; see LATIN PLURALS
and the individual words. 2. Many
are from Latin fourth-declension
words, whose Latin plural is *-us*
(pronounced ūs); but the English
plural *-uses* is almost always preferred, as *prospectuses*; -ūs is occasionally seen as a plural of *hiatus* and
a few of the rarer words, e.g. *lusus,
meatus*. Words of this class, which
must never have plural in *-i*, are
*afflatus, apparatus, conspectus, hiatus,
impetus, lusus, meatus, nexus, plexus,*

*prospectus, saltus, senatus, status.* 3. Some are from Latin third-declension neuters, whose plurals are of various forms in *-a*; so *corpus, genus, opus,* make *corpora, genera, opera,* which are almost always preferred in English to *-uses.* 4. *Callus, octopus, platypus, polypus,* and *virus,* nouns variously abnormal in Latin, can all have plural *-uses*; for any alternatives see the words. 5. Some English nouns in *-us* are in Latin not nouns, but verbs etc.; so *ignoramus, mandamus, mittimus, non possumus*; for these, as for the dative plural *omnibus* and the ablative plural *rebus,* the only possible plural is the English *-uses.*

**usage, use, user.** Those who write *usage* or *user* when they mean no more than *use* must be presumed to do so for one of two bad reasons: that they prefer either the longer word to the shorter (see LONG VARIANTS) or the unusual one to the common (see WORKING AND STYLISH WORDS). *Usage* implies a manner of using (e.g. *harsh usage*), especially of habitual or customary practice creating a right or standard (*modern English usage*). An example of its misuse is *There is a serious shortage of X-ray films due to increasing usage in all countries. User* is a legal word for *use* (exercise of a right) and should be left to the lawyers.

**use, n.** The forms *What is the use of complaining?,* and *There is no use in complaining,* are current and uncriticized. The forms *It is no use complaining* (or *to complain*), and *Complaining* (or *To complain*) *is no use,* are still more current, but much criticized, and the critics would have us correct them by inserting *of* (*is of no use*). General adoption of their *of* is at this time of day past praying for. We might of course take refuge instead in *useless,* which would do well enough if we could remember to say it, and thought it worth while. But most of us would like to be allowed their *It is no use bothering about such things,* if only on the footing of a STURDY INDEFENSIBLE. The OED admits without comment *use* in this sense 'with ellipsis of prep.', dat-

ing it from 1820 and quoting Shelley *Alas! It is no use to say 'I'm poor',* and Newman *From their thinking it no use doing good unless it is talked about.*

**use, v.,** makes *-sable*; see MUTE E. Pronounce *ūz*; but *used,* which is *ūzd* in general senses, is *ūst* in the senses was accustomed, and (as adj.) accustomed. As an intransitive verb, meaning be wont to, *use* is now confined to the past tense: we may say *He used to live in London,* but not, as we might once have done, *he uses to live in London.* (*Be merciful unto me as Thou usest to do unto those that love Thy name*). The proper negative form is therefore *He used not to* (or, colloquially, he usedn't to); but *He didn't use to* should be regarded rather as an archaism than as the vulgarism, like *He didn't ought to* (see OUGHT 2), it is generally thought to be in England, though not in U.S.

**usherette.** See FEMININE DESIGNATIONS s.f.

**usual.** Of the pronunciations, *ū'zhl* is slipshod, *ū'zhooal* normal, and the more precise *ū'zūal,* which might once have been called pedantic, is gaining ground.

**usurp.** *Eden is of course aware of these Russian schemings. He is I am sure advising his visitors to stop their attempts to usurp us from the Middle East.* This use of *usurp* (called by the OED 'rare') betrays either ignorance or a hankering after WARDOUR ST. English. Today's idiom requires *usurp our place in.*

**utilitarian.** See HEDONIST.

**utilize.** If differentiation were possible between *utilize* and *use* it would be that *utilize* has the special meaning of make good use of, especially of something that was not intended for the purpose but will serve. But this distinction has disappeared beyond recall; *utilize* is now ordinarily treated as a LONG VARIANT of *use. A form is enclosed herewith for favour of your utilization* is an example of the pretentious diction that prefers the long word.

# V

**vacation** is in America the ordinary word for what we call a *holiday*. In Britain it is not so used except (often abbreviated to *vac.*) for the intervals between terms in the Law Courts and Universities. The corresponding word for Parliament is *recess*.

**vacuity, -uousness.** The first is the usual word; the second may reasonably be chosen when a noun is wanted for *vacuous* as applied to the face, eyes, expression, etc.; see -TY AND -NESS.

**vacuum.** Pl. *-ums* popular, *-ua* scientific.

**vade-mecum.** Pronounce *vā'dĕ mē'cŭm*. See LATIN PHRASES.

**vagary.** Pronounce *văgār'ĭ*; the OED gives this pronunciation only, and among its verse quotations requiring it are lines from Milton, Gay, and the *Ingoldsby Legends*.

**vagina.** Always pronounced *văjī'na*; but the adjective either *văjī'nal* or *vă'jĭnal*. See FALSE QUANTITY.

**vainness.** See VANITY.

**valance, valence.** The first is the spelling of the drapery; the second (usually *valency*) that of the scientific term.

**-valent.** Those who have occasion to use unusual *-valent* compounds such as *bivalent, divalent, trivalent* and *polyvalent* do not seem to be agreed whether to stress the penultimate and make the *a* long or to stress the antepenultimate and make the *a* short. The latter is recommended on the analogy of *equi'vălent* and *ambi'vălent*. See RECESSIVE ACCENT.

**valet.** Pronounce both noun and verb *vă'lĕt*; the word is sufficiently naturalized to drop the alternative *vă'lā* for the noun. The verb makes *-eted, -eting*, see -T-, -TT-.

**valise.** Except in military use as the official term for the cylindrical canvas cover in which a soldier carries his bed-roll and other kit, the word is not now used in England, but survives in America.

**Valkyrie.** This is the prevailing spelling in modern English; pl. *-s*. The pronunciation shown in verse, and suggested by the formerly common spelling *Valkery*, is *vă'lkĭrĭ*; but *vălkĭ'rĭ* and *valkē'rĭ* are often heard.

**valour, valorous.** For spelling see -OUR AND -OR, and -OUR- AND -OR-.

**value,** n. *What value will our Second Chamber be to us if it is not to exercise such control?* An interesting specimen of ANALOGY. *What good will it be?* is unexceptionable; *What use will it be?* is condemned by some, but a plea has been put in for it in USE, n.; *What value will it be?* is ruled out, because no instinct tells us, as it does about *Of what use*, that *Of what value* is a piece of pedantry. *Is no good* is idiomatic and gives a false impression of being grammatical (false because the *no* shows that *good* is not here an adjective); *is no use* is idiomatic but not grammatical; and *is no value* is neither.

**valve.** For the preference of *valvular* as the adjective over possible formations such as *valval* and *valvar*, see -ULAR.

**Van Dyck, Vandyke, vandyke.** The painter's name, originally *Van Dyck*, was anglicized into (*Sir Anthony*) *Vandyke*; the derived word (noun and verb) should be, and usually is, *vandyke*; the painter or a picture of his may properly be called by either the first or preferably the second form.

**vanguard.** See AVANT-GARDE.

**vanity, vainness.** 1. Now that *vanity* is seldom used in its old sense of futility, waste of time, without any of its modern implication of conceit (*I looked on all the work that my hands had wrought . . . and behold all was vanity and vexation of spirit and there was no profit under the sun*), *vainness* ('rare', COD) is sometimes called on to fill the gap.
2. The appurtenances of this wicked world that the candidate for Confirmation is required to renounce together with the devil and all his works are its pomps and vanity, not, as often misquoted, vanities.

**vapid.** Of its nouns, *vapidness* is usually better than *vapidity* (in strong contrast with the nouns of *rapid*), except when the sense is a vapid remark; then *-ity* prevails, and still more the plural *-ities*; see -TY AND -NESS.

**vapour** and its belongings. For the spelling of the word itself see -OUR AND -OR. Allied words are best spelt: *vapourer, vapourings, vapourish, vapourless, vapoury*; but *vaporific, vaporize (-zation, -zer), vaporous (-osity)*; for the principle see -OUR- AND -OR-.

**variability, -bleness.** Both are in current use, without any clear difference of sense or application. This is unusual (see -TY AND -NESS); but, while -ity would be expected to prevail, -ness probably persists owing to the familiar 'with whom is no variableness, neither shadow of turning' (James i. 17).

**variance.** *It is utterly at variance from the habit of Chaucer.* Idiom demands *with*, not *from*.

**variant,** n., as compared with *variation* and *variety*, is the least ambiguous name for a thing that varies or differs from others of its kind; for it is concrete only, while the others are much more often abstract. *Variation* is seldom concrete except in the musical sense, and *variety* seldom except as the classifying name for a plant, animal, mineral, etc., that diverges from the characteristics of its species. It is worth while to help on the differentiation by preferring *variant* in all suitable contexts.

**variorum,** when used as a noun, has pl. *-ms*; see -UM. The word is a compendious way of saying *editio cum notis variorum*, and means an edition of a work that contains the notes of various commentators on it. *Variorum* is a genitive plural, not the neuter nominative singular that the bookseller took it to be when he offered *a good variorum edition including variora from MSS. in the British Museum.*

**various** as a pronoun. ANALOGY has played tricks with the word and persuaded many people that they can turn it at will, as *several, few, many, divers, certain, some,* and other words are turned, from an adjective into a pronoun. In the OED article, published in 1916, there is no hint of such a use, which was apparently thought too illiterate to be even worth condemnation; nor is there in the 1933 Supp. The COD in earlier editions gave examples of it, calling it 'vulg.', but withdrew even this amount of recognition in 1964. The OID also ignores it. From this we should perhaps infer that the authority of the Oxford dictionaries is still behind the opinion expressed in *Modern English Usage* that 'to write *various of them* etc. is no better than to write *different of them, diverse of them*, or *numerous* or *innumerable of them*'. Nevertheless, there are still writers who persist in doing it, and *various* may eventually be found to have forced its way into the company of those other adjectives that have established themselves as quasi-pronouns. Recent examples are: *But various of his county opponents rate him the most difficult of the off-spin bowlers to play on good wickets. / There are various of last season's books by no means out of print. / The two ministers chiefly concerned, Mr. Heath and Mr. Sandys, have been paying private visits to various of the Commonwealth representatives. / Various of the sonnets are interconnected by words, by images, and by ideas.*

**varlet.** Now, outside the historical novel, a PEDANTIC-HUMOUR word.

**varsity** is perilous stuff for those who are not familiar with universities to deal in; it plays them just the tricks that any English slang plays the foreigner. Thinking that to say the word shows intimacy with the undergraduate's (or the Englishman's) characteristic language, they naturally put it into places where it would never occur to him, and reveal themselves not as natives, but as foreigners. Their only safe course is to avoid it altogether; it has been falling into disrepute for some time, and, apart from its occasional use as a battle-cry by spectators of a sport, it is no longer tolerated unless linked adjectivally to a few nouns with which it has long been associated, and written *'Varsity*, e.g. *'Varsity match*.

**vase.** Pronounce *vahz*. *Vawz*, once considered correct, went out of fashion towards the end of the 19th c. The earlier pronunciation *vās* is still current in U.S.

**vastly.** In contexts of measure or comparison, where it means by much, by a great deal, as *is vastly improved*, *a vastly larger audience*, *v.* is still in regular use. Where the notion of measure is wanting, and it means no more than much or to a great degree, as in *I should vastly like to know*, *is vastly popular*, it was fashionable in the 18th c., but is now a WARDOUR ST. affectation. It has been supplanted by a series of even less appropriate adverbs. See TERRIBLY.

**vaticinate** makes *-tor*; see -OR. The verb, formerly equivalent to *prophesy*, now usually connotes contempt, and means rather to play the prophet, to be a Cassandra, cf. *pontificate*; *vaticination* is similarly limited.

**-ve(d), -ves,** etc., from words in -f and -fe. Corresponding to the change of sound discussed in -TH AND -DH that takes place in the plural etc. of words ending in -th, like *truth*, there is one both of sound and of spelling in many words ending in -f or -fe, which become -ves, -ved, -vish, etc. As the change is far from regular, and sometimes in doubt, an alphabetical list follows of the chief words about which some doubt may exist, showing changes in the plural of the noun and in the parts of the verb and in some derivatives (d). When alternatives are given the first is better.

beef. Pl. *beeves* oxen, *beefs* kinds of beef; d *beefy*.

calf. Pl. *calves*; vb. *calve*; d *calfish*, *calves-foot* or *calfs-foot*.

elf. Pl. *elves*; d *elfin*, *elfish*, *elvish*.

handkerchief. Pl. *-chiefs*.

hoof. Pl. *hoofs*, *hooves*; vb. *hoof*, *hoofed*, *hooved*; d *hoofy*.

knife. Pl. *knives*; vb. *knife*, *knive*, *knifed*, *knived*.

leaf. Pl. *leaves*; vb. *leaf*, *leave*, *leaved*, *leafed*; d *leafy*.

life. Pl. *lives*; vb. *live*, *-lived*; d *liven*, *lifer*.

loaf. Pl. *loaves*; vb. *loaf*, *loave*; d *loafy*.

oaf. Pl. *oafs*, *oaves*; d *oafish*.

proof. Vb. *prove*, but *proof* for to make waterproof.

roof. No *v* forms.

scarf. Pl. *scarves*, *scarfs*, vb. *scarfed*.

scurf. d *scurfy* having scurf, *scurvy* contemptible etc.

self. Pl. *selves*; d *selfish*.

sheaf. Pl. *sheaves*; vb. *sheave*, *sheaf*, *sheaved*; d *sheafy*.

shelf. Pl. *shelves*; vb. *shelve*, *shelved*; d *shelfy*, *shelvy*.

staff. Pl. *staffs*. (arch. and mus.) *staves*.

turf. Pl. *turfs*, *turves*; vb. *turf*; d *turfen*, *turfy*.

wharf. Pl. *wharfs*, *wharves*; d *wharfage*, *wharfinger*.

wife. Pl. *wives*; vb. *wive*; d *-wifed*, *-vived*, *-wifery*.

wolf. Pl. *wolves*; vb. *wolf*; d *wolfish*, *wolvish*.

**veld(t).** The modern form is *veld*, but the *-dt* may still be found in English use, and has the advantage of not disguising the sound, which is *fĕlt*.

**velleity, volition.** *Volition* in its widest sense means will-power. In a narrower but more usual sense it means an exercise of will-power for a specific purpose—a choice or resolution or determination. *Velleity* (now rare) is an abstract and passive preference. It is properly used either in direct opposition to volition or, when volition is understood in its widest sense, as equivalent to that inactive form of it which is sometimes called 'mere volition'. The man in Browning —'And I think I rather ... woe is me! —Yes, rather should see him than not see, If lifting a hand would seat him there Before me in the empty chair

Tonight'—is expressing a velleity, but not in the ordinary sense a volition. And the OED quotes from Bentham: 'In your Lordship will is volition, clothed and armed with power—in me, it is bare inert velleity'.

**venal, venial.** These words are so like in appearance that they are sometimes confused in spite of their being so unlike in meaning; *Venal* mercenary; *venial* excusable.

**vend** makes *vendible*; see -ABLE 2. *Vendor* and *vender* are both in use with the differentiation that *-or* is better when the contrast or relation between seller and buyer is prominent, and *-er* when purveyor or dealer is all that is meant. But *-or*, the invariable legal form, is tending to displace *-er* for all purposes.

**venery.** The existence of homonyms, one synonymous with hunting, the other with sexual indulgence, makes it necessary to provide against ambiguity in using either—the more since neither of them is now an everyday expression.

**vengeance.** See AVENGE.

**venison.** The standard pronunciation is *věn'zn*, but *věn'izn* seems likely to supplant it. Cf. MEDICINE and see PRONUNCIATION I.

**venturesome, venturous.** See ADVENTUROUS.

**venue.** Pronounce *vě'nū*. This term, formerly common in fencing (obs.— OED) and still used in law as the place appointed for a jury trial (esp. *lay*, and *change, the v.*), has become something of a VOGUE WORD for what used to be called a rendezvous or meeting-place, e.g. for races etc. The following quotation, in which it means merely place without the meeting- (or stage?), shows it undergoing the loss of character to which vogue words are liable: *One of our most distinguished actresses acquired one of these coastal landmarks* [a lighthouse] *a good many years ago, and I believe the quietude of* the interior provided a much appreciated venue for her dramatic work.

**veranda(h).** OED gives the *-da* form first, and there is indeed no reason for the *-h*; the adjective is best written *veranda'd*, see -ED AND 'D.

**verbal.** *The object of the provision was to apply it to all contracts, whether in writing or verbal.* / *The British Embassy have made both written and verbal protests to the Soviet Foreign Office.* The primary meaning of *verbal* is consisting of words. Written contracts and protests consist of words no less than spoken ones, and we have had for more than 300 years another adjective—*oral*—with which to distinguish the spoken word from the written. To give *verbal* that narrower sense was therefore both quite unnecessary and also a possible cause of ambiguity unless it is expressly contrasted with the written word, as in the above examples. But its use with this meaning is very common; the OED recognizes it without any deprecatory comment, and gives examples from the 16th c. The COD however calls it 'loose', and the supersession of *oral* is not yet so complete that those whose care for the niceties of language leads them to prefer it need fear a charge of pedantry.

**verbatim.** Pronounce *-ā-* not *-ah-*. See LATIN PHRASES.

**verbless sentences.** A grammarian might say that a verbless sentence was a contradiction in terms; but, for the purpose of this article, the definition of a sentence is that which the OED calls 'in popular use often, such a portion of a composition or utterance as extends from one full stop to another' (See SENTENCE.)

The verbless sentence is a device for enlivening the written word by approximating it to the spoken. There is nothing new about it. Tacitus, for one, was much given to it. What is new is its vogue with English journalists and other writers, and it may be worth while to attempt some analysis of the purposes it is intended to serve

**1.** Transitional. A verbless sentence may contain a summary comment on what has gone before: *True, no doubt. | So far so good. | Of course not.* Or it may introduce what is to follow: *The practical conclusion? | Finally on one small point. | Lastly the poetry of metaphor.* This is the most common literary use; all these examples are taken from the TLS.

**2.** Afterthought. The use of a full stop instead of lighter punctuation may suggest a pause for reflection. Among living novelists E. M. Forster, I. Compton Burnett, and Angus Wilson have done this. And C. P. Snow. | *Some lines might have been written by Auden himself. Well almost. | He thought as much as he observed. More in fact. | Mr. Laughton shuffles about the stage, apple-cheeked and rosy, looking very like Mr. Boffin in fancy dress. Yet not altogether like Mr. Boffin, neither.*

**3.** Dramatic climax. *The winter seas, endlessly hammering, endlessly probing for a weakness, had found one. The cement. | The intruder was no gay young man, but a grey-haired naval captain with one eye and one arm. Nelson. | Unless something is done soon, Oxford, the home of lost causes, will lose the last cause of all. Oxford itself. | We shall face difficulties as we always have done. As a united nation.*

**4.** Comment, especially if arch or strident or intended to surprise. *We solved the whole thing by appointing a Royal Commission. A neat solution. Clever us. | I could make my own survey on this new social phenomenon. A sort of Kinsey Report. | From Mr. K. down to the smallest party unit, the goal is simply 'Produce or die.' Recipe à la Danton. | At the end of the book he goes down the pit and describes the agony of work at the coal face. Brilliant. Searing. | Did someone whisper 'Who the hell is Jack Paar?' Incredible. Crazy. | I used to eat at a restaurant where she sang. She was vital. Fascinating. Tremendous.*

**5.** Pictorial. *The courteous inquisitor of television eyed me across his plain brown deal table. Politely. Courteously. Unaggressively. | Here silence and beauty were absolute. No aeroplane. Not even tree. | Eel Pie Island is like the Deep South. The same feeling of soft dereliction. The Thames green and dulled —a New Orleans bayou? The moon a silver magnolia. | And now the copse is thinned out. No badgers. No tramps. | It is an entire streetful of shops. Complete with side arcades. And a restaurant. And two snack bars. All piled on top of another. A whole civilization all to itself. Practically a State.* This last quotation is aptly described by a reviewer of the book from which it is taken as an example of the 'verbless convulsive'. The classic exponent of that idiom was Alfred Jingle.

**6.** Aggressive. *The particular dynamism of the publishing group which this book concerns springs, of course, from the rumbustious school of journalism it nurtured. Defying the conventions. Hastening the inevitable in social change. Cocking a snook at the hoary traditions and pomposities of our times. Fighting the taboos.*

Lastly here are a few examples, from the many that might be given, of verbless sentences that do not lend themselves to classification but are, it seems, merely the product of a writer's conviction that the more staccato the style the livelier the effect. *Now we are getting on to weaker ground. And a script. Alas, the script. | He receives no official praise or reproof. A free-lance. | So it will be a miracle if we get our restoration. Undoubtedly. | She makes sure the conversation is gay, witty, and light. But business. | Some see the League trying to become independent. Unlikely. | He hasn't got the proper mind for legal technicalities. Too much commonsense.*

Since the verbless sentence is freely employed by some good writers (as well as extravagantly by many less good ones) it must be classed as modern English usage. That grammarians may deny it the right to be called a sentence has nothing to do with its merits. It must be judged by its success in affecting the reader in the way the writer intended. Used sparingly and with discrimination, the device can no doubt be an effective medium of emphasis, intimacy, and rhetoric. Overdone, as it is in the sprightlier sort of modern journalism, it gets on a reader's nerves, offending against the principle of good writing immortalized in Flaubert's aphorism 'L'auteur, dans son œuvre, doit être comme Dieu dans l'univers, présent partout et visible nulle part.'

The same is true of the similar device of promoting a dependent clause to independent status by the use of a full stop, and so giving it greater importance. *Moreover the leaves are disproportionately few. As though the frangipani kept all its energies for flowering. | They demand long years of accurate study, even when the student has the necessary aptitude for such things. Which three students out of every four have not. | These are the days when taxes are breaking up the big estates. When the ascendancy of the large landowner is over. When you can only keep your stately home by admitting the public at half-a-crown a time. | So grows a tree. On which one day a child may swing or lovers carve their names.* This also is frowned on by some pedagogues, jealous for the integrity of the sentence as defined by grammarians. But it should be judged by the same test as the verbless sentence. Does it come off?

**verbs in -ie, -y, and -ye,** sometimes give trouble in the spelling of inflexions and derivatives. The following rules apply to the normally formed parts only, and are merely concerned with the question whether -y-, -ie-, or -ye-, is to be used in the part wanted; they are not concerned with parts of entirely different formation such as *flew, lay, applicable, liar*.

1. -ay: *plays, played, playing, player, playable*, is the form for all except *lay, pay*, and *say*, and their compounds (*inlay, repay, gainsay*, etc.), which use -*aid* instead of -*ayed*. *Allay, assay, belay, delay*, and *essay*, do not follow *lay* and *say*, but use -*ayed*.

2. -ey: *conveys, conveyed, conveying, conveyer, conveyable.* All follow this type, except that *purvey, survey*, have

*purveyor, surveyor*, and *convey* has *conveyor* for the machine.

3. -ie: *ties, tied, tying*; all except *hie* follow the type so far, and also for the most part, in having no -*er*, -*or*, or -*able* forms in common use.

4. -oy: *destroys, destroyed, destroying, destroyer, destroyable*; no exceptions.

5. -uy: *buys, guyed, buying, buyer, buyable.*

6. -y after consonant: *tries, tried, trying, trier, triable; denies, denied, denying, denier, deniable; copies, copied, copying, copier, copiable.* Neither number of syllables, place of accent, nor difference between ȳ and y̆, affects the spelling. But see DRY.

7. -ye: *dyes, dyed, dyeing, dyer, dyable; dyeing* is so spelt to avoid confusion with *dying* from *die* (cf. *singeing*); *eying* has five quotations in the OED as against two for *eyeing*, but the latter is now more usual, no doubt because it preserves more obviously the connexion with *eye*.

**verbs in -ize.** See -IZE, -ISE IN VERBS and NEW VERBS IN -IZE.

**verbum sap.** (sc. *sapienti sat est*), a word is enough to the wise. Also *verb. sap., verbum sat, sat verbum*, or at full length. Ostensibly an apology for not explaining at greater length, or a hint that the less said the better, but more often in fact a way of soliciting attention to what has been said as weightier than it seems.

**verdigris.** The orthodox pronunciation is -ĭs, the popular -ēs; -*gris* is derived not from *grease*, though the notion that it is probably accounts for the prevalent -ēs, but from Greece (green of Greece). As the true origin no more requires -ĭs than the false, there seems no reason why the -ēs of the majority should not be accepted by the minority.

**veridical.** Apart from its use in spiritualism for a phenomenon that has stood up to scientific tests, mostly a PEDANTIC-HUMOUR word whose vogue has now waned.

**verily** will not be forgotten as long as the Gospels are read. But as a current word, apart from its occasional appearances as a stylistic¹ ornament, and its legitimate use in the dialogue of historical novels, it is now perhaps confined to one single phrase (and even that with a touch of WARDOUR STREET)—*I verily believe*, which has the special meaning, 'It is almost incredible, yet facts surprise me into the belief'.

**veritable,** in its modern use, is probably to be classed as a journalistic GALLICISM, and its function is, when one contemplates an exaggeration, to say compendiously, but seldom truthfully, 'I assure you I am not exaggerating': *a veritable hail of slates* etc. It is a pity that the early 19th c. could not leave well alone; for the OED records that by about 1650 the word was dead, but the early 19th c. revived it. Would it had not! Its appearance in a description has always the effect of taking down the reader's interest a peg or two, both as being a FORMAL WORD, and as the now familiar herald of a strained top note. The adverb, which could equally well be spared, does to adjectives the same service, or disservice, as the adjective to nouns (*veritably portentous* etc.). It is also used with verbs as a supposed improvement on the various natural adverbs, as in: *If this is to be the last word, we shall find ourselves thrown back into a hopeless impasse, and there will veritably be no way of reforming our Parliamentary institutions* (indeed? actually? really? positively? absolutely? in very truth?).

**vermin.** The plural form *-ns* is not now used: the word is a COLLECTIVE meaning either all the creatures entitled to the name, or any particular species or set of them, or some of them. It is treated usually as a plural (*these v.; the v. are an incessant torment; v. infest everything*), but sometimes as singular (*this v.* = these rascals etc.), and occasionally has *a* both in the collective sense (*a v. that I hope to reduce the numbers of*) and as denoting an individual (*such a v. as you*).

**vernacular.** For v., idiom, slang, etc., see JARGON.

**verruca.** Pl. *-cae* (-sē).

**vers libre.** Versification or verses in which different metres are mingled, or prosodical restrictions disregarded, or variable rhythm substituted for definite metre.
The French phrase is still in general use; but there seems to be no good reason why 'free verse' should not be preferred. As a name for the writers custom has established *verslibrist,* as queer a fish in English waters as *belletrist.*

**vertigo.** The pronunciation in accordance with the Latin quantity is *vertĭ'gō,* but the OED gives *ver'tĭgō* precedence, and so do more recent dictionaries. See FALSE QUANTITY. It is worth remark, however, that all the OED verse quotations (Jonson, Swift, Fletcher, Wither) show *vertĭ'gō* (or *-ĕ'gō*).

**very** with passive participles. The legitimacy of this, or at least the line limiting its idiomatic use, is an old and not very easy puzzle. It will at once be admitted that *I was much tired* needs correction by the substitution of *very* for *much*, whereas, in *I was very inconvenienced by it, much* or *very much* should be substituted for *very*. [Throughout the rest of this article *much* must be read as meaning *much* or *very much*.] And it may be said generally that the critics of *very* have a way of going too far and damning the laudable; they fail to recognize that *very* and *much* are complementary, each being suited to places in which the other is unnatural or wrong. Here is part of a newspaper letter: *Sir, When the 'Westminster Gazette' can write and publish 'the "Common Cause" is very affronted',* it seems time for some-one to raise a gentle protest. '*Very much affronted*', or '*highly affronted*', if you *like, but surely not 'very affronted'.* Try another tense: one cannot say *'your language very affronts me'*; but *I can say (with truth) 'your language very much affronts me'.* How reasonable it sounds: see DIFFERENT for a similar argument. But it proves too much. Similarly I cannot say *This very tires me*; and yet, dissimilarly, I can say *I am very tired*; and with *affronted* itself I can say *He wore a very affronted look.* And *His look was very affronted* is, if not a likely expression, at least better than *His look was much affronted.*

The points that have to be taken into account are: (1) Has the p.p. passed into a true adjective in common use, as *tired* and *celebrated* have, and *inconvenienced* has not? (2) Is it used attributively (*a ——— damaged reputation*), or predicatively (*the car is ———damaged*)? (3) Is the noun to which the p.p. belongs the name of the person or thing on which the verbal action is exercised (*he was ——— surprised at the question*), or that of something else (*his expression was ——— surprised*)? (4) Is its participial or verbal (as opposed to adjectival) character unbetrayed by e.g. a telltale preposition such as *by*?

A word that is in form a p.p., if it is to be qualified by *very* instead of *much*, must be able to say Yes to (1), *or* to the first part of (2), *or* to the second part of (3), *and* to (4). That is, *He is very celebrated* is right (1), but *Attic taste is very celebrated by the poets* is wrong (4); *A very harassed official appeared* is right (2), but *The Government, very harassed, withdrew the motion* is wrong (2), and still more *The Government, very harassed by questions, withdrew* etc. (4); *His tone was very annoyed* is right (3), but *You seem very annoyed* is wrong (3), and still more *He was very annoyed by the interruption* (4).

All this amounts substantially to no more than that a participle (in -ed) that *is* a participle requires *much*, while a participle that is an adjective prefers *very*; but the bare rule is not very intelligible without some such expansion as has been given. Moreover, the process by which a participle becomes an adjective is gradual;

whether any particular one has passed the barrier must often be a matter of opinion, and some readers may disagree with a refusal to allow adjectival status to some of the participles denied it in this article. It will be noticed that of the following examples only one (the first) shows the wrong use of *much* for *very*; the rest are of *very* for *much*. The disproportion is natural, since *much* is the proper (or at least the admissible) adverb for all participles except the few that (like *tired*, and *limited*, and *pleased*) are in such common use as adjectives that avoidance of *very* can only be attributed to a misguided pedantry. *Opera and theatre engagements are also* much *limited and by no means easy to get.* | *We should be* very *surprised if the Liberal agents ever received the alleged 'warning'.* | *Sir Alfred said that the hostility of the Arabs was* very *exaggerated in this country.* | *A friend in Cornwall tells me that listeners there are* very *annoyed because . . .* | *We are not* very *concerned about these subtle distinctions.* | *Both parties are* very *jealous, and* very *afraid of each other* (*afraid*, and other purely predicative adjectives, rank with the p.p.). | *Your mind seems* very *exercised just now as to whether . . .* | *The peasant deputies consider themselves* very *aggrieved.* | *When the husband returned, he found her manner towards him* very *changed.*

**vesica.** Pronounce *vĕsī'ka*; but the derived words *vĕ'sĭkl* (*vesical* and *vesicle*), *vĕ'sĭkāt*, etc.; cf. *doctrinal* in FALSE QUANTITY.

**vest.** The older meanings robe, tunic, or collectively clothes (= *vesture*), are preserved in poetic or archaic use; as a synonym for a man's waistcoat it is chiefly a shop word in Britain but general in U.S. and other English-speaking countries. In Britain it is the name given by both sexes to the garment (in U.S. etc. an *undershirt*) that is worn next the skin on the upper part of the body.

**vet.** *Veterinary*, pronounced as it should be with the stress on the first syllable, is an example of the trouble we sometimes make for ourselves by our fondness for the RECESSIVE ACCENT: *vetnary* is the almost inevitable result, and that is slovenly speech. The CURTAILED WORD *vet* is thus doubly welcome: it has given us not only a convenient name for the surgeon but also a useful colloquial verb that enables us to pack into one syllable the notions of scrutinize, check, criticize, revise, and eventually 'O.K.' —now the recognized word for investigations into the reliability of persons whom it is proposed to employ on secret government work.

**veto.** Pl. *-oes*; see -O(E)S I.

**via, viâ.** In *via media, via* is the Latin nominative, and must not have *-â*. In its use as a preposition meaning 'by way of' or 'passing through' it is the Latin ablative, which may be distinguished by a circumflex accent but is better without one. As both forms are pronounced *vī'ă* (not *vē'a*, as is sometimes heard for the second; see LATIN PHRASES) and there is never any risk of confusion, it is idle to retain the accent, but italics are still usual. *Via* can only be properly used of the route; to apply it to the means of transport is a vulgarism. *The luggage is being sent via London*, but not *The luggage is being sent via rail*; and still less *May I send a message via you?*

**viable (-bility), credible.** *If the city centre is no longer viable for traffic a new road south of the river and the meadow becomes feasible.* There is no dictionary authority for this use of *viable* in the sense of passable, as though it came from Latin *via*. *Viable* is formed from the French *vie*: *viability* means the capacity of a newly created organism, a new-born child for instance, or a country becoming independent, to maintain its separate existence. As VOGUE WORDS they are having a wonderful time. Nothing is *durable, workable, lasting, effective, practicable* any longer: *viable* must always be the word. *What is the alternative? I do not pretend to know the answer, but who can*

doubt that *no viable answer is possible unless and until the Commonwealth is strong and united within itself?* | *A hotel had been proposed for the Portman Square site, but it was found that, to be economically viable, it would have to be a very tall building and would overshadow the adjacent square.*| *Comedy after all is an assertion of the viability of private judgement.* In the first of these quotations *no viable answer is possible* seems to be a pretentious and obscure way of saying *no alternative will work.* In the second *to be economically viable* clearly means no more than what we used to call *to pay*, but now, it seems, must say *to pay off* (see PHRASAL VERBS). In the third what the writer meant is anyone's guess.

*Credible* has lately invaded the field previously held by *viable*, though not with quite the same meaning. 'Is it viable?' means 'Will it work?' 'Is it credible?' means 'Do people believe that it will work?' This being its significance, *credible* naturally appeared first in company with *nuclear deterrent*; the all-important question about a weapon designed not to win a war but to prevent one must obviously be whether a potential enemy believes that it will work. *British industry could provide, I am certain, a completely c. deterrent system from our own resources.* | *Without Skybolt, the European manned bomber simply will not provide a c. basis for an independent nuclear policy.* Even for this purpose the word was not a very good choice; *convincing* would have been better. *Credible* is now showing symptoms of the SLIPSHOD EXTENSION that is the occupational disease of all VOGUE WORDS. 'In the past month or so' writes a journalist in 1963 'I have seen the word applied to almost everything from the threat of a strike to a campaign for selling more bicycles.'

**viaticum.** Pl. *-ms*, *-ca*, see -UM; but the plural is rare, and e.g. in the PEDANTIC HUMOUR uses one's provisions etc. are one's *viaticum*.

**vicar.** Vicar and rector, as designations of a parish priest, do not reflect any difference in function; he is a *rector* if his parish is one where the incumbent used to retain the tithes, a *vicar* if it is one where they had been lost to him by having been 'appropriated' to a monastery or other religious corporation or 'impropriated' to a lay person or corporation. Since the passing of the Tithe Act 1936 tithes have no longer been payable to any parish priest, but the designation *rector* is preserved where it previously existed; in all other parishes the incumbent is *vicar*.

**vice**, prep., prefix, and abbreviated noun. The preposition is pronounced *vī'sĕ*, and means in the place of (esp. in the sense succeeding to), being, like PACE, the ablative of a Latin noun followed by an English noun regarded as in the genitive (*appointed Secretary vice Mr. Jones deceased*).

The prefix is the same word treated as an adverb compounded with English nouns such as *chancellor, president, chairman, admiral,* but meaning deputy, and pronounced *vīs*.

The noun is the prefix used without its second element, but with the aid of context, as a CURTAILED WORD for some of its compounds, e.g. for *vice-chairman* and *vice-president*, but not for *vice-admiral*. The ablative of the Latin noun also survives in the phrase *pro hac vice* (for this turn) where it is usually pronounced *vēchĕ*, and in *vice versa*, where it is *vīsĕ*. A monosyllabic pronunciation in this last phrase is a solecism often committed by people who ought to know better.

**vicegerent, viceregent.** The first is a word of very wide application, including anyone who exercises authority committed or supposed to be committed to him by another, from the Pope as the Vicar of Christ on earth or the regent of a sovereign State to the clerk running an office during his employer's holiday. *Viceregent*, on the

other hand, is defined in the OED as 'One who acts in the place of a regent'; but from the quotations given it would appear that that is rather what it ought to mean than what it does. A regent is a particular kind of vicegerent, viz. a sovereign's. But *viceregent* is sometimes used in error for *vicegerent*, and sometimes used pleonastically for *regent* (which word includes the notion of vice-), so that it seems to have no right to exist, and may be classed among SUPERFLUOUS WORDS. It has a familiar look because of our choice of *viceregal* rather than *viceroyal* as the adjective of *viceroy*, and that no doubt accounts for its tendency to intrude.

**vicious circle.** For this as a term of logic see ARGUING IN A CIRCLE. *There is a vicious circle in which starvation produces Bolshevism, and Bolshevism in its turn feeds on starvation.* What, then, produces starvation, and on what does starvation feed? The writer can no doubt retort with truth that nothing (i.e. no food) produces starvation, and that starvation feeds on nothing; but he will have proved his wit at the expense of his logic. Such blunders in stating the elements of a vicious circle are not uncommon. As a term of economics *vicious spiral* has been coined on the analogy of *vicious circle* for action and reaction that intensify each other (e.g. rising prices and rising wages), a more suitable metaphor for something that is not closed, as a circle is.

**vide.** Pronounce *vī'dĕ*; literally 'see' (imperative). It is properly used in referring readers to a passage in which they will find a proof or illustration of what has been stated, and should be followed by something in the nature of chapter and verse, or at least by the name of a book or author. But for this purpose *see* will usually do as well as *vide*, and *see above* (or *below*) as well as *vide supra* (or *infra*), and *above* and *below* are superfluous if a page is mentioned. There is, however, no convenient English equivalent of *quod vide* (q.v.). *Vide* has, like RE, been taken over by the illiterate, and is used by them 'in extended senses with an incongruity of which the following is a comparatively mild specimen: *Numbers count for nothing—vide the Coalition—it is the principles that tell.*

**videlicet** in its full form is now rare except in PEDANTIC HUMOUR, the abbreviation *viz.* being used instead; see VIZ. for meaning.

**view** forms part of three idioms each equivalent to a preposition, and each liable to be confused in meaning or in form with the others. These are *in v. of*, *with a v. to*, and *with the v. of*. *In view of* means taking into account, or not forgetting, or considering, and is followed by a noun expressing external circumstances that exist or must be expected: *In v. of these facts, we have no alternative*; *In v. of his having promised amendment*; *In v. of the Judgement to come*. *With a view to* means calculating upon or contemplating as a desired result, and is followed by a verbal noun or a gerund: *With a v. to diminution of waste*, or *to diminishing waste*. It may sometimes be found with an infinitive (*with a v. to diminish waste*), but this construction ('vulg.'—COD) can hardly be called idiomatic. *With the view of* (*with the v. of proving his sanity*) has the same meaning as *with a v. to*, which is more usual and preferred by good writers. It will be observed that in the first phrase *v.* means sight, in the second eye, and in the third purpose. The confusion that produces blunders consists in giving the first the meaning of the others or vice versa, and neglecting the correspondences *a* and *to*, *the* and *of*, in the second and third. After each of the following quotations a correction, or a statement that it is right, is added: *This may be interesting* in view of *the fact that the atmosphere has been reeking with pugilism for some time* (right). | *I will ask your readers to accept a few further criticisms on matters of detail*, in view of *ultimately finding a workable solution* (read *with a v. to*.) | *He, like Basil, had scorned to order his* *life* with a view of *longevity or spurious youth* (read *to* for *of*). | *My company has been approached by several firms* with a view of *overcoming the difficulty* (read *to* for *of*). | *They have been selected* with a view to illustrate *both the thought and action of the writer's life* (read *illustrating* for *illustrate*). | *The question of reducing the cost of bread production,* with the view *both* to *preventing the price of the loaf from rising and* of *arresting any increase in the subsidy, is under consideration* (ELEGANT VARIATION again? read *of* for *to* or, better, *a* for *the* and *to* for *of*). For *point of view* see POINT.

**vigour, -gorous.** For spellings, see -OUR AND -OR, -OUR- AND -OR-.

**villain, villein.** The retention of the second form for the word meaning serf is a useful piece of DIFFERENTIATION, and the OED accordingly gives it in a separate article, though it states that 'the tendency to use the form *villain* [in this sense] has increased in recent years'. This tendency looks like PRIDE OF KNOWLEDGE, the man in the street who is familiar with the two forms having to be shown that he has been under a delusion all this time.

**vindictive** has become so generally restricted to the notion of personal thirst for revenge or desire to hurt that the phrases in which it meant punitive and not revengeful or cruel are now obsolete: *v.* damages are *exemplary*, and *v.* justice is *retributive*.

**viola.** The flower is *vī'ola*, the instrument *vēō'lă*.

**violin.** See FIDDLE.

**violoncello.** So spelt (not *-lin-*); pl. *-os*, see -O(E)S 6. For pronunciation, *vĕŏlŏnchĕ'lō* is the approximation to the Italian; *vīŏlŏnsĕ'lō*, which the OED puts first, is the complete anglicization; and *vīolonchĕ'lō* is the usual compromise, which, having in its favour both *violin* and *'cello* (*chĕ'lō*), has properly prevailed on the rare occasions when the word is spoken in full.

**virement** is a French word introduced into England by British Treasury officials early in the 20th c. In France it had dubious associations with financial sharp practice, but with us it is wholly respectable; it has become a standard term of British public finance meaning a government department's duly authorized use of money for a purpose other than that for which it was voted by parliament. The word is naturalized enough to be no longer italicized in official documents, but not yet enough to have shed its French pronunciation; the fully anglicized *vĭrmĕnt* is, however, increasingly heard and will no doubt prevail. The back-formation verb *vire*, freely used orally but not yet considered proper to be set down in print, is always pronounced to rhyme with *wire*.

**virile.** The accent is always on the first syllable, but the pronunciation of both *i*s has varied between *ī* and *ĭ*; OED puts first *vĭ'rĭl* and this is now usual in Britain (U.S. *vĭrĭl*). See -ILE. The proper sense is 'having the qualities of a male adult', but the emphasis is on *male*, and, though *vigorous* can often be substituted for *v.* without affecting the required meaning, *v.* must not be substituted for *vigorous* where the notion male is out of place, as in: *VIRILE AT 93: Despite her great age, Mrs. Jones is fairly virile, and performs all her own household work.* Perhaps the reporter associated *v.* with *viridis* green, not *vir* man, and was thinking of a green old age.

**virtu.** So spelt (not *ver-*); pronounce *vertōō'*.

**virtue.** *To make a v. of necessity* is one of the maltreated phrases illustrated in IRRELEVANT ALLUSION, being often applied to the simple doing of what one must, irrespective of the grace with which one does it.

**virtuoso.** Pl. *-si* (-sē).

**virus.** Pl. *-uses.* See -US and MICRO-ORGANISMS.

**visa.** See VISÉ.

**visage.** See COUNTENANCE.

**viscount.** For *V. Smith* and *Lord Smith*, see TITLES. Of the two forms of the rank-name, *viscounty* and *viscountcy*, the first is both much older and of better formation.

**visé** (*vē'zā*), in French a p.p. = 'examined' or 'endorsed', was formerly used in English for what is done to passports—*visés, visé'd, viséing.* It has now been replaced by the French noun *visa* for all purposes—*visas, visa'd, visaing.*

**visibility, visibleness.** See -TY AND -NESS. The second was formerly in more frequent use than most *-ness* words with predominant partners in *-ty.* The modern use of *visibility* to mean the effect of the weather on the power of seeing things at a distance might have been expected to establish a useful differentiation. But it has not; the result has been that the more familiar word is driving *visibleness* out.

**visible, visual.** *Visible* means capable of being seen; *visual* means pertaining to seeing. The visual arts are concerned with the production of the beautiful in visible form, visually appreciated. This differentiation is sometimes obscured by the misuse of *visual* for *visible*, for which indeed dictionary authority can be found. But the differentiation is worth preserving. For instance the wrong word is used in the descriptive phrase *Diagnosis by visual symptoms*; the method of diagnosis is visual, but the symptoms are visible.

**vision**, in the sense of statesmanlike foresight or political sagacity, enjoyed a noticeable vogue some years ago. 'Where there is no vision the people perish' (Prov. xxix. 18) is perhaps what makes the word tempting to politicians who wish to be mysteriously impressive; at any rate they are much given to imputing lack of *v.* to their opponents and implying possession of it by themselves when they are at a loss for more definite matter. But they now seem to be getting tired of this word, and tend to use REALISM instead. See VOGUE WORDS.

**visit, visitation.** 1. *Visitation*, once a formal word for visiting, as in the Prayer Book Service for the Visitation of the Sick, is now little used except for official visits of inspection, especially ecclesiastical, by someone in authority, and for an affliction attributed to divine or other supernatural agency. 2. For *visit up with* see PHRASAL VERBS.

**visor** etc. *Visor* and *vizor* pronounced *vīz-*; *vizard* and *visard* pronounced *viz-*. The -ard forms are not etymologically significant, being merely corruptions, but they differ in meaning by being restricted to the sense mask (lit. and fig.), whereas the -or forms have also, and chiefly, the sense movable helmet-front.

**vitamin.** *Vīt-* is the better pronunciation, in conformity with other words derived from *vita*, but seems unlikely to hold its own against the more popular *vĭt-*.

**viva, vivat, vive** (pronounce *vē'-vah, vī'văt, vēv*) are the Italian, Latin, and French, for 'long live ——!'; the first two can be used as nouns also, with plural *-s*; the last needs the addition of *la* to make it one. The verbs have plurals (*vivano, vivant, vivent*) for use with plural subjects—a fact forgotten in: *Cries of 'Vive les Anglais' attended us till we were inside the hotel.* | *Triumphal arches were prepared; 'Bienvenue à nos libérateurs'—'Vive les Alliés'—such were the words variously devised in illuminations and in posters.* The expression *on the qui vive* (= alert) originates in the sentry's challenge *qui vive?—long live who?*, i.e. *whose side are you on?*

**viva voce.** Pronounce *vīvă vō'sĭ*. See LATIN PHRASES. Often shortened colloquially into the CURTAILED WORD *viva*, which is used both as noun and as transitive verb (past and p.p. *viva'd*)=examine in viva voce. See also QUIZ.

**vividity.** *A theme worthy of poetry . . .; here it is handled with occasional vividity and general inconsequence. V.'s* ugliness is no doubt its misfortune rather than its fault; but it is as natural to prefer *vividness* to it as to choose the one of two otherwise equal applicants who does not squint.

**viz, sc(il)., i.e.** Full forms *videlicet, scilicet, id est*. The meanings of the second and third are explained in their dictionary places, but are here for convenience summarized and related to *viz*. The three are so close to one another that the less appropriate is often chosen. *Viz*, as is suggested by its usual spoken representative *namely*, introduces especially the items that compose what has been expressed as a whole (*For three good reasons*, viz 1 . . ., 2 . . ., 3 . . .) or a more particular statement of what has been vaguely described (*My only means of earning,* viz *my fiddle*). Sc. or *scil*. is in learned rather than popular use, is for instance commoner in notes on classical texts than elsewhere, and has as its most characteristic function the introducing of some word that has been not expressed, but left to be 'understood'; so *His performance failed to satisfy* (sc. himself),=not, as might be guessed, other people. What I.E. does is not so much to particularize like *viz*, or supply omissions like *scil*., as to interpret by paraphrasing a previous expression that may mislead or be obscure: *Now you are for it,* i.e. punishment; *The answer is in the negative*, i.e. is No; *Than that he should offend* (i.e. harm) *one of these little ones.*

Even the above examples suffice to show that choice may sometimes be difficult; it does not follow that it is not worth making rightly. The writing of *viz* rather than *viz.* depends partly on the principle stated in PERIOD IN ABBREVIATIONS, but partly also on the fact that z is not the letter, but the old symbol of contraction for the *-et* of *videlicet*; but *viz.* is the prevalent form.

**vizard.** See VISOR.

**vocabulary.** See GLOSSARY.

**vocation.** See AVOCATION.

**vogue words.** Every now and then a word emerges from obscurity, or even from nothingness or a merely potential and not actual existence, into sudden popularity. It is often, but not necessarily, one that by no means explains itself to the average man, who has to find out its meaning as best he can. His wrestlings with it have usually some effect upon it; it does not mean quite what it ought to, but to make up for that it means some things that it ought not to, by the time he has done with it. (See also SLIPSHOD EXTENSIONS.) Ready acceptance of vogue words seems to some people the sign of an alert mind; to others it stands for the herd instinct and lack of individuality. The title of this article is perhaps enough to show that the second view is here taken; on the whole, the better the writer, or at any rate the sounder his style, the less will he be found to indulge in the vogue word. It is unnecessary here to discuss in detail the specimens that will be given; many of them are to be found in their dictionary places, and they will here be roughly classified only. The reason for collecting them under a common heading is that young writers may not even be aware, about some of them, that their loose use is corrupting the vocabulary, and that when they are not chosen as significant words but gatecrash as CLICHÉS they are repulsive to the old and the well-read. Many, it should be added—perhaps most—are vogue words in particular senses only, and are unobjectionable, though liable now to ambiguity, in the senses that belonged to them before they attained their vogue.

1. Old vogue words. As the phrase implies, the popularity enjoyed by vogue words is ordinarily brief. Some of those listed here may have had their day before this book is in print. But for a few it is curiously persistent. *Individual* and *nice* may be instanced; the first now past its vogue but linger-ing in its vogue sense as a nuisance; the second long since established in a loose and general sense instead of its earlier and now infrequent precise one.

2. Words owing their vogue to the ease with which they can be substituted for any of several different and more precise words, saving the trouble of choosing the right: *alternative*; *amenity*; *appropriate* (adj.); *emergency*; *factor*; *framework*; *image*; *impact*; *implement* (vb.); *intensive*; *involve*; *issue*; *level*; *major*; *overall*; *realistic*.

3. Words owing their vogue to the joy of showing that one has acquired them: *allergic*; *ambience*; *ambivalent*; *catalyst*; *complex*; *climate* (fig.); *equate*; *global*; *idiosyncrasy*; *protagonist*; *repercussion*; *seminal*; *streamlined*, and many POPULARIZED TECHNICALITIES.

4. Words taken up merely as novel variants on their predecessors: *adumbrate* for *sketch*; *blueprint* for *plan*; *breakthrough* for *achievement*; *built-in* for *solid*; *ceiling* for *limit*; *claim* for *assert*; *integrate* for *combine*; *intrigue* for *interest*; *liquidate* for *destroy*; *reaction* for *opinion*; *optimistic* for *hopeful*; *redundant* for *superfluous*; *rewarding* for *satisfying*; *significant* for *important*; *sabotage* for *wreck*; *target* for *objective*; *smear* for *calumny*; *viable* for *workable*.

5. Words owing their vogue to some occasion: *acid test*; *coexistence*; *iron curtain*; *psychological moment*; *wind of change*.

6. Words of rhetorical appeal: *archetypal*; *challenging*; *dedicated*; *fabulous*; *fantastic*; *massive*; *overtones*; *sensational*; *unthinkable*.

**volcano.** Pl. *-oes*; see -O(E)S 1.

**volition.** See VELLEITY.

**vouch.** See AVOUCH.

**vulgarization.** Many words depend for their legitimate effect upon rarity; when blundering hands are laid upon them and they are exhibited in unsuitable places, they are vulgarized. *Save* (prep.) and *ere* were formerly seldom seen in prose, but, when seen, they then consorted well with any passage of definitely elevated style, lending to it and receiving from it the dignity that was proper to them. Things are now so different that the elevated style shuns them as tawdry ornament; it says what the man in the street says, *before* and *except*, and leaves *ere* and *save* to those writers who have not yet ceased to find them beautiful—which is naturally confusing, and an injury to the language. The fate of *awful* (see TERRIBLY) is of rather earlier date, but is still remembered, and *weird* and *ghastly* have, almost in our own century, been robbed of all their weirdness and ghastliness. There is little in common between the LYRICS of Pindar and those of a modern 'musical', or between the *cartoons* of Raphael and the strip variety of our popular press, or between the SATIRES of Juvenal and much of what goes by that name today, or between the SAGA of Grettir the Strong and that of Flook. One would like to represent to the gossip writers and advertisers that they are desecrating the words EPIC, *glamorous*, and FABULOUS by applying them to cocktail parties, cosmetics, and striptease girls; but they would probably be as indignant at the notion that their touch pollutes as the writer would be who was told that he was injuring *faerie* and *evanish* and *mystic* and *optimistic* and *unthinkable* and *replica* by selecting them in honourable preference to *fairy* and *vanish* and *mysterious* and *hopeful* and *incredible* and *copy*. Vulgarization of words that should not be in common use robs some of their aroma, others of their substance, others again of their precision; but nobody likes to be told that the best service he can do to a favourite word is to leave it alone, and perhaps the less said here on this matter the better. See also WORSENED WORDS.

# W

**wag(g)on.** The OED gives precedence to *wagon*, but concludes its note on the two forms with: 'In Great Britain *waggon* is still very commonly used; in the U.S. it is rare'. It has now become rare in Britain too; *wagon* is the ordinary spelling.

**waistcoat.** The pronunciation *wĕ's-kut* was once regarded as correct, but the speak-as-you-spell movement has now made *wās(t)kōt* the ordinary one. See PRONUNCIATION 1.

**wait.** The transitive use, as in *w. one's opportunity*, *w. the result*, *w. another's convenience* or *arrival*, is good English, but is described by the OED as 'now rare' and as being 'superseded' by *await* and *wait for*. The assignment of the intransitive uses to *wait* and of the transitive to *await* is a natural differentiation, and may be expected to continue; see also AWAIT. *Wait* retains a quasi-transitive use colloquially in such a phrase as *don't wait lunch for me*.

**waive.** The broad distinction between *wave* and *waive*, viz. that to wave is proper to physical motion and to waive is not, is now generally observed. But confusion, arising especially from the assumption that the two forms are mere spelling variants, still occurs, and is confirmed by the fact that senses certainly belonging to *waive* used frequently to be spelt *wave*. The following example is typical of this confusion: *The problem of feeding the peoples of the Central Empires is a very serious and anxious one, and we cannot* waive *it aside as though it were no concern of ours*. To *waive* is not a derivative, confined to certain senses, of to *wave*, but a derivative of *waif*, meaning to make waif or abandon. To wave aside or away is one method of waiving, but to *waive aside* or *away* is no better English than to *abandon aside* or to *relinquish away*.

**wake.** See AWAKE.

**wale, weal, wheal.** For the mark left on flesh by a cane etc., the original word appears to have been *wale*; this was confused with *wheal*, properly a pimple or pustule, and *weal*, now the ordinary spelling, was a wrong correction of the mistake.

**walled-up object.** *I shut and locked him in* is permissible English; *I scolded and sent him to bed* is not. In the first, *in* is common to *shut* and *locked*; *him* is therefore not walled up between *locked* and a word that is the private property of *locked*. In the second, *to bed* is peculiar to *sent*, and therefore *him*, enclosed between *sent* and *sent*'s appurtenance *to bed*, is not available as object to *scolded*; it is necessary to say *I scolded him and sent him to bed*, though *I scolded and punished him* requires only one *him*. If it is said that the agitated disciplinarian cannot be expected to decide when her boy is two *hims*, and when not, the plea may at once be admitted. It is not in hasty colloquial use that such lapses are wicked, and the examples chosen were the simplest possible in order that the grammatical point might be unmistakable. But in print it is another matter. The string of quotations following shows how common this slovenliness is, and no more need be said of them than that for nearly all the cure is to release the walled-up words, place them as object to the unencumbered verb (which usually comes first), and fill their now empty place with a pronoun, *it*, *them*, etc.; this is done in brackets after the first quotation, and any change not according to this simple formula is shown for later ones: *An earnest agitation for increasing and rendering* that force *more efficient* (read *for increasing that force and rendering it* . . .)./ *It is for its spirited reconstructions of various marches and battles that we counsel the reader to buy and make* the book *his own.* / *I hope the Ministry will also avail itself of the same effective machinery if not to reform, then to make* the Church of England *a free Church in a free country.* / *He had to count, trim, press, and pack* the furs *into bales* (read *and press the furs, and pack them*); or, of course, the omission of *into bales* would put all right). / *They had definitely beaten and knocked* one of their opponents *out of the war.* / *A season in Opposition will invigorate and restore* them *to health* (read *invigorate* them and . . .). / *The wish to hear a sermon which will soothe or 'buck* you *up', according to the needs of the moment* (read *soothe you or* . . .).

The great majority of such mistakes are of that form; one or two are added in which the principle infringed is the same, but some slight variation of detail occurs: *We were not a little proud of the manner in which we transported to and maintained* our Army *in South Africa.* This is the old type, complicated by the well-meant but disastrous *to*; read *in which we transported our army to S.A. and maintained it there.*

*I trust you will kindly grant me a little space to express, in my own and in the* name *of those elements in Russia whom I have the honour to represent, our indignation at* . . . The walled-up noun here (*name*) is governed not by a verb, but by a preposition; read *in my own name and in the name of the elements.* . . .

*The fourteen chapters explore the belief in immortality in primitive and in the various* civilizations *of antiquity taken in order.* Like the preceding; read *in primitive civilizations and in those of antiquity.*

**wall-eye** should be so written, not as separate words; its proper meaning is the state of being wall-eyed, i.e. of having both eyes, or one, abnormally light-coloured, whether because the iris is very pale, or because the whites are disproportionately large, or because a squint exposes one white excessively. *Wall-eye* is formed from *wall-eyed*, not vice versa.

**waltz, valse.** The first, described in 1825 as 'the name of a riotous and indecent German dance', is the form that has established itself as the ordinary English.

**want.** *No man can say what* is wanted to be done *in regard to the military affairs of a nation till* . . . For this ugly construction, see DOUBLE PASSIVE. For confusion between *want* and *need* see NEED 3.

**Wardour Street.** 'The name of a street in London mainly occupied by dealers in antique and imitation-antique furniture'—OED. It is now occupied mainly by the film industry, but the old flavour hangs about the name. As Wardour Street itself offered to those who lived in modern houses the opportunity of picking up an antique or two that might be conspicuous for good or ill among their surroundings, so this article offers to those who write modern English a selection of oddments calculated to establish (in the eyes of some readers) their claim to be persons of taste and writers of beautiful English. And even as it is said of some dealers in the rare and exquisite that they have a secret joy when their treasures find no purchaser and are left on their hands, so the present collector, though he has himself no practical use for his articles of *virtu*, yet shows them without commendation for fear they should be carried off and unworthily housed. Those marked with an asterisk are illustrated in the paragraph that follows.

ANENT\*; archaic or coined *a-* compounds like *aplenty*\*; *as touching*\*; AUGHT\*; *belike*; ERE; *erst(while)*; ETHIC DATIVE with *me* or *you*; FOREBEARS, n.; *haply*\*; HITHER; HOWBEIT; *maugre*; *methinks*; *more by* TOKEN; *oft*; *of yore*; past participles in *-en* as *knitten*, *litten*; PERCHANCE; PROVEN; *rede*; SANS\*; SAVE (prep. or conj.); SHALL as in *you shall find*; *suffer* (allow); some *there* compounds (e.g. *-from*); THITHER; *to wit*\*; *trow*; VARLET; *ween*; (WELL)-NIGH\*; *what time* = when; some WHERE COMPOUNDS (e.g. *-through*); *whit*\*; *withal*; *wot*.

*So much has been written about the selfishness and stupidity of women anent the servant clause of the Insurance Bill. | In both inclusion and exclusion the British Davis Cup selectors have supplied talking points aplenty. | There are University matches which one approaches with some apprehension as touching the standard of cricket to be expected. | I have never written aught conflicting with that theory of State function. |*

*There is nothing to be done about it but surrender to the spell of a power which one may haply be pardoned for imagining to be a voice from another sphere. | The prospect of Parks and Illingworth batting nos. 7 and 8 is nigh delirious. | Batting sans spirit, sans skill, sans everything, in the face of some fine off-break bowling. | Like many other eminent scientific men—Huxley, to wit —Sir Ray Lankester has a cultivated taste. | The platform, the golf club, the bridge table, in no whit less than the factory and the workshop, must relax their claims.*

The words in small capitals are further commented upon in their dictionary places. See also ARCHAISMS and INCONGRUOUS VOCABULARY.

**-ward(s).** Words ending with *-ward(s)* may most of them be used as adverbs, adjectives, or nouns. The *-s* is usually present in the adverb, and absent in the adjective; the noun, which is rather an absolute use of the adjective, tends to follow it in being without *-s*; *moving eastwards*; *the eastward position*; *looking to the eastward(s)*. This usage prevails especially with the words made of a noun + *-ward(s)*, but is also generally true of the older words in which the first part is adverbial, such as *downward*. Some words, however, have peculiarities; see AFTERWARDS, FORWARDS, ONWARDS, TOWARDS.

**warmonger,** formerly a mercenary soldier, has become a term of invective, much used in modern methods of open diplomacy, and representing perhaps the lowest point yet reached in the degradation of the once respectable word *monger*. Since the 16th c., says the OED, this suffix has been used chiefly for one who carries on a petty or disreputable traffic; and, though *iron-*, *fish-*, and *coster-* survive to remind us that *monger* used to be a blameless word, popular opinion now regards it as more suitable for such compounds as *gossip-*, *whore-*, and *scandal-*.

**warn** is ordinarily used transitively. But its intransitive use in the sense of giving a warning at large is now common in journalism. (*The Chancellor warned that more drastic measures might have to be taken.* | *The BOAC warned that more flights will have to be cancelled if the strike goes on.*)

**warp**, n. See WOOF.

**wash.** The rivals *wash-hand-basin*, *washing-basin*, and *wash-basin*, are all in themselves justifiable, but it would be well to be rid of two of them. The first is obviously cumbrous, and there is no reason for preferring the longer second to the shorter third, especially with (*Moab is my*) *wash-pot*, *wash-house*, and *washtub*, waiting to welcome *wash-basin*, which is recommended.

**washing.** For *take in one another's w.*, see WORN-OUT HUMOUR.

**wassail** (now archaic) was originally a greeting, 'be of good health'. The pronunciation *wŏ'sl* has more support than *wă'sl*.

**wastage** is properly used of loss caused by wastefulness, decay, leakage, etc., or, in a staff, by death or resignation. It would be well if it were confined to this meaning instead of being used, as it habitually is, as a LONG VARIANT of *waste*.

**waste** (vb.). *It is important that as many documents as possible should be wasted at the earliest permissible date*, says an instruction issued by H.M. Stationery Office to government departments. This use of *w.* in the sense of convert into waste paper is a publishers' term, well understood inside their establishments, however puzzling to the uninitiated.

**wastrel.** The sense spendthrift or ne'er-do-well, now the most frequent one, is a comparatively modern development, the older senses being a piece of waste land, and a flawed or spoilt piece of workmanship.

**watchword.** For synonymy see SIGN.

**watershed.** The original meaning of the word, whether or not it is an anglicization of German *Wasserscheide* (lit waterparting), was the line of high land dividing the waters that flow in one direction from those that flow in the other, called in America a *divide*. The older of us were taught that that was its meaning, and that the senses sometimes given to it of river-basin and catchment area and drainage-slope were mere ignorant guesses due to confusion with the familiar word *shed*. Such classics as Lyell and Darwin and Geikie are all quoted for the correct sense; and that being so it is lamentable that the mistaken senses should have found acceptance with those who could appreciate the risks of ambiguity. Yet Huxley proposed that *water-parting* should be introduced to do *watershed*'s work, and that *watershed* should be allowed to mean what the ignorant thought it meant. The inevitable result is that now one has no idea, unless context happens to suffice without aid from the word itself, which meaning it has in any particular place. The old sense should have been rigidly maintained. OED quotations from Lyell and Geikie follow to make the old use clear, and an extract from a newspaper shows the modern misuse: (Lyell) *The crests or watersheds of the Alps and Jura are about eighty miles apart.* | (Geikie) *The watershed of a country or continent is thus a line which divides the flow of the brooks and rivers on two opposite slopes.* | (Newspaper) *The Seine, between its source in the Côte d'Or and the capital, has many tributaries, and when there is bad weather in the watershed of each of these an excessive flow is bound to be the result.*

On the other hand the figurative use of watershed now in journalistic favour may help to preserve its proper meaning. *This space of four years* [*1914–18*] *was a w. of history.* | *If we have crossed the w. in our relations with the Soviet Union, Mr. Butler's powers will find full scope at the Foreign Office.*

**waxen.** See -EN ADJECTIVES.

**way.** 1. For 'at the parting of the ways' see PARTING. 2. *Under way* (not *weigh*) is the right phrase for in motion; it has nothing to do with the anchor's being aweigh. Strictly a vessel is under way when she is not at anchor or made fast or aground; she may be under way and yet have no way on her. 3. *See* one's *way to*. *We hope that the Government will see their way* of giving effect to *this suggestion*. What has happened? The writer doubtless knows the idiomatic phrase as well as the rest of us, but finding himself saying 'will see their way *to* give effect *to*' has shied at the two *to*s; but he should have abandoned instead of mutilating his phrase; see OUT OF THE FRYING-PAN, and CAST-IRON IDIOM. 4. For *way* as an aphetic form of *away* (*In one practical respect Soviet society is way ahead of ours*) see AWAY.

**-ways.** See -WISE.

**we.** 1. Case. 2. National, editorial, and generic use. 3. Personal use.

1. Case. Use of *us* for *we* has been illustrated under US 1; the converse is seen in: *Whether the Committee's suggestions are dictated by Patriotism, Political expediency, or . . ., is not* for we outside mortals *to decide*.

2. National etc. uses: *We* may mean I and another or others, or the average man, or this newspaper, or this nation, or several other things. The newspaper editor occasionally forgets that he must not mix up his editorial with his national *we*. *But still*, we *are distrusted by Germany, and* we *are loth, by explaining how our acts ought to be interpreted, to put her in a more invidious position*. The first *we* is certainly England, the second is probably the newspaper. See I 2, and OUR 2, for similar confusions of different senses that are legitimate apart, but not together.

3. Personal use. Writers of books and articles should not use *we* in circumstances where the collective anonymity of the editorial of a newspaper is out of place. An author may, taking the reader with him, say *we have seen how thus and thus . . .*, but he ought not, meaning *I*, to say *we believe thus and thus*; nor is there any sound reason why, even though anonymous, he

should say *the present writer* or *your reviewer*, expressions which betray his individuality no less and no more than the use of the singular pronoun. Modern writers are showing a disposition to be bolder than was formerly fashionable in the use of *I* and *me*, and the practice deserves encouragement. It might well be imitated by the many scientific writers who, perhaps out of misplaced modesty, are given to describing their experiments in a perpetually passive voice (*such-and-such a thing was done*), a trick that becomes wearisome by repetition, and makes the reader long for the author to break the monotony by saying boldly *I did such-and-such a thing*.

**weal.** See WALE.

**wear, gear.** *Gear* in the sense of wearing apparel is archaic, though still in use for equipment that may include clothes. It has, however, lingered in *headgear* as a generic word for what is worn on the head, though *-wear* is now the natural partner of *foot-*, *neck-*, and *under-*.

**weather gage.** The nautical *gage* (relative position in regard to the wind) is the same word as the landsman's *gauge*.

**weave.** Ordinary p.p. *woven*; see also WOVE.

**weazen.** See WIZENED.

**web.** See WOOF.

**wed** is a poetic or rhetorical synonym for *marry*; and the established past and p.p. is *wedded*; but it is noticeable that the need of brevity in newspaper headings has brought into trivial use both the verb instead of *marry* (*STAR WEDS THIRD HUSBAND*), and the short instead of the long p.p. (*SUICIDE OF WED TEEN-AGERS*) see HEADLINE LANGUAGE.

**ween.** A WARDOUR-STREET word.

**weft.** See WOOF.

**weigh.** For *under* see *w.*, WAY.

**weird.** A word ruined by VULGARIZATION.

**Welch.** See WELSH

**well.** 1. *As well as.* 2. The preliminary *well.*

1. *As well as.* It is time for someone to come to the rescue of the phrase *as well as*, which is being cruelly treated. Grammatically, the point is that *as well as* is a conjunction and not a preposition. Or, to put it in a less abstract way, its strict meaning is not *besides*, but *and not only.* Or, to proceed by illustration, English requires not *You were there as well as me* (as it would if the phrase were a preposition and meant *besides*), but *You were there as well as I* (since the phrase is a conjunction and means *and not only*). The abuses occur, however, not in simple sentences like this with a common noun or pronoun following *as well as.* Indeed, it is usually not possible in these to tell whether the construction intended is right or wrong; in *They killed women as well as men, men* may be rightly meant to be governed by *killed*, or wrongly meant to be governed by *as well as* = besides; only the writer, and very likely not he, can say. They occur in places where the part of a verb chosen reveals the grammar: *The Territorial officer still has to put his hand in his pocket as well as giving his time.* Read *give*; it depends on *has to*; or else substitute *besides.* / *Its authoritative reports would help to build up an informed public opinion as well as guiding the Government.* Read *guide*; it depends on *would*; or else substitute *besides.* / *His death leaves a gap as well as creating a by-election in Ross and Cromarty.* Read *creates*; it is parallel to *leaves*; or else substitute *besides.* / *What should be made into cheap meals is now being used by dog-biscuit and other animal food makers as a basis of their wares*, as well as converting it *into manure.* Read *converted* for *converting it*, continuing the construction of *being.* A more obviously illiterate sentence than the rest.

2. What the OED calls the 'preliminary or resumptive' use of *well*, tantamount to a plea for a moment's grace to think what to say, is for most of us a reflex response to the stimulus of any question. Those taking part in broadcast discussions would add to the pleasure of listeners if they would try to curb it, though it would be unreasonable to expect them to do without it altogether. A preliminary word, just by way of an introductory noise, is a natural feature of the spoken language. *I say, lo, look* (by itself or with *here* or *you*), *marry, listen, pray, why*: all these have had their turns of duty at this. *Now* or *And now* is the broadcaster's favourite way of indicating that he is passing to a new topic, as it serves the parson to show that he has finished his sermon and is starting to pray again. *Look* is much used as the first word of a telephone call—a curiously unsuitable word, one would have said, for beginning a conversation with someone who is out of sight. See also MEANINGLESS WORDS.

**well and well-.** In combinations of a participle and *well* there is often a doubt whether the two parts should be hyphened or left separate. The danger of wrong hyphens is greater than that of wrong separation; e.g. to write *His courage is well-known* (where *well known* is the only tolerable form) is much worse than to write *His well known courage*, which, though unusual, is justifiable. Some help will be found under HYPHENS; and it may be here repeated that if a participle with *well* is attributive (*a well-aimed stroke*) the hyphen is often used, though not necessary; but if the participle is predicative (*the stroke was well aimed*) the hyphen is wrong. Similarly in such phrases as *well off*: *They are not well off*, but *Well-off people cannot judge.* These are not arbitrary rules; they follow from acceptance of the principle that hyphens should not be used except when a reader needs their help.

**well-nigh.** See WARDOUR STREET. *Archaeology had strengthened its hold on art, and went well-nigh to strangling it.* The natural English would have been *and came near strangling it*, or *and nearly strangled it.* But if the writer was bent on displaying his antique, he should at least have said *and well-nigh strangled it*. The use of *well-nigh* is purely adverbial; i.e., it needs a following verb or adjective or noun to attach itself to; *well nigh worn to pieces*, and *well nigh dead*, says Shakespeare, and *well nigh half the angelic name*, says Milton. To say *come well-nigh to* is to put the antique in an incongruous frame.

**well-read.** See INTRANSITIVE P.P.

**Welsh, Welch.** The established modern spelling is *-sh*, except in the official names of the Welch Regiment and the Royal Welch Fusiliers; the Welsh Guards are so spelt. *Welsh* is also the only spelling of the verb; its derivation is uncertain. For *Welsh rabbit* see TRUE AND FALSE ETYMOLOGY.

**wen.** For *the Great W.*, see SOBRI-QUETS.

**were.** For the subjunctive uses in the singular, as *If I were you*, *Were he alive*, *It were futile*, some of which are more inconsistent than others with the writing of natural English, see SUBJUNC-TIVES.

**werewolf, werw-.** The first is recommended; it is the more familiar, it suggests the usual pronunciation, and it dates back to Old English.

**westerly.** See EASTERLY.

**wh.** Whether this is pronounced as plain *w* or as *hw* in words such as *where, whether, nowhere*, etc. is a matter of locality or nationality or education. The aspirated sound is natural to the Scots, the Irish, and the Americans; in England (except in the extreme north) *w* was long the normal pronunciation, but *hw* is gaining ground under the influence of the speak-as-you-spell movement.

**wharf.** For plural etc., see -VE(D).

**wharfinger.** Pronounce *-jer*.

**what** is a word of peculiar interest, because the small problems that it

poses for writers are such as on the one hand yield pretty readily to analysis, and on the other hand demand a slightly more expert analysis than they are likely to get from those who think they can write well enough without stopping to learn grammar.

1. Wrong number attraction.   2. *What* singular and *what* plural. 3. One *what* in two cases. 4. *What* resumed by (*and, but*) *which*. 5. Miscellaneous.

1. **Wrong number attraction.** In each of the examples to be given it is beyond question that *what* starts as a singular pronoun (= that which, or a thing that), because a singular verb follows it; but in each also the next verb, belonging to the *that* of *that which*, or to the *a thing* of *a thing that*, is not singular but plural. This is due to the influence of a complement in the plural, and the grammatical name for such influence is *attraction*; all the quotations are on the pattern *What is said are words*, instead of *What is said is words*. Whether attraction of verb to complement is idiomatic in English has already been discussed in NUMBER 1; it is here assumed that it is not, and that therefore in the quotations that follow, if the singular *is* is to stand (as to which see below), the roman-type verb should have its number changed from plural to singular: *What is of absorbing and permanent interest* are *the strange metamorphoses which this fear underwent.* | *What is required* are *houses at rents that the people can pay.* | *What is required* are *three bedrooms, a good large living-room* . . . | *What is wanted to meet it* are *proposals which are practical.* | *What is needed* are *a few recognized British financial corporations.* | *What is of more importance in the official statement of profits* are *the following figures.*

2. *What* singular and *what* plural. In each of the above quotations, the writer made it plain, by giving *what* a singular verb, that he conceived *what* there as a singular pronoun. But the word itself can equally well be plural: *I have few books, and what there* are do *not help me*. So arises

another problem concerning the number of verbs after *what*, and this second one naturally gets mixed up with the first, in which some of the examples could be mended as well by giving *what* a plural verb as by giving the complement a singular one. In dealing with this other problem, however, we will ignore the complication that the number of a verb may be affected by the 'attraction' of the complement, and consider only the question when a verb governed by *what* should be treated as singular and when as plural.

First comes a particular form of sentence in which plural *what* is better than singular, or in other words in which its verb should be plural. These are sentences in which *what*, if resolved, comes out as *the ——s that*, *——s* standing for a plural noun actually present in the complement. After each quotation a correction is first given if it is desirable, and in any case the resolution that justifies the plural: *We have been invited to abandon what* seems *to us to be the most valuable parts of our Constitution* (read *seem*; abandon the parts of our Constitution that seem). / *The Manchester City Council, for what* was *doubtless good and sufficient reasons, decided not to take any part* (read *were*; for reasons that were). / *It is a diatribe against M. Loucheur and M. Clementel, but the personal aspect is of little importance to English readers; what* are *important* are *the criticisms of the operation of protective duties in France* (The criticisms that; but *What is important is* would have been better, *what is* standing for *the thing that is*, in contrast with *aspect*). / *They specially approved what to Liberals* was *the most reactionary and disastrous parts of it* (read *were*; approved the parts of it that to Liberals were). / *Confidence being inspired by the production of what* appears *to be bars or bricks of solid gold* (read *appear*; production of bars or bricks that appear).

But resolution of *what* often presents us not with a noun found in the complement, but with some other noun of wider meaning, or again with the still vaguer *that which*. A writer should make the resolution and act on it without allowing the number of the complement to force a plural verb on him if the most natural representative of *what* is *that which* or *the thing that*. In several of the following quotations the necessary courage has been lacking; corrections and resolutions are given as before: *No other speaker has his peculiar power of bringing imagination to play on what* seems, *until he speaks, to be familiar platitudes* (read *seem*; on sayings that seem)./ *Instead of the stupid agitation now going on in South Wales, what* are *needed* are *regular working and higher outputs* (read *what is needed is*; the thing that is needed—rather than things, as opp. *agitation*). / *What* are *wanted* are *not small cottages, but the larger houses with modern conveniences that are now demanded by the working classes* (read *what is wanted is*; the thing that is wanted—rather than the buildings that are). / *What* provoke *men's curiosity* are *mysteries, mysteries of motive or stratagem; astute or daring plots* (read *provokes . . . is*; that which provokes—rather than the things that provoke)./ *In order to reduce this material to utility and assimilate it, what* are *required* are *faith and confidence, and willingness to work* (read *what is required is*; but the qualities that are required justifies the plurals, though it does not make them idiomatic).

It will be observed that there is more room for difference of opinion on this set of examples than on either those in 1 or the previous set in 2, and probably many readers will refuse to accept the decisions given; but if it is realized that there are problems of number after *what*, and that solutions of them are possible, that is sufficient.

**3.** One *what* in two cases. For the general question whether in a language that, like English, has shed nearly all its case-forms the grammatical notion of case still deserves respect, see CASES 3 D. It is here assumed that it does, to the extent that no word, even if it has not different forms such as *I* and *me* for the subjective and objective uses, ought to be so placed that it has, without being repeated, to be taken twice over, first in one and then in the other case. The word *what* is peculiarly liable to such treatment. There are two chief ways of sparing grammatically-minded readers this outrage on their susceptibilities: sometimes a second *what* should be inserted; sometimes it is better to convert a verb to the other voice, so that *what* becomes either object, or subject, or both. Corrections are given in brackets; to correct Pater, from whom the last example comes, is perhaps impudence, but grammar is no respecter of persons: *This is pure ignorance of what the House is* and its *work consists of* (and what its). / *But it is not folly to give it what it had for centuries* and was only artificially taken from it *by force rather more than a hundred years ago* (what belonged to it for). / *Mr. —— tells us not to worry about Relativity or anything so brain-tangling, but to concentrate on what surrounds us,* and we *can weigh and measure* (and can be weighed and measured). / *Impossible to separate later legend from original evidence as to what he was,* and said, and *how he said it* (and what he said).

**4.** *What* resumed by *and which* or *but which. Francis Turner Palgrave, whose name is inseparably connected with what is probably the best,* and which *certainly has proved the most popular, of English anthologies* (what is probably the best, and has certainly proved). / *It is an instructive conspectus of views on what can hardly be described as a 'burning question',* but which *certainly interests many Irishmen* (but certainly interests). / *We are merely remembering what happened to our arboreal ancestors,* and which *has been stamped by cerebral changes into the heredity of the race* (and has been stamped). A want of faith either in the staying power of *what* (which has a good second wind and can do the two laps without turning a hair), or in the reader's possession of common sense, has led to this thrusting in of *which* as a sort of relay to take up the running. These sentences are not English; nothing can represent *what*—except

indeed *what.* That is to say, it would be English, though hardly idiomatic English, to insert a second *what* in the place of the impossible *which* in each. If the reader will try the effect, he will find that the second *what*, though permissible, sometimes makes ambiguous what without it is plain; in the last example, for instance, 'what happened' and 'what has been stamped' might be different things, whereas 'what happened, and has been stamped' is clearly one and the same thing. The reason why *which* has been called 'impossible' is that *what* and *which* are of different grammatical values, *which* being a simple relative pronoun, while *what* (= that which, or a thing that) is a combination of antecedent and relative. The second verb needs the antecedent-relative just as much as the first, if *but* or *and* is inserted; if neither *but* nor *and* is present, *which* will sometimes be possible, and so omission of *and* would be another cure for the last example.

Two specimens are added in which the remedy of simply omitting *which* or substituting for it a repeated *what* is not possible without further change. The difficulty is due to the superstition against PREPOSITION AT END, and vanishes with it. *I can never be certain that I am receiving what I want and for which I am paying.* Read *what I want and am paying for.* / *But now we have a Privy Councillor and an ex-Minister engaged daily in saying and doing what he frankly admits is illegal,* and for which *he could be severely punished.* Read *and what he could be severely punished for.* The repetition of *what* is required because the relative contained in the first *what* is subjective, and that in the second objective; see 3.

**5.** Miscellaneous. Some writers with an excessive zeal for correctness seem to think that *what* is inelegant and *that which* is an improvement on it. But *hold fast to that which is good* is no better English than *hold fast to what is good*; in fact the first has now a slightly old-fashioned air. On the other hand the beautiful conciseness belonging to *what* as antecedent-relative seems to lure the unwary into experiments in further concision. They must remember that both parts of it, the antecedent (*that* or *those*) and the relative (*which*), demand their share of attention: *What I am concerned in the present article is to show that not only theory but practice support the unrestricted exercise of the prerogative.* Read *concerned to do*; otherwise the *which* in *what* is without government. | *What my friend paid less than a pound a day for last year he had to pay two guineas a day at a minor Brighton establishment last Easter.* Read *two guineas a day for*; otherwise the *that* in *what* is without government. | *Entering the church with feelings different from what he had ever entered a church before, he could with difficulty restrain his emotions.* Read *entered a church with*; otherwise the *which* of *what* is without government.

For *what* used as an 'interrogative expletive' see MEANINGLESS WORDS.

**what ever, whatever.** The various uses are complicated, and cannot be all set out, for readers who are not specialists in grammar, without elaborate explanations that would demand too much space. This article will avoid all technicalities except what are needed in dealing with two or three common mistakes.

1. The interrogative use. 2 The antecedent-relative use. 3. The concessive use.

1. The interrogative use. *What ever can it mean? What ever shall we do?* For the status of this, see EVER. It should never appear in print except when familiar dialogue is being reproduced, and should then be in two separate words, differing in this from all other uses. Three examples follow in which both these rules are disregarded; in the second of them we have an indirect instead of a direct question, but the same rules hold: *Which is pretty, but whatever can it mean?* | *Whatever you mean by 'patriotic' education I do not know, but Lord Roberts's use of the term is plain enough.* | *And, considering that 180,000 actually arrived in the country, whatever was the cost?*

2. The antecedent-relative use. *What ever* in this use is an emphatic form of *what* as antecedent-relative (see WHAT); i.e., while *what* means that which or the (thing, things) that, *whatever* means all that or any(thing etc.) that. The point ignored in the quotations below is that, since *whatever* contains in itself the relative (*that* or *which*) as well as the demonstrative or antecedent (*all, any,* etc.), another relative cannot grammatically be inserted after it; *whatever* (or *whatsoever*) means not *any*, but *any that*, and *whatever that* is as absurd as *any that that*. *His cynical advice shows that whatever concession to Democracy that may seem to be involved in his words, may not be of permanent inconvenience.* | *Keep close in touch with Him in whatsoever creed or form that brings you nearest to Him.* | *They use in the shell, the gun—in whatever component, big or small,* upon which *their attention is concentrated—the essence of all that matters.* In the first two, omit the roman-type *that*; in the third, which has gone wrong, as often happens, owing to the PREPOSITION-AT-END superstition, get rid of *which* by rewriting *in whatever component, big or small, their attention is concentrated upon.*

3. The concessive use. *Whatever one does, you are not satisfied*; *I am safe now, whatever happens*; *Whatever you do, don't lie.* These are concessive clauses, short for Though one does A or B or C, Though this or that or the other happens, Though you do anything else. They differ from the *whatever* clauses dealt with above in being adverbial, *whatever* meaning not *all* or *any that* (*that* beginning an adjectival clause), but *though all* or *any*. The difference is not a matter of hairsplitting; *Whatever he has done he repents* may mean (a) He is one of the irresolute people who always wish they had done something different, or (b) Though he may be a great offender, repentance should count for something; *whatever* antecedent-relative gives (a) and *whatever* concessive gives (b). In practice it should be noticed that proper punctuation distinguishes

the two, the (a) meaning not having the two clauses parted by a comma, since *whatever* belongs to and is part of both, and the (b) meaning having them so parted, since *whatever* belongs wholly to one clause. In the following sentence the reader is led by the wrong comma after *have* to mistake the *whatever* clause for a concessive and adverbial one: *He has no reason to be displeased with this sequel to his effort, and, whatever responsibility he may have, he will no doubt accept gladly*. This should run: 'and whatever responsibility he may have he will no doubt accept gladly'.

**wheal.** See WALE.

**wheaten.** See -EN ADJECTIVES.

**whence, whither.** The value of these subordinates of *where* for lucidity and conciseness seems so obvious that no one who appreciates those qualities can see such help being discarded without a pang of regret. Why is it that *whence* substitutes apparently so clumsy as *where . . . from*, and *where . . . to*, can be preferred? It is surely because the genius of the language actually likes the PREPOSITION AT END that wiseacres have conspired to discourage, and thinks 'Where are you coming to?' more quickly comprehensible in moments of threatened collision than 'Whither are you coming?' We who incline to weep over *whence* and *whither* must console ourselves by reflecting that in the less literal or secondary senses the words are still with us for a time; 'Whither are we tending?', and 'Whence comes it that . . .?', are as yet safe against *where . . . to* and *where . . . from*; and the poets may be trusted to provide our old friends with a dignified retirement in which they may even exercise all their ancient rights. But we shall do well to shun all attempts at restoration, and in particular to eschew the notion (see FORMAL WORDS) that the writer's duty is to translate the *where . . . from* or *where . . . to* of speech into *whence* and *whither* in print. On the other hand, let us not be ultra-modernists and assume that *whence* and *whither*, even in their primary

senses, are dead and buried; that must be the view of the journalist who writes: *The Irregulars have been compelled to withdraw their line from Clonmel*, to where *it is believed they transferred their headquarters when they had to flee from Limerick*. If *whither* was too antiquated, the alternative was 'to which'; but occasions arise now and then, as in this sentence, to which *whence* and *whither* are, even for the practical purposes of plain speech, more appropriate than any equivalent. They should be allowed to stand on their own feet: not even the examples that can be found in the Psalms and elsewhere justify the use today of the tautology *from whence*. *The whole middle of the pitch needs to be dug up and the turf sent back to the creek from whence it came*.

**whenever,** the right form for the ordinary conjunction, should not be used instead of the colloquial *when ever* (*When ever will you be ready?*), for which see EVER.

**where- compounds.** A small number of these are still in free general use, though chiefly in limited applications, with little or no taint of archaism; these are *whereabouts* (as purely local adv. and n.), *whereas* (in contrasts), *wherever*, *wherefore* (as noun plural in *whys and wherefores*), *whereupon* (in narratives), and *wherewithal* (as noun). The many others—*whereabout, whereat, whereby, wherefore* (adv. and conj.), *wherefrom, wherein, whereof, whereon, wheresoever, wherethrough, whereto, wherewith,* and a few more—have given way, though to different degrees, in both the interrogative and the relative uses either to the preposition with *what* and *which* and *that* (*whereof*=of what?, what . . . of?, of which, that . . . of), or to some synonym (*wherefore* = why); resort to them generally suggests that the writer has a tendency either to FORMAL WORDS or to PEDANTIC HUMOUR.

**wherever, where ever.** As WHENEVER.

**wherewithal.** The noun, as was mentioned in WHERE- COMPOUNDS, has survived in common use (*but I haven't got the w.*), no doubt because the quaintness of it has struck the popular fancy. But the noun should remember that it is after all only a courtesy noun, not a noun in its own right; it means just 'with which', but seems to have forgotten this in: *They* [France's purchases] *have been merely the* wherewithal with which *to start business again.*

**whether.** 1. For *w.* and *that* after *doubt(ful)*, see DOUBT(FUL). 2. *W.* or *no(t)*. *Whether he was there or was not there* easily yields by ellipsis *Whether he was there or not*, and that by transposition *Whether or not he was there*. *Whether or no he was there* is not so easily accounted for, since *no*, unlike *not*, is not ordinarily an adverb (see NO). In fact the origin of the idiom is uncertain; but the fact remains that *whether or not* is 'less frequent' (OED) than *whether or no*—especially, perhaps, when the *or* follows *whether* immediately : *Whether or no he did it* ; *whether he did it or not*. Whichever form is used, such a doubling of the alternative as the following should be carefully avoided : *But clearly, whether or not peers will* or will not *have to be made depends upon the number of the Die-Hards.* Omit either *or not* or *or will not.* 3. *Whether* is often repeated as a clearer pointer than a bare *or* to an alternative that forms a separate sentence. *I cannot remember whether they were lowered into the street or whether there was a window opening out at the back.* 4. For the misuse of *As to whether* see AS 3.

**which.** Relative pronouns are as troublesome to the inexpert but conscientious writer as they are useful to everyone, which is saying much. About *which*, in particular, problems are many, and some of them complicated; that the reader may not be frightened by an article of too portentous length, the two that require most space are deferred, and will be found

in the separate articles WHICH, THAT, WHO, and WHICH WITH AND OR BUT. Reference should also be made to the article THAT, REL. PRON. The points to be treated here can be disposed of with more certainty and at less length, under the headings: 1. Relative instead of demonstrative. 2. One relative in two cases. 3. One relative for main and subordinate verbs. 4. Break-away from relative. 5. Confused construction. 6. Late position. 7. Commas. 8. *In which to.*

1. Relative instead of demonstrative. The type is: *He lost his temper, which proving fatal to him.* The essence of a relative is to do two things at once, to play the part of a noun in a sentence and to convert that sentence into a subordinate clause. *He lost his temper; this proved fatal*; these can be made into one sentence (a) by changing the demonstrative *this* into the relative *which*, or (b) by changing the verb *proved* to the participle *proving*; one or the other, not both as in the false type above. Actual examples of the blunder, with corrections in brackets, are: *Surely what applies to games should also apply to racing, the leaders* of which being *the very people from whom an example might well be looked for* (read *of this* or *of the latter*; or else *are* for being). / *Persons who would prefer to live in a land flowing with milk and honey if such could be obtained without undue exertion,* but, failing which, *are content to live in squalor, filth, and misery* (read *failing that*; or else *failing which they* for *but failing which*). / *The World Scout principle—namely, of bringing into an Order of the young the boys of different races,* by which means not only educating *the children in scouting, but . . .* (read *by this means*; or else *we should not only educate* for *not only educating*).

2. One relative in two cases. See WHAT 3 for this question; in all the following extracts, a single *which* is once objective and once subjective. The cure is either to insert a second *which* in the second clause, or to convert one of the two verbs into the same voice as the other, e.g. in the first

example 'and others to study': *Mr. Roche is practising a definite system, which he is able to describe, and could be studied by others.* | *He went up to a pew in the gallery, which brought him under a coloured window which he loved and always quieted his spirit.* | *The queer piece, which a few find dull, but to most is irresistible in its appeal.* | *Shakespearian words and phrases which the author has heard, and believes can be heard still, along this part of the Avon valley.*

3. One relative for main and subordinate verbs (or verb and preposition). The following sentence is provided with three endings, A, B, C, with each of which it should be read successively: *This standard figure is called Bogey, which if you have beaten (A) you are a good player, (B) you are apt to mention, (C) is sometimes mentioned.* In A the grammar is unexceptionable, *which* being the object of *have beaten*, and having no second job as a pronoun (though as relative it attaches to *Bogey* the clause that is also attached by *if* to *you are* etc.). In modern use, however, this arrangement is rare, being usually changed to 'if you have beaten which'. In B we come to questionable grammar, *which* being object first to *have beaten* and then to *mention*; English that is both easy and educated usually avoids this by making *which* object only to *mention*, and providing *have beaten* with another—*which, if you have beaten it, you are apt to mention.* Meeting the B form, we incline to ask whether the writer has used it because he knows no better, or because he knows better than we do and likes to show it. Grammatically, it must be regarded as an ellipsis, and to that extent irregular, but many ellipses are idiomatic; this particular kind is perhaps less justified by idiom than noticeable as irregular. About C there are no such doubts; it is indefensible, the *which* having not only to serve twice (with *have beaten*, and with *is mentioned*), but to change its case in transit; see 2.

Illustrations follow of B and C; A, being both legitimate and unusual, and having been introduced only for purposes of comparison, need not be quoted for.

B, doubtfully advisable

*With a fire in her hold which he managed to keep in control, although unable to extinguish* (add *it?*). | *Mr. Masterman was a little troubled by the spirit of his past, which, if he had not evoked, no one would have remembered* (evoked *it?*). | *And it was doubtless from Weldon that he borrowed the phrase which his use of has made so famous* (of *it?* or *his use of which has made it?*). This last is no more ungrammatical, though certainly more repulsive, than the others.

C, undoubtedly wrong

*The programme is divided up into a series of walks, which, if the industrious sightseer can undertake, will supply him with a good everyday knowledge of Paris* (undertake them). | *In general the wife manages to establish a status which needs no legal proviso or trade union rule to protect* (protect it; *or* which it needs; *or* to protect which needs no . . . rule).

4. Break-away from relative. *He shows himself extremely zealous against practices in some of which he had greatly indulged, and was himself an example of* their *ill effects.* | *It imposes a problem which we either solve or perish.* Both of these are strictly ungrammatical. In the first, which is the easier to deal with, it will be noticed that in sense the third part (*and was* to the end) is clearly coupled by *and* not to the first part or main sentence (*He shows* to *practices*), but to the second part or relative clause (*in some* to *indulged*). Nevertheless, by the use of *their* it has been definitely broken away from connexion with *which*, and become grammatically, but illogically, a second main sentence with, for subject, the *He* that begins the first. There are two possible correct versions of the second and third parts, (a) *some of which he had greatly indulged in and himself exemplified the ill effects of,* or (b) *in some of which he had greatly indulged, and of the ill effects of (some of) which he was himself an example*; (a) will be repudiated, perhaps more justifiably

than usual, by those who condemn final prepositions; (b) fails to give the precise sense, whether the bracketed *some of* is inserted or not; some will therefore prefer the break-away, which is not an uncommon construction, to both (a) and (b).

The other example (*It imposes a problem which we either solve or perish*) is, owing to Lord Grey's 'The nations must learn or perish', of a now very popular pattern. The break-away depends on the nature of *either . . . or* alternatives, in which whatever stands before *either* must be common to both the *either* and the *or* groups. *Either we solve this or we perish* can therefore become *We either solve this or perish*, but cannot become *This we either solve or perish*, because *this* is peculiar to the *either* group—otherwise the full form would be *Either we solve this or we perish this*. With *this* as object the escape is easy—to put *this* after *solve* —; with *which* as object that is not tolerable (*we either solve which or perish*), and strict grammar requires us to introduce into the *or* group something that can take *which* as object—*a problem which we either solve or perish by not solving, either solve or are destroyed by*, etc. Even those who ordinarily are prepared to treat *either* with proper respect (see EITHER 5) may perhaps allow themselves the popular form; if not, 'A problem which if we do not solve we perish' (see 3 A) is worth considering.

**5.** Confused construction. *He may be expected to make a determined bid for the dual rôle* which *is his right and duty as Prime Minister to occupy*. In that sentence, is *which* subject to *is*, or object to *occupy*? It is in fact, of course, the latter, *occupy* having no other object, and not being able to do without one; but the writer has effectually put us off the track by dropping the *it* that should have parted *which* from *is*. *To occupy which is his right* becomes, when *which* is given its normal place, *which it is his right to occupy*. This mistake is very common, and will be found fully discussed under IT 1.

**6.** Late position. In the examples, which are arranged as a climax, the distance between *which* and its antecedent is shown by the roman type. Grammar has nothing to say on the subject, but common sense protests against abuse of this freedom. *She is wonderful in her brilliant* sketch *of that querulous, foolish little old lady* which *she does so well.* | *The whole art of clinching is explained in this little* book *from the concentrated harvest of wisdom in* which *we present some specimens to our readers.* | *Nothing has more contributed to dispelling this illusion than the* camera, *the remarkable and convincing evidence it has been possible to obtain with* which *has enormously added to the knowledge of the habits of animals*.

**7.** Commas. The plea is made elsewhere in this book (THAT, REL. 1; WHICH, THAT, WHO, 9) that defining and non-defining relative clauses ought to be readily distinguishable by the use of *that* for the former and of *which* etc. for the latter. But since most writers continue to use *which* for both it is important to have another means of distinguishing. A comma preceding *which* shows that the *which*-clause is non-defining, and the absence of such a comma shows that it is defining. *He declares that the men were treated like beasts throughout the voyage, and he gives the worst description of the general mismanagement* which *was most conspicuous*. There is no comma before *which*, and therefore the clause must be presumed to be a defining one; i.e., it limits the sense of *the general mismanagement* by excluding from it such parts as were less conspicuous; the most conspicuous part of the mismanagement is described as very bad indeed—that is what we are told. Or is it not so, and are we to understand rather that the whole of the general mismanagement is described as very bad, and moreover that it was conspicuous? Surely the latter is meant; but the absence of the comma forbids us to take it so. The difference between the two senses (or the sense and the nonsense) is not here of great importance, but is at least perfectly clear, and the importance of not misinterpreting will vary infinitely elsewhere. That right interpretation should depend on a mere comma is a pity, but, until *that* and *which* are differentiated, so it must be, and writers must see their commas safely through the press.

**8.** *In* etc. *which to*. Examples: *England is, however, the last country in which to say so.* | *I have no money with which to buy food*. The current English for the second is indisputably *I have no money to buy food with*; and there can hardly be a doubt that this has been formalized into the other by the influence of the PREPOSITION-AT-END superstition. No one need hesitate about going back to nature and saying *to buy food with*. And even for the first 'the last country to say so in' is here recommended, though the very light word *so* happens to make with the other very light word *in* an uncomfortably weak ending; much more is 'a good land to live in' superior to 'a good land in which to live'. A confessedly amateur guess at the genesis of these constructions may possibly throw light. The assumption underlying the *in which to* form is obviously that there is an elliptical relative clause —'This is a good land in which (one is) to live'. The amateur guess is that there is no relative clause in the case at all, and that the form *to live in* originated in an adverbial infinitive attached to the adjective *good*. He is a hard man to beat; how hard? why, to beat; what Greek grammars call an epexegetic (or explanatory) infinitive. *It is a good land to in-habit* is precisely parallel, and *to live-in* is precisely the same as *to in-habit*. If this account is true, the unpleasant form 'in which to live' might be dismissed as a grammarians' mistaken pedantry.

**which, that, who.** 1. General. 2. *Which* for *that*. 3. *Which* after superlative etc. 4. *Which* in *It is . . . that . . . .* 5. *Which* as relative adverb. 6. Elegant variation. 7. *That* for *which*. 8. *Which*,

*who*, and nouns of multitude. 9. *Who* and *that*.

**1.** General. If the evidence of a first-class writer who was no purist or pedant counts for anything, Lord Morley's opinion exhibited below should make it worth while to master the differences between *which* and the other relatives. The extracts are from an article in the *Westminster Gazette* of 3rd Oct. 1923 by Miss Hulda Friedrichs: 'In 1920 Messrs. Macmillan published a new edition of Lord Morley's works . . . He was determined to make it a carefully revised edition, and made one or two attempts at revising it himself. . . . He then asked me whether I would care to help him, and explained what my part of the work would be. It sounded rather dull, for he was particularly keen on having the word *which*, wherever there was the possibility, exchanged for *that*. . . . He was always ready and very willing to go with me through the notes I had jotted down while going through a book page by page, "which" hunting and looking out for other errors.'

Let it be stated broadly, before coming to particular dangers, that: (A) of *which* and *that*, *which* is appropriate to non-defining and *that* to defining clauses; (B) of *which* and *who*, *which* belongs to things, and *who* to persons; (C) of *who* and *that*, *who* suits particular persons, and *that* generic persons. (A) *The river, which here is tidal, is dangerous*, but *The river that flows through London is the Thames*. (B) *The crews, which consisted of Lascars, mutinied*, but *Six Welshmen, who formed the crew, were drowned*. (C) *You who are a walking dictionary*, but *He is a man that is never at a loss*. To substitute for the relative used in any of those six examples either of the others, if the principles maintained in this book are correct, would be a change for the worse; and, roughly speaking, the erroneous uses (if they are so) illustrated below are traceable to neglect or rejection of A, B, and C.

**2.** *Which* for *that*. The importance and convenience of using *that* as the regular token of the defining clause has been fully illustrated under THAT, REL. PRON., and no more need be said here on that general point.

**3.** *Which* after superlative etc. When the antecedent of a defining clause includes a word of exclusive meaning, such as a superlative, an ordinal numeral, or 'the few', the use of *which* instead of *that* (or *who* as second best) is bad enough to be almost a solecism even in the present undiscriminating practice. The question between *that* and *who* in such places will be touched upon in para. 9; but at least *which* should be expelled from the following extracts. By rule B *who* is better than *which*, and by rule C *that* is better than *who*: *All three will always be ranked among the foremost physical theorists and experimenters which Great Britain has produced.* / *Had the two men of greatest genius in the respective spheres, which the British Navy has ever produced, had their way. . . .* / *He was a true musical poet—perhaps, with one exception, the most gifted which England has ever produced.* / *One of the few composers of the first rank which England has* produced.

Besides the particular type here described there are others in which for various reasons *which* is wrong, but whether *who* or *that* should replace it is doubtful; these will be dealt with in para. 9.

**4.** *Which* in *It is . . . that . . .* The constructions exemplified in simple forms by *It was the war that caused it, It was yesterday that we came,* are often difficult to analyse grammatically or account for. The difficulty need not concern us here; one thing can be confidently said about them, namely that they require *that* and not *which*—*that* the defining relative (*It was Jones that did it,* the clause defining *It*) or *that* the conjunction (*It is with grief that I learn . . .*). In the three examples, *that* should replace *to which, in which,* and *which*: *It is to the State, and to the State alone,* to which *we must turn to acquire the transfer of freeholds compulsorily, expeditiously, and cheaply.* / *It is in the relation between*

*motive, action, and result in a given chain of historical causation,* in which *history consists.* / *So once again East is West, and it is shown that it is not only the Japanese* which *have the imitative instinct strongly developed.*

**5.** *Which* as relative adverb. The curious but idiomatic use of *that* in this construction is explained in THAT, REL. PRON. 3, where it is added that here *that* is equivalent to *which* with a preposition, and that *which* by itself is unsuitable for similar treatment. The clauses are defining, attached to such words, expressed or implied, as *way, extent, time, place. That* should be substituted in each example for *which*. *In England the furthest north* which *I have heard the nightingale was near Doncaster.* / *Parliament will be dissolved not later than Monday week—the earliest moment, that is to say,* which *it has ever been seriously considered possible for the dissolution to take place.* / *He made a good 'legend' during his lifetime in a way* which *very few actors have done.* / *Before railway working was 'sped up' to the extent* which *it is at present, continuous work of this character was no great strain.*

**6.** Elegant variation. I was surprised many years ago when a very well-known writer gave me his notion of the relation between *which* and *that*: When it struck him that there was too much *which* about, he resorted to *that* for a relief. So he said; it was doubtless only a flippant evasion, not a truthful account of his own practice, but still a tacit confession that he followed instinct without bothering about principles. Of the unskilled writer's method it would be a true enough account; here is a specimen: *Governments find themselves almost compelled by previous and ill-informed pledges to do things* which *are unwise and to refrain from doing things* that *are necessary.* The two relative clauses are exactly parallel, and the change from *which* to *that* is ELEGANT VARIATION at its worst. When two relative clauses are not parallel, but one of them depends on the other, it is not such a simple matter; as is stated in REPETITION (Dependent Sequences), there is a reasonable objection to one *which*-clause, or one *that*-clause, depending on another. Two examples will show the effect (a) of scorning consequences and risking repetition, and (b) of trying elegant variation; neither is satisfactory: (a) *Surely the reductio ad absurdum of tariffs is found in a German treaty with Switzerland* which *contains a clause* which *deserves to remain famous*; (b) *The task is to evolve an effective system* that *shall not imperil the self-governing principle* which *is the corner-stone of the Empire*. The repetition is easily avoided in (a) by the change of *which contains* to *containing* (*which* in both places ought to have been *that*, but that is here irrelevant). In (b) the absence of a comma shows that the *which* is meant as a defining relative and should therefore be *that*; but, as a non-defining clause would here give a hardly distinguishable sense, the escape is to use one and keep *which*, merely inserting the necessary comma. The reader may like another example to play with: . . . *was recalled to the passer-by in Pall-Mall by Foley's fine bronze statue of the War Minister who deeply cared for the private soldier*, which *stood in front of the now destroyed War Office*, that *has very recently given place to the palatial premises of the Royal Automobile Club.*

**7.** *That* for *which*. After all these intrusions of *which* into the place of *that*, it must be recorded that retaliation is not quite unknown; but it is rare. *In the island of South Uist, that I have come from, there is not one single tree.* | *A really happy party was the Chiverton family, that had a carriage to itself and almost filled it.* The justification of *that* in these would require that there should be several South Uists from one of which I have come, and several Chiverton families of which one only had a carriage to itself; but even those suppositions are precluded by the commas. Other examples are given in THAT, REL. PRON. 2.

**8.** *Which, who*, and nouns of multitude. Words like *section, union, world*, sometimes mean all the persons com-

posing a section etc.; idiom then allows us to regard them as grammatically singular or plural as we prefer, but not to pass from one to the other; see NUMBER 6. Now a section, if we elect to treat it as singular, is a thing; but, if we make it plural, it is persons, and by rule B in para. 1 above *which* belongs to things, whereas *who* belongs to persons. Three examples that accordingly need correction are: *There was a strong section* which *were in favour of inserting the miners' 5s. and 2s., as the debate proved* (*which* is required by the preceding *was*, but in turn requires a second *was* instead of *were*). | *All the world* who *is directly interested in railway projects will have paid a visit to the Brussels Exhibition* (*who* is possible, but only if *are* follows; otherwise not indeed *which* since the clause is defining, but *that* is required). | *The Canon is writing in justification of the Christian Social Union*, which, *he tells us*, are *tired of the present state of things* (*which is*, or *who are, tired*).

**9.** *Who* and *that*. It would be satisfactory if the same clear division of functions that can be confidently recommended for *that* and *which*, viz. between defining and non-defining clauses, could be established also for *that* and *who*; this would give us *that* for all defining clauses whether qualifying persons or things, and *who* for persons but *which* for things in all non-defining. But at present there is much more reluctance to apply *that* to a person than to a thing. Politeness plays a great part in idiom, and to write *The ladies that were present*, or *The general that most distinguished himself*, is perhaps felt to be a sort of slight, depriving them of their humanity as one deprives a man of his gentility by writing him Mr. instead of Esq. At any rate the necessarily defining *that* is displaced by the not necessarily defining *who* where the relative refers to a particular person or persons, but holds its own better when the person is a type or generic. In *It was you that did it*, the *It* defined is the doer—a type, not an individual; and such antecedents as *all*, *no one*, *a man*, ask for nothing better than *that*. Expressions in which we may prefer *that* without being suspected of pedantry are: *The most impartial critic that could be found*; *The only man that I know of*; *Anyone that knows anything knows this*; *It was you that said so*; *Who is it that talks about moral geography*|? To increase by degrees the range of *that* referring to persons is a worthy object for the reformer of idiom, but violent attempts are doomed to failure. Accordingly, in the following sentences, all exhibiting a wrong *which*, *that* should be sparingly preferred to *who*, though it is in all of them strictly legitimate: *The greater proportion of Consols are held by* persons or corporations which *never place them on the market* (*that*, the only relative applicable to both persons and things, is here specially suitable). | *They are harassing an enemy which is moving in the open* (who). | *Among other distinguished visitors which the Crawfords had at Rome was Longfellow* (that). | *A woman who is devoted to the many dear and noble friends, famous in art, science, and literature, which she possesses* (whom).

10. For the omission of the relative in defining clauses see THAT, REL. PRO. 4.

**which with and or but.** It is well known that *and which* and *but which* are kittle cattle, so well known that the more timid writers avoid the dangers associated with them by keeping clear of them altogether—a method that may be inglorious, but is effectual and usually not difficult. Others, less pusillanimous or more ignorant, put their trust in a rule of thumb and take the risks. That rule is that *and which* or *but which* should be used only if another *which* has preceded. It is not true; *and-which* clauses may be legitimate without a preceding *which*, and its natural though illogical corollary— that *and which* is always legitimate if another *which* has preceded—induces a false security that begets many blunders. On the other hand, it probably

saves many more bad *and-which*s than it produces. Anyone who asks no more of a rule of thumb than that it should save him the trouble of working out his problems separately, and take him right more often than it takes him wrong, should abandon the present article at this point.

Those for whom such a rule is not good enough may be encouraged to proceed by a few sentences in which it has not averted disaster: *A special measure of support and sympathy should be extended to the Navy and Admiralty*, which *have certainly never been more in need of it*, and to which *they have never been more entitled than today*. | *After a search for several days he found a firm* which *had a large quantity of them* and which *they had no use for.* | *A period in* which *at times the most ungenerous ideas and the most ignoble aims have strutted across the stage*, and which *have promptly been exploited by unscrupulous journalists and politicians.* True, it is easy to see the flaw in all these, viz. that the two *which*s have not the same antecedent, and to say that common sense is to be expected of those who apply rules; but then rules of thumb are meant just for those who have not enough common sense to do without them, and ought to be made foolproof.

Here, on the other hand, are examples in which there is no preceding *which*, and yet *and* (or *but*) *which* is blameless: *Mandates issued, which the member is bound blindly and implicitly to obey, to vote and to argue for, though contrary to the clearest conviction of his judgment and conscience—these are things* utterly unknown to the laws of this land, and which *arise from a fundamental mistake of the whole order and tenor of our Constitution* (Burke). | *Another natural prejudice*, of most extensive prevalence, and which *had a great share in producing the errors fallen into by the ancients in their physical inquiries, was this* (J. S. Mill). | *In the case of calls* within the London area, but which *require more than three pennies, the same procedure is followed.* | *The naked-eye comet* discovered by Mr. Brooks in the summer, and which *was visible in the early evening a few weeks since, has now reappeared.*

The first of these is from Burke, the second from Mill, and the other two from the most ordinary modern writing. Supporters of the rule of thumb will find it more difficult to appeal here to common sense, and will perhaps say instead that, no matter who wrote them, they are wrong. It will be maintained below that they are right. The rule of thumb fails, as such rules are apt to do, for want of essential qualifications or exceptions. The first qualification needed is that the *which* that has preceded must belong to the same antecedent as the one that is to be attached by *and* or *but*; our set of wrong examples, as we have seen, do not conform to that qualification. The next amendment is both more important and, to the lovers of simple easy rules, more discouraging: the 'another *which*' that was to be the test must be understood as meaning 'a clause or expression of the same grammatical value as the coming *which*-clause'. Now what is of the same grammatical value as a *which*-clause is either another *which*-clause or its equivalent, and its equivalent may be an adjective or participle with its belongings (*utterly unknown to the laws of this land, discovered by Mr. B. in the summer*), or an adjectival phrase (*of most extensive prevalence*; *within the London area*); for before these there might be inserted *which are, which was*, etc., without any effect on the meaning. And, secondly, what is of the same grammatical value as the *which*-clause that is coming must be an expression that agrees with it in being of the defining, or of the non-defining, kind; i.e., two defining expressions may be linked by *and* or *but*, and so may two non-defining, but a defining and a non-defining must not.

A defining expression is one that is inserted for the purpose of enabling the reader to identify the thing to which it is attached by answering ————? If the Burke quotation had stopped short at

*things* (*Mandates . . . are things.*), we should have said No doubt they are things, but what sort of things? We cannot tell what sort of things Burke has in mind till the expressions meaning 'unknown to law' and 'arising from mistake' identify them for us. Both expressions are therefore of the defining kind, and legitimately linked by *and*; whether *which* occurs in both, or only in one, is of no importance. In that example there can, owing to the vagueness of the antecedent *things*, be no sort of doubt that the expressions are defining. Often there is no such comfortable certainty. In the Mill sentence, for instance, 'another natural prejudice' is not a vague description like *things*, demanding definition before we know where we are with it. If the sentence had run simply *Another natural prejudice was this*, we should not have suspected a lacuna. We cannot be sure whether the two expressions were defining, meaning *Another natural, widespread, and fatal prejudice was this*, or non-defining, meaning *Another natural prejudice—and it was a widespread and fatal one—was this*. It is clear, however, that, whether 'of most extensive prevalence' is defining or non-defining, 'which had a great share' is the same, and the *and which* is legitimate. It was because it is not always possible to say whether clauses and expressions of the kind we are considering are defining or non-defining that the phrase 'inserted for the purpose of enabling the reader to identify' was so worded; the difference is often, though not always, a matter of the writer's intention.

After these explanations the rule, as now amended, can be set down: *And which* or *but which* should not be used unless the coming *which*-clause has been preceded by a clause or expression of the same grammatical value as itself. And a reasonable addition to this is the warning that, though the linking of a relative clause to a really parallel expression that is not a relative clause is logically and grammatically permissible, it has often an ungainly effect and is not unlikely to convict the writer of carelessness. If he had foreseen that a relative clause was to come (and not to foresee is carelessness), he could usually have paved the way for it by throwing his first expression into the same form.

It may possibly be noticed by persons who have read other parts of the book that so far *that* has not been mentioned in this article, though both defining and non-defining clauses have been in question. That is so; it has been assumed, to suit the large number of people for whom the relative *that* hardly exists in print, that *which* is the only relative. In what follows, which will consist largely of bad *and-which* or *but-which* clauses with corrections, the assumption will be, on the contrary, that *which* and *that* are, with some special exceptions, respectively the non-defining and the defining relatives.

Quotations will be arranged, with a view to their serving a practical purpose, in groups according to the particular cure that is most appropriate to each, and not according to the fault that necessitates a change. But, in order that any doubts about the latter may be resolved, an index letter appended to each quotation will refer to the following table:

A. No preceding parallel clause or equivalent
B. Different antecedents
C. Defining and non-defining expressions linked
D. *Which* instead of *that*
E. *What* preceding (see WHAT 4)
F. Right but ungainly

There is often room for difference of opinion either about the fault found or about the remedy offered. In some of the quotations the relative pronoun *who*, or the relative adverb *where* (= *at* or *in which*) plays a part instead of *which* or *that*, but this need cause no difficulty.

### CURE BY USING DEMONSTRATIVE INSTEAD OF RELATIVE

*I have also much Russian literature on that subject*, but from which, *out of respect to certain English prejudices,*

*I forbear to quote* (C; from this). | *The tunnel will be closed daily for several hours whilst the work is in progress,* and which *is expected to take two years* (A; and this). | *At one time there was a drop of something over 35 per cent.,* but from which *point there has been a recovery* (A; from that). | *In the next act—Athens during the Trojan War—we meet Diogenes, and are entertained by many clever allusions to ancient Greek mythology,* and where *our millionaire tourist falls in love with Helen of Troy* (A; and there). | *Motor-car accessories have been taxed in America, in the belief that the 5 per cent. would be absorbed by the makers or dealers,* but which *in reality is being passed on to the consumer* (A; but in reality it).

### CURE BY OMITTING THE RELATIVE

*A book the contributors to which come from many different countries,* and who *are writing under conditions which necessarily impose some restrictions upon them* (A; and are writing). | *How different from hers is Saint Augustine's, whose 'Confessions' are the first autobiography,* and which *have this to distinguish them from all other autobiographies, that they are addressed directly to God* (A; and have). | *An effort in this direction is, I believe, under consideration,* and which, *if given effect to, should be greatly in the interest of effectual unity* (A; and, if). | *The first peer was Attorney General in the first Reform Government,* and who *developed into what Greville calls 'a Radical of considerable vehemence'* (A; and developed).

### CURE BY USING *THAT* IN THE FIRST EXPRESSION ONLY

*This does not include the amount payable in respect of the buildings and improvements* erected and provided during the past year, and which *were not the property of the company vendors* (D; that were erected . . . and were not). | *I have carefully noted the earnest and sagacious advice* constantly given in your columns to the Ottoman Government, and which *may be summed up in the phrase 'Put your house in order'* (D; that has constantly been given . . . and may). | *The 'Matin' details the policy agreed upon at yesterday's meeting of the Cabinet, and which the French Government will pursue in dealing with the grave problem of Reparations* (D; that was agreed upon . . . and will be pursued by). When both expressions are defining, if the first is not a clause, the unfortunate result follows that the second requires a *that* far removed from its antecedent, on which point see THAT, REL. PRON. 4; correction may entail a change of voice or some other detail, as shown above.

## CURE BY OMITTING THE *AND* OR *BUT*

Vastly the greater number of mistakes, whether they are of the worse or the more venial kinds, can be treated thus, but the number of examples need not be correspondingly greater: *Again, take Pascal, the praise of whom in Sainte-Beuve never rings true,* and who *sees in the 'Pensées' which Pascal crowded into his short life mainly attacks on Papal Catholicism* (B; *whom* is Pascal, but *who* is Sainte-Beuve; *who* without the *and* does not go far enough; it should really be *for he sees*). | *He has attempted to give an account of certain events of which, without doubt, the enemy knew the true version,* and which *version is utterly at variance with everything that fell from my hon. friend* (B; the writer has tried to mend things by putting in the second *version,* but failed; omit that as well as *and*).|*His Majesty then took up the case of the Dartmoor Shepherd, who had been three times in the Church Army,* and whose *officers had failed to produce any lasting results upon the shepherd* (B).| *Large crowds congregated in the vicinity of the Dublin Mansion House last night, where the James Connolly anniversary concert was to have been held,* and which *was proclaimed* (A; read *but was*). | *So he sent him what he spoke of to Forster as a 'severe rating',* but which *was in reality the mildest of remonstrances* (E).

## CURE BY USING *THAT* (OR ELLIPSIS) IN FIRST EXPRESSION AND *WHICH* IN SECOND

*The class to which I belong* and which *has made great sacrifices will not be sufferers under the new plan* (C; class I belong to, which).|*No one can fail to be struck by the immense improvement which they have wrought in the condition of the people,* and which *often is quite irrespective of the number of actual converts* (C; improvement they have . . ., which is). | *The Pan-German papers are calling for the resignation of Herr von Kühlmann in consequence of the speeches which he has just made in the Reichstag,* and in which *he admitted that it was impossible for Germany to win by force of arms* (D; speeches that he . . ., in which).

## CURE BY ADVANCING THE *WHICH* (OR *WHO*) INTO THE FIRST EXPRESSION

If this is done the *which* after *and* etc. may be omitted or retained as seems best: *The enormous wire nets, marked by long lines of floating barrels and buoys,* and which *reach to the bottom of the sea, were pointed out to me* (F; which are marked . . ., and reach).| *Mr. Corbett's Nelson is a very great commander, bountifully endowed with that indispensable gift, a sound 'imagination',* but who *scorned to rely upon mere uncorroborated insight* (F; who was bountifully . . ., but scorned).| *Hallam, that most impassive of writers,* and whose *Liberalism would at the present day be regarded as tepid, tells us that* . . . (A; who was the most . . ., and whose). |*A Byzantine cross, reported to be valued at £250,000,* and which *belonged to a church in the province of Aquila, has to be returned to the parish priest* (F; which is reported . . ., and which). Anyone who has lasted out to this point may like to finish up with a few specimens of exceptional interest or difficulty, to be dealt with according to taste:

With what difficulty had any of these men to contend, save that eternal and mechanical one of want of means and lack of capital, *and of which* thousands of young lawyers, young doctors, young soldiers and sailors, of inventors, manufacturers, shopkeepers, have to complain?— *Thackeray.* / Nothing would gratify, or serve the purpose of, our enemies so much as would a panic in the capital of the Empire, as a result of their murderous aircraft attacks, *and which* might involve serious national consequences. / An amendment setting forth that the Government's action is in accordance with the strict Constitutional practice of the country and is *the only method by which* the will of the people as expressed by the majority of the elected representatives of the House of Commons can be made effective, *and among the good consequences of which* will be that the absolute veto of an unrepresentative and hereditary Chamber will for ever cease to exist. / But the review contains several criticisms which are uncalled-for, incorrect, *and to which* I wish to take exception. / Dealings are allowed in securities in such cases as those *where* negotiations between buyer and seller had been in course before the close of the House, *but which* were not completed by three o'clock. / Mr. De Haviland made a preliminary test with consummate success, *and which* was all the more impressive as the craft went through it in a casual way. / It is precisely in those trades *in which* unionism is the strongest *that* we have the most stability *and in which* we have made the greatest advance. / I got him to play in one of the charity matches at Lord's, many of which were held during the war, *and by which means* we raised a good deal of money.

**whichever, which ever.** See EVER.

**while** (or less commonly **whilst**) is a conjunction of the kind called strong or subordinating, i.e. one that attaches a clause to a word or a sentence, not a weak or coordinating conjunction that joins two things of equal grammatical value; it is comparable, that is, with *if*

and *although*, not with *and* and *or*. The distinction is of some importance to what follows. Nothing, perhaps, is more characteristic of the flabbier kind of journalese than certain uses of *while*, especially that which is described by the OED as 'colourless'; see (1) below. The proper use of *while* as a strong conjunction may be either temporal (= during the time that) or non-temporal (= whereas or though). *While she spoke, the tears were running down* | *While this is true of some it is not true of all.* It also admits of ellipsis. *While walking on the road he was run over* | *While convinced you are wrong I am sure you are sincere.* But writers are often tempted into incorrect ellipsis of two kinds (*a*) disregard of the full form, (*b*) wrongly attached participle. Examples of (*a*) are: *But while being in agreement with Sir Max Waechter's main thesis, I am bound to confess my opinion that he . . .* (the full form is not *while I am being*, but *while I am*, which should be used without ellipsis). | *We abide by that generous gesture, and while being prepared to remit all that our Allies owe to us . . . we ask only that they should. . . .* Omit *being.* | An example of (*b*) is: *While willing to sincerely sympathize with those who would suffer by such an order,* they *can only console themselves with the thought how lucky they have been that the fortunes of war have not affected them sooner* (the full form would be not *while they are willing*, but *while I am* or *we are willing*, so that *willing* is wrongly attached; read *while we are willing*). | For other examples see UNATTACHED PARTICIPLES.

The stages of degradation of *while* from a strong conjunction to a weak one may be exhibited thus:

(1) Strong conjunction playing the part of weak, i.e. introducing what is in grammar a subordinate clause but is in sense a coordinate sentence; the 'colourless' use, = *and*: *White outfought Ritchie in nearly every round, and the latter bled profusely, while both his eyes were nearly closed at the end.* Grammatically this use is unexceptionable, but it wantonly

emasculates a useful word; in this example it serves merely for ELEGANT VARIATION to avoid repeating *and*. Moreover, the temporal sense that lurks in *while* may lead those who so use it into the absurdity of seeming to say that two events occurred, or will occur, simultaneously which cannot possibly do so. *The early morning will be rather cold while afternoon temperatures will rise to the seasonal average.* Why not just *but*?

(2) Weak conjunction (= *and*) masquerading as strong. In these cases the interrogative form of the *while* sentence precludes the defence of grammatical correctness: *There is surely in this record a plain hint to the twin-Protectionist members for the City, Mr. Balfour and Sir Frederick Banbury; while was it not Disraeli who in 1842 admiringly traced the close connexion of the Tory Party with Free Trade principles? | We can only console ourselves with the thought that the German people are also 'slaves' on this showing; whilst what are we to think of a House of Lords which permitted this Slavery Act to become law?*

(3) Weak conjunction, not pretending to be anything else, but merely serving as a FORMAL WORD or ELEGANT VARIATION for *and*, with complete abandonment of the strong conjunction character: *Archbishops, bishops, and earls were allowed eight dishes; lords, abbots, and deans six; while mere burgesses, or other 'substantious' men, whether spiritual or temporal, no more than three. | The initial meridian to be that of Greenwich, while the descriptive text to be in the language of the nation concerned.*

**while**, vb. In the expression *while the time away* that is now the standard spelling. But *wile* was formerly not uncommon; it was used by Dickens. Skeat says that *while* was probably used by confusion with *wile*; the OED on the other hand calls *wile* a substitute for *while*.

**whilom.** For the adverbial use (*the wistful eyes that w. glanced down*), see WARDOUR STREET and POETICISMS; for the adjectival (*a w. medical man*), LATE.

**whilst.** See WHILE.

**whin.** See FURZE.

**whir(r).** The second r is now usual, in the noun as well as in the verb.

**whisky, -ey.** The first is the standard form but the Irish variety is usually spelt *-ey*.

**whit.** See WARDOUR STREET.

**Whit.** The forms recommended are *Whit Sunday*. *Whit Monday*, etc., *Whit-week*, *Whitsuntide*. The adjective is *Whit* (i.e. white), and the word *Whitsun* is a curtailment of *Whit Sunday*, used in the forms *Whitsun Monday*, *Whitsun week*. It is true that *Whit* with other words than *Sunday* is merely a further curtailment of *Whit Sunday*; but, as *Whit Monday* is now established, it is better to prefer *Whit* to *Whitsun* wherever the latter is not, as in *Whitsuntide*, too firmly in possession to be evicted. It must be remembered, however, that *Whitsun Week* and *Whitsunday* are the Prayer Book forms, and the Scottish term day is also written *Whitsunday* (or *Whitsun Day*); so that the advice given above can be neglected without danger. The *Whitsun* forms owe their survival partly also to the mistaken derivation (denounced with some asperity by Skeat as 'a specimen of English popular etymology') from German *Pfingsten* = pentecost.

**white(n)**, vb. See -EN VERBS. For the noun meaning prepared chalk the old word, still in use, is *whiting*; but it is being ousted by *whitening*, perhaps partly because the verb is now to *whiten* instead of to *white*, and partly for distinction from the fish whiting.

**whither.** See WHENCE.

**whiz(z).** See -Z-, -ZZ-.

**who and whom.** 1. Miscellaneous questions of case. 2. *Young Ferdinand, who(m) they suppose is drown'd.* 3. *Who(m)* defining and non-defining. 4. *And* or *but who(m)*. 5. Person and number of *who(m)*. 6. Personification. 7. *Who(m)* and participle. 8. Late position.

1. Miscellaneous questions of case. *Who* being subjective and *whom* objec-

tive, and English-speakers being very little conversant with case-forms, confusions are sure to occur. One is of importance as being extraordinarily common, and is taken by itself in Sec. 2; the others can be quickly disposed of here.

In talk *who* is constantly used for the objective case, especially when an interrogative is governed by a verb or preposition that follows it, as in *Who did you meet there? Who did you hear that from?* This colloquialism is indeed so common that it is invading printed matter. When a weekly journal, always scrupulous about its English, chooses *Who will the Opposition oppose?* as the heading of an article, and when a book reviewer in *The Times* writes *Who are such conspectuses really for?*, we must presume the choice to have been made deliberately, to avoid any suspicion of pedantry; and, if a writer given to the VERBLESS SENTENCES now so popular writes *Who from?* as one, his flouting of grammar will be more in keeping with modern idiom than *Whom from?* or *From whom?* But *who's* invasion of the province of *whom* has not gone so far in indirect questions as in direct, and we may reasonably suspect such sentences as the following to be due to carelessness rather than a splendid defiance of grammar: *When the Queen asks her retiring Prime Minister about who she should summon to head the government he wants to be able to offer her an unequivocal answer. | Before the intrusion of television into politics, politicians knew pretty well who they were speaking at.*

The opposite mistake of a wrong *whom* may sometimes be found in indirect questions, especially when it is preceded by a transitive verb or a preposition. Even Henry James could write *He had an air of being but vaguely aware of whom Miss Chancellor might be.* Other examples: *Speculation is still rife as to whom will captain the English side to Australia.| The French-Canadian, who had learned whom the visitors were, tried to apologize to Prince Albert.*

The relative *who* now and then slips

in for *whom*, giving the educated reader a shock; so: *The play, however, does not turn on Posket but on his stepson, Cis, who, though he is 19, his mother passes off as 14, and accordingly dresses, the year being 1885, in Etons. |As Mr. Bevin reminds those who in other circumstances* we should call *his followers, the agreement provided for....* That is a mistake that should not occur in print; nor should the making of one *whom* serve two clauses of which the first requires it as the object, and the second as subject. This practice is untidy enough with words that, like *which* and *that*, have only one form for both cases (see CASES 3 D, THAT, REL. PRON. 5, WHICH 2), but is still worse with *who* and *whom*. The correct form should be inserted in the second clause when a different case is wanted: *He ran upstairs and kissed two children* whom *he only faintly recognized, and yet* were *certainly his own. | But there has emerged to the final a Spaniard, in Señor Alonso,* whom *few people* would have supposed to have *a good chance a fortnight ago but* is delighting *the advocates of the older style by the beauty and rhythm of his strokes.*

For the formula *whom failing* see ABSOLUTE CONSTRUCTION; and for *than whom* see THAN 6.

**2.** *Young Ferdinand, whom they suppose is drown'd*—Tempest III. iii. 92. It was said above that the question between *who* and *whom* illustrated by this Shakespeare quotation is of importance. That is because the *whom* form is now so prevalent that there is danger of its becoming one of those STURDY INDEFENSIBLES of which the fewer we have the better, and of good writers taking to it under the hypnotism of repetition. We have not come to that pass yet; good writers usually keep clear of it, but by no means always, and it is high time for emphatic protests. What makes people write *whom* in such sentences? In the Shakespeare the preceding words are 'while I visit', so that *Ferdinand* is objective; the relative, which should be *who* as subject to *is drown'd*, may have become *whom* by attraction to the case of *Ferdinand*; or by confusion with another way of putting the thing —*whom they suppose (to be) drown'd*; or again a writer may have a general impression that, with *who* and *whom* to choose between, it is usually safer to play *whom* except where an immediately following verb decides at once for *who*. Any of these influences may be at work, but none of them can avail as a defence against the plain fact that the relative is the subject of its clause. Nor can Shakespeare's authority protect the modern solecist; did not the Revisers, in an analogous case, correct the *whom* of a more familiar and sacred sentence (*But whom say ye that I am?*—Matt. xvi. 15) into conformity with modern usage? Of the examples that follow (more than one from contemporary writers held in high esteem), the earlier show easily intelligible *whoms*, because an active verb follows that could be supposed by a very careless person to be governing it, but in the later ones a passive verb or something equivalent puts that explanation out of court: *The German people, whom Hitler had determined should not survive defeat, did survive.* / *Lord Montgomery liked to choose his own subordinates and to have around him only men* whom *he knew respected him.* / *Your reviewer,* whom *I suspect does not like this book.* / *Banquo's issue* whom *the witches predicted would be king.* / *The girl* whom *you wish was a boy.* / *There was a big man* whom *I think was a hotelier from Pnompenh.* / *Bateman could not imagine* whom *it was that he passed off as his nephew.* / *Among others* whom *it is hoped will be among the guests* are ... / *If South Africa were to be expelled from the Commonwealth the first to suffer would be those* whom *it would least be intended should suffer.* / *J. I. M. Stewart must now be quite as well known as Michael Innes,* whom *he also is.*

After reading these we can perhaps conclude that the decisive influence is probably the vague impression beforehand that *whom* is more likely to be right; but it need hardly be said that

slapdash procedure of that kind deserves no mercy when it fails. That every *whom* in those quotations ought grammatically to be *who* is beyond question, and to prove it is waste of time since the offenders themselves would admit the offence; they commit it because they prefer gambling on probabilities to working out a certainty.

As, however, an unsound proof is worse than no proof at all, discrediting, when itself discredited, the truth that depends on it, one argument sometimes brought against this use of *whom* should be abandoned. This is that the necessity of the correct form (whether *who* or *whom*) is shown when it is realized that the words between *who(m)* and what decides its case are parenthetic—*Ferdinand who (they suppose) is drown'd*. It is true that that analysis is much more often possible than impossible; it is even sometimes, though rarely, probable. But it is often impossible, as in: *Jones, who I never thought was in the running, has won*. That sentence is built up thus. Jones has won; I never thought that Jones was in the running: Jones, I never thought (that) who was in the running, has won: Jones, who I never thought was in the running, has won. No parenthesis there; nor, surely, in most examples where it is logically possible. A single live example of the impossible parenthesis is: *Cambridge's Vice-Chancellor lumped all these interesting and inspiring folk together as 'foreigners and others',* whom *he did not intend should desecrate Cambridge by their presence on a Sunday. Whom* should be *who*, not on the parenthesis argument, since 'he did not intend' cannot be parenthetic, but because the object of *did not intend* is the clause 'that who should desecrate'. The argument from parenthesis is unsound, unless indeed its champions are prepared to support it seriously by the analogy of 'You are a beauty, I don't think', where the essential main statement is playfully dressed up as a parenthesis. But it is as true that *who* is the only right case in the quoted sentences as it would be if the parenthesis argument were unassailable.

3. *Who(m)* defining and non-defining. As has been suggested in WHICH, THAT, WHO 9, the thing to aim at is the establishment of *that* as the universal defining relative, with *which* and *who(m)* as the non-defining for things and persons respectively. That consummation will not be brought about just yet; but we contribute our little towards it every time we write *The greatest poet* that *ever lived*, or *The man* that *I found confronting me*, instead of using *who* and *whom*. Failing the use of *that* as the only defining relative, it is particularly important to see that *who* defining shall not have a comma before it, and *who* non-defining shall. *Readers of the 'Westminster',* who *are also readers at the great Bloomsbury institution, will be able to admire the new decorations for themselves.* Those wrong commas (see WHICH 7) make the sentence imply that all readers of the 'Westminster' frequent the British Museum. For the omission of the relative in defining clauses see THAT, REL. PRON. 4.

4. *And* or *but who(m)*. The use of these is naturally attended by the same dangers as that of *and which*. These have been fully discussed under WHICH WITH AND OR BUT, and nothing need here be added beyond a few specimens containing *who(m)*; the letters appended refer to the table of faults given in that article: *Alfred Beasley was examined as to a meeting at which Mrs. Pankhurst was present* and a note of whose *speech he had taken* (A). | *A letter speaks of the sorrows of children* which *their parents are powerless to assuage,* and *who have little experience of the joys of childhood* (B). | *The working classes,* for long in enjoyment *of all the blessings of 'Tariff Reform',* and *who are therefore fully competent to appreciate their value, are moving with a startling rapidity towards Socialism* (F). | *We should be glad of further assistance to pay the cost of putting up relatives of men* who *live in the provinces,* and *to whom we like to extend invitations to come and stay near them for a few days at a time* (B).

In this last, the antecedent of *who* is *men,* but that of *whom* is *relatives.*

5. Person and number of *who(m)*. *To me, who* has *also a copy of it, it seems a somewhat trivial fragment.* Read *have*; the relatives take the person of their antecedents; the Lord's Prayer and the Collects, with *which art, who shewest,* and scores of other examples, are overwhelming evidence that *who* is not a third-person word, but a word of whichever person is appropriate.

The relatives take also the number of their antecedents—a rule broken in: *The death of Dr. Clifford removes one of the few* Free Churchmen whose *work had given* him *a national reputation.* The antecedent of *whose* is not *one,* but *Churchmen,* whereas the use of *him* instead of *them* shows that the writer assigned *whose* to *one*; read either *removes a Churchman whose work had given him,* or *removes one of the few Churchmen whose work has given them.* See NUMBER 5.

6. Personification. *Who(m)* must be ventured on in personifications only with great caution. It will be admitted that in the following *who* is intolerable, and *which* the right word: *The joint operation for 'pinching out' the little kingdom of Serbia,* who *had the audacity to play in the Balkan Peninsula a part analogous to that which the little kingdom of Piedmont had played in the old days in Italy.* Yet, if we had had *little Serbia* instead of *the little kingdom of Serbia,* who might have passed. Again, when we say that a ship has lost *her* rudder, we personify; yet, though *She had lost her rudder* is good English, *The ship,* who had lost her rudder is not, nor even *The Arethusa, who* etc.; both these can do with *her,* but not with *who*; possibly *Arethusa, who* (and the naval writers drop the *the* with ships' names) is blameless; if so, it is because the name standing alone emphasizes personification, which must not be half-hearted or dubious if *who* is to follow. See PERSONIFICATION.

7. *Who(m)* and participle. *I have been particularly struck by the un-selfishness of the majority of sons and daughters,* many of whom even re-

maining *unmarried because they lacked the wherewithal to do more than help their parents.* The mistake has been treated under WHICH 1. Read *many of them remaining,* or *many of whom remain.*

8. Late position. Like *which* (see WHICH 6), *who*(*m*) must not be unduly delayed. Ambiguity may result, as in *The alderman was a famous old socialist warrior, now living peaceably with his grand-daughter and her husband, who accepted as a mark of honour the nomination of him by the local businessmen's club as citizen of the year.* Here the antecedent of *who* is the alderman, not the husband as the reader might be momentarily misled into thinking.

**whodunit** is a FACETIOUS FORMATION whose felicity quickly won it a place in the dictionaries, though not yet of a higher status than slang. 'Detective or mystery story'—COD.

**whoever** etc. 1. Forms. 2. *Who ever, whoever.* 3. Case.

1. Forms. Subjective: *whoever, whosoever* (emphatic), *who-e'er* (poetic), *whoso* (archaic), *whosoe'er* (poetic). Objective: *whomever* (rare), *whoever* (colloq.), *whomsoever* (literary), *whomsoe'er* (poetic), *whomso* (archaic). Possessive: *whose ever, whoever's* (colloq.), *whosesoever* (archaic).

2. *Who ever, whoever.* See EVER. In print, when the common colloquialism of an emphasizing *ever* is used, it must be separate from *who.* It is illiterate to write *But whoever could have supposed that the business interests which are threatened would not have organized to resist?* Correct to *But who ever could* or, better, to *But who could ever,* etc.

3. Case. 'For whoever was responsible for that deliberate lie there can be no forgiveness.' The reviewer who quotes these words does so after saying 'His views on . . . are by an accident ungrammatical, but vigorous'. Obviously there is nothing ungrammatical in the sentence unless *whoever* is so, and we must conclude that the reviewer would have written *whom-ever* or

*whomsoever,* and that the subjective case therefore requires defence. The defence is not difficult, and *whom-ever* would be wrong, as it is for instance in *The slaves of the lamp render faithful service to whomsoever holds the talisman.* This, the ordinary use of the pronoun, should be distinguished from (a) the incorrect interrogative use mentioned in para. 2 and (b) the concessive use as in *Whoever consents, I refuse.* Apart from these, *whoever* is a relative that resembles *what* in containing its antecedent in itself; as *what* = that which, so *whoever* = any person who. The *that* and the *which* of *what* may or may not be in the same case, and similarly the *any person* and the *who* of *whoever* are often in different cases. But the case of *whoever* is that of the *who,* not that of the *any person,* that is, it is decided by the relative clause, not by the main sentence.

**whom.** See WHO.

**whose.** 1. General. 2. *Whose* = of which.

1. General. The word is naturally liable to some of the same misuses as *who,* which need not be here discussed separately; see WHO AND WHOM, 3–6. It may tempt writers into unconscious use of the figure of speech called SYNECDOCHE (taking the part as equivalent to the whole) as in *The women who had been killed in the riots would be like Grace, whose skin was a pleasing coffee-colour and wore pink and made such good pancakes.* Or *whose* may be made to serve in two clauses requiring different cases: *The whole scheme may be likened to the good intentions of the dear old lady whose concern for the goldfish led her to put hot water into their bowl one winter's day, and was grievously surprised when they died. | There was nothing to show for all those years, if one was a woman whose great desire was to have a child and was nearing 40.* Even in a single clause a writer may forget that he has written *whose* and end as though it had been *whom*: *Pip Thompson, an intelligent young man whose hatred of his*

*father (a general turned banker) has turned into a rebel.*

2. *Whose* = of which. A literary critic observes of an author: 'His style is clear and flexible; yet it still needs a little clarifying—weeding out "whose" as a relative pronoun of the inanimate, and the like'. If one knows neither who the author nor who the critic is, one cannot help suspecting that the flexibility commended may owe something to the condemned *whose*; in the starch that stiffens English style one of the most effective ingredients is the rule that *whose* shall refer only to persons. To ask a man to write flexible English, but forbid him *whose* 'as a relative pronoun of the inanimate', is like sending a soldier on 'active' service and insisting that his tunic collar shall be tight and high; activity and stocks do not agree. If the reader will glance at the specimens of 'late position' given in WHICH 6, he will see how cumbrous a late-placed relative is. Now insistence on *of which* instead of *whose* accounts for more late-placed relatives than anything else; *whose* would often replace not only *of which*, but *in which*. Even the WHICH 6 specimens just referred to, though they were selected long before the present article was designed, supply illustrations of that; 'This book, from the-concentrated-harvest-of-wisdom-in-which we' would become 'This book, from whose concentrated harvest of wisdom we'; 'The camera, the-remarkable-and-convincing-evidence-it-has-been-possible-to-obtain-with-which-has' would become 'The camera, whose remarkable and convincing evidence has'. To take everyday samples instead of such monstrosities, would not 'Courts whose jurisdiction', and 'a game of whose rules it is ignorant' be clear improvements in the following?—*The civilians managed to retain their practice in* Courts the jurisdiction of which *was not based on the Common Law.* / *In Whistler* v. *Ruskin—the subject of a most entertaining paper—we have the law standing as umpire in a* game of the rules of which *it is quite ignorant.* Of course they would, and

of the convenience of *whose* = *of* etc. *which* there can really be no question; nor is the risk of ambiguity worth considering, so rare is it in comparison with that of artificial clumsiness. The tabooing of *whose* inanimate is on a level with that of the PREPOSITION AT END; both are great aids to flexibility; both are well established in older as well as in colloquial English. *My thought, Whose murder yet is but fantastical* (Macbeth), and *The fruit Of that forbidden tree whose mortal taste Brought death into the world* (Paradise Lost), are merely the first instances that come to mind. The Milton happens to be a little out of the ordinary in that *whose* is not a mere possessive, but an objective genitive; but that even such a use is not obsolete is shown by the following from a newspaper: *Sir William Harcourt thrice refused an earldom,* whose acceptance *he feared might be a barrier to his son's political career.*

Let us, in the name of common sense, prohibit the prohibition of *whose* inanimate; good writing is surely difficult enough without the forbidding of things that have historical grammar, and present intelligibility, and obvious convenience, on their side, and lack only—starch.

**wide.** 1. For the distinction between *w.* and *broad*, which is of considerable idiomatic importance, see BROAD. 2. *Wide(ly)*. It should be remembered that there are many positions in which, though *widely* is grammatically possible, *wide* is the idiomatic form; see UNIDIOMATIC -LY for other such adjectives; *yawn wide, aim wide, wide apart, wide awake, open* one's *eyes wide, is widespread*, are all usually better than *widely apart*, etc., and there are many more.

**wide(-)awake.** He is *wide awake*; A very *wide-awake* person; He was wearing *a wideawake*.

**wig.** See PERIWIG.

**wight.** A WARDOUR STREET word.

**wild.** 1. Hyphens etc. On the principle that hyphens should not be used except when needed to prevent ambiguity (see HYPHENS), *wild* in the sense of not domesticated or cultivated (*w. ass, w. rose*) should not ordinarily be given one; but when the two words are used together attributively they must be hyphened or consolidated (*wild-goose chase, wildcat strike*), and it is possible, though unlikely, that even when used substantively they might have to be so written to make plain that the creature referred to was a goose or cat *ferae naturae* and not an unruly domestic one. 2. *Wildly* is the adverb. In such phrases as *gone w., run w., blow w., wild* is a predicative adjective. See UNIDIOMATIC -LY.

**wile.** See WHILE, vb.

**will,** n. Phrases like *the will to power*, in which a noun is tacked on to *will* by *to*, have come from Germany and been allowed to sojourn amongst us for a time; but there is a stronger case for their deportation and repatriation than that against many human aliens. *Will to power* may perhaps claim a prescriptive right to stay, but for the rest it may now be hoped that our philosophers, if they really do require the meaning of them, will at least dress it in English clothes. Meanwhile, *GROWING WILL TO RECONSTRUCTION*, says a newspaper headline.

**will,** vb. 1. Forms. 2. *Will* and *shall*. 3. The 'presumptive' and the 'habitual' *will*.

1. Forms. There is a verb *to will*, conjugated regularly throughout—*will, willest, wills, willed, willedst, willing*; it means to intend so far as one has power that so-and-so shall come about, the so-and-so being expressed by a noun or a *that*-clause or an infinitive with *to*: *You willed his death, that he should die, to kill him.* The much commoner auxiliary verb has none of the above forms except *will*, and on the other hand has *wilt* and *would* and *would(e)st*; it has also none of the above constructions, but is followed by an infinitive without *to*: *He will die, Would it be true?*

2. *Will* and *shall*. The meaning of this auxiliary is curiously complicated by a partial exchange of functions with *shall*, the work of merely giving future and conditional forms to other verbs being divided between certain persons of *shall* and certain persons of *will*, while the parts of each not so employed retain something of the senses of ordering or permitting (*shall*) and intending (*will*) that originally belonged to the stems. These might be called respectively the 'plain' and the 'coloured' functions of the auxiliary.

Here the English of the English differs from the English of those who are not English. The idiom of the former may be roughly summarized thus: that in the first person *shall* is the 'plain' auxiliary and *will* the 'coloured', and in the second and third persons it is the other way about. 'I shall see him tomorrow' implies no more than that that event will occur; 'I will see him tomorrow' implies that I intend to do so. Conversely 'You (or they) will see him tomorrow' implies no more than that that event will occur; 'You (or they) shall see him tomorrow' implies promise or permission. That bare summary gives a very incomplete picture; dividing lines are blurred and broached, as for instance by the emphatic *shall* used in the first person ('You surely won't do that'; 'Indeed I *shall*') and the use of *will* in the second and third persons in giving formal orders ('You will proceed at full speed to . . .'; 'The company will attack at dawn'). But in its essentials it has long been regarded as the shibboleth of the Englishman's allocation of duties to these auxiliaries, especially the use of *shall* and *should* as the 'plain' ones in the first person. The increasing number of those who cannot 'frame to pronounce it right', reveals the power wielded by Americans and Scots and others who are not English. That power need not be grudged them, and it is perhaps presumption to take for granted that *shibboleth* is better than *sibboleth*. But there are still Englishmen who are convinced that their *shall* and *will* endow their speech with a delicate precision that could not be attained without them and serve more important purposes than that of a race-label.

The idiom has been illustrated under SHALL; a small selection is here added of various common wrong forms, with references to the sections of *shall*:

See SHALL, 1

If we add too much to these demands we *will* be in grave danger of getting nothing. / Read *shall* for *will*.

See SHALL, 2

If it were true, the Germans would be right and we *would* be wrong. / Read *should* for the second *would*.

See SHALL, 3

We *would* like to bring together two extracts dealing with the effects of the Budget on land./Read *should* for *would*. (But reasons are given in SHALL for regarding this cause as lost.)

See SHALL, 4

The Gold Medal of the Royal Astronomical Society will go to a foreign astronomer, when this evening the President of the Society *will* present it to Professor Max Wolf. Read *presents* for *will present*.

See SHALL, 5

In a very few years we *shall* not remember, and *will* scarcely care to inquire, what companies were included. Read *shall* for *will*.

See SHALL, 6

It is intended that the exterior scenes in no fewer than four different pictures *will* be taken before they return. Read *shall* for *will*.

**3.** Two idiomatic uses of *will* are worth special mention. These are (a) the inferential or presumptive *will*, as in *You will have heard the news*; *That will be the telephone call I have been expecting*; and (b) the habitual *will*, as in *He will* (or *would*) *sit for hours without speaking*; *Accidents will happen.*

**wind,** verbs. The verb meaning to twist, etc.(*wind, wound*) is unconnected with that meaning to exhaust the breath or to scent (*wĭnd, wĭnded*), or to blow the horn, etc. (*wĭnd, wĭnded*, or *wound*). The past form *wound* for the word meaning to blow (the horn) is by confusion with that meaning to twist, etc.

**windward(s).** See -WARD(S).

**wine.** Except for 'wines' made from fruit other than grapes (*cowslip w., currant w.*, etc.), the word is now almost wholly a generic term (*white w., red w.*, etc.) and is not appended to the names of particular wines. *Port-wine* is the only survivor of an idiom no longer in polite use, and it too is passing.

**wink.** See BLINK.

**-wise, -ways.** 1. The ending *-ways*, or occasionally *-way*, is often used indifferently with *-wise*, and is very seldom the only form without one in *-wise* by its side; perhaps *always* is the only such word that has never had one, though with some others, e.g. *edge-* and *side-*, the *wise* form is now rare. 2. In a few established words, *-wise* is alone, e.g. in *clockwise, coastwise, likewise, otherwise, sunwise*. 3. In other established words both forms are (or were once) used, as *breadth-, broad-, end-, least-, length-, long-, no-, slant-*, though not in all cases with equal propriety. 4. In words made for the occasion from nouns, as in *Use it clubwise* or *pokerwise*, *Go crabwise* or *frogwise*, *Worn cloakwise* or *broochwise* or *chainwise*, *Placed studwise* or *fencewise*, *-wise* is now much the commoner.

**wish.** *Those wishing tickets should apply to the Secretary.* / *Do you wish some more vegetables, Madam?* This revival of an obsolete use of *wish* by way of a GENTEELISM for *want* is not to be encouraged. Modern usage recognizes *wish* with a direct object in the sense of *wish for* only when there is an indirect object too: *I wish the plan success.* / *I wish you all happiness.* / *I wish*

*him the suffering he has caused to others.*
Hence the modern colloquialism, *This
is a nice job that has been wished on me.*
See also WANT.

**wishful.** In 1881 Skeat said of this
word that, once common, it had been
almost supplanted by *wistful*; and
added that it was a reasonable inference that the latter was a corruption
of the former. It is true that *wishful* in
its original sense now gives the reader
a slight shock as he comes to it: *We
should recommend a perusal of the whole
article to those wishful to understand
the real nature of the conflict.* But the
word, has gained a new, and vigorous
lease of life in partnership with *thinking* as a term of disparagement for
those whose wishes are father to their
thoughts.

**wistaria.** So spelt and so pronounced
(*-ārĭa*), not, as one often hears, *-ērĭa*.

**wit,** n. See HUMOUR. That the two
are different names for the same thing
may still be a popular belief; but
literary critics at least should not allow
themselves to identify the two, as in:
*It is to be doubted whether the author's
gifts really do include that of humour.
Two jests do not make a wit.*

**wit,** vb. Except for the phrase *to wit*
and the derived adverbs *wittingly* and
*unwittingly,* this obsolete verb is now
remembered mainly in the archaic *God
wot,* especially in T. E. Brown's line
*A garden is a lovesome thing, God wot.*
Hence the FACETIOUS FORMATION *godwottery,* another word for what Stevenson called *tushery.* See ARCHAISMS and
WARDOUR STREET.

**witch-.** See WYCH-.

**with.** I. Writers who have become
conscious of the ill effect of AS *to* and
*in the* CASE *of,* casting about for a substitute that shall enable them still to
pull something forward to the beginning of a sentence ('The modern journalistic craving for immediate intelligibility' said Henry Bradley), have hit
upon *with.* It was a useful discovery,
but has become so popular that *with* is
sometimes found displacing *of* or some

really appropriate preposition—a trick
that should be avoided: *With pipes, as
with tobacco, William Bragge was one
of the most successful collectors.* | *With the
former class the heroes are frequently
dismissed as bad men.* Read *of pipes,
of tobacco, by the former class.* 2. The
popular expression *with it,* meaning
abreast of the times, is probably
ephemeral, like all juvenile slang.

**withal.** See WARDOUR STREET.

**withe, withy.** Both spellings and both
pronunciations (the monosyllabic as
well as the disyllabic) are in use.
To confound those who condemn the
monosyllable as a novelty or an
ignorance, there is the plural *withs* in
the A.V. of Judges xvi. 7 But probably
*withy,* pl. *-ies,* is the best form for
modern purposes, obviating uncertainty.

**withhold** not *withold.*

**without.** I. *W.* = outside. 2. *W.* =
unless. 3. *Without* ... *or without.* ... 4.
*Without hardly.* 5. Negative confusion.
  I. *W.* = outside. Both as adverb
(*listening to the wind without*; *clean
within and without*), as preposition (*is
without the pale of civilization*), the
word retains this meaning; but it is no
longer for all styles, having now a
literary or WARDOUR STREET sound that
may be very incongruous or even ambiguous. Most of us, as children, must
have wondered why it should be
thought worth mentioning that the
green hill far away was without a city
wall.
  2. *W.* = unless. *No high efficiency
can be secured* without we first secure
the hearty cooperation of the 30,000,000
or so workers. The use is good old
English, but bad modern English—
one of the things that many people say,
but few write; it should be left to
conscious stylists who can rely on their
revivals not being taken for vulgarisms.
  3. *Without* ... *or without.* ... *It can
be done without any fear of his knowing
it,* or without *other evil consequences.*
The well meant repetition of *without*
is not merely needless, but wrong.
See OR 4.

**4.** *Without hardly.* Example: *The introduction of the vast new refineries has been brought about quickly, silently, and effectively, and* without the surrounding community hardly being aware *of what was happening.* Again, like **2**, a common colloquialism, but, unlike it, one that should never appear outside speech or printed talk, and Trollope cannot be excused for writing *It seemed to her as though she had neglected some duty in allowing Crosbie's conduct to have passed away* without hardly a word *of comment on it between her and Lily.* The English for *without hardly* is *almost without.*

**5.** Negative confusion. Like all negative and virtually negative words, *without* often figures in such absurdities as: *It is* not safe *for any young lady to walk along the Spaniards-road on a Sunday evening by herself* without having *unpleasant remarks spoken as she passes along.*

**wizened, wizen, weazen.** All three forms are or have been used as adjectives, but the first is now usual. The -en of *wizen* and *weazen* is a p.p. termination, as is also the -ed of *wizened.*

**woman.** See FEMALE, WOMAN and FEMININE DESIGNATIONS. *Womankind,* not *womenkind,* for the whole sex or women in general; but *womenkind* or more commonly *womenfolk* for one's female relatives, etc.

**womanly.** See FEMALE, FEMININE.

**wonder.** For *I shouldn't w. if it didn't rain,* see NOT 4, and STURDY INDEFENSIBLES.

**wont(ed).** *Wont,* the pp. of an obsolete verb *won* (meaning to be accustomed to) is still used as a predicative adjective (*as he was wont to do*). The participial origin of the word having been forgotten, *wont* was used as a verb (*Talbot whom we wont to fear*) and given a past participle of its own, *wonted*—an experience similar to that of *hoise* and HOIST. In its other parts the verb is no longer used, but *wonted* survives as an attributive adjective (*he showed his wonted skill*). There is also a noun *wont* (*as was his*

*wont*), a partner in the SIAMESE TWIN *use and wont.* The traditional pronunciation *wŭnt,* useful as differentiating this word from the abbreviation of *will not,* has been preserved in U.S., but in England (though not in Scotland) *wōnt* has displaced it.

**wood.** *Woodbine,* not *-bind,* is the established form, especially with Shakspeare and Milton to maintain it. *To-morrow to fresh woods,* not *fields*; see MISQUOTATION.

**woof, warp, web, weft.** The *warp* is a set of parallel threads stretched out vertically; the threads woven horizontally between these are the *woof* or *weft*; and the fabric that results is the *web.*

**wool** makes, in British spelling, *woollen, woolly,* and in American *woolen, woolly.*

**word-patronage.** Under SUPERIORITY, the tendency has been mentioned to take out one's words and look at them and to apologize for expressions that either need no apology or should not have been used. To pat oneself on the back instead of apologizing for one's word is a contrary manifestation of the same weakness, viz. self-consciousness. It is rare, but perhaps deserves this little article all to itself: *That is a contingency which has been adumbrated (to revive a word which has been rather neglected of late); but this is one more case in which we must be content to wait and see./ He died believing he had solved the riddle of the universe. Engagement, to borrow that recently fashionable word, could scarcely go further. / To use an inevitable and expressive vulgarism, it 'took him out of himself'.*

**work,** vb. The decline of the form *wrought* is so manifest, yet so far from complete, that it is impossible to say from year to year where idiom still requires it and where it is already archaic. A few sentences with blanks for *wrought* or *worked* will illustrate. As the direction of progress is clear, prudence counsels falling in with it in good time. *A contemporary who——in* *brass. These things have——together for good. She——upon his feelings. This——infinite mischief. They have ——their will, Conscience——within him. Payment by weight of coal——. He——his audience into fury. When they were sufficiently——up.* But *overwrought* does not mean the same as *overworked*, and *wrought* remains unchallenged in *wrought-iron*.

**workaday** is now displaced, wholly in the noun use, and for the most part as an adjective, by the normal *workday*, of which it is regarded as a slipshod pronunciation to be used only as a genial unbending. To the extent that it survives (e.g. *This workaday world*) it differs from *workday* in connoting the humdrum element in work—'the trivial round, the common task'. Or it may serve, like *pedestrian*, as a patronizing term of disparagement— *his workaday talents*.

**working and stylish words.** No one, unless he has happened upon this article at a very early stage of his acquaintance with this book, will suppose that the word *stylish* is meant to be laudatory. Nor is it; but neither is this selection of stylish words to be taken for a blacklist of out-and-out undesirables. Many of them are stylish only when they are used in certain senses, being themselves in other senses working words; e.g., VIABLE is a working word to apply to a newly formed organism, though nothing if not stylish when used to indicate that a political programme is practicable; *initiate* is a working word for formal admission to an office or society, though stylish as a mere synonym for *begin*; DEEM is a working word in the sense in which lawyers use it, though stylish for *think*. Others again, such as *bodeful* and *dwell* and *perchance*, lose their unhappy stylish air when they are in surroundings of their own kind, where they are not conspicuous like an escaped canary among the sparrows.

What is to be deprecated is the notion that one can improve one's style by using stylish words, or that important occasions necessarily demand important words. The motorist before the magistrate does not improve his chances of acquittal by saying *I observed that I should not impede her progress* when he means *I saw that I should not get in her way.* The words in the list below, like hundreds of others, have, either in certain senses or generally, plain homely natural companions; the writer who prefers the stylish word for no better reason than that he thinks it stylish, so far from improving his style, makes it stuffy, or pretentious, or incongruous. About the words in small capitals remarks bearing on the present subject will be found in their dictionary places:

| STYLISH | WORKING |
|---|---|
| ameliorate | improve |
| ANGLE, vb. | fish |
| beverage | drink |
| BODEFUL | ominous |
| catarrh | cold |
| comestibles | eatables, food |
| DEEM | think |
| DESCRIPTION | kind, sort |
| dubiety | doubt |
| DWELL | live |
| edifice | building |
| ENVISAGE | foresee |
| FEASIBLE | possible |
| FUNCTION vb. | work |
| IMPLEMENT vb. | carry out |
| INITIATE | begin |
| INTEGRATE | combine |
| LIQUIDATE | do away with |
| partake | share |
| PERCHANCE | perhaps |
| peruse | read |
| reside | live |
| SLUMBER | sleep |
| terminate | end |
| VIABLE | workable |

See also GENTEELISMS and FORMAL WORDS.

**world.** *All the w. and his wife* is like the Psalmist; it has been young and now is old; see WORN-OUT HUMOUR.

**worn-out humour.** 'We are not amused.' So Queen Victoria baldly

stated a fact that was disconcerting to
someone. Yet the thing was very likely
amusing in its nature; it did not amuse
the person whose amusement mattered,
that was all. The writer's Queen Victoria is his public, and he would do
well to keep a bust of the old Queen
on his desk with the legend 'We are
not amused' hanging from it. His
public will not be amused if he serves
it up the small facetiae that it remembers long ago to have taken delight in.
We recognize this about anecdotes,
avoid putting on our friends the depressing duty of simulating surprise,
and sort our stock into chestnuts and
still possibles. Anecdotes are our
pounds, and we take care of them; but
of the phrases that are our pence we
are more neglectful. Of the specimens
of worn-out humour exhibited below
nearly all have had point and liveliness
in their time; but with every year that
they remain current the proportion of
readers who are 'not amused' to those
who find them fresh and new inex-
orably rises.

Such grammatical oddities as *muchly*;
such jests as *May all your troubles be
little ones*; such allusions as the *Law
of the Medes and Persians*; such parodies as *to—or not to*; such quotations
as *On—intent* or *single blessedness* or
*blushing honours*; such oxymorons as
the *gentle art* of doing something
ungentle or the *tender mercies* of a
martinet; such polysyllabic uncouth-
ness as calling a person an *individual*
or an old maid an *unappropriated
blessing*; such needless euphemisms as
*in an interesting condition* or *depart
this life*; such meioses as *the herring
pond* or *epithets the reverse of complimentary* or such hyperboles as *all the
world and his wife*; such playful archaisms as *hight* or *yclept*; such legalisms
as *the said——, the same*, and *this
deponent*; such shiftings of application as *innocent of aitches*; or *discuss
a roast fowl* or *be too previous*; such
unassimilated gallicisms as *return to
our muttons* or *give furiously to think*;
such metonymies as the *leather* and the
*ribbons* for ball and reins; such syl-
lepses as *in muddy gumboots and a*

*shocking temper*; such sobriquets as
*sky pilot* and *trick cyclist*; such arch
appellations as *my better half* or *my old
woman*; such threadbare circumlocu-
tions as *well endowed with this world's
goods* or *looked upon the wine when it
was red*; such happy thoughts as *taking
in each other's washing*—with all these
we, i.e. the average adult, not only are
not amused; we feel a bitterness, possibly because they remind us of the lost
youth in which we could be tickled
with a straw, against the scribbler who
has reckoned on our having tastes so
primitive. See also BATTERED ORNA-
MENTS, CLICHÉ, and IRRELEVANT ALLU-
SION.

**worsened words.** Changes in the
meaning of words, and still more in
their emotional content, often reflect
changes of opinion about the value of
what they stand for. Occasionally a
pejorative word becomes commendatory—BAROQUE and PRESTIGE for instance—but much more often it is the
other way about. *Imperialism* was defined at the end of the 19th c. as 'a
greater pride in empire, a larger patriotism'; in the middle of the 20th it was
described as 'a word so charged with
hostile meaning as to be almost useless.' *Colonialism* has a different significance today from what it had when
we prided ourselves on carrying the
white man's burden. The degradation
of BOURGEOIS and *capitalist* is notorious,
and *class*, as a social term, has become
unmentionable (see GROUP). *Appeasement* has suffered a similar change:
there is no hint in the OED definitions
of anything discreditable or humiliat-
ing about that word. No one thought
any the worse of Aeneas for letting
Cerberus have his usual sop. Under
the influence of EUPHEMISM the blameless word EPITHET is becoming ob-
jurgatory, and the second world war
has taken the virtue out of *collaborator*.
Not until the 1933 supplement
appeared did the OED contain any
suggestion that *academic* could be
used in a derogatory sense, reflecting
the view, widely held today, that
acquiring knowledge is a waste of

time unless it is a means of material advancement. *Hypothesis* seems to be going the same way. 'Today', says Medawar, 'the pejorative use of *hypothesis* (*Evolution is only an* h.; *It's only an* h. *that smoking causes lung cancer*) is a sign of semi-literacy.' Some old people may regret that a *pop singer* no longer means quite what it did in the days when Santley and Sims Reeves sang in St. James's Hall, and that a pop concert is no longer likely to consist of 'a series of masses and fugues and "ops" by Bach interwoven with Spohr and Beethoven'. Possibly the waning of our conviction that all progress must be beneficial will bring a converse experience to that now highly pejorative word REACTIONARY. It must also be some change in our national character —perhaps our greater reserve and selfconsciousness—that explains why many words, once reputable, are now seldom used by educated people except on a note of parody: such are *buxom, gallant, genteel, intrepid, sinful, virtuous, winsome.*

This article is concerned only with a few modern examples of a process that goes on ceaselessly, imperceptibly, and uncontrollably. Looking farther back, we should find many more words that have had similar experiences. A small sample is: CANDID, COMMONPLACE, *crafty*, GARBLE, EGREGIOUS, *idiot*, INDIFFERENT, *knave, leer, lewd,* -MONGER compounds, OFFICIOUS, PLAUSIBLE, *prevent, purlieu, respectable, rhetoric,* SENSUAL, *silly,* SPECIOUS, WORTHY, VILLAIN.

**worser.** Of this double comparative the OED says: 'Common in the 16th and 17th cc. as a variant of *worse* in all its senses. In modern use it is partly a literary survival, especially in phrases like the *worser part, sort, half,* and partly dialect and vulgar,' illustrating the last from Dickens: *Your poor dear wife as you uses worser nor a dog.* For the analogous *lesser* see LESS.

**worth, worth while. 1.** In certain uses great confusion prevails, which can be cleared up with the aid of grammar.

The important fact is that the adjective *worth* requires what is most easily described as an object. It is meaningless to say *This is worth,* but sense to say *This is worth sixpence,* or *This is worth saying* (i.e. the necessary expenditure of words), or *This is worth while* (i.e. the necessary expenditure of time). But one such object satisfies its requirements, so that *This is worth while saying,* with the separate objects *while* and *saying,* is ungrammatical.

**2.** *Worth* with two objects. There is no need to give more than two examples of this mistake; all that is necessary to set them right is to omit *while,* leaving *worth* with one object only—*turning* in the first and *pursuing* in the second. *A spare captain, to take charge of any prize that might be worth while turning into a raider. / Was not that a line worth while pursuing?*

**3.** *Worth* with no object. The commonest cause of this mistake is the anticipatory *it*; see IT 1 and 2. As is there explained, the function of the anticipatory *it* is to herald a deferred subject. It follows that such a sentence as *It is worth saying this* is equivalent to 'It, namely saying this, is worth'; *worth* is left without an object, and nonsense results. Such sentences, if they are short and simple, can be corrected in one of two ways: either by getting rid of the *it* or by inserting *while. It is worth quoting the 'Echo de Paris'* can be rewritten either *The 'Echo de Paris' is worth quoting* or *It is worth while to quote etc. It is worth dwelling on this method of approach* can be rewritten either *This method of approach is worth dwelling on* or *It is worth while to dwell on etc.* For longer and more complex sentences the insertion of *while* may be the only remedy, as in *It is worth recalling Lord Salisbury's declaration in 1885 that, if she yielded to pressure, we would consider ourselves released from our obligations.*

**3.** If an initial *it* is not anticipatory but an ordinary pronoun the case is of course quite different. In *I have said this because it is worth saying* 'it' is a pronoun whose antecedent is 'this' and *worth* is rightly given only one object—*saying*; to write *it is worth while saying* would give it two. The following quotation shows the risk of confusion: *In your excellent account of the late Miss——there is one omission, and it is worth filling it up*. Here the first *it* is anticipatory, meaning 'filling it up' and the second is the ordinary pronoun, meaning 'the omission'. If the second is left out the first takes over its job; the sentence becomes *It (i.e. the omission) is worth filling up*, and all is well.

**4.** It is true that in some sentences in which an anticipatory *it* is used there are two possible views of what *It* stands for. *It is worth while remarking that the Greek National Anthem is really a very interesting and harmonious air. | It is worth while pointing out that out of an electorate of nearly fourteen and a half millions no fewer than four and a quarter million votes were recorded in 1912 for the Socialist candidates.* Does the first quotation mean that 'it' (i.e. remarking this fact about the anthem) is worth while, or that 'it' (i.e. this fact about the anthem) is worth while remarking? Does the second mean that 'it' (i.e. pointing out this fact about the election) is worth while or that 'it' (i.e. this fact about the election) is worth while pointing out? If the former, the sentences are grammatically correct; if the latter, they are not. But the meaning is substantially the same in either case, and we may be content to give the writer the benefit of the doubt.

**5.** A complication is sometimes introduced by a relative clause. *The Chinese Labour Corps and its organization was one of the side issues of the war which is well worth while to hear about.* The skeleton of this, before subordination by the relative, is: *The Corps was a side issue; to hear about this* (issue) *is worth while*, or *it is worth while to hear about this*. Subordination by the relative should give accordingly either *to hear about which is worth while*, or *which it is worth while to hear about*. But the writer has taken that anticipatory *it*,

(= to hear about which) for the ordinary pronoun *it* (= this issue), and has therefore left it out because he supposes it to mean only the same thing as the *which* that is to connect the clause; the result is that his *which* is both subject to *is* (which he has deprived of its *it*) and object to *about*. Correct grammar would be either *which is well worth hearing about*, or *which it is well worth while to hear about*, or *which it is well worth while hearing about*.

**6.** Although the use of the gerund with *worth while* is not incorrect (*w. w. hearing about*), the infinitive (*w. w. to hear about*) is more idiomatic.

**worth-while.** This attributive-adjective compound extracted from the phrase 'is worth while' (*a worth-while experiment* from *the experiment was worth while*) is not without its uses; but, having been seized upon as a VOGUE WORD, it is fast losing all precision of meaning: *That motherhood is a full-time job all worth-while mothers will readily admit. | An attractive programme of worth-while topics has been arranged for discussion.* Like *rewarding* in a similar sense, it is a popular word with schoolmasters.

**worthy. 1.** As an adjective meaning possessed of estimable qualities, *w.* is a WORSENED WORD, seldom used except patronizingly.

**2.** The construction in which *w.* was treated like *worth* and *like*, governing a noun (*in words worthy the occasion*; *a deed worthy remembrance*, without *of*), is now rare, and appropriate only in exalted contexts.

**3.** The suffix -*worthy* is described by the OED as 'an adjective employed as a second element in a number of compounds of which only a few have come into regular use'. The large number that have not survived, some dating from the 12th c., include *death-, faith-, fame-, fault-, honour-, labour-, laugh-, mark-, name-, paint-, sale-, scorn-, song-, teach-, thank-, wonder-*, and *worship-*. Of the few that have established themselves, *blame-* dates from the 14th c., *note-* and *praise-* from the 16th, *battle-, road-* and *sea-* from the 19th and *air-* from the 20th. New formations are now mostly jocular, as when a piece of information is pronounced *newsworthy*, or something suitable for broadcasting on the BBC Third Programme said to be *third-worthy*, or a parliamentary candidate called *Westminster-worthy*, or the name for a new type of car is chosen as being *sales-worthy*.

**would.** See WILL (vb.) and SHALL AND WILL.

**wove,** p.p., instead of the usual *woven*, is chiefly in commercial terms, as *wove paper, hard-wove fabrics, wire-wove*.

**wrack.** See RACK AND RUIN.

**wrapt, wrapped, rapt.** See RAPT for the confusion between the English adjective made from Latin *raptus* and the p.p. of *wrap*. It is as well that the form *wrapt* seems to be disappearing, so that writers will have to make up their minds between *rapt* and *wrapped*.

**wrath, wrathful, wroth.** It is very desirable that differentiation should be clearly established, though all smack of WARDOUR STREET. The OED should be consulted on the history of these words; but it may safely be said (1) that many people ignore the existence of *wroth* and treat *wrath* as both noun and adjective, pronouncing it always *rawth*, and (2) that the useful arrangement would be for *wrath* to be noun only = anger and pronounced *rawth*, and for *wroth* to be the adjective = angry and pronounced *rŏth*. This does not put *wrathful* out of use; it is the attributive adjective, and *wroth* is the predicative: *A wrathful god*, but *God was wroth*. For *wroth*, the pronunciation *rŏth* is better not only than *rawth*, but also than *rŏth*, because much more easily distinguishable from the *rawth* of *wrath*.

**wreath.** Pronounce the plural -dhz; see -TH AND -DH for this, and for *wreathe*, vb.

**write.** 1. *W*. with personal object. 2. *Writ large*.

1. *W*. with personal object. In *I will write you the result*, there are two objects, *the result* (direct object), and *you* (indirect object). In literary English an indirect object is used after *write* only if there is also a direct object, but the direct object may be used without an indirect; that is, *I will write the result*, and *I will write you the result*, are idiomatic, but *I will write you soon*, or *about it*, is not. If a direct object is wanting, the person written to must be introduced by *to*: *I will write to you about it*. In commercial use *We wrote you yesterday*; *Please write us at your convenience*, etc., are established but avoided elsewhere. The following from a novel is to be condemned: *The Lady Henrietta, she who was to keep him out of Arcadia, and who believed him to be in Cannes or Mentone*, wrote him *regularly through his bankers, and once in a while* he wrote her. U.S. idiom is different; there the use of the indirect object without the direct (I will write you soon) is standard usage.

2. *Writ large.* The famous line *New Presbyter is but old Priest writ large* (Milton, Sonnet On the New Forcers of Conscience . . .) owes its fame to its double sense; *priest* and *presbyter* being derived alike from Greek πρεσβύτερος, the second word is literally a larger writing of the first; metaphorically, a presbyter turns out to be a priest, only more so. Nowadays, whenever a reform disappoints, the new state is said to be the old writ large; but, as circumstances seldom allow the literal sense as well as the other, some wrong is done to the inventor of the phrase by blunting its point.

**wrong** is one of the words whose adverbial use should be remembered; *did his sum wrong, guessed wrong*, etc., are better than with *wrongly*, but *a wrongly done sum*. See UNIDIOMATIC -LY.

**wroth.** See WRATH.

**wrought.** See WORK.

**wych-, wich-, witch-,** in *w.-elm* etc. The first and third forms are those usually seen, though the second is nearest to the earliest spelling *wice* (= drooping). Of the current forms *wych-* has the real advantage of not suggesting connexion with witches, and is recommended.

# X

**-x,** as French plural. It is still usual, in varying degrees, to write -x instead of the English -s in the plurals of words in -eau and -eu borrowed from French, the pronunciation being -z, as in English plurals. It is to be hoped that some day all of these that are in familiar English use will be anglicized with -s; but a list of the chief words, here given with an English plural in order that the reader may judge of their looks, is admittedly forbidding: adieus; beaus; châteaus; flambeaus; plateaus; portmanteaus; rouleaus; tableaus; trousseaus. The fact, however, that *purlieu*, which has all the air of a French word without being one, looks right with the plural -s (*purlieus*) because we are used to it suggests that courage with the others might soon be rewarded. For all of them except *beau, chateau,* and *tableau* the COD admits a plural in -s. as an alternative. Phrases such as *feux de joie* and *jeux d'esprit* would naturally keep their French -x, and so would any single words whose anglicization was so far from accomplished that the plural was still pronounced like the singular, without the sibilant; that is hardly true of any of the list above; we say not 'bō like Brummell', but 'bōz like Brummell', and 'portmantōz are out of fashion', not 'portmantō. . . .'

**-xion, -xive.** About certain nouns, especially *connexion, deflection, inflexion,* and *reflection,* there is a doubt whether they should be spelt with *-xion* or *-ction,* and the adjectives in *-ive* are also concerned. The forms *connexion, deflexion, inflexion,* and *reflexion,* are all called by the OED the 'etymological spellings'. In the first place, each is derived from an actual Latin noun in -xio, the change to English -ction being due partly to the influence of the verbs *connect* and *de-, in-, re-, flect,* and partly to that of the multitude of English nouns in -tion; and secondly, a vast majority of nouns in -ion were formed from the p.p. stem and not from the present stem of Latin verbs, so that *flecto flex-,* and *necto nex-,* would be expected to use *flex-* and *nex-* as the basis of their -ion nouns. As a few Latin nouns in -io were nevertheless formed from present stems, e.g. *oblivio,* the philological lapse is of no great importance. It may be well to retain the x in *connexion, deflexion* and *inflexion,* in which it has by no means gone out of use, though the earlier *connexive* has been displaced by *connective.* For *reflection* and *reflexion, reflective,* and *reflexive,* with which attempts at differentiation have had unequal success, see the separate articles.

# Y

**-y.** For the suffix used in making adjectives from nouns (*racy* etc.), as it affects spelling, see -EY AND -Y IN ADJECTIVES. For the diminutive suffix (*bookie, doggie,* etc.), see -EY, -IE, -Y, IN PET NAMES.

**y and i** were in older English writing freely interchanged; that general liberty has long been abandoned, but there are still a few words in which usage varies or mistakes are common; they are, in the spelling here recommended as the better, though not in all cases (e.g. tire) the prevailing one in Britain: cider, CIPHER; GYPSY; Libya(n); LICHGATE; Mytilene; pygmy; SIBYL; SILLABUB; SILVAN; siphon; SIREN; STILE (in hedge) and style (manner); STYMIE; TILER; TIRE (of wheel); tiro; WYCH-elm.

In *Libya, sibyl,* and *Mytilene,* the right spelling is indisputable, but with the same sound in successive syllables it is difficult to remember which is i and which y; even those who have read Herodotus and Thucydides often have to visualize the Greek words before they feel safe. *Lydia* makes them pause over *Libya*, and, as a name in current use, *Sybil*, is commoner than *Sibil*. In *cypher, lychgate, syllabub, sylvan, syphon, syren*, and *tyro*, the intrusive y is probably due to a vague feeling that an unEnglish-looking word is all the better for a little aggravation of its unEnglishness. In *tyler* and *tyre* differentiation may have been at work, but without need; and on *tyre* it may be added that, in some people's opinion, to say that *tire* is the American spelling is a sufficient reason for our using *tyre*. On the contrary, agreement between English and American spelling is much to be desired wherever it is practicable. In *gipsy* and *pigmy*, we have dissimilation, again without need; for if *invisibility* can carry five *i*s, these can surely do with two *y*s. But the fact that *stymie* has prevailed over both *stimy* and *stymy* illustrates the power of dissimilation; and it may be guessed that the y starts in the oftener used *stymied*, in which the necessary i of the second syllable tends to produce y in the first; with this compare the greater frequency of the correct y in *gypsies* and *pygmies* than in *gypsy* and *pygmy*. On the words in the list that are in small capitals, further remarks will be found in the separate articles.

**yankee.** The mistaken practice of applying this term to any citizen of the U.S.A. dies hard in Britain. Yankee was originally the nickname of those who lived in New England; at its most comprehensive a hundred years later it was extended contemptuously by the confederates during the civil war to all soldiers of the federal armies. The derivation of the word, like those of JAZZ and O.K., has been the subject of much discussion and little agreement.

**yclept.** See WORN-OUT HUMOUR

**ye.** This archaic form of *the*, now only in jocular use, should properly be pronounced *the*, not *ye*, the *y* being not our letter, but a representation of the obsolete single letter (þ, called *thorn*) now replaced by *th*.

**year.** Phrases such as *last year, next year*, may be either nouns or adverbs (*Next year may be warmer*; *We may have warmer weather next year*); they should not be both at once, as in: *The quinquennium ending and including last year*. The 'last year' that the quinquennium included was a noun; the 'last year' that the quinquennium ended was an adverb; indeed, far from the quinquennium's ending the year, the year ended the quinquennium. It is the same kind of mistake as making one word serve twice in two different cases, for which see CASES 3 D. For the use of an apostrophe in e.g. *Five years' imprisonment* see POSSESSIVE PUZZLES 6, and for *The first time in years* see IN I.

**yellow-(h)ammer.** It cannot be said with safety either that the *h* is due to ignorant assimilation by popular etymology to *hammer*, or that the absence of *h* is mere *h*-dropping. Each form has an etymological theory on its side, and OED says that both forms 'are historically justifiable'. The only reason for resisting the prevalent *h* is thus removed.

**yeoman.** *Yeoman service* and *yeoman's service* are both idiomatic, but the greater ease of the former has naturally made it the usual form.

**yester-.** Other combinations than *yesterday* are incongruous except in verse or in designedly poetic prose. It is true that *yestereve* is shorter than *yesterday evening*, but the saving of space is paid for by the proof that one has no literary sense.

**yet. I.** As an adversative adverb or conjunction (= nevertheless, for all that) *yet* is particularly liable to two misuses. One of these—its tendency to inspire foolish inversions—has been specially treated in INVERSION under the heading *Yet, Especially, Rather*. The other is what may be called its illogical pregnant use. When *yet* is used to point a contrast, the opposition between the fact it introduces and that which has gone before should be direct and clear. Examples of failure in this respect must necessarily be of some length; some simpler specimens of a rather similar kind will be found under BUT 3. In each of those that follow it will be noticed that the particular fact with which the *Yet* sentence is in contrast is by no means the essential contents of the previous sentence, but has to be got out of it at the cost of some thought. *We confess to being surprised at the line taken by the railwaymen at Crewe with reference to Colonel Yorke's conclusion that the Shrewsbury disaster occurred through the engine-driver having momentarily fallen asleep. Yet at a meeting the Crewe railwaymen are very indignant at the suggestion, and denounce Colonel Yorke as an Army officer who does not understand the real working of railways.* Here the *Yet* fact is that the men are indignant. What is that in contrast with? Apparently with the correctness of Colonel Yorke's conclusion; but, though many other things not in contrast with their indignation can be got out of the sentence, the correctness of the conclusion is inferable only from the newspaper's surprise at the men's indignation at the conclusion. If *yet* were omitted, the second sentence would come in logically enough as an explanation of what the men's 'line' referred to had actually been.

*Sir, I doubt if sufficient attention has been drawn to the injustice of throwing on the landlord in whose house they happen to be resident the cost of a large additional insurance benefit for those who are sick. Yet, under Clause 51, a sick tenant would be able to live rent free for a year at the expense of his or her landlord.* This is a less glaring case. The essence of the *Yet* sentence is that a tenant has power to injure a landlord. What is that in contrast with? With the fact that justice should protect landlords; that is, not with the main sentence preceding, which is a statement of why the writer is writing, but with a mere inference from a noun that occurs in it, viz. 'injustice'. As in the

first example, the logical work of the second sentence is to explain the nature of a noun contained in the first, but an explanation is presented in the guise of contrast; the sentence would do its work properly if *yet* were omitted. No doubt the contrast that was at the back of the writer's mind, but did not come to the front of it, was between the greatness of the injustice and the lack of public interest in it.

2. *Yet* in its temporal sense may be ambiguous: its old sense of 'now as before', 'still' has lasted longer in Scotland and Ireland than in England. To a Scotsman or Irishman the natural meaning of 'Is it raining yet?' is 'Is it still raining?' To an Englishman it is 'Has it yet begun to rain?'

**yiddish** is not a kind of Hebrew, but a kind of German. It is the language used by Jews in Europe and America, consisting mainly of German (orig. from the Middle Rhine area) with admixture of Balto-Slavic or Hebrew words and written in Hebrew characters.

**yon.** See WARDOUR STREET.

**you, your, yours.** 1. For the interjectory *you know* see MEANINGLESS WORDS. 2. For the impersonal use of *you* (= one) see ONE 8. 3. For misuse of *yours* in place of *your*, see ABSOLUTE POSSESSIVES. 4. For epistolary uses, see LETTER FORMS.

# Z

**zeugma.** Pl. *-as* or *-ata*. See SYLLEPSIS AND ZEUGMA.

**zinc,** n. and vb. Inflexions and derivatives give trouble with spelling and pronunciation, especially those in which an *e, i,* or *y* follows the *c,* and so converts its natural pronunciation to *s*. *Zinced* and *zincing* are clearly anomalous; verbs ending *c* normally make these inflexions *-ck-* so that there shall be no mistake about the pronunciation —e.g. *trafficking, picnicked.* But in compounds and derivatives the *c* with an *s* sound may be tolerated, e.g. *physicking* but *physicist*. The forms recommended are accordingly *zincked*, *zincking*, *zincky* (or *zink-*), but *zincic*, *zinciferous*, *zincify*, etc. See -C -CK.

**zingaro.** Fem. *-ara*; pl. *-ari*. Pronounce with stress on the first syllable and a short *a*.

**zither(n).** See CITHERN.

**zoology.** Pronounce *zŏŏ'logy*. The convenience of the CURTAILED WORD *Zoo* is no doubt the reason, but is not an excuse, for the very common vulgarism *zōōŏ'logy*.

**-z-, -zz-.** In *buz(z)*, *fiz(z)*, *friz(z)*, *quiz*, and *whiz(z)*, there is no need for a second z, and when it appears it is doubtless due to the influence of inflected forms like *buzzer*, *quizzed*, and *whizzing*, in which it serves to show that ĭ and ŭ, not ī and ū, are the sounds. But in fact the behaviour of these words is capricious. *Quiz* alone firmly dispenses with the second z. *Friz(z)* (of hair) and *whiz(z)* vacillate. *Buzz*, *fizz*, and *frizz* (of the noise from a frying-pan) retain that form, perhaps because of the onomatopoeic suggestion of the double z. For the same reason the newcomer, *jazz*, is unlikely to change to one.

PRINTED IN GREAT BRITAIN
AT THE UNIVERSITY PRESS, OXFORD
BY VIVIAN RIDLER
PRINTER TO THE UNIVERSITY